ExoticsCon 2023

Building Exotics Excellence. One City. One Conference.

Boston, Massachusetts, USA
20 - 24 August 2023

Volume 1 of 2

ISBN: 979-8-3313-1554-2

Printed from e-media with permission by:

Curran Associates, Inc.
57 Morehouse Lane
Red Hook, NY 12571

CURRAN ASSOCIATES INC.
proceedings
.com

Some format issues inherent in the e-media version may also appear in this print version.

Printed with permission by Curran Associates, Inc. (2025)

For permission requests, please contact Association of Reptilian & Amphibian Veterinarians
at the address below.

Association of Reptilian & Amphibian Veterinarians
5003 SW 41st Blvd
Gainesville FL 32608-4930
USA

Phone: 480-703-4941

info@arav.org

Additional copies of this publication are available from:

Curran Associates, Inc.
57 Morehouse Lane
Red Hook, NY 12571 USA
Phone: 845-758-0400
Fax: 845-758-2633
Email: curran@proceedings.com
Web: www.proceedings.com

EXOTICSCON

Building Exotics Excellence. One City. One Conference.

THE WESTIN BOSTON SEAPORT DISTRICT
BOSTON, MASSACHUSETTS | AUGUST 20 - AUGUST 24

2023 PROCEEDINGS

PRESENTED BY

AAV ASSOCIATION OF AVIAN VETERINARIANS

AEMV ASSOCIATION OF EXOTIC MAMMAL VETERINARIANS

ARAV ASSOCIATION OF REPTILE AND AMPHIBIAN VETERINARIANS

EXOTICSCON.ORG

Welcome to ExoticsCon 2023

On behalf of the Board of Directors for the Association of Avian Veterinarians (AAV), Association of Exotic Mammal Veterinarians (AEMV), and the Association of Reptile and Amphibian Veterinarians (ARAV), we welcome you to Boston, Massachusetts.

The Trilateral Conference Planning Committee is excited to bring you quality education and amazing networking opportunities. We hope you'll make the most of the program the committee worked tirelessly to assemble. From educational opportunities like Hands-On Workshops and lecture sessions, to special and social events, we know there is something for everyone.

We have the privilege to work with various professionals to plan this event and would like to acknowledge the following groups and individuals who helped make this conference possible:

- **Auction chairs,** for undertaking this monumental task and insuring the future of all the organizations' research grant programs:
    - o AAV: Jonathan Stasiek and Jackie Bucio
    - o AEMV: Victoria Bender, DVM and Colleen Roman, DVM
    - o ARAV: Marianne Caron, DVM, Emily Thulson, DVM, and Caitlin Cunningham, DVM
- **The *volunteer* abstract and proceedings reviewers,** for ensuring high-quality, accurate, and evidence-based, scientific programing and proceedings.
- **Session chairs,** for their time and expertise to ensure the conference proceeds smoothly and attendees have the best educational experience possible.
- **Hands-On Workshop instructors,** who develop exciting hands-on experiences to convey new clinical techniques for attendees to implement in their practices.
- **Presenters,** for their time, energy, expertise, and willingness to share their knowledge and further the profession with their passion for exotic animals.
- **Executive directors,** who provide leadership and guidance through the conference planning process, and for their devotion to our organizations and adopting our missions and passions as their own:
    - o AAV: Robert Groskin, DVM
    - o AEMV: Marisa Hackemann
    - o ARAV: Annie Trout, M.Ed.
- **Conference planning team with MJB Management, LLC,** for their help with budgets, registration, A/V, and other logistics and details impacting the conference:
    - o Therese Lombardo, CMP
    - o Jarisa Santos, CMP

In addition, thank you to the **lead conference planner Marisa Hackemann, MJB Management, LLC, CEO**, for taking on and tackling the overwhelming task of coordinating, aligning, creating compromise, and promoting the individual and collective needs of three devoted and passionate associations to create and shape a record-setting conference.

Last, but certainly not least, **we thank our members, attendees, sponsors and exhibitors,** for their faith and loyalty as well as their contributions to our associations. AAV, AEMV, and ARAV exist because of you and your generous contributions. Over the years, you have helped ExoticsCon grow in both size and scope. Your feedback continues to shape future planning and we look forward to your feedback. We hope you enjoy ExoticsCon 2023!

Sincerely,
Byron De La Navarre, DVM, *AAV Conference Chair*
Anna Osofsky, DVM, DABVP (Avian Practice), *AAV*
Stephanie Lamb, DVM, DABVP (Avian Practice), *AAV*
Todd Driggers, DVM, *AEMV Conference Chair*
Sara Gardhouse, DVM, Dipl ACZM, DABVP (Exotic Companion Mammal Practice), *AEMV*
Melissa Kling, DVM, *AEMV*
La'Toya Latney, DVM, DipECZM (Zoo Health Management), DABVP (Reptile and Amphibian Practice), *ARAV Conference Chair*
James E. Bogan, DVM, DABVP (Canine and Feline Practice), DABVP (Reptile and Amphibian Practice), CertAqV, *ARAV*
Sean Perry, DVM, PhD, *ARAV*

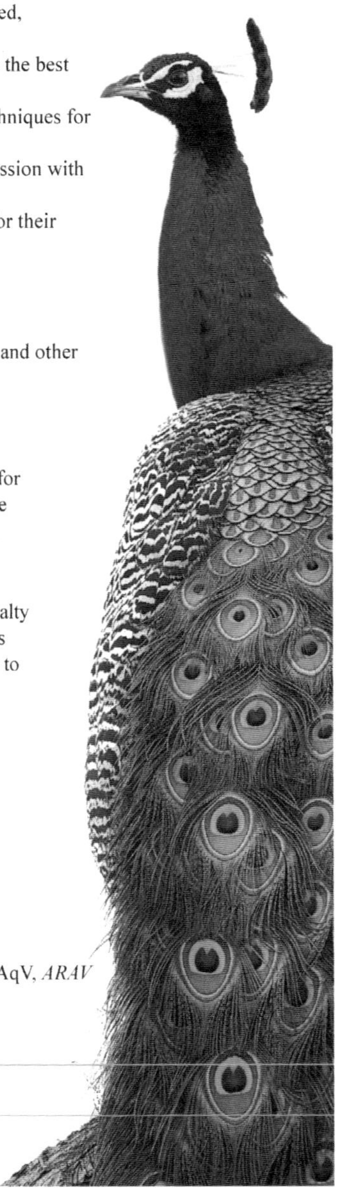

Introduction

We invite you to enjoy the 2023 ExoticsCon Proceedings. Within the proceedings, you will find a wealth of information related to the variety of lectures presented at this year's conference.

AAV, AEMV, and ARAV recognize and thank the following committees and staff for their time and hard work in assuring a quality program and publication:

ASSOCIATION OF AVIAN VETERINARIANS AAV

AEMV ASSOCIATION OF EXOTIC MAMMAL VETERINARIANS

ARAV ASSOCIATION OF REPTILE AND AMPHIBIAN VETERINARIANS

Continuing Education Accreditation

The American Association of Veterinary State Boards (AAVSB) Registry of Approved Continuing Education (RACE) committee has reviewed and approved the ExoticsCon 2023 program as meeting the standards adopted by the AAVSB. Please contact the AAVSB RACE program at race@aavsb.org or 877-698-8482 if you have any questions regarding Continuing Education for this program.

Conference Sponsors

Platinum Sponsors:

Lafeber's EmerAid Life Saving Nutrition™ A DIVISION OF LAFEBER

OXBOW ANIMAL HEALTH EST. 1980

Moichor

COLORADO EXOTIC ANIMAL HOSPITAL

ARIZONA EXOTIC ANIMAL HOSPITAL

Disclaimer

The material appearing in this publication comes exclusively from the authors and contributors identified in each manuscript. The techniques and procedures discussed reflect the personal knowledge and experience of the authors and contributors, and demonstrate their views of the methods that may be used for these medical procedures. The procedures demonstrated do not incorporate all known techniques, are not exclusive, and other techniques and technology may also be available. Any questions or requests for additional information concerning any of the manuscripts should be addressed directly to the authors.

AAV, AEMV, and ARAV do not research, review, or otherwise verify any of the information contained in this publication. Opinions expressed in this publication are those of the authors and contributors and not necessarily those of AAV, AEMV, or ARAV. AAV, AEMV, and ARAV are not responsible for errors or for opinions expressed in this publication.

AAV, AEMV, and ARAV expressly disclaim any warranties or guarantees, expressed or implied, and shall not be liable for damages of any kind in connection with the material, information, techniques, or procedures set forth in this publication.

Publication Terms

Conference proceedings manuscripts are accepted as full manuscripts or as scientific abstracts. Scientific abstracts are accepted if the author(s) express(es) intent to submit their work for consideration for future publication in a peer-reviewed scientific journal. Several distinguished veterinary journals, including the *Journal of the American Veterinary Medical Association* and the *American Journal of Veterinary Research* may reject work that has been previously published of 250 words or more and included additional content such as figures, tables, illustrations, and references. Scientific abstracts published in the ExoticsCon proceedings are still required to contain data and meaningful information to the reader.

For permission to reproduce or order copies of this publication, contact the respective association office:

- AAV Office, www.aav.org, PO Box 9, Teaneck, New Jersey 07666, USA, email pubs@aav.org.
- AEMV Office, www.aemv.org, MJB Management, LLC, PO Box 1868, Mt. Juliet, Tennessee 37121, email info@aemv.org.
- ARAV Office, www.arav.org, MJB Management, LLC, PO Box 1868, Mt. Juliet, Tennessee 37121, email info@arav.org.

For information on ExoticsCon educational conferences, visit https://www.exoticscon.org. For information on membership and other association programs visit the association websites listed above.

Table of Contents

Association of Avian Veterinarians

Section 1: AAV Fundamentals Session

Section 2: AAV Plenary Session

Section 3: AAV Cutting Edge Session

Section 4: AAV Wildlife, Anesthesia, and Surgery Session

Section 5: AAV Masterclasses Track 1

Section 6: AAV Masterclasses Track 2

Section 7: AAV Masterclasses Track 3

Section 8: AAV Masterclasses Track 4

Section 9: AAV Special Panel on Geriatrics

Section 10: AAV Potpourri Session

Section 11: AAV "I'll Take Topics with the Acronym AI for 500 Please, Alex" Session

Section 12: AAV Potpourri II Session

Association of Exotic Mammal Veterinarians

Section 13: AEMV Fundamentals Session

Section 14: AEMV Plenary Session

Section 15: AEMV Therapeutics Session

Section 16: AEMV Dentistry Session

Section 17: AEMV Respiratory System Session

Section 18: AEMV Cardiology Session

Section 19: AEMV Epidemiology Session

Section 20: AEMV Masterclasses Track 1

Section 21: AEMV Masterclasses Track 2

Section 22: AEMV Masterclasses Track 3

Section 23: AEMV Masterclasses Track 4

Section 24: AEMV Gastrointestinal System Session

Section 25: AEMV Anesthesia and Analgesia Session

Section 26: AEMV Neurology Session

Section 27: AEMV Dermatology Session

Section 28: AEMV Ophthalmology Session

Section 29: AEMV Pharmacokinetics Session

Section 30: AEMV Oncology Session

Section 31: AEMV Urogenital System Session

Section 32: AEMV Mammal Medley Session

Section 33: AEMV Posters

Association of Reptile and Amphibian Veterinarians

Section 34: ARAV Plenary Session

Section 35: ARAV Novel Therapies and New Perspectives Session

Section 36: ARAV Nutrition and Diet Composition Session

Section 37: ARAV Masterclasses Track 1

Section 38: ARAV Masterclasses Track 2

Section 39: ARAV Infectious Disease Session

Section 40: ARAV Neoplasia Session

Section 41: ARAV Clinical Pathology and Reproduction Session

Section 42: ARAV Surgery and Imaging Session

Section 43: ARAV Critical Care and Conservation Session

Section 44: ARAV Potpourri Session

Section 45: ARAV Posters

Hybrid Sessions

Section 46: Hybrid Session 1

Section 47: Hybrid Potpourri Session

Section 48: Hybrid Special Session

Section 49: Hybrid Marla Lichtenberger Emergency and Critical Care Session

Section 50: Hybrid Session 2

Section 51: Hybrid Session 3

Technician Sessions

Section 52: Technician Session 1

Section 53: Technician Session 2

Section 54: Technician Tutor Session

Section 55: Technician Emergencies Session

Author Index

Section 1

AAV Fundamentals Session

Cheryl Greenacre, DVM, DABVP (Avian Practice), DABVP (Exotic Companion Mammal Practice)
Session Chair

Fundamental Skills in Avian Medicine

Laura Neuhaus, DVM

Session #3190

Affiliation: From Grafton Animal Hospital, Community Veterinary Partners, 449 Grafton Drive, Yorktown, VA 23693, USA.

Abstract: There are various skills that are needed when first starting to work with avian species. This presentation describes some of the most common techniques. Topics include blood collection, subcutaneous injections, intramuscular injections, intravenous and intraosseous catheters, crop tubing, abdominocentesis, and nasal flush.

Introduction

When first starting to practice avian medicine there are a variety of skills that are necessary. The purpose of this discussion is to further describe some of commonly utilized techniques. For skills not listed, there are a variety of resources available to continue learning. These include but are not limited to continuing education at conferences such as ExoticsCon, the Education Portal and the Golden Egg on the Association of Avian Veterinarians' website (http://aav.org), the Avian Veterinary Clinical Competency Program (https://aav.org/page/avccp), and LafeberVet (http://lafeber.com/vet).

Blood Collection

Blood samples permit the practitioner to perform a wide variety of diagnostic tests. The number and type of diagnostics that can be run, however, can be limited by small patient sizes and blood volume. While blood volume is dependent on the species, a rough estimate is 10% of the body weight.[1] In a healthy bird, a clinician can safely collect 10% of this amount, or 1% of the body weight (for example 1 milliliter in a 100 gram cockatiel).[1] In severely ill birds, smaller sample sizes should be taken to avoid further compromising the patient.[1]

A variety of different sites can be used for blood collection. The right jugular vein is a frequently used site due to the ability to collect a large volume of blood quickly from even small patients.[1,2] To collect a sample, use a small amount of alcohol to expose the overlying aptera (present in most species) then apply pressure to the thoracic inlet to occlude and help stabilize the vessel.[1,2] The disadvantage of this site is that the surrounding large subcutaneous space predisposes the area to hematoma formation.[1]

The basilic vein is another common site for blood collection. For this collection, the patient is placed in dorsal recumbency and the wing is gently extended.[1] Pressure is then applied to the humeral area to occlude the vein prior to collection. Hematomas are common in this area due to movement of the wing.[1]

The medial metatarsal vein is the author's preferred site for blood collection. The patient can be maintained upright during collection, the leg is easily restrained, and hematomas are typically minimal due to protection from the surrounding muscles.[1,2] Due to the smaller size of this vessel, use is typically reserved for medium to large species. When collecting from smaller species, an insulin syringe can help to reduce negative pressure and vessel collapse. To collect a sample, locate the vessel in the region of the tibiotarsus-tarsometatarsal joint and apply pressure proximally over the tibiotarsus to occlude it.

Subcutaneous Injections

The subcutaneous route is one of the most common locations for fluid administration.[3] Fluids are warmed prior to administration and can be given in the intrascapular area, inguinal region, flank, or the axilla with the inguinal region being most common.[2,3] The feathers are parted with a small amount of alcohol and the needle is introduced at a shallow angle with the bevel up.[4] As fluids are injected, the subcutaneous space will expand and the needle can be moved into the middle of the pocket to keep it from inadvertently slipping out.[4] Be careful not to place the needle too deeply, however, as accidental administration into the air sacs can occur with fatal consequences.[4]

Intramuscular Injections

Intramuscular injections are commonly performed for administration of various therapeutic agents. The most common site used is the pectoral musculature.[2,5] The needle is introduced at an acute angle on either side of the keel in the mid to caudal portion of the muscles to avoid the larger vessels located cranially.[2] The plunger is withdrawn prior to administration of the medication to ensure a vessel has not been inadvertently entered.[2] It is important to note that not all medications are suited for intramuscular injection (e.g., Baytril).

Intravenous and Intraosseous Catheters

Intravascular fluids are given to patients that are severely debilitated or undergoing anesthesia. Patients that are more alert may become tangled in fluid lines or remove the catheter which could lead to fatal hemorrhage.[2,3,5] When intravascular fluids are indicated, either an intravenous or intraosseous catheter can be placed. To place a catheter, a local anesthetic, sedation or anesthesia may be needed to reduce patient stress and movement.[5] In the case of an intravenous catheter, placement sites are the same as those for phlebotomy. These catheters can then be secured using tape alone when using the medial metatarsal vein and sutured when using the basilic and jugular vein.[5]

Intraosseous catheters are useful for small or collapsed patients where intravenous access is not possible.[2,5] Intraosseous catheters are placed in the ulna or tibiotarsal bone using either an intraosseous or hypodermic needle.[2,3,5] While hypodermic needles are more readily available, they will frequently become obstructed with bone during insertion and may need to be replaced.[2,3] To place an ulnar catheter, the carpus is flexed and the dorsal condyle of the carpus is identified. The needle is rotated into the distal ulna and directed proximally.[2,5] When placing a tibiotarsal catheter, flex the stifle and insert the needle at or to either side of the insertion of the patellar tendon then rotate and direct the needle distally through the cnemial crest.[2,5] Catheter position for both sites can be checked with radiographs or in the case of ulnar catheters, by visualization of the basilic vein blanching when flushed.[5] Intraosseous catheters that are to be maintained in the awake patient can be secured with tape in addition to a figure-8 style bandage in the case of ulnar catheters.[5] Unlike intravenous catheters, fluids administered through an intraosseous catheter need to be given under pressure either by hand or using a syringe pump.[2]

Ingluvial Gavage

Ingluvial gavage is performed for a variety of reasons including to administer medications, provide nutrition, and collect diagnostic samples. Most commonly either metal gavage needles or red rubber catheters are utilized. Care must be taken when using soft tubes to ensure that the patient does not bite and ingest the material.[2] For this reason, metal tubes are often preferred in psittacine birds and softer red rubber catheters in other species such as Anseriformes and Galliformes.

When placing the tube or gavage needle, the patient is held with the head and neck extended. Lubricate the tube or gavage needle prior to placement to reduce the risk of accidental trauma to the oropharynx and thin walled esophagus.[2,5,6] Next, pass the tube or gavage needle from the left side of the mouth down the right side of the oral cavity. Prior to administration of food, water, or medications, the clinician should palpate for the tip of the tube or gavage needle in the crop to confirm the location.[2,3,6] While administering the syringe contents, the oral cavity should be monitored for regurgitation and the tube removed if seen.[2]

A variety of products are available on the market for nutritional support depending on the species and clinician preference. Up to 50 ml/kg of formula can be given though reduced amounts should be used when first starting gavage feeding.[6] Gavage feeding should be the last treatment performed to reduce the risk of regurgitation secondary to stress and feedings can be administered two to four times per day.[6]

An ingluvial lavage can be performed using the same technique. Rather than food, warm water is infused with smaller amounts than those used for gavage feeding to take into account current crop contents.[2] After infusing water into the crop, gently massage it and aspirate the fluid back into the syringe.[2] Do not forcibly aspirate the fluid back into the syringe as this can cause damage to the crop.[2]

Nasal Flush

Nasal flushes are used for both therapeutic and diagnostic purposes. To perform a nasal flush, position the patient so that the head is directed towards the floor, place a syringe containing warm saline against the nare, and flush the fluid through the nares.[7] The fluid expressed can be collected in a sterile container or on a swab for cytology and culture and sensitivity.

Abdominocentesis

Abdominocentesis is performed in birds with coelomic effusion to help relieve pressure on the air sacs and lungs as well as to collect samples for diagnostics.[8] The skin over the abdomen is disinfected and a 23- to 27-gauge needle attached to a syringe is introduced along the midline.[3] If introduced lateral to midline, the air sacs may be inadvertently entered.[3] A fluid extension set and stopcock may be added to aid in removal of larger volumes of fluid.[8] Be aware that when removing large amounts of fluid, protein shifts and changes in abdominal pressure can cause serious or fatal side effects.[3,8]

References

1. Campbell TW, Grant KR. *Exotic Animal Hematology and Cytology.* 5th Ed. Hoboken, NJ: Wiley-Blackwell; 2022.

2. Longley LA. *Anesthesia of Exotic Pets.* Philadelphia, PA: Elsevier; 2008.

3. Harrison GJ, Lightfoot TL. *Clinical Avian Medicine.* Palm Beach, FL: Spix Publishing, Inc.; 2006.

4. LafeberVet. Subcutaneous fluids in birds. Available at https://lafeber.com/vet/subcutaneous-fluids-in-birds/. Accessed March 20, 2023.

5. Speer BL. *Current Therapy in Avian Medicine and Surgery.* St. Louis, MO: Elsevier, Inc.; 2016.

6. LafeberVet. Tube feeding birds. December 9, 2010. Available at https://lafeber.com/vet/tube-feeding-birds/. Accessed March 20, 2023.

7. LafeberVet. Avian respiratory emergencies: an approach to the dyspneic bird. Available at https://lafeber.com/vet/respiratory-emergencies/. Accessed March 20, 2023.

8. LafeberVet. Abdominocentesis in birds. Available at https://lafeber.com/vet/abdominocentesis-in-birds/. Accessed March 20, 2023.

Fundamentals of Avian Orthopedics

Seth C. Oster, DVM

Session #3124

Affiliation: From the Department of Clinical Sciences, Auburn University, College of Veterinary Medicine, 1220 Wire Road, Auburn, AL 36849, USA and the Southeastern Raptor Center, Auburn University, College of Veterinary Medicine, 1350 Pratt-Carden Dr, Auburn, AL 36849, USA.

Abstract: Fundamentals of Avian Orthopedics is a 1-hour masterclass designed to update new and experienced clinicians with the fundamentals of, triaging, diagnosing and treating orthopedic injuries in avian patients. The course begins with a discussion of initial case presentation and management, including triage and short-term stabilization of injuries. The discussion then focuses on guiding the selection of conservative versus surgical management and exploring those methodologies.

Introduction

Orthopedic injuries affecting birds include fractures, degenerative joint disease, and developmental abnormalities. Avian cortical bone is thinner and more brittle than their mammal counterparts. This is due to the pneumatic medullary cavity and increased mineral content. In this author's experience, fractures are the most common orthopedic injury that presents in practice.

Fractures

Fractures commonly present as an emergency after acute trauma. Clinical signs include wing droops, lameness, instability of long bones, and crepitation on palpation. The principles of managing fractures in avian patients are similar to those of mammalian orthopedics. These include initial stabilization and triage, fracture assessment, and conservative or surgical stabilization.

Initial triage consists of supporting normal respiration and circulation, as many traumatic injuries such as fractures can leave the patient in shock. This can include fluid therapy, supplemental oxygen, and heat based on the patient's initial assessment. The remaining elements of triage include stabilization of the fracture site, usually through external coaptation, and providing analgesic support. In addition, antimicrobial therapy and local cleaning and disinfection may be required if the bones fragments are exposed (compound fracture).

External coaptation of fractures varies based on bones involved. A figure 8 bandage can be used to stabilize fractures involving the radius, ulna, or metacarpal bones. A body bandage can be used to stabilize fractures of the clavicle, coracoid, and scapula. Fractures of the humerus are stabilized with a combination of a figure 8 and a body bandage. Fractures of the tibiotarsus or tarsometatarsus can be stabilized using a roll gauze splint with side bars of ridged material such as aluminum or thermoplastic tape for support.

Analgesic support for orthopedic injuries is important from both a healing and ethical standpoint. Multimodal pain support is required to reach adequate levels of analgesia in an acute fracture. Based on available pharmacokinetic data a combination of meloxicam and buprenorphine SR are this author's preferred combination.

Radiographs are required to evaluate most fractures suitability for surgery and to guide the selection of the appropriate surgical fixation method. Due to the pain and potential for further injury while properly positioning a bird for diagnostic radiographs, sedation or anesthesia is required. Orthogonal views of the affected limbs should be used to assess the type and location of the fracture, and the morphology and the positioning of the fragments.

Positioning is key to diagnostic radiographs. Having a few items can make positioning easier. Foam wedges or pads can be very useful to elevate heads or synsacrum to get proper alignment for lateral views. Foam can also be useful in separating legs in a lateral view. White tape or masking tape can be used to secure limb in position with minimal damage to feathers. Bird Boards have also been used to aid in positioning. These devices are made of acrylic and are more appropriate for larger birds. Most of these devices have straps or ties to secure wings and legs and an acrylic plate to secure the head in place. This author prefers the use of foam and tape.

A properly positioned ventrodorsal view of the coelom should, if possible, include the thoracic inlet cranially and legs caudally. The patient's dorsum should be laid flat on the table with the keel pointed straight towards the ceiling. This should result in the keel being superimposed directly over the thoracic vertebrae in the resulting imaging.[1] Both legs and wings should be extended equally to prevent rotation of the body. Wings should also be extended equally to prevent rotation of the body. When using tape to restrain, wing tape should touch the wing over the distal radius/ulna. Leg tape should contact the leg over the distal tarsometatarsus. The keel should be aligned with the + created with the light of the beam and the beam should be centered over the caudal portion of the keel. The entire coelomic cavity should be fit within a single image in all but the largest bird. Labeling with a metal R or L is of course part of standard procedure.

When positioning for a lateral view of the coelomic cavity, both wings should be fully extended dorsally. It is important to not superimpose the wings directly on top of each other. The down wing is commonly place more cranial to allow easy identification on the resulting image. Tape can again be placed over the distal radius/ulna. In large birds, a foam pad or wedge between the wings can prevent overextending wings. Legs should be extended ventrocaudally. Again, avoid superimposing legs by moving the down leg more cranially. Tape can again be placed around the distal tarsometatarsus. In most birds, the hips are narrower than the shoulders. This will cause most birds to rotate their hips towards the downside, twisting the coelom. Placing a small foam wedge under the hips will help prevent this rotation and make a well aligned image. To confirm correct positioning, the heads of the humerus should superimpose with each other as well as the head of the femur on the final image.

Lateral images of the wings are designed to have the beam base from the ventral aspect of the wing through the dorsal aspect. To position this view, place the patient on the table in dorsal recumbency. Extend the wing that you wish to view until it reaches the natural end of its range of motion. In this position, it is best to use tape passing over the metacarpus to make sure both the elbow and carpus are fully extended. In larger birds, tape can be added over the mid radius and ulna to spread the pressure of the tape out. Center the radiographic tube over the mid radius and ulna if you can fit the entire wing in the field. In larger birds, it may take two or three images to get the entire wing radiographed.

The caudal-cranial view of the wing is another often used position. This positioning starts with the patient in dorsal recumbency on the table. Feet should be towards the restrainer and the head should point away from the restrainer. While keeping the patient's head flat on the table, the restrainer grasps the legs and tail in a glove hand and raises them towards the ceiling. Be aware this technique has an increased risk of regurgitation. The other hand holds the primary feathers of the wing and extends the wing out from the body. The beam should pass from the caudal aspect of the wing through the cranial aspect. Be sure to avoid getting fingers in the view. The beam can be centered over the proximal radius and ulna. In large birds, multiple views may be required to get the entire wing imaged. If the image is properly positioned, the radius and ulna will be super imposed in the resulting image.

Lateral images of the legs can be accomplished by placing the patient in the same position as the lateral whole-body images. In the case of the legs, the down leg should be moved cranially (example, the right lateral view of the legs should have the right leg moved cranially). The toes should be spread out and taped in position to prevent superimposition. The beam should be centered over the proximal tibiotarsus and the entire leg from the femur to the toes should be included in the image if possible. Large birds or long-legged birds may require multiple images.

A ventrodorsal view of the synsacrum is achieved by placing the patient in dorsal recumbency. The legs are fully extended and taped in place similar to the positioning to the ventrodorsal images of the whole body. Digits should be spread out and taped in place to prevent superimposition. The beam should be centered over the proximal tibiotarsus and the entire leg from the femur to the toes should be included. In larger birds or long-legged birds, multiple images maybe required.

The H-view (oblique view, perhaps) is a recent addition to the toolkit of avian practitioners. It is used to examine the coracoid, clavicle, and scapula without these bones being super imposed upon one another. This image is acquired by placing the patient in the same position that would be used for a standard whole body ventrodorsal view. It is important that the patient's head and feet are aligned with the length of the table. This technique requires a machine that will allow for the head of the radiographic unit to rotate to a 30-45° angle. The exact angle depends on the species of bird being examined. This practitioner routinely starts at 30° for most patients and increases the angle if there is still superimposition of the bones of interest. Rotate the head of the unit to the desired angle and then move the unit caudally in relation to the patient. The beam should be aimed at the shoulders of the patient passing through it in a caudoventral-craniodorsal angle. As this beam is at an angle to the patient, traditional measurement for the determination of technique can be difficult. Technique set based on ventrodorsal measurements of the patient tend to result in overexposed images. This practitioner uses a ventro-dorsal measurement and decreases the technique by 1 centimeter in most cases.

Surgical correction of fractures in birds can be done with a variety of methods including external fixators (EXT), intramedullary pins (IM), EXF-IM tie-ins, ring fixators, and bone plates. Each of these methods has their own benefits and contraindications. External fixators are relatively easy to place as pins can be placed through incision or percutaneous. Reduction of the fracture can be achieved either close or open depending on the surgeon's experience and overall displacement of the distal segment. This type of fixation is ideal for controlling rotation and compressional forces. It also provides adequate stabilization against bending forces. It is also possible to combine this method with an intramedullary pin. By tying the two fixation methods together, this provided increased stabilization.

Intramedullary pins are another common method of fracture repair in birds. The placement of this fixation is more difficult, as the surgeon must plan where the pin will exit the bone. In some bones, such as the ulna or humerus, the anatomy of the bones allows the pin to exit without interfering with a joint surface. In other bones, such as the metacarpus or tarsometatarsus, it is not possible to place an IM pin without damaging the joint. Place of the pin is commonly done by entering the fracture site through an incision and retrograding the pin into place. This type of fixation method is superior at stabilization against bending forces but has minimal affect to rotational or compressional forces.

In both surgical and conservative management of fracture, appropriate follow up care is key to recover. Patients should have limited activity and use of the affected limb until evidence of bony healing is present on radiographs. Avian bone healing is often quicker than mammalian bone healing. This author often observes bridging bony callus formation on radiographs 4 weeks after the injury. Ideally, return to normal activity is delayed until there is evidence of such healing. Once stabilized, pain from a fracture is significantly reduced. In most cases, this author continues multimodal analgesic support for 1 week after a surgical procedure. After that, the analgesic plan is decreased to an NSAID alone, usually for another week. Injuries to the pelvic limb can cause offloading of weight of the injuried limb and potentially cause pododermatitis on the contralateral limb. Both feet should be monitored for development of pododermatitis and changes in substrate, perching materials, or use of protective bandaging should be used if it is noted.

Regarding injuries of the wing, it is important to prevent contraction of the patagial membrane or loss of range of motion of the joints of the wing. This author's preferred protocol is passive range of motion (PROM) therapy performed with either sedation or anesthesia depending on the patient's temperament. Warm packing the affected joints or tissue for 5-10 minutes prior to start of a PROM session can be of value as well. While sedated or anesthetized, the bandage is removed and measurements of ROM of both elbow and carpus are taken prior to start of the session. Then the wing is gently extended and flexed, with particular attention paid to achieving full extensions. This may require the practitioner to apply gentle pressure during this process. It is important for the practitioner to support any areas that still are unstable during this process. This process is continued for 5 minutes or until full ROM is achieved. The bandaging can then be replaced and the patient returned to the enclosure. Patients with extensive contracture may require additional analgesia after PROM, in which case this author will usually resume use of along acting opioid for 24 hours. These physiotherapy sessions are performed up to 3 times a week based on patient needs and tolerance of the procedure.

Degenerative Joint Disease

Degenerative joint disease is commonly seen in older psittacine patients but can be seen in all species and younger birds when there is previous trauma or injury to joints. Clinical signs depending on the limb affected, but include lameness, loss of flight, loss of range of motion of joint, falling or difficulty ambulating, and lethargy. In some cases, this author's patients have demonstrated feather destructive behavior around the affected joints. Diagnosis is made by the characteristic radiographic lesions of sclerosis of subchondral bone, deceased joint space, and osteophyte formation.

Treatment includes analgesic support and husbandry changes. The most common analgesic used with degenerative joint disease is meloxicam due to the ability to administer orally at home and the relative safety of prolonged use. In addition, joint supplements such omega-3 fatty acids and glucosamine have been reported to improve clinical signs. Husbandry changes will include wider perches to allow easier stability and reducing access to vertical height to minimize the chance of falls.

Developmental Abnormality

Perosis, commonly called slipped tendon, is the most common developmental abnormality seen in young growing birds. This is commonly seen in poultry, waterfowl, cranes, ratites, and other long-legged birds. The cause of this deformity is due to cartilage deformity in the hock joint leading to medial luxation of the gastrocnemius tendon. This results in swelling of the hock joint, lateral deviation of the foot of the affected leg, and lameness in the affected leg. The cause is associated with diets high in protein, manganese deficiency, choline deficiency, and biotin deficiency. Treatment of this condition primarily involves surgical fixation of the tendon in the normal anatomical position. This type of intervention is best applied as early as possible to minimize any deformation of the tibiotarsus or tarsometatarsus. In addition, supplementing manganese, choline, and biotin along with reducing the protein content of the diet is recommended.

References

1. Darrow B, Bennett RA. Avian Orthopedics. In: Bennet RA, Pye GW, eds. *Surgery of Exotic Animals.* Hoboken, NJ: Wiley-Blackwell; 2022:112-153.

2. Calvo Carrasco D. Fracture management in avian species. *Vet Clin North Am Exot Anim Pract.* 2019;22: 223-238.

3. Hawkins H, Guzman D, Beaufrere H, et al. Birds. In: Carpenter JW, ed. *Exotic Animal Formulary.* 5th Ed. St. Louis, MO: Elsevier; 2018:167-375.

4. Guzman DSM, Ceulemans SM, Beaufrère H, et al. Evaluation of the thermal antinociceptive effects of a sustained-release buprenorphine formulation after intramuscular administration to American kestrels (*Falco sparverius*). *J Avian Med Surg.* 2018;32:1-7.

5. Harcourt-Brown NH. Orthopedic conditions that affect the avian pelvic limb. *Vet Clin North Am Exot Anim Pract.* 2002;5:49-81.

6. Hoybers Y, Bosmans T, Risselada M, et al. General anesthesia for the surgical repair of a tarsometatarsal fracture in a Harris's hawk (*Parabuteo unicinctus*). *Vlaams Diergeneeskd Tijdschr* 2008;77:309-3014.

7. Longley L. Avian anaesthesia. In: *Anaesthesia of Exotic Pets.* Elsevier, 2008;129–170.

8. Taiyari H, Abu J. External skeletal fixator to stabilize the orthopedic conditions in avian species: a systematic review of case reports and case series. *Pertanika J Sci Technol.* 2022;30:1815-1839.

9. Hatt JM, Christen C, Sandmeier P. Clinical application of an external fixator in the repair of bone fractures in 28 birds. *Vet Rec.* 2007;160:188-194.

10. Yoon H-Y, Fox DB, Jeong S-W. Tibiotarsal and ulnar fracture repair in a great horned owl (*Bubo virginianus*). *J Vet Clin.* 2008;25:218-220.

11. Naldo J, Saggese M. Radiography. In: Samour J, ed. *Avian Medicine.* 3rd Ed. St. Louis, MO: Elsevier; 2016:130-153.

12. Krautwald-Junghanns M, Pees M. Imaging techniques. In: Tully TN, Dorrestein GM, Jones AK, Cooper JE, eds. *Handbook of Avian Medicine.* St. Louis, MO: Elsevier; 2009:85-100.

13. Krautwald-Junghanns M. In: Coles B, ed. *Essentials of Avian Medicine and Surgery.* 2nd Ed. Ames, IA: Blackwell;1997:56-102

14. Silverman S, Tell L. *Radiology of Birds: An Atlas of Normal Anatomy and Positioning.* St. Louis, MO: Elsevier Saunders; 2010.

Nutrition of Psittacine Birds for the Avian Practitioner

Stephanie Lamb, DVM, DABVP (Avian Practice)

Session #3138

Affiliation: From Arizona Exotic Animal Hospital, 744 N Center St #101, Mesa, AZ 85201, USA.

Abstract: Optimal nutrition is critical to maintain health and well-being. Currently dietary recommendations for psittacine birds are derived from research in other species, several studies on specific nutrients in individual psittacine species and first-hand experience in aviculture. This lecture is geared towards providing the veterinary practitioner with the basics for understanding psittacine nutrition using available scientific literature.

Introduction

To determine if a diet is healthy and appropriate for a psittacine bird one must look at the nutrients that are required by the individual. Macronutrients, including proteins, fats and carbohydrates, must be evaluated. Micronutrients including vitamins and minerals must also be critiqued. Not all micronutrients have been studied in psittacine birds; however, there is information available on certain vitamins and minerals that can help guide our understanding of nutrition in psittacine birds and allow us to make appropriate recommendations for certain species. All nutrients must be provided within certain ranges or levels in the diet to provide adequate nutrition.

Water

The most important nutrient that is often forgotten is water. It is required for many functions including maintenance of appropriate cellular structure, transport nutrients throughout the vascular system, provide cushioning to joints, maintain body temperatures, and encourage normal passage of excretory products from the body. Generally, younger animals, larger birds, and those living in warmer environments have higher water needs. Studies in parrots show that they require 2.4% of their body weight per day as water when they are under thermoneutral conditions. If they are in hot environments though this can increase by 12 times to account for evaporative losses of water.[1,2]

Energy

Energy requirements of the individual bird can be determined by doing simple mathematical calculations. The basal metabolic rate (BMR), also known as the resting energy requirement (RER), is calculated according to the following equation.[1]

$$\text{BMR/RER} = 73.6 \times \text{BW (kg)}^{0.73} \text{ OR } 0.697 \times \text{BW (g)}^{0.705}$$

Once the BMR is established, this value is then multiplied by a factor based on the bird's energy needs for its activity level, health status, or age to calculate the metabolic energy requirement (Table 1).[1]

Table 1. Metabolic energy requirements of birds at different health status, activity level, and life stage.

Health status, activity level, life stage	Multiplication factor
Healthy juvenile	BMR × 2-3
Sick juvenile	BMR × 3-4
Perching, resting	BMR × 1
Perching, alert	BMR × 1.9-2.1
Preening	BMR × 1.6-2.3
Singing, grooming, bathing	BMR × 2.9
Mild stress	BMR × 1.2
Moderate stress	BMR × 1.5
Major stress, sepsis	BMR × 1
Gliding flight	BMR × 2
Sustained flight	BMR × 11
Aerial flight	BMR × 2.7-5.7

An example of this can be illustrated by calculating the energy requirements of an adult, female African grey parrot that is not laying eggs but is overweight. The bird presents with a weight of 450 grams and a body condition score of 7/9. The bird's BMR is calculated as $78 \times (0.45)^{0.75} = 42.8$ kcal. Since the parrot is overweight the perching resting rate (BMR × 1) or mild stress rate (BMR × 1.2) can be used to determine metabolic energy requirement. For this patient that would be a metabolic energy requirement that equals 42.8-51.4 kcal of food per day. The caloric content of the diet offered is then calculated to determine the volume of food that is needed to meet caloric needs.

Macronutrients

Carbohydrates

Carbohydrates are important as energy sources, function for storage within cells and provide cellular structure. They are also known as sugars and are made up of monomers, known as monosaccharides, that combine to make larger polymers, known as disaccharides and polysaccharides. Examples of monosaccharides are glucose, galactose and fructose. When these combine to make larger molecules the different linkages between the monomers create the overall larger structural form. The two main types of linkages are alpha and beta linkages and different enzymes are required to break these links down to release individual monomer units. Beta linkages are often broken down by microbes that can be present in the gastrointestinal tract, particularly in species that have a cecum or ceca. Psittacine birds have either absent or vestigial ceca which means they do little to no hind gut fermentation of carbohydrates with beta linkages. Absorption of carbohydrates occurs across intestinal microvilli and some frugivorous and nectivorous species have exceptionally long intestinal microvilli to aid in the absorption of free sugars found in their food.[2]

Protein

A wide range of dietary protein levels, from 6-31%, has been described for psittacine birds.[1-6] This is a wide range and considering the needs of individual species of birds, studies have shown variability. When sourced from high-quality ingredients, 11% dietary protein was adequate for maintenance in adult cockatiels.[1] Free-ranging cockatiels will select seeds containing 8.8-14% protein.[7] Interestingly, a nectarivore like the lorikeet can be maintained on 3% dietary protein, but the source of protein must be a good-quality source.[1] Experiments in African grey parrots indicate dietary protein must be 10-15% to maintain plasma protein levels.[2]

Studies in cockatiel chicks have offered variable levels of protein with different outcomes. Chicks were severely stunted when given only 5% dietary protein. When fed 10-15% dietary protein, chicks demonstrated slow growth and were slightly stunted. Optimal growth was associated with 20% protein with 0.8% lysine; however, when dietary protein levels reached 25%, the chicks continued to grow well but exhibited aggression. Even higher levels of 35% protein caused growth to slow and continued aggression.[8,9] It has been theorized that larger psittacine birds may require higher protein levels during growth. In one study evaluating the crop contents of wild scarlet macaw chicks being parent fed, 18-26% protein was present.[4] In studies where Amazon parrots were given free choice food selection, they chose diets that contained about 15.5% protein; however, it was not determined if this was a healthy level.[5]

Protein malnutrition in the bird can manifest as weight loss, immunodeficiency, and a loss of cell integrity and feather structure. Essential amino acids found in keratin include histidine, isoleucine, lysine, methionine, threonine, tryptophan and valine. A deficiency in any one of these may result in poor feather condition. Methionine deficiency during growth or molting results in dark, horizontal stress marks on the feathers. In addition, this amino acid deficiency can lead to a loss of structural proteins in feathers that help make the blue coloration observed. With a loss of this structural protein those feathers that are green lose their blue structural component and appear yellow.[10] A deficiency of this particular protein is also one of the components that can lead to fatty liver syndrome.[11]

Excess dietary protein has been studied. Many have had concerns that too much protein could lead to gout. A relationship between excess dietary protein and gout has been proposed in psittacine birds; however, research does not support this assumption. When African grey parrots were offered 30% dietary protein, this level was not associated with an increase in uric acid levels.[3] In cockatiels fed 11%, 20%, 35% and even 70% dietary protein, uric acid levels did increase as protein levels increased but gout was not induced. This is likely because of the increased breakdown of protein into the nitrogenous waste product, uric acid but it does not necessarily correlate with kidney damage. These studies also showed an increase in metabolic enzymes, ALT and AST, as protein levels increased but this is likely because of the increased demand for processing the protein that was entering the body. One negative finding in these studies was that the incidence of lipogranulomas within the liver increased as protein levels increased.[7]

Lipids

Published ranges for dietary fat levels in the diets of psittacine birds are between 4-20% showing there is quite a bit of variability; however, for maintenance purposes many sources recommend the levels be lower around 4-5%.[1,4,6,9] Studies show that wild living birds and birds fed free choice select much higher levels of dietary fat. Studies with scarlet macaws show that parents will feed their young food items containing 21-31% fat.[4] When fed free choice, captive Amazon parrots will select food items that contain 20.6% fat.[5] The benefits or harms of these levels of fats were not evaluated in these observational studies and further investigation is needed.

There are essential fatty acids that need to be in bird diets. Although birds possess desaturase enzymes that can form double bonds at the delta 5, 6 and 9 positions of fatty acid chains, they do require linoleic acid (n-6) and linolenic acid (n-3) in the diet.[2,6,12,13] There are some studies in psittacine birds that support the benefits of omega-3 fatty acid supplementation.[14-16] For instance, a custom-formulated diet containing 3% fish oil led to a reduction in triglyceride and cholesterol levels.[14] In addition, a postmortem study comparing the incidence of atherosclerosis with linolenic acid levels in breast and adipose tissue found that birds without atherosclerosis had higher tissue levels of linolenic acid.[15]

Micronutrients

Vitamins

Many sources recommend a level of vitamin A between 3,000-8,000 IU/kg of feed.[13,17] Adult cockatiels given 2,000 IU vitamin A per kg feed did not have any issues but when given 10,000 IU/kg the birds displayed hyper-excitability, modified vocalizations and an impaired antibody response. Higher vitamin A levels (100,000 IU/kg) also led to more vocalizations and impaired antibody response as well as deteriorating feather quality and pancreatitis.[17] Lorikeets fed 20,000 to 45,000 IU/kg of vitamin A in the diet developed liver damage, skin and feather changes, and splenic hemosiderosis.[18] Hypovitaminosis A has been well described in psittacine birds with the development of squamous metaplasia occurring in epithelium of the respiratory, renal and reproductive systems when the nutrient is chronically under supplemented.[10] A study in cockatiels suggested that vitamin A can be stored in the liver for up to 2 years, therefore adverse effects are more profound in young, growing birds of those with severe nutritional deficiencies.[17]

Recommended dietary vitamin D levels in bird diets range from 500 to 2,000 IU/kg of the diet.[10,13] Budgerigars have specifically been suggested to receive 10 IU per day according to one source.[3] Vitamin D deficiencies will lead to thin-shelled eggs, poor reproductive success and rickets. Excesses can be problematic as well with anorexia, lameness, lethargy, diarrhea, polyuria and metastatic mineralization seen.[10] A study looking at various levels of calcium and vitamin D in budgerigars diets found that metastatic mineralization was more dependent upon calcium levels and not effected by vitamin D levels. They showed that vitamin D could range from 500 to 2,000 IU/kg of the diet and may be able to go up to 3,300 IU/kg of the diet without adverse effects.[19] Growing macaws are thought to be more susceptible to hypervitaminosis D and historically there have been clinical reports of nutritionally imbalanced batches of commercial hand-feeding formulas containing excess levels of vitamin D and causing pathology in birds. In one report, a juvenile cockatoo and a blue and gold macaw chick fed vitamin D at 26,790 IU/kg feed developed hypercalcemia, hyperphosphatemia, hyperuricemia and renomegaly.[20]

Vitamin E is poorly studied in psittacine birds; however, suggested dietary levels range between 11-50 ppm in the diet.[6,13] Deficiencies in vitamin E can lead to encephalomalacia, exudative diathesis, muscular dystrophy, and poor reproduction.[6,9,10] Vitamin E levels that exceed requirement levels 100 fold can manifest as decreased growth, anemia, decreased bone mineralization, as well as reduced storage of vitamin A within the liver.

Vitamin K is also poorly studied in psittacine birds. Approximately 1 ppm has been suggested in the diet.[13] Deficiencies can lead to clotting disorders and skeletal abnormalities.[6] Excesses of this nutrient are said to need to be 1000 times the requirement to manifest as problems, such as anemia and interference with vitamins B and C.

Several of the B vitamins, including B_1, B_3, B_6, B_7, B_9 and choline, have not been well studied in psittacine birds. For these nutrients, we extrapolate information from other birds like chickens to supply us with information on what is believed to be appropriate. Vitamin B_1 (thiamin) is thought to be needed at 2-4 ppm in diet and deficiencies could result in polyneuritis, polioencephalomalacia, paresis, paralysis, opisthotonos, and convulsions.[9,10,13] Vitamin B_3 (niacin) is suggested to be at 39-50 ppm in the diet and deficiencies can lead to scaley dermatitis and hock disorders in chickens.[9,10,13] Vitamin B_6 (pyridoxine) is required at 4-6 ppm in the diet of chickens and if supplied lower than this it could cause a reduced appetite, convulsions, dermatitis and anemia.[9,10,13] Excesses have been reported to cause problems in other birds including gyrfalcons having green urates, regurgitation, hyporexia and death within 24-36 hours and pigeons having liver degeneration and death with 24-48 hours of getting excessive levels.[21] Vitamin B_7 (biotin) is thought to be needed at 0.25 ppm in the diet and if it is at lower levels problems like perosis, fatty liver and kidney syndrome, dermatitis, swollen eyelids, and mandibular swellings can occur in those species it has been studied in.[10,13] Vitamin B_9 (folic acid) is suggested to be at 1-1.5 ppm in the diet and deficiencies can result in anemia, slow growth, poor feathering, and reduced egg production and hatchability.[9,10,13] Choline (sometimes referred to as vitamin B_4) is thought to be required at 1500 ppm of the diet and deficiencies could play a role in liver disease, atherosclerosis, neurologic disorders and in cockatiels unpigmented feathers.[10,13] All of these nutrients do need further investigation in psittacine birds.

There are studies on vitamin B_2 (riboflavin) in cockatiels. If birds lack this nutrient in the diet they can develop achromatosis of the feathers or a failure of development of the normal pigment in the feathers. Cockatiels will also have a reduced number of eggs in the clutch and reduced hatchability. [6] This nutrient is supposed to be at 4-6 ppm in the diet.[9,13]

Vitamin B_5 (pantothenic acid) also has some limited studies in cockatiels. When this nutrient was not available, cockatiel chicks would die 7-10 days post hatching. When fed low levels of pantothenic acid, chicks would die within 3-3.5 weeks and display an absence of body feathers but wing and tail feathers were present.[6] This nutrient is suggested to be at 11-20 ppm in the diet.[9,13]

Vitamin B_{12} (cyanocobalamin) is suggested to be at 0.01 ppm in the diet.[13] There has been documentation of a macaw developing anemia when it was not receiving adequate levels of this nutrient.[22]

Vitamin C is not required in the diet of psittacine birds because they are able to synthesize this nutrient on their own.[6,9] It is produced within the kidneys of psittacine birds and therefore, may be required in birds suffering from kidney disease (Orosz, personal communication). Addition of this nutrient to the diet must be done with caution in those species who have a propensity towards the development of iron storage disease as vitamin C will enhance the absorption of iron.[1]

Minerals

Several studies have evaluated the calcium needs of different psittacine birds. Recommended adult maintenance levels of this nutrient range from 0.3-1.1%.[4,6,9,13] In cockatiels, 0.35% dietary calcium was found to be adequate for maintaining appropriate shell thickness and egg production.[6] A study evaluating various levels of calcium and vitamin D in budgerigar diets found that metastatic mineralization was more dependent upon calcium levels and not affected by vitamin D. Although both nestling and adult budgerigars did well with 0.3% dietary calcium, concentrations of 0.7% and 1.5% calcium were toxic. The most severe metastatic mineralization lesions were seen in chicks that died spontaneously.[19]

Maintenance dietary phosphorous levels range between 0.3-0.4% of the diet.[13] The budgerigars in the aforementioned study were fed 0.7% dietary phosphorous. This level was well tolerated with 0.3% calcium.[19] Deficiencies in phosphorous can result in rickets, weakness, thin eggshells, and a depressed appetite.[10]

Recommended dietary magnesium levels range between 500-667 ppm in the diet.[9,13] Low levels of this nutrient can result in poor egg production or result in neuromuscular disorders.[10] It has been suggested that seizures occur in African grey parrots with hypomagnesemia; however, a case report investigating this problem looked at plasma levels of magnesium and not dietary nutrient levels.[23]

There are several minerals that have been poorly studied in psittacine birds and recommendations for these nutrients are based on data for other granivores. Potassium, sodium, chloride, manganese, zinc, copper, selenium, iron, and iodine recommendations are in need of more studies in psittacine birds. Potassium is suggested to be at 0.3-0.4% of the diet and deficiencies could result in retarded growth, reduced egg production with thin shells and muscle weakness.[9,10,13] Sodium is recommended at 0.1-0.15% of the diet and low levels could result in decreased egg production, poor growth, softened bones, decreased cardiac output, decreased plasma volume and decreased gonadal inactivity.[6,9,10,13] Chloride is thought to need to be at 0.1-0.15% of the diet. Deficiencies could lead to poor growth, dehydration, hemoconcentration and mortality.[6,9,10,13] Manganese is recommended to be at 60-75 ppm in the diet and deficiencies in this could lead to decreased egg production and shell strength and in chickens it is one of the nutrient deficiencies that leads to perosis.[6,9,13] Zinc is suggested to be at 40-50 ppm in the diet and deficiencies could result in retarded growth, reduced egg production and hatchability, frayed feathers and poor wound healing.[6,9,10,13] Copper is suggested to be in the diet at a level of 6-8 ppm and deficiencies can result in anemia, cardiac hypertrophy, bone deformities, and dissecting aneurysms of the aorta in waterfowl.[6,9,13] Excesses of copper could result in kidney or neurologic disorders. Selenium is believed to be required at 0.1 ppm and deficiencies can result in poor growth and chick mortality, white muscle disease, liver necrosis and pancreatic fibrosis.[6,9,10,13,24] Excesses of this nutrient can cause impaired hatching success.[10]

Recommended dietary iron levels are extrapolated from what is known in other species and ranges from 60-80 ppm.[9,13] This particular nutrient bears important consideration for those individuals or species that are prone to iron storage disease. Granivorous psittacine birds are not generally susceptible to iron storage disease, but disease has been documented in lorikeets, blue and gold macaws and African grey parrots.[1,25,26,27] Based on what is known for other species it is recommended that those individuals who are prone to this disease do not get more than 50 ppm of iron in the diet.[1] Deficiencies in iron can result in anemia.[10]

Suggested dietary iodine levels range between 0.3-0.4 ppm; and deficiencies can result in low thyroid levels, goiters, and poor growth.[9,10,13] Budgerigars on iodine-deficient diets have been found to develop goiters.[28]

Dietary Recommendations

Although many nutrients have not been specifically studied in psittacine birds, based on our current level of understanding, the general recommendations extrapolated from other granivorous species seem to meet the needs of psittacine birds. The formulated diets currently on the market for psittacine birds do meet the requirements set forth for other granivores. Therefore, formulated diets for psittacine birds are generally recommended as the base of the psittacine diet. The ratio of formulated diets in relation to other food items is debated with suggestions ranging from 60-80% of the diet. The remainder of the diet should consist of fresh items, like vegetables, fruits, grains, beans, nuts and seeds.[1] Some aviculturists are proponents of all-natural diets that do not contain formulated foods but balancing these diets can be tricky and difficult for the average owner. In addition, birds can preferentially select items out of a mix, which can create nutritional imbalances.[5]

The eclectus parrot is thought to need a higher level of vitamin A in the diet and some anecdotal reports suggest this species may be more sensitive to high levels of pellets or formulated foods in the diet. Color variety cockatiels, budgerigars, lovebirds, and parrotlets may also be sensitive to too high proportions of pellets in the diet as well and have been seen anecdotally to develop kidney disease.[20] For them it is recommended that they receive no more than 50-60% of their diet as pellets. These observations are from aviculture and in-depth studies are needed to further specify the legitimacy of these concerns.

An all-seed diet has been historically used for captive psittacine birds due to observations of what they consume in the wild; however, these diets are not currently recommended as they can contain excessive fat, low levels of calcium, phosphorus, sodium, zinc, iron, lysine, and vitamin A.[5] Problems noted in previous sections regarding these nutrients have been seen in psittacine birds kept on all seed diets.

Nutritional Strategies

There are certain differences among species that must be considered because a variety of nutritional strategies are employed by psittacine birds. Generally, most psittacine birds are classified as florivores. These are animals that consume a plant-based diet. Among these, there are granivores that consume grains, frugivores which consume fruit, and nectarivores, which consume nectars. Some birds are classified as exhibiting a mix of these nutritional strategies. There are also some omnivorous psittacine birds.[2] Many macaws, queen of Bavaria conures and Palm cockatoos are believed to require higher levels of dietary fat and fiber.[29] Lories and lorikeets are specialists that consume nectar and pollen along with some seeds, fruits, buds and insects.[1] When compared to other psittacine birds, lorikeets have lower protein needs and possibility lower vitamin A requirements as well.[1,18] Aviculturists have said African grey parrots need a higher protein and fat diet, although good scientific references for this do not exist. African grey parrots in the wild have been found to forage from 14 different tree species and will consume different parts of these trees. They have been seen to eat the seeds and arils of four of these tree species, the fruits of seven to nine of these trees, the flowers of two of these trees, and the buds of one of these trees.[30] They also have been observed to eat the vegetative parts of an aquatic plant of the *Lemma* species.[31,32] Some of the birds classified as omnivores include the Eastern rosella, Major Mitchell cockatoo, and the sulphur-crested cockatoo. These individuals still consume seeds, fruits, and flowers like other birds but they also will consume insects.[2] It is important to note that although these may be the items birds are consuming in the wild, a wild life and a captive life are vastly different in terms of energy expenditure and risks to the birds so caution must be exercised when looking to the wild diet to dictate the captive diet.

Additional psittacine species have been studied in the wild to observe what food items they are selecting. Hyacinths consume acuri and bocaiuva palm nuts. *Calyptorhyncus* genus cockatoos eat seeds from the cones of sheoak and bulloak genera coniferous trees.[29] Austral parakeets have been observed to eat pollen, seeds, honeydew, lichens, and parasitic mistletoe.[29] Cockatiels have been documented to eat 17 species of native grasses and herbaceous dwelling plants in their native environment along with imported cereal grains.[29]

Nutritional Additives

Understanding of the importance of nutritional additives and supplements, such as prebiotics and probiotics, is poor in psittacine nutrition at the time of this writing; however, it is important to at least acknowledge that there are prebiotics, probiotics, and dietary textures that contribute to an individual's nutrient basis and health status. Knowledge about the gut microbiome is a hot topic in human medicine and surely its importance will likely be further elucidated in time with birds as well.

Conclusions

Further work is needed to investigate various aspects of nutrition for different species of psittacine birds. With time, more knowledge of individual nutrients in different species of psittacine birds will be gained, which will help practitioners.

References

1. Koutsos E, Gelis S, Echols MS. Advances in nutrition and nutritional therapy. In: Speer BL, ed. *Current Therapy in Avian Medicine and Surgery.* St. Louis, MO: Elsevier; 2016:142-176.

2. Koutsos EA, Matson KD, Klasing KC. Nutrition of birds in the order Psittaciformes: A review. *J Avian Med Surg*. 2001;15(4):257-275.

3. Harper EJ, Skinner ND. Clinical nutrition of small psittacines and passerine. *Sem Exot Pet Med.* 1998;7: 116-127.

4. Brightsmith DJ, Mcdonald D, Matsafuji D, Bailey CA. Nutritional content of the diets of free-living scarlet macaw chicks in southeastern Peru. *J Avian Med Surg*. 2010;24:9-23.

5. Brightsmith DJ. Nutritional levels of diets fed to captive Amazon parrots: does mixing seed, produce, and pellets provide a healthy diet? *J Avian Med Surg*. 2012;26:149-160.

6. Roudybush TE. Nutrition. In: Altmann RB, Clubb SL, Dorrestein GM, Quesenberry K, eds. *Avian Medicine and Surgery.* Philadelphia, PA: WB Saunders; 1997:27-53.

7. Koutsos EA, Smith J, Woods LW, Klasing KC. Adult cockatiels (*Nymphicus hollandicus*) metabolically adapt to high protein diets. *J Nutr*. 2001;131:2014-2020.

8. Roudybush TS. Food and water interrelations and the protein requirements for growth of an altricial bird, the cockatiel (*Nymphicus hollandicus*). *J Nutr*. 1986;116:552-559.

9. Brue RN. Nutrition. In: Ritchie BW, Harrison GJ, Harrison LR, eds. *Avian Medicine: Principles and Application.* Lake Worth, FL: Wingers Publishing Inc; 1994:63-95.

10. Macwhirter P. Malnutrition. In: Ritchie BW, Harrison GJ, Harrison LR, eds. *Avian Medicine: Principles and Application.* Lake Worth, FL: Wingers Publishing Inc; 1994:842-861.

11. Lumeij JT. Hepatology. In: Ritchie BW, Harrison GJ, Harrison LR, eds. *Avian Medicine: Principles and Application.* Lake Worth, FL: Wingers Publishing Inc; 1994:522-537.

12. Strunk A. Psittacine nutrition: focus on lipid metabolism and obesity. *Proc Annu Conf Am Board Vet Pract.* 2012.

13. Kollias GV, Kollias HW. Feeding passerine and psittacine birds. In: Hand MS, Thatcher CD, Remillard RL, Roudebush P, eds. *Small Animal Clinical Nutrition,* 4th Ed. Marceline, MO: Walsworth Publishing Co; 2000:979-991.

14. Heinze CR, Hawkins MG, Gillies LA, et al. Effect of dietary omega-3 fatty acids on red blood cell lipid composition and plasma metabolites in the cockatiel, *Nymphicus hollandicus*. *J Anim Sci*. 2012;90: 3068-3079.

15. Bavelaar FJ, Beynen AC. Severity of atherosclerosis in parrots in relation to the intake of alpha-linolenic acid. *Avian Dis*. 2003;47:566-577.

16. Petzinger C, Hetley JJ, Cornejo J, et al. Dietary modification of omega-3 fatty acids for birds with atherosclerosis. *J Am Vet Med Assoc*. 2010;236:523-528.

17. Koutsos EA, Tell LA, Woods LW, Klasing KC. Adult cockatiels (*Nymphicus hollandicus*) at maintenance are more sensitive to diets containing excess vitamin A than to vitamin A deficient diets. *J Nutr*. 2003;133: 1898-1902.

18. Park F. Vitamin A toxicosis in a lorikeet flock. *Vet Clin North Am Exot Anim Pract*. 2006;9:495-502.

19. Phalen DN. Renal failure associated with metastatic mineralization of the kidneys in birds. *Proc Annu Conf Assoc Avian Vet*. 2009.

20. Echols MS. Avian renal disease: part I and II. *Proc Annu Conf Am Board Vet Pract*. 2012.

21. Samour J, Perlman J, Kinne J, et al. Vitamin B_6 (pyridoxine hydrochloride) toxicosis in falcons. *J Zoo Wildl Med*. 2016;47:601-608.

22. Tully TN, Beaufrere H, Sirninger JR, et al. Suspected diet-related vitamin B_{12} anemia in a hyacinth macaw (*Andorhynchus hyacinthinus*) *Proc Annu Conf Assoc Avian Vet*. 2010:317-320.

23. Kirchgessner MS, Tully TN, Nevarez J, et al. Magnesium therapy in a hypocalcemia African grey parrot (*Psittacus erithacus*). *J Avian Med Surg*. 2012;26:17-21.

24. Schmidt RE, Reavill DR, Phalen DN. *Pathology of Pet and Aviary Birds*, 2nd Ed. Ames, IA: Wiley Blackwell; 2016.

25. Rasidi EK, Xie S, Hsu C. Iron storage disease in a blue and gold macaw (*Ara ararauna*). *Proc ExoticsCon*. 2019;384.

26. Lamb S, Reavill D, Biswell E. Pathology in practice: hepatic iron overload in a Congo African grey parrot (*Psittacus erithacus*). *J Am Vet Med Assoc*. 2022;259:S2. https://doi.org/10.2460/javma.21.04.0209.

27. O'Connor MR, Garner MM. Iron storage disease in African grey parrots (*Psittacus erithacus*) exposed to a carnivorous diet. *J Zoo Wildl Med*. 2018;49:172-177.

28. Loukopoulos P, Bautista AC, Puschner B, et al. An outbreak of thyroid hyperplasia (goiter) with high mortality in budgerigars (*Melopsittacus undulatus*). *J Vet Diagn Invest*. 2015;27:18-24.

29. Toft CA, Wright TF. *Parrots of the Wild*. Oakland, CA: University of California Press; 2015.

30. Tamungang SA, Ajayi SS. Diversity of food of the grey parrot (*Psittacus erithacus*) in Korup National Park, Cameroon. *Bull African Bird Club*. 2003;10:33-36.

31. Fotso R. Survey status of the distribution and utilization of the grey parrot in Cameroon. Secretariat CITES; 1998.

32. May DL. The behavior of African grey parrots in the rainforest of the Central African Republic. *PsittaScene*. 1996;8:8-9.

Section 2

AAV Plenary Session

Anna Osofsky, DVM, DABVP (Avian Practice), *AAV President*
Session Chair

Use of Intravenous Lipid Emulsion Therapy for the Treatment of Acute Poisonings in Birds

Lauren V. Powers, DVM, DABVP (Avian Practice), DABVP (Exotic Companion Mammal Practice), and
Laura A. Stern, DVM, DABVT

Session #3436

Affiliation: From Avian and Exotic Pet Service, Carolina Veterinary Specialists, 12117 Statesville Road, Huntersville, NC 28078, USA (Powers), The ASPCA Animal Poison Control Center, 1806A Fox Drive, Champaign, IL 61820, USA (Stern).

Note: Abstract format selected by author(s). See Front Matter for description of conference proceedings manuscripts submitted as scientific abstracts.

Abstract: A 4-year-old male white bellied caique (*Pionites leucogaster*) was presented for cannabis (water pipe resin) ingestion that was observed 2 hours prior. On presentation the bird was stuporous and hyperesthetic. The bird began having seizures responsive to intramuscular midazolam approximately 7 hours after admission. The bird's condition progressively worsened over the following 3 hours although bradycardia was not observed. An intravenous catheter was placed and intravenous lipid emulsion (ILE; Intralipid 20%, Baxter Healthcare, Deerfield, IL, USA) was administered at 16 ml/kg/h with an estimated total volume of 62 ml/kg. A rapid favorable response to treatment was observed. Monitoring for lipemia was not performed due to patient size. The bird was discharged the following day and remains well at the time of this writing.

ILE has been used to treat lipophilic toxicoses in a variety of animal species including dogs and cats,[2-5,14] horses,[6] inland bearded dragons,[7] rabbits,[8] guinea pigs,[9] and several avian species.[10-12] ILE has previously been used as a successful therapy for tetrahydrocannabinol (THC) toxicosis in dogs and humans.[1,2,14] From 2018-2022, the ASPCA Animal Poison Control center received 38 calls regarding birds (excluding poultry and waterfowl) exposed to cannabis, representing 1.8% of total bird cases.[15]

ILE has occasionally been associated with significant adverse effects,[3-5] including a dog with THC toxicosis that developed acute respiratory distress syndrome (ARDS).[2] The comparatively high dosage of ILE used in this caique (62 ml/kg vs. 9.5-20.1 ml/kg suggested for dogs[16-19] and 16.5-30 ml/kg used in other avian species[10,11]) was well-tolerated in this bird.

References

1. Aksel G, Güneysel Ö, Taşyürek T, et al. Intravenous lipid emulsion therapy for acute synthetic cannabinoid intoxication: Clinical experience in four cases. *Case Rep Emerg Med.* 2015:1-5.

2. Botha H, Jennings SH, Press SA, et al. Suspected acute respiratory distress syndrome associated with the use of intravenous lipid emulsion therapy in a dog: a case report. *Front Vet Sci.* 2019;6:1-6.

3. Romans CW, Day TK, Smith JJ. Oxidative red blood cell damage associated with propofol and intravenous lipid emulsion therapy in a dog treated for 5-fluorouracil toxicosis. *J Vet Emerg Crit Care.* 2020;30(4): 481-486.

4. Seitz MA, Burkitt-Creedon JM. Persistent gross lipemia and suspected corneal lipidosis following intravenous lipid therapy in a cat with permethrin toxicosis. *J Vet Emerg Crit Care.* 2016;26(6):804-808.

5. Epstein SE, Hopper K, Farrell KS. Manual plasma exchange to treat an accidental overdose of intravenous lipid emulsion in a dog with baclofen toxicosis. *J Am Vet Med Assoc.* 2022;260(6):650-656.

6. Vieitez V, Gómez de Segura IÁ, Martin-Cuervo M, et al. Successful use of lipid emulsion to resuscitate a foal after intravenous lidocaine induced cardiovascular collapse. *Equine Vet J.* 2017;49(6):767-769.

7. DeMel D, Gleeson M, Schachterle K, et al. Successful treatment of ivermectin overdose in a bearded dragon (*Pogona vitticeps*) using gastric lavage and intravenous lipid emulsion. *J Vet Emerg Crit Care.* 2022;32(5):680-684.

8. Branco SEMT, Mattoso CRS, Botelho AFM, et al. Intravenous lipid emulsion treatment in rabbits with ivermectin toxicosis. *J Vet Emerg Crit Care.* 2021;31(3):340-350.

9. Ebel JJ, Gannon K, Kraenzlin M. The use of intralipid emulsion therapy to treat status epilepticus in a guinea pig (*Cavia porcellus*) secondary to severe ivermectin toxicity. *J Vet Emerg Crit Care.* 2023;33(1):107-111.

10. Schmidt LK, Keller KA, Tonozzi C, et al. Intralipid emulsion therapy for the treatment of suspected toxicity in 2 avian species. *J Avian Med Surg.* 2022;36(4):394-399.

11. Lubian E, Capitelli R, Nappi S, et al. Use of intralipid emulsion therapy to treat suspected oleander toxicosis in a domestic goose (*Anser anser domesticus*). *J Exot Pet Med.* 2021;39:78-80.

12. Barron HW, Kehoe S, Bast R, et al. Use of intravenous lipid emulsion therapy as a novel treatment for brevetoxicosis in double-crested cormorants (*Phalacrocorax auritus*). *Proc Annu Conf Assoc Avian Vet.* 2019:151.

13. Amissah RQ, Vogt NA, Chen C, et al. Prevalence and characteristics of cannabis-induced toxicoses in pets: Results from a survey of veterinarians in North America. *PLoS One.* 2022;17(4):e0261909.

14. Williams K, Wells RJ, McLean MK. Suspected synthetic cannabinoid toxicosis in a dog. *J Vet Emerg Crit Care.* 2015;25(6):739-744.

15. American Society for the Prevention of Cruelty to Animals. Unpublished data. AnTox^{TM}. ASPCA Animal Poison Control Center; 2023.

16. Epstein SE, Hollingsworth SR. Ivermectin-induced blindness treated with intravenous lipid therapy in a dog. *J Vet Emerg Crit Care.* 2013;23(1):58-62.

17. Herring JM, McMichael MA, Corsi R, et al. Intravenous lipid emulsion therapy in three cases of canine naproxen overdose. *J Vet Emerg Crit Care.* 2015;25(5):672-678.

18. Royle K, Bandt C. Intravenous lipid emulsion to treat suspected cocaine toxicosis in a dog. *Can Vet J.* 2020;61(1):49-52.

19. Wright HM, Chen AV, Talcott PA, et al. Intravenous fat emulsion as treatment for ivermectin toxicosis in three dogs homozygous for the ABCB1-1Δ gene mutation. *J Vet Emerg Crit Care.* 2011;21(6):666-672.

Agency-Based Parrot-to-Parrot Video Calling as Enrichment for Birds

Rebecca Kleinberger, PhD,
Ilyena Hirskyj-Douglas, PhD,
Jennifer Cunha, JD,
Susan Clubb, DVM

Session #3447

Affiliation: From Northeastern University, 360 Huntington Ave, Boston, MA 02115, USA (Kleinberger), Massachusetts Institute of Technology, 77 Massachusetts Ave, Cambridge, MA 02139, USA (Kleinberger), University of Glasgow, Glasgow G12 8QQ, UK (Hirskyj-Douglas), Parrot Kindergarten, Inc., 601 Heritage Dr, Ste 424, Jupiter, FL 33458, USA (Cunha), Rainforest Clinic Veterinary Clinic for Birds and Exotics, 3319 E Road, Loxahatchee, FL 33470, USA (Clubb).

Abstract: Over 20 million parrots are kept as pets in the United States, often lacking appropriate stimuli to meet their high social, cognitive, and emotional needs. An approach was developed based on strong ethics and animal agency to allow parrots to engage in video-calling other parrots. A pilot experiment was conducted with feedback from bird experts and a 3-month study with 18 pet birds was run to evaluate the potential, value and usability of a parrot-to-parrot video-calling system. The system was evaluated with regard to perception, agency, engagement, and overall perceived benefits. Results show that every bird used the system (147 corroborated bird-triggered calls), that most of them exhibited high motivation and intentionality, and that all caretakers reported perceived benefits, including learning to forage or even fly by watching others.

Introduction

Parrots are a highly social species of birds that often suffer from lack of companionship and species-specific socialization in captivity.[1] This can result in stereotypical behaviors such as self-mutilation, feather-destructive behavior, and aggression, which can be indicative of poor psychological well-being.[2] Various approaches have been recommended to alleviate these issues, including environmental enrichment, positive reinforcement-based behavior modification, and the use of restraints or collars.[3] While enrichment and training are important for parrots, none of these approaches replace the importance of socialization, particularly with other parrots. In addition, co-housing parrots with other birds or arranging in-person meetups theoretically may improve bird well-being,[4] but there are many drawbacks, including the practical difficulties of arranging in-person social gatherings for birds, and the high rate of fatal transmissible diseases such as avian ganglioneuritis, which inflicts 43% of the parrot population and calls into question the ethics of parrots' proximal contact and socialization. In recent decades, the human-computer interaction (HCI) community has worked on providing tools, approaches, and applications to help support the well-being and socialization of captive animals. This paper focuses on the development of agency-based parrot-to-parrot calling devices to help meet parrots' social needs. The aim is not to assess parrot's intelligence but to provide them with at-home enrichment.

The contributions of this project are 1) an ethical framework and novel protocol for bird-triggered video calls, 2) an 18-bird longitudinal study over 3 months assessing the potential of tablet-based video calling for parrots, 3) analyses and insights into how birds use their new agency and engagement in the video-calling process, 4) interviews and feedback from the human caretakers and 5) new perspectives and a design proposal for future parrot-centered video-calling systems. The Ethics Committee approved this project for Animal (number EA 01/22) and Human Use as Subjects (number 300210172) at Glasgow University.

Background

Assessing the use of tablets for birds requires determining if parrots have the biologic and physiologic ability to perceive and understand screen-based stimuli and the feasibility of using off the shelf devices. Although parrot vision differs from humans as well as within parrot species, decades of research suggests that they can make sense of screen-based stimuli, with early previous work pigeons pecking at targets.[5] Although perceiving single dots on screens is different from complex scenes, some birds also appear to make sense of moving images despite reduced color spectrum and flicker.[6] Standard validation tests have been used to compare birds' attention to the object of interest on the screen with objects in real life.[7] In this study, sense-making is assessed by measuring parrots' attention to the other bird appearing on screen, moving off screen, or moving within the screen in the initial exposure to video calls in Phase 1.

Research in the HCI and animal-computer interaction (ACI) communities have focused on developing tools to provide enrichment and socialization to captive animals, including pets left home alone for long periods of time. Such research and technology development includes investigating how animals can use technology to interact with each other as well as humans.[8] Interactivity and agency are important considerations to ensure ethical and meaningful animal experience, and both animal consent and capacity are necessary for animals to express agency. To measure animal intention, an economics-based approach has been used to assess how highly an animal values a proposed option, often implemented in the context of operant conditioning (e.g., how much will an animal pay with behavior for a certain outcome).[9] The current study implements corroboration tasks to increase the pay and measure the value of consent from participating birds by requiring them not only to ring a bell but also to touch the photo of a bird to call.

Interactive systems that give animals agency in shaping their surroundings have shown potential benefits in reducing stereotypies and improving behavior.[10] Researchers have developed choice-based technology for environmental control for animals, such as cockatoos[11] and macaws[12] to choose food, toys, music, and even call their caretakers.[13] The perception of self-agency can be beneficial to animals, and there is potential for interactive tools that maintain animal control and agency. The concept of competency is central to understanding agency. Here, the authors assessed agency quantitatively by counting the number of outgoing calls made independently by birds during a specific phase. Engagement can also be assessed quantitatively through time-on-task, and in this research, the time-on-call was used as a method for measuring engagement. Measured positive engagement provides further validation of the birds' agency and intent around making the video calls.

Materials and Methods

Following the expert survey and pilot, a longitudinal at-home study on 18 pet parrots was conducted over a 3-month period. The study was conducted in two phases. The first phase, called "Meet-and-Greet", focused on establishing the association between 1) touching a bell and 2) touching the image of another bird on the screen to trigger a call. The second phase called "Open Calls" focused on measuring the birds' social interaction through pairwise video calls. The participants were grouped according to their time zone, availability, species, and bird size, and each bird was individually supported. The study involved home-based parrots, over 1 year old, with no known behavioral issues, and that were comfortable with looking at screens and touching objects.

Data was gathered from video submissions and diary inputs. Parrots' interaction times with the bell and tablet, as well their engagement times as caller and callee birds via the recorded video, were measured. Although human participants were instructed to stop calls when birds exhibited disengagement behaviors, the end call time was confirmed by an expert parrot behaviorist to verify if the birds showed earlier signs of disengagement that may have gone unnoticed. Eight such cases were adjusted in the analysis, and caretakers were notified to improve their understanding of disengagement behaviors.

To assess whether parrot participants perceived the other bird on the screen and potentially made sense of the video calls, meet-and-greet sessions (Phase 1) were examined since they represented the first introduction to video calling. The examination focused on changes in behavior between instances when the other bird was present versus absent from the screen, looking for signs of visual and auditory perception and interactive behaviors. Specifically, the distance the bird chose to stand from the screen, presence of timely responses to the other bird either appearing on or leaving the screen, clear indications that the bird was following the other bird on the screen, and timely responses to auditory stimuli were noted.

More details about the methods, ethics, setup, training protocol for both humans and parrots, and analysis methods can be found in Kleinberger, et al.[13]

Results

Based on the principles of animal behavior and cognition and ACI/HCI, the data analysis method aimed to evaluate parrots' calls based on four variables: perception, agency, engagement and perceived benefits.

Two authors independently labeled each video, with an inter-rater reliability of 89%. Regarding sense-making assessment, 50% of the birds stood close to the screen. Fourteen out of 18 (78%) birds exhibited clear timely responses to the other bird appearing on screen, six of which came closer, while three reacted with head bobbing behaviors. Five out of 18 (28%) parrots reacted when the other bird moved off-screen. Five out of 18 (28%) birds appeared to follow the other birds on the screen.

To evaluate whether birds made use of their newfound agency during Phase 2, for each bird and each session, measurements were made of how many times the bird triggered an outgoing call (up to two total outgoing calls per bird per session), including the distribution and variability between and within groups. Birds' individual social scores were measured as the ratio between the number of calls triggered and the total potential number of calls over the eight sessions. Pearson's correlation coefficient was used to test the linear relationships between the number of calls the parrots made and the number of calls they received. All the birds participated in video calls and were observed to make outgoing calls. One key finding was the strong correlation between the number of calls made by a bird and the number of calls it received. There are various possible explanations for this connection, such as increased familiarity with other birds or the voice of the other caretaker, which may lead to rapport and more attention from caretakers. Another interesting finding was which birds initiated the calls.

Aside from the number of outgoing calls triggered, the study also examined quantitatively the engagement of the birds through the calls, a potential marker of their level of interest, through how long they were engaged in the calls they themselves triggered versus those they received using the corrected duration of calls. Correlation between use of their agency (individual social scores) and their engagement during calls was determined. A significant correlation between high social scores and low disengagement from calls was observed.

Various factors may influence engagement, including size, age, household composition, and human daily availability; however, no significant correlation was found between bird size and engagement. Additional longitudinal validation of the study and co-specific grouping are needed to establish the effects of specific factors. Regardless of individual factors, the more a bird uses its agency, the more likely it is to stay engaged during the calls it triggers.

In both the pilot and full study, the human's role in the process was deemed crucial; from training the birds in the calling system to providing continued presence and encouragement during the calls. The video call system aimed to enhance the birds in their regular life context, where humans already play a significant part. It required careful monitoring and balance as caretakers' enthusiasm and desire for the birds to engage could cause stress for the parrots. This emphasizes the importance of caretaker training and adherence to protocols in any future systems centered around birds' agency.

Bird Caregiver Interview and Survey

The study found that all caretakers believed the experience was beneficial for their birds and for themselves. The caretakers reported various new observed behaviors supporting a beneficial outcome to the birds, including more confidence, calmer behavior, new foraging behaviors and new flying behaviors. One caregiver said, "She came alive during the calls." Another reported, "He would clearly ring the bell, and very decisively select a friend" and another said, "After the calls with [another participant bird], she started flying more. I think she realized that flying is an okay behavior. [...] I'm very happy about this!"

No negative behavior was reported by human participants. In addition, caretakers reported interactive-mirroring behaviors between the birds during calls, including both birds playing with toys, foraging, head-bobbing, dancing, waving, saying hello, vocalizing, learning new sounds and singing together. Those observations were further supported by analysis of the video recordings.

Human participants also reported having learned more about their birds and feeling better able to recognize behaviors. The sessions also provided bonding time between humans and birds, mediated by technology. The improved understanding of the caretakers for their animals may provide a healthy ground for interspecies bonding.

Discussion

Results of the perception analysis suggest that birds exhibited a diverse range of sense-making responses triggered by visual and auditory stimuli, including reacting to various sound sources from the correspondent's side, the presence of another bird on the screen, following the bird on-screen or reacting to their correspondent disappearing.

A significant correlation between the number of outgoing calls and the number of received calls was observed, which led to the development of a social score for each bird and group. This connection may be explained by increased familiarity with the other bird leading to rapport and increased special attention from caretakers. It was also found that the role of the human in the loop was instrumental, from initial training in the calling system to their continued presence and encouragement during the calls. Caretaker enthusiasm and desire for their bird to engage could be stressful for a parrot, emphasizing the importance of caretaker training and protocol adherence.

Screen-based parrot-to-parrot video-calling could be used for social enrichment, but further longitudinal validation and co-specific grouping are needed to establish the effects of specific factors. In addition, groups of two birds may allow for more and easier enrichment opportunities through socialization, as coordinating schedules was easier. Overall, the study highlights the potential of screen-based video-calling for parrots and the importance of considering their agency and needs when developing such systems.

The study required a large time commitment from caretakers and synchronizing schedules, which may make it difficult to sustain the calls long-term without support and structure provided by the research team. The limited range of the touchscreen device-housed cameras for video calls made the viewing experience inconsistent in some settings. Future work could investigate whether the birds would react similarly to pre-recorded videos and if ergonomic improvement on the tablet app display could better support their experience. While exposing birds to live, interactive stimuli could be argued as more natural and appropriate for their species identity, video calls offer more flexibility and ease of usage. There are ethical concerns about deceiving animals in playback experiments and birds' cognitive processing was not evaluated in this study. Further research could investigate whether grouping birds with others of the same species affects interaction levels, whether different populations of parrot species find video calls more or less meaningful, and optimal call duration and frequency.

Conclusions

Following a pilot experiment and expert survey, a 3-month study was conducted with 18 pet birds to evaluate the potential, value and usability of a parrot-to-parrot video-calling system. The system was assessed in terms of perception, agency, motivation, engagement strength and overall perceived benefits. With 147 corroborated bird-triggered calls, results show that every bird used the system, that most of them exhibited high engagement, and that all caretakers reported perceived benefits. The role of the caretaker appeared central in the birds' interaction with technology. The experience appeared to not only provide benefits to the parrots through bird-specific interactions and behavior learning, but also to the human participants by learning more about their animal companions. Leveraging internet calls ultimately allowed the birds to bond socially with other birds while also intimately reconnecting with their human companion.

References

1. Bradshaw GA, Yenkosky JP, McCarthy E. Avian affective dysregulation: psychiatric models and treatment for parrots in captivity. *Proc Annu Conf Assoc Avian Vet.* 2009.

2. Ebisawa K, Nakayama S, Pai C, et al. Prevalence and risk factors for feather-damaging behavior in psittacine birds: Analysis of a Japanese nationwide survey. *PLoS ONE.* 2021;16(7):e0254610.

3. Duncan B. Avian intelligence, parrot behavior and toys. *AFA Watchbird.* 1997;24(4):13-16.

4. German Federal Ministry of Food, Agriculture and Consumers. 1995. German Minimum Requirements for the Keeping of Parrots. Available at: http://www.thebirdschool.com/caring-for-parrots/parrot-cages-parrot-cages/german-minimum-requirements-for-the-keeping-of-parrots/. Accessed September 10, 2022.

5. Gibson BM, Wasserman EA, Frei L, Miller K. Recent advances in operant conditioning technology: a versatile and affordable computerized touchscreen system. *Behav Res Methods Instrum Comput.* 2004;36(2): 355-362.

6. Pepperberg IM, Wilkes SR. Lack of referential vocal learning from LCD video by grey parrots (*Psittacus erithacus*). *Interact Stud.* 2004;5(1):75-97.

7. Lea SEG, Dittrich WH. What do birds see in moving video images. In: Fagot J, ed. *Picture Perception in Animals.* New York, NY: Taylor & Francis Routledge; 2000:143-180.

8. Hirskyj-Douglas I, Lucero A. On the internet, nobody knows you're a dog...unless you're another dog. *Proc Conf Human Factors Comput Syst.* 2019;1-12.

9. Duncan IJH. Measuring preferences and the strength of preferences. *Poult Sci.* 1992;71(4):658-663.

10. Grandin T, Curtis SE, Widowski TM, Thurmon JC. Electro-immobilization versus mechanical restraint in an avoid-avoid choice test for ewes. *J Anim Sci.* 1986;62(6):1469-1480.

11. Cunha J, Rhoads C. Use of a tablet-based communication board and subsequent choice and behavioral correspondences in a Goffin's cockatoo (*Cacatua goffiana*). *Proc Conf Anim-Comput Interact.* 2020;1-9.

12. Kleinberger R, Harrington AHK, Yu L, et al. Interspecies interactions mediated by technology: an avian case study at the zoo. *Proc Conf Human Factors Comput Syst.* 2020;1-12.

13. Kleinberger R, Cunha J, Vemuri M, Hirskyj-Douglas I. Birds of a feather video-flock together: agency-based parrot-to-parrot video-calling system for interspecies ethical enrichment. *Proc Conf Human Factors Comput Syst.* doi.org/10.1145/3544548.3581166.

Section 3

AAV Cutting Edge Session

Jennifer Graham, DVM, DABVP (Avian Practice), DABVP (Exotic Companion Mammal Practice), Dipl ACZM
Session Chair

The Attending Veterinarian's Role in Compliance with Animal Welfare Standards in Birds

Susan Clubb, DVM

Session #3433

Affiliation: From Rainforest Clinic for Birds and Exotics, and Hurricane Aviaries, Inc., 3319 E Road, Loxahatchee, FL 33470, USA.

Abstract: On February 21, 2023 the U.S. Department of Agriculture (USDA) Animal and Plant Health Inspection Service (APHIS), published the final rule, Standards for Birds Not Bred for Research Under the Animal Welfare Act (9 CFR Parts 1, 2, and 3 Docket No. APHIS 2020-0068). The rule is effective March 23, 2023. Current Animal Welfare Act (AWA) licensees and registrants must comply with this rule by August 21, 2023. New AWA licensees and registrants, must comply with this rule by February 21, 2024.

Introduction

In 2002, Congress amended the definition of animal in the AWA by limiting the exclusion of birds to only those bred for use in research. In 2020, in response to lawsuits from animal welfare organizations, the US Court of Appeals for the District of Columbia ruled that USDA must publish a proposed rule by Feb 22, 2022, and a final rule no more than a year later. The final rule was 237 pages. The most important points for attending veterinarians are summarized here.

At least five requirements for attending veterinarians were removed between publication of the proposed rule and the final rule. An attending veterinarian must be a graduate of an accredited school of veterinary medicine, but a state license is not required, and the attending veterinarian could even reside in a different state. Ideally the attending veterinarian would be experienced in avian medicine. Board certification is not required. The attending veterinarian must make at least the first inspection. There is no required frequency of subsequent inspections, but annually is recommended. If the attending veterinarian lives far from the facility, s/he can work in conjunction with a surrogate local veterinarian who has less experience. The attending veterinarian can work with the aviculturist by telemedicine between visits to the facility. Regulations, forms and more can be found at www.aphis.usda.gov/animalwelfare.

Part 1 - Definitions

Bird - Any member of the class Aves, excluding eggs.

Carrier - Operator of any airline, railroad, motor carrier, shipping line or other enterprise engaged in transporting animals for hire. (Except anyone transporting a migratory bird, MBTA).

Exhibitor - Any person public or private, exhibiting any animals including carnivals, circuses, animal acts, free flight shows, zoos, educational exhibits whether operated for profit or not.

Exemptions - Retail pet stores, pigeon races, household pets, county fairs or livestock shows, falconry, bird fancier shows, exhibitions intended to advance agriculture.

Farm animals - Poultry (chickens, turkeys, swans, partridges, guinea fowl, peafowl, ducks, geese, pigeons and doves, grouse, pheasants and quail.

Intermediate handler - Any person who receives custody of birds in connection with transportation in commerce.

Pet animal - Includes canaries, cockatiels, lovebirds and budgerigar parakeets. This term excludes exotic or wild animals. See complete list on the USDA website.

Retail pet store - A place of business of residence at which the seller, buyer and animals available for sale are physically present so buyer may observe.

Weaned - A bird that has become accustomed to take food and has done so, without supplemental feeding ... for a period of 5 days.

Also Exempt – Any person who sells 200 or fewer birds 250 grams or less, or sells eight or fewer birds more than 250 grams determined by average adult weight of the species.

Any person maintaining four or fewer raptors for exhibition, and holds a valid U.S. Fish and Wildlife Service (USFWS) permit.

License types

Class A- Breeder - Sells birds bred at the licensed facility for pets or exhibition.
Class B - Dealer - Buys and sells birds not bred by the licensee.
Class C - Exhibitor - Any person public or private, exhibiting any animals including carnivals, circuses, animal acts, free flight shows, zoos, educational exhibits whether operated for profit or not.

Registration types

Carrier (Exp - airlines)
Intermediate carrier
Research facility

Application can be made on the USDA APHIS Animal Care website.

Part 2 - Regulations

2.40. Attending veterinarian and adequate veterinary care

The role of an attending veterinarian is much more comprehensive than providing veterinary medical care to sick animals. They also provide important guidance to help licensees and registrants keep their animals healthy. In general, an attending veterinarian's duties include:

- Providing each individual animal veterinary care that ensures its health and well-being.

- Working with the facility to develop a program that covers all aspects of animal health: communication with the veterinarian, disease monitoring and prevention, husbandry, nutrition, provision of emergency care, employee knowledge, and humane euthanasia.

- A written program of veterinary care is required for all facilities with a part-time attending veterinarian or consultant arrangement. The attending veterinarian is not required to sign the program of veterinary care.

- Conducting regular visits to the premises. There is no minimum frequency of visits for birds. APHIS recommends annually, but this is at the discretion of the attending veterinarian. Visits should occur as often as necessary to ensure that the attending veterinarian is aware of the conditions of the animals and facilities; and has the information needed to ensure the provision of adequate veterinary care and to oversee the adequacy of animal care and use. Additional site visits may be necessary for disease outbreaks; ill or injured animals; training of caretakers, or for the proper introduction of new animals.

- Animals in need of care may be taken to the clinic or practice of the attending veterinarian. These visits do not replace the need for regularly scheduled visits to the premises of the licensee or registrant.

- The attending veterinarian is also responsible for overseeing and approving additional practices for certain covered species and at research facilities.

2.35. Recordkeeping requirements

New information collection requirements created by the regulation include bird identification records, environmental enrichment plan records, cleaning and sanitation records, consignment documents and certifications for shipments of birds. Individual animal identification is not specifically required for birds. The birds should be on their acquisition records. If the birds are transported they must be identified at the time of delivery by a label (attached to the primary enclosure with the number of birds, species, distinctive physical features and identifying marks), or by either a leg or wing band applied to each bird, or by microchip

2.50. Time and method of identification

When one or more birds are confined in a primary enclosure, the birds shall be identified by:

1) A label attached to the primary enclosure which shall bear a description of the birds included in the primary enclosure: number of birds, species, any distinctive feathers or identifying marks.

2) A leg or wing band.

3) A transponder placed in a standardized location (facility must have a compatible reader).

2.75. Breeders, exhibitors and dealers

They will be required to maintain records of all birds. See regulations for specific details.

Part 3 - Standards

New facility standards are extensive. While the attending veterinarian should be familiar with those standards, they are not expected to evaluate compliance with those standards; however, some of the facility standards require approval or consultation by the attending veterinarian.

Space requirements

Both part-time and full-time attending *veterinarians* must consult with the facility to ensure that the space in all enclosures housing birds is adequate and allows for normal postural and social adjustments. Inadequate space may be indicated by evidence of malnutrition, poor condition, debility, stress or abnormal behavior patterns.

The normal postural and social adjustments of a bird may be restricted: When the attending *veterinarian* determines that making species-typical postural or social adjustments, such as dust-bathing, foraging, or running, would be detrimental to the bird's good health and well-being. The attending veterinarian must document the reason and recommended duration for the restriction and make such records available for review by an APHIS inspector.

§ 3.154 Environment enhancement to promote psychological well-being

Dealers, exhibitors, and research facilities must develop, document, and follow a species appropriate plan for environment enhancement adequate to promote the psychological well-being of birds. The plan must be approved by the *attending veterinarian* and must be in accordance with the regulations in this subpart and with currently accepted professional standards as cited in appropriate professional journals or reference guides. This plan must be made available to APHIS upon request, and, in the case of research facilities, to officials of any pertinent funding agency. The plan, at a minimum, must address each of the following:

(1) Social grouping. The environment enhancement plan must include specific provisions to address the social needs of species of birds known to exist in social groups in nature. Such specific provisions must be in accordance with currently accepted professional standards as cited in appropriate professional journals or reference guides. The plan may provide for the following exceptions:

(a) If a bird exhibits vicious or overly aggressive behavior, or is debilitated as a result of age or other conditions (e.g., arthritis), it can be housed separately;

(b) Birds that have or are suspected of having a contagious disease must be isolated from healthy animals in the colony as directed by the *attending veterinarian*. When an entire group or room of birds is known to have been or believed to be exposed to an infectious agent, the group may be kept intact during the process of diagnosis, treatment, and control.

(c) Birds may not be housed with other species of birds or animals unless they are compatible, do not prevent access to food, water, or shelter by individual animals, and are not known to be hazardous to the health and well-being of each other. Compatibility of birds must be determined in accordance with generally accepted professional practices and actual observations as directed by the *attending veterinarian*, to ensure that the birds are in fact compatible. Individually housed social species of birds must be able to see and hear birds of their own or compatible species unless the attending veterinarian determines that it would endanger their health, safety, or well-being. If individually housed social species of birds are unable to see and hear birds of their own or compatible species then special attention regarding enhancement to their environment must be provided.

(2) Environmental enrichment. The physical environment in the primary enclosures must be enriched by materials or activities that would provide the birds with the means to express non injurious species-typical activities. Species differences should be considered when determining the type or methods of enrichment. Examples of environmental enrichments include providing perches, swings, mirrors, and other increased cage complexities; providing objects to manipulate; varied food items; using foraging or task-oriented feeding methods; and providing interaction with the care giver or other familiar and knowledgeable person consistent with personnel safety precautions.

(3) Special considerations. Certain birds must be provided special attention regarding enhancement of their environment, based on the needs of the individual species and/or individual bird and in accordance with the instructions of the *attending veterinarian*. Birds requiring special attention are the following:

- Nestlings, chicks, or fledglings;
- Those that show signs of being in psychological distress through behavior or appearance;
- Those used in research for which the Institutional Animal Care and Use Committee (IACUC)-approved protocol requires restricted activity; and
- Individually housed social species of birds that are unable to see and hear birds of their own or compatible species.
- The *attending veterinarian* may exempt an individual bird from participation in the environment enhancement plan because of its health or condition, or in consideration of its well-being. The basis of the exemption must be recorded by the attending veterinarian for each exempted bird. Unless the basis for the exemption is a permanent condition, the exemption must be reviewed at least every 30 days by the *attending veterinarian*.

(4) Restraint devices. Birds must not be permitted to be maintained in restraint devices unless required for health reasons as determined by the *attending veterinarian* or by a research proposal approved by the IACUC at research facilities. Any restraining actions must be for the shortest period possible.

§ 3.155 Feeding.

(a) The diet for birds must be appropriate for the species, size, age, and condition of the bird. The food must be wholesome, palatable to the birds, and free of contamination. It must be of sufficient quantity and nutritive value to maintain a healthy condition and weight range of the bird and to meet its normal daily nutritional requirements. Birds must be fed at least once a day except as directed by the *attending veterinarian*, normal fasts, or other professionally accepted practices. If birds are maintained in group housing, measures appropriate for the species must be taken to ensure that all the birds receive a sufficient quantity of food.

§ 3.156 Watering.

Potable water must be provided in sufficient quantity to every bird housed at the facility, unless restricted by the *attending veterinarian*

§ 3.160 Compatibility and separation.

(a) Socially dependent birds, such as clutch-mates, must be housed in social groups, except where the *attending veterinarian* exempts an individual bird because of its health or condition, or in consideration of its well-being, or for specific management needs, or where such social grouping is not in accordance with a research proposal and the proposal has been approved by the research facility IACUC.

Birds that have or are suspected of having a contagious disease or communicable condition must be separated from healthy animals that are susceptible to the disease as directed by the *attending veterinarian.*

Transportation and Shipping

Carriers and intermediate handlers may accept un-weaned birds for transport as long as the *attending veterinarian* approves the un-weaned birds can be transported safely, provides written instructions for the conditions of transport, and is signed within 10 days of shipment. The attending veterinarian must approve that the un-weaned birds can be transported safely and provide written instructions of transport, and must sign those instructions within 10 days of the shipment. The veterinarian is not required to physically see the birds, this can be done virtually.

§ 3.164 Food and water requirements.

(a) All weaned birds must be offered food and potable water within 4 hours before being transported in commerce, unless the _attending veterinarian_ approves a delay or a delay is in accordance with professionally accepted standards.

(b) Dealers, exhibitors, research facilities, and operators of auction sales must provide potable water to all weaned birds transported in their own primary conveyance at least every 12 hours after such transportation is initiated, except for birds which, according to professionally accepted standards or under the direction of the _attending veterinarian_, require watering or feeding more or less frequently.

(c) Carriers and intermediate handlers must provide potable water to all live, weaned birds at least every 12 hours after accepting them for transportation in commerce, except for birds which, according to professionally accepted standards or under the direction of the _attending veterinarian,_ require watering or feeding more or less frequently.

(d) A sufficient quantity of food and water or other source of hydration must accompany the bird to provide food and water for such bird during period of transport, except as directed by _veterinary_ treatment and other professionally accepted standards.

§ 3.165 Care in transit.

(a) Surface transportation (ground and water). During surface transportation, any person subject to the Animal Welfare regulations in this part and parts 1 and 2 of this subchapter transporting birds in commerce must ensure that the operator of the conveyance, or a person accompanying the operator, visually observes the birds as frequently as circumstances may allow, but not less than once every 4 hours, to ensure that the birds are receiving sufficient air for normal breathing, that climatic and environmental conditions are being maintained in accordance with the requirements in § 3.168, and that all other applicable standards are met. The regulated person must ensure that the operator or person accompanying the operator determines whether any of the birds are in physical distress and obtains _any veterinary_ care needed for the birds as soon as possible.

(b) Air transportation. When transported by air, live birds must be visually observed by the carrier as frequently as circumstances may allow, but not less than once every 4 hours, if the animal cargo space is accessible during flight. If the animal cargo space is not accessible during flight, the carrier must visually observe the live birds whenever they are loaded and unloaded and whenever the bird cargo space is otherwise accessible to ensure that they are receiving sufficient air for normal breathing, that climatic and environmental conditions are being maintained in accordance with the requirements in § 3.168, and that all other applicable standards are met. The carrier must determine whether any such live birds are in physical distress and arrange for any needed veterinary care as soon as possible.

(c) Prohibition on the transport of ill, injured, or distressed birds. Any person subject to the Animal Welfare regulations in this part and parts 1 and 2 of this subchapter may not transport in commerce birds that are ill, injured, or in physical distress, except to receive _veterinary_ care for the condition.

Determining if a license is needed by a breeder

The following groups and species are considered to be small pet bird species (250 grams or less as determined by average adult weight of the species). A breeder must be licensed if they sell more than 200 small birds.

- Finches and finch-like birds from the family Fringillidae*
- Cockatiels
- Parrotlets
- Lovebirds
- Caiques
- Senegals and other *Poicephalus* parrots
- Mini-macaws, 250 grams or less by average adult weight of species
- Conures
- Pionus
- Lories and lorikeets
- Parakeets and other parakeet-like birds, such as but not limited to ringnecks, monks, lineolated, kakarikis, and rosellas
- Doves and pigeons in the order Columbiformes 250 grams or less by average adult weight of species
- Poultry, for use as pets, that are 250 grams or less by average adult weight of species

The following groups and species are considered to be <u>large pet bird species</u> (more than 250 grams as determined by average adult weight of the species). A breeder must be licensed if they sell more than eight large birds.

- Cockatoos
- African greys
- Eclectus
- Macaws
- Amazons
- Toucans
- Hornbills
- Turacos

- Kookaburras
- Doves and pigeons in the order Columbiformes more than 250 grams by average adult weight of species
- Poultry, for use as pets, more than 250 grams by average adult weight of species

References

1. 9 CFR Parts 1, 2, and 3 (Docket No. APHIS 2020-0068).

Effects of Atorvastatin and Rosuvastatin on Blood Lipids in Quaker Parrots (*Myiopsitta monachus*)

Hugues Beaufrère, DVM, PhD, Dipl ACZM, DABVP (Avian Practice), DipECZM (Avian),
Trinita Barboza, DVM, DVSc,
Alysha Burnett,
Ken D. Stark, PhD,
R. Darren Wood, DVM, DVSc, DACVP

Session #3146

Affiliation: From the Department of Medicine and Epidemiology, School of Veterinary Medicine, University of California–Davis, Davis, CA 95616, USA (Beaufrère), the Department of Clinical Sciences, Cummings School of Veterinary Medicine, Tufts University, North Grafton, MA 01536, USA (Barboza), the Department of Pathobiology, Ontario Veterinary College, University of Guelph, Guelph, ON N1G 2W1, Canada (Wood), the Department of Kinesiology and Health Studies, University of Waterloo, Waterloo, ON N2L 3G1, Canada (Stark).

Note: Abstract format selected by author(s). See Front Matter for description of conference proceedings manuscripts submitted as scientific abstracts.

Abstract: Statin drugs are the most effective class of hypolipidemic and anti-atherosclerotic drugs, with atorvastatin and rosuvastatin being the most effective. While the use of statins would be a tremendous asset in the treatment of dyslipidemia and lipid-accumulation disorders in birds, there are limited data available in psittacine species.

Two consecutive randomized cross-over trials on quaker parrots were performed to study the effect of these two statins. Ten birds were used in an initial balanced cross-over experiment with five oral treatments for 2 weeks each (control, atorvastatin 10 mg/kg q12h and q24h, rosuvastatin 10 mg/kg q12h and q24h). Plasma lipidomics and lipoprotein profiling were performed after each treatment. Twelve birds were used in a second experiment consisting of two parallel cross-over studies, each on six birds either fed their regular diet or a 0.3% cholesterol diet and receiving two oral treatments for 2 weeks each (control, atorvastatin 20 mg/kg q12h). Plasma lipidomics, lipoprotein profiles, and β-Hydroxy-β-methylglutaryl-CoA reductase activity (HMG-CoA) were subsequently measured. Results were analyzed with serial linear mixed models and trends were assessed graphically.

No statistically significant effect of any statin treatment was detected on plasma lipids, lipoproteins, creatine phosphokinase, and HMG-CoA. In the first trial, all the rosuvastatin treatments led to some non-significant decreases in several triacylglycerol species while in the second trial, this was observed only in birds on atorvastatin 20 mg/kg q12h being fed their regular diet. These results suggest that quaker parrots may require much higher doses of statin drugs to show significant and clinically useful lipid-lowering effects.

Acknowledgments: Funding for this study was provided by the Morris Animal Foundation (grant D19ZO-301) and the OVC Pet Trust Foundation. This publication has not been reviewed or endorsed by the Morris Animal Foundation and the views expressed do not necessarily reflect the views of the Foundation, its officers, directors, affiliates, or agents. We also thank Hagen Inc. for providing and manufacturing the experimental diets. The authors thank the Analytical Facility of Bioactive Molecules, the Hospital for Sick Children, Toronto, Canada for performing metabolomics analysis.

Pharmacokinetics of Single-Dose Oral Atorvastatin and its Metabolites in Orange-Winged Amazon Parrots (*Amazona amazonica*)

Nicole Mikoni, DVM,
David Sanchez-Migallon Guzman, LV, MS, DipECZM (Avian), DipECZM (Small Mammal), Dipl ACZM,
Heather Knych, DVM, PhD, DACVCP,
Hugues Beaufrère, Dr med vet, PhD, DABVP (Avian Practice), DipECZM (Avian), Dipl ACZM

Session #3149

Affiliation: From the Veterinary Medical Teaching Hospital (Mikoni), Department of Medicine and Epidemiology (Guzman, Beaufrère), K.L. Maddy Equine Analytical Pharmacology Laboratory (Knych), School of Veterinary Medicine, University of California Davis, Davis, CA, 95616, USA.

Note: Abstract format selected by author(s). See Front Matter for description of conference proceedings manuscripts submitted as scientific abstracts.

Abstract: Dyslipidemias are common in captive psittacine species and have been associated with lipid-accumulation disorders such as atherosclerosis and hepatic lipidosis.[1-5] Amazon parrots are especially prone to atherosclerosis.[2,3] Use of statins is considered a mainstay of human hyperlipidemia and atherosclerosis treatment.[6] Previous studies evaluating atorvastatin (10 mg/kg q24h for 30 days) in hypercholesterolemic Hispaniolan Amazon parrots did not find significant hypolipidemic effects at studied dosages;[7] however, evaluation of statin drug pharmacokinetics in psittacine species is limited and could help determine dosages required for hypolipidemic effect. This study reports the pharmacokinetics of a compounded oral atorvastatin (10 mg/ml) administered via gavage to orange-winged Amazon parrots. A pilot study using escalating doses from 20-80 mg/kg was performed on six birds. A follow-up study using eight birds (four males/four females, varying ages) was then conducted using a single dose of 20 mg/kg. For both studies, plasma concentrations of atorvastatin and the two main active metabolites (para- and ortho-hydroxyatorvastatin) were measured in each bird by LC-MS/MS at 10 different time points over 30 hours. Pharmacokinetic analysis was performed using non-compartmental analysis. Mean atorvastatin half-life, T_{max}, and C_{max} were 5.96 hours, 1.6 hours, and 82.6 ng/ml, respectively. For para-hydroxyatorvastatin, the half-life, T_{max}, and C_{max} were 6.46 hours, 5 hours, and 34.1 ng/ml, respectively, and 5.58 hours, 3.38 hours, and 7.35 ng/ml for ortho-hydroxyatorvastatin. These results provide pharmacokinetic data on atorvastatin and its main metabolites in orange-winged Amazon parrots that can be used to inform therapeutic recommendations for clinical patients and plan future pharmacodynamic studies.

Acknowledgments: This study was supported by the UC Davis Center for Companion Animal Health and the Richard M. Schubot Parrot Wellness and Welfare Program.

References

1. Beaufrère H, Reavill D, Heatley J, et al. Lipid-related lesions in quaker parrots (*Myiopsitta monachus*). *Vet Pathol*. 2019;56(2):282-288.

2. Beaufrère H, Ammersbach M, Reavill D, et al. Prevalence of and risk factors associated with atherosclerosis in psittacine birds. *J Am Vet Med Assoc*. 2013;242(12):1696-1704.

3. Beaufrère H, Cray C, Ammersbach M, et al. Association of plasma lipid levels with atherosclerosis prevalence in Psittaciformes. *J Avian Med Surg.* 2014;28(3):225-231.

4. Martín-Castillo A, Castells MT, Adánez G, et al. Effect of atorvastatin and diet on non-alcoholic fatty liver disease activity score in hyperlipidemic chickens. *Biomed Pharmacother.* 2010;64(4):275-281.

5. Adánez G, Castells MT, García Pérez B, et al. Effects of atorvastatin on progression-regression of renal injury in hyperlipidemic chickens. *Histol Histopathol.* 2008;23(9):1131-1142.

6. McKenney JM, Roth EM. Statins. In: Ballantyne CM, ed. *Clinical Lipidology: A Companion to Braunwald's Heart Disease E-Book.* Elsevier; 2015:227-256.

7. Robertson JA, Guzman DSM, Graham JL, et al. Evaluation of orally administered atorvastatin on plasma lipid and biochemistry profiles in hypercholesterolemic Hispaniolan Amazon parrots (*Amazona ventralis*). *J Avian Med Surg.* 2020;34(1):32-40.

Clinicopathologic, Gross Necropsy, and Histopathologic Effects of Repeated Glucocorticoid Administration in Domestic Pigeons (*Columba livia*)

Olivia A. Petritz, DVM, Dipl ACZM,
Mandy Womble, DVM, PhD,
Erika Gruber, DVM, DACVP (Clinical),
Andrea E. Thomson, BS,
Tara M. Harrison, DVM, MPVM, Dipl ACZM, Dipl ACVPM,
DipECZM (Zoo Health Management), CVA

Session #3430

Affiliation: From the Department of Clinical Sciences (Petritz, Thomson, Harrison), Population Health and Pathobiology (Womble, Gruber), North Carolina State University, College of Veterinary Medicine, Raleigh, NC 27607, USA.

Note: Abstract format selected by author(s). See Front Matter for description of conference proceedings manuscripts submitted as scientific abstracts.

Abstract: Steroids are commonly used in chemotherapy protocols yet the safety of these medications in any avian species, a reportedly steroid-sensitive taxon, has not been evaluated. Previous publications have focused on the avian pituitary-adrenocortical axis and its response to steroid administration, rather than the safety or side effects of long-term oral steroid administration.[1,2] Eleven adult unknown sex Carneau pigeons (*Columba livia*) aged 7-9 years of age were randomly assigned to a control (n = 5) or treatment group (n = 6). All treatment birds received prednisolone 1 mg/kg PO q24h for 29 days, while the control group received a similar volume of saline PO q24h. Complete blood counts and body weights were performed once weekly. Plasma biochemistries, *Aspergillosis* antibody, and protein electrophoresis were performed prior to the start of the study and at the end of the treatment course on all birds. At the end of the treatment period, the birds were humanely euthanized, and complete necropsies including histopathology were performed. No adverse clinical signs were noted in any bird throughout the study period. There was no statistically significant difference in the white blood cell count, absolute lymphocyte numbers, or *Aspergillosis* antibodies between treatment and control groups at day 0 compared to day 29. Post-mortem examination of all birds in both groups was largely unremarkable. There was no evidence of infectious disease on gross or histologic postmortem examination. Overall, no adverse clinical, clinicopathologic, or histopathologic changes were noted in the treated birds compared to the control birds following repeated administration of oral prednisolone.

References

1. Westerhof I, Van den Brom WE, Mol JA, et al. Sensitivity of the hypothalamic-pituitary-adrenal system of pigeons (*Columba livia domestica*) to suppression by dexamethasone, cortisol, and prednisolone. *Avian Dis.* 1994;38:435-445.

2. Westerhof I, Pellicaan CHP. Effects of different application routes of glucocorticoids on the pituitary-adrenocortical axis in pigeons (*Columba livia domestica*). *J Avian Med Surg.* 1995;9:175-181.

Effects of Capromorelin, Mirtazapine, and Cyproheptadine on Food Intake in Budgerigars (*Melopsittacus undulatus*)

Christoph Mans, Dr med vet, Dipl ACZM,
Caroline E. Titel, BS,
Grayson A. Doss, DVM, Dipl ACZM

Session #3387

Affiliation: From the School of Veterinary Medicine, University of Wisconsin-Madison, 2015 Linden Drive, Madison, WI 53706, USA.

Note: Abstract format selected by author(s). See Front Matter for description of conference proceedings manuscripts submitted as scientific abstracts.

Abstract: Hyporexia or anorexia frequently occur in sick and/or hospitalized psittacine birds. Reduced caloric intake in these species can lead to a variety of detrimental effects and prolong hospitalization. Several medications have been used to stimulate appetite in dogs and cats, but only benzodiazepines have been evaluated in psittacine species.[1,2] In a series of randomized, blinded, placebo-controlled complete crossover studies, the effects of mirtazapine (1 and 5 mg/kg), capromorelin (10 and 40 mg/kg), and cyproheptadine (0.5 and 2.5 mg/kg) on food intake were evaluated in healthy budgerigars (*Melopsittacus undulatus*, n = 12). After administration of a single dose of each drug via crop gavage, food intake was measured in hourly intervals for the first 8 hours and then one measurement for the 8- to 12-hour period. The birds were monitored by video recording for potential adverse effects. No clinically relevant or statistically significant effects on food intake were found following the administration of mirtazapine and cyproheptadine, regardless of the dose administered. Following administration of capromorelin at either dose, food intake was significantly increased for the first 8 hours compared to the control treatment. A dose-dependent difference in capromorelin efficacy was not found. Regurgitation was observed in 42% and 92% of birds following administration of capromorelin at 10 and 40 mg/kg, respectively, while gavage of the control treatment resulted in regurgitation in 25% of birds. Therefore, capromorelin at 10 mg/kg should be considered for future studies, and lower doses may be effective in increasing food intake with potentially reduced adverse effects.

References

1. Martel A, Doss G, Berg C, Mans C. Effects of midazolam on food intake in budgerigars. *J Avian Med Surg.* 2022;36:53-57.

2. Scagnelli A, Titel C, Doss G, Mans C. Effects of midazolam and lorazepam on food intake in budgerigars. *J Exot Pet Med.* 2022;41:42-45.

Use of a Continuous Glucose Monitor in a Diabetic Toco Toucan (*Ramphastos toco*)

Marcie Logsdon, DVM

Session #3426

Affiliation: From the Exotics and Wildlife Department, Washington State University Veterinary Teaching Hospital, 100 Ott Rd, Pullman, WA 99164, USA.

Note: Abstract format selected by author(s). See Front Matter for description of conference proceedings manuscripts submitted as scientific abstracts.

Abstract: An 11-year-old male toco toucan (*Ramphastos toco*) previously diagnosed with diabetes mellitus[1] presented for acute necrosis of the tongue. On examination the patient was severely emaciated at 512 grams with a body condition score (BCS) of 1/5. Discussion with the owner revealed that the toucan's overall condition was declining despite glipizide administration and dietary management. Hospitalization and insulin trial were initiated. Initial blood glucose (BG) reading was 1790 mg/dl[2] (Demand xPAND Plus Chemistry Analyzer, Siemens Medical Solutions, Malvern, PA, USA). Due to cost and sample volume requirements of the analyzer, attempts were made to use a glucometer for monitoring. This was unsatisfactory as the glucometer (not validated for birds) maxed out at 750 mg/dl[3] and provided inconsistent results if dilutions were attempted. It also required frequent restraint and blood draws for sampling. In pursuit of a less invasive monitoring system continuous glucose monitor (CGM)[4] (FreeStyle Libre 2, Abbott, Abbott Park, IL, USA) was placed on the lateral aspect of the left distal leg in the flexor muscles caudal to the tibiotarsus. The monitor was well tolerated by the patient and took continuous readings, recording every 5 minutes, for 4 days at which point the patient was euthanized due to overall decline. At four timepoints blood was drawn and compared with the clinical pathology lab. The CGM values were comparable, measuring 6-28 mg/dl higher with greatest deviation happening at levels ≥ 500 mg/dl. Given the physiologic lag between blood and interstitial glucose levels, this was felt to be adequate alignment for aiding in clinical decisions.

References

1. Rae M. Endocrine disease in pet birds. *Semin Avian Exot Pet Med.* 1995;4:32-38.

2. International Species Information System. ISIS Physiological Reference Intervals for Captive Wildlife: A CD-ROM Resource. Ramphastos_toco_Males_only__All_ages_combined_Conventional_American_units__2013_CD.html.

3. Zoeits (2017). AlphaTrak 2 Product Insert. Available at: https://www.zoetisus.com/content/_assets/docs/vmips/package-inserts/alphatrak_2.pdf. Accessed 4/24/23.

4. Wiedmeyer CE, Johnson PJ, Cohn LA, Meadows RL. Evaluation of a continuous glucose monitoring system for use in dogs, cats, and horses. *J Am Vet Med Assoc.* 2003;223(7):987-992.

Radiographic Measurements of Internal Organs in Red Hybrid Hens and Their Evolution During Egg Formation

Thomas Coutant, Dr med vet, IPSAV,
Ciro Cococcetta, Dr med vet,
Albert Phouratsamay, Dr med vet,
Chloé Touzet, Dr med vet,
Graham Zoller, Dr med vet, IPSAV, DipECZM (Avian),
Minh Huynh, Dr med vet, DipECZM (Avian), Dipl ACZM

Session #3121

Affiliation: From Service des Nouveaux Animaux de Compagnie (Coutant, Cococcetta, Phouratsamay, Zoller,* Huynh); Service d'imagerie médicale (Touzet), Centre Hospitalier Vétérinaire Fregis, 43 Avenue Aristide Briand, 94110 Arcueil, France.

*Present address: Clinique Vétérinaire Only Vet, 7 rue Jean Zay, 69800 Saint-Priest.

Note: Abstract format selected by author(s). See Front Matter for description of conference proceedings manuscripts submitted as scientific abstracts.

Abstract: There is no study that describes radiographic measurements of internal organs in backyard hens. Moreover, the effect of egg laying on these measurements has not been studied in avian species even though it could affect radiographic measurements, depending on the stage of egg formation. This issue is of particular concern in laying hens since they can lay on a daily basis. Thirteen laying adult hens and three non-laying juvenile Rhode Island Red hybrid hens were used to provide preliminary radiographic reference values for internal organ measurement in clinically healthy chickens as previously described in psittacine birds.[1] Whole body radiographs of five adult hens were taken every 2 hours during a 24-hour period to study the evolution of these measurements throughout egg formation. The only reliable organ measurements were heart and liver width on ventrodorsal views, which were significantly greater in adult compared to juvenile hens ($P = 0.024$). Among the different possible ratios, only heart width:total was significantly different between age classes, being greater in juvenile hens ($P = 0.024$). Hepatic silhouette width and ratios with coelomic width were found to increase over time ($P < 0.02$) but did not follow egg formation since liver width does not decrease immediately after the oviposition. Eggshell thickness was found to be a promising parameter to evaluate the stage of egg formation. This study demonstrates that egg formation should be considered when trying to establish reference ranges for internal organ radiographic measurements in laying hens in future studies.

References

1. Geerinckx L, Van der Vekens E, Saunders JH, et al. Literature review of radiographic measurements of internal organs in Psittaciformes. *J Exot Pet Med.* 2019;28(1):9.

Use of a Tablet-Based Communication Board and Subsequent Choice and Behavioral Correspondences in a Goffin's Cockatoo (*Cacatua goffiniana*)

Jennifer Cunha, JD,
Rebecca Kleinberger, PhD,
Susan Clubb, DVM

Session #3441

Affiliation: From Parrot Kindergarten, Inc., 601 Heritage Dr, Ste 424, Jupiter, FL 33458, USA (Cunha), Northeastern University, 360 Huntington Ave, Boston, MA 02115, USA (Kleinberger), Rainforest Clinic for Birds and Exotics, 3319 E Road, Loxahatchee, FL 33470, USA (Clubb).

Abstract: An enculturated individual is one who learns the content of a culture and assimilates it into their practices. An enculturated Goffin's cockatoo (*Cacatua goffiniana*) was taught to use a commercially available, android-based communication board. Through associative conditioning, the subject learned to press pictures representing items in categories for foods, beverages, activities, objects, and interactions. Over the course of 22 days, communication opportunities were tracked. Each of cockatoo's responses to the question "What do you want?" was corroborated by follow-up questions and body language verification (voluntary transfer to hand and other locations, acceptance or rejection of request execution). In all 33 reviewed communication opportunities, the researcher responded to the request with a corroboration task, and 26 of the opportunities involved two corroboration tasks. The interactions comprised 81 request and follow-up question tasks, and the cockatoo's response-positive corroboration consistency rate was 95%. Inter-observer agreement was independently collected on 30% of the reviewed communication opportunities with a score of 97.5% fidelity. In addition, the cockatoo used the communication board for the following requests: 42% - interactive activity; 33% - food/beverage; 18% - extra-familial interaction; 6% - object. These results show positive interactions with a communication board and warrant further replication with additional birds.

Introduction

Many animals have demonstrated the ability to use a variety of symbols to communicate. Since 1953 research has tracked a variety of species' abilities to use symbols to represent choice requests. Rats press levers to request food[1]or to remove aversive stimuli.[2] Horses have also demonstrated the ability to request the application or removal of blankets, and their choices were corroborated by the environmental temperature.[3]

Computer interfaces have also been used by researchers to communicate with non-human animals. Keyboard interfaces have been developed for dolphins to request objects, events, and locations.[4] A bonobo utilizes a touch-screen computer for lexigram use to communicate requests and ideas.[5]

Birds have demonstrated an impressive level of cognitive processing in wild and experimental settings. Corvids communicate information across generations[6] and have solved complex problems, outperforming 7- to 10-year-old children.[7]

Goffin's cockatoos evidence a limited form of the inference-through-exclusion.[8]They have also used and manufactured tools,[9] and can outperform 7-year-old children in puzzle tasks.[10] An African grey parrot in the Pepperberg laboratory, learned and spoke in context more than 100 words,[11] identified shapes and numbers by name,[12] understood the concept of zero,[13] and the ordinality of numbers.[14]

Choice and control over environment is a primary reinforcer among animals. Brent and Weaver (1996) discovered that greater degrees of environment control led to beneficial effects on the behavior of singly-housed animals.[15] Control is a vital part of psychological wellbeing, and some researchers urge that animals should even be allowed to choose their preferred enrichment objects and activities.[16]

Some parrots exhibit challenging behaviors. Nearly 40% of cockatoos and African grey parrots engage in feather destructive behavior, a form of self-injury.[17] Many also display aggressive behavior and excessive vocalization. Although the origin of these behaviors is complex, some behaviors may result from captive environments that lack choice and control. Some researchers have found that captive parrots that are free-roaming for 4-10 hours per day display lower levels of feather destructive behavior.

Augmentative Alternative Communication and Choice-Making

Augmentative alternative communication (AAC) is a graphic-symbol based communication method that has been developed for use by some special-needs individuals.[18] One type is comprised of a touchscreen computer board that allows the user to touch a picture to communicate requests and participate socially with others. Some special needs individuals exhibit challenging behaviors including self-injury, aggression toward others, damage to the physical environment, and tantrums.[19] AAC devices aid in social interaction, cognitive development, increase independence and control over the individual's environment through choice making, and decrease challenging behavior.[19]

In a meta-analysis of twenty-four single subject case studies, AAC training intervention was found to decrease these behaviors as individuals developed the ability to communicate requests using pictures as an alternative to the challenging behavior, and their quality of life was enhanced through the facilitation of communication.[19]

With millions of animals in home-settings, facilitating an accessible means of communication via commercially available AAC apps on a tablet can open up pathways to choice requests and enhance quality of life.

This study analyzed whether a Goffin's cockatoo can be trained to use an AAC device through associative conditioning and observed behavioral corroborations and communication consistency as evidence of intention matching the symbols requested.

Methods

This study involved an enculturated female Goffin's cockatoo that participated in previous research on symbol-representation. The bird was housed with other cockatoos and had constant access to food and water, regular mental enrichment training, and access to engagement with natural foraging materials. It was trained on word and picture symbols based on associative conditioning. It was also trained to use the symbols for choice tasks that are generalized to other objects and concepts, including toys, food and beverage items, and activities. The bird has seemingly learned to use the symbols on the CommBoard app on a tablet. Historically, the bird scored 70% accuracy on a match-to-sample paradigm using color cards and its use of the symbols, and 70% on sound-text correspondence tasks. This project followed applicable international and national guidelines for the care and use of animals. The training method involved voluntary training sessions three-to-five times per week for 20-45 minutes per session, and the use of positive reinforcement.

In this project, a Samsung Galaxy Tab A tablet with a protective case and the CommBoards AAC Speech Assist for Android, purchased in the Playstore, were used. The picture stimuli were derived from in-app pictures, Google images, and photos taken by the researcher. The bird was trained using varied reinforcers such as sunflower seeds and pine nuts. The researcher taught the bird to ring a toy bell to request access to a speech board. The training procedure included frustration avoidance techniques and pexigram training using a higher value (pine nut) and a lower value (sunflower seed) reinforcer, and if it got two answers wrong in a row, it was invited to engage in a simpler target touch task for a treat, to maintain momentum and lower frustration in learning.

Results

Inter-observer agreement (IOA) data were collected during observations of each of the communication opportunities. For 30% (10) of the 33 included communication opportunities, each researcher scored the recorded videos independently. IOA was calculated for each intervention step through point-by-point agreement by dividing the total agreements by the total agreements plus disagreements and multiplying by 100%. For the steps collected with event recording, point-by-point agreement was calculated by comparing each recorded response. IOA averaged 97.5% between the two researchers.

The 33 included communication opportunities comprised 81 cockatoo requests and researcher follow-up questions on the tablet. In all interactions, the cockatoo ultimately selected a pexigram that it corroborated. The cockatoo had a 94% corroboration rate on the initial request-corroboration task questions. Corroboration of the tasks comprised the following: follow-up questions (e.g., repeated question or yes-no question) - 85%; body language corroboration - 97%; two forms of corroboration - follow-up questions and body language corroboration - were observed in 79% of the communication opportunities.

The cockatoo's requests comprised: interactive activities (e.g., tracing, play outside) - 44%; food or beverage (e.g., treat, tea) - 36%; interactions with other humans - 12%; and object requests - 8%.

The bird's incorrect answers comprised a submenu item that was directly under the menu selection and may have been an inadvertent press and another request that was declined and not corroborated. In addition, the two excluded interactions involved corroboration attempts which were declined by the cockatoo: the cockatoo pushed away the offered hand for voluntary transport based on the requests in the two excluded sessions.

Some additional observations were made. Although the cockatoo was not specifically trained in-depth on the use of the want pexigram, in one session, it spontaneously began pressing "want - treat," which mirrored the question format ("What do you want?") The cockatoo also was not trained on the "to drink" menu, and spontaneously touched the picture of a glass with the label "To Drink," requested tea by picking a picture of a teacup with a label "Tea," and corroborated with a "yes" response. The researcher equidistantly offered the bird a cup of water and a cup of tea, and it drank five gulps of warm tea, rejecting the water.

Finally, most communication opportunities concluded with a yes-no corroboration question (i.e., "Do you want a treat? Yes or no.") In some later communication opportunities, the cockatoo selected a request picture and then immediately touched "yes" (e.g., treat-yes).

Discussion

Goffin's cockatoos have demonstrated remarkable skill on cognitive tasks, including puzzle tests [12], inference through exclusion [26], and tool manufacturing [2].

The goal of this pilot project was to see whether an enculturated Goffin's cockatoo could learn to touch pictures on a communication board to make requests, and whether after selecting the picture, the request selection would be corroborated by consistency in follow-up question responses and body language (accept/reject of selection, voluntary transfer on to hand to execute request). The two corroboration types were selected because of the bird's training in the use of "yes" and "no" symbols, its prior testing results indicating basic knowledge of yes and no symbols, and its history of body language (accepting/refusing objects, dropping food items, or refusing to engage/walking away from offered activities).

The subject cockatoo in these communication opportunities demonstrated high levels of consistency (95%) between its selection requests and its corroboration tasks. The IOA evidenced a high level of fidelity (97.5%) between the two researchers' observations on video-recorded requests and corroboration tasks.

The cockatoo most often requested play and learning activities and favorite food items, but also engaged in unexpected generalization of requests to non-associated pictures. This study demonstrates the potential for using AAC devices and communication training as a behavioral intervention for challenging behaviors in animals. The study is significant because it demonstrates the potential of using tablet devices and AAC apps to train non-human animals in communication.

Conclusions

This study investigated whether a Goffin's cockatoo's communication board selections after associative training would be positively corroborated with follow-up questions and body language. Specifically, the research comprised two questions: 1) can an enculturated Goffin's cockatoo be trained through associative conditioning and positive reinforcement training to select pictures on a commercially available tablet communication board? and 2) are the picture requests positively corroborated by follow-up questions and body language?

Following associative conditioning, the cockatoo was presented with the board by making requests across 33 analyzed sessions. The bird demonstrated a 95% corroboration rate with high IOA fidelity on instances of clear selection, indicating that it was able to use the tablet communication board app to communicate requests.

References

1. Davey GC, Cleland, GG. Food anticipation and lever-directed activities in rats. *Learn Motiv.* 1984;15(1):12-36. doi.org/10.1016/0023-9690(84)90014-6.

2. Barker DJ, Sanabria F, Lasswell A, et al. Brief light as a practical aversive stimulus for the albino rat. *Behav Brain Res.* 2010;214(2):402-408. doi.org/10.1016/j.bbr.2010.06.020.

3. Mejdell C, Buvik T, Jørgensen G, Bøe K. Horses can learn to use symbols to communicate their preferences. *Appl Anim Behav Sci.* 2016;184.

4. Herzing D. Interfaces and keyboards for human-dolphin communication: what have we learned? *Anim Behav Cogn.* 2016;3:243-254. doi:10.12966/abc.04.11.2016.

5. Rafaelle P. Speaking bonobo. Smithsonian Magazine. 2006. Available at: www.smithsonianmag.com/science-nature/speaking-bonobo-134931541. Accessed June 1, 2023.

6. Marzluff JM, Walls J, Cornell HN, et al. Lasting recognition of threatening people by wild American crows. *Anim Behav.* 2010;79(3):699-707. doi.org/10.1016/j.anbehav.2009.12.022.

7. Logan CJ, Jelbert SA, Breen AJ, et al. Modifications to the Aesop's Fable Paradigm Change New Caledonian Crow Performances. *PLoS ONE.* 2014;9(7):e103049. doi.org/10.1371/journal.pone.0103049.

8. O'Hara M, Auersperg AMI, Bugnyar T, Huber L. Inference by exclusion in Goffin cockatoos (*Cacatua goffini*). *PLoS ONE.* 2015;(10)8:e0134894. doi:10.1371/journal.pone.0134894.

9. Auersperg AM, Szabo B, von Bayern AM, Kacelnik A. Spontaneous innovation in tool manufacture and use in a Goffin's cockatoo. *Curr Biol.* 2012;22(21):R903-R904. doi:10.1016/j.cub.2012.09.002

10. Habl C, Auersperg AMI. The keybox: Shape-frame fitting during tool use in Goffin's cockatoos (*Cacatua goffiniana*). *PLoS ONE.* 2017;12(11):e0186859. doi.org/10.1371/journal.pone.0186859.

11. Pepperberg IM. Functional vocalizations by an African grey parrot (*Psittacus erithacus*). *Zeitschrift Für Tierpsychologie.* 1981;55(2):139-160.

12. Pepperberg IM. Acquisition of the same/different concept by an African grey parrot (*Psittacus erithacus*): learning with respect to categories of color, shape, and material. *Anim Learn Behav.* 1987;15(4):423-432. doi.org/10.3758/BF03205051.

13. Pepperberg IM. Grey parrot (*Psittacus erithacus*) numerical abilities: addition and further experiments on a zero-like concept. *J Comp Psychol.* 2006;120(1):1-11. doi.org/10.1037/0735-7036.120.1.1.

14. Pepperberg IM, Carey S. Grey parrot number acquisition: the inference of cardinal value from ordinal position on the numeral list. *Cognition.* 2012;125(2):219-232. doi.org/10.1016/j.cognition.2012.07.003

15. Brent L, Weaver O. The physiological and behavioral effects of radio music on singly housed baboons. *J Med Primatol.* 1996;25(5):370-374. doi.org/10.1111/j.16000684.1996.tb00031.x.

16. Washburn DA. The four Cs of psychological wellbeing: lessons from three decades of computer-based environmental enrichment. *Anim Behav Cogn.* 2015;2(3):218-232. doi.org/10.12966/abc.08.02.2015.

17. Jayson S, Williams D, Wood J. Prevalence and risk factors of feather plucking in African grey parrots (*Psittacus erithacus erithacus* and *Psittacus erithacus timneh*) and cockatoo (*Cacatua* spp.). *J Exot Pet Med.* 2014;23:250-257. doi:10.1053/j.jepm.2014.06.012.

18. Drager K, Light J, McNaughton D. Effects of AAC interventions on communication and language for young children with complex communication needs. *J Pediatr Rehabil Med.* 2010;3(4):303-310. doi.org/10.3233/PRM-2010-0141.

19. Ganz JB, Earles-Vollrath TL, Heath AK, et al. A meta-analysis of single case research studies on aided augmentative and alternative communication systems with individuals with autism spectrum disorders. *J Autism Dev Disord.* 2012;42(1):60-74. doi.org/10.1007/s10803-011-1212-2.

Section 4

AAV Wildlife, Anesthesia, and Surgery Session

Seth Oster, DVM;
Sayrah Gilbert, DVM
Session Chairs

Juvenile Wild Galah Wasting Syndrome

Bob Doneley, BVSc, FANZCVS (Avian Medicine), and
David Phalen, DVM, PhD

Session #3269

Affiliation: From the School of Veterinary Science, The University of Queensland, The UQ Gatton Campus, Gatton, Queensland 4343 Australia (Doneley), Sydney School of Veterinary Science, University of Sydney, 425 Werombi Rd, Camden 2570 Australia (Phalen).

Note: Abstract format selected by author(s). See Front Matter for description of conference proceedings manuscripts submitted as scientific abstracts.

Abstract: Over the last 20 years, the authors have observed a wasting syndrome in wild juvenile galahs (*Eolophus roseicapilla*) and little corellas (*Cacatua sanguinea*) in eastern Australia. The syndrome is characterized by mucoid, watery, or tarry diarrhea and emaciation. *Macrorhabdus ornithogaster* (MO) and/or a *Spironucleus* sp. may be present in the feces. Few birds respond to treatment and most die. To determine if there was a correlation between MO or *Spironucleus* infection and the wasting syndrome, 21 wild galahs and corellas were allocated into two groups: those with possible wasting syndrome (n = 14; group 1) or those with other causes for presentation (e.g., road trauma; n = 7; group 2). Wet preparations of feces were examined at 400x magnification. All birds were euthanized, necropsied, and samples were collected for PCR and histopathology. The mean weight of group 1 birds was significantly less than those in group 2. MO and *Spironucleus* were identified in a significantly greater proportion of the birds in group 1 than in group 2. PCR proved to be more sensitive than direct fecal examination for the detection of infection with MO and *Spironucleus* (Table 1). A chronic lymphoplasmacytic enteritis with blunting and fusing of the duodenal villi was found in group 1 birds, but not in group 2 birds. In conclusion, a correlation between wasting disease and infection with either or both MO and *Spironucleus* was found. Therefore, treatment for both is indicated. Duodenal lesions in group 1 birds were most consistent with those caused by a recent enteric viral infection, suggesting the involvement of a third pathogen contributing to this syndrome. The intestinal lesions may be sufficiently severe to cause a maldigestion or malabsorption syndrome and treatment of this may be required to improve treatment outcomes.

Table 1. Weight and faecal examination results of galahs and corellas.

	Group 1		Group 2	
Mean weight (galahs)	205 g (n = 8)		305 g (n = 5)	
Mean weight (corellas)	238 g (n = 6)		350 g (n = 2)	
	Number	%	Number	%
Direct fecal examination: *Macrorhabdus*	3/12	25.0	0/7	0.0
PCR: *Macrorhabdus*	11/14	78.6	4/7	57.1
Direct fecal examination: *Spironucleus*	1/11	0.09	0/7	0.0
PCR: *Spironucleus*	7/12	58.3	2/6	33.0

Retrospective Postmortem Evaluation of Cardiac Valvular Disease in Eagles

Kristina S. Vuong, Dr med vet,
Michael Jones, Dr med vet, DABVP (Avian Practice),
Linden E. Craig, Dr med vet, PhD, DACVP

Session #3321

Affiliation: From the Department of Small Animal Clinical Sciences (Vuong, Jones), and the Department of Biomedical and Diagnostic Sciences, University of Tennessee, College of Veterinary Medicine, 2407 River Drive, Knoxville, TN 37996, USA (Craig).

Abstract: Cardiovascular disease in birds of prey is underreported compared to their companion avian counterparts.[1,2] Literature regarding valvular lesions in birds of prey is scarce and include singular reports of left atrioventricular valvular endocarditis in an adult, free-ranging, male bald eagle (*Haliaeetus leucocephalus*)[3] and aortic valvular endocarditis in an adult, free-ranging, female red-tailed hawk (*Buteo jamaicensis*).[4] The purpose of this presentation is to describe the prevalence, signalment, gross necropsy findings, and histologic lesions of valvular lesions in eagles. In this retrospective study, necropsy reports of 24 free-ranging and captive eagles were evaluated over a 15-year period (July 3, 2006 - February 28, 2021). Five bald eagles (*Haliaeetus leucocephalus*) and one golden eagle (*Aquila chysaetos*) met the inclusion criteria demonstrating gross and histologic cardiac valvular lesions (25% prevalence). All six birds were captive adults (> 1 year of age) with the majority having a known age equal to or greater than 25 years (5/6; 83.3%). Four of the affected birds were female and two were male. The aortic and left atrioventricular valves were equally affected. Myxomatous valvular degeneration was observed in four eagles and endocarditis including *Staphylococcus aureus* vegetative endocarditis was observed in two birds. Acute or chronic cerebral infarcts were present in all six of the birds, which contributed to their death. Valvular cardiac disease should be considered as a differential diagnosis in eagles exhibiting respiratory distress, neurologic signs, syncope, and/or sudden death. This presentation contributes to the limited literature regarding cardiac valvular disease and secondary cerebral infarcts in these avian species.

Introduction

Limited data is available about cardiac diseases in birds of prey despite their prevalence as avian patients.[1,2] Cardiomyopathies, endocarditis, atherosclerosis, and pathogenic arrhythmias (e.g., supraventricular premature complexes, second degree atrioventricular blocks) have been described in raptor species.[1,3,4] A retrospective study of postmortem examinations conducted on 75 captive birds of prey found that 24% of the birds had cardiac lesions, including myofiber degeneration, myocarditis, endocarditis, pericarditis, atherosclerosis, and thrombi.[5] These lesions can have infectious (e.g., West Nile virus, bornavirus, sarcocystis), nutritional, toxic, inflammatory, or idiopathic origins.[6] Information pertaining to cardiac valvular diseases in birds of prey are scarce, and secondary cerebral infarcts have not been reported in these species.

The most common valvular diseases described in other veterinary species, such as companion birds and dogs, include vegetative endocarditis and myxomatous degeneration of the left atrioventricular valve.[7-9] Myxomatous valvular degeneration is the non-inflammatory deposition of myxomatous connective tissue (e.g., glycosaminoglycans) within a disorganized extracellular matrix.[6, 10-13] Right-sided congestive heart failure is more common in avian species.[6,9,14] During the initial stages of right-sided congestive heart failure, the right ventricle undergoes concentric hypertrophy in response to pressure overload.[14] Dilation of the right ventricle occurs in response to volume overload, leading to regurgitation of blood through the spiral right atrioventricular valve in birds.[7,14] Birds are more prone to pulmonary hypertension due to the limited ability of the pulmonary capillaries to dilate within non-expansile lungs.[14] Right-sided congestive heart failure often progresses to biventricular heart failure.[6] Clinical signs of cardiac valvular disease are typically subclinical but can progress to dyspnea, syncope, and arrhythmias.[1] The prognosis becomes guarded if secondary congestive heart failure develops.[7]

Bald eagles (*Haliaeetus leucocephalus*) are often admitted to rehabilitation centers and veterinary teaching hospitals in North America.[15] The objective of this retrospective study was to evaluate the prevalence, gross necropsy findings, and histopathologic lesions of cardiac valvular disease in eagles.

Materials and Methods

Necropsy reports completed by the pathology service at the University of Tennessee, College of Veterinary Medicine (Knoxville, TN, USA) were searched to identify eagles with cardiac valvular lesions. Necropsies were evaluated over a 15-year period (July 3, 2006 - February 28, 2021). The pathology archival database was surveyed for the term "valve" within the necropsy findings of eagles during this period. Only cases with gross and/or microscopic valve lesions were included in this retrospective study. The database included necropsies conducted on free ranging eagles brought to the facility by good Samaritans and captive, non-releasable eagles housed at the American Eagle Foundation (Pigeon Forge, TN, USA). Information regarding signalment, diet, housing, and previous medical information was recorded when available. Captive eagles that resided at the American Eagle Foundation were housed in approximately 50-meter × 100-meter outdoor enclosures. The diet consisted of commercially sourced quail, rats, mice, and trout, and the birds also had continuous access to water. The outdoor enclosures were large enough for birds to fly short distances. No extra exercise regimens were enforced on the eagles. Annual physical examinations with complete blood counts, plasma biochemistry panels, West Nile virus vaccinations (1 ml intramuscularly; Vetera West Nile Equine Vaccine, Boehringer Ingelheim Vetmedica Inc, Saint Joseph, MO, USA), and ivermectin injections (0.4 mg/kg intramuscularly; Durvet, Blue Springs, MO, USA) were completed by the Avian and Zoological Medicine service at the University of Tennessee.

Results

A total of 24 eagles had necropsies completed between July 3, 2006 - February 28, 2021. Six of the eagles (25%, 95% confidence interval: 8.9-58.9% of the eagles) were reported to have cardiac valvular lesions. Five bald eagles and one golden eagle (*Aquila chrysaetos*) met the inclusion criteria. All six birds were captive adults (> 1 year of age), with the majority having a known age of equal to or greater than 25 years (83%; 5/6). Four of the affected birds were female and two were male. The main clinical signs noted in the study birds prior to death were respiratory (dyspnea, inspiratory and expiratory wheezes), neurologic (seizures, torticollis, nystagmus, head tremors), or cardiovascular in origin (syncope). Three cases were found dead in their enclosures, two were euthanized due to a poor prognosis, and one died despite aggressive supportive care with intraosseous fluids, oxygen support, and antibiotic treatment (ceftiofur 10 mg/kg IM; Excede, Zoetis, Parsippany, NJ, USA).

The most common heart valves to have pathologic changes were the left atrioventricular valve (cases 1, 2, 4, 5) and aortic valve (cases 1, 2, 3, 6). The valvular lesions were characterized as either myxomatous valvular degeneration (cases 1, 3, 4, 6) or endocarditis (cases 2, 5). Thickened myxomatous valves with chondroid metaplasia were present in the left atrioventricular valves of cases 1 and 4 and the aortic valves of cases 1, 3, and 6. Valvular mineralization was present in cases 3, 4, and 6. The cause of the valvular changes was unknown in most cases. Heterophilic endocarditis affecting the left atrioventricular and aortic valves was present in case 2, which also was diagnosed with a pulmonary adenocarcinoma that had metastasized to the kidneys, liver, and pleura. Left atrioventricular vegetative endocarditis secondary to a *Staphylococcus aureus* infection was present in case 5. Acute or chronic cerebral infarcts were present in all cases. The valvular lesions and cerebral infarcts were diagnosed as the cause of death in 83.3% (5/6) of the cases. Pulmonary adenocarcinoma with multisystemic metastasis was the main reason case 2 was euthanized.

Discussion

Myxomatous valvular degeneration (MVD) was the most prevalent cardiac valvular lesion identified in this retrospective study of eagles. Valves grossly appear smooth and thickened due to the non-inflammatory deposition of myxomatous connective tissue.[6,10-13] Myxomatous valvular degeneration is rarely reported in avian species but there are descriptions of MVD affecting the left atrioventricular valve in an Indian ringneck parakeet (*Psittacula krameria*),[9] ostrich (*Struthio camelus*),[16] and Amazon parrot (*Amazona aestiva*),[17] as well as the right atrioventricular valve in an African black-footed penguin (*Spheniscus demersus*).[18] Conversely, MVD is the most prevalent cardiovascular disease in dogs, often progressing to left atrioventricular valve regurgitation and left-sided congestive heart failure.[8,11,12,19] The left atrioventricular valve is the most commonly affected valve in the dog (62%), with 32.5% of cases involving both the left and right atrioventricular valves.[8,11,13] The pathophysiology of MVD in dogs is unclear but there is a predisposition in small to medium breeds, older animals (> 7 years), and males.[8,11-13] Myxomatous valvular degeneration is rarely reported in other common animal species treated by veterinarians.[12,19] There are histopathologic similarities between the structure of the left atrioventricular valve in humans and dogs.[12] On scanning electron microscopy, there is an uneven distribution of endothelial damage, especially on the edge of the valvular leaflets in dogs with MVD.[19] Myxomatous valvular degeneration in humans affects the left atrioventricular valve and aortic valve, similar to the results we found in the six eagles.[20,21] The cause in humans is also unclear but is correlated to age (>65 years), genetics, and is more common in females.[12,20] The prevalence of MVD in our study (16.6%, 95% confidence interval: 0.1-31.5; 4/24) was higher than that of small breed dogs (14.4%)[22] and humans (2.5%).[20] No conclusions can be made given the limited sample size from this study.

Endocarditis was present in two eagles in this study. Vegetative left atrioventricular valve endocarditis with multisystemic necrotizing vasculitis was present in case 5. *Staphylococcus aureus* cultured from blood and the vegetative valve was thought to have been spread hematogenously from a previous traumatic skin wound. Bacteria, including *Staphylococcus* species, *Streptococcus* species, *Pasteurella multocida*, *E. coli*, and *Pseudomonas* species, are commonly cultured in avian vegetative endocarditis with the most common isolate being *Staphylococcus aureus*.[2] These bacteria originate from acute traumatic wounds and focal infections such as salpingitis, pododermatitis, or gut translocation.[1,3,7] The lesions in this case were consistent with endocarditis lesions in other species.[10] The most commonly affected valves in birds are the left atrioventricular and aortic valves.[23] Clinical signs associated with vegetative endocarditis include lethargy, decreased appetite, and sudden death.[3] Bacterial thrombi from vegetative lesions can embolize to multiple organs including the lung, liver, spleen, kidney, pancreas, and brain. Vegetative endocarditis has been described most frequently in poultry,[24] with additional reports of non-hemolytic, coagulase positive *Staphylococcus* species on the aortic and left atrioventricular valve in an emu (*Dromaius novaehollandiae*),[23] *Streptococcus* species on both atrioventricular valves in a Waldrapp ibis (*Geronticus eremita*),[25] *Staphylococcus aureus* on the left atrioventricular valves of mallard ducks (*Anas platyrhynchos*),[26] *Enterobacter cloacae* on the left atrioventricular valve of a blue and gold macaw (*Ara ararauna*),[27] and *Staphylococcus aureus* on the left atrioventricular valve of a cockatoo (*Cacatua alba*).[28] In birds of prey, left atrioventricular valvular endocarditis caused from suspected *Pasteurella multocida* with disseminated thromboembolic lesions was described in an adult free-ranging male bald eagle.[3] Suspected staphylococcal aortic valvular endocarditis with multisystemic thrombi was reported in a free-ranging female red-tailed hawk (*Buteo jamaicensis*) with an acute left proximal radial fracture.[4] Diagnosis of the underlying cause of vegetative endocarditis is made through blood culture and echocardiogram but prognosis is poor despite long-term antibiotic therapy.[29]

The acute cerebral infarctions in cases 1, 3, 4, and 6 were attributed to a turbulence-induced thromboembolus produced from the degenerate left atrioventricular and/or aortic valves. Case 3 had an acute cerebellar infarct in addition to a chronic cerebral infarct that likely resulted from endocarditis and left atrioventricular valve regurgitation, which were diagnosed through echocardiographic imaging 5 months prior to necropsy. The relationship between valvular disease and cardioembolic cerebral infarctions in birds is not well established.[14] There is a high risk of a cerebrovascular event in humans with valvular heart disease, especially if the person has concurrent atrial fibrillation, myocardial infarction, prosthetic valves, and dilated cardiomyopathy.[12,19,30-32] Mitral valve prolapse is seen with left atrioventricular MVD in humans and surgical reconstruction increases the risk of thromboemboli.[19,33] There are no reports that MVD increases the risk of a thromboembolism in animal species (dogs) commonly treated by veterinarians despite observing platelets adhered to the exposed subendothelial matrix under scanning electron microscopy.[11,12,19] The differences observed between the risk of cardioembolic cerebral infarctions in dogs and humans remains unknown but are thought to involve differences in platelet function.[19] Avian thrombocytes have longer clotting times and have less tight aggregates that may result from differences in the GPIb*alpha* receptor.[34] Vascular encephalopathies in dogs have been reported to be associated with atherosclerosis secondary to hypothyroidism, parasitic emboli from *Dirofilaria immitis*, bacterial vegetative endocarditis, sepsis, neoplasia, coagulopathies, and trauma.[35,36] Thrombi move through the middle cerebral artery in both dogs and humans and result in weakness, syncope, altered mentation, and seizures, as observed in the eagles.[31,35]

Thromboemboli are rarely reported in birds. Fat and egg yolk emboli in psittacine birds and have been associated with hepatic lipidosis and atherosclerosis.[37] Avian thrombocytes have different aggregation properties and are less able to form shear-resistant thrombi.[14] Atherosclerosis is a progressive, degenerative disease that leads to the dystrophic calcification, chondroid metaplasia, necrosis, collagen, lipid, foam cell and fibrin deposition, luminal stenosis, and decreased elasticity of arteries;[2,10,38] however, there was no evidence of atherosclerosis in the six eagles.

The left atrioventricular and aortic endocarditis found in case 2 were unexpected findings. One cause of illness in this bird was pulmonary adenocarcinoma with widespread metastasis to the kidneys, liver, and pleura. The cerebral infarct resulted in the neurologic signs, which contributed to the decision to euthanize this animal and was more chronic than the identified left atrioventricular and aortic endocarditis. Primary respiratory tract neoplasms (e.g., pulmonary adenocarcinomas) are typically malignant and rare, affecting approximately 0.1-0.2% of birds.[39-42] Pulmonary carcinoma causing progressive weakness and paraparesis has been reported in a great horned owl (*Bubo virginianus*),[43] Moluccan cockatoo (*Cacatua moluccensis*),[39] hyacinth macaw (*Anodorhynchus hyacinthinus*),[41] African Grey parrot (*Psittacus erithacus*),[44] and blue and gold macaw (*Ara ararauna*).[45] These neoplasms often metastasize to the air sacs, pneumatized humerus, vertebrae, and spinal cord.[39,41,43-45] Multiple thromboembolic events secondary to a nonbacterial thrombotic endocarditis was observed in a human with concurrent pulmonary adenocarcinoma.[46] The nonbacterial endocarditis was attributed to a paraneoplastic hypercoagulation in the human patient.[46] We speculate that the same pathogenesis was involved in the eagle identified as case 2 in this study.

The cause of the valvular lesions in this retrospective study was undetermined, but age is a suspected factor in the MVD cases. Valvular disease was not encountered on any wild eagle necropsies. Captive birds of prey are susceptible to cardiac disease due to their higher fat diet, less active lifestyle, and ability to reach geriatric ages.[1] Five of the six eagles were > 25 years old. Females may have a higher prevalence of cardiac disease given the mobilization of cholesterol, low density lipoproteins, and triglycerides during ovulation (egg laying).[1,47] Limitations of this study include a potential geographic bias as well as similar lifestyles of the captive eagles at this one facility.

Acknowledgments: We thank the American Eagle Foundation for their contributions to the health and wellbeing of these non-releasable eagles.

References

1. Oster SC, Pariaut R. Cardiac disease of raptors. *J Avian Med Surg.* 2021;35(4):382-389.

2. Jones MP. Vascular diseases in birds of prey. *J Exot Pet Med.* 2013;22(1):348-357.

3. Jessup DA. Valvular endocarditis and bacteremia in a bald eagle. *Mod Vet Pract.* 1980;61(1):49-51.

4. Lemon MJ, Pack L, Forzán MJ. Valvular endocarditis and septic thrombosis associated with a radial fracture in a red-tailed hawk (*Buteo jamaicensis*). *Can Vet J.* 2012;53(1):79-82.

5. Cooper JE, Pomerance A. Cardiac lesions in birds of prey. *J Comp Pathol.* 1982;92(1):161-168.

6. Knafo SE, Rapoport G, Williams J, et al. Cardiomyopathy and right-sided congestive heart failure in red-tailed hawk (*Buteo jamaicensis*). *J Avian Med Surg.* 2011;25(1):32-39.

7. Strunk A, Wilson GH. Avian cardiology. *Vet Clin North Am Exot Anim Pract.* 2003;6(1):1-28.

8. Borgarelli M, Buchanan JW. Historical review, epidemiology and natural history of degenerative mitral valve disease. *J Vet Cardiol.* 2012;14(1):93-101.

9. Oglesbee BL, Lehmkuhl L. Congestive heart failure associated with myxomatous degeneration of the left atrioventricular valve in a parakeet. *J Am Vet Med Assoc.* 2001;218(3):376-380.

10. Schmidt RE, Reavill DR, Phalen DN. Cardiovascular system. In: Schmidt RE, Reavill DR, Phalen DN, eds. *Pathology of Pet and Aviary Birds.* 2nd Ed. Hoboken, NJ: John Wiley & Sons, Inc; 2015:1-19.

11. Keene BW, Atkins CE, Bonagura JD, et al. ACVIM consensus guidelines for the diagnosis and treatment of myxomatous mitral valve disease in dogs. *J Vet Intern Med.* 2019;33(1):1127-1140.

12. Aupperle H, Disatian S. Pathology, protein expression and signaling in myxomatous mitral valve degeneration: comparison of dogs and humans. *J Vet Cardiol.* 2012;14(1):59-71.

13. Fox PR. Pathology of myxomatous mitral valve disease in the dog. *J Vet Cardiol.* 2012;14(1):103-126.

14. Fitzgerald BC, Beaufrère H. Cardiology. In: Speer BL, ed. *Current Therapy in Avian Medicine and Surgery.* St. Louis, MO: Elsevier; 2016:252-328.

15. Deem S, Terrell S, Forrester D. A retrospective study of morbidity and mortality of raptors in Florida: 1988-1994. *J Zoo Wildl Med.* 1998;29(1):160-164.

16. Kubba MAG, Al-Azreg SA. Endocardiosis and congestive heart failure in a captive ostrich (*Struthio camelus*). *Open Vet J.* 2013;3(2):121-125.

17. Kagohara A, dos Santos Filho M, Marinho JPM, et al. Myxomatous degeneration of the left atrioventricular valve in a true parrot (*Amazona aestiva*): a case report. *Braz J Vet Med.* 2020;42(1):e104720-e104720.

18. McNaughton A, Frasa S Jr, Mishra N, et al. Valvular dysplasia and congestive heart failure in a juvenile African penguin (*Spheniscus demersus*). *J Zoo Wildl Med.* 2014;45:987-990.

19. Corcoran BM, Black A, Anderson H, et al. Identification of surface morphologic changes in the mitral valve leaflets and chordae tendineae of dogs with myxomatous degeneration. *Am J Vet Res.* 2004;65(2):198-206.

20. Lung B, Vahanian A. Epidemiology of valvular heart disease in the adult. *Nat Rev Cardiol.* 2011;8(1): 162-172.

21. Shipton B, Wahba H. Valvular heart disease: review and update. *Am Fam Physician.* 2001;63(11):2201-2208.

22. Serfass P, Chetboul V, Sampedrano CC, et al. Retrospective study of 942 small-sized dogs: prevalence of left apical systolic heart murmur and left-sided heart failure, critical effects of breed and sex. *J Vet Cardiol.* 2006;8(1):11-18.

23. Randolph JF, Moise S, Graham DL, et al. Bacterial endocarditis and thromboembolism of a pelvic limb in an emu. *J Am Vet Med Assoc.* 1984;185(11):1409-1410.

24. Chadfield MS, Christensen JP, Christensen H, et al. Characterization of streptococci and enterococci associated with septicemia in broiler parents with a high prevalence of endocarditis. *Avian Pathol.* 2004;33(6): 610-617.

25. Greenwood AG, Marshall J, Tinsley EGF. Vegetative endocarditis in a Waldrapp ibis. *Avian Pathol.* 1996;25(1):387-391.

26. Wobeser G, Kost W. Starvation, Staphylococcosis, and vitamin A deficiency among mallards overwintering in Saskatchewan. *J Wildl Dis.* 1992;28(2):215-222.

27. Isaza R, Buergelt C, Kollias GV. Bacteremia and vegetative endocarditis associated with a heart murmur in a blue-and-gold macaw. *Avian Dis.* 1992;36(1):1112-1116.

28. Steinagel A, Quesenberry K, Donovan T. Vegetative endocarditis due to Staphylococcus aureus in an umbrella cockatoo (*Cacatua alba*). *J Avian Med Surg.* 2019;33(4):419-426.

29. MacDonald K. Infective endocarditis in dogs: diagnosis and therapy. *Vet Clin North Am Small Anim Pract.* 2010;40(1):665-684.

30. Fleming HA, Bailey SM. Mitral valve disease, systemic embolism and anticoagulants. *Postgrad Med J.* 1971;47(1):599-604.

31. Arbox A, Alió J. Acute cardioembolic cerebral infarction: answers to clinical questions. *Curr Cardiol Rev.* 2012;8(1):54-67.

32. Petty GW, Khandheria BK, Whisnant JP, et al. Predictors of cerebrovascular events and death among patients with valvular heart disease. *Stroke.* 2000;31(1):2628-2635.

33. Cohn LH, Couper GS, Aranki SF, et al. Long-term results of mitral valve reconstruction for regurgitation of the myxomatous mitral valve. *J Thorac Cardiovasc Surg.* 1994;107(1):143-151.

34. Ribeiro AM, Zepeda-Mendoza L, Bertelsen MF, et al. A refined model of the genomic basis for phenotypic variation in vertebrate hemostasis. *BMC Evol Biol.* 2015;15(1):1-8.

35. Cook LB, Coates JR, Dewey CW, et al. Vascular encephalopathy associated with bacterial endocarditis in four dogs. *J Am Anim Hosp Assoc.* 2005;41(1):252-258.

36. Garosi L, McConnell JF, Platt SR, et al. Results of diagnostic investigations and long-term outcome of 33 dogs with brain infarction (2000-2004). *J Vet Intern Med.* 2005;19(1):725-731.

37. Zwart P, Samour J. The avian respiratory system and its non-infectious disorders: a review. *J Exot Pet Med.* 2021;37(1):39-50.

38. Lujan-Vega C, Keel K, Barker CM, et al. Evaluation of atherosclerotic lesions and risk factors of atherosclerosis in raptors in northern California. *J Avian Med Surg.* 2021;35(3):295-304.

39. Jones MP, Orosz SE, Richman LK, et al. Pulmonary carcinoma with metastasis in a Moluccan cockatoo (*Cacatua moluccensis*). *J Avian Med Surg.* 2001;15(2):107-113.

40. Reavill DR, Tumors of pet birds. *Vet Clin North Am Exot Anim Pract.* 2004;7(1):537-560.

41. Baumgartner WA, Guzman DSM, Hollibush S, et al. Bronchogenic adenocarcinoma in a Hyacinth macaw (*Anodorhynchus hyacinthinus*). *J Avian Med Surg.* 2008;22(3):218-225.

42. Garner MM. Overview of tumors: section II. In: Harrison GJ, Lightfoot TL, eds. *Clinical Avian Medicine.* Palm Beach, FL: Spix Publishing; 2006:566-571.

43. Rettenmund C, Sladky KK, Rodriguez D, et al. Pulmonary carcinoma in a great horned owl (*Bubo virginianus*). *J Zoo Wildl Med.* 2010;41(1):77-82.

44. André JP, Delverdier M. Primary bronchial carcinoma with osseous metastasis in an African grey parrot (*Psittacus erithacus*). *J Avian Med Surg.* 1999;13(3):180-186.

45. Fredholm DV, Carpenter JW, Shumacher LL, et al. Pulmonary adenocarcinoma with osseous metastasis and secondary paresis in a blue and gold macaw (*Ara ararauna*). *J Zoo Wildl Med.* 2012;43(4):909-913.

46. Kaufmann CC, Wessely E, Huber K. Non-bacterial thrombotic endocarditis in the context of pulmonary adenocarcinoma: a case report. *Eur Heart J.* 2020;4(1):1-5.

47. Beaufrère H, Ammersbach M, Reavill DR, et al. Prevalence of and risk factors associated with atherosclerosis in psittacine birds. *J Am Vet Med Assoc.* 2013;242(12):1696-1703.

Can Packed Cell Volume and Total Solids Guide Wildlife Rehabilitation Decisions for Birds?

Zoey Lex, BS,
Laken Russell, DVM,
Corinne Mayer, DVM,
Jordan Padlo, CVT,
Tithipong Plangsangmas, DVM,
Mark Mitchell, DVM, MS, PhD, DipECZM (Herpetology)

Session #3372

Affiliation: From the Department of Veterinary Clinical Sciences, School of Veterinary Medicine, Louisiana State University, Skip Bertman Drive, Baton Rouge, LA 70803, USA.

Note: Abstract format selected by author(s). See Front Matter for description of conference proceedingsmanuscripts submitted as scientific abstracts.

Abstract: Wildlife are being presented to veterinarians because of negative encounters with humans. Many of these cases are critical and require medical and surgical management; however, because of limited resources, it is important to develop triage protocols to ensure resources are applied to cases with the best chance for success. The Louisiana State University Wildlife Hospital of Louisiana (LSU-WHL) currently receives over 1,000 wildlife cases a year, primarily avian species (68.8%, 2021). As a donation-based institution, it is imperative that resources are used wisely. Because of this, there is a need to understand how economical diagnostics could better evaluate prognosis in these species. Packed cell volumes (PCV) and total solids (TS) are examples of inexpensive physiologic data that could assist veterinarians with triage. The purpose of this study was to evaluate if PCV and TS measurements obtained on intake could be used as significant prognostic indicators for avian cases presented to the LSU-WHL. The hypotheses for this study were that PCV and TS would be predictive for case outcome, with cut-off values being useful in deciding the prognosis for a case. A cross-sectional study was performed using LSU-WHL records from 2020-2022. Avian cases that had a PCV and TS recorded on intake before treatment and a final disposition were included. Birds with PCV $< 20\%$ were always found to have a negative outcome (died or euthanized); TS <1.8 g/dl were also found to be a negative predictor of outcome. These results suggest that these inexpensive tests are useful to triage avian cases.

Use of Trimethoprim Sulfamethoxazole in the Treatment of Brevetoxicosis in Seabirds and Shorebirds

Maria E. Passarelli, VMD

Session #3123

Affiliation: From Save Our Seabirds, 1708 Ken Thompson Pkwy, Sarasota, FL 34236, USA.

Note: Abstract format selected by author(s). See Front Matter for description of conference proceedings manuscripts submitted as scientific abstracts.

Abstract: The toxic dinoflagellate *Karenia brevis* blooms frequently in the warm waters of the Gulf of Mexico. These harmful algal blooms can lead to significant morbidity and mortality in marine wildlife populations of the Florida Gulf Coast. In the rehabilitation setting, treatment has traditionally focused on the primary effects of the toxin;[1,2] however, brevetoxins can also significantly alter the immune response of the patient,[3,4] which may lead to secondary complications such as infection.[5] We examined the success of a broad-spectrum antibiotic treatment protocol in comparison to a supportive care-only protocol in the treatment of brevetoxicosis in 86 seabird and shorebird patients. The species most commonly affected were double-crested cormorants (*Nannopterum auritum*), laughing gulls (*Leucophaeus atricilla*), and sanderlings (*Calidris alba*), although other species were also affected. The supportive care-only group received fluid therapy and deworming as needed with fenbendazole 20 mg/kg q24h for 3 days. The treatment group was administered trimethoprim sulfamethoxazole 30 mg/kg q12h for 7 days, fenbendazole 20 mg/kg q24h for 3 days, and fluid therapy. In the supportive care-only group, 39 out of 56 died (69.6%) and 17 out of 56 survived to release (30.4%). In the treatment group, 8 out of 30 died (26.7%) and 22 out of 30 survived to release (73.3%). This study demonstrates that the addition of a broad-spectrum antibiotic to a red tide treatment protocol may help to counteract secondary infection due to brevetoxin-induced immunocompromise in avian species.

References

1. Fauquier DA, Flewelling LJ, Maucher JM, et al. Brevetoxicosis in seabirds naturally exposed to *Karenia brevis* blooms along the central west coast of Florida. *J Wildl Dis.* 2013;49:246-260.

2. Perrault JR, Barron HW, Malinowski CR, et al. Use of intravenous lipid emulsion therapy as a novel treatment for brevetoxicosis in sea turtles. *Sci Rep.* 2021;11:1-12.

3. Gebhard E, Levin M, Bogomolni A, De Guise S. Immunomodulatory effects of brevetoxin (PbTx-3) upon in vitro exposure in bottlenose dolphins (*Tursiops truncatus*). *Harmful Algae.* 2015;44:54-62.

4. Walsh CJ, Butawan M, Yordy J, et al. Sublethal red tide toxin exposure in free-ranging manatees (*Trichechus manatus*) affects the immune system through reduced lymphocyte proliferation responses, inflammation, and oxidative stress. *Aquat Toxicol.* 2015;161:73-84.

5. Niedringhaus KD, Shender LA, DiNuovo A, et al. Mortality in common (*Sterna hirundo*) and sandwich (*Thalasseus sandvicensis*) terns associated with Bisgaard taxon 40 infection on Marco Island, Florida, USA. *J Comp Pathol.* 2021;184:12-18.

Pharmacokinetics of Hydromorphone Hydrochloride and its Metabolite in Great Horned Owls (*Bubo virginianus*)

Mariana Sosa-Higareda, MVZ,
David Sanchez-Migallon Guzman, LV, MS, DipECZM (Avian), DipECZM (Small Mammal), Dipl ACZM,
Heather Knych, DVM,
Michelle G. Hawkins, VMD, DABVP (Avian Practice)

Session #3307

AAV House Officer Manuscript Competition Winner

Affiliation: From William T. Pritchard Veterinary Medical Teaching Hospital, School of Veterinary Medicine, University of California-Davis, Davis, CA 95616, USA (Sosa-Higareda), Department of Medicine and Epidemiology, School of Veterinary Medicine, University of California-Davis, Davis, CA 95616, USA (Sanchez-Migallon Guzman, Hawkins), Maddy Equine Analytical Pharmacology Laboratory, School of Veterinary Medicine, University of California-Davis, Davis, CA 95616, USA (Knych).

Note: Abstract format selected by author(s). See Front Matter for description of conference proceedings manuscripts submitted as scientific abstracts.

Abstract: Hydromorphone hydrochloride is a semisynthetic full μ-opioid more potent than morphine for acute pain.[1] A recent study demonstrated that hydromorphone has thermal antinociceptive effects in great horned owls following IM administration of 0.3 and 0.6 mg/kg for 3 and 1.5 hours respectively.[2] The objective of this study was to determine the pharmacokinetic parameters of hydromorphone and its metabolite, hydromorphone-3-glucuronide (H3G) after a single IV and IM dose in great horned owls (*Bubo virginianus*). Six healthy adult great horned owls were administered a single dose of hydromorphone (0.6 mg/kg) either IM or IV with a 6-week washout period between experiments. Blood samples were collected at 5 minutes and 0.5, 1.5, 2, 3, 6, 9, and 12 hours after drug administration. Plasma hydromorphone and H3G concentrations were determined with liquid chromatography–tandem mass spectrometry, and pharmacokinetic parameters were calculated with a non-compartmental model. Hydromorphone had high IM bioavailability 130%, with rapid elimination after administration and a large volume of distribution. Mean C_{max} was 225.46 ± 0.21 ng /ml at 13 minutes after IM injection. Mean $t_{1/2}$ were 1.62 ± 0.36 and 1.35 ± 0.59 hours after IM and IV administration, respectively. The metabolite H3G was readily measured after IM and IV administration. Mean C_{max} was 41.14 ± 12.0 ng /ml and $t_{1/2}$ 4.65 ± 3.37 hours for the IM route. Hydromorphone was well tolerated in all birds. This study is the first to document the presence of the metabolite H3G in avian species, suggesting a hydromorphone metabolism similar to that of mammals.

References

1. Murray A, Hagen NA. Hydromorphone. *J Pain Manag.* 2005;29(5):57-66. doi:10.1016/j.jpainsym-man.2005.01.007

2. Monopoli MR, Sanchez-Migallon Guzman D, Paul-Murphy Joanne, et al. Evaluation of the thermal antinociceptive effects of hydromorphone hydrochloride in great horned owls (*Bubo virginianus*). Proc *ExoticsCon*; 2022:61.

Computed Tomography-Guided Surgical Correction of Air Sac Rupture in a Scarlet Macaw Combined with Pleurodesis-Like Procedure

Lauren Thielen, DVM, DABVP (Avian Practice), and
Scott Echols, DVM, DABVP (Avian Practice)

Session #3467

Affiliation: From Cornell University Hospital for Animals, Ithaca, NY 14853, USA (Thielen), Texas Avian & Exotic Hospital, Grapevine, TX 76051, USA (Thielen), Mobile Avian Surgical Services, Oakley, CA 94561, USA (Echols), Parrish Creek Veterinary Hospital & Diagnostic Center, Centerville, UT 84014, USA (Echols).

Note: Abstract format selected by author(s). See Front Matter for description of conference proceedings manuscripts submitted as scientific abstracts.

Abstract: An 11-year-old male scarlet macaw (*Ara macao*) presented with a 2-year progressive history of chronic air-sac rupture. A large accumulation of subcutaneous air, distending a space of 15 × 6 centimeters, along the ventrocaudal left side of the neck extending onto the dorsal head, was appreciated as a ruptured left cervical air sac. A computed tomography (CT) scan was performed to locate a rent location for surgical planning. CT findings diagnosed the open communication between the cervical and interclavicular air sacs via two individual rents. Upon surgical investigation, the largest rent (7 millimeters) was identified by visual inspection; however, the smaller rent was present deep to the initial rent and located based on CT findings. Closure of the small rent (2 millimeters) was made possible by performing surgery through the larger rent, followed by closure of the large rent. The day after surgery, although dramatically reduced, a mild to moderate amount of subcutaneous air was present. A pleurodesis-like procedure was performed by injecting doxycycline hyclate into the air sac-subcutaneous space, to generate a reaction to the skin and the air sac tissues and keep it collapsed.[1] An intravenous catheter was used to deflate the air and then inject the doxycycline, and then maintained as an air sac cannula for 1 week to prevent subcutaneous air recurrence. The parrot is currently symptom-free 1 year post operation. Despite correction of air sac rupture being described in previous literature, this is the first report of a combined CT-guided surgical correction and pleurodesis-like procedure.[2]

References

1. Antinoff N. Attempted pleurodesis for an air sac rupture in an Amazon parrot. *Proc Annu Conf Assoc Avian Vet.* 2008.

2. Browning GR, Eshar D, Tucker-Mohl K, Berke K. Diagnosis and surgical repair of a chronic ruptured cervical air sac in a double yellow-headed Amazon parrot (*Amazona ochrocephala oratrix*). *J Exot Pet Med.* 2019;29:45-50.

Hypothermia During General Anesthesia Under Isoflurane, Sevoflurane and Desflurane in Hispaniolan Amazon Parrots (*Amazona ventralis*)

Hugo Gonzalez-Jassi, MVZ,
Thomas N. Tully Jr., DVM, MS, DABVP (Avian Practice), DipECZM (Avian),
Jeannette Cremer, Dr med vet, DACVAA, CVA,
Noelia Diaz Falcón, DVM, MS,
Chin-Chi Liu, MS, MApStat, PhD,
Patricia Queiroz-Williams, DVM, MS

Session #3459

Affiliation: From the Department of Veterinary Clinical Sciences, College of Veterinary Medicine, Oklahoma State University, 2065 W. Farm Rd., Stillwater, OK 74078, USA (Gonzalez-Jassi), Department of Veterinary Clinical Sciences, School of Veterinary Medicine, Louisiana State University, Skip Bertman Dr., Baton Rouge, LA 70803, USA (Tully Jr, Cremer, Diaz Falcón, Chin-Chi Liu, Queiroz-Williams).

Note: Abstract format selected by author(s). See Front Matter for description of conference proceedings manuscripts submitted as scientific abstracts.

Abstract: Hypothermia caused by general anesthesia using isoflurane (iso), sevoflurane (sevo), and desflurane (des) was evaluated in Hispaniolan Amazon parrots (HAP). Twelve adult parrots were premedicated with butorphanol (0.5 mg/kg IM) and randomly assigned to three groups: iso, sevo, and des. Anesthesia was induced via face mask (4% iso, 6% sevo and 12% des) followed by orotracheal intubation and maintenance with 1.4-2% iso, 2.4-3% sevo, and 8.5-9.2% des. Each parrot received each inhalant on different occasions (7-day washout period). Hypothermia was defined as body temperature (BT) below 100 °F.[1] External heat support was provided when BT dropped to 99.5 °F. Time for heat loss from 102 °F to 99.5 °F (T1), time to first increase in BT from 99.5 °F (T2) and time from initial heat support to 102 °F (T3) were statistically analyzed. Heart rate, respiratory rate, oxygen saturation, end-tidal carbon dioxide, and non-invasive blood pressure were also analyzed. All three inhalants rapidly caused hypothermia (T1 Mean ± SD in mins iso:13.92 ± 8.58, sevo:11.42 ± 4.14 and des:12.58 ± 4.62). There were no statistically significant differences among inhalants over T1 and T3, but T2 had a significant difference ($P = 0.0412$) between des and sevo. Sevoflurane caused transitory apnea and respiratory depression. When hypothermic, 10 animals had a systolic arterial pressure below 100 mmHg (iso:4, sevo:3 and des:3). Two types of bradyarrhythmia were identified with all inhalants. Hypothermic HAP under general anesthesia using iso, sevo and des, had similar complications. When heat support is provided, the inhalant used does not play a major role in the time to correct this complication.

References

1. Rembert MS, Smith JA, Hosgood G, et al. Comparison of traditional thermal support devices with the forced-air warmer system in anesthetized Hispaniolan Amazon parrots (*Amazona ventralis*). *J Avian Med Surg*. 2001;15(3):187-193.

Device Composition Using a 3-Way Stop Cock for Monitoring of Capnography in Small Birds

Jungha Lee, DVM,
Donhwi Shin, DVM,
Seongchan Yeon, DVM, PhD,
Inhyung Lee, DVM, PhD,
Won-gyun Son, DVM, PhD

Session #3209

Affiliation: From the Department of Veterinary Clinical Sciences, College of Veterinary Medicine,Seoul National University, Seoul 08826, Korea.

Note: Abstract format selected by author(s). See Front Matter for description of conference proceedings manuscripts submitted as scientific abstracts.

Abstract: In small birds, perioperative monitoring of vital signs is challenging due to their size. End-tidal carbon dioxide ($EtCO_2$) monitoring is especially difficult due to relatively small tidal volume and large dead space of airway devices. In this study, without specialized equipment minimizing dead space, a low-cost device composition using a 3-way stop cock for intraoperative capnography monitoring was evaluated in small birds (eight species) weighing 70-260 grams. Medical records were reviewed for nine birds receiving general anesthesia for bone or beak fracture repair. The birds were premedicated with none, sedatives (medetomidine, midazolam, alfaxalone) and/or analgesics (hydromorphone, ketamine, tramadol) intramuscularly (IM). After mask induction, intubation was performed using an intravenous catheter (14- to 22-gauge), and was maintained with isoflurane (end-tidal concentration 2.0-2.5%) for 25-100 minutes. Monitoring of heart rate, oxygen saturation, temperature, $EtCO_2$ was attempted, but temperature and $EtCO_2$ monitoring was only possible due to the small size of the patients. To minimize dead space, an $EtCO_2$ sampling line was connected to the intubated catheter on the opposite side via 3-way stop cock for effective sampling of exhalation gas. The other way that a 3-way stop cock was connected to the breathing tube was through an extension tube. Extubation was performed after surgery and the patients recovered uneventfully except for one case; hypoperfusion and hypovolemic shock was estimated to be the reason of death in a cachexic patient. Complications associated with intubation and ventilation were not observed during peri-anesthetic period.

Comparison of Non-Controlled (Dexmedetomidine-Nalbuphine) and Partially Controlled (Dexmedetomidine-Butorphanol) Sedation Protocols in Chickens (*Gallus gallus domesticus*)

Greta Doden, DVM,
Kaitlyn Crocker, DVM,
Hannah Kunzman, DVM,
Julie Fithian, DVM,
João Brandão, LMV, MS, DipECZM (Avian), Dipl ACZM

Session #3298

Affiliation: From the Department of Veterinary Clinical Sciences, College of Veterinary Medicine, Oklahoma State University, 2065 W. Farm Rd., Stillwater, OK 74078, USA.

Note: Abstract format selected by author(s). See Front Matter for description of conference proceedings manuscripts submitted as scientific abstracts.

Abstract: Typical avian sedation protocols include controlled drugs, creating regulatory challenges for veterinarians and field researchers.[1] Recent avian studies have investigated the use of dexmedetomidine, a non-controlled α2-agonist.[2,3] Butorphanol is a commonly used κ-agonist/μ-antagonist opioid and a Schedule IV controlled substance. Nalbuphine, also a κ-agonist/μ-antagonist opioid, is currently not controlled. Few studies have assessed nalbuphine in avian species.[4-7] An effective, non-controlled sedation protocol would be particularly useful in field situations. The purpose of this study was to determine the effective sedation dosage and compare the sedation quality between non-controlled nalbuphine-dexmedetomidine (ND) and partially controlled butorphanol-dexmedetomidine (BD) protocols. Effective dose 99% of each protocol was estimated using an Up-and-Down method in 20 healthy, adult Buff Orpington hens (*Gallus gallus domesticus*). Ten hens were subsequently included in a blinded, randomized, complete crossover design. Each animal received 2 mg/kg butorphanol and 50 μg/kg dexmedetomidine or 12.5 mg/kg nalbuphine and 20 μg/kg dexmedetomidine intramuscularly. Cardiorespiratory parameters and sedation score were evaluated by a blinded observer every 10 minutes until recovery. A lower score indicated effective sedation. Both protocols induced sedation. Heart and respiratory rates did not differ between treatments, but heart rates were significantly different from baseline. BD induced lower scores for recumbency at 10, 40, and 60 minutes; muscle tone at 30, 40, and 90 minutes; and radiographic positioning at up to 30 minutes. Although BD provided lower scores at most time points, the ND combination also induced sedation and may be a useful alternative to controlled substances for light to moderate sedation in avian patients.

References

1. González MS, Adami C. Psittacine sedation and anesthesia. *Vet Clin North Am Exot Anim Pract*. 2022;25: 113-114.

2. Mumm LE, Mans C. Evaluation of dexmedetomidine-midazolam sedation in budgerigars (*Melopsittacus undulatus*). *J Am Vet Med Assoc*. 2022;260(10):1194-1199.

3. Sha J, Kongara K, Singh P, et al. Pharmacokinetics and pharmacodynamics of butorphanol and dexmedetomidine after intranasal administration in broiler chickens (*Gallus gallus domesticus*). *Vet Sci*. 2022;9(212):1-13.

4. Guzman DSM, KuKanich B, Keuler NS, et al. Antinociceptive effects of nalbuphine hydrochloride in Hispaniolan Amazon parrots (*Amazona ventralis*). *Am J Vet Res.* 2011;72:736-740.

5. Keller DL, Guzman DS-M, Klauer JM, et al. Pharmacokinetics of nalbuphine hydrochloride after intravenous and intramuscular administration to Hispaniolan Amazon parrots (*Amazona ventralis*). *Am J Vet Res.* 2011;72(6):741-745.

6. Guzman DSM, Braun JM, Steagall PVM, et al. Antinociceptive effects of long-acting nalbuphine decanoate after intramuscular administration to Hispaniolan Amazon parrots (*Amazona ventralis*). *Am J Vet Res.* 2013;74(2):196-200.

7. Guzman DS-M, KuKanich B, Heath TD, et al. Pharmacokinetics of long-acting nalbuphine decanoate after intramuscular administration to Hispaniolan Amazon parrots (*Amazona ventralis*). *Am J Vet Res.* 2013;74(2):191-195.

Section 5

AAV Masterclasses
Track 1

Chistal Pollock, DVM, DABVP (Avian Practice);
Michael Bignall, DVM
Session Chairs

Avian Zoonoses

Cheryl B. Greenacre, DVM, DABVP (Avian Practice), DABVP (Exotic Companion Mammal Practice)

Session #3347

Affiliation: From College of Veterinary Medicine, University of Tennessee, Knoxville, TN 37996, USA (Emeritus). Current address: Office of Research, University Research Animal Resources, University of Georgia, 501 D.W. Brooks Drive, Room 3504, Athens, GA 30602, USA.

Abstract: There are many diseases that can be spread from birds to humans, the most well-known being chlamydiosis. There are other less notable zoonotic diseases from birds including those that can spread directly (salmonellosis, campylobacteriosis, yersiniosis, influenza A, Newcastle disease, by mosquito (viral encephalitides) or indirectly (histoplasmosis and cryptococcosis). An overview of typical clinical signs, diagnostic and treatment options for birds are provided for common avian zoonotic pathogens.

Introduction

Diseases that can be spread from birds to humans are collectively called avian zoonoses. These diseases can be categorized by causative agent, in this case, bacterial and chlamydial, viral, fungal, and parasitic. Although the reported incidence of disease from avian zoonotic pathogens is low, and most if not all can be prevented by proper sanitation and hygiene practices and/or proper use of personal protective equipment (PPE), it is best to be familiar with the clinical signs, and diagnostic and therapeutic options available to manage these diseases. There are many resources available for further information on avian zoonoses including books, published articles, websites, and free informational pamphlets from governmental websites.[1-21]

Avian Chlamydiosis

Taxonomy and nomenclature

The disease psittacosis goes by the names parrot fever, ornithosis, chlamydiosis, and chlamydophilosis.[1-17,19-21] The term psittacosis refers to the disease in people originating from a parrot (a psittacine bird), whereas the term ornithosis refers to the disease in people originating from any species of bird. The term chlamydiosis is a more generic term and refers to infection in any animal or person with any organism in the *Chlamydia* or *Chlamydophila* (previous name) genus. The term avian chlamydiosis is used to specify an infection in a bird with any *Chlamydia* sp., including the recently identified *Chlamydia gallinacea* (turkeys and chickens) and *Chlamydia avium* (pigeons). Many changes in the nomenclature of organisms in the Chlamydiaciae family have occurred over the years. For this discussion, the organism will be called by its current name, *Chlamydia psittaci*. Note that *Chlamydia psittaci* should not be confused with a related organism in people, *Chlamydia trachomatis*, a sexually transmitted disease of people, or another related organism, *Chlamydia pneumoniae*, a common respiratory pathogen of people.

History

In the late 1800s psittacosis was recognized as a clinical entity and was called pseudotyphus. In 1892, 51 cases and 16 deaths occurred in Paris, France. In 1929-1930 a pandemic occurred involving 12 countries and approximately 800 cases. This outbreak stimulated research into the disease and it was found to be associated with parrots, particularly those recently imported from South America. Immediately prior to the pandemic a large shipment of birds from South America was distributed worldwide for sale during the Christmas season. The infective stage, known as elementary bodies, were first described independently by three different researchers, Leventhal, Cole and Lillie in 1930, which lead to the alternate name of LCL bodies. In 1945 the term chlamydia, from the Greek word chalmus for cloak, was introduced. As late as 1965 the organism was thought to be a virus. It wasn't until the organism was observed with electron microscopy to have bacterial RNA, ribosomes and cell wall structures that it was proven to be an obligate intracellular bacterial organism (a bacteria that, like viruses and the Rickettsia, is incapable of extracellular living). The *Chlamydia* sp. were lumped in taxonomically with the Rickettsia until 1966 when Page validated the placement of the organism into its own family, Chlamydiaceae.

Epidemiology

The organism has been found in 130 species of birds worldwide and a variety of mammals, including humans. There are numerous potential avian species that may act as a source of infection for people. The most common source of infection (70% of all cases in the 1980's) is exposure to a recently acquired psittacine bird, most often a budgerigar, cockatiel or lovebird. Birds smuggled across the United States borders as well as those raised in captivity can potentially be infected with avian chlamydiosis. Other birds can be a potential source of infection, such as domestic or wild pigeons, passerines (soft-billed birds) or poultry.

The general prevalence of chlamydial infection in captive birds is thought to be less than 5%, and in wild birds less than 1%, but may increase dramatically in birds stressed by shipping, crowding, chilling and breeding. If an outbreak of psittacosis occurs in a facility with multiple birds or a facility that buys and sells birds, then the potential for rapid spread to other birds and people is high. Common high-traffic situations are pet stores, bird breeding facilities, and flea markets. Often it is difficult to later identify the seller of birds at a flea market.

People can become infected with psittacosis even after brief, passing exposure to infected birds or environments and may not present with an obvious bird-related exposure history. In the 1980s, 43% of the psittacosis cases reported in humans involved bird fanciers or pet bird owners. Others at vocational risk include pigeon fanciers. Others at occupational risk include pet store employees, veterinarians, veterinary technicians, laboratory workers, workers in avian quarantine stations, farmers, wildlife rehabilitators, zoo workers and employees of poultry slaughtering and processing plants (usually turkey).[20,21] Occasionally exposure to wild pigeon roosts is a source of infection to the general public.

The distribution of this disease in birds and people is worldwide. From 1988 to 1998 the Centers for Disease Control and Prevention (CDC) received reports of 813 human cases in the United States, but since psittacosis is only a notifiable disease (physicians are not required to report it) and diagnosis is sometimes missed, the actual number of cases is probably much higher. During the 1980s, 70% of the reported cases were associated with exposure to infected pet birds. The prevalence of disease in birds is considered to be low (< 5%), but can reach 100% during an outbreak. In birds, the morbidity during an aviary outbreak is generally up to 50-80% and mortality is up to 10-30%. An excellent source of yearly documented cases in humans is the CDC's Morbidity and Mortality Weekly Report. Through the CDC there is a graph in PowerPoint form of reported human psittacosis cases in the United States between 1971 and 2001 at this website: http://www.cdc.gov/epo/dphsi/annsum/1999/slides/psitgraph.ppt.

Pathogenesis

The *C. psittaci* organism is transmitted by either inhalation or ingestion of the spore-like elementary body phase of the organism. Shedding in birds can be activated by stress, such as shipping, crowding, chilling and breeding. Person to person transmission has been suggested, but never proven. Those individuals that are immunosuppressed are more susceptible to the disease and its effects.

The organism *C. psittaci* is relatively resistant, surviving in the soil for 3 months or within a bird dropping for up to 1 month. *C. psittaci* is a dimorphic organism, meaning it exists in two different forms called phases, the infectious phase and the replicating phase. In general, the *C. psittaci* organism enters the host via inhalation or ingestion and replicates in the host's cells. The infectious form is small (0.2-0.6 microns) and called an elementary body. After the elementary body is inhaled or ingested it is endocytosed by a host cell where it transforms into a reticulate body. The reticulate body is larger (1.5 microns) than the elementary body and is the replicating, or metabolically active, form of the organism inside the host's cell. The reticulate body uses nutrients from the host cell and undergoes multiple rounds of binary division before releasing multiple elementary bodies from the cell when it ruptures. Two or more days pass between the time the host cell is infected and the elementary bodies are released. These infectious elementary bodies then infect other of the host's cells or are released into the environment via feces, nasal secretions, sputum, blood or infected tissues. The elementary body is metabolically inactive and resistant to environmental forces, so it can survive long enough to be inhaled or ingested by another host.

The *C. psittaci* organism is a documented hazard among laboratory personnel and appropriate precautions should be taken when working with this organism or birds naturally or experimentally infected with this organism. *C. psittaci* may be present in tissues, feces, nasal secretions and blood of birds and present in tissues, sputum and blood of humans. The organism easily aerosolizes making it easy to inhale the infective phase of the organism. Persons performing a necropsy of any bird suspected of having avian chlamydiosis should wear gloves and wet the carcass with soapy water to decrease aerosolization or organisms, as well as practice Animal Biosafety Level 2 practices, containment equipment and facilities and respiratory protection.

Clinical signs

Parrots

There are many species of parrots and each species, even each individual, may respond differently to being infected with *C. psittaci*. For example, some birds present quite ill whereas others exhibit very subtle signs of disease. Generally, the parrot with psittacosis presents with depression, lethargy, anorexia, dyspnea, nasal or ocular discharge, conjunctivitis, and biliverdinuria. Rarely, birds present comatose, which has been observed in sensitive species such as macaws. Commonly both the spleen and liver are enlarged.

Pigeons and passerine birds

Pigeons and passerine birds seem to exhibit little if any clinical signs of disease while infected with the *C. psittaci* (or *C. avium*) organism and therefore are sometimes referred to as asymptomatic carriers of the disease.

Turkeys and chickens

Turkeys and chickens seem to exhibit little if any clinical signs of disease while infected with the *Chlamydia* sp. and therefore are sometimes referred to as asymptomatic carriers of the disease.

Diagnosis

Suggestive

Consistent clinical signs and radiographs showing splenomegaly, ± hepatomegaly are suggestive. A CBC showing a heterophilic, monocytic leukocytosis and a mild non-regenerative anemia are also suggestive. A plasma electrophoresis may be suggestive of either acute or chronic disease.

More specific

There are many tests available for use in birds including tests to detect antibodies in the serum (elementary body assay [EBA] and immunofluoescent antibody [IFA]) and test to detect antigen in the feces or blood (enzyme linked immunofluorescent antibody assay [ELISA] and polymerase chain reaction [PCR]). It is best to perform a panel of three tests including PCR of blood, PCR of feces and IFA of serum. Also, a fluorescent antibody (FA) test can be performed on tissue such as liver tissue from a biopsy or necropsy. For legal purposes, cell culture from the feces is the best test, but the organism does not consistently grow, and shedding of the organism in the feces is intermittent, and there is also risk to laboratory personnel when the organism is grown in the laboratory. There is one commercial laboratory, Texas Medical Diagnostic Laboratory, offering culture.[1]

The following definitions have been accepted by the American Veterinary Medical Association (AVMA) and the Association of Avian Veterinarians and can be found in the Compendium of Measures to Control *Chlamydia psittaci* Infection Among Humans (Psittacosis) and Pet Birds (Avian Chlamydiosis), 2017 by the National Association of State Public Health Veterinarians (NASPHV).[1]

A confirmed case of avian chlamydial infection is defined on the basis of at least one of the following:

- Isolation of *C. psittaci* from a clinical specimen.
- Identification of chlamydial DNA by use of in situ hybridization to detect chlamydial DNA followed by specific *C. psittaci* DNA detection using PCR-based testing of in situ hybridization-positive tissues or secondary *C. psittaci*-specific DNA probes in combination with characteristic pathology. The commercial antibodies used for immunohistochemistry staining cross-react with non-chlamydial epitopes and are not diagnostic.
- A fourfold or greater change in serologic titer in two specimens from the bird obtained at least 2 weeks apart and assayed simultaneously at the same laboratory.
- Identification of suggestive intracellular bacteria within diseased cells in smears or tissues (e.g., liver, conjunctival, spleen, respiratory secretions) stained with Gimenez or Macchiavello stain in combination with detecting *C. psittaci* DNA in the same tissue sample using a *C. psittaci*-specific PCR-based detection assay.

A suspected case of avian chlamydial infection is defined as a compatible illness and at least one of the following:

- Identification of chlamydial nucleic acid by PCR-based testing in conjunctival, choanal, or cloacal swabs; blood; or feces.

- Chlamydiaceae antigen (fluorescent antibody) in feces, a cloacal swab specimen, or respiratory tract or ocular exudates. The commercial antibodies used for fluorescent antibody staining cross-react with non-chlamydial epitopes and are not diagnostic.

- Is epidemiologically linked to a confirmed case in a human or bird.

There may be cause for concern and further investigation if one of the following occurs:

- Identification of chlamydial nucleic acid by PCR-based testing in conjunctival, choanal, or cloacal swabs; blood; or feces in a healthy-appearing bird, or

- Compatible illness with positive results from a nonstandardized test or a new investigational test.

Treatment

Treatment of birds should be supervised by a licensed veterinarian. The treatment of choice in birds is oral doxycycline, since it is absorbed better and eliminated slower than other tetracyclines. Care should be taken to observe for signs of toxicosis while treating such as lethargy, anorexia, biliverdinuria. If it occurs, the doxycycline should be discontinued and the bird offered supportive care until recovered, and then a lower dose can be attempted. Doxycycline monohydrate or calcium syrup formulations are administered at a dose of 40-50 mg/kg/day for most species, although a lower dose of 25 mg/kg/day is used in African grey parrots, Goffin's cockatoos, blue and gold macaws and green-winged macaws.[13,14]

A medicated feed can be administered to budgies made of one part steel cut oat groats and three parts small-hulled millet (measured by volume) and 5-6 milliliters of sunflower oil per kilogram of feed, then mixing to coat seeds, then adding and mixing in 300 milligrams of doxycycline hyclate (from capsules) per kilogram of oat/seed/oil mix. Make fresh daily.

It has been shown through pharmacologic studies that alternatively medicated water can be offered to cockatiels at 200-400 milligrams of doxcycline hyclate per liter of drinking water, or can be offered to Goffin's cockatoos at 400-600 milligrams per liter of drinking water, or can be offered to African grey parrots at 800 mg per liter of drinking water.[13,14]

The intramuscular formulation of doxycycline called Vibravenos (Pfizer, London) is available in Europe and used at a dose of 75-100 mg/kg body weight every 7 days for four or more doses. Realize the intravenous (IV) formulation of doxycycline hyclate is not acceptable for intramuscular use, but for IV use only.

The injectable, long-acting oxytetracycline (LA-200, Pfizer, Exton, PN) has been administered in some species (Goffin's cockatoos, blue-fronted and orange-winged Amazon parrots and blue and gold macaws) subcutaneously at a dose of 75 mg/kg body weight every 3 days, but is only used short term in birds that aren't eating initially because it causes irritation at the site of injection.

There is also a chlortetracycline medicated feed (mash or pellet) available; however, it must be the only source of food and birds may not accept it very well.

Prevention and control

The following is an outline of prevention and control information provided in the Compendium of Measures to Control *Chlamydia psittaci* Infection Among Humans (Psittacosis) and Birds (Avian Chlamydiosis), 2017.[1]

- Protect persons at risk.
- Those people cleaning cages or handling infected birds, should wear protective clothing, gloves, a disposable surgical cap and a respirator.
- Maintain accurate records of all bird-related transactions to aid in identifying sources of infection and potentially exposed persons.
- Records should include date of purchase, species of bird, source of birds, leg band numbers, and describe any illness or deaths of birds.
- Avoid purchasing or selling birds that have signs of avian chlamydiosis.
- Isolate newly acquired birds.
    - New birds should be quarantined at least 30-45 days.
- Test birds before they are to be boarded or sold on consignment.
- Practice preventive husbandry.
- Do not stack cages. Keep cages and bowls clean and disinfect before adding new birds or using the cage for a different bird.
- Prevent the spread of infection.
    - Isolate birds and treat.
- Use disinfection measures.
- Chlamydial organisms can be killed with most commonly used disinfectants including a 1:100 dilution bleach solution for 10 minutes, 1% Lysol®, 70% isopropyl alcohol, or 1:1000 dilution of Roccal®. Gloves and respirators should be worn in an avian outbreak in any areas that have been exposed to the positive bird or its fecal matter (remember dried fecal matter is still contagious for several months). It is important to clean and remove all organic matter first and then disinfect.

Specific considerations

In most states avian chlamydiosis is reportable by veterinarians. In most states human chlamydiosis caused by organisms other than *C. trachomatis* is reportable by physicians (*C. trachomatis* is reportable by physicians in all states).

Salmonellosis

Taxonomy and nomenclature

The *Salmonella* genus comprises gram-negative rod-shaped bacteria. The serotypes that penetrate the mucosal barrier better are usually the pathogen, whereas the other serotypes are usually the secondary pathogens. Genus, species and variety are depicted in Table 1.

Table 1. *Salmonella* genus, species and variety.

Genus	Species	Variety
Salmonella	*typhimurium*	Many (e.g., var. Copenhagen)

Epidemiology

Salmonella has been reported in many avian families including Galliformes, Columbiformes, Psittaciformes, Passeriformes, Strigiformes, and Falconiformes. Generally, most birds with a cecum, such as Galliformes, commonly have *Salmonella* spp. in their gastrointestinal (GI) tract, but if it is isolated from any other tissue in these birds it is considered significant. Usually birds without cecae, such as Psittaciformes, do not have *Salmonella* sp. in their GI tract, and if cultured from the feces it is significant.

Pathogenesis

Transmission is usually through the fecal oral route, but may be from the aerosol route, after exposure to clinically or subclinically infected birds (i.e. asymptomatic carriers are common). Humans developing salmonellosis from birds usually have a history of being exposed to ducks (especially young children during Easter time), gulls or poultry. Infection is either acute, peracute, chronic or subclinical.

The incidence of salmonellosis in birds and people varies greatly with the host species and serotype. Factors that increase the likelihood of disease are exposure to high numbers of organisms, susceptibility of the host, and the serotype of the organism. Lories and African grey parrots seem to be highly susceptible to salmonellosis, even to non-invasive serotypes. *Salmonella* spp. can survive up to 28 months in avian feces.

Clinical signs

Clinical signs can include depression, lethargy, anorexia, weight loss, diarrhea (pasty vent), lameness, abscess formation and death. In people the clinical signs are primarily a hemorrhagic diarrhea and death. People under 5 years of age, the elderly, or those that are immunosuppressed are more likely to develop severe clinical signs and death.

Diagnosis

Salmonella is best grown on MacConkey's agar. *Salmonella* isolated from tissues other than the GI tract of any bird is significant, as well as if it is isolated from the feces or GI tract of birds without a cecum.

Treatment

Antibiotic sensitivity varies with the serotype isolated, therefore, use antibiotics known to include a gram-negative spectrum and change antibiotics accordingly when results of the sensitivity return. Birds whose normal fecal flora is known to have *Salmonella* sp. (those with cecae such as chickens) should not be treated because it is unnecessary, not possible to completely eliminate, and creates resistant organisms.

Prevention and control

Educate the public regarding the risk of salmonellosis from Easter ducks and chicks, waterfowl at parks that children feed and poultry. Free posters and downloadable handouts are available from the CDC website.

Specific considerations

In most states salmonellosis is reportable by veterinarians.

Mycobacteriosis

Taxonomy and nomenclature

Mycobacteriosis has been diagnosed in many species including Galliformes, Anseriformes, Gruiformes, Charadriiformes, Columbiformes, Falconiformes, Strigiformes, Ciconiiformes, Passeriformes and Psittaciformes. Domestic fowl are relatively resistant to infection and seem to be the primary host for many avian infections, whereas psittacine birds seem to be highly susceptible. Genus, species, and common hosts are depicted in Table 2.

Table 2. *Mycobacterium* genus, species and common hosts.

Genus	Species	Common host
Mycobacterium	*tuberculosis*	People
Mycobacterium	*avium*	Birds
Mycobacterium	*bovis*	Birds and people

Epidemiology and pathogenesis

Transmission is through inhalation or oral inoculation. Because of the long incubation period it is sometimes difficult to determine the source of people or bird exposure. There have been cases where *Mycobacterium avium* was isolated from both the people and a bird in a household, and it is unknown who had it first. The reported incidence of mycobacteriosis in wild populations varies between 4-40%.

Clinical signs

Avian mycobacteriosis generally causes chronic weight loss and unthriftiness. Radiographically or at necropsy granulomatous masses of the liver and intestines are most common. People generally develop tubercles in the lung. Gulls will present with mycobacteria related osteomyelitis in the limbs. Rarely, birds will present with a cutaneous form.

Diagnosis

Radiographs are suggestive. A significant monocytic leukocytosis is usually present. Demonstration of acid-fast organisms in feces is not definitive, since other non-pathogenic acid-fast organisms can be in the feces. Demonstration of acid-fast organisms in tissues is highly suggestive. A FITES acid-fast test is preferable to the standard acid-fast test since it is less destructive to the organism, especially for delicate mycobacteria species such as *M. genovense* or *M. kansasii*. Usually, *M. avian* is associated with large numbers of organisms in birds, whereas *M. tuberculosis* and *bovis* are not. The organism is difficult and dangerous to culture. The intradermal tuberculin test used in many mammals is not reliable in birds. There is a PCR test available.

Treatment

Treatment should be discouraged and euthanasia should be the first option discussed due to the difficulty of treating the *M. avium* in people. Veterinarians should probably have clients sign a statement outlining the human risks of treatment and to consult their physician. Various regimens have been described including the use of ethambutol which renders the cell wall permeable to other medications such as chlorambucil, azithromycin, enrofloxacin, among others.

Yersiniosis

Taxonomy and nomenclature:

Genus, species, common hosts, and geographic area for yersiniosis are depicted in Table 3.

Table 3. *Yersinia* genus, species, common hosts and geographic areas.

Genus	Species	Common host	Geographic area
Yersinia	*pseudotuberculosis*	People and animals	Northern Europe
Yersinia	*enterolitica*	People and animals	Worldwide

Epidemiology

Yersinia pseudotuberculosis has been reported in many species, whereas *Y. enterolitica* has been reported in geese, ducks, robins, canaries, pigeons and gulls. Some wild species are known to be refractory. The reservoir of *Yersinia* spp. is wild rodents and birds. In the author's experience toucans fed wild caught rodents at an aviary became infected and died of yersiniosis. Control rodents in an aviary to prevent spread of disease from rodents to birds.

Pathogenesis

Transmission is through the oral route from exposure to infected feces or urine.

Clinical signs

Clinical signs are usually of acute fatal sepsis, but clinical signs can include depression, lethargy, weight loss, diarrhea, respiratory distress and incoordination.

Diagnosis

Antemortem diagnosis is difficult. The organism is difficult to isolate in the laboratory.

Treatment

Treatment is difficult to start in time to be of benefit, but broad spectrum bacteriocidal antibiotics may be of benefit if started early in the course of the disease.

Campylobacteriosis

Taxonomy and nomenclature

Campylobater are gram-negative curved or spiral rods that cannot be distinguished by light microscopy from *Vibrio* spp. or *Helicobacter* spp. Genus, species and common hosts are depicted in Table 4.

Table 4. *Campylobacter* genus, species and common hosts.

Genus	Species	Common host
Campylobacter	*fetus*	Birds
Campylobacter	*intestinalis*	Birds and people
Campylobacter	*jejuni*	Birds and people

Epidemiology and pathogenesis

Campylobacter jejuni is the most common isolate from birds. The common bird host species are chickens, turkeys, pheasants, geese, pigeons, robins, finches, canaries, ducks, sparrows, starlings and blackbirds. Transmission is through the fecal-oral route and from contaminated or infected vectors. Eating contaminated poultry is a reported source of infection for humans.

Clinical signs

Clinical signs include lethargy, diarrhea and emaciation. Fledgling mortality can be high.

Diagnosis

Identification of curved rods on a fecal Gram's stain is suggestive. Culture or electron microscopy is definitive.

Treatment

Treatment for campylobacteriosis varies, but usually involves tetracyclines. Birds whose normal fecal flora is known to have *Campylobacter* sp. (those with cecae such as chickens) should not be treated as it is unnecessary and creates resistant organisms.

Influenza A

Overview and current update

There are two forms of avian influenza (AI): low pathogenic AI (LPAI) and high pathogenic AI (HPAI). Most AI encountered in the United States are LPAI, generally causing a mild respiratory disease.[18] There is a LPAI vaccine available, and is used in commercial poultry settings. Conversely, HPAI causes severe disease, is associated with a high mortality, and is a foreign animal disease in the United States with no available vaccine. The combined value of the U.S. production from broilers, eggs, turkeys, and the value of sales from chickens in 2020 was $35.5 billion. During the 2014/2015 HPAI outbreak, the U.S. Department of Agriculture (USDA) Animal and Plant Health Inspection Service (APHIS) spent $1 billion to eradicate HPAI and provide indemnity (payment) to those that lost birds and were National Poultry Improvement Plan (NPIP) certified. In that outbreak, about 49 million birds died or were depopulated and the United States lost 10% of its egg laying birds. HPAI is a threat to the U.S. economy, a threat to birds (backyard and commercial poultry, zoo birds, wild birds), and a possible threat to humans. Although a few cases have been shown to pass from birds to humans, and from birds to mammals (rats, mice, stone martens, civets, ferrets, pigs, cats, tigers, leopards, dogs, horses, coyotes, fox, raccoon, skunk, bear, dolphins, and more recently, harbor seals in the Netherlands, and harbor and gray seals in New England), the virus easily mutates, and could mutate into a strain that passes easily from person to person, creating a potential pandemic situation. The recent H5N1 clade 2.3.4.4b Eurasian strain that spread from Europe has now infected wild and domestic birds in North America, and South America, with devasting results. This strain is more pathogenic to wild birds (bald eagles, black vultures, waterfowl, pelicans, penguins, gulls) than past strains with high mortality events in many species and is trending towards becoming endemic in these new areas. The USDA is actively pursuing vaccine development and may change tactic in the near future to control HPAI from test and cull to vaccination.

Taxonomy and nomenclature

Influenza virus is in the Orthomyxoviridae family. There are three types of influenza, types A, B and C, but only type A is found in birds. There are two influenza A subtypes based on surface proteins, H for hemagglutinin, and N for neuraminidase. There are 15 known H subtypes and nine known N subtypes. All subtypes are found in birds, but only H1, H2, H3 and N1 and N2 are in humans. All avian virulent strains to date have been H5 or H7 subtype, but most H5 or H7 isolates have been of low virulence.

History

In 1918-1919 there was an influenza A pandemic in people called the Spanish flu that was found to be H1N1. This outbreak may have killed 20-50 million people worldwide, and 50% were young, healthy adults. Greater than 500,000 deaths were attributed to the flu in the United States.

In 1957-1958 the Asian flu, H2N2, caused approximately 70,000 human deaths in the United States. In 1968-1969, the Hong Kong flu, H3N2, caused 34,000 human deaths in the United States and still circulates today. In 1961, influenza A H5N1 was first isolated from terns in South Africa. In 1997, influenza A was shown to pass from birds to humans in an outbreak in Hong Kong where 18 people became sick and six died that had contact with infected birds. This outbreak was controlled by killing 1.5 million chickens.

Since December 2003, an Asian HPAI H5N1 virus has resulted in high mortality in poultry and wild birds in Asia, the Middle East, Europe, and Africa. In 2003, influenza A H5N1 was isolated from two human cases in Hong Kong, one of which died. Then the H5N1 subtype was confirmed in turkey poults in Cambodia, China

(Hong Kong had a single positive peregrine falcon), Indonesia, Japan, Laos, South Korea, Thailand and Viet Nam. Human deaths of influenza A H5N1 were reported in Thailand and Viet Nam after contact with infected birds or their excretions (nasal, saliva or feces). This H5N1 subtype isolated from humans has been genetically sequenced and all genes were of bird origin, but it was found to be resistant to two antiviral drugs, amantadine and rimantadine. It was still sensitive to oseltamivir and zanamivir. Also in 2003, influenza A H7N7 was isolated from poultry workers in the Netherlands. Over 80 cases were reported and one died, a veterinarian. People in the Netherlands outbreak exhibited eye infections with some respiratory infections. Another case occurred in 2003 involving one child in Hong Kong with influenza A H9N2 that recovered. In Canada in 2003, H7N3 was isolated.

In 2022, (January), HPAI (H5N1) was identified in the United States, initially in wild ducks in South Carolina and North Carolina. By March of 2022, HPAI had spread along the eastern seaboard in wild birds (AL, CT, DE, FL, GA, IA, IN, KS, KY, MA, MD, ME, MO, NE, NH, NJ, NY, NC, SC, TN, VA), and along the Atlantic and Mississippi migration flyways in commercial and backyard poultry (CT, DE, IA, IL, IN, KS, KY, MD, ME, MI, MO, NY, SD, VA, WI). Waterfowl act as a reservoir host and usually do not show any signs of infection, but the current outbreak is highly pathogenic to waterfowl and wild birds as well. The current outbreak is associated with migrating birds from Europe where the Eurasian strain of HPAI (H5N1) is essentially becoming endemic. Updated information can be found on the USDA-APHIS website regarding wild bird confirmations, and commercial and backyard flock confirmations, at the following websites, respectively:

- https://www.aphis.usda.gov/aphis/ourfocus/animalhealth/animal-disease-information/avian/avian-influenza/hpai-2022/2022-hpai-commercial-backyard-flocks
- https://www.aphis.usda.gov/aphis/ourfocus/animalhealth/animal-disease-information/avian/avian-influenza/hpai-2022/2022-hpai-wild-birds

Epidemiology and pathogenesis

Wild birds can be the natural hosts for influenza A, especially waterfowl, and with previous strains these bird species did not generally exhibit illness from infection, but with the currently circulating H5N1 wild birds including ducks, vultures, etc. are experiencing high mortality events. Domestic birds, particularly chickens are very susceptible to high mortality with influenza A. Animal to human transmission has occurred but is unusual at the current time, but a mutation in the virus could change this at any time. The chance for influenza A to develop a mutation that will allow it to be transmitted from human to human is very real and may happen in the near future. The virus is not efficient yet at human to human transmission.

The World Health Organization (WHO), as well as other organizations, are monitoring for human gene development. Currently, the goal is to stop the virus from spreading by culling any birds infected or exposed to pathogenic strains (H5 or H7). The OIE, an international organization based in France, and formed in 1928 by 28 countries, and now comprises 165 countries, is a clearinghouse for reported cases of animal disease from each of these countries. Their goal is to respond to multi-country outbreaks in a rapid manner.

Clinical signs

The current circulating HPAI (H5N1) is a foreign animal disease in the United States and is causing high mortality in both wild and domestic bird species. If clinical signs present prior to death, they are quite variable, so if you witness high avian mortality in any bird species, contact your state veterinarian immediately. AI is a highly contagious reportable foreign animal disease. You do not need to know or prove what is causing the high mortality event before you call them. The clinical signs of HPAI can be quite variable and may include: oculonasal discharge, infraorbital sinusitis, dyspnea, tachypnea, cyanosis, edema of the head, comb (swollen comb is common in chickens), and wattle, subcutaneous ecchymotic hemorrhages of the legs and feet, blood-tinged oral and nasal discharges, green diarrhea, torticollis, opisthotonos, and ataxia. Non-specific lesions are seen at gross necropsy, but a friable liver and/or splenic necrosis is commonly encountered.

Diagnosis

Many tests are available including the agar gel immunodiffusion (AGID) test, ELISA and RT-PCR, but these tests must be performed by a certified laboratory (usually state diagnostic laboratories or contact your state or USDA veterinarian). For PCR swabs it is recommended to obtain both a choanal and a cloacal sample (it can be on same swab), even in waterfowl. Not all birds develop demonstrable antibodies on the ELISA. Also available are hemagglutination and hemagglutination inhibition tests.

Treatment

Birds with H5 of H7 in the United States are required to be depopulated. The 2003 influenza A H5N1 isolated from people in Hong Kong, was genetically sequenced and found to be resistant to two antiviral drugs, amantadine and rimantadine. It was still sensitive to oseltamivir and zanamivir.

Prevention and control

A killed vaccine is available for the low pathogenic strains in the United States. There is a vaccine for the high pathogenic strains, but they are not currently used in the United States. Depopulation is the current recommended action taken by the USDA during an outbreak. The USDA is actively pursuing vaccine development and may change tactic in the near future to control HPAI from test and cull to vaccination.

Specific considerations

HPAI is a reportable foreign animal disease. Refer to the USDA website for current information on this rapidly changing outbreak. Indemnity from the federal government is available for birds that died after a diagnosis was made and if the flock is NPIP certified.

Newcastle Disease

Overview

Virulent Newcastle disease (VND) is a foreign animal disease and millions of dollars are spent to eradicate all exposed birds when there is an outbreak in the United States. VND affects many species of birds. The causative agent is a paramyxovirus. Technically it is considered a zoonotic disease since it can cause conjunctivitis in humans.

Taxonomy and nomenclature

Newcastle disease is caused by a paramyxovirus of which there are four pathotypes, lentogenic, mesogenic, neurotropic velogenic and viscerotropic velogenic (VVND). The latter, a foreign animal disease goes by many names and abbreviations including VVND, exotic Newcastle disease (END), and more recently as VND.

History

The non-exotic form of Newcastle disease was common before the use of vaccines for those forms. There is a vaccine available for exotic Newcastle disease, but its use is currently not allowed in the United States because the tactic to control the disease in this country is to test and cull any positive birds. In the early 1970s there was an outbreak of VND and millions of dollars were spent to destroy chickens in the outbreak to control the disease. In 2002, there was another outbreak of VND and again millions were spent to destroy infected and surrounding flocks to control the disease.

Epidemiology

The disease is spread by direct contact with viral particles from aerosolization of respiratory secretions or feces, or from fecal contaminated food, water or litter. Recovered birds are believed to shed the virus indefinitely.

Clinical signs

The clinical signs vary with the pathotype but include depression, diarrhea, anorexia, ruffled feathers, oculonasal discharge, conjunctivitis, dyspnea, ataxia, muscle tremors, paralysis and death. In people Newcastle disease causes a mild, acute granular conjunctivitis, general malaise and sinusitis that resolves within 7-20 days.

Diagnosis

Avian serum can be sent to a laboratory for PMV 1, 2 and 3 antibody testing. Other tests are available including the hemagglutination inhibition test, the ELISA, AGID, PCR, and virus isolation.

Treatment

For pet birds with mild lentogenic or mesogenic forms of Newcastle disease offer supportive care. Any bird, pet or otherwise that is positive for VND in the United States is to be reported to federal authorities and culled.

Prevention and control

Vaccines are available for control of lentogenic and mesogenic forms of Newcastle disease in the United States. VND is a foreign animal disease and millions are spent to eradicate all exposed birds when there is an outbreak in the United States (one was in California in 2002-2003; 3.5 million birds were culled). The latest outbreak in California started May, 2018 and was declared ended on June 1, 2020. At least 1.2 million birds were euthanized or depopulated. The cost is still undetermined, but over $100 million were spent to eradicate.

Specific considerations

Quarantine stations for birds entering the United States were developed to keep VND out, therefore any bird entering U.S. quarantine is tested serologically for VND (and also avian influenza). They are not tested for other diseases, not even chlamydiosis, although they are provided chlortetracycline impregnated food while in quarantine to prevent shedding of any possible chlamydial organisms to quarantine personnel while housed.

Viral Encephalitides

Taxonomy and nomenclature

Eastern and Western equine encephalomyelitis virus are known as EEE and WEE, respectively.

Epidemiology

Birds are the known reservoirs for these diseases. Many species of birds are involved. The virus is transmitted through a mosquito bite or in pheasants it has been shown to be transmitted through the broken skin created by pecking.

The mosquitos known to transmit EEE virus in the United States and Canada include *Culiseta melanura*, *Aedes* spp. and *Coqiullettidia* spp. Interestingly, *Culiseta melanura* feeds mainly on birds and rarely on people. The *A. sollicitans* and *A. vexans* mosquitos feed on both birds and people, but are less likely to be infected with the virus. Bird mites are known to be potential vectors. The mosquitos known to transmit WEE virus include *Culex tarsalis* in the western United States and Canada, and *Culiseta melanura* in the eastern United States.

People are an accidental host to the virus and if infected with EEE can have up to 80% mortality, and if infected with WEE can have up to 5-15% mortality. In Canada, a direct connection has been shown between the prevalence of the virus in wild birds and the number of human cases. During the 1962 outbreak in Saskatchewan, Canada the WEE virus was isolated from 22% of the wild bird population.

Clinical signs

Birds may show no clinical signs to ataxia, trembling, weakness, paralysis and death. People show varying degrees of neurologic signs and death.

Diagnosis

Antemortem diagnosis is difficult. Histopathologic lesions of lymphoplasmacytic encephalitis is suggestive.

Treatment

There is no specific treatment for viral encephalitis other than supportive care.

Rabies

All warm-blooded animals, including birds are susceptible to rabies virus. Avian infections are rare, but have been reported in chickens, waterfowl and raptors. A passive hemagglutination test in wild birds was seropositive in 23.1% of 65 predatory birds and 2.9% of 278 non-predatory birds. While no case has yet been documented of avian to human transmission of the rabies virus, preexposure rabies prophylaxis and preventing trauma while handling raptors is recommended.

West Nile Virus

Taxonomy and nomenclature

The causative organism is a flavivirus.

History

West Nile virus (WNV) is endemic in other countries, but in the late 1990's it was found within the eastern United States and has since spread across the country.

Epidemiology

Crows, jays and raptors and horses are susceptible species, whereas poultry are considered resistant. The WNV virus is spread by mosquitos. If people or dogs are affected they are usually older or immunosuppressed.

Clinical signs

Clinical signs range from none in resistant species such as poultry to neurologic signs (ataxia, circling, head tilt, and seizuring) and death in susceptible species. A CBC is usually normal or a lymphocytosis is present.

Diagnosis

There is a serum antibody test available.

Treatment

Treatment consists of supportive care. Recently in people the use of alpha interferon has seemed to result in better success.

Prevention and control

This disease has spread throughout the United States, so it is too late to prevent the disease in this country. There is a conditionally licensed vaccine available for use in horses that is currently being used intramuscularly in birds at the same or a reduced dose. Studies are underway evaluating the antibody response and protectiveness of this vaccine in some species of birds. Another vaccine specifically for use in birds is being developed.

Histoplasmosis

Taxonomy and nomenclature

Histoplasma capsulatum is an imperfect dimorphic fungus with a worldwide distribution.

Epidemiology and pathogenesis

Transmission is through the inhalation of highly infective spores. There is a wide host range including birds, mammals, and humans. The fungus proliferates particularly well in soil that is enriched with nitrogen from bird droppings or within accumulations of bird droppings (e.g., pigeon roosting sites are a common source of infection).

Clinical signs

In immunocompetent humans the infection is common, benign and self-limiting. In immunosuppressed humans, severe respiratory disease can result in chronic cavitary lesions or disseminated disease.

Diagnosis

Antemortem diagnosis is apparently difficult in humans since it mimics other less serious respiratory diseases, but thoracic radiographs and culture of sputum are recommended, as well as an antigen skin test to indicate sensitivity to the organism.

Treatment

Antifungals are used in the attempted treatment in people.

Cryptococcosis

Taxonomy and nomenclature

Cryptococcus neoformans is a saprophytic yeast-like fungus.

Epidemiology and pathogenesis

Cryptococcus infections have been reported in many avian species including swallows, pigeons, green winged macaw, canaries, thick-billed parrot, African grey parrot, triton cockatoo and Mollucan cockatoo. Cryptococcosis in people from exposure to infected birds is well documented and is particularly common after exposure to infected pigeon droppings which are high in nitrogen and encourage growth. The infectivity of *Cryptococcus* is thought to be lower in avians due to their higher core body temperature and the poor growth of this organism at higher temperatures such as 40 °C.

Clinical signs

Most birds show no clinical signs. Clinical signs in birds vary with the location of the growth of fungus but can include depression, weakness, anorexia, weight loss, acute diarrhea, incoordination, blindness, dyspnea, nasal discharge, oral masses and death. Generally, the respiratory lesions are the most common. Clinical signs in people usually include a meningitis and/or respiratory disease.

Diagnosis

Cryptococcal infections are associated with a gelatinous or granulomatous lesion usually in the air sacs. The organism has been isolated form intraocular sites, bone marrow, pectoral muscles, neck fascia, meninges, hepatic veins, air sacs, trachea, pararenal areas and lungs.

Treatment

Antifungal including amphotericin-B and flucytocine are recommended, and the success rate in people with this combination is 68%.

Summary

Several diseases can be transmitted from birds to people, but most are treatable if diagnosed early. This early diagnosis is enhanced by good communication between veterinarians and their clients and proper regulatory officials. Because avian veterinarians work in a field involving recently imported, recently domesticated or wild birds, they are exposed to disease entities that are not commonly encountered by the average veterinarian and should therefore be more vigilant in educating themselves and others to these diseases as well as keeping an open mind when observing clinical signs in their patients.

References

1. Balsamo G, Maxted AM, Midla JW, et.al. Compendium of measures to control *Chlamydia psittaci* infection among humans (psittacosis) and pet birds (avian chlamydiosis). *J Avian Med Surg.* 2017;31(3):262-282.

2. Chu J, Yarrarapu SNS, Vaqar S, et al. Psittacosis. In: StatPearls [Internet]. Treasure Island, FL: StatPearls Publishing; 2023.

3. Human psittacosis linked to a bird distributor in Mississippi, Massachusetts and Tennessee. *Morbidity and Mortality Weekly Report.* 1992, 41(42);794-797.

4. Chin J. *Control of Communicable Diseases Manual.* 17th Ed. Washington, DC: American Public Health Association; 2000.

5. Centers for Disease Control and Prevention. Epidemiologic Notes and Reports Psittacosis at a Turkey Processing Plant, North Carolina. 1989. Available at: http://www.cdc.gov/mmwr/preview/mmwrhtml/00001662.htm. Accessed June 1, 2023.

6. Centers for Disease Control and Prevention. Psittacosis in the United States. 1979. Available at: http://www.cdc.gov/mmwr/preview/mmwrhtml/00014688.htm. Accessed June 1, 2023.

7. Everett KD, Bush RM, Anderson AA. Emended description of the order Chlamydiales, proposal of Parachlamydiaceae fam. nov. and Simkaniaceae fam. nov., each containing one monotypic genus, revised taxonomy of the family Chlamydiaceae, including a new genus and five new species, and standards for the identification of organisms. *Int J Syst Bacteriol.* 1999;49(2):415-440.

8. Everett KD, Anderson AA. Identification of nine species of the Chlamydiaceae using PCR-RFLP. *Int J Syst Bacteriol.* 1999;49(2):803-813.

9. Centers for Disease Control and Prevention. The CDC's Division of Bacterial and Mycotic Diseases Disease Information Sheet on Psittacosis. Available at: http://www.cdc.gov/ncidod/dbmd/diseaseinfo/psittacosis_t.htm. Accessed June 1, 2023.

10. Centers for Disease Control and Prevention. Facts About Disease Case Definitions. Available at: https://www.cdc.gov/pneumonia/atypical/psittacosis/index.html. Accessed June 1, 2023.

11. Centers for Disease Control and Prevention. Morbidity and Mortality Weekly Report. Notice to Readers: Changes in National Notifiable Diseases List and Data Presentation. Available at: https://www.cdc.gov/mmwr/preview/mmwrhtml/mm5101a5.htm. Accessed June 1, 2023.

12. Federation of American Scientists. CDC Bioterrorism Agents. Available at: https://biosecurity.fas.org/resource/documents/CDC_Bioterrorism_Agents.pdf#:~:text=CDC%20Bioterrorism%20Agents%20Category%20A%20%C2%BE%20Anthrax%20%28Bacillus,Lassa%2C%20Machupo%5D%29%20Category%20B%20%C2%BE%20Brucellosis%20%28Brucella%20species%29. Accessed June 1, 2023.

13. Powers L, Flammer K, Papich M. Preliminary investigation of doxycycline plasma concentration in cockatiels (*Nymphicus hollandicus*) after administration by injection or in water or feed. *J Avian Med Surg.* 2000;4:23-30.

14. Flammer K, Whitt-Smith D, Papich M. Plasma concentrations of doxycycline in selected psittacine birds when administered in water for potential treatment of *Chlamydophila psittaci* infection. *J Avian Med Surg.* 2001;15:276-282.

15. Raso T, Junior AB, Pinto AA. Evidence of *Chlamydophila psittaci* infection in captive Amazon parrots in Brazil. *J Zoo Wildl Med.* 2002;33(2):118-121.

16. Ritchie BW, Dreesan DW. Avian zoonoses: proven and potential diseases. Part I. Bacteria and parasitic diseases. *Comp Cont Educ.* 1988;10(4):484-491.

17. Ritchie BW, Dreesan DW. Avian zoonoses: proven and potential diseases. Part II. Viral, fungal and miscellaneous diseases. *Comp Cont Educ.* 1988;10(6):688-695.

18. World Organisation for Animal Health. New OIE web portal on avian influenza. 2016. Available at: https://www.woah.org/en/new-oie-web-portal-on-avian-influenza/#:~:text=The%20new%20OIE%20portal%20on%20avian%20influenza%20brings,actions%20taken%20by%20the%20OIE%20and%20its%20partners. Accessed June 1, 2023.

19. Sachse K, Laroucau K. Two more species of *Chlamydia*-does it make a difference? *Pathog Dis.* 2015;73(1):1-3. doi.org/10.1093/femspd/ftu008.

20. Guo W, Li J, Kaltenboeck B, et al. *Chlamydia gallinacea,* not *C. psittaci*, is the endemic chlamydial species in chicken (*Gallus gallus*). *Sci Rep.* 2016;6:19638.

21. Centers for Disease Control and Prevention. CDC Multistate Psittacosis Outbreak among Poultry Plant Workers, 2018. Available at: https://www.cdc.gov/pneumonia/atypical/psittacosis/surveillance-reporting/outbreaks/2018-poultry-multistate-investigation.html. Accessed June 1, 2023.

Birds' Use of Technology for Enrichment and Socialization

Ilyena Hirskyj-Douglas, PhD,
Rebecca Kleinberger, PhD,
Jennifer Cunha, JD,
Susan Clubb, DVM

Session #3444

Affiliation: From University of Glasgow, Glasgow G12 8QQ, UK (Hirskyj-Douglas), Northeastern University, 360 Huntington Ave, Boston, MA 02115, USA (Kleinberger), Parrot Kindergarten, Inc., 601 Heritage Dr, Ste 424, Jupiter, FL 33458, USA (Cunha), Rainforest Clinic for Birds and Exotics, 3319 E Road, Loxahatchee, FL 33470, USA (Clubb).

Abstract: Technology has changed drastically in the past decade, creating new opportunities for enrichment and interactive devices for parrots and their caregivers. While recent advances have been made in parrot husbandry and care practices, the physical and mental health issues in captive parrots persist. While parrots in people's homes often do use technologies, such as tablets and smartphones, these technologies are designed by and for humans, who have vastly different physiology and cognitive processes than what parrots need. In this workshop, veterinarians and other experts in the parrot industry are invited to speculate on how technology can support parrots and caregivers to improve everyday lives and welfare, and increase sense of agency. In this workshop, the future of parrot technology is envisioned to critically explore and interrogate what emerging technologies can, and should, look like for birds. Using these visions, central themes of technology are distilled and important features of these systems are listed to create an expert-guided framework of key features that parrot-technology needs to address.

Introduction

Recent research has demonstrated that parrots (psittacine birds) are capable of high-level cognitive processing, including inference-by-exclusion,[1] counting,[2] music preference,[3] symbol use, and abstract concept processing.[4] Such cognitive stimulation further appears to have positive outcomes in terms of mental and physical health suggesting potential to use these tasks for enrichment. While research on non-human primates spans over a century, comparative research into avian cognition is much more recent and upcoming.

Comparing parrots to non-human primates (*Homo sapiens*), replications conducted on certain corvids (Corvidae) and parrots yield striking results. Güntürkün, Ströckens, Scarf and Colombo compared the performance of primates, corvids and pigeons (Columbidae) across a variety of tasks including short-term memory, object permanence, abstract numerical competence, and orthographic processing.[5] They discovered that pigeons and corvids demonstrated similarly high levels of performance when compared to non-human primates. Likewise, Goffin's cockatoos (*Cacatua goffiniana*), African grey parrots (*Psittacus erithacus*), and certain species of corvids have out-performed 7- to 10-year-old children and adult gorillas on some complex cognitive tests such as reasoning, memory, and mental manipulation tasks.[6-8]

Stereotypy and Enrichment

Perhaps due to their cognitive ability and unmet need for stimulation, captive birds often display stereotypic behaviors. Stereotypies are abnormal repetitive, unvarying, and apparently functionless behaviors that are often performed by captive and domesticated animals housed in barren environments.[9] They are believed to be in response to life in a captive environment. In birds, stereotypies include a variety of oral and locomotive stereotypies including feather-destruction , pacing, excessive vocalizations and hanging on cage bars.

Stereotypy rates in parrots are reported as high as 42% in cockatoos (*Cacatua galerita*) and 40% in African grey parrots[10] giving indication towards other parrot species. Some research has suggested a relationship between lack of enrichment and stereotypy.[9] For example, locomotive stereotypy (i.e., route-tracing and cage-hanging) may be due to lack of space and physical complexity and oral stereotypy may be related to lack of foraging enrichment. Orange-wing Amazon parrots (*Amazona amazonica*) raised in a barren environment had much higher rates of stereotypy than those raised with locomotion- and foraging-based enrichment environments.[11] Further, once the barren-environment raised parrots were relocated to enriched environments, their rates of stereotypy decreased.

Contemporary researchers have suggested a need for higher levels of cognitive enrichment for captive animals, beyond traditional methods of environmental stimulation.[11-13] Enrichment is traditionally composed of foraging opportunities, toys, variety in diet, destructible materials, and passive sensory stimulation (such as the radio or television). While these toys and devices may elicit natural behaviors, research suggests that these may not provide adequate mental stimulation with varied learning, motor, attentional and sensory components. This is attributed to the lack of cognitive challenges that increase in difficulty upon mastery of tasks, as may be demanded in wild settings.[12] These researchers argue that enhanced enrichment using non-traditional modalities including complex toys, puzzles, and even interactive technology including learning games or social interactions could additionally be utilized to meet the cognitive needs of captive animals.

Technology Enrichment

To meet the growing complex needs of animals, researchers have begun exploring how different technologies can support an animal's life and wellbeing. Nicky, et al. shows the growing relevance of interactive digital applications for captive primates and insists on the importance of giving control of the technologies to the animal.[14] Drawing from this, researchers have now deployed several technologies with animals in zoos, and various pets that aim to give animals agency by recognizing that they are sentient beings that benefit from controlling their environment.[15-17] Agency can be defined as the propensity of an animal to engage actively with the environment with the main purpose of gathering knowledge and enhancing its skills for future use.[18]

Likewise, offering parrots a similar kind of agency in shaping their environment, especially in turning systems on or off, may yield substantial benefits. Technological improvement in terms of sensing and recognition opens the door to various sorts of technologies for parrots.[16,19] Thus, while such systems should not be used in isolation to replace human care or co-specific interactions, there is potential for technologies to enrich, and connect birds to each other, their humans and extend their environment around them.

In terms of technological innovation, the field of animal-computer interactions has produced inspiring previous work in solo enrichment systems. Technology for animals is a growing billion-dollar industry.[17]

In the context of zoo-housed animals, computers and tablets have been used to provide primates with a variety of enriching applications, from face-matching games to "Tinder for orangutans" used in breeding research.[20]

In those examples, the main role of the human volunteers was to hold the devices. Keepers have expressed the need for enrichment systems that are hands-off since tablets often have to be held by caregivers, which is time-consuming and can cause safety issues. With the orangutans, frustration was observed because they were not able to hold the devices themselves.

Some of the most interesting works involving animals and digital technologies are the ones that acknowledge the key role played by humans in animal environments. Previous inspiring work from the human-computer interaction (HCI) community involving animals has acknowledged and curated the human's role and behavior to appeal to the specific animal or species, from a challenging play partner,[15] to an empathetic audience.[21] Such approaches have the potential to lead to higher levels of animal-human interaction, playfulness, and energy expenditure especially when the animal is provided with increased agency through the digitally mediated interspecies interaction.

Workshop Organization

The purpose of this workshop is to explore the design of technology that can improve the welfare of parrots. Although there have been efforts to develop technology for animals, most current technologies are not adapted to the unique needs of birds. Therefore, it is essential to develop technology based on parrot needs, with a good user interface and user experience for birds. The workshop is divided into three parts:

1. Overview of animal-computer interaction (ACI) and previous work on innovative experience design for parrots

   In this 30-minute overview, the field of ACI is introduced and inspiring previous work on technology for parrots is presented both by authors and other practitioners. This section proposes key considerations when working with animals and technology including agency, control, affordances, consent, and ethics.

2. Group design exercise

   In this 45-minute interactive section, participants gather in groups of four and are given a prompt to create an ideation around what they see the future of parrot-computer interfaces looking like. The method of design fiction is used to help move from current technology constraints to envision, explore and criticize future parrot-technologies. For this task, researchers support the groups and provide additional information on current technologies and take key notes on overarching themes that are present within the designs. The designs are focused on imaging technologies for parrot-to-parrot interaction, parrot-human interaction and parrots using technologies in general.

3. Interactive session with audience members

   After 45 minutes, all groups gather and present their thoughts on key features of their designs, any ethical concerns they have, any key opportunities that technology can take and any criticisms of future technologies. These thoughts are gathered and visualized in real-time to present key findings of future parrot-computer interfaces and how the future of technology can support parrots and things to be concerned about. This section is interactive with the audience to help identify the main contexts, needs, and senses of parrots where technology can be leveraged. Potential contexts are introduced in which technology can be used and audience feedback is sought through live polling that participants can access via their phones to help address the main needs and requirements that parrots have. The aim is to collect and reflect together quantitative and qualitative data based on participants observation and expertise and on the above exercise. The topics covered in this section include:

- Identifying and ranking main needs and requirements from parrots
- Identifying and ranking potential technologies that could be leveraged
- Identifying risks and ethical issues linked to using such technologies
- Identifying ways to alleviate those risks
- Reflecting on the role of human caretakers in deploying technologies for parrots
- Identifying rigorous ways to assess and evaluate the value/benefits of the systems on birds

Ethical Statement

All data collected is anonymized and following the conference copies of the overall findings are sent to participants desiring them. In encouraging the development of technologies for parrots, an emphasis is placed on not only ethical design features to create ease in interacting with the device but also in the deployment and training of animals to engage with the technologies in a force-free, uncoerced manner.

Outcomes

In this workshop, ACI is introduced and examples of currently available ACI technologies are provided. Participants are encouraged to develop innovative and practical technology solutions that benefit parrots and other animals.

This workshop gathers information from veterinarians and other avian experts regarding the ways technology might be developed to benefit parrots, as well as the unique physiology of parrots that need to be considered in technology. This is accomplished in several ways including via an online survey as well as with small group discussions.

With this information, data are presented to technology experts and developers to encourage the design and deployment of ethical technologies for animals. As a result of this information, the future will include improved welfare conditions, lower stereotypies, and higher rates of health for parrots in captivity.

Conclusions

The workshop looks at how technologies can support parrots, bringing together experts in parrot husbandry and veterinary science to explore the possibilities and limitations of future parrot-computer interfaces. Through group ideation and the design fiction method, participants are able to overcome current technological constraints and imagine potential technologies to support parrots and their caregivers. The workshop investigates key features of the designs, ethical concerns, opportunities for technology, and critiques of future technologies for parrots exploring how they can support the birds themselves, veterinarians and their human caregivers. The real-time visualization of the ideas provides a clear picture for attendees of the central themes that emerge from the workshop, and the expert-guided framework of key features for parrot technology serves as a valuable resource for researchers and practitioners in the field on the current state of the art technologies for animal welfare. The workshop emphasizes the importance of collaboration and interdisciplinary approaches to address the complex challenges in parrot care and welfare. The insights gained from this workshop will inform future research and development of technologies to support parrots and their caregivers, and ultimately contribute to improving the well-being of captive parrots.

References

1. O'Hara M, Auersperg AMI, Bugnyar T, Huber L. Inference by exclusion in Goffin cockatoos (*Cacatua goffini*). *PLoS ONE.* 2015;(10)8:e0134894. doi:10.1371/journal.pone.0134894.

2. Pepperberg IM, Carey S. Grey parrot number acquisition: the inference of cardinal value from ordinal position on the numeral list. *Cognition.* 2012;125:219-232.

3. Copping, J. The Business Insider. Parrots are Picky When it Comes to Music. Available at: https://www.businessinsider.com/parrots-have-specific-music-tastes-2012-12. Accessed June 1, 2023.

4. Pepperberg IM, Koepke A, Livingston P, et al. Reasoning by inference: further studies on exclusion in grey parrots (*Psittacus erithacus*). *J Comp Psychol.* 2013;127(3):272-281. doi.org/10.1037/a0031641.

5. Güntürkün O, Ströckens F, Scarf D, Colombo M. Apes, feathered apes, and pigeons: differences and similarities. *Curr Opin Behav Sci.* 2017;16:35-40.

6. Pepperberg IM, Gray SL, Lesser JS, Hartsfield LA. Piagetian liquid conservation in grey parrots (*Psittacus erithacus*). *J Comp Psychol.* 2017;131(4):370-383. doi.org/10.1037/com0000085.

7. Pepperberg IM, Gray SL, Mody S, et al. Logical reasoning by a Grey parrot? A case study of the disjunctive syllogism. *Behaviour.* 2019;156(5-8):409-445. doi.org/10.1163/1568539X-00003528.

8. Pepperberg IM, Pailin H. Evolution of mechanisms underlying visual working memory manipulation: when "bird-brain" is a compliment. *Proc Annu Conf Vision Sci Soc.* 2017.

9. Garner J, Meehan C, Famula T, Mench J. Genetic, environmental, and neighbor effects on the severity of stereotypies and feather picking in orange-winged Amazon parrots (*Amazona amazonica*): an epidemiological study. *Appl Anim Behav Sci.* 2006;96(1):153-168. doi:10.1016/j.applanim.2005.09.009.

10. Jayson S, Williams D, Wood J. Prevalence and risk factors of feather plucking in African grey parrots (*Psittacus erithacus erithacus* and *Psittacus erithacus timneh*) and cockatoo (*Cacatua* spp.) *J Exot Pet Med.* 2014;23:250-257. doi:10.1053/j.jepm.2014.06.012.

11. Meehan C, Garner J, Mench J. Environmental enrichment and development of cage stereotypy in orange-winged Amazon parrots (*Amazona amazonica*). *Dev Psychobiol.* 2004;44(4):209-218.

12. Bennett A, Bailoo J, Dutton M, et al. A model quantitative assessment tool for nonhuman primate environmental enrichment plans. *bioRxiv.* 2018;341206. doi.org/10.1101/341206.

13. Washburn DA. The four Cs of psychological wellbeing: lessons from three decades of computer-based environmental enrichment. *Anim Behav Cogn.* 2015;2(3):218-232. doi.org/10.12966/abc.08.02.2015.

14. Kim-McCormack NNE, Smith CL, Behie AM. Is interactive technology a relevant and effective enrichment for captive great apes? *Appl Anim Behav Sci.* 2016;185:1-8.

15. Hirskyj-Douglas I, Lucero A. On the internet, nobody knows you're a dog...unless you're another dog. *Proc Conf Human Factors Comput Syst.* 2019;1-12.

16. Cunha J, Rhoads C. Use of a tablet-based communication board and subsequent choice and behavioral correspondences in a Goffin's cockatoo (*Cacatua goffiana*). *Proc Conf Animal-Comput Interact.* 2020;1-9.

17. Kleinberger R, Cunha J, Vemuri M, Hirskyj-Douglas I. Birds of a feather video-flock together: agency-based parrot-to-parrot video-calling system for interspecies ethical enrichment. *Proc Conf Human Factors Comput Syst.* 2023. doi.org/10.1145/3544548.3581166.

18. Spinka M, Wemelsfelder F. Environmental challenge and animal agency. *Animal Welfare.* Wallingford, UK: CAB International; 2011:27-44.

19. Scheel B. Designing digital enrichment for orangutans. *Proc Conference Animal-Comput Interact.* 2018. doi.org/10.1145/3295598.3295603.

20. Westerlaken M, Gualeni S. Situated knowledges through game design: a transformative exercise with ants. *Proc Conf Philos Comput Games.* 2016.

21. Webber S, Marcus C, Sherwen S, et al. Kinecting with orangutans: zoo visitors' empathetic responses to animals' use of interactive technology. *Proc Conf Human Factors Comput Syst.* 2017:6075-6088. doi.org/10.1145/3025453.3025729.

Sedation in Birds

Christoph Mans, Dr med vet, Dipl ACZM, and
Grayson A. Doss, DVM, Dipl ACZM

Session #3486

Affiliation: From the School of Veterinary Medicine, University of Wisconsin-Madison, 2015 Linden Drive, Madison, WI 53706, USA.

Abstract: Sedation techniques provide a means of reducing physiologic stress in birds undergoing non-painful clinical procedures. In addition, sedation in birds provides easy restraint, increases the safety of many clinical procedures (e.g., blood collection, radiography, ultrasonography), and allows for a complete examination, which would otherwise only be achieved under general anesthesia. Using safe and effective sedative protocols in pet birds provides substantial benefits to the patients as well as the veterinarian and staff. Sedation should be considered for a variety of clinical procedures in avian patients.

Introduction

Sedation of dogs and cats in veterinary practice is a daily routine for a variety of procedures, such as radiographs and ultrasonography, or other non-painful but potentially stressful procedures. However, historically for avian patients, either manual restraint of conscious birds or general anesthesia is typically performed to complete most clinical procedures.[1] General anesthesia predisposes birds to cardiovascular and respiratory depression and may cause aspiration of gastric or crop contents and hypothermia. In contrast, manual restraint in conscious birds is simple to perform but can have negative consequences, including stress to the bird and/or handler, negative conditioning to the clinic environment (e.g., the person restraining or the towel used for restraint), hyperthermia, and the predisposition of trauma to the handler and/or bird. Several recent studies demonstrated that manual restraint of birds causes increased body temperature and respiratory rate.[1] In sick, old, or very stressed birds, acute collapse, and death secondary to manual restraint have been reported. Therefore, sedation techniques provide a useful alternative for reducing physiologic stress in birds undergoing non-painful clinical procedures. Furthermore, sedation in birds provides easier restraint, increases the safety of many clinical procedures (e.g., blood collection, radiography, ultrasonography), and allows for a complete examination, which would otherwise only be achieved under general anesthesia. Using safe and effective sedative protocols in pet birds provides substantial benefits to the patients as well as the veterinarian and staff and should be considered for a variety of clinical procedures.

Route of Drug Administration

Historically, sedative drugs have been most administered by intramuscular injection in birds. Recently, the intranasal administration of sedative drugs in birds has gained increased attention. Intranasal administration of midazolam in Hispaniolan Amazon parrots, ring-necked parakeets, budgerigars, and canary finches resulted in rapid onset of dose-dependent sedation, which is completely reversible with intranasal flumazenil.[1] Intranasal drug administration offers an alternative, non-invasive technique for drug administration in birds (Table 1). It is characterized by its ease of administration, high bioavailability, rapid onset of action, and reduced pain compared to intramuscular administration. Elevation of muscle enzymes in biochemistry panels secondary to intramuscular drug administration is avoided if intranasal administration is used instead. In addition, clients perceive the intranasal route as non-invasive, compared to intramuscular injection, which leads to better client compliance in cases in which sedation is recommended. The time of onset to sedation is rapid, typically within 3-5 minutes, which is comparable to intramuscular administration;[1-4] however, limitations of intranasal administration include incomplete drug delivery due to sneezing during administration, physiologically narrowed nostrils (e.g., cockatoos), or upper respiratory disease (e.g., blocked or stenotic nostrils). In some cases, in larger birds (e.g., macaws), the drug volume can also limit the effectiveness and produce excessive sneezing, therefore leading to incomplete drug delivery. Higher concentrated drugs (e.g., midazolam 50 mg/ml, Zoopharm, Windsor, CO) are available, but intramuscular administration might be more feasible in these cases.[1]

Table 1. Comparison of intramuscular and intranasal drug administration in pet birds.

	Intramuscular administration	Intranasal administration
Pro	Reliable drug delivery Faster administration	Non-invasive and not painful Higher client acceptance Reduced risk of needle stick injury No changes in the biochemical panel
Contra	Invasive Potentially painful Possible post-injection hemorrhage Changes in the biochemical panel	Risk of incomplete drug delivery Difficult to administer larger volumes May cause sneezing More difficult in birds with feathered nares

Drugs Used for Sedation

Midazolam

Midazolam is currently the most used drug for the sedation of pet birds and has a wide margin of safety. Midazolam has sedative, muscle-relaxing, anxiolytic, amnestic, and appetite-stimulating properties in birds.[1,5,6] Standard concentrated forms of injectable midazolam (midazolam hydrochloride (1 and 5 mg/ml); Hospira Inc, Lake Forest, IL) or a more concentrated form (50 mg/ml, Zoopharm, Windsor, CO) can be administered intranasally and intramuscular, without side effects.[1-3] Dosages commonly used in pet birds range from 0.5-3 mg/kg.[7] At the University of Wisconsin, we routinely use 2 mg/kg of midazolam in pet birds if administered intranasally and as the sole sedative agent. In smaller birds, such as finches or budgerigars, we routinely use 4-6 mg/kg of midazolam if administered alone (which is rarely indicated).

Diazepam

Diazepam is of similar efficacy as midazolam in birds following intranasal administration but has a slightly longer onset time and duration of action.[3,4] While less commonly used, diazepam represents a suitable alternative in cases in which midazolam might not be available. The intramuscular administration of diazepam should be avoided if possible due to the potential for delayed absorption and muscle irritation.[5] Dosages commonly used in pet birds range from 0.2-2 mg/kg if used as a sole sedative agent. Dosages as high as 10-15 mg/kg have been administered to finches and budgerigars without significant side effects, but this is not recommended in clinical settings, and instead, the benzodiazepines should be combined with butorphanol or dexmedetomidine.[3,4]

Butorphanol

Butorphanol is currently the most used opioid analgesic in birds. Besides its analgesic effects, butorphanol has sedative effects, which are potentiated by benzodiazepines (i.e., midazolam, diazepam). The combined administration of midazolam and butorphanol is recommended in birds for which midazolam alone provides only an insufficient level of sedation or where deeper sedation is needed for certain clinical procedures such as radiographic positioning. Butorphanol can be given intramuscularly as well as intranasally, drawn into a single syringe in combination with midazolam. No adverse effects of intranasal administration of butorphanol at a dose range of 1-3 mg/kg were seen in psittacine birds.[7] At the University of Wisconsin, we routinely use butorphanol at 2-3 mg/kg combined with 2-3 mg/kg of midazolam administered intranasally or by intramuscular injection in pet birds.

Dexmedetomidine

Historically, alpha$_2$-agonists have not been commonly used for sedation in birds; however, recently its safety and efficacy have been reported in budgerigars.[8] If combined with midazolam (3 mg/kg), dexmedetomidine (10 or 40 μg/kg IM) provided dose-dependent, moderate to deep sedation without obvious adverse effects. At the higher dexmedetomidine dose, positioning for radiographs was feasible. Reversal was quick following flumazenil and atipamezole administration.[8] The authors recommend 20 μg/kg IM as a starting dose in small psittacine and passerine birds instead of midazolam-butorphanol protocols, due to the superior sedation quality achieved. In birds with known or suspected cardiac disease, this medication should be avoided.

Reversal of Sedation

Reversal of sedation in pet birds will depend on the patient and the purpose of sedation. Sedation performed to facilitate physical examination, and diagnostic sample collection should always be reversed to have the patient return to normal behavior and food intake as soon as possible. It is important not to discharge sedated patients, as owners are often concerned about caring for a partially sedated bird, which might be ataxic, sleepy or refuse to eat. Birds that were sedated for e-collar or bandage placement or control of seizures should not be reversed. In these cases, birds should be carefully monitored, and reversal should be considered if the level of sedation is perceived as too deep or the duration of sedation is prolonged and might interfere with physiological behavior, particularly food intake.

Flumazenil

Flumazenil is a benzodiazepine antagonist and is used to reverse the sedative effects of midazolam and diazepam in birds.[1,2,7] The injectable form of flumazenil (flumazenil hydrochloride (0.1 mg/ml), Abaxis Pharmaceutical Products, Schaumburg, IL) can be administered intranasally, intramuscularly, or intravenously, without side effects. The recommended dosages range from 0.01-0.1 mg/kg. Alternatively, a flumazenil to midazolam ratio of 13:1 has been recommended, but it has been shown that lower doses of flumazenil achieve complete recovery from midazolam-induced sedation in birds.[1] The author prefers to administer 0.05 mg/kg initially. If the reversal is deemed unsatisfactory 10 minutes after drug administration, then the same amount of flumazenil can be administered repeatedly. Recovery from sedation is usually complete within 10-15 minutes. Reversal with flumazenil is recommended in all outpatient cases, to avoid injuries from falling off perches and negative client perceptions of interacting with their abnormally behaving bird. Therefore, it is important to ensure that complete reversal has occurred prior to discharge from the hospital by assessing complete recovery of the sensory and motor function (e.g., approaching the bird, rotating finger/perch to evaluate balance).

Atipamezole

Sedation protocols using dexmedetomidine or medetomidine should always be reversed using atipamezole. Atipamezole should be dosed at 10× the dose of dexmedetomidine in milligrams or 5× the dose of medetomidine in milligrams. Redose together with flumazenil if recovery is considered insufficient.

Step-by-Step Procedure

1. Educate the bird owners, if present during the sedation event, about what to expect regarding the sedation of their bird. Sedation using published protocols is safe in healthy pet birds, but birds will appear sleepy, which owners are not used to. Inform them that in some cases, regurgitation might occur, particularly in macaws.

2. Manually restrain the bird using a towel and immobilize the head.

3. Administer the sedative drugs either by intramuscular injection or intranasally. For intranasal administration, administer into the nostrils over 5-10 seconds and split the total dose between both nostrils.

4. Release the bird from manual restraint and allow 7-10 minutes to reach the maximum sedative effect. Do not allow the bird to remain high off the ground; instead, place it on the floor or in its carrier.

5. Manually restrain the sedated bird. Monitor respiration throughout the restraint period, as well as heart rate. Perform necessary clinical procedures (e.g., physical examination, blood collection, radiographs).

6. If reversing sedation, administer flumazenil IM or intranasally (split total volume between nostrils).

7. Place the bird on the floor or in its carrier and monitor recovery. Do not allow the bird to remain high off the ground until it has completely recovered to avoid trauma from falling. Most birds will recover within 10-15 minutes. Assess recovery before discharging the bird. Administer a second dose of reversal flumazenil if recovery is incomplete. Cockatoos often recover slowly and, in most cases, need a second dose. Do not discharge an incompletely recovered bird.

Useful Tips for Sedation of Birds

Below are tips for use during sedation of birds.

- Each bird requires an individual assessment prior to choosing a suitable sedative drug protocol. Birds which have no previous experience with manual restraint may require less or no sedation compared to birds that had previous negative experiences and become readily stressed or fearful in a veterinary clinic environment.

- Macaws and grey parrots usually require a combination of midazolam and butorphanol to achieve a sufficiently deep level of sedation suitable to perform a variety of clinical procedures.

- Smaller psittacines, passerines and other non-psittacines often need higher doses of midazolam-butorphanol and often the induced sedation is not satisfactory. Therefore, consider using dexmedetomidine-midazolam protocols instead.

- Even though a bird might not appear sufficiently deeply sedated while under manual restraint for a clinical procedure, it might become more sedated once stimulation is discontinued, making it unsuitable to be discharged from the hospital. Therefore, every bird, which is intended to be discharged shortly after completion of the procedures performed under sedation, should receive flumazenil reversal.

- Cockatoos often do not recover completely after the initial dose of flumazenil. A second dose of flumazenil is typically necessary.

- Record the sedation protocol used and the achieved level of sedation, indication for sedation and adverse effects in the patient's medical record, to better guide sedation protocol selection for future visits.

References

1. Doss GA, Mans C. Avian sedation. *J Avian Med Surg.* 2021;35:253-268.

2. Mans C, Guzman DSM, Lahner LL, et al. Sedation and physiologic response to manual restraint after intranasal administration of midazolam in Hispaniolan Amazon parrots (*Amazona ventralis*). *J Avian Med Surg.* 2012;26:130-139.

3. Sadegh AB. Comparison of intranasal administration of xylazine, diazepam, and midazolam in budgerigars (*Melopsittacus undulatus*): clinical evaluation. *J Zoo Wildl Med.* 2013;44:241-244.

4. Vesal N, Zare P. Clinical evaluation of intranasal benzodiazepines, alpha-agonists and their antagonists in canaries. *Vet Anaesth Analg.* 2006;33:143-148.

5. Scagnelli A, Titel C, Doss G, Mans C. Effects of midazolam and lorazepam on food intake in budgerigars. *J Exot Pet Med.* 2022;41:42-45.

6. Martel A, Doss G, Berg C, Mans C. Effects of midazolam on food intake in budgerigars. *J Avian Med Surg.* 2022;36:53-57.

7. Mans C, Braun J, Feeny M. Intranasal sediation with midazolam or midazolam-butorphanol in psittacine birds: a retrospective evaluation of 114 cases (2010-2012). *Proc ICARE.* 2013.

8. Mumm L, Mans C. Evaluation of dexmedetomidine-midazolam sedation in budgerigars. *J Am Vet Med Assoc.* 2022;260:1194-1199.

Section 6

AAV Masterclasses Track 2

Michael Huffaker, DVM;
Seth Oster, DVM
Session Chairs

Avian Wound Management

Heather Wilson Barron, DVM, DABVP (Avian Practice), and
Rebecca Duerr, DVM, MPVM, PhD

Session #3220

Affiliation: From Loggerhead Marinelife Center, 14200 US HWY 1, Juno Beach, FL 33478, USA (Barron), International Bird Rescue, 4369 Cordelia Rd, Fairfield, CA 94534, USA (Duerr).

Abstract: The purpose of this masterclass is to provide in-depth understanding of wound management of birds. Techniques are applicable to all bird species and case examples are given. The course includes details on practical applications of things like bandaging, with clear guidelines on how and when to use bioactive, hydrocolloid, hydrogel, and other dressings. Further discussion involves topical medications, antibiotic impregnated beads, regional limb perfusion, bioactive glass, and surgical management options like autografting, advancement flaps, and closure of severely contaminated wounds when appropriate due to species concerns. In addition, vacuum-assisted closure is discussed. Videos are used to ensure understanding of more complex procedures. Poll questions are used to ensure audience engagement and understanding.

Introduction

By definition, a wound disrupts the normal skin barrier, which is the body's first line of defense against invading microorganisms. Wounds are a common and debilitating problem in domestic and companion birds, as well as being a common reason that wild birds are presented for care. Understanding and delivering modern, appropriate wound care for birds can present a unique challenge for the avian practitioner. Some of the special anatomic features, physiology, or behavior of the myriad of bird species seen can engender a need for creative thinking. When wounds do not heal in a timely manner, there is an increase in discomfort and disease that interferes with health and quality of life, and in wild species that present in an extremely debilitated state, or with unfamiliar behavioral or physiologic needs, delays in healing can lead to the development of new problems and eventual euthanasia. There is a vast array of options for wound treatment and it is important that the avian practitioner remains current with this knowledge to offer the best patient care. Wound care in companion birds is well-covered in other sources, hence the options presented here largely focus on techniques used by the authors in their roles as wild bird veterinarians.

Avian Integument

One thing that clinicians may appreciate when attempting to suture bird skin is that it is less elastic than mammalian skin, drier, and much thinner (only ~4-10 cells thick), although grossly there is much variation due to species. For example, duck skin seems thick, fatty, and friable while songbird skin may resemble tough wet tissue paper. Other variations include air cells as in pelicans, dermal ornamentations such as wattles, combs, and beards, species-dependent anatomic locations of subcutaneous fat deposits, and changes due to seasonal transitions or loss of body condition.

The other thing that is obviously unique to birds is the presence of feathers. Be very conservative when removing feathers from wild birds, especially flight or tail feathers, lest you inadvertently prolong the bird's captivity.

Some species may not be able to stay in captivity long enough to regrow feathers before developing sometimes-lethal, captivity-related problems. Larger species may take three or more years to molt their entire set of flight feathers, for example.

The dermis is the site of feather development and is also quite thin. Therefore, be cautious in the application of topical medications as they may be more easily absorbed transdermally. Further, caution should always be used in the application of any ointments or creams in birds topically (unless well-bandaged) because these products may damage the plumage and render it ineffective in keeping the bird warm and waterproof. This holds special importance during wound care in wild birds because some species require rapid transition into water-based housing, and if dirtied by topical products, birds may require washing. Birds have thin skin and limited soft tissue covering the distal extremities and thus a reduced vascular supply to these areas, which means wounds in this area may be slow to heal and simple closure of skin defects may be difficult. Birds also lack myeloperoxidase and tend to have very caseous pus and thus drains are typically not used, although because birds are so diverse, there are exceptions to every rule. One author (RD) has successfully used drains on several species with persistent liquid pus production from wounds, particularly those involving aquatic birds with severe fishing hook or line injuries (including pelicans, shorebirds, gulls, geese, others). Finally, birds lack biliverdin reductase and hence bruising in avian skin and muscle may appear green.

Tissues deep to the skin vary by anatomic location and may be merely exposed and subject to desiccation or infection, or the deeper tissues may be involved in the injury itself. Muscle, tendons, ligaments, bones, nerves, and other tissues may be directly involved, and the fate of these tissues may drive the outcome of the case. With wild birds that must have appropriate function and athleticism for their wild lifestyle after recovery, permanent disability may become an indication for euthanasia. Once in a good nutritional plane, most wild birds have an excellent ability to heal even severe wounds with appropriate care, as long as the injury has not ruined a non-expendable structure such as a joint vital for flight, or a toe needed for killing prey.

Anesthesia and Analgesia

Anesthesia or sedation may be necessary for adequate wound care. Analgesics are an important part of treatment and general anesthesia may be indicated. One author (HB) generally prefers buprenorphine or hydromorphone for analgesia in most species of wild birds. Avian species responses to these drugs are highly variable, even with closely related species The other author (RD) commonly uses butorphanol, meloxicam, and gabapentin, again with awareness of species differences and an eye out for beneficial and adverse responses to therapy. There are concerns regarding the use of NSAIDS in poorly studied taxonomic groups of birds, hence staying abreast of recent developments in avian pharmacology is prudent and more work in this area is needed. Clinical presentation including species considerations may help dictate anesthetic protocols. The patient should be ideally warm and stable prior to anesthesia or aggressive wound management. Until then, simple bandaging may provide temporary fracture stabilization and wound coverage.

Wound Care

Basic care for most wounds necessitates fully identifying the nature of the injury. Begin with gross removal of debris from the wound, and carefully pluck a narrow margin of healthy skin around the edge. Pluck towards the wound whenever possible to avoid creating a larger defect by tearing the skin or hold flat-tipped forceps on the skin at the base of each feather to reduce tension. Follow the wound margin to fully identify the extent of the lesion. This can be quite time consuming in aquatic birds with dense plumage, as necrotic skin may be covered with relatively normal-looking feathers. Do not remove flight (remiges) or tail (rectrices) feathers if at all possible, as it is preferred to allow soft tissues to heal with a feather in the follicle.

Once gross debris has been manually removed, follow with lavage of large amounts of warm sterile physiologic fluid with or without dilute chlorhexidine, betadine, colloidal silver solutions, or Tris EDTA as appropriate. Irrigation pressure is ideally around 8-10 PSI to adequately clean and debride the wound. Lavage is contraindicated in wounds that include an opening into the bird's air sac system. If uncertain whether flushing a wound will risk drowning the bird, it is helpful to have the bird anesthetized and intubated, and have the anesthetist give the bird a nice big breath to check for leakage of air out the wound. If the wound does communicate with the respiratory system, cleansing the area repeatedly with sterile gauze moistened with sterile flushing fluids may suffice.

Debridement may be done as needed surgically or with wet to dry bandages over the first few days. Lavage may be done once daily and then gradually less often. Wet-to-dry bandages can be very helpful in removal of debris and in moistening dried out jerky-type tissue that needs removal. Coagulation necrosis is common in wounded avian tissue, and these necrotic tissues must be removed for full healing to occur. Sterile hydrogel-impregnated gauze is also useful for softening necrotic tissues prior to debridement. During debridement it may become apparent that an important structure is involved in the wound. Prognosis and outcomes are highly dependent on what structures are involved. There are some tendons that can be lost without a perceptible effect on function but losing a large bundle of tendons that completely control, for example, flexion or extension of the hock joint, is not compatible with being a releasable (or even comfortable in captivity) wild bird. Quality of life must be assessed for captive birds with similar loss of function. The outcome of these cases is usually dependent on what function has been lost, but it is often worth attempting treatment to see what functionality is retained.

Muscle tissue in birds has a tremendous capacity for repair. Large wounds in muscle, even when a large portion of the muscle is essentially destroyed, may heal well and result in a full return to function. Tendons are, however, more delicate, and tend to desiccate, die, and rupture if allowed to remain exposed for any length of time. Freshly exposed tendons that are shiny and white should be prioritized for treatment, as they are still healthy and need help to stay that way. As might be expected from a slippery structure designed to move freely and not adhere to surrounding tissue, tendon can be difficult to heal over with second intention healing and requires primary closure for best healing. See Duerr[1] for information on the use of autografts in providing coverage when locally available skin is not adequate. Tendon that has been exposed for long periods that looks yellowish or tan or has fibrinous debris covering it has a poor prognosis. Even if the unhealthy tendon appears intact and continues to function for days to weeks, it is not unexpected that it may rupture after a few weeks in care. Dead tendons can continue to perform their mechanical functions until rupture occurs. These necrotic tendons typically feel like firm tubular structures on palpation and appear denser than surrounding soft tissue on radiography. Careful regular palpation of tissues surrounding wounds can identify these problems prior to occurrence. In the author's (RD) experience, if the tendon feels thickened but springy in texture, it may merely be inflamed and may recover; monitor it regularly to assess whether the inflamed zone is increasing or decreasing in size. But if it is firm with any sort of lumpy character, it may be dead or dying and may require removal or may suddenly extrude itself from a nearby wound being treated when the area is manipulated.

One advantage that birds have over mammals is that bone exposure in wounds is usually much less concerning, and with appropriate care these areas heal very well. See Duerr[1] for repair of degloving injuries to the neck and scalp, and Duerr[2] for information regarding repair of keratin damage exposing pelican mandible, which is also applicable to bone exposure in any species near the feathered skin-bill keratin border. Open joints are always of serious concern and healing that results in a fully functional joint is uncommon.

Skull, synsacrum, or other bone may be laid bare by skin damage. Small areas no larger than the skin is thick do not usually require treatment, but larger exposures usually do. In wild birds these injuries may not present until chronic. Left to its own devices, the skin at the edges of the defect adheres to the bone and begins to granulate in place, while cells are unable to cross the defect because it cannot heal from below. A raised edge of granulating skin surrounding a flat plain of eroding bone is often the result. In a radiograph that catches a cross section of these areas on long bones, it may be apparent that the exposed area is eroding while the inside of the medullary cavity is becoming reinforced. This at least means that the bone is rarely penetrated by the erosion, but the site is a perpetual source of discomfort from the tissue layers being stuck together (which may impede movement such as flight) and the open wound remains a source of potential infection. Exposed skull from degloving injuries in ducks and geese tends to develop a leathery scar on the bone and wound contracture may cause the palpebral fissure to become displaced from the globe, leading to visual impairment. One author's (RD) preferred resolution for these injuries is to meticulously debride the surface of the bone and associated crevices by scraping with a scalpel blade, then removing the granulated edge to create a fresh skin margin, undermining the adjacent tissue and then closing the gap across the bone with as little tension as possible. If this is occurring on an exposed skull, this closure becomes an advancement graft from the cranial neck skin, which is placed onto the crown like a hood. Excess tension or inadequate debris removal may result in graft necrosis or dehiscence. Ducks and geese that have been hit by cars may present with macerated skin with areas of exposure of the synsacrum. These areas often granulate well with minimal treatment after simple debris removal. Periodic debridement may be useful; autografts are usually not needed.

Topical medications

The use of topical antibiotics is indicated when treating infected wounds. Ointments may initially help to keep tissues moist but can delay wound healing by impeding the migration of fibroblasts into the area. These agents most often are indicated when applied to small, recent wounds for short periods of time. Silver sulfadiazine cream is effective against most gram-positive and gram-negative bacteria as well as most fungi. It is often used for thermal burns, as in electrocution cases, because of its ability to penetrate eschar and necrotic tissue.[3,4] But it may have a negative effect on fibroblasts and thus may impede wound contraction and so is used infrequently by an author (HB).[5,6] Medical grade honey and silver impregnated hydrogels have been documented to be effective in a variety of avian wounds and are generally both authors' empiric preferences.[5] Honey and granular sugar also have uses in reducing the gross contamination of wounds that are producing a lot of discharge and debris; however, be aware that hyperosmotic dressings may cause tissue such as exposed tendons or bare bone to desiccate. Whenever possible, bone and tendon should be covered with vital tissue, even if it means only partially closing a wound.

Some species of birds tend to quickly close skin wounds over pockets of contaminated debris; hence keeping an abscessed area open for dressings and debridement may be best while a wound heals from below. Routine palpation of deeper tissues as a wound heals allows monitoring for abscess or granuloma formation. In birds, the skin should remain largely loose in relation to deeper tissues—once the skin has healed, an area of adhesion may indicate a problem such as a pocket of debris.

Petrolatum impregnated dressing works well to keep the tissues moist as they heal, although oily dressings of any kind are not suitable for use in species of birds that will require waterproof plumage in the very near future (loons, grebes, diving ducks, and seabirds to name a few). All wild birds must be waterproof prior to release lest they become soaked and chilled the first time it rains. Washing a wild bird simply for ointment removal adds stress to an already-stressed patient.

Vacuum-Assisted Closure

Once hemorrhage is controlled and a clean wound bed established, vacuum assisted closure may reduce wound healing time. In one study in humans, vacuum-assisted closure (VAC)-treated wounds decreased in size by 78% compared to a 30% size reduction in wounds treated with wet-to-dry bandages over the same time frame. There are few publications describing the adaptations for this technique in birds.[6] The VAC is usually left in place until a healthy bed of granulation tissue is noted. Bandage changes generally start out being done more frequently and then may stretch out to once every few days once the wound is progressing well. Aquatic species may need to be alternated in water and dry docking with the VAC if waterproofing isn't possible. Antimicrobials commonly used by the author (HB) with the VAC include water-soluble products, such as honey or colloidal silver hydrogels. Do not use topical antimicrobials that will impair the VAC, such as creams or ointments.

External Coaptation and Physical Therapy

Keep a bandage box with appropriately sized bandage materials for small patients available to help facilitate rapid application in emergent situations. A figure-of-8 bandage may be indicated in birds for temporary stabilization of fractures of the radius, ulna or both; fractures or luxations of the elbow or carpus; or soft tissue injuries to the brachium or antebrachium.[7] Generally, cast padding and an outer layer of vet wrap will suffice for most wild birds. Usually no more than 2-3 circuits around are needed; otherwise the bandage is needlessly bulky. A body wrap is indicated for shoulder or humeral injuries. A correctly placed bandage will have the primary feathers sitting **under** the secondary feathers, as seen in the normal perching pose. Be sure to downsize bandaging materials appropriately for small patients.

Generally physical therapy is indicated in wing wraps to prevent patagial membrane contracture and keep a normal range of motion (ROM) in the joints. Physical therapy (PT) may begin as soon as 3 days post-admission or fixation of the limb. Initially, PT should be done daily or every other day. If ROM is good, this may be extended to every third day. Once contracture has occurred, it is more difficult to reverse it. So, it is better to do PT more often. Initially, it may be necessary to do PT under general multi-modal anesthesia with injectable pain medications plus inhalant anesthesia. After some initial healing, it may be possible to do PT awake in sedated animals given injectable analgesic drugs. Eventually, it may be possible to do PT using oral pain medication with or without sedation. The author (HB) tries to minimize the use of intramuscular (IM) injectable medications where possible and puts medications into food items to avoid the pain and trauma that repeated IM injections may cause. Generally, external coaptation devices are removed between 10-21 days if they are the only method of fixation. Joints should never be bandaged for longer than necessary and timely PT is always required for a good outcome. In species with calmer demeanors, the author (HB) may keep full coaptation on for shorter time periods, gradually reduce, and try strict cage rest with anxiolytics and analgesics to cut down coaptation time.

Functions of tendons and joints are assessed by manipulation of the reciprocal apparatus of the wings and legs. For example, in the wings, the carpus is depressed caudally while the elbow is depressed cranially. These combined forces should cause the elbow and carpus to pivot open and spread the wing utilizing the tendons primarily. See Freshman and Duerr[8] for more information on developing physical therapy plans for wild birds.

Goniometer measurements should be used to assess ROM. Heat packing prior to PT exercises may facilitate better ROM. While there is some anecdotal information that low level laser therapy may also be effective, there are very few published studies demonstrating clear efficacy in birds.[9] In one study in green iguanas, laser therapy did not result in substantial improvement in histologic evidence of wound healing, compared with results for traditional wound management.[10]

References

1. Duerr RS. Surgical repair of keel lesions and lacerations in aquatic birds. *Proc ExoticsCon.* 2016;83-88.

2. Duerr RS. Successful surgical management of pouch and bill injuries in pelicans. *Proc ExoticsCon.* 2018;363-367.

3. Grunkemeyer V. Management of avian skin wounds. *Proc Annu Conf Assoc Avian Vet.* 2009;57-62.

4. Burke HF, Swaim SF, Amalsadvala T. Review of wound management in raptors. *J Avian Med Surg.* 2002;16:180-191.

5. Mans C, Sunohara-Neilson J, Higginson G, et al. Wound management in a trumpeter swan using honey and a sustained release ionic silver hydrogel. *Exot DVM.* 2006;8(5):21-23.

6. Knapp-Hoch H, de Matos R. Clinical technique: negative pressure wound therapy-general principles and use in avian species. *J Exotic Pet Med.* 2014;23(1):56-66.

7. Duerr RS, Purdin GJ, eds. *Topics in Wildlife Medicine: Orthopedics.* Vol. 4. St. Paul, MN: National Wildlife Rehabilitator's Association; 2017.

8. Freshman J, Duerr R. Physical therapy. In: *Topics in Wildlife Medicine: Orthopedics.* Vol. 4. St. Paul, MN: National Wildlife Rehabilitator's Association; 2017:6-14.

9. Hippólito AG, Stanski G, Fernandes LL, et al. Evaluation of topical medications, lavage solutions, and laser therapy for treating open wounds induced in blue-fronted Amazon parrots, *Amazona aestiva. J Exot Pet Med.* 2019;29(0):1-9.

10. Cusack LM, Mayer J, Cutler DC, et al. Gross and histologic evaluation of effects of photobiomodulation, silver sulfadiazine, and a topical antimicrobial product on experimentally induced full-thickness skin wounds in green iguanas (*Iguana iguana*). *Am J Vet Res.* 2018;79(4):465-473.

Vulture Medicine and Conservation

John Chitty, BVetMed, CertZooMed, MRCVS,
Michael Jones, DVM, DABVP (Avian Practice),
Julia Ponder, DVM, MPH

Session #3212

Affiliation: From Wombourne, Allington, Salisbury, SP4 0DD, UK (Chitty), The American Eagle Foundation, 2957 Veterans Blvd.,Pigeon Forge, TN 37863, USA (Jones), College of Veterinary Medicine, University of Minnesota, 1365 Gortner Avenue, St. Paul, MN 55108, USA (Ponder).

Abstract: Old World and New World vulture species face threats worldwide and are of major conservation significance. Many of these species have previously been extremely common but have suffered major population crashes. Veterinarians working with these species—whether in zoos, rehabilitation, or as part of conservation projects—need an understanding of the unique biology of these birds to provide successful treatment and outcomes. Their intelligence and complex social structures require an understanding of their behavior and behavioral diseases they may experience. Differences are highlighted by the sensitivity of some of these species to commonly used drugs, especially non-steroidal anti-inflammatory drugs (NSAIDs). This masterclass provides information about the natural history as well as rearing and husbandry of Old and New World vultures encountered in managed care, handling and examination, behavioral conditions (i.e., self-mutilation syndromes), common conditions, diagnostic techniques and clinical pathology, therapeutics, analgesia, anesthesia and surgical techniques, all while highlighting similarities and differences between vultures and other raptors.

Vulture Taxonomy

Taxonomy and classification of raptors has gone through several significant changes in the last decade. Currently, four avian lists are available; however, they may not completely agree, which some would argue is detrimental to avian conservation.[1] In particular, species level taxonomy is often the basis for decisions for conservation prioritization.[1] The four lists available include the International Ornithological Community (IOC) World Birds List, the Howard and Moore Checklist of the Birds of the World, the eBird/Clements Checklist of Birds of the World, and the Handbook of the Birds of the World and Birdlife International Digital Checklist of the Birds of the World. These lists differ not only in the scientific names they use but also in the groups of individual organisms to which the scientific name refers.[1] Previously, diurnal birds of prey were included in the Order Falconiformes under which five families were noted to include Sagittariidae (Secretary birds), Pandionidae (ospreys), Accipitridae (Old World Vultures (OWVs), harriers, hawks, eagles, and buzzards), Falconidae (falcons and caracaras) and the family Cathartidae (New World vultures). Nocturnal birds of prey were listed under the Order Strigiformes with two families, the Strigidae (most owls) and Tytonidae (barn and bay owls). Subsequently, diurnal birds of prey were divided into three different Orders including Falconiformes, Accipitriformes, and Cariamiformes to include the five families listed previously but with the addition of the family Cariamidae (Seriemas). Strigiformes remained the same. At one point, Cathartidae were considered part of the Order Ciconiformes along with storks. Currently, while raptor classifications are still in a flux, there appears to be some agreement that there are five taxonomic Orders of birds of prey.[1] These include the Orders Acciptirformes, Falconiformes, Cariamiformes and Cathartiformes for diurnal birds of prey, and Strigiformes for nocturnal birds of prey. Interestingly, OWVs are classified with Accipitriformes despite their natural history as obligate scavengers compared to all other member of the Order.

Vulture Conservation

Worldwide, vulture populations are under pressure with 61% of vulture species threatened with extinction. Of the 16 African-Eurasian species of OWVs, eight are critically endangered (four in Africa and four in Asia), two are endangered, one is vulnerable and three are near-threatened (IUCN Redlist). Most of the seven New World vulture species are categorized as least concern although the California condor (*Gymnogyps californianus*) is critically endangered and the Andean condor is in rapid decline. While turkey vulture and black vulture populations are on the rise, most of the rest are in decline.[2]

For some Old OWV species, the current population problems can be traced to the 1990's, when three species of *Gyps* vultures experienced 95% population reductions over the course of one decade through exposure to diclofenac, a non-steroidal anti-inflammatory drug used by veterinarians in treating cattle in Southeast Asia. The exquisite sensitivity of vultures to this drug as well as the communal feeding strategies led to massive die-offs throughout the region.

There are many reasons for the on-going vulture population declines, with poisoning and persecution at the top of most (if not, all) lists. As the only known obligate scavengers outside of the ocean, vultures have a high-risk feeding strategy that increases the likelihood of exposure to contaminants and toxicants. They often feed communally, increasing the risk that significant numbers of individuals will be impacted by a single carcass. In addition, they are long-lived, high trophic level feeders, which increases the risk for bioaccumulation of toxicants. These shared characteristics of vulture populations put them at high risk for poisoning, whether the exposure is intentional or unintentional.[3] Banning of the use of diclofenac for cattle in several counties has led to some populations stabilizing and recovering. In Pakistan, a population of colony of long-billed vultures was shown to have a significant decline in reproductive parameters in the three years before the 2006 ban of diclofenac and an increase in the same parameters for the two years after the ban.[4]

For OWVs, the number one threat to vultures is probably unintentional poisoning secondary to deliberate poisonings of carnivores to protect game and livestock. These poisonings may be efforts to reduce predators of game animals or in retaliation for carnivore predation on livestock.[5] Additionally, vultures have been intentionally poisoned to prevent their signalling poachers of the presence of an elephant carcass, as well as for collection of body parts for belief-based use and trade. Between September 2019 and March 2020 across Guinea-Bissau, West Africa, greater than 2000 hooded vultures were killed by poisoning with methiocarb, a carbamate pesticide, and their heads were collected for cultural use.[6]

New World vulture populations also face many problems. The main threats for California Condors, a highly endangered population with approximately 300 free-flying individuals, are lead poisoning from spent hunting ammunition; utility line-related mortality (largely mitigated) and vulnerability to catastrophic losses secondary to forest fires and changing climate. One study demonstrated the risk of exposure to second-generation anticoagulant rodenticides (SGAR) with 42% of condors tested being positive for residues and evidence of prolonged clotting time in 16% of free-flying condors.[7] For Andean condors, persecution is the greatest population threat with poisoning (sodium monofluoroacetate and carbofuran) on top of the list with shooting second.[8] Other threats faced by vulture populations solidly reflect the intersections with human activity and include collision and electrocution at energy infrastructures; reduced food supplies; habitat destruction and degradation; and human disturbances.

While many of the threats are based in beliefs that vultures cause problems for humans, they need to be balanced by the ecologic, economic and social case for conservation. Vultures provide many ecosystem services. They clean up carcasses through scavenging, promoting the flow of energy through the food web and preventing disease outbreaks. They act as sentinels for other predators in locating food sources. And their presence can contribute to ecotourism.[9]

Literature Review of Infectious Diseases in Vulture Species

Given their natural history as obligate scavengers that consume carcasses in various stages of decomposition, vultures are exposed to many infectious disease-causing organisms. In a systemic global review of available scientific literature, infectious diseases account for 16% of causes of morbidity and mortality in free-living vulture species worldwide.[10] Other reports of opportunistic infectious diseases occurring in vultures in managed care are available but are also limited in number, suggesting a gap in our understanding of infectious disease in vulture species world-wide. In many instances, susceptibility to infectious diseases may be a consequence of co-morbidities (i.e., toxin exposure, malnutrition) that inhibit an appropriate immune response in affected vultures.

Fungal diseases

Ives et al (2022)[10] reported fungal diseases, as the most commonly reported infectious disease in free-living Old World and New World vultures. *Aspergillus* spp. was the most commonly reported in the literature; however, *Candida* spp. affected more individual free-living vultures and was also reported more in nestling and fledglings.[10] Chege et al. (2013)[11] reported on opportunistic *Aspergillus* spp. infection and mixed bacterial infection in Cape vultures (*Gyps coprotheres*) in managed care.

Bacterial diseases

Bacteria reported to cause disease in vulture species include *Mycoplasma* spp., *Mycobacterium* spp., *Escherichia coli*, *Salmonella* spp., and *Staphylococcus* spp.

Several reports describe disease caused by *Mycobacterium* spp. For example, granulomatous disease associated with *Mycobacteria* spp. has been described in an adult male white-rumped vultures (*Gyps bengalensis*) in Pakistan.[12] In addition, upper respiratory tract disease associated with *Mycobacteria* spp. has also been described in other vultures.

It is difficult to determine the significance of *Mycoplasma* spp. infections in raptors as the presence of the organism does not always correlate with a disease process; however, Panagala (1993)[13] described a new species, *Mycoplasma corogypsi* sp. associated with footpad abscesses in a black vulture (*Coragyps atratus*). Similarly, Ruder et al (2009)[14] were the first to report polyarthritis diagnosed in association with *Mycoplasma corogypsi* in a black vulture and documented an acute fibrino-heterophilic polyarthritis. The vulture also suffered from lead toxicosis. Oaks et al (2004)[15] described mild heterophilic inflammation in the trachea and bronchi associated with a novel *Mycoplasma* spp. in an Oriental white-backed vulture. This vulture was found dead and demonstrated renal failure and visceral gout related to diclofenac intoxication.[15]

Other bacterial species such as *Salmonella* spp. and *E. coli*, while reported as occurring in vulture species, appear to be associated more with intense agricultural practices, antimicrobial resistance and concerns of carrier or reservoir status in vultures and may not represent actual diseases in these species. Interestingly, it is suspected that the microbiome of vultures may make them more resistant to enteric diseases.

Viral diseases

Viral diseases reported in vulture species world-wide include West Nile virus (WNV), Influenza A, and poxvirus. The Centers for Disease Control and Prevention reports detection of WNV in all bird species from 1999 to 2016. In this report, WNV was detected in a free-living black vulture, a turkey vulture and a cinereus vulture in managed care.[16] Plaza et al (2020)[17] reviewed recent literature and reported the presence of WNV in California condors, bearded vultures (*Gypaetus barbatus*), and a cinereous vulture (*Aegypius monachus*). In this report, the presence of inflammation and hemorrhage produce by WNV was only reported in a California condor.[17] The other species in which WNV was detected either had no disease condition reported or had concomitant conditions such as oral lesions associated with *Candida* spp. in a cinereous vulture and footpad abscesses in a black vulture.

Highly pathogenic avian influenza A (HPAI), specifically the Eurasian H5N1 lineage, has caused severe disease and large-scale die-offs in vultures in the United States, especially in Georgia, Kentucky, New Jersey, and Tennessee over the past 1 to 2 years, The largest die-off of 700 black vultures occurred at a Sanctuary in Georgia. As of April 13, 2023, the United States Fish & Wildlife Service reported the deaths of 18 California condors in the Arizona/Utah flock; HPAI has been confirmed for six of those condors. Certainly, this is a major concern for the population of condors should the disease spread to other regions, especially to California. To protect the species, biologist are considering capturing wild condors to protect them from infection until a vaccine can be developed.[18] Interested parties can follow detection of HPAI in the United States by reviewing the U.S. Department of Agriculture Animal and Plant Health Inspection Service site.[19] Certainly, HPAI infections in vultures have been described in other countries many years prior to 2022. Vultures are most likely to become infected by consuming infected migratory bird species (waterfowl and shorebird), and cannibalism of dead infected vultures may also contribute to the spread of this disease. Characteristic neurologic clinical signs associated with avian influenza are similar to those reported in other raptor species along with pathologic lesions demonstrating multiorgan necrosis.

Poxvirus that affects raptors are distributed worldwide, and infections are likely underreported in vulture species. This author (MPJ) and others in the United States previously noted what appeared to be poxvirus infections in turkey vultures (*Cathartes aura*). Poxvirus also has been detected in a turkey vulture in Chile by Gomez et al (2014)[20] and by Mora-Carreño et al (2019)[21] in Venezuela. The raised and sometimes ulcerated appearance of poxvirus lesions noted on unfeathered areas of the feet and the head of vultures are similar to lesions noted in other raptor and avian species.

Parasitic and protozoal diseases

Similar to other raptors, parasitic and protozoal diseases are reported in vulture species. Martinez-Herrero, et al (2014 and 2019)[22,23] documented oropharyngeal trichomonosis with a novel Trichomonas spp., *Trichomonas gypaetinii*, in Griffon (*Gyps fulvus*), Egyptian (*Neophron percnopterus*), and Cinereous vultures. The reported lesions were similar to those reported in other raptor species and trichomonad trophozoites were detected from wet mount preparations of the lesions.

In general, hemoparasites are considered non-lethal at lower levels of infection in most avian species. Webb SL et al (2005)[24] performed a survey of blood parasites in black vultures and turkey vultures from South Carolina, USA and noted that while blood parasites have been reported in black vultures, no blood protozoans were observed on blood smears of black vultures with the exception of a microfilaria found in one black vulture. In another study, Yabsley et al. (2018)[25] opportunistically obtained blood and/or tissue samples from 162 turkey vultures and 95 black vultures to identify haemosporidian parasites in these species. Similar to the previously noted study, no parasites were detected in the black vultures by blood smears or PCR testing of their samples.

Yabsley et al. (2018)[25] also provide a brief review of hemoparasites documented in New World and OWV species. *Hemoproteus elani*, *Haemoproteus janovyi*, *Leucocytozoon toddi* and *Plasmodium fallax* have been documented in OWVs, and *Haemoproteus catharti*, *Leucocytozoon toddi* and *Plasmodium elongatum* have been reported in New World vultures. Yabsley et al. (2018)[25] further asserts that there are numerous records of *Haemoproteus* spp., *Plasmodium* spp., and *Leucocytozoon* spp. that have yet to be morphologically or genetically characterized in New World vultures.

New World Vulture Rehabilitation and Ethics

Vultures are frequently seen in wildlife rehabilitation facilities. As with any injured or ill animal admitted for treatment, they present with a wide variety of issues, with miscellaneous trauma and toxicities being two of the more common presenting problems. While the goal of wildlife rehabilitation is always the treatment and release back to the wild of the patient, there are a series of decisions from triage through final outcome that determine the bird's future and require thoughtful and ethical considerations that prioritizes the bird's welfare. There are three possible outcomes for a vulture admitted to a wildlife rehabilitation facility: successful treatment and release, euthanasia and placement in permanent captivity. While the veterinarian's role in clinical treatment and release is the most obvious, the DVM also plays a key role in each of the other two outcomes. With the high bar of a full return to normal being required for release, the veterinarian's experience is helpful in determining a prognosis. When that bar cannot reasonably be expected to be reached, the veterinarian's access to humane and acceptable methods of euthanasia is a critical support for wildlife rehabilitators.

Vultures present some unique concerns when considering the option of placement in permanent captivity. As with all animals, it is important to consider the physical, affective and environmental states for the bird when determining if it is a candidate for placement. Physically, the bird's injuries must be completely healed, the illness/injury should be expected not to progress (no future complications), the bird should have no imbalance issues (such as those associated with amputations, pelvic/leg issues or nerve damage), and the bird's permanent injury must be compatible with its captive purpose. In the United States, a wild native bird cannot be kept in permanent captivity unless it is deemed non-releasable by a veterinarian, who must also attest to the fact that the bird meets the other conditions listed above.

Links to U.S. Fish and Wildlife Service legal requirements and documents

- Guidance for veterinary certification of condition that renders migratory bird/eagle as non-releasable https://www.nwrawildlife.org/resource/resmgr/symp_2020/handouts/Handout_Template_Vet_Letter.pdf

- Euthanasia requirements (50 CFR 21.76(e)(4)(iii)) https://www.ecfr.gov/current/title-50/chapter-I/subchapter-B/part-21/subpart-C/section-21.76#p-21.76(e)(4)(iii)

- Requirement of commitment of veterinarian to provide medical care of a placed bird (50 CFR 21.76(e)(4)(iii)(A)) https://www.ecfr.gov/current/title-50/chapter-I/subchapter-B/part-21/subpart-C/section-21.76#p-21.76(e)(4)(iii)(A)

Beyond these physical requirements, the consideration of the affective or mental state of a vulture is a key consideration for placement. Vultures are highly social animals, extremely intelligent and very long-lived. While young vultures are the most likely to adapt well to captivity, they also require considerable social learning from conspecifics. Single vultures housed alone present a welfare consideration as it is likely detrimental to their mental state. These social animals tend toward neophobia when imprinted or taken out of their social context, which impacts the ability to adapt. It can require extensive knowledge, time, resources and skills to integrate them into an established community. Confidence-building is key to avoiding behavioral issues, which are often seen as aggression, which covers for neophobia. Conspecifics are critical for this confidence-building. With enough space, non-imprinted adults may integrate well into a display situation as long as there is no need for constant management or medical treatment.

Finally, when considering placement, the longevity of vultures must be taken into consideration. It is extraordinarily rare that a captive vulture will have the same curator/handler or housing situation for its whole captive life and there is a significant probability that there will be need for extended management of chronic disease as the bird ages. All of these issues should be taken into consideration by the veterinarian when determining whether to sign the official statement saying the bird is appropriate for captive placement. The question is not "can the animal be placed," but "should the animal be placed?"

Husbandry and Medicine of Old World Vultures

Enclosure design and construction

The size and weight of OWVs dictates several recommended design parameters. In addition, they are generally social birds so aviaries should be large enough to hold multiple birds (they are also suitable for mixed vulture species exhibits) with sufficient space and nesting areas to allow pairs to break away from the main group in breeding season.

Height is important and should be a minimum of 3 meters (10 feet). Overall dimensions for a pair of birds can be in the range of 8.5-10 meters (25-33 feet) long and 6-8 meters (20-24 feet) wide, although larger enclosures are possible. A mixture of covered and uncovered areas is recommended, as birds will often deliberately sit unprotected in the rain and will actively sunbathe with wings outstretched but need the option of shelter from rain and summer sun.

A suitable enclosure will ideally be positioned to allow at least 2-3 hours of winter sun daily, with protection from prevailing winds in the form of a solid barrier wall. Heating is not generally necessary provided adequate shelter is available. The provision of mild heat to nesting areas is a possibility, particularly in Europe where some species will breed during the winter months (November to March).

In winter months, birds should be observed regularly for wing drooping and lameness or discoloration of the feet-peripheral vascular changes are common (especially in older birds) and can result in loss of digits or wing-tip edema syndrome. It is likely these may be linked to underlying cardiovascular or atherosclerotic conditions but may be triggered in cold weather.

Enclosure materials should be robust with consideration given to potential for injury. Chicken wire is totally unacceptable. Welded mesh is a superior enclosure medium and should be of sufficient gauge (40-50 millimeters) to avoid abrasion.

Bathing, preening and sunbathing constitute a significant proportion of the daily activity of OWVs. Fresh water is essential, and should be available in a bath at least 1 meter in diameter and 10-15 centimeters deep. Vultures will often only bathe in or drink water that is changed virtually every day.

The type of enclosure substrate used is variable and can consist of sand, pea gravel, grass or a combination of these. Woodchips should be avoided, particularly in enclosed or confined areas due to risk of aspergillosis although, in general, OWVs are fairly resistant to this infection. They do not appear overly prone to geosedimentation problems when housed on sand and rarely deliberately eat stones.

Perch sizes should be varied. Although OWVs can land and sit on surprisingly narrow perches, an enclosure should contain a number of perches 10-15 centimeters in diameter sufficient for the number of birds enclosed. Care is required to ensure perch positions do not allow feces to reach water baths or feeding areas, and perches should be spaced throughout the enclosure with sufficient in sheltered areas that subordinate individuals are not pushed out of the shelter. Large vultures kept on small, hard perches are more prone to bumblefoot as are those where a significant part of the flooring is concrete. Astroturf™ or similar artificial materials can aid weight dispersal on perches, although some vultures will destroy this material rapidly.

Old World Vultures should not be tethered, even when being used regularly in flying demonstrations. They do not tolerate well the actual tethering, or the equipment used, and will often remove items attached to their legs. Even where equipment is tolerated, tethered birds often become depressed and fail to thrive. Where part of an educational programme, demonstration birds should therefore be flown directly from an enclosure, without equipment.

Feeding

The gregarious feeding behaviour of many OWVs needs to be taken into account for successful captive management. Interactions between birds during feeding is important for general social interaction, the establishment of a colony pecking order and for pair formation.

Food type and presentation is important, and ideally consists of large animal parts (e.g., jointed beef) or small to medium sized mammals (e.g., rabbit). It is important that bone is given and that large bones are smashed to enable swallowing. Nutritional osteodystrophy has been recorded in these birds when fed meat alone or when given bone in a form they cannot ingest. Calcium supplementation also is recommended for breeding and growing birds in the form of a calcium and vitamin D_3 powder (e.g., Nutrobal: Vetark, Winchester, UK).

Feeding regimes should attempt to mimic wild conditions. At least one fasting day per week is recommended and should follow a sizeable feed the previous day. It is not unusual for vultures to still exhibit a protruding crop more than 24 hours after a large feed. Within reason, carcass remains should be left in aviaries for the day. Fossicking of carcass remains and the consumption of small food pieces around a feeding area is an important behavioural characteristic and is an important activity for young or subordinate birds held within a colony.

Care should be taken that birds do not become obese. If food is regularly found uneaten after more than a day, then overfeeding should be suspected and body condition monitored.

Behavioral characteristics and breeding

Females generally take main responsibility for nest building, although males will also bring material and add to the nest. Nest areas should have a moderate depth (5-10 centimeters or 2-4 inches) of substrate, such as a mixture of sand and wood shavings upon which the birds can build a nest. Care should be taken that nesting material is not rotten as aspergillosis may result from inappropriate material.

The nest, varying between 50 centimeters and 180 centimeters in diameter, is made of sticks and twigs. In the wild, some OWVs are tree-nesting species (e.g., *Gyps africanus*, *G. bengalensis*, *Aegypius*); however, the option for providing large trees for captive birds is often limited. Captive birds are traditionally offered two main nesting areas: platforms and caves. The use of caves by tree-nesting OWVs in captivity is worthy of note, given that they tend to avoid cliffs and/or cliff cavities for nesting in the wild.

In captivity, courtship displays vary from phlegmatic or non-existent to preening and head-turning (especially *Aegypius* and *Torgos*). Mating takes place either in the nest or on a preferred perch nearby. Copulation can be prolonged (compared with other avian species) and lasts between 10 and 30 seconds. The male will often bite the nape of the female's neck during copulation.

Among the species, OWVs generally lay one egg (rarely two). The incubation period ranges between 42 (*Neophron*) and up to 58 days (larger *Gyps*). Both the male and female will perform incubation duties. While OWVs will breed successfully as a single pair kept in an enclosure, the colonial nesting nature of some species in the wild makes it possible for a pair of birds to breed in an enclosure with many specimens. Interference from other birds at the nest and during incubation is limited, as breeding birds are defensive of the small area surrounding the nest.

Restraint

Compared with other raptors the talons of OWVs are relatively weak for their size; however, all large OWVs have a powerful grip, and although significant injury to a keeper is rare, due caution must be exercised. In any event, the handler should primarily restrain the head due to the strength of the bite and sharpness of the bill.

In general, two people are required to capture and restrain a large vulture. The first person should secure the head and neck immediately prior to the second person securing the wings, body and legs. A vulture that is aware of the fact it is to be captured will usually vomit, and this reaction is likely to continue during restraint. Care must be taken that the bird does not aspirate regurgitated food. Where handling is planned, the bird should be starved for 24 hours before the event to ensure an empty crop. Similarly positive reinforcement training of young birds to accept handling will make the whole less stressful for all concerned, especially as repeated handlings over many days (e.g., for long term veterinary treatment) can result in loss of body condition due to repeated regurgitation.

Blood sampling

Ulnar or medial metatarsal veins are suitable. The right jugular vein may also be used but generally is more difficult due to the strength of these birds and difficulties in head and neck restraint.

If sampling zoo birds, it is generally advised to fast birds for 24 hours pre-sampling to reduce post-prandial effects, especially when looking at renal and lipid parameters. Otherwise, blood values appear similar to other raptor species.

Imaging

Imaging is frequently required as part of a clinical investigation. Radiography is the mainstay of this but computed tomography should be considered especially when looking at suspected osteoarthritis of coxo-femoral joints as these are quite difficult to adequately evaluate using conventional radiography. New World Vulture species appear prone to laying down considerable amounts of new bone around osteoarthritic joints.

Anesthesia

OWVs should be anesthetised using isoflurane with induction by restraint and open mask. Breath holding is common making stabilization difficult on occasion. They should therefore be intubated and placed on mechanical ventilation as soon as possible after induction.

Pain management

Old World Vultures are sensitive to the renal side-effects of NSAIDs and this author (JRC) will avoid their use until the post-anaesthetic period when blood pressure is likely to have been restored. Fluid therapy should be given during all procedures in OWVs to assist with renal perfusion.

Analgesia in the pre- and peri-operative period should be based on opiate use, typically buprenorphine at 0.5 mg/kg IM q12h. For longer term management (e.g., for osteoarthritis) meloxicam may be used but this author (JRC) will generally use this at lower dosage and frequencies compared to other avian species. A dose of 0.2 mg/kg PO q24h generally appears effective and well-tolerated. Use of management techniques to assist movement, and complementary agents (e.g., glucosamine or poly-sulphated glycosaminoglycans) may assist in reducing NSAID use.

Fenbendazole

Old World Vultures have been shown to be sensitive to the hepatotoxic and radiomimetic side-effects of fenbendazole. The drug may be used for deworming OWV species, but lower doses should be used (generally 20 mg/kg PO q24h for 3-5 days) and the suspension should always be well-shaken both before dispensing from a larger bottle, and before dosing in order to reduce the chances of accidental overdosage due to settling out of the suspension.

Therapeutic dosing

Due to the risks of regurgitation, oral drugs should generally be given hidden in part of the daily ration (e.g., in body cavity of day-old cockerel or rat) and not by direct administration into the esophagus or crop. Handling interventions for birds on medication should be performed before feeding/ dosing rather than after. For debilitated birds, placement of a proventricular feeding tube should be considered where repeated oral dosing and/or supportive feeding is needed.

References

1. McClure CJ, Lepage D, Dunn L, et al. Towards reconciliation of the four world bird lists: hotspots of disagreement in taxonomy of raptors. *Proc Royal Soc. B.* 2020;287(1929):20200683.

2. Safford R, Andevski J, Botha A, et al. Vulture conservation: the case for urgent action. *Bird Conserv Int.* 2019;29(1):1-9.

3. Ogada DL, Keesing F, Virani MZ. Dropping dead: causes and consequences of vulture population declines worldwide. *Ann NY Acad Sci.* 2012;1249(1):57-71.

4. Chaudhry MJI, Ogada DL, Malik RN, et al. First evidence that populations of the critically endangered long-billed vulture *Gyps indicus* in Pakistan have increased following the ban of the toxic veterinary drug diclofenac in south Asia. *Bird Conserv Int.* 2012;22(4):389-397.

5. Ogada D, Shaw P, Beyers RL, et al. Another continental vulture crisis: Africa's vultures collapsing toward extinction. *Conserv Lett.* 2016;9(2):89-97.

6. Henriques M, Buij R, Monteiro H, et al. Deliberate poisoning of Africa's vultures. *Science.* 2020;370(6514): 304-304.

7. Herring G, Eagles-Smith CA, Wolstenholme R, et al. Collateral damage: anticoagulant rodenticides pose threats to California condors. *Environ Pollut.* 2022;311:119925.

8. Restrepo-Cardona JS, Parrado MA, Vargas FH, et al. Anthropogenic threats to the vulnerable Andean condor in northern South America. *PLoS One.* 2022;17(12):e0278331.

9. Carucci T, Whitehouse-Tedd K, Yarnell RW, et al. Ecosystem services and disservices associated with vultures: a systematic review and evidence assessment. *Ecosyst Serv.* 2022;56:101447.

10. Ives AM, Brenn-White M, Buckley JY, et al. A global review of causes of morbidity and mortality in free-living vultures. *EcoHealth.* 2022;19:40-54.

11. Chege S, Howlett J, Al Qassimi M, et al. Opportunistic infection of *Aspergillosis* and bacteria in captive Cape vultures (*Gyps copreotheres*). *Asian Pac J Trop Biomed.* 2013;3(5):401-406.

12. Gilbert M, Virani MZ, Watson RT, et al. Breeding and mortality of Oriental white-backed vulture *Gyps bengalensis* in Punjab Province, Pakistan. *Bird Conserv Int.* 2002;12:311-326.

13. Panagala VS, Stingfellow JS, Dybvig K, et al. *Mycoplasma corogypsi* sp. nov., a new species from the footpad abscess of a black vulture, *Coragyps atratus*. *Int J Syst Bacteriol.* 1993;43:585-590.

14. Ruder MG, Feldman SH, Wünschmann A, et al. Association of *Mycoplasma corogypsi* and polyarthritis in a black vulture (*Coragyps atratus*) in Virginia. *J Wildl Dis.* 2009;45(3):808-816.

15. Oaks JL, Donahoe SL, Rurangirwa FR, et al. Identification of a novel *Mycoplasma* species from an Oriental White-backed vulture (*Gyps bengalensis*). *J Clin Microbiol.* 2004;42(12):5909-5912.

16. Centers for Disease Control and Prevention. West Nile Virus and Dead Birds. Available at: https://www.cdc.gov/westnile/dead-birds/index.html. Accessed May 8, 2023.

17. Plaza PI, Blanco G, Lambertucci SA. Implications of bacterial, viral, and mycotic microorganisms in vultures for wildlife conservation, ecosystem services and public health. *Ibis.* 2020;162:1109-1124.

18. U.S. Fish and Wildlife Service. California Condor Recovery Program. Available at: https://www.fws.gov/program/california-condor-recovery/southwest-california-condor-flock-hpai-information-updates-2023. Accessed April 17, 2023.

19. U.S. Department of Agriculture Animal and Plant Health Inspection Service. 2022-2023 Detections of Highly Pathogenic Avian Influenza in Wild Birds. Available at: https://www.aphis.usda.gov/aphis/ourfocus/animalhealth/animal-disease-information/avian/avian-influenza/hpai-2022/2022-hpai-wild-birds. Accessed May 8, 2023.

20. Gomez O, Lopez R, Naveda-Rodriguez A. Avian pox in turkey vulture (*Cathartes aura*). *Vulture News.* 2014;66:42-48.

21. Mora-Carreño M, Guerra-Correa C, Moroni M, Paredes E. Avian pox in a turkey vulture (*Cathartes aura*) from northern Chile. *Austral J Vet Sci.* 2019;51:41-43.

22. Martinez-Herrero MC, Sansano-Maestre J, López Márquez I, et al. Genetic characterization of oropharyngeal trichomonad isolates from wild birds indicates that genotype is associated with host species, diet and presence of pathognomonic lesions. *Avian Pathol.* 2014;43(6):535-546.

23. Martinez-Herrero MC, González-González F, López-Márquez I, et al. Oropharyngeal trichomonosis due to *Trichomonas gypaetinii* in a cinereous vulture (*Aegypius monachus*) fledgling in Spain. *J Wildl Dis.* 2019;55(1):153-157.

24. Webb SL, Fedynich AM, Yeltatzie SK, et al. Survey of blood parasites in black vultures and turkey vultures from South Carolina. *Southeast Nat.* 2005;4(2):355-360.

25. Yabsley MJ, Vanstreels RET, Martinsen ES, et al. Parasitaemia data and molecular characterization of *Haemoproteus catharti* from New World vultures (Cathartidae) reveals a novel clade of *Haemosporida*. *Malar J.* 2018;17:12.

Avian Orthopedic Surgery: From Theory to Practice

Shangzhe Xie, BSc/BVM, MVS, PhD, DABVP (Avian Practice), DACCM

Session #3104

Affiliation: From Mandai Wildlife Group, 80 Mandai Lake Road Singapore 729826.

Abstract: Basic principles of avian orthopedics are well-established and well-described in several review papers and textbook chapters. The first part of this masterclass covers thebasics of avian orthopedic surgery including the suitability of each fracture fixation method (e.g., intramedullary pin, external skeletal fixator, tie-in fixator, coaptation) for different fracture locations and types. Given the wide range of wing and limb sizes, shapes and conformations, some principles cannot simply be extrapolated across all avian species. The second part of the masterclass utilizes real case studies to discuss the need for adjustment of these techniques and methods for patients of various avian species, including Malagasy turtle dove, honey buzzard, olive-backed sunbird, sun conure, rainbow lorikeet and Vietnamese pheasant. In particular, the honey buzzard case includes a discussion on the level of physical therapy and rehabilitation required for cases to regain full flight and ability to thrive back in the wild, as backed by post-release tracking data.

Introduction

Avian bones are the epitome of oxymorons: they are lightweight, yet strong; they are easy to break, yet quick to heal; they are simple in structure, with components similar to mammalian bones, yet complex in function, with physiologic integration into other organ systems such as the respiratory and reproductive systems; the cortices are thin, yet can withstand the forces associated with flight. The powerful flight muscles that provide some protection to the bone can also cause rotational deformity of fragments of long bones after they are fractured. These unique features of avian bone result in fracture presentations that may differ from mammalian fractures. For example, compared to mammalian fractures, avian fractures are more frequently open and comminuted due to the lack of soft tissue protection.

Clinically Relevant Avian Musculoskeletal Anatomy and Physiology

Very few fractures are orthopedic emergencies. Taking time to stabilize the patient from life-threatening injuries resulting from the same traumatic event that caused the fracture will improve overall survival and outcome of fracture repair attempts.

It is important to understand the clinically relevant anatomic and physiologic details of each patient's fracture, which will differ according to patient species, size, fracture location and type etc. The periosteal, medullary, metaphyseal and epiphyseal blood vessels supply blood to avian bone. The periosteal blood supply is often the most important component in callus formation, especially in pneumatic bones such as the humerus and femur.[1] Therefore, preserving as much of the soft tissue around the bone as possible during fracture repair will ensure that bone healing occurs at the optimal rate. Bone fragments in a comminuted fracture that have been displaced far from soft tissue and therefore lost vascular supply are potentially likely to become sequestra.[2] Management of sequestra depends on their size and location. A large sequestrum that may be important in contributing to the overall stability of the repaired bone could be better left in place until a stable callus has formed,[1] unless it starts acting as a nidus for infection. Smaller sequestra that serve no function in the bone healing should be removed.[2]

Avian bones can heal in the combination of a few ways:

1. Primary healing through the Haversian system when there is good fixation and bone apposition. There is usually minimal callus formation if this is the main way of healing, however fracture healing time will be greater than with secondary healing.

2. Secondary healing, of which there are two types:

   a. Endosteal callus formation between well-aligned bone fragments, which is often the most realistic objective of fracture repair.

   b. Periosteal callus formation when fractures are not aligned, with movement at the fracture site.[1]

Most fracture healing is a combination of these two.

The type of healing affects the rate of healing; however, the general rate of healing is faster in a bird than in a mammal for a similar fracture type. Depending on factors such as age, size, weight, activity level, fracture location, and specific bone (i.e., humerus, ulna, metatarsus etc.), fracture healing may be as short as 2 weeks and as long as 9-18 weeks. In general, more distal bones heal more slowly than those close to the body. Clinical stability of a fracture (i.e. a firm fibrous callus) at 2-3 weeks often precedes radiographic evidence of bone healing, which often takes 4-8 weeks.[1] Therefore, although radiographs taken at 10-14 days post-op are important for evaluation of adequacy and integrity of fixation, evidence of osteomyelitis, sequestrum formation, and other issues, the need for such radiographs should be assessed. For example, if the callus is palpable and solid and return to function of the patient is progressing as expected, then there is less need for stressful restraint and anesthesia for early radiographic reassessment.

Ligaments surrounding avian joints and tendons that are immobilized for long periods of time can become contracted, resulting in a permanently reduced range of motion.[1,2] Similarly, fracture callus formation and ligament/tendon adhesions can also affect the joint range of motion. As formation of hematomas and cartilaginous bridges delays fracture healing, there should be balance between excessive movement at the fracture site versus maintenance of the range of motion through early and regular physiotherapy. Without considering and addressing these factors, complete healing of the bone may not correlate with complete return to flight or weight-bearing capabilities of the patient.

Principles of Fracture Management

Many good resources exist that describe different types of surgical techniques used to repair avian fractures, including the most appropriate fixation type for each anatomic location.[1,2] In addition, considerable assessment and judgement must be made by the clinician in choosing a management option. The objectives of fracture management are to achieve alignment of the bone fragments with minimal loss of bone length and alignment, while minimizing soft tissue damage. Depending on the type of patient, full return of function may be important, especially for patients that are intended for release back into the wild, or into a large flight aviary. The skill and experience of the clinician with different techniques may also be an important factor, as surgical fixation requires skills that are developed through training and experience. Lastly, the financial cost of repair should also be considered, as an appropriately charged surgical fracture repair will cost a lot more than external coaptation alone. However, the number of follow-up revisits required for maintenance of a splint and the higher likelihood of failure may also contribute significantly to the overall cost.

External coaptation

After assessment of a fracture, including well-positioned radiographs, external coaptation is usually the first form of temporary stabilization of the fracture. This is important to prevent further damage to the soft tissues from excessive movement of mobile bone fragments. Good bone alignment can sometimes be achieved with this alone, and it can be an appropriate sole treatment for some fractures and patients. It is also the only option for patients that are too small for surgery. Bone alignment is best assessed on radiographs, so radiographs after external coaptation should be taken wherever possible. Examples of external coaptation techniques include figure of eight bandages for antebrachial and metacarpal fractures; wing to body wraps for humeral fractures; Robert Jones bandages, Altman splints, Schroeder-Thomas splints, or stirrup splints for tibiotarsal and tarsometatarsal fractures.[1] Pressure necrosis is a likelihood from bandages that are applied too tightly, so frequent monitoring is required when coaptation is selected. Some patients also tolerate bandages poorly, and the frequent movement and self-traumatization associated with them may also diminish the likelihood of a favorable outcome.

Surgical Repair Techniques

Intramedullary pins

Intramedullary (IM) pins can be inserted normograde or retrograde, depending on the nature of the fracture. Retrograde insertion is generally more straightforward as the fracture fragments can be visualized directly and manipulated into position; however, unless the fracture was already open, the surgical approach should be well-planned to ensure minimal soft tissue disruption, including avoiding joint penetration wherever possible. Anatomic guides such as *Avian Surgical Anatomy: Thoracic and Pelvic Limbs*[3] are useful in planning the most direct approach based on the fracture location. Normograde pin placement requires practice and generally works best for fractures that are not excessively displaced. IM pins are very good for maintaining alignment and length of the bone, but lack rotational stability.[1,2] Stack pinning using multiple smaller pins of the humerus and femur can provide further rotational stability, but in general, combining an IM pin with an external skeletal fixator (i.e., tie-in fixation, or with external coaptation is utilized for this purpose).[1]

External skeletal fixator

External skeletal fixators (ESF) are light, strong, inexpensive and adaptable to many fracture types ESFs also allow easier access to the soft tissue wounds in open fractures, if those need to be managed by secondary intention healing. ESFs may be the preferred fixator type if the anatomy precludes placement of an IM pin (e.g., if the pin has to go through joints, potentially affecting long-term function). Type 1 ESF, with pins connected on one side of the limb only may be the only type possible in some anatomic locations due to the lack of space, compared to Type II ESFs, with pins through the bone and connected on both sides of the limb.[1] Maintaining alignment of the bones until the ESF is completely placed and stabilized can be challenging because the bone fragments are not held in place unlike in IM pinning, so a surgical assistant or the use of temporary jigs is essential.

IM pin-ESF tie-in fixator

Combining an IM pin and ESF overcomes the disadvantages of each. It may not be suitable for the tarsometatarsus as there is no marrow cavity into which to insert an IM pin, but is otherwise a very versatile technique. Technically, this is just a combination of each of the two techniques with some minor adjustments (e.g., a smaller IM pin, 50-65% of the medullary cavity) to allow the cross pins for the ESF to be placed within the same cavity.[1] A connecting bar can be either a separate pin, or the protruding end of the IM pin can be bent back 180° to act as the connecting bar.

Bone plates

Bone plating is not as well-described as the above techniques in literature, and is not as well-practiced as IM pinning. The cortices of avian bones were traditionally thought to be too thin to engage the thread of bone screws. Also, given how well IM pins and tie-in fixators work for the majority of avian fractures, the theoretical advantages (e.g., quicker return to function, superior rotational stability) of bone plates may not result in a better outcome for avian fractures.

Case Studies

Although the methods discussed above form a sound basis for managing many avian fractures, these methods have historically been applied to a limited range of avian species (e.g., raptor and psittacine species). Some of these methods may not be readily extrapolated to some avian species and factors such as life history and patient size, among others, need to be considered for each individual patient, especially if the alternative to an imperfect fracture repair is euthanasia. A few case studies are presented here, demonstrating a combination of uncommon fractures in common species, common fractures in uncommon species, typical fracture management plan with setbacks etc. The common theme with these cases is that assessment by the clinician of the overall patient circumstances resulted in modification of standard methodology to create a good outcome for the patient.

Malagasy turtle dove (*Nesoenas picturatus*)

A Malagasy turtle dove suffered a right comminuted humeral fracture, which was repaired with IM pinning and figure of 8 bandage. As the fracture was comminuted and the patient was small, surgical repair was not straightforward and bone length was not fully restored; however, with staged physiotherapy, starting with passive range of motion exercises and progressing to increasingly large cages for flight exercises, the patient regained sufficient flight capabilities for release into a large flight aviary after approximately 8 weeks.

Oriental honey buzzard (*Pernis ptilorhynchus*)

A wild oriental honey buzzard presented with an oblique fracture of the left humerus that was repaired with IM pinning and figure of 8 bandage. Aesthetic complications precluded the additional surgical time needed to add an ESF tie-in. Bone alignment was good, but not perfect, and healing was slightly delayed, likely due to excessive mobility due to the temperament of the patient and the need to restrain it regularly for supportive feeding. The patient was finally released 16 weeks post-surgery, with 10 weeks of active rehabilitation, including passive range of motion exercises, flight exercises in large flight cages and creance flying.

Olive-backed sunbird (*Cinnyris jugularis*)

An olive-backed sunbird presented with a left radius/ulna fracture. Due to the small size of the patient, surgical fixation was not possible, and a tape splint was applied. The splint was removed 2 weeks later, by which time a bony callus could be palpated. However, the alignment of both bones was not perfect. The patient was placed in a small cage while waiting to see if it could be placed permanently in captivity, but it seemed to start flying well in this space during this time. A decision was made to move the patient to a larger space to further assess its flight capabilities and it continued to demonstrate good flight. Eventually, 6 weeks post presentation, it was released into the wild in a location with abundant food resources.

Sun conure (*Aratinga solstitialis*)

A sun conure suffered a traumatic injury resulting in concurrent closed, mid-shaft fractures of the right femur and right tibiotarsus. IM pins were placed in both the right femur and right tibiotarsus. Both pins exited at the stifle joint and were connected just in front of the stifle joint externally using acrylic. Both pins were removed 2 weeks post-surgery. The right leg was fully functional 3 weeks post-surgery (and no apparent morbidity to the stifle joint).

Rainbow lorikeet (*Trichoglossus moluccanus*)

A rainbow lorikeet was presented acutely with left leg lameness. Radiographic assessment revealed a midshaft, mildly displaced femoral fracture. The patient developed respiratory problems during this time that made it an unsuitable anaesthetic candidate. As the focus shifted to management of its respiratory problems, the femoral fracture healed during this 2-week period, with the left leg returning to full function without any form of fixation, even though the bone alignment was not perfect.

Vietnamese pheasant (*Lophura edwardsi*)

A juvenile Vietnamese pheasant presented acutely with left leg lameness. Radiographic assessment revealed a midshaft, mildly displaced left femoral fracture. As time was taken to plan for potential surgical fixation while avoiding damage to the growth plates, the patient's lameness improved significantly over the next 3 days. Due to this rapid improvement, a decision was made to forego surgery altogether. This patient is still being monitored at the time of writing but is expected to return to full function.

References

1. Doneley B. *Avian Medicine and Surgery in Practice: Companion and Aviary Birds*. 2nd Ed. Boca Raton, FL; CRC Press; 2016.

2. Redig PT, Ponder J. Orthopedic surgery. In: Samour J, ed. *Avian Medicine.* 3rd Ed. St. Louis, MO: Elsevier; 2016:312-358.

3. Orosz SE, Ensley PK, Haynes CJ. *Avian Surgical Anatomy: Thoracic and Pelvic Limbs*. Philadelphia, PA: WB Saunders; 1992.

Section 7

AAV Masterclasses Track 3

Shangzhe Xie, BSc/BVMS, MVS, PhD,
DABVP (Avian Practice), DACCM;
Hannah Attarian, DVM
Session Chairs

Management of Reproductive and Behavioral Diseases of Captive Psittacine Birds

Amber Lee, BVSc, DABVP (Avian Practice)

Session #3186

Affiliation: From The Unusual Pet Vets, Peninsula, 210 Karingal Drive, Frankston, VIC, 3199, Australia.

Abstract: Captive psittacine birds commonly present for reproductive problems; often including a spectrum of medical, surgical and behavioral disorders. This stems from a birds' strong innate drive to reproduce, and oftentimes the conditions in captivity are perfect. There are often links between behavior changes and the onset of reproductive activity, of which many owners misunderstand or are unaware of their significance. Similarly, management of many of these conditions requires both medical therapy and behavioral modification to ensure that the bird has no further problems in the future. This masterclass covers the importance of the avian behavioral exam, how to recognize and diagnosis reproductive disease, and case examples for management and treatment options. The medical and surgical management of egg binding, dystocia and cloacal prolapse are highlighted.

Introduction

Reproductive disorders are a common occurrence in avian practice. Birds frequently present for evaluation of chronic egg laying, egg binding or dystocia, egg yolk coelomitis, gonadal neoplasia and cloacal prolapse. Avian veterinarians must have a solid understanding of reproductive anatomy, physiology and normal breeding behaviors to be able to diagnose and treat reproductive disease appropriately. Similarly, it is important for veterinarians to be able to recognise the role that environmental, nutritional and behavioral modification play in the management of these conditions.

There can often be an association between the development of these reproductive diseases and behavioral disorders for many avian patients. These can include territorial aggression, inappropriate bonding, redirected aggression, attention seeking behavior, excessive screaming, masturbation and feather destructive behaviors. Understanding these problem behaviors requires detailed history acquisition to ascertain the bird's hand-rearing status or degree of early learning and development, its current household relationships as well as its health, diet, husbandry and environmental cues.[1]

Behavior as Part of a History and Why It is Important

Pet birds lack domesticity, and as a result require the need for daily maintenance behaviors of their wild parents or grandparents. Often, the husbandry provided and the human relationships that result are not appropriate or sufficient for the engagement of these normal behaviors.[2] When owners fail to understand normal parrot behavior, they anthropomorphise and expect the bird to fit in with their human expectations. This then leads to the lack of appropriate learning opportunities for the bird to engage in normal flock interactions such as conflict resolution, self-maintenance and other social behaviors. This can lead to the development of many behavioral problems including the inappropriate pair bonding that occurs with many birds and their owners. Most bird owners are not cognizant of the fact that physical touching, cuddling or sharing food with their bird is perceived by their bird as mating behaviors.

Captive psittacine birds, particularly those that are hand-reared often do not have the opportunities to learn normal social behaviors. If a bird is not appropriately socialized, attention-seeking behaviors such as begging and screaming or displacement behaviors such as biting, phobias and feather destructive behavior can result. As these antisocial behaviors develop, the bird often becomes more isolated, and therefore the behaviors can worsen as the bird is positively or negatively reinforced.

Changes in behavior can indicate underlying reproductive disease. For example, broody behavior without oviposition may indicate cystic ovarian changes.[3] Veterinarians can work with owners to modify any unwanted or problem behaviors that could lead to conditions such as chronic egg laying, dystocia, coelomitis and feather destructive behavior.

One of the most common causes of relinquishment of birds, especially for larger species like macaws and cockatoos, is for behavioral reasons.[4] When a bird is relinquished to a rescue the effects are threefold. First, there is disruption for the bird. Second, there is disappointment for the owner. Third, there is a loss of clientele and income for the veterinarian. Finally, it is important to consider difficulties handling many birds in the consultation room, the use of Fear Free techniques for the welfare of the bird, and the safety of the staff.

Normal Sexual Behaviors of Wild Psittacine Species and How They Become Problems in Captivity

It is imperative to understand normal sexual behaviors of wild psittacine birds if veterinarians are going to be able to prevent the development of problem sexual behaviors in captive counterparts. It is important to be able to recognize normal sexual behaviors, environmental triggers of wild parrots, and how they occur or are expressed in the home setting.[5]

Reproductive behaviors of wild psittacine birds include pair bonding, courtship regurgitation, nest seeking, nest building, territorial aggression and copulation. The display of these behaviors is regulated by environment (e.g., photoperiod, rainfall, food supply, presence of a nest, presence of a mate), hormonal and social cues (e.g., pair bond formation and social hierarchy of the flock).[6] The following table illustrates the above reproductive stimuli and addresses management of these factors from the perspective of pet bird owners (Table 1). Prevention, rather than intervention, is key as is understanding the unique natural history and species-specific differences in breeding strategies exhibited by captive parrots.

Table 1. Therapeutic considerations for treatment of reproductive behaviors.[1,2,5,26-28]

Stimulus	Explanation	Management
Environmental		
Presence of a nest site Presence of nest material Maintenance of a territory	Presence and maintenance of a nest site in the wild helps stimulate reproductive drive.	Do not provide a nest or happy hut. Deny access to drawers, underneath furniture, other dark cavities. Deny access to the newspaper at the bottom of the cage, move the paper to underneath the grate. Frequently rearrange perches and cage furniture to create some (gentle) environmental instability. Provide opportunities for exercise, especially flight.
Long day length	Increasing day length can signal the start of breeding season for many species.	Ensure birds gets at least 12 hours of uninterrupted sleep. Ideally, sleeps in a room that no one uses after the bird goes to bed.
Rainfall	Rain can indicate that the availability of fresh food is about to increase, which is ideal for reproduction.	Rainfall cannot be controlled but owners can limit the amount of shower time or misting a bird receives.
Loss of eggs	Females may try to replace eggs lost from clutch.	Once a hen has started laying a clutch, allow it to brood the clutch for at least 3 weeks (instead immediately removing eggs).
Nutritional		
High energy diet	A high-energy diet facilitates egg production and chick rearing.	Birds fed a seed-only diet or predominantly seed diet should be converted to a pellet-based diet with low energy vegetables and only a small amount of fruit, seed and nuts.
Easily accessible food	Accessible food makes feeding a mate or chicks easier. Wild birds spend 60-80% of time foraging, but most captive birds meet nutritional requirements in less than 20 minutes over the course of the day.	Birds should be required to work and expend energy (forage) for their food rather than receive it freely. Dry food can be mixed with recycled cat litter and fruit and vegetables can be wrapped in paper and cardboard to increase difficulty of procurement. Various foraging toys are available for purchase or can be made by hand by owners.
Social		
Interactions with a real or perceived mate	In the wild, courtship behavior between a bonded pair is a strong driver for nest building and reproduction. In captivity, a bird may pair-bond with another bird, a human or an inanimate object.	If the bird is bonded to another bird, denying visual and auditory contact with that bird may reduce reproductive drive in the long term, but is likely to cause significant distress and therefore is generally not recommended. If the bird is bonded to a person, reduce the reproductive behaviors performed by that person. Keep all physical contact to the head, avoid touching the back or under the wings. Do not share food with the bird or feed the bird directly from your mouth. Do not allow the bird to masturbate on you. Foster platonic interactions, such as clicker training, as an alternative to sexual interactions. Recommendations for different species vary but keep birds of the same sex of the same or similar species together. Allow other members of the household the opportunity to interact positively with the bird. If the bird is bonded to an object (e.g., food bowl, mirror, bell), consider removing that object. Do not purchase unweaned baby birds! These birds are more likely to be poorly socialized and develop reproductive problems.
Innate		
Genetic	Some species, or individuals within a species are more prone to excessive reproductive drive.	Small parrots including cockatiels, budgerigar and lovebirds. Seasonal aggression of Amazon parrots. Peach-faced lovebirds tuck long pieces of nest material into the feathers of their back for transportation.

Formation of pair bonds

The formation of pair bonds is not well understood but appears to be the result of increased associative interactions, including spending more time together, allopreening, and reduced aggression towards pair mate than to non-mates.[7] Many pair mates remain together almost constantly, performing many daily activities together, with some species exhibiting social synchrony (mimicking each other's behaviors). The main advantage of forming a pair bond for psittacine birds is in the ability to undertake the huge task of hatching and rearing altricial offspring, that would not be possible for an individual bird. Often these pair bonds exist year-round due to the advantages that pair bonds provide including increasing social standing of the pair within the flock.

Vocalization

Unique vocalizing and mating calls of various species signal their sexual readiness. Male cockatiels can easily be led to start singing their mating call with encouragement from owner vocalizations (author's personal experience). Female cockatiels will cluck softly or chatter when ready to mate. Pair bonded captive psittacine birds will often vocalise when their caregiver leaves the room or house. Their wild counterparts will contact call to keep in communication with a mate, so excessive screaming may be escalation of contact call if the bird is distressed by their mate's absence.

Courting behavior

Psittacine birds exhibit several different moves in performing courtship behavior including hopping, bowing, strutting or tail wagging. Many species show pinning of the pupils (rapid dilation and constriction) or blushing of facial skin areas (macaws and Palm cockatoos). Cockatoos will display movements of the crest feathers on the head. Pair bonded birds will bow their head to solicit preening from a mate. Some species usually remain physically distant and only come together for breeding. Thus, attention paid by owners and excessive physical contact in the form of petting (preening) can encourage chronic reproductive behavior.

Nest seeking

Nest cavities are often a limiting factor for breeding for many Neotropic species, and in captivity the provision of a cage (nest), nesting material (newspaper), or free ranging through the house allows cavity-seeking behavior (e.g., under furniture, dark spots or corners). Fleece huts are a particular problem as many owners think that their birds require a nest to sleep in at night and inadvertently provide, a dark cozy place for egg laying or nesting behaviors.

Copulation

The drive to mate is strong for pair bonded birds. In captive birds, this desire may result in masturbation on the owner, or on perches or toys within the cage.

Laying eggs

Generally, larger species lay smaller clutches compared to smaller species. For cockatiels, budgerigars and lovebirds, the development of chronic egg laying in captivity is very common. Some species share incubation duties (e.g., cockatoos), while female macaws incubate and it is the male's job to ensure she is fed through regurgitation. Thus, excessive regurgitation may present a problem in captivity as the bird tries to feed their owner as their perceived mate.

Aggression

Territorial aggression towards non-mates for protection of the nest or guarding of a mate is common in wild psittacine birds. In captivity this may lead to cage aggression, mate-related aggression towards competing human family members, or even redirected aggression towards a favored human if that person is approached by another person. This is commonly seen in Amazon parrots, cockatoos, (miniature) macaws, grey parrots, conures, quaker parrots and lovebirds.[8]

Feather destructive behavior

Chronic hormone stimulation and inappropriate bonding can be an underlying cause for feather destructive behavior (FDB) in cockatoos, macaws, conures, and grey-cheeked parakeets.[3] Sexual frustration can result from elongated photoperiod, high-fat diet, loss of a mate or inappropriate mate selection, or the presence of mating pairs in the environment. Some reproductive disorders can lead to pain or neuropathy which may result in FDB.

Female Reproductive Diseases

Disorders of the reproductive tract in captive psittacine birds involve a complete spectrum from mild and self-limiting to severe and life-threatening. Disease can involve any part of the reproductive tract, and many diseases can present with non-specific signs so it is important to determine whether the reproductive tract is involved from history, clinical signs and diagnostics. The following is a review of the general approach to diagnosing reproductive disease and general treatment, with a specific focus on egg binding, dystocia and cloaca prolapse.

History

Confirmation of the sex of the bird is important. Changes in behavior can come with puberty (6 months to 4 years depending on species) with changes including screaming, irritability, preference of a caregiver, territorial aggression, and overt sexual behaviors (regurgitation, masturbation). Nest seeking or building, along with recent oviposition are indicators of reproductive activity. Changes in reproductive cycle with oviposition and then cessation of lay may suggest egg binding or dystocia along with the laying of abnormal or soft-shelled eggs.

Clinical signs

Often signs are non-specific including lethargy, dyspnoea, coelomic distension, lameness, or feather destructive behavior. Specifically, there may be a palpable egg or mass in coelom, straining, tenesmus, tail wagging, wide-based stance or drooped wings. Depending on the chronicity of disease, the bird may present severely emaciated and weak. With hypocalcemia there maybe pathologic fractures or tremors. Cloacal prolapse can occur secondary to straining from dystocia, space-occupying masses or behaviorally in male cockatoos.

Diagnosis

A diagnosis can be difficult; therefore, it is important to obtain clues from history, physical examination and diagnostics. Blood work can be completely normal. A complete blood cell count may show anemia (sometimes non-regenerative), or evidence of infection or inflammation. Biochemical changes include elevation of alkaline phosphatase, elevated creatine phosphokinase, hyperglobulinemia, hypercalcemia or hypocalcemia, hypercholesterolemia with lipemic serum.[9,10] Radiography often will help to detect an intact egg or shell fragments, loss of serosal detail (e.g., free fluid), mass effect in the region of the reproductive tract, and medullary hyperostosis.

Ultrasonography provides a more sensitive assessment of the reproductive tract and can help identify neoplasms, unmineralized eggs, ectopic ova, free fluid, and cystic tissue associated with the ovaries or oviduct. Ultrasound-guided coelomocentesis can be both therapeutic to help relieve dyspnoea and diagnostic as aspirated fluid can be submitted for cytology and culture and sensitivity. The use of endoscopy can further help elucidate the type of reproductive disease present with direct visualization of reproductive organs and biopsy of lesion at the same time. For definitive diagnosis of some conditions, exploratory laparotomy may be the only way to confirm and/or treat the disease.

Treatment

Generally, treatment is dependent upon the diagnosis (or suspected diagnosis) and the condition of the patient. While birds may present acutely, often disease may actually be chronic in nature, and the patient must be stabilized and any secondary complications treated. Once the reproductive disease has been identified and treatment initiated, the caregivers of the bird must be counselled on environmental, behavioral and dietary modifications to eliminate the risk of future egg laying or reproductive stimulation. This includes altering the relationship to include less intimacy; less petting, sitting on the shoulder, physical touch or sharing of food. Positive reinforcement training is recommended to get the bird to step-up readily, and only use the hand to transfer to and from the cage with the use of stations and play stands in the environment. Any high fat or high sugar foods should be eliminated from the diet, as well as addressing photoperiod (12-14 hours of darkness) and removal of any potential nesting sites or nesting material.

The pharmacologic use of hormonal therapy for control of chronic hormone stimulation and reproductive disease has been largely through the use of gonadotropin releasing hormone (GnRH) agonists leuprolide acetate (e.g., Lucrin; AbbVie Pty Ltd, Mascot NSW 2020) or deslorelin acetate (e.g., Suprelorin; Virbac (Australia) Pty Limited, Milperra NSW 2214). Previously used agents included medroxyprogesterone acetate, human chorionic gonadotropin and testosterone have since fallen out of use due to severe and potentially fatal side effects.[3] GnRH agonists initially stimulate the secretion of luteinizing hormone (LH) and follicle-stimulating hormone (FSH) from the pituitary gland, which results in oestrus and ovulation as well as the production of GnRH receptors.[11] When administered long term or as sustained release they lead to the downregulation of GnRH receptors via negative feedback.[12]

Leuprolide acetate is a short-acting depot formulation that has been studied in several species.[13-17] Length of efficacy varies between species with the average length of cessation of egg laying in cockatiels being 12-19 days[13] and the effect shorter in budgerigars at the same dose, only lasting for 12-14 days.[16] There is a wide range of doses being recommended from 400-1,000 µg/kg IM every 2-3 weeks.[12] There is also anecdotal evidence that the efficacy of Lupron reduces with long-term usage.

Deslorelin acetate is a longer-acting, subcutaneous, controlled release implant that is available in a 4.7-mg or 9.4-mg dose. Deslorelin acetate, like leuprolide acetate has been used successfully to reduce reproductive behaviors and egg laying in birds, but additional uses have been identified in the management of ovarian neoplasia in cockatiels,[18,19] and budgerigars with Sertoli cell tumors.[16] There is also evidence in chickens that deslorelin acetate has chemoprotective properties and prevents against the development of ovarian neoplasia.[20] The only prospective studies of deslorelin acetate have been done in chickens,[21] quail[22-24] and pigeons.[25] The length of treatment varies with species and the type of disease being treated. Recommended dosing intervals are difficult to ascertain, and indicators such as return of reproductive behaviors, egg laying or both, have been used as indicators for reimplantation.[12] In psittacine birds, the most common dosing frequency is every 3 months, with a range of 2-5 months in addition to behavioral and environmental modifications.[12]

Egg binding and dystocia

Birds presenting with egg binding or dystocia can present in various stages of shock. Patient stability will determine when further imaging (e.g., radiographs or ultrasound) is performed to locate the egg and determine its position. If the bird is obstructed (i.e., dystocia) and not able to pass feces, then immediate stabilization and removal of the egg must occur.

In general, provide supportive care with supplemental heat 84-90 °F (29-32 °C), fluid therapy, analgesics, parenteral nutrition and parenteral calcium. Antibiotics can be provided if there is suspicion of ruptured oviduct (i.e., egg no longer within oviduct, secondary coelomitis). Some eggs can pass with supportive care alone, however if the patient declines in condition, then further intervention is warranted. The use of prostaglandin, oxytocin and hormone therapy can be considered if there is no suspicion of oviduct obstruction, adhesion or perforation and the uterovaginal sphincter is patent.[3] Parenteral calcium must be administered prior as these treatments potentiate muscle contraction.

If these treatments are ineffective then gentle massage and manipulation of the lower coelom under anesthesia to facilitate vaginal opening can be attempted. Once the uterovaginal sphincter is dilated, then direct pressure is applied above and partially behind the egg to move it caudally into oviposition. Ovocentesis is then performed if the above treatments have failed. Transcloacal ovocentesis is preferred over transcoelomic due to the risk of trauma. Ideally, crown the egg to visualise through the sphincter and then use a needle and syringe to collapse and drain the contents. A vaginal speculum or lone star retractor can be used to help open the lips of the vent and identify the uterovaginal sphincter. If the egg is not visualized, then it is likely not in the cloaca and the procedure should be converted to a coeliotomy. Prepare owners that many hens may lay another egg within the 72 hours following the passage of an egg. In our practice, we usually administer leuprolide acetate at the time of egg removal, and then wait a few days before administering long-acting deslorelin acetate. The owner is then counselled on environmental, behavior and nutritional factors to help prevent further egg laying.

Cloacal prolapse

Prolapse of cloacal tissues, vagina, uterus or oviduct are a common emergency presentation and can result from dystocia, normal oviposition, manual delivery of an egg, malnutrition, obesity and general debilitation from chronic egg laying.[26,27] General prolapse treatment principles follow.

- Prolapsed tissue must be protected from further damage and desiccation.
- Supportive care is provided to the patient, which may include shock therapy, fluid resuscitation, antibiotics, analgesia and nutritional support.
- Resolution of the prolapse is based upon identification and treatment of the underlying cause.

At presentation, the stability of the patient must be carefully evaluated as well as history acquisition to suggest the potential for underlying reproductive disease as the cause for cloacal prolapse. If the patient is bright, alert and responsive (with a smaller prolapse), initial interventions may include warmth, fluids and gentle cleansing and lubrication of prolapsed tissues. For those patients presenting in poor condition due to septicemia, blood loss or dystocia, more aggressive emergency and shock therapy measures are required.

If the prolapse is small and no egg or mass is palpated, then it can be replaced into the cloaca with stay sutures, while further diagnostics and definitive therapy is planned. Two lateral horizontal mattress sutures are placed individually either side of the vent, while a cotton-tipped applicator is used to hold the replaced tissue within the cloaca. The resulting opening must be small enough that the tissue does not re-prolapse but large enough that urofeces can still pass normally.

For those prolapses where a large amount of tissue is present, or there are signs of devitalization, replacement of necrotic tissue is not recommended due to risk of coelomitis and sepsis which can result in death. These cases require further advanced coelomic surgical resolution (i.e., salpingohysterectomy) and necessitate referral to an avian specialist. If an egg is present within the prolapsed tissue, it may be gently manipulated out of oviductal tissue after flushing and lubrication. The egg should crown through the opening of the uterovaginal sphincter. If the egg is unintentionally passed through the wall of the oviduct or cloaca, then the defect must be repaired.

Cloacal prolapses can occur in cases with chronic hormone stimulation and/or masturbation as a result of chronic straining. It is important that all other underlying causes of prolapse are ruled out in these cases. In chronic cases, the cloacal wall and sphincter may become flaccid and this loss of tone leads to soiling of skin/feathers with malodorous urofeces and repeated prolapse. These cases require reconstructive surgery with cloacopexy or ventplasty as well medical and behavioral therapy. Diagnosis of underlying reproductive disease in cloacal prolapse cases is imperative as in many cases if the bird resumes oviposition, then the prolapse is likely to recur. Initial diagnostic work-up can include complete blood cell count, biochemistry panel and radiographs in two views. The use of ultrasound may be superior to radiographs in visualizing eggs or egg material (non-mineralized) within the oviduct or coelom (egg yolk coelomitis) as well as aide in the diagnosis of salpingitis or metritis and differentiate oviductal masses from caudal coelomic masses. The examination of the cloaca itself can be helpful. This can be achieved with manual eversion with a lubricated cotton-tipped applicator in larger birds. Cloacoscopy is ideal to evaluate for underlying masses, and to collect mucosal biopsies.

While it is important to protect and replace the prolapsed tissues, it is imperative in the management of prolapses that the underlying disease process is identified and addressed. Then all attempts to decrease hormone stimulation and discourage reproductive behavior must also occur to prevent recurrence. These include the aforementioned husbandry, behavioral,[28] environmental and hormonal therapies.

Male Reproductive Diseases

Diseases of the male reproductive tract include infectious or neoplastic etiologies. Orchitis is inflammation associated with the testicular tissue and can be due to local disease, extension from the cloaca or hematogenous spread of an infection. Clinical signs can be vague with decreased fertility, lethargy, decreased appetite or pain.

The most common neoplasms of the male gonads include Sertoli cell tumors, seminomas, Leydig (interstitial) cell tumors, adenocarcinoma, leiomyosarcoma, and lymphosarcoma. These can present with the above signs, but due to space-occupying effects also present with lameness, paresis, or paralysis of the hindlimbs and coelomic distension.

Diagnosis is as above for with female reproductive diseases, with ultrasound or computed tomography scan likely providing greater detail and more accurate images over convention radiography. Coelioscopy can provide visualization of the testes and the ability to biopsy affected tissues.

Treatment is aimed at the underlying disease with antibiotics based on culture and sensitivity for orchitis. For testicular neoplasms the use of GnRH agonists may aide in reduction of size, along with regular coelomocentesis for fluid accumulation, and general supportive care. Removal of neoplasms is ideal, but difficult in many small avian patients and there is limited information regarding chemotherapeutic options.[3]

References

1. Doneley B. Reproduction. In: Chitty J, Monks D, eds. *BSAVA Manual of Avian Practice: A Foundation Manual.* Gloucester, UK: British Small Animal Veterinary Association; 2018:54-67.

2. Stelow E. Avian behavior consultation for exotic pet practitioners. *Vet Clin North Am Exot Anim Pract.* 2021;24(1):103-117. doi:10.1016/j.cvex.2020.09.006.

3. Scagnelli AM, Tully TN. Reproductive disorders in parrots. *Vet Clin North Am Exot Anim Pract.* 2017;20(2):485-507. doi:10.1016/j.cvex.2016.11.012.

4. Clubb S. Parrot relinquishment in the US: why are birds losing their homes? *Proc Annu Conf Assoc Avian Vet.* 2011:279-284.

5. Van Sant F. Problem sexual behaviors of companion parrots. In: Luescher AU, ed. *Manual of Parrot Behavior.* Ames, IA: Blackwell Publishing; 2006:233-245.

6. Seibert LM, Sung W. Psittacines. In: Tynes VV, ed. *Behavior of Exotic Pets.* Hoboken, NJ: Wiley-Blackwell; 2010:1-11.

7. Spoon TR. Parrot reproductive behavior, or who associates, who mates, and who cares? In: Luescher AU, ed. *Manual of Parrot Behavior.* Ames, IA: Blackwell Publishing; 2006:67-82.

8. Welle KR, Luescherm Andrew U. Aggressive behavior in pet birds. In: Luescher AU, ed. *Manual of Parrot Behavior.* Ames, IA: Blackwell Publishing; 2006:211-217.

9. Bowles HL. Reproductive diseases of pet bird species. *Vet Clin North Am Exot Anim Pract.* 2002;5(3):489-506. doi:10.1016/S1094-9194(02)00008-7.

10. Joyner KL. Theriogenology. In: Ritchie BW, Harrison GJ, Harrison LR, eds. *Avian Medicine: Principles and Application.* Lake Worth, FL: Wingers Publishing; 1994:748-775.

11. Johnson JG. Therapeutic review: deslorelin acetate subcutaneous implant. *J Exot Pet Med.* 2013;22(1): 82-84. doi:10.1053/j.jepm.2012.12.013.

12. Petriz OA. Advancements in methods for decreasing reproductive success. In: Speer BL, ed. *Current Therapy in Avian Medicine.* St. Louis, MO: Elsevier; 2016:446-454.

13. Millam J, Finney H. Leuprolide acetate can reversibly prevent egg laying in cockatiels. *Zoo Biol.* 1994;13:149-155.

14. Millam JR. Reproductive management of captive parrots. *Vet Clin North Am Exot Anim Pract.* 1999:93-110.

15. de Wit M, Westerhof I, Pefold L. Effect of leuprolide acetate on avian reproduction. *Proc Annu Conf Assoc Avian Vet.* 2004:73-74.

16. Straub J, Zenker I. First experience in hormonal treatment of Sertoli cell tumors in budgerigars (*M. undulatus*) with absorbable extended-release GnRH chips (Suprelorin). *Proc ICARE.* 2013:299-301.

17. Klaphake E, Fecteau K, DeWit M, et al. Effects of leuprolide acetate on selected blood and fecal sex hormones in Hispaniolan Amazon parrots (*Amazona ventralis*). *J Avian Med Surg.* 2016;23(4):253-262.

18. Keller KA, Beaufrère H, Brandão J, et al. Long-term management of ovarian neoplasia in two cockatiels (*Nymphicus hollandicus*). *J Avian Med Surg.* 2013;27(1):44-52. doi:10.1647/2012-006.

19. Nemetz L. Deslorelin acetate long-term suppression of ovarian carcinoma in a cockatiel (*Nymphicus hollandicus*). *Proc Annu Conf Assoc Avian Vet.* 2012:37-42.

20. de Matos R. Investigation of the chemopreventative effects of deslorelin in domestic chickens with a high prevalence of ovarian cancer. *Proc ICARE.* 2013:90.

21. Noonan B, Johnson P, de Matos R: Evaluation of egg-laying suppression effects of the GnRH agonist deslorelin in domestic chickens. *Proc Annu Conf Assoc Avian Vet.* 2012:321.

22. Petritz OA, Sanchez-Migallon Guzman D, Paul-Murphy J, et al. Evaluation of the efficacy and safety of single administration of 4.7-mg deslorelin acetate implants on egg production and plasma sex hormones in Japanese quail (*Coturnix coturnix japonica*). *Am J Vet Res.* 2013;74:316-323.

23. Petritz O, Guzman DS, Hawkins M et al: Evaluation of deslorelin acetate implant dosage on egg production and plasma progesterone in Japanese quail (*Coturnix coturnix japonica*). *Proc Annu Conf Assoc Avian Vet.* 2013:17.

24. Schmidt F, Legler M, Einspanier A et al: Influence of the GnRH slow release agonist deslorelin on the gonadal activity of Japanese quail (*Coturnix coturnix japonica*). *Proc ICARE.* 2013:501-502.

25. Cowan ML, Martin GB, Monks DJ, et al. Inhibition of the reproductive system by deslorelin in male and female pigeons (*Columba livia*). *J Avian Med Surg.* 2014;28:102-108.

26. Lennox A, Lee A. Cloacal, uterine and rectal prolapse. In: Chitty J, Monks D, eds. *BSAVA Manual of Avian Practice: A Foundation Manual.* Gloucester, UK: British Small Animal Veterinary Association; 2018: 358-365.

27. Curd S. Egg retention. In: Chitty J, Monks D, eds. *BSAVA Manual of Avian Practice A Foundation Manual.* Gloucester, UK: British Small Animal Veterinary Association; 2018:350-3537.

28. van Zeeland YRR, Friedman SG, Bergman L. Behavior. In: Speer BL, ed. *Current Therapy in Avian Medicine and Surgery.* St. Louis, MO: Elsevier; 2016:177-251.

Veterinary Care of Breeding Collections of Birds of Prey and Psittacine Species: Tips and Tricks to Help Your Client

Michael Lierz, DZooMed, DipECZM (Wildlife Population Health), DipECPVS

Session #3108

Affiliation: From Clinic for Birds, Reptiles, Amphibians and Fish, Faculty of Veterinary Medicine, Justus Liebig University Giessen, Frankfurter Str. 114, D-35392 Giessen.

Abstract: Captive breeding has contributed to the successful restoration of many species of birds of prey and psittacine birds. Avicultural breeding techniques include double clutching, direct fostering, cross fostering, hatch and switch, hacking, imprinting and assisted reproduction techniques such as semen collection and artificial insemination; however reproductive failure is a common occurrence. Failures are mainly related to management problems, including hygiene measures, food quality issues, breeding collection structure, or to individual health issues of breeding birds. These issues may result in non-egg laying females, low quality eggs or infertile eggs due to male infertility. Therefore, veterinary care of breeding collections is important. This masterclass provides an overview of veterinary involvement in birds of prey and psittacine bird breeding projects. Insights into the most common management problems are provided, basic aviculture techniques are explained and state-of-the-art as well as cutting-edge assisted reproduction techniques in avian reproduction are described. Information from this veterinary field allows the veterinarian to expand expertise and service to clients. The latest developments in prophylactic health measures, preparation of birds and cages, semen collection, evaluation and cryopreservation techniques, artificial insemination and ex-ovo fertilization are covered.

Manuscript previously published in: Lierz M. Veterinary care of breeding flocks of birds of prey and psittacine species: tips and tricks to help your client. Proc ExoticsCon. 2022;57-62.

Introduction

Many raptor and psittacine species are endangered or threatened with extinction and species conservation programs exist. In addition, commercial breeding is common. Reproduction attempts in birds of prey or psittacine birds in captivity regularly fail or the reproduction rate is low. The reason for this is multifaceted. Management-related problems are the most common reason, followed by diseases and infertility. To solve problems in a breeding collection, a detailed investigation is necessary. The veterinarian plays the most important role in this process as s/he has the capability to perform all necessary detailed investigations. A veterinarian in this role needs to understand how to manage a breeding collection and which aspects are of special importance. This masterclass provides a basic introduction. For more detail the reader is directed to further literature.[1-16]

Basics of Bird of Prey Reproduction

Birds of prey are monogamous and reach their sexual maturity usually between 2-6 years. Pairs are very territorial and the bond to their nesting site is strong. The male is around one third smaller than the female and provides the food for the brooding female. Males brood to a much lesser extend compared to the female. Birds of prey have usually a sibling sexual imprint starting around day 8 after hatch. Purely human hand-reared single birds of prey are unable to reproduce naturally, but can be stimulated by man and artificially inseminated. This is usually used for to produce species hybrids. Birds of prey hand-reared in groups may be used for natural breeding and sexual stimulation by humans, but parent-reared birds are best for natural reproduction.

Birds of prey usually lay two to four eggs in one clutch. By removing the eggs before the clutch is completed, the female may be stimulated to lay more eggs (egg pulling). If the eggs are removed within 2-3 weeks after clutching, a second clutch will be laid.

Breeding birds should be kept in flying cages with a nesting site. In cases where assisted reproduction techniques are used, imprinted semen donors can be kept by falconry methods.

Basics of Psittacine Bird Reproduction

Compared to birds of prey, psittacine birds have multiple reproductive strategies, from strongly monogamic, to group nesting and polygamic. Imprinting birds for reproduction purposes is not used in this avian group as imprinting causes usually behavioral problems preventing natural breeding or even nesting. Also artificial insemination is not routinely used as semen collection is much more difficult in this group compared to birds of prey. Breeders usually use double clutching to increase the amount of offspring. Psittacine reproduction usually relies on natural pair breeding.

Assisted Reproduction Techniques

In falconry outside Germany, hybrid falcons are commonly used. Those hybrids are mixed breeds of two different falcon species, which under natural circumstances, do not reproduce together. Therefore, artificial insemination techniques are used. Breeding birds are imprinted to humans and when they reach sexual maturity the breeder can stimulate the bird. Semen is collected from the males by massage methods or by voluntary semen donation. This semen is transferred into the cloaca or the oviduct of a female immediately prior to egg laying of the first egg and shortly after each subsequent egg laid.[12,16]

In psittacine species, artificial insemination techniques have only recently been introduced, as semen collection is more difficult compared to birds of prey. These techniques are usually used in species conservation programs, but commercial breeders begun to include them in the management. More and more parrot owners ask for species hybrids to produce uncommon colors, but there is really no argument to support this change.

In several psittacine or bird of prey breeding facilities, eggs are removed and placed in an incubator. Natural incubation in captivity can be risky, especially with inexperienced females as they might leave the nest or damage the eggs. Artificial incubation of eggs is not easy and each egg needs to be treated individually. Apart from the temperature (37.2-37.4 °C), the humidity is an important key factor. Each egg should lose between 14-16% weight during the complete incubation period. Therefore, the egg needs to be weighed and humidity adjusted to reach those loss rates. If alteration is insufficient, the eggshell needs to be manipulated to increase or decrease the water loss through the shell. In addition, vitality of the embryo should be monitored, either by candling or by a device that is able to monitor the heart rate (Buddy ® Digital Egg monitor).[11]

Chicks break the eggshell from inside (i.e., pipping) at the end of the species-specific incubation period. The total incubation period is highly variable between species. After pipping it still takes 48-50 hours until hatch is completed and during this time no manipulations should be done. Chicks are then fed at the earliest 6 hours after hatch and usually are placed back to the parents between day 6-10 of life (birds of prey), or purely hand reared, which must be done in groups in psittacine birds to reduce risk of behavioral problems later in life.

Veterinary Considerations of Bird of Prey and Psittacine Breeding

The process of breeding birds includes many risk factors that need to be considered. Pairs are regularly in large cages with no direct control of the individual. Especially during the reproduction period, disturbances usually interfere with breeding activity. In birds of prey, imprinted single breeding birds are much better to control from the health point of view; however, repeated inseminations do carry an increased risk of diseases of the reproductive organs. Finally, reproductive behavior and egg laying is an increased burden to the birds. In Europe, psittacine and raptorial birds are usually kept in outside aviaries where those birds have contact with free- ranging populations of various birds. Also, partners are usually placed together without an intensive health check. Breeders regularly place eggs from other breeders in their incubator or place chicks of different sources together, even to the same pair. All this significantly increases the risk of infectious disease.

Problems in Reproduction

Infertile clutches

The most common problem in bird of prey and parrot breeding facilities is infertile eggs. Pair behavior is not always ideal and despite the owner not seeing a problem because birds demonstrate display behaviour, copulation does not occur successfully. Psittacine birds are highly social birds with a complex partner selection system. Forced pairing might regularly lead to insufficient partner acceptance. In birds of prey, the male, as the smaller partner, must approach the larger female and must fly onto her back. Females might interpret this as aggression and hurt the male. Compared to nature, males cannot escape aggressive behavior because they are in an aviary. Therefore males must approach the female carefully, offer her food and stimulate her for copulation. Even after an attack from the female, males need to be brave enough to try copulations repeatedly. Many males fail this, offer food, demonstrate reproductive behavior, but do not try to copulate again after one poor experience. Therefore, the male represents the key factor in natural reproducing pairs. As food-offer behavior from the male to the female is important, males must be supported by the breeder. During reproduction period, the pair should be fed several times during the day with high-quality food.

In psittacine birds as well as birds of prey, infertility of the male is a common reason for poor breeding performance. Infections (e.g., *Salmonella* sp., *Chlamydia* sp., *Mycoplasma* sp.) involving the reproductive tract, environmental toxic agents (e.g., wood protecting paint, heavy metals, pesticides, mycotoxins), medications (e.g., coccidiostats given in animal feed and consumed by raptors) or vitamin deficiencies (e.g., vitamin E, A) are possible causes. In cases where infertility of a male is suspected, a detailed spermatologic investigation (e.g., motility, pH, live/death ratio, spermatozoa morphology) should be performed. The semen of falcons contains so-called round bodies (RB), which are a normal structure, but not found in all avian species. In falcons the number of RB can be high despite that their function is presently unknown.[5] Reference values are available for the different falcon species,[4] but not for all birds of prey species and a very rare in psittacine birds.[2] If reference values are not available, the comparison of ejaculate parameters from the investigated male with proven breeding males is useful. To date, spermatologic investigations can be supported by computer assisted semen analysis (CASA) which provides additional very useful information.[7] Functional spermatologic tests, such as the perivitelline membrane penetration test,[9] are very helpful in investigating infertile males that demonstrate motile semen. Electron microscopic investigations of individual spermatozoa[15] assist in the determination of the reason for infertility and might be used to complete the evaluation of the semen.

In addition, a detailed endoscopic investigation of the reproductive tract is recommended. It should include a testicle biopsy in case the spermatologic values or the endoscopic appearance of the testicle are altered.

Finally, infertile clutches occur after artificial insemination with contaminated semen. This is especially true when urate or fecal contamination decreases spermatozoa vitality significantly.

Reduced egg quality

Alterations of the eggshell occur if the female faces stressful events. These stresses might come from outside or from disturbed pair behavior. Thin eggshells might also be caused by calcium deficiencies. Causes include low calcium content or imbalanced calcium:phosphorus ratios in the food, vitamin-D deficiency and intestinal or kidney disorders. Diseases of the female reproductive tract, such as oophoritis, salpingitis or cloacitis need to be considered.

Reduced hatching rate

A reduced hatching rate is caused by incubation management failure. Incorrect temperatures (calibrate the thermometer!) and not controlling the weight loss of each egg are causes. In addition, poor incubation hygiene and malpositioning of embryos in the eggs also reduce hatching rate.

An egg's content is a closed environment. Once the egg is laid, nothing can be added or removed. Therefore, feeding management of the parents, especially the female, is of major importance for a satisfactory hatching rate. The egg follicle starts growing approximately 6 weeks prior to ovulation and the egg yolk contains most of the nutritional ingredients for the growing embryo. High quality, vitamin-enriched (E, B_2, D_3) food must be supplied months before the first reproductive activity is visible. This is in many cases not considered by the breeder.

Vertically transmitted pathogens need to be considered. Infected females often are clinically healthy, but carriers of pathogens and their vertically infected embryos may die. The same occurs with environmental toxic agents (e.g., pesticides, mycotoxins), which may not affect the female clinically, but are transmitted via the egg yolk and affect the embryo severely.

The investigation of a decreased hatching rate involves a detailed assessment of the breeding and feeding management and a postmortem examination of the egg, including evaluation of the time of death of the embryo. Common reasons for early embryonic death are inadequate egg transport, egg storage prior to incubation, severely inappropriate incubation temperature, microbiologic egg contamination or environmental toxic agents. Late embryonic death usually occurs after malnutrition of the parents, vertically transmitted pathogens, failure in incubation management such as deficient egg turning, slightly wrong incubation temperature, insufficient or increased egg weight loss.

The incubator needs to be tested and calibrated so all electronics and mechanisms work correctly. Swabs for microbiologic investigations should be taken to assess incubation hygiene. Investigation of the egg and embryos that failed to hatch should involve a microbiologic examination of the egg yolk. The egg yolk membrane and the trachea of the embryo should be taken to investigate for mycoplasma, the eggshell for detection of *Salmonella*, liver and kidney of the embryo should be stored for virologic and molecular biologic examination and the complete embryo should then be placed in formalin and sent for histopathologic examination.

Veterinary Management of Breeding Birds

Breeding birds demand special attention, not only because of their high value, but also due to their increased health risk (see above). On a regular basis (e.g., every 3-6 months) a fecal sample should be taken and examined for the presence parasites and *Salmonella* spp. At least once per year the birds should be evaluated in detail. This involves a swab for detection of *Chlamydia* sp. and a stress test[3] to assess the function of the respiratory tract. In cases where the stress test demonstrates alterations, an endoscopic examination of the celomic cavity of the bird should be performed. A blood sample should be taken to investigate the antibody level against paramyxovirus-1 and avian influenza virus H5 and H7, blood chemistry to investigate liver and kidney function, hematology to detect chronic diseases and determination of the vitamin and trace element level, especially if a nutritional imbalance is suspected. In psittacine birds, samples to check for circo-, polyoma-, borna-, herpes-, and adenovirus should be taken, including antibody screening, especially for antibodies directed against avian bornavirus. Treatments (e.g., against parasites) or prophylactic measures such as vaccinations (e.g., in birds of prey: paramyxovirus-1 [PMV-1], avian influenza [AIV, H5/H7] and West Nile virus) can be done at that time. In birds of prey, it has been demonstrated in vaccination and challenge studies, that vaccination against H5 using inactivated vaccine protects the birds completely from clinical disease[10] and partial protection against West Nile virus can be achieved.

Conclusions

Breeding collections of birds of prey and psittacine birds in captivity are complex systems. The basics of management and behavior of the birds are not fully understood by many owners. Reproductive success rates of 40-50% (hatched chicks to laid eggs) are considered successful; however, this is well below natural reproduction rate (> 80%), not to mention pairs that do not lay at all. The role of a veterinarian in a breeding facility is critical. Increasing success rate is usually possible. The veterinarian plays a key role in increasing the success of such facilities and the health of the breeding birds.

References

1. Bailey T, Lierz M. Veterinary aspects of bird of prey reproduction. *Vet Clin North Am Exot Anim Pract.* 2017;20(2):455-483.

2. Bublat AD, Fischer S, Bruslund H, et al. Seasonal and genera-specific variations in semen availability and semen characteristics in large parrots. *Theriogenology.* 2017;5(91):82-89.

3. Chitty J, Lierz M, eds. *BSAVA Manual of Raptors, Pigeons and Passerine Birds.* Gloucester, UK: British Small Animal Veterinary Association; 2008.

4. Fischer D, Garcia De La Fuente J, Wehrend A, Lierz M. Semen quality and semen characteristics of large falcons with special emphasis on fertility rate after artificial insemination. *Reprod Domest Anim.* 2011;46:14.

5. Fischer D, Garcia De La Fuente J, Wehrend A, Lierz M. Celibate birds needs assistance-spermatology and artificial insemination in falcons. *Proc 4th Int Giessen Grad Centre Life Sci.* 2011;127.

6. Fischer D, Neumann D, Purchase C, et al. The use of semen evaluation and assisted reproduction in Spix´s macaws in terms of species conservation. *Zoo Biol.* 2014;33(3):234-244.

7. Fischer D, Neumann D, Wehrend A, Lierz M. Comparison of conventional and computer-assisted semen analysis in cockatiels (*Nymphicus hollandicus*) and evaluation of different insemination dosages for artificial insemination. *Theriogenology.* 2014;82(4):613-620.

8. Fischer D, Angenvoort J, Ziegler U, et al. DNA vaccines encoding the envelop protein of West Nile virus lineages 1 or 2 administered intramascularly, via electroporation and with recombinant virus protein induce partial protection in large falcons (*Falco* spp.). *Vet Res.* 2015;46:87.

9. Krohn J, Meinecke-Tillmann S, Schneider H, et al. Technical note: investigations on the accomplishment of the avian perivitelline membrane-penetration-assay. *Reprod Domest Anim.* 2012;33.

10. Lierz M, Hafez HM, Klopfleisch R, et al. Protection and virus shedding of falcons vaccinated against highly pathogenic avian influenza A virus (H5N1). *Emerg Infect Dis.* 2007;13(11):1667-1674.

11. Lierz M, Gooss O, Hafez HM. Non-invasive heart rate measurement using a digital egg monitor in chicken and turkey embryos. *J Avian Med Surg.* 2006;20(3):141-146.

12. Lierz M. Raptors: reproductive disease, incubation and artificial insemination. In: Chitty J, Lierz M, eds. *BSAVA Manual of Raptors, Pigeons and Passerine Birds.* Gloucester, UK: British Small Animal Veterinary Association; 2008:235-249.

13. Lierz M, Reinschmidt M, Müller H, et al. A novel method for semen collection and artificial insemination in large parrots (Psittaciformes). *Sci Rep.* 2013;3:2066.

14. Lierz M. Reproduction: advancements in methods for improving reproductive success. In: Speer B, ed. *Current Therapy in Avian Medicine and Surgery.* St. Louis, MO: Elsevier; 2016:433-445.

15. Schneider H, Fischer D, Bergmann M, et al. Investigations on crypreservation of psittacine semen. *Proc ICARE.* 2013;386-387.

16. Temple SA. Artificial insemination with imprinted birds of prey. *Nature.* 1972;237:287-288.

Cloacal Prolapses in Cockatoos

Brynn McCleery, DVM, DABVP (Avian Practice)

Session #3136

Affiliation: From Red Bank Veterinary Hospital, 197 Hance Avenue, Tinton Falls, NJ 07722, USA.

Abstract: Cloacal prolapse is the most common reported disease condition of the cloaca in companion parrots with cockatoo species over-represented. Cloacal anatomy, causes of prolapse, signalment, clinical signs, and diagnostic techniques utilized in cases of cloacal prolapse are covered in this masterclass. The focus of the masterclass is on medical and surgical management options for cases of chronic, recurrent cloacal prolapses.

Introduction

Chronic cloacal prolapse of psittacine birds is a common and frustrating disease process for both owners and veterinarians. Cockatoos are overrepresented amongst birds suffering from cloacal prolapse, and prolapsing in this species may be due to a hypersexual or idiopathic syndrome, making treatment more challenging and recurrence more likely. Chronic cloacal prolapses in cockatoos almost always must be treated with a combination of medical, behavioral, and surgical therapies.

Cloacal Anatomy

The cloaca serves as a common termination point for the gastrointestinal, urinary, and reproductive tracts in birds.[1-4] It consists of three compartments (in oral to aboral positioning): the coprodeum, urodeum, and proctodeum.[1-4] The most proximal compartment, the coprodeum, is the largest of the three and connects with the rectum.[1,3,4] In psittacine birds, the rectum attaches to the left side of the coprodeum at an approximately 80-90 degree angle.[3] It is separated from the urodeum via the sphincter-like coprourodeal fold, which is everted from the vent during defecation to prevent fecal contamination of the other cloacal compartments.[1-4] The urodeum is the middle compartment and the smallest of the three.[1,3] It contains the termination of the ureters in the dorsolateral walls.[1-4] The urodeum also connects to the reproductive system of males and females.[1,3,4] In female birds, the oviduct has a rosette-like opening into the urodeum on the left ventrolateral wall, while in males the ductus deferens exits into the ventrolateral wall bilaterally on symmetric, raised papillae.[1,2] The dorsal semicircular uroproctodeal fold separates the urodeum from the most distal compartment, the proctodeum. In juvenile birds, the dorsal wall of the proctodeum contains the opening of the Bursa of Fabricius, an important side of B-lymphocyte differentiation.[1,3,4] This opening regresses into an ostium after the involution of the cloacal bursa as birds mature.[1,3] In some avian species, such as ratites and waterfowl, a copulatory structure is also present within the proctodeum.[2] Vasa parrots (*Coracopsis vasa*) can form an eversion of the cloacal for copulation.

The muscular folds between the chambers of the cloaca create both a physical and functional separation and facilitate many of the important functions of the cloaca. In addition to defecation, as mentioned above, the folds assist with urination, retroperistalsis of urine into the colon for water and solute resorption, copulation, oviposition, and exposure of the Bursa of Fabricius to environmental antigens.[3,4]

The vent is composed of dorsal and ventral lips with right and left commissures and serves as the outer opening of the cloaca, analogous to the anus of mammals.[3-5] Opening and closing of the vent is controlled by the sphincter, transverse, and levator muscles of the cloaca.[4]

Vascular supply to the cloaca is provided by the pudendal artery and vein. The pudendal nerve provides innervation, along with the connexus caudalis and caudal coxal nerves. This pudendal nerve travels parallel to the ureter to the dorsolateral cloacal wall where a plexus is formed.[1,3]

Cloacal Prolapse

Cloacal prolapse is a common disorder reported in psittacine birds and can occur from primary disease within the cloaca or secondary to diseases involving structures that terminate in or near the cloaca.[5-7] Cloacal prolapses may occur in conjunction with prolapses of the oviduct or rectum.[3]

Causes

Cloacal prolapse is ultimately caused by excessive straining and/or loss of normal vent sphincter tone. Confirmed or theorized causes of cloacal prolapse in psittacine birds include: [3,5-9]

Primary: infectious or inflammatory cloacitis (bacterial including mycobacteriosis, fungal [pathogenic yeasts, especially *Candida albicans*[3]], viral [Herpesvirus], parasitic [*Sarcocystis* spp., *Cryptosporidium* avian genotype V[19]]), trauma, neoplasia (adenocarcinoma, leiomyoma, lymphosarcoma, squamous cell carcinoma, fibrosarcoma), hematoma, stricture (especially after treatment for papillomatous masses), adenomatous polyps, viral induced or nonviral papillomatous masses[18]

Secondary: chronic behavioral maladaptation, inappropriate bird-human bond, dystocia, other reproductive diseases (salpingitis, metritis, cystic ovaries), gastrointestinal disease (megacloaca, infectious enteritis, heavy metal gastroenteritis, giardiasis, helminths), cloacolithiasis, ureterolithiasis, neurogenic or neuromuscular deficits resulting in reduced sphincter tone or cloacal atony (from chronic vent trauma or spinal cord disease), coelomic masses/space occupying lesions (organomegaly, tumors, granulomas), ascites

Cloacal prolapse from excessive straining associated with secondary causes is considered more common than primary cloacal disease.[5]

In a study of > 500 cloacal biopsies from avian patients, the three most common diagnoses included: cloacitis (40%), cloacal papillomas (20%), and neoplasia (14%).[5] Herpesvirus is one of the more commonly reported causes of cloacitis in new world parrots, especially green-winged macaws.[3,5] Bacterial cloacitis has been reported due to *Pasteurella* spp., *Salmonella* spp., *E. coli*, and *Staphylococcus* spp.[3] Cloacal prolapses suspected to be from reproductive, behavioral, or idiopathic causes may be less likely to be biopsied compared to other cloacal diseases, and the results of the above retrospective are unlikely to be reflective of the most common causes of cloacal prolapse.

Reproductive tract disease is the most common cause of prolapsed tissue in female birds and can include infection or inflammation (salpingitis, metritis), neoplasia, and recent oviposition. Nutritional deficiencies, obesity, and debilitation from chronic egg laying are risk factors associated with prolapse in these birds.[5] In male birds, masturbation and other sexual behaviors are a common cause of cloacal prolapsing.[10]

Syndromes of cloacal prolapse in cockatoos

Coprodeal prolapse syndrome has been reported cockatoos (and a few other species). This is sometimes considered a single syndrome, while other authors have suggested two separate syndomes.[3,6,11,12]

1. *Secondary to excess hormone stimulation and hypersexual behavior in imprinted parrots (Hypersexual):* This syndrome is reported as more common in cockatoos and affects both males and females. Chronic hormone stimulation leads to anatomic changes to the cloaca and vent.

2. *Progressive idiopathic coprodeal prolapse and straining (Idiopathic):* This syndrome generally starts around sexual maturity (5-8 years in most cockatoos), but the birds do not display normal sexual behaviors—which is considered the distinguishing feature from the other syndrome. This syndrome is most reported in cockatoos (including umbrella and Moluccan cockatoos) but has also been reported in African grey parrots and blue and gold macaws. Males are more commonly affected. Speculated risk factors include living in a single parrot household, being raised without conspecifics, holding feces for prolonged periods, inappropriate bond to one human, and a prolonged weaning period.

Prevalence and signalment

In a retrospective study of companion parrots, prevalence of cloacal disease overall was 3.8%. Prolapse (cloacal, oviductal, and rectal) was diagnosed most often (65.1%), followed by masses and cloacoliths. Cloacitis was identified in approximately one-third of cases (34.9%) and was considered a secondary disease condition in each instance.[4]

Although cloacal prolapse has been reported in a multitude of psittacine bird species, cockatoos are overrepresented.[5] Of the 61.5% of parrots with cloacal prolapse in the study mentioned above, 60.7% were cockatoos, despite cockatoos making up only 10% of the parrot population evaluated. All of the cockatoos that had prolapses were umbrella cockatoos (*Cacatua alba*). Of the cockatoos included in the study, 94.4% were presented for prolapse (cloacal, oviductal, rectal). Sex distribution of the cockatoos with prolapses was 67.9% female, 25% male, and 7.1% of unknown sex. Almost half of the birds in the study (46.4%) with cloacal disease had a history of inappropriate sexual behaviors (excluding egg laying); 76.9% were cockatoos and 84.6% were females. This contradicts previous reports of male cockatoos being overrepresented for cloacal prolapse. The authors theorized that females may have more commonly prolapsed secondary to hypersexual behavior and increased hormones while males had prolapses more commonly caused by idiopathic prolapse syndrome. Sexual behavior was commonly noted in cockatoos and females; however, male cockatoos and a single African grey parrot made up the group of birds for which no underlying cause of prolapse was identified.[4]

Cockatoos were also one of the most common bird groups for which cloacal biopsies were submitted to a pathology lab, along with Amazons and macaws. Cloacitis was commonly diagnosed on histopathologic diagnosis of the biopsy samples from cockatoos. In that study, prolapse was rare with only 10/712 (1.4%) samples reported to be from prolapsed tissue; however, underreporting may have affected those numbers due to the retrospective nature of the study.[13]

Clinical Signs

Clinical signs associated with cloacal prolapse can vary immensely. Birds may show no other apparent signs of disease aside from the prolapse or may present in critical condition.[5] Severity of symptoms depends on the length of time the prolapse has been present, the amount of tissue prolapsed, and the health of the prolapsed tissue.

Symptoms reported by the owners or seen at presentation can include a permanent or intermittent prolapse of variable size, diarrhea, reduced fecal production, tenesmus, dyschezia, haematochezia, malodorous droppings, flatulence, alterations in respiratory rate and/or effort, self-mutilation of the vent, a soiled pericloacal area/vent gleet, infertility, change in perching posture, and pelvic limb paresis or paralysis.[4,5,8] The owners may also report recent oviposition.

Sequelae of Prolapse

The prolapsed tissue is at risk of trauma and subsequent hemorrhage and/or inflammation, infection (cloacitis), desiccation, ischemia, and necrosis.[6,14] Cloaca prolapses can lead to impaired excretion of urinary and gastrointestinal waste products, ascending urogenital infections, cloacitis, enteritis, impaired water resorption, dehydration, egg binding, dystocia, oviductal impaction, gastrointestinal intussusception, and cloacal rent formation.[6,9] With chronic prolapses especially, there may be loss of cloacal wall and vent sphincter muscle tone.[9]

Vegetative endocarditis due to *Staphylococcus aureus* was also reported in an umbrella cockatoo secondary to suspected translocation from the owner's skin during digital reduction of a chronic cloacal prolapse over the course of 1 year.[15]

Diagnosis

A full diagnostic work-up is recommended in cases of cloacal prolapse to try and identify the underlying cause and to guide therapy. Even if chronic hormone stimulation and masturbation is suspected, other causes of prolapse should be ruled out. Unfortunately, in many cases, a cause may not be identified.[7,9]

Physical exam findings may help elucidate a cause of the prolapse, especially if a coelomic mass is palpated or effusion is suspected. During physical examination, the prolapse should be evaluated and cloacal tissue may need to be differentiated from prolapsed rectum or oviduct. Prolapsed cloacal tissue should be carefully evaluated for any areas of discoloration, necrosis, ulceration, or other trauma. The prolapse itself, as well as the feathers around the vent, may be heavily soiled with feces, urates, and blood and should be gently cleaned with sterile saline for evaluation.

Physical examination should include cloacal evaluation. On awake exam, the vent and a small portion of the proctodeum can be evaluated by gentle eversion using a cotton tipped applicator. Further cloacal evaluation can be performed via indirect visualization with an otoscope and sterile cone; however, this technique is still limited and can cause trauma without sedation or anesthesia.[3,5] The preferred method of cloacal evaluation is through cloacoscopy.[5] Saline-infused cloacoscopy is preferred due to improved visualization and maneuverability.[3,5,6,16] This procedure can be performed with a 2.7 or 1.9 mm rigid telescope with a diagnostic sheath for sterile saline insufflation.[3,5,16] Cloacal lesions should be biopsied for histopathology and/or culture. Care must be taken to only sample superficial tissue (avoid full thickness biopsies) and to avoid pertinent anatomic structures.[3]

A minimum database of a complete blood cell count and biochemistry panel is recommended for patients presenting with a cloacal prolapse.[5,6] Blood work will help evaluate for anemia, systemic infection, renal disease, liver disease, and changes associated with ovulation and egg laying.[5] Protein electrophoresis may also be indicated. Fecal evaluation should be performed including fecal cytology (evaluated especially for sporulated rods), fecal parasite evaluation (direct, floatation), and possibly fecal culture.[6] Malodorous fecal production with gas may be associated with *Clostridium* spp. or *E. coli* cloacitis.[3-5]

Orthogonal radiographs are recommended to evaluate for reproductive changes in females (e.g., eggs, polyostotic hyperostosis or physiologic osteomyelosclerosis), organomegaly, and coelomic masses. Contrast gastrointestinal radiographs or a cloacal contrast study may help delineate diseases of the GI tract causing secondary prolapse or cloacal luminal masses.[5] Alternatively, fluoroscopy can be utilized to assess GI motility and evaluate for luminal abnormalities/masses. Advanced imaging (such as abdominal ultrasound, CT, or MRI) may also be indicated. Ultrasound and CT are superior for the detection of unmineralized eggs or egg products, changes to the oviduct (salpingitis, neoplasia), and differentiation of coelomic masses compared to radiographs.

Additional testing may be beneficial in some cases including specific pathogen testing (e.g., avian bornavirus, avian antiganglioside antibody, herpesvirus) and coelioscopy. Biopsy for cytology or histopathology is required for definitive diagnosis of cloacal or coelomic masses and may be performed with ultrasound guidance, endoscopically, or surgically.

Treatment

If left untreated, cloacal prolapse can potentially be fatal.[6] Initial treatment should be centered around supportive care and may include providing appropriate warmth, fluid resuscitation, and cleaning of the prolapsed tissue with subsequent lubrication to present further desiccation or damage.[5]

Although most chronic cloacal prolapses in cockatoos are mild (and often intermittent), some patients may present with severe prolapses and be in shock secondary to blood loss/hypovolemia, underlying disease, compression of coelomic nerves or vasculature (secondary to a space occupying lesion, mass, or egg), and/or septicemia. These patients should receive emergency management including shock therapy.[5]

Appropriate therapeutic intervention will depend on the underlying cause of the prolapse. Although many theorized or reported causes of cloacal prolapse were listed above for psittacine birds, many of these causes have not been identified in cockatoos or are considered extremely rare. For this reason, specific management of papillomatosis, cloacolithiasis, and ureterolithiasis are not covered in these proceedings. Diagnosis and treatment of cloacitis should be guided by biopsy and culture results. Similarly, if neoplasia is diagnosed within the cloaca or coelom, treatment will depend on diagnostic results (e.g., advanced imaging, cytology, histopathology) and may include complete surgical debridement vs. debulking and adjunct therapies (e.g., chemotherapy, radiotherapy).

Summary of treatment goals

- *Protect the tissue:* Prolapsed tissue must be protected against trauma and desiccation. The tissue should be gently cleaned with sterile saline and lubricated. Hypertonic saline or dextrose can be applied if the tissue appears edematous.

- *Patient stabilization:* Supportive care including warmth, fluid resuscitation, analgesia (topical anesthetics, systemic opioids, nonsteroidal anti-inflammatory medication), antibiotics if indicated for primary or secondary infection, and nutritional support should be provided prior to diagnostics or any surgical therapies.

- *Reduction:* Prolapse reduction can be performed digitally with a gloved hand, utilizing cotton-tipped applicators as needed. Tissue must be replaced in normal anatomic positioning. Transverse interrupted or horizontal mattress vent sutures may be needed to maintain position of the tissue during diagnostic work-up and definitive therapy.

- *Resolution:* This will be based on identification and treatment of the cause of prolapse. It is important to inform owners that a simple reduction with transverse sutures is unlikely to solve the problem in most cases.

Prolapse related to egg laying

In cases of confirmed egg binding or dystocia, the egg should be removed manually or through transcloacal ovocentesis under heavy sedation or general anesthesia. If the egg is located proximally, cannot be safely removed from prolapsed tissue, or the tissue is necrotic, surgery is indicated and may include a salpingohysterectomy and/or partial cloacal resection. Replacement of necrotic tissue puts the bird at risk of disease and death secondary to coelomitis and septicemia.[5] Procedures for ovocentesis and salpingohysterectomy have described in detail elsewhere.[17]

Hormone therapy can be utilized in birds with a suspected or confirmed reproductive cause of cloacal prolapse (e.g., recent egg binding or dystocia, history of oviposition prior to prolapse). Long-acting gonadotropin releasing hormone analogues (e.g., leuprolide acetate, deslorelin acetate) are used commonly.[12] Although these medications initially stimulate release of leutinizing hormone (LH) and follicle stimulating hormone (FSH) from the anterior pituitary gland, downregulation of GnRH receptors in cells producing these hormones follows through a negative feedback mechanism on both the hypothalamus and pituitary gland.[5] Chronic use of GnRH agonist medication may cause the medications to be less effective over time, and an increased dose or dosing frequency may be required.[12]

Surgical Intervention

Many cockatoos will present with chronic, intermittent, or persistent prolapses. These cases typically require surgical intervention. Each of these surgical procedures only addresses the clinical signs associated with prolapsing. The underlying cause must still be identified and addressed or recurrence is likely. These procedures can often be performed concurrently to address chronic changes to the anatomy (e.g., atony of the cloaca, atony of the vent).

Transverse sutures

After prolapse reduction, transverse simple interrupted or horizontal mattress sutures can be used to reduce the vent opening to prevent prolapse recurrence while treating the underlying cause and to allow time for reduction of tissue inflammation. The vent opening must be reduced to a size small enough to prevent prolapse recurrence but large enough to allow normal passage of feces.[3,5] It is rare for simple reduction as a sole therapeutic modality to provide permanent resolution of cloacal prolapses, and owners should be properly informed. In chronic cases, the cloacal wall and vent sphincter may be damaged, leading to fecal incontinence, secondary infections, and repeated prolapsing without more aggressive surgical intervention (i.e., reconstructive surgeries).

Purse strings in the vent are not recommended due to possible damage to the vent sphincter or its innervation and impedance of dorsal and ventral movements of the vent during defecation, predisposing to obstipation.[3,9] Placement and periodic replacement of stainless steel transverse sutures has also been reported.[3,6]

Cloacotomy

Direct incision into the cloaca through a coelomic approach can be performed for removal or biopsy of masses or abnormal tissue.[5] This is performed through a ventral midline approach to the caudal coelom.[17] Avoidance of important anatomic structures, including the ureters and reproductive tract openings, is essential.[5]

Cloacopexy

This procedure creates a permanent adhesion between the cloaca and body wall to help prevent further prolapse. Several approaches have been reported in psittacine birds. Intracoelomic approaches involve performing a coeliotomy via a vertical or horizontal skin incision made over the cloaca and reflection or removal of the fat pad over the ventral cloaca.[17] A cotton-tipped applicator placed in the cloaca prior to the below procedures can help confirm placement of the cloaca within the coelom.

- *Percutaneous:* Multiple (two to three) sutures are placed blindly through the skin and cloacal wall into a cotton-tipped applicator passed retrograde into the cloaca from the vent. The cotton-tipped applicator is removed from the vent and the cotton is teased off the sutures, which are then tightened. This procedure is more likely to cause inadvertent perforation of other structures (ureter, rectum, duodenum, pancreas), cause GI or ureteral entrapment, or cause functional colonic obstruction (if sutures placed extremely cranial) than an open procedure and is not recommended by the author.[6,9]

- *Circumcostal:* Sutures are placed around the most distal rib (at the junction of the sternal and vertebral portions) and passed full thickness through the craniolateral cloacal wall. This procedure is performed bilaterally and can be combined with a ventral body wall cloacopexy during closure.[6,9,17]

- *Sternal:* Anchoring to the pubic bones and sternum has also been reported. For sternal cloacopexy, two sutures are passed through the cartilaginous border of the sternum, then full thickness through the cloaca.[17]

- *Incisional:* The serosal surface can be incised on the coprodeum (exposing the subserosa) and sutured to an incision in the abdominal wall, similar to a gastropexy procedure in canine patients.[6,17] To perform this procedure, a 2- to 5-mm incision is made in the serosa of the coprodeum longitudinally, approximately 5 to 10 millimeters from the midline. An opposite incision is created in the paramedial body wall at a position that results in slight inversion of the vent when the cloaca is held up to it. Sutures are preplaced (three to four) then tied. During closure, a ventral midline cloacopexy can also be performed.[17]

- *Ventral midline*: This can be performed on its own or with one of the above cloacopexy procedures. To encourage adhesion formation, the ventral surface of the cloaca should be scarified or incised. Full thickness bites through the cloaca are taken during closure.

Possible and reported complications of cloacopexy include adhesion formation, colonic entrapment (reported in two cockatoos),[9] infection, intraoperative trauma to the cloaca/oviduct GI tract, failure or tearing of the cloacopexy site, and obstructions.[5,9,17]

Traditional ventplasty

Ventplasty reduces the vent size permanently in an effort to prevent cloacal prolapse.[10,17] This procedure should be reserved for chronic cloacal prolapses where cloacal muscles and supporting structures are permanently stretched and nonfunctional, leading to vent atony.[6,10] The amount of tissue resected depends on the extent of the vent dilation. One section can be removed in cases of mild to moderate dilation, while two sections are typically removed in cases of severely dilated vents using a traditional approach.[10,17] The goal is to return to vent to a more normal diameter by reducing the opening by approximately 25-30%.[6,10] Triangular incisions are made in both approaches with the base of the triangle along the leading edge of the vent with the point away from the vent.[6,17] Excision of the epidermis and dermis is performed in each approach with effort made to spare the sphincter and transverse cloacal muscle, if visible. The dermis can typically be bluntly dissected from the underlying muscular and submucosal layers. After resection of the triangular piece of tissue, gentle manual apposition of the new epidermal edges will form the vent diameter. If insufficient, additional epidermal and dermal tissue can be resected as needed until the desired vent diameter is achieved. During closure, a simple interrupted pattern is used to appose the submucosa, dermis, and epidermis in separate layers using absorbable sutures.[6,17]

- *Singular incision:* A single incision over the cranioventral aspect of the vent can be performed in cases of mild to moderate dilation.[17]

- *Bilateral approach:* Bilateral triangular wedge resections on the left and right lateral aspects of the vent can be performed in more severe cases.

It is important to note that ventplasty alone is likely to fail if the underlying cause of straining is not identified and addressed (as with cloacopexy alone).[10] Furthermore, prolapse of tissue through a more narrowed opening puts the tissue at increased risk of vascular compromise, inflammation, and necrosis. In females, egg laying must be controlled through surgery (salpingohysterectomy), behavioral modification, or hormone therapy, otherwise dystocia or rupture of the ventplasty sutures may occur.[10]

Asymmetric ventplasty[6]

During this procedure, the mucocutaneous junction of the vent is superficially incised in a C-shape on the right side with the incision encompassing approximately 50% of the vent circumference. A width of 1 to 2 millimeters of mucosa is removed with tenotomy scissors, and the mucosa is subsequently apposed with horizontal mattress sutures (5-0 PDS) placed on the vent sphincter in order to maximize contact between the dorsal and ventral vent lips. Skin edges are oversewn with simple interrupted sutures (6-0 PDS) to prevent inward curling. Finally, an additional tension-relieving mattress suture (4-0 PDS) is placed cranial to the cloacal sphincter as the luminal edge of the incision and removed 2 weeks later. The tension-relieving suture helps protect the area most susceptible to stretching leading to dehiscence.

This procedure reduces incision tension secondary to midline straining that occurs post-operatively. The asymmetric positioning creates a physical barrier to coprodeal tissue during straining but still allows defecation in parrots due to left-sided positioning of the rectal opening into the coprodeum. As with traditional ventplasty, this procedure is used in chronic cases associated with chronic reproductive conditions or idiopathic prolapsing leading to cloacal atony.

Reported acute complications include dehiscence, transitory straining, transitory mild fecal staining, and high velocity and lateralized fecal output. Possible long-term complications include dystocia, cloacolith formation or cloacitis secondary to fecal/urate stagnation.

Colopexy[7]

Colopexy was reported as a treatment for chronic recurrent colocloacal prolapse in an 8-year-old male sulphur-crested cockatoo (*Cacatua galerita*). The prolapses often occurred in association with hypersexual behaviors including masturbation. A ventral U-shaped skin incision was made, and the distal colon was sutured to the left abdominal wall using the same technique reported in dogs and cats. The seromuscular layer of the colon was incised longitudinally along the antimesenteric side. A corresponding incision was made in a craniocaudal direction in the opposing left transverse abdominal muscle. The dorsal and ventral edges of the seromuscular incision were sutured to the corresponding edges of the abdominal wall incision with simple continuous sutures with the colon retracted cranially under slight tension. Although technically more challenging than cloacopexy due to the relatively thin colon wall and risk of perforation, this procedure provides cranially directed tension on the cloaca with the intestines fixed in their normal anatomic position with the pexy site on the left body side, thereby reducing the risk of intestinal entrapment.

Possible complications of colopexy include infection/coelomitis perforation or suture penetration of the lumen, fistula formation, and stricture.

Environmental Changes

Environmental factors need to be addressed to help prevent recurrence of chronic cloacal prolapse in cockatoos, as the vast majority of patients are exhibiting excessive reproductive behaviors (e.g., hypersexual, chronic egg laying) or have idiopathic prolapse with environmental/behavioral risk factors. Information regarding the diet, housing, reproductive status, exposure to other birds, acquisition, and behavior at home should be obtained from the owner.[5]

Environmental changes and enrichment should include dietary improvement (e.g., conversion to a formulated diet, limiting high-calorie foods, switching to meal feeding if obese), increasing exercise, and encouraging foraging.

If a reproductive cause is identified or confirmed, special attention must be placed on at-home reduction of hormonal stimulation and discouragement of reproductive behavior. This can include reduced photoperiod, removal of nesting material, removal or preventing access to actual or perceived nesting sites, separation from other birds in the home, removal of toys stimulating masturbation or other reproductive behaviors, and dietary changes to reduce high fat, warm, or mushy foods (mimics courtship feeding).[5,12] The owner-bird bond should be evaluated for behavior contributing to reproductive activity/hormonal stimulation, such as petting below the head/neck (mimics mating), kissing (mimics courtship feeding), and cuddling (mimics nesting).

Applied behavioral analysis should be utilized to identify and correct undesirable behaviors that may be contributing to prolapsing, or the prolapsing itself if intermittent and the patient appears to have some control.

Behavioral Pharmacology

Behavioral pharmacology may be beneficial in some patients. Clomipramine hydrochloride has been anecdotally reported to contract the vent orifice and help prevent cloacal prolapse; however, these claims have not been validated in scientific studies.[20] Clomipramine is a tricyclic antidepressant and serotonin-norepinephrine reuptake inhibitor (SNRI) that has been used for the treatment of obsessive-compulsive disorders in humans.[21] It was evaluated in a double-blind, placebo-controlled trial in cockatoos and was showed to reduce feather picking.[22] Pharmacokinetic studies in avian species are lacking, and dosing is based on extrapolation or progressive titration from low doses. Side effects have been reported and are typically mild in birds including drowsiness, regurgitation, transient ataxia, and suspected hallucinations (from paradoxical anxiety). More severe extrapyramidal side effects were reported in a blue and gold macaw treated concurrently with haloperidol.[21] The benefit of clomipramine or other behavioral modification medications in cockatoos with chronic cloacal prolapse is unknown.

Prognosis

Management of chronic cloacal prolapse in cockatoos requires a dedicated owner. Full resolution can be difficult to achieve and often requires a combination of medical, surgical, environmental, and behavioral interventions. Even with surgical management, recurrence is possible, and repeated surgical procedures may be necessary.

References

1. Graham JE, Tell LA, Lamm MG, Lowenstine LJ. Megacloaca in a Moluccan cockatoo (*Cacatua moluccensis*). *J Avian Med Surg.* 2004;18;41-49.

2. Sanmeier P. Anatomy and physiology. In: Chitty J, Monks D, eds. *BSAVA Manual of Avian Practice: A Foundation Manual.* Gloucester, UK: British Small Animal Veterinary Association; 2018;14-34.

3. Taylor WM. Clinical significance of the avian cloaca: interrelationships with the kidneys and the hindgut. In: Speer BL, ed. *Current Therapy in Avian Medicine and Surgery.* St Louis, MO: Elsevier; 2016:329-344.

4. Gill KS, Helmer PJ. Clocal diseases in companion parrots: a retrospective study of 43 cases (2012-2018). *J Avian Med Surg.* 2020;34(4):364-370.

5. Lennox A, Lee A. Cloacal, uterine, and rectal prolapse. In: Chitty J, Monks D, eds. *BSAVA Manual of Avian Practice: A Foundation Manual.* Gloucester, UK: British Small Animal Veterinary Association. 2018; 358-365.

6. Zaheer O, Beaufrere H, Jajou S, Gardhouse S. Asymmetrical cloacoplasty for the treatment of chronic cloacal prolapse in Psittaciformes: a case series. *J Avian Med Surg.* 2020;34(2):172-180.

7. van Zeeland YRA, Schoemaker NJ, van Sluijs FK. Incisional colopexy for treatment of chronic, recurrent colocloacal prolapse in a sulphur-credted cockatoo (*Cacatua galerita*). *Vet Surg.* 2014;43:882-887.

8. Detweiler DA, Carpenter JW, Kraft SL et al. Radiographic diagnosis: avian cloacal adenocarcinoma. *Vet Radiol Ultrasound.* 2000;41(6):539-541.

9. Radlinsky MG, Carpenter JW, Mison MB, Armbrust LJ. Colonic entrapment after cloacopexy in two psittacine birds. *J Avian Med Surg.* 2004;18(3):175-182.

10. Echols MS. Basic surgery. In: Chitty J, Monks D, eds. *BSAVA Manual of Avian Practice: A Foundation Manual.* Gloucester, UK: British Small Animal Veterinary Association. 2018;242-268.

11. Beaufrere H, Taylor WM. Case series: long-term management in five umbrella cockatoos with idiopathic cloacal prolapse syndrome. *Proc Annu Conf Eur Assoc Avian Vet.* 2009:163-169.

12. van Zeeland YRA, Friedman SG, Bergman L. Behavior. In: Speer BL, ed. *Current Therapy in Avian Medicine and Surgery.* St Louis, MO: Elsevier; 2016:177-251.

13. Lee A, Lennox A, Reavill D. Diseases of the cloaca: a review of 712 cloacal biopsies. *Proc Annu Conf Assoc Avian Vet.* 2014:51-60.

14. Dutton TAG, Forbes NA, Carrasco DC. Cloacal prolapse in raptors: review of 16 cases. *J Avian Med Surg.* 2016;30(2):133-140.

15. Steinagel A, Quesenberry K, Donovan T. Vegetative endocarditis due to *Staphylococcus aureus* in an umbrella cockatoo (*Cacatua alba*). *J Avian Med Surg.* 2019;33(4):419-426.

16. Taylor M. Examining the avian cloaca using saline infusion cloacoscopy. *Exot DVM.* 2001;3(3):77-79.

17. Rubin JA, Runge JJ, Mison M, et al. Surgery. In: Speer BL, ed. *Current Therapy in Avian Medicine and Surgery.* St Louis, MO: Elsevier; 2016:631-667.

18. Gartrell BD, Morgan KJ, Hoe L, et al. Cloacal papillomatosis in the absence herpesvirus and papillomavirus in a sulphur-crested cockatoo (*Cacatua galerita*). *N Z Vet J.* 2009;57(4):241-243.

19. Curtiss JB, Leone AM, Wellehan JFX, et al. Renal and cloacal cryptosporidiosis (*Cryptosporidium* avian genotype V) in a Major Mitchell's cockatoo (*Lophochroa leadbeteri*). *J Zoo Wildl Med.* 2016;46(4): 934-935.

20. Bowles H, Lichtenberger M, Lennox A. Emergency and critical care of pet birds. *Vet Clin North Am Exot Anim Pract.* 2007;10:345-394.

21. Starkey SR, Morrisey JK, Hickman HD, et al. Extrapyramidal side effects in a blue and gold macaw (*Ara ararauna*) treated with haloperidol and clomipramine. *J Avian Med Surg.* 2008;22(3):234-239.

22. Seibert LM, Crowell-Davis SL, Wilson GH, Ritchie BW. Placebo-controlled clomipramine trial for the treatment of feather picking disorder in cockatoos. *J Am Anim Hosp Assoc.* 2004;40(4):261-269.

How to Perform a Salpingohysterectomy

Cheryl B. Greenacre, DVM, DABVP (Avian Practice), DABVP (Exotic Companion Mammal Practice)

Session #3343

Affiliation: From College of Veterinary Medicine, University of Tennessee, Knoxville, TN 37996, USA (Emeritus). Current address: Office of Research, University Research Animal Resources, University of Georgia, 501 D.W. Brooks Drive, Room 3504, Athens, GA 30602, USA.

Abstract: The Association of Avian Veterinarians routinely offers a high-quality instructional hands-on laboratory on coelomic surgery of birds that includes instruction on how to perform a salpingohysterectomy. For various reasons, including cost, popularity of the laboratory with limited space, or timing, not everyone who has an interest in learning this topic can benefit from the laboratory. Although hands-on training is the best method to learn a surgical procedure, the following description provides step-by-step instruction of how to perform a salpingohysterectomy in various sizes and species of birds. A description of common clinical signs and diagnosis of reproductive disease is included as well as pathology. It is recommended that these skills be practiced on a cadaver and enroll in laboratories or consult a mentor for perfecting your technique before performing the surgery, including learning microsurgical techniques for delicate tissue handling.

Introduction

Salpingohysterectomy (SHE) refers to the surgical removal of the entire oviduct and uterus (i.e., shell gland). Female birds commonly present for reproductive disease necessitating medical or surgical intervention. If oviductal disease cannot be medically managed, then removal of the oviduct and uterus can be performed. A SHE can be performed to manage dystocia (i.e., egg binding), egg impaction, cystic right oviduct, chronic egg laying, chronic infection, and neoplasia. The following is a description of important anatomic landmarks, indication for the need for a salpingohysterectomy and step-by-step instruction on how to perform a salpingohysterectomy. There are many resources available to clinicians including published articles, webinars, websites, and a newly published book, *Surgery of Exotic Animals*.[1-7]

Anatomy

The gonads of birds are intracoelomic, therefore, to determine the gender of a bird, surgical sexing, blood testing, or observation of external or behavioral characteristics must be performed. Most parrots are not obviously sexually dimorphic. To determine the gender of a parrot, surgical sexing or blood sexing must be performed. An exception would be the eclectus parrot that at a few days of age the female will grow red feathers and the male will grow green feathers. Surgical sexing involves visualizing the gonads and reproductive tract via a rigid endoscope placed into the abdominal or caudal thoracic air sac. Blood sexing involves evaluating 0.2 ml of blood via PCR that tests for a heterogamete (female is ZW) or homogamete (male is ZZ).

The female reproductive tract consists of an infundibulum, magnum, isthmus, uterus (i.e., shell gland) and very short vagina.[6] The ovary is not typically surgically removed from birds due to its very close attachment via the mesovarium to the aorta and dorsal body wall making it prone to fatal hemorrhage. Therefore, a salpingohysterectomy is performed in birds. Usually this involves only removing the left oviduct since the right oviduct is rudimentary in most birds and does not develop. Occasionally, birds can develop a cystic right oviduct, which is most common in pet chickens but can occur in psittacine birds as well, and would also need to be removed.[1-7]

The oviduct is suspended within the dorsal part of the intestinal peritoneal cavity by the dorsal and ventral suspensory ligaments and extends from the cranial division of the left kidney to the cloaca. Blood supply to the oviduct consists of the cranial, accessory cranial, middle, and caudal, and oviductal arteries. These vessels are usually easily visible within the suspensory ligament.[4-6]

Before the Procedure

The bird should be in the best health possible for the surgical procedure. Correct any underlying problems (e.g., such as dehydration, infection, inflammation, negative energy balance). Decrease the size of the reproductive tissues several weeks prior to surgery, if possible, by administering leuprolide acetate or deslorelin. A chicken, which is a food animal, cannot be given hormones in the United States because they are a prohibited drug in any food-animal species (e.g., chicken or turkey), regardless of their use.

Once the patient is in the best possible health, consider premedication with an opiate pain reliever such as butorphanol about 15 minutes or more before anesthetic induction and midazolam or other injectable anesthetic or sedative to relax the patient and use less anesthetic gases.[2,3] Begin mask induction with isoflurane or sevoflurane. An uncuffed endotracheal tube is placed after all monitoring equipment is placed (e.g., Doppler, ECG, pulse oximeter, respiratory monitor, blood pressure cuff). Ideally, an intravenous or intraosseous catheter is placed and fluids are given. Thermal support is important. A heated surgical table can be used as well as a forced heated air blanket. For larger birds, a thermometer can be used in the cloaca.

Procedure

Preparation

The entire ventrum and/or left side of the coelom is plucked and the edges of the proposed sterile area are taped off with white tape to keep the feathers out of the surgical site. The area is aseptically prepared with chlorhexidine scrub and sterile saline (alcohol can drop the body temperature precipitously). For most psittacine birds, a left lateral incision is preferred that extends from 1 cm cranial to the left pubic (pin) bone cranially to about 1 cm dorsal to the uncinate process of the last two ribs. For large birds such as poultry a ventral midlne incision following the linea ventrally is preferred.[4,5] Some perform a T-shaped incision by making a ventral midline incision and then a 90-degree angle to the left. Castro-Viejo or eyelid retractors, or a Lone Star retractor can be used for better visualization. The author uses bipolar radiocautery with a Harrison tip to cut through the skin and the coelom wall musculature. Long, balanced, microsurgical instruments (ring-tip forceps, needle holders, scissors) are used for delicate manipulation of fine tissues, and good quality 5x magnifying loupes. Purchase good-quality tools for this job. The uterus is identified as a whitish-pink, longitudinally striated tissue that can be gently retracted with ring-tip forceps and followed cranially to the flimsy cranial end of the salpinx near the left ovary. Bipolar radiocautery is used to cut and coagulate through the fascia and use small to medium surgical clips to ligate vessels before cutting (usually two clips stay with the bird and one clip goes with the tissue being removed). Work your way from cranial to caudal ligating and cutting vessels until the juncture of the uterus to the cloaca is reached where it is ligated with 5-0 or appropriate sized polydioxanone suture depending on the size of the bird.[4,5] The same suture is used to close the coelomic musculature in a simple continuous pattern and the skin is also closed with the same suture in a continuous Ford-interlocking pattern. During recovery from anesthesia, the patient is given meloxicam, and then repeated butorphanol injections as needed approximately every 4 hours. There are no pain scales developed for birds. The patient usually eats within a few hours of surgery. Care is taken not to give fluids subcutaneously near the surgical site, so they are given in the right inguinal region post operatively. Sutures are removed in about 14 days.

Reproductive Diseases

Egg-related coelomitis or peritonitis

Egg-related coelomitis or peritonitis is acute or chronic inflammation or infection involving all or part of the reproductive structures. It can be septic or not. It is common for chickens to have some degree of egg-related coelomitis or peritonitis, and mild cases are commonly encountered at necropsy in production hens. Generally, chickens tolerate mild coelomitis or peritonitis better than parrots. Causes usually involve retrograde flow of shelled or shell-less eggs from the oviduct back into the coelomic cavity due to oviductal bacterial infection, oviductal impaction, or abnormal confirmation of the oviduct. Heavy production hens or those with inadequate calcium in their diet can have calcium depletion and uterine inertia leading to retrograde flow of egg material. Bacterial infections commonly involve *E. coli* migrating up the oviduct from the vent.[1]

Clinical signs of egg-related coelomitis or peritonitis can be obvious (e.g., a sudden drop or cessation in egg laying) or can be vague and seem unrelated to the reproductive tract (e.g., lethargy, partial anorexia, weight loss, and lameness). Physical examination findings can include a thin bird, increased respiratory and heart rates, crackles of lower respiratory tract, a large doughy abdominal region, fluctuant coelomic fluid, and a lameness not explained by other obvious reasons.

Diagnosis is based on typical clinical signs, ultrasound, and radiographs. Endoscopy is possible but may be hampered or risky due to coelomic fluid and limited air sac space. Radiographically, a large variety of findings can be encountered such as the ground glass appearance of coelomic fluid, multiple or single-shelled or shell-less eggs inside or outside the oviduct, hernia or expanded coelomic wall, thickened air sacs, and masses in the caudal thoracic and abdominal air sac area. An ultrasound is helpful in determining if the egg material is within or outside the oviduct. A CBC and chemistry profile including total and ionized calcium can be performed to determine the degree of infection or inflammation, amount of dehydration, if there is concurrent liver or kidney disease, or if calcium depletion is present.

Treatment can involve fluids, antibiotics, non-steroidal anti-inflammatory drugs (NSAID), butorphanol, and/or surgery. Sometimes it is difficult to ascertain whether surgery will provide a better outcome than antibiotics and pain relievers alone. Generally, the more severe the egg-related coelomitis or peritonitis then the more likely surgery will provide a better outcome. A few clients chose no surgery for their hens with what appeared to be a moderate egg-related coelomitis or peritonitis (i.e., lameness, large fluctuant coelomic cavity, cessation of egg laying, but no shelled eggs in the coelomic cavity) and it cleared up with just oral antibiotics and meloxicam. Oviductal impaction should be resolved with surgery to remove the egg material before it adheres to the oviduct. Parenteral calcium is given for oviductal inertia due to calcium depletion. There are minimal to no adverse effects from administering one calcium injection after drawing blood to perform an ionized calcium while you are waiting for the results.

Egg binding or dystocia

Egg binding describes a condition in which a fully formed egg in the uterus cannot be laid. There are many causes for this condition including too large of an egg or a misshapen egg, inability of uterus to contract sufficiently from low blood calcium, or other causes of uterine paralysis. Clinical signs usually include straining to lay the egg (which sometimes owners think is straining to defecate), dyspnea, fluffed appearance, anorexia and sitting at the bottom of the cage. Diagnosis can be made on history, palpation on physical examination (including digital cloacal examination), ultrasound, and radiographs showing an egg or eggs. Radiographs can be taken without anesthesia in a cardboard box because positioning is not necessary to determine the presence of a calcified egg. It cannot be radiographically determined if the egg is in the uterus or out loose in the coelomic cavity. In larger birds, an ultrasound may help to determine this. Total and ionized calcium are helpful, as is a CBC and chemistry profile, but egg binding is a true emergency and injectable calcium, subcutaneous fluids with or without 2.5% dextrose, and a warm incubator are needed immediately. The primary cause of egg binding is usually hypocalcemia from insufficient calcium in the diet, such as a seed diet. Oxytocin, or if available, vasotocin, or prostaglandin E2 alpha can be given, but only if the egg is deemed able to come out.[2,3] If the egg is too large or irregular to come out naturally, or if the egg is putting pressure on the kidneys and has progressed to shock, then the egg should be imploded via ovocentesis as soon as possible. Implosion involves creating a negative pressure within the egg to collapse the egg in on itself, not crushing the egg. Prevention of egg laying in the future involves removing all sexual stimuli such as mirrors, decreasing daylight (10 hours of light and 14 hours of dark daily), and removing the presence of a male. Leuprolide acetate, a GnRH agonist, can also be given monthly to non-food animal species, or a salpingohysterectomy can be performed.

References

1. Saif YM. *Diseases of Poultry.* 12th Ed. Ames, IA: Blackwell Publishing; 2008.
2. Crespo R, Petritz OA. Backyard poultry, gamebirds, and waterfowl. In: Carpenter JW, Harms CA. *Carpenter's Exotic Animal Formulary.* 6th Ed. St. Louis, MO: Elsevier; 2023:444-495.
3. Guzman DS, Beaufrere H, Welle KR, et al. Birds. In: Carpenter JW, Harms CA. *Carpenter's Exotic Animal Formulary.* 6th Ed. St. Louis, MO: Elsevier; 2023:222-443.
4. Mehler SJ, Bennett RA. Avian reproductive procedures. In: Bennett RA, Pye GW. *Surgery of Exotic Animals.* Hoboken, NJ: Wiley-Blackwell; 2022:163-174.
5. Bennett RA, Harrison GW. Soft tissue surgery. In: Ritchie BW, Harrison GW, Harrison LR. *Avian Medicine: Principles and Application.* Lake Worth, FL: Wingers Publishing; 1996:1096-1136.
6. Joyner KL. Theriogenology. In: Ritchie BW, Harrison GW, Harrison LR. *Avian Medicine: Principles and Application.* Lake Worth, FL: Wingers Publishing; 1996:748-804.
7. Espinosa RA. Cystic right oviduct in poultry. Merck Veterinary Manual. Available at: https://www.merckvetmanual.com/poultry/disorders-of-the-reproductive-system/cystic-right-oviduct-in-poultry?query=cystic%20right%20oviduct%20poultry. Accessed April 4, 2023.

Egg Binding in Small Psittacine Species: Tips, Tricks, Treatment and Outcomes

Ariana Finkelstein, DVM, and
Laurel Harris, DVM

Session #3455

Affiliation: From Sebastian Animal and Bird Hospital, 948 US 1, Sebastian, FL 32958, USA (Finkelstein), Wasatch Exotic Pet Care, 892 Fort Union Blvd, Cottonwood Heights, UT 84121, USA (Harris).

Abstract: This presentation covers different presentations of egg binding in small psittacine species (e.g., budgerigar, cockatiel). It addresses the complexity of the egg binding process and recommended treatments. Pathophysiology and nutrition are addressed. Outcomes and possible scenarios are discussed. Discussions with the pet owner on reasonable expectations also are addressed. This is more of a how-to or in-our-experience discussion than a scientific presentation. Questions are addressed during the lecture. Videos are provided.

Introduction

Egg binding can present differently in small psittacine birds. Egg binding is a complex process. Pathophysiology (e.g., hypothalamus-pituitary-ovarian axis [HPG] is upregulated to have increased estrogenic activity), nutrition (e.g., excessive food, especially high-fat food sources), and behavior play a large part in egg binding cases.

Egg Binding

New to a practice, small birds often present on an emergency basis. They are fluffed in the bottom of the cage, not eating or drinking and often not able to pass urates and feces. Often the owner is unaware of how long their pet has been in distress.

On presentation, these patients are fragile. They should be managed as quickly and handled as minimally as possible. Sometimes a quick weight, subcutaneous fluids, brief examination and placement in an incubator is the starting point. A good history from the owner and discussion of all potential outcomes and scenarios can be done while the bird rests. Once this is accomplished, pain management (e.g., premedication with meloxicam or gabapentin, especially if the bird is in shock) is an important next step. Opioids can drop blood pressure too. It is important to remember that unless there is oviductal pathology, the painfulness of this condition is temporary. Sedation (butorphanol ± midazolam), lube around the cloaca and potentially a calcium injection can be performed as the next step. After some time, gentle manipulation may be attempted if the egg has not passed. If at this stage there is still no relief, oxygenation, radiology, and/or ultrasound should be the next step to potentially guide you in the manipulation process. Oxytocin or prostaglandin E2 can be considered but they are not as effective in avian patients as in mammalian patients. If the egg is still not progressing, general anesthesia to attempt oviposition and manipulation from the vent is the next step. With a technician watching the patient's breathing closely, place one or two fingers above the egg and apply constant resistance while waiting for the cloaca to stretch. Active pushing or manipulation is not required, just firm constant pressure.

In the event that the cloaca is not still functionally able to oviposit normally or the bird has hormonally cycled out of the laying phase (meaning an old egg,), then consideration is the avian equivalent of a closed mammalian cervix and you have to collapse the egg. These eggs are sometimes also adhered to the oviduct and are always more problematic.

In the event that this fails consideration for ovocentesis, surgical removal may be the next step. The potential for complications from both of these different treatments would need to be addressed and discussed at this time.

Once the bird is stable and the egg has been removed, nutrition, proper light cycle, and behavior modification will all be necessary to help downregulate the HPG axis to decrease estrogenic activity and potentially prevent future occurrences.

Excessive egg producers and small psittacine birds with egg binding issues are among the scariest presentations that require triage on a seasonal basis. Management of these cases provides unique challenges.

The Avian Uropygial Gland in Health and Disease

Lauren V. Powers, DVM, DABVP (Avian Practice),
DABVP (Exotic Companion Mammal Practice)

Session #3437

Affiliation: From Avian and Exotic Pet Service, Carolina Veterinary Specialists, 12117 Statesville Road, Huntersville, NC 28078, USA.

Abstract: The uropygial (preen) gland is a unique structure to birds and is easily overlooked on the routine physical examination. Uropygial gland disease is relatively uncommon in psittacine birds and can be challenging to manage. The most common conditions of the uropygial gland in psittacine birds are impaction and neoplasia. Surgery is often curative for advanced or aggressive disease but can pose significant surgical challenges.

Introduction

The uropygial (preen) gland is a unique anatomic structure to birds. The gland is situated on the dorsal rump and produces an oily to waxy substance that is placed on the feathers by the bird's beak during autogrooming. The gland serves to maintain feather condition and may have additional purposes in birds.

Anatomy

The skin in birds is glandless apart from the uropygial gland, glands of the ear canal, and pericloacal glands. The uropygial gland is the only compact gland in the avian integument.[1,2]

Presence and size

The gland is generally larger in aquatic birds compared to terrestrial species.[2-4] Some birds lack a uropygial gland. The uropygial gland is absent in the following avian species.

Large flightless birds

The uropygial gland is absent in the ostrich (*Struthio camelus*), emu (*Dromaius novaehollandiae*), rhea (e.g., *Rhea americana*), and Southern cassowary (*Casuarius casuarius*).[2]

Columbiformes

The uropygial gland is absent or naked (lacking tuft feathers) in columbiform birds.[2,5] In the rock dove (*Columba livia*), the gland is present but indistinctly bilobed and the papilla is large and naked.[2] The gland is absent in crowned pigeons (e.g., *Goura victoria*).[2]

Psittaciformes

The uropygial gland is absent in all Amazon parrots and pionus parrots examined to date, non-*Ara* blue macaws (hyacinth macaw [*Anodorhynchus hyacinthinus*], Lear's macaw [*Anodorhynchus leari*], Spix's macaw [*Cyanopsitta spixii*]) and some *Brotegeris* species examined to date (e.g., grey cheeked parakeet [*Brotogeris pyrrhopterus*]).[2]

Topography and macroscopic anatomy

The uropygial gland is generally situated dorsal to the last sacral and first caudal vertebrae but is occasionally found more caudally such as over the pygostyle.[2,6] The gland lies dorsal and medial to the levator caudae and lateralis caudae muscles of the tail.[2,6] It is embedded in a cushion of fat and has deep fibrous attachments between the ventral capsule and the connective tissue covering the rectrice follicles and caudal vertebrae.[1] The gland is covered by skin that is often feathered.[2,7]

The uropygial gland is bilobed in nearly all avian species examined to date and has a characteristic heart or pear shape.[1,8] The two lobes are divided by the interlobular septum. In the chicken and all psittacine species evaluated in one report,[1] there is one duct for each of the two lobes with each excretory duct exiting through a nipplelike papilla to a single external orifice on each side; however, some avian species (generally those with no primary storage cavity) have up to 18 orifices.[1,2] The papilla itself is normally bare apart from a small tuft of feathers at the tip, termed the circulus uropygialis or uropygial wick.[1,2,8] These feathers are either modified down or modified semiplume feathers.[2] The number and appearance of these feathers varies widely across avian species.[2]

The blood supply to the uropygial gland arises from the median caudal (coccygeal) artery,[1,6] which gives rise to four or more paired caudal segmental arteries.[6] One of these caudal segmental arteries is a specialized uropygial gland artery that then divides into two or more branches that vascularize the gland. Branches from adjacent caudal segmental arteries may also feed the gland while other branches feed the lateral caudae and levator caudae muscles.[6] There may be anastomoses between the branches of the uropygial gland along the dorsal surface.[6] The gland lacks striated muscle of its own, and movement of the gland is through movement of the tail or rectrices.[1] The glandular uropygialis nerve arises from the caudalis dorsalis nerve that also innervates the skeletal muscles of the tail.[1]

Microscopic anatomy

The gland is covered by a dense external connective tissue capsule comprised of collagenous, non-elastic fibers and may contain some smooth muscle fibers.[1,6] The interlobar septum is comprised of similar connective tissue but may also possess elastic fibers.[1,6] The septum is perforated in some avian species (e.g., many passerine species).[1] The capsule can be deeply pigmented with melanin granules in some psittacine species.[1]

The inner aspect of the gland is comprised of glandular tubules where secretions are generated and various cavities that store the secretions.[1] The presence of storage cavities is highly variable, even in psittacine species.[1,8] For example, a larger central primary cavity is present in the sulfur-crested cockatoo (*Cacatua sulphurea*) but not in the cockatiel (*Nymphicus hollandicus*).[1] Although many avian species with a compact-type papilla possess smooth muscle layers in the papilla, the cockatiel was the only psittacine bird in which discrete muscle fibers were detected in the papilla in one report.[1]

Secretions

The uropygial gland is a compound tubuloalveolar holocrine gland with a lipoid sebaceous secretion containing cell fragments and secretory granules.[1,3,8] Chemical substances found in the secretions include lipids, fatty acids (both volatile and non-volatile), aliphatic monoester waxes, and long-chain monohydroxy wax-alcohols. Secretions also contain 7-dehydrocholesterol, a vitamin D_3 precursor.[1,9]

Flora

The normal flora of the uropygial gland has been investigated in several avian species but no psittacine species.[10] For example, Enterococcaceae, Clostridiales Family X1, Clostridiaceae 1, Comamonadaceae, Enterobacteriaceae, and Micrococcaceae were identified in the preen glands of free-living house sparrows (*Passer domesticus*) in one report.[10]

Function

It is widely accepted that the uropygial gland is involved with maintaining feather condition in those species that possess them; however, the understanding of the full functions of the gland is still in its infancy.[2,8,11,12] Postulated functions of the uropygial gland include those listed below.

Waterproofing and in-flight drag reduction

A waterproofing function of the uropygial gland is well-supported in the literature but the exact mechanism remains uncertain.[2,12,13] Postulated mechanisms include direct waterproofing by providing plumage with a covering layer of lipids and indirect waterproofing through the general maintenance of feather structure.[2]

Antimicrobiosis

Preen oil contains compounds that inhibit bacterial and fungal growth and has been shown to reduce growth of keratinase-containing, feather-degrading bacteria such as *Bacillus licheniformis*.[2,11,12,14]

Intraspecific communication and predator deterrence

Preen gland oil shows some level of hormonal control and exhibits individual, sex, and age-specific patterns. Support for the hypothesis that the uropygial gland produces pheromones is mixed.[2,12,15] An intact olfactory system appears to be critical to courtship and mating in mallard ducks (*Anas platyrhynchos*).[2]

Feather coloration

Preen gland oil likely does not significantly alter coloration or ultraviolet light reflectance in birds but the evidence is mixed.[2] For example, preen gland oil in the greater flamingo (*Phoenicopterus roseus*) contains carotenoids which make the pink plumage more intense.[2]

Vitamin D_3 synthesis

Preen gland oil contains 7-dehydrocholesterol, a vitamin D_3 precursor; however, evidence of a definitive role in vitamin D_3 synthesis through exposure to ultraviolet light is dated and mixed. The presence of the gland was not essential for calcium metabolism altered by rickets in birds.[9,12]

Clinical Evaluation

Physical examination

The uropygial gland should always be visually inspected and palpated during the physical examination in species that possess one.[7,16] A diseased preen gland is easy to miss during the exam. The gland is comparatively large in grey parrots (*Psittacus erithacus*) and is often larger in juvenile birds compared to adults.[17] This large size may be mistaken for impaction or infection.[17] Routine expression of the gland during the exam can be painful and injurious and is generally not advised.

Evidence for uropygial gland disease includes signs of localized pain or discomfort during autopreening or the physical examination; excessive enlargement or cystic protrusion of one or both lobes; asymmetry of the lobes; crust, ulcerated skin, or masses on or around the gland; feather loss or damage on or around the gland; automutilation of the gland and/or surrounding tissues; presence of blood on the surrounding feathers or beak, and abnormal (e.g., hemorrhagic) secretions from the duct(s).[18] Many birds with preen gland disease do not display overt signs of pain or discomfort and disease.

Clinical pathology and imaging

While routine diagnostic testing such as hematology and clinical biochemistry are generally not specifically diagnostic for diseases of the uropygial gland, clinical pathology evaluation is important in assessing the overall health of the bird, particularly in the presurgical patient.

The diagnostic strength of routine radiography is generally considered to be low for preen gland diseases although the uropygial gland can often be observed on the lateral-lateral radiographic projection. Computed tomography can be useful to evaluate the uropygial gland and evaluate for evidence of distant metastatic disease in cases where neoplasia is confirmed or suspected.[4]

Microbiologic evaluation

The microbiologic flora of secretions from the uropygial gland and excised tissue specimens can be evaluated by cytology, Gram stain and other biological stains, culture and sensitivity, PCR, and next generation sequencing.

Cytology and histopathology

Cytologic evaluation can be performed on uropygial gland secretions, impression smears from ulcerated masses, and fine needle aspirates of solid masses. Masses within or around the uropygial gland can be carefully biopsied and tissue samples submitted for histopathology. Caution is warranted when collecting biopsy samples that involve the papilla. Excised masses and in toto uropygial glands should also be evaluated by histopathology.[7,20] Masses can be biopsied prior to more aggressive surgery such as mass excision or uropygialectomy. A small dermal biopsy punch can be used for sample collection.[7]

Diseases

Disease of the uropygial gland is not commonly diagnosed in companion birds, possibly in part due to the difficulty in visualizing the gland without handling and restraint and failure to inspect the gland during the routine physical examination. The most recognized clinical conditions of the preen gland are impaction and neoplasia.

Impaction

Impaction is often a sequela to squamous metaplasia and hyperkeratosis associated hypovitaminosis A and can also be secondary to uropygial gland adenitis.[4,7,20] Impaction of the uropygial gland can be unilateral or bilateral. Clinical signs include excessive enlargement of one or both lobes, cystic protrusions, evidence of localized pain or discomfort, and inability to manually express the gland. Caution is warranted with attempts to express the gland as the gland may rupture internally or externally with excessive force. In severe cases of impaction, the gland may rupture spontaneously[7] but this author has found this to be uncommon.

In one report, a king penguin (*Aptenodytes patagonicus*) had uropygial impaction and abscessation and died 6 months later with bacterial septicemia and myiasis of the preen gland.[4] In the same report, 21 gentoo penguins (*Pygoscelis papua*) had uropygial gland disease (all impactions) with an annual incidence from 2.6 to 11.5%. No uropygial gland disease was seen in macaroni penguins (*Eudyptes chrysolophus*) or rockhopper penguins (*Eudyptes chrysocome*).[4]

Neoplasia

The most reported neoplasms of the uropygial gland are squamous cell carcinoma and adenocarcinoma,[16,18,20] although other neoplasms have been reported including adenoma,[20] epithelioma,[19] and melanoma.[21,22] Squamous cell carcinoma often appears as areas of crust or ulceration and may progress to solid masses.[18] Neoplastic lesions generally advance or enlarge over time. Deeper tissues are frequently involved with more aggressive neoplasms.[18] Histopathology is necessary for a definitive diagnosis. Distant metastasis has been reported.[21]

Infection

Viral (e.g., poxvirus[4]), bacterial, and mycotic (e.g., *Candida albicans*[23]) causes of uropygial gland adenitis have been reported. Infected glands are often enlarged or thickened and can be abscessed.[4,16] Thick, pasty, fetid, or hemorrhagic glandular secretions may be present.[23]

Treatment

Conservative management

In most cases of simple impaction, conservative management is the preferred option. For birds that are not displaying signs of localized pain or discomfort, benign neglect with periodic monitoring is a reasonable option. Nutritional deficiencies (e.g., hypovitaminosis A) should be corrected. Warm compresses can be applied periodically to the gland if this is possible.[4,7] Careful attempts at manually expressing the gland can be attempted but caution is warranted. Applying direct pressure to the gland can be painful and risks rupturing the gland. Recurrence is common after manual expression.

Non-surgical treatment of impaction and infection

Cannulation (catheterization)

Cannulation of the uropygial gland duct is possible in larger birds.[24] Cannulation should be performed using appropriate local and/or systemic analgesia and sedation or general anesthesia.[24] Flushing with warm sterile saline by catheterization was ineffective in all cases of impaction in gentoo penguins in which it was attempted.[4]

Antimicrobial therapy

In cases of uropygial gland infection or abscessation, antibiotics and/or antifungal agents can be used. The selection of suitable antibiotics ideally is based on results of culture and sensitivity testing. Antibiotics are unlikely to resolve impactions when used alone.

Analgesic therapy

Appropriate analgesic therapy should be used in cases with evidence of localized pain or birds with conditions of the preen gland likely to be painful.

Non-surgical treatment of neoplasia

Cryotherapy

Cryotherapy can be attempted but it can be difficult to freeze tumors completely and recurrence is likely.[18] Injury to surrounding tissues is challenging to avoid. Cryotherapy of the uropygial gland is not advised.

Radiation therapy

The successful use of Strontium-90 (S-90) β-ray plesiotherapy for the treatment of uropygial gland neoplasia has been reported in birds including cockatiels and budgerigars (*Melopsittacus undulatus*).[20,25] The depth of penetration is shallow (often cited as 5 millimeters or less)[18,25] and so larger masses should be debulked prior to treatment. The gland may need to be divided into sections for treatment in larger birds. Single and repeated fractions of 100 Gray (Gy) have been reported.[20,25]

Chemotherapy, electrochemotherapy

There are no reports of the use of systemic chemotherapy alone for the treatment of uropygial gland neoplasia in birds. The successful use of electrochemotherapy has been reported in birds.[26,27] Electrochemotherapy consists of intratumoral application of a chemotherapeutic agent such as cisplastin, carboplatin, or bleomycin followed by electroporation through localized treatment with electrical pulses with either plate or needle electrodes with the patient under general anesthesia. Two or more treatment sessions are often needed for a successful outcome.

Immunotherapy

Immunotherapy is a treatment for canine melanomas that utilizes the human tyrosinase enzyme to stimulate the host immune system against tyrosinase enzyme-producing cells.[23] Immunotherapy with a canine melanoma vaccine (Oncept, Merial, Duluth, GA, USA) was reported in a black-footed penguin (*Spheniscus demersus*) with melanoma of the uropygial gland.[23] The penguin was euthanized after the second dose and had extensive metastatic disease.[23]

Surgical treatment

Surgery should be considered for cases of uropygial gland neoplasia and for impaction, infection, or abscessation if the impaction recurs, if there is a chronic non-responsive infection, or if the gland has ruptured. Surgical options for impaction and infection include surgical debridement and partial or complete uropygialectomy. Smaller isolated neoplastic lesions can often be completely excised with primary closure or healing by second intention.[16,18] Larger, more aggressive masses and those involving the papilla may require a partial or complete uropygialectomy. Very large or aggressive masses may require amputation of the tail. Potential surgical complications include hemorrhage, incision dehiscence, infection, incomplete surgical margins, and vascular metastasis.[20,27] Birds should be monitored postoperatively for picking at the surgical site and a neck collar or brace should be considered if this occurs.

Surgical treatment options

Surgical debridement

An incision (or incisions) is made directly over the affected lobe(s) and the contents of the primary cavity are expressed or otherwise drained and the lumen is irrigated with warm sterile saline.[28] The incision(s) can be left open or closed routinely. Attempts can be made to flush and/or cannulate the duct(s) intraoperatively.[28] In gentoo penguins this approach resulted in fewer repeat episodes of impaction and was the recommended treatment for uropygial gland impactions in captive penguins.[4]

Complete uropygialectomy

Complete excision (extirpation) is indicated for advanced cases of impaction or infection that have not responded to medical management and with large and aggressive neoplasms.[19,27] Complete excision can be challenging and significant hemorrhage is possible. Radiosurgical equipment or a surgical (CO_2) laser,[30] should be considered for proper hemostasis. Magnification using surgical loupes or an operating microscope and proper illumination are strongly advised as the borders of the gland can be difficult to identify, particularly with chronic or advanced disease (Nemetz L, personal communication, March 2021). Local anesthesia with lidocaine or bupivacaine is recommended. The patient is placed in lateral or semi-ventral recumbency. A towel can be placed under the tail and the rectrices can be securely taped. A circumferential skin incision is made around the gland.[17,31] Or, alternatively, the skin over the dorsal aspect of the gland can be incised transversely and carefully dissected free from the underlying gland and reflected cranially. This technique may improve the likelihood of successful primary closure (Nemetz L, personal communication, April 2023). The soft tissues underlying the gland are then carefully undermined.[31] The gland has deep soft tissue facial attachments to the rectrice follicles and caudal vertebrae. Care must be taken not to damage the blood supply to the rectrice follicles.[17] Closure is routine.[17,31] If the remaining defect is too large to close by primary closure it can be allowed to heal by second intention. A hydroactive or hydrocolloid dressing can be used. Although a drain can be placed postoperatively,[17] drains are generally not necessary and may result in irritation or inflammation.[17]

Partial uropygialectomy

Partial uropygialectomy with physical removal or cauterization of remaining glandular tissue of one or both lobes can be considered for severe or persistent impactions or infections unresponsive to conservative treatment (Nemetz L, personal communication, March 2021). Partial excision avoids the risks of serious hemorrhage and damage to surrounding important tissues such as the rectrice follicles.[32,33] Partial excision is not recommended for large and aggressive neoplasms (Nemetz, personal communication, March 2021). As with complete excision, magnification and illumination are strongly advised and local anesthesia is recommended. For a unilateral approach, the dorsal wall and sides of the affected lobe are excised. For a bilateral approach, a circumferential skin incision is made that incorporates the papilla and the dorsal and lateral walls of both lobes. The remaining contents of the affected lobe(s) are evacuated. Deeper biopsy samples can be collected at this time. The remaining glandular walls are then manually debrided using a bone curette or similar device. The walls can be chemically cauterized using silver nitrate applicator sticks. Caution is warranted with the use of silver nitrate and the area should be copiously lavaged after application. Hemorrhage is controlled with direct pressure and/or chemical or thermal cauterization. The incision is left open and the area is packed with antibiotic ointment or cream and allowed to heal by second intention. Postoperative antibiotics and analgesics are recommended. Repeated treatment sessions are often required for complete resolution, particularly if removal or cauterization of remaining glandular tissue is unadequate. If repeated treatments are unsuccessful, complete excision can be then performed. The gland is often significantly smaller after partial excision and easier to fully excise (Nemetz L, personal communication, March 2021).

Tail amputation

Very large or aggressive neoplasms may require amputation of the tail. In one report of a 33-year-old male blue and gold macaw (*Ara ararauna*) with uropygial gland adenocarcinoma, a tail amputation consisting of transection of musculature to the vertebral column, disarticulation between the second and third caudal vertebrae, and routine closure was performed but the bird died of an undetermined cause approximately 4 months after surgery.[29]

Acknowledgments: The author extends deep appreciation and gratitude to Larry Nemetz, MS, DVM for advisement on clinical and surgical management of uropygial gland disease and for contributions to this manuscript.

Disclosure Statement: The author has no financial or commercial conflicts or interests to disclose.

References

1. Jacob J, Ziswiler V. The uropygial gland. In: Farner DS, King JR, Parkes KC, eds. *Avian Biology*. New York: Academic Press; 1982:199-324.

2. Johnston DW. Morphological atlas of the avian uropygial gland. *Bull Br Mus Nat Hist (Zool)*. 1988;54(5): 199-259.

3. Baumel JJ, Raikow RJ. Integumentum speciale. In: Baumel JJ, Kings AS, Breazile JE, et al, eds. *Handbook of Avian Anatomy: Nomina Anatomica Avium*. 2nd Ed. Cambridge, MA: Nuttall Ornithological Club; 1993:22-44.

4. Keeble EJ. The incidence and treatment of uropygial gland disease in a mixed species exhibit of captive penguins. *Proc Int Conf Dis Zoo Wild Anim*. 2007.

5. Chiale MC, Carril J, Montalti D, et al. The uropygial gland of the eared dove and its evolutionary history within the Columbiformes (Aves). *J Ornithol*. 2019;160:1171-1181.

6. Atalgin H, Kürtül İ. Arterial vascularization of the uropygial glands (gl. uropygialis) in the Japanese quail (*Coturnix coturnix japonica*) and silver polish (*Gallus gallus domesticus*). *Anat Histol Embryol*. 2008;37(3):177-180.

7. Abou-Zahr T. Avian dermatology. *Vet Clin North Am Exot Anim Pract*. 2023;26(2):327-346.

8. Carril J, Chiale MC, Barbeito CG. The uropygial gland of the monk parakeet *Myiopsitta monachus*: Histology, morphogenesis, and evolution within Psittaciformes (Aves). *Evol Dev*. 2020;22(3):269-282.

9. Salibian A, Montalti D. Physiological and biochemical aspects of the avian uropygial gland. *Braz J Biol*. 2009; 69: 437-446.

10. Grieves LA, Bottini CL, Gloor GB, et al. Uropygial gland microbiota differ between free-living and captive songbirds. *Sci Rep.* 2022;12(1):18283.

11. Fülöp A, Czirják GÁ, Pap PL, et al. Feather-degrading bacteria, uropygial gland size and feather quality in house sparrows *Passer domesticus*. *Ibis.* 2016;158(2):362-370.

12. Moreno-Rueda G. Preen oil and bird fitness: a critical review of the evidence. *Biol Rev.* 2017;92(4): 2131-2143.

13. Møller AP, Laursen K. Function of the uropygial gland in eiders (*Somateria mollissima*). *Avian Res.* 2019;10(1):1-6.

14. Bodawatta KH, Schierbech SK, Petersen NR, et al. Great tit (*Parus major*) uropygial gland microbiomes and their potential defensive roles. *Front Microbiol.* 2020;28;11:1735.

15. Galván I, Møller AP. Odor transmission and olfaction: the tuft of the uropygial gland and olfactory ability in birds. *The Condor.* 2013;115(4):693-699.

16. Howard L, Lamberski N, Newton AL. Uropygial gland inflammation, neoplasia and surgical resection in six northern carmine bee-eaters (*Merops nubicus nubicus*). *Proc Annu Conf Am Assoc Zoo Vet.* 2006.

17. Wissman MA. Diseases of the uropygial gland. Available at: http://www.exoticpetvet.net/avian/uropygial.html. Accessed April 9, 2023.

18. Quesenberry K. Neoplastic diseases in birds - treatment options. *Proc Conf Pacific Vet.* 2018.

19. Ambar N, Eshar D, Njaa BL. Intraluminal uropygial gland epithelioma in an African grey parrot (*Psittacus erithacus*). *J Avian Med Surg.* 2022;35(4):433-439.

20. Nemetz LP, Broome M. Strontium-90 therapy for uropygial neoplasia. *Proc Annu Conf Assoc Avian Vet.* 2004.

21. Reavill D. Pet bird oncology. *Proc Conf Assoc Avian Vet Australasian Comm Unusual Exot Pet.* 2011.

22. Steeil J, Schumacher J, Ramsay EC, et al. Application and limitation of advanced diagnostic and therapeutic modalities in zoologic species with malignant melanoma. *Proc Annu Conf Am Assoc Zoo Vet.* 2013.

23. Hines RS, Sharkey P, Friday RB. Itraconazole treatment of pulmonary, ocular, and uropygeal aspergillosis and candidiasis in birds – data from five clinical cases and controls. *Proc Annu Conf Am Assoc Zoo Vet.* 1990;364-370.

24. Faux CM, Logsdon ML. Impacted uropygial (preen) gland. In: Orsini JA, Grenager NS, de Lahunta A, eds. *Comparative Veterinary Anatomy: A Clinical Approach.* London, UK: Elsevier; 2022.

25. Pignon C, Azuma C, Mayer J, et al. Radiation therapy of uropygial gland carcinoma in psittacine species. *Proc Annu Conf Assoc Avian Vet.* 2011.

26. Racnik J, Svara T, Zadravec M, et al. Electrochemotherapy with bleomycin in the treatment of squamous cell carcinoma of the uropygial gland in a cockatiel (*Nymphicus hollandicus*). *J Exot Pet Med.* 2019;29: 217-221.

27. Hirata SB, Brunner CHM, Santos SV. Electrochemotherapy for the treatment of fibroma in red-billed curassow (*Crax blumenbachii*). *Proc Conf Am Assoc Zoo Vet, Eur Assoc Zoo Wildl Vet, Int Zoo Wildl Vet.* 2016.

28. Hochleithner M, Hochleithner C, Fuchs A. Surgical treatment of impacted uropygial glands. *ExoticDVM.* 1999;1(1):23-25.

29. Robertson J, Guzman DS, Sinnott D, et al. Modified tail amputation technique in a blue and gold macaw (*Ara ararauna*) with uropygial gland adenocarcinoma. *J Avian Med Surg.* 2020;34(1):57-64.

30. Van Sant F, Kureshi M. Successful use of a surgical (CO_2) laser in squamous cell carcinoma in companion psittacines. *Proc Symp Southwest Vet.* 2019.

31. Guzman DS. Avian soft tissue surgery. *Vet Clin North Am Exot Anim Pract.* 2016;19(1):133-157.

32. Jawad HS, Idris LH, Bakar ZB, et al. Partial ablation of uropygial gland effect on carcass characteristics of Akar Putra chicken. *Poult Sci.* 2016;95(8):1966-1971.

33. Hasan SAJ. Uropygialectomy method overview. *Dairy Vet Sci J.* 2017;2(4):555592.

Section 8

AAV Masterclasses Track 4

Brian Speer, DVM, DABVP (Avian Practice), DECZM (Avian);
Anthony Pilny, DVM, DABVP (Avian Practice)
Session Chairs

Avian Bornavirus: A Global Perspective on Pathology, Current Diagnostics, Research and Treatment

Sharman Hoppes, DVM, DABVP (Avian Practice),
Michael Lierz, DZooMed, DipECZM (Wildlife Population Health), DipECPVS,
Helene Pendl, Dr med vet

Session #3173

Affiliation: From Texas Avian & Exotic Hospital, Grapevine, TX 76051, USA, Texas A&M University, College Station, TX 77843, USA (Hoppes), Clinic for Birds, Reptiles, Amphibians and Fish, Justus Liebig University, Giessen, Germany (Lierz), Pendl Lab, Zug, Switzerland (Pendl).

Abstract: Avian bornavirus (ABV) is a neurotropic virus affecting parrots and other species of birds worldwide including North America, South America, Europe, South Africa, Asia and Australia. ABV is the cause of proventricular dilation disease also known as avian ganglioneuritis. The diagnosis, treatment and management of this disease both in the individual pet bird and in flocks has been a source of frustration to veterinarians, aviculturists and owners for decades. This masterclass is an international collaboration discussing pathology, immunology, current research and treatment for birds exposed and infected with avian bornavirus. Current research on transmission and susceptivity is reviewed. Treatment options are discussed along with management and prevention protocols.

Introduction

Proventricular dilatation disease (PDD) or avian ganglioneuritis (AG) was first recognized in the late 1970s in macaws imported into North America and Europe.[1-12] Initially diagnosed in macaws (*Ara* spp.) it was later identified in other parrot species. It is also known as Macaw wasting disease, neuropathic ganglioneuritis, lymphoplasmacytic ganglioneuritis, infiltrative splanchnic neuropathy, and psittacine encephalomyelitis. The nomenclature of proventricular dilatation disease (PDD) and avian bornavirus (ABV) has evolved over the years. The recommended names now are psittacine bornavirus (PaBV) for infected psittacine birds and either avian ganglioneuritis or avian bornaviral ganglioneurits for the clinically diseased state or PDD bird. This presentation uses avian ganglioneuritis (AG). AG better defines the disease process and takes the focus off the proventriculus as other organ systems can be affected.

The disease has been identified in over 80 species of psittacine and non-psittacine birds. In parrots, African grey parrots (*Psittacus erithacus*), macaws (*Ara* species), Amazon parrots (*Amazona* species), and cockatoos (*Cacatua* species) are the most common species affected.[3] The disease is minimally represented in Quaker parrots (*Myiopsitta monachus*) and lovebird species (*Agapornis* species).[3] It has been reported in captive psittacine birds worldwide. AG/PDD has historically been considered an infectious disease based on its spread through aviaries. But the etiologic agent proved elusive for over thirty years. Then in 2008, two different research groups Honkavuori and colleagues and Kistler and colleagues detected a novel bornavirus in parrots with PDD.[13,14] During this same time, Ouyang and colleagues detected ABV Nucleoprotein (N-protein) in the brains of birds with AG/PDD.[15]

Bornaviruses

The bornaviruses are neurotropic, negative-encoded, single-stranded, nonsegmented RNA viruses and are members of the family Bornaviridae of the order Mononegavirales. The replication of and transcription of bornaviruses takes place in the cell nuclei of nerve cells. Currently, the family comprises the three genera Orthobornavirus, Carbovirus and Cultervirus. Due to the discovery of various new bornaviruses during the past 15 years, the taxonomy and nomenclature of the family has undergone several revisions.[16-18] Borna disease virus (BDV) causes encephalitis in horse, sheep, and other domestic mammals in Europe.[3] BDV has also been detected in wild mallards and corvids in Scandinavia, and ostriches in Israel.[4,5,19] Experimental infections of chickens showed that 1-day-old chicks inoculated intracerebrally with brain homogenates from rabbits with Borna disease virus developed paralysis of legs and wings.[20] There are currently two BDV genotypes. BoDV-1 infects mammals: horses, sheep and the white toothed shrew (its natural reservoir). BoVD-2 was isolated from a horse.[21] A new mammalian bornavirus, variegated squirrel bornavirus 1 (VSBV-1) has been associated with a fatal encephalitis in humans.[22]

Avian Bornavirus

Currently there are 15 genetically diverse Bornavirus genotypes comprising six distinct viral species identified within the genus Orthobornavirus, which can occur in avian species. Eight genotypes are found in parrot populations worldwide that share 91 to 100% of their nucleotide sequences within genotypes and 68 to 85% between genotypes.[17] The two species known to cause AG/PDD in psittacine birds are psittaciform bornavirus 1 and 2 and include PaBV genotypes 1-8. PaBV-2 and PaBV-4 are the most reported genotypes in parrots.[17]

It has been speculated that the diversity of bornavirus genotypes in captive parrots compared to free-living waterfowl is the result of the spread of bornaviruses across species during the unregulated housing and transport associated with wild captures decades ago.[23] Genotypes for bornaviruses in parrots do not appear to be clearly linked to geographic regions unlike bornaviral genotypes identified in waterfowl.[24] Avian bornavirus does not appear to be zoonotic.

Other known ABV's includes passeriform bornavirus 1 (CNBV-1,2,3 canary) and (MuBV-1 munia finch), passeriform bornavirus-2 (EsBV-1, estrildid finch bornavirus 1) and waterbird bornavirus 1 (ABBV-1, aquatic bird bornavirus-1).[17]

Epidemiology of Avian Bornavirus

Avian bornavirus infections have been described in a wide range of families within the order Psittaciformes. It is assumed that all psittacine species are susceptible to some degree of infection. The viruses can be found in captive psittacine populations worldwide.[1-12] This information on the distribution of ABV is based on sampling of birds in aviaries, veterinary clinics, and information gained by pathologists. Few actual prevalence studies have been published. In the most extensive study, Heffels-Redmann et al.[25] found 23% of 1442 captive psittacine birds from several European countries were positive for markers of avian bornavirus infection (viral RNA detection and/or bornavirus-reactive antibodies), suggesting a wide distribution of these viruses.[25] In captive psittacine birds, PaBV-2 and PaBV-4 are the most commonly reported genotypes.[3,16,21] In contrast, all other known psittacine bornaviruses (PaBV-1, PaBV-3 and PaBV-5 to PaBV-8), have been detected in less than ten birds each.[24]

Mixed infections with two psittacine bornaviruses, such as PaBV-2 and PaBV-4, have been described rarely. Very little data is available on avian bornaviruses in free-ranging psittacine populations, but it has been reported to be present in South America and Australia.[26,27]

Transmission

Transmission of avian bornaviruses has only been studied for parrot bornaviruses and canary bornavirus. Direct transmission between birds does not occur easily. The virus is shed by urine and feces, but infection via oral and nasal route were not successful in an experimental study compared to parenteral application routes which consistently induce infection. This is supported by proof of horizontal transmission via wounds.[28] Vertical transmission seems to play a major role in virus epidemiology, but the influencing factors are not clear. It seems that carrier birds infected at a young age transmit the virus more easily via vertical transmission compared to birds infected as adults.[29] In summary, introducing the virus into collections is more likely via carrier birds rather than by indirect introduction through contaminated accessories, feed or humans. Introduction of untested birds and especially the release of untested birds back into wild populations in a species conservation program should be avoided.

Pathology and Pathophysiology of Avian Borna Virus Related Disease

The non-cytopathic nature of Bornaviruses

Bornaviruses cause a persistent noncytolytic infection. Persistence is achieved by tight regulation of viral replication involving the X, N, and P proteins. A central role seems to be played by the X protein, which inhibits cellular apoptosis and neuronal degeneration via protective effects on mitochondria. This is also likely to be true for avian bornaviruses, as they are detected in many tissues without concurrent signs of tissue damage.[16]

Wounds as a possible entry point of transmission

Traumatic skin wound transmission has recently been proposed for PaBV-2,[30] PaBV-4[28] in parrots and variegated squirrel bornavirus-1 (VSBV-1) in humans. The presence of VSBV-1 in the oral cavity of infected variegated squirrels further supports the bite wound transmission hypothesis and prompts comparisons with rabies virus (RABV), another neurotropic negative-sense single-stranded RNA virus with a similar mode of spread to bornaviruses.[30]

Non-suppurative meningoencephalomyelitis and ganglioneuritis as the key lesion

Avian Borna virus related disease is based on non-suppurative inflammatory lesions in the central and peripheral nervous system. In contrast to bornavirus disease (BVD) in mammals and humans, the enteric nervous system is particularly affected. This is especially true in parrots, where the disease was first described, and has led to the misleading initial term of proventricular dilation disease (PDD). Although inflammatory lesions of nervous tissue have been demonstrated in correlation with canary and aquatic bornaviruses, psittacine borna viruses, namely PaBV-2 and PaBV-4, currently remain the most clinically relevant avian borna viruses.[16] Several successful experimental infections fulfilled Koch's postulates and confirmed PaBVs as the causative agent of PDD in psittacine birds.[31-38] Moreover, it appears that different parrot bornaviruses (PaBVs) cause differing clinical patterns with predominant neurologic or predominant gastrointestinal manifestations.[34] It remains a subject of intense investigation, how these non-cytopathic viruses induce clinical disease and which factors contribute to the high variability of clinical onset and outcome.

T-cell mediated immunopathogenesis

Based on the histopathologic findings of a mononuclear inflammation of nervous tissues, immune mediated mechanisms have long been suspected to contribute to the pathogenesis of AG/PDD. It seems to be mainly mediated by T-cells, as in BoDV-1 - onset of clinical neurologic signs is independent from virus load and distribution but is correlated to the presence of CD3+ T cells in the brain.[21]

For BoDV-1 infections, it is known that this cell-mediated response is driven mainly by CD4+ and CD8+ T cell subsets with CD8+ cells representing the effector cells targeting the viral nucleoprotein. Experimental studies in parrots suggest a similar pathogenesis for PaBV related disease, but parrot specific or cross-reactive tools for further immunophenotyping are lacking.[16,39,40]

Immunohistochemical investigations in brain sections of cockatiels experimentally infected with PaBV-2 confirm a predominance of T lymphocytes (CD3+) with lesser amounts of macrophages/microglia (IbA1+) and very few B-cells (PAX5+) in perivascular cuffs,[39] which is in alignment with immunophenotypic patterns seen BoDV-1 infected rodents/horses and VSBV-1 infected humans.[22,41,42]

In the gastrointestinal tract, inflammatory infiltrates were more frequent in the proventricular ganglia than in the small intestine, again consisting mainly of T lymphocytes, with lower levels of macrophages and B lymphocytes. Small intestinal ganglia showed a milder infiltration, consisting of a more uniform mixture of the three cell types.[39]

The route of PaBV-2 spread and inflammation

Recently, time-based studies in cockatiels (*Nymphicus hollandicus*) provided important insights into the spread of the virus and the development of lesions after intramuscular inoculation of parrot PaBV-2.[30,43] Based on the first time point of PaBV N-protein detection by IHC in histologic sections, viral spread in days post infection (dpi) could be characterized as an initial retrograde (centripetal) axonal migration from the injection site to the CNS (spinal cord 20 dpi, brain 25 dpi), followed by an anterograde (centrifugal) axonal migration to peripheral neural tissues (adrenal glands and ganglia in the gastrointestinal tract 30 dpi, epicardial ganglia, kidney 35 dpi) and, with a clear delay, to non-neural tissues (skin 80 dpi).

Mononuclear inflammatory lesions showed an identical time course and spread, appearing earliest in the brachial plexus at 20 dpi, in the thoracic and cervical spinal cord at 20-25 dpi (then spreading evenly to more caudal parts of the cord) and reaching the brain at 25 dpi, where the lesions remained confined to the central ventral parts of the brain until 40 dpi. The most severe lesions were found in the thalamic nuclei and midbrain, with the optic tectum involved at 35 dpi. The cerebrum, cerebellum, and the meninges showed no signs of inflammation before 40 dpi and were only mildly affected.

Consequently, defining the anatomic location of the inflammatory lesion in the central nervous system may allow conclusions to be drawn about the duration of the infection. Furthermore, there is a good chance that cerebrum and cerebellum (the most common samples taken from the brain) will be (falsely) negative, especially in the early stages of clinical disease, which can occur as early as 35 dpi. A similar timeline of virus spread has been discussed for embryonic brain tissue of cockatiels (*Nymphicus hollandicus*) experimentally infected with PaBV-4 and PaBV-2.[44]

Retrograde axonal transport to the brain has also been demonstrated for the spread of Borna disease virus 1 (BoDV-1) in experimentally infected rodents via nerve endings of the nasopharyngeal mucosa and the olfactory bulb and for Rabies virus (RABV) via motor neurons and spinal cord to the brainstem and limbic system.[30]

Clinical presentation from the perspective of pathology and pathophysiology

Factors affecting clinical presentation

Given that immunopathogenesis requires a functional immune system, it is reasonable to assume that any stressor that alters the host's immune competence will influence clinical onset, severity, duration and outcome. Examples of such stressors include age, genetic background, and immunosuppression due to co-infection with other pathogens or during certain life stages such as migration or breeding. In addition, viral dose and strain, route of infection, and degree of virus-host adaptation are likely to influence disease development. This explains, at least in part, the high variability in incubation period, clinical presentation, course, and outcome of parrot bornavirus infections.[16]

CNS lesions

Depending on the location in the CNS affected, different neurologic symptoms may be observed clinically. Cerebral perivascular cuffing and glial cell injury may cause seizures; optic lobe lesions result in cortical blindness; disruption of cerebellar cell layers results in fine motor deficits such as ataxia, intention tremor, incoordination, and proprioceptive abnormalities. In addition, cognitive deficits, and abnormalities in vocalization may be noted.[45]

Vagus nerve innervated organs as target tissues

Peripheral tissues mainly affected in histologic sections of AG (i.e., the anterior gut, the heart, the adrenal gland, and the kidneys) correlate with the innervation areas of the 10th cranial nerve (*N. vagus*) suggesting a pivotal role of this nerve for anterograde axonal migration of the virus from the brain to the periphery.[21]

Proventricular dilation

Comparison of the pathophysiology of gastroparetic disease events in humans and laboratory rodents, as well as proventricular dilations of other etiologies in birds, suggests that the dilation of the proventriculus (and sometimes other parts of the upper digestive tract) in AG is primarily due to a loss of nitrergic neurons in the autonomic enteric nervous system (ENS) and the vagus nerve plexus.[21] A simple causal relationship between the lymphocytic infiltrations and the destruction of nitrergic neurons is unlikely, because ganglioneuritis affects all neurons and not selectively nitrergic neurons. Rather, anatomic and neurophysiologic findings in laboratory rodents and poultry suggest that a complex interplay between direct inflammatory damage and indirect impairment of neuroimmunologic and neuroendocrine reflex arcs within the ENS and the gut-brain axis (communicating between the ENS and CNS) leads to the disruption of peristalsis and impairment of mucosal secretion and absorption. Together, these imbalances result in a net excess of cholinergic (stimulating, muscle-contracting) over nitrergic (inhibiting, muscle-relaxing) stimuli in the gastric muscle wall, leading to its permanent contraction and functional obstruction of the isthmus. The upstream proventriculus expands as a result of the downstream obstruction. This expansion is a reversible process in physiologic peristalsis, mediated by vagal pathways and mechanoreceptors in the gastric wall. With this reflex relaxation, the proventriculus regulates food flow and pressure in the downstream ventriculus and duodenum. If the gastro-duodenal contraction pattern is disturbed over a long period of time or if there is continuous contraction, a point of no return is passed and chronic (i.e., irreversible) dilation of the proventriculus with complete loss of function develops.[21] Clinical symptoms reflecting impaired proventricular or esophageal function include delayed crop emptying, regurgitation, anorexia, impaired GI transit, and dilatation and impaction of the upper GI tract. Passage of undigested food material in the feces and diarrhea are signs of impaired mucosal secretion and absorption and pancreatic activity (both of which depend on intact vagal innervation), leading to weight loss and starvation.[45]

Sudden death due to cardiac failure

In a retrospective study of the prevalence of cardiac lesions in naturally and experimentally ABV-infected birds with clinical signs of AG, epicardial ganglioneuritis was diagnosed as the most common finding in more than 66% of both groups, followed by myocarditis and endocarditis. PaBV-2 antigen was most frequently detected in the epicardial ganglia by IHC. These results suggest that heart failure caused by PaBV infection may be the underlying pathogenesis of sudden death in psittacine birds with AG.[46]

Blindness

Blindness is a common clinical feature of bornavirus disease in horses, sheep and rodents and has been associated with both central blindness[40] and degeneration of retinal neurons.[30] BoDV-1 spreads from the brain along the second cranial nerve (*N. opticus*) into the retina and causes immune-mediated retinopathy. A similar pathogenesis may be applicable for blindness and retinitis in avian bornavirus disease. Blindness has been reported in the course of bornaviral infections in parrots and canaries with occasional demonstration of mononuclear infiltrates within the retina. ABBV-2 has been detected within the retinas of infected ducks and ABBV-1 in the retinas of gulls.[21]

Feather destructive and auto-mutilative behavior

AG-associated peripheral neuritis has been suggested as a cause of feather picking and self-mutilation, possibly mediated by impaired sensory fiber signaling in the dorsal root ganglia of the spinal cord.[45] To date, this hypothesis lacks sufficient experimental evidence.

Autoimmunity as an alternative hypothesis for pathogenesis

A humoral autoimmune response to brain gangliosides has been proposed as an alternative pathogenesis for AG, based on the hypothesis that avian borna viruses induce the production of anti-ganglioside antibodies.[45,47-49] According to the original experiment, administration of purified gangliosides to cockatiels resulted in the development of clinical signs of AG. Based on these results it was claimed that AG was a suitable animal model for human Guillain-Barré syndrome (GBS) and that serologic testing for anti-ganglioside antibodies was a useful antemortem diagnostic for AG. Further statements suggested the presence of Campylobacter in the gut as a prerequisite for the development of AG and antibody-mediated phagocytosis of axons by intraganglionic and intracerebral macrophages recruited by CD4+ T cells as the cause of demyelination.[45] To date, despite attempts by other groups to substantiate the findings, these hypotheses still lack sufficient experimental evidence.[16,21,50,51]

Nevertheless, autoantibodies against brain antigens can be found in the sera of birds experimentally infected with PaBV-2. Similar to results from human VSBV-1 infection and human achalasia, these antibodies do not appear to play a central role in the pathogenesis and clinical course of AG and their significance in the disease process requires further investigation. In addition, future studies should also investigate the role of possible autoreactive T cells, a direction that has not yet been explored in AG.[51]

Clinical disease and influencing factors

Avian Bornavirus is a neurotropic virus affecting primarily psittacine birds. The symptoms include gastrointestinal signs such as dilatation of the proventriculus (PDD), regurgitation and maldigestion. Central nervous system signs include head shaking, ataxia and tremors, seizures as well as blindness and sudden death. Recently it was demonstrated that PaBV infection was more common in birds with neurologic disorders than in clinically healthy birds. The neurologic birds had higher antibody titers and higher levels of viral shedding.[52] The same study discussed that feather plucking was observed in birds with PaBV-infection at a higher prevalence than controls. Statistically the difference was not significant and feather plucking should be considered multi-factorial and studies determining that PaBV incites feather destructive behavior still need to be performed.

Virus strain appears to cause different disease patterns. and may influence the clinical outcome of infection.[34] Whether this is really linked to the genotype or the virus strain within the same genotype still needs to be evaluated. As described above the status of the immune system seems to play a major role in the clinical outcome of an infection. Therefore it is not surprising that under experimental conditions, cockatiels infected during their first week of life (immature immune system) developed into carriers without clinical symptoms, whereas cockatiels infected as adults developed severe clinical signs.[53]

In non- psittaciform species clinical symptoms are rarely described. ABBV-1 associated clinical signs in waterfowl included esophageal dilation and encephalitis.

Laboratory diagnosis

When testing for avian bornavirus, it needs to be determined if the diagnostic focus is detection of an infected (PaBV positive) bird, or if the goal is to determine if a clinical symptom is caused by avian bornavirus. Healthy infected birds are common. Therefore, PaBV positive birds may present with clinical signs similar to ABV infection that are not caused by ABV infection. Other potential differential diagnoses should be ruled out. Demonstration of the infection is important for introducing birds into a collection, releasing it to the wild, pairing, pre-purchase exam or as a potential diagnosis for a disease.

Ideally, testing a bird for avian bornavirus should include serology and PCR testing. Some birds shed the virus without seroconverting and in birds which seroconverted shedding may be irregular. So, performing both tests provides the highest accuracy. Detection of the virus is usually done by PCR using a pooled choanal and cloacal swab. In deceased birds, the brain or retina is the best material for viral detection. Feces, feather and blood are not ideal sample materials and are not preferable. Not all PCR-protocols detect all genotypes of avian bornaviruses, therefore it should be mentioned to the lab which genotype to focus on. In suspected bornaviral clinical cases with a negative PCR, additional protocols involving other bornavirus genotypes might be necessary. Sampling, storage and sample transport are very important as viral RNA degrades quickly leading to false negative laboratory results. Gold standard for the detection of anti-ABV-antibodies is the immunofluorescence assay.[54] Alternatively, ELISA or Western Blot test are described but seem to have a lower sensitivity, especially in cases with a low titer.

A bird should be considered infected if both PCR and serology provide a positive result. In cases where only the PCR is positive, the bird can be retested 4-6 weeks later, to confirm infection or exclude contamination of the sample. In addition, a seroconversion of the bird might be detected.[55] If the bird seroconverted or the PCR detects bornaviral-RNA repeatedly, the bird should be considered positive. Seropositive birds with a negative PCR should be considered infected if the titer is high. Low titers might become negative in following tests and clearance of an infection cannot be ruled out, although very unlikely.

Histopathology of tissues detects lesions typical for bornaviral disease. Tissue biopsies might be helpful in live birds although crop biopsies have been shown to have a low sensitivity. Therefore, demonstration of infection and ruling out other differential diagnoses is preferable in clinical affected cases. However, it has been shown that high antibody titers often correlate with clinical symptoms and therefore may be more useful for clinical diagnosis.

The detection of antiganglioside antibodies in clinically affected PaBV infected birds has been proposed. This theory may have some merit, but proof for the occurrence of antiganglioside- antibodies in all and only affected birds is lacking, as well as a validation of the tests. As a direct test for clinical disease, further investigations are necessary.

Treatment Protocols

Although currently, AG/PDD is considered an incurable disease, there are treatments to help diminish clinical signs and prolong life. Based on the observed histopathologic lesions, that are inflammatory in nature, nonsteroidal anti-inflammatory drugs (NSAID's) are widely utilized. Anecdotally this treatment appears beneficial. Effectiveness is difficult to measure, however, with no controlled studies. For example, the effectiveness of treatment may depend on the time of onset of treatment and the stage of the disease when initiated. Most clinicians believe the preferential and selective COX-2 inhibitors improve and extend the quality of life of AG/PDD birds. For birds manifesting neurologic deficits, Gabapentin may be helpful.

Drugs used to treat AG/PDD

The main stay of AG/PDD treatment has been the nonsteroidal anti-inflammatory drugs (NSAIDs) that inhibit cyclooxygenase (COX) isozymes. Celecoxib, meloxicam and robinacoxib are commonly utilized to treat AG. Celecoxib is believed to be more selective to COX-2 enzymes than meloxicam and therefore less likely to affect the GI tract and kidney and was the initial NSAID utilized for AG/PDD.[56] Dalhausen and colleagues administered celecoxib at 10 mg/kg PO q24h to treat birds with clinically diagnosed PDD and birds showed marked clinical improvement. Unfortunately, there were no control groups provided so the significance of this report is unclear. Currently, Celecoxib is administered at 20-30 mg/kg PO q12-24h. Anecdotally the COX-2 inhibitors appear to reduce clinical signs of disease and are utilized by veterinarians to treat AG worldwide. Anecdotally most practitioners also report beneficial effects in AG/PDD birds using oral and injectable Meloxicam. The dose for meloxicam for AG/PDD birds is 1-4 mg/kg IM or PO q12-24h.

Robinacoxib is also a COX-2 selective and COX-1 sparing NSAID. It is recommended at 1-2 mg/kg IM q3days or 10 mg/kg IM weekly. After 4 weeks of weekly treatment decrease to 2-10 mg/kg IM monthly.[45,57] Anecdotally, robinacoxib appears to be especially effective in parrots presenting with neurologic clinical signs.

Cyclosporine is a calcineurin inhibitor that selectively suppresses T-cell function that has anecdotally lessened symptoms in birds who presented with end stage AG/PDD who had not responded to either meloxicam or celecoxib (this was prior to the availability of robinacoxib injections). Cyclosporine at 10 mg/kg PO q12h along with meloxicam at 0.5 mg/kg PO q12h was given to severely affected birds who were unresponsive to NSAID alone. The addition of cyclosporine appeared to reduce the clinical signs of disease and prolonged their life. Since cyclosporine is an immunosuppressive drug, the birds were on antibiotics and antifungals prophylactically.[58]

Ribavirin is a ribonucleic analog that stops viral RNA synthesis and readily kills mammalian bornavirus and ABV in tissue culture.[59] Pharmacokinetic studies of ribavirin in birds show that it is excreted rapidly and that appropriate mammalian doses are insufficient to reach and maintain therapeutic levels in birds. Reuter and colleagues[60] demonstrated a synergistic effect with Ribavirin and interferon-a (IFN-a) against PaBV in avian cells, suggesting that a combination of ribavirin with type 1 IFNs may be more effective in vivo. More studies are needed on the use of ribavirin with and without IFNs.

Favipiravir (T-705): In Japan, researchers performed a study looking at favipiravir (T-705) an antiviral similar to ribavirin. Favipiravir was found to be a potent inhibitor of both BoDV-1 and PaBV-4 in vitro.[61] Further studies are needed to determine if T-705 would be effective in vivo.

Gabapentin is a gamma-Amino butyric acid (GABA) agonist. It is used to relieve neuropathic pain in mammals and birds. Although there have been no controlled studies in birds, anecdotally there have been positive clinical responses. The dose of 10-25 mg/kg PO q12h is recommended for central or peripheral nervous system signs such as seizures or ataxia. Gabapentin 50 mg/kg PO has been used in birds that self- mutilate.[45,57]

Adjunctive therapies for birds with PDD include drugs such as cisapride and metoclopramide to improve gastrointestinal tract transit. Antibiotics or antifungals may be necessary for treatment of secondary infections. Essential fatty acids such as flax seed oil (0.1 ml/kg PO daily or Vet Omega 0.22-0.44 ml/kg PO daily) are natural anti-inflammatories and help support kidney function. Reproductive stress has been reported to exacerbate clinical signs, so in some cases GNRH agonists such as Leuprolide acetate or the deslorelin implants have been used to reduce reproductive hormones. Anecdotally, stress has been deemed a factor in the development of clinical signs. Owners should avoid overcrowding of birds, practice good hygiene, and provide nutritionally adequate diets.

Vaccination

RNA viruses lack the proofreading enzymatic mechanisms that are present in DNA viruses, leading to increased mutation rates and increased changes to their genetic code, making a vaccine potentially more challenging. Several studies looking at vaccination options have been performed,[37,38,62] but to date, no ABV/PABV vaccines are commercially available. The studies have employed either recombinant bornavirus N protein or viral vector vaccines expressing bornavirus N and P proteins. The viral vector vaccines have shown promise, but more research and study in the area of vaccines is needed. In particular, the vaccination trials to date did not reach a sterile immunity and infectious virus was still detectable in vaccinated birds. As vaccinated birds seroconvert, PaBV-antibody test will not be able to differentiate between antibodies due to a field infection or due to vaccination.

Control

AG/PDD is an incurable and fatal disease. With limited treatment options and no vaccine available, control must be sought by other methods. Fortunately, ABV is not an environmentally stable virus. It is easily destroyed with soap or dilute bleach and is inactivated by ultra-violet light and desiccation. Bornaviruses can be inactivated with chlorine-containing disinfectants, and research suggests that allowing infected cages to dry thoroughly for 8 and 24 hours reduced the amount of infectious virus by 48% and 86%, respectively. Bornaviruses become rapidly less infective with heat treatment at 56 °C (or 132.8 °F).[63-65]

Controlling the spread of ABV in homes and in aviaries involves good hygiene, testing, and good quarantine procedures. In an aviary setting, the goal of such measures is the eradication of the virus from a flock and the prevention of its reintroduction. An eradication program is based on repeated rT-PCR testing of all birds combined with serology, followed by separation and ultimately removal of all positive birds from the flock. Newly introduced birds from flocks with a known positive or unknown status would be tested and kept in quarantine until deemed negative. With the challenges in testing and the unknown mechanism of transmission, this could be a long process (i.e., taking years). Along with practicing good hygiene, testing and quarantine measures, a nutritious diet should be provided.

The following are guidelines to control PaBV in aviaries and homes.

- Good hygiene and housing birds outdoors when temperature permits.
- For pets, reduce reproductive stress via diet and husbandry modifications and use of GNRH agonists. Avoid overcrowding and provide good nutrition.
- Isolation of all new, sick, or PaBV-positive birds, separation of birds from a flock when they are PaBV positive.
- Existing birds and newly presenting birds should be tested using both serology and PCR, and birds should be separated based on the results of this testing. There should be at least three negative tests at 3- to 6-week intervals before considering a bird negative.
- With the likelihood of vertical transmission, chicks from infected birds should be incubator hatched, hand-raised, kept separate from other non-infected birds and monitored for development of disease.
- Due to the intermittent shedding and inconclusive serology results, testing and separating birds may require years to obtain PaBV-negative aviaries, depending on the willingness of the owner.
- Healthy birds that test positive with either PCR or serology testing should not be euthanized.

References

1. Daoust PY, Julian RJ, Yason CV, et al. Proventricular impaction associated with nonsuppurative encephalomyelitis and ganglioneuritis in two Canada geese. *J Wildl Dis.* 1991;27(3):513-517.
2. Delnatte P, Ojkic D, Delay J, et al. Pathology and diagnosis of avian bornavirus infection in wild Canada geese (*Branta canadensis*), trumpeter swans (*Cygnus buccinator*) and mute swans (*Cygnus olor*) in Canada: a retrospective study. *Avian Pathol.* 2013;42(2):114-128.
3. Hoppes S, Gray PL, Payne S, et al. The isolation, pathogenesis, diagnosis, transmission, and control of avian bornavirus and proventricular dilatation disease. *Vet Clin North Am Exot Anim Pract.* 2010;13(3):495-508.
4. Payne SL, Delnatte P, Guo J, et al. Birds and bornaviruses. *Anim Health Res Rev.* 2012;13(2):145-156.
5. Perpiñán D, Fernández-Bellon H, López C, et al. Lymphoplasmacytic myenteric, subepicardial, and pulmonary ganglioneuritis in four nonpsittacine birds. *J Avian Med Surg.* 2007;21(3):210-214.
6. Weissenbock H, Sekulin K, Bakonyi T, et al. Novel avian bornavirus in a nonpsittacine species (Canary; *Serinus canaria*) with enteric ganglioneuritis and encephalitis. *J Virol.* 2009;83(21):11367-11371.
7. Avian Bornavirus Research Forum. *Proc Annu Conf Assoc Avian Vet.* 2014.
8. Berg M, Johansson M, Montell H, et al. Wild birds as a possible natural reservoir of Borna disease virus. *Epidemiol Infect.* 2001;127(1):173-178.

9. Payne S, Covaleda L, Jianhua G, et al. Detection and characterization of a distinct bornavirus lineage from healthy Canada geese (*Branta canadensis*). *J Virol.* 2011;85(22):12053-12056.

10. Shivaprasad H. Proventricular dilatation disease in a peregrine falcon (*Falco peregrinus*). *Proc Annu Conf Assoc Avian Vet.* 2005;107-108.

11. Payne S, Guo J, Tizard I. Bornaviruses. In: Hambrick J. Gammon LT, eds. *Ducks: Habitat, Behavior and Diseases.* New York, NY: Nova Scientific Publishers; 2013:1-20.

12. Smith D, Berkvens C, Kummrow M, et al. Identification of avian bornavirus in the brains of Canada geese (*Branta canadensis*) and trumpeter swans (*Cygnus buccinator*) with non-suppurative encephalitis. *Proc Wildl Dis Assoc.* 2010.

13. Honkavuori KS, Shivaprasad H, Williams BL, et al. Novel borna virus in psittacine birds with proventricular dilatation disease. *Emerg Infect Dis.* 2008;14(12):1883-1886.

14. Kistler AL, Gancz A, Clubb S, et al. Recovery of divergent avian bornaviruses from cases of proventricular dilatation disease: identification of a candidate etiologic agent. *Virol J.* 2008;5(1):1-15.

15. Ouyang N, Storts R, Tian Y, et al. Histopathology and the detection of avian bornavirus in the nervous system of birds diagnosed with proventricular dilatation disease. *Avian Pathol.* 2009;38(5):393-401.

16. Rubbenstroth D. Avian bornavirus research-a comprehensive review. *Viruses.* 2022;14(7):1513.

17. Kuhn JH, Dürrwald R, Bào Y, et al. Taxonomic reorganization of the family Bornaviridae. *Arch Virol.* 2015;160:621-632.

18. Rubbenstroth D, Briese T, Dürrwald R, et al. ICTV virus taxonomy profile: Bornaviridae. *J Gen Virol.* 2021;102(7):001613.

19. Malkinson M. Borna disease in ostriches. *Vet Rec.* 1993;132:304.

20. Ludwig H, Kraft W, Kao M, et al. Borna virus infection (Borna disease) in naturally and experimentally infected animals: its significance for research and practice. *Tieraerztl Prax.* 1985;13(4):421-453.

21. Tizard I, Shivaprasad H, Guo J, et al. The pathogenesis of proventricular dilatation disease. *Anim Health Res Rev.* 2016;17(2):110-126.

22. Hoffmann B, Tappe D, Höper D, et al. A variegated squirrel bornavirus associated with fatal human encephalitis. *New Engl J Med.* 2015;373(2):154-162.

23. Fluck A, Enderlein D, Piepenbring A, et al. Correlation of avian bornavirus-specific antibodies and viral ribonucleic acid shedding with neurological signs and feather-damaging behaviour in psittacine birds. *Vet Rec.* 2019;184(15):476-476.

24. Rubbenstroth D, Schmidt V, Rinder M, et al. Phylogenetic analysis supports horizontal transmission as a driving force of the spread of avian bornaviruses. *PLoS ONE.* 2016;11(8):e0160936.

25. Heffels-Redmann U, Enderlein D, Herzog S, et al. Occurrence of avian bornavirus infection in captive psittacines in various European countries and its association with proventricular dilatation disease. *Avian Pathol.* 2011;40(4):419-426.

26. Encinas-Nagel N, Enderlein D, Piepenbring A, et al. Avian bornavirus in free-ranging psittacine birds, Brazil. *Emerg Infect Dis.* 2014;20(12):2103.

27. Sutherland M, Phalen DN, Herzog S, et al. Detection of Immunoreactivity to Psittaciform Bornavirus in the Serum of a Wild Cacatuid in Victoria, Australia. *J Wildl Dis.* 2021;57(2):471-473.

28. Heckmann J, Enderlein D, Gartner AM, et al. Wounds as the portal of entrance for parrot bornavirus 4 (pabv-4) and retrograde axonal transport in experimentally infected cockatiels (*Nymphicus hollandicus*). *Avian Dis.* 2020;64(3):247-253.

29. Link J, Herzog S, Gartner AM, et al. Factors Influencing Vertical Transmission of Psittacine Bornavirus in Cockatiels (*Nymphicus hollandicus*). *Viruses.* 2022;14(12):2721.

30. Leal de Araujo J, Rech RR, Heatley JJ, et al. From nerves to brain to gastrointestinal tract: A time-based study of parrot bornavirus 2 (PaBV-2) pathogenesis in cockatiels (*Nymphicus hollandicus*). *PLoS ONE.* 2017;12(11):e0187797.

31. Piepenbring AK, Enderlein D, Herzog S, et al. Pathogenesis of avian bornavirus in experimentally infected cockatiels. *Emerg Infect Dis.* 2012;18(2):234-241.

32. Rubbenstroth D, Brosinski K, Rinder M, et al. No contact transmission of avian bornavirus in experimentally infected cockatiels (*Nymphicus hollandicus*) and domestic canaries (*Serinus canaria forma domestica*). *Vet Microbiol.* 2014;172(1-2):146-156.

33. Payne S, Shivaprasad HL, Mirhosseini N, et al. Unusual and severe lesions of proventricular dilatation disease in cockatiels (*Nymphicus hollandicus*) acting as healthy carriers of avian bornavirus (ABV) and subsequently infected with a virulent strain of ABV. *Avian Pathol.* 2011;40 (1):15-22.

34. Piepenbring AK, Enderlein D, Herzog S, et al. Parrot Bornavirus (PaBV)-2 isolate causes different disease patterns in cockatiels than PaBV-4. *Avian Pathol.* 2016;45(2):156-168.

35. Gray P, Hoppes S, Suchodolski P, et al. Use of avian bornavirus isolates to induce proventricular dilatation disease in conures. *Emerg Infect Dis.* 2010;16(3):473-479.

36. Mirhosseini N, Gray PL, Hoppes S, et al. Proventricular dilatation disease in cockatiels (*Nymphicus hollandicus*) after infection with a genotype 2 avian bornavirus. *J Avian Med Surg.* 2011;25(3):199-204.

37. Olbert M, Römer-Oberdörfer A, Herden C, et al. Viral vector vaccines expressing nucleoprotein and phosphoprotein genes of avian bornaviruses ameliorate homologous challenge infections in cockatiels and common canaries. *Sci Rep.* 2016;6(1):1-14.

38. Runge S, Olbert M, Herden C, et al. Viral vector vaccines protect cockatiels from inflammatory lesions after heterologous parrot bornavirus 2 challenge infection. *Vaccine.* 2017;35(4):557-563.

39. Leal de Araújo J, Rech RR, Rodrigues-Hoffmann A, et al. Immunophenotype of the inflammatory response in the central and enteric nervous systems of cockatiels (*Nymphicus hollandicus*) experimentally infected with parrot bornavirus 2. *Vet Pathol.* 2022;59(3):493-497.

40. Herden C, Briese T, Lipkin W, et al. Bornaviridae. In: Knipe D, Howley P, Griffin D, et al., eds. Fields Virology. Philadelphia, PA: Wolters Kluwer Health Lippincott Williams & Wilkins; 2013:1124-1150.

41. Tappe D, Schmidt-Chanasit J, Rauch J, et al. Immunopathology of fatal human variegated squirrel bornavirus 1 encephalitis, Germany, 2011-2013. *Emerg Infect Dis.* 2019;25(6):1058.

42. Caplazi P, Ehrensperger F. Spontaneous Borna disease in sheep and horses: immunophenotyping of inflammatory cells and detection of MHC-I and MHC-II antigen expression in Borna encephalitis lesions. *Vet Immunol Immunopathol.* 1998;61(2-4):203-220.

43. Leal de Araujo J, Rodrigues-Hoffmann A, Giaretta PR, et al. Distribution of viral antigen and inflammatory lesions in the central nervous system of cockatiels (*Nymphicus hollandicus*) experimentally infected with parrot bornavirus 2. *Vet Pathol.* 2019;56(1):106-117.

44. Wuest E, Malberg S, Petzold J, et al. Experimental infection of embryonic cells and embryonated eggs of cockatiels (*Nymphicus hollandicus*) with two parrot bornavirus isolates (PaBV-4 and PaBV-2). *Viruses.* 2022;14(9):1984.

45. Rossi G, Dahlhausen RD, Galosi L, et al. Avian ganglioneuritis in clinical practice. *Vet Clin North Am Exot Anim Pract.* 2018;21(1):33-67.

46. Leal de Araujo J, Hameed S, Tizard I, et al. Cardiac lesions of natural and experimental infection by parrot bornaviruses. *J Comp Pathol.* 2020;174:104-112.

47. Pesaro S, Crosta L, Bertoni P, et al. Anti-ganglioside antibodies production as a theory of PDD pathogenesis. *Proc Euro Assoc Avian Vet.* 2009;89.

48. Rossi G, Ceccherelli R, Crosta L, et al. Anti-ganglioside specific auto-antibodies in ganglia of PDD-affected parrots. *Proc Eur Assoc Avian Vet.* 2011;177-178.

49. Rossi G, Crosta L, Pesaro S. Parrot proventricular dilation disease. *Vet Rec.* 2008;163(10):310.

50. Nobach D, Müller J, Tappe D, et al. Update on immunopathology of bornavirus infections in humans and animals. *Adv Virus Res.* 2020;107:159-222.

51. Leal de Araujo J, Tizard I, Guo J, et al. Are anti-ganglioside antibodies associated with proventricular dilatation disease in birds? *PeerJ.* 2017;5:e3144.

52. Fluck A, Enderlein D, Piepenbring A, et al. Avian bornavirus infection in psittacine birds with neurological signs and feather plucking. *Proc ICARE.* 2013;388.

53. Gartner AM, Link J, Bücking B, et al. Age-dependent development and clinical characteristics of an experimental parrot bornavirus-4 (PaBV-4) infection in cockatiels (*Nymphicus hollandicus*). *Avian Pathol.* 2021;5(2):138-150.

54. Herzog S, Enderlein D, Heffels-Redmann U, et al. Indirect immunofluorescence assay for intra vitam diagnosis of avian bornavirus infection in psittacine birds. *J Clin Microbiol.* 2010;48(6):2282-2284.

55. Lierz M. Avian bornavirus and proventricular dilatation disease. In: Speer B, ed. *Current Veterinary Therapy in Avian Medicine and Surgery.* St. Louis, MO: Elsevier; 2016:28-46.

56. Dahlhausen RD, Aldred S, Colaizzi E. Resolution of clinical proventricular dilatation disease by cyclooxygenase 2 inhibition. *Proc Annu Conf Assoc Avian Vet.* 2002;9-12.

57. Clubb S, Meyer M. Clinical management of psittacine birds affected with proventricular dilatation disease. *Proc Annu Conf Assoc Avian Vet.* 2006;85-90.

58. Gancz AY, Elbaz D, Farnoushi Y, et al. Clinical recovery from proventricular dilatation disease following treatment with cyclosporine A in an African grey parrot (*Psittacus erithacus*). *Proc DVG Tagung über Vogelkrankheiten.* 2012;130-133.

59. Musser JM, Heatley JJ, Koinis AV, et al. Ribavirin inhibits parrot bornavirus 4 replication in cell culture. *PLoS ONE.* 2015;10(7):e0134080.

60. Reuter A, Horie M, Höper D, et al. Synergistic antiviral activity of ribavirin and interferon-α against parrot bornaviruses in avian cells. *J Gen Virol.* 2016;97(9):2096-2103.

61. Tokunaga T, Yamamoto Y, Sakai M, et al. Antiviral activity of favipiravir (T-705) against mammalian and avian bornaviruses. *Antiviral Res.* 2017;143:237-245.

62. Rall I, Amann R, Malberg S, et al. Recombinant modified vaccinia virus Ankara (MVA) vaccines efficiently protect cockatiels against parrot bornavirus infection and proventricular dilatation disease. *Viruses.* 2019;11(12):1130.

63. Boatright-Horowitz SL. Avian bornaviral ganglioneuritis: current debates and unanswered questions. *Vet Med Int.* 1-9; 2020. [Review article]; doi: https://doi.org/10.1155/2020/6563723.

64. Kranz JB. A study on the chemical and physical inactivation of avian bornavirus. Doctoral dissertation. College Station, TX: Texas A&M University; 2014.

65. Kranz JB, Escandon P, Musser JM. Environmental stability of avian Bornavirus: pH and drying. *Proc ExoticsCon.* 2015;89.

Practical Applications of Fluoroscopy in Avian Practice

Rachel Baden, DVM, DABVP (Avian Practice), and
Brian Speer, DVM, DABVP (Avian Practice), DipECZM (Avian)

Session #3217

Affiliation: From the Medical Center for Birds, 3807 Main Street, Oakley, CA 94561, USA.

Abstract: Fluoroscopy is an under-utilized imaging modality that has numerous clinical applications in avian practice. This masterclass will briefly review the physics and methods behind fluoroscopy and its use in avian clinical practice. Real clinical cases and video images will be used to illustrate its potential applicability in diagnostically shaping cases. Although the primary focus will be on diseases of the gastrointestinal tract, other systems and disease conditions will be included to best show the potential broad value and scope of this imaging modality as an addition to the diagnostic tools for clinical practice.

Physics of Radiology

While this masterclass does not spend much time on the physics involved in fluoroscopy, it can be useful to recall the basic principles and techniques, as these may inform interpretation of the resulting images. Both radiographs and fluoroscopic images are generated by x-rays, which are part of the electromagnetic spectrum. In particular, x-rays are given off by electron interactions outside of the atomic nucleus, and the energy of a given x-ray is dependent upon the energy of the original electron and its subsequent photonic energy. Since they are on the electromagnetic spectrum, x-rays (energetic photons) are simultaneously both a particle and a wave, like visible light.[1]

In fluoroscopy, x-rays are produced when high-speed electrons strike a metal x-ray tube. To generate electrons, electrical current is run through the x-ray tube filament, similar to that of a lightbulb, creating heat that causes electrons to leave the filament and form an electron cloud. The number of electrons in the cloud is directly related to the current running through the filament, which is controlled by the milliamperage (mA). Increasing mA is similar to increasing the wattage of a lightbulb. The milliampere-second (mAs) describes the total amount of radiation produced in a period of time.

Once produced, the electrons in the electron cloud are accelerated into metallic target, driven by a voltage differential between the filament and the target. The voltage difference is controlled by the kilovoltage peak (kVp); increasing kVp increases the voltage differential which more strongly attracts electrons to the target, producing higher velocity electrons and leading to higher-energy x-rays.

An important distinction to make is between the kiloelectron volt (keV, the unit of energy for electromagnetic radiation) and kVp, the kilovoltage peak. The first is a unit that describes the amount of energy of an electron or photon, and the second is the maximum voltage difference between the x-ray tube generating electrons and the target.

When the electrons strike the metallic anode (often tungsten, with current configurations), they are suddenly slowed down (Bremsstrahlung process) creating multiple photonic energy wavelengths. The creation of an x-ray image, ultimately, involves the x-rays (photons) that then pass through a patient (or don't) and strike a detector, the x-ray plate. One of the reasons that x-rays can produce an interpretable image is because of the difference between radiation exposure and absorption. For a given exposure to radiation, different tissues may absorb different amounts; for example, bone absorbs much more radiation than soft tissue, leading to a different appearance on radiography and fluoroscopy between the two tissues.

Fluoroscopy of Different Organ Systems

Gastrointestinal tract

Evaluation of the size, shape, and function of the gastrointestinal organs is likely the most common use of fluoroscopy in veterinary medicine, and can be done either with or without contrast. For contrast fluoroscopy, a barium sulfate suspension is anecdotally used most often, but iodine-based media (iohexol) has also been reported.[2-4] Some of the advantages of fluoroscopic assessment of the GI over standard radiography include the ability to monitor contractions of the crop and ventriculus, and evaluate intestinal filling and peristalsis.[5] With the addition of contrast, radiolucent intraluminal foreign objects, mural irregularities, or abnormal diverticula can be identified.[2,3,6] Additionally, contrast can be used to measure overall gastrointestinal transit time, as a way of evaluating overall motility, and has been assessed in several avian species to date.[4,7-11] Contrast-enhanced intestinal loops can also assist in locating specific segments of the tract as well as mass effects relative to other organs in the abdomen.[12]

One of the major advantages of fluoroscopy-based contrast GI study compared to radiography is the reduction in intrusiveness and risk. With radiographs, in order to get a properly positioned and interpretable film without causing undue stress to the patient, sedation and/or anesthesia is often required, and the patient is placed in lateral recumbency; both of these things can be dangerous with a crop full of liquid contrast medium, as regurgitation and subsequent aspiration are possible. Fluoroscopy avoids the need for chemical restraint and lateral recumbency both, as the patient can be standing on a perch or in a radiolucent container for the duration of the study.[13] In addition, sedation has been shown to artificially slow GI motility and prolong transit time.[4]

Cardiovascular system

Protocols for contrast-based fluoroscopy of the cardiovascular system have been reported in humans and other mammals, and are beginning to be reported in birds as well. There are two main methods, nonselective and selective contrast fluoroscopy, with the latter not (yet) published as being used in birds. The difference between the two is the selective fluoroscopy, also called digital subtraction angiography, involves digital processing of the images, where the non-contrast images are digitally subtracted from the views containing contrast, to enhance visualization of the important structures and reduce interference from other organs.[14]

Several reports of fluoroscopic angiography exist in birds, including the diagnosis of infundibular pulmonic stenosis in a Moluccan cockatoo (*Cacatua moluccensis*), a right coronary artery aneurysm in an umbrella cockatoo (*Cacatua alba*), and cardiac chamber dilation in a whooping swan (*Cygnus cygnus*).[15-17] One study has established a protocol with iodinated contrast medium for fluoroscopic angiography in Hispaniolan Amazon parrots (*Amazona ventralis*), but the literature on avian cardiovascular fluoroscopy remains sparse.[18]

Other organs

Reports on the use of fluoroscopy to evaluate other organ systems in birds are few, but include guidance for placement of a wire stent for tracheal stenosis in an eclectus (*Eclectus roratus*) and introduction of a tube into the ventriculus for foreign body removal in a honey buzzard (*Pernis apivorus*).[19,20] One study has investigated the kinematics of bipedal locomotion in common quail (*Coturnix coturnix*) using fluoroscopic video analysis.[21]

References

1. Thrall D, Widmer W. Radiation protection and physics of diagnostic radiology. In: Thrall D, ed. *Textbook of Veterinary Diagnostic Radiology.* Philadelphia, PA: Elsevier Saunders; 2018:2-22.

2. Degernes LA, Wolf KN, Zombeck, DJ, et al. Ventricular diverticula formation in captive parakeet auklets (*Aethia psittacula*) secondary to foreign body ingestion. *J Zoo Wildl Med.* 2012;43(4):889-97.

3. Cotton RJ III, Divers SJ. Endoscopic removal of gastrointestinal foreign bodies in two African grey parrots (*Psittacus erithacus*) and a hyacinth macaw (*Anodorhynchus hyacinthinus*). *J Avian Med Surg.* 2017;31(4):335-343.

4. Martel A, Mans C, Doss GA, Williams JM. Effects of midazolam and midazolam-butorphanol on gastrointestinal transit time and motility in cockatiels (*Nymphicus hollandicus*). *J Avian Med Surg.* 2018;32(4):286-293.

5. Keller D, Honkavuori KS, Briese T, et al. Proventricular dilatation disease associated with avian bornavirus in a scarlet macaw (*Ara macao*). *J Vet Diagn Invest.* 2010;22(6):961-965.

6. Laniesse D, Beaufrère H, Mackenzie S, et al. Perforating foreign body in the ventriculus of a pet pigeon (*Columba livia domestica*). *J Am Vet Med Assoc.* 2018;253(12):1610-1616.

7. Doss GA, Williams JM, Mans C. Determination of gastrointestinal transit times in barred owls (*Strix varia)* by contrast fluoroscopy. *J Avian Med Surg.* 2017;31(2):123-127.

8. Doss GA, Williams JM, Mans C. Contrast fluoroscopic evaluation of gastrointestinal transit times with and without the use of falconry hoods in red-tailed hawks (*Buteo jamaicensis*). *J Am Vet Med Assoc.* 2017;251(9):1064-1069.

9. Beaufrère H, Nevarez J, Taylor WM, et al. Fluoroscopic study of the normal gastrointestinal motility and measurements in the Hispaniolan Amazon parrot (*Amazona ventralis*). *Vet Radiol Ultrasound.* 2010;51(4):441-6.

10. Vink-Nooteboom M, Lumeij JT, Wolvekamp WTC. Radiography and image-intensified fluoroscopy of barium passage through the gastrointestinal tract in six healthy Amazon parrots (*Amazona aestiva*). *Vet Radiol Ultrasound.* 2003;44(1):43-48.

11. Kubiak M, Forbes NA. Fluoroscopic evaluation of gastrointestinal transit time in African grey parrots. *Vet Rec.* 2012;171(22):563.

12. Murphy B, Fitzgerald B, Olsen G, Speer B. Diagnosis and surgical repair of an acute abdominal wall hernia and partial cloacal strangulation in a yellow-naped Amazon parrot (*Amazona auropalliata*). *J Avian Med Surg*. 2018;32(4):336-341.

13. Dennison S. Common imaging modalities: selecting the right modality and getting the most out of your imaging. In: Speer B, Van Zeeland Y, eds. *Current Therapy in Avian Medicine and Surgery.* Vol II. St. Louis, MO: Elsevier; 2023.

14. Beaufrère H, Pariaut R, Rodriguez D, Tully TN. Avian vascular imaging: a review. *J Avian Med Surg*. 2010;24(3):174-84.

15. Vink-Nooteboom M, Schoemaker NJ, Kik MJ, et al. Clinical diagnosis of aneurysm of the right coronary artery in a white cockatoo (*Cacatua alba*). *J Small Anim Pract*. 1998;39:533-537.

16. Zoller G, Guzman DS-M, Summa N, et al. Infundibular pulmonic stenosis in a Moluccan cockatoo (*Cacatua moluccensis*). *J Avian Med Surg*. 2017;31(1):53-61.

17. Fischer I, Christen C, Scharf G, Hatt JM. Cardiomegaly in a whooper swan (*Cygnus cygnus*). *Vet Rec*. 2005;156:178-182.

18. Aller TL, Guzman DS-M, Stern JA, et al. Evaluation of a fluoroscopic angiography protocol in Hispaniolan Amazon parrots (*Amazona ventralis*). *J Avian Med Surg*. 2022;36(2):178-186.

19. Mejia-Fava J, Holmes SP, Radlinsky, MA, et al. Use of a nitinol wire stent for management of severe tracheal stenosis in an eclectus parrot (*Eclectus roratus*). *J Avian Med Surg*. 2015;29(3):238-249.

20. Lumeij JT, Wolvekamp WT, Bron-Dietz GM, Schotman AJ. An unusual case of lead poisoning in a honey buzzard (*Pernis apivorus*). *Vet Q*. 1985;7(2):165-8.

21. Ogihara N, Oku T, Andradaet E, et al. Planar covariation of limb elevation angles during bipedal locomotion in common quails (*Coturnix coturnix*). *J Exp Biol*. 2014;217(22):3968-3973.

Recognition, Assessment and Management of Pain Improves Avian Welfare

Michelle G. Hawkins VMD, DABVP (Avian Practice), and
Joanne Paul-Murphy DVM, Dipl ACZM, Dipl ACAW

Session #3166

Affiliation: From the Department of Medicine and Epidemiology, School of Veterinary Medicine, University of CA, Davis, Davis, CA 95616, USA.

Abstract: This masterclass provides an overview of current information regarding avian pain. The review includes recognition of avian pain, and the current evidence-based clinical analgesic use in birds. Recognition and assessment of behavioral changes associated with pain in birds are crucial tools towards pain management. The field of avian analgesia has expanded dramatically, identifying more options for alleviating acute and chronic pain for many species. Opioids, nonsteroidal anti-inflammatory drugs (NSAIDs), local anesthetics, and drugs including gabapentin, amantadine, and cannabinoids are being used for birds. Analgesic combinations acting at different points in the nociceptive system provide greater pain relief, reduce the risk of adverse effects and improve welfare of the patient.

Avian Analgesia and Welfare

There is still much to be determined to create criteria to recognize and assess pain in birds because this is a highly diverse class of species. Our team recently published two papers[1,2] that will be helpful. Both publications provide information on the physiology of avian pain, observed behavioral and physiologic alterations with pain, and how different sources and degrees of pain affect behavior. Species vary widely in their response to pain, and individual animals of the same species can have different responses to different types of pain. Layered on top of that, the individual animal's response to pain under different conditions, for example a bird under direct vs remote observation, or a bird under different environmental conditions such as home or the clinical setting. The World Organization for Animal Health (OIE) defines an animal as having good welfare if it is "healthy, comfortable, well nourished, safe, able to express innate behavior, and if it is not suffering from unpleasant states such as pain, fear and distress". At the heart of welfare definitions is the understanding that the welfare of a bird belongs to the bird.[3] The definition of welfare includes the bird's psychological well-being. Applying these concepts to recognition and management of pain often occurs without conscious effort when the clinical staff have experience and advanced training. Knowledge of the normal appearance and behavior of the species, and the individual bird involved, is essential for recognition of pain. When a bird experiences pain, one or more of the normal behavior typically shift or are absent. Given the overall reduction in ability to express pain via facial expression in birds, in contrast to mammals, evaluation of the avian patient should instead focus on the whole individual, and variations in positioning of the tail, wings, and crest (if present) may provide useful information.[1] In most circumstances, pain is assessed based on a bird's clinical appearance and overall behavior. This approach can be unreliable; however, it is usually effective in detecting severe pain in many species. It is also effective when pain is localized to one limb causing lameness or wing droop, or to a specific body area resulting in a marked behavioral response if that area is palpated.[4] Assessing pain is going to be highly subjective and based largely on preconceived ideas about the appearance and behavior of birds in response to pain.[4] A team approach can reduce subjectivity, and requires communication and cooperation among all involved; veterinarians, veterinary technicians, animal care staff and owners will advance the welfare of the bird in pain. Accurate assessment of avian pain can be confounded by the effect of the observer on the bird's behavior (especially in terms of wildlife species), effects of some analgesics or sedatives that can alter behavior, and concurrent physiologic conditions (anesthetic recovery, hypothermia, and others.) that could make a bird appear painful despite adequate analgesia.[2] The intensity of a bird's pain needs to be recognized to aid in the selection of the correct analgesic. Differences in the source (visceral, somatic, neuropathic), chronicity (acute vs. chronic), and degree (mild, moderate, severe) of pain can lead to differences in pain response in birds. If the wrong analgesic or the wrong dose is selected, it could result in suboptimal treatment.

The options to provide reduction of pain and analgesia are expanding for birds as the field of analgesia for all vertebrate species is advanced. Management of pain has numerous components with administration of analgesic drugs being an important tool. Published dosages of many analgesic drugs for birds are determined by pharmacokinetics (PK) studies using a limited number of animals and a restricted set of species. Pharmacodynamic studies to determine effectiveness is lacking for most analgesic drugs currently used to treat birds; however, the information provides a critically helpful reference point to extrapolate from, whether it be to another species or a different type of painful condition. Successive observations are critical to assess the effect of treatment and determine if the bird's condition is improving or declining. Having a structure for assessment, shared by the care team, is always helpful when monitoring throughout a period of time, especially when several people are involved. Additionally, as a clinical team shares observations and evaluations of birds' responses to analgesic treatments, the team's proficiency to make accurate assessments will improve. Published pain scales and assessment schemes are useful as a starting point to adopt the aspects that resonate with the observers. Generalized methods of quantifying pain such as visual analogue or numeric rating pain scales can be applied as a tool for evaluation of an avian patient's pain while awaiting development of more avian-specific pain scales.[2]

Specific Analgesia

Balanced analgesia

Combinations of drugs acting at different points in the nociceptive system are more effective and less toxic than either drug given alone. For example, opioids act centrally to limit nociceptive input whereas NSAIDS act peripherally to decrease inflammation, limiting nociceptive information that enters the CNS. Synergy has been demonstrated in laboratory animals and is now being tested in the clinical environment.

Opioids

Analgesia is species- and intra-species variable at equipotent dosages. Opioids are used for moderate to severe pain, such as traumatic or surgical pain. Opioids reversibly bind to specific receptors in the central and peripheral nervous system. These drugs are categorized as full agonists, partial agonists, mixed agonist/antagonists, or antagonists based on their ability to induce an analgesic response once bound to a specific receptor. The action of opioid drugs on these receptors activates G-proteins, leading to a reduction in transmission of nerve impulses and inhibition of neurotransmitter release.

Agonist drugs have a linear dose–response curve that may be titrated to reach the desired effect, whereas the agonist/antagonist drugs may reach a ceiling effect after which increasing the dose does not appear to provide additional analgesia. During anesthesia, opioids are used to provide perioperative analgesia that may reduce the concentrations of volatile anesthetics (i.e., gas anesthesia-sparing effects) while maximizing analgesia. The most common adverse effects reported with opioids are cardiac and/or respiratory depression. In many cases, these drugs may be reversed with antagonists, which will also terminate analgesia. The application and dosages of few opioid formulations have been scientifically evaluated and clinically applied in birds.

Few studies have been conducted regarding the distribution, quantity, and function of each opioid receptor type in birds, and although marked species variability is assumed, further studies are necessary.[5] There are small differences in receptor densities between cockatiels and pigeons but these may not be sufficient to make functional changes.[6] In conditioning experiments pigeons were able to differentiate μ from κ receptor agonists. Isoflurane-sparing effects of μ and κ agonists are quite variable in mammals and it is likely that similar variability will be evident in birds.

Morphine (MOR) is a μ-receptor agonist that has been infrequently used in avian medicine because early studies in domestic fowl yielded conflicting results and suggested very high dosages were required. The PK of MOR at 2 mg/kg IV in the domestic chickens found that they cleared MOR at a higher rate than humans or cats and plasma MOR analgesic concentrations were similar to humans for only approximately 2 hours. Following 2 mg/kg IV MOR, chickens had an increased time to finish the obstacle course and exhibited sedation, confounding assessment of its analgesic effects.

Hydromorphone HCl (HYDRO) is a μ-opioid receptor agonist with an approximate potency 5-7 times that of MOR. Hydromorphone showed a dose-responsive thermal antinociceptive effect when administered at 0.1, 0.3, and 0.6 mg/kg IM to American kestrel (Falco sparverius; AMKE) for up to 6 hours.[7] Severe sedation was detected in four birds after 0.6 mg/kg HYDRO, and when the sedated birds were handled, they appeared agitated for a few seconds before resuming sedated behaviors. Hydromorphone PK following IM and IV administration in AMKEs found high bioavailability (75%) and rapid elimination.[8] A similar thermal nociception study in cockatiels (*Nymphicus hollandicus*) administered the same dosages showed no significant effect on the thermal threshold, with only a mild sedative effect.[9] Thermal antinociceptive effects of HYDRO were evaluated following IM administration to orange-winged Amazon parrots (*Amazona amazonica*; OWAP).[10] Hydromorphone at 1 and 2 mg/kg significantly increased thermal foot withdrawal responses compared with the response after saline administration, and plasma concentrations were maintained above 1 ng/ml for 6 hours in OWAP receiving 1 mg/kg HYDRO IV and IM, respectively. Significant agitation was also observed for 1 and 2 mg/kg HYDRO doses. This kind of agitation in response to an opioid in normal birds may not be present in birds with pain. Other significant adverse effects after administration of 1 and 2 mg/kg HYDRO included nausea-like behavior, ataxia and pupillary constriction. The presence of adverse effects warrants consideration of lower doses for clinical use in this species.

Methadone (MET) is a μ-opioid receptor agonist that also is a non-competitive inhibitor of NMDA receptors. Methadone can reduce re-uptake of norepinephrine and serotonin, which may contribute to its analgesic effects. Due to its additional mechanisms of action, MET potentially may be more efficacious than other μ-agonists, particularly for neuropathic or chronic pain. In hens, MET at 6 mg/kg IM decreased the MAC_{iso} by 30% but it also produced mild respiratory acidosis, decreased heart rate, and increased systemic blood pressure, with occasional cardiac arrhythmias. The IM bioavailability was approximately 80%. [11] In OWAP, 6 mg/kg IM MET increased thermal thresholds significantly; lower doses (1 and 3 mg/kg) did not result in a significant effect. Methadone caused significant agitation and nausea-like behavior when compared to saline administration. The IM mean bioavailability was 85.7 %. While 6 mg/kg MET IM appears to have desired analgesic effects, the presence of agitation, nausea-like behavior and ataxia warrants consideration of lower doses for clinical use in OWAP.

Fentanyl (FEN) is a short-acting μ-opioid receptor agonist often administered as a continuous rate infusion (CRI) or as a transdermal patch. Early work evaluating FEN in psittacine birds was challenging to interpret. Fentanyl administered as an IV constant rate infusion (CRI) in red-tailed hawks (*Buteo jamaicensis;* RTHA) reduced the minimum anesthetic concentration of isoflurane 31-55% in a dose-related manner, without significant effects on heart rate, blood pressure, $PaCO_2$, or PaO_2. In Hispaniolan Amazon parrots (*Amazona ventralis*; HAPs), FEN administered as an IV CRI reduced $MAC_{isoflurane}$ similarly to the RTHAs; however, approximately 20 times higher dosing was necessary to achieve the same effects with significant effects detected in HR and BP suggesting that close monitoring is advised.[12] In several breeds of chickens, the plasma concentrations of FEN were evaluated following application of 25 μg/hour transdermal FEN patches.[13] There was significant inter-individual difference in maximum plasma FEN concentrations attributed to absorption variability, but all chickens achieved human target plasma concentrations within 2-4 hours and had a rapid decrease in concentrations following patch removal. The FEN transdermal patch resulted in plasma concentrations within the range considered therapeutic in people for 72 hours.[13]

Butorphanol (BUT) is a mixed κ-opioid receptor agonist and μ-opioid receptor antagonist opioid drug. This is one of the most commonly used opioid analgesics in birds, despite many studies suggesting it requires frequent dosing. There is evidence that BUT does not produce dose-related respiratory depression. Butorphanol (1 mg/kg) produced an isoflurane-sparing effect in cockatoos (*Cacatua* spp.) and grey parrots, *Psittacus erithacus*), but not in blue-fronted Amazons (*Amazona aestiva*), indicating species variability in response to the drug itself or to the dose administered. When 4 mg/kg IV of BUT was administered rapidly, there were significant arrhythmias in all birds, significant decreases in heart rates and arterial blood pressures, and one bird died.[14] Other studies have shown efficacy of BUT administered 2-6 mg/kg IM without significant anesthetic or cardiopulmonary changes. Oral administration had very limited bioavailability (5.9%) in HAPs and is not recommended. In contrast, hyperesthesia, hyperalgesia and agitation were noted in male AMKE receiving 6 mg/kg; thus BUT is not recommended in this species for pain management. Likewise, when a dose of 2 mg/kg IV BUT was administered to chickens, its terminal half-life was shorter than in humans, dogs, cattle, rabbits, and owls. Chickens finished an obstacle course faster after 2 mg/kg BUT injection, but concentrations only lasted 2 hours. All these data confirm frequent BUT dosing is necessary in birds. This may be impractical because of lack of personnel to provide frequent dosing and the stress of frequent handling on the patient. In common peafowl, the PK of butorphanol were assessed after administration via an osmotic pump, a novel delivery method.[15] Plasma concentration reached 60 ng/ml after 24 hours in most birds, and most birds stayed above this threshold for 7 days, then decreased rapidly after implant removal. This delivery method holds promise for pain management in larger birds with the limitation that it will likely require deep sedation or general anesthesia for implantation and removal.

Buprenorphine HCl (BUP) has a slow onset of action with longer duration than other opioid drugs and a complex pharmacological profile. It is thought to act as a partial μ agonist but its κ-receptor activities are less well defined. Several studies suggest that BUP demonstrates κ-receptor agonist/antagonist activity. Buprenorphine has unusual receptor binding characteristics as a result of slow drug dissociation from opioid receptors. Buprenorphine 0.1 mg/kg IM in grey parrots did not demonstrate any antinociceptive effect but pigeons administered 0.25 and 0.5 mg/kg IM BUP had increased withdrawal thresholds from a noxious electrical stimulus for 2 and 5 hours, respectively. In AMKEs, BUP did cause a significant dose-dependent thermal antinociceptive response for 6 hours compared with controls. At 0.6 mg/kg, a mild sedative effect was appreciated. The PK of 0.6 mg/kg IM and IV BUP was also evaluated in AMKEs and high bioavailability (94.8%) after IM dosing was identified. In a PK study in AMKEs, SC and IM administration of 1.8 mg/kg BUP SR-LAB™ (Wildlife Pharmaceuticals, Windsor, CO, USA) were both characterized by rapid absorption and mean plasma concentrations remained above the target concentrations for 48 hours after administration.[16] The elimination half-life was approximately 9 times longer than those reported in AMKEs receiving 0.6 mg/kg IM standard BUP. A follow-up study evaluating the thermal antinociceptive effects of this formulation in the same species concluded that 1.8 mg/kg IM requires administration at least every 24 hours to manage pain but might have a longer duration of action depending on the severity and type of pain, adjunctive therapy, and the individual response.[17] In RTHAs, the PK of two doses of the concentrated formulation of BUP (Simbadol®, Zoetis, Kalamazoo, MI, USA) were evaluated after SC administration.[18] Plasma concentrations were maintained above 1 ng/ml for at least 24 and 48 hours, respectively. Adverse effects were not noted in any birds. The use of this concentrated BUP formulation might help reduce stress related to repeated handling and administration of treatments while providing consistent pain management.

In contrast, no significant difference was identified in thermal withdrawal threshold using the same dosages of BUP in cockatiels and no sedative effects were observed at any dosage.[19] In OWAPs, the thermal antinociceptive effects of the concentrated formulation of BUP (Simbadol®) after 2 mg/kg SC administration was evaluated. Concentrated BUP resulted in a small but significant increase in thermal threshold, while lower doses did not result in any significant effect. These results provide evidence of weak efficacy at lower doses for pain management in this species and highlight the differences in efficacy. Further studies are needed to fully determine the analgesic and adverse effects of buprenorphine in psittacines and warrants evaluation in other species.

Opioid-like medications

Tramadol HCl (TRAM) is an analgesic drug often used due to its ability to be dosed orally. It is a synthetic μ-opioid receptor agonist that appears to have variable efficacy depending on the species. The efficacy is thought to be primarily related to the production of the O-desmethyl-tramadol metabolite (M1). This metabolite acts centrally over the opioid receptor and is thought to have almost 200 times higher affinity for this receptor than the parent drug. Tramadol also has effects on the k and δ- opioid receptors, as well as inhibiting reuptake of norepinephrine and serotonin. The conversion to the M1 metabolite is variable among species, but it is produced in the bird species studied thus far. Pharmacokinetic profiles have been reported for a multitude of avian species, and important differences exist with PK parameters between species. Oral bioavailability of 11 mg/kg TRAM PO was higher (98%) for American bald eagles achieved concentrations in the human analgesic range for 10 hours; M1 plasma concentrations reached human ranges only in two eagles at much earlier time points. In RTHAs administered 11 mg/kg PO, analgesic thresholds for humans were maintained for approximately 4hrs.[20] In HAPs, doses of 10 or 20 mg/kg PO did not result in significant antinociception to thermal stimuli and human analgesic thresholds were only reached after dosing with 30 mg/kg orally or 5 mg/kg IV.[21] Transient bradycardia was noted after IV administration. In adult peafowl gavaged with 7.5 mg/kg, M1 concentrations remained above the human therapeutic threshold for 12-24 hours.[22] This underscores the importance of assaying for pharmacologically active metabolites in these studies. In AMKEs, 5 mg/kg PO TRAM significantly increased thermal thresholds for 9 hours when compared with baseline thresholds, while 15 and 30 mg/kg PO resulted in less antinociceptive effect.[8] In Muscovy ducks (*Cairina moschata domestica*), 30 mg/kg PO TRAM was effective in improving gait in a temporary arthritis model and assessed by ground-reactive forces measured by a pressure-sensitive walkway (PSW) system.[23] In this study, Muscovy ducks readily produced the M1 metabolite and maintain it well above concentrations known to be therapeutic for humans and other avian species. In African penguins (*Spheniscus demersus*) following 10 mg/kg PO administration, the elimination half-life was 7.3 hours, significantly longer than other species reported.[24] Tramadol and M1 concentrations were consistent with therapeutic concentrations in humans for 12 hours for tramadol and 36 hours for M1.

Tapentadol (TAP) is an analgesic agent which acts as both a μ-opioid receptor agonist and a norepinephrine reuptake inhibitor. It is a common therapeutic agent in human medicine for management of acute and chronic pain and is beginning to be investigated for use in veterinary medicine. The PK of TAP were evaluated in HAPs following a single 30 mg/kg PO of a compounded 5 mg/ml TAP suspension.[25] Low oral bioavailability and short $t_{1/2}$ of this drug in HAPs suggest that this drug might be of limited clinical use in this species.

Non-steroidal anti-inflammatory drugs (NSAIDs)

NSAIDs are used to relieve musculoskeletal and visceral pain, acute pain and chronic pain such as in osteoarthritis. The pharmacologic activity of NSAIDs has been reviewed elsewhere and the mechanism of action is similar when administered to birds. The relative expression of COX-1 and COX-2 enzymes varies between species and both enzymes are important in pain and inflammation. The application and dosages of NSAID formulations continues to be evaluated and clinically applied in birds. The intention of recently developed NSAIDs has been to spare COX-1 and emphasize COX-2 inhibition with the goal of providing analgesia and suppressing inflammation without inhibiting physiologically important prostaglandins. The common NSAIDs used in avian medicine include meloxicam (MEL), carprofen, celecoxib, robenacoxib and piroxicam, each with a distinct COX-1/COX-2 ratio and differing reports of effectiveness and toxicity in birds.

The most common adverse effects of NSAIDs include effects on the gastrointestinal system, renal system, and coagulation. NSAIDs have been implicated in mammals with an increased risk of myocardial infarction and delays in bone healing, but these effects have not been substantiated in birds. These adverse effects are often dose dependent and associated with chronic administration. Originally it was hypothesized that the adverse renal effects of NSAIDs were linked primarily to COX-1 inhibition but COX-2 metabolites have been implicated in maintenance of renal blood flow, mediation of renin release, and regulation of sodium excretion. Therefore, in conditions of relative intravascular volume depletion and/or renal hypoperfusion such as dehydration, hemorrhage, hemodynamic compromise, heart failure, and renal disease, interference with COX-2 activity can have significant deleterious effects on renal blood flow and glomerular filtration rate. The PK of NSAIDs is typically measured from plasma samples but it is important to remember that tissue kinetics may differ significantly, with drugs that have short plasma half-lives being active in tissues long after the drug has disappeared from the plasma. In mammals this has been evaluated using tissue cages where it is possible to measure the effect and duration of the NSAID by quantifying the effect in tissue PGs. A single report describes the use of this model in turkeys for evaluation of sodium salicylate.[26]

Meloxicam (MEL) is the most widely used anti-inflammatory medication today. Meloxicam is a COX-2-selective NSAID that demonstrates COX-1 selectivity at higher doses. Single dose MEL PK studies have been evaluated in a plethora of avian species; references for these can be requested from the authors. Differences in dosage, bioavailability, clearance and elimination half-lives likely contribute to the clinical differences seen with the use of this NSAID. For example, the MEL elimination half-life in chickens and pigeons is three times as long as that of ducks, ostriches and turkeys.[27] Wyandotte hens had greater AUC, longer $t_{1/2}$ and MEL persisted longer in their egg yolks compared with white leghorn hens, suggesting their PO dosing intervals may be every 24 hours and every 12 hours in white leghorns. Based on these results, a 2-week withdrawal time should be adequate to avoid drug residues in eggs meant for consumption.[28,29] Meloxicam was also identified in 2/3 unfertilized eggs of captive bearded vultures after breeding failure, but the implication is unclear.[30] Following 1 mg/kg MEL PO the elimination half-life in grey parrots was twice that of HAPs with significantly lower oral bioavailability (38.1% ± 3.6%).[31] In cockatiels, only 11% oral bioavailability following 1 mg/kg meloxicam PO was found with and elimination half-life of less than 1 hour.[32] The highly variable bioavailability in avian species[33-35] is quite different than mammals where it is usually 85% or greater.[28,29,32] A possible explanation could be variability in proventricular pH among species. In a multiple dose PK study performed in grey parrots evaluating three different dosages, plasma concentrations were greater than effective concentrations described for other avian species. Evidence of moderate drug accumulation was found with both IM and PO routes of administration.[36] The large variability in PK between avian species is one of the main issues with analgesic therapy in birds, as cross-species allometric scaling is not reliable and therefore PK/PD analysis is required in every species.[27] A sustained-release formulation of MEL (MEL SR, Wildlife Pharmaceuticals, Windsor, CO, USA) has been evaluated in two species- HAPs[37] and American flamingos.[38] Its use cannot be recommended due to huge PK variability (parrots) and lack of difference from standard MEL (flamingos).

There are very few studies evaluating the PDs of MEL in birds. In lame chickens, MEL significantly increased the thermal threshold compared to saline.[39] Administration of MEL at 0.5 mg/kg IM or 2 mg/kg IM reduced clinical signs associated with experimental systemic inflammation induced by *E. coli* LPS in broiler chickens, but dexamethasone was ultimately preferred by the authors for its clinical effects of reducing the acute inflammation.[40]

Since higher doses of MEL can block COX-1 enzyme activity, this drug has the potential to induce the adverse effects of gastrointestinal, renal and cardiac toxicity seen in mammals, yet reports of intoxication at doses used in avian practice are minimal. Numerous studies evaluating the adverse effects of MEL in different species[41-45] with no significant adverse effects were reported at doses in therapeutic range; however, 2/9 AMKEs had gastric ulceration after a dose of 20 mg/kg.[44] Plasma and urine N-acetyl-β-d-glucosaminidase activities were significantly increased in grey parrots at 1.6 mg/kg MEL administered PO q24h for 7 days.[36] Elimination half-lives of MEL are highly variable and those species with rapid elimination are less likely to have adverse effects.[32] Anecdotal reports of visceral gout and gastrointestinal ulceration in brown pelicans, using dosages commonly used in other species, prompted a PK study using 0.2 mg/kg PO which found the MEL elimination $t_{1/2}$ was 36 hours, the highest reported to date in any species, heightening suspicion that adverse effects, including renal toxicity, may be attributed to this.[46] The renal effects of NSAIDs have been of particular concern in the Asian vultures, but MEL may be the exception because of its overall short elimination half-life (< 1 hour); however, recently MEL was detected in eggs of captive bearded vultures (*Gypaetus barbatus*).[30] Despite MEL's large therapeutic range and relative safety compared with other NSAIDs, these new data demonstrate that dose extrapolation is not a suitable method for MEL dose determination in avian species. Species-specific PK, efficacy and toxicity studies are necessary to determine appropriate dosages for different species.

The newer NSAID coxibs celecoxib, mavacoxib, robenacoxib are COX-2 selective as opposed to MEL which is COX-2 preferential, inhibiting both COX-1 and COX-2. Celecoxib in humans is used for acute pain, osteoarthritis, rheumatoid arthritis and spondylitis. Its use in veterinary medicine is off label in the United States; it is not approved for use in any veterinary species. Celecoxib 10 mg/kg PO q24h has been used to reduce inflammation associated with avian ganglioneuritis.[47] Anecdotally there are suggestions of very high doses (30-40 mg/kg) PO q12h, then long term 15-20 mg/kg PO q12h for GIT signs, and 60-80 mg/kg divided PO q12h for CNS involvement with ABV.[48] Mavacoxib administered orally has a long mean half-life of 15.5 days in dogs, allowing for monthly dosing after two loading doses administered 1 week apart. [49] Plasma concentrations of 0.4 µg/ml mavacoxib were associated with efficacy in reducing osteoarthritis in dogs.[50]

The single dose PK and oral bioavailability of celecoxib 10 mg/kg PO and mavacoxib 4 mg/kg PO were evaluated in the cockatiel.[32] Mavacoxib was completely absorbed (bioavailability = 111%) but celecoxib only had a mean PO bioavailability of 56%.[32] Bioavailability of the coxibs is often directly associated with feeding status; the presence of food has been shown to increase absorption of both NSAIDs.[51] The cockatiels in this study were fed ad libitum. A further study in cockatiels with ad libitum feeding and additional gavage feedings during the study showed that changes in body temperature, mentation, and posture were less pronounced with a single dose of mavacoxib 4 mg/kg PO compared to celecoxib 10 mg/kg at the tested dosages.[52] This is likely associated with the higher oral bioavailability of mavacoxib in cockatiels compared to celecoxib and MEL. Plasma concentrations of 0.4 µg/ml mavacoxib, effective in osteoarthritic dogs[53] were maintained for 1,000 hours in the cockatiels[32,52] and Caribbean flamingos maintained plasma concentrations > 0.4 µg/ml for 120 hours after a single dose of 6 mg/kg.[54] Elimination half-lives in cockatiels (135 hours)[32] and Caribbean flamingos (*Phoenicopterus ruber*, 74.5 hours)[54] were not as long as in dogs but are still much longer than most NSAID elimination half-lives in other bird species and prompts additional studies to determine dosing regimens.[54]

Piroxicam is a nonselective NSAID used for its anti-inflammatory properties and its value as a chemopreventive and antitumor agent. It has a much higher potency against COX-1 than COX-2. In chickens, intramuscular administration of 1-2 mg/kg of piroxicam twice daily for 4 days failed to result in gastrointestinal lesions consistent with toxicity.[55] Piroxicam has been used in cranes for more than 20 years to treat a variety of conditions including trauma, inflammation, and postoperative pain with minimal observed side effects.[56] A PK study in brolga cranes (*Grus rubicunda*) evaluated 0.5 mg/kg and 1 mg/kg PO, finding that daily administration might be appropriate for this species.[56]

Scavenging bird species such as vultures and eagles may be exposed to NSAIDs when they feed on livestock carcasses. The NSAID diclofenac played the major role in the catastrophic decline in vulture populations on the Indian sub-continent. Subsequently, other NSAIDs have also been identified as toxic to vultures. Ketoprofen and nimesulfide, a sulfonamide NSAID, caused visceral gout and death in vultures whereas MEL and tolfenamic acid (TA) were reported to be safe.[57-61]

Other NSAIDs must be identified to reduce the use of those toxic to vultures and other scavengers. Tolfenamic acid is an NSAID belonging to the fenamates group. This NSAID inhibits both COX-1 and COX-2 receptor activity and exerts potent anti-tumor activity by both COX-dependent and independent pathways. Tolfenamic acid was detected in 20% Eurasian griffon vulture (*Gyps fulvus*) nestling samples, [62] suggesting it is being used more frequently by veterinarians in the region. A study evaluating the safety of TA in three *Gyps* species, all of which are susceptible to diclofenac, was performed with 3.5 mg/kg TA PO via gavage and two birds died with elevated uric acid levels and severe visceral gout; the remainder showed no adverse clinical or biochemical signs at the same or even higher dosages, suggesting the deaths may have been an anomaly due to the high dose used and the high ambient temperatures at the time of the experiment.[58] Tolfenamic acid is likely to be safe to *Gyps* vultures at concentrations encountered by wild birds and could therefore be promoted as a safe alternative to toxic NSAIDs.[58] Otherwise the pharmacokinetics of TA in chukar partridge (*Alectoris chukar*),[63] Japanese quail (*Coturnix japonica*)[64] and geese (*Anser cygnoides*)[65] have been reported. The $t_{1/2\lambda z}$ of TA administered PO in chukar partridges, Japanese quail and geese was similar to MEL.[63,65] These short $t_{1/2\lambda z}$ are important in preventing drug accumulation and delaying toxic effects in repeated administrations in birds.[66] Bioavailability after a 2 mg/kg PO dose was 88%, 78%, and 76% for the IM, SC, and oral routes, respectively in geese, and 57% in Japanese quail suggesting relatively good absorption in these species. Currently there are no specific dosing recommendations for TA in birds but in mammals and reptiles 2-4 mg/kg PO is used.[67,68] While the few studies to date are promising, additional studies evaluating clinical efficacy, multi-dose pharmacokinetics, safety, and residue data of TA should be conducted before widespread use of TA in birds at mammalian doses.

Prostaglandin receptor antagonist

Grapiprant (GRAP) is a prostaglandin E2 (PGE_2) receptor antagonist approved for use in dogs in the United States. Grapiprant binds to the EP4 receptor, blocking the sensitization of sensory neurons and stimulation of inflammation, mediated by the PGE_2 receptor. It does not rely on inhibition of COX enzymes and preserves functions associated with COX enzymes pathways. Oral administration of GRAP at 30 mg/kg to fasted RTHAs achieved plasma concentrations approximately 100 times the canine therapeutic concentrations with no adverse effects identified. It is still unknown whether the dog concentrations are analgesic in birds or toxic in hawks after multi-dosing.[69] Grapiprant administered at the same dosage to the same RTHAs with food had significantly lower plasma concentrations compared to when fasted, with delayed absorption and similar half-lives (Hawkins MG, personal communication, April 1, 2023). Due to the concern that target plasma concentrations may not be maintained if this dosage was administered every 24 hours versus every 12 hours, multi-dose modeling simulations showed that the concentrations troughs are still above the target plasma concentrations in dogs for analgesia (Hawkins MG, personal communication, January 9, 2022). Further studies evaluating the PK/PD of multi-dose GRAP with food in RTHAs are underway.

Gabapentin

Gabapentin is an analog of the neurotransmitter gaba-aminobutyric acid (GABA) that binds to voltage-gated calcium channels acting presynaptically to decrease the release of excitatory neurotransmitters, especially glutamate. In humans and animals, it is best used as an adjunct to other analgesics and seems to be most effective for neuropathic pain. Some formulations of gabapentin contain xylitol and these should not be used in birds until further safety evidence is available. Oral bioavailability was found to be high (> 80%) in HAPs.[70] The terminal $t_{1/2}$ has been in the range of 3-5 hours and so gabapentin should be administered 2-3 times per day to maintain adequate plasma concentrations. In great horned owls (*Bubo virginianus*) following oral administration of 11 mg/kg PO, plasma gabapentin concentrations were maintained above 2 μg/ml for 9.6 hours.[71] In Caribbean flamingos (*Phoenicopterus ruber ruber),* gabapentin dosed at 25 mg/kg PO maintained plasma concentrations above the human therapeutic range for approximately 12 hours; however, dosages as high as 37 mg/kg PO q12h may be needed in some individuals.

Amantadine

Amantadine is an antiviral medication that increases dopamine in the central nervous system and weakly inhibits the NMDA receptors that contribute to chronic pain. Amantadine is synergistic with NSAIDs, opioids, and gabapentin/pregabalin. In humans, amantadine has been shown to decrease opioid use in post-operative patients and aid in treating neuropathic pain. Amantadine administered to dogs at 3-5 mg/kg PO q24h with meloxicam showed significant improvement in management of osteoarthritis pain refractory to meloxicam alone. In OWAPs, the single and multiple dose PK profile was evaluated following 10 mg/kg PO once and 5 mg/kg PO q24h for 7 days.[72] No obvious adverse effects have been observed. Based on these results, once-daily administration of amantadine at 5 mg/kg PO in OWAPs can maintain above target plasma concentrations with weak accumulation. For long term administration, it is recommended to check plasma concentrations over time. The use of amantadine in chickens, ducks, and turkeys is prohibited in the United States by the Federal Drug Administration.

Cannabidiol

Cannabidiol (CBD) oil is the primary non-psychoactive cannabinoid derived from the cannabis plant (*Cannabis sativa*). The pharmacokinetics of a CBD oral formulation (50 mg/ml; Canna Companion, Mill Creek, WA, USA) was evaluated in HAPs. The highly variable results and short half-life of the drug in HAPs, even at high doses, suggests that this formulation would need to be administered at least three to four times daily. Evaluation of other formulations and in other species is recommended to further characterize the analgesic potential of CBD in birds.

References

1. Mikoni NA, Guzman DSM, Fausak E, et al. Recognition and assessment of pain-related behaviors in avian species: an integrative review. *J Avian Med Surg.* 2022;6:153-172.
2. Mikoni NA, Sanchez-Migallon Guzman D, Paul-Murphy J. Pain recognition and assessment in birds. *Vet Clin North Am Exot Anim Pract.* 2023;26:65-81.
3. Paul-Murphy J. Foundations in avian welfare In: Speer BL, ed. *Current Therapy in Avian Medicine and Surgery*. St. Louis, MO: Elsevier; 2015;669-677.
4. National Research Council (US). *Recognition and Alleviation of Pain in Laboratory Animals.* National Academies Press; 2010.
5. Duhamelle A, Raiwet DL, Langlois I, et al. Preliminary findings of structure and expression of opioid receptor genes in a peregrine falcon (*Falco peregrinus*), a snowy owl (*Bubo scandiacus*), and a blue-fronted Amazon parrot (*Amazona aestiva*). *J Avian Med Surg.* 2018;32:173-184.

6. Fousse SL, Golsen BM, Sanchez-Migallon Guzman D, et al. Varying expression of mu and kappa opioid receptors in cockatiels (*Nymphicus hollandicus*) and domestic pigeons (*Columba livia domestica*). *Front Genet.* 2020;11.

7. Guzman DS, Drazenovich TL, Olsen GH, et al. Evaluation of thermal antinociceptive effects after intramuscular administration of hydromorphone hydrochloride to American kestrels (*Falco sparverius*). *Am J Vet Res.* 2013;74:817-822.

8. Guzman DS, Drazenovich TL, Olsen GH, et al. Evaluation of thermal antinociceptive effects after oral administration of tramadol hydrochloride to American kestrels (*Falco sparverius*). *Am J Vet Res.* 2014;75: 117-123.

9. Houck EL, Guzman DS-M, Beaufrère H, et al. Evaluation of the thermal antinociceptive effects and pharmacokinetics of hydromorphone hydrochloride after intramuscular administration to cockatiels (*Nymphicus hollandicus*). *Am J Vet Res.* 2018;79:820-827.

10. Sanchez-Migallon Guzman D, Douglas JM, Beaufrère H, et al. Evaluation of the thermal antinociceptive effects of hydromorphone hydrochloride after intramuscular administration to orange-winged Amazon parrots (*Amazona amazonica*). *Am J. Vet Res.* 2020;81:775-782.

11. Escobar A, Barletta M, Pypendop BH, et al. Pharmacokinetics and pharmacodynamics of methadone administered intravenously and intramuscularly to isoflurane-anesthetized chickens. *Am J. Vet Res.* 2021;82: 181-188.

12. Hawkins MG, Pascoe PJ, DiMaio Knych HK, et al. Effects of three fentanyl plasma concentrations on the minimum alveolar concentration of isoflurane in Hispaniolan Amazon parrots (*Amazona ventralis*). *Am J. Vet Res.* 2018;79:600-605.

13. Delaski KM, Gehring R, Heffron BT, et al. Plasma concentrations of fentanyl achieved with transdermal application in chickens. *J Avian Med Surg.* 2017;31:6-15.

14. Escobar A, Valadão CA, Brosnan RJ, et al. Cardiopulmonary effects of butorphanol in sevoflurane-anesthetized guineafowl (*Numida meleagris*). *Vet Anaesthes Analg.* 2014;41:284-289.

15. Clancy MM, KuKanich B, Sykes JM. Pharmacokinetics of butorphanol delivered with an osmotic pump during a seven-day period in common peafowl (*Pavo cristatus*). *Am J Vet Res.* 2015;76:1070-1076.

16. Guzman DS-M, Knych HK, Olsen GH, et al. Pharmacokinetics of a sustained release formulation of buprenorphine after intramuscular and subcutaneous administration to American kestrels (*Falco sparverius*). *J Avian Med Surg.* 2017;31:102-107.

17. Guzman DS, Ceulemans SM, Beaufrère H, et al. Evaluation of the thermal antinociceptive effects of a sustained-release buprenorphine formulation after intramuscular administration to American kestrels (*Falco sparverius*). *J Avian Med Surg.* 2018;32:1-7.

18. Gleeson MD, Guzman DSM, Knych HK, et al. Pharmacokinetics of a concentrated buprenorphine formulation in red-tailed hawks (*Buteo jamaicensis*). *Am J Vet Res.* 2018;79:13-20.

19. Guzman DS-M, Houck EL, Knych HKD, et al. Evaluation of the thermal antinociceptive effects and pharmacokinetics after intramuscular administration of buprenorphine hydrochloride to cockatiels (*Nymphicus hollandicus*). *Am J Vet Res.* 2018;79:1239-1245.

20. Souza MJ, Martin-Jimenez T, Jones MP, et al. Pharmacokinetics of oral tramadol in red-tailed hawks (*Buteo jamaicensis*). *J Vet Pharmacol Ther.* 2011;34:86-88.

21. Sanchez-Migallon Guzman D, Souza MJ, Braun JM, et al. Antinociceptive effects after oral administration of tramadol hydrochloride in Hispaniolan Amazon parrots (*Amazona ventralis*). *Am J Vet Res.* 2012;73: 1148-1152.

22. Black PA, Cox SK, Macek M, et al. Pharmacokinetics of tramadol hydrochloride and its metabolite O-desmethyltramadol in peafowl (*Pavo cristatus*). *J Zoo Wildl Med.* 2010;41:671-676.

23. Bailey RS, Sheldon JD, Allender MC, et al. Analgesic efficacy of tramadol compared with meloxicam in ducks (*Cairina moschata domestica*) evaluated by ground-reactive forces. *J Avian Med Surg.* 2019;33: 133-140.

24. Kilburn JJ, Cox SK, Kottyan J, et al. Pharmacokinetics of tramadol and its primary metabolite O-desmethyltramadol in African penguins (*Spheniscus demersus*). *J Zoo Wildl Med.* 2014;45:93-99.

25. Duvall A, Tully Jr TN, Carpenter JW, et al. Pilot study of a single dose of orally administered tapentadol suspension in Hispaniolan Amazon parrots (*Amazona ventralis*). *J Avian Med Surg.* 2021;35:45-50.

26. Cramer K, Schmidt V, Richter A, et al. Investigations on the acute, carrageenan-induced inflammatory reaction and pharmacology of orally administered sodium salicylate in turkeys. *Berl Munch Tierarztl Wochenschr.* 2015;128:240-251.

27. Baert K, De Backer P. Comparative pharmacokinetics of three non-steroidal anti-inflammatory drugs in five bird species. *Comp Biochem Physiol C Toxicol Pharmacol.* 2003;134:25-33.

28. Souza MJ, Bailey J, White M, et al. Pharmacokinetics and egg residues of meloxicam after multiple day oral dosing in domestic chickens. *J Avian Med Surg.* 2018;32:8-12.

29. Souza MJ, Gerhardt LE, Shannon L, et al. Breed differences in the pharmacokinetics of orally administered meloxicam in domestic chickens (*Gallus domesticus*). *J Am Vet Med Assoc.* 2021;259:84-87.

30. Zorrilla I, Richards NL, Benítez JR, et al. Case study: detection of two nonsteroidal anti-inflammatory drugs (NSAIDs) in the eggs of captive-reared bearded vultures at a breeding center in southern Spain. *J Wildl Rehabil.* 2018;38.

31. Montesinos A, Ardiaca M, Gilabert JA, et al. Pharmacokinetics of meloxicam after intravenous, intramuscular and oral administration of a single dose to African grey parrots (*Psittacus erithacus*). *J Vet Pharmacol Ther.* 2017;40:279-284.

32. Dhondt L, Devreese M, Croubels S, et al. Comparative population pharmacokinetics and absolute oral bioavailability of COX-2 selective inhibitors celecoxib, mavacoxib and meloxicam in cockatiels (*Nymphicus hollandicus*). *Sci Rep.* 2017;7:12043.

33. Montesinos A, Ardiaca M, Diez-Delgado I. Pharmacokinetics of parenteral and oral meloxicam in African grey parrots (*Psittacus erithacus erithacus*). *Proc Annu Conf Eur Assoc Avian Vet.* 2011;129-134.

34. Molter CM, Court MH, Cole GA, et al. Pharmacokinetics of meloxicam after intravenous, intramuscular, and oral administration of a single dose to Hispaniolan Amazon parrots (*Amazona ventralis*). *Am J Vet Res.* 2013;74:375-380.

35. Lacasse C, Gamble KC, Boothe DM. Pharmacokinetics of a single dose of intravenous and oral meloxicam in red-tailed hawks (*Buteo jamaicensis*) and great horned owls (*Bubo virginianus*). *J Avian Med Surg.* 2013;27:204-210.

36. Montesinos A, Encinas T, Ardiaca M, et al. Pharmacokinetics of meloxicam during multiple oral or intramuscular dose administration to African grey parrots (*Psittacus erithacus*). *Am J Vet Res.* 2019;80:201-207.

37. Guzman DS, Court MH, Zhu Z, et al. Pharmacokinetics of a sustained-release formulation of meloxicam after subcutaneous administration to Hispaniolan Amazon parrots (*Amazona ventralis*). *J Avian Med Surg.* 2017;31:219-224.

38. Sim RR, Cox SK. Pharmacokinetics of a sustained-release formulation of meloxicam after subcutaneous administration to American flamingos (*Phoenicopterus ruber*). *J Zoo Wildl Med.* 2018;49:839-843.

39. Hothersall B, Caplen G, Parker RM, et al. Thermal nociceptive threshold testing detects altered sensory processing in broiler chickens with spontaneous lameness. *PloS One.* 2014;9:e97883.

40. Nakhaee P, Mosleh N, Nazifi S, et al. Comparative effects of meloxicam and dexamethasone in chickens with experimental systemic inflammation: clinical outcome and cardiovascular parameters. *Comp Clin Pathol.* 2021;30:681-691.

41. Montesinos A, Ardiaca M, Juan-Sallés C, et al. Effects of meloxicam on hematologic and plasma biochemical analyte values and results of histologic examination of kidney biopsy specimens of African grey parrots (*Psittacus erithacus*). *J Avian Med Surg.* 2015;29:1-8.

42. Miller KA, Hill NJ, Carrasco SE, et al. Pharmacokinetics and safety of intramuscular meloxicam in zebra finches (*Taeniopygia guttata*). *J Am Assoc Lab Anim Sci.* 2019;58:589-593.

43. Desmarchelier M, Troncy E, Fitzgerald G, et al. Analgesic effects of meloxicam administration on postoperative orthopedic pain in domestic pigeons (*Columba livia*). *Am J Vet Res.* 2012;73:361-367.

44. Summa NM, Guzman DS-M, Larrat S, et al. Evaluation of high dosages of oral meloxicam in American kestrels (*Falco sparverius*). *J Avian Med Surg.* 2017;31:108-116, 109.

45. Dijkstra B, Guzman DS, Gustavsen K, et al. Renal, gastrointestinal, and hemostatic effects of oral administration of meloxicam to Hispaniolan Amazon parrots (*Amazona ventralis*). *Am J Vet Res.* 2015;76:308-317.

46. Horgan MD, Knych HK, Siksay SE, et al. Pharmacokinetics of a single dose of oral meloxicam in rehabilitated wild brown pelicans (*Pelecanus occidentalis*). *J Avian Med Surg.* 2020;34:329-337.

47. Hoppes SM, Tizard I, Shivaprasad HL. Avian bornavirus and proventricular dilatation disease. *Vet Clin North Am Exot Anim Pract.* 2013;339-355.

48. Rossi G, Dahlhausen RD, Galosi L, et al. Avian ganglioneuritis in clinical practice. *Vet Clin North Am Exot Anim Pract.* 2018;21:33-67.

49. Lees P, Pelligand L, Elliott J, et al. Pharmacokinetics, pharmacodynamics, toxicology and therapeutics of mavacoxib in the dog: a review. *J Vet Pharmacol Ther.* 2015;38:1-14.

50. Payne-Johnson M, Becskei C, Chaudhry Y, et al. Comparative efficacy and safety of mavacoxib and carprofen in the treatment of canine osteoarthritis. *Vet Rec.* 2014.

51. Cox SR, Lesman SP, Boucher JF, et al. The pharmacokinetics of mavacoxib, a long-acting COX-2 inhibitor, in young adult laboratory dogs. *J Vet Pharmacol Ther.* 2010;33:461-470.

52. Gasthuys E, Houben R, Haesendonck R, et al. Development of an in vivo lipopolysaccharide inflammation model to study the pharmacodynamics of COX-2 inhibitors celecoxib, mavacoxib, and meloxicam in cockatiels (*Nymphicus hollandicus*). *J Avian Med Surg.* 2019;33:349-360.

53. Cox SR, Liao S, Payne-Johnson M, et al. Population pharmacokinetics of mavacoxib in osteoarthritic dogs. *J Vet Pharmacol Ther.* 2011;34:1-11.

54. Huckins GL, Carpenter JW, Dias S, et al. Pharmacokinetics of oral mavacoxib in Caribbean flamingoes (*Phoenicopterus ruber ruber*) *J Zoo Wildl Med.* 2020;51:53-58.

55. Awan AF, Ashraf M, Malik A, et al. Toxicity study of piroxicam in broilers. *Lat Am J Pharm.* 2012;31.

56. Keiper NL, Cox SK, Doss GA, et al. Pharmacokinetics of piroxicam in cranes (Family Gruidae). *J Zoo Wildl Med.* 2017;48:886-890, 885.

57. Galligan TH, Mallord JW, Prakash VM, et al. Trends in the availability of the vulture-toxic drug, diclofenac, and other NSAIDs in South Asia, as revealed by covert pharmacy surveys. *Bird Conserv Intl.* 2021;31: 337-353.

58. Chandramohan S, Mallord JW, Mathesh K, et al. Experimental safety testing shows that the NSAID tolfenamic acid is not toxic to *Gyps* vultures in India at concentrations likely to be encountered in cattle carcasses. *Sci Total Environ.* 2022;809:152088.

59. Galligan TH, Green RE, Wolter K, et al. The non-steroidal anti-inflammatory drug nimesulide kills *Gyps* vultures at concentrations found in the muscle of treated cattle. *Sci Total Environ.* 2022;807:150788.

60. Nambirajan K, Muralidharan S, Ashimkumar AR, et al. Nimesulide poisoning in white-rumped vulture *Gyps bengalensis* in Gujarat, India. *Environ Sci Pollut Res.* 2021:1-7.

61. Stokstad E. Vultures face new toxic threat. *Science.* 2021;373:1187-1187.

62. Gómez-Ramírez P, Blanco G, García-Fernández AJ. Validation of multi-residue method for quantification of antibiotics and NSAIDs in avian scavengers by using small amounts of plasma in HPLC-MS-TOF. *Int J Environ Res Public Health.* 2020;17.

63. Cetin G, Corum O, Corum D, et al. Pharmacokinetics of intravenous meloxicam, ketoprofen and tolfenamic acid in chukar partridge (*Alectoris chukar*). *Br Poult Sci.* 2021:1-7.

64. Turk E, Tekeli IO, Corum O, et al. Pharmacokinetics of meloxicam, carprofen, and tolfenamic acid after intramuscular and oral administration in Japanese quails (*Coturnix coturnix japonica*). *J Vet Pharmacol Ther.* 2021;44:388-396.

65. Turk E, Tekeli IO, Durna Corum D, et al. Pharmacokinetics of tolfenamic acid after different administration routes in geese (*Anser cygnoides*). *J Vet Pharmacol Ther.* 2021;44(3):381-387.

66. Naidoo V, Wolter K, Cromarty AD, et al. The pharmacokinetics of meloxicam in vultures. *J Vet Pharmacol Ther.* 2008;31:128-134.

67. Corum O, Atik O, Durna Corum D, et al. Pharmacokinetics of tolfenamic acid in red-eared slider turtles (*Trachemys scripta elegans*). *Vet Anaesth Analg.* 2019;46:699-706.

68. Tekeli IO, Turk E, Durna Corum D, et al. Effect of dose on the intravenous pharmacokinetics of tolfenamic acid in goats. *J Vet Pharmacol Ther.* 2020;43:435-439.

69. Rodriguez P, Paul-Murphy JR, Knych HK, et al. Pharmacokinetics of grapiprant administered to red-tailed hawks (*Buteo jamaicensis*) after food was withheld for 24 hours. *Am J Vet Res.* 2021;82:912-919.

70. Baine K, Jones MP, Cox S, et al. Pharmacokinetics of compounded intravenous and oral gabapentin in Hispaniolan Amazon parrots (*Amazona ventralis*). *J Avian Med Surg.* 2015;29:165-173.

71. Yaw TJ, Zaffarano BA, Gall A, et al. Pharmacokinetic properties of a single administration of oral gabapentin in the great horned owl (*Bubo virginianus*). *J Zoo Wildl Med.* 2015;46:547-552.

72. Berg KJ, Sanchez-Migallon Guzman D, Knych HK, et al. Pharmacokinetics of amantadine after oral administration of single and multiple doses to orange-winged Amazon parrots (*Amazona amazonica*). *Am J Vet Res.* 2020;81:651-655.

Section 9

AAV Special Panel on Geriatrics

Shangzhe Xie, BSc/BVMS, MVS, PhD, DABVP (Avian Practice), DACCM
Session Chair

Avian Geriatric Diseases

Sharman Hoppes, DVM, DABVP (Avian Practice),
Scott Echols, DVM, DABVP (Avian Practice),
Yvonne van Zeeland, DVM, MVR, PhD, DipECZM (Avian),
DipECZM (Small Mammal), CPBC,
Susan Orosz, PhD, DVM, DABVP (Avian Practice), DipECZM (Avian)

Session #3282

Affiliation: From Texas Avian & Exotic Hospital, Grapevine, TX 76051, USA, (Hoppes), Medical Center for Birds, 3807 Main Street, Oakley, CA 94561, USA (Echols), Parrish Creek Veterinary Hospital and Diagnostic Center Centerville, UT 84014, USA (Echols), Echols Veterinary Services, Cottonwood Heights, UT 84093, USA (Echols), Scarlet Imaging, PO Box 57515, Murray, UT 84157, USA (Echols), Division of Zoological Medicine, Department of Clinical Sciences of Companion Animals, Faculty of Veterinary Medicine, Utrecht University, Utrecht, The Netherlands (van Zeeland), Bird and Exotic Pet Wellness Center, Toledo, OH 43623, USA (Orosz).

Abstract: As a result of improved psittacine husbandry and nutrition, and advances in the diagnosis and treatment of diseases, life expectancy of captive parrots has substantially increased over the years. This in turn has led to a growing number of elderly birds seen in veterinary practices and associated with it a higher incidence of geriatric-related diseases such as cataracts, cardiovascular disease, chronic lung, hepatic or renal disease, spinal and joint disease, cystic ovarian disease, and neoplasia. In this masterclass, four of the most common conditions seen in aging psittacine birds (i.e., atherosclerosis and cardiac disease, osteoarthritis, renal disease and neoplastic conditions of the reproductive system) are discussed including their clinical presentation, diagnostics, and the preventive and therapeutic management. A panelist team, consisting of specialists in avian medicine that work in private practice and academia present their approach to each of these conditions from their respective backgrounds to help provide a broad and international perspective on the approach to these conditions. This session provides practitioners with guidelines for immediate implementation in practice, as well as insight into some of the more advanced, novel, or future options to optimize care for the aging psittacine bird.

Introduction

Avian geriatric medicine comprises the care, treatment, and management of the elderly bird. This presentation focuses primarily on pet parrots. In the past, few pet birds lived to be geriatric due to poor husbandry, nutrition, and infectious diseases. Over the years, avian veterinarians, avian industry, and the pet bird owner have gained knowledge in husbandry, nutrition, medicine, and surgery, hence resulting in an increased life expectancy of companion birds. Nevertheless, despite their increased longevity, few parrots outlive their owners. Despite life spans of up to 60 to 80 years having been documented (dependent on the species), most parrots do not live that long as a result of a sedentary lifestyle, over abundant nutrition, and possible genetic factors.

Organ Changes Over Time

Physiologic changes that occur in geriatric pet birds have not been well studied. Similarly, data from poultry cannot be easily extrapolated because studies are based on reproductive capacity for production and not on a bird's actual potential life span.

In humans and small animals, there are a number of documented age-related issues. In a study looking at young versus older dogs (> 10 years), older dogs had significantly lower venous oxygen saturation and lower plasma glucose after exercise, and a slower hematopoietic response to acute anemia compared to young dogs.[1]

In humans over 80 years of age, a hypercoagulable shift can be noted, which is associated with a higher incidence of pulmonary thromboembolism. Similar changes occur in some species of small mammals. The thoracic cage becomes more rigid, the lungs less elastic, and respiratory muscle strength decreases by 25%. Respiratory diseases are generally more common and more difficult to treat in elderly patients of all species.[1] In addition, both older humans and older parrots are prone to atherosclerosis, which leads to hypertension and cardiac disease.[2]

In mammals, glomerular filtration rate, urine concentrating and diluting ability and creatinine clearance all diminish with age, resulting in decreased renal function.[2] Similarly, a decrease in liver function is common in geriatric mammals. Geriatric mammalian patients can have up to 50% age related reduction in liver mass. Less information is available on the effects of age on the avian kidneys and liver, but we do know that older avian patients have an increased incidence of renal and liver disease overall, of which many are chronic in nature.[1,2]

Osteoarthritis is common in geriatric patients of all species, including older companion birds. Older animals lose lean body muscle mass which is replaced by fat. Aging changes in the musculoskeletal system, most notably the cells and extracellular matrix of joint tissues, likely increase the susceptibility of older individuals to osteoarthritis, especially when other risk factors such as joint injury, obesity, sedentary lifestyle, or anatomic features that affect joint mechanics are present.

Aging also affects the reproductive system, leading to changes in gonadal tissues, decreased production of sperm, declining stores of primary oocytes, and overall reduced fertility as a result of changing hormone levels. Preliminary comparisons of reproductive aging of female hens, quail, and budgerigars with mammals of similar size show that birds age slightly to much slower than their mammalian counterparts.[3,4] Nevertheless, certain reproductive diseases are seen more frequently in older hens (e.g., cystic ovarian disease, neoplasms).

Cognitive changes have been documented in many mammalian mammals; however, effects of aging on avian cognition remain largely unknown. There are anecdotal accounts from owners and veterinarians that suggest senility may occur. For example, changes in behavior and cognition have been reported in birds with cardiac disease or atherosclerosis.

Geriatric Diseases

Defining geriatric age in parrots is challenging. The life span of many parrots has been documented to be 60 to 80 years in larger parrots and 30 to 40 years in smaller parrots; however, due to malnutrition or over abundant feeding and obesity, combined with the sedentary lifestyles of most pet birds, many parrots develop old-age diseases much earlier than expected. Since many owners still view their bird as a teenager at age 20, they are hardly prepared for their bird to be diagnosed with cardiac disease or atherosclerosis.

The list of common geriatric diseases in the bird is similar to the list for mammals, including humans. Osteoarthritis, cataracts, metabolic bone disease (osteoporosis), reproductive disease, renal disease, hepatic disease, atherosclerosis, cardiac disease, and neoplasia are amongst the diseases commonly seen in the elderly patient. In this masterclass osteoarthritis, metabolic bone disease, renal disease, atherosclerosis and cardiac disease, and reproductive disease, most notably neoplastic conditions of the reproductive system, are discussed.

Advanced Imaging

New imaging technology, most notably computed tomography (CT) and to a lesser degree magnetic resonance imaging (MRI), has been instrumental in identifying many geriatric diseases. That statement does not discount the benefits of high-resolution radiograph and ultrasound studies. This presentation focuses on the use of newer imaging technology to identify the diseases discussed in this masterclass.

Newer high-resolution flat-panel CT scanners and 3T and greater MRI often are required to identify fine details necessary for accurate diagnoses in small patients. While older CT and MRI technology may be acceptable for identifying large masses and gross changes that may otherwise be identified via physical examination and radiographs and/or ultrasound, high resolution and fine detail obtained from newer scanners are often required to make accurate diagnoses.

Contrast studies for both CT and MRI improve the diagnosis of normal versus abnormal tissue. Without contrast, CT (ionizing radiation) is only capable of differentiating five basic densities (same as with radiographs): air, fat, soft tissue, bone and metal. Properly performed contrast studies provide greater differentiation of soft tissue structures. This allows the reviewer to visualize blood vessels, different organs, masses and other structures that would otherwise appear as one density. Contrast studies in MRI (radiofrequency and magnetization) also help to better differentiate soft tissue structures and highlight pathology.

Osteoarthritis

The prevalent form of arthritis in birds is osteoarthritis (OA). Osteoarthritis is a common cause of lameness and chronic pain, particularly in psittacine birds over 15 years of age and in birds that have experienced trauma. It also occurs at a younger age in waterfowl and poultry kept as pets, especially those kept on packed earth, concrete or other hard surfaces. Bumblefoot, ulcerative dermatitis, and callus formation may be secondary effects of OA. Osteoarthritis is caused by damage to chondrocytes. This damage or stress on the cartilage may be related to trauma, obesity, normal wear on abnormal joints or abnormal wear on normal joints. Osteoarthritis is a progressive disease; unless the cycle can be interrupted, further degradation of the cartilage causes continued damage to the joint.

Joint pain is the essential problem associated with OA. The synovium and subchondral bone are the major sources of OA pain, but pain can also arise from periarticular muscles and ligaments. Clinical signs can vary, depending on the location and severity of the disease. Birds may present with lameness and a history of decreased activity, inability or reluctance to perch, falling off the perch, or altered preference for perches. In addition, a flighted bird may not want to fly or cannot fly as well as it used to fly. Some birds may demonstrate a poor grip response, which may mimic more commonly recognized disorders such as leg fractures, renal disease, nerve trauma, heavy metal poisoning and more. Other, less obvious signs of OA include swollen joints, decreased range of motion, self-injurious behavior such as feather destructive behavior or mutilation, and excessive vocalization or biting/aggression.

Diagnosis is based on clinical signs, and findings from the physical examination and radiographs or CT. Radiographic lesions include narrowing of the joint space, sclerosis of subchondral bone, misalignment of the joint, and osteophyte formation. Such changes are more readily recognized using a CT scan, as this is a more sensitive tool for diagnosing osteoarthritis than radiology. Commonly affected joints include the phalangeal joints, the tibiotarsus-tarsometatarsal joint and stifle joint. Joints of the thoracic limb appear less commonly affected than the pelvic limb, probably due to many companion birds not being flighted.[2] The shoulder appears to be the most commonly identified wing joint with OA based on CT evaluation (MSE). Another, possibly most underdiagnosed region is in the spine. While high-resolution CT is often required for diagnosis, one author (MSE) would argue that spinal arthritis is fairly common in aged birds.

Although not formally adopted in avian medicine, the Kellgren and Lawrence system for radiographic classification of osteoarthritis and the more recently published CT arthritis classification system used in humans may serve as a guide in birds.[5-7] The tenets of Kellgren and Lawrence system for radiographic OA are described in Table 1.

Table 1. Kellgren and Lawrence system for radiographic osteoarthritis.

Grade	Description
Grade 0 (none)	Absence of radiographic changes of osteoarthritis
Grade 1 (doubtful)	Questionable joint space narrowing and possible osteophytic lipping
Grade 2 (minimal)	Possible joint space narrowing and clearly defined osteophytes
Grade 3 (moderate)	Definite joint space narrowing with some sclerosis and possibly deformed bone ends and moderate multiple osteophytes
Grade 4 (severe)	Severely narrowed joint space, severe sclerosis and definite deformity of bone ends with large osteophytes

The CT scoring system used in humans describes osteophytes sizes that may be too large for most bird species. The basic concepts mostly involve evaluation of the joint space, presence of osteophytes and identifying cysts and can be applied to avian patients.[7]

A multimodal treatment plan is recommended to provide the best outcome for the patient. Additionally, use of multiple drugs may allow for decreased drug dosages, tapered to the individual drug to prevent toxicity (which may be especially important given the increased incidence of renal or liver disease at old age). Management of OA incorporates both medical and nonmedical modalities to achieve therapeutic goals. Medical options include the (combined) use of nonsteroidal anti-inflammatory drugs (NSAIDs), chondroprotectants, and possibly opioids. The most common NSAID utilized in avian medicine is meloxicam, a COX-2 inhibitor, which is commonly used at a dose of 1-1.6 mg/kg PO q12-24h . Celebrex can be used at 20 mg/kg q12-24h for up to several months. There are potential side effects of NSAIDs such as renal ischemia or gastric ulceration. NSAIDs should be used with caution, with the goal to use the lowest therapeutic dose possible to reduce the risk of adverse effects. In addition, gastric protectants (e.g., sucralfate) or antacids (e.g., omeprazole, ranitidine, cimetidine) may be added in patients in which long-term treatment with NSAIDs is indicated. Chondroprotectants, including glucosamine, glycosaminoglycan and chondroitin can also be useful for treatment of OA, as these have been shown useful to slow the progression of arthritis and pain in dogs. Glucosamine has been shown to promote production and repair of cartilage and has anecdotally been used in birds at dosages of 20-40 mg/kg q12h or 35 mg/kg PO q24-48h. Adequan, a polysulfated glycosaminoglycan, has also been used in birds with OA without apparent side effects when used in dosages of 5 mg/kg IM once weekly for 4 weeks, then once monthly. As a note of caution, fatal hemorrhage has been reported in two raptors, one psittacine and one coraciiforme after an intramuscular injection of polysulfated glycosaminoglycans.[8] For chondroitin sulfate, which promotes water retention, elasticity of cartilage and inhibits breakdown of cartilage, no known dosages exist in birds.

Opioids may be needed for acute exacerbations of pain due to OA, or for birds that initially do not respond adequately to NSAIDs. Depending on the species, tramadol at dosages of 11 mg/kg PO q6-12h up to 30 mg/kg PO q12h) or butorphanol (0.5-3 mg/kg IM q4h) may be used for a short period of time until the anti-inflammatory effects of the NSAIDs take effect. Gabapentin (10-25 mg/kg PO q8-12h) is also being used more in avian medicine for both neuropathic and chronic pain. Gabapentin is a lipophilic structural analogue of the inhibitory neurotransmitter GABA. This drug was initially developed as an anticonvulsant and is now indicated for the treatment of seizures and neuropathic pain. In mammals, gabapentin has been shown to be highly effective as an adjunctive therapy (i.e., combined with opioids or NSAIDs).

In addition to aforementioned medications, omega-3 fatty acids, ginger and turmeric have been used in birds with OA.[9] Omega-3 fatty acids (e.g., VetOmega; 0.22-0.44 mL/kg PO daily) have been shown in many species to be anti-inflammatory and renal protective (an added bonus when using NSAIDs which can be damaging to the kidneys in high doses or in renal compromised or dehydrated patients). Ginger and turmeric have similar anti-inflammatory effects, but not the side effects of NSAIDs. Ginger (gingerols), which inhibits platelet cyclo-oxygenase synthase, thereby leading to a decrease of thromboxanes and leukotrienes, is suggested to have greater COX-1 inhibition than aspirin as well as being a potent inhibitor of COX-2.[10] It appears to have hepatoprotective properties, which can be helpful for older birds that are suspected of or diagnosed with liver disease. Turmeric is considered to have anticarcinogenic properties.[11] White willow bark is the original source of aspirin; however, it does not inhibit platelet aggregation. The active ingredients are converted from salicylic acid to acetylsalicylic acid. As such, this herb has also been implemented in the treatment of osteoarthritis, muscle pain, and gout.[12]

Non-medical treatment modalities for OA include husbandry changes, weight management, an exercise plan, feeding a healthy diet rich in omega-3 fatty acids, and physical and/or laser therapy. An exercise plan in the captive avian patient can be challenging, particularly in non-flighted birds. In flighted birds, allowing them to fly or encouraging flight is the best form of exercise, provided that it can be done safely. If this is not possible, encouraging climbing and walking in birds, or even repeatedly stepping up and down for several times during the day can be exercise for parrots. Foraging for food can also be used to encourage exercise (e.g., by scatter feeding or placing foraging boxes on opposite sides of the cage or enclosure). Moreover, providing foraging opportunities can entertain and distract the bird from its discomfort, thereby diminishing feather destructive behavior and mutilation.

Physical therapy is important for any patient with OA as it encourages blood flow to the joints and cartilage and strengthens the muscles which helps to support the joints. Should the bird be overweight, weight loss is essential as studies have identified obesity as a risk factor for OA in many species. Weight management includes transitioning the bird slowly to a healthy pelleted diet with the addition of omega-3 fatty acids. Husbandry changes can dramatically affect the life of a bird with OA. Cages should be equipped with adequately sized perches and platforms for the bird to rest on. Perches may need to be lowered to prevent trauma if a patient is frequently falling and can be wrapped with self-adhesive bandage or felt to provide better grip. For smaller birds, hutches can be provided as an alternative means of support. In severe cases, the bird may need to be moved to a box or flat surface environment with soft bedding or towels to prevent bumblefoot. To prevent the bird from getting its nails caught in toys, floor or bars of the enclosure, nails should be trimmed and dull. The toenails should be trimmed and dull to prevent them from getting caught in toys or the bars of the cage.

Metabolic Bone Disease

Nutritional secondary hyperparathyroidism, or metabolic bone disease, occurs commonly in elderly birds that are on a poor diet and housed indoors. Diets deficient in calcium and vitamin D and high in phosphorus (i.e., all seed diets) are risk factors. The decreased bone density that results from the vitamin/mineral imbalance increases the risk for pathologic fractures during restraint or following a fall,[13]

Clinical signs include weakness, ataxia, inability to perch, and lameness as a result of (pathologic) fractures. Reproductively active hens may present with egg binding or cloacal prolapse.

Diagnosis is based on the bird's history, which typically reveals a sedentary bird on a poor diet, and findings during the physical exam. In addition, the bird may present with decreased blood calcium (plasma and ionized), increased phosphorus, and decreased vitamin D (25-hydroxycholecalciferol) plasma levels, while radiographs or CT may reveal decreased bone density and/or pathologic fractures.

While bone density evaluation using radiographs is quite subjective (pixel intensity), calibrated CT scans can be used to measure Hounsfield Units which is a measurement of voxel density. Most common DICOM viewers have measurement tools which allow the viewer to select a region to measure density. Numerous bone density studies have been published, mostly relative to poultry species.[14-16] At this time, there are no normal established bone density parameters for captive companion birds. Measurements can be taken between normal-appearing bone and abnormal (usually more radiolucent) bone in different birds of the same species. Bone health goes beyond the scope of this discussion and is a complex balance between developmental nutrition, long term nutrition, physical activity, sunlight exposure (importance varies highly between species), reproductive activity, sex and presence of certain diseases.

Some generalizations can be made when evaluating bone density. Birds that spend most of their time perched and do not climb around much will have poor relative bone density along the proximal femurs and at the stifle joint. Birds that are non or poor flyers tend to have low relative bone density at the proximal humeri and coracoids. The combination of limited flying and climbing may additionally result in poor relative density along the notarium and synsacrum. Focal or regional osteoporosis and can be associated with a variety of health concerns including disuse atrophy (as can be seen with certain orthopedic and neurologic diseases), localized bone cancer and infections, arthritis and more.

Reproductive activity and subsequent physiologic (and abnormal) hyperostosis can make bone density determination challenging. Hyperostosis is a physiologically normal phenomenon in hens that allows for rapid mobilization of calcium and other bone minerals to move from the cortex into the medullary cavity. By having these dense elements in the medullary space, minerals can be rapidly pulled from the medulla into the blood stream and used to make the eggshell. Short term, this physiologic process results in minor mineral depletion from the cortex that is (under normal circumstances and with a good diet) replenished once the hen is no longer reproductively active. Long term, this process results in cortex thinning and osteoporosis with increased risk of fracture development.

Radiographically, medullary hyperostosis makes the medulla appear more dense than normal. While the overall bone density (cortex and medulla combined) may be normal on a volume bases, the cortex thickness and density often suffer with chronic medullary hyperostosis. As has been reported in poultry, this may result in micro (common in the spine, costrochondral junctions and more) and macro fractures.[17] Micro CT is often needed to visualize small fractures; however, secondary features such as arthritis and poor relative density can be seen with high-resolution clinical CT scans. Specific to spinal osteoporosis, one author (MSE) has noticed a tendency towards increased incidences of intervertebral disk disease, spondylosis and other spinal column diseases as viewed on high-resolution clinical CT scans. This statement requires further investigation to prove a true correlation between osteoporosis and spinal disease.

Treatment includes dietary modifications (converting to a diet appropriate for the species), providing ultraviolet lighting (indoor) or, ideally time outdoors, increasing physical activity, reducing reproductive activity (especially if inappropriate or chronic), managing underlying disease (obesity, arthritis, etc), and providing supplemental omega-3 fatty acids. Omega-3 fatty acids added to the diet has been shown to increase bone mineral density and/or improve markers of bone health in multiple bird models.[18,19] Physical activity appears to be necessary for omega-3 fatty acids to have maximum benefit to bone health. Calcium supplementation and modifying the cage to prevent falling are also advised. In the case of an existing fracture, strict cage rest is recommended whereby climbing should be minimized and any potential for falling be avoided. Research has shown that birds may benefit from both oral and UV-B delivered vitamin D_3. Regardless, proper diet, exercise and managing the other factors described above are still required for proper bone health. Limited research also suggests African grey parrots (*Psittacus erithacus*) have a greater dependence on UV-B light for maintaining adequate serum calcium levels than do *Amazona* spp.[13] This suggests that species differences may account for unique UV light needs.

Renal Disease

Birds can develop renal disease at any age, but it is more common in the older bird. Often, there is a history of poor diet and inappropriate husbandry. Renal disease may also be due to infectious diseases, toxin exposure, poor genetics, neoplasia and more. Diets deficient in vitamin A, high in phosphorus, calcium or vitamin D, or just long-term generalized malnutrition from an all-seed diet can lead to renal disease. Numerous viruses can infect and affect the kidneys including polyoma virus, psittacine herpes virus, and possibly also avian borna virus (ABV), which has been found in the kidney, is shed in the birds' urine, and may induce polyuria/polydipsia (PU/PD), and lead to elevated blood uric acid levels. Chronic bacterial infections or septicemia can potentially lead to (chronic) nephritis, tubular dilatation, glomerulopathies and more, which in turn can lead to chronic renal insufficiency or renal failure.[20]

Clinical signs of renal disease include reduced or poor appetite, weight loss, PU/PD, depression, regurgitation, and (unilateral) lameness or paresis. Owners often report that the bird is less active and quieter. In some cases, articular or visceral gout may develop. Visceral gout is the accumulation of uric acid crystals on the internal organs (e.g., heart, kidney, spleen), and is often seen in advanced disease, and is usually fatal. Articular gout is the accumulation of uric acid crystals in the joints and tendons, particularly in the feet and hocks and is more commonly seen as a side effect of chronic renal disease. Uric acid crystals irritate surrounding tissues and cause severe inflammation and pain. The prognosis for gout is usually grave.

Diagnosis of renal disease includes blood work, which reveals persistent hyperuricemia that may be accompanied by hyperphosphatemia, hypocalcemia, and changes in electrolytes. A CBC may reveal leukocytosis and (non-regenerative) anemia. Due to the relative lack of sensitivity of uric acid plasma concentrations, diagnosing kidney disease is often difficult in avian patients. In well hydrated birds, persistently elevated blood uric acid and urinary GGT are moderate indicators of some forms of renal disease. Newer biomarkers that could prove useful include N-acetyl-b-d-glucosaminidase (NAG), which can be measured in the blood or urine to assess tubular (dys-)function, and symmetric dimethylarginine (SDMA), which can be measured in the blood to assess glomerular filtration rates.[21] However, further studies are needed to assess their value for diagnosing renal disease in birds. Radiographs or CT may reveal renal enlargement and mineralization and altered uptake of iodinated contrast media by the kidneys. Endoscopic examination can be performed to visualize the kidneys and collect biopsies for histopathologic examination, which may help determine the etiologic process and prognosis.

High-resolution CT with contrast is particularly helpful in identifying renal lesions, especially prior to the development of laboratory abnormalities. Kirk reported 47 avian cases with CT identifiable renal disease. Only six of those 47 cases (12.8%) had lab abnormalities (hyperuricemia) supporting renal disease. Eight of those cases had histology performed, all of which had histologically confirmed renal disease. Of those eight histology confirmed cases, only two had hyperuricemia.[22] CT with contrast not only provides shape, size, density and vasculature of the kidneys, a properly performed study will also show renal contrast clearance giving functional information. Areas of non or poor contrast uptake and distribution indicate abnormally functioning renal tissue. These areas of poor contrast processing may be due to infectious disease, cancer, degenerative processes, cysts and more and aid in targeting tissue selection for biopsy.

Treatment for renal disease includes fluid therapy and nutritional support. Allopurinol (10 mg/kg PO q24h), colchicine (0.01-0.2 mg/kg PO q12-24h), and omega-3 fatty acids (VetOmega 0.22-0.44 ml/kg PO q24h) are used as anti-inflammatory agents. Use caution as allopurinol has been shown to have a low margin of safety in red tailed hawks (*Buteo jamaicensis*) and is generally not recommended for *Buteo* species.[23] Butorphanol (0.5-3 mg/kg IM q4h) or tramadol (11 mg/kg PO q6-12h to 30 mg/kg PO q12h) are important if gout is present. In addition, birds may benefit from treatment with anti-emetics, antacids or gastric protectants to stimulate appetite and reduce uricemia-associated gastric upset. Changes in the diet and husbandry may be indicated in birds with renal disease.[20] In certain countries, diets formulated specifically for geriatric birds might be available. Older birds, especially cockatiels and budgerigars, on poor diets should be converted very slowly to a formulated diet. They may develop PU/PD, lethargy and elevated uric acid levels when placed suddenly on a formulated diet. This may be due to reduced renal and hepatic function from a life-long deficiency in vitamin A and certain amino acids.[20]

Rehmania is a nutraceutical that significantly improved renal function in rats with chronic nephritis,[24] and may therefore also be worthwhile to consider in the treatment of birds with chronic kidney disease. One of the authors (SEO) has used it for chronic renal disease in a number of species along with boosting omega fatty acids (balanced 3-6-9 PUFAs) and fluids, and found patients to benefit from this treatment.

Reproductive Disease and Neoplasia

Common reproductive diseases in pet birds include cystic ovaries, salpingitis, egg yolk peritonitis, egg binding, hernias and neoplasia. Many of these can occur in any reproductively active hen, but they tend to occur more commonly as the bird ages. Ovarian neoplasia is reported with increasing frequency in pet birds, and may be associated with egg binding, cystic ovarian disease, medullary hyperostosis, abdominal hernias, oviductal impaction, ascites, and general malaise. One study by the U.S. Department of Agriculture Animal and Plant Health Inspection Service of mature fowl noted that 38% of the birds were condemned because of neoplastic disease, most of which related to the genital tract. Granulosa cell tumors and ovarian adenocarcinomas are most frequently reported, but carcinomas, leiomyosarcomas/leiomyomas, adenomas, teratomas, dysgerminomas, fibrosarcomas, lipomas and lymphomatosis have all been reported in bird ovaries.[13] Oviductal tumors are less common than ovarian neoplasia and include adenocarcinomas/adenomas, adenomatous hyperplasia, carcinoma, and carcinomatosis.[25]

Testicular neoplasia has been reported in numerous bird species and includes Sertoli and interstitial cell tumors, seminomas, teratomas, and lymphoproliferative diseases. Sertoli cell tumors are most common, especially in budgerigars, and are typically benign neoplasms.[25] Nevertheless they can result in clinical signs as a result of their size, and production of hormones. Reported neoplasms of the epididymis and ductus deferens include leiomyosarcoma and carcinoma.[25]

Many reproductive tract neoplasms are functional and can cause increased plasma hormone levels. Granulosa cell tumors have been reported in older female budgerigars (*Melopsittacus undulatus*) and may be functional, resulting in hyperestrogenism. Similarly, Sertoli-cell tumors are functional and can lead to feminization, leading to a change in cere color from blue to brown in the budgerigar. Polyostotic (medullary) hyperostosis may occur as a paraneoplastic syndrome with functional ovarian, oviductal or testicular neoplasms.

Clinical signs include reduced appetite, depression, lethargy, ascites, dyspnea, straining, cloacal prolapse, paresis/lameness, and poor or altered reproductive performance. Birds may present with feather destructive behavior or nesting behavior. Ceres may become hypertrophied or change color in budgerigars.

Diagnostics include hematology, biochemistry profile, radiography, CT, magnetic resonance imaging (MRI), ultrasound, endoscopy and biopsy. Hematology is often non-specific, though it may reveal a leukocytosis and anemia. The biochemistry profile may reveal elevations in AST, creatine phosphokinase (CPK), total protein, (total) calcium, cholesterol, triglycerides, and lipemia. Imaging may reveal an enlarged ovary, testicle or oviduct, hepatomegaly, ascites, and/or mineralization of the medulla of the long bones. Exploratory celiotomy may confirm the diagnosis.

Advanced imaging, such as high-resolution contrast CT and MRI, can be especially beneficial when evaluating the reproductive tract. The ovary should appear as a grape cluster with each follicle taking up contrast in a similar manner. In reproductively active hens, active follicles (F series follicles) will appear enlarged. Enlarged 'F' follicles should all vary sequentially in size from F1 (largest and closest to being released into the infundibulum) to F2, F3, F4, F5 approaching the size of the small inactive ovarian follicles. Birds that lay only two eggs in a clutch, will only have F1 and F2 follicles visible at the time of egg production. Birds that are continuous or large clutch layers will often have visible F1-F5 follicles. All follicles should be smooth spheres with a relatively homogenous vascular pattern (that evenly transitions from the capsule to the center of the follicle). Some species will have two ovaries (and sometimes two oviducts).

Large follicles that are either too big to be a normal yolk for the species or are of similar size and do not appear to be sequentially getting smaller are abnormal. Ovarian cysts often have a thin rim of contrast uptake and no visible contrast in the remainder of the mass. Both abnormalities may be associated with cancer and chronic inflammation. Inflamed, infected and neoplastic follicles may have an irregular vascular pattern and uneven surface. Follicles or cysts found separate from the ovary and oviduct may indicate ectopic ovulation, coelomitis, or neoplastic spread (such as carcinomatosis). Ovarian cancer is sometimes associated with invasion into the cranial renal division (usually left, but may involve either side).

The oviduct can vary significantly from indistinguishable to massive and occupy the bulk of the caudal (usually left) coelom. As hens become reproductively active, the oviduct enlarges, folds upon itself, and becomes highly vascular. Eggs in various stages of development (follicles to shelled eggs) may be visualized within the oviduct.

A developing egg found outside the oviduct is abnormal and may represent reverse peristalsis, oviductal rupture or sometimes failure of the infundibulum to catch the follicle. If there is a fresh oviductal tear, contrast may be helpful to visualize the point of rupture. Oviductal impaction, hyperplasia and metritis often result in a thickened and vascular wall with a non-contrast enhancing lumen. Oviductal cysts, masses, abnormal anatomy, etc. should all be noted.

The testicles are also highly variable in size relative to the bird's reproductive status. Testicles are often slightly asymmetrical with the left testicle usually being larger. Testicular shapes also vary between species and should all have a smooth surface and homogenous vascular pattern. Any variations (e.g., cysts, avascular areas or mixed vascularity, irregular surfaces, shapes) should be noted and may indicate some form of testicular disease.

Treatment options include dietary modifications (if needed), antibiotics (in case of suspected bacterial involvement), anti-inflammatories or analgesics (e.g., NSAIDs and/or opioids), and GnRH agonists such as leuprolide acetate (700-800 µg/kg IM; Lupron Depot, AbbVie, Chicago, IL, USA) or deslorelin acetate implants (Suprelorin, Virbac AH, Inc. Fort Worth, TX, USA).[26] For severe reproductive disease that is unresponsive to medical management, a partial or complete ovariectomy, and in the case of oviductal neoplasia, salpingohysterectomy may be the only feasible option, despite this carrying a more guarded prognosis in an elderly, debilitated bird. Dependent on the neoplasia involved, adjunct chemotherapy or radiation therapy may be considered following consultation with a veterinary oncologist. For testicular tumors, surgical removal of the affected testicle is the treatment of choice and carries a good prognosis if metastasis is not present.

Atherosclerosis

Atherosclerosis is a chronic inflammatory and degenerative condition of the arterial wall. The mineralization and plaque formation that occurs results in narrowing of the lumen and decreased blood flow. In birds, changes associated with atherosclerosis are commonly seen in the cardiac outflow tract, aorta, brachiocephalic trunk, pulmonary arteries, and main cerebral vasculature. Lesions may also occur in the peripheral vasculature, most notably the coronary, carotid, celiac, cranial mesenteric, and ischiatic arteries. Atherosclerosis is likely the most common cardiovascular disease in middle aged to older pet birds. Reported incidence has been as high as 91% in some species. Risk factors include a high-fat diet, low omega-3 fatty acid diet, sedentary lifestyle, age, and there may be a genetic predisposition. Females are overrepresented, as are Amazon and African grey parrots.[27]

Clinical signs depend on the site of the lesion but may include sudden death, dyspnea, depression, lethargy, ascites, ataxia, tremors, seizures and paresis. Owners may report their birds act disoriented or are exhibiting extension and rigidity of a leg or having difficulty perching. Similarly, self-injurious behaviors and wing web dermatitis may be noted.

Antemortem diagnosis of atherosclerosis is challenging. Diagnostics include radiography, contrast-enhanced CT imaging, and measuring cholesterol and triglyceride levels. Radiographs or CT may reveal an increased density and size of the aortic arch and/or tortuous or mineralized vessels. Some birds will have elevated cholesterol and triglycerides levels and lipemic blood.[27] Direct correlation between cholesterol and triglyceride levels and atherosclerosis development in birds is weak.

High-resolution CT with contrast can be very beneficial in identifying atherosclerotic lesions and calculating blood flow in larger vessels. While vascular mineralization can often be seen with radiographs, high-resolution CT can also better define fatty changes within the vasculature. Fatty atheromas typically are less dense and have a mixed density compared to the arterial wall. These may be seen with or without contrast. The addition of contrast fills the arterial lumen allowing one to calculate rough percentage of patency through regions where atheromas are present. Such lesions may have no evidence of vascular tortuosity or mineralization as has been discussed with radiographs.

Treatment is challenging and includes dietary modifications (formulated diet with vegetables and fruit), weight loss, and omega-3 fatty acids. Omega-3 fatty acids (VetOmega 0.22-0.44 ml/kg PO q24h) have been advocated to reduce cholesterol and inflammation, and seem to be correlated with a lower incidence of atherosclerosis in multiple bird species, thereby suggesting supplementation may have benefit in managing this disease. Isoxsuprine (10 mg/kg PO q12-24h (Bi-Coastal Pharm International LLC, Red Bank, NJ, USA), a peripheral vasodilator that causes vascular smooth muscle relaxation, has previously been used anecdotally with some success in birds, but is currently no longer available. Other blood pressure lowering drugs such as ACE-inhibitors may therefore be considered. In addition, pentoxifylline, a vasoactive drug that improves blood flow by reducing blood viscosity, has been used with good success in many atherosclerotic patients by one of the authors (YvZ). Statins are a group of lipid-lowering drugs used in humans with atherosclerosis. Statins have been used in birds, but thus far there is little information to support their use. A study investigating rosuvastatin (Crestor, AstraZeneca Pharmaceuticals LP, Wilmington, DE, USA; 10 mg/kg and 25 mg/kg PO) did not yield detectable plasma concentrations.[28] Similarly, atorvastatin did not yield detectible changes in the lipid profile following daily dosing of 10 mg/kg PO for 30 days.[29] Further studies would be required to determine whether higher or more frequent dosing would be effective.

One of the authors (SEO) uses Chinese ginseng to help reduce cholesterol in the serum and liver. This is based on a study using hypercholesterolemic chickens that showed Chinese/Wisconsin ginseng reduces HMG-CoA reductase and cholesterol 7α-hydroxylase activities, which in turn resulted in lowering of cholesterol.[30] Anecdotally, a nutraceutical/vitamin combination developed by Dr. Cathy Johnson-Delaney has been suggested to be useful for parrots as a dietary supplement. Her formula consists of 1000 mg L-carnitine, 1000 mg taurine, 1000 IU vitamin E, and 400 mg CoQ10 that is placed in 30 ml of a balanced omega 3-6-9 fatty acid oil (such as VetOmega). For some patients, 10 mg of policosanol is added as a cholesterol reducer.[31]

Hawthorn leaf with flower is another nutraceutical that can be considered for patients with atherosclerotic disease. This herbal acts as a positive inotrope, increasing coronary and myocardial perfusion, and decreasing vascular resistance, and is also hypocholesterolemic as it upregulates LDL receptors in the liver.[32]

Cardiac Disease

Cardiac disease is being diagnosed more frequently in geriatric birds. In the past, diagnosis was commonly only made following postmortem examination. One study found heart disease in 11% of avian necropsies.[33] With advancing diagnostic capabilities like digital imaging, CT scan and ultrasound, cardiovascular disease is increasingly diagnosed antemortem. Major predisposing factors for cardiac disease in birds include poor diet, obesity and a sedentary lifestyle. Common diseases of the aging cardiovascular system include cardiomyopathy, valvular disease and atherosclerosis. Vegetative lesions of the aortic and mitral valves may occur secondary to chronic infections such as hepatitis, reproductive infections and pododermatitis.[33] Congestive heart failure is associated with fluid retention resulting from sustained inadequate cardiac output. Pulmonary edema predominates with left sided heart disease and hepatomegaly and ascites predominate in right sided heart disease.

Clinical signs include periodic or sustained weakness, syncope, lethargy, increased respiratory effort, hepatomegaly and ascites. The physical exam may reveal brady-/tachycardia, cardiac arrhythmias, a heart murmur, ascites and jugular distention, weakened pulse, tachypnea/dyspnea or slow respiratory recovery following restraint (3-5 minutes is normal). There may also be muffled heart sounds from pericardial effusion, or hepatomegaly.[33]

Diagnostics include CT imaging, radiography, electrocardiography, and cardiac ultrasound. Whole body radiographs can reveal an enlarged cardiac silhouette, the presence of tortuous vessels, hepatomegaly and/or ascites. Radiographic detection of an enlarged cardiac silhouette with muffled heart sounds is suggestive of pericardial effusion. Ultrasound is the standard in non-invasive cardiac evaluation. The probe can be directed cranially through the liver to get a four-chamber view of the heart. Chamber size, wall thickness, blood flow and valvular defects may be seen. CT allows visualization of the heart and the blood vessels and is more sensitive in evaluating vasculature than radiology. Additionally, CT with contrast can be used to measure chamber and vasculature size and calculate chamber volumes. The electrocardiogram (ECG) characterizes cardiac arrhythmias and conduction disturbances, but results may be affected by sedation and anesthesia. Indirect blood pressure can be measured using a Doppler flow detector placed over the superficial ulnar artery or cranial tibial artery and a cuff and sphygmomanometer.[34] Studies have shown consistently that this method does not agree with direct methods and may be of limited clinical use, especially in small avian patients.[34] Results suggested that indirect monitoring of a single bird during a single cuff placement such as during an anesthetic procedure or an emergency may be useful to detect changes within an event.[34] Direct blood pressure may also be performed in the anesthetized patient.

Management of a patient with congestive heart failure first aims at stabilization of the patient through provision of supportive care and oxygen therapy. Furosemide (1-10 mg/kg PO q4-24h) or coelomocentesis can be used to alleviate signs of congestion by promoting diuresis and eliminating excess fluids. Following stabilization, an ACE-inhibitor such as enalapril (1.25-5 mg/kg PO q12-24h) and positive inotrope such as pimobendan (0.1-10 mg/kg PO q12-24h) can be used to improve cardiac function by reducing afterload and contractility, respectively.[33] Consultation with a veterinary cardiologist is recommended, when possible, to optimize the medical treatment.

Omega-3 fatty acids (VetOmega at 0.22-0.44 ml/kg PO daily) have been advocated for cardiac patients and used to reduce cholesterol and inflammation. Dandelion may be used as a nutraceutical, in either the root or leaf form, or both. In rats, the diuretic effects of the leaves, often described as the herb, appeared comparable to furosemide, while the root was far less effective. Unlike furosemide, dandelion does not lead to hypokalemia, as the high potassium content of the leaf replaces the loss during diuresis.[35]

Long-term care may include exercise restriction, cage rest with appropriate housing, and conversion to a healthier diet, along with weight loss, if the bird is obese. Because many cardiac patients will need to be on long-term medication, positive reinforcement training for targeting and syringe training may aid in administration of medications. Medications should be formulated or compounded to be as palatable and concentrated as possible to provide for better acceptance and reduced-volume dosing. Nevertheless, the long-term prognosis of heart failure in birds is guarded to grave.

References

1. McMichael TM. Emergency and care of the geriatric patient. *Proc Conf Int Vet Emerg Crit Care.* 2012.

2. Lightfoot TL. Geriatric psittacine medicine. *Vet Clin North Am Exot Anim Prac.* 2010;13:27-45.

3. Holmes DJ, Thomson SL, Wu J, et al. Reproductive aging in female birds. *Exp Gerontol.* 2003;38:751-756.

4. Abdul-Rahman I, Jeffcoate I, Obese FY. Age-related changes in the gross anatomy of the reproductive organs, and associated steroid hormone profiles in male and female guinea fowls (*Numida meleagris*). *Vet Anim Sci.* 2018;6:41-49.

5. Kellgren J, Lawrence J. Radiological assessment of osteo-arthrosis. *Ann Rheum Dis.* 1957;16(4):494-502.

6. Kohn M, Sassoon A, Fernando N. Classifications in brief: Kellgren-Lawrence classification of osteoarthritis. *Clin Orthop Relat Res.* 2016;474(8):1886-1893.

7. Gielis WP, Weinans H, Nap FJ, et al. Scoring osteoarthritis reliably in large joints and the spine using whole-body CT: osteoarthritis computed tomography-score (OACT-Score) *J Pers Med.* 2021;11:5.

8. Anderson K, Garner MM, Reed HH, et al. Hemorrhagic diathesis in avian species following intramuscular administration of polysulfated glycosaminoglycan. *J Zoo Wildl Med.* 2013;44(1):93-99.

9. Frech TM, Clegg DO. The utility of nutraceuticals in the treatment of osteoarthritis. *Curr Rheumatol Rep.* 2007;9:25-30.

10. Ginger: *Zingiber officinale* Roscoe. In: Blumenthal M, ed. *The ABC Clinical Guide to Herbs*. Austin, TX: American Botanical Council; 2003:176.

11. Rao CV, Rivenson A, Simi B, et al. Chemoprevention of colon carcinogenesis by dietary curcumin, a naturally occurring plant phenolic compound. *Cancer Res.* 1995; 55:259266.

12. Beer AM, Wegener T. Willow bark extract (*Salicis cortex*) for gonarthrosis and coxarthrosis-results of a cohort study with a control group. *Phytomed.* 2008;15:907-913.

13. Stanford M. Calcium metabolism. In: Harrison GJ, Lightfoot TL, eds. *Clinical Avian Medicine*. Lake Worth, FL: Spix Publishing; 2006:141-152.

14. Aguado E, Pascaretti-Grizon F, Goyenvalle E, et al. Bone mass and bone quality are altered by hypoactivity in the chicken. *PLoS ONE.* 2015;10(1):e0116763.

15. Oviedo-Rondon EO, Mente PL, Arellano C, et al. Influence of gait on bone strength in turkeys with leg defects. *Poult Sci.* 2018;97(8):2621-2637.

16. Martinez-Cummer MA, Heck R. Leeson S. Use of axial X-ray microcomputed tomography to assess three-dimensional trabecular microarchitecture and bone mineral density in single comb white leghorn hens. *Poult Sci.* 2006;85(4):706-711.

17. Webster AB. Welfare implications of avian osteoporosis. *Poult Sci.* 2004;83(2):184-192.

18. Tompkins YH, Chen C, Sweeney KM, et al. The effects of maternal fish oil supplementation rich in n-3 PUFA on offspring-broiler growth performance, body composition and bone microstructure. *PLoS ONE.* 2022;17(8):e0273025.

19. Tarlton JF, Wilkins LJ, Toscano MJ, et al. Reduced bone breakage and increased bone strength in free range laying hens fed omega-3 polyunsaturated fatty acid supplemented diets. *Bone.* 2013;52(2):578-586.

20. Echols S. Evaluating and treating the kidneys. In: Harrison GJ, Lightfoot TL, eds. *Clinical Avian Medicine.* Lake Worth, FL: Spix Publishing; 2006:451-492.

21. Alberton S, Vergneau-Grosset C, Summa N. Advances in exotic animal clinical pathology. *Vet Clin North Am Exot Anim Pract.* 2019;22:367-385.

22. Kirk N, Echols MS. Comparison of contrast CT renal lesions and uric acid values in avian cases. *Proc ExoticsCon.* 2022;21.

23. Poffers J, Lumeih JT, Timmermans-Sprang EPM, et al. Further studies on the use of allopurinol to reduce plasma uric acid concentrations in the red-tailed hawk (*Buteo jamaicensis*) hyperuricaemic model. *Avian Pathol.* 2002;31(6):567-572.

24. Su ZZ, He YY, Chen G. Clinical and experimental study on effects of man-shen-ling oral liquid in the treatment of 100 cases of chronic nephritis. *Zhongguo Zhong Xi Yi Jie He Za Zhi.* 1993;13:259-272.

25. Lightfoot TL. Clinical avian neoplasia and oncology. In: Harrison GJ, Lightfoot TL, eds. *Clinical Avian Medicine.* Lake Worth, FL: Spix Publishing; 2006:560-566.

26. Ottinger MA. Neuroendocrine regulation of reproduction in birds and clinical applications of GnRH analogues in birds and mammals. *Semin Avian Exotic Pet Med.* 2002;11(2):71-79.

27. Beaufrere H. Avian atherosclerosis: parrots and beyond. *J Exotic Pet Med.* 2014;22(4):336-347.

28. Beaufrere H, Papich MG, Nevarez J, et al. Pharmacokinetics of orally administered Rosuvastatin in Hispaniolan Amazon parrots (*Amazon ventralis*). *Proc Annu Conf Assoc Avian Vet.* 2014:25.

29. Robertson JA, Guzman DS, Graham JL, et al. Evaluation of orally administered atorvastatin on plasma lipid and biochemistry profiles in hypercholesterolemic Hispaniolan Amazon parrots (*Amazona ventralis*). *J Avian Med Surg.* 2020;34:32-40.

30. Qureshi AA, Abuirmeileh N, Din ZZ, et al. Suppression of cholesterogenesis and reduction of LDL cholesterol by dietary ginseng and its fractions in chicken liver. *Atherosclerosis.* 1983;48:81-94.

31. Orosz SE, Johnson-Delaney CA. Application of nutritional supplements and herbs in avian practice. *Proc Annu Conf Assoc Avian Vet.* 2007.

32. Upton R. *Hawthorn Leaf with Flower: Crataegus spp.* Santa Cruz, CA: American Herbal Pharmacopoeia and Therapeutic Compendium; 1999.

33. Fitzgerald BC, Beaufrere H. Cardiology. In: Speer B, ed. *Current Therapy in Avian Medicine and Surgery.* St. Louis, MO: Elsevier; 2016;252-328.

34. Johnston MS, Davidowski LA, Rao S, et al. Precision of repeated, Doppler-derived indirect blood pressure measurements in conscience psittacine birds. *J Avian Med Surg.* 2011;25:83-90.

35. Leung Ay, Foster S. *Encyclopedia of Common Natural Ingredients Used in Food, Drugs, and Cosmetics.* 2nd Ed. New York, NY: John Wiley & Sons; 1996.

Section 10

AAV Potpourri Session

Jessica Magnotti, DVM
Session Chair

Cryptosporidiosis in Cockatiels and Other Psittacine Birds

Susan L. Clubb, DVM

Session #3155

Affiliation: From Rainforest Clinic for Birds and Exotics, Inc., 3319 E Rd, Loxahatchee, FL 33470, USA.

Abstract: Cryptosporidium, sometimes informally called crypto, is a genus of apicomplexan parasitic alveolates that can cause a respiratory gastrointestinal illness (cryptosporidiosis) that primarily involves watery diarrhea (intestinal cryptosporidiosis), sometimes with a persistent cough (respiratory cryptosporidiosis) in humans.[1,2] Cryptosporidiosis is also widespread in many species of mammals, birds, reptiles and fish. This paper addresses intestinal cryptosporidiosis in psittacine birds.

Introduction

Cryptosporidiosis, caused by parasite species of the genus *Cryptosporidium,* is a major diarrheal disease in both people and animals globally, with *C. hominis* and *C. parvum* the main species infecting humans. Environmentally robust oocysts which are shed in high numbers in the feces of infected individuals are resistant to disinfectants, including levels of chlorine normally used in drinking water. As a result, *Cryptosporidium* is a major cause of waterborne and foodborne outbreaks.[1,2]

Intestinal cryptosporidiosis is typically an acute, short-term infection, can be recurrent through reinfection in immunocompetent hosts, and become severe or life-threatening in immunocompromised individuals. In humans, it remains in the lower intestine and may remain for up to 5 weeks. The parasite is transmitted by environmentally hardy cysts (oocysts) that, once ingested, exist in the small intestine and result in an infection of intestinal epithelium. Impaired intestinal epithelial barrier function and increased permeability are commonly associated with diarrheal diseases caused by enteric infections.[3,4]

Species delimitation in *Cryptosporidium* is largely based on four criteria: a description of oocyst morphology, a description of natural host specificity (where possible), genetic characterizations, and compliance with the International Code of Zoological Nomenclature (ICZN). There are now at least 44 validated species of *Cryptosporidium* (many of which were originally identified as genotypes and later described as species), including six avian *Cryptosporidium* spp. (*Cryptosporidium ornithophilus, Cryptosporidium proventriculi, Cryptosporidium avium, Cryptosporidium galli, Cryptosporidium baileyi* and *C. meleagridis*).[1,5]

Zoonotic potential of *C. meleagridis* or *C. baileyi* has been reported in domestic birds. *Cryptosporidium* has been detected in cockatiels worldwide, but the species or genotype of isolates usually remain unclear.[1,6,7] One study documents the first discovery of *C. meleagridis* and *C. baileyi* in cockatiels and suggests that pet birds may play an important role in the epidemiology of cryptosporidiosis.[6] Therefore, accurately identifying isolates is important for understanding the epizootiology of *Cryptosporidium* infection in birds and for the control of human cryptosporidiosis.

Studies in numerous countries find *Cryptosporidium* as a prevalent parasite in psittacine species. A study of a Japanese pet shop found *Cryptosporidium* in 27 of 98 (14.8%) of psittacine birds.[5]

A new species of *Cryptosporidium*, *Cryptosporidium proventriculi*, has been reported in cockatiels. Transmission studies failed to infect chickens, budgerigars and mice. Experimentally, cockatiels shed 4000 to 60,000 oocysts per gram feces for 30 days post inoculation; however, no infected cockatiels showed clinical signs.[7]

No vaccine and few treatments are currently available, which has greatly limited control of this disease to date. Livestock are both an important reservoir and source of human infections, and some authors have suggested pet birds may represent a source of human infections.

An outbreak of *Cryptosporidium* was reported in a psittacine nursery in 1996.[8] In this study it was demonstrated that *Cryptosporidium* often is missed on necropsy and histopathology if the tissues are not collected very shortly after death of affected birds. Postmortem autolysis in which the intestinal epithelium can preclude detection by histopathology. Diagnostic techniques were contrasted, but at that time PCR did not exist. In the author's opinion PCR is the most reliable diagnostic test. In the 1996 outbreak, oral paromomycin 100 mg/kg q12h for 7 days provided rapid and effective therapy and was used (by the author) subsequently with predictably excellent results.[8]

An Outbreak of *Cryptosporidium* in Handfed Cockatiels (2019-2020)

Because of low wholesale and retail prices, many breeders in the late 2000s reduced or ceased breeding of cockatiels resulting in a nation-wide shortage. In an attempt to gear up production, some breeders turned to utilizing imported cockatiels for breeding stock. In this case the new stock was batch tested for *Cryptosporidium* upon arrival at the aviary quarantine, and it was not detected; however, an outbreak occurred when these birds started to produce.

Chicks were removed from the nest at approximately 3-4 weeks of age. Young cockatiels exhibited a disease syndrome showing poor growth, diarrhea, crop stasis, high mortality, and secondary or concurrent infections with gram-negative bacteria and yeast. Tests for viral pathogens were negative. Cryptosporidiosis was diagnosed by pathology and PCR testing of feces.

To determine how widespread *Cryptosporidium*, and later *Helicobacter*, were in breeding stock and chicks, 281 clutches of cockatiels were swabbed (pooled fecal swabs) for PCR when removed from the nest. Results are provided in Table 1.

Table 1. Clutches of cockatiels tested by PCR of pooled feces for *Cryptosporidium* and *Helicobacter*.[a]

Clutches tested	Organism	Number positive	Percent positive
281	*Cryptosporidium*[b]	99/281	35%
64	Also tested for *Helicobacter*	20/64	31%

[a]*Cryptosporidium* were identified by sequencing as *C. parvum* and *C. meleagridis*.
[b]217 clutches only tested for *Cryptosporidium*.

Because *Cryptosporidium* is typically considered a waterborne organism, the well, concentrated filtered water, and the reverse osmosis system membranes were tested by PCR with negative results. It was assumed that the organism was introduced by the imported birds, and either missed on initial screening, or potentiated by breeding activity.

At the time of the 2019-2020 outbreak paromomycin was not available commercially in the United States. It is currently available from compounding pharmacies. The only drug available for human use in the United States, Alinia (nitazoxanide), was cost prohibitive, so an alternative was necessary.

Many husbandry interventions, medications, even herbal and alternative treatments were tried. Finally, a feed additive, an acidified clay product, was added to the handrearing formula at the rate of 1 level teaspoon in 5 pounds of dry handrearing formula. The morbidity and mortality rates declined rapidly and the population stabilized. Unfortunately post treatment testing was not performed; however, when the supplement ran out, health issues returned. When the supplement was available again, mortality and morbidity and mortality rates again returned to normal levels.

Clinical *Cryptosporidium* Cases in Adult Psittacine Birds

While cryptosporidiosis is typically considered self limiting, it appears that it can cause disease, and possibly chronic illness, in adult psittacine birds.

Cryptosporidium was diagnosed by PCR of feces in 14 of 39 (35.8%), in clinical cases in psittacine birds. Most of the cases were in adult pet birds in home environments. Cases were selected based on chronic gastrointestinal disease, intermittent diarrhea, mal-odor of feces, or recurrent clostridium infections. In many cases birds were tested simultaneously for *Cryptosporidium* and *Helicobacter*. Most birds were swabbed (choanal and cloacal) and the swab also introduced into the feces. Swabs were stored under refrigeration until shipped overnight to the laboratory for PCR testing.

In the author's opinion, *Helicobacter* infections can show clinical presentation similar to *Cryptosporidium*, so in most cases testing for both was elected. The case positive rate for *Helicobacter* was five positive of 35 total cases or 14.2% (Table 2).

Birds that tested positive were treated with compounded paromomycin, 100 mg/kg, q12h for 7 days. Most owners reported clinical improvement. In five cases, post therapeutic testing was negative for *Cryptosporidium*.

Table 2. Incidence and signalment for *Cryptosporidium* and *Helicobacter* in psittacine birds tested by PCR.

Species	Age (years)	Presentation	*Cryptosporidium*	*Helicobacter*	
African grey parrot	0.5	Diarrhea	Negative	Negative	
Baby blue throat macaw	0.5	Delayed weaning	Negative	Not tested	
Blue-headed pionus	1	New birds, diarrhea	Negative	Not tested	
Cockatiel	2	Exposure	Negative	Negative	
Cockatiel	2	Exposure	Negative	Negative	
Eclectus parrot	2	Proventricular dilatation disease confirmed	Negative	Negative	
Black-headed caique	3	Chronic illness, second opinion	Negative	Negative	
Cockatiel	5	Chronic intestinal issues	Positive	Negative	*
White belly caique	6	Export exam	Negative	Negative	
Eclectus parrot	7	Export exam	Negative	Negative	
Blue and gold macaw	12	Weight loss, lethargy	Negative	Positive	
Senegal parrot	12	Anemia	Positive	Negative	
Blue and gold macaw	12	Chronic intestinal issues	Positive	Negative	
Blue and gold macaw	12	Bloody stools	Positive	Negative	
Cockatiel	17	Diarrhea	Negative	Negative	
Cockatiel	17	Abnormal stools	Positive	Negative	
Yellow-naped Amazon	18	Rehoming, health screen	Negative	Negative	
Senegal	19	Chronic intestinal disorder	Negative	Negative	
Yellow-crowned Amazon	19	Recurrent intestinal infections	Positive	Negative	*
Catalina hybrid macaw	19	Weight loss, lethargy	Positive	Not tested	
African grey parrot	20	Falling, anemic	Negative	Negative	
Blue and gold macaw	21	Smelly stools, *Clostridium*	Positive	Not tested	*
Eclectus parrot	22	Chronic liver disease	Negative	Negative	
Palm cockatoo	22	Chronic regurgitation	Positive	Negative	
Hyacinth macaw	22	Breeder, diarrhea	Positive	Negative	
Double yellow-headed Amazon	24	Chronic crop problems	Negative	Negative	
Cockatiel	26	Old bird, chronic gastrointestinal issues	Positive	Negative	
Scarlet macaw	27	Proventricular dilatation disease suspect	Negative	Positive	
Severe	28	Undigested seeds in stool	Positive	Negative	*
Harlequin hybrid macaw	29	Chronic gastrointestinal disorder, clostridium	Negative	Positive	
Blue-fronted Amazon	32	Sick, anemic	Negative	Negative	
Yellow-naped Amazon	37	Noise when defecates, cloacitis	Negative	Positive	
Yellow-naped Amazon	38	Chronic gastrointestinal issues, recurring infections	Negative	Positive	
Orange-winged Amazon	44	Papillomas, fecal accumulation around vent	Negative	Negative	
Timneh grey parrot	NA	Abnormal stools	Negative	Negative	
Red-lored Amazon	NA	Proventricular dilatation disease suspect, regurgitation	Negative	Negative	
Alexandrine parakeet	NA	Old bird, chronic illness	Negative	Negative	
Yellow-naped Amazon	NA	Old pair, purchased for breeders	Positive	Negative	
Number of positive tests		14/39	5/35		
Percent positive tests		35.8	14.2		

*Retest after paromomycin treatment: negative.

Discussion

Cryptosporidium has world-wide distribution in surface water and water systems, and is the most common waterborne disease in humans and many other species.[3,4] Likewise reports of cryptosporidiosis shows a world-wide distribution in many avian species, including cockatiels and other psittacine birds. While it is primarily self-limiting in immunocompetent humans, moderate to severe watery diarrhea, crampy abdominal pain, low grade fever, dehydration, weight loss and fatigue typically last 2-4 weeks. As few as 2-10 oocysts can initiate infection. Immunocompromised humans can be chronically infected. Re-infection can occur typically with reduced virulence.[3,4] While not documented in this study, it appears that cryptosporidiosis may not be self-limiting in many psittacine birds.

Cryptosporidium oocysts are extremely resistant in the environment. They are highly resistant to chlorine disinfection levels used in water systems. The required chlorine levels generally preclude the use of chlorine as a reliable control method. With high enough concentrations and contact time, oocysts will be inactivated by chlorine dioxide and ozone treatment. Ultraviolet light at relatively low doses will inactivate *Cryptosporidium*. Boiling at least 1 minute (3 minutes at 200 m altitude) or filtration with a 1 micrometer filter, will also decontaminate drinking water. Ammonia can inactivate *Cryptosporidium*, but requires an 18-hour exposure time. Hydrogen peroxide is more effective than chlorine. Alcohol-based hand sanitizers are ineffective. Outside the body, oocysts can remain infectious for 6-8 months.

In clinical cases, owners should be advised that *Cryptosporidium* can be a zoonotic disease and aggressive disinfection is required. Porous items such as perches and toys should be discarded or baked in the oven. Cages should be scrubbed to remove all organic material. UV light may be the most practical means of disinfection of cages, potential fomites such as equipment, and rooms.

Testing for *Cryptosporidium* and *Helicobacter* by PCR of feces, should be considered. This testing is especially warranted in cases of gastrointestinal disease where conventional therapy is ineffective.

References

1. Ryan UM, Feng Y, Fever R, Xiao L. Taxonomy and molecular epidemiology of *Cryptosporidium* and *Giardia* – a 50 year perspective (1971-2021). *Intl J Parasitol.* 2021;51(13-14):1099-1119.

2. Zahedi A, Ryan U. *Cryptosporidium* – an update with an emphasis on foodborne and waterborne transmission. *Res Vet Sci.* 2020;132:500-512.

3. Wikimedia Foundation. *Cryptosporidium.* In: Wikipedia. 2023. https://en.wikipedia.org/wiki/Cryptosporidium.

4. Centers for Disease Control and Prevention. Enteric Diseases Epidemiology Branch. 2022. https://www.cdc.gov/ncezid/dfwed/edeb/index.html.

5. Akira A, Marcelo N, Meireles V. *Cryptosporidium* infections in birds - a review. Infecção por *Cryptosporidium* em aves - uma revisão. *Parasitol Res.* 2010;106(6):1491-1497. doi: 10.1007/s00436-010-1810-5.

6. Abe N, Iseki M. Identification of *Cryptosporidium* isolated from cockatiels by direct sequencing of the PCR-amplified small subunit ribosomal RNA gene. *Parasitol Res.* 2004;92(6):523-526. doi: 10.1007/s00436-004-1082-z.

7. Holubova N, Zikmundova V, Limpouchova Z, et al. *Cryptosporidium proventriculi* sp. n. (Apicomplexa: Cryptosporidiidae) in Psittaciformes birds. *European J Protistol.* 2019;69:70-87.

8. Clubb S, Cray C, Greiner E, Latimer K. Cryptosporidiosis in a psittacine nursery. *Proc Annu Conf Assoc Avian Vet.* 1996.

Captive Monk Parakeets (*Myiopsitta monachus*) Allowed to Choose Sodium Intake Consume Less Sodium than Their Counterparts

EV Voltura, PhD,
Christopher A. Bailey, PhD,
J. Jill Heatley, DVM, MS, DABVP (Avian Practice), Dipl ACZM

Session #3401

Affiliation: From Schubot Center for Avian Health (Voltura, Heatley), the Department of Entomology (Voltura), the Department of Poultry Science, College of Agrilife and Life Sciences (Bailey), Clinical Small Animal Sciences Department, School of Veterinary Medicine and Biomedical Sciences (Heatley), Texas A&M University, 400 Bizzell St, College Station, TX 77843, USA.

Note: Abstract format selected by author(s). See Front Matter for description of conference proceedings manuscripts submitted as scientific abstracts.

Abstract: Many of the nutrient requirements of Psittaciformes, including sodium (Na), have been derived from poultry nutrition studies. This derivation is likely inappropriate because species in the Psittaciformes and Galliformes differ vastly from one another. This difference is supported by the fact that some free-living parrots appear to thrive on diets containing much lower sodium (Na) than what Psittaciformes are currently thought to require as a minimum level. We sought to determine whether captive monk parakeets (*Myiopsitta monachus*) fed a low Na diet would remain as healthy, reproduce effectively, and produce offspring similarly to those fed a high Na diet. Sixteen adult pairs were fed a custom-formulated diet containing either 280 or 1500 ppm [Na] and offered an aqueous Na solution containing either 500 or 1200 ppm [Na]. Feed, water, Na solution intake, and reproductive output were measured. Total Na intake was calculated. Health checks were conducted pre- and post-treatment (e.g., body weight, body condition score, venous biochemistries). Low Na diet pairs consumed less Na (i.e., 400 ppm [Na]) than high Na diet pairs (i.e., 1500+ ppm [Na]) as well as less water and total aqueous solution. Although venous [Na^+] of blood decreased over the study, the change was small and deemed clinically irrelevant. No other significant differences were detected in measures of health or reproductive output between treatments. Monk parakeets' Na requirements appear to be significantly lower than 1500 ppm [Na]. Additional research is needed to determine whether feeding parrots high Na diets over long periods of time negatively impacts their health.

Figure-of-8? More Like, Figure-I-Hate!

Renée Schott, DVM, CWR

Session #3443

Affiliation: From Wildlife Rehabilitation Center of Minnesota, 2530 Dale St. N, Roseville, MN 55113, USA.

Note: Abstract format selected by author(s). See Front Matter for description of conference proceedings manuscripts submitted as scientific abstracts.

Abstract: Figure-of-8 wraps are commonly used for various fractures of the avian wing; however, this author and others, anecdotally, have noted it is one of the most incorrectly applied bandages.[1] In addition, many peer-reviewed sources show improperly applied figure-of-8 wraps.[2-5] Complications of improperly placed figure-of-8 wraps are common due to the difficulty of placing this looping and winding wrap, which may take several tries to achieve correct placement, along with immense species anatomy variation.[2,6] While it has occasionally been recommended as a temporary fixation in addition to a body wrap for a humeral fracture, it is no longer recommended because it tends to distract the humeral fracture further.[6] Other complications of this wrap include over-tightening effect when using a self-adhesive bandage,[6] placing wrap caudal to elbow joint,[2,6] hyper-flexion of the carpus and/or elbow,[6,7] wrap placed too loose,[7] and patagial damage.[2,7-9] More studies are needed comparing the figure-of-8 wrap to a wing-to-body wrap when dealing with distal wing fractures.

References

1. Tully T. Simple avian surgical procedures. DVM360. Available at: https://www.dvm360.com/view/simple-avian-surgical-procedures-proceedings. Accessed June 1, 2023.

2. Darrow B, Bennett RA. Avian orthopedics. In: Bennett RA, Pye GW, eds. *Surgery of Exotic Animals*. John Wiley & Sons; 2021:112-153.

3. Kayikci C, Kuşcu Y, Durmuş A, Aslan L. Fractures and treatment methods in wild avians. *Van Vet J*. 2019;30(2):115-119.

4. Pollock C. Wing wrap placement in birds. LafeberVet. Available at: https://lafeber.com/vet/wing-wrap-placement-in-birds/. Accessed June 1, 2023.

5. Department of Conservation; New Zealand Government. Wildlife Health: Avian Bandaging Techniques (Figure-of-8 Module). Available at: https://www.doc.govt.nz/globalassets/system/training-courses/wildlife-health-7/index.html#/. Accessed June 1, 2023.

6. Graham JE, Doss GA, Beaufrère H. *Exotic Animal Emergency and Critical Care Medicine*. Hoboken, NJ: Wiley-Blackwell; 2021.

7. Wade L. Treating angel wing deformity: a sling for the wing. *Today's Vet Pract*. 2022;68-74.

8. Gull JM, Saveraid TC, Szabo D, Hatt J-M. Evaluation of three miniplate systems for fracture stabilization in pigeons. *J Avian Med Surg*. 2012;26(4):203-212. doi:10.1647/2010-062R1.1

9. Duerr RS, Purdin GJ. *Topics in Wildlife Medicine: Orthopedics*. St Cloud, MN: National Wildlife Rehabilitators Association; 2017.

Novel 3D Imaging of the Psittacine Beak and a Note on Papillae and Beak Trimming Procedures

Emily J. Lessner, PhD,
Scott Echols, DVM, DABVP (Avian Practice),
Joanne R. Paul-Murphy, DVM, Dipl ACZM, Dipl ACAW,
Brian L Speer, DVM, DABVP (Avian Practice), DipECZM (Avian),
Casey M. Holliday, PhD

Session #3165

Affiliation: From Denver Museum of Nature and Science, 2001 Colorado Blvd., Denver, CO 80205, USA (Lessner), Medical Center for Birds, 3807 Main Street, Oakley, CA 94561, USA (Echols, Speer), Scarlet Imaging, PO Box 57515, Murray, UT 84157, USA (Echols), Department of Medicine and Epidemiology, School of Veterinary Medicine, University of California-Davis, Davis, CA 95616, USA (Paul-Murphy), University of Missouri School of Medicine, 1 Hospital Drive, Columbia, MO 65212, USA (Holliday).

Note: Abstract format selected by author(s). See Front Matter for description of conference proceedings manuscripts submitted as scientific abstracts.

Abstract: The avian beak is a complex organ complete with bone, neurovascular tissue and a keratinized covering called the rhampotheca. Beak trimming is a common procedure described in avian species and may be used for corrective, cosmetic and sometimes behavioral modification purposes- often with little regard to the underlying neurovascular structures. Previous publications describe papillae extending from the premaxillary bone into the upper beak rhampotheca providing sensory input;[17,28] however, no recommendations as to how beak trimming can be safely performed without damaging these sensory papillae have been accurately defined. In this study, we evaluated the upper and lower beaks of two deceased grey parrots (*Psittacus erithacus*). In one bird, we used a novel stain (Scarlet Tissue Stain, Scarlet Imaging, Murray, UT, USA) and micro computed tomography (CT) to visualize the papillae in the upper and lower beaks. In a second grey parrot, we hand isolated the upper and lower beak dermal papillae and used high resolution photography plus traditional paraffin histology, and staining using a modified Masson's trichrome. The utilized processes clearly identified the papillae three-dimensionally (CT) and two-dimensionally (photography and histology) and in both cases demonstrated associated nerves. This allowed us to determine the approximate lengths of the papillae at the rostral tomia for the upper and lower beaks. Based on these findings, the authors recommend lateral radiographs of the bird's head and beak prior to considering performing beak trims to identify the location of the underlying bone relative to the overlying rhampotheca. Based on this preliminary study and specifically in grey parrots, the authors recommend the rostral tomia (upper and lower beaks) should not be trimmed closer than 8-10 mm from the underlying bone. Further work will be needed to support these recommendations and possibly provide guidelines for other species.

References

1. Bailleul AM, Witmer LM, Holliday CM. Cranial joint histology in the mallard duck (*Anas platyrhynchos*): new insights on avian cranial kinesis. *J Anat.* 2017;230(3):444-460.

2. Baumel JJ. Handbook of avian anatomy: *nomina anatomica avium. Publications of the Nuttall Ornithological Club.* 1993;23.

3. Berkhoudt H. Taste buds in the bill of the mallard (*Anas platyrhynchos L.*). *Neth J Zool.* 1976;27(3): 310-331.

4. Berkhoudt H. Touch and taste in the mallard (*Anas platyrhynchos L.*). *Rijksuniversiteit te Leiden.* 1980.

5. Bolze GV. Anordnung und bau der herbstschen korperchen in limicolenschnabeln im zusammenhang mit nahrungsfindung. *Zool Anz.* 1968;181:313-355.

6. Crole MR, Soley JT. Comparative distribution and arrangement of Herbst corpuscles in the oropharynx of the ostrich (*Struthio camelus*) and emu (*Dromaius novaehollandiae*). *Anat Rec.* 2014;297(7):1338-1348.

7. Cunningham SJ, Alley MR, Castro I, et al. Bill morphology of ibises suggests a remote-tactile sensory system for prey detection. *The Auk.* 2010;127(2):308-316.

8. Cunningham SJ, Corfield JR, Iwaniuk AN, et al. The anatomy of the bill tip of kiwi and associated somatosensory regions of the brain: comparisons with shorebirds. *PLoS One.* 2013;8(11): e80036.

9. Demery ZP, Chappell J, Martin GR. Vision, touch and object manipulation in Senegal parrots *Poicephalus senegalus*. *Proc Royal Soc B: Biol Sci.* 2011;278(1725):3687-3693.

10. Dubbeldam JL. The sensory trigeminal system in birds: input, organization and effects of peripheral damage. A review. *Arch Physiol Biochem.* 1998;106(5):338-345.

11. Dubbeldam JL, De Bakker MA, Bout RG. The composition of trigeminal nerve branches in normal adult chickens and after debeaking at different ages. *J Anat.* 1995;186(3):619.

12. Genbrugge A, Adriaens D, De Kegel B, et al. Structural tissue organization in the beak of Java and Darwin's finches. *J Anat.* 2012;221(5):383-393.

13. Gentle MJ, Breward J. The bill tip organ of the chicken (*Gallus gallus* var. *domesticus*). *J Anat.* 1986;145:79.

14. Gignac PM, Kley NJ, Clarke JA, et al. Diffusible iodine-based contrast-enhanced computed tomography (diceCT): an emerging tool for rapid, high-resolution, 3-D imaging of metazoan soft tissues. *J Anat.* 2016;228(6):889-909.

15. Gottschaldt KM. Structure and function of avian somatosensory receptors. In: King AS, McLelland J, eds. *Form and Function in Birds.* 1985;375-461.

16. Gottschaldt KM, Lausmann S. The peripheral morphological basis of tactile sensibility in the beak of geese. *Cell Tissue Res.* 1974;153(4):477-496.

17. Goujon DE. An apparatus of tactile corpuscles situated in the beaks of parrots. *J Anat Physiol Norm Pathol.* 1869;6:449-455.

18. Gustafson LA, Cheng HW, Garner JP, et al. The effects of different bill-trimming methods on the well-being of Pekin ducks. *Poult Sci.* 2007;86(9):1831-1839.

19. Halata Z, Grim M. Sensory nerve endings in the beak skin of Japanese quail. *Anat Embryol.* 1993;187(2): 131-138.

20. Heppleston PB. Anatomical observations on the bill of the oystercatcher (*Haematopus ostralegus occidentalis*) in relation to feeding behaviour. *J Zool.* 1970;161(2):519-524.

21. Hoerschelmann H. Strukturen der Schnabelkammer bei Schnepfenvögeln (Charadriidae und Scolopacidae). *Zeitschrift für wissenchaftliche Zool J.* 1972;185:105-121.

22. Homberger DG, Brush AH. Functional-morphological and biochemical correlations of the keratinized structures in the African grey parrot, *Psittacus erithacus* (Aves). *Zoomorphol.* 1986;106(2):103-114.

23. Kuenzel WJ. Neurobiological basis of sensory perception: welfare implications of beak trimming. *Poult Sci.* 2007;86(6):1273-1282.

24. Lederer RJ. The role of avian rictal bristles. *The Wilson Bulletin*, 1972;193-197.

25. Lucas AM, Stettenheim PR. Avian anatomy: integument. Washington, DC: US Department of Agriculture; *Agricult. Handbook 362.* 1972.

26. Lüdicke M. Aufbau und Abnutzung der Hörnzähne und Hornwülste des Vogelschnabels. *Zeitschrift für Morphologie und Ökologie der Tiere*, 1940;37(2):155-201.

27. Mench J, Paul-Murphy J, Klasing K, Cussen V. True parrots (Psittacidae). In: Yeates J, ed. *Companion Animal Care and Welfare: The UFAW Companion Animal Handbook*; 2018;338-354.

28. Menzel R, Lüdicke M. Funktionell-anatomische und autoradiographische Untersuchungen am Schnabelhorn von Papageien (Psittaci). *Zool Jahrb Anat.* 1974;93:175-218.

29. Nebel S, Jackson DL, Elner RW. Functional association of bill morphology and foraging behaviour in calidrid sandpipers. *Anim Biol.* 2005;55(3):235-243.

30. Orosz SE, Bradshaw GA. Avian neuroanatomy revisited: from clinical principles to avian cognition. *Vet Clin North Am Exot Anim Pract.* 2007;10(3):775-802.

31. Schneider ER, Mastrotto M, Laursen WJ, et al. Neuronal mechanism for acute mechanosensitivity in tactile-foraging waterfowl. *Proc Natl Acad Sci.* 2014;111(41):14941-14946.

32. Soliman SA, Madkour FA. A comparative analysis of the organization of the sensory units in the beak of duck and quail. *Histol Cytol Embryol.* 2017;1(4):1-16.

33. Soons J, Herrel A, Genbrugge A, et al. Multi-layered bird beaks: a finite-element approach towards the role of keratin in stress dissipation. *J R Soc Interface.* 2012;9(73):1787-1796.

34. Speer B, Echols S. Beak deformities: form, function and methodology for medical treatment. *Assoc Latinoamericana Vet Fauna Silv.* 2016.

35. Speer B, Powers LV. Anatomy and disorders of the beak and oral cavity of birds. *Vet Clin North Am Exot Anim Pract.* 2016;19(3):707-736.

36. Speer BL, Fecchio R, Holliday CM. The beak: form, function, and problems. In: Speer B, van Zeeland Y, eds. *Current Therapy in Avian Medicine and Surgery,* Vol 2. St. Louis, MO: Elsevier; 2023.

37. Stettenheim PR. The integumentary morphology of modern birds—an overview. *Am Zool.* 2000;40(4): 461-477.

38. Struthers S, Classen HL, Gomis S, Schwean-Lardner K. The effect of beak tissue sloughing and post-treatment beak shape on the productivity of infrared beak-treated layer pullets and hens. *Poult Sci.* 2019;98(9): 3637-3646.

39. Van Hemert C, Handel CM, O'brien DM. Stable isotopes identify dietary changes associated with beak deformities in black-capped chickadees (*Poecile atricapillus*). *The Auk.* 2012;129(3):460-466.

40. Wild JM. The avian somatosensory system: a comparative view. In: Scanes CG, ed. *Sturkie's Avian Physiology*, 6th Ed. New York NY: Academic Press; 2015:55-69.

Novel Use of Antifungal Nail Lacquer to Treat *Candida* Bill Infections in Ducks

Glenn H. Olsen, DVM, MS, PhD

Session #3290

Affiliation: From the US Geological Survey, Eastern Ecological Science Center (formerly Patuxent Wildlife Research Center), 12302 Beech Forest Road, Laurel, MD 20708, USA.

Abstract: Captive wild ducks in the research colony at the U.S. Geological Survey's Eastern Ecological Science Center in Laurel, Maryland frequently contract *Candida* infections on the upper and lower bills. Various standard treatments were attempted with mixed success. Several of the surf scoters suffered extreme erosions of the bill leading to euthanasia prior to using this new treatment. For the past 3 years a human medical product, ciclopirox 8% nail lacquer, has been used to treat the *Candida* successfully. The treatment requires careful trimming and debriding of the dead tissue before applying the ciclopirox, then keeping the ducks caged away from water for 15 minutes to allow the product to dry. Standard treatment is twice weekly. For persistent cases oral fluconazole is added to the treatment regime. Common eider, surf scoter, white-winged scoter, black scoter, long-tailed ducks, and harlequin ducks have been treated successfully with ciclopirox and no adverse reactions were observed.

Introduction

The bill or beak of birds can be affected by fungal pathogens, namely *Candida albicans*, *C. parapsilosis*, *C. krusei*, *C. tropicalis*, *C. glabrata*, and *Aspergillus*.[1-3] Presenting signs are described as white caseous exudate or plaques on the bill or oropharynx.[1,3] *Candida* species is a common cause of fungal infections in birds including flamingoes, raptors, parrots, pigeons, and poultry.[5-7] These *Candida* infections often are associated with antibiotic therapy, malnutrition, poor sanitation, other concurrent illnesses, or immunocompromised birds.[1,5]

Treatments have involved the use of topical and systemic antifungal medications. Topical applications of clotrimazole cream (Lotrim, Schering, Kenilworth, NJ, USA), amphotericin B ointment (Fungizone, Bristol-Myers Squibb Co, Princeton, NJ, USA), nystatin ointment (Mycostatin topical, Bristol-Myer Squibb Co, Princeton, NJ, USA), miconazole gel (Daktarin Oral Gel, Jansen-Cilag Ltd, Wycombe, Buckinghamshire, UK), or chlorhexidine (Nolvasan Solution, Zoetis, Inc, Kalamazoo, MI, USA) have all been used to treat *Candida* infections.[1,2,5] Systemic antifungal therapy used against *Candida* have included itraconazole (Sporonox, Janssen Pharmaceutical Products LP, Titusville, NJ, USA), nystatin (Mycostatin oral, Bristol-Myer Squibb Co, Princeton, NJ, USA), or fluconazole (Diflucan, Pfizer, Inc, New York, NY, USA).[1,5]

Clinically Relevant Data

The sea duck colony at the U.S. Geological Survey, Eastern Ecological Science Center (EESC) in Laurel, Maryland houses approximately 125 ducks of six species used for research and education. The sea ducks at the EESC include common eider (*Somateria mollissima*), black scoter (*Melanitta nigra*), surf scoter (*M. perspicillata*), white-winged scoter (*M. fusca*), long-tailed ducks (*Clangula hyemalis*), and harlequin ducks (*Histrionicus histrionicus*). In addition, two non-sea duck species are maintained, lesser scaup (*Aythya affinis*), also used for research, and wood ducks (*Aix sponsa*), used primarily for education by our visitor center. The EESC holds the first captive breeding record for surf scoters. The research with the captive sea ducks revolves around some physiologic studies such as hearing acuity[9] and vision,[10] and also, avian influenza.[11] The EESC Institutional Animal Care and Use Committee (IACUC) annually approves the standard operating procedures for the duck colony husbandry and the veterinary medical care of the ducks. The information reported in this paper resulted from several years of attempting to and successfully treating *Candida* infections of the bill and oropharynx of the captive ducks in the colony.

Animal husbandry staff often first find signs of *Candida* as an elongated bill or misshaped bill. Sometimes the lesions appear more in the oropharynx as white plaques or flaky exudate. Diagnosis has been made by fungal cultures submitted to Antech Laboratories. Cultures consistently yield *Candida* species, but not every duck was tested. Often the diagnosis was made on the signs of the lesions alone. Early treatments were with nystatin, clotrimazlole, or chlorhexidine topically (Table 1). Sometimes itraconazole was used orally (10 mg/kg twice daily), but often there was no resolution of the bill lesions with any of these treatments. Lesions could result in severe crusting and deformity of the upper bill in the surf scoters especially. No birds were euthanized because of these lesions, but a group of surf scoters were chosen for a terminal avian influenza study partly due to the bill deformities. Even with the bill deformities the birds otherwise were in good body condition.

Ciclopirox (Ciclopirox Topical Solution 8% Nail Lacquer, Perrigo, Bronx, NY, USA) was at first just used alone, with some initial success (Table 1), but some cases did not resolve, and one surf scoter male continued to get worse. In an effort to help this bird, fluconazole at 20 mg/kg orally twice weekly (Fluconazole Tablets USP 50 mg, Rising Health LLC, Saddle Brook, NJ, USA) was added to the treatment regime. This bird responded and improved markedly within 1 month. The current treatment involves handling ducks twice weekly to debride the tissues infected on the bill and oropharynx to remove any white plaques and debris, then to give fluconazole tablet (15-25 mg/kg), and finally to paint the affected areas with the ciclopirox nail lacquer. Birds are held out of the water for 15 minutes to allow the ciclopirox time to dry. Most ducks resolve in 2-3 months, mean days from onset of treatment to resolution is 108 (range 46-283 days).

Table 1. Select results from several years of treating *Candida* species bill and oropharynx infections in sea ducks at the U.S. Geological Survey Eastern Ecological Science Center, Laurel, Maryland USA. Clotrimazole, chlorhexidine, nystatin, and ciclopirox all applied topically to lesions. Fluconazole given orally as a tablet.

Date	Species[a]	Band #	Sex	Age	Treatment used[b]	Date resolved	Length of treatment	Comments
16 Sep 2016	SUSC	635	F	13	Clotrimazole Chlorhexidine	Not resolved		Discontinued 11 Jan 2017
30 Oct 2017	COEI	196	M	4 mo	Nystatin	20 Dec 2017	51	Treatment 1X or 2X day
6 Oct 2017	COEI	169	M	4 mo	Nystatin	Not resolved		Treatment 1X or 2X day
14 Nov 2018	COEI	164	M	1	Nystatin	Not resolved		Treatment 2X day
7 Sep 2018	SUSC	F-461	M	2	Ciclopirox	Not resolved		Treatment 1X week
12 Feb 2021	SUSC	F-461	M	5	Ciclopirox + Fluconazole	22 Nov 2021	283	Severe case Trt 2X week
24 Sep 2020	WWSC	157	M	6	Ciclopirox	Not resolved		Treatment 1X week
1 Nov 2021	WWSC	157	M	7	Ciclopirox + Fluconazole	28 Oct 2021	211	Treatment 2X week
7 Dec 2018	WWSC	181	F	1	Ciclopirox	Not resolved		Treatment 3X month
1 Nov 2021	WWSC	181	F	4	Ciclopirox + Fluconazole	2 Aug 2021	123	Treatment 2X week
20 Jun 2019	COEI	166	F	2	Ciclopirox	10 Oct 2019	112	Treatment 1X week
29 Apr 2021	HARD	285	M	5	Ciclopirox + Fluconazole	21 Jun 2021	53	Treatment 2X week
17 Mar 2022	HARD	548	M	16	Ciclopirox + Fluconazole	12 May 2022	46	Treatment 2X week
28 Sep 2022	BLSC	137	M	16+	Ciclopirox + Fluconazole	6 Feb 2023	125	Treatment 2X week
29 Sep 2022	LTDU	754	F	5	Ciclopirox + Fluconazole	12 Dec 2022	54	Treatment 2X week
26 Dec 2022	SUSC	176	F	10	Ciclopirox + Fluconazole	16 Mar 2023	80	Treatment 2X week

[a]Species: BLSC = black scoter (*Melanitta nigra*), COEI = common eider (*Somateria mollissima*), LTDU = long-tailed duck (*Clangula hyemalis*), HARD = harlequin duck (*Histrionicus histrionicus*), SUSC = surf scoter (*M. perspicillata*), WWSC = white-winged scoter (*M. fusca*).
[b]Treatment used: Chlorhexidine (Nolvasan Solution, Zoetis Inc, Kalamazoo, MI, USA), Ciclopirox (Ciclopirox Topical Solution 8% Nail Lacquer, Perrigo, Bronx, NY, USA), Clotrimazole (Lotrim Cream, Schering, Kenilworth, NJ, USA), Fluconazole (15-25 mg/kg, PO, twice weekly, Fluconazole Tablets USP 50 mg, Rising Health LLC, Saddle Brook, NJ, USA), Nystatin (Mycostatin topical, Bristol-Myer Squibb Co, Princeton, NJ, USA).

Conclusions

Candida infections, though not common, occur on the bill and oropharynx of many avian species. The author of this paper has diagnosed and treated the condition in several psittacine species. Generally, the available topical products cited in the introduction worked well in these other species; however, in ducks, especially sea ducks, the birds are almost constantly on water. Even restraining the birds in cages for up to 30 minutes after treatment with one of the other topical products failed to resolve the problem in ducks. The advantage of the ciclopirox is that after 15 minutes it is dry to the touch and persists, not being easily washed away in water.

Many ducks with mild cases resolved with just the ciclopirox (one example is listed in Table 1); however, adding fluconazole orally to the treatment regime seems to improve outcome, from a clinical perspective. The treatment frequency of twice weekly is dictated by the species. The sea ducks are all captive reared, except the black scoter, which was wild caught. Even so, they do not tolerate frequent handling. For long-term treatments such as required to control *Candida* infections, twice weekly has worked well, with one surf scoter receiving treatments over a 9-month period. The addition of fluconazole at 15-25 mg/kg twice weekly is similar to one published recommendation for psittacine birds of 20 mg/kg PO q48h.[12]

Acknowledgments: The author thanks veterinary technician, Carlyn Caldwell, for assistance provided and the leader of our animal husbandry unit, Sharon Peregoy, for support in preparing this manuscript.

Disclaimer: Any use of trade, product, or firm names is for descriptive purposes only and does not imply endorsement by the U.S. Government.

References

1. Olsen GH. Oral biology and beak disorders of birds. *Vet Clin North Am Exot Anim Pract.* 2005:281-291.

2. Samour JH, Naldo JL. Diagnosis and therapeutic management of candidiasis in falcons in Saudi Arabia. *J Avian Med Surg.* 2002;16(2):129-132.

3. Donnelly KA, Wellehan JFX, Quesenberry K. Gastrointestinal disease associated with non-*alibicans Candida* species in six birds. *J Avian Med Surg.* 2019;33(4):413-418.

4. Hoefer HL. Diseases of the gastrointestinal tract. In: Altman RB, Clubb SL, Dorrestein GM, et al, eds, *Avian Medicine and Surgery.* Philadelphia, PA: WB Saunders; 1997:419-453.

5. Rayment KM, Garner MM, Drees R, et al. Atypical candidiasis of the gnathotheca in a lesser flamingo (*Phoeniconaias minor*). *J Avian Med Surg.* 2021;35(4):457-463.

6. Silvanose C. Infectious diseases: candidiasis. In: Samour JH, ed. *Avian Medicine.* 3rd Ed. St. Louis, MO: Elsevier; 2016:472-473.

7. Crespo R, Franca MS, Fenton H, Shivaprasad HL. Galliformes and Columbiformes. In: Terio KA, McAloose D, St Leger J, eds. *Pathology of Wildlife and Zoo Animals.* Cambridge, MA: Academic Press; 2018: 747-773.

8. Ciclopirox Topical Solution, 8%. Bronx, NY; Perrigo. Drug insert.

9. Crowell SE, Wells-Berlin AM, Carr CE, et al. A comparison of auditory brainstem responses across diving bird species. *J Comp Physiol A.* 2015;201:803-815.

10. Cantlay JC, Bond AL, Wells-Berlin AM, et al. Ineffectiveness of light emitting diodes as underwater deterrents for long-tailed ducks *Clangula hyemalis*. *Global Ecol Conserv.* 2020: doi.org/10.1016/j.gecco.2020.e01102.

11. Prosser DJ, Schley H, Simmons N, et al. A lesser scaup (*Aythya affinis*) naturally infected with Eurasian 2.3.4.4 highly pathogenic H5N1 avian influenza virus – movement ecology and host factors. *Transbound Emerg Dis.* 2022;69(5):e2653-e2660. doi.org/10.1111/tbed.14614.

12. Hawkins MG, Guzman DSM, Beaufrere H, et al. Birds. In: Carpenter JW, ed. *Exotic Animal Formulary.* St. Louis, MO: Elsevier; 2018:167-375.

Assessment of Novel Serologic Assays in Avian Species

Carolyn Cray, PhD,
Adriana Callico, MS,
A. Leclerc, DVM, DipECZM (Zoo Health Management),
Ellen Bronson, Dr med vet, Dipl ACZM

Session #3227

Affiliation: From Division of Comparative Pathology, Department of Pathology & Laboratory Medicine, University of Miami Miller School of Medicine, P.O. Box 016960 (R-46), Miami, FL 33101, USA (Cray, Callico), Zoo De Beauval, 41110 Saint-Aignan-sur-Cher, France (Leclerc), The Maryland Zoo in Baltimore, 1 Safari Place, Baltimore, MD 21217, USA (Bronson).

Note: Abstract format selected by author(s). See Front Matter for description of conference proceedings manuscripts submitted as scientific abstracts.

Abstract: While antemortem serodiagnostics of avian aspergillosis remains problematic, recent advances in gliotoxin detection as well as quantitation of hydroxybutyrate and the use of protein electrophoresis have been documented for use especially in penguin species.[1-3] While gliotoxin detection is highly specific, the sensitivity as a routine element of a screening panel is questionable. In a review of 345 samples submitted for analysis, less than 5% were positive for this marker. With the goal of improving antibody detection, additional studies were undertaken to examine options for ELISA and Western blot technologies. For the ELISA, Afmp1p, a recombinant antigen was used as previously described in falcons.[4] More than 260 penguin samples were examined and 29% were positive. With limited case follow up, these were inclusive of penguins with confirmed and suspect disease and those with known exposure. In some cases, antibody levels were followed through repeated measures and limited results suggest that titers may have prognostic value. Lastly, a novel implementation of a commercially available Western blot for antibody detection[5] was examined using samples from African penguins from an institution in France and from several parks in the United States. In this method, the presence of antibody is scored by the reactivity to four different *Aspergillus* antigens. While greater reactivity was observed in penguins with confirmed infection (over presumed normal penguins), differences were found between France and U.S. samples suggesting that institutional differences may be present. Additional studies need to be conducted to address the application of these methods in other avian species.

Acknowledgments: Thank you to the clinicians who shared clinical information related to submitted samples.

References

1. Cray C, Watson T, Arheart KL. Serosurvey and diagnostic application of antibody titers to *Aspergillus* in avian species. *Avian Dis.* 2009;53:491-494.

2. Desoubeaux G, Rodriguez M, Bronson E, et al. Application of 3-hydroxybutyrate measurement and plasma protein electrophoresis in the diagnosis of aspergillosis in African penguins (*Spheniscus demersus*). *J Zoo Wildl Med.* 2018;49(3):696-703.

3. Reidy LR, Desoubeaux G, Cardenas J, et al. Detection of gliotoxin but not bis(methyl) gliotoxin in plasma from birds with confirmed and probable aspergillosis. *J Zoo Wildl Med.* 2022;53(1):60-69.

4. Wernery U, Tsang C, Hebel C, et al. Serodiagnosis of aspergillosis in falcons (*Falco* spp.) by an Afmp1p-based enzyme-linked immunosorbent assay. *Mycoses.* 2018;61:600-609.

5. Oliva A, Flori P, Hennequin C, et al. Evaluation of the *Aspergillus* Western blot IgG kit for the diagnosis of chronic aspergillosis. *J Clin Microbiol.* 2015;53:248-254.

Effects of a Single Dose of Gabapentin in Adult Domestic Pigeons (*Columba livia domestica*)

Laura Burns, DVM,
Kelsey Trumpp, DVM,
Connor Jacobson, DVM,
Stephen Divers, BVetMed, DZooMed, DipECZM, Dipl ACZM, FRCVS

Session #3375

Affiliation: From the Department of Small Animal Medicine and Surgery (Zoological Medicine), College of Veterinary Medicine, University of Georgia, Athens, GA 30602, USA.

Note: Abstract format selected by author(s). See Front Matter for description of conference proceedings manuscripts submitted as scientific abstracts.

Abstract: Avian species are commonly presented for routine wellness evaluations and clinical disease. Handling for physical examinations can lead to increased stress that worsens clinical signs. Gabapentin has been used in other species as a safe and effective treatment to reduce stress for transportation and physical examination.[1-3] This study aimed to evaluate the effects of gabapentin on physiologic parameters (heart rate and respiratory rate) and stress-associated behaviors of pigeons. The hypothesis was that gabapentin administered once would improve physiologic parameters, improve tractability, and reduce stress-associated behaviors. As part of a blinded, randomized, crossover study, twenty domestic pigeons received 50 mg/kg of oral gabapentin and an equal volume oral saline control. All animals were subjected to an 8-day washout period between treatments. Each pigeon was assessed prior to any handling and then assessed at 1 hour and 2 hours post-treatment. Individual behaviors and physiologic parameters were evaluated before and after handling at each of these time points. A modified human intruder test, sedation scale (1-5) and resistance score (0-3) were also developed and compared between time points.[1,4,5] Average heart rate was decreased by 30 beats per minute 2 hours post treatment. Resistance scores were 0.35 points lower than control 2 hours post treatment. Attempts to capture for examination were lower by 0.49 attempts 2 hours following treatment compared to the control. Mild sedation was noted in 20% of pigeons receiving gabapentin. Results from this study indicate that gabapentin may be beneficial when handling pigeons and future studies should be considered in other avian species.

References

1. Van Haaften KA, Eichstadt Forsythe LR, Stelow EA, Bain MJ. Effects of a single preappointment dose of gabapentin on signs of stress in cats during transportation and veterinary examination. *J Am Vet Med Assoc.* 2017;251:1175-1181.

2. Bleuer-Elsner S, Medam T, Masson S. Effects of a single oral dose of gabapentin on storm phobia in dogs: A double-blind, placebo-controlled crossover trial. *Vet Rec.* 2021;e453.

3. Conway R, Burton M, Mama K et al. Behavioral and physiologic effects of a single dose of oral gabapentin in rabbits (*Oryctolagus cuniculus*). *Proc ExoticsCon.* 2022;183-184.

4. Heatley JJ, Cary J, Kingsley L, et al. Midazolam sedates passeriformes for field sampling but affects multiple venous blood analytes. *Vet Med (Auckl).* 2015;6:61-69.

5. Bethell EJ, Cassidy LC, Brockhausen RR, Pfefferle D. Toward a standardized test of fearful temperament in primates: a sensitive alternative to the human intruder task for laboratory-housed rhesus macaques (*Macaca mulatta*). *Front Psychol.* 2019;10.

Effect of Capromorelin (Entyce®) on Appetite and Weight Gain of Domestic Pigeons (*Columba livia domestica*)

Kelsey Trumpp, DVM,
Laura Burns, DVM,
Connor Jacobson, DVM,
Stephen J. Divers, BVetMed, DZooMed, DipECZM (Herpetology),
DipECZM (Zoo Health Management), Dipl ACZM, FRCVS

Session #3179

AAV House Officer Manuscript Competition Winner

Affiliation: From the Department of Small Animal Medicine and Surgery (Zoological Medicine), University of Georgia College of Veterinary Medicine, 501 D. W. Brooks Drive, Athens, GA 30602, USA.

Note: Abstract format selected by author(s). See Front Matter for description of conference proceedings manuscripts submitted as scientific abstracts.

Abstract: Weight loss and decreased appetite are commonly encountered sequela of disease and stress in avian patients. There is currently minimal information in the literature regarding appetite stimulation in birds.[1,2] Capromorelin (Entyce®, Elanco Animal Health, Greenfield, IN 46140, USA) is a potent agonist of the growth hormone secretagogue receptor (GHS-R), which increases food consumption via direct stimulation of the hunger centers of the hypothalamus. This drug is approved by the U.S. Food and Drug Administration for use as an appetite stimulant in dogs and has been shown to also increase food consumption in New Zealand white rabbits and broiler chickens.[3-9] Twenty adult domestic pigeons, housed in groups of five, were involved in a randomized, placebo-controlled study to investigate the effect of capromorelin on appetite and weight gain in birds. Each pigeon was randomly assigned to the control or treatment group. The birds were individually weighed and given either oral water (control) or capromorelin at 12 mg/kg once daily for the duration of the 6-day study period. Total food consumed was recorded in grams per cage each day. Pigeons given Entyce® consumed a greater percentage of food daily and gained more weight over the course of the study period than those that were given the same volume of water. No adverse side effects were noted in any birds in this study. Entyce® shows promise as an appetite stimulant in pigeons and further investigation into its use in additional avian taxa is warranted.

References

1. Scagnelli A, Titel C, Doss G, Mans C. Effects of midazolam and lorazepam on food intake in budgerigars (*Melopsittacus undulatus*). *J Exot Pet Med.* 2022;41:42-45.

2. Martel A, Berg C, Doss G, Mans C. Effects of midazolam on food intake in budgerigars (*Melopsittacus undulatus*). *J Avian Med Surg.* 2022;36(1):53-57.

3. Rhodes L, Zollers B, Wofford JA, Heinen E. Capromorelin: a ghrelin receptor agonist and novel therapy for stimulation of appetite in dogs. *Vet Med Sci.* 2018;4(1):3-16.

4. Zollers B, Rhodes L, Heinen E. Capromorelin oral solution (Entyce®) increases food consumption and body weight when administered for 4 consecutive days to healthy adult beagle dogs in a randomized, masked, placebo controlled study. *BMC Vet Res*. 2016;13(1):1-5.

5. Zollers B, Wofford JA, Heinen E, et al. A prospective, randomized, masked, placebo-controlled clinical study of capromorelin in dogs with reduced appetite. *J Vet Intern Med*. 2016;30(6):1851-1857.

6. Zollers B, Rhodes L, Smith RG. Capromorelin increases food consumption, body weight, growth hormone, and sustained insulin-like growth factor 1 concentrations when administered to healthy adult beagle dogs. *J Vet Pharmacol Ther*. 2017;40(2):140-147.

7. Zollers B, Huebner M, Armintrout G, et al. Evaluation of the safety in dogs of long-term, daily oral administration of capromorelin, a novel drug for stimulation of appetite. *J Vet Pharmacol Ther*. 2017;40(3):248-255.

8. Draper JM, Savson DJ, Lavin ES, et al. Comparison of effects of capromorelin and mirtazapine on appetite in New Zealand white rabbits (*Oryctolagus cuniculus*). *J Am Assoc Lab Anim Sci*. 2022;61(5):495-505.

9. Ceron-Romero N, Taofeek N, Thomas A, et al. Capromorelin, a ghrelin receptor agonist, increases feed intake and body weight gain in broiler chickens (*Gallus gallus domesticus*). *Poult Sci*. 2021;100(8):101204.

Successful Conservative Management of a Neuro-Compressive Unstable Synsacral Fracture in a Yellow-Headed Amazon Parrot (*Amazona oratri*)

Lauren Thielen, DVM, DABVP (Avian Practice),
Ian Porter, DVM, Dipl ACVR,
Laura St Clair, DVM,
Nicola Di Girolamo, DVM, PhD, DipECZM (Herpetology), Dipl ACZM, Dipl ACVPM

Session #3473

Affiliation: From the Department of Clinical Sciences, College of Veterinary Medicine, Cornell University, 930 Campus Road, Ithaca, NY 14853, USA.

Note: Abstract format selected by author(s). See Front Matter for description of conference proceedings manuscripts submitted as scientific abstracts.

Abstract: Although conservative management for synsacral fractures in birds is suggested in the literature, case reports describing the conservative management and long-term outcomes have not been reported.[1] This case report describes successful conservative management of an acute, traumatic, neuro-compressive synsacral fracture in a parrot. A 32-year-old female Amazon parrot (*Amazona oratri)* was presented with increased respiratory effort and rate, depressed mentation, and an inability to stand after an acute fall. On initial assessment, cloacal tone was absent and proprioceptive deficits were noted in both hind legs. The patient had superficial sensation, grip of the feet, and movement of the legs. A contrast-enhanced computed tomography (CT) scan was performed and found an unstable complete linear circumferential fracture of the mid-caudal aspect of the synsacrum. The fracture resulted in ventral displacement and a step deformity attenuating approximately 40% of the vertebral canal diameter. Moderate pulmonary consolidation secondary to trauma was noted. Conservative management was elected due to the size of the synsacrum and the risk for surgical complications. The patient was hospitalized to provide analgesia, supportive care and multiple time daily cloacal stimulation to defecate. After 8 days in hospital, the patient began to develop cloacal tone, and was able to perch on a rolled towel. Strict cage rest was performed for 6 weeks, in a confined space devoid of perches. In the first week following discharge the patient returned to normal defecation and function of both hind limbs. A repeat CT scan performed 8 months after the trauma showed adequate healing with a moderate amount of smooth, bridging bone crossing the fracture site. Conservative management could be considered as a treatment option for similar synsacral fractures.

References

1. Powers LV, Bergman RL. Surgical spinal stabilization in an umbrella cockatoo (*Cacatua alba*). *Proc Annu Conf Assoc Avian Vet.* 2005.

Effects of Benzodiazepines on Food Intake in Budgerigars (*Melopsittacus undulatus*)

Christoph Mans, Dr med vet, Dipl ACZM,
Caroline E. Titel, BS,
Grayson A. Doss, DVM, Dipl ACZM

Session #3454

Affiliation: From the School of Veterinary Medicine, University of Wisconsin-Madison, 2015 Linden Drive, Madison, WI 53706 USA.

Note: Abstract format selected by author(s). See Front Matter for description of conference proceedings manuscripts submitted as scientific abstracts.

Abstract: Reduced food intake is a common concern in sick and hospitalized birds, and insufficient caloric intake can lead to a variety of detrimental effects and prolong hospitalization. Currently, insufficient food intake in pet birds is predominately managed by veterinarians by providing nutritional support via crop or proventricular gavage, depending on the species. Several medications have been used to stimulate appetite in dogs and cats, including benzodiazepines. In psittacine birds, our knowledge about the effects of benzodiazepines is currently limited to single-dose studies (1 mg/kg) of midazolam and lorazepam.[1-2] The purpose of this study was to evaluate two different doses of midazolam (1 mg/kg and 0.5 mg/kg IM once) and diazepam (1 mg/kg IM once) in a randomized, controlled, blinded, complete crossover study. Food intake was measured in hourly intervals for the first 8 hours and then one measurement for the 8- to 12-hour period. The birds were monitored by video recording for potential adverse effects. Both midazolam and diazepam at 1 mg/kg resulted in a significant increase in food intake compared to the control treatment. At 0.5 mg/kg, midazolam-induced food intake was less, and sedation behavior was of shorter duration compared to midazolam at 1 mg/kg. Results of this study indicate that diazepam has similar positive effects on food intake in budgerigars compared to midazolam, and lowering the dose of midazolam will result in a decrease in its appetite-stimulating as well as sedative effects. Either benzodiazepine can be used clinically as a short-acting appetite stimulant in budgerigars.

References

1. Martel A, Doss G, Berg C, Mans C. Effects of midazolam on food intake in budgerigars. *J Avian Med Surg.* 2022;36:53-57.

2. Scagnelli A, Titel C, Doss G, Mans C. Effects of midazolam and lorazepam on food intake in budgerigars. *J Exot Pet Med.* 2022;41:42-45.

Section 11

AAV "I'll Take Topics with the Acronym AI for 500 Please, Alex" Session

Nick Kirk, DVM, DABVP (Avian Practice)
Session Chair

Highly Pathogenic Avian Influenza in Raptors

Dana Franzen-Klein, DVM, MS,
Annette Ahlmann, DVM, MS, CWR,
Victoria Hall, DVM, MS, Dipl ACVPM

Session #3340

Affiliation: From The Raptor Center at the University of Minnesota, St. Paul, MN 55108, USA.

Note: Abstract format selected by author(s). See Front Matter for description of conference proceedings manuscripts submitted as scientific abstracts.

Abstract: The strain of highly pathogenic avian influenza (HPAI, strain EA/AM 2.3.4.4. H5N1) that arrived in the United States in early 2022 is behaving very differently from previous HPAI outbreaks. The outbreak is persisting through more than one migration season, and there is a high rate of transmission and clinical disease among wildlife in all four of the migratory flyways. From March 28-December 31, 2022, The Raptor Center at the University of Minnesota screened 1051 individuals for HPAI via polymerase chain reaction (PCR) testing. Of those tested, 215 individuals representing 12 species were confirmed HPAI positive. Great horned owls (*Bubo virginianus*, n = 92) were the most affected species, followed by bald eagles (*Haliaeetus leucoceophalus*, n = 48) and red-tailed hawks (*Buteo jamaicensis*, n = 39). Clinical signs were characterized and video examples of the various presentations of HPAI in raptors will be presented. Most birds admitted with the virus were exhibiting severe neurologic signs. One great horned owl recovered from the infection and was released to the wild. In addition, the utility of a point-of-care avian influenza test (Zoetis FluDETECT®) was assessed by comparing the results to PCR. Paired FluDETECT® and PCR tests have been run on 101 individuals representing 10 species to date, and sampling is ongoing. Based on preliminary data analysis, the FluDETECT® test has a high false negative rate compared to PCR results, making this diagnostic of limited value in a raptor rehabilitation setting.

Acknowledgments: The authors thank Dr. Irene Bueno and Dr. Carol Cardona for their contributions to this work, and Dr. Arno Wuenschmann for pathology results.

References

1. Bertran K, Busquets N, Abad FX, et al. Highly (H5N1) and low (H7N2) pathogenic avian influenza virus infection in falcons via nasochoanal route and ingestion of experimentally infected prey. *PLoS ONE*. 2012;7(3):e32107.

2. Caliendo V, Leijten L, van de Bildt MWG, et al. Pathology and virology of natural highly pathogenic avian influenza H5N8 infection in wild common buzzards (*Buteo buteo*). *Sci Rep*. 2022;12:920.

3. Hall JS, Ip HS, Franson C, et al. Experimental infection of a North American raptor, American kestrel (*Falco sparverius*), with highly pathogenic avian influenza virus (H5N1). *PLoS ONE*. 2009;4(10):e7555.

4. Nemeth NM, Ruder MG, Poulson RL, et al. Bald eagle mortality and nest failure due to clade 2.3.4.4 highly pathogenic H5N1 influenza a virus. *Sci Rep*. 2023;13:191.

5. Shearn-Bochsler VI, Knowles S, Ip H. Lethal infection of wild raptors with highly pathogenic avian influenza H5N8 and H5N2 viruses in the USA, 2014-15. *J Wildl Dis*. 2019;55(1):164-168.

6. Uno Y, Soda K, Tomioka Y, et al. Pathogenicity of clade 2.3.2.1 H5N1 highly pathogenic avian influenza virus in American kestrel (*Falco sparverius*). *Avian Pathol*. 2020;49(5):515-525.

7. U.S. Department of Agriculture. Final Report for the 2014-2015 Outbreak of Highly Pathogenic Avian Influenza (HPAI) in the United States. Revised August 11, 2016. Washington, DC: Animal and Plant Health Inspection Service. https://www.aphis.usda.gov/animal_health/emergency_management/downloads/hpai/2015-hpai-final-report.pdf.

8. van den Brand JMA, Krone O, Wolf PU, et al. Host-specific exposure and fatal neurologic disease in wild raptors from highly pathogenic avian influenza virus H5N1 during the 2006 outbreak in Germany. *Vet Res*. 2015;46:24.

Quarantined Emu (*Dromaius novaehollandiae*) Infected with Highly Pathogenic Avian Influenza H5 Survives, Despite Previous Fulminant Encephalitis Vaccine-Induced Rhabdomyolysis

April Romagnano, PhD, DVM, DABVP (Avian Practice)

Session #3383

Affiliation: From the Avian & Exotic Clinic of Palm City, Palm City, FL 34990, USA.

Note: Abstract format selected by author(s). See Front Matter for description of conference proceedings manuscripts submitted as scientific abstracts.

Abstract: This presentation describes a male emus' recovery from vaccine-induced rhabdomyolysis despite contracting highly pathogenic avian influenza (HPAI), as confirmed by the U.S. Department of Agriculture's (USDA) National Veterinary Services Laboratory, from a positive feral Egyptian goose. Avian influenza (AI) is being disseminated throughout Florida by migratory and feral birds, particularly Egyptian geese. The farm that maintained this emu reported two dead birds, a feral Egyptian goose and a rooster, to Florida's Department of Agriculture and Consumer Services (FDACS). The birds tested positive for AI. The farm was quarantined and over 60 ducks, poultry and guinea fowl were euthanized. FDACS sagely choose not to euthanize one black swan, two Sebastopol geese and two adult emus, including the male reported here. Thanks to FDACS, the farm experienced no further AI deaths and all bordering neighbors were unaffected. AI, which is a major threat to the poultry industry and animal health officials, has strains that vary in severity and pathogenicity. Those most likely strains to kill poultry are labeled highly pathogenic. When HPAI flocks are identified, FDACS collaborates with USDA to eliminate, control and contain the virus. The emu reported here, previously cared for by the farm's large animal veterinarians, was referred weeks after encephalitis vaccination and AI exposure. Down and moribund on examination, its swollen right leg and thigh was severely bruised and leathery at the vaccination site. Bloodwork confirmed severe rhabdomyolysis. The bird was treated diligently by its owners with parenteral fluids and medications. Daily handfeeding addressed anorexia. Intense daily physical therapy was used to get the emu up, including manual and ceiling-built-in support-track sling walking, which continued for many months. The emu recovered safely in a blanket-covered hay-bale enclosure, where it eventually ate, drank, defecated, urinated, stood, walked and finally ran on its own.

Artificial Intelligence Assisted Cell Morphology Assessment in Avian Blood Smear

Kaori Davis, DVM, PhD, DACVP (Anatomic and Clinical Pathology),
Liam Cattell, MEng, DPhil,
Bianca Murphy, DVM, DAVBP (Avian Practice),
Elyssa Dimaano,
Dan Huang,
Thanh Le, MS, PhD

Session #3415

Affiliation: From Moichor, 555 De Haro St Suite 220, San Francisco, CA 94107, USA.

Note: Abstract format selected by author(s). See Front Matter for description of conference proceedings manuscripts submitted as scientific abstracts.

Abstract: The complete blood count (CBC) is a foundational assay to ascertain the overall health status and detect various conditions, including inflammation, infection and neoplasia. The alterations in numerical parameters of the CBC, such as shifts in leukocyte absolute and differential counts are informative, although they can be non-specific. Morphologic assessment of leukocyte populations further aids in interpreting the leukogram. Morphologic assessment is typically performed manually; however, manual morphologic assessment has limitations as it is time and labor-intensive, making it only available in selected patients. In this study, we evaluated the effectiveness of cutting-edge artificial intelligence (AI) and computer vision techniques for analyzing leukocyte morphology in avian hematology samples, specifically in 500 Congo African grey parrots (*Psittacus erithacus*). Among them, 238 were clinically healthy, 196 had various diseases, and the health status and clinical history of 66 animals were unknown. AI-generated counts of morphologic abnormalities, including band heterophils and toxic heterophils, were compared to manual microscopic counts conducted by blinded clinical pathologists. Within-run, between-run, and between-observer coefficients of variation were significantly lower when counted by AI when compared to the manual microscopic count. There was a good agreement between the automated and manual methods. This study demonstrated automated techniques to screen CBC morphologic parameters in a high throughput manner.

Biosecurity and Quarantine Procedures During a Highly Pathogenic Avian Influenza Outbreak

Dana Franzen-Klein, DVM, MS,
Annette Ahlmann, DVM, MS, CWR,
Victoria Hall, DVM, MS, Dipl ACVPM

Session #3341

Affiliation: From The Raptor Center at the University of Minnesota, St. Paul, MN 55108, USA.

Note: Abstract format selected by author(s). See Front Matter for description of conference proceedings manuscripts submitted as scientific abstracts.

Abstract: The 2022 highly pathogenic avian influenza (HPAI) outbreak presented an extraordinary challenge to wildlife rehabilitators - how do they complete their work while preventing the spread of a highly contagious foreign animal disease? From March 28-December 31, 2022, The Raptor Center at the University of Minnesota (TRC) admitted 215 HPAI positive raptors into our wildlife hospital. Despite the high volume of positive cases, TRC was able to use biosecurity principles and quarantine procedures combined with polymerase chain reaction (PCR) testing to effectively prevent disease transmission in the wildlife hospital and protect the center's educational ambassadors from this deadly virus. PCR testing was conducted at admission and at each stage of the quarantine process to safely move over 175 individuals from quarantine into the main patient population. This data allows us to better understand incubation periods and associated disease transmission risk in a raptor rehabilitation setting. Affordable PCR testing with timely results is not readily available to the majority of wildlife rehabilitators and veterinarians. TRC has collected detailed clinical signs data and paired that with PCR results to help guide the use of clinical signs in place of PCR testing as an indicator of possible HPAI infection in a raptor patient. TRC has used the PCR and clinical data collected during the spring of 2022 to adjust our biosecurity and quarantine protocols to ensure sustainability as this outbreak continues. Biosecurity logistics, quarantine time frames, triage parameters, and the data that support our practices and recommendations are discussed.

Acknowledgments: The authors thank Dr. Irene Bueno and Dr. Carol Cardona for their contributions to the development of our biosecurity methods.

References

1. Bertran K, Busquets N, Abad FX, et al. Highly (H5N1) and low (H7N2) pathogenic avian influenza virus infection in falcons via nasochoanal route and ingestion of experimentally infected prey. *PLoS ONE*. 2012;7(3):e32107.

2. Hall JS, Ip HS, Franson C, et al. Experimental infection of a North American raptor, American kestrel (*Falco sparverius*), with highly pathogenic avian influenza virus (H5N1). *PLoS ONE*. 2009;4(10):e7555.

3. Stull JW, Bjorvik E, Bub J, et al. 2018 AAHA infection control, prevention and biosecurity guidelines. *J Am Anim Hosp Assoc*. 2018;54:6.

Avian Influenza: Practical Ways to Protect the Backyard Flock

Brynn McCleery, DVM, DABVP (Avian Practice)

Session #3134

Affiliation: From Red Bank Veterinary Hospital, 197 Hance Avenue, Tinton Falls, NJ 07722, USA.

Abstract: This session briefly reviews the origin of the current highly pathogenic avian influenza (HPAI) outbreak and its significance to wildlife, poultry, and humans. Clinical signs of disease and testing are reviewed. The focus of the session is on precautions to take for backyard poultry to help prevent infection with HPAI.

Introduction

The deadliest outbreak of avian influenza virus (AIV) in United States and European history began in 2020 and is ongoing. This outbreak of H5N1 AIV has been associated with enormous loss of poultry, as well as increased outbreaks and death of wild birds and sporadic spillover into mammals, including a handful of human cases. In the United States, >490 backyard poultry flocks have been affected.

Avian Influenza Viruses

The avian influenza viruses (AIVs) belong to the family Orthomyxoviridae.[1,2] They are all influenza type A viruses, which are single stranded, negative-sense ribonucleic acid (RNA) viruses with segmented genomes that code for 10 viral proteins.[2,3] The 10 viral proteins are encoded by 8 segmented genomic strands.[3] Compared to the other 2 types of influenza viruses, influenza type A has the most genetic variation and the broadest host range.[3] Viral naming is based on the influenza type, place of isolation, strain, designation, and year of isolation. Further classification into subtypes is based on the hemagglutinin (HA) and neuraminidase (NA) surface glycoproteins.[2] HA mediates viral attachment and entry into the host cell by binding to sialic acid receptors on the host cell surface and is responsible for neutralizing host antibodies. NA catalyzes the cleavage of virus particles from the sialic acid sites on the host cell, thereby allowing for spread of the virus within the host.[2,3] Many combinations of HA and NA proteins are possible to create different subtypes of AIV. There are 18 known HA subtypes and 11 NA subtypes.[3,4] Related viruses within a subtype are referred to as a lineage.

Influenza viruses have a high capacity to change due to both genetic drift (e.g., mutations) and genetic shift (e.g., reassortment through exchange of gene segments). Antigenic alterations occur frequently in these viruses, particularly in the HA and NA sites, and are a mechanism through which the virus adapts to the host and survives.[2] Antigenic drift is characterized by smaller variations secondary to accumulated amino acid changes. Mutations that avoid antibody neutralization are selected for over time. Antigenic drift tends to occur every year to few years and is the underlying reason why the annual flu vaccine is constantly changing for humans.[2] Reassortment of recombination of the 8 gene segments can lead to larger variations (i.e., antigenic shift) which may result in new viruses for which a host population has no immunity. Antigenic shift is what typically leads to influenza pandemics.[2]

Pathogenicity of the AIVs is classified as high (HPAI) or low (LPAI) based on the ability of the strain to produce disease and mortality in chickens.[1,3] AVIs have been recognized as a cause of disease in birds for almost 150 years. HPAI was first described in 1878 and referred to as "fowl plague." A viral cause was identified in 1901, but the link to influenza was not made until 1955.[1]

Natural infections of LPAI typically cause no clinical signs or tissue lesions in wild birds. LPAI infections can be asymptomatic in poultry or cause mild symptoms.[1,3,4] H5 and H7 subtypes have the potential to mutate or evolve into HPAI viruses. However, not all viruses of these subtypes cause HPAI.[1] HPAI are rarely isolated from wild birds, and when they do occur in wild birds, the infections typically occur in the vicinity of outbreaks of HPAI in poultry.[1] Instead, the proposed mechanism for emergence of HPAI viruses is that these strains occur only after the viruses have crossed from wild birds to poultry, where they mutate and adapt.[3,4]

Natural Hosts

AIVs infect over 100 different bird species globally and are endemic in some wild birds. Natural reservoirs of the viruses include ducks, swans, geese, terns, shorebirds (including storks, plovers, and sandpipers), and gulls.[3,4] Wild birds, especially dabbling ducks (mallards, pintails, and widgeons), may be infected in the gastrointestinal and respiratory tracts with no symptoms. Waterfowl congregate in huge flocks, often on lakes, annually prior to migration, at which time AIVs spread rapidly through the population. In one study evaluating infection rate of ducks in Alberta, Canada prior to southern migration, > 60% of juvenile ducks were estimated to be infected.[1] Previous surveillance studies reported prevalence rates of LPAI in wild birds of around 15% in ducks and geese and 2% in other species.[1]

Aberrant Hosts

Since LPAI naturally circulates in wild birds, spillover can occur into aberrant hosts in contact with these birds or their feces. Aberrant hosts include poultry (chickens, turkeys, pheasants, quail, domestic ducks, geese, and guinea fowl), pigs, wild mammals, and humans. After transmission to poultry, LPAI may lose its adaptation to wild birds preventing reinfection. Further mutations can occur leading to new strains of HPAI due to exchange of genetic material with other AIVs. Unfortunately, as with the current outbreak, H5 and H7 infections in poultry can spill back into wild birds resulting in further geographic spread as the birds migrate.[4]

Even in areas where AIV outbreaks occur with regularity, such as Minnesota, variation in virus subtype, number of outbreaks seen each year, and seasonal relationship of outbreaks suggest that the epizootics are a result of new primary introductions from wild birds and that AIVs are not enzootic in poultry species.[1]

Transmission

Shedding occurs primarily in the feces, as well as through oral and nasal secretions. Transmission is through the oral-fecal route and can occur primarily from direct contact with infected birds or shared environments (including contaminated water) or secondarily through mechanical transfer of infective feces to susceptible birds.[1,2]

In the USA, spread to poultry typically occurs indirectly through contaminated feed, clothing, and equipment, and less frequently from direct contact with wild birds as poultry are more commonly reared in bird-proof confinement. In other countries, birds may be reared in ways that allow contact with wild birds such as ducks reared on ponds and turkeys reared on range. These birds are at increased risk of infection, especially if situated on a migratory waterfowl route.[1] HPAI are extremely infectious with rapid flock-to-flock spread reported.

Clinical Symptoms

LPAI infections can be asymptomatic in poultry or cause mild symptoms and are most commonly associated with a mild primary respiratory disease, fluffed feathers, reduced egg production, and low mortality.[1,3,4] HPAI, on the other hand, can cause acute high mortality (up to 90-100% in chickens, often within 48 hours).[1] Symptoms of HPAI include acute death, severe depression, anorexia, neurologic disease (twisted neck, swimming in circles, tremors), lethargy, reduced appetite, decreased egg production, irregular eggs (e.g., soft, thin shelled, misshapen), dyspnea, nasal discharge, coughing, sneezing, diarrhea, and swelling of the head, comb, eyelids, wattles, and legs with purple discoloration.[3]

Impact

HPAI is associated with enormous economic losses in poultry and high numbers of bird loss, through death and culling ("stamping out") as part of control policies.[1,5] HPAI is considered a foreign animal disease in the United States. Internationally, HPAI is also reportable to the OIE. Although reportable in poultry, backyard flocks with no sales are considered non-poultry on an international level.

Previous Outbreaks

In 1996, H5N1 HPAI (A/goose/Guangdong/1/1996) reinfected wild birds after spilling over into poultry. This panzootic was unprecedented. The outbreak was contained through culling all the poultry in Hong Kong, but the AIVs continued to circulate in healthy duck populations in the surrounding areas. This strain of HPAI is now considered endemic in 6 countries worldwide, including Bangladesh, China, Egypt, Indonesia, India, and Vietnam.[3]

In 2014-2015, an outbreak of HPAI subtype H5Nx occurred in poultry in the USA and Canada. HPAI H5N8 spread through the flyways initially leading to infection in North America from migratory birds originating in Asia. First identified over the winter and last detected in poultry in June 2015, the outbreak affected 211 commercial flocks and 21 backyard flocks, leading to $3.3 billion in economic losses and the loss of 50 million birds. The outbreak affected 21 states. Fortunately, no human cases were detected. Three mortality events were also reported in wild birds, affecting snow geese and ringed-necked ducks. Multiple captive raptor deaths were reported after being fed meat from infected wild waterfowl. Subtypes involved included H5N8, H5N1, and H5N2.[6,7]

Current Outbreak

The current 2021 H5N1 outbreak is a different strain than the 2014 outbreak. The virus has spun off several variants within clade 2.3.4.4b making it even more successful. The first case was identified in an American wigeon duck in South Carolina. The infection travelled from western Europe to eastern Canada and has been detected in all four flyways. Most early infections centered around the intersection of the Central and Mississippi migratory flyways. There has been wide geographic spread in wild birds and poultry worldwide with sporadic spillover into a variety of mammalian species.[4]

Disease has been variable among species. The current outbreak affects layers more than broilers and turkeys.[4] In wild birds, swans, geese, waterbirds (e.g., pelicans, mergansers), raptors, and scavengers are all considered highly susceptible. The current outbreak has been associated with a higher wild bird mortality than that of 2014, as well as with a wider diversity of species affected. In addition to wild birds, HPAI has been detected in a variety of wild mammals (e.g., harbor seals, grey seals, foxes, skunks, bobcats, lynx, Eurasian otters, Eurasian polecats, European badger, domestic pigs, ferret, common raccoon dog, opossums, bears, farmed mink, sea lions).[4,5] A small number of human infections have been reported, all of which were associated with recent poultry exposure. Overall, symptoms in human cases have been mild. No cases of human-to-human transmission have occurred.[4]

Statistics (at the time of writing)

Global

- > 11,300 animal outbreaks of H5N1 reported by 73 member states to the World Organization of Animal Health since January 2022[4]
- Largest AIV epidemic thus far in European history (2398 outbreaks in poultry, 46 million birds culled, 168 detections in captive birds, 2733 die off events in wild birds, 36 European countries affected)[5]
- Humans: Jan 2022-Mar 2023, 10 cases in humans from 7 countries (1 in the USA)[4]

USA (as of March 2023)

- 815 confirmed poultry flocks: 323 commercial, 492 backyard (in 47 states)[4,8]
- Almost 6,500 wild birds (in 49 states)[4]
- 58.65 million birds affected[4,8]
- Iowa: two massive egg laying operations culled > 5 million birds in single incidents[8]
- Mammals: 144 mammals (in 23 states)[4]

Summary

The above statistics make this the deadliest AIV outbreak in USA and European history. While most previous outbreaks were halted by change in season, the current H5N1 outbreak made it through the summer and some authorities are suggesting that the observed persistence of the virus in wild birds since the 2020-2021 epidemic wave may indicate that this strain has become endemic in wild birds.[5]

Prevention

In some ways, preventing infection in smaller, backyard flocks is easier than in large, commercial flocks due to reduced farm-to-farm contact, equipment, and transport of birds; however, in other ways, prevention can be more difficult due to increased free range flocks and a reduced understanding of proper biosecurity practices in backyard flock owners.

A line of separation should be placed around the flock and may include the walls of a coop/barn, fence around the run, or fence around the yard. Consider the risks of anything crossing the line—such as other birds, other animals, and humans—and evaluate ways to reduce those risks. Consider both structural biosecurity (e.g., the enclosure the birds are housed in) and operational biosecurity (e.g., practices and procedures followed by the people caring for the birds).

Basic biosecurity principles

Basic biosecurity principles should be followed and include:

- Isolate newly acquired birds
- Become familiar with signs of illness; quarantine sick birds; seek veterinary care and testing
- Clean and disinfect enclosures and equipment regularly
- Limit visitors: only allow people caring for the flock (family, friends) interact with the birds, make sure all participants follow biosecurity principles, avoid having people with separate flocks interact with the birds
- Wash hands before and after contacting birds
- Consider a foot bath (remove debris first) or dedicated pair of shoes (rubber boot covers work well)
- Change clothes or wear protective outer garments when working with poultry

Specific prevention for AIVs

The biggest risk to the backyard flock is from waterfowl, which are less likely to encounter backyard flocks unless there is a pond or large amount of acreage. Songbirds and other visitors to the backyard (corvids, pigeons/doves) are less likely to be infected with AIVs including HPAI; however, due to the variety of species that can be infected, including wild mammals, care should be taken to avoid contact of all wild animals with the flock. Farm-to-farm transmission is even more common, and owners of backyard flocks should do their best to eliminate contact with other flocks during the outbreak (e.g., follow all in/all out procedures and not buy/adopt new birds from various sources, not share equipment).

Prevent cross-contamination with other flocks:

- Avoid having people with separate flocks interact with the backyard flock
- Do not share equipment
- Avoid use of soiled egg cases, do not reuse cardboard egg cases between different flocks

Prevent contact with free-ranging wild birds/animals:

- Prevent nesting/shelter seeking
    - Build structures with shorter eaves or no overhangs/ledges
    - Fill small burrows, tunnels, or holes with gravel

- Repair any holes or tears in barn/cool walls
- Exclusionary screens/netting: utilize small gauge (1-inch opening excludes most bird species)

- Reduce water access
  - Remove standing water (event small pools) through proper grading or filling areas where water stands < 24-48 hours after rain
  - Ponds: If present, consider using a boat, bird deterrent balls, or predator decoys to discourage wild birds.
  - Do not offer bird baths around the coop or run

- Reduce food sources
  - Mow grass frequently (< 3 inches)
  - Keep fruit trees and other edible plants father away from the coop or run
  - Remove fallen fruit
  - Avoid grass, weeds, and shrubbery next to the coop/run to prevent wildlife seeking shelter
  - Do not feed wild birds/wildlife near poultry (i.e., no bird seed, suet, corn, etc.)
  - Prevent access of wildlife to stored feed (lidded container, sweep up spills)
  - Keep compost and garbage away from poultry and prevent wildlife access

- Reduce fecal contamination
  - Shade cloth over the top of the run can prevent wild bird droppings from falling into the run area
  - If covering the entire area is not desirable, place food/waterers under cover to prevent fecal contamination from wild birds
  - Do not store extra equipment outside. Cleaned equipment should be stored in a garage/barn/basement until needed.
  - Large troughs/swimming pools: Cover with a tarp/lid when not in use. Drain and refill frequently to keep clean.

APHIS is an excellent resource for owners (www. aphis.usda.gov/animalhealth/defendtheflock).

Vaccination is not allowed in the United States at this time due to impacts on international trade as vaccinated birds cannot be distinguished from infected birds through testing. Vaccination is also complicated by timing for commercial poultry: birds would require two doses of vaccine with a withdrawal period after the vaccine prior to slaughter, which is not feasible.

Testing

In poultry with suspected HPAI, testing should be performed by a USDA APHIS accredited veterinarian. RT-PCR is recommended. Appropriate samples include an oropharyngeal swab in gallinaceous poultry or a cloacal swab in waterfowl.[2] Possible results include not detected or non-negative. A non-negative sample with clinical signs consistent with HPAI is considered a presumptive positive. If detected on PCR, virus isolation and growth of the virus in an embryonated chicken egg may be performed.[2] Serology is not recommended as mortality occurs acutely before antibody production. After reporting a presumptive positive flock, the flock is placed under quarantine with control and surveillance zones established. Depopulation and a fallow period (where repopulation is not allowed) will follow.

Testing of wild bird populations is performed for surveillance purposes only. Surveillance of AIVs in wild birds began in the mid 1970s.[1] The USDA began monitoring migratory birds in different flyways between 1998-2005.[2] Surveillance efforts have increased due to the current outbreak. Eradication of HPAI from wild birds is not being attempted. Instead, attempts to limit spillover to domestic species through preventive measures are recommended.

References

1. Alexander DJ. A review of avian influenza in different bird species. *Vet Microbiol.* 2000;74:3-13.

2. Thomas JK, Noppenberger J. Avian influenza: a review. *Am J Health Syst Pharm.* 2007;64:149-165.

3. Nunez IV, Ross TM. A review of H5Nx avian influenza viruses. *Ther Adv Vaccines Immunother.* 2019;7:1-15.

4. Centers for Disease Control and Prevention. Technical report: highly pathogenic avian influenza A (H5N1) viruses. Available at: https://www.cdc.gov/flu/avianflu/spotlights/2022-2023/h5n1-technical-report.htm#print. Accessed March 25, 2023.

5. Adlhoch C, Fusaro A, Gonzalez JL, et al. Avian influenza overview March-June 2022. *Eur Food Safety Author J.* 2022;20(8):7415.

6. U.S. Department of Agriculture. Final report for the 2014-2015 outbreak of highly pathogenic avian influenza (HPAI) in the United States. Available at: https://www.aphis.usda.gov/animal_health/emergency_management/downloads/hpai/2015-hpai-final-report.pdf. Accessed March 25, 2023.

7. Greene JL. Update on the highly-pathogenic avian influenza outbreak of 2014-2015. Available at: https://nationalaglawcenter.org/wp-content/uploads/assets/crs/R44114.pdf. Accessed March 25, 2023.

8. Animal and Plant Health Inspection Service. 2022-2023 confirmations of highly pathogenic avian influenza in commercial and backyard flocks. Available at: https://www.aphis.usda.gov/aphis/ourfocus/animalhealth/animal-disease-information/avian/avian-influenza/hpai-2022/2022-hpai-commercial-backyard-flocks Accessed March 25, 2023.

Measuring Disease Prevalence in Barred Owls (*Strix varia*) Presented to a Louisiana Wildlife Hospital (2014-2021)

Aspen La-Riene Settle, BSc, and
Mark A. Mitchell, DVM, MS, PhD, DipECZM (Herpetology)

Session #3487

Affiliation: From Louisiana State University School of Veterinary Medicine, Baton Rouge, LA 70803, USA.

Note: Abstract format selected by author(s). See Front Matter for description of conference proceedings manuscripts submitted as scientific abstracts.

Abstract: The Wildlife Hospital of Louisiana (WHL) sees an annual average of about 1,600 patients. Of that caseload, approximately 15% are raptors. This retrospective study catalogued the presentation of barred owls (*Strix varia*), the most numerous raptor patients presented to the WHL. Electronic records were categorized according to patient outcome (6/431 dead on arrival, 220/431 euthanized within 24 hours, 22/431 died within 24 hours, 22/431 died after 24 hours, and 161/431 released). Primary diagnoses of orphaning, trauma, neurologic disease, or unknown were recorded for each outcome, as well as classification of trauma type and existence of comorbidities (e.g., ballistic trauma, vehicular trauma, comminuted fractures, open fractures, ocular damage, luxation, and parasitism). Three-hundred thirty-two (77%) of all owls presented with evidence of trauma. Metal opacities outside of the gastrointestinal tract, entry or exit wounds, or at least one identified puncture wound combined with a fracture warranted classification of ballistic trauma, which was the case for six birds. Vehicular trauma was the most frequent suspected cause of trauma, which affected 119 (35%) birds. Of those presenting with trauma, comminuted fractures were identified 24 times, open fractures 82 times, ocular damage 93 times, luxation 21 times, and parasitism 28 times. Further results, including analysis of comorbidity impact on release rate, will be discussed during the lecture. The authors use this data to analyze the efficiency of the WHL triage assessment protocols, the success of the team's rehabilitation and release efforts, and the danger to these animals of anthropomorphic activity and encroachment.

Section 12

AAV Potpourri II Session

Rachel Baden, DVM, DABVP (Avian Practice)
Session Chair

How to Maximize Foraging Opportunities for Parrots: A Case Study

Mandy Beekmans, MSc,
Claudia Vinke, PhD,
Nico Schoemaker, DVM, PhD, DipECZM (Small Mammal), DipECZM (Avian),
Yvonne van Zeeland, DVM, MVR, PhD, DipECZM (Avian), DipECZM (Small Mammal), CPBC

Session #3275

Affiliation: From the Division of Zoological Medicine, Department of Clinical Sciences (Beekmans, Schoemaker, van Zeeland), Division of Animals in Science and Society, Department of Animal Population Health, Faculty of Veterinary Medicine, Utrecht University, Yalelaan 108, 3584 CM, Utrecht, The Netherlands (Vinke).

Note: Abstract format selected by author(s). See Front Matter for description of conference proceedings manuscripts submitted as scientific abstracts.

Abstract: Foraging enrichment is considered an effective tool to stimulate species-typical foraging behavior and to reduce abnormal (repetitive) behavior;[1-3] however, for parrots in captivity, even when using foraging enrichment it remains difficult to approximate foraging times of wild parrots.[4,5] As such current strategies may be insufficient to meet the parrots' foraging requirements. Moreover, currently available enrichment often stimulates a single aspect of foraging, while the behavior in fact comprises of a multitude of different activities including flying, searching, selecting and procuring, opening, extracting, processing, and finally the actual consumption of food.[2-4] We therefore designed a foraging enrichment for parrots, which consisted of two components that respectively stimulated the various appetitive (flying, searching, selecting, procuring) and consumptive behaviors (opening, extracting, processing), and studied its effect on foraging times in 12 grey parrots (*Psittacus erithacus*). The two-component enrichment was (tentatively) able to increase daily foraging times from 2 h/day (control condition) to 4 h/day, while the separate components increased daily foraging times to 3 h/day. Hence, through stimulation of both appetitive and consumptive behaviors, the two-component enrichment was able to increase daily foraging times to produce natural time budgets in captivity. While further studies are needed to investigate the importance of individual foraging activities for parrots, and evaluate whether naturalized time budgets can more effectively eliminate abnormal behaviors, findings of this study emphasize that efficacy of foraging enrichment can be increased by incorporating and combining multiple foraging activities typical for the species.

References

1. Meehan CL, Millam JR, Mench JA. Foraging opportunity and increased physical complexity both prevent and reduce psychogenic feather picking by young Amazon parrots. *Appl Anim Behav Sci.* 2003;80(1):71-85.

2. Lumeij JT, Hommers CJ. Foraging 'enrichment'as treatment for pterotillomania. *Appl Anim Behav Sci.* 2008;111(1-2):85-94.

3. Fangmeier ML, Burns AL, Melfi, VA, et al. Foraging enrichment alleviates oral repetitive behaviors in captive red-tailed black cockatoos (*Calyptorhynchus banksii*). *Zoo Biol.* 2020;39(1):3-12.

4. Rozek JC, Danner LM, Stucky PA, et al. Over-sized pellets naturalize foraging time of captive orange-winged Amazon parrots (*Amazona amazonica*). *Appl Anim Behav Sci.* 2010;125(1-2):80-87.

5. van Zeeland YRA, Schoemaker NJ, Ravesteijn MM, et al. Efficacy of foraging enrichments to increase foraging time in grey parrots (*Psittacus erithacus erithacus*). *Appl Anim Behav Sci.* 2013;149(1-4):87-102.

Desert Rose (*Adenium obesum*) Acute Toxicity and Treatment in 33-Year-Old Blue and Gold Macaw

April Romagnano, PhD, DVM, DABVP (Avian Practice),
Kian Martin,
Elliott Leonard,
Stephanie Monteith, CNA,
Lisa Smith, CVT,
Scott Martin, DVM, MS

Session #3380

Affiliation: From the Avian & Exotic Clinic of Palm City, Palm City, FL 34990, USA (Romagnano, Martin, Leonard, Monteith, Smith, Martin), Martin Animal Clinic Palm City, FL 34990, USA (Romagnano, Martin, Leonard, Monteith, Martin).

Note: Abstract format selected by author(s). See Front Matter for description of conference proceedings manuscripts submitted as scientific abstracts.

Abstract: This presentation reports the first case of desert rose *(Adenium obesum)* acute toxicity and successful treatment in an avian species. Desert rose *(Adenium obesum)*, of the family Apocynaceae, is a cardioactive steroid and cardiac glycoside plant. The entire plant produces a sap that is a cardiac toxin from its roots to its flowers. Historically, its sap was used to create poison arrows to hunt large game and fish in Africa. Desert rose poisoning has been reported in humans, dogs, cats and horses.[1] A 33-year-old male blue and gold macaw ate an entire desert rose flower and base from an indoor houseplant. After ingestion, the bird fell to the floor, ran to its owners who witnessed the bird foaming at the mouth, collapse, bluing of the cheeks and seizing, which rendered it moribund. The owners placed a call to the veterinary emergency phone and drove 2 hours to the clinic with the car's air-conditioner on high directed towards the bird's face. The macaw presented moribund, with dark-blue cheeks, rigid limbs and a severe bradycardia. It was weighed and then placed in a low heat and high oxygen incubator. Treatment included atropine, parenteral fluids, dextrose, calcium-gluconate, vitamins, anti-inflammatories, antibiotics and activated charcoal gavage. Initial bloodwork revealed a severe hyperkalemia > 10 mmol/L which was treated with fluids and furosemide. Diagnostics and supportive treatment continued for 3 weeks, including tube-feeding, sucralfate and hyperbaric oxygen. Constant incubator oxygen coupled with multiple hyperbaric oxygen treatments significantly improved perfusion, decreased inflammation and facilitated recovery. Presently the bird is bright and active with normal bloodwork.

References

1. Botelho AFM, Pierezan F, Soto-Blanco B, Martins Melo M. A review of cardiac glycosides: structure, toxicokinetics, clinical signs, diagnosis and antineoplastic potential. *Toxicon.* 2019;158:63-68.

Patient-Patient Owner Relationships and Their Influence on Budgerigar Husbandry Systems

Ruediger T. Korbel, Prof Dr med vet habil, DipECZM (Avian),
Cécile Danner, DVM,
Katrin Drasch, Dr phil,
Johannes Kallenbach,
Monika Rinder Prof Dr med vet, DVM,
Nicole J. Saam, Prof Dr phil

Session #3327

Affiliation: From the Center for Clinical Veterinary Medicine, Clinic for Birds, Small Mammals, Reptiles and Ornamental Fish, LMU University of Munich, Oberschleissheim, Germany (Korbel, Danner, Rinder), Institute for Sociology, Chair for Methods of Empirical Sociological Studies, Friedrich-Alexander-University of Erlangen-Nuremberg, Germany (Drasch, Kallenbach, Saam).

Note: Abstract format selected by author(s). See Front Matter for description of conference proceedings manuscripts submitted as scientific abstracts.

Abstract: According to the 1966 Veterinary Surgeons Act (an act of the Parliament of the United Kingdom), central tasks within the profession of veterinary medicine are: (1) diagnosis of disorders and injuries including diagnostics, (2) advisory (for the patient owner) based on diagnostics, (3) medical and (4) surgical treatment of animals. Mullan and Fawcett defined veterinary treatment as "interventions to prevent disease and suffering, diagnose and manage or cure disease in animals."[1] According to this definition, including prevention of suffering, everything must be undertaken to assure maximum animal welfare. Within the complex set of relations that may be described as a clinician-companion animal-client triangle (CCC-triangle), in routine daily clinical work, not only in pet bird medicine but also for state authority veterinary, commercially kept poultry veterinary work methods to date are based more on subjective approaches (i.e., relying on general unspecified experience within clinical daily routine) without validated standardized protocols.

Introduction

The housing, care, feeding and environmental demands of birds can vary greatly depending on the species and are often not easy to meet while in human care. In the present study, a husbandry index was developed to assess budgerigar husbandry and to examine the relevance of the human-bird relationship to budgerigar husbandry in captivity.

Husbandry Index

As a first step, a husbandry index for noncommercially kept budgerigars from private households (Budgerigar Husbandry Index BHIg-14) was developed, taking into account current valid husbandry recommendations and guidelines. The additive weighted index consists of 14 housing dimensions: housing environment, grating, perches, floor covering/floor substrate, nesting material, environmental enrichment, light, temperature, food, cleaning/hygiene, free flight, possibility of separation, socialization and bathing. It measures, as a preliminary stage of an animal welfare index, the quality of the existing husbandry conditions. The index was queried together with the owner-bird-relationship-scale (OBRS),[2] a psychometric scale that makes the human-bird relationship measurable in four factors in a standardized online questionnaire for private budgie owners in Germany. The evaluation was based on 1536 online questionnaires and was carried out using multivariate analysis methods.

Results

Overall study results reveal good housing conditions could be determined and confirmed objectively in the survey. Marked deficits were found in owners who kept exactly two or only one budgerigar, in sociologic groups with a low monthly net household income and in owners with a commercial background. Within the framework of the multivariate linear regression method, all four factors of the human-bird relationship described, namely the bird as human; the bird as meaning of life; empathy, mindfulness and respect; and the relationship of the bird-bird owner showed a highly significant positive correlation with the quality of budgerigar husbandry in human care. In addition, connections in relation to the behavior of the owners to gain knowledge, the number of budgerigars kept and other socio-demographic characteristics could be identified.

References

1. Mullan S, Fawcett A. *Veterinary Ethics: Navigating Tough Cases.* Sheffield, UK: 5m Publishing; 2017.

2. Burmeister A-K, Drasch K, Rinder M, et al. Development and application of the owner-bird relationship scale (OBRS) to assess the relation of humans to their pet birds. *Front Vet Sci.* 2020;7. https://doi.org/10.3389/fvets.2020.575221.

Clinical Signs Associated with Elevated Copper Levels in Mixed Species Captive Birds: A Case Series

Dawn Daugherty, DVM, MPH, CPH, and
M. Scott Echols, DVM, DABVP (Avian Practice)

Session #3248

Affiliation: From Parrish Creek Veterinary Hospital and Diagnostic Center, Centerville, UT 84014, USA (Daugherty, Echols), Echols Veterinary Services, Cottonwood Heights, UT 84093, USA (Echols), Scarlet Imaging, Murray, UT 84157, USA (Echols).

Note: Abstract format selected by author(s). See Front Matter for description of conference proceedings manuscripts submitted as scientific abstracts.

Abstract: While normal values of copper have been established at 0.07-0.19 ppm in Hispaniolan Amazon parrots (*Amazona ventralis*), there is little information on the effects of copper toxicity in captive birds.[1] Gastrointestinal distress, hemolytic anemia, liver disease and death are reported in mammals. The poultry literature describes cases of severe acute copper toxicity resulting in anorexia and death with ulcerations and congestion in the ventriculus and high levels of copper in various tissues.[2] Regardless, there is very little information on low level or chronic copper toxicity in birds. This case series examines six birds at the time of writing, (one cockatoo, one macaw, one Pionus parrot, one African gray parrot, two Amazon parrots) with variable clinical signs, recognizable metal in the proventriculus and/or ventriculus on standard projection radiographs and/or high-resolution computed tomography and elevated blood copper levels between 0.34-1.52 ppm. All birds were treated conservatively to remove the radiographically detected gastrointestinal metal and all birds returned to normal upon successful treatment. The case series describes clinical signs, diagnostics used, treatments and outcomes of birds entered in the study. The purpose of the study is to find associations between elevated copper levels and clinical signs in captive birds.

References

1. Osofsky A, Jowett PLH, Hosgood G, Tully TN. Determination of normal blood concentrations of lead, zinc, copper, and iron in Hispaniolan Amazon parrots (*Amazona ventralis*). *J Avian Med Surg.* 2001;15:31-36.

2. Giergiel M, Durkalec MM, Nawrocka A, et al. Ingestion of bedding material as a cause of acute copper sulfate poisoning in turkey poults. *Poult Sci.* 2019;98:707-711.

Determining Fecal Microbiome Using Next-Generation Genomic Sequencing in Cockatiels (*Nymphicus hollandicus*) Fed Different Diets

Nicole Sands, DVM,
Shachar Malka, DVM, DABVP (Avian Practice), DACEPM,
Gina Vecere, DVM, DACEPM,
Margaret Lee, DVM,
Jonathan Stockman, DVM, DACVIM (Nutrition),
Janina Krumbeck, PhD

Session #3425

Affiliation: From Long Island Bird and Exotics Veterinary Clinic, 333 Great Neck Road, Great Neck, NY 11021, USA (Sands, Malka, Vecere, Lee), Department of Clinical Veterinary Sciences, LIU College of Veterinary Medicine, Long Island University, 720 Northern Boulevard, Brookville, NY 11548, USA (Stockman), MiDOG LLC, 17062 Murphy Ave., Irvine, CA 92614, USA (Krumbeck).

Note: Abstract format selected by author(s). See Front Matter for description of conference proceedings manuscripts submitted as scientific abstracts.

Abstract: Formulated diets provide balanced nutrition with essential vitamins and minerals, and therefore are often recommended as the majority of a bird's diet by veterinarians.[1-4] The goal of this study is to characterize the normal fecal microbiome of client-owned cockatiels (*Nymphicus hollandicus*) fed a commercial seed mix as compared to cockatiels fed commercial brands of formulated pellets. Fecal samples from 34 healthy client-owned cockatiels, 15 eating a generic commercially available seed diet and 19 eating a formulated pellet diet, were collected and analyzed by next-generation DNA sequencing. samples were collected according to the manufacturer's instructions using a swab collection kit provided by the NGS technology diagnostic center and processed as previously described.[5] Next-generation DNA sequencing has been shown to provide practical applications for characterizing microbiomes and providing better identification of microbes as compared to traditional culture methods.[5-8] A total of 179 different bacterial species representing 94 genera were detected. The most abundant bacterial species in the pellet-fed group were *Lactobacillus alvi*, *Ureaplasma* spp., and *Lactobacillus salivarius* while *Ureaplasma* spp., *Lactobacillus alvi*, and *Lactobacillus crispatus* were the most abundant species in the seed-fed group. For fungi, 244 fungal species representing 156 genera were detected. *Cladosporium* spp., *Fusarium* spp., and *Alternaria* spp. were the most abundant in the pellet-fed group, and *Cladosporium* spp., *Cladosporium lignicola/sphaerospermum*, and *Meyerozyma carpophila/guilliermondii* in the seed-fed group. These results illustrate how various diets affect the fecal microbiome of cockatiels and should be taken into consideration by clinicians when identifying a disease state. Elaboration of how these results compare to each other are provided.

References

1. Brightsmith D. Nutritional levels of diets fed to captive Amazon parrots: does mixing seed, produce, and pellets provide a healthy diet? *J Avian Med Surg.* 2012;26(3):149-160.

2. Ullrey D, Allen ME, Baer DJ. Formulated diets versus seed mixtures for psittacines. *J Nutr.* 1991;121(11): 193-205.

3. Harrison G, Lightfoot T. *Clinical Avian Medicine.* Palm Beach, FL: Spix Publishing; 2006:63-94.

4. Samour J. *Avian Medicine.* St. Louis, MO: Elsevier Saunders; 2016:25-30.

5. Tang S, Prem A, Tjokrosurjo J, et al. The canine skin and ear microbiome: a comprehensive survey of pathogens implicated in canine skin and ear infections using a novel next-generation-sequencing-based assay. *Vet Microbiol.* 2020;247:108764.

6. Krumbeck J, Turner DD, Diesel A, et al. Skin microbiota of quaker parrots (*Myiopsitta monachus*) with normal feathering or feather loss via next-generation sequencing technology. *J Exot Pet Med.* 2022;42: 26-34.

7. Vecere G, Malka S, Holden N, et al. Comparison of ear canal microbiome in rabbits with and without otitis externa using next generation DNA sequencing. *J Exot Pet Med.* 2022;42:35-41.

8. Melgarejo T, Oakley BB, Krumbeck JA, et al. Assessment of bacterial and fungal populations in urine from clinically healthy dogs using next-generation sequencing. *J Vet Intern Med.* 2021;35(3):1416-1426.

Military Macaw with Chronic Active *Klebsiella* Stomatitis Diagnosed by Next-Generation Sequencing

April Romagnano, PhD, DVM, DABVP (Avian Practice), and
Kian Martin

Session #3468

Affiliation: From the Avian & Exotic Clinic of Palm City, Palm City, FL 34990, USA (Romagnano, Martin), Martin Animal Clinic, Palm City, FL 34990, USA (Romagnano, Martin).

Note: Abstract format selected by author(s). See Front Matter for description of conference proceedings manuscripts submitted as scientific abstracts.

Abstract: This presentation reports a chronic oral proliferative stomatitis case in a 27-year-old male military macaw that progressed and evolved over a 6-year period. The oral growths were pale tan to dark brown, raised, and ranged from 0.2-1.3 centimeters. They were smooth to roughened, firm and irregularly shaped, often with a head and pedunculated stalk seated in an inflamed circular base surrounded by normal oral mucosa. These growths caused chronic-active stomatitis, complicated by rhinitis that initially started in 2017 as a proliferative ear growth referral case. Despite extensive diagnostics, various treatments-including antimicrobials based on culture + sensitivity, nebulization and hyperbaric-oxygen, and repeated surgical and cautery removal, the growths recurred every 4-6 weeks impeding mastication. Histopathology had repeatedly shown locally extensive, chronic active, proliferative heterophilic stomatitis with marked hyperkeratosis which evolved into chronic mixed mononuclear and heterophilic inflammation. Early histopathologic examination suggested a bacterial component. Numerous cultures revealed gram-negative and gram-positive bacteria with *Klebsiella* sp., *Pseudomonas* sp., and *Staphylococcus* sp. predominating. PCRs from multiple labs were negative for herpesvirus, papillomavirus, poxvirus, circovirus, chlamydia and *Aspergillosis*, *Candida* and *Malassezia*. Bloodwork had repeatedly shown an elevated WBC with a heterophilia, low bile acid, elevated beta-globulins, and negative *Aspergillosis* AB and galactomannan. MiSEQ next-generation sequencing of an oral mass from late in the disease process identified no viruses, but a predominance of *Klebsiella* sp. and particularly *K. pneumoniae*. Based on the compilation of all the findings and diagnostics in this case, it is believed that *Klebsiella* played a primary role in the disease process throughout this still-thriving bird's history.

Fourteen Cases of Naturally Infected Birds with Avian Bornavirus: Usefulness of Histologic Examination of Skin Lesions and Virus Localization of Feather Damaging Behavior for Early Diagnosis

Reiko Soga, DVM,
Souichi Yoshino,
Naoyuki Aihara,
Junichi Kamiie,
Akiko Makino,
Keizo Tomonaga

Session #3200

Affiliation: From Azabu University Veterinary Medicine and Pathology (Yoshino, Aihara, Kamiie), Kyoto University Medical and Biological Research Institute (Makino, Tomonaga), Grow-Wing Animal Hospital (Soga).

Abstract: An outbreak of avian bornavirus (ABV) disease was observed in a domestic breeding farm for pet Psittaciformes. Some individuals infected with ABV are known to exhibit feather destructive behavior. In this study, we report the results of a pathologic study of the defeathered skin area in ABV-positive individuals found to be defeathered.

Materials and Methods

Two live (i.e., surviving) birds and 14 dead birds with feather shedding and confirmed bornavirus infection were studied. Surviving birds had a clinical follow-up by multiple fecal PaBV PCR tests, and skin biopsies were pathologically examined. Dead birds were submitted for pathologic autopsy, and sampling of brain tissue, digestive tract, and defeathered skin area. Subsequently, histopathologic examination and PaBV immunostaining was performed.

Discussion and Conclusions

In infected birds that exhibited defeathering, infection and inflammation of the peripheral nerves surrounding the feather follicle was observed, as well as infection of the feather pulp. It was suggested that viral infection of the cutaneous peripheral nerves or feather pulp may be involved in defeathering. PaBV antigen was detected in skin peripheral nerves by immunostaining two live biopsied birds. All 14 dead birds were PaBV-positive in the brain and 14 were also PaBV-positive in the skin. Four of the 10 birds whose intestines could be examined had no bowel lesions.

Despite a negative fecal PCR test, ABV infected birds present with feather destructive behavior. Early PCR fecal testing is often negative. Therefore, if the diagnosis of ABV and feather destructive behavior can be confirmed early by anti-ABV antibody immunostaining, treatment will be successful and life may be prolonged.

References

1. Horie M, Veda K, Veda A, et al. Detection of avian bornavirus 5 RNA in *Eclectus roratus* with feather picking disorder. *Microbiol Immunol.* 2012;56(5):346-349 .

2. Kolb SJ. Feather Destruction Disorder. Available at: https://www.vin.com/ce/TECH281-0120_VSPN.htm. Accessed May 2, 2023.

3. Antinoff N. Feather damaging…for a reason. *Pacific Vet Conf.* 2019.

4. Sassa Y, Horie M, Fujino K, et al. Molecular epidemiology of avian bornavirus from pet birds in Japan. *Virus Genes.* 2013;47(1):173-177.

5. Philadelpho NA, Rubbenstroth D, Guimarães MB, et al. Survey of bornaviruses in pet psittacines in Brazil reveals a novel parrot bornavirus. *Vet Microbiol.* 2014;174(3-4):584-590.

6. Horie M. Parrot bornavirus infection: correlation with neurological signs and feather picking. *Vet Rec.* 2019;184(15):473-475.

Antemortem Diagnosis and Successful Long-Term Management of Disseminated Xanthogranulomatous Disease in an Eclectus Parrot (*Eclectus roratus*)

Christoph Mans, Dr med vet, Dipl ACZM,
Neta Ambar, DVM,
David Gasper, DVM, PhD, DACVP

Session #3476

Affiliation: From the School of Veterinary Medicine, University of Wisconsin-Madison, 2015 Linden Drive, Madison, WI 53706, USA.

Note: Abstract format selected by author(s). See Front Matter for description of conference proceedings manuscripts submitted as scientific abstracts.

Abstract: A 12-year-old male eclectus parrot was referred for evaluation of coelomic distention. A computed tomography scan and blood work revealed coelomic effusion with free coelomic mineral attenuating material and severe elevation of plasma bile acids. Coelomic effusion was consistent with macrophagic inflammation with abundant intracellular lipids. Initial treatment with meloxicam showed minimal improvement. Disseminated xanthogranulomatous inflammation was suspected based on imaging and laboratory results, which were consistent with those previously reported.[1-2] Biopsies of the liver and intracoelomic masses confirmed this diagnosis. Treatment with prednisolone at 1 mg/kg/day for 6 months, followed by 0.5 mg/kg/day for 3 months, was initiated. Clinical improvement was assessed based on owner evaluation, plasma bile acid levels, and diagnostic imaging. After 2 months of prednisolone treatment, the bird's behavior and appetite improved, which persisted throughout treatment and at follow-up 17 months after the cessation of steroid therapy. Bile acid levels were normal 10 months after the cessation of therapy. Diagnostic imaging showed minimal coelomic effusion at 10 months post-cessation of therapy, with progressive expansion of the air sacs and static to improved dystrophic mineral foci. This is the first report describing an antemortem diagnosis and treatment of disseminated coelomic xanthogranulomatous disease in a psittacine species, with an objectively documented response to prednisolone therapy.

References

1. Hanson ME, Donovan TA, Quesenberry K, et al. Imaging features of disseminated xanthogranulomatous inflammation in eclectus parrots (*Eclectus roratus*). *Vet Radiol Ultrasound.* 2020;61:409-416.

2. Donovan TA, Garner MM, Phalen D, Reavill D, Monette S, et al. Disseminated coelomic xanthogranulomatosis in eclectus parrots (*Eclectus roratus*) and budgerigars (*Melopsittacus undulatus*). Vet Pathol. 2022;59(1):143-151.

Prevalence, Distribution, and Risk Factor Correlates of Adipocytic Tumors in Psittaciformes: 648 Cases (1998-2018)

Lucyanne Megan, DVM,
David Sanchez-Migallon Guzman, LV, MS, DipECZM (Avian), DipECZM (Small Mammal), Dipl ACZM,
Kevin Keel, DVM, PhD, DACVP,
Hugues Beaufrère, Dr med vet, PhD, DABVP (Avian Practice), DipECZM (Avian), Dipl ACZM

Session #3361

Affiliation: From the William R. Pritchard Veterinary Medical Teaching Hospital, University California-Davis, Davis, CA 95616, USA (Megan), Department of Medicine and Epidemiology, School of Veterinary Medicine, University of California-Davis, Davis, CA 95616, USA (Guzman, Beaufrère), Department of Pathology, Microbiology, and Immunology, School of Veterinary Medicine, University of California-Davis, Davis, CA 95616, USA (Keel).

Note: Abstract format selected by author(s). See Front Matter for description of conference proceedings manuscripts submitted as scientific abstracts.

Abstract: Adipocytic tumors are mesenchymal tumors that are commonly reported in psittacine birds; however, large-scale studies evaluating their prevalence and associated risk factors are lacking. A retrospective study of adipocytic tumors in psittacine birds was performed by reviewing pathology submissions from the UC Davis Drury Reavill Pathology Database, containing 26,013 submissions from psittacine birds. Age, sex, genus, anatomic distribution, and pathologic diagnosis were collected for each case when available. Four-hundred fifty cases of lipoma, 129 cases of myelolipoma, 35 cases of hemangiolipoma, and 31 cases of liposarcoma were identified. Adipocytic tumors were identified in 27 genera. Lipomas were most commonly identified in *Amazona* (130/450), *Nymphicus* (65/450), and *Melopsittacus* (41/450) spp., while liposarcomas were most commonly identified in *Nymphicus* (6/31), *Myiopsitta* (5/31), and *Melopsittacus* (4/31) spp. The mean ages at diagnosis of lipoma and liposarcoma were 16.7 years (range 0.3-46 years) and 15.7 years (range 1-41 years), respectively. Of the psittacine birds diagnosed with lipoma, 35.3% (159/450) were female, 34% (153/450) were male, and 30.7% (138/450) were of unknown sex. Of those diagnosed with liposarcoma, 45.2% (14/31) were female, 25.8% (8/31) were male, and 29% (9/31) were of unknown sex. Anatomic location was specified in 87.7% (568/648) of cases of adipocytic tumors. The most common sites of lipoma formation were the wing and chest, and the most common site of liposarcoma formation was the wing. The prevalence of adipocytic tumors was determined as a function of age, sex, and genus. Risk factors and association with other comorbidities, such as atherosclerosis, hepatic lipidosis, and xanthoma, also were evaluated.

Radiation Therapy for the Treatment of Squamous Cell Carcinoma in a Barn Owl (*Tyto alba*)

Tracey K. Ritzman, DVM, DABVP (Avian Practice), DABVP (Exotic Companion Mammal), and
Tara M. Harrison, DVM, MPVM, Dipl ACZM, DACVPM, DipECZM (Zoo Health Management), CVA

Session #3130

Affiliation: From Carolina Raptor Center, 6000 Sample Road, Huntersville, NC 28078, USA (*Ritzman*), Carolinas Veterinary Care Clinic, 10110 Northcross Center, Ste 100, Huntersville, NC 28078, USA (*Ritzman*), North Carolina State University College of Veterinary Medicine, 1010 William Moore Drive, Raleigh, NC 27607, USA (*Harrison*).

Note: Abstract format selected by author(s). See Front Matter for description of conference proceedings manuscripts submitted as scientific abstracts.

Abstract: This presentation describes the collaborative medical and surgical treatment of an 18-year-old female captive barn owl with squamous cell carcinoma neoplasia of the thoracic limb (wing). Surgical, medical therapy, and radiation treatments, were utilized to treat the bird, and preserve the function of the wing. The tumor originated from the soft tissues of the right medial ulna and was initially treated with surgical debulking. Biopsy samples collected at the time of the surgery were submitted for histopathology which confirmed the diagnosis of squamous cell carcinoma. Medical therapy after surgery consisted of analgesia, systemic antibiotic therapy, bandaging, when necessary, physical therapy for the wing, and monitoring of the affected site. The tumor recurred 6 months after the initial debulking surgery. Palliative radiation therapy was utilized to treat the tumor and preserve wing function. Radiotherapy was effective at reducing the tumor size and delayed progression of disease, thus extending the life expectancy of this bird with a good quality of life. Radiotherapy for this patient was performed manually with single 12e-electron beam with 6 X 6 cone, prescribed to 90% IDL. A total of six (6) weekly fractions of 6 Gy was delivered for six consecutive weeks at the North Carolina State University College of Veterinary Medicine. The owl tolerated the radiation therapy very well and there were minimal side effects in this patient. For this patient there was delayed wound healing and delayed feather growth at the site. Other potential side effects of radiotherapy in general can include thermal injury at the treatment site and the inherent risk of the repeated general anesthesia sessions required for the treatments. This case report illustrates the positive benefits and feasibility of radiotherapy for avian patients with neoplasia.

References

1. Adair JE, Riggs GL. Treatment of recurrent multifocal cutaneous squamous cell carcinoma in an African grey parrot (*Psittacus erithacus*) with intralesional carboplatin in Poloxamer 407. *J Avian Med Surg.* 2022;36(2):200-205.

2. Alexander AB, Griffin L, Johnston MS. Radiation therapy of periorbital lymphoma in a blue- and -gold macaw (*Ara ararauna*). *J Avian Med Surg.* 2017;31(1):39-46.

3. Cutler DC, Shiomitsu K, Liu CC, Nevarez JG. Comparison of calculated radiation therapy versus actual radiation delivery in military macaws (*Ara militaris*). *J Avian Med Surg.* 2016;30(1):1-7.

4. Ferrell ST, Marlar AB, Garner M, et al. Intralesional cisplatin chemotherapy and topical cryotherapy for the control of choanal squamous cell carcinoma in an African penguin (*Spheniscus demersus*). *J Zoo Wildl Med.* 2006;37(4):539-540.

5. Guzman DS. Avian soft tissue surgery. *Vet Clin North Am Exot Anim Pract.* 2016;19(1):133-157.

6. Guthrie AL, Gonzalez-Angulo C, Wigle WL, et al. Radiation therapy of a malignant melanoma in a thick-billed parrot (*Rhynchopsittica pachyrhyncha*). *J Avian Med Surg.* 2010;24(4):299-307.

7. Hinkle D, Eschar D, Ambar N, et al. Successful radiation treatment of undertail of fibrosarcoma in a major Mitchell's cockatoo (*Lophochroa leadbeateri*). *J Am Vet Med Assoc.* 2022;260(12):1-4.

8. Kent MS. Principals and application of radiation therapy in exotic animals. *Vet Clin North Am Exot Anim Pract.* 2017;20(1):255-270.

9. Murtaugh RJ, Ringler DJ, Petrak ML. Squamous cell carcinoma in the esophagus of an Amazon parrot. *J Am Vet Med Assoc.* 1986;188(8):872-873.

10. Ramis A, Gibert X, Majó N, Grifols J. Metastatic oral squamous cell carcinoma in a Montagu's harrier (*Circus pigargus*). *J Vet Diagn Invest.* 1999;11(2):191-194.

11. Rettenmund CL, Newton AL, Calle PP. Uropygial gland squamous cell carcinoma in chinstrap (*Pygoscelis antarcticus*) and gentoo (*Pygoscelis papua*) penguins at the Wildlife Conservation Society's Central Park Zoo. *J Zoo Wildl Med.* 2015;46(1):113-119.

12. Rocha CMDS, Wilson TM, Sousa DER, et al. Squamous cell carcinoma of the footpad with systemic metastasis in a captive crowned solitary eagle (*Buteogallus coronatus*). *J Vet Med Sci.* 2020;82(11):1627-1630.

13. Staudenmaier AM, Simone-Freilicher EA, Noonan BP, Ewing PJ. Antemortem diagnosis of esophageal squamous cell carcinoma in a blue-fronted Amazon parrot (*Amazona aestiva*) and a mealy Amazon parrot (*Amazona farinosa*). *J Am Vet Med Assoc.* 2019;254(11):1324-1328.

14. Suedmeyer WK, Henry C, McCaw D, Boucher M. Attempted photodynamic therapy against patagial squamous cell carcinoma in an African rose-ringed parakeet (*Psittacula krameri*). *J Zoo Wildl Med.* 2007;38(4):597-600.

15. Swisher SD, Phillips KL, Tobias JR, et al. External beam radiation therapy of squamous cell carcinoma in the beak of an African grey parrot (*Psittacus timneh*). *J Avian Med Surg.* 2016;30(3):250-256.

16. Weinstock D, Correa MT, Rives DV, et al. Histopathology and epidemiology of condemnations in broiler chickens due to squamous cell carcinoma in North Carolina. *Avian Dis.* 1995;39(4):676-686.

Analyzing Baseline and Post-Restraint Plasma Corticosterone Levels in Hispaniolan Amazon Parrots (*Amazona ventralis*)

Shelby N. Parks, BS,
Thomas N. Tully, Jr. DVM, MS, DABVP (Avian), DipECZM (Avian),
Aspen L. Settle, BS,
Christine R. Lattin, PhD

Session #3171

Affiliation: From the Department of Veterinary Clinical Sciences, School of Veterinary Medicine, Louisiana State University, 1909 Skip Bertman Dr., Baton Rouge, LA 70803, USA (Parks, Tully, Settle), Department of Biological Sciences, Louisiana State University, 202 Life Science Building, Baton Rouge, LA 70803, USA (Lattin).

Note: Abstract format selected by author(s). See Front Matter for description of conference proceedings manuscripts submitted as scientific abstracts.

Abstract: The purpose of this study was to measure baseline plasma corticosterone levels in 22 Hispaniolan Amazon parrots (*Amazona ventralis*) and assess the effects of handling and restraint on corticosterone levels over 1 hour. To do this, each parrot was removed from its cage and wrapped in a towel for restraint, similar to that performed in a clinical setting. Once restrained, an initial baseline blood sample was collected in < 3 minutes,[1] after which blood samples were taken every 15 minutes for 1 hour (a total of five blood samples). To determine concentrations of plasma corticosterone, an enzyme-linked immunoassay was used. This assay was validated for Hispaniolan Amazon parrots by assessing parallelism of diluted parrot plasma samples with a corticosterone standard curve. On average, parrots showed a significant increase in corticosterone between baseline samples and all subsequent post-restraint time points. Females, on average, displayed significantly higher corticosterone levels than males after 30, 45, and 60 minutes of restraint. The plasma corticosterone concentrations measured in parrots with feather destructive behavior were compared to those without the behavior, and no significant differences were found between the groups at baseline or any post-restraint time points. Understanding the physiologic stress response in companion psittacine birds during routine handling will allow clinicians to better evaluate how this may affect the patient's condition and diagnostic test results.

References

1. Romero LM, Reed JM. Collecting baseline corticosterone samples in the field: is under 3 min good enough? *Comp Biochem Physiol A Mol Integr Physiol.* 2005;140(1):73-79.

Section 13

ASSOCIATION OF EXOTIC MAMMAL VETERINARIANS

AEMV Fundamentals Session

Jeffery Toraason, DVM
Session Chair

Dermatologic Diseases of Exotic Companion Mammals

Cheryl B. Greenacre, DVM, DABVP (Avian Practice), DABVP (Exotic Companion Mammal Practice)

Session #3351

Affiliation: From College of Veterinary Medicine, University of Tennessee, Knoxville, TN 37996, USA (Emeritus), Clinical Laboratory Animal Veterinarian, Office of Research, University Research Animal Resources, University of Georgia, 501 D.W. Brooks Drive, Room 3504, Athens, GA 30602, USA.

Abstract: Exotic companion mammals, including rabbits, rodents, ferrets, hedgehogs, and sugar gliders, can present with a variety of dermatologic disorders due to various etiologies including bacterial, viral, parasitic, neoplastic, traumatic, and other causes. In some cases, the treatment is straightforward and proven, while in other cases the treatment options are unproven or palliative. This masterclass describes in detail the diagnostic and therapeutic techniques used to manage selected dermatologic diseases in various exotic companion mammal species.

Introduction

Dermatologic diseases of exotic companion mammals (ECMs) can be categorized by their cause. Bacterial dermatologic diseases include rabbit syphilis, secondarily infected pododermatitis, and abscesses. Viral dermatologic diseases include papilloma of rabbits or ferrets. Ringworm is the most common fungal dermatologic disease of ECMs. Various parasites cause dermatologic disease in ECMs including ear mites of ferrets and rabbits, fur mites of rabbits, and species-specific mites and lice of mice, rats, guinea pigs and hedgehogs. There are various forms of cutaneous neoplasia that affect ECMs. Lastly, some diseases are not as easily categorized such as idiopathic ulcerative dermatitis in mice. There are many excellent sources available for information on diseases of exotic companion mammals including books, articles, and websites.[1-24]

Bacterial Dermatologic Diseases

Rabbit syphilis (*Treponema cuniculi*)

T. cuniculi is a sexually transmitted disease of rabbits causing crusting and erosive lesions of the skin and mucocutaneous junctions of the perineum, periocular, perioral areas, and sometimes the base of the ear. This disease is not zoonotic; human syphilis is caused by *T. palladium*. The incubation period is long, approximately 3-6 weeks. This can be a self-limiting disease. Diagnosis is based on typical clinical signs, ELISA, dark field microscopy demonstrating spiral organisms on wet mounts or stains of fluids expressed from active lesions, or rapid plasma regain (RPR) card test.[13] Treatment is procaine penicillin G with benzathine given subcutaneously only every 7 days for three treatments. Never give rabbits oral penicillin antibiotics as it causes fatal enteritis. Separate affected or positive rabbits so there is no direct contact between rabbits and do not breed infected rabbits. Rabbits can transmit the disease without obvious clinical signs, therefore quarantine all new rabbits.

Ulcerative pododermatitis (and secondarily infected pododermatitis)

Ulcerative pododermatitis (sore hock) in rabbits can be caused by obesity, genetically sparse hair cover over the plantar surface of the hock (i.e., the plantar surface of the proximal metatarsus), dirty and wet bedding, mobility issues, or trauma from wire cage flooring with no substrate or carpet causing shearing forces on the hair. Normally the hair provides the cushion for the plantar surface. Mild pododermatitis usually presents as alopecia and hyperemia of the intact skin. Moderate pododermatitis involves breakdown of the skin with erosions, scabs, proliferative tissue, swelling or infection of skin. Severe pododermatitis with osteomyelitis can easily occur since there is very little soft tissue coverage in this area. Guinea pigs present with this same disease, but the cause is usually unsanitary conditions either past or present, and it is very difficult to treat. Changes in husbandry, such as providing a flat place to stand (kitchen cutting boards work well), removing wire substrate, limiting or eliminating time on carpet, providing towels or hay, or making a bunny bootie to bandage or somehow cover and protect the proximal metatarsus are helpful. It is more difficult to bandage guinea pigs, chinchillas, and rats, therefore soft and/or deep bedding may be an alternative. The prognosis is good for mild pododermatitis. Moderate to severe cases may require surgery to remove abscessed areas, multiple days flushing an infected joint, and antibiotics (based on aerobic and anaerobic culture). The prognosis is guarded to grave for moderate and severe pododermatitis, respectively. Metronidazole is typically used in rabbits if *Fusobacterium necrophorum* or similar is cultured. Consider for deep, severe ulcerative pododermatitis, surgical debridement, and from the deep tissue, run a comprehensive PCR as cultures can often miss slow growing bacteria. These tests can give you a good analysis of antimicrobials to use. Guinea pigs require improved substrate, soaking feet in warm water to increase circulation, frequent cleaning and moisturizing of feet, and long-term antibiotics. As this condition is painful, consider analgesics such as NSAIDs. Laser therapy may be helpful in stimulating circulation.

Abscesses

Many abscesses in rabbits and guinea pigs are dental related, but other common sites include the skin, lungs, or tympanic bulla(e). Ideally, complete surgical removal of an abscess and its capsule is recommended but rarely possible due to location. Other treatment options such as marsupialization and secondary healing, or primary closure with implanted antibiotic impregnated beads can be used, although in some cases the beads may further act like foreign material. A full dental examination with imaging is necessary before determining if there are infected teeth needing extraction. Drains, or simply just lancing and draining are not recommended because rabbits and guinea pigs have heterophils instead of neutrophils. Heterophils lack the enzyme myeloperoxidase needed to liquefy pus. Therefore, if an abscess is present, expect the pus to be firm, fibrous, tenacious, and the abscess will have a thick lining or capsule. Aerobic and anaerobic culture of the full thickness of the abscess wall is the best method to grow the primary organisms causing disease. Culturing the necrotic pus, rather than the lining will result in no or less growth. Note that most rabbits do not develop a leukocytosis with abscess formation. After marsupialization, the area is cleaned and flushed with dilute iodine solution and a topical antibiotic ointment, cream or honey can be applied, but ensure the rabbit does not ingest these topicals while grooming. Packing the area with gauze, with or without a tie-over bandage works well.

Antibiotic choice in rabbits and guinea pigs is limited due to the fact that orally administered penicillins, macrolides (like erythromycin, lincomycin), and cephalosporins can quickly cause a fatal diarrhea from destruction of gram-positive normal flora and overgrowth of gram-negative and anaerobic (such as *Clostridia* sp. bacteria and release of endotoxins. Antibiotics that are generally used in rabbits and guinea pigs include the fluoroquinolones, metronidazole, sulfa drugs, and aminoglycosides. Of course, it is best to select antibiotic therapy not only on what is safe to give to rabbits and guinea pigs but what is indicated by culture and sensitivity or microbiome testing using next-generation sequencing. Even with the use of the safer antibiotics, diarrhea can occur at any time due to dysbiosis.

The best methodology to assess the full extent of an abscess is computed tomography (CT). A radiograph can be helpful in determining if there is underlying bony lysis suggestive of osteomyelitis, but it is not as sensitive as CT. Abscess formation in the tympanic bulla(e) can sometimes be difficult to diagnose. The best diagnostic tool in this case is a radiograph or CT of the skull to evaluate the wall thickness, content, shape and size of the tympanic bulla. A CT is far more sensitive than radiographs in identifying bulla osteitis. The osseus portion of the tympanic bulla of rabbits is generally very thin, less than 1 millimeter in thickness. Any increased thickness, irregular margin, or bony lysis of the tympanic bulla wall, or presence of material in the cavity of the tympanic bulla, is indicative of otitis media. Culture of the exudates in the external ear canal will likely grow many contaminants. If the abscess is within the tympanic bulla and the tympanic membrane is intact, then only an aggressive approach, such as a tympanic bulla osteotomy will allow access to the contents or lining for culture. A tympanic bulla osteotomy (usually a lateral approach in rabbits) also provides a method for flushing/removing the pus.[14] Long-term (i.e., months long) parenteral antibiotic therapy is a possible treatment for rabbits with otitis media, but rarely cures the infection due to its protected location. Many protocols call for giving enrofloxacin orally and giving penicillin G procaine with benzathine subcutaneously every 5 to 7 days. Dosages can be found in Carpenter's Formulary.[15] The benzathine formulation of penicillin provides longer action. Antibiotic therapy based on culture is best. Flushing, cleaning and instilling antibiotics into the external ear canal does little to resolve the infection and rarely removes the all the exudates, but it may make the patient more comfortable and scratch at the ears less.

Antibiotic impregnated polymethylmethacrylate or medical grade calcium sulfate beads can be placed in areas of infection, such as an abscess, after thorough cleaning and debriding. The placement of beads can be delayed until the tissue has undergone some second intention healing before using first intention healing methods when closing beads into the tissue. Antibiotic elutes from the beads into the nearby tissue and is effective for 2-3 millimeters around each bead. Approximately 5% of the antibiotic present is released into nearby tissue in the first 24 hours, but systemic levels are low. The exact point at which no further elution occurs is not known, but beads impregnated with gentamycin had completed eluted by 37 days after placement into tissue. Many factors affect the rate of elution including bacterial or debris contamination at the site, blood supply to nearby tissue, the number of beads placed, the amount of antibiotic in the bead, the bead's proximity to the affected tissue, and the appropriateness of the antibiotic chosen for the pathogen present. The beads can be removed later or left in place. Antibiotics that work well in beads are those that are heat stable to withstand steam sterilization and those that come in a powdered form, such as ceftiofur, cephazolin, and gentamycin. Penicillin, ampicillin, amoxicillin, and clindamycin have also been used in the beads. As always, assess pain using facial grimace scales and address observed pain or potential pain with appropriate analgesics, either NSAIDs or opioids as needed.[16,17] Dosages can be found in the latest edition of Carpenter's Formulary.[15]

Viral Dermatologic Diseases

Papilloma

In ferrets, papillomas are dry, proliferative lesions on the plantar surfaces of the feet. In rabbits, there is an oral form of papilloma that is found within the oral cavity and is usually self-limiting. Rabbits also occasionally present with rectal papillomas that are not self-limiting and usually need to be treated. The papilloma virus of rabbits that affects the face (Shope papilloma virus) causes proliferative, keratinized, horn-like projections on the face that have been shown to predispose the rabbit to squamous cell carcinoma later. Rabbit facial papilloma is spread by mosquitos feeding on infected cottontail rabbits in the area, so it is recommended to house rabbits indoors. Rabbit oral papillomas are usually monitored until they resolve several months later. Seldom do they interfere with mastication. The rabbit rectal papillomas can be monitored or if associated with tenesmus are best treated by weekly silver nitrate applications to the mass while the rabbit is under anesthesia. Immediately after applying the silver nitrate, deactivate the chemical burning by rinsing with water. The rabbit facial papilloma needs to be surgically removed to prevent later squamous cell carcinoma, but this is often difficult because there is little spare tissue for adequate closure after surgical resection of the masses. Prognosis is good, except for rabbits that get squamous cell carcinoma secondary to the facial form.[12]

Fungal Dermatologic Diseases

Dermatophytosis

Clinical signs of dermatophytosis (ringworm) in small mammals are very similar in appearance to what is seen in a dog or cat. The most common exotic animal species presenting with ringworm are guinea pigs and chinchillas, although any small mammal species can present with the disease and spread it to humans. Generally *Trichophyton* sp. are more common than *Microsporum* sp. A dermatophyte test medium (DTM) fungal culture or PCR can be performed. Cytology or biopsy could also be performed, but is more invasive. Topical miconazole or terbinafine is typically used daily for 3 months or at least 1 month past resolution of clinical signs. Griseofulvin orally is described in the older literature but is generally avoided now because of hepatic side effects. Discourage high humidity environments.

Of special note is *Trichophyton benhamiae*, previously known as either *Anthroderma benhamiae* or *Trichophyton mentagrophytes* var. *benhamiae,* a dermatophyte described more commonly in Europe and Asia, but also the United States from guinea pigs, and less commonly from other small mammals. The disease in people is more severe than that of *T. mentagrophytes*.[18,19] It is also more difficult to treat in guinea pigs and requires oral terbinafine.

Parasitic Dermatologic Diseases

Fleas

The common cat flea, *Ctenocephalides felis*, can severely affect ferrets causing life threatening anemia. Clinical signs include pruritis, erythematous dermatitis, alopecia, evidence of fleas and flea dirt. There is a licensed product for ferrets to treat fleas, Advantage II (Bayer) for ferrets that contains imidacloprid (9.10%) and pyriproxyfen (0.46%). Rabbits can also be affected by the common cat flea but remember to NEVER use fipronil in rabbits as it will cause seizures and death in rabbits. Many use imidacloprid in rabbits as a choice for flea treatment. including Advantage Multi (Bayer) consisting of imidacloprid and moxidectin which is now licensed for use in dogs, cats, and ferrets.

Mites and lice

Common mites of exotic companion mammals are generally species specific. Fur mites generally cause dandruff and patchy alopecia with little associated pruritis. The rabbit fur mite (*Cheyletiella parasitovorax*) is considered zoonotic, causing a mild hyperemia and pruritis in humans. The ear mite, *Otodectes cyanotis*, is commonly found in ferrets and can spread between cats and ferrets, therefore it is important to treat all susceptible pets in the household. The presence of black ear wax in ferrets does not necessarily denote ear mite infestation since ferret ear wax is normally dark; however, ear mites are readily visible, and many ferrets will scratch their ears violently as you try to clean them if ear mites are present. Shake the swab on a dark surface. Mites will be moving white spots while also present on microscopic examination. The rabbit ear mite, *Psoroptes cuniculi*, can cause such an accumulation of dry crusts that an upright ear can be weighted down into the lop-ear position. *Demodex* spp. are commonly demonstrated in hamsters greater than 2 years of age often with concomitant Cushing's disease, or any form of immunosuppression. Hamsters are generally exposed to the mites when young, but do not exhibit clinical signs until immunocompromised. Rabbits and guinea pigs with intense pruritis should be checked for their respective mites since it would be the most common differential. African pygmy hedgehogs kept as pets in the United States can present with *Chorioptes* sp. mites resulting in quill loss, skin crusting and flaking. The mouse louse (*Polyplax serrata*) is common in wild mice, causing anemia, debilitation, and intense pruritis. If it is found in pet mice, there is usually exposure to wild mice. The spined rat louse (*Polyplax spinulosa*) is rare, but can also cause pruritis, irritability and anemia. The lice of guinea pigs (*Gliracola porcelli, Gyropus ovalis*) are commonly seen in pet guinea pigs, and if heavy in number can cause pruritis, rough hair coat and alopecia.

Mites are generally treated with multiple doses of ivermectin or even better one dose of selamectin. Do not attempt to remove the debris from the ears of rabbits with ear mites prior to treatment as the crusts will fall off on their own days after treatment.

Cuterebra, maggots, ticks

Rabbits that graze outside or are fed fresh cut grass from outside can get *Cuterebra* spp. bot larva subcutaneously. The *Cuterebra* is ingested and migrates out of the oral cavity or esophagus to a subcutaneous site and is distinguished by the presence of a breathing hole in the skin. *Cuterebra* larvae are removed surgically and generally the wound is left open to heal by second intention. Maggots are seen mostly in rabbits housed outdoors, usually in unsanitary conditions. Maggots can be killed by topical, oral, or rectal application of nitenpyram (Capstar) that kills the maggots, which can then be cleaned out of the affected area. It may be necessary to clip the matted, soiled hair, and the underlying, affected skin may need antimicrobial cleansing and care. NSAIDs and/or analgesics may be necessary for comfort. A variety of ticks can infest domestic rabbits housed outdoors or wild rabbits and are removed manually. The Centers for Disease Control and Prevention website has a summary and explanation of diseases that are transmitted by ticks.[20]

Neoplasia

Mast cell tumor

In ferrets, mast cell tumors (MCT) are very common and appear as round, red to pink, erosive skin lesions that may or may not be raised, that may worsen or spontaneously resolve. Unlike dogs, mast cell tumors in ferrets are not known to metastasize. A slow healing traumatic wound may resemble a mast cell tumor in a ferret, or rarely, a skin tumor of another type. If the mass causes irritation or excessive bleeding due to its location, then removal is recommended; however, benign epitheliomas are more common than MCT, and any skin mass should be removed and submitted for histopathology.

Adrenal disease

Adrenal disease is common in ferrets over 3 years of age, but adrenal neoplasias have been diagnosed as early as 10 months. Clinical signs include bilaterally symmetric alopecia, some with petechiation; rat tail that includes alopecia, comedones; thinning, dry haircoat; muscle atrophy, pendulous abdomen and splenomegaly (extramedullary hematopoiesis). Clinical signs may also include vulvar enlargement even in spayed females, prostatic hyperplasia, often with cysts, and subsequent urethral blockage in males, pruritis, aggressiveness, return to male behavior in neutered males, or bone marrow suppression in either males or females. Diagnosis is based on clinical signs, a hormone panel and/or ultrasound. Treatment of adrenal gland disease in ferrets used to involve surgical removal of the gland, but now hormonal therapy is the treatment of choice with deslorelin implants (Suprelorin-F, Virbac), which are conditionally licensed for use in ferrets, It blocks sex steroid production thereby reducing clinical signs, but adrenal neoplasia may continue to grow. The deslorelin implant are also used preventively starting at 1 year of age to prevent the development of adrenal tumors in spayed and neutered ferrets.

Mammary fibroadenoma and fibrosarcoma

Mammary fibroadenoma and fibrosarcoma in rats and mice can present with varying incidence and types of tumors depending on the strain of the rat as well as age, sex, diet, caloric intake and environment. Unfortunately, the genetic background of pet rats is unknown. The most common tumor of rats is the mammary gland tumor, described as an incidence of 30-57% in female Sprague-Dawley rats, 50-90% in female aged rats, and 16% in male rats.[22] Most rat mammary tumors are fibroadenomas (prolactin sensitive), with about 10% being adenocarcinomas (estrogen sensitive).[21] In mice, it is the opposite, 90% are fibroadenocarcinoma and 10% are fibroadenomas. Mammary fibroadenomas in rats are easily removed surgically, but often recur in nearby mammary tissue. Some inguinal masses are more difficult to remove if associated with the urethra. Mammary tissue can occur anywhere along the shoulder, neck, ventrum, flank or tail base. One study showed that rats ovariectomized at 90 days of age had significantly lower incidence of mammary tumors than intact female rats, 2/47 versus 24/29, respectively, but there is no evidence that spaying an adult rat prevents the occurrence, or recurrence, of fibroadenoma or fibroadenocarcinoma.[23] Because 90% of rat mammary tumors are fibroadenoma, and prolactin sensitive, the administration of anti-estrogen drugs such as tamoxifen, has not been shown to be helpful unless the tumor is a fibroadenocarcinoma as is found in mice or 10% of rats. Energy consumption restriction has been shown to decrease tumor incidence in mice.[22]

Ventral abdominal scent gland neoplasia

The ventral abdominal scent gland can develop a tumor in gerbils. It is important to know the appearance of the normal gerbil ventral abdominal gland compared to a cancerous one. Benign adenomas of the ventral abdominal gland have been reported to be their third most common spontaneously occurring tumor. Surgical removal of the mass if the gland is abnormal. Prognosis after complete surgical removal is good.

Other Considerations

Sugar gliders

Adult male sugar gliders possess a frontal scent gland on their forehead and gular scent glands in the ventral cervical area which should not be mistaken for an abnormality. It is common for sugar gliders to traumatize themselves with stress, post-surgery, due tolocal irritation, or other unknown reasons. Male sugar gliders will sometimes chew the penis, which can be amputated distal to the urethral opening at the fork with no untoward effects. This behavior may be brought on by impacted cloacal glands. A low dose of acepromazine post-operatively is helpful in curbing self-mutilatiomutilation.[24] Also, pain relievers such as NSAIDs and opioids are used for pain and gabapentin is sometimes helpful.

Cystic ovary, bilateral alopecia

The incidence of cystic ovary(ies) in intact female guinea pigs over 5 years of age is approximately 80%. Generally, the left ovary is involved, but it can be the right or both. The ovarian cyst can commonly reach 5 × 5 centimeters in size. Clinical signs include lethargy, painful abdomen, partial anorexia, enlarged abdomen, enlarged nipples, and bilaterally symmetric alopecia along the flanks and ventrum. Palpation or radiograph or ultrasound demonstrates the large cyst. Some veterinarians give leuprolide acetate, others recommend a complete spay procedure, while others suggest that only an ovariectomy is needed and that it is a quicker surgery. The author prefers ovariectomy if the ultrasound shows that the uterus is normal otherwise a complete ovariohysterectomy is performed since an abnormal uterus in a guinea pig is most indicative of a concomitant uterine leiomyoma/ leiomyosarcoma (common). In the author's experience hormonal therapy is not curative and the disease recurs.

Urine scald

Rabbit urine is very alkaline, with a pH of approximately 8 and it very easily irritates the thin inguinal skin. When rabbits urinate they lift up slightly on their rear legs to direct the flow of urine caudally. Any interference with this motion can result in the skin being exposed to urine including: arthritis (spine, hips, knees, hocks), pododermatitis, disc disease, obesity, neurologic disorders such as encephalitozooanosis, or weakness and debilitation.

Hyperkeratosis

Sometimes excess keratin is seen where the fifth toenail would be if guinea pigs had five toes on the front feet (i.e., they only have four toes on the front feet). It is similar to a chestnut in a horse. It can be trimmed just as a toenail since it is just excess keratin.

Porphyrin tears

Chromodacryorrhea, or red tears occur in rats and other rodents due to stress causing excess Harderian gland secretions which contain red porphyrins that look like blood. Since the porphyrin fluoresces under ultraviolet light a simple black light can help distinguish it from blood.

Hairless varieties

There are hairless varieties of guinea pigs, mice, and rats.

Tattoos

One commercial breeder of ferrets places two dot tattoos in the right ear signifying that it has been spayed or neutered and demusked. Occasionally, show or research rabbits will present with a tattoo in the ear used for identification.

Ulcerative dermatitis of mice

Idiopathic ulcerative dermatitis of mice is known to occur in the C57BL/6 strain of mice or in mice that have C57BL/6 in their background. The cause is not known but it is known to be a pruritic condition secondary to a vasculitis. The lesions are usually about the thorax and head and can be differentiated from barbering that is associated with a non-pruritic alopecia about the face and whiskers.[11] Trimming the nails is imperative. Other treatments, such as omega-3 fatty acids, topical antibacterial treatments, or even steroids have some success.

References

1. d'Ovidio D, Santoro D. Dermatologic diseases of ferrets. In: Quesenberry KE, Orcutt CJ, Mans C, Carpenter JW. *Ferrets, Rabbits, and Rodents: Clinical Medicine and Surgery.* 4th Ed. St. Louis, MO: Elsevier; 2021:109-116.

2. Varga M, Paterson S. Dermatologic diseases of rabbits. In: Quesenberry KE, Orcutt CJ, Mans C, Carpenter JW. *Ferrets, Rabbits, and Rodents: Clinical Medicine and Surgery.* 4th Ed. St. Louis, MO: Elsevier; 2020:220-232.

3. Graham JE. Rabbit wound management. *Vet Clin North Am Exot Anim Pract.* 2004;(7):37-55.

4. Mickelson MA, Mans C, Colopy SA. Principles of wound management and wound healing in exotic pets. *Vet Clin North Am Exot Anim Pract.* 2016;(19):33-53.

5. Langlois I. Wound management in rodents. *Vet Clin North Am Exot Anim Pract.* 2004;(7):141-167.

6. Pilny A, Hess L. Ferrets: wound healing and therapy. *Vet Clin North Am Exot Anim Pract.* 2004;(7): 105-121.

7. Hernandez-Divers S. Principles of wound management of small mammals: hedgehogs, prairie dogs, and sugar gliders. *Vet Clin North Am Exot Anim Pract.* 2004;(7):1-18.

8. White SD. Rabbit dermatology. *Vet Clin North Am Exot Anim Pract.* 2023;26(2):347-357.

9. Halck ML, Schoemaker NJ, von Zeeland YRA. Ferret dermatology. *Vet Clin Norh Am Exot Anim Pract.* 2023;26(2):359-381.

10. Sarvi J, Eshar D. Rodent dermatology. *Vet Clin North Am Exot Anim Pract.* 2023;26(2):383-408.

11. Frolich J. Rats and mice. In: Quesenberry KE, Orcutt CJ, Mans C, Carpenter JW, eds. *Ferrets, Rabbits, and Rodents: Clinical Medicine and Surgery.* 4th Ed. St. Louis, MO: Elsevier; 2021:345-367.

12. Varga M, Paterson S. Dermatologic diseases of rabbits. In: Quesenberry KE, Orcutt CJ, Mans C, Carpenter JW, eds. *Ferrets, Rabbits, and Rodents: Clinical Medicine and Surgery.* 4th Ed. St. Louis, MO: Elsevier; 2021;220-232.

13. Harkness JE, Turner PV, VandeWoude S, Wheler C, eds. *Harkness and Wagner's Biology and Medicine of Rabbits and Rodents.* 5th Ed., Ames, IA: Wiley-Blackwell; 2010.

14. Chow EP, Bennett A, Whittington JK. Total ear canal ablation and left bulla osteotomy for treatment of otitis externa and media in a rabbit. *J Am Vet Med Assoc.* 2011;239(2):228-232.

15. Fisher P, Graham JE. Rabbits. In: Carpenter JW, Harms CA, eds. *Carpenter's Exotic Animal Formulary.* 6th Ed., 2023;575-626.

16. Miller AL, Clarkson JM, Quigley C, et al. Evaluating pain and analgesia effectiveness following routine castration in rabbits using behavior and facial expressions. *Front Vet Sci.* 2022;9:782486. doi: 10.3389/fvets.2022.782486.

17. Keating SCJ, Thomas AA, Flecknell PA, Leach MC. Evaluation of EMLA cream for preventing pain during tattooing of rabbits: changes in physiological, behavioural and facial expression responses. PLoS ONE 20 12;7(9):e44437. doi:10.1371/journal.pone.0044437.

18. Drouot S, Mignon B, Fratti M, et al. Pets as the main source of two zoonotic species of *Trichophyton mentagrophytes* complex in Switzerland, *Arthroderma vanbreuseghemii* and *Arthroderma benhamiae*. *Vet Dermatol.* 2009;20(1):13-18.

19. Bartosch T, Frank A, Gunther C, et al. *Trichophyton benhamiae* and *Trichophyton mentagrophytes* target guinea pigs in a mixed small animal stock. *Med Mycol Case Reports.* 2019;23:37-42.

20. Centers for Disease Control and Prevention. Diseases Transmitted by Tick. Available at: https://www.cdc.gov/ticks/diseases/index.html. Accessed April 7, 2023.

21. Frohlich J. Rats and mice. In: Quesenberry KE, Orcutt CJ, Mans C, Carpenter JW, eds. *Ferrets, Rabbits, and Rodents: Clinical Medicine and Surgery.* 4th Ed. St. Louis, MO: Elsevier; 2021:345-367.

22. Greenacre CB. Spontaneous tumors of small mammals. *Vet Clin North Am Exot Anim Pract.* 2004;7(3):627-651. doi: 10.1016/j.cvex.2004.04.009. PMID: 15296867.

23. Hotchkiss C. Effect of surgical removal of subcutaneous tumors on survival of rats. *J AmVet Med Assoc.* 1995;206(10):1575-1579.

24. Doss G, Johnson-Delaney C. Sugar gliders. In: Carpenter JW, Harms CA, eds. *Carpenter's Exotic Animal Formulary.* 6th Ed. 2023;496-510.

Respiratory Diseases of Exotic Companion Mammals

Cheryl B. Greenacre, DVM, DABVP (Avian Practice), DABVP (Exotic Companion Mammal Practice)

Session #3349

Affiliation: From College of Veterinary Medicine, University of Tennessee, Knoxville, TN 37996, USA (Emeritus), Clinical Laboratory Animal Veterinarian, Office of Research, University Research Animal Resources, University of Georgia, 501 D.W. Brooks Drive, Room 3504, Athens, GA 30602, USA.

Abstract: Exotic companion mammals, including rabbits, rodents, ferrets, hedgehogs, and sugar gliders, can present with a variety of respiratory disorders due to various causes including bacterial, viral, parasitic, neoplastic, traumatic, and other causes. In some cases, the treatment is straightforward and proven, while in other cases the treatment options are unproven or palliative. This masterclass describes in detail the diagnostic and therapeutic techniques used to manage these respiratory diseases in various exotic companion mammal species.

Introduction

Respiratory diseases of exotic companion mammals (ECMs) can be categorized by their cause. Common bacterial respiratory diseases include mycoplasmosis, bordetellosis, and pasteurellosis. Causes of viral respiratory diseases include Sendai virus, sialodacryoadenitis virus, calicivirus, distemper, influenza, and SARS-CoV-2. Fungal respiratory disease of ECMs include pneumocystosis. The most common parasite to cause secondary respiratory disease would be heartworm disease in ferrets. Lastly, various forms of neoplasia affect ECMs including thymoma of rabbits and metastasis of various forms of neoplasia to the lungs. There are many excellent sources available for information on diseases of exotic companion mammals including books, articles, and websites.[1-18]

Bacterial Respiratory Diseases

Mycoplasmosis

Mycoplasma pulmonis is a gram-negative bacteria that causes variable clinical signs based on exposure dose, strain of bacteria, and strain or species of animal, but can usually be described as a chronic bronchopneumonia that spreads from the hilus. The disease in rats is more severe than as seen in mice.[4,5] Older animals are more commonly affected. Rats show signs of dyspnea, rales, oculonasal discharge, and porphyrin tears. The mouse and rat have a single left lung lobe so if an infection is on the left side then the entire lobe/left side can be affected instead of being compartmentalized as it could be on the right where there are four lobes. The rat shows the typical gross cobblestone lung describing the lung's surface appearance. Diagnosis is via ELISA, RIA, culture, and characteristic lesions on gross and histopathologic examination of tissues. Treatment is usually enrofloxacin and doxycycline given concurrently for 21 days since the bacteria is intracellular; however, mycoplasmosis is not cured and recurrence is common. Tylosin has also been used parenterally and nebulized. In rats, clinical signs will often return after signs of stress, despite treatment. Remember that pleuritis can be very painful, therefore pain relief should be addressed using opiates if needed.

Bordetellosis

Rabbits can carry *Bordetella* sp. as normal flora, but guinea pigs exposed to rabbits can be infected with *Bordetella* sp. and become clinically ill, especially young guinea pigs.[7] Some shipments of guinea pigs in the pet trade may have a high incidence of bordetellosis and may have already received prophylactic antibiotics. Coinfection with *Chlamydia caviae* may also be present and exacerbate clinical signs. Clinical signs in guinea pigs include a purulent bronchopneumonia. Diagnosis consists of an ELISA test for antibodies, radiographs, history of being housed with rabbits, or culture and sensitivity. Commonly, treatment consists of either enrofloxacin, trimethoprim sulfa, or chloramphenicol.

Pasteurellosis

Pasteurellosis is a very common cause of disease in rabbits and manifests in many different tissues[6,8] The causative organism of bacterial pneumonia in rabbits is usually (55%) *Pasteurella multocida* but can be due to other bacteria such as *Bordetella* sp., *Pseudomonas* sp., and *Staphylococcus* sp., or a combination of bacteria. Nasal or sinus infections can also include *Moraxella* sp., or *E. coli*. Clinical signs of pasteurellosis in rabbits include upper or lower respiratory signs, neurologic signs such as vestibular disease, and abscessation in skin or soft tissue. Diagnosis is based on culture/sensitivity and PCR of a nasal swab. A CBC may show a leukocytosis, but surprisingly, abscesses in rabbits do not raise the WBC. Treatment is based on culture and sensitivity, but until the results return, enrofloxacin is recommended if pasteurellosis is suspected. For severe sinus/nasal infections a rhinotomy or rhinostomy may be necessary to remove debris and instill local antibiotics. Remember that pasteurellosis can only be treated, not cured. It may recur during times of stress.

Viral Respiratory Diseases

Sendai virus infection

The highly contagious Sendai virus (SV), a paramyxovirus, is transmitted by aerosol. Mice are the natural host, but rats, hamsters and guinea pigs are also susceptible. Clinical signs in mice include hunched posture, erect hair coat, weight loss, dyspnea with chattering, and ocular discharge (crusts). Weanling and younger mice suffer more severe clinical signs than older mice and may even die. Rats have a low prevalence of clinical signs but can show a rhinitis that progresses to bronchitis and focal alveolitis. In hamsters most infections are subclinical. Diagnosis is by serologic tests (IFA, ELISA), RT-PCR, and characteristic lesions on histopathologic examination of tissues. There is no treatment, but the virus is highly labile, so general hygiene helps. In a truly closed mouse population (i.e., no new animals, no new births), the infection will burn out in about 4-6 weeks.[1]

Sialodacryoadenitis virus

Sialodacryoadenitis virus (SDAV) is caused by rat coronavirus. Clinical signs of SDAV in rats is variable including asymptomatic infection to a transient (about 1-2 weeks) sialodacryoadentitis. Cervical swelling from edema and inflammation of submaxillary salivary glands is a commonly seen clinical manifestation. At necropsy, necrosis of the epithelium in salivary and lacrimal glands is seen. Diagnosis is via serologic tests (ELISA), and characteristic lesions on immunohistochemical staining of tissues. There is no treatment.

Rabbit hemorrhagic disease

Rabbit hemorrhagic disease (RHD), with the most common current strain circulating known as RHDV-2, is a highly contagious foreign animal disease in the United States caused by a calicivirus that is associated with near 100% mortality in both wild (*Sylvilagus* spp.) and domestic (*Oryctolagus cuniculus*) rabbits. Epistaxis and death are the classic hallmarks of RHD, but not all rabbits will show epistaxis and some may only present with GI stasis, respiratory disease, or death. The disease is now spreading across the United States from west to east in both the wild and domestic rabbit populations. If you suspect RHD immediately contact your state veterinarian.

As of March, 2020, RHDV-2 was identified in wild rabbit species (*Sylvilagus audubonii* and *Lepus californicus*) for the first time in New Mexico and within weeks RHDV-2 spread to wild and domestic rabbits in the western United States and has been found sporadically in domestic rabbits in the eastern half of the country. The virus is highly stable and can remain actively infectious on surfaces for > 3.5 months. The loss of wild prey species in these states will have significant ecological impacts on other species that feed on rabbits and on co-prey species that will see an increase in predation.

Currently there is a killed vaccine manufactured in the United States that has been granted Emergency Use Authorization. It is made by Medgene and the state veterinarian must be contacted in order to use it and to provide specific rules for its use in the state. Administration is 0.5 milliliters subcutaneously and it requires a booster at 21 days, and annually.

Resources

USDA Emerging Risk Notice:
https://www.aphis.usda.gov/animal_health/downloads/rhdv2.pdf

USDA APHIS RHD Factsheet:
https://www.aphis.usda.gov/publications/animal_health/fs-rhdv2.pdf

USDA General Guidance for Cleaning and Disinfection of RHDV Contaminated Premises:
https://www.aphis.usda.gov/animal_health/downloads/rhdv-cleaning-guidance.pdf

USDA RHDV2 Vaccine FAQs:
https://www.aphis.usda.gov/aphis/ourfocus/wildlifedamage/programs/nwdp/nwdp-rabbit-hdv

Distemper

Most literature describes canine distemper virus, a paramyxovirus, as causing a nearly 100% fatal disease in ferrets, but recent information suggests better outcomes if treatment is begun in the early stage, hence the strong recommendation for vaccination. There are two vaccines licensed for use in ferrets in the United States, Purevax Ferret Distemper Vaccine® (Merial) and Neovac FD® (Neotech). Due to supply issues in the past some have used Nobivac Puppy DPv® (Merck Animal Health), but this is not allowed if there is a licensed product available. Clinical signs include an opaque oculo-nasal discharge, fever, and hyperemic, crusty footpads and chin. Crusty chin and foot pads are almost pathognomonic for canine distemper viral disease in the ferret, although both of these occur late in the clinical disease. Diagnosis is based the clinical signs and/or FA or PCR on conjunctival swab. Titers can be performed as well as histopathology. Influenza and SARS CoV-2 can mimic the early stages of distemper. Euthanasia is recommended, but for some low pathogenic strains or if the disease is caught early, there are reports of survival with antibiotics for secondary infection, vitamin A and C, CDV hyperimmune serum IV, interferon, vaccinate in the face of the outbreak and supportive care.

Influenza

Influenza virus A and B of the Orthomyxoviridae family, causes influenza in ferrets, the same as in humans.[1,9] The clinical signs include serous oculonasal discharge, fever, sneezing and coughing. Ferrets are used extensively in flu research and vaccine development. The incubation period is approximately 48 hours. PCR testing is available. Rarely, the disease can progress to pneumonia or sinusitis with secondary bacterial infection. The clinical course of the disease is 7-14 days and rarely is any treatment necessary other than supportive care. Occasionally, diphenhydramine may be necessary. Secondary bacterial infections usually respond well to oral antibiotics such as amoxicillin with clauvulinic acid, but cultures should be performed. The virus can be transmitted to the humans in the household: and the humans transmit it to the ferrets. It is recommended that all humans working with or owning ferrets receive influenza vaccination.

SARS-CoV-2

Ferrets, mink, tigers and several other mammals can become infected and develop clinical signs with SARS-CoV-2 (COVID-19). It was proven that ferrets can become infected, shed, and transmit to other ferrets SARS CoV-1 (i.e., 2003 outbreak). Recent studies show they have the appropriate ACE2 receptor for SARS-CoV-2. There are several reports of mink to human transmission of the virus, but to date no ferret to human transmission has been documented.[10-12] Since it is thought that SARS-CoV-2 can be shed for at least a week without evidence of clinical signs of disease and is potentially fatal to people, it is prudent to be proactive and develop a protocol for how to handle a ferret with possible SARS-CoV-2 in the hospital before it arrives. Clinical signs in ferrets include respiratory disease, fever, diarrhea, vomiting, and death. Idexx and other companies offer testing (PCR swab of oropharynx and rectum) for SARS-CoV-2 in animals, but it is important to coordinate any testing with the state veterinarian even if the test is being sent to another state.

Fungal Respiratory Diseases

Pneumocystosis

Although fungal pneumonias such as histoplasmosis, blastomycosis, coccidioidomycosis, and cryptococcosis have been reported in ferrets, they are rare.[14] It is important to be aware of *Pneumocystis carinii*, a ubiquitous organism that was initially classified as a protozoan, but now has been reclassified as a fungus and interestingly, it seems to respond to antibiotics. Generally, infection with *P. carinii* in the ferret and humans is asymptomatic, but clinical signs can develop in immunocompromised individuals. It can cause a slowly progressive pneumonia, which should be kept in mind when treating ferrets with chemotherapy for lymphoma. Diagnosis is based on serologic tests (ELISA), PCR, and characteristic lesions on immunohistochemical or silver staining of tissues or tracheal wash. Treatment is with trimethoprim-sulfa. There has never been animal to human transmission documented and humans are thought to acquire their infections from other humans when young.[13] Ferrets are commonly used in pneumocystosis research because they mimic the disease in humans so well.

Parasitic Respiratory Diseases

Heartworm

Dirofilaria immitis, a filarial parasite, can infect ferrets and cause weakness, lethargy, depression, dyspnea, cyanosis, anorexia, cough, and pale mucous membranes. The low worm burdens make diagnosis or screening difficult, but traditional canine antigen kits can be used. Definitive diagnosis involves ultrasound of the heart demonstrating the worms. In severe cases typical radiographic lesions can be seen. The FDA recently approved Advantage Multi® for Cats (Bayer, 0.4 ml formulation) for use in ferrets as a heartworm preventative.[15] It is a once a month topical of imidacloprid and moxidectin. Also recently approved is Midamox for Cats (Norbrook Laboratories) which is also a topical heartworm preventative. There are a few anecdotal reports of treatment of heartworms with Imiticide® (Meriel, Athens, GA) at the cat-dosing regimen. Note that heartworm(s) may outlive the ferret!

Neoplasia

Metastasis to the lungs

Some forms of cancer are more prone to metastasize to the lungs than others. The most encountered thoracic metastasis in exotic companion mammals is uterine adenocarcinoma in the rabbit. Intact female rabbits over 4 years of age have a very high incidence of uterine adenocarcinoma and can be as high as 40-50% in some breeds including the tan, French silver, Havana, Dutch, and in the author's experience, the Silver Marten.[16] If clinical signs are present, they include anorexia, depression, hematuria, dysuria or milk production. Diagnosis is based on palpation, radiographs, or ultrasound that suggests an enlarged, possibly fluid filled uterus, sometimes with masses or cystic endometrial hyperplasia. Treatment consists of ovariohysterectomy a radiograph should be performedprior to surgery to rule out metastasis to lung or bone.

Thymoma

It is not uncommon for adult and older rabbits to present with thymoma.[2,3] Clinical signs include respiratory difficulty, swelling of the head, exophthalmos, and prolapse of the third eyelids. The first sign seen is usually exercise intolerance and visibility of the sclera at the medial canthus. If the heart is muffled on one side compared to the other on thoracic auscultation, this provides further suspicion of thymoma. Radiographs or computed tomography (CT) will show a cranial mediastinal mass that is elevating the thoracic trachea. An ultrasound guided needle aspirate is useful for diagnosis and can help distinguish thymoma from other mediastinal masses such as lymphoma. Treatment options include surgery or radiation therapy, but radiation, using a very precise linear accelerator, is the treatment of choice due to less side effects. The mean response time to radiation therapy is about 8 months. Judicious steroid therapy has also been beneficial. If fluid filled thymic cysts are identified at ultrasound, then aspiration of the fluid will immediately help with improved breathing and lessening the anesthetic risk for future anesthetic episodes.

Other Respiratory Diseases

Pseudo-odontoma

Trauma to upper incisor reserve crown(s) from chewing on cage has been considered a possible factor in the formation of pseudo-odontomas. The mass decreases the nasal cavity lumen making respiration difficult for obligate nasal breathers. Clinical signs include upper respiratory noise and dyspnea. Although not a true odontoma, it is a proliferative growth, likely in response to trauma. Radiographs demonstrate the lesion and decreased nasal cavity lumen size. A CT scan will further delineate the extent of the mass and may be instrumental in deciding surgical excision. Treatment may include removal of the maxillary incisors, along with adherent tissues. If full restoration of the nasal airway cannot be restored via mass and tooth removal, a nasal cavity stent may need to be installed.

Dilated cardiomyopathy

Dilated cardiomyopathy (DCM) has been described in ferrets and rabbits and is treated similarly to dogs with DCM, including pimobendan, furosemide, spironolactone, and enalapril, except the lower dose of enalapril is used to prevent creating hypotension. Dosages can be found in Carpenter's Formulary.[18] Although ferret diets should have adequate amounts of taurine, this can be supplemented. Diagnosis is also similar to dogs and cats via echocardiogram and can be followed up with repeat echocardiograms. Complete echocardiographic measurements have been published for ferrets.[17]

Trauma

Ferrets present with trauma from either falling, having something heavy fall on them or being caught in a door for example, or from being bitten, shaken, or crushed by a dog. Be sure to perform auscultation and radiographs to rule out a pneumothorax, and if present perform thoracocentesis to remove the air. Hemorrhage within the thorax is also possible. Radiographs and thoracic ultrasonography are warranted.

References

1. Perpinan D. Respiratory diseases of ferrets. In: Quesenberry KE, Orcutt CJ, Mans C, Carpenter JW, eds. *Ferrets, Rabbits, and Rodents: Clinical Medicine and Surgery.* 4th Ed. St. Louis, MO: Elsevier; 2021: 71-76.

2. Lennox AM, Mancinelli E. Respiratory diseases of rabbits. In: Quesenberry KE, Orcutt CJ, Mans C, Carpenter JW, eds. *Ferrets, Rabbits, and Rodents: Clinical Medicine and Surgery.* 4th Ed. St. Louis, MO: Elsevier; 2021:188-200.

3. Quesenberry KE, Pilny AA, St-Vincent RS. Lymphoreticular disorders, thymoma, and other neoplastic diseases. In: Quesenberry KE, Orcutt CJ, Mans C, Carpenter JW, eds. *Ferrets, Rabbits, and Rodents: Clinical Medicine and Surgery.* 4th Ed. St. Louis, MO: Elsevier; 2021:258-269.

4. Frohlich J. Rats and mice. In: Quesenberry KE, Orcutt CJ, Mans C, Carpenter JW, eds. *Ferrets, Rabbits, and Rodents: Clinical Medicine and Surgery.* 4th Ed. St. Louis, MO: Elsevier; 2021:345-367.

5. Fox JG, Anderson LC, Loew FM, Quimby FW. *Laboratory Animal Medicine.* 2nd Ed. San Diego, CA: Elsevier; 2002.

6. Varga Smith M. *Textbook of Rabbit Medicine.* 3rd Ed. St. Louis, MO: Elsevier; 2022.

7. Garcia MA, Barcelo AM, Nadeu CB, Jekl V. Respiratory diseases in guinea pigs, chinchillas, and degus. *Vet Clin North Am Exot Anim Pract.* 2021;24(2):419-457.

8. Jekl V. Respiratory diseases of rabbits. *Vet Clin North Am Exot Anim Pract.* 2021;24(2):459-482.

9. Lennox A. Respiratory diseases of ferrets. *Vet Clin North Am Exot Anim Pract.* 2021;24(2):483-493.

10. Oude Munnink BB, Sikkema RS, Nieuwenhuijse DF, et al. Transmission of SARS-CoV-2 on mink farms between humans and mink and back to humans. *Science.* 2021;371(6525):172-177. doi: 10.1126/science.abe5901. PMID: 33172935; PMCID: PMC7857398.

11. Hammer AS, Quaade ML, Rasmussen TB, et al. SARS-CoV-2 transmission between mink (*Neovison vison*) and humans, Denmark. *Emerg Infect Dis.* 2021;27(2):547-551. doi: 10.3201/eid2702.203794. PMID: 33207152; PMCID: PMC7853580.

12. Rabalski L, Kosinski M, Mazur-Panasiuk N, et al. Zoonotic spillover of SARS-CoV-2: mink-adapted virus in humans. *Clin Microbiol Infect.* 2022;28(3):451. doi: doi: 10.1016/j.cmi.2021.12.001.

13. Centers for Disease Control and Prevention. Pneumocystis pneumonia. Available at: https://www.cdc.gov/fungal/diseases/pneumocystis-pneumonia/index.html. Accessed April 5, 2023.

14. Greenacre CB, Dowling M, Nobrega-Lee M. Diagnosis and treatment of histoplasmosis in a group of four domestic ferrets (*Mustela putorious furo*) and a review of histoplasmosis. *J Exot Pet Med.* 2019;29: 194-201.

15. Drugs.com. Advantage Multi (imidacloprid + moxidectin) Topical Solution. Available at: https://drugs.com/vet/advantage-multi-imidacloprid-moxidectin-topical-solution.html. Accessed April 5, 2023.

16. Di Giralamo N, Salleri P. Disorders of the urinary and reproductive systems. In: Quesenberry KE, Orcutt CJ, Mans C, Carpenter JW, eds. *Ferrets, Rabbits, and Rodents: Clinical Medicine and Surgery.* 4th Ed. St. Louis, MO: Elsevier; 2021:201-219.

17. Dudás-Györki Z, Szabó Z, Manczur F, Vörös K. Echocardiographic and electrocardiographic examination of clinically healthy, conscious ferrets. *J Small Anim Pract.* 2011;52:18-25.

18. Applegate J, Harms C. Ferrets. In: Carpenter JW, Harms CA, eds. *Carpenter's Exotic Animal Formulary.* 6th Ed. St. Louis, MO: Elsevier; 2023:626-657.

Point-of-Care Ultrasonography in Exotic Animals

Cathy A. Johnson-Delaney, DVM

Session #3266

Affiliation: From Washington Ferret Rescue & Shelter, 6124 Evergreen Way, Everett, WA 98203, USA, NW Zoological Supply, 3614 Smith Ave, Everett, WA 98201, USA.

Abstract: Point-of-care ultrasound (POCUS) and abdominal-focused-assessment with sonography for trauma (AFAST), and thoracic-focused-assessment with sonography for trauma (TFAST) techniques can be performed in exotic patients such as the ferret, rabbit, guinea pig, small rodent, hedgehog and sugar glider. This POCUS can quickly look for fluids in the body cavities and be used as screening for other diseases such as masses, organomegaly, and uroliths. Sonographic uses for aspirations, and needle biopsies will be discussed. POCUS to assess the heart can facilitate initiation of treatment, although it may not take the place of a detailed echocardiograph. Ultrasound is also useful to visualize gut motility. Because of differences in anatomy, scanning positions for different organs are used. In rabbits, guinea pigs and chinchillas, paralumbar scanning is useful for kidneys and the reproductive tract. Ferret positioning differs from those used in dogs and cats. Assessment in hedgehogs requires getting the probe onto the ventrum. Small rodents are better scanned with a probe with a small footprint and high frequency for more detail. Sugar gliders and other marsupials with pouches can be scanned from within the pouch. The handheld scanner that projects onto a cellphone or tablet computer makes point of care scanning practical to add to any examination.

Introduction

Wouldn't veterinarians all love to be able to use a Star Trek medical tricorder and scanner right now? In some ways, use of a point-of-care ultrasound (POCUS) device, particularly one that is a handheld unit transmitting wirelessly to a cell phone, tablet or computer currently provides some of that ability. While performing a physical examination is still a starting point for the hands-on examination, results can be inconsistent if the animal has a tense body, even after administration of a sedative. In many animals, all the organs cannot be fully palpated because other organs are in the way, such as a cecum in a rabbit or guinea pig making palpation of the rest of the abdomen difficult. Being able to observe organs, fluids, and even function (e.g., heart) while having the animal on the exam table or cageside, can expedite care timelines, improve outcomes, and even give clients peace-of-mind that their pet is receiving a thorough examination. Ultrasound units are now in a price range and of improved technology so that every practice can afford to have one available.

Instrumentation Basics

While many practices have large console units with multiple probes, portable units are available that have multiple probes. The probe, or transducer, should be selected by the size of the footprint that is the point-of-contact with the animal, and the power of the transducer (i.e., will it be able to penetrate the depth of the patient or to the organ you want to examine). Most small animal probes that range from 7.5-12 MHz work well for all exotics. The type of transducer is also important. A convex, sector probe or a linear probe will give different images, with different levels of detail. All machines allow tuning of the image by increasing the depth of the focus, increasing the gain (blackening the picture) or lessening the gain (whitening the image). Many of the newer wireless point-of-care probes allow adjustments on the tablet or smart phone screen, thus there is no need to turn knobs or move sliders. The recommendation is to set aside some time with the machine and read the user's manual to become familiar with the menus and features. Newer POCUS machines are user friendly and adjustments can be made quickly, as well as freezing the frame, taking pictures, saving the images with many programs saving to the cloud for future downloads and analysis.

A few basic terms are necessary to know, as are relative tissue densities. Knowing these can be helpful when pathology is encountered. Essentially, the more sound waves come back to the transducer, the whiter the image will be. Hyperechoic shows white on the image, and the opposite, hypoechoic will have some waves returning, showing up in varying levels of grey. Anechoic means all the waves go through, usually liquid, and so will show up as black. There are also things to consider like the density of the tissue. Not all sound waves return at the same rate, so again, different densities and different shades of gray will be seen. Without spending a lot of time on the physics of this, there are some easy rules to follow. Start with the bladder. Urine should be black (anechoic), except if the rabbit has sludge, in cases of hematuria, or when stones are present. But it is still a good place to start your scan to tune your picture to what makes sense to you. Then slide cranially and laterally to the kidney; the medullary should be fairly anechoic, the parenchyma, cortex increasingly dense allowing to tune for depth, and gain (the strength of the beams desired to go through). Once these two organs look like they should, then for the most part, additional changes may not be necessary unless there are some serious pathologies present altering normal tissue.

Remember that there will be scatter or artefacts from tissues such as bone, or in the lungs due to the gas. Artifacts can also be from gas and material in the stomach and gastrointestinal tracts. Uroliths, or choleliths will also create scatter as beams cannot penetrate and will bounce off and may not return in a straight line to the transducer. Practice scanning to literally see around these artifacts where they are supposed to be present and to be alerted when an artifact is picked up where it is not supposed to be present. For more detailed information about the physics, please consult any ultrasonography textbook.[1]

The modes to be familiar with and those that can help the most are:

1. B-mode: Brightness mode. This is used the most.

2. M-mode: Motion mode. This is used for echocardiography. These tracings record depth on the vertical axis and time on the horizontal axis. It allows cardiac chamber and wall measurements and quantitative evaluation of valve and wall motion with time.

3. Color Doppler: This allows detection of movement with flow away from the probe being blue and flow towards the probe being red. It is useful to evaluate regurgitation of blood around valves in the heart, blood supply into a mass, and any organ.

4. Pulse-wave Doppler: This is used primarily in cardiac evaluation. The sound waves are transmitted in pulses similar to B-mode, but with tracking showing echoes arising from moving blood, showing the vessel's depth. It is shown on a split screen similar to that when using M-mode, although this allows for determination of blood flow.

A chart listing the order of increasing echogenicity of body tissues and substances can be helpful in interpreting images (Table 1).

Table 1. Anechoic to hyperechoic echogenicity.

Echogenicity	Example tissue or substance
Lowest (anechoic)	Bile, urine, serum/effusion within cysts
	Renal medulla
	Muscle
	Renal cortex
	Liver
	Storage fat
	Spleen
	Prostate
	Renal sinus
	Structural fat, vessel walls, sometimes intestinal walls
Highest (hyperechoic)	Bone, gas, organ boundaries (including sometimes intestinal walls)

Source: Nyland TG, Mattoon JS, Wisner ER. Physical principles, instrumentation, and safety of diagnostic ultrasound. In: Nyland TG, Mattoon JS, eds. *Veterinary Diagnostic Ultrasound.* Philadelphia, PA: WB Saunders; 1995:3-18.

There are many other types of imaging variations that the operator needs to be familiar with including mirroring. It takes practice to freeze images that will be diagnostic, and in all the types of measurements including tissue volumes. Most machines allow recording of images that may allow returning to the images to do the measurements. This way the animal does not have to be present and that can also allow the practitioner to go back through all the images and movies to do a more in-depth evaluation. Many of the new point-of-care wireless ultrasound systems save the images to the cloud so the veterinarian can not only return to the exam, make measurements, and further evaluations, but also create a report with images for the animal's medical record. A copy of the images can also be generated for the owner. Cloud-based imaging storage also allows practitioners to share cases with others for aid in evaluation.

Ultrasound is extremely useful for needle aspiration of fluid-filled tissues, fine needle aspirations or actual tissue biopsies. Urinary catheter placements can also be checked. Post-operative internal scans can show if there is hemorrhage or fluid seepage, or if other complications are arising such as dehiscence with leakage and even early peritonitis is starting.

The bottom line is that the practitioner needs to be familiar with the gross anatomy and imaging of the normal animal so imaging of abnormalities will seem obvious. Having normal organ images handy as reference whether from a text or from prior scans can be helpful.[2]

Principles of Quick Imaging

An animal needs to be in an appropriate position to be comfortable, while allowing a scan through the abdomen's four quadrants. Consider the cranial quadrant containing the liver, spleen, stomach and associated lymph nodes. The left quadrant contains the left kidney, adrenal, retroperitoneal space. From the same probe position, scan the caudal abdomen including the urinary bladder and reproductive tract. The right quadrant contains the right kidney adrenal, pancreas, and associated lymph nodes of the mesentery. In rabbits and rodents with large cecums, move the probe ventrally to look at the bladder and reproductive tracts. Within 3-5 minutes answers may be apparent. This essentially is the abdominal- focused-assessment with sonography for trauma (AFAST) technique to tell if there is free fluid in the abdomen.

The thoracic-focused-assessment with sonography for trauma (TFAST) technique can be applied to the thoracic area to quickly look for free fluid, major cardiac dysfunction, even air or fluid within the lungs. For many species, the ability to scan the cranial thoracic area can rule out masses versus enlarged lymph nodes. [3-10]

Application for Use in Exotic Pets

Unfortunately, POCUS techniques have been described in detail for dogs and cats but have not been published on application for use in exotic pets.[3-7] It is possible to apply these techniques, but given the varied anatomy of the different mammals, birds and herpetological species, the practitioner needs guidelines for the best positioning of the probe. Practitioners need to (1) have knowledge of the normal anatomy of the different species; (2) practice scans on normal, healthy individuals of different species (in small animals, and particularly in those with little cutaneous fat, a stand-off may be needed); and (3) practice scans on ill animals. Many of these scans should be following administration of an analgesic and/or a mild sedative. If an animal is being anesthetized for radiographs or other diagnostics, while the animal is being induced or recovered, quickly check some organs or body cavities. For example, every male ferret over the age of 3 years, quickly scan the bladder and prostate during the regular physical examination. POCUS can be incorporated into every geriatric animal examination. The examination fee can be adjusted to just include this 3- to 5-minute scan.

Remember that gas, air, bone, and hair can introduce artifacts. Uroliths or choleliths, tumors or other structures will cause artifacts and difficulties imaging organs that are masked by these. This is why POCUS can be tailored to any animal to look at almost any organ.[8-10] Many veterinarians learned ultrasound as a rather static imaging system with specified probe locations and angles; however, this applied to dogs and cats. Be creative with imaging. It is also helpful to image an animal before going to surgery or necropsy and compare the ultrasonographic image to what is actually present. Again, practice is important!

Ferrets

Ferrets can be easily imaged completely with the ferret in dorsal recumbency, or even from the scruffed, vertically held position. The author uses a two-handed palpation/scanning technique as it is easy to feel each organ and stabilize it for scanning. A systematic scan beginning with the bladder, moving craniad to left kidney, spleen, left adrenal, stomach/pyloric area, liver, gallbladder, right kidney, right adrenal, intestines and back to the bladder, prostate (male). Females: check for uterine stump. Special attention should be paid throughout the abdomen and thorax to lymph nodes. The heart can be done from this position, as the probe can be moved laterally. M-mode and pulse wave doppler exam can be done. The mediastinum should be checked. The lungs themselves can be evaluated, particularly if there is cardiac disease or evidence of potential neoplasia.[11]

Rabbits, guinea pigs, chinchillas and other animals with large cecums

From para/lumbar fossa area view kidneys, and possibly some of the reproductive tract. From the caudal ventrum view the bladder directly and use the bladder as a standoff to view the reproductive tract. The liver is viewed moving the probe ventrally and tracing the liver silhouette. The heart is imaged from the lateral thoracic areas, realizing rib artifact. The cranial mediastinum can be imaged as well from the lateral body wall. This is also the site for ultrasound-guided needle biopsy in rabbits.

Ultrasonography is especially useful to assess liver lobe torsion. The lobe will be more echogenic than other liver lobes, and using color doppler, it will show no blood flow.

Other small animals

Hedgehogs

Sedation is almost always needed to be able to conduct imaging. All images come from under the mantle. In lateral recumbency, image positions are close to those described for a dog, although the mantle must be displaced so the probe can be on the abdominal or thoracic body wall.

Marsupials

Female marsupials such as sugar gliders, opossums and wallabies may have the caudal abdomen imaged from inside the pouch. The abdominal wall is easily accessible. The reproductive tract can be confusing unless the practitioner is familiar with the different anatomy; however, the uterus, vaginal canals, bladder, and ovaries can be imaged. Fluid within any of these or enlargements may signify pathology, although it must be assessed if the dam has just been bred or has passed a fetus.

Rodents

The probe size may image almost all internal structures from the ventrum of the abdomen or the lumbar fossa areas. Hamster cheek pouches can be checked by imaging over the pouch.

Conclusions

Principles of POCUS can be used in exotic pet practice, although positioning of the probe may be different depending on species. The practitioner needs to be familiar with normal anatomy, as well as be able to recognize abnormalities. The development of handheld or portable ultrasound probes makes it easy for all exotic practitioners to do ultrasonography. In addition, it has become economically feasible to own the equipment.

References

1. Nyland TG, Mattoon JS, Wisner ER. Physical principles, instrumentation, and safety of diagnostic ultrasound. In: Nyland TG, Mattoon JS, eds. *Veterinary Diagnostic Ultrasound.* Philadelphia, PA: WB Saunders; 1995:3-18.

2. Krautwald-Junghanns ME, Pees M, Reese S, Tully R. *Diagnostic Imaging of Exotic Pets.* Hannover, DE: Schlutersche Verlagsgesellschaft mbH & Co; 2011.

3. Boysen SR, Rozanski EA, Tidwell AS, et al. Evaluation of a focused assessment with sonography for trauma protocol to detect free abdominal fluid in dogs involved in motor vehicle accidents. *J Am Vet Med Assoc.* 2004;225(8):1198-1204.

4. FASTVET. Available at: http://www.FASTvet.com. Accessed April 2, 2022.

5. Lisciandro GR, Lisciandro SC. Global FAST for patient monitoring and staging in dogs and cats. *Vet Clin North Am Small Anim Pract.* 2021;51:1315-1333.

6. Lisciandro GR. AFAST target-organ approach and fluid scoring system in dogs and cats. *Vet Clin North Am Small Anim Pract.* 2021;51:1217-1231.

7. McMurray J, Boysen S, Chalhoub S. Focused assessment with sonography in nontraumatized dogs and cats in the emergency and critical care setting. *J Vet Emerg Crit Care.* 2016;26(1):64-73.

8. Lisciandro GR. Abdominal and thoracic focused assessment with sonography for trauma, triage and monitoring in small animals. *J Vet Emerg Crit Care.* 2011;21(2):104-122.

9. Boysen SR, Lisciandro GR. The use of ultrasound in the emergency room (AFAST and TFAST). *Vet Clin North Am Small Anim Pract.* 2013;43(4):773-797.

10. Lisciandro GR. *Point-of-Care Ultrasound Techniques for the Small Animal Practitioner.* 2nd Ed. Ames, IA: Wiley-Blackwell; 2021.

11. Johnson-Delaney CA. Ultrasonography in ferret practice. In: Lewington JH, ed. *Ferret Husbandry, Medicine and Surgery.* 2nd Ed. Sydney, ASTL: Elsevier Saunders; 2007:417-429.

Section 14

AEMV Plenary Session

Sara Gardhouse, DVM, Dipl ACZM, *AEMV President*
Session Chair

Effects of Capromorelin and Mirtazapine on Food Intake in Chinchillas

Christoph Mans, Dr med vet, Dipl ACZM, and
Emily Hamilton, BS

Session #3376

Recipient of the 2020 AEMV Research Grant

Affiliation: From the School of Veterinary Medicine, University of Wisconsin-Madison, 2015 Linden Drive, Madison, WI 53706, USA.

Note: Abstract format selected by author(s). See Front Matter for description of conference proceedings manuscripts submitted as scientific abstracts.

Abstract: Reduced food intake due to underlying disease processes or following anesthesia or sedation is a common problem in chinchillas and other small mammals. In dogs and cats, drug-induced appetite stimulation is routinely prescribed by veterinarians to promote an increase in voluntary food intake;[1-4] however, information on such drugs is limited in small companion mammals. In several randomized, blinded, complete crossover studies, the effects of capromorelin (Entyce™) and mirtazapine were evaluated for their appetite-stimulating effects in chinchillas. Capromorelin was evaluated at 3 and 10 mg/kg PO q24h in normorexic chinchillas and at 10 mg/kg PO q24h in chinchillas with experimentally induced post-sedation hyporexia. Mirtazapine was evaluated at 5 and 20 mg/kg PO q24h in normorexic chinchillas and at 20 mg/kg PO q24h in a study in chinchillas with experimentally induced post-sedation hyporexia. No adverse effects were noted with either drug. Capromorelin significantly increased food intake in normorexic chinchillas and significantly attenuated the sedation-induced reduction in food intake compared to the control. In contrast, mirtazapine did not have any significant or clinically relevant impact on food intake compared to the control. The findings of these studies suggest that capromorelin, but not mirtazapine is an effective appetite stimulant in healthy chinchillas. Administration of capromorelin should be considered for anorexic or hyporexic chinchillas to increase voluntary food intake, if dental disease has been ruled out.

References

1. Rhodes L, Zollers B, Wofford JA, et al. Capromorelin: A ghrelin receptor agonist and novel therapy for stimulation of appetite in dogs. *Vet Med Sci.* 2018;4:3-16.

2. Zollers B, Rhodes L, Heinen E. Capromorelin oral solution (ENTYCE®) increases food consumption and body weight when administered for 4 consecutive days to healthy adult beagle dogs in a randomized, masked, placebo controlled study. *BMC Vet Res.* 2017;13:1-5.

3. Zollers B, Huebner M, Armintrout G, et al. Evaluation of the safety in dogs of long-term, daily oral administration of capromorelin, a novel drug for stimulation of appetite. *J Vet Pharmacol Ther.* 2017;40:248-255.

4. Quimby JM, Gustafson DL, Samber BJ, et al. Studies on the pharmacokinetics and pharmacodynamics of mirtazapine in healthy young cats. *J Vet Pharmacol Ther.* 2011;34:388-396.

Investigation of the Genetic Basis of Cystine Urolithiasis in Ferrets (*Mustela putorius furo*)

Nicola Di Girolamo, DMV, PhD, DipECZM (Herpetology), Dipl ACZM, Dipl ACVPM,
Udaya DeSilva, BVSc, PhD,
Gill Blake, BS,
João Brandão, LMV, MS, DipECZM (Avian), Dipl ACZM,
Holly Carter, BVetMed,
Vondelle McLaughlin

Session #3485

Recipient of the 2019 AEMV Research Grant

Affiliation: From Exotic Pet Service, Department of Clinical Sciences, College of Veterinary Medicine, Cornell University, 930 Campus Road, 14853, Ithaca, NY 14853, USA (Di Girolamo), Department of Animal & Food Sciences, Ferguson College of Agriculture, Oklahoma State University, Stillwater, OK 74078, USA (DeSilva, Blake), Department of Veterinary Clinical Sciences, College of Veterinary Medicine, Oklahoma State University, Stillwater, OK 74078, USA (Brandao), Evergreen Avian and Exotic Animal Hospital, Kirkland, WA 98034, USA (Carter), Washington Ferret Rescue & Shelter, 6124 Evergreen Way, Everett, WA 98203, USA (McLaughlin).

Note: Abstract format selected by author(s). See Front Matter for description of conference proceedings manuscripts submitted as scientific abstracts.

Abstract: Cystine urolithiasis has risen from 15% to 93% prevalence in North American ferrets in just 10 years.[1] The purpose of this study was to evaluate the presence of a genetic cause of cystine urolithiasis in ferrets, a disorder that has been proven to be of genetic origin in humans, dogs and cats. The authors have investigated genes SLC3A1 and SLC7A9 that were implicated in cystinuria in other mammals. Both genes are fairly large in size. SLC3A1 consists of 10 exons spread over 32.5 mb in the ferret genome while SLC7A9 consists of 12 exons covering 27.4 mb in the genome. Oral swabs for DNA collection were obtained from 46 ferrets, including 30 Marshall ferrets, 14 that had urolithiasis, and 16 that did not have urolithiasis, and 16 non-Marshall ferrets. PCR primers were designed to amplify all exons from both genes and all exons from several ferrets were sequence analyzed. All sequences were compared against the ferret reference sequence at NCBI and they were compared against each other to identify possible mutations. Up to this point no mutations were detected among the samples that were tested; however, the limited amount of DNA from the affected ferrets currently sampled, somehow deters from fully confirming the absence of mutations.

References

1. Hanak EB, Di Girolamo N, DeSilva U, et al. Variation in mineral types of uroliths from ferrets (*Mustela putorius furo*) submitted for analysis in North America, Europe, or Asia over an 8-year period. *J Am Vet Med Assoc.* 2021;259(7):757-763.

Section 15

AEMV Therapeutics Session

Todd Driggers, DVM
Session Chair

Bee Venom Therapy to Treat a Spindle Cell Sarcoma in a Guinea Pig

Jörg Mayer, MSc, DABVP (Exotic Companion Mammal Practice), DipECZM (Small Mammal), Dipl ACZM, and
Laura Burns, DVM

Session #3181

Affiliation: From the Department of Small Animal Medicine and Surgery, College of Veterinary Medicine, University of Georgia, Athens, GA 30602, USA.

Note: Abstract format selected by author(s). See Front Matter for description of conference proceedings manuscripts submitted as scientific abstracts.

Abstract: A 5-year-old guinea pig was presented with a significant swelling in the neck area. The mass was surgically excised, and histology revealed a well-differentiated spindle cell sarcoma which was then incompletely excised. The mass rapidly regrew and was excised again a few weeks later. The patient was presented for an oncologic consultation to explore different treatment options. Medical oncology and radiation therapy options were deemed unsuitable due to the low likelihood of providing significant relief and the potential for significant side effects. The decision was made to use bee venom therapy since evidence exists that this therapy can be helpful when traditional oncology options have not been successful or have been exhausted.[1] In human medicine the activation of the stimulator of interferon genes (STING)-controlled innate immune pathway is a promising therapeutic strategy that induces tumor regression with long-term antitumor immunity.[2] Bee venom is a potential STING agonist which can be used in cases such as this one. A single bee was captured from the teaching apiary and placed in an empty glass tube and held to the site of the tumor until a sting was achieved. No side effects of the venom therapy were noted. The animal was presented 14 days later and no growth was observed. A second venom injection was performed at that time to initiate the STING response.

References

1. Yang W, Zhang Y, Yang G, et al. Anti-PD-1 immunotherapy and bee venom for relapsed and refractory liposarcoma: a case report. *Front Oncol.* 2021;11:668992. doi: 10.3389/fonc.2021.668992.

2. Luo K, Li N, Ye W, et al. Activation of stimulation of interferon genes (STING) signal and cancer immunotherapy. *Molecules.* 2020;27(14):4638:1-17.

Use of Photobiomodulation in Rabbit Abscesses

Molly Varga Smith, BVetMed, DZooMed, MRCVS

Session #3128

Affiliation: From Rutland House Veterinary Referrals, Rutland House Veterinary Hospital, Abbotsfield Road, St Helens, Merseyside, WA9 4HU, UK.

Abstract: Photobiomodulation is the use of various wavelengths of light to bring about effects at a cellular and molecular level when applied to the skin.[1] Lasers and both red and blue wavelength light emitting diodes can be used for this purpose.[2] The biologic effects of different wavelengths of light can vary. While rabbits have been used as experimental models for investigating the use of various types of photobiomodulation,[3] there are few reports of its use in clinical cases involving this species. This case report describes the author's experience using photobiomodulation in an extensive rabbit abscess.

Introduction

Treatment of abscesses in rabbits is notoriously difficult. Abscesses in this species present three issues that can limit the effectiveness of treatments that are effective in other species. Rabbit pus is thick and does not flow, meaning that lancing and flushing has a limited effect. Rabbit abscesses are usually surrounded by a thick fibrous capsule that limits access of antibiotics to the interior. Finally, many abscesses in rabbits are not diagnosed until they are extensive, meaning that removal in toto can be difficult or impossible. While surgical exploration and removal or marsupialisation remains the cornerstone of care, the issue of dealing with the infected tissue that is left still needs to be addressed. Topical therapies such as hydrocolloid wound gel or honey have been used with variable success; however, neither of these addresses the comfort of the wound bed. This case report describes the author's experience using photobiomodulation in an extensive rabbit abscess.

Signalment

The patient was an unvaccinated intact male rabbit aged 3 years and 10 months. The animal was kept indoors.

History

The rabbit was presented to a referral center with a complaint of unexplained weight loss, severe lethargy and gastrointestinal stasis. The owner noted weight loss and reduced appetite over the preceding 2-3 weeks. The owner also reported polydipsia and polyuria. There was no history of previous trauma, and the rabbit was kept alone, with no other pets in the household.

Presenting signs

On examination the rabbit was quiet, alert and responsive, with a mild to moderate degree of dehydration (7%). There were no gut sounds audible on any abdominal quadrant, and on palpation the abdomen was slightly tense, with areas of nodular swelling palpable under the skin of the ventral abdomen and thorax. The rabbit also was covered in urine around its perineum and ventrum. The skin was excoriated and foul-smelling.

Dermatologic signs

The rabbit had urine scalding on its perineum and ventrum. The underlying skin was wet and inflamed with areas of necrosis and evidence of purulent exudate underneath. On subsequent examination under sedation, once the fur had been clipped away, most of the ventrum was found to be affected by a chain of abscesses extending under the skin from near the thoracic inlet to the groin.

Further Investigations

A full hematologic and biochemical profile was run to evaluate the reported polydipsia, polyuria and weight loss. Results indicated that there was a degree of renal compromise, which had likely caused the polydispsia and polyuria and was sufficient to account for the weight loss. The abscess formation was believed to result from the wet, scalded skin, which had become necrotic.

Treatment Plan

Renal parameters were stabilized medically. The abscesses and skin damage were addressed surgically. Because of the extent of the abscess area, it was not possible to undertake full removal. Samples (including portions of the abscess wall) were taken for culture and sensitivity testing. The portions of the fibrous abscess capsule that could not be removed were marsupialized and topical treatment with dressings was initiated.

The use of a gel-based photosensitiser (Phovia™ Vetoquinol, France) photobiomodulation with blue-light emitting LED was considered as part of the treatment plan. This treatment was applied to the open areas of the abscess capsules and 2 × 2-minute LED activation cycles were undertaken on a twice weekly basis. In between treatments the areas were packed with medical grade honey and covered with non-adherent dressings kept in place with self-adhesive bandage. Pending the culture and sensitivity results, antibiosis was initiated with oral enrofloxacin. Directly post-surgery tramadol (15 mg/kg)[4] and meloxicam (0.6 mg/kg)[5] were used as analgesia, with the tramadol being stopped after 5 days.

Culture results demonstrated growth of coagulase negative *Staphylococcus* and alpha hemolytic *Streptococcus* under enriched conditions. Both isolates were sensitive to enrofloxacin, so this treatment was continued. Both bacteria could potentially be normal skin or respiratory flora.

Resolution

The abscess cavities progressed to full resolution within 4 weeks of treatment. This resolution is faster than would be expected if treatment were limited to standard options.

Discussion

Photobiomodulation is the use of various wavelengths of light to bring about effects at a cellular and molecular level when applied to the skin.[1,2] Lasers and both red and blue wavelength light emitting diodes can be used for this purpose. The biologic effects of different wavelengths of light can vary. While rabbits have been used as experimental models for investigating the use of various types of photobiomodulation,[3] there are few reports of its use in clinical cases involving this species.

The challenges surrounding treatment of rabbit abscesses include frequent late diagnosis resulting in extensive invasion of local tissues, compromised blood supply to the infected tissues, bacterial colonization of the abscess capsule, the presence of biofilms which protect the bacteria present and the tendency of rabbits to produce excessive amounts of purulent exudate that is tenacious and hard to clear. These factors have meant that clinically ongoing wound care can present a significant challenge, often necessitating repeated visits and sometimes even repeated anesthetic interventions.

Because of the almost immunologically privileged status of the inside of abscess capsule, the body's attempt to wall-off infection and a lack of blood supply to the capsule interior, antibiotics alone are ineffective as penetration to the relevant tissues is near impossible. This means that surgical treatment of abscesses is preferable. The ideal situation would be removal of the abscess within its capsule in its entirety. The extent of many abscesses at the time of diagnosis can prevent total excision, and removal of as much of the abscess as possible, marsupialisation of the remaining capsule to the skin, and subsequent topical treatment is a common outcome. The first issue this presents is the fact that the open area of tissue is painful and also open to further wound contamination. sThese issues can be mitigated by suturing a dressing over the open portion of the abscess for the first few days. The advantage of this is that it allows a bed of granulation tissue to form, and the dressing gives some protection to the wound. Another advantage is that an absorbent dressing will take care of any wound exudate. Once the dressing is removed, there is essentially an open, infected wound still present and regular topical wound care is necessary to achieve resolution. It depends on the individual rabbit and the owner whether this is successful or not. Typically, ongoing care of an extensive abscess can last for 6-8 weeks and sometimes longer, begging the question of whether it is fair to both the animals and the owners. Clearly a better option needs to be explored.

Photobiomodulation presents several advantages in this situation. The modality has shown to be effective in the treatment of deep pyoderma in dogs[6] so its use in open infected areas is proven. There are also published reports in rats and suggested uses in rabbits.[7,8] Photobiomodulation will help to eradicate microrganisms within a wound, without causing microbial resistance and without its efficacy being affected by the presence of biofilm.[2] Light can hasten wound healing by promoting fibrous tissue formation and epidermal and endothelial cell proliferation. Light (depending on the wavelength) can have effects at both cellular and molecular levels, altering protein synthesis, affecting cell migration and survival and providing anti-inflammatory action. Light has been shown to reduce inflammation and pain in skin wounds and confer more rapid restoration of function.[9]

Our experience using photobiomodulation (in this case a combination of Phovia gel™ and a blue LED lamp) in an extensive rabbit abscess wound post-surgically has been positive. The wounds themselves appear more comfortable immediately after the LED therapy, the treatment appears to be well tolerated with no observed side effects, and the amount of purulent discharge produced appeared reduced in comparison to other similar wounds. Over a period of time, wound contraction appeared more rapid in comparison and the time to full resolution was shorter than for other comparable wounds in the author's experience.

The limitations of photobiomodulation include a lack of standardized optical dosimetry parameters and a poor understanding of the details of its underlying mechanism. In this case we used consistent 2-minute cycles (applied twice at each visit) with the LED lamp approximately 5-7 centimeters away from the wound; however, this distance was not measured, and the angle of incidence could certainly have been variable. In subsequent cases we have been more consistent and have had ongoing positive results in a series of both rabbit as well as other exotic pet cases. While the evidence presented here is anecdotal, photobiomodulation did appear to shorten time to resolution and certainly improved perceived pain levels in this patient during the treatment period. In the longer term a prospective study with some standardized criteria would contribute to the evidence base for the use of Phovia gel™ in exotic patients.

Acknowledgments: The author thanks the owners of the rabbits discussed in this paper, as well as the clinical staff of Rutland House Veterinary Hospital Exotics Service. The author also thanks Dr. Sue Paterson for her support in using photobiomodulation.

Disclosures

The author has presented this paper previously at Vetoquinol Photobiomodulation Congress in 2022 and has been received an award for this work from the European College of Veterinary Dermatologists. She has also written papers and webinars for Vetoquinol and received payment for them.

References

1. Ailionie LM, Litscher G. Probiotics, photobiomodulation and disease management: controversies and challenges. *Int J Mol Sci.* 2021;22(9):4942 https://doi.org/10.3390/ijms22094942.

2. Heiskanen V, Hamblin M. Photobiomodulation: lasers vs light emitting diodes. *Photochem Photobiol Sci.* 2018;17(8):1003-1017: DOI:10.1039/c8pp90049c.

3. Li Y, Zhang J, Yanfeng L, et al. The histopathological investigation of red and blue light emitting diodes on treating skin wounds in Japanese big ear white rabbit. *PLoS ONE* 2016;11:e0157898.

4. Souza MJ, Greenacre CB, Cox SK. Pharmacodynamics of orally administered tramadol in domestic rabbits (*Oryctolagus cuniculus*). *Am J Vet Res.* 2008;69(8):979-982.

5. Fiorello CV, Divers SJ. Rabbits. In: Carpenter JW, Marion CJ eds. *Exotic Animal Formulary.* 4th Ed. St Louis, MO: Elsevier; 2012:517-559.

6. Marchegiani A, Spaterna A, Cerquetta M. Current applications and future perspectives of fluorescent light energy biomodulation in veterinary medicine. *Vet Sci.* 2021;25;8(2):20. doi:10.3390/vetsci8020020.

7. Sabater Gonzalez M, Mayer J. Technological advances in wound treatment of exotic pets. *Vet Clin North Am Exot Anim Pract.* 2019;22(3):451-470.

8. Gong J, Park H, Lee J, et al. Effect of photodynamic therapy on multispecies biofilms including *Streptococcus mutans*, *Lactobacillus casei* and *Candida albicans*. *Photobiomodul Photomed Laser Surg.* 2019;37(5):282-287. https://doi.org/10.1089/photob.2018.4571.

9. Yadav A, Gupta A. Non-invasive red and near-infra-red wavelength-induced photobiomodulation: promoting cutaneous wound healing. *Photoderm Photoimmunol Photomed.* 2016;33(1):4-13. https://doi.org/10.111/phpp.12282.

Methylprednisolone Epidural Treatment of Bladder Atony Secondary to Traumatic Spinal Injury

Nicolle Lofgren, VMD, DACEPM, and
Andrew Burton, DVM, Dipl ACVS (Small Animal)

Session #3396

Affiliation: From Mountain West Veterinary Specialists, 908 N Main St, Layton, UT 80401, USA.

Note: Abstract format selected by author(s). See Front Matter for description of conference proceedings manuscripts submitted as scientific abstracts.

Abstract: A 2-year-old male castrated ferret presented to an emergency clinic for inability to urinate for several days. On examination the ferret's bladder was firmly distended and the patient was lethargic. No arrhythmia was noted. Radiographs and ultrasound revealed no evidence of bladder stones, and normal sized adrenal glands and prostate. Bloodwork and urinalysis were normal. A computed tomography scan revealed right-sided disc extrusion at L 5-6 with probable associated cord and cauda equina compression and chronic malunion fractures of L 7 and sacrum with severe neural canal stenosis. The patient was started on pain control, a urethral dilator and antibiotics, and was sedated for a steroid epidural of methylprednisolone at 1 mg/kg for suspected trauma-secondary bladder atony. The patient's bladder could not be manually expressed and an indwelling urinary catheter was placed. After 1 week the patient's bladder could be manually expressed and the catheter was removed. A second steroid epidural was given 14 days after the first. The patient urinated unassisted 4 days after the second epidural administration. A final dose was given 18 days later. The patient has been able to urinate unassisted since receiving the second dose but is unable to fully express its bladder. The patient is maintained on daily manual bladder expressions and has urinalysis performed monthly. In dogs, a single epidural steroid injection with methylprednisolone improved clinical signs of lumbo-sacral stenosis in 84.4% of patients.[1] This is the first report of steroid epidurals to treatment of symptoms due to vertebral trauma in a ferret.

References

1. Gomes SA, Lowrie M, Targett M. Single dose epidural methylprednisolone as a treatment and predictor of outcome following subsequent decompressive surgery in degenerative lumbosacral stenosis with foraminal stenosis. *Vet J.* 2020;257:105451. doi.org/10.1016/j.tvjl.2020.105451.

Treatment of Type II Diabetes in a Sugar Glider

Nicolle Lofgren, VMD, DACEPM

Session #3381

Affiliation: From Mountain West Veterinary Specialists, 908 N Main St, Layton, UT 80401, USA.

Note: Abstract format selected by author(s). See Front Matter for description of conference proceedings manuscripts submitted as scientific abstracts.

Abstract: A 3.5-year-old intact male sugar glider presented to an emergency hospital obtunded and with a day history of lethargy and anorexia after ingesting steak. The patient weighed 80 grams on presentation and had polyuria and polydipsia (PU/PD). The diagnostic work-up, including blood chemistry, urinalysis and CT imaging showed elevated liver enzymes and renal markers (ALT 505U/L, AST 513U/L, ALP 724 U/L and a BUN of 136 mg/dl with a creatinine of < 0.1 mg/dl, ketonuria 1+, and hyperglycemia of 935 mg/dl.) Patient size precluded some testing including insulin levels. No gross abnormalities of the liver, gall bladder or GI were seen on a computed tomography scan. All values improved steadily after insulin administration. The patient was started on fluids, antibiotics, GI protectants and Novolin® N insulin 100 U/ml diluted 1:10 into sterile water at each administration. The glider was maintained on Glycobalance wet diet from Royal Canin®. The insulin dose was increased based on a glucose curve over several days until the glucose came down to 180 mg/dl at 1.25 U/kg subcutaneously (SC) in the morning and 2.5 U/kg SC in the evening. Despite resolution of clinical signs and normalized liver values, the patient has not been able to discontinue insulin and is suspected to have permanent pancreatic damage from suspected acute pancreatitis causing diabetes.[1] This report details the use of diluted insulin as a treatment for type II diabetes secondary to suspected triaditis in a sugar glider. As far as the author is aware, this is the first report of insulin use in a sugar glider.

References

1. Hart PA, Bradley D, Conwell DL, et al. Diabetes following acute pancreatitis. *Lancet Gastroenterol Hepatol.* 2021;6(8):668-675. doi: 10.1016/S2468-1253(21)00019-4. PMID: 34089654; PMCID: PMC8277724.

Effects of Metronidazole and Tinidazole on Food Intake in Chinchillas (*Chinchilla lanigera*)

Christoph Mans, Dr med vet, Dipl ACZM

Session #3456

Affiliation: From the School of Veterinary Medicine, University of Wisconsin-Madison, 2015 Linden Drive, Madison, WI 53706, USA.

Note: Abstract format selected by author(s). See Front Matter for description of conference proceedings manuscripts submitted as scientific abstracts.

Abstract: The use of compounded metronidazole suspensions in chinchillas is controversial due to the documented negative effects on food intake;[1-2] however, commercial metronidazole suspensions marketed for veterinary and human patients may differ in their impact on food intake.[3] In a randomized, controlled, complete-crossover study, two commercial metronidazole suspensions were administered at 20 mg/kg or 10 mg/kg q12h for 3 days, and food intake was recorded. At 20 mg/kg PO q12h, both commercial products induced a reduction in food intake within 24 hours after starting the treatment. At 10 mg/kg PO q12h, the reduction in food intake was only ~ 50% of that of the 20 mg/kg q12h treatments, suggesting that the negative effect of metronidazole on food intake in chinchillas is dose-dependent. Tinidazole has the same antimicrobial spectrum as metronidazole and a superior efficacy against *Giardia* spp.[4] Its effects on food intake were evaluated at 100 mg/kg and 200 mg/kg PO single dose and 20 mg/kg PO q12h for 5 days. A significant self-limiting reduction in food intake was recorded at 100 mg/kg and 200 mg/kg. In contrast, the repeated administration of tinidazole at 20 mg/kg q12h did not result in any clinically relevant or statistically significant reduction in food intake compared to the control treatment. In conclusion, metronidazole-induced hyporexia in chinchillas appears to be dose-dependent, and the 10 mg/kg PO q12h dose may result in fewer adverse effects. Tinidazole at 20 mg/kg PO q12h can be considered a safe alternative to metronidazole in chinchillas.

References

1. Thomas L, Doss G, Mans C. Metronidazole benzoate induced anorexia in two healthy chinchillas (*Chinchilla lanigera*). *J Exot Pet Med.* 2021;36:52.

2. Mans C, Fink D, Giammarco H, Ciarrocchi C. Effects of compounded metronidazole and metronidazole benzoate oral suspensions on food intake in healthy chinchillas (*Chinchilla lanigera*). *J Exot Pet Med.* 2021;36:75-79.

3. Mans C, Fink D. Effects of commercial metronidazole and metronidazole benzoate suspensions on food intake in chinchillas. *J Small Anim Pract.* 2021;62:174-177.

4. Tournade C, Fink D, Williams S, Mans C. Effects of orally administered tinidazole on food intake in chinchillas (*Chinchilla lanigera*). *J Am Assoc Lab Anim Sci.* 2021;60:587-591.

Section 16

AEMV Dentistry Session

Teresa Bousquet, DVM
Session Chair

Dental Examination Findings and Feasibility of Dental Radiography in Sedated Four-Toed Hedgehogs (*Atelerix albiventris*)

Grayson A. Doss, DVM, Dipl ACZM,
You Na Jeon, DVM,
Christopher Snyder, DVM, DAVDC

Session #3187

Affiliation: From the Department of Surgical Sciences, University of Wisconsin-Madison, School of Veterinary Medicine, Madison, WI 53706, USA.

Note: Abstract format selected by author(s). See Front Matter for description of conference proceedings manuscripts submitted as scientific abstracts.

Abstract: While oral and dental disorders are commonly reported in hedgehogs[1], normal oral exam findings for these species are poorly described. Additionally, while dental radiology is commonly used in other pet species, its use has not been detailed in hedgehogs. This study evaluated the feasibility of sedation for a complete oral examination and dental radiography in a group of 12 healthy, adult, captive-bred, four-toed hedgehogs (seven male, five female). Hedgehogs were sedated with a combination of alfaxalone (5 mg/kg) and midazolam (1 mg/kg) SC and an oral examination performed. The dental formula, documentation of calculus, missing teeth, fractures, mobility, and microdontia was recorded. Periodontal probing evaluating the sulcus depth for each tooth was performed and a gingivitis score assigned, if applicable. Dental radiography was performed using commercial phosphor dental imaging plates. Intraoral images of maxillary and mandibular arcades as well as skull images were collected. Sedation was sufficient for performing most of the oral examination, except inspection of the tonsils, soft palate, and pharynx, and permitted intraoral radiography in all hedgehogs. The dental formula for each hedgehog varied but most animals (7/12, 58%) had I 3/2, C 1/1, P 3/2, M 3/3. Most animals (9/12, 75%) had dental abnormalities, which were most noted in the maxillary arcade (9/12, 75%). The most noted oral abnormality was calculi, seen in 7/12 (58%) animals. Gingivitis was present in 5/12 (42%) hedgehogs and was most associated with the maxillary incisors. Calculus accumulation, missing teeth, and periodontal disease were common findings in this group of hedgehogs.

References

1. Doss GA, Carpenter JW. African pygmy hedgehogs. In: Quesenberry K, Mans C, Orcutt C, eds. *Ferrets, Rabbits and Rodents: Clinical Medicine and Surgery.* 4th Ed. St. Louis, MO: Elsevier Saunders; 2021: 401-415.

Surgical Approach for Treatment of Odontogenic Retromasseteric Abscesses in Fifteen Pet Rabbits

Vittorio Capello, DVM, DipECZM (Small Mammal),
DABVP (Exotic Companion Mammal Practice)

Session #3242

Affiliation: From Clinica Veterinaria S. Siro, via Lampugnano 99, 20151, Milano, Italy, Clinica Veterinaria Gran Sasso, via Donatello 26, 20131, Milano, Italy.

Note: Abstract format selected by author(s). See Front Matter for description of conference proceedings manuscripts submitted as scientific abstracts.

Abstract: Among the different kinds and topographic positions of odontogenic abscesses, the most challenging affecting the mandible is the retromasseteric.[1] Scarsely reported,[2-4] it originates from perforation of the thin bone of the masseteric fossa, following elongation and periapical infection of the reserve crown of the 4th (sometimes also the 5th) mandibular cheek tooth.[1] Computed tomography is necessary for accurate diagnosis, prognosis and a surgical plan.[1] The surgical treatment of a retromasseteric abscess, associated with underlying bone lysis of the masseteric fossa, was performed in 15 pet rabbits following exposure and blunt dissection of the zygomatic muscle and of the three layers of the masseter muscle; surgical debridement of the abscess; extraoral luxation and extraction of the diseased cheek tooth/teeth avoiding iatrogenic fracture of the mandible; marsupialization of the muscle layers and skin, and accurate local postoperative treatment, until healing by second intention. The postoperative long-term follow-up (1 year, or longer) was positive in all rabbits but one, which died 2 months after surgery for unrelated reasons. In the first case, the surgical reconstruction of the muscle layers was followed by relapse, requiring surgical review and marsupialization. In three cases, the postoperative period was complicated by an intraoral/extraoral fistula that required intraoral review and gingival suture in one case, while the fistula healed after 8 and 12 weeks in the other two cases. The surgical treatment of the retromasseteric abscess is challenging, implies a long and complex postoperative local treatment, but carries a good prognosis.

References

1. Capello V. Surgical treatment of facial abscesses and facial surgery in pet rabbits. *Vet Clin North Am Exot Anim Pract.* 2016;19:799-823.

2. Harcourt-Brown F, Chitty J. Facial abscesses. In: Harcourt-Brown F, Chitty J, eds. *Rabbit Surgery, Dentistry and Imaging.* Gloucester, UK: British Small Animal Veterinary Association; 2013:395-422.

3. Böhmer E. Abscesses. In: Böhmer E. *Dentistry in Rabbits and Rodents.* Hoboken, NJ: Wiley-Blackwell; 2015:213-241.

4. Varga M. Abscesses. In: Varga M. *Rabbit Medicine.* 2nd Ed. Oxford, UK: Butterworth- Heinemann; 2014:249-270.

Rostral Subtotal Mandibulectomy for Treatment of Extensive Osteomyelitis in Seven Pet Rabbits

Vittorio Capello, DVM, DipECZM (Small Mammal), DABVP (Exotic Companion Mammal Practice)

Session #3241

Affiliation: From Clinica Veterinaria S. Siro, vial Lampugnano 99, 20151, Milano, Italy, Clinica Veterinaria Gran Sasso, via Donatello 26, 20131, Milano, Italy.

Note: Abstract format selected by author(s). See Front Matter for description of conference proceedings manuscripts submitted as scientific abstracts.

Abstract: In cases of severe or extreme complications following advanced dental disease, pet rabbits may be diagnosed with extensive osteomyelitis of the mandible.[1-5] When the supporting bone is completely lytic, the diseased part of the mandible should be removed en-bloc, because thorough surgical debridement would not be sufficient to stop the infection and to elicit healing by second intention.[1] If the masseteric part of the affected mandible is not involved, unilateral rostral (partial) mandibulectomy[1,5,6] can be a feasible therapeutic option.[1] Computed tomography is necessary to make accurate diagnosis, prognosis and a surgical plan.[1] Mandibulectomy has been reported in rabbits for treatment of oral tumors as well, but the surgical technique and the follow-up are not reported in detail.[6,7] Unilateral rostral mandibulectomy was performed in seven pet rabbits following exposure of the osteomyelitic bone, separation of the mandibular symphysis, lateral buccotomy, transverse osteotomy of the body of the mandible cranial to the masseteric fossa; suture of the buccal mucosa, suture of the muscles overlying the defect, and suture of the skin. The postoperative follow-up was positive in three rabbits in the long-term (> 2 years); one was lost at follow-up after positive mid-term follow-up (3 months); two died after positive follow-up in the mid-term (3 months); one died intraoperatively. Unilateral rostral mandibulectomy is a challenging surgical procedure implying long and complex postoperative period and adaptation to dedicated nutrition,[1] but can be considered a salvage procedure versus euthanasia, for selected patients.

References

1. Capello V. Surgical treatment of facial abscesses and facial surgery in pet rabbits. *Vet Clin North Am Exot Anim Pract.* 2016;19:799-823.

2. Böhmer E. Abscesses. In: Böhmer E. *Dentistry in Rabbits and Rodents.* Hoboken, NJ: Wiley-Blackwell; 2015:213-241.

3. Böhmer E. Other changes of the jaw. In: Böhmer E. *Dentistry in Rabbits and Rodents.* Hoboken, NJ: Wiley-Blackwell; 2015:254-259.

4. Harcourt-Brown F, Chitty J. Facial abscesses. In: Harcourt-Brown F, Chitty J, eds. *Rabbit Surgery, Dentistry and Imaging.* Gloucester, UK: British Small Animal Veterinary Association; 2013:395-422.

5. Risi E, Sauvaget S, Boutoille F et al. Five successful cases of partial mandibulectomy and their medical followup in rabbits (*Oryctolagus cuniculus*) suffering from mandibular abscess or tumor. *Proc Annu Conf Assoc Exot Mamm Vet.* 2012;72.

6. Verstraete F. Mandibulectomy and maxillectomy. *Vet Clin North Am Small Anim Pract.* 2005;35:1009-1039.

7. Miwa Y, Nakata M, Takimoto H, et al. Spontaneous oral tumours in 18 rabbits (2005-2015). *J Small Anim Pract.* 2021;62:156-160.

Surgical Approach and Techniques for Treatment of Odontogenic Retrobulbar Abscesses in Nine Pet Rabbits

Vittorio Capello, DVM, DipECZM (Small Mammal), DABVP (Exotic Companion Mammal Practice)

Session #3247

Affiliation: From Clinica Veterinaria S. Siro, via Lampugnano 99, 20151, Milano, Italy, From Clinica Veterinaria Gran Sasso, via Donatello 26, 20131, Milano, Italy.

Note: Abstract format selected by author(s). See Front Matter for description of conference proceedings manuscripts submitted as scientific abstracts.

Abstract: Among the many different kinds and topographic positions of odontogenic abscesses in pet rabbits, the most challenging affecting the maxilla is the retrobulbar abscess. It originates from perforation of the alveolar bulla after elongation and periapical infection of the reserve crown of maxillary cheek teeth 3-6.[1-3] The most common presenting clinical sign is unilateral exophthalmos, associated with ocular disease or panophthalmitis.[1] Computed tomography and/or magnetic resonance imaging are necessary to make accurate diagnosis and a treatment plan.[1] Three different surgical approaches are possible for treatment of a retrobulbar abscess to save the eye globe.[1,4,5] Nine pet rabbits cases are presented which have been treated for a retrobulbar abscess and related exophthalmos. Four, with the lateral approach to the retrobulbar space dorsal to the zygomatic process; two, with the lateral approach to the alveolar bulla and the retrobulbar space after partial ostectomy of the zygomatic process; three, with an endoscopic-guided intraoral approach to the alveolar bulla and the retrobulbar abscess after extraction of selected maxillary cheek teeth. The approach depends on the individual case and findings from advanced diagnostic imaging. In seven cases the abscess was successfully treated saving the eye; in two cases the panophtalmitis did not resolve, and enucleation was necessary. Six rabbits among the seven had a positive follow-up longer than 1 year. Historically, the retrobulbar abscess was treated routinely with enucleation of the eye globe. The surgical treatment of the retrobulbar abscess is challenging but feasible, which allows saving the affected, exophthalmic eye in most cases.

References

1. Capello V. Surgical treatment of facial abscesses and facial surgery in pet rabbits. *Vet Clin North Am Exot Anim Pract.* 2016;3:799-823.

2. Harcourt-Brown F, Chitty J. Facial abscesses. In: Harcourt-Brown F, Chitty J, eds. *Rabbit Surgery, Dentistry and Imaging.* Gloucester, UK: British Small Animal Veterinary Association; 2013:395-422.

3. Böhmer E. Abscesses. In: Böhmer E. *Dentistry in Rabbits and Rodents.* Hoboken, NJ: Wiley-Blackwell; 2015:213-241.

4. Ward ML. Diagnosis and management of a retrobulbar abscess of periapical origin in a domestic rabbit. *Vet Clin North Am Exot Anim Pract.* 2006;3:657-665.

5. Martìnez-Jimenéz D, Hernàndez-Divers SJ, Dietrich UM, et al. Endosurgical treatment of a retrobulbar abscess in a rabbit. *J Am Vet Med Assoc.* 2007;230:868-872.

Section 17

AEMV Respiratory System Session

Dan Johnson, DVM, DABVP (Exotic Companion Mammal Practice)
Session Chair

Characterization of the Rabbit (*Oryctolagus*) Sinus Microbiome Using Next-Generation DNA Sequencing

Linda Siperstein, DVM,
Janina A. Krumbeck, PhD,
Elizabeth Marie Rush, BS, DVM, Dipl ACZM

Session #3346

Affiliation: From Dr. Sip Exotics Veterinary House Calls, 2625 Alcatraz Avenue Suite 325, Berkeley, CA 94705, USA (Siperstein), MiDOG LLC, 14762 Bentley Cir, Tustin, CA 92780, USA (Krumbeck), Antech Imaging Services, Staff Imaging Specialist, Zoo, Wildlife and Exotic Species, 315 Great View Circle, Hoover, AL 35226, USA (Rush).

Note: Abstract format selected by author(s). See Front Matter for description of conference proceedings manuscripts submitted as scientific abstracts.

Abstract: Upper respiratory infections in rabbits[1] commonly present special challenges for veterinary practitioners. Cultures may not grow relevant organisms, while next-generation DNA sequencing (NGS)[2] can reveal organisms not previously identified or of unknown significance. The goal of this study was to characterize the healthy rabbit sinus microbiome, while also considering the possibility of regional differences. Samples were collected following a standard protocol swabbing both nostrils of 25 clinically healthy rabbits on the West Coast, and 24 rabbits on the East Coast of the United States. Samples were analyzed to profile the bacterial and fungal microbiome using NGS. In total, 186 different bacterial species were detected, nine of these representing at least 1% of the microbiome. On average, 37.5 ± 18 different bacterial species were detected per individual. Specifically, the most abundant bacterial species were *Pseudomonadales* sp. (12.7% ± 17), *Moraxella cuniculi* (11.9% ± 18), and *Neisseriaceae* sp. (8.5% ± 9). Even though these were all considered healthy rabbits, *Mycoplasma* sp. was detected in 37 out of 49 samples with an average abundance of 4.7% ± 9. The mycobiome was more diverse than the bacteriome with 401 different species detected, and on average 48.3 ± 22 per individual. The most abundant fungus was *Cladosporium* sp. (38.1% ± 25), followed by *Alternaria* sp. (5.5% ± 8), and *Pleosporales* sp. (3.8% ± 10). Twenty-three fungi were more abundant in the West Coast rabbits, and only three species were more common in East Coast rabbits. This is the first study describing these differences in pet rabbits.

References

1. Jekl V. Respiratory disorders in rabbits. *Vet Clin North Am Exot Anim Pract.* 2021;24:459-482.

2. Tang S, Prem A, Tjokrosurjo J, et al. The canine skin and ear microbiome: a comprehensive survey of pathogens implicated in canine skin and ear infections using a novel next-generation-sequencing-based assay. *Vet Microbiol.* 2020;247:108764. doi: 10.1016/j.vetmic.2020.108764.

Diagnosis and Treatment of Chronic Bacterial, Fungal, and Algal Rhinitis in a Rabbit (*Oryctolagus cuniculus*)

Gina Vecere, DVM,
Shachar Malka, DVM, DABVP (Avian Practice),
Michael Garner, DVM, DACVP

Session #3392

Affiliation: From the Long Island Bird and Exotics Veterinary Clinic, 333 Great Neck Road, Great Neck, NY 11021, USA (Vecere, Malka), Northwest ZooPath, 654 West Main Street, Monroe, WA 98272, USA (Garner).

Note: Abstract format selected by author(s). See Front Matter for description of conference proceedings manuscripts submitted as scientific abstracts.

Abstract: A 6-year-old male castrated rabbit (*Oryctolagus cuniculus*) was presented for sneezing and purulent nasal discharge. Computed tomography and rhinoscopy were performed for diagnostic and therapeutic purposes. Severe destruction of the nasal turbinates was noted. Endoscopic debridement and sample collection was performed, and samples were collected for traditional aerobic, anaerobic, and fungal culture, next-generation DNA sequencing, cytologic, and histologic evaluation. Next-generation DNA sequencing identified *Pasteurella multocida*, *Bordetella bronchiseptica/parapertussis/pertussis*, and *Cladosporium* sp. Cytologic examination revealed heterophilic inflammation with rare fungal hyphae and extracellular bacteria. Histologically, severe suppurative rhinitis with intralesional bacteria and algae was noted. The samples were positive by PCR for *Aspergillus* spp. and *Prototheca* spp. The patient was treated with a combination of orbifloxacin, azithromycin, and itraconazole orally, as well as nasal flushes using dilute amphotericin B. Serial rhinoscopies were performed and the rabbit's clinical disease improved over 3 months; however, approximately 1 month after nasal amphotericin B application, the rabbit developed progressive azotemia of 3 months duration despite daily fluid therapy and was subsequently euthanized. There are no reports of rhinitis in rabbits associated with *Prototheca* spp. although the disease has been reported in goats, horses, and cats.[1-3] Amphotericin B is potentially nephrotoxic; in this report, azotemia was suspected to be associated with nasal application of amphotericin B.[4,5] This case highlights the presence of fungal and algal rhinitis in conjunction with severe chronic bacterial rhinitis. It also shows the treatment success and failure of nasal amphotericin B.

References

1. Schoniger S, Roschanski N, Rosler U, et al. *Prototheca* species and *Pithomyces chartarum* as causative agents of rhinitis and/or sinusitis in horses. *J Comp Pathol.* 2016;155(2-3):121-125.

2. Riet-Correa F, Silva do Carmo PM, Uzal FA. Protothecosis and chlorellosis in sheep and goats: a review. *J Vet Diagn Invest.* 2021;33(2):283-287.

3. Maboni G, Elbert JA, Stilwell JM, et al. Genomic and pathologic findings for *Prototheca cutis* infection in cat. *Emerg Infect Dis.* 2021;27(3):979-982.

4. Plumb DC. Amphotericin B. *Plumb's Veterinary Drugs.* Available at: https://app.plumbs.com/drug-monograph/IbDAylzFlePROD. Accessed February 22, 2023.

5. Hosny KM, Alhakamy NA. Nasal gel loaded with amphotericin nanotransferosomes as antifungal treatment for fungal sinusitis. *Pharmaceutics.* 2020;13(1):35.

Incidence of Nasal Septum Deviation and Comorbidities in Domestic Rabbits (*Oryctolagus cuniculus*)

Len Donato, VMD, DABVP (Avian Practice), and
Kaleigh Tomasson, VMD

Session #3382

Affiliation: From Radnor Veterinary Hospital, 107 North Aberdeen Avenue, Wayne, PA 19087, USA (Donato, Tomasson), Department of Clinical Sciences and Advanced Medicine, The University of Pennsylvania School of Veterinary Medicine, 3900 Spruce Street, Philadelphia, PA 19104, USA (Donato).

Note: Abstract format selected by author(s). See Front Matter for description of conference proceedings manuscripts submitted as scientific abstracts.

Abstract: Fifty-five rabbits of various signalments underwent computed tomography (CT) of the skull, from February 2019 to January 2023, to diagnose the underlying cause of their clinical signs. Twenty- three rabbits presented with rhinitis characterized by sneezing, nasal discharge or increased upper respiratory noise. Twenty rabbits presented with malocclusion or any palpable or visual tooth root abscess. Thirteen rabbits presented with a head tilt, eight rabbits presented with exophthalmos, and six rabbits presented with other clinical signs. Several rabbits had multiple clinical signs. Each rabbit was placed under general anesthesia and positioned in sternal recumbency in the CT unit (Vimago HU, Epica Animal Health, Duncan, SC, USA). Nasal septum deviation, rhinitis, dental disease, otitis media/interna, sinusitis, and retrobulbar abscess were the abnormalities most observed. These abnormalities have been reported in the literature.[1-8] Thirty-five (64%) rabbits had a deviated nasal septum, 31 (56%) rabbits had rhinitis, 27 (49%) rabbits had dental disease, 20 (36%) rabbits had otitis media/interna, 12 (22%) rabbits had sinusitis, 8 (15%) rabbits had a retrobulbar abscess and 15% of rabbits had a specific pathology to that patient that did not fall into any of these categories. Multiple rabbits had a few comorbidities therefore numbers exceed total patients. The rabbits that had nasal septum deviation were then grouped based on the comorbidities of rhinitis, dental disease, otitis media/interna, and sinusitis. Of the 35 rabbits that had nasal septum deviation, 17 (49%) had rhinitis, 14 (40%) had dental disease, 13 (37%) had otitis media/ interna, and 4 (11%) had sinusitis. It is important to note that multiple rabbits had multiple comorbidities, so the numbers exceed the total patients with nasal septum deviation.

References

1. Lennox A. Rhinotomy and rhinostomy for surgical treatment of chronic rhinitis in two rabbits. *J Exot Pet Med.* 2013;22(4):383-392.

2. Lennox A, Mancinelli E. Respiratory disease. In: Quesenberry K, Orcutt C, Mans C, Carpenter JW, eds. *Ferrets, Rabbits, and Rodents: Clinical Medicine and Surgery.* 4th Ed. St. Louis, MO: Elsevier; 2020: 188-200.

3. Salihoglu M, Cekin E, Altundag A, et al. Examination versus subjective nasal obstruction in the evaluation of the nasal septal deviation. *Rhinology.* 2014;52(2):122-126.

4. Teixeira J, Certal V, Chang E, et al. Nasal septal deviations: a systematic review of classification systems. *Plast Surg Int.* 2016;1:7089123.

5. Chavadaki J, Raghu K, Vi Patel. A retrospective study of establishment of association between deviated nasal septum, sinusitis, and chronic dacryocystitis. *Indian J Otolaryngol Head Neck Surg.* 2020;72(1):70-73.

6. Ji X, Fu H, Song A. Study on the correlation between chronic sinusitis with nasal septum deviation. *J Otolaryngol.* 2015;29(12):1103-1104.

7. Lifeng L, Zang H, Han, D, et al. Impact of varying types of nasal septal deviation on nasal airflow pattern and warming function: a computational fluid dynamics analysis. *Ear Nose Throat J.* 2019;100(6):283-289.

8. Van Egmond MM, Rovers, MM, Tillema, AH, et al. Septoplasty for nasal obstruction due to a deviated nasal septum in adults: a systematic review. *Rhinology.* 2018;56(3);195-208.

Outcomes for Rabbit Rhinostomy and Rhinotomy for Severe Rhinitis Based on Anatomic Location of Lesions

Angela M. Lennox, DVM, DABVP (Avian Practice), DABVP (Exotic Companion Mammal Practice), DipECZM (Small Mammal)

Session #3193

Affiliation: From Avian and Exotic Animal Clinic, 9330 Waldemar Road, Indianapolis, IN 46268 USA.

Note: Abstract format selected by author(s). See Front Matter for description of conference proceedings manuscripts submitted as scientific abstracts.

Abstract: Rhinotomy and rhinostomy are indicated for cases of severe rhinitis not responsive to aggressive medical management, which includes antimicrobials delivered parenterally, and by nebulization based on culture and sensitivity results.[1-4] The severity of disease is variable, from varying degrees of filling of the nasal cavity with fluid and presence of caseous abscess material to osteomyelitis affecting turbinates and supporting bone. Cases collected thus far have been for rabbits with severe rhinitis. Data from a limited number of cases suggests location of disease within the nasal cavity is associated with overall prognosis, with infection of the caudal nasal cavity, in particular the nasopharynx associated with guarded to poor prognosis.

References

1. Berglof A, Norlander T, Feinstein R, et al. Association of bronchopneumonia with sinusitis due to Bordetella bronchiseptica in an experimental rabbit model. *Am J Rhinol.* 2000;14(2):125-130.

2. Rougier S, Galland D, Boucher S, et al. Epidemiology and susceptibility of pathogenic bacteria responsible for upper respiratory tract infections in pet rabbits. *Vet Microbiol.* 2006;115(103):192-198.

3. Lennox AM. Rhinotomy and rhinostomy for surgical treatment of chronic rhinitis in two rabbits. *J Exot Pet Med.* 2013;22(4):383-392.

4. Wright L, Mans C. Lateral rhinostomy for treatment of severe chronic rhinosinusitis in two rabbits. *J Am Vet Med Assoc.* 2018;252(1):103-107.

Respiratory Distress Caused by Neoplastic Upper Respiratory Tract Obstructions in Two Pet Rats (*Rattus norvegicus domestica*)

Rachael Figueroa, DVM,
Miranda J. Sadar, DVM, Dipl ACZM,
Chad Frank, DVM, MS, DACVP,
Megan A. McCarthy, DVM, DACVP,
David Sanchez-Migallon Guzman, LV, MS, DipECZM (Avian),
DipECZM (Small Mammal), Dipl ACZM

Session #3360

Affiliation: From Colorado State University, 300 West Drake Road, Fort Collins, CO 80523, USA (Figueroa, Sadar, Frank), ·IDEXX Laboratories, Inc, 1 IDEXX Drive, Westbrook, ME 04092, USA (McCarthy), University of California, Davis, 1 Garrod Drive, Davis, CA 95616, USA (Guzman).

Note: Abstract format selected by author(s). See Front Matter for description of conference proceedings manuscripts submitted as scientific abstracts.

Abstract: Respiratory disease is one of the primary reasons for pet owners to seek veterinary attention for their rats. While chronic respiratory disease is most often responsible for lower respiratory signs in pet rats, consideration should be given to other causes including upper respiratory obstructive disease, which can be the result of neoplasia, trauma, degeneration, infection, or inflammation. This case series describes two pet fancy rats (*Rattus norvegicus domestica*) presenting with a multi-day history of progressive respiratory signs refractory to empiric medical management. Complete blood cell count and plasma biochemistry panels were performed in both cases and were unremarkable. Survey radiographs were performed under sedation in both cases. In the first case, no cause for the clinical signs could be identified, in part due to suboptimal radiographic positioning. In the second case, cervical tracheal luminal narrowing and increased soft tissue opacity along the walls of the trachea were identified. On necropsy, the first case had a squamous cell carcinoma originating from the Zymbal's gland causing obstruction of the larynx. The second case had an intra-luminal tracheal mass obstructing the airway, most consistent with a soft tissue sarcoma on histopathology. Empiric management of respiratory signs with antimicrobials, anti-inflammatories, and supportive care may be indicated given the high prevalence of infectious respiratory disease in the pet rat population. Diagnostic testing, including radiographs, should be considered due to the potential for other respiratory diseases, such as obstructive etiologies including neoplasia.

Fluoroscopic Balloon Dilation of Nasopharyngeal Stenosis Associated with Epiglottic Chronic Dislocation in a Pet Rabbit

Imal Khelik, DVM, and
Nicola Di Girolamo, DVM, PhD, DipECZM (Herpetology), Dipl ACZM, Dipl ACVPM

Session #3309

Affiliation: From Zoological Medicine Service, College of Veterinary Medicine, Oklahoma State University, 208 S McFarland St, Stillwater, OK 74078, USA (Khelik, Di Girolamo), Aurora Animal Hospital 20250 E Smoky Hill Rd, Centennial, CO 80015, USA (Khelik), Exotic Pet Service, Department of Clinical Sciences, College of Veterinary Medicine, Cornell University, 930 Campus Road, Ithaca, NY 14853, USA (Di Girolamo).

Note: Abstract format selected by author(s). See Front Matter for description of conference proceedings manuscripts submitted as scientific abstracts.

Abstract: A 4-year-old, male neutered, rabbit (*Oryctolagus* sp.) was presented for evaluation of a 3-month history of sneezing with nasal discharge. Mild upper respiratory stertor was appreciated with nasal flaring. A computed tomography scan showed a stenosis of the caudal nasopharynx and the dislocation of the glottis from its normal position over the soft palate. Airway fluoroscopy was performed with the patient in a large, clear, anesthesia chamber that restricted patient movements and revealed that the patient was open mouth breathing and the air was flowing from the mouth directly through the trachea, thanks to the dislocation of the epiglottis. The rabbit was scheduled for a balloon dilation of the nasopharyngeal stenosis. Once anesthetized, endoscopy with a rigid otoscope confirmed again that the glottis was abnormally open. The rabbit was then intubated and placed in right lateral recumbency for balloon dilation under fluoroscopic guidance.[1] A guide wire was introduced into the left naris and fed into the gastric fundus. Then, a balloon catheter was passed over the wire and centered in the caudal nasopharynx extending into the pharynx. The balloon was dilated several times. Following removal of the balloon and wire, fluoroscopy revealed no stenosis of the caudal nasopharynx as previously described. Balloon dilations were repeated for two more times, after 3 and 5 weeks, as advised by the internal medicine specialist. The rabbit respiratory effort had improved after each procedure. On an email recheck 18 months from the first examination the owner confirmed the rabbit has been healthy and not showing any respiratory signs.

References

1. Pollack SZ, Chapman PS, Klag A. Balloon dilation for the treatment of nasopharyngeal stenosis in seven cats. *JFMS Open Rep.* 2017;3(2):2055116917729987.

Clinical and Computed Tomographic Findings of Rabbits with Pulmonary Emphysema

Athinodoros Athinodorou, DVM, MVetSci (ConsMed), PGCertVetEd, CertAVP (ZooMed), PGCertACP, FHEA, MRCVS,
Jenna Richardson, BVM&S, DipECZM (Small Mammal), FHEA, MRCVS,
Tobias Schwarz, MA, Dr med vet, FRSB, DVR, Dipl ECVDI, Dipl ACVR, FRCVS,
Nicolas Israeliantz, LdoVet, MRCVS

Session #3219

Affiliation: From Hospital for Small Animals, Royal (Dick) School of Veterinary Studies, Rabbit & Exotic Animal Practice and Diagnostic Imaging Service, The University of Edinburgh, Scotland, UK.

Note: Abstract format selected by author(s). See Front Matter for description of conference proceedings manuscripts submitted as scientific abstracts.

Abstract: Pulmonary emphysema (PE) is a rarely reported and poorly understood condition in pet rabbits.[1,2] This retrospective study investigates clinical and computed tomographic (CT) findings of PE in domestic rabbits. Seven-hundred twenty-six archived thoracic CT studies of 529 rabbits were reviewed with their medical records for clinical data. Studies were scored based on the classification and lobar location of PE. Seventy-four CT studies of 58 rabbits (10.7% of the population) were positive for PE, which commonly affected the left cranial (52/74) and right cranial (38/74) lung lobes. Identified pulmonary changes included diffuse (12/75), bullous (2/75) and peripheral (61/75) emphysema. Of the rabbits that underwent follow-up CT evaluation, PE remained stable in 7/13 and progressed in 6/13. No cases demonstrated improvement. Patient presentation reasons included upper (19.3%) and lower (15.6%) respiratory tract disease. Naso-ocular discharge (14 %), and dyspnea (26.3%) were the most common clinical signs. Respiratory distress was manifested with indications of cyanosis (seven rabbits, 12.3%) and open-mouth breathing (two rabbits, 3.5%) which were considered poor prognostic factors. Twenty-six rabbits (44.8%) were asymptomatic. Euthanasia resulting from respiratory disease was the outcome in 10/58 rabbits (17.3%). Study findings conclude that pulmonary emphysema is a relatively common condition in rabbits, readily detected with advanced diagnostic imaging. The most common imaging finding is hypoattenuation of the pulmonary parenchyma, which can be focal, multifocal, or diffuse. While frequently asymptomatic, clinical signs can include naso-ocular discharge, and dyspnea. Pulmonary emphysema should be considered as a differential diagnosis of respiratory disease in rabbits.

References

1. Cooper TK, Griffith JW, Chroneos ZC, et al. Spontaneous lung lesions in aging laboratory rabbits (*Oryctolagus cuniculus*). *Vet Pathol.* 2017;54(1):178-187.

2. Guillerit F, Gros L, Touzet C, et al. Computed tomographic findings in 5 rabbits with pulmonary emphysematous bullae. *Proc ICARE.* 2022;160.

Section 18

AEMV Cardiology Session

Charly Pignon, DVM, DECZM (Small Mammal)
Session Chair

Point-of-Care Cardiac Troponin I (cTnI) Assay Evaluation in New Zealand White Rabbits (*Oryctolagus cuniculus*)

Sarah Ozawa, DVM, Dipl ACZM,
Kursten Pierce, DVM, DACVIM (Cardiology),
Andrea Thomson, RVT, BA,
Maria Boggess, BS,
Nina Moiseiwitsch, BS,
Ashley Brown, PhD,
Hiroyuki Mochizuki, BVMS, PhD, DACVP

Session #3301

Affiliation: From the Department of Clinical Sciences, College of Veterinary Medicine, North Carolina State University (Ozawa, Pierce, Thomson), Department of Population Health and Pathobiology, College of Veterinary Medicine, North Carolina State University (Boggess, Mochizuki), Joint Department of Biomedical Engineering, UNC-Chapel Hill and North Carolina State University (Moiseiwitsch, Brown).

Note: Abstract format selected by author(s). See Front Matter for description of conference proceedings manuscripts submitted as scientific abstracts.

Abstract: Cardiac troponin I (cTnI) is a cardiac-specific biomarker that is released into peripheral circulation secondary to cardiomyocyte injury.[1] It is an important biomarker in humans for disease states such as myocardial infarction. Rabbits have been shown to develop increased cTnI concentrations following administration of cardiotoxins, perfusion injuries, sepsis and myocardial insult.[2-4] While rabbit-specific ELISAs have been developed to evaluate for cTnI, they are not commercially available. Despite the high homology of cTnI between mammalian species, different commercially available assays have different detection capabilities. The objective of this study is to validate a point-of-care cTnI assay (i-STAT cTnI, Abbot Point of Care Inc, Union City, CA) in healthy adult rabbits and to establish a reference interval for cTnI in this breed. Blood was collected from 40 rabbits using manual restraint and placed in lithium heparin tubes for analysis on the point-of-care machine. Cardiac tissue and skeletal muscle were harvested from a single rabbit and resuspended in phosphate-buffered saline and whole rabbit blood to ensure the assay detected cardiac tissue but not skeletal muscle. Pilot results from 19 rabbits revealed a median cTnI of 0.01 ng/ml (range 0-0.04 ng/ml). Validation of this assay has implications for laboratory animal medicine and for the detection of myocardial injury in pet rabbits.

References

1. Langhorn R, Willesen JL. Cardiac troponins in dogs and cats. *J Vet Int Med.* 2016;30(1):36-50.
2. Adamcova M, Popelova-Lencova O, Jirkovsky E, et al. Cardiac troponins—translational biomarkers in cardiology: theory and practice of cardiac troponin high-sensitivity assays. *Biofactors.* 2016;42(2):133-148.
3. Tümer K, Özdemİr H, Eröksüz H. Evaluation of cardiac troponin I in serum and myocardium of rabbits with experimentally induced polymicrobial sepsis. *Exp Anim.* 2020;69(1):54-61.
4. Shaty MH, Arif IS, Al-Ezzi MI, Hanna DB. Metformin attenuate fibrosis in both acute and chronic doxorubicin cardiotoxicity in rabbits. *J Pharm Sci Res.* 2018;10(6):1559-1565.

Epicardial Pacemaker Implantation and Outcome in a Pet Ferret (*Mustela putorius furo*)

Katrina Ramsell, PhD, DVM

Session #3506

Affiliation: From NW Exotic Pet Vet, 2425 Southwest Cedar Hills Blvd, Beaverton, OR 97005, USA.

Note: Abstract format selected by author(s). See Front Matter for description of conference proceedings manuscripts submitted as scientific abstracts.

Abstract: A 5-year-old, spayed female pet ferret (*Mustela putorius furo*) was found in lateral recumbency with increased respiratory effort, severe bradycardia, dehydrated, and hypothermic. The animal was administered antibiotics and furosemide for suspected congestive heart failure. The ferret had a similar episode approximately 2 weeks later also in addition to hypoglycemia. It was diagnosed with third-degree atrioventricular block (AVB) and suspected insulinoma. The ferret was referred to the Oregon State University Veterinary Teaching Hospital 4 days later for placement of an epicardial pacemaker and insulinoma surgery. A pre-operative echocardiogram revealed moderate left atrial dilation, mild left ventricular eccentric hypertrophy and mild aortic insufficiency. The ferret had normal systolic function and bradycardia, with a heart rate of 60 beats per minute (bpm). Pacemaker implantation was recommended and performed via a transdiaphragmatic approach. The heart was temporarily paced intraoperatively at 80 bpm by contacting the epicardium with an endocardial pacing lead through the pericardiotomy. A permanent epicardial pacemaker (Medtronic pulse generator; Adapa ADSR01) was implanted then the temporary pacing lead was removed. Three pancreatic nodules were excised immediately following pacemaker implantation. The ferret recovered from surgery uneventfully and a post-operative ECG confirmed a ventricular paced rhythm at 140 bpm. At a 3-week post-operative recheck examination, the ferret was bright and alert, and had been clinically doing well since surgery. An ECG revealed third-degree AVB with a paced rhythm at 140 bpm. Pacemaker settings were adjusted during the device interrogation and battery life was estimated at 5.5-8 years. Approximately 6 months after pacemaker implantation, the ferret was exhibiting clinical signs attributed to hypoglycemia and was started on diazoxide. Prednisolone was added to help maintain normoglycemia approximately 1 year later. The ferret was rechecked 2.5 years after its pacemaker surgery. The animal was bright and alert and had been doing well clinically. Repeat ECG and pacemaker interrogation indicated correct pacemaker function and an estimated 2.5-4 years of battery life left. The ferret developed severe periodontal disease and swelling of the face and head. A severe infection of *Bacteroides pyogenes* and *Pasteurella canis* developed and was likely the cause of death. The ferret's heart rate was 140 bpm and the pacemaker appeared to be working well right before death. Necropsy results revealed severe necrosuppurative cellulitis, lymphoma and an islet cell carcinoma. Histopathology results showed the ferret had mild only myocardial fibrosis and left- or right-sided cardiac insufficiency was not present. The author has observed third degree AVB in many ferrets over the last 20 years and medical intervention often has resulted in minimal to no improvement. Epicardial pacemakers have been placed in a small number of ferrets with varying degrees of success.[1-2] The ferret in this case report did not have any complications associated with the epicardial pacemaker implantation. The pacemaker likely prolonged survival considering the ferret's comorbidities. Epicardial pacemakers should be considered a viable option for treating ferrets with high grade AVB.

Acknowledgments: The author thanks Dr. Alex Alvarez, Ldo Vet, MS, DACVS (Small Animal), Dr. Katy Townsend, BVSc (Hons), MS, DACVS (Small Animal), and the Small Animal Surgery, Cardiology, and Anesthesiology services at the Carlson College of Veterinary Medicine, Oregon State University.

References

1. Alvarez-Sanchez A, Townsend KL, LeBlanc NL, et al. Epicardial pacemaker implantation and outcome in eight pet ferrets (*Mustela putorius furo*). *J Vet Cardiol.* 2022;41:44-56.

2. Sanchez-Migallon Guzman D, Mayer J, Melidone R, et al. Pacemaker implantation in a ferret (Mustela putorius furo) with third-degree atrioventricular block. *Vet Clin North Am Exot Anim Pract.* 2006;9: 677-687.

Section 19

AEMV Epidemiology Session

Claire Vergneau-Grosset, DVM, IPSAV, CES, Dipl ACZM;
Marian Allison, DVM
Session Chairs

Assessment of Novel Detection Methods for Antibody to *Encephalitozoon cuniculi* in Rabbits

Carolyn Cray, PhD,
Adriana Callico, MS,
Ann Gottenger, BS

Session #3225

Affiliation: From Division of Comparative Pathology, Department of Pathology & Laboratory Medicine, University of Miami Miller School of Medicine, P.O. Box 016960 (R-46), Miami, FL 33101, USA (Cray, Callico, Gottenger), The Ohio State University, 829 Bethel Road, Columbus, OH 43214, USA (Gottenger).

Note: Abstract format selected by author(s). See Front Matter for description of conference proceedings manuscripts submitted as scientific abstracts.

Abstract: Current serologic assays for detection of antibody to *Encephalitozoon cuniculi* in rabbits in the United States involve tissue culture grown reagents, which include a multitude of different antigens.[1,2] In the present study, an ELISA has been implemented using a recombinant protein produced for a polar tube protein (PTP2) which is part of the specific organelle that enables host cell invasion.[3] Initial validation work has been undertaken to determine significance and applicability of IgM and IgG titers using samples from 108 patients submitted to the clinical laboratory. In a group of rabbits (n = 25) that were strongly presumptive for *E. cuniculi* infection and demonstrated negative results on the traditional ELISA, the PTP2 ELISA detected IgM antibody in 10/25 and IgG antibody in 1/25 patients. Of the rabbits categorized as possibly infected with *E. cuniculi* by submitting clinicians that were seronegative by traditional ELISA methods, the PTP2 ELISA detected IgM antibody in 12/44 and IgG antibody in 3/25 animals. In healthy rabbits with no clinical signs, 3/39 and 2/39 samples demonstrated PTP2 IgM and IgG reactivity, respectively. Overall, the data suggest the PTP2 ELISA may be more reliable in the detection of infection in suspect and presumed infected rabbits versus the traditional ELISA. Additional preliminary data supports that the PTP2 antibody levels persist with mild decreases observed in some animals with resolution or partial resolution of clinical signs. More studies are underway to understand the final applications of this ELISA in the diagnosis and prognosis of *E. cuniculi* infection.

Acknowledgments: Thank you to the clients who shared clinical information on the patients included in this study.

References

1. Cray C, McKenny S, Perritt E, et al. Utility of IgM titers with IgG and C-reactive protein quantitation in the diagnosis of suspected *Encephalitozoon cuniculi* infection in rabbits. *J Exot Pet Med.* 2015;24(3):356-360.

2. Desoubeaux G, Piqueras MDC, Pantin A, et al. Application of mass spectrometry to elucidate the pathophysiology of *Encephalitozoon cuniculi* infection in rabbits. *PLoS ONE.* 2017;12(7):p.e0177961.

3. Xu Y, Weiss LM. The microsporidian polar tube: a highly specialised invasion organelle. *Int J Parasitol.* 2005;35(9):941-953.

Review of an International Survey on *Encephalitozoon cuniculi* Infection in Rabbits

Carolyn Cray, PhD,
Fabiano Montiani-Ferreira, BMV, MCV, PhD,
Ann Gottenger, BS,
Franz Riegler Mello, DVM,
Jennifer Graham, DVM, DABVP (Avian Practice), DABVP (Exotic Companion Mammal Practice), Dipl ACZM,
Jean-Francois Quinton, DVM, DipECZM, EBVS,
Anja Joachim, PhD, DipEVPC, EBVS,
Emma Keeble, BVSc, DipZooMed (Mammalian), MRCVS,
Jaume Martorell Monserrat

Session #3229

Affiliation: From Division of Comparative Pathology, Department of Pathology & Laboratory Medicine, University of Miami Miller School of Medicine, P.O. Box 016960 (R-46), Miami, FL 33101, USA (Cray, Montiani-Ferreira, Gottenger, Graham), Departamento de Medicina Veterinária, Universidade Federal do Paraná, Rua dos Funcionários, 1540, 80035-050 Curitiba-PR, Brazil (Montiani-Ferreira, Riegler Mello), The Ohio State University, 829 Bethel Road, Columbus, OH 43214, USA (Gottenger), Graham Veterinary Consulting, Madison, AL 35756, USA (Graham), Centre Hospitalier Veterinaire ADVETIA, Velizy-Villacoublay, France (Quinton), University of Veterinary Medicine, Vienna, Austria (Joachim), University of Edinburgh, The Royal (Dick) School of Veterinary Studies, Midlothian, EH25 9RG, Scotland (Keeble), Universtat Autonoma de Barcelona, Department of Animal Medicine and Surgery, 08193 Bellaterra, Spain (Martorell Monserrat).

Note: Abstract format selected by author(s). See Front Matter for description of conference proceedings manuscripts submitted as scientific abstracts.

Abstract: In late 2022, a worldwide online survey was undertaken to investigate different experiences that may aid to better understand *E. cuniculi* (ECUN) infection in rabbits and build a consensus statement. Over 3.5 months, 339 individuals responded, including small animal clinicians (n = 71), exotic animal clinicians (n = 245), veterinary technicians (n = 19), biologists (n = 1) and others associated with animal welfare and rescue (n = 3). Comparisons of responses were made using Chi-Square tests. Respondents represented the United States (n = 163) and outside the United States (n = 176) with the latter mostly from Europe and the United Kingdom (n = 115). While non-U.S. respondents treated more rabbits, U.S. respondents reported a higher incidence of suspect ECUN cases. For diagnosis, physical exam, serology, and imaging were the most common tools. Lop, dwarf, and crossbred rabbits were the most common breeds examined. Most clinicians believe there is no breed or age predilection for ECUN disease and neurologic signs are the most common presentation. Regardless of location, ECUN and/or otitis were considered the top three diagnoses for head tilt. Other causes included tumors and, in the U.S., a significantly higher number of cases of meningitis were considered (p = 0.028). Fenbendazole, fluids, and meloxicam were the top three treatments used regardless of location, although antibiotics, meclizine and oxibendazole were significantly more used inside the United States (p = 0.003, 0.0001, 0.004, respectively). Albendazole, steroids, and analgesics were significantly more common outside the United States (p = 0.003, 0.001, 0.007, respectively). Bone marrow complications from the use of fenbendazole were reported as rare. Hematologic monitoring was performed more frequently in the United States. Rare associations were seen between vaccination and ECUN infection.

Acknowledgments: Thank you to the participants and to the Association of Exotic Mammal Veterinarians for their help in promoting this survey.

Epidemiology of Odontogenic Abscesses in Client-Owned Rabbits: One Hundred Sixteen Cases (2011-2022)

Ivana Levy, DVM, and
Christoph Mans, Dr med vet, Dipl ACZM

Session #3358

Affiliation: From the University of Wisconsin-Madison, School of Veterinary Medicine, 2015 Linden Drive, Madison, WI 53706, USA.

Note: Abstract format selected by author(s). See Front Matter for description of conference proceedings manuscripts submitted as scientific abstracts.

Abstract: The medical record database of a veterinary teaching hospital was searched from 2011-2022 for records of rabbits diagnosed with odontogenic abscesses by computed tomography. One hundred and sixteen client-owned rabbits with odontogenic abscesses were included. Retrobulbar abscesses were excluded from this study and evaluated separately. Relevant findings included that 25% of cases had more than one oral quadrant affected by an odontogenic abscess. Bacterial culture (aerobic, anaerobic, or both) was performed in 40% of cases, and the most common aerobic isolates were *Streptococcus intermedius* (21%) and alpha-hemolytic *Streptococcus* spp. (13%). Common anaerobic bacterial isolates included *Bacteroides pyogenes* (27%), *Bacteroides fragilis* (15%) and *Fusobacterium nucleatum* (15%). About 45% of cases received long-term systemic antibiotics as the primary treatment. Another 30% of cases were euthanized, sent home with palliative care, or died within 48 hours. In 15% of the cases, repeated abscess packing protocol[1] with antibiotic-soaked gauze was performed in conjunction with long-term systemic antibiotics, which resulted in greater than 80% long-term resolution. Other surgical procedures performed in 10% of cases included tooth extraction, marsupialization, or lancing and draining. The study highlights the prevalence of diffuse dental disease and the difficulty in predicting appropriate antimicrobial treatment without culture and susceptibility testing. As previously reported,[1] the abscess packing protocol combined with appropriate systemic antibiotic therapy results in long-term resolution without recurrence of odontogenic abscesses in the majority of affected rabbits.

References

1. Taylor WM, Beaufrere H, Mans, C, et al. Long-term outcome of treatment of dental abscesses with a wound-packing technique in pet rabbits: 13 cases (1998-2007). *J Am Vet Med Assoc.* 2010;237:1444-1449.

Epidemiology of Retrobulbar Abscesses in Client-Owned Rabbits: Twenty-One Cases (2011-2022)

Ivana Levy, DVM, and
Christoph Mans, Dr med vet, Dipl ACZM

Session #3398

Affiliation: From the University of Wisconsin-Madison Veterinary Care, 2015 Linden Drive, Madison, WI 53706, USA.

Note: Abstract format selected by author(s). See Front Matter for description of conference proceedings manuscripts submitted as scientific abstracts.

Abstract: The medical record database of a veterinary teaching hospital was searched from 2011-2022 for records of rabbits diagnosed with retrobulbar abscesses by computed tomography. Data reviewed included age, breed, presenting complaint, association with an odontogenic abscess, aerobic and anaerobic culture results, treatment, and outcome. Twenty-one rabbits were included. The primary presenting complaint was exophthalmos (19/21, 90%). Dental disease (14/21, 67%) was a common co-morbidity on computed tomography. Most cases (15/21, 71%) were associated with an odontogenic abscess. Bacteria culture (aerobic, anaerobic, or both) was performed in 11/21 (52%) cases, and the most common aerobic isolate was *Streptococcus intermedius* (5/11, 45%). The most common anaerobic isolate was *Fusobacterium nucleatum* (2/7, 29%). There was no growth on 1/11 (9%) aerobic and 4/7 (57%) anaerobic bacterial cultures. Surgical treatment combined with long-term systemic antibiotic therapy was performed in 9/21 (43%) cases and included intraoral tooth extraction (4/9, 44%) versus extraoral peribulbar abscess lancing with either abscess packing with antibiotic-soaked gauze (3/9, 33%) or drainage (2/9, 22%). Clinical sign resolution with no recurrence for at least 6 months occurred in 7/9 (78%) surgically treated cases. Due to poor prognosis, euthanasia was performed or recommended in 7/21 (33%) cases. Medical treatment with long-term systemic antibiotic therapy was performed in 5/21 (24%) cases, and 3/5 (60%) resolved. Based on the data from this study, retrobulbar abscesses in rabbits carry a guarded prognosis, and the most effective treatment appears to be surgical intervention with long-term systemic antibiotic therapy.[1-6]

References

1. Capello V. Surgical treatment of facial abscesses and facial surgery in rabbits. *Vet Clin North Am Exot Anim Pract.* 2016;19:799-823.

2. Harcourt-Brown, F, Chitty J, eds. *BSAVA Manual of Rabbit Surgery, Dentistry and Imaging*, Gloucester, UK: British Small Animal Veterinary Association; 2013.

3. Lennox AM, Legendre LF. Small mammal dentistry. In: Quesenberry KE, Orcutt CJ, Mans C, Carpenter JW, eds. *Ferrets, Rabbits, and Rodents: Clinical Medicine and Surgery.* 4th Ed. St. Louis, MO: Elsevier; 2020:514-535.

4. Martínez-Jiménez D, Hernàndez-Divers SJ, Dietrich UM, et al. Endosurgical treatment of a retrobulbar abscess in a rabbit. *J Am Vet Med Assoc.* 2007;230:868-872.

5. Thomas A, Kazakos G, Pardali D, et al. Surgical management of orbital abscesses in domestic rabbits (*Oryctolagus cuniculus*): a report of seven cases. *J Hellenic Vet Med Soc.* 2020;71:2251-2260.

6. Ward M. Diagnosis and management of a retrobulbar abscess of periapical origin in a domestic rabbit. *Vet Clin North Am Exot Anim Pract.* 2006;9:657-665.

Rabbit Hemorrhagic Disease in New England: Vaccination, Demographics, and Opportunity

Peter A. Sojka, DVM, DABVP (Avian Practice)

Session #3418

Affiliation: From Pieper Memorial Veterinary Hospital, 730 Randolph Road, Middletown, CT 06457, USA.

Note: Abstract format selected by author(s). See Front Matter for description of conference proceedings manuscripts submitted as scientific abstracts.

Abstract: With the release of an American-made rabbit hemorrhagic disease virus (RHDV-2) vaccine in October 2021, it is now possible to vaccinate client-owned European rabbits (*Oryctolagus cuniculus*) in nearly all of the United States under an emergency use authorization by the United States Department of Agriculture (USDA). At the time of manuscript submission, Connecticut and Rhode Island allow for RHDV-2 vaccination, but the nearby remaining New England states of Massachusetts, New Hampshire, Vermont, and Maine do not. Vaccination of rabbits in Connecticut began prior to the first reported case in the state in September 2022. The author's clinic has vaccinated 176 Connecticut resident rabbits since February 2022, 36% of which did not have a primary care veterinarian at the time of vaccination and 62% of which were surgically altered. The average age of the vaccinated group was 36.7 months with a range of 2-132 months. The lack of recommended or required vaccinations for rabbits in North America makes establishing a valid veterinary-client-patient relationship (VCPR) early in life challenging.[1] A recent study has shown that clients are knowledgeable about RHDV-2, but for households with more than five rabbits, vaccination can be cost-prohibitive.[2] Regular vaccination clinics throughout the year have facilitated the diagnosis of chronic or occult diseases in rabbits, including dilated cardiomyopathy. RHDV-2 vaccination can also offer veterinarians the opportunity to establish care with new rabbit patients and institute a preventive care plan including sterilization, nutritional support, and behavior consultation.

References

1. Sutherland M, Baron H, Llinas J. Recommended health care and disease-prevention programs for herds/flocks of exotic animals. *Vet Clin North Am Exot Anim Pract.* 2021;24:697-737.

2. Shapiro HG, Ruder MG, Pienaar EF. Down the rabbit hole: domestic rabbit owners' perceptions of rabbit hemorrhagic disease virus 2. *Ecohealth.* 2022;19(4):487-501.

Preliminary Clinical Pathology Assessment of Vaccination for Rabbit Hemorrhagic Disease Virus in Rabbits

Carolyn Cray, PhD,
Chris Griffin, DVM, DABVP (Avian Practice),
Salina Locke, DVM, DABVP (Avian Practice),
Andressa Lopes Grego, DVM,
Fabiano Montiani-Ferreira, BMV, MCV, PhD

Session #3228

Affiliation: From Division of Comparative Pathology, Department of Pathology & Laboratory Medicine, University of Miami Miller School of Medicine, P.O. Box 016960 (R-46), Miami, FL 33101, USA (Cray), Griffin Avian & Exotic Veterinary Hospital, 2100 Ln St, Kannapolis, NC 28083, USA (Griffin), Avian and Exotic Animal Care, 8711 Fidelity Blvd., Raleigh, NC 27617, USA(Locke), Departamento de Medicina Veterinária, Universidade Federal do Paraná, Rua dos Funcionários, 1540, 80035-050 Curitiba-PR; Brazil (Lopes Grego, Montiani-Ferreira).

Note: Abstract format selected by author(s). See Front Matter for description of conference proceedings manuscripts submitted as scientific abstracts.

Abstract: With the spread of rabbit hemorrhagic disease virus type 2 (RHDV2) in North America, Emergency Use Authorization was issued for a killed baculovirus vector-based vaccine (Medgene Labs, Brookings, SD, USA).[1-3] The vaccine is recommended to be used in a two-dose regimen with the second dose provided at 21 days after the initial dose. An efficacy study conducted by the company found excellent protection in the vaccinated group with minor safety concerns. The goal was to assess changes in routine clinical pathology parameters including hematology, biochemistry, and C-reactive protein (CRP) levels. A total of 29 patients participated in the study with the median age of 2 years (CI: 1-3) including 12 female and 17 male animals in good health. Animals were examined on day 0, 7, 21, and 49. Owners did not report any clinical signs such as pain, fever, swelling, or inappetence. No clinical abnormalities were observed on physical exam. Data were examined for normality and on that basis, RM ANOVA or Friedman's test were performed on the repeated measurements. No significant difference was found in CRP levels although a few rabbits displayed elevations by day 21. Total protein was significantly increased at this time point ($p = 0.02$). In addition, there was a trend towards significantly higher levels of BUN by day 49 ($p = 0.02$) with a significant and positive correlation between time (in weeks) and BUN levels ($r = 0.223$), although all values were within normal limits. Additional studies are examining seroconversion post vaccination and assessing any changes in *E. cuniculi* titers.

Acknowledgments: Thank you to the staff of the Avian & Wildlife Laboratory for their assistance in the analysis of the samples.

References

1. Ambagala A, Schwantje H, Laurendeau S, et al. Incursions of rabbit haemorrhagic disease virus 2 in Canada-clinical, molecular and epidemiological investigation. *Transbound Emerg Dis.* 2021;68:1711-1720.

2. U.S. Department of Agriculture. Animal and Plant Health Inspection Service. Rabbit hemorrhagic disease interactive map. 2021. Available at: https://www.aphis.usda.gov/aphis/maps/animal-health/rhd. Accessed March 30, 2023.

3. U.S. Department of Agriculture. Animal and Plant Health Inspection Service. Rabbit hemorrhagic disease virus, type 2. 2020. Available at: https://www.aphis.usda.gov/animal_health/downloads/rhdv2.pdf. Accessed March 30, 2023.

Myxomatosis Caused by the California Myxoma Virus: Description and Diagnosis

Hilary Stern, DVM, DABVP (Exotic Companion Mammal Practice)

Session #3119

Affiliation: From the Exotic Pet Clinic of Santa Cruz, Scotts Valley, CA 95066, USA.

Note: Abstract format selected by author(s). See Front Matter for description of conference proceedings manuscripts submitted as scientific abstracts.

Abstract: Myxomatosis is a disease of European rabbits (*Oryctolagus cuniculus*) caused by the myxoma virus, a poxvirus naturally found in two American cottontail rabbit species, *Sylvilagus brasiliensis* and *Sylvilagus bachmani.*[1-10] This virus, which causes only mild disease in adult cottontail rabbits, creates an often-lethal disease in European (domestic) rabbits. During the 1950s strains of myxoma virus from the common tapeti (*Sylvilagus brasiliensis*) in South America were intentionally transferred to Australia and Europe, where myxomatosis became a widespread and well-studied disease. Much less is known about the strains of myxoma virus carried by the brush rabbit (*Sylvilagus bachmani*), which is endemic to coastal California and Oregon in the United States and to Baja California in Mexico. Highly virulent with near 100% mortality, and a reportable disease at the federal and state levels, Californian myxoma virus produces fever, edema of the eyelids and genitals, and sudden death. The cutaneous nodules from which the disease myxomatosis gets its name are not observed with the Californian MSW or MSD strains. The peracute and less florid symptoms of Californian myxomatosis can make recognition and diagnosis challenging. This prospective study of pet rabbits with naturally occurring infections documented clinical signs, radiographs, complete blood counts, chemistry panels, real-time PCR, histopathology, and immunohistochemistry for each patient. Real-time PCR of eyelid biopsies was found to be a reliable method of diagnosis, and pancytopenia and hypocalcemia were often observed. Epistaxis was also noted, and cases should not be confused with RHDV2.

References

1. Fenner, F. Deliberate introduction of the European rabbit, *Oryctolagus cuniculus*, into Australia. *Revue Sci Et Technique De L'oie.* 2010;29(1):103-111.

2. Kerr PJ, Donnelly TM. Viral infections of rabbits. *Vet Clin North Am Exot Anim Pract.* 2013;16(2):437-468.

3. Kerr PJ, Ghedin E, DePasse JV, et al. Evolutionary history and attenuation of myxoma virus on two continents. *PLOS Pathog.* 2012;8(10):e1002950.

4. Kessel JF, Prouty CC, Meyer JW. Occurrence of infectious myxomatosis in southern California. *Proc Soc Exp Biol Med.* 1931;28(4):413-414.

5. Luna RML. First report of myxomatosis in Mexico. *J Wildl Dis.* 2000;36(3):580-583.

6. Marshall ID, Regnery DC, Grodhaus G. Studies in the epidemiology of myxomatosis in California I: observations on two outbreaks of myxomatosis in coastal California and the recovery of myxoma virus from a brush rabbit (*Sylvilagus bachmani*). *Am J Hyg.* 1963;77(2):195-204.

7. Patton NM, Holmes HT. Myxomatosis in domestic rabbits in Oregon. *J Am Vet Med Assoc.* 1977;171(6): 560-562.

8. Regnery DC, Miller JH. A myxoma virus epizootic in a brush rabbit population. *J Wildl Dis.* 1972;8(4): 327-331.

9. Silvers L, Inglis B, Labudovic A, et al. Virulence and pathogenesis of the MSW and MSD strains of Californian myxoma virus in European rabbits with genetic resistance to myxomatosis compared to rabbits with no genetic resistance. *Virology.* 2006;348(1);72-83.

10. Snook TS, White SD, Hawkins MG, et al. Skin diseases in pet rabbits: a retrospective study of 334 cases seen at the University of California at Davis, USA (1984-2004). *Vet Dermatol.* 2013;24(6):613-e148.

Canine Distemper Outbreak in Ferrets in North America, 2022

Cathy A. Johnson-Delaney, DVM

Session #3265

Affiliation: From Washington Ferret Rescue & Shelter, 6124 Evergreen Way, Everett, WA 98203, USA, NW Zoological Supply, 3614 Smith Ave, Everett, WA 98201, USA.

Note: Abstract format selected by author(s). See Front Matter for description of conference proceedings manuscripts submitted as scientific abstracts.

Abstract: Ferrets supplied to the pet trade in North America largely come from one source. In September 2022, severe illness and mortality began appearing in individual pet stores and distributors. One store chain isolated canine distemper from euthanized ferrets, and diagnostic work began by the breeder; however, the breeder acknowledged a change in the company manufacturing the vaccine months prior to the outbreak. As no cases were being seen at the facility, no additional changes were made. Kits would be vaccinated at approximately 7 weeks of age with a canine distemper virus (CDV) modified live vaccine (MLV) primarily marketed for mink. The U.S. Department of Agriculture's Animal and Plant Health Inspection Service had approved the biologic for use in ferrets both with safety and efficacy data in November of 1969, although the study data are no longer available.[1] Kits are then shipped within a week of vaccination. Clinical signs and mortality began within 3-14 days after shipment. Cases were seen in the United States and Canada, and involved different distributors and stores. Cases were still being seen in November. The breeder then halted shipments of ferrets until further investigation could be completed. The strain of CDV isolated is that from the vaccine, not a wild-type. Outbreaks from MLV vaccine have occurred in farmed mustelids in the past.[2] Distemper in ferrets does not present or progress like in dogs, and can be successfully treated.[3,4] Clinical signs included anorexia, dehydration, diarrhea, severe ocular and nasal discharge, coughing, pneumonia and death. Time from the start of signs to death was within 72 hours, as 8-week-old kits cannot survive without eating and drinking for more than a day. Treatment of individuals and shipments included vaccinating in the face of the outbreak, vitamin A, vitamin C, interferon, hyperimmune serum, nebulizing, keeping discharges clear, broad spectrum antibiotics, famotidine, hand feeding, fluid therapy, diphenhydramine, NSAIDs and analgesics.[4] Monitoring of recovered ferrets includes repeated bloodwork, serum electrophoresis and titers. Ferrets do not seem to develop chronic neurologic disease as seen in dogs. The breeder has now changed the protocol and will be releasing ferrets to pet stores that have had a two-vaccine series, as well as the cohort having a proven protective titer. This means that ferrets in the stores will be 11-14 weeks old, and considerably larger than previously sold kits.

Acknowledgments: The author thanks Alex and Denise Orleans, NW Zoological Supply, Peter Reid, COO, Marshall Pet Products, and the Association of Exotic Mammal Veterinarians.

References

1. U.S. Department of Agriculture. Animal and Plant Health Inspection Service. Product Summary 245-14500: Mink Distemper Vaccine, Modified Live Virus. Available at: https://www.aphis.usda.gov/aphis/ourfocus/animalhealth/veterinary-biologics/product-summaries/vet-label-data/eb97d4da-b4b7-44f1-9f44-7762c5464084. Accessed March 24, 2023.

2. Ek-Kommonen C, Rudbäck E, Anttila M, et al. Canine distemper of vaccine origin in European mink, *Mustela lutreola*--a case report. *Vet Microbiol.* 2003;92(3):289-293.

3. Richardson J, Perpiñan D. Disorders of the respiratory system. In: Johnson-Delaney CA, ed. *Ferret Medicine and Surgery.* Boca Raton, FL: CRC Press; 2017:311-324.

4. Johnson-Delaney CA. Rescue centre medicine. In: In: Johnson-Delaney CA, ed. *Ferret Medicine and Surgery.* Boca Raton, FL: CRC Press; 2017:437-448.

Section 20

AEMV Masterclasses Track 1

Laurel Harris, DVM;
Salina Locke, DVM, DABVP (Avian Practice)
Session Chairs

Ear Disease in Pet Rabbits: Diagnostic Accuracy, Prognosis, Options and Strategies for Surgical Treatment

Vittorio Capello, DVM, DipECZM (Small Mammal), DABVP (Exotic Companion Mammal Practice)

Session #3126

Affiliation: From Clinica Veterinaria S. Siro, vial Lampugnano 99, 20151, Milano, Italy, Clinica Veterinaria Gran Sasso, via Donatello 26, 20131, Milano, Italy.

Note: Abstract format selected by author(s). See Front Matter for description of conference proceedings manuscripts submitted as scientific abstracts.

Abstract: Ear disease of pet rabbits is a complex syndrome, frequently encountered in practice.[1] Lop-eared rabbits are mostly affected, because the anatomy of the cartilaginous ear canal is abnormally different in this breed. The many different clinical aspects of this syndrome include: the involvement of either the external ear canal or the middle ear (the tympanic bulla), or both; when affected, the tympanic bulla may either be associated or not with bony changes; the otitis externa and/or media may be unilateral or bilateral; and may either be associated or not with clinical signs and symptoms.[1] All these variables make diagnosis and staging of the disease challenging.[1,2] Prognosis and treatment options may represent a dilemma for the rabbit practitioner. This 2-hour masterclass includes a brief review of the anatomy of the rabbit's ear, and is focused on the discussion of the challenging aspects of this complex disease from the practical point of view, including the clinical evaluation; the diagnostic imaging modalities for accurate diagnosis; the abnormal radiologic findings at computed tomography;[1-3] the staging of the disease affecting the middle and the external ear;[1,2] the prognostic elements; and the therapeutic approach tailored to selected patients. The masterclass visually reviews the approaches and techniques for the surgical treatment, including the lateral ear canal resection (LECR);[1] the partial ear canal resection (PECR);[1,4-6] the flushing of the tympanic bulla;[1] the minimal lateral bulla osteotomy (LBO);[1] and the standard lateral bulla osteotomy.[1,4-6] Postoperative management and follow-up are visually illustrated and discussed.

References

1. Capello V, Mancinelli E, Lennox A. *Ear Surgery of Pet Rabbits.* Available at: http://www.ebooksdynamic.vet/ebook.html. Accessed May 1, 2023.

2. Richarson J, Longo M, Liuti T et al. Computed tomographic grading of middle ear disease in domestic rabbits (*Oryctolagus cuniculus*). *J Small Anim Pract.* 2013;54:325-330.

3. De Matos R, Ruby J, Van Hatten RA, et al. Computed tomographic features of clinical and subclinical middle ear disease in domestic rabbits (*Oryctolagus cuniculus*): 88 cases (2007-2014). *J Am Vet Med Assoc.* 2015;246:336-343.

4. Eatwell K, Mancinelli E, Hedley E, et al. Partial ear canal ablation and lateral bulla osteotomy in rabbits. *J Small Anim Pract.* 2013;54:325-330.

5. Csomos R, Boschner G, Hardie R et al. Surgical management of ear disease in rabbits. *Vet Clin North Am Exot Anim Pract.* 2016;19:189-204.

6. Chitty J, Raftery A. Ear and sinus surgery. In: Harcourt-Brown F, Chitty J, eds. *Rabbit Surgery, Dentistry and Imaging.* Gloucester, UK: British Small Animal Veterinary Association; 2013:212-232.

A Comprehensive Introduction to Exotic Animal Endoscopy: Why You Need Endoscopy in Your Exotic Practice

Stephen J. Divers, BVetMed, DZooMed, Dipl ACZM, DipECZM (Herpetology), DipECZM (Zoo Health Management), FRCVS, and
Scott Stahl, DVM, DABVP (Avian Practice)

Session #3700

Affiliation: From the Department of Small Animal Medicine and Surgery (Zoological Medicine), College of Veterinary Medicine, University of Georgia, 2200 College Station Road, Athens, GA 30602, USA.

Abstract: A definitive diagnosis is important to maximize treatment success, and many exotic animal cases are unsuccessfully managed simply because of a failure to identify the specific problem. It is therefore clear that tissue samples offer the most expedient means to a diagnosis in many cases, and endoscopy typically offers the least invasive ante-mortem method to collect such material. Endoscopy is entrenched as a primary diagnostic tool for evaluation and sampling from the gastro-intestinal tract, respiratory tract, and body cavities of humans and small animals. So why is this not the case in exotic pets? Is it due to a lack of available equipment, poor understanding of the benefits in birds, reptiles and exotic mammals, or deficiencies in training and comfort with the techniques? This masterclass aims to confront and dispel these obstacles.

Introduction

A definitive diagnosis is important to maximize treatment success, and many exotic animal cases are unsuccessfully managed simply because of a failure to identify the specific problem. Definitive diagnosis relies upon the demonstration of a host pathologic response (by histopathology, less reliably cytology, or using paired rising antibody titers), and the causative agent (by microbiology, parasitology, or toxicology). There are few reliable serologic tests available for most exotic pet diseases, and those that are available require a minimum of 2 weeks to demonstrate the necessary two-fold increase in titers. It is therefore clear that tissue samples offer the most expedient means to a diagnosis in many cases, and endoscopy offers the least invasive ante-mortem method to collect such material.

Endoscopy is entrenched as a primary diagnostic tool for evaluation and sampling from the gastro-intestinal tract, respiratory tract, and body cavities of humans and small animals. Indeed, most of us would consider it exceptional for a human to undergo joint, tracheal, gastric, bladder, abdominal, or thoracic evaluation by anything other than endoscopy. So why is it that endoscopy is not the generally accepted primary modality for evaluating gastric disease in ferrets, bladder disease in rabbits, liver disease in birds, or renal disease in reptiles? Why do the generally available diagnostic approaches to exotic pets differ so obviously from the recommended endoscopic diagnostics for humans and small animals? Is it due to lack of equipment? Is it due to lack of training or confidence? This masterclass aims to confront and dispel some of these obstacles. Once you start employing endoscopy in your exotic animal practice you will never look back.

Instrumentation

Given the variation in size, species-specific anatomy, and variety of procedures that may be performed, a selection of different endoscopes and instruments may be required; however, when dealing with most exotic pets, the 2.7-mm system offers flexibility and diagnostic power for an affordable investment (Table 1).[1] The 2.7-mm system is most widespread in practice, offers greatest versatility, and can be expanded as individual practice caseload dictates. This system offers several advantages including single-entry procedures, ports for gas or fluid infusion, and an operating channel for the introduction of 1.7-mm (5-Fr) instruments (Fig 1). In addition, the 1.9-mm telescope with integrated sheath, and the 1-mm semi-rigid miniscope are extremely useful for smaller species.

Table 1. Initial endoscopic system for zoological companion animals.

Equipment description	Primary indications
Visualization and documentation	
Tele Pack+ (monitor, light source, camera, digital capture)	Required for all endoscopy procedures.
Rigid Telescopes & Endoscopes	
2.7-mm × 18-cm telescope, 30° oblique with separate operating sheath (14.5-Fr, 4.8 mm) or 2.7-mm × 14.5-cm telescope, 30° oblique with integrated sheath (9.5-Fr, 3.2 mm)	Preferred for most birds, reptiles and small mammals between 100 grams and 5 kilograms.
Flexible instruments for use with rigid telescopes	
1.7-mm biopsy forceps 1.7-mm single-action scissors 1.7-mm remote injection needle 1.7-mm grasping/retrieval forceps	For use with 2.7-mm system.
Insufflation and infusion	
CO_2 endoflator and tubing connected to the sheath port	Used for laparoscopy in mammals and coelioscopy in reptiles.
Sterile saline suspended above with intravenous drip line to the sheath port	Used for sterile saline infusion for GI, cloaca, bladder, and coelom of small chelonians < 200 grams.

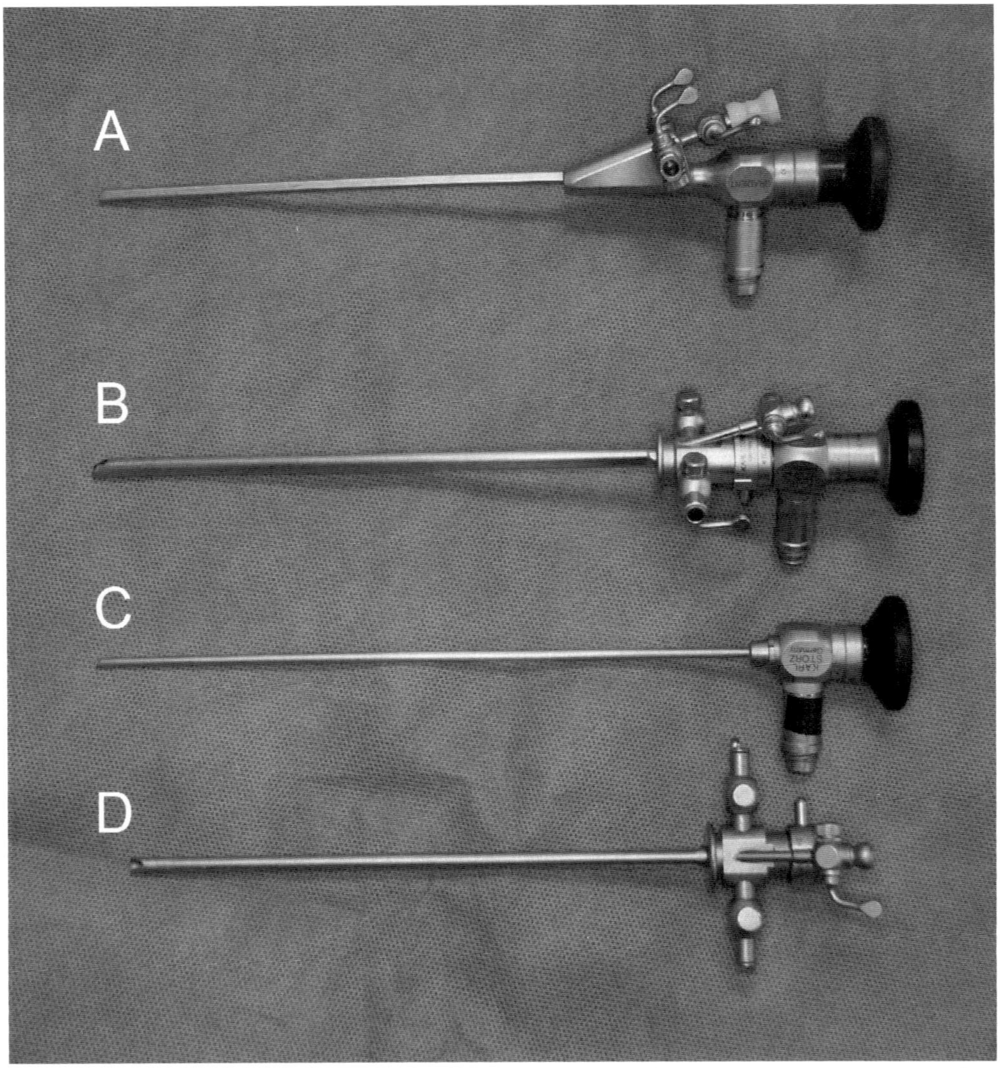

Figure 1. 2.7-mm telescope options. (A) 2.7-mm telescope with integrated sheath (total diameter 3.3 mm, 9.5-Fr), (B) 2.7-mm telescope housed inside the operating sheath (total diameter 4.8 mm, 14.5-Fr); (C,D) Separate 2.7-mm telescope and operating sheath.
Source: Dr. Stephen Divers.

The telescope is connected via a fiber-optic light guide cable to a light source. An endovideo camera, connected to the eyepiece of the telescope, is an integral part of the endoscopy system and greatly facilitates the surgeon's performance. Cameras, available in both PAL and NTSC formats, can vary dramatically in cost from budget single-chip cameras to 3-chip digital 4K models. The fact is that any camera, however cheap, will greatly improve performance compared to using the eyepiece, and will facilitate photo-documentation. The TelePack+ system combines an LED light source, HD camera, and touchscreen monitor with digital image/video storage.

Operating room set-up is important, and the monitor should be positioned directly in front of the endoscopist, with instruments within easy reach. CO_2 insufflation is required for mammal laparoscopy and reptile coelioscopy, but not in birds. The use of saline infusion can be especially helpful when dealing with a hollow viscus (e.g., gastro-intestinal tract, bladder, cloaca).

The 1.7-mm (5-Fr) grasping forceps are useful for manipulating tissues, debridement and retrieving foreign objects or parasites. The fine aspiration/injection needle can be used for the aspiration of fluid from cystic structures where biopsy may be contraindicated due to post-sampling leakage. The needle can also be used for irrigation, and drug administration. The flexible biopsy forceps are used to harvest tissue samples for histopathology and microbiology. The small sample size usually permits the collection of several biopsies for multiple laboratory tests, and serial biopsies to monitor disease progression over time to assess response to treatment. To take a tissue sample, the biopsy forceps are inserted down the operating channel and into the field of view. It is much easier to advance and manipulate the sheath-telescope-instrument as a single device than to try and keep the sheath-telescope still and independently move the biopsy forceps back and forth. With the biopsy forceps held open, the sheath-telescope-instrument is advanced to the tissue of interest and when tissue enters the biopsy cup, the handle is released. These instruments are delicate and the biopsy handle is only required to open the biopsy jaws. The handle's spring mechanism is usually sufficient to take a soft tissue biopsy without additional manual pressure, as long as the instrument is sharp. Clamping down on the handle will damage the forceps and increase biopsy crush artifact. Some organs may be protected by a more fibrous membrane. The fixed blade of the scissors is inserted at a shallow angle through the membrane, and the sheath-telescope-scissors are advanced as a single unit, cutting the membrane as they proceed. The scissors can then be replaced by the biopsy forceps to take a sample through the capsular incision.

Biopsies are delicate and measure approximately 1.3 mm^3 in size when collected using a 1.7-mm instrument. Handling and other histologic artifacts are reduced by gently dislodging the tissue from the biopsy cups into a small volume of sterile saline, before then decanting into a biopsy cassette, and submitting in 10% neutral buffered formalin. Picking the biopsy out of the instrument using a needle will cause severe damage, and even moistened cotton-tipped applicators have been shown to cause tissue alteration. Biopsies for microbiology are best submitted in sterile saline for immediate processing. Alternatively, if mailing to a laboratory, they should be submitted in appropriate transport media. For the submission of samples for toxicology or parasitology, it is wise to consult with the laboratory prior to sample collection.

Birds

Avian veterinarians have been performing endoscopic assessments of the avian reproductive tract for many years, and despite the advent of DNA probes for sex identification of many species, clinicians involved with breeders, wholesalers, or retailers may still be asked to perform surgical sexing and evaluate the health or disease of the reproductive system. The ability to exploit the airsac system of birds enables the endoscopist to visualize most, if not all, of the major organs of clinical interest including liver, lung, airsac, heart, kidney, adrenal, spleen, pancreas, reproductive and intestinal tracts. In addition, the oral approach permits examination of the buccal cavity, esophagus, crop, proventriculus, ventriculus, glottis and trachea down to the level of the syrinx. The vent approach permits examination of the cloaca, and openings to the shell gland and ureters.[2]

There are four approaches to the coelom; left (most indicated), right (common), ventral (uncommon), and interclavicular (rarely indicated). Physical examination, diagnostic imaging (including radiography or computed tomography, and ultrasonography), and clinical pathology should be conducted to identify the most appropriate approach. For example, the spleen is best visualized via a left approach, the psittacine pancreas from the right, and both liver lobes can only be seen from a ventral approach. Most approaches to the avian coelom involve entry into and examination from within the airsacs. Therefore, it is important to keep sheath ports closed to avoid anesthetic compromise. In addition, left and right approaches should be avoided in birds with ascites because fluid leakage into the airsac system is almost unavoidable; however, in such cases, a ventral midline approach is practical.

The most employed procedure involves a left approach into the airsac system because male and female reproductive organs can always be seen (only a few species have bilateral ovaries). The bird is positioned in right lateral recumbency with wings secured dorsad over the bird's back using self-adhesive tape. The left pelvic limb is pulled craniad and secured to the neck, again using self-adhesive bandage, to expose the left flank. The entry site is located immediately behind the last rib, and just ventral to the flexor cruris medialis muscle as it courses from caudal stifle to ischium (Fig 2). Very few feathers, if any, need to be plucked prior to aseptic preparation of the area. Following a 2- to 4-mm skin incision, straight hemostats, directed in a slight craniodorsal direction, are used to bluntly dissect between the thin subcutaneous tissues and enter the left caudal thoracic airsac. The hemostats are replaced by the sheathed telescope, and correct position within the caudal thoracic airsac is confirmed by the identification of lung (straight ahead), cranial thoracic airsac (left), abdominal airsac (right), caudal edge of liver and proventriculus (ventral), and ribs and intercostal muscles (dorsal).

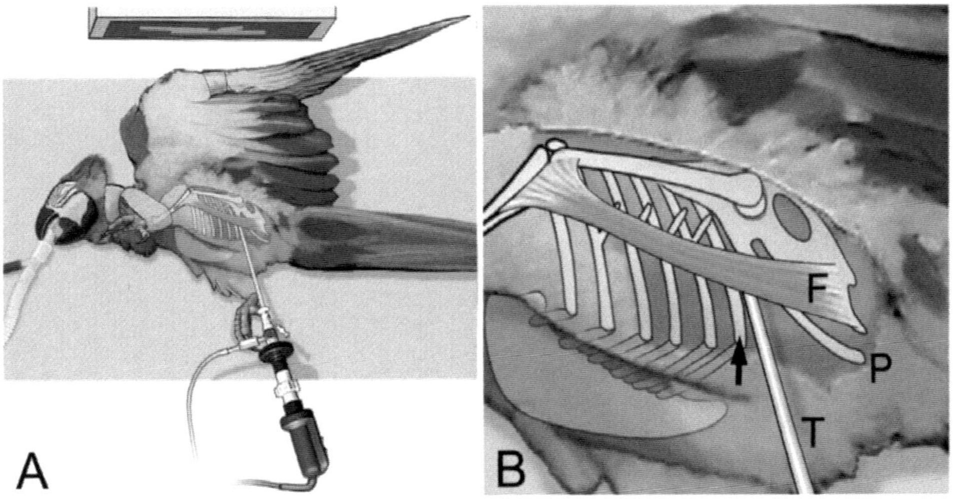

Figure 2. (A) Macaw in right lateral recumbency having left coelioscopic approach undertaken. (B) Close up of the telescope (T) entering the bird just ventral to the flexor cruris medialis muscle (F), just caudal to the last rib (black arrow), and well cranial of the pubis bone (P).
Source: Dr. Stephen Divers.

Exploration of adjacent airsacs is accomplished by pressing the tip of the telescope against the airsac membrane and advancing the telescope in a sweeping motion until the airsac membranes are breached. Normal membranes are transparent and tissues in the adjacent airsac can be visualized and avoided. Great care is required when breaking through thickened, opaque airsacs because vision is impaired and visceral trauma can occur if the telescope is blindly advanced. Lung, liver, heart and associated great vessels can be examined from the cranial thoracic airsac, while urogenital, intestinal, splenic, adrenal, and associated vasculature can be visualized from the abdominal airsac (Fig 3). There is no need to repair the small holes punctured in the airsac membranes as they generally heal within 5-10 days. Therefore, either a single absorbable (e.g., poliglecaprone 25) suture that incorporates both muscle and skin, or a two-layer closure is recommended.

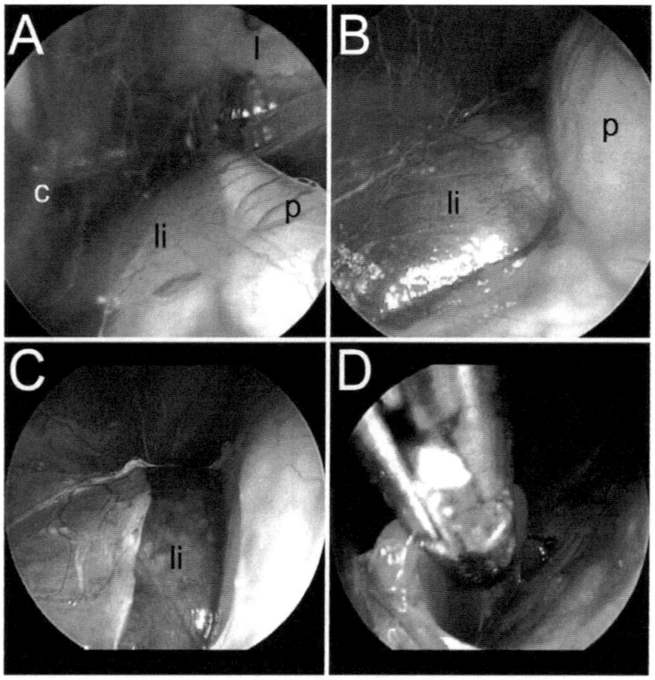

Figure 3. Liver examination and biopsy from within the left caudal thoracic airsac. (A) View of the normal liver (li), proventriculus (p), lung (l), and cranial thoracic airsac (c) in an Amazon parrot. (B) View of the swollen and discolored liver (li), and distended proventriculus in a black palm cockatoo with chlamydophilosis. (C) In preparation for biopsy, the ventral floor of caudal thoracic airsac and the hepatoperitoneal membranes have been incised using 1.7-mm scissors to expose the liver parenchyma (li). (D) 1.7-mm biopsy forceps are used to collect a sample of liver.
Source: Dr. Stephen Divers.

The right approach is essentially the same as previously described. Of particular note is the asymmetrical location of the psittacine pancreas, which can be most easily accessed from the right abdominal airsac in most species by following the duodenum (most caudal intestinal loop) caudoventrad.

The ventral approach provides excellent access to both left and right liver lobes. In cases of ascites, the ventral approach is preferred because the telescope can enter the hepatoperitoneal cavity without entering the airsac system. The bird is positioned in dorsal recumbency and following aseptic preparation, entry is made in the ventral midline, just caudal to the keel. The hepatoperitoneal cavity is divided into left and right sides and the midline membrane can be perforated as previously described.

Reptiles

Common endoscopic procedures in reptiles include evaluation of the respiratory tract, gastrointestinal tract (including the cloaca and bladder), and coelom, including visceral biopsy.[3]

Tracheoscopy and pulmonoscopy

Depending upon the size of the reptile, a 2.7-mm, 1.9-mm, or 1.0-mm endoscope can be used to examine the glottis and trachea of most lizards, chelonians, and snakes. It is always preferable to use a protection sheath with the finer telescopes but the increased diameter may be prohibitive. Rigid scopes of sufficient length can usually be directed into the lungs of lizards by careful manipulation of the lizard's head, neck and body. For large snakes and chelonians, fine diameter flexible endoscopes or videoscopes will permit a deeper examination into the lungs.

The reptile is positioned in dorsal or sternal recumbency, with head and neck extended. A mouth gag is recommended to guard against a lightly anesthetized reptile from biting down on the endoscope. The endoscope should be inserted through the glottis and gently advanced under direct visual control to avoid damaging the mucosa. Complete tracheal occlusion for several minutes is well tolerated by anesthetized reptiles, and there is no need to provide an alternative airway.

A tracheal approach to the chelonian lung is particularly difficult because of the narrow trachea and meandering primary bronchi. It is possible in the largest species (e.g., giant tortoises and sea turtles) using fine flexible endoscopes, but not in smaller specimens, and therefore transcarapacial and prefemoral approaches to the chelonian lung have been developed. A transcutaneous approach to the lung of snakes has also been described. This involves a mini-coeliotomy approach to the lung approximately 35-45% from snout-to-vent, on the right side (or either left or right side for boas and pythons because they have two lungs).

Stomatoscopy and gastroscopy

Sternal or dorsal recumbency with head and neck extended is preferred for stomatoscopy and gastroscopy. It is often helpful to have the reptile close to the table edge or raised above the table surface. The rigid telescope-sheath system can be used to examine the buccal cavity, esophagus, and stomach of small lizards, and chelonians. Flexible endoscopes are required to gain access to the stomach for most snakes, and larger reptiles of other orders. Air insufflation can be used to dilate the tract which provides good exposure for detecting gross lesions and foreign bodies. Warm saline irrigation provides superior clarity especially when examining for mucosal detail, although tracheal intubation is essential to avoid aspiration.

Cloacoscopy

While snakes can also be positioned in sternal or dorsal recumbency, chelonians and lizards are easier in dorsal. Rigid endoscopy using warm saline irrigation promotes unparalleled examination of the proctodeum, urodeum, and coprodeum (Fig 3). Detailed appraisal of the distal colon, cloacal mucosa, urogenital papillae, oviductal openings in females, and (when present) the urethral opening and bladder. With practice, the endoscope can be directed through the urethra into the thin-walled bladder in chelonians, and some lizards. It is also possible to enter the distal oviducts of reproductively active females. Cloacoscopy has been used to remove shell and egg material from the distal oviducts, and as a guide for the exteriorization of partially prolapsed tissue and resection. Mucosal biopsies should be taken with care especially from the colon due to its thin nature, and the risk of perforation.

Coelioscopy

As reptiles lack a true abdomen, the term coelioscopy is preferred over laparoscopy. Precise positioning for coelioscopy depends upon the reptile's body shape and confirmation, the structure(s) of interest, and the preferences of the endoscopist. Insufflation is essential for lizards and snakes and although the shell prevents coelomic collapse can still be helpful in chelonians. Typically CO_2 insufflation pressures of 3-5 mmHg are used, which are lower than those recommended for mammals (8-14 mmHg). In some situations, saline infusion may be preferred over gas. In the author's experience this is especially true when dealing with hatchlings and other very small reptiles, or when dealing with aquatic animals, where residual gas may have a negative effect on post-operative buoyancy. Most, if not all visceral structures can be evaluated and sampled if indicated (Figs 4 and 5).

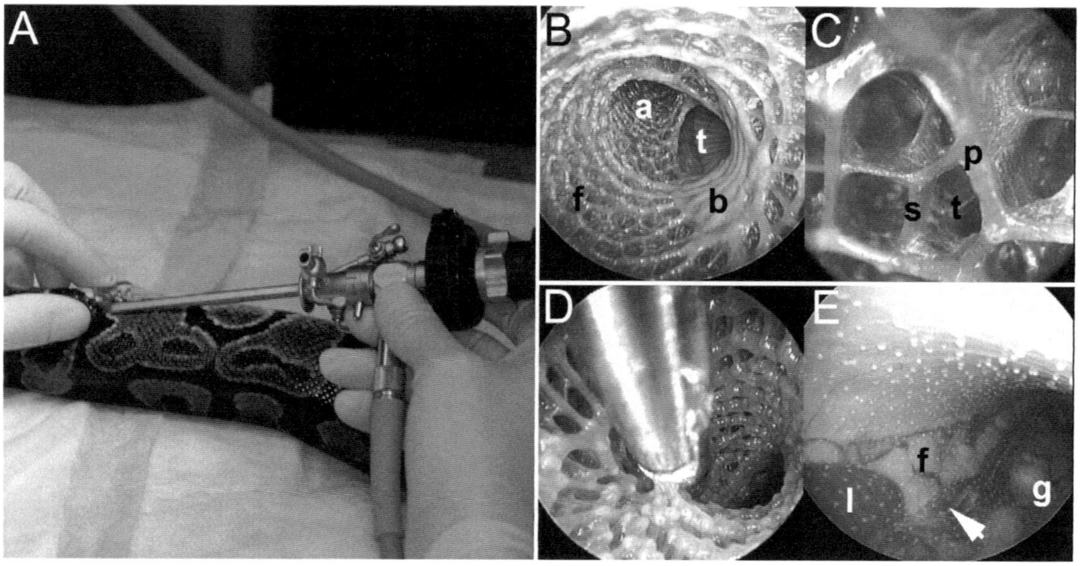

Figure 4. Ball python (*Python regius*) transcutaneous pulmonoscopy. (A) The sheathed 2.7-mm telescope inserted through the lateral body wall and introduced into the cranial lung of a research python that has not been draped to permit visualization of the snake's position and surgical entry site for photography (B) View of the cranial lung demonstrating the distal trachea (t), intrapulmonary bronchus (b), anterior lung lobe (a), and the faveolar lung tissue (f). (C) Close-up view of the faveolar region demonstrating the primary (p), secondary (s) and tertiary (t) septae of the anterior vascular lung. (D) Lung biopsy using the 1.7 mm biopsy forceps. (E) View of the thin, transparent, caudal airsac through which the caudal edge of the liver (l), fat body (f), gastro-intestinal tract (g) and caudal vena cava (arrow) can be seen.
Source: Dr. Stephen Divers.

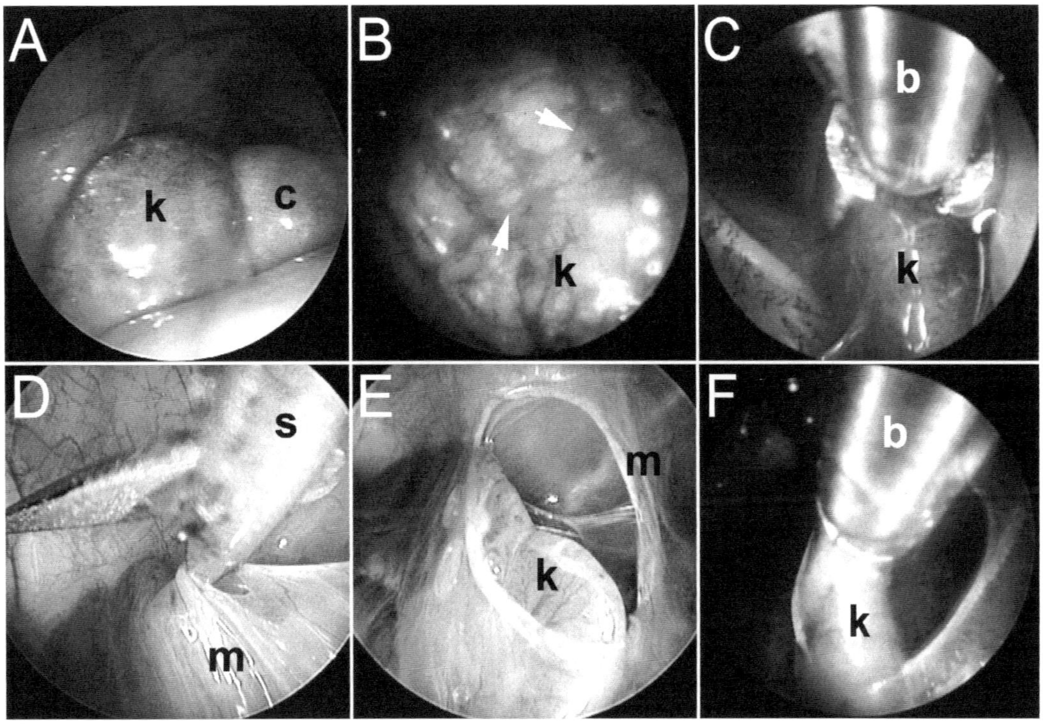

Figure 5. Endoscopic pathology and kidney biopsy. (A) Enlarged kidney (k) with associated renal cyst (c) in a green iguana (*Iguana iguana*) that presented with anorexia and cachexia. Radiography and ultrasonography confirmed renomegaly, while plasma biochemistry revealed reversed calcium:phosphorus ratio and elevated uric acid. Endoscopic biopsy confirmed the diagnosis of glomerulonephrosis with renal gout and mineralization (left lateral coelioscopy). (B) Abnormal kidney (k) with fibrous bands (arrows) in a female Greek tortoise (*Testudo gareca*) that presented with anorexia, lethargy, and polyuria. Diagnostic imaging and clinicopathology were unremarkable; however, renal biopsy confirmed the diagnosis as severe tubulonephrosis (left prefemoral approach). (C) Biopsy of an iguanid kidney (k) using 1.7-mm biopsy forceps (b) (right lateral coelioscopy). (D) Incising the caudodorsal coelomic membrane (m) of a yellow-bellied slider (*Trachemys scripta scripta*) using 1.7-mm endoscopy scissors (s) in order to gain access to the retrocoelomic kidney (left prefemoral approach). (E) View of the incised coelomic membrane (m) revealing the kidney (k) (left prefemoral approach). (F) Biopsy forceps (b) being passed through the incision to collect a biopsy from the kidney (k) (left prefemoral approach).
Source: Dr. Stephen Divers.

Small Mammals

Probably the most common endoscopic procedures performed in small mammals include stomatoscopy, otoscopy, laryngotracheoscopy, rhinoscopy, gastroscopy, cystoscopy and laparoscopy.[4]

Dental disease is undoubtedly one of the most common presentations for rabbits and rodents, and stomatoscopy under general anesthesia ensures a far more detailed examination than can be achieved in the conscious animal in the examination room (Fig 6). Injectable anesthetic agents may be used, and although this does not negate the need for supplying oxygen via nasal line or mask, it does reduce staff exposure to inhalant agents. The animal is positioned in sternal recumbency with the head supported and mouth held open using a rodent/rabbit table retractor restrainer and cheek spreaders (Sontec Instruments Inc., Centennial, CO 80112, USA). The 2.7-mm telescope are preferred because the 30° angle provides a better view of the occlusal surfaces with the light guide cable down, the view is angled towards the maxillary arcades, and with the light guide cable up the mandibular teeth are preferentially seen. The telescope can also be used periodically to evaluate the intra-operative progress of premolar or molar extractions, or to examine the cavity left following extraction. Indeed, the telescope has also been used to target flushing and antimicrobials into dental cavities via the oral cavity. These techniques have been used to successfully treat retrobulbar abscesses in rabbits via the oral cavity, thereby avoiding enucleation. Although rare, soft tissue masses may be biopsied using the 1.7-mm biopsy forceps, while foreign bodies may be removed using retrieval forceps.

Otoscopy is performed with the anesthetized animal in sternal or lateral recumbency. In cases of severe disease, superficial exudates and debris should be gently removed before employing sterile saline infusion and working within a fluid environment. Abnormalities can be sampled for histopathology and microbiology using the biopsy forceps, which tends to provide more precise results than introducing a culturette down the vertical canal. Depending upon the size of the animal, diameter of the scope, and nature of the aural disease, it is often possible to examine the vertical and horizontal canals down to the tympanum. Endoscopic myringotomy is a practical option in rabbits, ferrets and guinea pigs.

Intubation is more difficult in rodents and rabbits, and the endoscope can serve as a useful aid for intubation prior to any prolonged diagnostic or surgical procedure. The endoscope can be used as a laryngoscope to provide visualization of the glottis to aid direct intubation. The author favors inserting the endoscope into the endotracheal tube, passing the endoscope through the glottis and then advancing the tube off the endoscope and into the trachea.

Gastrointestinal disease is especially common in ferrets; however, there has been a noticeable absence of endoscopy in the pursuit of definitive diagnoses. This is unfortunate because ferrets and other small mammals > 1 kg can often accommodate the smaller flexible gastroscopes, while the stomach can often be reached using telescopes in animals < 1 kg. The ability to confirm the presence of gastric ulceration and collect mucosal biopsies for cultures and histology should be considered routine in the investigation of gastric diseases in ferrets. Unfortunately, the stomach of small herbivores is never empty which can make endoscopic evaluation near impossible.

Hematuria is not uncommon in rabbits and rodents, and can be related to diseases affecting the urinary or reproductive systems. With the animal in dorsal recumbency and the perineum close to the table edge, a small 30° sheathed telescope can be inserted through the vulva and into the vagina. Using sterile saline infusion, it is certainly possible to evaluate the vagina, urethra, bladder, and the surface of the cervices of animals as small as 500 grams.

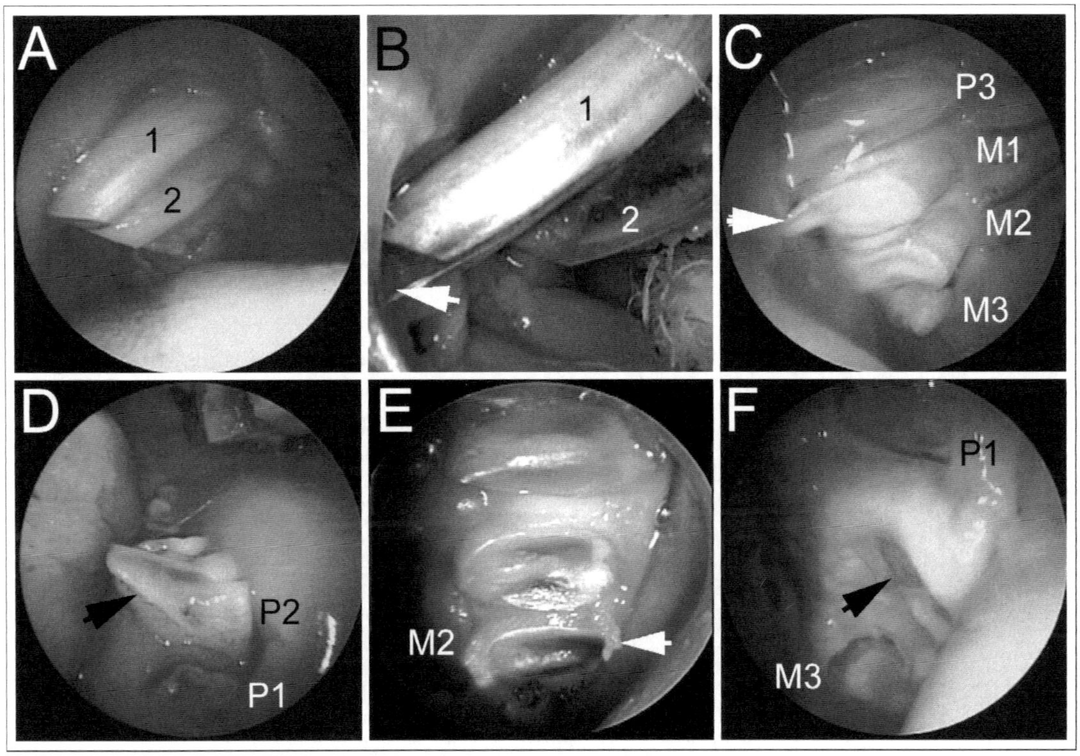

Figure 6. Endoscopic dental pathology in rabbits. (A) Abnormally long right maxillary premolars 1 and 2 due to inadequate dietary forage. (B) Close-up of more severe elongation of the right maxillary premolars 1 and 2. In this case the teeth have caused severe ulceration in the buccal mucosa (arrow). (C) Mild elongation of right maxillary premolar 3, molar 1, and molar 2; molar 3 appears normal. Note that molar 1 also has a small sharp spur associated with its buccal border (arrow). This spur would be buried in the buccal mucosa and would be impossible to visualize without endoscopy and lateral retraction of the mucosa. (D) Gross elongation of left mandibular premolar 2 with associated lingual spur (arrow), and fracture or severe wear of premolar 1 to the level of the gingival. (E) Caseous exudate emanating from around the left maxillary molar 2 following the application of pressure using a dental probe. (F) End stage dental disease. In this view of the left maxillary arcade only premolar 1 and molar 3 are clearly visible, premolar 2 and molar 2 are missing, while part of molar 1 (arrow) can be seen emerging from swollen gingiva that may well indicate an underlying dental abscess.
Source: Dr. Stephen Divers.

Single-entry laparoscopy has been used most extensively for the collection of visceral biopsies from rodents and rabbits. In general, a 3- to 4-mm surgical approach is made through the umbilicus or at some other convenient point along the linea alba. Following insertion of the sheathed telescope, a mattress suture can be tied to create an airtight seal; however, this is seldom necessary if the incision through the linea alba is small. For larger rabbits and rodents CO_2 insufflation is required; however, for small rodents it is often possible to simply attach a syringe containing air to one of the sheath ports, and manually inject air into the abdomen. For larger animals multiple-entry techniques using 2 and 3 mm human pediatric instruments are preferred and provide greater opportunities for endosurgery. Indeed, laparoscopic ovariectomy is now the author's sterilization method of choice for rabbits because it involves less tissue manipulation and results in less post-operative discomfort, and faster return to normal feeding and behavior.

For procedures involving dorsolateral structures (e.g., ovariectomy), the telescope is inserted caudal to the umbilicus with two additional cannulae placed 2-3 cm cranial and caudal to the telescope, along the linea alba (Fig 7). In this way the instruments can be directed to one lateral. When required to examine the other lateral, the surgeon simply moves around the table. It is often helpful to have a second slave monitor located on the other side of the operating table rather than move the entire endoscopy tower.

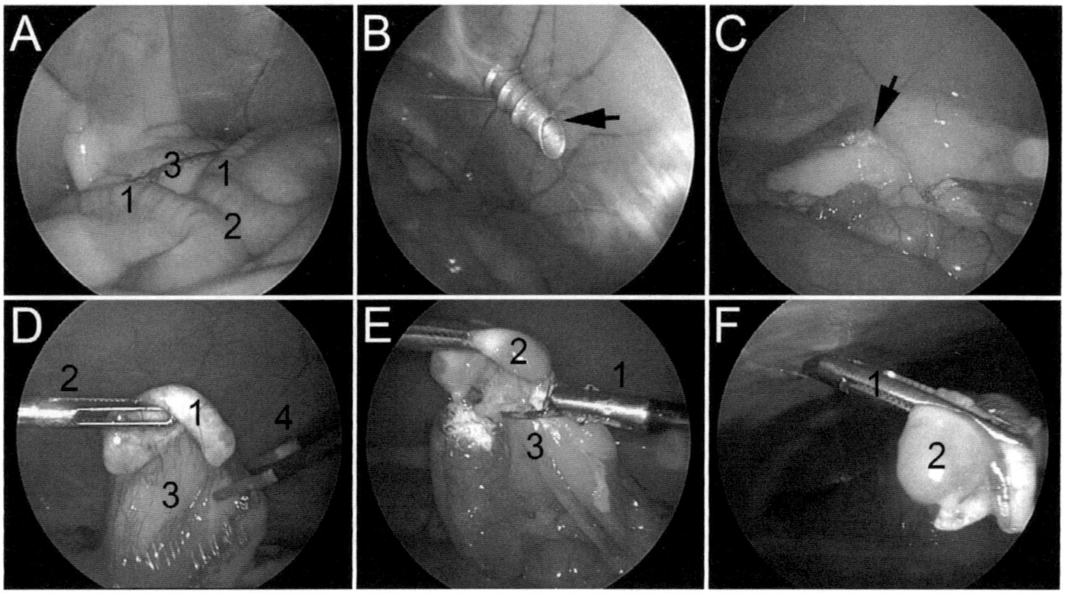

Figure 7. Endoscopic oophorectomy in a rabbit. (A) Initial view of the caudal abdomen revealing normal uteri (1), vagina (2), and bladder (3). (B) Placement of a 3.5-mm graphite/plastic endotip (threaded) cannula (arrow) under endoscopic direction. (C) View of the right ovary (arrow) at 30° lateral tilt. (D) View of the ovary (1) elevated using grasping forceps (2) to reveal the vasculature within the mesovarium (3). The bipolar forceps (4) are advanced to coagulate the vessels prior to transection in this large rabbit. (E) Monopolar scissors (1) being used to coagulate and cut to dissect the ovary (1) free from the mesovarium (3). (F) The cannula has been slid up the shaft of the grasping forceps (1) and is no longer visible. The forceps and ovary (2) are gently withdrawn through the hole in the body wall.
Source: Dr. Stephen Divers.

For access to the liver, gastrointestinal tract, spleen, pancreas, and bladder, the telescope is again passed through the linea alba or umbilicus, but the instruments are inserted transversely across the abdomen rather than longitudinally. In this manner, the telescope and instruments can be advanced into the cranial abdomen for access to the liver, stomach and intestinal tract, liver, spleen and pancreas, or advanced caudally towards the large intestine and bladder. It is important to aspirate all abdominal gas following surgery as the presence of residual gas is a source of post-operative discomfort. Cannula holes are closed using a single suture.

Summary

It is impossible to provide specialist level care to humans or small animals without endoscopy, and the same is true for exotic companion animals. This masterclass highlights the value of endoscopy in zoological companion animal practice, and the need for practitioners to embrace this modality. A more complete (16 hours) on-line, on-demand course is available at least annually through the University of Georgia (https://vet.uga.edu/events/ or email vetmedce@uga.edu). Those that complete the on-line course are eligible for a 2-day intensive practical course undertaking various diagnostic and endosurgical procedures in birds, reptiles and mammals.

References

1. Divers SJ. Diagnostic and surgical endoscopy equipment. In: Divers SJ, Stahl SJ, eds. *Mader's Reptile and Amphibian Medicine and Surgery*. 3rd Ed. Elsevier; 2019:589-599.

2. Divers SJ. Avian diagnostic endoscopy. Review. *Vet Clin North Am Exot Anim Pract*. 2010;13(2):187-202. doi:10.1016/j.cvex.2010.01.002.

3. Divers SJ. Diagnostic endoscopy. In: Divers SJ, Stahl SJ, eds. *Mader's Reptile and Amphibian Medicine and Surgery*. 3rd Ed. Elsevier; 2019:604-614.

4. Divers SJ, Sladakovic I. Exotic mammal diagnostic and surgical endoscopy. In: Quesenberry KE, Orcutt CJ, Mans C, Carpenter JW, eds. *Rabbits, Ferrets and Rodents: Clinical Medicine and Surgery*. 4th Ed. Elsevier; 2021:498-513.

AEMV
ASSOCIATION OF EXOTIC
MAMMAL VETERINARIANS

Section 21

AEMV Masterclasses Track 2

Lauren Powers, DVM, DABVP (Avian Practice), DABVP (Exotic Companion Mammal Practice);
Cathy Johnson-Delaney, DVM
Session Chairs

Diagnosis and Management of Pseudo-Odontoma in Squirrel-Like Rodents

Dan H. Johnson, DVM, DABVP (Exotic Companion Mammal Practice)

Session #3416

Affiliation: From Avian and Exotic Animal Care, 8711 Fidelity Boulevard, Raleigh, NC 27617, USA.

Abstract: Pseudo-odontoma is dysplasia of the incisor apex that causes obstructive upper airway disease in members of the squirrel family. For practitioners in the United States, this means prairie dogs and grey squirrels primarily, but most sciuromorphs appear susceptible.[2] Although all four incisors are typically affected, only maxillary incisor lesions result in decreased nasal airflow. Since squirrel-like rodents are obligate nasal breathers, pseudo-odontoma causes debilitating disease.

Rodent Dental Terminology

Teeth that grow throughout life and never develop roots are termed elodont while those that do not continuously grow and have roots are termed anelodont. The root of an elodont tooth is more accurately termed the apex, and it contains the periapical bud. Within the periapical bud is germinal tissue responsible for the continual growth of elodont teeth.

Damage to the apical bud can result in the formation of an apical mass (elodontoma) comprised of continuously developing odontogenic tissue and alveolar bone.[1] A benign tumor-like lesion such as this, composed of an overgrowth of mature tissue that normally occurs in the affected part of the body but with disorganization and often one element predominating, is termed a hamartoma. Elodontomas are similar histologically to odontomas, the most common type of odontogenic hamartoma in man. Thus, early reports of apical masses of the incisors in squirrels and similar species often referred to these lesions as odontomas; however, since scuiromorphs have elodont dentition, the term elodontoma was proposed instead. A further distinction between lesions that do not contain evidence of dysplasia (i.e., elodontoma) and those that do has led to the more appropriate term pseudo-odontoma being applied to the particular dysplastic, non-neoplastic dental disease of scuiromorphs. For clinical purposes, pseudo-odontoma, elodontoma, odontoma all refer to the same thing.

Etiology and Pathophysiology

Etiology is unknown; however, trauma to germinal tissue at the incisor apex is believed to be a factor. Trauma to the incisor is thought to disrupt epithelial cords (EC) in the apical bud of the tooth, separating the developing tooth from its germinal cells and thereby altering tooth development. At the same time, impact may also cause trabecular bone to tear the EC into several epithelial islands (EI), which now have no connection to the developing tooth. Each daughter EI, due to continuously proliferating (elodont) nature, continues to form dental hard tissue, resulting in a haphazardly arranged mass of odontogenic hard and soft tissue.

Chewing on cage bars, falling from a height, improper trimming of maloccluded incisors, or even local infection in the vicinity of the apical bud are all considered to be possible sources of damage to the apical bud and surrounding alveolar bone. All four incisors are usually affected, but only the maxillary incisor lesions typically result in clinical signs. As pseudo-odontomas grow and impact the tissues around them, there is secondary inflammation. The pseudo-odontoma and associated swelling eventually intrudes into nasal airways, resulting in reduced nasal air passage and eventually complete obstruction. Because prairie dogs and squirrels are obligate nasal breathers, the situation becomes life threatening.

Signalment

Pseudo-odontoma typically affects adult prairie dogs and squirrels between 2 and 6 years of age. It has also been reported in various other species, including the rabbit, guinea pig, degu, red backed vole, rat, and mouse.

Prevalence

According to one author, pseudo-odontoma represents about 25% of prairie dog presentations for veterinary care and about 10% of all neoplasms in prairie dogs. Information concerning the prevalence in other sciuromorphs is not reported.

Clinical Signs

Clinical signs include rhinitis and congestion of the nasal passages, sneezing, a particular type of snoring (often referred to as reverse sneezing), nasal discharge, stridor, pronounced inspiratory effort, open mouth breathing, reduced activity, reduced food intake, weight loss, and emaciation.

Physical Examination Findings

Physical examination may reveal missing or stunted upper incisors. The labial surface of the upper incisors may have a ribbed, washboard appearance. The occlusal surface of affected incisors is typically flat or notched rather than chiseled to a point. Oral examination may reveal a firm swelling or mass of the hard palate caused by the deformed apices. Occasionally, the mass has ruptured through the palate. There may be unilateral or bilateral nasal discharge, and unilateral or bilateral reduction in airflow. Nasal airflow can be assessed using a down feather, glass slide, or wisp of hair.

Diagnosis

The diagnosis is usually based on clinical presentation and radiographic findings. Radiographic images are good, but CT imaging is even better. Radiographic signs of pseudo-odontoma include scalloped dorsal surfaces to the upper incisors on lateral view, notched or flat incisor occlusal surfaces, incisors worn to the gumline, mineral opacity at the incisor apices, and/or apical masses. Typically, both upper and both lower incisors are affected. Lesions are usually visible on radiograph and CT long before they result in clinical signs. A complete blood count and biochemical panel allows evaluation of the general health status of a patient. Bacterial culture and sensitivity are important in cases of periapical infections and abscessation to allow a more targeted antibiotic treatment. Histopathology is indicated for a definitive diagnosis.

Medical Treatment

Symptomatic treatments including non-steroidal anti-inflammatory drugs (NSAIDs), corticosteroids, antibiotics, decongestants, and antihistamines may provide temporary relief but are generally unrewarding.

Surgical Treatment

In the late 1990s, the first prairie dogs to present with pseudo-odontoma were treated by creating a permanent rhinostomy with a stent, forming a kind of blowhole in the bridge of the nose; however, this blowhole technique is not curative. The stent requires constant maintenance and is not recommended as a final solution, but it can provide an alternate airway for animals that are too debilitated to undergo surgical extraction. Rhinostomy is useful for palliative care and for getting a pseudo-odontoma patient into good enough condition for curative surgery. Temporary rhinostomy is also part of the recovery strategy in most cases of surgical extraction (see below)

Surgical extraction of a maxillary pseudo-odontoma provides permanent improvement to nasal airflow. The author prefers to remove only the mass that is causing the greatest obstruction. Pseudo-odontoma removal is complicated by the fact that apical masses are usually too large to be extracted through the alveolar canal. Extraction is traumatic, painful, and surgical bleeding can be severe. By removing only one pseudo-odontoma, surgical trauma, blood loss, anesthetic time and risk are cut by half. Restoring normal airflow to one nasal airway is usually enough to normalize respiration and provide a good quality of life to the patient. Plus, by leaving one maxillary incisor, the two mandibular incisors may be maintained by wearing against the single remaining upper incisor. The maxillary pseudo-odontoma on the opposite side can always be removed later, if necessary.

Perioperative Assessment

Prairie dogs and squirrels with pseudo-odontoma present on a continuum of severity from asymptomatic to extreme dyspnea and debilitation. Pseudo-odontoma extraction is a major surgery that requires vigor and a will to live for recovery. Patients that are still able to eat, drink, and sleep because breathing is not severely compromised are better candidates for extraction, while those with severe clinical signs (e.g., open mouth breathing, lethargy, weight loss) are better candidates for rhinostomy (blowhole) followed by several weeks of aggressive supportive care prior to extraction.

If pseudo-odontoma can be identified before a large mass has developed at the incisor apex, then removal can be almost as easy as a normal incisor extraction. But after pseudo-odontoma has progressed to a large apical mass, then the affected tooth cannot be removed in routine fashion because the alveolus is too narrow. Either the incisor crown will break off leaving the pseudo-odontoma behind, or extraction will fracture the delicate bones of the alveolus, the floor of the nasal cavity and incisive bones. After a sizable apical mass has developed, surgical extraction is indicated.

The nasal passage with the least airflow is almost always the higher priority for pseudo-odontoma extraction. Airflow through each naris is compared to determine which pseudo-odontoma is causing the greater interruption in nasal airflow. This can be accomplished by comparing condensation from both nares on a microscope slide or mirror, with the aid of a small tuft of fur, or using a down feather. The skull is imaged via radiograph or CT and measured to determine where the rhinotomy incision should be located. Note that the pseudo-odontoma that appears largest on imaging is *usually but not always* the one causing the greatest obstruction to airflow. By not disturbing the nasal passage with better airflow, the patient will ideally retain partial use of it during the immediate post operative period when the operative nasal passage is blocked by swelling and debris.

Anesthesia

For patients with mild clinical signs, induction of anesthesia is routine. The author uses ketamine 10 mg/kg + midazolam 0.25 mg/kg + buprenorphine 0.03 mg/kg IM and then masks the patient with inhalant anesthetic. For patients with moderate to severe nasal obstruction and open mouth breathing, induction of anesthesia is complicated by the fact that the patient must be conscious to breathe through its mouth. Instead, these patients are placed in an induction chamber so as not to prevent open mouth breathing. As soon as possible, the patient is intubated using a 1.9-mm semi-flexible fiberscope (MDS Inc, Valrico, FL) and a 2.0-mm uncuffed endotracheal tube. Intubation is essential to securing the airway as intranasal hemorrhage is inevitable, will create further airway obstruction, and risks aspiration.

Surgical Extraction

Both lateral and dorsal rhinotomy approaches for pseudo-odontoma removal are described. An approach through the soft palate is also described but carries with it some risk of causing oronasal fistula. Surgical removal via rhinotomy approach as described by Bennet is the technique used successfully by the author in a variety of squirrel-like rodents.

A skin incision is made using a 5-mm biopsy punch, the nasal bone directly over the pseudo-odontoma is perforated using ¼-inch IM bone pin, and the rhinotomy is enlarged with a rongeur, entering the nasal passage with the least nasal airflow. Once the soft tissues and thin layer of bone overlying the mass are removed, the pseudo-odontoma lesion can be appreciated through the rhinotomy incision. The incisor crown and pseudo-odontoma are removed intact or in sections. The crown is elevated in routine fashion while the pseudo-odontoma is loosened and (usually) extracted through the rhinotomy incision. Bleeding is significant but controllable with topical phenylephrine (OTC nasal spray) and hemostatic sponges (Gelfoam, Vetspon). Occasionally, the mass extends across midline and extraction creates a shunt between the left and right nasal passages, which is important to keep in mind when assessing nasal airflow later (see below).

Temporary Rhinostomy

The rhinotomy incision is kept open by placing a stent fashioned from the hub of a red rubber catheter. Holes to anchor the stent are pre-drilled in the nasal bones with a 25-ga needle, and three to four sutures of 4-0 nylon are placed to fasten the stent. Note that the swaged needle may need to be bent into a tighter curvature to permit passage through the narrow stent opening. The rhinostomy is usually maintained for 1-3 weeks to provide a patent airway and a means for flushing and debriding the nasal passage during the initial post-operative period. The author instills several drops of a custom nasal flush solution [sterile saline (15 cc), amikacin 50 mg/ml injectable (3 cc), phenylephrine 1% (1.5 cc), and acetylcysteine 20% (0.75 cc) keep refrigerated] in the nares and stent opening several times a day to assist in post-op sloughing, auto-debridement, and to prevent infection. Initially, the stent will need to be cleaned with an ear loop curette several times daily to keep it patent. Because rhinostomy is the path of least resistance for air, the nares tend to become plugged with debris and must be kept clean with warm, wet gauze several times a day, as well.

Post-Operative Supportive Care

Patients are hospitalized for supportive care and observation for 3-5 days or more following extraction. Fluid support usually includes warm LRS 20-30 ml/kg SC q8h. Nutritional support usually consists of omnivore hand feeding formula (Oxbow or Lafeber) 20 ml/kg PO q8h along with 10-12 ml/kg of oral fluids (Pedialyte). Additional buprenorphine injections are indicated every 8 hours during the first 24 hours post operation. Patients also receive meloxicam 0.75 mg/kg PO q12h and tramadol 10 mg/kg PO q8h for 5-7 days post-operatively. The author covers for infection either with SMZ-TMP 30 mg/kg or chloramphenicol 40 mg/kg PO q12h for 10-14 days post operation. Patients are provided supplemental heat (e.g., 75-100 watt heat lamp) at one end of the hospital cage to prevent torpor.

Assessment Prior to Stent Removal

Beginning 5-10 days after surgery, breathing through the nasal passages is assessed. A few drops of nasal flush or saline are instilled into the nares and stent opening. The nares are cleaned and debrided with warm, wet gauze, and then air is alternately suctioned and forced through them with a 5-cc syringe (i.e., as if the patient was sniffling and sneezing, but it is unable because the stent bypasses the nares). Once the nasal airways are clear and moist, temporarily plug the stent opening with a wet, lubed cotton plug (or similar) and let the animal breath through its nose. Monitor for open mouth breathing. Measure airflow at each naris with a feather, mirror, or glass slide as before. Once the patient can breathe normally through the operative nasal passage, the stent can be removed. Rhinostomy closure is optional. The opening usually becomes plugged with mucus and dried tissue fluid within 24 to 48 hours and closes by second intension within 10-14 days.

Prognosis and Outcome

In the author's experience, complete removal of one pseudo-odontoma provides a return to normal function and quality of life for most patients. The remaining pseudo-odontoma will likely continue to develop but will not impair airflow on the operative side in most cases.

The prognosis is best if pseudo-odontoma is detected before there are any signs, or while only one nasal passage is affected. Outcome is best if the patient is presented before open mouth breathing and secondary debilitation (i.e., lethargy, difficulty eating, weight loss, dehydration, etc.). It is best to catch the disease in the very early stages, when both nares are functional, and all four incisors are still growing and wearing normally. Removal of pseudo odontoma before it becomes clinical means the patient has been able to eat, drink, and sleep normally and is in much better condition. Patients still in early stages of disease will have better reserve airflow during the immediate post operative period, when there is swelling and discharge. The patient will be stronger and more resilient to stress of surgery.

Removal after a patient has reached the stage of open mouth breathing means the patient will be compromised. It will have been unable to eat, sleep, or drink normally, will be in a negative energy balance. It will have minimal reserves and less resilience to the stresses of anesthesia and surgery, and thus will have a worse prognosis for survival.

Take-Home Points

- Intervene as early as possible.
- Remove only the incisor/pseudo-odontoma complex from the most severely affected side.
- Leave rhinotomy site open during the initial post-operative period to provide an essential route for breathing and topical medication.

References

1. Smith M, Dodd JR, Hobson HP, et al. Clinical techniques: surgical removal of elodontomas in the black-tailed prairie dog (*Cynomys ludovicianus*) and Eastern fox squirrel (*Sciurus niger*). *J Exotic Pet Med.* 2013;22(3):258-264.

2. Pelizzone I, Vitolo GD, D'acierno M, et al. Lateral approach for excision of maxillary incisor pseudo-odontoma in prairie dogs (*Cynomys ludovicianus*). *In Vivo.* 2016;30(1):61-68.

Managing the African Pygmy Hedgehog Patient

Grayson A. Doss, DVM, Dipl ACZM

Session #3164

Affiliation: From the Department of Surgical Sciences, University of Wisconsin-Madison, School of Veterinary Medicine, Madison, WI 53706, USA.

Abstract: The four-toed (*Atelerix albiventris*) or African pygmy hedgehog is a nocturnal insectivore, grouped in the order Eulipotyphla along with shrews and voles. Four-toed hedgehogs should be housed in well-ventilated enclosures and provided exercise wheels. Supplemental heat may be necessary to prevent pet hedgehogs from entering a torporous state. Diet consists of a high protein, moderate fat kibble as the base, along with additional supplemented food items like vegetables and live insects. Physical examination and diagnostic sample collection is challenging in awake hedgehogs due to an inclination to roll into a protective ball. Chemical immobilization is necessary for a complete physical examination in hedgehogs. Common disorders include hemorrhagic vulvar discharge, neoplasia, parasitic dermatologic disease, neurologic disease and oral cavity disease.

Introduction

Four-toed (*Atelerix albiventris*) or African pygmy hedgehogs are popular pets due to their inquisitive, often convivial nature; however, because of their unique anatomy and behavior, hedgehogs can be challenging patients. The lack of prospective research in this species forces veterinarians to rely on anecdotal information, particularly regarding therapeutics, which could lead to difficulties with diagnostics and ineffective or even dangerous treatment strategies.

Natural History and Anatomy

The four-toed hedgehog (*Atelerix albiventris*), also known as the African pygmy or central African hedgehog is a nocturnal insectivore originating from central and eastern Africa. Hedgehogs are the family Erinaceidae and were previously classified in the now obsolete order Insectivora. They are now grouped in the order Eulipotyphla that includes other small insectivorous mammals like shrews and moles. The two most familiar hedgehog species are the four-toed hedgehog and the European hedgehog (*Erinaceus europaeus*).

Four-toed hedgehogs are very active invertebrate predators and travel long distances at night to search for food. They are territorial, solitary animals and can enter a state of torpor in the wild during harsh environmental conditions.

The mantle is the formidable covering of the crown (head) and dorsum consisting of thousands of keratinaceous spines. There are four major muscles that function to roll the hedgehog into an impenetrable ball of spikes[1]: the *caudodorsalis* m. and the *frontodorsalis* m. start the action by rolling up the spines covering the rump and head, respectively. The *panniculus carnosis* m. then rolls up the dorsum of the hedgehog and the *orbicularis* m. cinches the mantle closed around the animal, like a purse-string. The mantle should be moved dorsally when performing lateral radiographic views of the thorax so as to not interfere with the lung and abdominal fields. The mantle spines are not barbed and have a basal bulb connecting the spine to the follicle that makes removal of healthy spines difficult. Teeth are brachydont and the first maxillary incisor teeth have a considerable gap between them. The adult dental formula is 2 (I 3/2, C 1/1, P 3/2, M 3/3) = 36, although variations have been reported.[1] The tail is short and spineless.

Sexing is easily accomplished by examining the anogenital distance. In females, the vulvar opening is located immediately cranial to the anus. In males, the conspicuous preputial opening is located mid-abdomen, a great distance from the anus. Males have dramatic accessory sex glands (particularly the seminal vesicles) which often fill most of the caudodorsal abdomen and can be mistaken for an abdominal mass when palpating dorsal to the urinary bladder. Normal size of male accessory sex glands has been reported.[2] The testicles are mobile and sit in a para-anal recess rather than a scrotal sac. Females have a tightly curled, bicornuate uterus that has a ram's horns appearance. The mesometrium is short and the ovaries are small and often hidden within adipose tissue.

Husbandry

Housing should be well-ventilated and cleaned regularly, as hedgehogs produce copious amounts of soft feces. A large, high-walled, smooth-sided plastic storage container is easy to move and sanitize. Because of their active nature and propensity for obesity in captivity, a solid surface exercise wheel should be provided and cleaned daily if frequently used due to rapid build-up of feces. Hedgehogs quickly learn to drink from either standing water or bottles. Bedding used for hedgehog enclosures varies from aspen shavings to artificial fleece material. Recycled paper bedding is easily cleaned, produces minimal dust, and a thick layer provides hedgehogs with opportunities to dig and burrow.

Insectivorous mammals are traditionally fed diets that are high in protein and have moderate fat content. Commercial hedgehog kibble or a high-quality weight management dry cat food is often used for the base diet. Vegetables may also be offered. Hedgehogs should be supplemented with live, low fat content invertebrates for enrichment.

Physical Examination, Chemical Restraint and Analgesia

Performing a complete physical examination in hedgehogs without chemical restraint is impossible, given their shy nature and propensity to curl into a protected ball. Stress level of the individual animal and poor habituation to handling probably play the largest roles as to why hedgehogs may remain curled up, sometimes for long periods. Even hedgehogs that are used to handling and rarely ball-up for their owners may become shy and non-approachable in a strange environment. While held suspended, holding the hedgehog with its ventrum facing the ground and then tilting the balled-up hedgehog head forward may coax tractable hedgehogs to unroll. For more stubborn individuals, placing them in a shallow (< 2 cm depth) dish of warm water will often cause them to rapidly unroll to try and exit the water. During this attempted escape from the water, a cursory abdominal palpation may be performed. Scruffing can be performed but is likely stressful. Inhalant gas anesthesia with isoflurane or sevoflurane is the most common method of chemical restraint in four-toed hedgehogs. Induction is via use of a small anesthetic chamber and then via facemask. It is difficult to visualize the glottis in this species given their relatively long oropharynx. Intubation requires small endotracheal tubes (1-2 mm I.D.) and displacement of the epiglottis from its normal position dorsal to the soft palate. A small supraglottic airway device designed for rabbits (v-gel R1, Docsinnovent Ltd, London, UK) can provide improved airway support during general anesthesia of adult animals when compared to a facemask and permits positive pressure ventilation.[3]

There are few studies examining injectable anesthetic agents in this species, although use of various combinations of benzodiazepines, $alpha_2$-agonists, opioids, and ketamine have been anecdotally reported.[4] Administration of midazolam (0.5 mg/kg SC, IM; West-ward, Eatontown, NJ, USA) and butorphanol (0.3-0.5 mg/kg SC, IM; Torbugesic, Fort Dodge Animal Health, New York, NY, USA) may result in mild sedation which may permit scruffing. A combination of alfaxalone (3-5 mg/kg) and midazolam (1 mg/kg) administered deep SC into the mantle results in heavy sedation to light anesthesia useful for examination or basic, non-painful procedures like diagnostic imaging.[5] It is partially reversed with flumazenil (0.05 mg/kg, deep SC). Subcutaneous tiletamine-zolazepam (Telazol, Zoetis, Inc., Kalamazoo, MI, USA) produces dose-dependent heavy sedation (10 mg/kg) to light anesthesia (30 mg/kg).[6] Compared with other injectable sedation options (e.g., alfaxalone-midazolam), recoveries with tiletamine-zolazepam are noticeably prolonged, especially at the higher dose.

While there is little information regarding analgesia in hedgehogs, a recent study[7] determined that a single, deep SC dose of buprenorphine hydrochloride administered into the mantle at 0.03 and 0.05 mg/kg provided analgesia for 36- and 48-hours, respectively, without apparent side effects. Methadone (1-1.5 mg/kg SC q1-2h) and hydromorphone (0.15-0.3 mg/kg SC q4-6h) may be used for short-term analgesia.[8]

Common Conditions and Presentations

Neoplasia

Neoplasia occurs commonly in older (> 2 years) adult African pygmy hedgehogs and is well reported in the scientific literature. Major organ systems affected include integumentary, hemolymphatic and alimentary.[9] Oral squamous cell carcinoma is one of the most commonly reported neoplasias in African pygmy hedgehogs.[9,10] Often, more than one type of malignant neoplasia has been reported in a single individual at the time of presentation. There are no studied oncologic treatments in pet hedgehogs.

Hemorrhagic vulvar discharge

At the author's institution, pet hedgehogs are commonly presented for hemorrhagic vulvar discharge. Urinalysis and abdominal ultrasound are useful tools for differentiating between two major differentials for hemorrhagic vulvar discharge, cystitis and uterine disease. Antibiotic treatment for cystitis should be based on urine bacterial culture and sensitivity. Uterine disease, often affecting the endometrium, is commonly reported in the literature,[11,12] with clinical signs including vaginal bleeding, hematuria, and weight loss. Recommended treatment for uterine disease is ovariohysterectomy.

Acariasis

Infestation with *Caparinia tripilis* is common in pet hedgehogs and can result in one or a combination of signs including spine loss, seborrhea, pruritus, or skin erythema. Diagnosis is through skin scrapings of affected areas. Both mite sexes have three long setae on the third pair of legs and there are trilobite projections on both sides of the posterior end of male mites.[13] Selamectin (15-30 mg/kg topically q14-30d; Revolution, Zoetis Inc, Kalamazoo, MI, USA) or ivermectin (0.2-0.4 mg/kg SC or topically q14d; Ivomec, Boehringer Ingelheim, St Joseph, MO, USA) may be used for treatment.

Oral disease

A complete oral examination under anesthesia is very important for this species. Gingival masses, gingival hyperplasia and recession, periodontitis, gingivitis, calculus formation, dental fractures and abscesses may be noted during an oral exam. Antimicrobial therapy may be used for cases of gingivitis.[14] Dental prophylaxis may be performed as in ferrets or small animals.

Obesity

Obesity is a common problem in captive hedgehogs. Healthy hedgehogs should be able to roll up completely, without any portion of the ventrum protruding. Obese hedgehogs may have large axillary or rump fat deposits or a decreased ability to roll up. Treatment includes eliminating high-fat foods, rationing the main diet, and increasing exercise. Hiding food in the substrate or disseminating it throughout the enclosure substrate can help promote weight loss through foraging and exercise. Weight reduction should be gradual to help prevent hepatic lipidosis, with owners monitoring their pet's weight with an accurate scale.

Neurologic signs

In the author's experience, pet hedgehogs are often presented for ataxia. Wobbly hedgehog syndrome is a disorder of the white matter of the central nervous system leading to often progressive neurologic signs. The etiology of the disease or effective treatment options remain unknown. While wobbly hedgehog syndrome is an attractive differential, the author commonly encounters patients that have advanced systemic disease and are severely weak and stumble while ambulating, resulting in an abnormal gait that may be misinterpreted as true ataxia. Hedgehogs inadvertently allowed to enter a state of torpor may also present in a similar fashion, although the prognosis is much more favorable. Depending on the home environment, supplemental heating is often beneficial in preventing torpor by keeping temperatures in a preferred optimum range of 75-85 °F (24-29 °C). In southern African hedgehogs (*Atelerix frontalis*), torpor occurred in most animals at environmental temperatures ≤ 68 °F (20 °C).[15]

References

1. Heatley JJ. Hedgehogs. In: Mitchell MA, Tully Jr TN, eds. *Manual of Exotic Pet Practice.* St. Louis, MO: Elsevier Saunders; 2009:433-455.

2. Thielen L, Di Girolamo N, Gilbert K, et al. Computed tomographic size and appearance of accessory sex glands and testicles in male African pygmy hedgehogs (*Atelerix albiventris*). *Proc ExoticsCon.* 2021;192.

3. Huckins G, Doss GA, Ferreira TH. Feasibility of supraglottic airway device use during maintenance of inhalant anesthesia in healthy African pygmy hedgehogs (*Atelerix albiventris*). *Vet Anaesth Analg.* 2021;48: 517-523.

4. Helmer P, Carpenter J. Hedgehogs. In: Carpenter J, ed. *Exotic Animal Formulary.* 5th Ed. St. Louis, MO: Elsevier Saunders; 2018:443-458.

5. Hawkins SJ, Doss GA, Mans C. Evaluation of subcutaneous alfaxalone-midazolam and ketamine-midazolam sedation protocols in African pygmy hedgehogs (*Atelerix albiventris*). *J Am Vet Med Assoc.* 2020;257: 820-825.

6. Hausmann KE, Doss GA, Mans C. Subcutaneous tiletamine-zolazepam immobilization and effect of flumazenil reversal in African pygmy hedgehogs (*Atelerix albiventris*). *J Exot Pet Med.* 2020;36:1-5.

7. Doss GA, Mans, C. Antinociceptive efficacy and safety of subcutaneous buprenorphine hydrochloride in African pygmy hedgehogs (*Atelerix albiventris*). *J Am Vet Med Assoc.* 2020;257:618-623.

8. Doss G, Peterson M, Mans C. Efficacy and safety of subcutaneous methadone and hydromorphone in African pygmy hedgehogs (*Atelerix albiventris*). *Proc ExoticsCon.* 2022;168.

9. Raymond JT, Garner MM. Spontaneous tumours in captive African hedgehogs (*Atelerix albiventris*): a retrospective study. *J Comp Pathol.* 2001;124:128-133.

10. Gardhouse S, Eshar D. Retrospective study of disease occurrence in captive African pygmy hedgehogs (*Atelerix albiventris*). *Isr J Vet Med.* 2015;70:32-36.

11. Chambers JK, Shiga T, Takimoto H, et al. Proliferative lesions of the endometrium of 50 four-toed hedgehogs (*Atelerix albiventris*). *Vet Pathol.* 2018;55:562-571.

12. Mikaelian I, Reavill DR. Spontaneous proliferative lesions and tumors of the uterus of captive African hedgehogs (*Atelerix albiventris*). *J Zoo Wildl Med.* 2004;35:216-220.

13. Kim DH, Oh DS, Ahn KS, Shin SS. An outbreak of *Caparinia tripilis* in a colony of African pygmy hedgehogs (*Atelerix albiventris*) from Korea. *Korean J Parasitol.* 2012;50:151.

14. Lennox AM, Miwa Y. Anatomy and disorders of the oral cavity of miscellaneous exotic companion mammals. *Vet Clin North Am Exot Anim Pract.* 2016;19:929-945.

15. Hallam SL, Mzilikazi N. Heterothermy in the southern African hedgehog, *Atelerix frontalis*. *J Comp Physiol B.* 2011;181:437-445.

Rabbit Nephrectomy: Case Management and Surgical Techniques

Emi Knafo, DVM, Dipl ACZM

Session #3122

Affiliation: From University of Tennessee College of Veterinary Medicine, Small Animal Clinical Sciences, Zoological Medicine, 2407 River Rd, Knoxville, TN 37996, USA.

Abstract: Nephrectomy, which may be indicated in cases of neoplasia, hydronephrosis, severe pyelonephritis, polycystic disease, and severe nephropathies is uncommonly performed in exotic pet practice. In rabbits, the nephrectomy surgical technique is not well described in the veterinary literature, and only a few case reports describe single-case management. This masterclass describes approaches to case work-up, surgical procedure details, and recommended post-operative management strategies.

Nephrectomy: Indications, Contraindications, and Considerations

Rabbit surgery can present unique challenges compared to other mammals due to airway anatomy, differences in tissue handling, prey species physiology, and the related anesthetic risks. Nephrectomy, the surgical removal of a kidney, is indicated if there is extensive, irreversible renal disease, which could negatively affect the patient if left in place. Indications include neoplasm, abscess, hydronephrosis, or pyelonephritis that is not responsive to antibiosis or causes uncontrollable pain, urine leakage or hemorrhage.[1]

Primary renal tumors have been rarely reported in rabbits, with incidences ranging from 0.066% to 0.64%.[2] These tumors include benign embryonal nephroma, renal carcinoma, renal adenocarcinoma, nephroblastoma, and kidney hamartoma.[3] The most common renal tumor reported in rabbits is nephroblastoma, a primary renal tumor of embryologic origin. Upper urinary tract transitional cell carcinomas result from malignant changes of transitional epithelial cells lining the urinary tract from the renal calyces to the ureteral orifices.[2]

The major contraindication to nephrectomy is significantly impaired function of the contralateral kidney. Pre-surgical diagnostics are aimed at assessing and confirming adequate renal function in the contralateral kidney. Prior to any surgery or anesthesia, a thorough physical exam and diagnostic work up should be performed on each patient. Baseline blood work (CBC, chemistry), abdominal imaging, and thoracic radiographs are generally recommended, especially in geriatric rabbits (> 6 years) to evaluate for any subclinical comorbidities. An excretory pyelogram or a glomerular filtration study should be considered to confirm adequate function of the unaffected kidney. Radiographs are useful to diagnose kidney and bladder stones, whereas ultrasonography may be more useful to assess renal architecture and ureters. Stabilization of sick rabbits prior to surgery is essential to minimize risks of complications. Rabbits do not need to be fasted as they cannot vomit, though removing food 1-2 hours prior to anesthesia (or rinsing the mouth) can help minimize food within the oral cavity.

Airway control is strongly recommended, either by way of direct intubation or use of a supraglottic airway device. The organs should be handled gently and examined without unnecessary manipulation. Regular lavage with sterile saline as well as the use of moistened cotton tipped applicators or gauze squares helps prevent drying of the serosa. A maintenance fluid rate of 100 -120 ml/kg/d is recommended, with adjustments as necessary to correct any azotemia or dehydration. Urine production, body weight, and cardiovascular function should be closely monitored to prevent volume overloading. Intraoperative fluid therapy during surgery at 5 - 10 ml/kg/h should be used.[1]

GI syndrome is one of the most common anesthesia related complications in rabbits, and therefore, they should be supported nutritionally if anorexic prior to and after surgery. Monitoring and correcting body temperatures are important as hypothermia is common in debilitated rabbits. Fluid therapy should be initiated prior to surgery by intravenous, intraosseous, or subcutaneous routes.

Nephrectomy technique

The patient is placed in dorsal recumbency and the entire ventral abdomen surgically prepared with an initial midline abdominal exploration incision made from sternum to the umbilicus. For renal surgery, a cranial abdominal approach should be used, with a midline incision initially from just caudal to sternum down to umbilicus.

The kidney is identified in the dorsal abdomen in the retroperitoneal space after shifting the gastrointestinal tract to either side to aid visualization. Lap sponges moistened with warm sterile saline can be used to keep viscera moist during dissection of the kidney. A Lone star retractor or other abdominal retractor is helpful to maximize exposure.

The affected kidney is bluntly separated from the parietal peritoneum using fine metzenbaum scissors and cotton tipped applicators. The kidney is gently lifted and rotated medially and the hilus, ureter, and renal artery and vein are identified.[1,3] The perinephric fat surrounding the vessels and ureter at the hilus of the kidney should be carefully dissected to permit individual ligation using 3-0 or 4-0 absorbable suture or vascular clips.[1] The ureter is followed and gently detached from the peritoneum along its length. The distal end of the ureter is ligated 1 cm from its insertion into the urinary bladder, transected above the ligature, and removed. If the surgeon has access to a vessel sealing device, this can be used to transect the vessels in addition to suture placement.

Postoperative considerations and complications

Fluid therapy, analgesia, and nutritional support are essential to reduce the risk of postoperative complications. The use of Elizabethan collars to prevent self-trauma to surgical wounds is often unnecessary, may be stressful for rabbits, and prevents cecotrophy. Postoperatively, a rabbit's appetite, fecal and urine output, heart rate, respiratory rate, and temperature should be monitored.

Closure of the linea, muscle, and/or subcutaneous layer in rabbits is best accomplished with 4-0 or 3-0 monofilament, absorbable suture material in a simple continuous pattern. Catgut, Vicryl, and other highly inflammatory suture materials should never be used in rabbits because of the tissue reaction it causes and the risk of adhesions. The holding layer of the abdominal wall is fascia, therefore it should always be incorporated in the sutures. Given the thin nature of each of these layers, it is common to close rabbit abdomens in two layers: linea and subcuticular. If the rabbit is larger or has a discernable subcutaneous layer, then a traditional three-layer closure can be accomplished. The skin is closed with continuous intradermal (or subcuticular) sutures using the same suture material as for the subcutaneous tissue. Tissue glue may be used to seal any small defects in skin closure. External sutures are not recommended as they may attract the rabbit's attention to the incision.

The most common complications after abdominal surgery are hypothermia, GI syndrome, adhesion formation, peritonitis, infection, seroma, and incisional dehiscence. Adhesion formation is common in rabbits and can interfere with intestinal motility, block a ureter, and cause pain. Prevention of adhesions is critical as the treatment is surgical breakdown, which is not always possible, and there is risk of recurrence. Viscera should be handled gently and should be kept moist using repeating irrigation. Contamination or chemical irritation (powder from gloves, urine contamination, and inappropriate suture materials) should be avoided as much as possible.[1]

Prolonged fluid therapy is necessary, especially after nephrectomy to support the remaining kidney. Antibiotics based on culture results should be considered if bacterial infections are suspected or contamination during surgery occurred. Monitoring urine output and urinalysis in the postoperative period is critical. Serial lab work to check BUN, creatinine, and urinalysis at 12, 24, and 48 hours post nephrectomy is ideal but not always possible. If blood work indicates stable renal values by 48 hours post-operatively, and the rabbit has otherwise recovered well after anesthesia (e.g., eating, defecating, urinating normally), then discharge from the hospital is appropriate. Recheck examination of rabbits after nephrectomy is recommended at 1 week for repeated full blood work, and then at 2 weeks for incision recheck and abbreviated blood work (renal panel or BUN, creatinine). Long-term follow up will depend on the individual case, but generally a physical exam every 3 months and renal panel is suggested. The recheck interval can be extended somewhat if the patient is stable and doing well clinically; however, since the rabbit has a single remaining kidney, examinations and blood work should not be any less frequent than every 6 months.

References

1. Szabo Z, Bradley K, Cahalane AK. Rabbit soft tissue surgery. *Vet Clin North Am Exot Anim Pract.* 2016; 159-188.

2. Rhody J. Unilateral nephrectomy for hydronephrosis in a pet rabbit. *Vet Clin North Am Exot Anim Pract.*2006;633-641.

3. Reavill D, Lennox A. Disease overview of the urinary tract in exotic companion mammals and tips on clinical management. *Vet Clin North Am Exot Anim Pract.* 2020;169-193.

The Path to Safe Rabbit Anesthesia

Nicola Di Girolamo, DMV, PhD, DipECZM (Herpetology), Dipl ACZM, Dipl ACVPM, and
Stephania Scarabelli, DMV, DECVAA, DACVAA

Session #3494

Affiliation: From the Department of Clinical Sciences, College of Veterinary Medicine, Cornell University, Ithaca, NY 14853, USA (Di Girolamo), Clinica Veterinaria Malpensa-Anicura, Via G. Marconi 27, Samarate, VA 21017, Italy (Scarabelli).

Abstract: Rabbit anesthesia is drastically different than anesthesia of domestic carnivores. Rabbits have a strikingly higher peri-anesthetic mortality as compared to dogs and cats. Difficulty selecting an ideal rabbit anesthesia lies in the fact that most published research studies include populations limited in numbers and procedures, making it difficult to generalize results to clinical patients. This presentation discusses available protocols and best practices to minimize the risk of anesthetic-related mortalities in rabbits. The veterinary practitioner should keep in mind that multi-modal analgesia is fundamental in rabbits as well as in any other living beings.

Introduction

Rabbit anesthesia is drastically different than anesthesia of domestic carnivores. Rabbits have a strikingly higher peri-anesthetic mortality as compared to dogs and cats.[1,2] Based on a large study including 8209 rabbit anesthetic procedures, the risk of anesthetic and sedation-related death in rabbits was 1.39% (95% CI 1.14-1.64%) within 48 hours of the procedure, almost ten times greater than dogs.[1] This indicated that approximately one rabbit out of 72 anesthetized or sedated died. When stratifying rabbits by healthy or diseased health status, healthy rabbits had a 0.73% risk of mortality and diseased rabbits had a 7.4% risk of mortality.[1] A more recent study found an even higher mortality rate within 72 hours of the procedure, equal to 4.8% (95% CI 0.025-0.086, six rabbits over 210 procedures).[2] Unfortunately while these studies are useful in order to identify this risk and to describe the risk to owners, there is limited clinical information that can be extrapolated reliably from the published reports to guide and improve future anesthetic events.

An additional difficulty selecting an ideal rabbit anesthesia lies in the fact that most published research studies include populations limited in numbers and procedures, making it difficult to generalize results to clinical patients. This presentation discusses available protocols and best practices to minimize the risk of anesthetic-related mortalities in rabbits. The veterinary practitioner should keep in mind that multi-modal analgesia is fundamental in rabbits as well as in any other living beings.

Preparation of the Patient

Intravenous catheter placement

Placement of an intravenous catheter should be performed in every anesthetic event to administer fluids, analgesic drugs and above all, emergency drugs if required. The marginal ear vein is the preferred site, but the cephalic and saphenous veins can be used as well. It is authors' opinion that in healthy animals, it is better to apply the intravenous catheter after premedication to avoid stress to the animal. If the patient is critical and intramuscular premedication is not appropriate, a racemic mixture of lidocaine and prilocaine (EMLA cream) can be applied

30 minutes before venipuncture to reduce pain and discomfort. If a catheter is already in place, it is important to check patency before anesthesia. It is key to perform intubation only once the rabbit has been properly premedicated and does not show any chewing or tongue movements. Pre-anesthetic medications are discussed later in these proceedings.

Intubation

Fasting is avoided in rabbits but, before intubation, it is fundamental to clean the rabbit's mouth to remove food residues that could be otherwise pushed inside the trachea and inhaled. To achieve endotracheal intubation (ETI), a suitable depth of anesthesia should be induced by means of premedication (including combinations of benzodiazepine, dissociative agents, opioids and/or alpha-2 agonists) and, if needed, induction agents (e.g., propofol or alfaxalone). Lidocaine (spray or the injectable preparation instilled topically) can be used to desensitize the larynx.

Regardless of the method used for ETI, the rabbit should be placed in sternal recumbency with the neck and head extended dorsally and cranially but not completely vertical. A 2- to 4.4-millimeter ID uncuffed endotracheal tube (ETT) should be available.

ETT in rabbits can be performed in different ways:

Endoscopically guided ETI. A 2.7-mm endoscope at 30 or 0 degrees can be used. If feasible, it can be inserted into the ETT and used as a guide to visualize the glottis and as a stylet to facilitate the ETT insertion. If ETT of smaller diameters a 'side by side', technique is used, just to simply visualize the glottis. Endoscopic intubation is the only technique currently in use by the authors because (1) it allows visualization of the glottis and therefore inappropriate transport of food material into the trachea with the ETT; (2) once mastered, it is fast and allows adaptation to different type of larynx of rabbits; and (3) it allows to identify alterations at the level of the glottis (e.g., laryngeal neoplasia).

Blind. The ETT is inserted into the mouth and the breath sounds of the rabbit or the presence of condensation on the ETT itself are used to guide the insertion.

Otoscope. For direct visualization of the larynx. The otoscope is inserted into the side of the patient's mouth, then moved centrally and used to elevate the soft palate and visualize the glottis. An ETT can be inserted through the cone of the otoscope or with a side-by-side technique.

In the authors' experience, endoscopically guided ETI allows insertion of a larger diameter ETT thus decreasing airway resistance and work of breathing for the rabbits and allowing more accurate and reliable monitoring of end tidal carbon dioxide ($EtCO_2$). Since a few years ago, a supraglottic airway device (V-GEL, Docsinnovent, Hemel Hempstead, United Kingdom) has been produced and marketed for use in rabbits. The device is made of a soft gel-like material and should provide a pressure seal around the airway and the esophagus. The authors had instances in which V-GELs resulted in poor rabbit oxygenation (low SPO_2) and dislocation from the front of the glottis of the rabbit, resulting in ventilation in the gastrointestinal tract. Until a properly sized, clinical-based study compares the safety and effectiveness of supraglottic airway devices in rabbits with standard endotracheal intubation, the authors' advice is to maintain use of endotracheal intubation.

Induction of general anesthesia (isoflurane, sevoflurane, propofol and alfaxalone)

Historically, induction with anesthetic gases has been considered safer in exotic small mammals than induction with injectable agents (such as propofol, etomidate or alfaxalone); however, there are several shortcomings associated with induction of anesthesia with anesthetic gases that need to be considered.

Isoflurane in rabbits causes apnea during induction and a severe cardiovascular depression at high concentrations.[3-5] These are two main reasons why the use of anesthetic gases and especially the widely used isoflurane should be tailored in rabbits. Considering anesthetic mortality in healthy patients, it is possible that isoflurane accounts for some of the deaths observed. When isoflurane was administered in rabbits for anesthetic induction, long periods of apnea (> 1 minute) occurred together with bradycardia, hypercapnia and hypoxaemia.[5] In another study, most animals struggled violently during induction with both isoflurane and sevoflurane.[3] The authors of these studies warned against the use of isoflurane, sevoflurane, and desflurane for anesthetic induction of rabbits.[3,5]

During anesthesia, reaching high isoflurane concentrations in rabbits results in a dose-dependent cardiopulmonary depression attributable to vasodilation and negative inotropy.[4] At an isoflurane concentration of 4.15% with mechanical ventilation, cardiovascular depression is severe.[4] The authors mention that use of unnecessarily high isoflurane concentrations in this species should be avoided.[4]

The widespread use of endoscopic intubation has allowed a more consistent use of injectable agents for induction. The window of time required to intubate the rabbit may be shorter as compared to anesthetic gasses (i.e., the rabbit may need to be intubated quicker), but there are several cardiovascular benefits associated with the use of different injectable agents for induction.

Among injectable agents, propofol is commonly used in rabbits, but recently alfaxalone has been licensed for use in this species as well. The main advantage of alfaxalone, compared with propofol, is that it can be used intramuscularly (even though it is not licensed for this route in rabbits yet). When used intravenously, a dose-dependent respiratory depression and apnea are commonly reported.[6,7] The Summary of Product Characteristics (SPC) of Alfaxan suggests that rabbits are pre-oxygenated before induction of general anesthesia and that oxygen is provided throughout the entire anesthetic procedure. Dosage is reported as 4 mg/kg in premedicated rabbits and 5 mg/kg in non-premedicated animals even if lower doses (1-2 mg/kg) are generally efficacious. Even though propofol is still the authors' agent of choice for induction of anesthesia in rabbits, apnea is common and therefore the clinician should be confident with endotracheal intubation. Respiratory depression is dose related and can be decreased with slow administration of the drug. Nevertheless, authors discourage inducing rabbits with propofol if endoscopic-guided intubation cannot be performed.

Constant rate infusions

Constant rate infusions (CRI) may be extremely advantageous in rabbits. A fentanyl CRI may reduce isoflurane MAC by approximately 60% in New Zealand white rabbits.[8] Furthermore, in rabbits anesthetized with isoflurane mean arterial blood pressure, systemic vascular resistance and cardiac output were significantly higher when different fentanyl concentrations were obtained through CRIs.[9]

Lidocaine CRI at 50 μg/kg/min resulted in the reduction of isoflurane MAC by 10.5%, while at 100 μg/kg/min CRI resulted in the reduction of MAC by 21.7%.[10] Lidocaine also induced significant decreases in arterial blood pressure and heart rate, but the variables were still within the reference intervals for rabbits.[10] When compared to buprenorphine for management of postoperative pain, a CRI of lidocaine provided better postoperative outcomes with respect to fecal output, food intake, and glucose concentrations, with similar pain scores.[11] Thus, lidocaine appeared to be a suitable alternative to buprenorphine for alleviating postoperative pain with minimal risk of anorexia and gastrointestinal ileus.[11]

Vasopressors

Based on a recent study, phenylephrine might be more indicated than dopamine to increase MAP in anesthetized rabbits.[12] Dopamine increased stroke index at the highest infusion rate of 30 μg/kg/min; however, cardiac output and mean arterial blood pressure remained unchanged from baseline values. Within the dose range of 5-30 μg/kg/min, dopamine was not an effective treatment for isoflurane-induced hypotension in rabbits. Phenylephrine instead, at a rate of 2 μg/kg/min increased mean arterial blood pressure to 62 mm Hg from the baseline value of 45 mm Hg. This was a result of an increase in systemic vascular resistance with a concomitant decrease in heart rate and no change in cardiac output. Phenylephrine was not effective at the dose of 0.125, 0.25, 0.5, 1.0 μg/kg/min only minimally effective at a dose of 2 μg/kg/min. Blood lactate concentration increased with time when rabbits received either treatment.[12]

Anesthetic drugs commonly used in rabbits

Several anesthetic protocols have been used and published in rabbits. Classes of drugs commonly employed include sedatives (midazolam, alpha-2 agonists, acepromazine), opioids (butorphanol, buprenorphine, fentanyl, methadone) and ketamine. These drugs should be used in different combinations and at different doses depending on the procedure performed, the required degree of sedation and analgesia and on the animal's temperament.

Alpha-2 agonists

Alpha-2 agonists (medetomidine and dexmedetomidine) are commonly used in rabbits due to their sedative and analgesic properties and their anesthetic-sparing effect; however, important side effects must be remembered, above all at cardiovascular level. Generally, there is an increase in systemic vascular resistance (SVR) and reflective reduction in heart rate followed by a centrally mediated bradycardia accompanied by normal or slightly increased/decreased SVR, depending on the species. In rabbits SVR remain increased during dexmedetomidine infusions.[13] Regarding the respiratory system, dexmedetomidine has been reported to affect the respiratory response to carbon dioxide in rabbits.[14] The effects of alpha-2 agonists on gastrointestinal (GI) motility have been extensively studied both in human and veterinary medicine leading to controversial results. A recent study suggests a delayed transit motility in rabbits administered ketamine + medetomidine as compared to rabbits receiving ketamine + midazolam, however the time to production of feces was identical in the two groups.[15] Other side effects of alpha-2 agonists include hyperglycemia and enhanced diuresis. The authors generally use dexmedetomidine at doses of 25-35 μg/kg IM + ketamine ± benzodiazepine ± opioids, either as a standalone anesthetic protocol for short and non-invasive procedures (e.g., dentals, castrations), or as a premedication before induction with propofol as needed, intubation and maintenance on isoflurane or sevoflurane.

Benzodiazepines

Benzodiazepines, particularly midazolam, can be used in rabbits. Midazolam can be given intramuscularly and side effects are very limited. In the authors' experience, midazolam can be added to alpha-2 agonists, in very nervous, healthy animals or can be given as only a sedative in critically ill patients. 0.5-1 mg/kg IM is the dose commonly used by the authors.

Ketamine

Ketamine can be used either for premedication or for induction of general anesthesia. Immobilization and analgesia are the main advantages offered by ketamine whereas poor muscle relaxation is the main disadvantage; the latter can be overcome by adding benzodiazepines or alpha-2 agonists. Several studies evaluated the use of ketamine in association with other drugs in rabbits.[16] An effective premedication consists of ketamine (4-5 mg/kg up to 10 mg/kg) with alpha-2 agonists and benzodiazepine all administered intramuscularly.

Opioids

Opioids are the mainstay for perioperative analgesia in humans, dogs and cats; their use in rabbits has always been controversial due to their effects on GI motility. Opioid receptors are highly expressed at the GI level and endogenous endorphins activation of mu-opioid receptors has been reported to inhibit GI motility.[17]

In rabbits, morphine has been reported to inhibit the small intestine's contractility in vitro[18] whereas buprenorphine is the opioid more commonly evaluated in clinical studies. Several studies have demonstrated a negative effect of buprenorphine on GI motility.[11,19,20]

The authors' drugs of choice are methadone at doses of 0.2-0.3 mg/kg SC or IM or hydromorphone at doses of 0.3-0.5 mg/kg SC or IM. In a recent study on healthy rabbits[20] methadone administration was associated with no negative effects on appetite and GI motility, compared to buprenorphine and hydromorphone.[20] Butorphanol 0.3-0.4 mg/kg is used to enhance the activity of an alpha-2 agonist or benzodiazepine when sedation is required for nonpainful procedures, such as radiography or computed tomography. Non-steroidal anti-inflammatory drugs and locoregional anesthetic techniques should always be considered in order to provide multimodal analgesia.

References

1. Brodbelt DC, Blissitt KJ, Hammond RA, et al. The risk of death: the confidential enquiry into perioperative small animal fatalities. *Vet Anaesth Analg.* 2008;35(5):365-373.

2. Lee HW, Machin H, Adami C. Peri-anaesthetic mortality and nonfatal gastrointestinal complications in pet rabbits: a retrospective study on 210 cases. *Vet Anaesth Analg.* 2018;45(4):520-528.

3. Flecknell PA, Roughan JV, Hedenqvist P. Induction of anaesthesia with sevoflurane and isoflurane in the rabbit. *Lab Anim.* 1999;33(1):41-46.

4. Barter LS, Epstein SE. Cardiopulmonary effects of three concentrations of isoflurane with or without mechanical ventilation and supramaximal noxious stimulation in New Zealand white rabbits. *Am J Vet Res.* 2013;74(10):1274-1280.

5. Hedenqvist P, Roughan JV, Antunes L, et al. Induction of anaesthesia with desflurane and isoflurane in the rabbit. *Lab Anim.* 2001;35(2):172-179.

6. Gil AG, Silván G, Villa A, Illera JC. Heart and respiratory rates and adrenal response to propofol or alfaxalone in rabbits. *Vet Rec.* 2012;170(17):444. doi:10.1136/vr.100573.

7. Grint NJ, Murison PJ. A comparison of ketamine-midazolam and ketamine-medetomidine combinations for induction of anaesthesia in rabbits. *Vet Anaesth Analg.* 2008;35(2):113-121.

8. Barter LS, Hawkins MG, Pypendop BH. Effects of fentanyl on isoflurane minimum alveolar concentration in New Zealand White rabbits (*Oryctolagus cuniculus*). *Am J Vet Res.* 2015;76(2):111-115.

9. Tearney CC, Barter LS, Pypendop BH. Cardiovascular effects of equipotent doses of isoflurane alone and isoflurane plus fentanyl in New Zealand white rabbits (*Oryctolagus cuniculus*). *Am J Vet Res.* 2015;76(7): 591-598.

10. Schnellbacher RW, Carpenter JW, Mason DE, et al. Effects of lidocaine administration via continuous rate infusion on the minimum alveolar concentration of isoflurane in New Zealand white rabbits (*Oryctolagus cuniculus*). *Am J Vet Res.* 2013;74(11):1377-1384.

11. Schnellbacher RW, Divers SJ, Comolli JR, et al. Effects of intravenous administration of lidocaine and buprenorphine on gastrointestinal tract motility and signs of pain in New Zealand white rabbits after ovariohysterectomy. *Am J Vet Res.* 2017;78(12):1359-1371.

12. Gosliga JM, Barter LS. Cardiovascular effects of dopamine hydrochloride and phenylephrine hydrochloride in healthy isoflurane-anesthetized New Zealand white rabbits (*Oryctolagus cuniculus*). *Am J Vet Res.* 2015;76(2):116-121.

13. Sazuka S, Matsuura N, Ichinohe T. Dexmedetomidine dose dependently decreases oral tissue blood flow during sevoflurane and propofol anesthesia in rabbits. *J Oral Maxillofac Surg.* 2012;70(8):1808-1814.

14. Chang C, Uchiyama A, Ma L, et al. A comparison of the effects on respiratory carbon dioxide response, arterial blood pressure, and heart rate of dexmedetomidine, propofol, and midazolam in sevoflurane-anesthetized rabbits. *Anesth Analg.* 2009;109(1):84-89.

15. Botman J, Hontoir F, Gustin P, et al. Postanaesthetic effects of ketamine-midazolam and ketamine-medetomidine on gastrointestinal transit time in rabbits anaesthetised with isoflurane. *Vet Rec.* 2020;186(8):249.

16. Grint NJ, Murison PJ. A comparison of ketamine-midazolam and ketamine-medetomidine combinations for induction of anaesthesia in rabbits. *Vet Anaesth Analg.* 2008;35(2):113-121.

17. Müller-Lissner S, Bassotti G, Coffin B, et al. Opioid-induced constipation and bowel dysfunction: a clinical guideline. *Pain Med.* 2017;18(10):1837-1863.

18. Feng ZY, Mao LG, Lu Y. Effect of morphine chloride on contractility of small intestinal muscle in vitro or in vivo and its mechanism. *Zhejiang Da Xue Xue Bao Yi Xue Ban.* 2008;37; 271-275.

19. Martin-Flores M, Singh B, Walsh CA, et al. Effects of buprenorphine, methylnaltrexone, and their combination on gastrointestinal transit in healthy New Zealand white rabbits. *J Am Assoc Lab Anim Sci.* 2017;56(2):155-159.

20. Pathak D, Di Girolamo N, Maranville R, et al. Effects of injectable analgesics on selected gastrointestinal physiologic parameters in rabbits. *Proc ExoticsCon.* 2021:342.

Understanding Timing and Techniques for Intestinal Surgery of Rabbits

Nicola Di Girolamo, DVM, PhD, DipECZM (Herpetology), Dipl ACZM, Dipl ACVPM

Session #3492

Affiliation: From Exotic Pet Service, Department of Clinical Sciences, College of Veterinary Medicine, Cornell University, 930 Campus Road, 14853, Ithaca, NY, 14853, USA.

Abstract: Rabbits often are presented for the sudden onset of anorexia, lack of defecation and different degrees of gas build-up in the gastrointestinal tract. Due to the anatomic inability of the rabbit to vomit, these conditions typically result in life-threatening gastric and enteric distension. Although in some cases medical treatment can be sufficient, there are conditions when surgical removal of a mechanical obstruction in the intestinal tract or of inflammation of the appendix is required. In this masterclass, the major surgical approaches to the intestinal tract in rabbits are discussed, including enterotomy, enterectomy and appendectomy, with a special focus on finding the correct timing for these interventions.

Introduction

Rabbits often are presented for the sudden onset of anorexia, lack of defecation and different degrees of gas build-up in the gastrointestinal (GI) tract.[1,2] Due to the anatomic inability of the rabbit to vomit, these conditions typically result in life-threatening gastric and enteric distension. Although in many cases medical treatment can be sufficient to resolve acute gastric distension, there are conditions when surgical removal of a mechanical obstruction in the intestinal tract is required. It is critical to distinguish early between cases with functional/paralytic ileum and cases with mechanical obstruction, and the veterinarian needs to be aware of the potential surgical options. In this masterclass, the focus is on the main surgical approaches to the intestinal tract in rabbits, including enterotomy, enterectomy and appendectomy.

Causes of Mechanical Obstruction

In a case series reporting outcomes of 84 events of gastric dilation in rabbits, 64 cases had an intestinal obstruction confirmed.[1] The other 20 cases recovered without surgery or, in four cases, were euthanized and the owners declined a postmortem examination. The striking majority of the events (49/64, 76%) were caused by pellets of compressed hair (trichobezoars) similar in size to a large fecal pellet (approximately 1 centimeter in size). The second most common cause for mechanical obstruction was the presence of an intestinal neoplasm (5/64, 8%).

Diagnostics

Diagnostic imaging is typically employed to discriminate rabbits with mechanical obstruction that requires surgery. Currently, the author's preferred way to locate the exact position of the obstruction and their size is computed tomography. Depending on size and location of the obstruction, radiographs repeated at 6- to 12-hour intervals are useful to determine the evolution of gas build-up in the GI tract of anorectic rabbits. Ultrasound is often diagnostic for observation of primary or secondary signs of GI obstruction.

Surgical Treatment

Peri-operative management and stomach decompression

Anesthetic management of patients with potential obstruction is critical. In the author's experience, rabbits suffering mechanical obstruction often have different degrees of metabolic alkalosis, and the anesthetic plan including fluid therapy should be corrected accordingly. The rabbit is pre-medicated, induced and endoscopically intubated. Once the rabbit is stable, the stomach is decompressed. A rubber tube of 18- to 24-Fr (larger size depending on the size of the rabbit), that has been previously modified by creation of multiple, elliptical (10 × 4 mm) holes is passed through the oral cavity into the stomach. The tube is attached to a 60-ml cone syringe in which the cone has been shortened to provide a larger diameter. This is critical as the ingesta in the stomach is often mixed with hair that can block a traditional syringe. Typically, 150-300 milliliters of ingesta mixed with hair can be removed by this method. Once the stomach is decompressed the rabbit is prepped for standard laparotomy.

Laparotomy, detection and isolation of affected intestinal tract surgical techniques

A 4- to 5-centimeter midline laparotomy incision is made through the skin and linea alba. The linea alba incision is performed with forceps rather than scalpel in order to avoid iatrogenic rupture of the GI tract. An auto-static retractor is positioned to maximize exposure of the abdomen. Without exteriorizing the entire intestinal tract, the stomach and the duodenum are gently palpated, progressing distally over the small intestine until the site of obstruction is detected. Only the affected intestinal section is then exteriorized. Wet gauze sponges are placed around the isolated intestinal section and either sterile hair clips are placed proximally and distally to the obstruction or a second surgeon may perform a manual compression. In the author's experience the use of stay sutures is contraindicated in the intestinal tract of the rabbit due to its fragility.

Enterotomy surgical technique

Once the affected tract is isolated, an incision, usually 4-5 millimeters in length, but dependent on the size of the foreign body, is performed on the proximal aspect of the foreign body. The foreign body is grasped with forceps and removed through the incision. Care is paid to avoid elongation of the surgical incision of the intestine at the moment of exteriorizing the foreign body. The intestinal tract is sutured with 5-0 or 6-0 PDS, in a simple interrupted pattern. Sutures are positioned approximately 1-2 millimeters from each other and typically four or five sutures are sufficient. The suture encloses 1 millimeter of intestinal margin. Including excessive intestinal margins would result in narrowing of the intestinal lumen and potential future stenosis. A leak test is performed administering 0.5-1 milliliter of saline with a syringe with a 25-gauge needle or smaller. The intestinal tract is flushed, the surgeon(s) change gloves and surgical instrument pack. The abdomen is explored for further foreign bodies or other abnormalities. The laparotomy incision is closed as per surgeon preference.

Enterectomy surgical technique

In cases of intestinal tract neoplasia, or in cases where the intestinal tract is severely damaged, or even perforated, due to an untreated foreign body, enterectomy needs to be considered. The preparation of the patient is identical to what is described above. After visualization of the affected tract, the surgeon needs to evaluate the diameter of the intestinal lumen proximally and distally to the pathologic intestinal section. In the author's experience rabbits with chronic subobstruction such as the one caused by neoplasm, often show an abnormal distension of the intestinal tract proximal to the affected section. In these cases, an incision perpendicular to the intestinal length is performed proximally to the affected section. An incision at 45 degrees is performed distally to the affected length. If the oblique incision is not sufficient in making intestinal tract of the same lumen diameter, one side of the proximal intestinal end is sutured in order to result in a lumen diameter similar to the one of the distal end. An end-to-end anastomosis is then performed by placing simple interrupted sutures (PDS 5-0, 6-0) at 1-2 millimeters from each other around the entire perimeter of the intestinal lumen. A leak test is performed administering 3-5 milliliters of saline with a syringe using a 25-gauge needle or smaller proximal to the anastomosis and by manipulation of the tract to expose any potential leakage. The intestinal tract is flushed, the surgeon changes gloves and the surgical instrument pack. The abdomen is explored for further foreign bodies or other abnormalities. The laparotomy incision is closed as per surgeon preference.

Appendectomy surgical technique

Appendicitis has been increasingly described in rabbits.[3,4] Since this condition may result in sepsis and death, in cases unresponsive to medical treatment, surgical options need to be considered, and appendectomy has been successfully performed. The approach to the intestinal tract is similar to the one described above for other intestinal surgery. In this case, the altered appendix is usually readily visualized following the incision on the linea alba. The ileocecal fold needs to be excised to isolate the rabbit appendix from the remainder of the intestinal tract. Excision of the ileocecal fold can be achieved with cautery or multiple sutures. Considering the large number and small size of the vessels, and the length of the ileocecal fold, bipolar radiosurgery is the best option for excision of this ligament. After excision of the ileocecal fold, one or two circumferential ligations are placed at the base of the appendix (leaving not more than 3 millimeters of appendix), followed by placement of an intestinal clamp distal to the sutures and subsequent excision of the appendix. The cecum is checked for potential leaking, and the abdomen is routinely lavaged and closed.

Post-operative care

Rabbits undergoing intestinal surgery require intensive post-operative monitoring. Assisted feeding is started 2-6 hours after surgery, typically with watery Emeraid Herbivore or Oxbow Critical Care. Fluid therapy is typically continued for the following 2-3 days. Pain medication and antibiotic treatment may be continued for 5-10 days depending on the case.

References

1. Harcourt-Brown FM. Gastric dilation and intestinal obstruction in 76 rabbits. *Vet Rec.* 2007;161(12): 409-414.

2. Di Girolamo N, Petrini D, Rota S, et al. Clinical features of acute necrotizing appendicitis in 8 pet rabbits. *Proc Annu Conf Assoc Exot Mamm Vet.* 2017;277.

3. Huynh M, Vilmouth S, Gonzalez MS, et al. Retrospective cohort study of gastrointestinal stasis in pet rabbits. *Vet Rec.* 2014;175:225.

4. Longo M, Thierry F, Eatwell K, et al. Ultrasound and computed tomography of sacculitis and appendicitis in a rabbit. *Vet Radiol Ultrasound.* 2018; doi: 10.1111/vru.12602.

Section 22

AEMV Masterclasses Track 3

Victoria Bender, DVM;
Mary Cate Miller, DVM
Session Chairs

Gross Lesion Recognition for the Small Mammal Clinician

Michael M. Garner, DVM, DACVP

Session #3254

Affiliation: From Northwest ZooPath, 654 West Main, Monroe, WA 98272, USA.

Note: Abstract format selected by author(s). See Front Matter for description of conference proceedings manuscripts submitted as scientific abstracts.

Note: Reprinted with permission. Garner MM. Gross lesion recognition for the small mammal clinician. *Proc ExoticsCon.* 2022;206.

Abstract: Small mammal veterinarians in clinical practice, by nature of their specialty, often have to select diagnostic procedures that will provide the best answers and guide the least expensive treatment options. A familiar course of action is establishing a viable list of differential diagnoses during clinical exam or necropsy based on the gross appearance of the lesions. Gross lesion recognition (and photography), like other imaging modalities, is a bit of an art form. Vast zones of grey may confront the investigator when considering what the lesion is, what it could be, and what to do with it. And nobody is good at all of it. A seasoned livestock pathologist could possibly clear an entire day's necropsies without the help of a single histology slide. A bee keeper would recognize immediately the maggot stages that parasitize his colony. A dog-cat pathologist in a busy private laboratory or a companion animal clinician may not have a clue about gross lesions in livestock or bees. Historically, our attempts at accurately diagnosing disease by gross lesion recognition have been a humbling experience, and that has been the impetus for all further diagnostic specialties. Gross lesions, or lack thereof, are the first visual indication of what may be wrong with the patient (or cadaver, as the case may be). These are the lumps, the effusions, the asymmetrical oddities, the discolorations, the odors of disease that (hopefully) stimulate a scientific thought process culminating in a list of differential diagnoses. The purpose of this 2-hour interactive workshop is to present clear and not-so-clear images of common and not-so-common gross lesions in live and necropsy specimens, and in a participatory manner, establish a differential diagnosis and means for establishing a definitive diagnosis.

Recommended Reading

1. King JM, Dodd DC, Roth L, Newson ME. The Necropsy Book. Charles Louis Davis Foundation, USA, 2002.

2. Garner MM. An overview of biopsy, cytology and necropsy techniques for clinical veterinarians. In: Aguilar RF, Hernandez SM, Divers SJ, Perpinan D, eds. *Atlas de Medicina de Animales Exoticos.* 2 da. Buenos Aires: Inter-Medica S.A. I.C.I.; 2010:437-443.

3. Garner MM. Biopsia y necropsia. In: Aguilar R, Hernandez-Divers SM, Hernandez-Divers SJ, eds. *Atlas De Medicina, Terapeutica y Patologia De Animales Exoticos.* Buenos Aires: Inter-Medica S.A. I.C.I.; 2005:361-375.

Rabbit Calcium Metabolism, Bladder Sludge, and Urolithiasis

Dan H. Johnson, DVM, DABVP (Exotic Companion Mammal Practice)

Session #3405

Affiliation: From Avian and Exotic Animal Care, 8711 Fidelity Boulevard, Raleigh, NC 27617, USA.

Abstract: Calcium is the most abundant mineral in the mammalian body, and the major component of bones and teeth. In addition, calcium is involved in a variety of vital physiologic processes including blood coagulation, muscle contraction, membrane permeability, nerve conduction, enzyme activity, and hormone release. Calcium balance is achieved by the interaction of various control mechanisms which regulate its absorption and excretion. In most mammals, the interaction of these regulatory mechanisms maintains serum calcium levels within a narrow range. In contrast, rabbits have adapted a unique strategy in which most dietary calcium is absorbed in the intestine, and the excess is excreted in the urine. In this species, calcium levels may vary within a wide range, and in direct proportion to the dietary calcium intake. This unique calcium metabolism has implications for the health and husbandry of the rabbit.

Calcium Metabolism in the Rabbit

Calcium metabolism in the rabbit differs in several respects from other mammals. In most species, calcium absorption is closely regulated to balance metabolic needs, and serum calcium concentrations are maintained within a narrow range, typically 1.25-1.6 mmol/L (5.0 – 6.4 mg/dl). Total serum calcium in the rabbit is 30-50% higher than in other mammals and varies over a wide range from 3.25-3.75 mmol/L (13-15 mg/dl). The intestinal absorption of calcium in most mammals involves primarily vitamin D_3-regulated active transport. In the rabbit, calcium is absorbed in direct proportion to the quantity ingested rather than in accordance with metabolic need, and calcium absorption is relatively independent of vitamin D.

Most mammalian species grow only one or two sets of teeth in their lifetime. In contrast, rabbit teeth are constantly erupting at a rate of approximately 2-2.4 mm/week. The rabbit's increased life-long demand for calcium compared to other mammals is met by its efficient intestinal calcium absorption. In addition, during normal dental wear, calcium is released from the teeth, swallowed, and reabsorbed from the intestine.

In the rabbit, urinary calcium excretion increases in parallel to dietary intake. The fraction of calcium that can be filtered out of the blood is higher than in other mammals. The fractional excretion of calcium for most mammals is less than 2%; the range in rabbits is 45%-60%. When the reabsorptive capacity of the kidney is reached, calcium precipitates as calcium carbonate in the alkaline urine of the rabbit, causing cloudy or sludgy urine. When metabolic demand for calcium is increased by growth, pregnancy, lactation, or metabolic disorders, less calcium is excreted, and the urine appears clear.

Hormonal Regulation of Calcium Metabolism

The main hormonal regulators of calcium homeostasis are parathyroid hormone (PTH), calcitonin and active vitamin D_3 (1,25 $(OH)_2D_3$). Other hormones including estrogen, testosterone, prolactin, growth hormone, glucagon, and gastrin, as well as corticosteroids and other minerals also play a role in calcium regulation.

Parathyroid hormone

Parathyroid hormone is the principal hormone involved in minute-to-minute fine regulation of blood calcium in mammals. It is secreted by the chief cells of the parathyroid gland in response to decreases in serum ionized calcium or 1,25 $(OH)_2D_3$ concentrations. In the kidney, PTH acts on the renal tubules to increase calcium reabsorption and promote phosphorous excretion. By increasing the activity of alpha-hydroxylase in renal tubules, PTH is also a major stimulator of renal 1,25 $(OH)_2D_3$ synthesis, which increases duodenal calcium absorption. In bone, PTH stimulates osteoclastic bone resorption thus increasing the release of calcium and phosphate into the blood. PTH levels have been shown to rise post-prandially, stimulating the release of gastrin and digestive hormones which are postulated to affect serum calcium concentrations. The result of these actions is a rise in serum calcium concentration, which inhibits further release of PTH by negative feedback. These actions also result in decreased urinary calcium excretion.

Rabbits exhibit a unique pattern of renal response to PTH. The rabbit's ionized calcium concentration is protected from hyper- and hypocalcemia by rapidly changing PTH and calcitonin secretion. Changes in PTH secretion are seen only at relatively high calcium concentrations, levels that are the physiologic norm for the rabbit. Despite having a high serum calcium concentration, rabbits have readily measurable levels of PTH which are dramatically reduced by infusion of calcium. This implies that the parathyroid gland and PTH actively contribute to calcium homeostasis in this species. In vitro perfusion studies of the rabbit nephron have demonstrated that PTH increases calcium transport in the cortical segment of the loop of Henle, but not in the medullary segment and causes an increase in calcium absorption in the connecting tubule. It is the distal nephron segments that play the key role in determining the calcium excretion in the final urine, and this content is regulated by the actions of PTH and calcitonin, and the presence of other minerals, particularly sodium.

Calcitonin

Calcitonin is produced by the C cells of the thyroid gland and released in response to increases in plasma calcium concentration. Release of calcitonin is also enhanced by estradiol, glucagon, and GI hormones including gastrin, cholecystokinin and secretin. Calcitonin levels increase post-prandially, especially following a high-calcium meal. Neonates and pregnant and lactating animals also demonstrate increased calcitonin levels. Calcitonin decreases PTH-stimulated osteoclastic bone resorption and lowers the cytosolic calcium concentration in bone cells, thus decreasing calcium efflux from the pool of labile bone calcium. The hormone may aid in sequestering calcium and phosphate in bone in a form in which it could be available for fasting needs. Renal tubular reabsorption of calcium and phosphate are decreased by calcitonin. In the GI tract, calcitonin causes decreased secretion of gastrin and gastric acid, and causes increased small bowel secretion of sodium, potassium, chloride, and water. No major effect has been shown on regulation of intestinal calcium absorption. The result of these actions is a lowering of serum calcium concentration.

There is marked species variability in the role of calcitonin in calcium homeostasis. In vitro studies with isolated rabbit nephrons have demonstrated specific calcitonin receptors in the kidney that are distinct from PTH receptors. The hormone increased net calcium flux in distal portions of the nephrons resulting in an overall calciuric effect in vitro. Still, the physiologic effects of calcitonin in the rabbit remain unclear. One group found that intramuscular injections of calcitonin decreased serum calcium concentration while another concluded that there was no change. No consistent effect has been demonstrated on urinary excretion of calcium in the rabbit.

Vitamin D

The precursors of active vitamin D_3 come from the diet or are synthesized in the epidermis from 7-dehydrocholesterol. The conversion of 7-dehydrocholesterol is catalyzed by ultraviolet irradiation from sunlight. Dietary precursors are absorbed in the proximal small intestine and modified by hydroxylation in the liver and kidneys. In the kidney, activation of the enzyme 1-alpha-hydroxylase results in the final conversion of 25-hydroycholicalciferol to 1,25 $(OH)_2D_3$. Synthesis of active vitamin D_3 is stimulated by PTH, and may also be increased by prolactin, estradiol, and growth hormone. Increased serum calcium and vitamin D concentrations will decrease the rate of conversion to active vitamin D_3. The main function of vitamin D is to regulate the absorption of enough calcium and phosphorus to ensure adequate concentrations for bone mineralization. In most mammals, vitamin D is the principal regulator of intestinal calcium and phosphorus absorption. The hormone is also necessary for osteoclastic bone resorption and mobilization of calcium from bone and increases renal tubular reabsorption of calcium and phosphorus. Vitamin D also decreases the formation and secretion of PTH by the parathyroid gland.

When rabbits are fed a diet deficient in vitamin D, serum PTH levels rise, but there is no detectable change in serum calcium levels. Chronically vitamin D deficient rabbits showed no change in net intestinal absorption of calcium or phosphorus as compared to controls, while urinary excretion of both minerals was greatly decreased. These results emphasize the vitamin D-independent nature of intestinal calcium absorption in the rabbit, as well as the importance of renal conservation of calcium and phosphorus in vitamin D deficient rabbits. Vitamin D also plays a role in insulin secretion and glucose tolerance in the rabbit, and helps maintain the transplacental calcium gradient in pregnant does.

Other hormonal regulators of calcium homeostasis

In humans, both estrogen and androgens are known to regulate renal calcium transport. This regulatory effect may partly account for the negative calcium balance that results from deficiency of these hormones. Estrogens also increase intestinal absorption of calcium, elevate total serum calcium concentration, and may influence the conversion to active vitamin D_3. Testosterone and progesterone both enhance reabsorption of calcium in the rabbit kidney. Testosterone has been shown to have a determinant effect in the formation of urinary stones in rats, while estrogen has a protective effect. The effects of gonadectomy on calcium homeostasis in the rabbit have yet to be determined. Sex steroids have been shown to play a role in decreasing the calcium content of atherosclerotic plaques in spayed rabbits fed a high-cholesterol diet.

Prostaglandins of the E series are potent activators of bone resorption. PGE1 causes increased synthesis of active vitamin D and subsequent hyperabsorption of calcium at the intestinal level. When administered intravenously to rabbits, PGE1 increased serum calcium concentration. PGE2 increases osteoclast numbers leading to increased bone resorption.

Exogenous steroid administration and endogenous glucocorticoids have been associated with decreased intestinal calcium absorption and increased renal calcium loss. Adrenal cortical steroids have actions that are antagonistic to vitamin D and PTH, and can affect responses of plasma calcium, phosphorus, PTH and calcitonin. Administration of dexamethasone to rabbits resulted in calcium retention in the wall of the duodenum. Suckling rabbits treated with methylprednisolone exhibited decreased intestinal calcium. Another group found decreased serum $1,25(OH)_2D_3$ concentration, decreased bone trabecular volume, and increased rate of bone turnover in methylprednisolone-treated rabbits.

Growth hormone (GH) is the principal hormone regulator of growth. It causes increased intestinal calcium absorption in humans and has been shown to produce hypercalciuria by an unknown mechanism. GH may also produce hypercalcemia, possibly by stimulating 1, $25(OH)2D_3$ production.

Several digestive and pancreatic hormones also affect mineral homeostasis. Insulin deficiency results in decreased intestinal calcium absorption in association with low levels of vitamin D. Pharmacologic doses of glucagon produce transient hypocalcemia, but the physiologic role of this hormone is not clear. The hypocalcemic effect of glucagon may depend on the presence of the thyroid gland in rabbits. Gastrin, secretin, and vasoactive intestinal peptide have been shown to alter extracellular calcium concentrations in animals, but their physiologic role has not been determined. It has been proposed that the association of the digestive enzymes and calcitonin in the post-prandial period may affect the hypocalcemic effects of these hormones. This may also explain the hypocalcemia seen in some cases of pancreatitis.

Urolithiasis and Bladder Sludge

Rabbit urine normally contains calcium-containing crystals, giving it a color and turbidity that varies from grapefruit juice to coffee with cream. Calcium sediment may combine with mucus, protein, and cellular debris, coming to rest in a dense layer along the ventral wall of the bladder. Bladder sludge can condense to a doughy consistency that resists going back into suspension. Bladder sludge may lead to cystitis, bladder distension, and other problems.

Etiology

Urolithiasis and sludgy urine are common in rabbits and may result from an exacerbation of normal physiologic calcium excretion.[1] When rabbits are fed a high calcium diet, urinary calcium excretion increases, but urine volume remains constant, increasing the likelihood of crystal aggregation and stone formation. The alkaline pH of rabbit urine also increases the risk of forming insoluble calcium precipitates. Genetic predisposition, dehydration, metabolic disorders, bacterial or parasitic infections, and nutritional imbalances may also be predisposing factors. Temporary obstruction of the kidney in rabbits recovering from hydronephrosis resulted in crystal aggregation and stone formation within a few weeks. This suggests that any condition causing urinary stasis (arthritis, *E. cuniculi* infection, spinal trauma or deformity, bacterial infection, lack of exercise, obesity) may also predispose the rabbit to urolithiasis.

As with many exotic animal diseases, it appears that most instances of bladder sludge are the result of improper husbandry. Calcium intake may not be as important to the etiology as fluid intake and exercise. Rabbits that become obese on a calcium-rich, protein-rich dry diet tend to move around less and to hold onto more concentrated urine. Lack of access to fresh water, water that tastes bad, or water that is not changed daily leads to chronic mild dehydration and concentration of the urine. Rabbits without adequate space to move are prone to obesity and are less likely to eliminate urine sediment. Those that have concurrent metabolic bone disease, arthritis, back problems, lameness, bladder atony or perineal pain may also be prone to bladder sludge.

Symptoms

Affected rabbits may be asymptomatic. Urine can vary from normal, to cloudy, to thick and muddy; it may appear darker than usual. Hematuria is possible and may be intermittent. Affected rabbits may urinate in an unusual area, at an unusual time, for longer than normal, or with increased frequency. Urinary incontinence and urine scald frequently occur. Straining, tail-bobbing, posturing, and vocalization are rare and may be mistaken for constipation. Affected rabbits may exhibit anorexia, weight loss, decreased stool production, GI stasis, lethargy, depression, hunched posture, bruxism, or otherwise indicate that they are in pain.

Diagnosis

Affected rabbits frequently show no symptoms and may urinate normally. Mild cases of bladder sludge may be discovered as an incidental finding on x-rays. Abdominal radiography will reveal a diffuse to solid mineral opacity in the bladder, often completely defining its borders. Palpation of the bladder may reveal a mass with a doughy consistency. Abdominal ultrasound and endoscopy are occasionally used. It is also important to assess the kidneys, ureters, and urethra. Bladder sludge is occasionally associated with stones. Urine culture is often negative.

Treatment

Mild cases of bladder sludge respond to increased fluid intake, exercise, and dietary changes. Increasing fluid intake will cause more urine to be produced, dilute the solutes in the urine, and enable more sludge to pass. Subcutaneous fluids can be administered during the initial phase of treatment, 20-30 ml/kg q8-12h, and owners can be taught how to continue treatment at home. Increase the amount of water the rabbit will drink by flavoring the water with fruit juice. Find out which fruit juice the rabbit likes best by giving it directly, via syringe or in a bowl. Add its favorite juice to the drinking water. Change the water daily and monitor the water intake. Gradually reduce the proportion of juice in the water to the point that water consumption drops, and then return the quantity of juice to its previous level.

Get affected rabbits up and keep them moving. Take rabbits out for walks on a harness. Let them exercise outdoors. Make them use the stairs. Bladder sludge occurs more commonly in obese, sedentary rabbits. Exercise will help the rabbit lose weight, produce more urine and urinate more often. Agitating the bladder helps to stir up sediment, move it out, and reduces its tendency to settle. Urinating more frequently and completely is likely to improve muscle tone to the bladder, as well.

Reduce or eliminate pellets in the diet. Dry pelleted feed may contribute to chronic dehydration and the production of concentrated urine. Removing alfalfa-based hay and pellets will reduce the calcium load, protein and energy intake, and limit weight gain. Timothy hay-based products are preferred. Water intake is greatest is rabbits fed a grass hay-only diet. Add green moisture rich foods to the diet to force even more water consumption. The calcium contained in some fresh greens is bound by oxalate and is less bioavailable than previously believed. Dandelion greens, which are high in calcium, actually have a diuretic effect. The benefit of increased moisture intake appears to outweigh the increase total calcium that greens add to the diet.

For mild cases of bladder sludge, and even in some cases where sediment appears dense, gentle manual expression of the bladder can be achieved without anesthesia. The author stands the patient on its hind limbs. The rabbit's chest is lifted and supported by the author's right hand, and its back is supported against the author's body. With the left hand the author gently agitates the bladder and expresses it. When expressing the bladder with the patient in this upright position, the heavy sediment in the bladder tends to flow out the urethra with the urine. Expressing the bladder may also help the bladder to return to its normal size and tone. This procedure can be repeated once or twice daily when the bladder is full. This awake procedure should not be utilized for rabbits that are in obvious pain.

For more severe cases, where urine sediment is extremely dense or the rabbit is in pain, the bladder should be catheterized and repeatedly flushed while the patient is anesthetized. The goal is to remove as much sediment as possible, to remove irritation and encourage the bladder to shrink to its normal size. The patient is premedicated with buprenorphine 0.02 mg/kg SC for pain, and midazolam 0.25 mg/kg IM to improve relaxation. Fluids are administered subcutaneously or via intravenous catheter. The patient is placed under general anesthesia with isoflurane. A urethral catheter is carefully placed. For bucks, a 3.5- or 5-fr catheter is recommended, and placing the catheter is routine. For does, a 5- or 8-fr catheter is advanced along the vaginal floor while simultaneously expressing the bladder. Once the catheter is placed, the bladder is repeated flushed with warm sterile saline and expressed until it is clean. Agitating the bladder will increase the output of sediment, however, the bladder should be handled gently as it is thin walled and prone to trauma. Cystotomy is an alternative to this method; however, surgery can usually be avoided.

Urolithiasis may occur anywhere along the urinary tract. It does not indicate an emergency unless a stone has lodged in the urethra, impairing urination, and distending the bladder. The bladder can be emptied by cystocentesis and, if possible, the stone should be flushed back into the bladder by urohydropropulsion. A cystotomy or urethrotomy is performed in routine fashion. If urethral blockage cannot be cleared, a urethrostomy or cystostomy can be established and maintained until another means of unblocking the urethra is found.

Post-operative pain meds are important for the bladder flushing procedure. Administer a second dose of buprenorphine 12 hours after the first and give meloxicam 0.2 mg/kg PO q24 h for 5-7 days. Hospitalize the patient for observation. Change the litter box frequently in order to monitor urination. If there is bladder atony, manually express the bladder several times daily. Antibiotics are generally not indicated for bladder sludge but may be indicated after the bladder flushing procedure. The patient is discharged once he is able to urinate comfortably and is eating and drinking normally.

Monitoring

The patient is rechecked and radiographed every 2-4 weeks until the problem has resolved. For mild cases improvement will be noted with fluids, diet change, and exercise alone. If conservative therapy fails, then offer manual expression and/or bladder flushing.

Dietary Considerations

The unusual calcium metabolism of rabbits makes it essential to maintain a diet that is well-balanced with appropriate calcium concentration, calcium to phosphorus ratio, and vitamin D content. If too little calcium is consumed, secondary hyperparathyroidism and bone resorption may result, while if too much calcium is present, the risk of urolithiasis and bladder sludge is increased. A dietary calcium level of 0.22% supports normal growth, but 0.35-0.4% calcium is required for optimal bone calcification and growth rates in young rabbits. There should be less than 25 μg (1,000 IU) cholecalciferol/kg food to prevent soft tissue mineralization. The pet rabbit's diet should be chosen carefully to prevent selective feeding. If the rabbit rejects pellets and whole grain, calcium intake may be insufficient. Most commercial rabbit feeds contain calcium from alfalfa hay and calcium carbonate, which are very absorbable. While many vegetables are high in calcium, they often contain calcium oxalate, which is not metabolized by rabbits, and thus cannot contribute to urinary sludge or stones.

The ideal diet for pet rabbits should contain at least 15-16% fiber, 12-13% protein, 1-3% fat, 0.6-1.0% calcium, calcium to phosphorus ratio of 1:1 - 2:1, and less than 1000 IU/kg vitamin D. A diet comprised of grass hay versus legume hay, a small portion of commercial pellets composed of timothy hay, fresh vegetables, and fresh drinking water has been recommended.

In the past, most of the emphasis concerning the prevention and treatment of bladder sludge and urolithiasis has been placed on the calcium content of the diet. Today more emphasis is placed on water intake, exercise, and preventing obesity. Most instances of the disease can be addressed by improving these three factors.

Acknowledgements: The information on rabbit calcium metabolism comes from an excellent review by the author's good friend and coworker of many years, Christine Eckermann-Ross, DVM, CVA, CVCH, CTPEP.[2]

References

1. Di Girolamo N, Selleri P. Disorders of the reproductive and urinary systems. In: Quesenberry KE, Orcutt CJ, Mans C, Carpenter JW, eds. *Ferrets, Rabbits, and Rodents: Clinical Medicine and Surgery.* 4th Ed. St. Louis, MO; Elsevier; 2020:201-219.

2. Eckermann-Ross C. Hormonal regulation and calcium metabolism in the rabbit. *Vet Clin North Am Exot Anim Pract.* 2008;11(1):139-152.

Thymomas in Rabbits

Zoltan Szabo, DVM, Dipl ACZM, DABVP (Exotic Companion Mammal Practice), DABVP (Avian Practice), GpCert(ExAP), MRCVS

Session #3176

Affiliation: From Concordia Pet Care, 5-7, Blue Pool Road, Happy Valley, Hong Kong.

Abstract: The thymus is found in the mediastinum, cranioventral to the heart, and in rabbits, it does not involute but persists for life. Thymomas are the most common mediastinal tumors in rabbits. The masterclass discusses the currently available diagnostic and treatment methods for thymomas, emphasizing the surgical approach. Multiple case studies are included with intraoperative videos to illustrate the most effective surgical methods and post-operative care to treat this common condition.

Introduction

Thymoma, a benign thymic mass, is the most common mediastinal neoplasia in rabbits. The thymus is a specialized primary lymphoid organ of the immune system. Within the thymus, thymus cell lymphocytes or T-cells mature. The thymus is located in the craniad part of the chest, in the cranial mediastinum, behind the sternum, and in front of the heart. Normal healthy rabbits can have a persistent large thymus. Fat can also be found in this region of the chest in obese rabbits.

The thymic mass could be thymic hyperplasia, thymoma, thymic lymphoma,or thymic carcinoma.

Thymic hyperplasia

In adult rabbits, the thymus can become hyperplastic and enlarge up to three or four times normal size, grossly resembling a tumor but with no neoplastic characteristics on histologic examination.[1]

Thymoma

Thymoma is a benign tumor originating from the epithelial components of the thymus. They are generally slow growing and rarely metastasize. Thymomas are more common in rabbits than thymic lymphoma or thymic carcinoma.[1]

Thymic lymphoma

Thymic lymphoma is a malignant neoplasm of the lymphoid tissue of the thymus, and it has a T-lymphocytic origin. The clinical presentation of thymic lymphoma is similar to that of thymoma; however, it can metastasize.[1]

Thymic carcinoma

Thymic carcinoma is a rare malignant neoplasm of the epithelial cells of the thymus, and it is likely to metastasize to other areas of the body.[1]

Sub-Categories

Thymomas in rabbits have been graded as lymphocyte-predominant, mixed lymphoepithelial, and epithelial-predominant types. Staging of thymomas in humans is based on the assessment of local disease, the presence of microscopic versus macroscopic invasion into mediastinal tissues, and metastasis. The stage of the thymoma is an essential factor in selecting a treatment protocol and has also been correlated with response and survival. Unfortunately, staging is not yet available for animals with thymoma.[1]

Signalment

Thymomas usually occur in middle-aged and older rabbits (3-10 years). There is no sex predilection. Some rabbits are asymptomatic, and the thymoma can be an incidental finding on thoracic radiographs.[2]

Clinical signs

One of the most common clinical signs is breathing difficulties; dyspnea and tachypnea appear in 80% of the rabbits with thymoma. Flaring of the nostrils while breathing is often visible. With auscultation, one can notice the absence of breath sounds in the cranial thorax and caudal displacement of the heart and ictus cordis.[3]

Bilateral exophthalmos, the protrusion of the eyes, appears in 50% of the rabbits with thymoma. The exophthalmos is caused by cranial vena cava syndrome (precaval syndrome) due to the pressure of the mass in the cranial thorax. Exophthalmos due to thymoma is always bilateral, and the eye can be pushed back into orbit. These signs help to distinguish thymomas from retrobulbar abscesses or tumors. Rabbits with compromised heart function or rabbits exhibiting extreme fear would also have these bulgy eyes. Typically, the bilateral exophthalmos can be exacerbated by changing the body position (wheelbarrow test) .[4]

Rabbits with thymoma may present with generalized hair loss, scaly skin, and dermatitis. This condition is associated with sebaceous adenitis causing exfoliative dermatitis. Sebaceous adenitis is a T-cell–mediated autoimmune reaction directed at the sebaceous glands and hair follicles. Lipid metabolism defects are also implicated. Affected rabbits exhibit nonpruritic, chronic exfoliative dermatitis associated with areas of alopecia and are prone to secondary dermatitis.[5]

Diagnosis

Radiology

Radiology is the most common method to diagnose a thymoma in rabbits, and it is often accidental finding in patients radiographed for another disease. Thoracic radiographs show an increased radiodensity suggestive of a mediastinal mass and dorsal deviation of the trachea due to a mass effect.[6]

Computed tomography (CT)

The CT is a sensitive and specific diagnostic modality for identifying thoracic masses. It determines the tumor's location, extent, and inner structure, which could help decide the treatment options. CT is also helpful in identifying metastases.[6]

Ultrasonography

Ultrasonographic scanning is important for assessing the cardiac size and function, distinguishing the heart from pericardial masses such as thymomas, and identifying other pulmonary masses and pulmonary or pericardial effusion. If a CT scan is not available, an ultrasonographic examination can also help determine the mass's architecture.[6]

Cytology

Cytology samples can be obtained by ultrasound guided fine needle aspiration under general anesthesia. The cytologic results of thymomas may vary because of the usually poor exfoliation of epithelial cells compared with lymphocytes and the variable amounts of lymphocytic proliferation. Results may show a mixed population of small to intermediate mature lymphocytes and varying numbers of epithelial cells with thymomas. In contrast, a finding of primarily intermediate to large lymphoblasts would support a diagnosis of lymphoma.[6]

Histology

Tissue biopsy samples are often needed for a definitive diagnosis of thoracic masses. Thoracic needle core biopsy can be a diagnostic option; however, it may yield necrotic or cystic material that prevents a definitive diagnosis, and the procedure is much riskier than ultrasound-guided fine-needle aspiration. Therefore, histology after thoracotomy should be performed for the most accurate diagnosis.[6]

Treatment

Surgery, chemotherapy, radiation therapy, and therapeutic aspiration are the primary treatment options for thymoma in people and animals.[7]

Aspiration

Thymomas can be primarily cystic or solid. In rabbits with large cystic components of thymomas, removing fluid from the cyst by ultrasound-guided aspiration may alleviate dyspnea and improve quality of life by decreasing lung compression. This is a palliative option if radiation or surgery is not feasible for financial or clinical reasons. The aspiration can be repeated at varying intervals, usually every 3 to 6 months, as clinical signs of dyspnea or tachypnea recur.[7]

Chemotherapy

There is limited information regarding the efficacy of chemotherapy in the management of thymomas in rabbits. Unfortunately, there are no published case series about successful treatment.

Prednisolone

Prednisolone is often used in rabbits with thymomas for its antineoplastic and anti-inflammatory effects. It is used specifically to reduce the lymphocytic component of the tumor. In addition, corticosteroids are potentially useful in rabbits undergoing radiation therapy to reduce the side effects associated with irradiating the thoracic organs, notably the lung, where the development of radiation pneumonitis is a concern.[8]

Surgical excision

Surgery is the treatment of choice for thymoma in all species when there is a solitary mass because it provides the best chance of removing the entire tumor and curing the patient. Surgical removal of benign, noninvasive thymomas in dogs and cats results in long-term disease control and survival. Surgery is also the mainstay of treatment for thymoma in humans. In humans, recurrences of thymoma are described in 10-30% of cases up to 10 years after surgical resection.[9]

Radiation therapy

In general, radiation therapy has three different roles in treating thymomas: pre- surgery, post-surgery, and more frequently as the sole treatment without surgical intervention. For large tumors, one low dose of preoperative radiation may decrease the size of the mass enough to make it more feasible to remove surgically.[2] In addition, the success rate of radiation therapy in rabbits treated with multiple-dose protocols does seem to favor this treatment over surgery. Therefore, radiation therapy should be considered the first treatment of choice wherever available.[10]

Complications of thymic radiation therapy in a rabbit have been reported, including pulmonary fibrosis and thrombosis of thoracic vessels. In addition, the repeated anesthetic episodes needed for radiation carry some risk, especially if cardiopulmonary dysfunction secondary to the mass effect of the thymoma is present. To ameliorate these side effects, non-steroidal anti-inflammatory drugs or prednisolone may be considered.[2]

References

1. Kunzel F, Hittmair KM, Hassan J, et al. Thymomas in rabbits: clinical evaluation, diagnosis, and treatment. *J Am Anim Hosp Assoc.* 2012;48(2): 97-104.

2. Dolera M, Malfassi L, Mazza G, et al., Feasibility for using hypofractionated stereotactic volumetric modulated arc radiotherapy (Vmat) with adaptive planning for treatment of thymoma in rabbits: 15 cases. *Vet Radiol Ultrasound.* 2016;57(3):313-320.

3. Jekl V. Respiratory disorders in rabbits. *Vet Clin North Am Exot Anim Pract.* 2021;24(2):459-482.

4. Kostolich M, Panciera RJ. Thymoma in a domestic rabbit. *Cornell Vet.* 1992;82(2):125-129.

5. Florizoone, K., Thymoma-associated exfoliative dermatitis in a rabbit. *Vet Dermatol.* 2005;16(4):281-284.

6. van Zeeland Y. Rabbit oncology: diseases, diagnostics, and therapeutics. *Vet Clin North Am Exot Anim Pract.* 2017;20(1):135-182.

7. Morrisey JK, McEntee M. Therapeutic options for thymoma in the rabbit. *Semin Avian Exot Pet Med.* 2005;14(3):175-181.

8. Palmer A, Wu CC, Miwa Y, et al. Outcomes and survival times of client-owned rabbits diagnosed with thymoma and treated with either prednisolone or radiotherapy, or left untreated. *J Exot Pet Med.* 2021;3835-43.

9. Clippinger TL, Bennett RA, Alleman AR, et al. Removal of a thymoma via median sternotomy in a rabbit with recurrent appendicular neurofibrosarcoma. *J Am Vet Med Assoc.* 1998;213(8):1140-1143.

10. McCready JE, Poirier VJ, Fleck A, et al. Adaptive radiation therapy using weekly hypofractionation for thymoma treatment: a retrospective study of 10 rabbits. *Vet Comp Oncol.* 2022;20(3):559-567.

Anesthesia-Related Morbidity and Mortality and Recent Advances in the Perioperative Management of Rabbits

Inga-Catalina Cruz Benedetti, DVM, MSc, DECVAA

Session #3216

Affiliation: From the Department of Clinical Sciences, Faculty of Veterinary Medicine, Université de Montréal, 3200 rue Sicotte, Saint-Hyacinthe (QC) J2S 2M2, Canada.

Abstract: Although rabbits are popular pets and one of the most frequently used laboratory animal species, sedation and anesthesia remain challenging. As a prey species, they are easily stressed. Therefore, many diagnostic procedures well-tolerated in other domestic animals are only possible under sedation or general anesthesia. High-stress levels and inadequately managed pain are factors thought to contribute to the high incidence of peri-anesthetic death of up to 4.8% reported in this species. Gastrointestinal complications have been observed in 38% of rabbits undergoing sedation and general anesthesia. This has been associated with increased body weight, repeated anesthetic events, pain and dose-dependent drug side-effects. This lecture provides a synthesis of the current scientific literature with regard to the underlying pathophysiologic mechanisms and proven drug effects attributed to perioperative mortality and morbidity in rabbits. Particular attention is given to the respiratory and cardiovascular system and the stress response. Results of multiple studies addressing stress reduction, sedation, anesthesia, and pain management, recently realized by our research team, are presented. For example, recommendations for the use of oral trazodone for preoperative stress reduction in rabbits, which were obtained from a series of pharmacologic and behavioral studies, are provided. An adaptation of the transverse abdominal plane block for rabbits, a block commonly used to reduce postoperative pain after abdominal surgery in humans which is widely used in other veterinary species, is presented. Recommendations for bupivacaine dosage obtained from a pharmacokinetic study are included.

Introduction

Rabbits (*Oryctolagus cuniculus*) are frequently used for animal research due to their size, docile nature, and relatively short life cycles[1,2] and they are becoming increasingly popular pets.[3] Veterinarians are therefore successively exposed to rabbit patients to whom they must provide a variety of diagnostic and surgical procedures. Rabbits, being prey species, are easily stressed. Therefore, many diagnostic procedures, well tolerated in other domestic animals, are only possible under sedation or general anesthesia. Although great advances have been made in veterinary anesthesia over the last decades, peri-anesthetic morbidity and mortality remain significantly higher in rabbits compared with other small animals with an incidence of perioperative mortality of 1.39-4.8% and morbidity (i.e., gastrointestinal complications) of 38% reported in recent studies.[4,5] In a large multi-center study, anesthetic records of 8209 rabbits undergoing sedation or general anesthesia were analyzed. In that study, they were able to determine an overall mortality rate of 1.39% within 48 hours following an anesthetic procedure, with a risk of 0.37% in healthy animals (American Society of Anesthesiologist Physical Status Classification System [ASA] I-II) and a risk of 7.37% in sick animals (ASA III-V). This mortality rate was eight-fold higher than that obtained for dogs in the same study. Of the 97 rabbits that deceased, 6 died during induction, 29 during the maintenance phase and 62 deaths occurred in the postoperative phase, with the highest prevalence occurring 0-3 hours and 6-12 hours following surgery. In rabbits, the cause of death remained unknown in 57% cases indicating suboptimal patient monitoring, 3% were classified as purely cardiovascular, 13% as purely respiratory, 22% as either cardiovascular or respiratory and 2% neurologic.[4] Although these findings may be considered partially unsatisfactory, they provide a basis for further explorations into rabbit morbidity and mortality and provide the basis for this lecture. Additionally, they highlight the need for better anesthetic and analgesic approaches in this species.

Identified Risk Factors

Unfortunately, the above-mentioned studies were not able to identify specific risk factors for peri-anesthetic mortality in rabbits.[4,5] Nonetheless, these studies were able to pinpoint some causes by demonstrating that the majority of clearly identified peri-anesthetic deaths were cardiorespiratory in nature.[4,5] In a follow-up article, the comparatively low use of endotracheal intubation and oxygen supplementation and superficial monitoring were highlighted.[6] In addition, the high surface-area to volume ratio which predisposes rabbits to perioperative hypothermia and the frequency of comorbidities (e.g., respiratory, digestive and other diseases causing fluid and electrolyte imbalances) were addressed.[6] Lastly the small size of rabbits and related challenges (i.e., fewer accessible veins and difficult intubation) and the fact that many veterinarians have less routine in rabbit anesthesia were named as risk factors.[6] The fact that most anesthetic-related deaths were of unknown cause, is an invitation to critically analyze anesthetic surveillance in rabbits. Regarding peri-anesthetic morbidity, another study identified an increased risk for the development of gastrointestinal complications in animals with increasing body weight.[5] The choice of anesthetic drugs, relative overdosing in overweight patients, pain, and repeated anesthetic events were all discussed for their potential to impact patient outcomes.

Temperament

As prey species, rabbits are susceptible to stress in unfamiliar environments, such as the veterinary hospital or the laboratory setting. This stress causes the activation of the hypothalamic-pituitary-adrenal-axes and in turn, the release of catecholamines and glucocorticoids from the adrenal gland. This increase in circulating catecholamines leads to tachypnoea, tachycardia, and hypertension thereby increasing the tissue oxygen demands. Consequently, this may lead to myocardial damage, renal ischemia, gastrointestinal stasis and inappetence as well as a decrease in body temperature with its respective complications.[7] In the context of anesthesia, this may enhance the occurrence of anesthetic complications as these animals will often have increased anesthetic requirements, making them susceptible to cardiorespiratory complications as higher drug dosages increase the potential for adverse drug effects.

As a prey species, rabbits also hide signs of illness until disease is well advanced. Therefore, subclinical disease might be overlooked.

Cardiovascular System

Rabbit hearts are susceptible to multiple diseases, which is the reason why they are extensively used as animal models of cardiovascular diseases. They are particularly popular in the study of atherosclerosis-related heart complications as rabbits are prone to develop myocardial infarction following ischemic damage which, in the clinical setting, can be caused by both hypotension and hypoxemia. As mentioned above, stress induces catecholamine release. Since the rabbit myocardium is particularly sensitive to the effects of catecholamines, preoperative stress may predispose them to fatal arrhythmias.[8,9] Intraoperative hypotension is a frequently observed anesthetic complication, particularly during inhalant anesthesia, that may cause tissue hypoperfusion.[10,11] All of these issues might contribute to frequently observed cardiac arrests.

Respiratory System

As previously mentioned, stress behavior makes rabbits hide illness. Subclinical respiratory disease such as that caused by *Pasteurella multocida*, bacterial pneumonia associated with lung abscessation, and pulmonary metastatic adenocarcinoma may remain unnoticed but provoke respiratory complications during anesthesia and in the postoperative period.

Rabbit anatomy and predisposition for laryngospasm makes endotracheal intubation difficult for inexperienced veterinarians.[12,13] Repeated attempts or rough introduction of the endotracheal tube can cause laryngeal edema and might cause hypoxemia, which in the worst case may trigger cardiorespiratory arrest. It is especially important to preoxygenate rabbits before anesthetic induction, to desensitize the larynx with lidocaine prior to intubation and to provide pauses in which animals are allowed to breathe oxygen via a facial mask in between intubation attempts. The use of supraglottic airway devices associated with monitoring of capnography and pulse oximetry might be a good alternative for short procedures.[14] It is noteworthy that induction with isoflurane may cause excitation and breath holding in rabbits.[15]

Gastrointestinal System

Anesthetic duration, pain, opioids and α_2-agonists have all been associated with gastrointestinal complications in rabbits.[16-18] In a recent study, gastrointestinal complications were observed in 38% of 210 anesthetic events.[5] In this study, 80% of rabbits having two anesthetic events presented gastrointestinal complications after both interventions. The authors believe that this might indicate some individuals could be more predisposed to this morbidity than others. Regarding opioid-induced ileus in rabbits, it has been suggested that opioid-induced ileus might be easier to treat than pain-induced ileus.[19] It is important to ensure adequate analgesia in all patients.

Monitoring

Patient monitoring can be a challenge during rabbit anesthesia because commercially available equipment is not always adapted to small patients nor validated in the species. Clinical monitoring must remain a priority. Monitoring and equipment pitfalls may include ECGs and pulse oximeters whose maximum count is too low, inappropriately sized non-invasive blood pressure cuffs, dampened invasive blood pressure curves and high equipment dead space causing carbon dioxide rebreathing. These criteria should be considered when buying novel equipment.

Postoperative monitoring is essential for patient outcomes, as it has been identified as a pivotal moment in both morbidity and mortality studies. During this period, the patient should receive supportive care in the form of active heating, oxygen supplementation, (re-) administration of reversal agents (e.g., flumazenil), analgesic drugs, if necessary, and be gavaged, or assisted in feeding. Only a normothermic and reactive patient should be left unsupervised following its return to consciousness.

Pain Management

Pain evaluation and pain management are essential in the perioperative period since pain or nociception can cause a stress response. Balanced analgesic protocols include a combination of systemically administered drugs, such as non-steroidal anti-inflammatory drugs, lidocaine, opioids and local anesthetic techniques. Loco-regional anesthetic techniques, which are considered the ideal analgesic technique, are capturing the veterinary medical stage with a rapid increase of available techniques after adaptation from humans to domestic animal species. The number of loco-regional techniques described in rabbits is also increasing. The author's laboratory has described the feasibility of the transversus abdominis plane block, an ultrasound guided locoregional anesthetic technique that consists of the administration of local anesthetic in four aliquots with two injections per hemiabdomen into the facial plane between the transversus abdominis and internal oblique muscle in female New Zealand white rabbits undergoing ovariohysterectomy.[20] This technique provides a sensory blockade of the abdominal muscles, subcutaneous tissues, skin and parietal peritoneum.[21]

Preoperative Stress Reduction

The decrease of stress in the pre- and perioperative period should be the highest priority in rabbit anesthesia due to the plethora of negative effects it entrains and should include adequate pain management. In absence of a perfect anxiolytic drug for rabbits, it is only logical that drugs, proven to be efficient in other domestic species, should be evaluated in this species. In recent studies, the author's team has evaluated the pharmacodynamics and pharmacokinetics of trazodone (Summit Veterinary Pharmacy LTD, Aurora, ON, Canada; 50 mg/ml) administered orally to six and eight juvenile, female, intact New Zealand white rabbits. We were able to show that compliance to handling was improved for up to 4 hours following the oral administration of 20 mg/kg trazodone, with a maximum effect occurring after 80 minutes.[22] In a behavioral study we observed that activity was increased in the first 2 hours after treatment and that rabbits were mostly exploring their enclosures during this time, which was considered an anxiety-free behavior. Maximum plasma concentration ($C_{max:}$ 6,592.6 ± 586.5 ng/ml) following oral trazodone administration was obtained within 30 minutes in all rabbits (T_{max} 0.3 ± 0.11 hours) and the mean half-life ($t_{1/2}$) was 3.38 ± 0.49 hours.[23] Based on our results an oral dose of 20 mg/kg trazodone administered 2 hours prior to the veterinary visit might be of clinical interest and beneficial to patient welfare.

References

1. Fan J, Chen Y, Yan H, et al. Principles and applications of rabbit models for atherosclerosis research. *J Atheroscler Thromb.* 2018;25:213-220.

2. Mapara M, Thomas BS, Bhat KM. Rabbits as an animal model for experimental research. *Dent Res J.* 2012;9:111-118.

3. American Veterinary Medical Association. *U.S. Pet Ownership and Demographics Sourcebook.* AVMA; 2018.

4. Brodbelt DC, Blissitt KJ, Hammond RA, et al. The risk of death: confidential enquiry into perioperative small animal fatalities. *Vet Anaesth Analg.* 2008;35: 365-373.

5. Lee HW, Machin H, Adami C. Peri-anaesethetic mortality and nonfatal gastrointestinal complications in pet rabbits: a retrospective study on 210 cases. *Vet Anaesth Analg.* 2018;45:520-528.

6. Brodbelt D. Perioperative mortality in small animal anaesthesia. *Vet J.* 2009;182:152-161.

7. Scarabelli S, Nardini G. Gastrointestinal vasculitis and necrosis in a rabbit under long-term meloxicam treatment. *J Exot Pet Med.* 2020;34:26.

8. Flecknell PA. *Laboratory Animal Anaesthesia.* 4th Ed. Amsterdam, Netherlands: Elsevier Academic Press; 2016.

9. Varga M. Anaesthesia and analgesia. In: Varga M, ed. *Textbook of Rabbit Medicine.* 2nd Ed. Toronto, Canada: Butterworth-Heinemann Elsevier Ltd; 2014:178-202.

10. Tearney CC, Barter LS, Pypendop BG. Cardiovascular effects of equipotent doses of isoflurane alone and isoflurane plus fentanyl in New Zealand white rabbits (*Oryctolagus cuniculus*). *Am J Vet Res.* 2015;76: 591-598.

11. Barter LS, Epstein SE. Cardiopulmonary effects of three concentrations of isoflurane with or without mechanical ventilation and supramaximal noxious stimulation in New Zealand white rabbits. *Am J Vet Res.* 2013;74:1274-1280.

12. Johnson-Delaney CA, Orosz SE. Rabbit respiratory system: clinical anatomy, physiology and disease. *Vet Clin North Am Exot Anim Pract.* 2011;14:257-266.

13. Fick TE, Schalm SW. A simple technique for endotracheal intubation in rabbits. *Lab Anim.* 1987;21: 265-266.

14. Engbers S, Larkin A, Rousset N, et al. Comparison of a supraglottic airway device (v-gel®) with blind orotracheal intubation in rabbits. *Front Vet Sci.* 2017;4:49.

15. Hedenqvist P, Roughan JV, Antunes L, et al. Induction of anaesthesia with desflurane and isoflurane in the rabbit. *Lab Anim.* 2001;35:172-179.

16. Goldschlager GB, Gillespie VL, Palme R, et al. Effects of multimodal analgesia with low dose buprenorphine and meloxicam on fecal glucocorticoid metabolites after surgery in New Zealand white rabbits (*Oryctolagus cuniculus*). *J Am Assoc Lab Anim Sci.* 2013;52:571-576.

17. Deflers H, Gandar F, Bolen G, et al. Influence of a single dose of buprenorphine on rabbit (*Oryctolagus cuniculus*) gastrointestinal motility. *Vet Anaesth Analg.* 2018;45:510-519.

18. Benato L, Rooney NJ, Murrell JCJVa, et al. Pain and analgesia in pet rabbits within the veterinary environment: a review. *Vet Anaesth Analg.* 2019;46:151-162.

19. Johnston MS. Clinical approaches to analgesia in ferrets and rabbits. *Semin Avian Exotic Pet Med.* 2005;14:229-235.

20. Garbin M, Violette D, Watanabe R, et al. Perioperative analgesic effects of an ultrasound-guided transversus abdominis plane block with bupivacaine in rabbits undergoing elective ovariohysterectomy: a pilot study. *Proc Assoc Vet Anaesth.* 2022;111.

21. Portela DA, Romano M, Briganti A. Retrospective clinical evaluation of ultrasound guided transverse abdominis plane block in dogs undergoing mastectomy. *Vet Anaesth Analg.* 2014;41:319-324.

22. Watanabe R, Gibert A, Benito J, et al. Evaluation of three doses of oral trazodone and its impact on handling, activity, and physiological parameters in rabbits: a prospective randomized blinded cross-over study. *Proc Assoc Vet Anaesth.* 2022;67.

23. Gibert A, Watanabe R, Desmarchelier M, et al. Pharmacokinetics and behavioral effects of a single oral dose of trazodone in rabbits. *Proc Assoc Vet Anaesth.* 2022;95.

Not One More Guinea Pig: A Guide to Safe Anesthesia and Surgery

Eva Stoffels, PhD, and
René Remie, PhD

Session #3271

Affiliation: From Marumoto Veterinary Practice, Hoevenweg 18, 7722 PN Dalfsen, The Netherlands (Stoffels), René Remie Surgical Skills Center (RRSSC), Operetteweg 27, 1323 VK Almere, The Netherlands (Remie).

Abstract: This work describes the authors' experiences with surgery of pet guinea pigs. To reduce mortality, the principles of good surgical practice must be observed. The preoperative assessment of the patient plays a pivotal role. Both visual observation and quantitative markers (blood lactate and urinary parameters) are of importance. Detailed anesthetic protocols, including monitoring, are provided for typical procedures (cutaneous, dental, abdominal). The most common causes of complications are discussed, and troubleshooting measures are proposed.

Introduction

Anesthesia and surgery of guinea pigs are still considered challenging. Complications are common; in one study, perioperative mortality of about 4% was reported.[1] It was unclear whether this pertained to all situations where anesthesia was required or only to specific types of surgery, but this number can be nevertheless considered surprisingly high.

The author performs two or three surgeries daily, amounting to about 500 in a year. About 70% of these are dental procedures, where the mortality risk is practically zero. The second most frequent procedures are non-visceral soft tissue surgeries (tumor excision, vasectomy), with similarly low complication rates. Finally, only laparotomic surgeries are associated with some risks; however, these can be significantly reduced. In the author's experience, the overall mortality rate lies well below one percent. In this work, the authors share experiences and successful protocols based on cases from the practice.

Many factors determine the success versus failure of surgery including the patient's initial condition, the procedure's invasiveness, the surgeon's skills, the duration and depth of anesthesia, the choice of drugs, and, eminently the adequate analgesia protocol. The principles of good surgical practice and typical anesthetic protocols for different types of surgery are presented.

Finally, the most common postoperative complications are discussed: the instantaneous (i.e., in the operating theater) and the delayed problems (i.e., in the recovery, up to several days after surgery). The underlying mechanisms are elucidated, and troubleshooting protocols are provided.

Patient Assessment

To do or not to do?

Even when carried out optimally, surgery still causes pain and tissue damage. In addition, the inevitable release of catecholamines, cortisol and pain mediators causes postoperative immunosuppression[2] and makes the patient prone to opportunistic infections. Miracle surgeries do not exist; there will always be adverse effects for the animal to cope with. The problems arise proportionally to the duration and depth of the anesthesia and the degree of tissue damage. Therefore, the surgeon should always perform the risk-benefit analysis and make a prudent decision.

Correct evaluation of the patient has a pivotal impact on the success of the surgery. The American Society of Anesthesiologists (ASA) classification[3] (I-V, where I refers to a healthy animal and V to a moribund one) applies to guinea pigs, although assigning the animal to the proper category may sometimes be challenging. Nevertheless, even before a thorough health examination, the surgeon must address the following questions.

- *Can the animal survive the surgery?* Surgery never warrants instant improvement. The animal must overcome the perioperative stress. Patients belonging to ASA IV and V categories are unlikely to do so. In such cases, humane endpoints should be considered. For example, animals with severe respiratory distress or metabolic acidosis should not undergo surgery unless they can be stabilized first.

- *Is surgery essential?* Non-essential (elective) surgeries should be performed solely on ASA I patients. Thus, the surgery should be postponed until the animal is disease-free. For example, an animal with (even slight) dyspnea, digestive problems, or a poor body condition should not be neutered. In addition, old animals (> 4 years) should not undergo elective surgeries because the postoperative immunosuppression may prove fatal (see Complications).

- *Does surgery actually solve the problem?* Answering this requires a correct and complete diagnosis. Multiple issues may present, but it is prudent to tackle only the ones that require intervention. For example, a female with cystic ovaria should not be spayed. Ovarian cysts are a symptom (of chronic stress[4]) rather than a disease. Surgery will definitely not solve the problem. The cause of chronic stress (e.g., cystitis) must be addressed with priority.

- *Will surgery improve the quality of life?* This question addresses the ethical side of veterinary medicine. Many procedures are technically feasible and may even prolong the animal's life, but they result in severe debilitation, handicap and/or pain. For example, a permanent nasogastric tube, gastric cannulation or even total intravenous nutrition via a jugular catheter are possible. The authors would rather consider humane endpoints if these are palliative measures.

- *Are there preexisting diseases?* A well-considered decision must be taken if the animal is known for previous heart, kidney or liver issues. For example, an animal with end-stage kidney disease should not undergo surgery because the chance of recovery and life expectancy are slim.

- *Are there less invasive treatment methods?* Surgery should not be considered the gold standard but rather the last resort. Depending on the animal's condition, less invasive methods should be attempted first. For example, small-sized bladder or ureter stones do not always require surgery but can be expelled using certain medications.[5] Many uroliths in females can be removed using a cystoscopic procedure, thus avoiding laparotomy.

In the following paragraphs, methods of perioperative assessment are described in detail.

Pain recognition

Pain has devastating physiologic impacts on guinea pigs and presumably on other prey animals. The type of nociception determines the severity of the physiologic response.

- *Somatic pain* is the result of integumentary damage, muscle and skeletal issues. The neural signal transmission is uncomplicated, via sensory fibers through the neospinothalamic tract and a single synapse in the thalamus, with little projections onto the brain stem and the limbic system.[6] Somatic pain is well-localized and rarely accompanied by autonomous responses. Guinea pigs tolerate this type of pain very well – even deep wounds do not alter the animal's behavior, appetite and digestion. With adequate pain management, the prognosis is usually favorable.

- *Visceral pain* is the result of stimulation of various nociceptors in the thoracic and abdominal cavity (e.g., pleura, peritoneum, urogenital system). It is transmitted through the archi- and paleospinothalamic tracts, with multiple synapses in the spinal cord and projections onto the brain stem (reticular formation and periaqueductal gray, and indirectly nucleus raphe and locus coeruleus).[6] This type of pain is not well-localized but rather experienced as general malaise. A remarkable autonomous response (due to catecholamine and glucocorticoid release) results in circulatory, digestive and other (electrolyte, thermoregulation) disturbances. Guinea pigs poorly tolerate visceral nociception. An animal suffering any grade of visceral pain can be automatically considered ASA III or higher.

- *Dental pain* sensation may be seen as a hybrid of the abovementioned categories. The trigeminal nerve transmits the nociception and follows a multi-synaptic pathway in the spinal cord.[6] There is a moderate autonomous response. The animals are visibly sick and often develop secondary digestive issues. Sadly, this is the most common type of pain in guinea pigs.

Guinea pigs are often tricky to decipher. Like other prey animals, they tend to hide their pain and discomfort. Simultaneously, they are intolerant to (especially visceral) pain. Thus, animals with obvious pain symptoms are in a much worse condition than the examiner might think. In addition, the communication of pain and discomfort is quite different from other species: there is (usually) no vocalization, and the behavioral changes (body language, facial grimaces) may be subtle. Several attempts have been made to standardize the facial expressions (e.g., orbital tightening, whisker and ear position) in response to certain pain stimuli in rats, mice and rabbits.[7] Similar signs can also be used to evaluate pain levels in guinea pigs, but this approach is unreliable as it does not accurately reflect the severity of pain.

Recognition of pain, or more generally, the poor condition of an animal, requires experience and skills. Features indicative of severe (usually visceral) pain and malaise are listed below.

- *Body temperature.* Critically ill guinea pigs usually have a low body temperature. They feel cold to the touch, and their rectal temperature may be 37 °C or lower. ASA IV or higher.

- *Bradycardia.* A heart rate lower than 150 bpm in a conscious animal indicates serious problems. Surgery should be postponed until the patient is stabilized.

- *Mastication.* Healthy guinea pigs masticate involuntarily. The absence of this reflex is a sign of severe pain (ASA IV or higher).

- *Anorexia.* An animal that eats less than usual but still shows interest in feeding may suffer light to moderate pain, such as pain due to dental issues. An animal that totally refuses to feed is in a critical condition (ASA IV or higher).

- *Posture and muscle tone.* Guinea pigs are usually alert and exploring. Critically ill animals appear depressed and tend to gaze downwards (Fig 1). There is remarkable muscle relaxation upon handling. This is called the sedative effect of pain, most likely caused by activating α_2 adrenergic receptors in the brain stem.[8] Surgery should be postponed until stabilization.

- *Body condition and fur condition.* Cachexia and an unkempt coat are always signs of chronic suffering. Animals in pain may lose weight even if they (apparently) feed normally. This is the catabolic effect of pain. They also stop grooming and often display alopecia and other skin conditions (the telogen effluvium effect). These patients (usually ASA II and III) can undergo surgery, but caution is required.

- *Stereotypic behavior.* Signs such as bruxism, head nod, twitches or tremors usually indicate pain. Not all animals display these signs.

Figure 1. Postural changes due to visceral pain. Above: an animal with enteritis (right) and healthy animals (left). Note that the sick guinea pig does not lift its head. Below: a moribund animal with end-stage kidney disease.

Quantitative markers

While the visual evaluation usually provides sufficient information, in some instances, one should resort to unbiased, quantitative criteria such as those obtained through blood and urine analysis.

The authors do not advocate routine blood sampling before surgery. It inflicts a substantial amount of stress, while the outcome is usually of little diagnostic value. For example, the electrolyte concentrations (Na^+, K^+, Cl^-) can deviate from the normal range in healthy animals, while they can lie in the normal range in terminal animals. The renal (urea, creatinine) and hepatic (bilirubin, ALT) parameters have some positive predictive value, but they are elevated only in end-stage diseases (ASA IV/V, not suitable for surgery). The same is valid for bicarbonate; it is affected only in terminal situations (e.g., shock), but otherwise, it lies in the normal range. There are a few exceptions that do have a high diagnostic value, which are listed below.

- *Lactate.* Lactate is the product of anaerobic glucose metabolism; lactate accumulation indicates inadequate oxygen supply to the tissue due to circulatory failure or inadequate ventilation. Blood lactate in healthy guinea pigs is 0.7-3 mmol/L (n = 20, measured using a Roche Accutrend Plus meter). Hyperlactatemia (4-10 mmol/L) occurs in life-threatening disorders (n = 13), such as paralytic ileus/tympany, enterotoxemia, (septic) shock and heart failure. Values above 10 mmol/L are only seen in terminal animals (n = 7). The authors' cut-off value is about 7 mmol/L. Patients with higher values are unlikely to survive anesthesia and surgery. Lactate can also be measured in urine; the values are about 50% higher than the blood values, but they can be helpful in decision-making.

- *Creatine phosphokinase (CPK).* The typical values are below 200 IU/L. Elevated CPK indicates rhabdomyolysis, which in turn suggests a severe systemic disease. Animals with CPK > 1000 IU/L are unlikely to survive the surgery.

- *Blood or urine lactate* is particularly valuable in preoperative assessment. Small amounts of blood can be collected from the ear vein, as described by Nemeth et al.[9] This is relatively uncomplicated and does not excessively stress the animal.

- *Urinalysis* should be performed routinely because it yields crucial information about the patient's condition. Urine can be obtained by spontaneous micturition or by applying gentle pressure to the caudal abdomen. The following values are most informative:
  - *Urinary pH.* This is probably the most critical parameter in patient evaluation. Normal guinea pig urine is strongly alkaline (pH 8-9). Lowered values indicate acidosis and are only seen in critical patients. Interestingly, respiratory acidosis (e.g., due to a lower respiratory tract disease) yields only moderate pH reduction (7-8). Acidic urine (pH 5-6.5) always indicates life-threatening metabolic acidosis. Such values are usually accompanied by strongly elevated blood or urine lactate. The guinea pig may still appear alert, but if left untreated, it will die within maximally 48 hours due to multiple organ failure. The cause of metabolic acidosis (enterotoxemia, septic shock, cardiogenic shock etc.) should be established and treated if possible. Guinea pigs with metabolic acidosis must not undergo surgery as the survival rates are negligible.
  - *Hemoglobin.* This must be differentiated from hematuria, as is common in cystitis or urolithiasis.[5] In terminal patients, urine may contain hemoglobin due to intravascular hemolysis; the corresponding pH is usually strongly acidic (pH 5). The blood samples from such animals are hemolytic. Stabilization can be attempted but is seldom successful. Euthanasia should be considered.

- *Specific gravity (SG).* Normal values, measured with a refractometer, should not exceed 1.020 g/ml. Since SG reflects the hydration state of the animal, it should be known before administering any fluids. Fluid replacement in animals that are actually not dehydrated can lead to fatal lung and brain edema. If SG is higher than 1.020 g/ml, fluids warmed up to the body temperature (38 °C) may be administered at 20 ml/kg through the subcutaneous route in the shoulder region. This is the easiest and safest method of administration of large volumes of liquids to guinea pigs. The resorption from this site is relatively fast (less than 5 minutes). The IV route (cephalic vein) is also possible, but catheter insertion requires sedation. The IV fluids must be infused very slowly to avoid overloading the circulation. Usually, 0.9% NaCl, NaCl with glucose (blood glucose should be determined before) or lactated Ringer's are used. In cases of metabolic acidosis, these solutions should be avoided because of their low pH. Instead, Plasmalyte (pH 7.5) with acid-base correction can be used (see Complications).
- *Nitrogen species.* Nitrate is an important marker molecule. In the urine of healthy guinea pigs, it does not exceed 300 mg/L; however, it is markedly elevated (500-900 mg/L) in animals fed a high-nitrate diet (greens). These animals usually suffer from (chronic) cystitis and urolithiasis. In animals fed a low-nitrate diet, elevated values result from increased inducible nitric oxide synthase (iNOS) activity; nitrate is a nitric oxide metabolite.[10] Increased iNOS activity is a sign of an inflammatory process; values higher than 1000 mg/L are alarming. Animals with high urinary nitrate can still undergo surgery but are prone to complications.

After the assessment, and if necessary, stabilize the patient. To do this, the surgeon must choose an adequate protocol.

Protocols

Various techniques of laboratory animal anesthesia have been described in the literature.[11] Good anesthetic practice comprises unconsciousness, adequate analgesia, muscle relaxation and autonomous block (in certain types of surgery). The stages of anesthesia (anesthetic depth), initially described by Guedel,[12] are depicted in Figure 2.

After testing the available methods including inhalation versus injectable anesthesia, various (pre) medications, intravenous versus intraperitoneal or subcutaneous administration routes, epidural versus infiltration nerve block, the author (ES) has selected a relatively simple protocol that includes injectable drugs and inhalation anesthesia. Depending on the invasiveness of the procedure, three variants (i.e., light, regular, and heavy) can be distinguished. Even in heavy protocols, the anesthetic depth should never exceed stage 3.2, as explained in Figure 2.

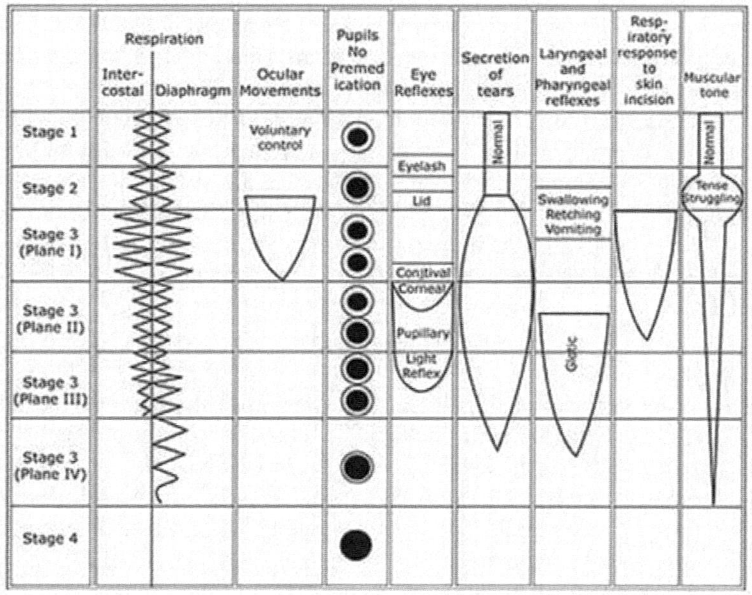

Figure 2. Stages of anesthesia after Guedel.12 Stages 3.1 or (maximally) 3.2 can be considered safe. Starting from 3.3, intercostal muscles start to get paralyzed. (Source: Guedel AE. Inhalation Anesthesia: A Fundamental Guide. New York, NY: The Macmillan Co.; 1937:14.)

General preparatory measures include:

- *Unlimited access to high-quality feed and water.* Guinea pigs must not be fasted before surgery, because this will increase the rate of gastrointestinal complications and significantly reduce survival. Anorexic animals should be fed about 1 hour before anesthesia and given parenteral fluids if necessary (see above). Just before induction, the mouth of the animal may be lavaged to remove food remains.

- *Preemptive administration of analgesics and antibiotics.* It is crucial to increase the threshold of sensory nerve firing even before the actual tissue damage has occurred. This will help to reduce postoperative hyperalgesia.[13] NSAIDs such as meloxicam (1 mg/kg SC) are usually sufficient. The author (ES) also advocates prophylactic use of antibiotics (e.g., enrofloxacin, 10-20 mg/kg SC) because this suppresses the intestinal pathogens and may moderate or eliminate post-surgical wound infections. Especially in unclean procedures, such as male neutering or vasectomy, dental procedures or abscess treatment, peri-operative use of antibiotics has undeniable benefits.

- *Other premedications.* Benzodiazepines, notably midazolam (1 mg/kg SC), have anxiolytic properties, cause anterograde amnesia and produce excellent muscle relaxation.[14] Preoperative administration of midazolam facilitates isoflurane induction. Midazolam also acts as an appetite stimulant postoperatively. Therefore, it is highly recommended for lengthy procedures (> 20 minutes) because it reduces the risk of secondary gastrointestinal disorders resulting from anorexia. It does increase the interval between the cessation of the general anesthesia and the time the animal fully regains consciousness. In critical patients, this may have adverse effects. If desired, midazolam can be antagonized with flumazenil.

Induction

After administering the drugs of choice, the animal is placed on a heated pad or surface (Fig 3) to assure a constant body temperature (about 38 °C). Unfortunately, general anesthesia impairs the thermoregulatory center, making the animals extremely prone to hypothermia. Hypothermia should be avoided at any cost because it leads to bradycardia and cardiac arrest.

Artificial tears can be applied to prevent cornea desiccation. Unlike in rabbits, rats or chinchillas, this step is not essential in guinea pigs, which do produce tears during isoflurane inhalation. Most guinea pigs keep their eyes closed while anesthetized with isoflurane.

Anesthesia is induced with isoflurane at 5% while the oxygen flow is maintained at 1.5 L/min. The gasses are administered through a nose mask (guinea pigs are obligate nasal breathers). The authors believe intubation is not beneficial unless the procedure requires mechanical ventilation. The specific anatomy of the nasopharynx, especially the very tight palatal ostium (Fig 4), limits access to the larynx.[15] Several intubation methods have been described,[16] but they usually require relatively deep anesthesia (with concomitantly higher complication rates, see Fig 2) and may cause trauma to the vulnerable tissue. Airway obstruction is seldom an issue in guinea pigs. These animals cannot vomit and usually do not regurgitate unless $\alpha 2$ adrenergic agonists (medetomidine, xylazine) have been administered. The patency of the airway (absence of congestion) must always be checked prior to surgery (see Patient Assessment).

The duration of induction with 5% isoflurane determines the eventual depth of anesthesia. Only brief immobilization is sufficient for light procedures (e.g., dental corrections). Keeping the animal in dorsal recumbency should be possible, usually achieved after 5 to 10 seconds inhalation. For standard surgical anesthesia, inhalation is sustained until the pedal reflex, and the eyelid reflex disappears (about 15-20 s); withdrawal reflexes in response to deep pain sensation (pinching the paw) may still be present. After prolonged inhalation (> 30 seconds), the breathing pattern may change from costal to abdominal, and the breathing rate may slow down; the anesthesia is already too deep (stage 3.3 or higher in Fig 2), and the isoflurane concentration must be reduced immediately.

The optimal maintenance dose of isoflurane is 1.5%. Anesthetic depth can be regulated with reasonable accuracy, but the working levels should never be higher than 2-2.5% because of the risk of respiratory depression. Isoflurane anesthesia is suitable for more lengthy procedures because, at 1.5%, it does not lead to tissue hypoxia and lactate accumulation (Fig 4).

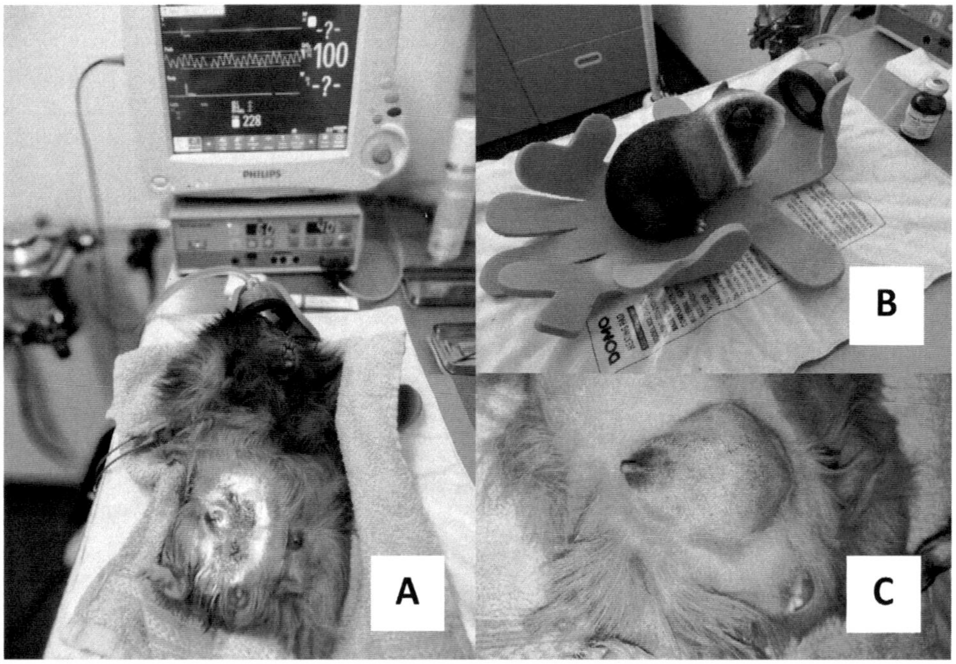

Figure 3. (A) A guinea pig on a heated pad, (B) fixed using a flexible holder. (C) In this case, a mammary tumor was excised.

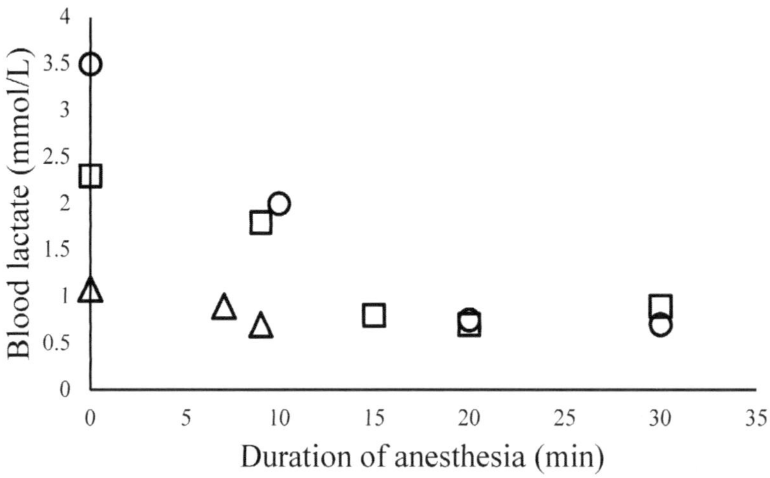

Figure 4. Blood lactate as a function of the exposure to 1.5% isoflurane in 1.5 L/min oxygen in three healthy animals.

Monitoring and troubleshooting

Generally, the deeper the anesthesia, the more careful patient monitoring is required. The most used options are pulse oximetry, ECG and capnography.

Pulse oximetry provides a very accurate and sensitive method of evaluating the animal's ventilation: it simultaneously records the oxygen saturation and pulse waveform. The sensor can be attached to the carpus or tarsus. Ears and tongue are not reliable, because of inadequate signal intensity. A problem arises in animals with pigmented skin; the technique is still applicable, but the signal quality deteriorates. If no waveform can be obtained, other monitoring techniques should be used. For example, capnography is possible without intubating the animal; however, a tightly fitting mask with an extra port is necessary. ECG is less suitable because it is not very sensitive to myocardial hypoxia. Besides, old or stressed animals may display aberrant ECGs, which are difficult to interpret.[17] ECG can still be used to detect the pulse in anesthetized animals, but it is not the first choice for monitoring rodents.

Optimally, the pulse oximeter displays a robust and stable signal, with a saturation between 90 and 100%. In dark pigmented animals, 80-90% is still acceptable. The waveform should be monophasic, and the rate should be between 160-220 bpm. Higher rates (230 bpm and more) suggest that the animal is merely immobilized but not fully anesthetized; this may be sufficient for minor procedures.

The most common problem is a drop in oxygen saturation to less than 80%. Most of the time, this usually occurs after induction and may be transient; however, measures should be taken if the animal fails to restore the normal saturation within minutes. For example, if the procedure is elective, one should consider abandoning it. Essential surgeries still must be carried out; doxapram (5-10 mg/kg SC or IP) can be used in such cases. Doxapram is an excellent respiratory stimulant which should always be available in the operating theater. The animal's head and thorax can be slightly tilted to alleviate the pressure on the diaphragm from the intestines.

Sudden changes in the pulse frequency and waveform are often indicative of problems. An increase in the heart rate suggests that the animal feels pain. Additional local anesthetics should be administered, and/or the isoflurane concentration can be temporarily increased. A decrease of the pulse frequency to less than 150 bpm accompanied by an irregular and weak waveform is an emergency situation, usually caused by a baroreflex following excessive catecholamine release (severe visceral pain) or hypovolemia (bleeding). Besides doxapram and lidocaine administration, atropine should be slowly titrated to effect in increments of 0.2 mg/kg (usually IP in laparotomic procedures). If hypovolemia due to blood loss is the cause, fluids should be administered (usually IP). The isoflurane concentration can be reduced to 0.5-1%, and more local anesthetics can be applied. The procedure should be finished as fast as possible.

Local anesthetics

Tissue damage is inherent to surgery. During tissue handling, inflammatory mediators (prostaglandins, cytokines) are released, triggering the nociceptive fibers to fire.[13] During general anesthesia, pain mediators and neurotransmitters accumulate, but pain transmission is temporarily blocked at the level of the brain stem (reticular formation). After termination of anesthesia, intense pain is experienced: even minor stimuli elicit a strong pain response (Fig 5). Postoperative pain slows the recovery and may cause the animal to auto-mutilate (e.g., chew on the sutures). Local anesthetics, especially lidocaine, provide a solution to this problem.[18] Lidocaine has very low toxicity in guinea pigs; it can be used for local infiltration at doses as high as 20 mg/kg. Adding epinephrine (e.g., xylocaine 2% with 5 µg/ml epinephrine) increases the duration of action of the lidocaine and may be suitable for most surgeries. Alternatively, longer acting anesthetics such as bupivacaine can be used (with caution).

Local anesthetics can be titrated to effect – additional amounts can always be administered during surgery, such as onto the incision site, into the periodontal ligament in dental procedures, or intraperitoneally during laparotomy. Lidocaine can also be used for epidural or paravertebral nerve blocks. However, this is much more difficult to titrate and adds complexity to the protocol, with no clear benefits.

Using local anesthetics allows us to keep the depth of anesthesia within safe limits, which results in much lower complication rates. Its advantages are enormous and should never be omitted in any surgery.

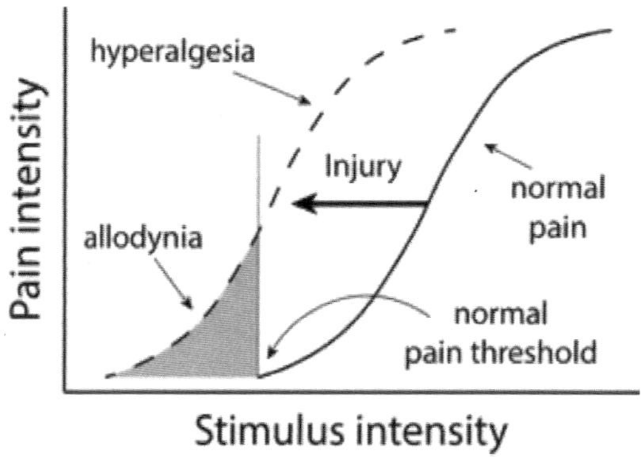

Figure 5. A schematic depiction of pain stimulus-pain response dependence. Tissue injury causes the curve to shift to the left; as a result, non-noxious stimuli are perceived as painful. (Source: Wikipedia).

Light, regular and heavy protocols

- *Light.* Short procedures (< 15 minutes) with minor tissue damage require no elaborate protocols or careful monitoring. There is no need to administer midazolam. The animal may even remain partly conscious (i.e., some reaction to pain stimuli may still be present, being a major advantage in dental procedures, which are often explorative). For example, while correcting the clinical crowns is painless because of their rudimentary innervation,[19] the actual sources of pain (e.g., wounds, abscesses, infected periodontal pockets, loose or fractured teeth) can be easily identified in this way. At the slightest pain reaction from the animal, lidocaine should be administered (to effect, 2-3 mg) directly into the periodontal space. The same light protocol can be applied to minor cutaneous surgery, like removing small skin tumors or opening abscesses other than odontogenic abscesses.

- *Regular.* Moderately invasive procedures that may cause substantial (somatic) pain require surgical anesthesia and monitoring. Therefore, the use of local anesthesia is always essential as part of a balanced, multi-modal analgesia/anesthesia plan When extensive tissue damage is expected, tramadol (20 mg/kg SC) should be added to the preoperative cocktail. Tramadol is relatively well-tolerated; unlike other opioids, it does not cause gastrointestinal complications. This is most likely due to its SSRI (selective serotonin reuptake inhibitor) properties.[20] Typical examples of regular procedures are male neutering or vasectomy, removing large tumors (e.g., mastectomy) and treating odontogenic abscesses.

- *Heavy.* This category contains all laparotomic procedures because they inevitably inflict visceral pain. The full preoperative cocktail should be administered, and preferably metamizole (100 mg/kg SC) should be given instead of meloxicam. Metamizole is an NSAID that is more effective in targeting visceral pain than meloxicam.[21] In the authors' opinion, administering midazolam to critical patients is not beneficial even if its action is reversed with flumazenil. Maropitant (2 mg/kg SC) can be added. Examples include urinary tract surgery including cystotomy or ureterotomy. During all laparotomies, special care must not be taken to traumatize the viscera (intestine), as this will cause postoperative ileus (see Complications).

Good Surgical Practice

The principles of good surgical practice were introduced in the XIX century by an eminent American surgeon, William Stewart Halsted. These principles can still be considered the gold standard in modern surgery.[22] The whole concept can be summarized in one sentence: the tissues should be treated respectfully. Therefore, in practice, the following must be observed:

- *Strict asepsis.* Thorough disinfection of the incision site can be performed using chlorhexidine solution - alcoholic on the skin, water-based on vulnerable surfaces such as mucosae. Povidone-iodide can be used as well, but not in combination with chlorhexidine. Contamination of the wound with iodine will impair wound healing. The authors prepare the sterile operation site using Press'n Seal cling film.[23] This provides an antibacterial and hydrophobic barrier that sticks to the skin. After disinfection of the film (alcoholic chlorhexidine), the adhesive spray is applied, and a sterile drape is fixed on the top (Fig 6). This double barrier provides good protection. Clamps should never be used because they cause painful wounds. Alternatively, other sterile clear drapes are commercially available.

- *Hemostasis.* Blood loss during surgery should be prevented. This is achieved by localization of major blood vessels and careful dissection to avoid incidentally damaging them. Certain surgeries carry a risk of excessive bleeding. Mastectomy is especially notorious because mammary gland tumors contain both large blood vessels and fine networks of capillaries. This can result in massive as well as diffuse bleeding. Doppler ultrasound examination should be done before surgery to identify vascular structures to minimize blood loss. After identifying the location of blood vessels, about 2-5 milliliters of 0.9% NaCl (with lidocaine) can be injected around the site. In this artificially created edema, blood vessels are visible during dissection and can be ligated before damage occurs. Blood loss during an average surgery is less than 0.3 milliliters. During excision of large tumors it should be less than 2 milliliters (2% of the blood volume). Bloodless surgery makes excessive administration of fluids redundant because the circulating volume is unaffected.

- *Minimizing tissue damage.* The surgeon has to avoid excessive grasping of the tissue with instruments. This particularly applies to guinea pig viscera, which do not tolerate handling. In laparotomic procedures, externalization of viscera should be avoided. For example, cystotomy should be performed through a small (< 2 cm) incision in the caudal linea alba, so that only the bladder is elevated out of the abdominal cavity (Fig 6). If internal organs must be externalized, it is imperative to keep the serosa moist using warm saline.

- *Anatomic reconstruction of tissues.* The anatomic structures (e.g., subcutis, muscles) should be restored to their original position. Tissues should be sutured layer-by-layer, the blood supply must be preserved, and tension on the sutures should be avoided. Good apposition of wound edges is crucial for optimal recovery and wound healing. Usually, the standard running (continuous) pattern with good edge apposition provides the best healing. It is superior to intracutaneous sutures because these provide less stability and may result in dehiscence. Be careful not to introduce foreign tissue into the peritoneum when suturing internal organs because this will elicit a strong inflammatory response. For example, after closing a hollow organ (bladder), the second inverting suture (Cushing) must be applied so that no mucosa or muscularis of the bladder is exposed (Fig 6). Failing to observe this will result in postoperative pain and if the animal survives, intraabdominal adhesions.

- *Reduction of dead space.* The space between structures (e.g., tissue layers) that are not attached to each other can fill up with blood, serum or exudate. This creates the perfect environment for bacterial growth and may lead to wound dehiscence. The same is valid for necrotic tissue left at the operation site. Dead space should be avoided by applying anchoring sutures between the layers. For example, while closing the abdominal wall (in three layers), the subcutis should be anchored to the muscular layer and the skin to the subcutis. Any necrotic tissue should be carefully removed.

The surgery should be completed in less than 30 minutes for the best result. Prolonged exposure to anesthesia (> 60 minutes) significantly increases the complication rates.

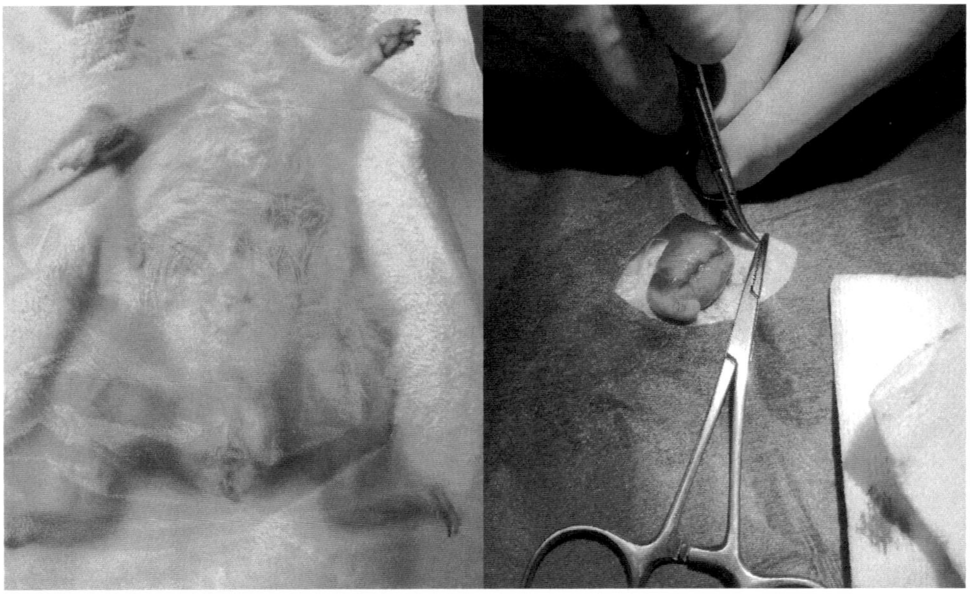

Figure 6. Left: the patient is covered with Press'n Seal plastic wrap. Right: after disinfection of the plastic, a sterile drape is placed on top of it. In this case, a large urolith was removed. Note that only the bladder was externalized through the incision and that an inverting suture pattern was applied.

Aftercare

Aftercare is a crucial part of the protocol: most complications occur not in the operating theater but in the recovery. Although problems can emerge up to 72 hours after the surgery, usually the first hours determine the outcome.

After termination of anesthesia, the animals should regain consciousness in less than 10 min. They should also start feeding within 15-20 minutes and urinating/defecating within 30 minutes. In older animals, or after prolonged and/or invasive procedures, one should add another 10-20 minutes. Obese animals may regain consciousness slower because isoflurane accumulates in adipose tissue during induction and maintenance and slowly dissociates from it. Sometimes "second anesthesia" can be observed – the guinea pig relapses after briefly regaining consciousness. This is not necessarily a pathologic sign. The animals should be left on the heated pad with the oxygen mask on, or in the incubator filled with oxygen, until they start to move around. During this period they should be closely watched. After light and regular procedures, the patients can be sent home when the recovery from anesthesia is fast and no alarming signs are observed. Usually the authors wait until the patient starts eating and defecating.

Postoperative analgesia is always necessary (NSAIDs such as meloxicam 0.5 up to 1 mg/kg PO q12h dependent on the severity of pain). In case of severe dental pain, tramadol (20 mg/kg PO q12h) and pregabalin (10-20 mg/kg PO q24h) are beneficial. Prophylactic antibiotics are also recommended for all procedures because of the postsurgical immunosuppression.[2] Enrofloxacin (10 mg/kg PO q12h for 5-7 days) is usually sufficient unless other recommendations (e.g., based on the antibiogram) exist. After heavy procedures, observing the animals for at least several hours is prudent. Broad-spectrum antibiotics should be given because animals are particularly prone to infections after invasive surgeries. The protocol for postoperative analgesia is usually more complicated. For example, meloxicam should be replaced with metamizole (100 mg/kg PO/SC q12h) or flunixin (5 mg/kg SC q12h), which is more suitable for combatting visceral pain.[21,24] The hydration state should be monitored, especially when flunixin is used. However, in the authors' experience, flunixin is well-tolerated by guinea pigs. If there is insufficient response to NSAIDs only, multimodal analgesia, including acetaminophen (100 mg/kg q6h PO) and tramadol (20 mg/kg q12h PO) can be applied. Prokinetics (metoclopramide SC and cisapride PO, both at 1 mg/kg q6h) may be added in case of little fecal output. If the animal does not eat properly, assisted feeding is recommended; wheat grass powder is a palatable and nutritious product that is ideal for assisted feeding of critical patients, but there are also other products available commercially. However, these measures are not always necessary. Certain types of abdominal surgeries (e.g., cystotomy) are well-tolerated; the patients can be dismissed the same day, and the owner can proceed with TLC at home.

Failure to regain full consciousness, or to start feeding, urinating and defecating within 1-2 hours, indicates complications. Therefore, such patients should always be hospitalized.

Complications

Although the complication rate is very low, the surgeon must always be prepared for emergencies. There are several types of complications in the operating theater or post-surgically.

In the operating theater

These complications are infrequent; the author (ES) experienced less than ten such cases in about ten years. In most cases, the animal's life was terminated deliberately (intraoperative euthanasia) because surgery could not solve the problem, and a humane endpoint decision was made.

The most common cause of death in the operating theater is cardiorespiratory depression. It manifests as a drastic drop in oxygen saturation and heart rate, followed by ventricular escape beats (extreme bradycardia with 50-100 bmp) and cardiac arrest. Reasons for this can be:

- A sudden catecholamine release (nociception/pain that the animal feels despite anesthesia) triggers a baroreflex. This can be anticipated and prevented by adequate monitoring and applying local anesthesia (see Monitoring and troubleshooting).

- The use of α_2 adrenergic agonists, especially in patients belonging to ASA III or higher, is likely to trigger a baroreflex.[8] Besides, α_2 agonists cause nausea and regurgitation and increase the risk of airway obstruction in non-intubated animals. For these reasons, the author (ES) never uses these drugs except for euthanasia.

- Hypovolemia due to excessive blood loss.

- Hypothermia. Anesthetized animals have impaired thermoregulation; a low body temperature potentiates endogenous acetylcholine[25] and may lead to bradycardia.

- Cardiac arrest due to preexisting cardiovascular diseases. This is usually due to the inadequate preoperative assessment of the patient.

Most of the abovementioned causes are due to procedural flaws. However, thorough patient assessment, adequate monitoring and observing the rules of good surgical practice allow the elimination of casualties in the operation theater.

In recovery

The general rule reads: the more profound the anesthesia, the more prolonged and more invasive the procedure, and the worse the condition of the animal before surgery, the higher the risk of complications in recovery. Typical postoperative issues are described below.

- The patient does not fully regain consciousness and dies within several hours. This is usually due to an incorrect preoperative assessment: unstable patients (ASA IV/V) should not undergo surgery unless preoperative stabilization is possible. Too deep anesthesia (> stage 3.3, Fig 2) and using α_2 agonists contribute to the risk. Alternatively, this can result from procedural flaws. Examples include inadequate suturing of internal organs, which leads to leakages (such as uroperitoneum), or septic shock due to a body cavity contamination with infectious material (abscess, fecal contamination). Treatment is unrewarding, and euthanasia should be considered.

- Simple postoperative ileus. The animals may appear awake and alert but refuse to feed and display signs of visceral pain (see Pain recognition). The fecal output is significantly reduced, and sometimes tympany is present. This type of complication is most often seen in non-visceral procedures carried out using α_2 agonists, even if the antidote (atipamezole) has been administered. The mechanism of gastrointestinal stasis is most likely the adrenergic-induced disruption of intestinal circulation,[8] which leads to temporary hypoxia. If there are no other complications (e.g., dysbiosis, metabolic acidosis), the patients can be treated with high doses of prokinetics (metoclopramide 1 mg/kg SC q3-4h, cisapride 1 mg/kg PO q3-4h), visceral analgesics (see Light, regular and heavy protocols), fluid replacement and TLC. Assisted feeding can be attempted, but only in the absence of tympany. If treated adequately, the condition usually resolves within several days.

- Complicated postoperative ileus. This life-threatening condition is usually a sequalae to complex laparotomic procedures that involve gut handling. The use of $\alpha 2$ agonists exacerbates the problem. The proposed mechanism is the inflammatory reaction to any injury (e.g., serosa laceration, blunt trauma, or desiccation) to the intestinal wall. An important mediator in this process is nitric oxide (NO),[26] resulting from the activation of inducible nitric oxide synthase (iNOS) by (among others) inflammatory cytokines and substance P. Nitric oxide is an inhibitory neurotransmitter in the myenteric plexus; activation of iNOS leads to paralytic ileus. The affected animals are soporose. There is no fecal output; metabolic acidosis (urinary pH < 7) and dehydration are often present. Untreated, the animals die within 24 to 48 hours. Although this condition may seem uncurable, there are still several treatment options. First, radical visceral analgesia should be applied: flunixin (5 mg/kg SC q12h),[24] preferably with fluid replacement therapy. Opioids should be avoided. Fluids with physiologic pH of 7.4 should be infused at 20 ml/kg q6h SC or IV (slowly). To correct metabolic acidosis, sodium bicarbonate is added to the fluid, at 1 mEq for every 0.5 of pH unit below 7 (e.g., 1 mEq for 6.5, 4 mEq for 5). Adding 3-5 mg lidocaine (without epinephrin) to the solution reduces the pain at the injection site and speeds up fluid absorption from the subcutis. The latter is due to vasodilatory properties of lidocaine.[27] Prokinetics are of crucial importance. Alizapride[28] (a relatively new dopamine agonist) is the drug of choice in cases of severe impairment of gut motility. Its efficacy in guinea pigs is higher than that of metoclopramide. The author (ES) has successfully used it in severe cases of paralytic ileus at 10-25 mg/kg SC q4-6h. Alizapride can be combined with cisapride (1 mg/kg PO q4-6h).

- Another novel treatment option is based on the inhibition of iNOS.[29] Aminoguanidine is a selective iNOS inhibitor; it can be administered preoperatively in anticipation of complications or postoperatively (20-40 mg/kg PO q12h). The combination of radical analgesia, prokinetics, and iNOS inhibition reduces mortality by 50-70%. Antibiotics should be given to prevent secondary dysbiosis.

- Dysbiosis. Perioperative immunosuppression increases the risk of infections. The intestinal flora of clinically healthy guinea pigs contains small numbers of (facultative) pathogens: coliforms, clostridia, *Enterococcus* spp., yeasts and protozoa. Shifts in gut microbiota composition can occur up to 72 h after surgery. The animals develop paralytic ileus (see above) or diarrhea and deteriorate quickly. Feces can be examined under a microscope for the presence of yeasts and protozoa. There is usually no time for bacterial culture and antibiogram. Therefore, broad-spectrum antibiotics (with activity against anaerobes) should be administered. For example, metronidazole (50 mg/kg q12h) is administered together with enrofloxacin (10 mg/kg q12h). If necessary, antifungal and antiprotozoal medications should be given. Adequate visceral analgesia and fluid replacement therapy are essential. The risk of postoperative dysbiosis increases with the increasing age of the patient. For this reason, older animals (4 years and more) should not undergo elective surgeries.

- Wound dehiscence and infection. This is a rare occurrence in guinea pigs: when the rules of good surgical practice are observed, there is practically no risk. Wound dehiscence due to auto-mutilation occurs mainly after the administration of ketamine. This NMDA antagonist causes neurologic disturbances[30] (e.g., agitation, hallucinations), so animals are more likely to gnaw on the sutures. Besides, if no local anesthetics have been applied, the animals that recover from general anesthesia face hyperalgesia. Too tight sutures resulting in disturbed blood supply and tissue hypoxia may also contribute to auto-mutilation. If the wounds are not critical (e.g., cutaneous), topical antimicrobial preparations and local anesthetics can be applied, and the wounds can be left to heal by secondary intention. Deep wounds with large amounts of necrotic material must be debrided surgically.

Conclusions

Anesthesia and surgery of guinea pigs can be safely carried out if specific rules are observed. The key to success is meticulous prescreening of the patient, followed by rigorous analgesia and minimalization of tissue trauma. Most complications can be entirely prevented, and even in emergencies, life-saving solutions are available.

References

1. Lennox AM. Mitigating risk factors for anesthetic death in exotic animals: what can we learn from canine and feline standards? *Proc ExoticsCon.* 2022;259-266.

2. Lachmann G, Von Haefen C, Kurth J, et al. Innate immunity recovers earlier than acquired immunity during severe postoperative immunosuppression. *Int J Med Sci.* 2018;15:1-9.

3. Portier K, Ida KK. The ASA Physical Status Classification: what is the evidence for recommending its use in veterinary anesthesia? A systematic review. *Front Vet Sci.* 2018;5:204.

4. Stoffels E. Quantitative stress assessment in guinea pigs. *Proc ExoticsCon.* 2023; In press.

5. Stoffels E. Lower urinary tract disease in guinea pigs as it relates to nutrition. *Proc ExoticsCon.* 2022; 267-283.

6. Neuroscience Online: An Open-Access Neuroscience Electronic Textbook. University of Texas. Available at: https://nba.uth.tmc.edu/neuroscience/. Accessed March 23, 2023.

7. National Centre for the Replacement, Refinement and Reduction of Animals in Research. Available at: https://www.nc3rs.org.uk/3rs-resources/grimace-scales. Accessed March 23, 2023.

8. Sinclair MD. A review of the physiological effects of alpha2-agonists related to the clinical use of medetomidine in small animal practice. *Can Vet J.* 2003;44:885-897.

9. Nemeth M, Pschernig E, Wallner B, et al. Non-invasive cortisol measurements as indicators of physiological stress responses in guinea pigs. *Peer J.* 2016;4:e1590.

10. Lundberg J, Weitzberg E, Gladwin M. The nitrate-nitrite-nitric oxide pathway in physiology and therapeutics. *Nat Rev Drug Discov.* 2008;7:156-167.

11. Flecknell PA. *Laboratory Animal Anesthesia.* London, UK: Academic Press; 2023.

12. Guedel AE. *Inhalation Anesthesia: A Fundamental Guide.* New York, NY: The Macmillan Co; 1937:14.

13. Dahl JB, Steen Møiniche S. Pre-emptive analgesia. *Br Med Bull.* 2005;71:13-27.

14. Sohn HM, Na HS, Lim D, et al. Immediate retrograde amnesia induced by midazolam: a prospective, non-randomized cohort study. *Int J Clin Pract.* 2021;75:e14402.

15. Adner M, Canning BJ, Meurs H, et al. Back to the future: re-establishing guinea pig in vivo asthma models. *Clin Sci.* 2020;134:1219-1242.

16. Nambiar MP, Gordon RK, Moran TS, et al. A simple method for accurate endotracheal placement of an intubation tube in guinea pigs to assess lung injury following chemical exposure. *Toxicol Mech Methods.* 2007;17:385-92.

17. Merentie M, Lipponen JA, Hedman M, et al. Mouse ECG findings in aging, with conduction system affecting drugs and in cardiac pathologies: development and validation of ECG analysis algorithm in mice. *Physiol Rep.* 2015;3:e12639.

18. Malamed SF. *Local Anesthesia.* St. Louis, MO: Elsevier; 2019.

19. Jayawardena CK, Takano Y. Nerve-epithelium association in the periodontal ligament of guinea pig teeth. *Arch Oral Biol.* 2006;51:587-595.

20. Sansone RA, Sansone LA. Tramadol: seizures, serotonin syndrome, and co-administered antidepressants. *Psychiatry.* 2009;6:17-21.

21. Ventura-Martinez R, Mares-Sánchez JJ, Avilés-Herrera J, et al. Antinociceptive synergy between metamizole and hesperidin in a model of visceral pain in mice. *Arch Med Res.* 2021;52:389-396.

22. Hunt GB. *BSAVA Manual of Canine and Feline Surgical Principles.* Gloucester, UK: BSAVA Publications; 2012;21:264-276.

23. Emmer KM, Celeste NA, Bidot WA, et al. Evaluation of the sterility of Press'n Seal cling film for use in rodent surgery. *J Am Assoc Lab Anim Sci.* 2019;58:235-239.

24. Lemonnier LC, Thorin C, Meurice A, et al. Comparison of flunixin meglumine, meloxicam and ketoprofen on mild visceral post-operative pain in horses. *Animals.* 2022;12:526.

25. Nagai M, Toda H, Iriki M. Hypothermia enhances acetylcholine-induced contraction of isolated rat ileum. *Jpn J Physiol.* 1987;37:545-549.

26. Moojen TM, Van Gulik TM, Hoek FJ, et al. Possible role of nitric oxide in postoperative ileus: a comparative study. *Neurogastroenterol Motil.* 1999;11:403-408.

27. Newton DJ, McLeod GA, Khan F, et al. Mechanisms influencing the vasoactive effects of lidocaine in human skin. *Anaesthesia.* 2007;62:146-150.

28. DrugBank Online. Alizapride. Available at: https://go.drugbank.com/drugs/DB01425. Accessed June 1, 2023.

29. Yanagida H, Sanders KM, Ward SM. Inactivation of inducible nitric oxide synthase protects intestinal pacemaker cells from postoperative damage. *J Physiol.* 2007;582:755-765.

30. Wu JL, Haberman RP, Gallagher M, et al. Probing for conditioned hallucinations through neural activation in a ketamine mouse model of schizophrenia. *Neurosci Bull.* 2020;36:937-941.

Section 23

AEMV Masterclasses Track 4

Colleen Roman, DVM
Session Chair

Palliative Treatment of Odontogenic Abscesses

Angela M. Lennox, DVM, DABVP (Avian Practice), DABVP (Exotic Companion Mammal Practice), DipECZM (Small Mammal), and
Lauren Powers, DVM, DABVP (Avian Practice), DABVP (Exotic Companion Mammal Practice)

Session #3192

Affiliation: From Avian and Exotic Animal Clinic 9330 Waldemar Road Indianapolis IN, 46268 USA (Lennox), Carolina Veterinary Specialists 12117 Statesville Rd, Huntersville, NC 28078, USA (Powers).

Abstract: Severe odontogenic abscesses typically do not respond well to medical management and may affect portions of the skull not amenable to complete excision. Some patients may be poor surgical candidates or experience difficult recoveries. For these, a palliative approach may be considered, with the goal of controlling discomfort and slowing progression of the disease process. Multiple case examples of palliative treatment of severe odontogenic abscesses are presented, with some patients surviving with good to excellent quality of life for 3 years or more.

Introduction

Odontogenic abscess can be a sequela of severe acquired dental disease or can develop in rabbits with more normal dentition via trauma or dental fractures. Abscesses in rabbits are unique in that they tend to form tough capsules with thick caseous pus and can produce marked widespread osteomyelitis.

Abscesses vary widely in severity, from single alveolar infections to those involving multiple teeth and associated bone. Odontogenic abscesses of the mandible can involve the more rostral portions, which provide easier surgical access, to the masseteric fossa covered with the large masseter muscle. Rostral maxillary abscesses are often technically more challenging. In particular are abscesses of the alveolar bulla, a unique bony structure containing the apices of the caudal maxillary cheek teeth (CHT) and associated with the retrobulbar space. Access to this site is via an enucleation site for the non-viable eye, from the oral cavity in cases where tooth loss allows adequate access, or through zygomatic arch resection.

Treatment

There are only a handful of retrospective studies describing abscess treatment outcomes.[1-3] An earlier study showed success with a simple technique of abscess lancing with implantation of antibiotic soaked gauze and suturing with weekly repeated surgeries. Other case reports feature a more aggressive approach with removal of the abscess, capsule and associated infected bone, with marsupialization for continuing debridement and healing by second intention. A recent study showed success with multiple cases of zygomatic arch resection to access the retrobulbar space. Development of guidelines for selecting the approach associate with greatest success would require a standardized staging system (number of teeth affected, extent of osteomyelitis, ease of access, mandibular vs. maxillary, secondary site involvement, e.g., nasal cavity) and retrospective data on outcomes, which is currently not available.

Attempts to surgically treat increasingly severe odontogenic abscesses have resulted in cases where complete cure is not achieved. The authors have identified multiple cases where outcomes can be best described as palliative, and several are associated with long periods of excellent quality of life, with symptom-free periods of several years.

References

1. Capello V. Surgical treatment of facial abscesses and facial surgery in pet rabbits. *Vet Clin North Am Exot Anim Pract.* 2016;19(3):799-823.

2. Capello V. Novel diagnostics and surgical techniques for treatment of difficult facial abscesses in pet rabbits. *Proc Annu Conf North Am Vet.* 2011;1685-1689.

3. Taylor WM, Beaufrere H, Mans C, Smith DA. Long term outcome of treatment of dental abscesses with a wound packing technique in pet rabbits: 13 cases (1998-2007). *J Am Vet Med Assoc.* 2010;237(12): 1444-1449.

The Gastrointestinal Tract of The Rabbit: Health and Disease

Dan H. Johnson, DVM, DABVP (Exotic Companion Mammal Practice)

Session #3413

Affiliation: From Avian and Exotic Animal Care, 8711 Fidelity Boulevard, Raleigh, NC 27617, USA.

Abstract: Rabbits are adapted to a coarse high-fiber diet. They require plant fiber to stimulate gastrointestinal motility and have a well-developed hindgut with a functional cecum and a colon capable of particle-dependent separation. Microbial fermentation within the cecum is the primary mechanism by which nutrients are released from ingested food. Rabbits produce and consume cecotropes which contain bacterial protein, amino acids, volatile fatty acids, and vitamins. Cecotrophy maximizes feed utilization by allowing rabbits to redigest and absorb previously undigested nutrients. Alteration of food and water intake, stress, pain, and other factors can disrupt the digestive process, alter gastrointestinal motility, and lead to dysbiosis, often with serious consequences for the rabbit.

Rabbit Natural History, Diet, and Feeding Behavior

Domestic rabbits are descended from the European rabbit, *Oryctolagus cuniculus*, and belong to the order Lagomorpha. The lagomorphs (rabbits, hares, and pikas) are characterized by the presence of a second small pair of upper incisors or peg teeth. Wild European rabbits eat a diet of grass and browse, which is high in fiber and low in energy.

Rabbits are selective feeders. They tend to consume the most nutrient-dense parts of a plant, allowing them to meet dietary requirements while minimizing the volume of food that must be consumed; however, they are adapted to a high-fiber diet. A large quantity of poorly digestible fiber is essential for normal gastrointestinal (GI) motility. Rabbits can consume large volumes of roughage without the need for a large gut like ruminants because of their rapid digestive transit time. As a prey animal, the rabbit's digestive strategy has evolved to permit efficient digestion of fibrous vegetation without the need to store large volumes of food within the body. Having a small gut is an advantage when swift evasion from predators is necessary.

Rabbits feed primarily in the early morning and night (crepuscular). They eat about 30 times a day, for 4-6 minutes at a sitting, consuming 2-8 grams at a time. Hunger is stimulated by a dry mouth and gastric contractions or a decrease in blood levels of metabolites such as glucose, amino acids, fatty acids, or lactic acid. Food selection is based upon smells and tactile information obtained from the sensitive vibrissae around the nose and lips. Cecotrophy occurs about 3-8 hours after feeding. When compared to other small herbivores, digestive transit time is fast (19 ± 10 hours). Like their wild counterparts, pet rabbits tend to void their droppings in a latrine.

Rabbits prefer feeds with sweet material (molasses or sugar) over unsweetened feeds. When low-fiber, alfalfa-based pellets are fed ad lib, obesity and chronic soft stool may develop. If pelleted food is restricted without supplementing the diet with hay or vegetables, fur pulling and gnawing on carpet, furniture, and wire may result. Boredom and destructive behavior can be prevented by supplementing the diet with fresh grass, grass hay, vegetables, and by providing safe items to gnaw on.

Rabbits have a relatively high metabolic rate and water intake (50-150 ml/kg/day). Water intake increases when food is withheld; after 3 days of food deprivation, water intake may increase by 6.5 times. Food consumption declines with water deprivation; after 3 days of deprivation, anorexia results.

Rabbit Anatomy and Physiology

Abdominal cavity

Rabbits have very large abdominal cavities. They are classified as true herbivores and hindgut fermenters. The gastrointestinal tract (GIT) is relatively long, and its contents make up 10-20% of body weight. The stomach and cecum are both very large and together make up 5-19% (average 10%) of total body weight. Because rabbits rely heavily on the diaphragm for inspiration, preanesthetic fasting is recommended to partially empty the GIT (not to prevent vomiting).

Stomach

Rabbits have a simple stomach which lacks specialized regions. It serves as reservoir for most of ingested feed and contains about 15% of alimentary tract ingesta (food, hair, and fecal pellets). The stomach is thin walled and often appears ruptured at necropsy due to gas accumulation/distention and autolysis. The cardia and pylorus are well developed. The rabbit is unable to vomit due to a massive muscular lower esophageal (cardiac) sphincter and a serrated mucosal rosette at the cardiac orifice. The pylorus is easily compressed by the duodenum, which exits at an acute angle; therefore, gastric distention or compression due to a trichobezoar, gas, or hepatomegaly contribute to pyloric compression and prevents emptying.

Hydrochloric acid and pepsin are secreted into the stomach to initiate digestion much the same as in other species. Gastric pH is more acidic than in most other species (1.0-2.0 postprandially; 3.0 following cecotrophy). This low pH effectively sterilizes ingesta and is one of the rabbit's first lines of defense against GI infection. The stomach pH of preweaned rabbits is much higher (5.0-6.5), which promotes the survival and passage of ingested bacteria, helping to establish normal large intestinal flora. Gastric transit time is approximately 3-6 hours; however, food, hair, and fluid remain in the stomach even after a 24-hour fast.

Small intestine

The rabbit's small intestine is relatively short and makes up about 12% of total GIT volume. The duodenum and jejunum have a relatively small lumen. At the terminal ileum, the lumen becomes even narrower and is a common site of foreign body impaction. Distal to this, at the ileocecocolic junction, the ileum dilates to become the rounded sacculus rotundus. Also called the ileocecal tonsil or ampulla ilei, the sacculus rotundus is unique to the rabbit and made up primarily of lymphoid tissue, giving it a honeycombed external appearance. The small intestine appears to play an important role in immune function. There are extensive aggregates of lymphoid tissue (Peyer's patches) throughout the intestinal tract, most prominently in the distal jejunum. Peyer's patches are absent from the duodenum and the first half of the ileum.

The small intestine is the primary site of digestion and absorption of amino acids, lipids, starches, simple sugars, and electrolytes. Nutrients present within ingested cecotropes (bacterial protein, amino acids, volatile fatty acids [VFAs], and vitamins) are absorbed across the small intestine. Calcium absorption in the rabbit is unusual in that it occurs via a passive, vitamin-D_3 independent mechanism. Bicarbonate is secreted into the duodenum to neutralize the stomach acid and is reabsorbed in the jejunum. Unlike in other species, pancreatic amylase is not a significant factor in digestion. Transit time through the small intestine is fast (10-20 minutes through jejunum, and 30-60 minutes through the ileum), which limits time for absorption in the small intestine. Vegetation is predominantly composed of lignin, cellulose, hemicellulose, oligosaccharides, pectins, and plant proteins that pass through the small intestine largely unaltered. Once ingesta leaves the small intestine, retrograde movement from the large intestine back into the ileum is inhibited by a valve-type mechanism at the ileocecocolic junction.

Large intestine

Rabbits have a well-developed hindgut with a functional cecum and a colon capable of particle-dependent separation. Fine particles and solutes are selectively retained by the proximal colon then delivered to the cecum for microbial fermentation, while indigestible fiber is formed into fecal pellets and expelled as rapidly as possible. The cecum acts as a fermentation vat where digestible fiber and starch are broken down by microbial action, and the colon sorts fiber particles and forms the feces. Like the small intestine, the large intestine features significant amounts of gut-associated lymphoid tissue (GALT); in fact the large intestinal GALT constitutes over 50% of total lymphoid tissue in the rabbit.

The cecum is the largest, most prominent organ in the abdominal cavity, having ten times the capacity of the stomach and containing about 40% of the ingesta. Loosely coiled into three folds, the cecum is in contact with most of the ventral abdominal wall. The cecum is thin-walled and possesses 18-22 haustra-like pouches delineated by a spiral, running constriction. It ends in a thick-walled, blind, pale, finger-like projection, the vermiform appendix, which contains abundant lymphatic tissue and has a secretory function (water and bicarbonate). The cecum receives short fibers and fluid selectively retained by the proximal colon; therefore, cecal contents are semi-fluid.

At the open end of the cecum, distal and adjacent to the sacculus rotundus, sits the bulbous ampulla coli, from which the proximal colon emerges. The colon has muscular bands running along its length and is functionally divided into two parts: the proximal colon, which has numerous distinct sacculations (haustra), and the distal colon, which has none. The proximal colon is the site of separation of ingesta into digestible and indigestible fractions, which is partly due to functional anatomy and partly due to colonic motility. The shorter, digestible particles tend to settle near the mucosa and are propelled retrograde, back into the cecum via a series of coordinated contractions. At the same time, the longer, coarse, indigestible particles tend to remain in the center of the colonic lumen and are directed aborally to the distal colon, where water and solutes are reabsorbed, and hard fecal pellets are formed then rapidly eliminated.

The proximal colon and distal colon are separated by the fusus coli, a thickened, muscular section of colon that is densely innervated, highly vascularized, and unique to lagomorphs. It acts as an intestinal pacemaker for the colon, regulating colonic contractions. The fusus coli (and thus, colonic motility) is influenced by hormones, such as aldosterone and prostaglandins, and by the course, indigestible fiber ("scratch factor") content of the ingesta. While long, indigestible fiber provides no nutritional value, it plays a key role as the stimulus for normal GI motility and for maintenance of normal cell regeneration, secretion, digestion, absorption, peristalsis, and excretion. Only one type of feces is excreted at a time, following a marked circadian rhythm, alternating between a hard-feces phase which coincides with feeding activity, and a cecotrope phase which usually coincides with rest four or more hours after feeding. This rhythm is influenced by diet, age, and reproductive status.

Hindgut flora and fermentation

Microbial fermentation is the primary mechanism by which nutrients are released from ingested food in the rabbit. Hindgut flora also serves as a defense against potential pathogens. GI microbes are established while the rabbit is young. Suckling rabbits maintain a stomach pH of 5.0-6.5. (Side note: they only nurse for 3-4 minutes a day) Rabbit pups produce an antimicrobial fatty acid (milk oil) following ingestion of doe's milk; rabbits fed the milk of other species do not produce milk oil and are therefore more susceptible to infection. Young rabbits have little or no GI microbes until about 3 weeks of age. From 3-6 weeks, as weaning takes place, the production of milk oil wanes and solids gradually replace milk. During this transitional period, as the production of milk oil ceases but before gastric pH drops to adult levels, bacteria and other microbes pass through the stomach to colonize the small intestine, cecum, and colon.

Studies to date, using culture, cytology, PCR techniques, and a variety of other methods, have indicated that the normal microbial inhabitants of the rabbit may include a variety of anaerobes, gram-positive and gram-negative facultative anaerobes, large ciliated protozoa, and hindgut-specific yeast of the Saccharomyces family, *Cyniclomyces guttalatulus*. *Bacteriodes* spp., strict anaerobes, are generally considered to be the predominant organisms within the cecum. Large anaerobic metachromatic staining bacteria (LAMB) are also present in large populations. Coliforms and *Clostridium* spp. may be present in the normal rabbit, but only in small numbers. In contrast with other mammals, lactobacilli are extremely rare. More recent work using molecular analysis has identified over 60 species of fungal organisms, with *Candida* species being found in the greatest numbers in all areas of the GI tract, and over 130 bacterial species, with anaerobes the most commonly found including those in the Order Clostridia. Of these, the most common organisms identified included *Clostridiales* and *Lachnospiraceae* species.

Cecal microorganisms are responsible for the breakdown of selectively retained materials. They produce VFAs (acetate, butyrate, and propionate) which provide up to 40% of the rabbit's maintenance energy requirement, vitamins, amino acids, enzymes, and microbial protein. VFAs are primarily absorbed across the cecal wall; other fermentation byproducts are expelled and reingested as cecotropes.

Cecotrophy

The production and consumption of cecotropes (soft feces, night feces, cecotrophs) contributes significantly to rabbit digestive efficiency. It maximizes feed utilization by allowing rabbits to redigest and absorb previously undigested nutrients. Considered a form of pseudorumination, cecotrophy makes bacterial fermentation products not absorbed in the cecum and colon available to the rabbit small intestine for absorption. Microbial protein makes up about 30% of rabbit total nitrogen intake and has high content of essential amino acids; thus, cecotrophy can greatly improve both the quantity and the biological value of the protein in the diet. Cecotrophy may have evolved as an adaptation to thrive in marginal or extreme habitats and during periods of poor food availability.

Cecotropes, appearing as clusters rather than as single pellets, are often mistaken for diarrhea when attached to fur around the anus. They contain high levels of VFAs, microbial protein, amino acids, enzymes, vitamins B and K, live microbes, sodium, potassium, and water. The microbial protein contained in cecotropes (mucin) can only be digested by lysozyme. As a cecotrope is formed, the colon secretes lysozyme and incorporates in into the pellet, breaking down mucin and making it available to the rabbit by the time the cecotrope reaches the small intestine. The large intestine also coats the cecotrope in a strong mucus envelope as a barrier to low pH in the stomach.

The cecum empties and cecotropes are produced once or twice a day. They are ingested directly (reflexively consumed) from the anus and swallowed without mastication. Ingestion is triggered by rectal sensation, VFA concentration (i.e., odor and taste) of cecotropes, and blood concentration of various metabolites and hormones. Amount consumed varies with dietary composition: fewer cecotropes are ingested when the diet is rich in protein than with a high-fiber diet. (Side note: Germ-free rabbits and cecotomized rabbits do not practice coprophagy) Nutrients present within cecotropes are made available to the rabbit as they pass through the stomach into the small intestine. The mucus envelope of a cecotrope resists stomach acid for up to 6 hr, and contents are maintained at a relatively neutral pH because of high levels of phosphate buffers it contains. Protected from stomach acid in this way, living bacteria, yeast, and protozoa can pass through the stomach and aid in the replenishment of hindgut microflora.

Gut motility

GI motility is influenced by the autonomic nervous system, certain hormones, and the nutrient content of the diet (especially indigestible fiber). The hormone motilin, secreted by the duodenum and jejunum stimulates GI smooth muscle; fat stimulates the release of motilin and carbohydrate inhibits it. Certain VFAs produced in the cecum directly stimulate cecocolic motility. Distention of the digestive tract by indigestible fiber also promotes increased GI motility.

Liver, gallbladder, and pancreas

The liver lies caudal to the diaphragm, mostly within the rib cage, and is divided into two major lobes: left and right. The left lobe his subdivided into lateral and medial lobes; the right lobe is divided into lateral, medial, quadrate, and caudate lobes. The caudate lobe has a narrow, stalk-like attachment to the dorsal hilar region of the liver and is prone to torsion.

The gallbladder lies deep between right medial and quadrate liver lobes. Like dogs, rabbits have separate bile and pancreatic duct openings in the duodenum; the bile duct empties into the proximal part of the duodenum. Like most non-mammals, rabbits secrete mainly biliverdin instead of bilirubin. Rabbits produce a large amount of bile, approximately 250 milliliters a day, which is seven times as much as a dog on a weight basis. Bilirubin and/or bile acid levels may be elevated with liver disease in the rabbit and can occur with bile obstruction. Unlike many other mammals, rabbits do not usually become jaundiced with liver disease.

The pancreas is closely associated with the duodenum; however, it is diffuse and difficult to differentiate from the surrounding mesentery. The pancreatic duct opens at the junction of the ascending and transverse duodenum. Ligation of the duct does not result in pancreatic insufficiency.

Nutrition

Improper nutrition is a predisposing factor for many diseases in the rabbit, including those of the GIT. An appropriate diet will provide a rabbit with sufficient fiber to promote normal GI motility, adequate digestible nutrients, proper substrates for cecal fermentation, and promote normal foraging behavior. Indigestible fiber is responsible for stimulating gut motility and controlling gut transit time. Low fiber diets tend to suppress voluntary food intake. If fiber is ground too finely, gut transit and cecal retention times increase, and the potential for dysbiosis grows. Indigestible fiber particles 0.5 millimeters and larger stimulate gut motility, whereas small particles < 0.3 millimeters lead to increased gut retention time, reduced gut motility, and enteritis. Long-stem fiber (hay and grass) promotes proper dental attrition and prevents behavioral vices. Total dietary fiber levels of 20-25% are recommended for rabbits. Ad lib hay and grass are recommended as a source of indigestible fiber, along with limited amounts of commercial pellets having a crude fiber level greater than 18%.

Protein is obtained directly from ingested foods and from cecotropes. A protein level of 12-16% is considered appropriate for maintenance, and 18-19% for pregnant/lactating does. Excessive dietary protein can alter cecal pH, resulting in cecal dysbiosis and overgrowth of pathogenic microorganisms.

Rabbits can utilize carbohydrates (simple sugars and starches) as an energy source, but care must be taken that excessive amounts are not fed. Because of the rabbit's rapid gut transit time (10-20 minutes through the jejunum, and 30-60 minutes through the ileum), excessive carbohydrates may pass undigested into the cecum and predispose rabbits (especially young animals) to enterotoxemia. High-energy grains such as corn or wheat should be avoided; oats are a better choice. Grains should not be finely processed, and fruits should be restricted or not provided. As selective feeders, rabbits may not eat a balanced diet if seed-mix-type rations are offered. Excessive intake of pelleted feed is associated with obesity, dental disease, gastroenteropathies, and/or metabolic issues, and boredom-associated behavioral problems. Therefore, limited intake of pelleted diets without seeds and dried fruits is preferred.

Fat provides energy and improved palatability. Levels are restricted to prevent obesity and hepatic lipidosis. Dietary fat levels of 2.5-4% are recommended.

Water should always be provided. Bottles are easier to keep clean than bowls, but drinking from a bowl is more natural for a rabbit. Water intake is approximately 10-12% of bodyweight per day. A rabbit can go several days without feed by relying on coprophagy but cannot go without water for more than 24 hours, and less in hot weather.

Gastrointestinal Diseases of the Rabbit

Gastrointestinal stasis

One of the most common presentations in pet rabbits, GI stasis indicates that one or more factors governing GI motility are out of order. GI stasis has no breed or gender predilection, and it can occur at any age. Affected rabbits will often initially stop eating pellets or hay, but will continue to eat greens and treats, followed later by complete anorexia. Fecal pellets may become scant, firm, small, or cease to be produced entirely. In cases with concurrent dysbiosis, there may be diarrhea and/or bloat. Initially, patients may be bright and alert, but with prolonged stasis may present depressed, lethargic, or shocky. Signs of pain include bruxism, hunched posture, failure to groom, and reluctance to move. Affected animals may also stretch out or roll in an attempt to relieve pain. GI stasis becomes life-threatening if left untreated.

Lack of sufficient fiber, either through dietary insufficiency or anorexia, leads to GI stasis. Proper hindgut fermentation and GI motility are dependent on the ingestion of large amounts of roughage, long-stemmed hay, and water. A diet containing inadequate course, indigestible fiber predisposes a rabbit to GI hypomotility and alters cecal fermentation, pH, and substrate production such that enteric microflora is affected. Because intestinal microflora depends on a steady flow of water and nutrients, and the digestive tract depends on fiber for normal motility, any event leading to inappetence or anorexia (pre-surgical fasting, sudden changes in the diet, concurrent illness, starvation) or dehydration (sipper malfunction, bad tasting water, careless mistake) can trigger GI stasis and/or concurrent dysbiosis. As nausea and GI discomfort lead to anorexia, fiber and water intake are further reduced, and the process becomes self-perpetuating. If prolonged, GI stasis often leads to hepatic lipidosis, dehydration, and other secondary complications.

Inadequate dietary fiber is the leading risk factor for GI stasis. Rabbits that do not receive enough fiber often suffer from subclinical GI hypomotility and are thus more susceptible to other risk factors for GI stasis, whereas those that have adequate fiber intake are more resistant. A diet of mostly pellets – especially pellet mixes that contain seeds, corn, and other high-energy ingredients – tends to promote GI hypomotility. Feeding cereal products (bread, crackers, and breakfast cereals) and sugary treats (fruits, yogurt drops, other treats) further increases the risk.

Stressful conditions have a negative effect on gut motility. GI motility is regulated in part by the autonomic nervous system; stress increases the adrenal output of epinephrine and inhibits peristalsis. Common causes of stress include dental disease (malocclusion, molar elongation, odontogenic abscesses), metabolic disease (renal disease, liver disease), pain (oral, trauma, postoperative, urolithiasis), anxiety (dyspnea, fear, fighting, lack of hide box), neoplasia, infection, parasitism, and environmental changes (boarding, new pets, unfamiliar noises). Other factors that can contribute to GI stasis include toxin ingestion, foreign material (scoopable cat litter, hair, carpet fiber), obesity, inactivity, confinement, and certain drugs (anesthetics, anticholinergics, opioids, antibiotics). Inappropriate antibiotics can damage enteric microflora, promote the growth of pathogens associated with GI stasis, and lead to antibiotic-associated enterotoxemia.

GI stasis should be suspected when there is decreased appetite and/or fecal production, and a recent history of inappropriate diet, illness, or stressful event. Abdominal palpation may reveal small, hard fecal pellets or absence of fecal pellets palpable in the colon. The cecum may be filled with gas, fluid, or firm, dry contents depending on the underlying cause. The normal stomach should be easily deformable, soft, and pliable; it should not remain pitted on compression. With complete GI stasis the stomach may be severely distended, hard, and nondeformable. The presence of firm ingesta in the stomach of a patient that has been anorectic for 1-3 days is compatible with the diagnosis of GI stasis. Differential diagnosis includes GI obstruction due to foreign body, volvulus, intussusception, or intestinal neoplasia, and any cause of anorexia and decreased fecal output (e.g., dental disease, metabolic disease, cardiac disease).

CBC, biochemistry, and urinalysis are often normal, but may be used to identify underlying causes of GI hypomotility and anorexia. PCV and TS may be elevated with dehydration. Liver parameters may be elevated in cases with hepatic lipidosis. Inflammatory leukogram may be seen with intestinal perforation.

Diagnosis is usually based upon history, clinical presentation, and radiography. Gastric contents (including hair) are normally present and visible radiographically, however, a distended stomach despite anorexia implies GI stasis. A halo of gas can be observed around the inspissated stomach contents in some cases of GI stasis, and there can be moderate to severe gas distension throughout the digestive tract, including the cecum. Small fecal balls or the absence of fecal balls in the colon is highly suggestive of hypomotility. Severe distention of the stomach with fluid and/or gas may indicate acute small intestinal obstruction, which is a surgical emergency (see below).

Medical management of GI stasis centers on basic supportive care including warmth, stress reduction, pain relief, fluid replacement and nutritional support. Affected rabbits should be housed in a dark, quiet space away from natural predators' noise and odors. Provide thermal support with incubators, heating pads, or radiant heat emitters; however, to avoid causing heat stress, ambient temperature should not exceed 80°F.

Anxiety can be safely reduced in rabbits with injectable midazolam (0.25-0.5 mg/kg SC or IM). Analgesics such as buprenorphine (0.02-0.05 mg/kg IM, SC q8-12h) or meloxicam (1 mg/kg IM, SC, PO q24h) are essential in most cases as intestinal pain decreases appetite and impairs GI motility.

Nausea and GI inflammation are managed with maropitant citrate (Cerenia) 1-2 mg/kg SC q24h.

Parental fluids are typically given at a rate of 25-35 ml/kg SC q8h. Warm all parental fluids prior to administration. Although most cases can be treated effectively with subcutaneous fluids, shocky, azotemic, and critically ill patients should receive intravenous fluids. Intravenous catheters can be placed in any site normally used for venipuncture, but the cephalic and saphenous veins are the most practical. In an emergency or when peripheral veins are collapsed, fluids can be administered intraosseously. Provide additional fluid support orally with oral electrolyte solutions such as Pedialyte (Abbott) at 15-20 ml/kg q8h. Fluids are continued and then gradually reduced until urine output and drinking return to normal.

Provide nutritional support by syringe-feeding gruel such as Oxbow Critical Care Herbivore, vegetable baby foods, or soaked, ground pellets, approximately 20-30 ml/kg q8h. For prolonged nutritional support, a nasogastric (NG) tube can be placed in a manner like that used for cats. Liquid enteral fluids (Ensure, Sustecal), Oxbow Critical Care Herbivore Fine Grind, and diluted vegetable baby food can be passed through an 8-fr or even a 5-fr NG tube. Rabbits should be tempted often with parsley and other fresh greens to see if they will eat on their own. Nutritional support is continued and then gradually reduced until fecal production and appetite return to normal.

Motility modifiers are indicated in cases of GI stasis, provided intestinal obstruction and perforation have been ruled out. The author usually begins prokinetic therapy with injectable medication in severe cases. Metoclopramide 0.5 mg/kg PO, SC, IM, and cisapride 0.5 mg/kg PO are administered q8-12h for 3-5 days, until appetite and fecal production return to normal. They may be used alone or in conjunction. Simethicone (20 mg/kg PO q8-12h) may be indicated in cases of gas distention.

Antibiotics and cholestyramine are indicated in some cases (see above). Because activity promotes GI motility, the patient should be allowed supervised time out of the cage to exercise. Access to a safe grazing area will provide additional fiber and enrichment.

The expected course and prognosis depend on severity and underlying cause. For mild cases with chronic symptoms due to inappropriate diet, the prognosis is generally excellent to good, and diet correction often leads to complete recovery. For moderately severe, acute cases which require hospitalization, the prognosis is good to fair; the rabbit usually improves after several days of intensive supportive care and is discharged for additional care at home. In advanced cases (i.e., those that went unnoticed for several days prior to presentation) the prognosis is much worse. Hepatic lipidosis, shock, and other complications frequently make treatment unrewarding. Surgical correction of gastric trichobezoar, if indicated, carries a better prognosis than for surgical treatment of acute intestinal blockage by hair (see below).

GI stasis may be avoided by providing a diet with adequate indigestible coarse fiber (grasses and long-stemmed hay) and low carbohydrate content. Rabbits should be allowed daily exercise to reduce the incidence of GI stasis and prevent obesity. Minimize changes in the daily routine that might cause stress and avoid sudden changes in the diet. Ensure that fresh water is always available and presented in a familiar manner (sipper vs. bowl). Use only broad-spectrum antibiotics in rabbits and discontinue if soft stools or other GI signs develop. Drugs that can depress GI motility should be used with caution. Avoid over-fasting rabbits prior to surgery, control perioperative pain and stress, and encourage patients to eat as soon as possible following surgery. Be certain that all postoperative patients are eating and passing feces prior to release.

Gastric trichobezoar

GI stasis in rabbits is commonly referred to as a hairball, wool block, or trichobezoar. As a result of their grooming habits, rabbits ingest a considerable amount of hair which dietary fiber normally removes. With GI stasis, fluid and fiber intake decrease and the gastric contents condense to form a semi-solid trichobezoar composed of hair and ingesta. Thus, the presence of a hairball is not a disease, but a consequence of chronic GI hypomotility. Surgery is not necessary in most cases. Fluid therapy and assisted feeding help to rehydrate the trichobezoar and facilitate its breakup. Note that papaya, pineapple juice, and proteolytic enzymes have been shown to be ineffective at dissolving hair.

Dysbiosis, enteritis, enterotoxemia, and diarrhea_

Disruption of the normal balance of microflora in the gut, usually with overgrowth of known or potential pathogens, is a common and serious problem in rabbit medicine. Dysbiosis occurs secondary to GI hypomotility, poor diet composition (low fiber, high carbohydrate, high protein), inappropriate antibiotic selection, exposure to pathogenic organisms or toxins, and increased glucocorticoid levels (iatrogenic or secondary to stress). Bacterial dysbiosis can lead to enteritis, enterotoxemia, GI stasis, and severe cases can be life-threatening.

Low dietary fiber is probably the most important risk factor, promoting dysbiosis, enteritis, and enterotoxaemia in several ways. First, low-fiber diets typically contain high concentrations of carbohydrate, which provide a ready source of fermentable products. This alters the large bowel ecology in a way that threatens favorable microorganisms and promotes bacterial pathogens (e.g., *E. coli* and *Clostridium*), and toxin production. Pathogens such as *E. coli* and *Clostridium* species proliferate in a carbohydrate-rich environment. Glucose, a byproduct of carbohydrate digestion, is necessary for *Clostridium* species to produce iota toxin. Second, low-fiber diets reduce hindgut motility, which alters fermentation and pH; even small alterations in cecal pH can lead to death of normal microorganisms and increases in pathogens. Similarly, a decrease in peristalsis in the small intestine (hypomotility) will allow normally small populations of potentially pathogenic bacteria to proliferate, causing a range of pathology, from mild diarrhea to death from enterotoxemia. Third, low-fiber diets hinder the ability of stomach acid to reach and kill potential pathogens. The low pH of the rabbit stomach (pH 1-2) essentially sterilizes ingesta, preventing bacterial infection of the GI tract. Coarse fiber particles normally form a loose ball of ingesta in the stomach, which can be penetrated by stomach acid, killing the bacteria within. However, with inadequate fiber intake, this ball of ingesta forms a dense, compacted mass that protects ingested bacteria from degradation by gastric acid.

Other factors involved in the development of dysbiosis, and enteritis include antibiotic administration and stress. Antibiotics including clindamycin, lincomycin, ampicillin, amoxicillin, amoxicillin-clavulanate, cephalsporins, many penicillins, and erythromycin can suppress normal flora and allow pathogens to proliferate. Stress is believed to cause epinephrine-mediated inhibition of gut motility, which, like other factors, has a detrimental effect on hindgut flora; it may also suppress the immune system and promote secondary infection through endogenous glucocorticoid release.

Clinical signs may include acute or chronic intermittent diarrhea, excessive gas (bloat), GI stasis, bloody stool, dehydration, shock, and/or death. Diagnosis is by history, clinical presentation, and fecal cytology. Bacterial culture and sensitivity may help to guide antibiotic therapy in chronic cases.

Because dysbiosis and enteritis often accompany GI stasis, the medical management for both is similar and involves basic supportive care (nutritional, fluid, and thermal support), and pain management (see above). Prokinetics may be indicated in cases with concurrent GI stasis. Antibiotics are recommended where secondary bacterial overgrowth has occurred. Broad-spectrum antibiotics such as trimethoprim-sulfamethoxazole (30 mg/kg PO q12h) or enrofloxacin (15 mg/kg PO q24h) should be selected. If antibiotic-associated enterotoxemia is suspected, it can be treated with metronidazole 20 mg/kg q12h (to treat *Clostridium*), and cholestyramine (binds bacterial toxins) 2 gm/20 cc water, divided q24h PO or by gavage. (Dose cited is for a typical rabbit). Probiotics (lactobacillus, yogurt) are of questionable efficacy, however transfaunation with cecotropes from a healthy individual should be considered.

Dysbiosis may be avoided by offering a diet high in course, indigestible fiber; by using only appropriate, broad-spectrum antibiotics (trimethoprim-sulfas, fluoroquinolones, chloramphenicol, aminoglycosides, azithromycin, and metronidazole); and by reducing stress (proper environment, adequate pain control, treatment for concurrent disease).

Acute GI dilation and bloat

Rabbits are unable to vomit because the cardia is well developed, and the duodenum exits the stomach at an angle such that the pylorus is easily compressed. Gastric emptying may be prevented by gastric distention or compression due to trichobezoar, gas, or hepatomegaly, and lead to the development of bloat. In the healthy digestive tract, a moderate amount of gas is normally produced by the fermentation process and eliminated by peristalsis; however, with GI stasis and/or bacterial dysbiosis there can be excessive gas production and without normal motility the stomach and intestines can overfill with gas and become distended. Affected rabbits should receive basic supportive care, analgesics, simethicone, and prokinetic agents (see above). Decompression by orogastric tube or transabdominal trocar may also be indicated.

GI obstruction

Obstruction of the GI tract is an acute, life-threatening condition. Pathogenesis, history, and clinical findings differ significantly from those of GI stasis (and the gastric hairball that may accompany it). In most cases, obstruction occurs because a compact mat of hair (trichobezoar) lodges in the small intestine. Mats of hair may be ingested from the coat during grooming, formed in the stomach, or ingested as cecal pellets containing compacted hair. Other potential foreign objects reported to acutely obstruct the intestinal tract include carpet or cloth fibers, locust beans, or plastic. Obstruction typically occurs in the proximal duodenum where the lumen narrows, the mid-duodenum, or the ileocecocolic junction. Acute obstruction can also occur secondary to neoplasia, herniation, torsion, or other causes, all of which cause physical obstruction of the gut lumen and are not related to functional GI stasis. Diagnosis is by severe, acute clinical presentation, typical gas pattern on radiographs, and exploratory surgery. Initial medical treatment may include analgesia, decompression (see above), and fluid therapy for shock. Emergency surgery is indicated in most cases. Prognosis is guarded to poor.

Cecal impaction and cecoliths

Altered cecocolonic motility, excessively fine-particle fiber, and excessive indigestible fiber (e.g., psyllium) can result in compaction and dehydration of cecal and colonic material and subsequent impaction or formation of abnormally hard lumps of fecal contents (cecoliths). Impacted material that breaks free can cause lower bowel obstruction, most commonly lodging at the distal end of the sacculated (proximal) colon, the fusus coli. Cecoliths are the most common cause of large bowel obstruction. Affected rabbits commonly have a chronic history of large intestinal symptoms including large, malformed fecal pellets, recurrent cecal impaction, abdominal pain, and anorexia. Diagnosis is by palpation of firm, doughy cecal material, and large-bowel obstruction pattern on x-rays. Treatment involves rehydration, analgesia, prokinetics, and careful enemas.

Mucoid enteropathy

Mucoid enteropathy affects young rabbits 7-14 weeks of age and is a major cause of morbidity and mortality. Clinical signs include anorexia, lethargy, weight loss, diarrhea, cecal impaction, and excessive mucus in the stool. Its cause is unknown, but it has been linked to improper diet, abnormally low cecal pH and resultant bacterial dysbiosis. The cecum and large intestine are rich in goblet cells. Dysbiosis causes inflammation of the cecum and colon, which increases goblet cell number and amount of mucus production. The presence of mucus is not pathognomonic, and the disease mucoid enteropathy is a descriptive term only, not an etiology. Treatment is as for bacterial dysbiosis (see above) and prevention is through a high-fiber, low-carbohydrate diet.

Clostridium spiroforme enterotoxemia

Iota-like toxin from *Clostridium spiroforme* may cause acute anorexia, depression, and watery, brown diarrhea that soils the perineum and legs. Diarrhea may contain blood or mucus. Mostly affects newly weaned rabbits 3-6 weeks of age, likely because of their underdeveloped population of normal GI flora and higher gastric pH. Treatment consists of aggressive supportive care, cholestyramine, and metronidazole (see above).

Enteropathogenic *E. coli*

Enteropathogenic *E. coli* (EPEC) causes diarrhea and mortality. It is a major cause of economic loss in the commercial rabbit industry and has not been reported in pet rabbits. Disease in neonatal rabbits is most common between 1 and 14 days of age. It causes watery, yellow diarrhea with morbidity and mortality rates within a litter approaching 100%. Older animals can also be affected; early treatment can give good results. Treat with trimethoprim-sulfamethoxazole (30 mg/kg PO q12h) or enrofloxacin (10 mg/kg PO q12h).

Proliferative enteritis

Proliferative enteritis is caused by an obligate intracellular bacterium *Lawsonia intracellularis*, which is Gram negative, curved to spiral-shaped, and found within intestinal epithelial cells. It causes acute diarrhea in weanling rabbits 2-4 months of age. It also affects many other species. Treat with chloramphenicol (30-50 mg/kg PO, SC q12h) for 7-14 days.

Tyzzer's disease

Tyzzer's disease is caused by *Clostridium piliforme* (formerly *Bacillus piliformis*), a motile gram-variable spore-forming obligate intracellular bacterium, which causes disease in many other small mammals. Stress of overcrowding, unsanitary conditions, high temperatures, or breeding may predispose rabbits to the disease. Clinical signs include watery diarrhea, depression, and death. Weanling rabbits experience acute disease, while older rabbits may develop chronic form resulting in chronic weight loss. Liver may show characteristic necrotic foci and myocardial lesions at necropsy. Treatment may be unsuccessful once clinical signs have started.

Oral papillomatosis

Oral papillomatosis is a benign disease caused by a papillomavirus, reported only in colonies of New Zealand white rabbits. Lesions consist of small white growths on the ventral surface of the tongue and rarely in other areas of the mouth. Papillomas vary in size and shape but are typically 1-3 millimeters and persist from several weeks to months. Lesions resolve without treatment.

Coronavirus

Rabbit enteric coronavirus is primarily a disease of laboratory rabbits, affecting animals 3-10 weeks of age, but also found in normal adults. Clinical signs include lethargy, diarrhea, abdominal swelling, and death. Coronavirus causes high morbidity and mortality. Diagnosis may be confirmed by isolating virus in feces or cecal contents.

Rotavirus

Rotavirus is ubiquitous and cosmopolitan. By itself, the virus is mildly pathogenic, associated with high morbidity but variable mortality. Coinfection with other microorganisms likely influences severity of clinical signs, which include anorexia, dehydration, and mucoid or greenish-yellow watery diarrhea in rabbits 30-80 days of age.

Mortality is usually low but may be as high as 80%. Diagnosis is based upon histopathology, virus isolation, and demonstration of antibodies. Prevention and control are by appropriate hygiene, reduction of stress, and treatment for concurrent disease.

Rabbit hemorrhagic disease virus

Caused by a calicivirus, rabbit hemorrhagic disease virus (RHDV) first emerged in China. RHDV 1 affects only the European rabbit, while RHDV2 has been crossing over to North American wild rabbits, and is endemic in China, Europe, Cuba, Australia, and New Zealand. Outbreaks have occurred in the Middle East, South America, Mexico, and the United States. Causing clinical disease only in rabbits over 2 months of age, RHDV replicates in the liver, resulting in severe hepatic necrosis and death from disseminated intravascular coagulation (DIC). Affected rabbits may die without premonitory signs, or may present with lethargy, depression, anorexia, fever, jaundice, constipation, or diarrhea. Some rabbits exhibit neurologic signs, and at the end stage of disease, tachypnea, cyanosis, and blood-tinged foamy nasal discharge are seen. Course of disease varies up to 2 weeks in some cases. In mild cases, rabbits may live and become carriers of the virus. Diagnosis is through identification of the virus. RHDV is reportable. A vaccine is available in endemic countries. There is now a vaccine available in the United States for RHDV2. The virus can be inactivated with bleach or formalin.

Coccidiosis

Coccidia are the most common parasite of the rabbit GI tract. Causes clinical illness mostly in young rabbits less than 6 months old; adult rabbits rarely show clinical signs. Infection does not equate to disease. Twelve *Eimeria* species are reported; all but *E. stiedae* (found in the liver) are found in the digestive tract.

Hepatic coccidiosis caused by *E. stiedae* is ubiquitous in commercial rabbitries. Many infections are asymptomatic; however, the disease may be fatal, especially in young rabbits. Heavy infections cause decreased liver function and bile obstruction. Clinical signs include anorexia, debilitation, diarrhea, and/or constipation; may cause increased liver enzymes and/or hepatomegaly. Diagnosis is by identification of oocysts in bile, on histopathology, or by fecal exam.

Intestinal coccidiosis infection is common in both young and adult rabbits. Most infections are subclinical; however, clinical signs vary widely depending on the rabbit's age, coccidia species, parasite burden, and host factors (age, stress, diet, etc.). Clinical signs are usually associated with poor husbandry or overcrowding, usually in rabbits less than 6 months of age, and include mild intermittent to severe diarrhea that may contain mucus or blood, weight loss, and dehydration. Animals with severe diarrhea may develop intussusception. Death is most often attributed to dehydration and secondary intestinal dysbiosis. Diagnosis is confirmed by histopathology and fecal exam. Treatment of individual rabbits is with sulfadimethoxine (15 mg/kg PO q12h) for 10 days or trimethoprim-sulfamethoxazole (30 mg/kg PO q12h) for 10 days. Outbreaks may be controlled by practicing good husbandry. Most healthy rabbits show no clinical signs, and rabbits that recover develop immunity that may be lifelong.

Cryptosporidiosis

Cryptosporidium parvum causes disease in young rabbits approximately 30-40 days of age. Clinical signs include diarrhea lasting 3-5 days, decreased appetite, depression, lethargy, weakness, and dehydration. Adults are not infected. The ileum and jejunum are most affected. Diagnosis may be made on histopathology, fecal exam (saline wet mount, floatation, or acid-fast stain), or fecal ELISA. There is no treatment.

Nematodes

Passalurus ambiguus is the common pinworm of rabbits. Infection is common and usually nonpathogenic. Adult worms are found in the lumen of the cecum and large intestine. Life cycle is direct, with eggs being ingested during cecotrophy. Diagnosis is made by identifying adult worms in feces or eggs on fecal exam. Treatment is not usually indicated, but fenbendazole (10-20 mg/kg PO, repeated in 10-14 days) is usually effective.

Hepatic lipidosis

Fatty liver disease is anecdotally reported in rabbits that go without food for several days or more. Incidence is probably greater in rabbits that are obese or on inappropriate diet. Diagnosis is through history, clinical presentation, and plasma biochemistry. Prevention is by instituting nutritional support (e.g., syringe feeding Oxbow Critical Care Herbivore) as soon as anorexia or decreased appetite is noticed.

Liver lobe torsion

The caudate lobe is most often affected in liver lobe torsion; however, torsion of the right lobe, quadrate lobe, and the posterior lobule of the left hepatic lobe also have been reported. Clinical signs of acute liver lobe torsion generally progress over 1-3 days. Affected rabbits are anorexic, weak, and depressed; they demonstrate cranial abdominal discomfort, and a firm, cranial abdominal mass may be palpable. Affected rabbits may appear jaundiced. Initial clinical presentation may be similar to GI stasis syndrome; however, affected rabbits will deteriorate rapidly, becoming obtunded, shocky, and hypothermic. Death may occur in 12-72 hours. Affected rabbits may exhibit anemia (PCV 16-17%) elevated liver enzymes. Diagnosis may be possible with radiography and/or ultrasound examination. Acutely ill rabbits may respond to analgesia, intravenous fluids, thermal support, and surgical removal of the affected liver lobe; however, postsurgical survival rate is low. Rabbits with chronic liver lobe torsion (e.g., poor doers with a history of recurrent GI stasis) are occasionally diagnosed, and surgical removal in these patients caries a better prognosis for recovery.

Aflatoxicosis

Aflatoxins produced by *Aspergillus flavus* and *A. parasiticus* cause liver and biliary damage in rabbits, which are among the most sensitive to aflatoxin of all species studied. Clinical signs include anorexia, dullness, weight loss, and jaundice; death occurs within 3-4 days of onset of illness. Treatment is removal of affected feed followed by appropriate support.

Neoplasia

Neoplasms of the GIT include adenocarcinoma and leiomyosarcoma of the stomach, leiomyoma and leiomyosarcoma of the intestine, papilloma of the sacculus rotundus, papilloma of the anorectal junction, and bile duct adenoma and carcinoma. Metastatic neoplasia, most commonly uterine adenocarcinoma, often involves the GIT. If diagnosed early, intestinal masses can be resected with good success. Rectal papillomas are benign, unrelated to oral or dermal papillomas, and removal is usually curative. *Eimeria stiedae* may be a causative factor in the development of bile duct neoplasia.

Husbandry and prevention of disease

Rabbits should receive an annual to semiannual complete physical examination. Animals should be weighed at every opportunity, and owners should be encouraged to weigh rabbits monthly. Diet and husbandry should be reviewed at each visit, and regular fecal examinations should be performed. Routine screening for disease (e.g., CBC, chemistry, urinalysis, radiographs) establishes an individual baseline and aids in the early detection of disease.

References

1. Donnelly TM, Vella D. Basic anatomy, physiology, and husbandry of rabbits. In: Quesenberry KE, Orcutt CJ, Mans C, Carpenter JW, eds. *Ferrets, Rabbits, and Rodents: Clinical Medicine and Surgery.* 4th Ed. St. Louis, MO: Elsevier; 2020:135-138.

2. Brandao J, Graham J, Quesenberry KE. Basic approach to veterinary care of rabbits. In: Quesenberry KE, Orcutt CJ, Mans C, Carpenter JW, eds. *Ferrets, Rabbits, and Rodents: Clinical Medicine and Surgery.* 4th Ed. St. Louis, MO: Elsevier; 2020:158-159.

3. Smith SM. Gastrointestinal physiology and nutrition of rabbits. In: Quesenberry KE, Orcutt CJ, Mans C, Carpenter JW, eds. *Ferrets, Rabbits, and Rodents: Clinical Medicine and Surgery.* 4th Ed. St. Louis, MO: Elsevier; 2020:162-173.

4. Oglesbee BL, Lord B. Gastrointestinal diseases of rabbits. In: Quesenberry KE, Orcutt CJ, Mans C, Carpenter JW, eds. *Ferrets, Rabbits, and Rodents: Clinical Medicine and Surgery.* 4th Ed. St. Louis, MO: Elsevier; 2020:174-187.

5. Meredith A. Biology, anatomy and physiology. In: Meredith A, Lord B, eds. *BSAVA Manual of Rabbit Medicine.* Gloucester, UK: British Small Animal Veterinary Association; 2014.

6. Prebble J. Nutrition and feeding. In: Meredith A, Lord B, eds. *BSAVA Manual of Rabbit Medicine.* Gloucester, UK: British Small Animal Veterinary Association; 2014:27-35.

7. Harcourt-Brown F. Digestive system disease. In: Meredith A, Lord B, eds. *BSAVA Manual of Rabbit Medicine.* Gloucester, UK: British Small Animal Veterinary Association; 2014:168-190.

Section 24

ASSOCIATION OF EXOTIC MAMMAL VETERINARIANS

AEMV Gastrointestinal System Session

Micah Kohles, DVM, MPA
Session Chair

Determining the Fecal Microbiome of Domestic Ferrets (*Mustela putorius furo*) using Next-Generation DNA Sequencing

Margaret Lee, DVM,
Shachar Malka, DVM, DABVP (Avian Practice),
Gina Vecere, DVM,
Nicole Sands, DVM,
Janina A. Krumbeck, PhD

Session #3482

Affiliation: From Long Island Bird and Exotics Veterinary Clinic, 333 Great Neck Rd, Great Neck, NY 11021, USA (Lee, Malka), MiDOG LLC, 14762 Bentley Circle, Tustin, CA 92780, USA (Krumbeck).

Note: Abstract format selected by author(s). See Front Matter for description of conference proceedings manuscripts submitted as scientific abstracts.

Abstract: The domestic ferret gastrointestinal tract is extremely reactive and frequently mounts a strong inflammatory response from a myriad of etiologies.[1,4] Therefore, ferrets are often presented with a nonspecific history of gastrointestinal signs, which can pose significant diagnostic challenges.[2] The purpose of this study is to compare the fecal microbiome of clinically healthy ferrets and ferrets with gastrointestinal disease using next-generation DNA sequencing. Fecal samples from 52 ferrets (27 clinically healthy individuals and 25 individuals with gastrointestinal signs) were collected and analyzed using next-generation DNA sequencing. The most abundant bacterial species in the diarrhea cases were *Clostridium perfringens* (27.5% ± 31), *Clostridium disporicum* (12.5% ± 19), and a species of the order *Clostridiales* (6.9% ± 15). In the clinically heathy samples, *Intestinibacter bartlettii* (14.4% ± 25), *Clostridium disporicum* (10.8% ± 19), and *Romboutsia* sp. (5.6% ± 10) were the most abundant species. The most abundant fungal species in the diarrhea group were *Saccharomyces cerevisiae* (38.1 ± 47), *Endocarpon adscendens* (3.5 ± 17), and *Lodderomyces elongisporus* (3.5 ± 16), and in the healthy group the most abundant species were *Saccharomyces cerevisiae* (26.5 ± 38), *Trichosporon asahii* (4.34 ± 9), and *Mucor plumbeus* (4.3 ± 18). These results are in agreement with a study that has reported that *Clostridium* spp. were isolated in healthy individuals; however, only two species were detected using traditional culture methods.[3] This study demonstrates that next-generation DNA sequencing is a more comprehensive diagnostic test and we hypothesize that it may elucidate on how the fecal microbiome changes in ferrets with gastrointestinal disease.

References

1. Lewington JH, ed. *Ferret Husbandry, Medicine and Surgery.* 2nd Ed. Edinburgh, UK: Elsevier Saunders; 2007:203-214.

2. Huynh M, Pignon C. Gastrointestinal disease in exotic small mammals. *J Exot Pet Med.* 2013;22(2):118-131.

3. Nizza S, Rando F, Fiorito F, et al. Fecal microbiota and antibiotic resistance in ferrets (*Mustela putorius furo*) from two captive breeding facilities in Italy. *Res Vet Sci.* 2014;96(3):426-428.

4. Lennox AM. Gastrointestinal diseases of the ferret. *Vet Clin North Am Exot Anim Pract.* 2005;8(2):213-225.

Surgical Treatment of Gastric Trichobezoars, Perigastric Abscess and Gastric Perforation in a Ferret (*Mustela putorius furo*) with Peritonitis

Patricia Muffat-es-Jacques, BVM&S,
Paul Raymond, DVM,
Roxane Chejfec, DVM,
Hadrien Balle, DVM,
Stéphanie Marotto, DVM,
Charly Pignon, DVM, DipECZM (Small Mammals)

Session #3211

Affiliation: From École Nationale Vétérinaire d›Alfort, Alfort Teaching Hospital, Exotics Medicine Service, Maisons-Alfort, F-94700, France.

Note: Abstract format selected by author(s). See Front Matter for description of conference proceedings manuscripts submitted as scientific abstracts.

Abstract: Ingestion of foreign bodies is frequent in ferrets and can occur at all ages as they remain inquisitive lifelong.[1] Complications, including formation of abscesses and the associated treatment has not been described in the literature. A 3-year-old intact male ferret was presented in compensated hypovolemic shock. Diarrhea, polydipsia and hyporexia were reported for 10 days. Gastric foreign bodies resembling trichobezoars along with gastric perforation and changes suggesting peritonitis were observed with ultrasonography. The laparotomy revealed a gastric perforation complicated by a perigastric abscess. The trichobezoars were removed via gastrotomy. The abscess was drained and the perforation was sutured. Due to the proximity of the abscess to the pylorus and extensive nature involving the lesser curvature, an omentalization of the abscess was performed. Post-operative care included supportive treatment, analgesia and antimicrobials. The ferret recovered well and was discharged five days post surgery. Resolution of signs and wound healing within 2 weeks. Two months after, one episode of vomiting and diarrhea occurred. Signs resolved with symptomatic treatment involving antibiotic, anti-emetic and gastroprotectants. Septic peritonitis is an emergency and animals have a guarded prognosis.[2,3] Omentalization was performed to minimize extensive tissue removal. Due to the proximity of the perforation to the pylorus a gastroduodenostomy such as Billroth I or II or Roux-en-Y gastrojejunostomy could have been required if the reconstruction led to pyloric stenosis;[4-6] however, little information regarding this surgery in ferrets is available. To the authors' knowledge, this is the first reported case of gastric perforation and perigastric abscess formation secondary to trichobezoars in a ferret, with successful surgical treatment.

References

1. Powers LV, Perpinàn D. Basic anatomy,physiology, and husbandry of ferrets. In: Quesenberry KE, Carpenter JW, eds. *Ferrets, Rabbits, and Rodents: Clinical Medicine and Surgery.* 4th Ed. St. Louis, MO: Elsevier; 2020:1-12.

2. Dayer T, Howard J, Spreng D. Septic peritonitis from pyloric and non-pyloric gastrointestinal perforation: prognostic factors in 44 dogs and 11 cats. *J Small Anim Pract.* 2013;54(12):625-629.

3. Bush M, Carno MA, St Germaine L, Hoffmann DE. The effect of time until surgical intervention on survival in dogs with secondary septic peritonitis. *Can Vet J.* 2016;57(12):1267-1273.

4. Jiang H, Li Y, Wang T. Comparación de las reconstrucciones Billroth I, Billroth II y Roux-en-Y después de la gastrectomía distal: una revisión sistemática y un metaanálisis de red. *Cirugía Española.* 2021;99(6): 412-420.

5. Nishizaki D, Ganeko R, Hoshino N, et al. Roux-en-Y versus Billroth-I reconstruction after distal gastrectomy for gastric cancer. *Cochrane Database Syst Rev.* 2021;9:1-25.

6. Barandun MA, Mulllins RA, Rytz U. Billroth II procedure for the treatment of spontaneous gastrointestinal perforation in two cats. *J Am Vet Med Assoc.* 2021;259(11):1325-1331.

A Retrospective Evaluation of Gastrointestinal Obstruction in 35 Rabbits (*Oryctolagus cuniculus*)

Amanda Steinagel, DVM, and
Barbara Oglesbee, DVM, DABVP (Avian Practice)

Session #3312

Affiliation: From MedVet Hilliard, 5230 Renner Road, Columbus, OH 43228, USA.

Note: Abstract format selected by author(s). See Front Matter for description of conference proceedings manuscripts submitted as scientific abstracts.

Abstract: Small intestinal obstruction also referred to as intestinal obstruction, gastric dilation, gastrointestinal obstruction or gastric stasis and dilation syndrome is an acute, rapidly progressive, and potentially life-threatening disorder of rabbits requiring immediate treatment.[1-7] Intestinal obstruction initiates rapid redistribution of fluid from the intravascular space into the gastrointestinal lumen proximal from the point of obstruction.[1,7-9] Depending on the location of the obstruction, fluid loss is rapid and substantial, leading to hypovolemic shock.[1,7-9] Both medical and surgical treatment of small intestinal obstruction in rabbits have been previously described.[1-9] The most common cause of gastrointestinal obstruction in the rabbit is a mat of compressed hair.[1-7] Orogastric decompression has been previously recommended as a part of both medical and pre-surgical treatment in rabbits with gastrointestinal obstructions;[1-7] however, none of these sources offer guidelines as to when orogastric decompression is indicated. This retrospective study was conducted to define parameters which may guide the decision to perform this procedure. The study reviewed the medical records of 35 rabbits, over a 3-year period, diagnosed with gastrointestinal obstruction and compares the complete blood count, serum chemistry, and radiographic findings. The results suggest that radiographic findings, when correlated with the severity of hyperglycemia and hyponatremia in rabbits with gastrointestinal obstruction, could support the need for performing orogastric decompression. The radiographic severity of gastric obstruction correlated with severe hyperglycemia (avg. 25.7 mmol/L; 463 mg/dl) and corresponding hyponatremia (Na $<$ 138 mmol/L; 138 mEq/L). The patients in this study were treated with either medical management or orogastric decompression and medical management, successfully, no gastrointestinal surgical procedures were completed. With the use of diagnostic tools, such as, radiographs and the appropriate clinicopathologic data, rabbits with gastrointestinal obstruction can be managed effectively with orogastric decompression and medical treatment.

References

1. Oglesbee BL, Lord B. Chapter 14: Gastrointestinal disease of rabbits. In: Quesenberry KE, Orcutt CJ, Mans C, Carpenter, JW eds. *Ferrets, Rabbits, and Rodents: Clinical Medicine and Surgery*. 4th Ed. Philadelphia, PA: Elsevier Saunders; 2021:174-187.

2. Oglesbee BL. Gastrointestinal dilatation (bloat). In: Oglesbee BL, ed. *Blackwell's Five-Minute Veterinary Consult: Small Mammal*. 2nd Ed. 2011;420-422.

3. Varga M. Digestive disorders. In: Varga M, ed. *The Textbook of Rabbit Medicine*. 2nd Ed. 2014;303-349.

4. DeCubellis J, Graham JE. Rabbits. In: Graham JE, Grayson AD, Beaufrere H eds. *Exotic Animal Emergency and Critical Care Medicine*. 1st Ed. Hoboken, NJ: John Wiley & Sons, Inc. 2021;238-283.

5. Harcourt-Brown FM. Gastric dilatation and intestinal obstruction in 76 rabbits. *Vet Rec.* 2007;161:409-414.

6. Harcourt-Brown TR. Management of acute gastric dilatation in rabbits. *J Exot Pet Med.* 2007;16:168-174.

7. Schulhmann B, Cope I. Medical treatment of 145 cases of gastric dilatation in rabbits. *Vet Rec.* 2014;175: 484-487.

8. R, Fehr M, Neumuller M, Thole M. Acid-base-balance status and blood gas analysis in rabbits with gastric stasis and gastric dilatation. *J Exot Pet Med.* 2020;32:18-26.

9. Harcourt-Brown FM, Harcourt-Brown S. Clinical value of blood glucose measurement in pet rabbits. *Vet Rec.* 2012;170:674.

Effect of Rabbit Gastrointestinal Stasis on the Fecal Microbiota of Pet Rabbits (*Oryctolagus cuniculus*)

Faith M. Rahic-Seggerman, BS,
Stephan Schmitz-Esser, PhD,
Cayla Beebe-Iske, PhD,
Jennifer Graham, DVM, DABVP (Avian Practice), DABVP (Exotic Companion Mammal Practice), Dipl ACZM,
Nicole Furst, DVM,
Micah R. Kohles, DVM, MPA

Session #3462

Affiliation: From the Department of Animal Science, Iowa State University, 1019 North University Boulevard, Ames, IA 50010, USA (Rahic-Seggerman, Schmitz-Esser), Omaha's Henry Doorly Zoo and Aquarium, 3701 South 10th Street, Omaha, NE 68107, USA (Beebe-Iske), Graham Veterinary Consulting, LLC, 119 Sally Lane, Madison, AL 35758, USA (Graham), VCA Abbott Animal Hospital, 21 E. Mountain St., Worcester, MA 01606, USA (Furst), Oxbow Animal Health, 11902 South 150th Street, Omaha, NE 68138, USA (Kohles).

Note: Abstract format selected by author(s). See Front Matter for description of conference proceedings manuscripts submitted as scientific abstracts.

Abstract: Rabbit gastrointestinal stasis (RGIS) is defined as reduced motility of any part of the digestive tract that can lead to impaction and death if left untreated.[1] This study aimed to describe the secondary effect of RGIS on the fecal microbiota of pet rabbits. Fecal samples from rabbits displaying one of three different health outcomes were obtained and the symptomology of the rabbits were recorded along with any medical intervention that took place. The three health outcomes were as follows: 1) healthy rabbits, 2) rabbits that displayed symptoms of RGIS that were treated and recovered, and 3) rabbits that displayed severe symptoms of RGIS and were euthanized. Fecal samples were analyzed using 16S rRNA and 18S rRNA amplicon sequencing. The 18S rRNA amplicon dataset revealed that the yeast *Cyniclomyces guttulatus* made up over 90% of the sequences. *Cyniclomyces guttulatus* is a known yeast symbiont of domestic rabbits.[2] In this dataset the relative abundance of four individual operational taxonomic units (OTU) were trending ($P < 0.05$, FDR < 0.1) towards being significantly altered due to the health status of the animal. In the 16S rRNA amplicon dataset there was no significant difference between health outcomes on an OTU level. Overall, the health status of the animal had a negligible effect on the community composition (Bray-Curtis) or the community structure (alpha diversity) of the samples, suggesting RGIS has little to no effect on the fecal microbiota.

References

1. Oglesbee BL, Lord B. Gastrointestinal diseases of rabbits. In: Quesenberry, KE, Orcutt CJ, Mans C, Carpenter JW, eds. *Ferrets, Rabbits, and Rodents: Clinical Medicine and Surgery.* St. Louis, MO: Elsevier; 2020:174-187.

2. Sioutas G, Evangelou K, Vlachavas A, et al. Deaths due to mixed infections with *Passalurus ambiguus*, *Eimeria* spp. and *Cyniclomyces guttulatus* in an industrial rabbit farm in Greece. *Pathogens.* 2021;10:756.

Computed Tomographic Diagnosis of Small Intestinal Obstruction Caused by Trichobezoars in Rabbits: Five Cases (2021-2022)

Nicola Di Girolamo, DVM, PhD, DipECZM (Herpetology), Dipl ACZM, Dipl ACVPM, and Christopher Tollefson, DVM, Dipl ACVR

Session #3480

Affiliation: From Exotic Pet Service, Department of Clinical Sciences, College of Veterinary Medicine, Cornell University, 930 Campus Road, 14853, Ithaca, NY 14853, USA (DiGirolamo), Radiology Section, Department of Clinical Sciences, College of Veterinary Medicine, Cornell University, 930 Campus Road, 14853, Ithaca, NY 14853, USA (Tollefson).

Note: Abstract format selected by author(s). See Front Matter for description of conference proceedings manuscripts submitted as scientific abstracts.

Abstract: Rabbits are sensitive to small intestinal obstructions, which typically result in a sudden onset of anorexia, lack of defecation and varying degrees of gas build-up in the stomach and intestinal tract.[1] There is a surprising lack of information on the way this condition is diagnosed with advanced diagnostic imaging. The purpose of this case series is to report features of small intestinal obstruction in rabbits using computed tomography (CT). All rabbits managed by a board-certified zoological medicine specialist between January 2021 and November 2022 with a surgical or postmortem diagnosis of intestinal obstruction that underwent CT scanning in the 24 hours before the diagnosis were included in this study. Five cases fulfilled the inclusion criteria. Of the five rabbits, one underwent radiographs followed by CT, while the four others underwent CT. In four of five rabbits, CT scan showed a distended proximal duodenum, followed by an area of mixed opacity and an empty ileum and jejunum. In one rabbit, most of the small intestinal tract was distended, with an area of mixed opacity at the level of the distal jejunum. In one rabbit, free abdominal fluid and gas was observed. All five rabbits received surgical treatment; four rabbits had a duodenal trichobezoar, and one had a jejunal trichobezoar. One of the four rabbits with duodenal obstruction also had free gastrointestinal content and gastric rupture. This rabbit died during surgery, while the other four rabbits recovered from surgery. The present case series highlights features of intestinal obstruction in rabbits observed upon CT.

References

1. Harcourt-Brown FM. Gastric dilation and intestinal obstruction in 76 rabbits. *Vet Rec.* 2007;161(12): 409-414.

Upper Gastrointestinal Flexible Endoscopy in Rabbits: A Pilot Study

Charly Pignon, DVM, DipECZM (Small Mammal), and
Lena Crescence, DVM

Session #3331

Affiliation: From Exotics Medicine Service, Ecole Nationale Vétérinaire d'Alfort, 7 avenue du Général De Gaulle, 94700 Maisons-Alfort, France.

Note: Abstract format selected by author(s). See Front Matter for description of conference proceedings manuscripts submitted as scientific abstracts.

Abstract: Gastrointestinal flexible endoscopy is a useful and common diagnostic technique performed in cats, dogs and ferrets.[1] It has been described in the literature that fasting a rabbit could lead to gastrointestinal stasis syndrome and without an empty stomach a gastroscopy could not be efficiently performed.[2-4] The goal of the pilot study was to test a fasting protocol to perform upper gastrointestinal flexible endoscopy in rabbits and to assess its safety. Two New Zealand White rabbits were fed with EmerAid IC Herbivore® (Lafeber, Cornell, IL, USA) for 24 or 48 hours, and then fasted for 6 hours. Under anesthesia a 5.6-millimeter and an 8.6-millimeter diameter flexible endoscope were used. The animals were monitored during the recovery period until their appetite returned to normal. This protocol was approved by an institutional ethics committee. The fasting protocols were well-tolerated, and rabbits returned promptly to normal appetite and fecal pellet production without showing discomfort. Too much food was still present in the stomach to fully explore its mucosa using the 24-hour gavage and the 6-hour fasting protocol. Good quality images were obtained from esophagus, stomach and duodenum using the 48-hour gavage and the 6-hour fasting protocol. In one of the rabbits, a perforation of the esophagus just cranial to the cardiac sphincter occurred 24 hours after the procedure after use of the 8.6-millimeter optic. No side effects were present with the use of the 5.6-millimeter optic, which would be the size recommended to perform upper gastrointestinal tract flexible endoscopy in rabbits.

References

1. Pignon C, Huynh M, Husnik R, Jekl V. Flexible gastrointestinal endoscopy in ferrets (*Mustela putorius furo*). *Vet Clin North Am Exot Anim Pract.* 2015;18(3):369-400.

2. Donnelly TM, Vella D. Basic anatomy, physiology, and husbandry of rabbits. In: Quesenberry KE, Orcutt CJ, Mans C, Carpenter JW, eds. *Ferrets, Rabbits, and Rodents: Clinical Medicine and Surgery.* 4th Ed. Philadelphia, PA: Elsevier Saunders; 2020:131-149.

3. Dayan MO, Besoluk K. Three-dimensional reconstruction of stomach and intestines in New Zealand white rabbits from computerized tomography images. *Isr J Vet Med.* 2011;66(3):108-113.

4. Nath SK, Das S, Kar O, et al. Topographical and biometrical anatomy of the digestive tract of white New Zealand rabbit (*Oryctolagus cuniculus*). *J Adv Vet Anim Res.* 2016;3(2):145-151.

Contrast Imaging of the Gastrointestinal Tract in Healthy Four-Toed Hedgehogs

Grayson A. Doss, DVM, Dipl ACZM,
Jessica Wild, DVM,
Seng Wai Yap, BVSc, MS, Dipl ACVR

Session #3188

Affiliation: From the Department of Surgical Sciences, University of Wisconsin-Madison, School of Veterinary Medicine, Madison, WI 53706, USA.

Note: Abstract format selected by author(s). See Front Matter for description of conference proceedings manuscripts submitted as scientific abstracts.

Abstract: Gastrointestinal (GI) diseases are commonly reported in pet hedgehogs.[1] Diagnostic imaging is often utilized to evaluate suspected GI disease; however, the paucity of studies evaluating radiographic GI anatomy of hedgehogs makes interpretation of radiographic GI changes challenging. Goals of this study were to develop a practical upper GI contrast protocol; describe the normal radiographic GI anatomy; and define GI transit times in nine healthy, adult hedgehogs. Following three-view survey images, anesthetized hedgehogs were placed in sternal recumbency on a rodent dental platform positioned at a 45° incline and administered barium sulfate suspension (60% wt/vol) via gastric gavage. Three-view radiographs were then collected at defined intervals following contrast administration (0 hours, 0.5 hours, 1 hour, 2 hours, 4 hours, and every 2 hours until contrast arrival in the region of the colon). Gastroesophageal reflux of barium occurred in all animals with no evidence of aspiration. Median (range) times to presence of contrast in the duodenum and gastric emptying were 0.5 hours (0.5-1 hours) and 4 hours (1-8 hours), respectively. Time to presence of contrast in the colon was 6 hours (4-8 hours). Survey measurements of the small (0.7 centimeters [0.6-0.8 centimeters]) and large (0.8 centimeters [0.4-1.3 centimeters]) intestinal diameter were similar. Post-contrast measurements of the small and large intestinal diameter were also similar. All barium was cleared by 24 hours post administration except in a single animal.

References

1. Doss GA, Carpenter JW. African pygmy hedgehogs. In: Quesenberry K, Mans C, Orcutt C, eds. *Ferrets, Rabbits, and Rodents: Clinical Medicine and Surgery.* 4th Ed. St. Louis, MO: Elsevier Saunders; 2020: 401-415.

Section 25

AEMV Anesthesia and Analgesia Session

Inge Catalina Cruz Benedetti, DVM
Session Chair

Locoregional Anesthesia in Small Mammals: A Clinical Approach

Giulia Bersanetti, DVM, MSc,
Elisa Silvia D'Urso, DVM, PhD, MRCVS,
Paolo Selleri, DVM, PhD, DipECZM (Herpetology), DipECZM (Small Mammal)

Session #3370

Affiliation: From Centro Veterinario Specialistico, Via Sandro Giovannini 53, Rome, Italy.

Note: Abstract format selected by author(s). See Front Matter for description of conference proceedings manuscripts submitted as scientific abstracts.

Abstract: Locoregional anesthesia is a useful aid for perioperative pain management and is widely described in dogs and cats.[1,2] Some techniques also have been reported in exotic species, including small mammals, but they are mostly sparsely described.[3] Furthermore, most studies report a cadaveric approach to locoregional blocks, which does not allow any clinical assessment as regard to both the effectiveness and possible side effects.[4–6] This lecture deals with a series of cases about the use of locoregional anesthesia in small mammals like rabbits, chinchillas, hedgehogs and meerkats which underwent surgical procedures. The aim is to describe the techniques applied, their effectiveness and their usefulness in a clinical setting. In these cases, the primary limitation of the applicability of the techniques as described in dogs and cats is related to the small size of the intended species. The use of ultrasound-guided or electrostimulation-guided approach was essential for the localization of the correct injection point in the reported cases. Examples include ultrasound-guided brachial plexus block in a rabbit for a forelimb amputation, electrostimulation-guided sciatic and femoral nerve block with a pre-iliac approach in a rabbit and in a hedgehog for hindlimb orthopaedic surgery, electrostimulation-guided lumbosacral plexus and femoral nerve block with pre-iliac approach in a meerkat and electrostimulation-guided brachial plexus block in a chinchilla for forelimb amputation. The use of locoregional anesthesia allowed provision of an adequate analgesic plane for the selected procedure, reducing the dose of systemic drugs for the anesthesia and provision of analgesia during the procedures and recovery.

References

1. Campoy L, Read M, Peralta S. Canine and feline local anesthetic and analgesic techniques. In: Grimm KA, Lamont LA, Tranquilli WJ, Greene SA, Robertson SA, eds. *Veterinary Anesthesia and Analgesia: The Fifth Edition of Lumb and Jones.* Chichester, UK: Wiley-Blackwell; 2015:827-856.

2. Otero P, Portela DA. *Manual of Small Animal Regional Anesthesia: Illustrated Anatomy for Nerve Stimulation and Ultrasound-Guided Nerve Blocks.* 2nd Ed. Buenos Aires, AR: Inter-Medica; 2018.

3. d'Ovidio D, Adami C. Locoregional anesthesia in exotic pets. *Vet Clin North Am Exot Anim Pract.* 2019;22(2):301-314.

4. Di Bella C, Pennasilico L, Staffieri F, et al. Ultrasound-guided lateral transversus abdominis plane (TAP) block in rabbits: a cadaveric study. *Animals.* 2021;11(7):1953.

5. Felisberto R, Flaherty D, Tayari H. Ultrasound-guided saphenous nerve block in rabbits (*Oryctolagus cuniculus*): a cadaveric study comparing two injectate volumes. *Animals.* 2022;12(5):624.

6. de Miguel Garcia C, Radkey DI, Hetzel S, Doss G. Injection techniques for auricular nerve blocks in the rabbit cadaver. *Vet Anaesth Analg.* 2020;47(2):274-279.

Efficacy of a New Single-Use Supraglottic Airway Device for Anesthesia in Four-Toed Hedgehogs (*Atelerix albiventris*)

Grayson A. Doss, DVM, Dipl ACZM,
Gail L. Huckins, DVM, Dipl ACZM,
Christoph Mans, Dr med vet, Dipl ACZM,
Tatiana H. Ferreira, DVM, MSc, PhD, DACVAA

Session #3183

Affiliation: From the Department of Surgical Sciences, University of Wisconsin-Madison, School of Veterinary Medicine, Madison, WI 53706, USA.

Note: Abstract format selected by author(s). See Front Matter for description of conference proceedings manuscripts submitted as scientific abstracts.

Abstract: Hedgehogs are routinely anesthetized for examinations and procedures, but endotracheal intubation is difficult and, therefore, uncommonly performed.[1] A multi-use supraglottic airway device (SGAD) for small rabbits was a suitable alternative to facemask use in this species but is no longer manufactured;[2] however, a single-use version is now available, and this current study evaluated the feasibility of this single-use SGAD designed for small rabbits (V-Gel ADVANCED R1, Docsinnovent, England, UK) in nine adult hedgehogs. Hedgehogs were maintained for the entire anesthetic period (30 minutes) using either a rabbit SGAD or a rodent nosecone in a randomized, complete cross-over study. Time to SGAD placement, ability to provide positive-pressure ventilation (PPV), and maintain a satisfactory anesthetic plane with the SGAD, capnography, pulse oximetry, and physiologic variables were assessed. Once the SGAD was placed, animals were randomly rotated into four different recumbencies to assess for dislodging of the device. The median (interquartile range) time to successful SGAD placement was 15 seconds (9-23 seconds). The SGAD permitted PPV at 10 cm H_2O in all animals. A blinded reviewer retrospectively evaluated rigid oropharyngoscopy videos recorded pre-and post- SGAD or nosecone, and no significant difference between treatments was found. Daily food intake and running wheel activity were assessed for 6 days post-anesthesia, and no significant differences were found between treatments. The use of the SGAD in hedgehogs appears effective and safe for anesthetic maintenance, allowing effective airway control and intermittent PPV, as compared to nosecone or facemask use.

References

1. Doss GA, de Miguel Garcia C. African pygmy hedgehog (*Atelerix albiventris*) and sugar glider (*Petaurus breviceps*) sedation and anesthesia. *Vet Clin North Am Exot Anim Pract.* 2021;25:257-272.

2. Huckins G, Doss GA, Ferreira TH. Feasibility of supraglottic airway device use during maintenance of inhalant anesthesia in healthy African pygmy hedgehogs (*Atelerix albiventris*). *Vet Anaesth Analg.* 2021;48: 517-523.

Development and Validation of a Pain Scale in Guinea Pigs

Charly Pignon, DVM, DipECZM (Small Mammal),
Floriane Benedetti, DVM,
Patricia Mufat-Jacques, DVM,
Caroline Gilbert, DVM, PhD, DipECAWBM

Session #3330

Affiliation: From Exotics Medicine Service, and Behavior Medicine Service, Ecole Nationale Vétérinaire d'Alfort, 7 avenue du Général De Gaulle, 94700 Maisons-Alfort, France.

Note: Abstract format selected by author(s). See Front Matter for description of conference proceedings manuscripts submitted as scientific abstracts.

Abstract: The goal of this study was to develop and validate a pain assessment scale for guinea pigs useful clinically for veterinarians. To achieve this goal, the scale was based on pre-existing pain scales in other species and created with the criteria described in the literature known to be indicators of pain in guinea pigs.[1-4] This scale was composed of 14 descriptors grouped in four categories: one related to appearance, one functional, one physiological and a behavioral component. During its use, the observation and manipulation of guinea pigs, a score was assigned for each item. The total score obtained, ranging from 0 to 26, is intended to give an indication of the intensity of pain. The experimental study for the validation of this scale was conducted on 45 adult guinea pigs presented at the Teaching Hospital of Alfort between March 2021 and March 2022. Statistical analysis showed that the scale had very good repeatability and intra-operator reproducibility. Agreement between pain intensities obtained with the scale and the expected pain was fair. The sensitivity of the scale was 52% and the specificity was 78%. The evaluation of the repeatability and reproducibility of some items highlighted those that needed to be clarified for the proper understanding of the scale. Further ongoing studies include modifications of these items and the use of a larger number of animals to improve this scale for clinical use.

References

1. Della Rocca G, Catanzaro A, Di Salvo A, Goldberg ME. Diagnosis of pain in small companion animals. *Am J Anim Vet Sci.* 2015;10(2):57-66.

2. Dunbar ML, David EM, Aline MR, Lofgren JL. Validation of a behavioral ethogram for assessing postoperative pain in guinea pigs (*Cavia porcellus*). *J Am Assoc Lab Anim Sci.* 2016;55(1):29-34.

3. Keown A, Farnworth M, Adams N. Attitudes towards perception and management of pain in rabbits and guinea pigs by a sample of veterinarians in New Zealand. *NZ Vet J.* 2011;59(6):305-310.

4. Konh DF, Martin TE, Foley PL, et al. Guidelines for the assessment and management of pain in rodents and rabbits. *J Am Assoc Lab Anim Sci.* 2007;46(2):13.

Section 26

AEMV Neurology Session

Jessica Robertson, DVM, Dipl ACZM
Session Chair

Treating Ferrets (*Musela putorius furo*) with Disseminated Idiopathic Myofasciitis

Katrina Ramsell, PhD, DVM

Session #3504

Affiliation: From NW Exotic Pet Vet, 2425 Southwest Cedar Hills Blvd, Beaverton, OR 97005, USA.

Abstract: Disseminated idiopathic myofasciitis (DIM) is a severe inflammatory condition that affects primarily muscles and surrounding connective tissues. DIM affects young ferrets and is characterized by a relatively acute onset of clinical signs and a general lack of response to conventional therapies. Affected ferrets usually have multiple, concurrent clinical signs including a severe, persistent, fluctuating fever, extreme lethargy, general paresis, dehydration and decreased appetite. Other clinical signs include enlarged lymph nodes or subcutaneous masses, abnormal stools, pain, tachypnea, tachycardia, heart murmur and serous nasal or ocular discharge. The duration of illness typically lasts weeks to months. Common diagnostic findings include neutrophilic leukocytosis, toxic neutrophils, mild to moderate anemia, hyperglycemia and hypoalbuminemia. Although a definitive diagnosis of DIM requires muscle biopsies, many ferrets are now being presumptively treated for DIM based on signalment, history and clinical signs considered classic for the disease. Although the current treatment protocol includes antibiotic therapy, prednisolone and medication(s) to control pain, the key drug for treating ferrets with DIM appears to be cyclophosphamide, and in many cases, injectable cyclophosphamide is more effective than oral. Supportive care (e.g., supplemental feedings, supplemental fluids and physical therapy) is also critical for recovery of ferrets with DIM. Although DIM was initially considered to be a fatal disease, many ferrets definitively diagnosed with DIM or suspected cases treated presumptively for DIM with the current treatment protocol have recovered and become long term survivors. DIM in ferrets should no longer be considered a hopeless and fatal disease.

Introduction

Disseminated idiopathic myofasciitis (DIM) is a disease that affects pet ferrets (*Mustela putorius furo*).[1-3] It is a severe inflammatory condition that affects primarily muscles and surrounding connective tissues.[1-3] There have been many suspected and confirmed cases since the disease was first recognized in 2003.[4] DIM typically affects young ferrets and is characterized by a relatively acute onset of clinical signs and a general lack of response to conventional therapies.[3-4] The previously published treatment protocol has been used to successfully treat many DIM ferrets, but the protocol has since been modified to improve patient care and safety to the caregivers.

Signalment and Clinical Signs

DIM generally affects ferrets 5 months to 24 months of age. Ferrets with DIM usually have multiple, concurrent clinical signs including a persistent, fluctuating, often severe fever (104-108 °F), extreme lethargy, general paresis, pain (especially in the lumbosacral region and hind legs), dehydration and decreased appetite. Most ferrets with DIM are depressed but cognizant. Other clinical signs include enlarged lymph nodes or subcutaneous masses, abnormal stools, tachypnea, tachycardia, heart murmur and serous nasal or ocular discharge. Despite the relatively acute to subacute onset of clinical signs in ferrets with DIM, with ferrets initially declining rapidly over a period of 12-36 hours, the duration of illness often lasts weeks to potentially several months.[3-4]

Diagnostic Results

Although the white blood cell count in a ferret with DIM may initially be in the normal range (2500-8000/µl, Antech Diagnostics), most ferrets with DIM eventually exhibit a significant leukocytosis, with a moderate to marked neutrophilia and an occasional left shift. Toxic neutrophils are often observed and a mild to moderate anemia is common in ferrets with DIM. Hyperglycemia and hypoalbuminemia are typical serologic changes in ferrets with DIM. Interestingly, creatine phosphokinase, a value typically associated with severe inflammation and muscle tissue necrosis, is rarely elevated.[3-4]

Antemortem biopsies of enlarged lymph nodes and muscle tissue from ferrets with DIM often reveal severe, widespread inflammation, often with suppurative changes. Postmortem microscopic findings include severe suppurative to pyogranulomatous inflammation in muscle tissue (i.e., skeletal, cardiac and smooth), with the esophagus being primarily affected. There is a multifocal distribution in muscle, fascia and surrounding adipose tissue with atrophy and fibrosis in severe cases. No infectious agents or neoplastic processes have been associated with DIM.[3-4] The pathogenesis of DIM is still poorly understood and there is no known definitive etiology.

Although a definitive antemortem diagnosis of DIM requires muscle biopsies, many ferrets are now being presumptively treated for DIM based on signalment, history and clinical signs, which are now considered classic for the disease. Many DIM ferrets are not stable enough, especially early in the disease process, to undergo anesthesia and surgery to obtain biopsies and some owners have financial constraints, so presumptive treatment in those cases is reasonable and often preferred over euthanasia.

Treatment

Since 2006, many confirmed and suspected cases of ferrets with DIM have improved with treatment and made a complete recovery. The original treatment protocol included cyclophosphamide, prednisolone and chloramphenicol.[3] The protocol has since been modified and has been used to treat many ferrets with great success. The current treatment protocol includes cyclophosphamide, prednisolone, antibiotic therapy and analgesics as well as supplemental care (e.g., supplemental feedings, supplemental fluids and other various supplemental therapies). The key drug for treating ferrets with DIM appears to be cyclophosphamide. The author does not know of any cases where a ferret with DIM has survived without treatment with cyclophosphamide. In some cases, injectable cyclophosphamide has been much more effective than oral preparations, but access to injectable cyclophosphamide may be limited for some practitioners. Tables 1 and 2 summarize the treatment protocol that is discussed below.

Current treatment protocol

1. Cyclophosphamide. 10 mg/kg SC to be given on day 1, day 14 and then every 14-28 days as needed. Some ferrets need treatments as frequently as every 10-14 days, especially early in the disease.

- Subcutaneous fluids (usually 45-60 ml/kg) given approximately 15 minutes prior to cyclophosphamide treatment can help prevent sterile cystitis (injectable cyclophosphamide can be injected into the subcutaneous fluid hump). Make sure the fluids are not leaking from the injection site before injecting the cyclophosphamide into the fluid hump.

- It is important to note that many ferrets begin to decline clinically (more lethargic and sometimes their fever spikes) as they are approaching the need for another dose of cyclophosphamide.

- Monitor for side effects, such as neutropenia, hemorrhagic cystitis, GI toxicity.
- The CBC should be monitored while a ferret is receiving cyclophosphamide treatments.
- It is not possible to predict how many total doses of cyclophosphamide a DIM ferret will need.
- A compounding pharmacy may be able to aliquot a bottle of sterile cyclophosphamide powder into individual 20 mg vials of injectable cyclophosphamide.
- Although oral cyclophosphamide is not palatable to ferrets and a suspension may be less effective than injectable cyclophosphamide, a compounding pharmacy can make a cyclophosphamide oral suspension.
- Partial tablets of cyclophosphamide should not be given as the drug is not evenly distributed throughout the tablet.

2. *Prednisolone.* 0.5-1 mg/kg PO q24h for 1-2 weeks, then PRN (may wean off).

- A bitter-free oral suspension is easily administered and can be obtained from a compounding pharmacy.
- Monitor for side effects, such as GI ulcers, muscle wasting, and other conditions.

3. *Antibiotics.* Amoxicillin, enrofloxacin PRN.

- Amoxicillin: 15 mg/kg PO q12h
- Enrofloxacin: 5 mg/kg PO q12h. Can also be given 10 mg/kg SC q24h. Put in fluids prior to giving. Preferably not intramuscularly as this could result in injection site necrosis and the muscles are already pathologic.
- Preferably not chloramphenicol (potentially toxic to humans), metronidazole (not palatable), or amoxicillin/clavulanic acid (can cause GI upset).

4. *Analgesics.* Buprenorphine and/or gabapentin PRN.

- Buprenorphine: 0.005-0.01 mg/kg (0.01-0.03 ml/kg or 0.3 mg/ml) SC or PO (monitor for sedation).
- Gabapentin: 1-2 mg/kg PO q12-24h (monitor for sedation).

5. *Supplemental fluids.* Almost always needed.

- Daily maintenance fluids can be given in clinic intravenously or subcutaneously or owners can give at home subcutaneously.
- Most ferrets with DIM are chronically dehydrated, especially if they have a persistent fever.
- Adding vitamin B complex to the bolus of subcutaneous fluids may be helpful.
- Be cautious not to fluid overload if significant hypoalbuminemia and/or anemia.

6. *Supplemental feedings.* Should be given 4-6 times a day if possible PRN.

- It is often painful for DIM ferrets to swallow, especially kibble, as the esophagus is often affected.
- Feedings of warmed, pureed high quality canned food or prescription food can be given.
- Avoid foods with peas, pea flour, pea fiber.

7. *Additional supplemental medical therapies.* May be helpful.

- GI protectants (e.g., sucralfate, famotidine) may be needed, especially as GI ulcers can occur.
- A multivitamin with iron may help with anemia (e.g., Pet Tinic).

8. *Supplemental physical therapies.* May be helpful.

- Physical therapy, water therapy, cold laser therapy, acupuncture.
- Remember to be gentle as ferrets with DIM are often painful.
- Love and TLC.

Table 1. Recommended medications for disseminated idiopathic myofasciitis (DIM) treatment protocol.

Drug	Suggested dose	Potential side effects
Cyclophosphamide[a-d]	10 mg/kg SC or PO on day 1, day 14, then every 14-28 days until recovery has been achieved. More frequent treatments may be needed.	Neutropenia Hemorrhagic cystitis GI toxicity
Prednisolone[e]	0.5-1.0 mg/kg PO q24h for 1-2 weeks, then PRN (wean off).	GI ulceration Muscle wasting Elevated liver enzymes Abdominal fat deposition Dermatologic effects PU/PD/Polyphagia
Antibiotics (examples)[f]:		
Amoxicillin	12.5 mg/kg PO q12h PRN	GI upset
Enrofloxacin	5 mg/kg PO q12h PRN	GI upset
Analgesics (examples):		
Buprenorphine	0.005-0.01 mg/kg SC, PO q8-12h PRN	Sedation
Gabapentin	1-2 mg/kg PO q12-24h PRN	Sedation

[a]Some ferrets need treatments of cyclophosphamide as frequently as every 10-14 days, especially early in the disease process. Many ferrets decline clinically as they approach the need for another dose of cyclophosphamide.
[b]Subcutaneous fluids given approximately 15 minutes prior to a subcutaneous cyclophosphamide treatment can help prevent sterile cystitis.
[c]Injectable cyclophosphamide can be injected into the subcutaneous fluid hump.
[d]The CBC should be monitored while a ferret is receiving cyclophosphamide treatments.
[e]A bitter-free oral prednisolone suspension can be obtained from a compounding pharmacy.
[f]Antibiotics are preferably *not*: chloramphenicol (toxic to humans), metronidazole (not palatable) or amoxicillin/clavulanic acid (tends to cause GI upset).

Table 2. Supplemental and supportive care for ferrets with disseminated idiopathic myofasciitis (DIM).

Supplemental treatment	Examples	Notes
Medications:		
GI protectants and antacids	Sucralfate 50-100 mg/kg PO q6-12h	For GI ulcers
	Famotidine 0.3 mg/kg PO, SC, IV q24h	For GI ulcers
	Pet Tinic 0.2 ml/kg PO q24h	For anemia
Multivitamin with iron		
Supplemental fluids: (intravenous or subcutaneous) Note: Vitamin B complex can be added to subcutaneous fluids (0.11 ml/kg) every 24 hours	LRS Normosol-R	Daily maintenance May need more, depending on dehydration, especially if febrile Do not fluid overload, especially if anemic and/or hypoalbuminemic
Supplemental feedings: High-quality canned food or prescription diets Critical care foods	Blue Buffalo canned cat food Hill's Science Diet a/d Critical Care Carnivore Care	Syringe feed/bowl every 4-6 hours. Can puree and warm canned food. Avoid foods with peas.
Supplemental therapies	Physical therapy Water therapy Cold laser therapy Acupuncture	DIM ferrets need to rest, but they tend to get stiff. Keep DIM ferrets clean. Lots of love and TLC.

Conclusions

The prognosis for ferrets with DIM that are treated appropriately is relatively good. The treatment appears to be most successful when it is started as early in the disease process as possible, but even ferrets with delayed treatment can still potentially recover. Ferret owners should be advised that recovery can take weeks to months, but many of the treatments, feedings and supportive care can be provided at home by the owner. Cyclophosphamide treatments and diagnostic tests should be done by the veterinarian. Although DIM was initially considered to be a fatal disease, many ferrets definitively diagnosed with DIM or suspected cases treated presumptively for DIM with the current treatment protocol have recovered and become long-term survivors. Although not all ferrets with DIM survive, usually because the treatment has been started too late or when cyclophosphamide is not included in the treatment regimen, most ferrets treated for the disease do eventually recover. DIM in ferrets should no longer be considered a hopeless and fatal disease.

Acknowledgments: The author thanks Dr. Michael Garner, Northwest ZooPath.

References

1. Garner MM, Ramsell K, Schoemaker, NJ, et al. Myofasciitis in the domestic ferret. *Vet Pathol.* 2007;44(1): 25-38.

2. Garner M, Ramsell K. Myofasciitis: an emerging fatal disease of the domestic ferret. *Exot DVM.* 2006;8(3): 23-25.

3. Ramsell K, Garner MM. Disseminated idiopathic myofasciitis in ferrets. *Vet Clin North Am Exot Anim Pract.* 2010;13(3):561-575.

4. Ramsell K. Disseminated idiopathic myositis (DIM) in ferrets. *Am Ferret Assoc.* 2019;1-5.

Focusing on the Myelin and Glia: Exploring the Pathophysiology of Wobbly Hedgehog Syndrome

Grayson A. Doss, DVM, Dipl ACZM,
Daniel Z. Radecki, PhD,
Julien Guevar, DVM, MVM, Dipl ECVN, MRCVS,
Andrea Pohly, DVM, DACVP,
Madelyn J. Reilly,
Arya Kethireddy,
Benjamin K. August,
Ian D. Duncan, BVMS, PhD,
Jayshree Samanta, MBBS, PhD

Session #3185

Affiliation: From the Departments of Surgical Sciences (Doss), Comparative Biosciences (Radecki, Reilly, Kethireddy, Samanta), Pathobiological Sciences (Pohly), and Medical Sciences (Duncan), University of Wisconsin-Madison, School of Veterinary Medicine, Madison, WI 53706, USA, the Department of Clinical Veterinary Sciences (Guevar), Vetsuisse Faculty, University of Bern, Bern, Switzerland, and the Electron Microscopy Facility (August), School of Medicine and Public Health, University of Wisconsin-Madison, Madison, WI 53706, USA.

Note: Abstract format selected by author(s). See Front Matter for description of conference proceedings manuscripts submitted as scientific abstracts.

Abstract: Wobbly hedgehog syndrome (WHS), a progressive, fatal neurologic disease, affects captive four-toed hedgehogs around the world.[1] Despite its high prevalence, the etiology of this debilitating disease remains unknown. Immunohistochemical evaluation of the CNS of hedgehogs with WHS has not been widely performed, particularly in regard to understanding myelin changes, leaving major gaps in our understanding of the etiopathogenesis. Previous studies have been retrospective in nature and relied upon immersion fixed CNS tissue, which can produce significant myelin artifact, making detailed evaluation of myelin pathology difficult to impossible. This prospective study evaluated the clinical signs, neurodiagnostic, and histochemistry and ultrastructure findings of the CNS in a group of eight hedgehogs with WHS. Client and breeder owned hedgehogs were presented for end-stage neurologic disease. Following a complete neurologic exam,[3] brainstem auditory-evoked responses (BAER)[4] were recorded and hedgehogs euthanized for confirmation of WHS and further evaluation of the CNS. Affected animals were presented with a variety of neurologic signs but all hedgehogs were ataxic or non-ambulatory. The BAER waveform morphology was abnormal in all animals, with varying levels of severity in waveform changes. Immunohistochemical and ultrastructural evaluation revealed notable demyelination and subsequent remyelination.

References

1. Díaz-Delgado J, Whitley DB, Storts RW, Heatley JJ, Hoppes S, Porter BF. The pathology of wobbly hedgehog syndrome. *Vet Pathol.* 2018;55:711-718.

2. Gardhouse S, Eshar D. Retrospective study of disease occurrence in captive African pygmy hedgehogs (*Atelerix albiventris*). *Isr J Vet Med.* 2015;70:32-36.

3. Berg CC, Doss GA, Guevar J. Neurologic examination of healthy adult African pygmy hedgehogs (*Atelerix albiventris*). *J Am Vet Med Assoc.* 2021;258:971-976.

4. Teubenbacher U, Doss GA, Guevar J. Brainstem auditory-evoked response in healthy African pygmy hedgehogs (*Atelerix albiventris*). *J Sm Anim Pract.* 2021;62:49-54.

Conservative Treatment of Multiple Intervertebral Disc Extrusions in a Lionhead Rabbit

Nicola Della Camera, DVM,
Serena Sola, DVM,
Cristian Falzone, DVM, DipECVN

Session #3222

Affiliation: Diagnostica Piccoli Animali, Veterinary Private Clinic, Zugliano (VI), Italy.

Note: Abstract format selected by author(s). See Front Matter for description of conference proceedings manuscripts submitted as scientific abstracts.

Abstract: A 4-year-old lionhead rabbit was presented to Diagnostica Piccoli Animali with an acute onset of abnormal hindlimb gait.[1,2] Neurologic examination revealed marked paraparesis with proprioception positioning and withdrawal reflexes decreased in both hindlimbs and worse on the left side. A lumbar myelopathy was suspected[3] and magnetic resonance imaging (MRI) of the thoraco-lumbosacral spine revealed an intervertebral L4-L5 disc extrusion, with subsequent mild left-sided spinal cord compression.[1,2,4-8] Based on the mild cord compression and neurologic signs, conservative treatment was elected and consisted of cage rest and non-steroidal anti-inflammatory drugs.[3,7,9,10] The rabbit improved weekly and fully recovered in 1 month. Clinical signs comparable to those previously described, albeit worse on the right side, were observed 2 months after the first diagnosis. A repeat MRI revealed the persistence of the previous disc herniation, with milder cord compression and a new L5-L6 intervertebral disc extrusion, with further mild cord compression, predominantly on the right side. Conservative treatment was administered as for the first episode and full recovery occurred within 1 month from the diagnosis.[7,9] Natural disc extrusions are reported as very rare in rabbits; this is the first case report of multiple lumbar disc herniations in one rabbit treated with conservative management, as well as the use of MRI for diagnosis and for monitoring the discs and the effectiveness of the conservative treatment.

References

1. Pamuk K. Imaging of thoracic and lumbar intervertebral disc herniation cases with MRI: a retrospective study. *Progr Nutr.* 2021;23(S2):e2021280.

2. Kozlov N, Pozyabin S, Bhattarai B, et al. Surgical management of intervertebral disc extrusion in a rabbit. *J Small Anim Pract.* 2022;63:570-570. https://doi.org/10.1111/jsap.13496.

3. Webb AA, Ngan S, Fowler D. Spinal cord injury II: prognostic indicators, standards of care, and clinical trials. *Can Vet J.* 2010;51(6):598-604.

4. Gomes SA, Volk HA, Packer RM, et al. Clinical and magnetic resonance imaging characteristics of thoracolumbar intervertebral disk extrusions and protrusions in large breed dogs. *Vet Radiol Ultrasound.* 2016;57(4):417-426.

5. da Costa RC, De Decker S, Lewis MJ, et al. Diagnostic imaging in intervertebral disc disease. *Front Vet Sci.* 2020;7:588338.

6. Meredith AL, Richardson J. Neurologic diseases of rabbits and rodents. *J Exot Pet Med.* 2015;24;21-33.

7. Olby NJ, Moore SA, Brisson B, et al. ACVIM consensus statement on diagnosis and management of acute canine thoracolumbar intervertebral disc extrusion. *J Vet Intern Med.* 2022;36(5):1570-1596.

8. Ashinsky BG, Gullbrand SE, Bonnevie ED, et al. Multiscale and multimodal structure-function analysis of intervertebral disc degeneration in a rabbit model. *Osteoarth Cartil.* 2019;27(12):1860-1869.

9. Moore SA, Tipold A, Olby NJ, et al. Current approaches to the management of acute thoracolumbar disc extrusion in dogs. *Front Vet Sci.* 2020;7:610.

10. Olby NJ, da Costa RC, Levine JM, et al. Prognostic factors in canine acute intervertebral disc disease. *Front Vet Sci.* 2020;7:596059.

Section 27

AEMV Dermatology Session

Jessica Robertson, DVM, Dipl ACZM
Session Chair

Clinical Diagnosis and Outcome of Pyrexic Rabbits with Cervicofacial Cellulitis: Six Cases (2014-2021)

Christoph Mans, Dr med vet, Dipl ACZM, and
Chelsey Tournade, DVM

Session #3474

Affiliation: From the School of Veterinary Medicine, University of Wisconsin-Madison, 2015 Linden Drive, Madison, WI 53706, USA.

Note: Abstract format selected by author(s). See Front Matter for description of conference proceedings manuscripts submitted as scientific abstracts.

Abstract: Cervicofacial cellulitis without odontogenic cause is a poorly described condition in rabbits.[1] A retrospective study was performed to evaluate the epidemiology, clinical management, and outcome of rabbits diagnosed with pyrexia and concurrent cervicofacial cellulitis. A total of six cases were included in this study. Rabbits presented with a median age of 6 years (8 months to 8 years) with a presenting complaint of anorexia or hyporexia. All rabbits had a rectal temperature > 40.2 °C (104.4 °F). Physical exam and contrast-enhanced computed tomography revealed unilateral submandibular and ipsilateral cervical diffuse soft tissue swelling in five of six rabbits. No antemortem evidence of periodontal or dental disease was found on the physical examination or computed tomography. Leukopenia was present in five of six rabbits, and four of six rabbits had a concurrent left shift and toxic heterophils. Bacterial cultures of aspirated subcutaneous soft tissue swelling grew *Escherichia coli*, *Pasteurella multocida*, *Granulicatella adiacens*, *Streptococcus species*, *Haemophilus species*, and *Bacteroides species*. Treatment was pursued in five rabbits, where all rabbits received supportive care, and four of five rabbits received systemic antimicrobial therapy. Three of five rabbits that received treatment were euthanized within 24 hours of hospitalization due to clinical decline. One rabbit was euthanized after 45 days due to the development of multifocal abscessation, and one rabbit developed diarrhea and presented dead on arrival 16 days after the completion of treatment and clinical resolution of cervicofacial cellulitis. In conclusion, cervicofacial cellulitis in pyrexic rabbits carries a poor long-term prognosis despite aggressive systemic antimicrobial therapy and supportive care.

References

1. Jenkins JR. Skin disorders of the rabbit. *Vet Clin North Am Exot Anim Pract.* 2001;4:543-563.

Treatment of Sebaceous Adenitis in Domestic Rabbits (*Oryctolagus cuniculus*)

Ian Hollinshead, VMD, and
Rebecca Ehrlich, VMD

Session #3472

Affiliation: From Radnor Veterinary Hospital, 107 North Aberdeen Avenue, Wayne, PA 19087, USA.

Note: Abstract format selected by author(s). See Front Matter for description of conference proceedings manuscripts submitted as scientific abstracts.

Abstract: Sebaceous adenitis is a disease described in domestic rabbits (*Oryctolagus cuniculus*) as a chronic, exfoliative dermatitis with areas of alopecia.[1,2] Diagnosis is primarily dependent on histopathology of skin biopsies and exclusion of other causes such as infection and thymoma-related exfoliative dermatitis.[3] Management of sebaceous adenitis in rabbits can be difficult for practitioners **as** there is no consensus on treatment, and the published literature is limited to a small number of cases, with different treatments resulting in variable success.[2,4,5] This study assessed the therapeutic effects of oral beta carotene supplementation with or without fatty acid supplementation in a group of rabbits. The animals (n = 5) ranged in age from 4-10 years old and were diagnosed with primary sebaceous adenitis through dermatohistopathology. Other etiologies were excluded by performing dermatological exams, radiology, hematology and biochemistry. Patients were administered a combination of fatty acids in the form of flax oil (Barlean's Flax Oil for Animals, 0.2 ml/kg/day) and beta carotene (Puritan's Pride, 10,000-12,500 IU/day) for a minimum of 8 weeks. One rabbit received only beta carotene due to owner compliance. After 2 months of therapy, all rabbits showed partial to complete resolution of clinical signs. Results of this report highlight a need to re-evaluate the poor prognosis that is associated with primary sebaceous adenitis in rabbits, and to provide a potential treatment plan for patients diagnosed with this condition.

References

1. Varga M, Paterson S. Dermatologic diseases of rabbits. In: Quesenberry KE, Orcutt CJ, Man C, Carpenter JW, eds. *Ferrets, Rabbits, and Rodents: Clinical Medicine and Surgery*. 4th Ed. St. Louis, MO: Elsevier; 2020:220-232.

2. Jassies-van der Lee A, Van Zeeland Y, Kik M, Schoemaker N. Successful treatment of sebaceous adenitis in a rabbit with ciclosporin and triglycerides. *Vet Dermatol.* 2009;20:67-71.

3. Benato L. Guide to skin diseases in rabbits. *In Pract.* 2019;41(10):488-497.

4. Kovalik M, Thoday KL, Eatwell K, van den Broek AHM. Successful treatment of idiopathic sebaceous adenitis in a lionhead rabbit. *J Exot Pet Med.* 2012;21(4):336-342.

5. White SD, Linder KE, Schultheiss P, et al. Sebaceous adenitis in four domestic rabbits (*Oryctolagus cuniculus*). *Vet Dermatol.* 2000;11:53-60.

Section 28

ASSOCIATION OF EXOTIC MAMMAL VETERINARIANS

AEMV Ophthalmology Session

Abigail Needleman, DVM
Session Chair

Retrospective Evaluation of Subconjunctival Enucleation Surgery in Rabbits: Eighteen Cases (2014-2022)

Christoph Mans, Dr med vet, Dipl ACZM,
Kara Hiebert, DVM,
Seth Eaton, VMD, DACVO

Session #3466

Affiliation: From the School of Veterinary Medicine, University of Wisconsin-Madison, 2015 Linden Drive, Madison, WI 53706, USA.

Note: Abstract format selected by author(s). See Front Matter for description of conference proceedings manuscripts submitted as scientific abstracts.

Abstract: Enucleation in pet rabbits is indicated for various ocular disorders, including glaucoma and phacoclastic uveitis;[1-3] however, a well-developed retrobulbar venous plexus in the orbit and the associated risk of potentially fatal hemorrhage substantially complicate enucleations in rabbits if the standard transpalpebral enucleation technique is used.[3] Therefore, the subconjunctival enucleation technique has been proposed as a safer technique in rabbits.[3] No information on this technique's short- or long-term complications has been published. A retrospective search of the medical records of a university veterinary teaching hospital was conducted for subconjunctival enucleations performed between 2014-2022, and records of rabbits with complete medical records and follow-up information were included in this study. A total of 20 enucleation procedures in 18 rabbits were included in this retrospective study. Intraoperative hemorrhage due to the nictitating membrane removal was the only intraoperative complication noted in the initial cases. In the subsequent surgeries, the nictitating membrane was not removed. Post-operative complications included serous to mucoid discharge from the surgical site in 7/20 surgeries and swelling at the surgery site in 2/20 surgeries. Post-operative swelling spontaneously resolved within 3 months of surgery; post-operative discharge spontaneously resolved within 10 months of surgery. In conclusion, the subconjunctival enucleation technique without removal of the third eyelid appears to be a safe and effective surgical technique in rabbits. Complications are limited but include serous to mucoid discharge in some cases, which is self-limiting.

References

1. Quesenberry KE, Orcutt CJ, Mans C, Carpenter JW, eds. *Ferrets, Rabbits, and Rodents: Clinical Medicine and Surgery.* 4th Ed. St. Louis, MO: Elsevier; 2020.

2. Harcourt-Brown F, Chitty J. *BSAVA Manual of Rabbit Surgery, Dentistry, and Imaging.* Gloucester, UK: British Small Animal Veterinary Association; 2014.

3. Diehl KA, McKinnon JA. Eye removal surgeries in exotic pets. *Vet Clin North Am Exot Anim Pract.* 2016;19(1):245-267.

Evaluation of the Schirmer Tear Test-I in Pet Rabbits

Claudia Bassi,
Maria Cristina Pincelli, DVM, GpCert (Ophthalmology),
Paolo Selleri, DVM, PhD, SpecPACS, DipECZM (Small Mammal), DipECZM (Herpetology),
Alessia Luciani, DVM

Session #3295

Affilation: From the Department of Veterinary Medicine, Teaching Veterinary Hospital, University of Teramo, SP18 64100, Piano D'Accio, Teramo, Italy (Bassi, Pincelli, Luciani), Centro Veterinario Specialistico, Via Sandro Giovannini 51-53, 00137 Rome, Italy (Selleri).

Note: Abstract format selected by author(s). See Front Matter for description of conference proceedings manuscripts submitted as scientific abstracts.

Abstract: This study provides physiologic values of the Schirmer tear test-I (STTI) in pet rabbits. The popularity of rabbits as pets and the limited literature available regarding the application of the STTI in this species[1-4] prompted an evaluation of tear production in pet rabbits. The study aimed to obtain standard reference values that can be applied during clinical examination to facilitate the diagnosis of eye diseases. The study included 52 rabbits, of which 33 were healthy. The animals evaluated were different breeds: Netherland dwarf rabbit, lionhead, Holland lop and mixed-breed. The rabbits previously underwent a general physical examination and an ophthalmologic examination. The parameters evaluated were menace response, palpebral reflex, corneal reflex, pupillary light reflexes, dazzle test, swinging flashlight test, STTI, fluorescein, and Jones' test. Several rabbits brought in as healthy presented asymptomatic eye diseases. The STTI analysis showed a significant difference between the values obtained from healthy eyes compared to diseased eyes (4.72 ± 0.21 e.s. mm\min versus 6.6 mm\min ± 0.70 e.s., $P < 0.01$) respectively, with a range of 2-10 mm\min for healthy ones. Six eyes of rabbits presented systemic pathologies but no ocular symptoms. Their values were within the average range of healthy eyes (4.6 mm\min ± 1.08 e.s.). These data can be used as reference values during the ophthalmologic examination of rabbits.

Acknowledgments: The authors thank Andrea Sigalotti, DVM, GpCert, (exAP) and Noemi Morara, DVM.

Disclosure: This research received no external funding.

References

1. Abrams KL, Brooks DE, Funk RS, Theran P. Evaluation of the Schirmer tear test in clinically normal rabbits. *Am J Vet Res.* 1990;51(12):1912-1913. PMID: 2085216.

2. Biricik HS, Oğuz H, Sindak N, et al. Evaluation of the Schirmer and phenol red thread tests for measuring tear secretion in rabbits. *Vet Rec.* 2005;156(15):485-487. doi: 10.1136/vr.156.15.485. PMID: 15828747.

3. Whittaker AL, Williams DL. Evaluation of lacrimation characteristics in clinically normal New Zealand white rabbits by using the Schirmer tear test I. *J Am Assoc Lab Anim Sci.* 2015;54(6):783-787. PMID: 26632789; PMCID: PMC4671795.

4. Corsi F, Arteaga K, Corsi F, et al. Clinical parameters obtained during tear film examination in domestic rabbits. *BMC Vet Res.* 2022;18(1):398. doi: 10.1186/s12917-022-03492-1. PMID: 36368996; PMCID: PMC9652783.

Section 29

ASSOCIATION OF EXOTIC MAMMAL VETERINARIANS

AEMV Pharmacokinetics Session

Abigail Needleman, DVM
Session Chair

Effects of a Single Oral Dose of Gabapentin on the Behavior and Mobility of New Zealand White Rabbits (*Oryctolagus cuniculus*)

Rachel E. Conway, DVM,
Marion Desmarchelier, DMV, MSc, Dipl ACZM, DipECZM (Zoo Health Management), Dipl ACVB,
Mollie Burton, MS, DVM,
Khursheed Mama, DVM, DACVAA,
Sangeeta Rao, BVSc, MVSc, PhD,
Lon V. Kendall, DVM, PhD, DACLAM,
Miranda J. Sadar, DVM, Dipl ACZM

Session #3449

Affiliation: From James L. Voss Veterinary Teaching Hospital, College of Veterinary Medicine and Biomedical Sciences, Colorado State University, 300 W Drake Rd, Fort Collins, CO 80523, USA (Conway), Department of Clinical Sciences, Faculté de médecine vétérinaire, Université de Montréal, 3200 rue Sicotte, St Hyacinthe, QC JS2 2M2 Canada (Desmarchelier), Laboratory Animal Resources and the Department of Microbiology, Immunology and Pathology, College of Veterinary Medicine and Biomedical Sciences, Colorado State University, 300 W Drake Rd Fort Collins, CO 80523, USA (Burton, Kendall), Department of Clinical Sciences, College of Veterinary Medicine and Biomedical Sciences, Colorado State University, 300 W Drake Rd, Fort Collins, CO 80523, USA (Mama, Rao, Sadar).

Note: Abstract format selected by author(s). See Front Matter for description of conference proceedings manuscripts submitted as scientific abstracts.

Abstract: Stress in rabbits (*Oryctolagus cuniculus*) can contribute to significant morbidity and mortality, including gastrointestinal hypomotility and poor anesthetic outcomes.[1,2] Gabapentin is an alpha-2 delta ligand commonly used as an anxiolytic in various species, but little information is available on its use in rabbits.[3] In this study five female intact, and three male intact, New Zealand white rabbits, aged 8-12 months, were administered a single oral dose of 25 mg/kg gabapentin. Effects on individual behavior, posture, and motor activity were evaluated using ANYMAZE software, a video tracking system, and behavioral ethograms. Rabbits were acclimated for 3 days, and baseline measurements (control group) were assessed 1 day prior to the first gabapentin administration. Treatment was performed three times daily to allow for gabapentin administration to each rabbit once a day for morning, midday, and late afternoon doses, with a 1-week washout period in between. Rabbits were assessed 2 hours after gabapentin administration. Data were analyzed as continuous with a linear mixed model ($P \leq 0.05$). There was a significant increase in frolicking behavior ($P < 0.01$) and a significant decrease in observing behavior ($P = 0.05$) between baseline and post-gabapentin administration. There was no significant change in other rabbit behaviors, postures, or motor patterns. These results indicate 25 mg/kg oral gabapentin increased play (frolicking), and decreased fear (observing), behaviors without significantly affecting motor patterns. This suggests that gabapentin reduced stress-associated behaviors without causing excessive sedation.[4-6] Oral gabapentin shows potential as a method to reduce fear associated with veterinary visits, transport, and other stressful events in rabbits.

References

1. Krall C, Glass S, Dancourt G, et al. Behavioural anxiety predisposes rabbits to intra-operative apnoea and cardiorespiratory instability. *Appl Anim Behav Sci.* 2019;221:1-8: e104875. doi.org/10.1016/j.applanim.2019.104875.

2. Oglesbee BL, Lord B. Gastrointestinal diseases of rabbits. In: Quesenberry K, Orcutt C, Mans CJ, Carpenter JW, eds. *Ferrets, Rabbits and Rodents: Clinical Medicine and Surgery.* 4th Ed. St. Louis, MO: Elsevier Saunders; 2020:174-187.

3. Crowell-Davis SL, Irimajiri M, Mattos de Souza Dantes L. Anticonvulsants and mood stabilizers. In: Crowell-Davis SL, Murray TF, Mattos de Souza Dantes, eds. *Veterinary Psychopharmacology.* Hoboken, NJ: Wiley and Sons; 2019:147-156.

4. Hansen LT, Bethelson H. The effect of environmental enrichment on the behaviour of caged rabbits (*Oryctolagus cuniculus*). *Appl Anim Behav Sci.* 2000;68:163-178. doi.org/10.1016/S0168-1591(00)00093-9.

5. Meijesser FM, Kersten AMP, Wiepkema PR, Metz JHM. An analysis of the open field performance of subadult rabbits. *Appl Anim Behav Sci.* 1989;24:147-155. doi.org/10/1016/0168-1591(89)90042-7.

6. Crowell-Davis S. Rabbit behavior. *Vet Clin North Am Exot Anim Pract.* 2021;24:53-62. doi:10.1016/j.cvex.2020.09.002.

Local and Systemic Antibiotic Levels of Subcutaneously Implanted Florfenicol Calcium Sulfate Beads in New Zealand White Rabbits (*Oryctolagus cuniculus*)

Megan Partyka, DVM,
Stephen J. Divers, BVetMed, DZooMed, DipECZM (Herpetology), DipECZM (Zoo Health Management), Dipl ACZM, FRCVS,
Laura Burns, DVM,
Kelsey M. Trumpp, DVM,
Crisanta Cruz-Espindola, MS,
Dawn M. Boothe, DVM, MS, PhD

Session #3460

Affiliation: From the Department of Small Animal Medicine and Surgery, College of Veterinary Medicine, University of Georgia, Athens, GA 30602, USA (Partyka, Divers, Burns, Trumpp), Department of Anatomy, Physiology and Pharmacology, College of Veterinary Medicine, Auburn University, Auburn, AL 36849, USA (Cruz-Espindola, Boothe), Veterinary Information Network, Davis, CA 95616, USA (Boothe).

Note: Abstract format selected by author(s). See Front Matter for description of conference proceedings manuscripts submitted as scientific abstracts.

Abstract: Abscesses in rabbits are very common and difficult to treat. Florfenicol (FF) has broad-spectrum activity and may be effective for the mixed aerobic-anaerobic gram-positive and negative bacterial infections involved in rabbit abscesses.[1-7] Local drug levels can be achieved with calcium sulfate beads (CSB), which are biodegradable, absorbable, and osteoconductive.[8,9] Ten FF CSB (5-mm, Kerrier beads, Palm Beach Gardens, FL, USA) and an ultrafiltration probe were implanted subcutaneously in five juvenile New Zealand white rabbits (*Oryctolagus cuniculus*). One control rabbit received CSB without FF. Blood was collected at 0, 3, 6, 12, 24, 48 hours and 7, 14, and 21 days. Interstitial fluid (ISF) was collected daily for 21 days. Plasma and ISF FF concentrations were measured by HPLC. Mean area under the curve was 16.63 ± 8.16 and 0.06 ± 0.06 h·µg/ml, mean maximum concentration was 0.83 ± 0.17 and 245.22 ± 222.77 µg/ml, mean time to maximum concentration was 2.90 ± 0.3 and 59.09 ± 40.08 h, mean disappearance half-life was 30.81 ± 16.89 and 27.34 ± 9.39 h, and mean residence time was 40.22 ± 15.85 and 73.66 ± 16.34 h for plasma and ISF respectively. Plasma concentrations were < 2 µg/ml at all timepoints and still present at day 7 in one rabbit. ISF concentrations were highest day 1-4, < 2 µg/ml day 9-13, and still present at day 21 in one rabbit. These findings support that FF CSB can achieve high local antibiotic concentrations, considered effective against most bacteria (MIC > 2 µg/ml) for 7-10 days, while maintaining low plasma concentrations in rabbits.[8]

References

1. Taylor WM, Beaufrère H, Mans C, Smith DA. Long-term outcome of treatment of dental abscesses with a wound-packing technique in pet rabbits: 13 cases (1998–2007). *J Am Vet Med Assoc.* 2010;237(12): 1444-1449.

2. Strocchi R, Orsini G, Iezzi G, et al. Bone regeneration with calcium sulfate: evidence for increased angiogenesis in rabbits. *J Oral Implantol.* 2002;28(6):273-278.

3. Ferreira TSP, Felizardo MR, Sena de Gobbi DD, et al. Virulence genes and antimicrobial resistance profiles of *Pasteurella multocida* strains isolated from rabbits in Brazil. *Sci World J.* 2012;685028.

4. Koc F, Ozturk M, Kadioglu Y, et al. Pharmacokinetics of florfenicol after intravenous and intramuscular administration in New Zealand white rabbits. *Res Vet Sci.* 2009;87(1):102-105.

5. Dowling PM. Chloramphenicol, thiamphenicol, and florfenicol. In: Giguère S, Prescott JF, Dowling PM, eds. *Antimicrobial Therapy in Veterinary Medicine.* 5th Ed. Hoboken, NJ: John Wiley & Sons; 2013: 269-277.

6. Chow EP. Surgical management of rabbit ear disease. *J Exot Pet Med.* 2011;20(3):182-187.

7. Capello V. Clinical technique: treatment of periapical infections in pet rabbits and rodents. *J Exot Pet Med.* 2008;17(2):124-131.

8. Pförringer D, Obermeier A, Kiokekli M, et al. Antimicrobial formulations of absorbable bone substitute materials as drug carriers based on calcium sulfate. *Antimicrob Agents Chemother.* 2016;60(7):3897-3905.

9. Nelson CL, McLaren SG, Skinner RA, et al. The treatment of experimental osteomyelitis by surgical debridement and the implantation of calcium sulfate tobramycin pellets. *J Orthop Res.* 2002;20(4):643-647.

Pharmacokinetics of Intramuscular and Oral Transmucosal Buprenorphine in the Domestic Ferret (*Mustela putoris furo*)

Kelsey M. Trumpp, DVM,
Stephen J. Divers, BVetMed, DZooMed, DipECZM (Herpetology), DipECZM (Zoo Health Management), Dipl ACZM, FRCVS,
Laura Burns, DVM,
Megan Partyka, DVM, Nia Chau, RVT, VTS (Exotic Companion Animal),
Stephen Harvey, DVM, DACLAM,
Heather Knych, DVM, PhD, DACVCP

Session #3180

Affiliation: From the Department of Small Animal Medicine and Surgery (Zoological Medicine), (Trumpp, Divers, Burns, Partyka, Chau) and the Department of Population Health, Laboratory Animal Medicine Program (Harvey), University of Georgia College of Veterinary Medicine, 501 D. W. Brooks Drive, Athens, GA 30602 USA, K.L. Maddy Equine Analytical Chemistry Laboratory, School of Veterinary Medicine, University of California, Davis, 620 West Health Science Drive, Davis, CA 95616 USA (Knych).

Note: Abstract format selected by author(s). See Front Matter for description of conference proceedings manuscripts submitted as scientific abstracts.

Abstract: Ferrets are a common companion animal in the United States.[1] While analgesia is an important component of veterinary practices, appropriate control of pain can be challenging. One difficulty of providing pain relief to zoological companion mammals is the lack of options that can be administered easily by non-veterinary staff outside of a hospital setting. While non-steroidal anti-inflammatory and neuropathic pain medications are commonly used in ferrets, non-injectable opioid options, as indicated for more moderate or severe pain, are limited. Buprenorphine administered via the oral transmucosal route (OTM) in other species, including domestic dogs and cats, has been shown to provide effective analgesia.[2-4] Eight female ferrets were involved in a 2-week prospective cross-over pharmacokinetic study. Each ferret received two treatments of 0.04 mg/kg buprenorphine (0.3 mg/ml) with a 7-10 day wash out period in between each treatment. Blood was collected at multiple time points for pharmacokinetic analysis. Mean maximum plasma concentrations were 5.03 and 1.00 ng/ml for the IM and OTM routes of administration, respectively. Mean times to maximum plasma concentration were 0.13 and 0.17 hours, and mean terminal half-lives were 2.77 and 0.715 hours for IM and OTM routes of administration, respectively. No adverse effects were observed, however mild sedation was noted in most animals regardless of route of administration. Based on effective buprenorphine plasma concentrations in dogs (0.6 ng/ml), OTM buprenorphine may only provide short-term analgesia to ferrets.[5] More work is needed to determine if a higher dose might be more appropriate for OTM buprenorphine in ferrets.

References

1. American Veterinary Medical Association. *U.S. Pet Ownership and Demographics Sourcebook.* AVMA; 2012.

2. Abbo LA, Ko JC, Maxwell LK, et al. Pharmacokinetics of buprenorphine following intravenous and oral transmucosal administration in dogs. *Vet Ther*. 2008;9(2):83-93.

3. Catbagan DL, Quimby JM, Mama KR, et al. Comparison of the efficacy and adverse effects of sustained-release buprenorphine hydrochloride following subcutaneous administration and buprenorphine hydrochloride following oral transmucosal administration in cats undergoing ovariohysterectomy. *Am J Vet Res.* 2011;72(4):461-466.

4. Robertson SA, Lascelles BD, Taylor PM, Sear JW. PK-PD modeling of buprenorphine in cats: intravenous and oral transmucosal administration 1. *J Vet Pharmacol.* 2005;28(5):453-460.

5. Ko JC, Freeman LJ, Barletta M, et al. Efficacy of oral transmucosal and intravenous administration of buprenorphine before surgery for postoperative analgesia in dogs undergoing ovariohysterectomy. *J Am Vet Med Assoc.* 2011;238(3):318-328.

Section 30

AEMV Oncology Session

Sarah Ozawa, DVM, Dipl ACZM
Session Chair

Palliative External Radiotherapy in a Pet Rabbit (*Oryctolagus cuniculus*) Diagnosed with an Abdominal Fibrosarcoma

Laëtitia Volait-Rosset, DVM, MVS, DipECZM (Small Mammal),
DABVP (Exotic Companion Mammal Practice),
Jenny Garcia, DVM, MVS,
Anna Skwarek, DVM,
Nanna Åkerlund Denneberg, DVM

Session #3329

Affiliation: From Evidensia Djurkliniken Roslagstull, Valhallavägen 12 a, 114 22 Stockholm, Sweden (Volait-Rosset, Garcia, Skwarek), AniCura Animal Hospital Jönköping, Hjortronvägen 8, 554 75 Jönköping, Sweden (Åkerlund Denneberg).

Note: Abstract format selected by author(s). See Front Matter for description of conference proceedings manuscripts submitted as scientific abstracts.

Abstract: A 9-year-old intact male lop-eared rabbit was presented for a tubular subcutaneous mass located on the mid-abdomen associated with a moderate loss of weight. The mass was palpated 3 months earlier and a cytological examination was unconclusive at this time. Ultrasonography revealed a 5-centimeter mass adherent to the rectus abdominis muscle with a heterogeneous echotexture. A computed tomography study confirmed the presence of infiltrative and suspected malignant subcutaneous masses located on the caudoventral abdominal wall. The lesions grossly were consistent with a fibrosarcoma or a rhabdomyosarcoma. A solitary nodule was also identified in the lungs. A surgical removal was performed, and the histological examination identified a fibrosarcoma with infiltrated margins. External radiotherapy treatment was given with a weekly fractionation scheme of 7 Gy, for 4 weeks (Philips RT 250 Orthovoltage Unit). Tangential technique was used to prevent the irradiation of viscera. The abdominal mass recurred one month after radiotherapy ended. A second aggressive surgical removal was performed. Histology confirmed the recurrence of a sarcoma with cancerous margins. Palliative medical treatment with meloxicam (1 mg/kg q12h PO) and gabapentin (15 mg/kg q12h PO) were prescribed. The patient was euthanized 3 months after the second surgery. Few reports have been published on fibrosarcoma in pet rabbit.[1-3] Hypofractionated radiotherapy protocols are not described for palliative treatment of fibrosarcoma in rabbits. We attempted to follow the guidelines for adjuvant radiotherapy in the management of feline fibrosarcoma. Despite the short remission of our patient, radiotherapy is an essential component of tumor control.[4,5]

References

1. Von Bomhard W, Goldschmidt MH, Shofer FS, et al. Cutaneous neoplasms in pet rabbits: a retrospective study. *Vet Pathol.* 2007;44:579-588.

2. Petterino C, Modesto P, Strata D. A case of interscapular fibrosarcoma in a dwarf rabbit (*Oryctolagus cuniculus*). *J Vet Diagn Invest.* 2009;21:900-905.

3. Thas I, Dorrestein GM, Cohen-Solal NA. Mandibular fibrosarcoma and bile duct adenoma in a pet rabbit (*Oryctolagus cuniculi*): a case report. *J Pathol.* 2014;4:32-40.

4. Lavra Zajc A, Harper A, Mason S, et al. Non-injection-site soft tissue sarcoma in cats: outcome following adjuvant radiotherapy. *J Feline Med Surg.* 2022;24(8):e183-e193.

5. McCready J, Poirier VJ, Fleck A, et al. Adaptive radiation therapy using weekly hypofractionation for thymoma treatment: a retrospective study of 10 rabbits. *Vet Comp Oncol.* 2022;20(3):559-567.

Mediastinal Lymphoma and Thoracic Lymphadenomegaly in a Geriatric Pet Rabbit (*Oryctolagus cuniculus*)

Margaret Lee, DVM, and
Shachar Malka, DVM, DABVP (Avian Practice)

Session #3484

Affiliation: From Long Island Bird and Exotics Veterinary Clinic, 333 Great Neck Rd, Great Neck, NY 11021, USA.

Note: Abstract format selected by author(s). See Front Matter for description of conference proceedings manuscripts submitted as scientific abstracts.

Abstract: Lymphoproliferative disorders are the second most common type of neoplasia in pet rabbits, of which most cases were diagnosed as generalized or multicentric lymphoma.[1-4] Lymphoma occurs in all age groups and like most tumors, are most commonly seen in older patients.[3,4] The clinical manifestations of lymphoma vary and generally correspond with the organ system affected.[2] Therefore, animals with mediastinal lymphoma may typically present for bilateral exophthalmia with or without respiratory distress.[2,3] Bilateral exophthalmos occurs due to the compression on the large veins by the tumor, thereby reducing venous return to the heart and leading to cranial vena cava syndrome.[3] A 12-year-old male neutered rabbit was presented with bilateral exophthalmos, mucopurulent ocular discharge and prolapsed nictitating membrane. On presentation, the blood chemistry profile was unremarkable, and the complete blood count revealed leukocytosis (17.6×10^3/µl), heterophilia (7392/µl), lymphocytosis (8096 /µl), monocytosis (1408 /µl) and basophilia (704 /µl). Computed tomography revealed mild left-sided atrophic rhinitis, right-sided Harderian gland enlargement, severe enlargement of the sternal and cranial mediastinal lymph nodes, and a large cranial mediastinal mass. Ultrasound-guided fine needle aspirates of the mediastinal mass were submitted for cytology. Based on cytologic analysis, final diagnosis of lymphoma was confirmed. Both radiation therapy and chemotherapy treatment options were discussed but declined by the owner. The patient was prescribed palliative oral prednisolone (0.5 mg/kg PO q12h). At the time of this report, the patient has been reportedly doing well with near resolution of exophthalmos.

References

1. Quesenberry KE, Pilny AA, St-Vincent RS. Lymphoreticular disorders, thymoma, and other neoplastic diseases. In: Quesenberry KE, Orcutt CJ, Mans C, Carpenter JW, eds. *Ferrets, Rabbits, and Rodents: Clinical Medicine and Surgery.* 4th Ed. St. Louis, MO: Elsevier Saunders; 2020;20:258-267.

2. Robertson JA, Guzman DS-M, Willcox JL, et al. Clinical and pathological findings of rabbits with lymphoma: 16 cases (1996-2019). *J Am Vet Med Assoc.* 2022;260(9):1-10.

3. van Zeeland Y. Rabbit oncology: diseases, diagnostics, and therapeutics. *Vet Clin North Am Exot Anim Pract.* 2017;20(1):135-182.

4. Bertram CA, Bertram B, Bartel A, et al. Neoplasia and tumor-like lesions in pet rabbits (*Oryctolagus cuniculus*): a retrospective analysis of cases between 1995 and 2019. *Vet Pathol.* 2021;58(5):901-911.

Diagnosis of Thymoma and Thymic Lymphoma in Rabbits by Flow Cytometry

Pin-Yu Chen, DVM, MS, and
Pin-Huan Yu, DVM, PhD

Session #3143

Affiliation: From Institute of Veterinary Clinical Sciences, School of Veterinary Medicine, National Taiwan University, No.1, Sec. 4 Roosevelt Road, Taipei, 10617, Taiwan (R.O.C).

Note: Abstract format selected by author(s). See Front Matter for description of conference proceedings manuscripts submitted as scientific abstracts.

Abstract: Thymoma and thymic lymphoma are the top two differential diagnoses of mediastinal masses in rabbits;[1] however, surgery and biopsy may be too invasive for rabbits and sometimes cytology alone is inconclusive.[2] In dogs and humans, flow cytometry has been used to differentiate mediastinal neoplasms routinely.[3] To achieve the goal of diagnosing thymoma and thymic lymphoma with a less invasive approach in rabbits, the potential of flow cytometry to distinguish these two neoplasms in clinical cases was investigated. Fine needle aspiration samples collected from patients with mediastinal masses went through cytology and flow cytometry examination within 48 hours. Tested results would be compared with histopathology diagnosis once patient passed away. A total of 23 cases were collected and four versions of antibody panels were developed. The final version of the antibody panel successfully diagnosed three thymomas, one B-cell lymphoma, and additionally one T-cell leukemia. Similar to those in dogs, higher ratios of CD4+ and CD8+ co-expressing lymphocytes (range = 28.4-43.8%, average = 36.43%) was noted in rabbit thymomas.[4] The current study demonstrated that flow cytometry is an effective tool for discriminating lymphoid neoplasms in rabbits. Nevertheless, more cases are still needed for establishing referable references. To our knowledge, this is the first study focused on diagnosing mediastinal masses via flow cytometry in rabbits. With more cases and research, we hope this technique could create practical benefits for rabbits suffering from lymphoid neoplasms and become a cornerstone for further research.

References

1. van Zeeland, Y. Rabbit oncology: diseases, diagnostics, and therapeutics. *Vet Clin North Am Exot Anim Pract.* 2017;20(1):135-182.

2. Atwater, SW, Powers BE, Park RD, et al. Thymoma in dogs: 23 cases (1980-1991). *J Am Vet Med Assoc.* 1994;205(7):1007-1013.

3. Carter BW, Marom EM, Detterbeck FC. Approaching the patient with an anterior mediastinal mass: a guide for clinicians. *J Thoracic Oncol.* 1994;9(9):S102-S109.

4. Lana S, Plaza S, Hampe K, et al. Diagnosis of mediastinal masses in dogs by flow cytometry. *J Vet Intern Med.* 2006;20(5):1161-1165.

Gastric Neuroendocrine Carcinoma (Carcinoid) in a Ferret (*Mustela putorius furo*)

Teresa Bousquet, DVM,
Maria Bravo-Araya, DVM, MSc, Dipl ACVM,
Jennifer L. Davies, DVM, MVSc, DACVP

Session #3114

Affiliation: From Park Veterinary Centre, 101 Broadway Blvd, Sherwood Park, Canada (Bousquet), University of Saskatchewan Prairie Diagnostic Services Inc. 52 Campus Drive, Saskatoon, Canada (Bravo-Araya), University of Calgary Faculty of Veterinary Medicine Diagnostic Services Unit and Department of Veterinary Clinical and Diagnostic Sciences, 11877 85th Street NW, Calgary, Canada (Davies).

Note: Abstract format selected by author(s). See Front Matter for description of conference proceedings manuscripts submitted as scientific abstracts.

Abstract: A 3-year-old spayed female domestic ferret (*Mustela putorius furo*) presented with a history of increased frequency of vomiting. No obvious cause for vomiting was noted on physical exam, and a complete blood cell count and serum biochemistry panel were unremarkable. Ultrasound examination revealed a 1.5- to 2-centimeter diameter mass, whose margins were confluent with the greater curvature of the stomach. Ultrasound guided fine needle aspiration revealed epithelial neoplasia with neuroendocrine-type appearance, and lymphoid inflammation. Surgical resection of the mass was discussed, but ultimately was declined due to the suspicion of metastasis, based on enlarged gastric lymph nodes also identified on ultrasound, and the lack of information about the prognosis for this condition in ferrets. The ferret was managed palliatively for 10 months, prior to being presented to an emergency clinic with a very large intra-abdominal mass and bicavitary effusion, at which time the ferret was euthanized. Postmortem examination revealed a 6 cm × 4 cm × 3 cm mass, involving ~75% of the greater curvature of the stomach, innumerable nodules throughout the omentum, mesentery, peritoneal surface of the diaphragm, and a 1-centimeter mass in the mesentery at the level of the ileocecocolic junction. Cytologic, histopathologic, histochemical and immunohistochemical examination supported a diagnosis of a well-differentiated gastric neuroendocrine carcinoma with lymph node and transcoelomic metastasis. This is the first case of a gastric neuroendocrine carcinoma described in a ferret. Gastric neuroendocrine carcinoma should be considered as a differential in cases of vomiting in a ferret.

References

1. Schoemaker N. Ferret oncology: diseases, diagnostics and therapeutics. *Vet Clin North Am Exot Anim Pract.* 2017;20:183-208.

2. Miwa Y, Kurosawa A, Ogawa H, et al. Neoplastic diseases in ferrets in Japan: a questionnaire study for 2000-2005. *J Vet Med Sci.* 2009;71:397-402.

3. Li X, Fox JG, Padrid PA. Neoplastic diseases in ferrets: 574 cases (1968-1997). *J Am Vet Med Assoc.* 1998;212:1402-1406.

4. Avallone G, Forlani A, Tecilla M, et al. Neoplastic diseases in the domestic ferret (*Mustela putorius furo*) in Italy: classification and tissue distribution of 856 cases (2000-2010). *BMC Vet Res*. 2016;12:1-8.

5. Antinoff N, Hahn K. Ferret oncology: diseases, diagnostics, and therapeutics. *Vet Clin North Am Exot Anim Pract.* 2004;7:579-625.

Index

I

J

K

L

P

U

V

W

ExoticsCon 2023

Building Exotics Excellence. One City. One Conference.

Boston, Massachusetts, USA
20 - 24 August 2023

Volume 2 of 2

ISBN: 979-8-3313-1554-2

Printed from e-media with permission by:

Curran Associates, Inc.
57 Morehouse Lane
Red Hook, NY 12571

Some format issues inherent in the e-media version may also appear in this print version.

Printed with permission by Curran Associates, Inc. (2025)

For permission requests, please contact Association of Reptilian & Amphibian Veterinarians at the address below.

Association of Reptilian & Amphibian Veterinarians
5003 SW 41st Blvd
Gainesville FL 32608-4930
USA

Phone: 480-703-4941

info@arav.org

Additional copies of this publication are available from:

Curran Associates, Inc.
57 Morehouse Lane
Red Hook, NY 12571 USA
Phone: 845-758-0400
Fax: 845-758-2633
Email: curran@proceedings.com
Web: www.proceedings.com

EXOTICSCON

Building Exotics Excellence. One City. One Conference.

THE WESTIN BOSTON SEAPORT DISTRICT
BOSTON, MASSACHUSETTS | AUGUST 20 - AUGUST 24

2023 PROCEEDINGS

PRESENTED BY

AAV ASSOCIATION OF AVIAN VETERINARIANS

AEMV ASSOCIATION OF EXOTIC MAMMAL VETERINARIANS

ARAV ASSOCIATION OF REPTILE AND AMPHIBIAN VETERINARIANS

EXOTICSCON.ORG

Welcome to ExoticsCon 2023

On behalf of the Board of Directors for the Association of Avian Veterinarians (AAV), Association of Exotic Mammal Veterinarians (AEMV), and the Association of Reptile and Amphibian Veterinarians (ARAV), we welcome you to Boston, Massachusetts.

The Trilateral Conference Planning Committee is excited to bring you quality education and amazing networking opportunities. We hope you'll make the most of the program the committee worked tirelessly to assemble. From educational opportunities like Hands-On Workshops and lecture sessions, to special and social events, we know there is something for everyone.

We have the privilege to work with various professionals to plan this event and would like to acknowledge the following groups and individuals who helped make this conference possible:

- **Auction chairs,** for undertaking this monumental task and insuring the future of all the organizations' research grant programs:
    - AAV: Jonathan Stasiek and Jackie Bucio
    - AEMV: Victoria Bender, DVM and Colleen Roman, DVM
    - ARAV: Marianne Caron, DVM, Emily Thulson, DVM, and Caitlin Cunningham, DVM
- **The *volunteer* abstract and proceedings reviewers,** for ensuring high-quality, accurate, and evidence-based, scientific programing and proceedings.
- **Session chairs,** for their time and expertise to ensure the conference proceeds smoothly and attendees have the best educational experience possible.
- **Hands-On Workshop instructors,** who develop exciting hands-on experiences to convey new clinical techniques for attendees to implement in their practices.
- **Presenters,** for their time, energy, expertise, and willingness to share their knowledge and further the profession with their passion for exotic animals.
- **Executive directors,** who provide leadership and guidance through the conference planning process, and for their devotion to our organizations and adopting our missions and passions as their own:
    - AAV: Robert Groskin, DVM
    - AEMV: Marisa Hackemann
    - ARAV: Annie Trout, M.Ed.
- **Conference planning team with MJB Management, LLC,** for their help with budgets, registration, A/V, and other logistics and details impacting the conference:
    - Therese Lombardo, CMP
    - Jarisa Santos, CMP

In addition, thank you to the **lead conference planner Marisa Hackemann, MJB Management, LLC, CEO**, for taking on and tackling the overwhelming task of coordinating, aligning, creating compromise, and promoting the individual and collective needs of three devoted and passionate associations to create and shape a record-setting conference.

Last, but certainly not least, **we thank our members, attendees, sponsors and exhibitors,** for their faith and loyalty as well as their contributions to our associations. AAV, AEMV, and ARAV exist because of you and your generous contributions. Over the years, you have helped ExoticsCon grow in both size and scope. Your feedback continues to shape future planning and we look forward to your feedback. We hope you enjoy ExoticsCon 2023!

Sincerely,
Byron De La Navarre, DVM, *AAV Conference Chair*
Anna Osofsky, DVM, DABVP (Avian Practice), *AAV*
Stephanie Lamb, DVM, DABVP (Avian Practice), *AAV*
Todd Driggers, DVM, *AEMV Conference Chair*
Sara Gardhouse, DVM, Dipl ACZM, DABVP (Exotic Companion Mammal Practice), *AEMV*
Melissa Kling, DVM, *AEMV*
La'Toya Latney, DVM, DipECZM (Zoo Health Management), DABVP (Reptile and Amphibian Practice), *ARAV Conference Chair*
James E. Bogan, DVM, DABVP (Canine and Feline Practice), DABVP (Reptile and Amphibian Practice), CertAqV, *ARAV*
Sean Perry, DVM, PhD, *ARAV*

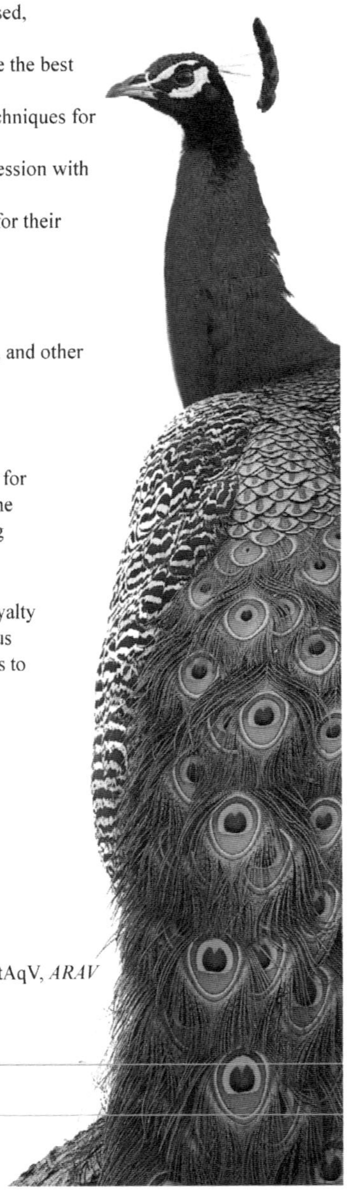

Introduction

We invite you to enjoy the 2023 ExoticsCon Proceedings. Within the proceedings, you will find a wealth of information related to the variety of lectures presented at this year's conference.

AAV, AEMV, and ARAV recognize and thank the following committees and staff for their time and hard work in assuring a quality program and publication:

ExoticsCon Trilateral Conference Committee

Byron De La Navarre, DVM, *AAV Conference Chair*
Anna Osofsky, DVM, DABVP (Avian Practice), *AAV*
Stephanie Lamb, DVM, DABVP (Avian Practice), *AAV*
Todd Driggers, DVM, *AEMV Conference Chair*
Sara Gardhouse, DVM, Dipl ACZM, DABVP (Exotic Companion Mammal Practice), *AEMV*
Melissa Kling, DVM, *AEMV*
La'Toya Latney, DVM, DipECZM (Zoo Health Management), DABVP (Reptile and Amphibian Practice), *ARAV Conference Chair*
James E. Bogan, Jr.,DVM, DABVP (Canine and Feline Practice), DABVP (Reptile and Amphibian Practice), CertAqV, *ARAV*
Sean Perry, DVM, PhD, *ARAV*

AAV Executive and Conference Committees

Anna Osofsky, DVM, DABVP (Avian Practice), *President, Conference Committee Member*
Stephanie Lamb, DVM, DABVP (Avian Practice), *President-Elect, Conference Committee Member*
Byron De La Navarre, DVM, *Immediate Past President, Conference Committee Chair*
Salina Locke, DVM, *Education Committee Co-Chair*
Julia Shakeri, DVM, DABVP (Avian Practice), *Education Committee Co-Chair*
Susan Tyson, MS, VMD, *AAV Hands-On Workshop Coordinator*
Christina Knight, DVM, *AAV Hands-On Workshop Coordinator*
Len Donato, VMD, DABVP (Avian Practice), *Treasurer*
Jen Riley, DVM, DABVP (Avian Practice), *Advisory Council Representative to the AAV Executive Board*
Robert Groskin, DVM, *Executive Director*

AEMV Executive Committee

Sara Gardhouse, DVM, Dipl ACZM, DABVP (Exotic Companion Mammal Practice), *President, Conference Committee Member*
Victoria Bender, DVM, *President-Elect*
Jessica Robertson, DVM, Dipl ACZM, *Vice President*
Eryn Hanak, DVM, *Secretary*
Melissa Kling, DVM, *Treasurer, Conference Committee Member*
Todd Driggers, DVM, *Immediate Past-President, Conference Committee Chair*
Cathy Johnson-Delaney, DVM, *Member-at-Large*
Carla Christman, DVM, *Member-at-Large*
Marisa Hackemann, *Executive Director*

ARAV Executive Committee

James E. Bogan, Jr., DVM, DABVP (Canine and Feline Practice), DABVP (Reptile and Amphibian Practice), CertAqV, *President, Conference Committee Member*
La'Toya Latney, DVM, DipECZM (Zoo Health Management), DABVP (Reptile and Amphibian Practice), *President-Elect, ARAV Conference Chair*
Sean M. Perry, DVM, PhD, *Vice President, Conference Committee Member*
Kourtney Grimm, DVM, *Secretary*
Stacey Leonatti Wilkinson, DVM, DABVP (Reptile and Amphibian Practice), *Treasurer*
Louisa M. Asseo, DVM, DABVP (Canine and Feline Practice), *Immediate Past President*
Elizabeth Hyde, DVM, *Member-at-Large*
Abby Needleman, DVM, *Member-at-Large*
Kelly Rockwell, DVM, *Member-at-Large*
Anna-Marie Ruoff, DVM, *Member-at-Large*
Annie Trout, M.Ed., *Executive Director*

ASSOCIATION OF AVIAN VETERINARIANS
AAV

AEMV
ASSOCIATION OF EXOTIC MAMMAL VETERINARIANS

ARAV
ASSOCIATION OF REPTILE AND AMPHIBIAN VETERINARIANS

Veterinary Review and Editing Committee

Natalie Antinoff
Gunther Antonissen
Joao Brandao
Bri Bryan
Sue Chen
John Chitty
Bob Dahlhausen
Byron de la Navarre
Bob Doneley
M. Scott Echols
Jesse Fallon
Jennifer Graham
Cheryl Greenacre
David Sanchez-Migallon Guzman
Michelle Hawkins
Cathy Johnson-Delaney
Stephanie Lamb
Isabelle Langlois
La'Toya Latney
Josh Llinas
Salina Locke
Elizabeth Mackey
An Martel
Brynn McCleery
Colin McDermott
Christina Miller
Mark Mitchell
Deborah Monks
Bianca Murphy
Anna Osofsky
Kristina Palmer
Olivia Petritz
Christal Pollock
Patrick Redig
Jaime Samour
Julia Shakeri
Marcy Souza
Brian Speer
Lauren Thielen
Tom Tully
Paul Welch
Kenneth Welle
Shangzhe Xie

Technical Editor

Charlotte Kirk Baer, MS

Conference Staff — MJB Management, LLC

Marisa Hackemann, *Chief Executive Officer*
Annie Trout, M.Ed., *Senior Vice President, Client Serivces*
Therese Lombardo, CMP, *Meeting Planner*
Jarisa Santos, CMP, *Meeting Planner*

Proceedings Layout and Design — BrightLizard, Inc.

Elizabeth Smith

Continuing Education Accreditation

The American Association of Veterinary State Boards (AAVSB) Registry of Approved Continuing Education (RACE) committee has reviewed and approved the ExoticsCon 2023 program as meeting the standards adopted by the AAVSB. Please contact the AAVSB RACE program at race@aavsb.org or 877-698-8482 if you have any questions regarding Continuing Education for this program.

Conference Sponsors

Platinum Sponsors:

Disclaimer

The material appearing in this publication comes exclusively from the authors and contributors identified in each manuscript. The techniques and procedures discussed reflect the personal knowledge and experience of the authors and contributors, and demonstrate their views of the methods that may be used for these medical procedures. The procedures demonstrated do not incorporate all known techniques, are not exclusive, and other techniques and technology may also be available. Any questions or requests for additional information concerning any of the manuscripts should be addressed directly to the authors.

AAV, AEMV, and ARAV do not research, review, or otherwise verify any of the information contained in this publication. Opinions expressed in this publication are those of the authors and contributors and not necessarily those of AAV, AEMV, or ARAV. AAV, AEMV, and ARAV are not responsible for errors or for opinions expressed in this publication.

AAV, AEMV, and ARAV expressly disclaim any warranties or guarantees, expressed or implied, and shall not be liable for damages of any kind in connection with the material, information, techniques, or procedures set forth in this publication.

Publication Terms

Conference proceedings manuscripts are accepted as full manuscripts or as scientific abstracts. Scientific abstracts are accepted if the author(s) express(es) intent to submit their work for consideration for future publication in a peer-reviewed scientific journal. Several distinguished veterinary journals, including the *Journal of the American Veterinary Medical Association* and the *American Journal of Veterinary Research* may reject work that has been previously published of 250 words or more and included additional content such as figures, tables, illustrations, and references. Scientific abstracts published in the ExoticsCon proceedings are still required to contain data and meaningful information to the reader.

For permission to reproduce or order copies of this publication, contact the respective association office:

- AAV Office, www.aav.org, PO Box 9, Teaneck, New Jersey 07666, USA, email pubs@aav.org.
- AEMV Office, www.aemv.org, MJB Management, LLC, PO Box 1868, Mt. Juliet, Tennessee 37121, email info@aemv.org.
- ARAV Office, www.arav.org, MJB Management, LLC, PO Box 1868, Mt. Juliet, Tennessee 37121, email info@arav.org.

For information on ExoticsCon educational conferences, visit https://www.exoticscon.org. For information on membership and other association programs visit the association websites listed above.

Table of Contents

Association of Avian Veterinarians

Section 1: AAV Fundamentals Session

Section 2: AAV Plenary Session

Section 3: AAV Cutting Edge Session

Section 4: AAV Wildlife, Anesthesia, and Surgery Session

Section 5: AAV Masterclasses Track 1

Section 6: AAV Masterclasses Track 2

Section 7: AAV Masterclasses Track 3

Section 8: AAV Masterclasses Track 4

Section 9: AAV Special Panel on Geriatrics

Section 10: AAV Potpourri Session

Section 11: AAV "I'll Take Topics with the Acronym AI for 500 Please, Alex" Session

Section 12: AAV Potpourri II Session

Association of Exotic Mammal Veterinarians

Section 13: AEMV Fundamentals Session

Section 14: AEMV Plenary Session

Section 15: AEMV Therapeutics Session

Section 16: AEMV Dentistry Session

Section 17: AEMV Respiratory System Session

Section 18: AEMV Cardiology Session

Section 19: AEMV Epidemiology Session

Section 20: AEMV Masterclasses Track 1

Section 21: AEMV Masterclasses Track 2

Section 22: AEMV Masterclasses Track 3

Section 23: AEMV Masterclasses Track 4

Section 24: AEMV Gastrointestinal System Session

Section 25: AEMV Anesthesia and Analgesia Session

Section 26: AEMV Neurology Session

Section 27: AEMV Dermatology Session

Section 28: AEMV Ophthalmology Session

Section 29: AEMV Pharmacokinetics Session

Section 30: AEMV Oncology Session

Section 31: AEMV Urogenital System Session

Section 32: AEMV Mammal Medley Session

Section 33: AEMV Posters

Association of Reptile and Amphibian Veterinarians

Section 34: ARAV Plenary Session

Section 35: ARAV Novel Therapies and New Perspectives Session

Section 36: ARAV Nutrition and Diet Composition Session

Section 37: ARAV Masterclasses Track 1

Section 38: ARAV Masterclasses Track 2

Section 39: ARAV Infectious Disease Session

Section 40: ARAV Neoplasia Session

Section 41: ARAV Clinical Pathology and Reproduction Session

Section 42: ARAV Surgery and Imaging Session

Section 43: ARAV Critical Care and Conservation Session

Section 44: ARAV Potpourri Session

Section 45: ARAV Posters

Hybrid Sessions

Section 46: Hybrid Session 1

Section 47: Hybrid Potpourri Session

Section 48: Hybrid Special Session

Section 49: Hybrid Marla Lichtenberger Emergency and Critical Care Session

Section 50: Hybrid Session 2

Section 51: Hybrid Session 3

Technician Sessions

Section 52: Technician Session 1

Section 53: Technician Session 2

Section 54: Technician Tutor Session

Section 55: Technician Emergencies Session

Author Index

Surgical Removal of a Ceruminous Adenocarcinoma with a Total Ear Canal Ablation and Bullae Osteotomy in a Domestic Ferret (*Mustela putorius furo*)

Kristin Britton, DVM

Session #3326

Affiliation: From NorthStar VETS, 315 Robbinsville-Allentown Road, Robbinsville, NJ 08691, USA.

Note: Abstract format selected by author(s). See Front Matter for description of conference proceedings manuscripts submitted as scientific abstracts.

Abstract: A 5-year-old female spayed ferret was treated by its veterinarian for an ear mass and infection for 2 months with topical medications with no resolution and was referred. On presentation the mass was irregular, infected, ulcerated, and filled the left ear canal. The ferret was concurrently diagnosed with insulinoma and managed with prednisolone and amoxicillin.[1-3] Eight weeks later an incisional biopsy was taken confirming a ceruminous gland carcinoma.[2-5] Six weeks later a pre-operative computed tomography scan confirmed a large soft tissue attenuating mass measuring 1.7cm x 1.8cm x 2.1cm in the left ear canal.[2,4,6] A surgical total ear canal ablation with bullae osteotomy (TECA-BO) was performed to remove the mass.[5-7] Histopathology confirmed a ceruminous gland adenocarcinoma with three mitotic figures in 10 hpf and minimal surgical margins.[2-5] Follow-up radiation therapy was not pursued by the owner. The patient recovered well with persistent facial nerve paralysis causing diminished palpebral reflex.[5,6] No obvious re-occurrence of the mass had been noted for 5 months. A small subcutaneous mass developed along the dorsal head after 5 months. Surgical removal and histopathology confirmed the mass was a ceruminous gland carcinoma. Two months later the patient developed a firm subcutaneous mass ventral to the AS pinna near the original TECA-BO surgical site. Nine months after the initial surgery the patient was humanely euthanized. TECA surgery should be considered a viable option in ferrets with ear canal disease or masses such as ceruminous gland adenocarcinoma.[5-7] Early diagnosis and removal is recommended.[5,6] Follow-up radiation should be considered due to re-occurrence.[2-4,6,7]

References

1. Schoemaker NJ, Van Zeeland YRA. Endocrine diseases of ferrets. In: Quesenberry KE, Orcutt CJ, Mans C, Carpenter JW, eds. *Ferrets, Rabbits, and Rodents: Clinical Medicine and Surgery*. 4th Ed. St. Louis, MO: Elsevier; 2022:92-108.

2. Williams BH, Wyre NR. Neoplasia in ferrets. In: Quesenberry KE, Orcutt CJ, Mans C, Carpenter JW, eds. *Ferrets, Rabbits, and Rodents: Clinical Medicine and Surgery*. 4th Ed. St. Louis, MO: Elsevier; 2022: 92-108.

3. Schoemaker NJ. Ferret oncology diseases, diagnostics, and therapeutics. *Vet Clin North Am Exot Anim Pract*. 2017;20:183-208.

4. Harrison TM, Kitchell BE. Principles and applications of medical oncology in exotic animals. *Vet Clin North Am Exot Anim Pract*. 2017;20:209-234.

5. Fox-Alvarez WA, Moreno AR, Bush J. Diagnosis and successful surgical removal of an aural ceruminous gland adenocarcinoma in a domestic ferret (*Mustela putorius furo*). *J Exot Pet Med.* 2015;24:350-355.

6. MacPhail C. Surgery of the ear. In: Fossum TW, ed. *Small Animal Surgery.* 5th Ed. Philadelphia, PA: Elsevier; 2019:302-330.

7. Steffey MA. Principles and applications of surgical oncology in exotic animals. *Vet Clin North Am Exot Anim Pract.* 2017;20:235-254.

Use of Electrochemotherapy for High Grade Scent Gland Adenoma in a Gerbil (*Meriones unguiculatus*)

Kimberly Lezama, DVM, and
Alicia McLaughlin, DVM, CertAqV

Session #3419

Affiliation: From Center for Bird and Exotic Animal Medicine, 11401 NE 195th St., Bothell, WA 98011, USA.

Note: Abstract format selected by author(s). See Front Matter for description of conference proceedings manuscripts submitted as scientific abstracts.

Abstract: A 1-year-old intact male gerbil (*Meriones unguiculatus*) presented to the hospital for a mass on the ventral scent gland that the owner had noticed 1 week prior. The patient was scheduled for surgical mass excision. An excisional biopsy of the mass with wide gross margins was performed and submitted for histopathology.[1,2] The mass was identified as a completely excised ventral marking gland (scent gland) epithelioma with narrow deep margins.[3-5] The gerbil recovered uneventfully, and owner was instructed to closely monitor for recurrence. One month later, there was regrowth of the mass, so the animal was scheduled for a second surgery. The mass and entirety of scent gland tissue and associated deep fascial planes were removed with grossly wide margins, and the tissues were submitted for histopathology.[1,7] The lesion was once again confirmed to have been completely excised with narrow deep margins. Following an oncology consultation, the patient received an intralesional dose of bleomycin while an electric pulse was applied to the surgical site (electrochemotherapy).[1,6] The patient recovered with minimal complications and the mass did not recur. The patient was subsequently euthanized 8 months later due to quality-of-life concerns from congestive heart failure, which could be secondary to the patient being obese, age related changes, or frequent incidences of short anesthesia for diagnostics and treatment over the course of several months.[2,7,8] Regardless of the outcome, this case demonstrates the potential benefit of intralesional electrochemotherapy for oncologic treatment of scent gland epitheliomas in gerbils.

References

1. Bennett RA, Pye GW. *Surgery of Exotic Animals.* 1st Ed. Hoboken, NJ: John Wiley & Sons; 2022.

2. Quesenberry KE, Orcutt CJ, Mans C, Carpenter JW, eds. *Ferrets, Rabbits, and Rodents Clinical Medicine and Surgery.* 4th Edition. St. Louis, MO: Elsevier; 2020.

3. Barthold SW, Griffey SM, Percy DH. *Pathology of Laboratory Rodents and Rabbits.* 4th Ed. Hoboken, NJ: John Wiley & Sons; 2016.

4. Otrocka-Domagała IW, Paździor-Czapula K, Fiedorwicz J, et al. Cutaneous and subcutaneous tumours of small pet mammals-retrospective study of 256 cases (2014-2021). *Animals.* 2022;12(8):965.

5. Turner PV, Brash ML, Smith DA. *Pathology of Small Mammal Pets.* 1st Ed. Hoboken, NJ: John Wiley & Sons; 2017.

6. Tellado M, Mir LM, Maglietti F. Veterinary guidelines for electrochemotherapy of superficial tumors. *Front Vet Sci.* 2022;9:868989.

7. Dutton M. Selected veterinary concerns of geriatric rats, mice, hamsters, and gerbils. *Vet Clin North Am Exot Anim Pract.* 2020;23:525-548.

8. Ventura LLA, Fortes NCL, Santiago HC, et al. Obesity-induced diet leads to weight gain, systemic metabolic alterations, tissue inflammation, hepatic steatosis, and oxidative stress in gerbils (*Meriones unguiculatus*). *PeerJ.* 2017;5:e2967.

Use of Chemotherapy in Syrian Hamsters (*Mesocricetus auratus*) with Epitheliotropic T-cell Lymphoma: Three Cases (2021-2022)

Shih Chieh Chan, DVM

Session #3137

Affiliation: From Zodiac Pet and Exotic Hospital, 101A, 1/F, Victoria Centre, No. 15 Watson Road, North Point, Hong Kong.

Note: Abstract format selected by author(s). See Front Matter for description of conference proceedings manuscripts submitted as scientific abstracts.

Abstract: Epitheliotropic T-cell lymphoma (ETCL) has been reported in Syrian hamsters (*Mesocricetus auratus*), although these cases focus on the pathologic findings without a description of treatment.[1-5] This case series includes three client-owned *M. auratus* diagnosed with spontaneous ETCL via full thickness skin biopsy or cytology. Individuals were treated with chemotherapy involving multiple agents, including prednisolone, chlorambucil, L-asparaginase, lomustine and cyclophosphamide. One of the hamsters showed prolonged survival time (276 days after diagnosis) compared to a previous report[1] and a period of total remission (33 days) with chemotherapy (prednisolone 0.5 mg/kg PO q24h + chlorambucil 0.4 mg/kg PO q24h). A single dose of lomustine (20 mg/m^2 PO) showed mild antineoplastic activity and no noticeable side effects. A single dose of L-asparaginase (400 U/kg SC) showed no apparent side effects. The other two hamsters that underwent chemotherapy without chlorambucil lived for 81 and 49 days after initial presentation. *Staphylococcus aureus* was cultured from cutaneous lesions in the end stages of all three cases. This finding is consistent with the research in human ETCL,[6-8] showing that *S. aureus* endotoxin is associated with the worsening of ETCLs. Early identification and chemotherapeutic intervention may lead to a period of remission and an improved quality of life for *M. auratus* with ETCL.

Acknowledgments: The author thanks Dr. Colin T McDermott and Dr. Nicole Wyre who contributed to this study by editing and providing professional advice regarding developing the treatment protocols.

References

1. HarveRG, Whitbread TJ, Cooper JE. Epidermotropic cutaneous T-cell lymphoma (*Mycosis fungoides*) in Syrian hamsters (*Mesocricetus auratus*). A report of six cases and the demonstration of T-cell specificity. *Vet Dermatol.* 1992;3(1):13-19.

2. Miles S. Hamster cutaneous lymphoma. *Vet Times.* 2017;47(26):6.

3. Otrocka-Domagała I, Pazdzior-Czapula K, Fiedorowicz J, et al. Cutaneous and subcutaneous tumours of small pet mammals-retrospective study of 256 cases (2014-2021). *Animals (Basel).* 2022;12(8):965.

4. Sarv J, Eshar D. Rodent dermatology. *Vet Clin North Am Exot Anim Pract.* 2023;26(2):383-408.

5. Saunders GK, Scott DW. Cutaneous lymphoma resembling mycosis fungoides in the Syrian hamster (*Mesocricetus auratus*). *Lab Anim Sci.* 1988;38(5):616-617.

6. Fujii K. Pathogenesis of cutaneous T cell lymphoma: involvement of *Staphylococcus aureus*. *J Dermatol.* 2022;49(2):202-209.

7. Krejsgaard T, Willerslev-Olsen A, Lindahl LM, et al. Staphylococcal enterotoxins stimulate lymphoma-associated immune dysregulation. *Blood.* 2014;124(5):761-770.

8. Willerslev-Olsen A, Krejsgaard T, Lindahl LM, et al. Staphylococcal enterotoxin A (SEA) stimulates STAT3 activation and IL-17 expression in cutaneous T-cell lymphoma. *Blood.* 2016;127(10):1287-1296.

Clinical, Imaging, and Pathologic Features of Hepatic Carcinosarcoma in a Sugar Glider (*Petaurus breviceps*)

Julianne E. McCready, DVM, DVSc, and
Brynn McCleery, DVM, DABVP (Avian Practice)

Session #3111

Affiliation: From the Department of Veterinary Clinical Sciences, College of Veterinary Medicine, Oklahoma State University, 2065 West Farm Road, Stillwater, OK 74078, USA (McCready), Avian & Exotics Department, Red Bank Veterinary Hospital, 197 Hance Avenue, Tinton Falls, NJ 07724, USA (McCleery).

Note: Abstract format selected by author(s). See Front Matter for description of conference proceedings manuscripts submitted as scientific abstracts.

Abstract: Hepatocellular tumors may be overrepresented in sugar gliders. In this case report, 5.5-year-old, neutered male sugar glider (*Petaurus breviceps*) presented for lethargy following the death of a cagemate. On physical examination, severe obesity was noted. A plasma biochemistry profile revealed severely elevated alkaline phosphatase and moderate hypercholesteremia. An abdominal ultrasound revealed a heterogenous liver and a right abdominal mass, which was determined to be hepatic in origin. A computed tomography scan was later performed, which revealed an enlarged liver with numerous ill-defined hypoattenuating regions and a heterogenous mass arising from the caudate lobe. A cytologic examination of the liver was consistent with low-grade mixed lipid and hydropic vacuolar change and mild to moderate mixed to mature lymphocytic infiltrate. The owner declined liver biopsies, and the patient was managed with silymarin and prednisolone. Hematology, biochemical analysis, and abdominal ultrasonography were monitored serially, revealing persistently elevated alkaline phosphatase and progressive heterogenous hepatomegaly and hepatic nodules. The sugar glider died over 2 years after original presentation. It had lost almost half of its original body weight at the time of death. Necropsy revealed a poorly differentiated hepatic neoplasm, with immunohistochemistry (neoplastic cells vimentin positive and pancytokeratin negative) most consistent with a carcinosarcoma. This report describes the presentation and management of a sugar glider that was ultimately diagnosed with hepatic carcinosarcoma. Although a diagnosis was not made antemortem in this case, hepatic biopsies could be considered for sugar gliders presenting with hepatic masses when cytology is unrewarding.[1-10]

References

1. Garner MM. Diseases of pet hedgehogs, chinchillas, and sugar gliders. *Proc Annu Conf Assoc Avian Vet.* 2011:405-412.

2. Gentz EJ, Richard MJ, Crawshaw G, et al. Neoplasia in sugar gliders (*Petaurus breviceps*): thirty-three cases. *Proc Annu Conf Am Assoc Zoo Vet.* 2003:132-134.

3. Lin YS, Wang TY, Lin JC, et al. Hepatic carcinosarcoma: clinicopathologic features and a review of the literature. *Ann Hepatol.* 2013;12(3):495-500. doi:10.1016/s1665-2681(19)31015-4.

4. Barrows M. Toxoplasmosis in a colony of sugar gliders (*Petaurus breviceps*). *Vet Clin North Am Exot Anim Pract.* 2006;9(3):617-623. doi:10.1016/j.cvex.2006.05.006.

5. Li J, Liang P, Zhang D, et al. Primary carcinosarcoma of the liver: imaging features and clinical findings in six cases and a review of the literature. *Cancer Imaging.* 2018;18(1):7. doi:10.1186/s40644-018-0141-0.

6. Kurita D, Mokuno Y, Matsubara H, et al. Primary hepatic carcinosarcoma with multimodal treatment. *Nagoya J Med Sci.* 2018;80(3):423-429. doi:10.18999/nagjms.80.3.423.

7. Gupta BN, Feldman DB. Carcinosarcoma in an opossum. *J Am Vet Med Assoc.* 1973;163(6):586-588.

8. Wang KY, Panciera DL, Al-Rukibat RK, Radi ZA. Accuracy of ultrasound-guided fine-needle aspiration of the liver and cytologic findings in dogs and cats: 97 Cases (1990-2000). *J Am Vet Med Assoc.* 2004;224(1): 75-78. doi:10.2460/javma.2004.224.75.

9. Yuschenkoff D, Graham J, Barton BA, Garner MM. Evaluation of the clinical presentation and histologic lesions of hepatic copper accumulation in sugar gliders (*Petaurus breviceps*). *J Exot Pet Med.* 2021;39: 51-56. doi:10.1053/j.jepm.2021.06.007.

10. Olson AR, Zimmerman AM, Sledge DG. Pathology in practice. *J Am Vet Med Assoc.* 2021;258(7): 1201-1203.

Optic Nerve Meningioma in a Pet Rat (*Rattus norvegicus*)

Julianne E. McCready, DVM, DVSc,
Sara M. Gardhouse, DVM, Dipl ACZM, DABVP (Exotic Companion Mammal Practice),
Karen Carlton, DVM,
Emily Rätsep, BVSc, DVSc, MRCVS, DACVP,
Ryan Appleby, DVM, Dipl ACVR

Session #3117

Affiliation: From the Department of Veterinary Clinical Sciences, College of Veterinary Medicine, Oklahoma State University, 2065 West Farm Road, Stillwater, OK 74078, USA (McCready), Evolution Veterinary Specialists, 34 Van Gordon Street Suite 160, Lakewood, CO 80228, USA (Gardhouse), Department of Pathobiology, University of Guelph, 50 Stone Road East, Guelph, ON N1G 2W1, Canada (Carlton, Rätsep), Department of Clinical Studies, Ontario Veterinary College, University of Guelph, 50 Stone Road East, Guelph, ON N1G 2W1, Canada (Appleby).

Note: Abstract format selected by author(s). See Front Matter for description of conference proceedings manuscripts submitted as scientific abstracts.

Abstract: Orbital neoplasms are not common in rats but have been reported.[1-9] This report describes the clinical findings, diagnostic work-up, attempted treatment, and pathologic findings of an optic nerve meningioma resulting in exophthalmos in a pet rat. A 2-year-old castrated male domestic rat (*Rattus norvegicus*) was evaluated for exophthalmos of the left eye. A computed tomography scan of the head revealed mildly increased vascularity of the left retrobulbar tissues but no distinct mass. Enucleation was planned but the rat arrested following anesthesia. Histopathology of the left retroorbital space revealed an unencapsulated mass encircling the optic nerve. The neoplasm was composed of round-to-spindle shaped cells forming sheets and vague streams. Neoplastic cells were strongly S100 positive, cytokeratin negative, and weakly to moderately vimentin positive. Results were most consistent with an optic nerve meningioma. A differential diagnosis of malignant peripheral nerve sheath tumor (PNST) was considered, but as PNSTs are comprised of a mixture of Schwann cells and endo- and epineural fibroblast cells, we would not expect such a high percentage of cells to express S100; only the Schwann cells should express this marker. A malignant Schwannoma is another consideration, but herringbone and Verocay body patterns were not appreciated, nor was there evidence of intra-neural growth. Optic nerve meningiomas have been rarely reported in laboratory rats, but, to the authors' knowledge, this is the first report in a pet rat. As computed tomography was of insufficient sensitivity to identify this mass, magnetic resonance imaging could be considered for rats presenting with exophthalmos.

References

1. Smith SF, Grieve KL. Exophthalmos in rats. *Lab Anim (NY).* 2008;37(6):247-249.

2. Miller PE, Teixeira LBC. Ocular tumors. In: Vail DM, Thamm DH, Liptak JM, eds. *Withrow and MacEwen's Small Animal Clinical Oncology.* 6th Ed. Elsevier; 2020:675-687. doi:10.1016/B978-0-323-59496-7.00032-3.

3. Mauldin EA, Deehr AJ, Hertzke D, Dubielzig RR. Canine orbital meningiomas: a review of 22 cases. *Vet Ophthalmol.* 2000;3(1):11-16. doi:10.1111/j.1463-5224.2000.00087.x.

4. Flaherty EH, Robinson NA, Pizzirani S, Pumphrey SA. Evaluation of cytology and histopathology for the diagnosis of canine orbital neoplasia: 112 cases (2004-2019) and review of the literature. *Vet Ophthalmol.* 2020;23(2):259-268. doi:10.1111/vop.12717.

5. Montoliu P, Añor S, Vidal E, Pumarola M. Histological and immunohistochemical study of 30 cases of canine meningioma. *J Comp Pathol.* 2006;135(4):200-207. doi:10.1016/j.jcpa.2006.06.006.

6. Yoshitomi K, Everitt JI, Boorman GA. Primary optic nerve meningiomas in F344 rats. *Vet Pathol.* 1991;28(1):79-81. doi:10.1177/030098589102800111.

7. Ramos MF, Baker J, Atzpodien EA, et al. Nonproliferative and proliferative lesions of the rat and mouse special sense organs (ocular [eye and glands], olfactory and otic). *J Toxicol Pathol.* 2018;31(3 Suppl):97S-214S. doi:10.1293/tox.31.97S.

8. Suzuki S, Uchida K, Nakayama H. The effects of tumor location on diagnostic criteria for canine malignant peripheral nerve sheath tumors (MPNSTs) and the markers for distinction between canine MPNSTs and canine perivascular wall tumors. *Vet Pathol.* 2014;51(4):722-736. doi:10.1177/0300985813501336.

9. Steinmetz A, Ellenberger K, März I, et al. Oculocardiac reflex in a dog caused by a choroidal melanoma with orbital extension. *J Am Anim Hosp Assoc.* 2012;48(1):66-70. doi:10.5326/JAAHA-MS-5685.

Spontaneous Primary Splenic Liposarcoma in a Guinea Pig (*Cavia porcellus*)

Miranda Govoni, DVM

Session #3288

Affiliation: From CVS, Centro Veterinario Specialistico, Via Sandro Giovannini 51/53, 00137, Rome, Italy, Happy Friends Veterinary Hospitals, Via Meucci 2, 24035, Curno, Bergamo and Via Comasina 155, 20843, Verano Brianza, Monza Brianza, Italy.

Note: Abstract format selected by author(s). See Front Matter for description of conference proceedings manuscripts submitted as scientific abstracts.

Abstract: Liposarcomas are rare malignant tumors of the adipose tissue which are described in humans,[1] dogs,[2,3] and companion exotic species.[4-6] Although spontaneous tumors in guinea pigs are relatively uncommon,[7,8] subcutaneous[9] and palpebral liposarcoma[10] were found. This case report describes for the first time a spontaneous primary splenic liposarcoma in a 6-year-old intact female guinea pig (*Cavia porcellus*), presented with a history of anorexia and abdominal distension. A 7-centimeter abdominal mass was palpated on physical examination. The animal was hospitalized and complete blood profile, serum biochemistry and full body radiographs were performed. Abdominal ultrasound demonstrated the presence of a soft tissue mass of splenic origin and an exploratory laparotomy was performed. Primary initial differential diagnoses were splenic hematoma, abscess, granuloma and neoplasm, such as hemangiosarcoma, which is rare but reported in guinea pigs.[11,12] The splenic mass occupied a large part of the abdominal cavity and compromised more than 90% of the organ, therefore splenectomy was performed and the lesion was submitted for histopathology and immunohistochemistry. The liver was diffusely yellowish and numerous 5- to 7-millimeter ovarian cysts were found. A few days after the surgery the animal's clinical conditions worsened and it was euthanized. The histopathologic features of the splenic mass were consistent with a liposarcoma, and tumor cells stained positive for vimentin and negative for cytokeratins and S-100. No metastatic lesions were found in other organs at the necropsy. Spontaneous splenic liposarcoma has not been described previously in guinea pigs, but should be considered as a differential diagnosis for abdominal masses.

References

1. The WHO Classification of Tumours Editorial Board. *WHO Classification of Tumours Soft Tissue and Bone Tumours.* 5th Ed. Lyon, France: IARC Press; 2020.

2. Gower KL, Liptak JM, Culp WT, et al. Splenic liposarcoma in dogs: 13 cases (2002-2012). *J Am Vet Med Assoc.* 2015;247(12):1404-1407.

3. Avallone G, Roccabianca P, Crippa L, et al. Histological classification and immunohistochemical evaluation of MDM2 and CDK4 expression in canine liposarcoma. *Vet Pathol.* 2016;53(4):773-780.

4. Gardhouse S, Eshar D, Fromstein J, et al. Diagnosis and successful surgical treatment of an unusual inguinal liposarcoma in a pet ferret (*Mustela putorius furo*). *Can Vet J.* 2013;54(8):739-742.

5. Doria-Torra G, Martínez J, Domingo M, et al. Liposarcoma in animals: literature review and case report in a domestic pig (*Sus scrofa*). *J Vet Diagn Invest.* 2015;27(2):196-202.

6. Tsai FY, Chang HM, Chang HK, et al. Case report: endometrial stromal sarcoma and liposarcoma in an African hedgehog (*Atelerix albiventris*). *Taiwan Vet J.* 2016;42:181-186.

7. Minarikova A, Hauptman K, Jeklova E, et al. Diseases in pet guinea pigs: a retrospective study in 1000 animals. *Vet Rec.* 2015;177(8):200.

8. Pignon C. Guinea pig. In: Quesenberry KE, Orcutt CJ, Mans C, et al, eds. *Ferrets, Rabbits and Rodents: Clinical Medicine and Surgery.* St. Louis, MO: Elsevier Saunders; 2021:270-297.

9. Sahoo A, Dwibedi B, Das J, et al. Spontaneous liposarcoma in a guinea pig (*Cavia porcellus*) and its surgical management. *Int J Lives Res.* 2017;7(3):193-196.

10. Quinton JF, Ollivier F, Dally C. A case of well-differentiated palpebral liposarcoma in a guinea pig (*Cavia porcellus*). *Vet Ophthalmol.* 2013;16(Suppl 1):155-159.

11. Thompson JJ, Burgmann PM, Brash ML, et al. Spontaneous splenic hemangiosarcoma in a guinea pig (*Cavia porcellus*). *J Exot Pet Med.* 2016;25:139-143.

12. Paoletti C, Rosas-Navarro J, Nicolier A, et al. Spontaneous splenic hemangiosarcoma in pet guinea pigs (*Cavia porcellus*): five cases (2007-2021). *J Exot Pet Med.* 2023;44:36-41.

Pheochromocytoma in an African Pygmy Hedgehog (*Atelerix albiventris*): A Case Report

Giulia Bersanetti, DVM, MSc,
Marco Masi, DVM, MSc,
Paolo Selleri, DVM, PhD, DipECZM (Herpetology), DipECZM (Small Mammal)

Session #3369

Affiliation: From Centro Veterinario Specialistico, Via Sandro Giovannini 53, Rome, Italy.

Note: Abstract format selected by author(s). See Front Matter for description of conference proceedings manuscripts submitted as scientific abstracts.

Abstract: A 6-year-old intact male African pygmy hedgehog was presented for syncope and severe bleeding from the oral cavity after the loss of two teeth. At the physical examination a severe parodontopathy associated with absence of all the right mandibular molars and premolars was identified. Blood biochemistry was within normal limits and complete blood count showed normocytic normochromic anaemia. Whole-body radiographs showed lumbar spondylosis. Abdominal ultrasonography showed right adrenal gland enlargement with multiple calcifications and enlargement of the prostate. Echocardiography showed moderate left ventricular hypertrophy. Blood pressure was not measured due to technical limitations. A whole-body computed tomography scan was then carried out and confirmed the presence of a right adrenal neoplasia without involvement of the caudal vena cava, multiple prostatic cysts and multifocal lumbar spondylosis. Concurrent presence of neoplasia on the left superficial gluteal muscle was visible. Ten days after initial presentation, a partial right adrenalectomy was performed. Partial excision of the neoplasia was carried out due to its close relation to the vena cava. Perioperative enrofloxacin (5 mg/kg SC q12h), meloxicam (0.5 mg/kg q12h) and buprenorphine (0.01 mg/kg IM q8h) were administered. Gabapentin (10 mg/kg PO q12h) was started upon discharge for the treatment of lumbar pain associated with the spondylosis. Histopathology of the excised adrenal gland tissue was consistent with pheochromocytoma. Within the next 5 months, clinical conditions deteriorated secondary to worsening of spondylosis-associated clinical signs. Euthanasia was elected, followed by post-mortem examination, which showed the presence of multiple neoplasias (nerve sheath tumor, fibrosarcoma and testicular tumor). This is the first report of in vivo diagnosis and surgical treatment of pheochromocytoma in an African pygmy hedgehog.[1-3]

References

1. Doss GA, Carpenter JW. African pigmy hedgehogs. In: Quesenberry K, Orcutt CJ, Mans C, Carpenter JW, eds. *Ferrets, Rabbits and Rodents: Clinical Medicine and Surgery*. 4th Ed. St. Louis, MO: Elsevier; 2020:401-415.

2. Johnson DH. Geriatric hedgehogs. *Vet Clin North Am Exot Anim Pract.* 2020;23(3):615-637.

3. Heatleyn JJ, Mauldin GE, Youn Cho D. A review of neoplasia in the captive African hedgehog (*Atelerix albiventris*). *Sem Avian Exot Pet Med.* 2005;14(3):182-192.

AEMV

ASSOCIATION OF EXOTIC MAMMAL VETERINARIANS

Section 31

AEMV Urogenital System Session

Carla Christman, DVM
Session Chair

Diagnosis and Treatment of Uroabdomen Due to Traumatic Urinary Bladder Rupture in a Pet Rabbit

Emma Monge, DVM,
Mathilde Charrasse, DVM,
Charly Pignon, DVM, DipECZM (Small Mammal)

Session #3230

Affiliation: From Zoological Medicine Department, Ecole Nationale Vétérinaire de Toulouse, Toulouse, France (Monge), Exotic Animal Service, CHUVA, Ecole Nationale Vétérinaire d'Alfort, Maisons-Alfort, France (Pignon), Veterinary Hospital Center Frégis, Gentilly, France (Charrasse), IHAP, Universite de Toulouse, INRAe, ENVT, Toulouse, France (Monge).

Note: Abstract format selected by author(s). See Front Matter for description of conference proceedings manuscripts submitted as scientific abstracts.

Abstract: An 8-month-old female spayed rabbit was presented for pelvic paresis and hematuria following a fall. Physical examination showed paraparesis. Whole body radiographs were unremarkable. Abdominal ultrasonography performed at admission, and 12 hours later, did not reveal any abnormality. The next day, pelvic paresis and hematuria were solved. Three days later, deterioration of general health and a major abdominal distension were observed. Abdominal ultrasound revealed a ruptured bladder associated with a severe effusion. Analysis of this fluid was consistent with uroabdomen. Biochemistry was compatible with acute renal failure. Supportive care with intravenous fluid allowed stabilization of the patient. A laparotomy with cystorrhaphy and abdominal lavage was performed. Recovery was uneventful. Effusion did not recollect during hospitalization and renal parameters returned to normal. On abdominal ultrasound rechecks on day 6, day 16 and day 30, the urinary bladder was intact. One year later, the rabbit was presented for routine health check and vaccination with no abnormalities found on physical examination and abdominal ultrasound. This report adds to the current literature[1-6] and describes successful surgical treatment and long-term survival of a pet rabbit with an uroabdomen due to traumatic urinary bladder rupture. This case highlights the importance of monitoring the patient after an abdominal trauma with the risk of urinary bladder leakage a few days after.

References

1. Bennett RA. Rabbit soft tissue surgery. In: Bennett RA, Pye GW, eds. *Surgery of Exotic Animals.* Hoboken, NJ: John Wiley & Sons, Inc; 2022:240-276.

2. Di Girolamo N, Selleri P. Disorders of the urinary and reproductive systems. In: Quesenberry, KE, Orcutt CJ, Mans C, Carpenter JW, eds. *Ferrets, Rabbits, and Rodents: Clinical Medicine and Surgery.* 4th Ed. St. Louis, MO: Elsevier Saunders; 2020:201-219.

3. Stafford JR, Bartges JW. A clinical review of pathophysiology, diagnosis, and treatment of uroabdomen in the dog and cat. *J Vet Emerg Crit Care.* 2013;23(2):216-229.

4. Balakrishann A, Drobatz KJ. Management of urinary tract emergencies in small animals. *Vet Clin North Am Small Anim Pract.* 2013;43(4):843-867.

5. Grimes JA, Fletcher JM, Schmiedt CW. Outcomes in dogs with uroabdomen: 43 cases (2006-2015). *J Am Vet Med Assoc.* 2018;252(1):92-97.

6. Hornsey SJ, Halfacree Z, Kulendra E, et al. Factors affecting survival to discharge in 53 cats diagnosed with uroabdomen: a single-centre retrospective analysis. *J Feline Med Surg.* 2021;23(2):115-120.

Surgical Management of a Case of Septate Bladder and Related Complications in a Dwarf Rabbit

Gianluca Deli, MRCVS, DVM, CertAVP(ZM), GPCert(ExAP), PgCert(EAS), GpAdvCert(SMP), PgCert(HE), FHEA, VetGDP Adviser,
Emanuele Lubian, DVM, GPCert(ExAP)

Session #3160

Affiliation: From Medivet Faygate, Tower Road, Faygate, Horsham, RH12 4SD, UK (Deli), Clinica MyPet, Viale D. Ranzoni, 10 20149 Milano, Italy (Lubian).

Abstract: With this clinical case we retrospectively analyze management of a case of septate bladder in a dwarf rabbit (*Oryctolagus cuniculus*) from diagnostic and therapeutic points-of-view. Septate bladder is a rare occurrence. Its identification and subsequent treatment required the synergistic use of imaging and laboratory diagnostics. Difficulties that most influenced its identification were the size of the patient, the absence of pathognomonic symptoms and the initial non-cooperation of the owner. The therapeutic approach was surgical, with resection of the membranous septum. No recurrences were documented in the 2 years following surgery. Complications that occurred were development of uroliths in various sections of the urinary tract. Currently the patient, under maintenance therapy and close medical supervision, has normal diuresis and has improved in general condition including weight gain.[1-11]

References

1. Appel S, Otto SJ, Weese JS. Cystotomy practices and complications among general small animal practitioners in Ontario, Canada. *Can Vet J.* 2012;53:303-310.

2. Brown C. Urolithiasis and cystotomy in the rabbit. *Lab Anim.* 2011;40:73-74.

3. Carpenter JW, Marion CJ, eds. *Exotic Animal Formulary.* 5th Ed. St. Louis, MO: Elsevier; 2018.

4. Cruciani B, Vachon C, Dunn M. Removal of lower urinary tract stones by percutaneous cystolithotomy: 68 cases (2012-2017). *Vet Surg.* 2020;49:138-147.

5. Harcourt-Brown F, Chitty J, eds. *BSAVA Manual of Rabbit Surgery, Dentistry and Imaging.* Quedgeley, UK: BSAVA Publications; 2016.

6. Kucera J, Koristkova T, Gottwaldova B, Jekl V. Calcium sulfate dihydrate urolithiasis in a pet rabbit. *J Am Vet Med Assoc.* 2017;250:534-537.

7. Martel-Arquette A, Mans C. Urolithiasis in chinchillas: 15 cases (2007 to 2011). *J Small Anim Pract.* 2016;57:260-264.

8. Meredith A, ed. *BSAVA Small Animal Formulary.* Part B: Exotic Pets. 9th Ed. Quedgeley,UK: BSAVA Publications; 2015.

9. Petrini D, Di Giuseppe M, Deli G, De Caro Carella C. Cystolithiasis in a Syrian hamster: a different outcome. *Open Vet J.* 2016;6:135-138.

10. Quesenberry KE, Orcutt CJ, Mans C, Carpenter JW, eds. *Ferrets, Rabbits, and Rodents: Clinical Medicine and Surgery.* 4th Ed. St. Louis, MO: Elsevier, 2020.

11. Szabo Z, Bradley K, Cahalane AK. Rabbit soft tissue surgery. *Vet Clin North Am Exot Anim Pract.* 2016;19:159-188.

Urine Specific Gravity and Urine Osmolality in Healthy Pet Rabbits (*Oryctolagus cuniculus*)

Kees Pellikaan,
Nico Schoemaker, DVM, PhD, DipECZM (Small Mammal), DipECZM (Avian)
Evert-Jan de Boer,
Lenie Boswijk-Appelhof,
Jos Vossen,
Yvonne van Zeeland, DVM, MVR, PhD, DipECZM (Avian), DipECZM (Small Mammal), CPBC

Session #3277

Affiliation: From Division of Zoological Medicine, Department of Clinical Science, Faculty of Veterinary Medicine, Utrecht University, Yalelaan 108, 3584 CM Utrecht, the Netherlands (Pellikaan, Schoemaker, van Zeeland), Dierenkliniek Wilhelminapark, Johan Willem Frisostraat 1, 3583 JR Utrecht, the Netherlands (de Boer); University Veterinary Diagnostic Laboratory, Faculty of Veterinary Medicine, Utrecht University, Yalelaan 104, 3584 CM Utrecht, the Netherlands (Boswijk-Appelhof, Vossen).

Note: Abstract format selected by author(s). See Front Matter for description of conference proceedings manuscripts submitted as scientific abstracts.

Abstract: Renal concentrating ability in mammals can be evaluated by determining either urine specific gravity (Usg) or urine osmolality (Uosm);[1-3] however, in rabbits, reference ranges are currently lacking for Uosm, and considered wide and as specific for Usg.[4-8] This study aimed to establish reference ranges for both parameters as well as study their correlation and the potential effects of gender, age and weight category of the rabbit, and the method and time of the sample collection.[9-12] Morning and evening urine samples as well as an additional blood sample were collected from 98 healthy privately owned rabbits of various breeds. Usg and Uosm were subsequently measured using respectively clinical refractometry and freezing-point osmometry. Findings of this study reveal a wide reference range for Usg (1.010-1.060) and Uosm (73-2181 mOsm/kg). A strong correlation was present between both parameters ($r = 0.93$). Mixed model analysis and logistic regressions revealed weight category to significantly influence both Usg and Uosm ($P < 0.001$), with dwarf breeds showing higher values for both parameters compared to other breeds. In contrast, gender or collection method had no significant effect, whereas time and age exerted a small but significant effect on Usg only ($P = 0.001$ and $P = 0.03$, respectively). Reference ranges were established for Uosm and Usg in healthy pet rabbits. When evaluating Usg or Uosm in rabbits, weight category likely needs to be taken into account; however, due to potential bias with other factors such as breed and origin of the rabbits further research is warranted prior to drawing definite conclusions.

References

1. Ayoub JA, Beaufrere H, Acierno MJ. Association between urine osmolality and specific gravity in dogs and the effect of commonly measured urine solutes on that association. *Am J Vet Res.* 2013;74:1542-1545.

2. Waldrop JE. Urinary electrolytes, solutes, and osmolality. *Vet Clin North Am Small Anim Pract.* 2008;38: 503-512.

3. Watson AD. Urine specific gravity in practice. *Aust Vet J.* 1998;76:392-398.

4. Melillo A. Rabbit clinical pathology. *J Exot Pet Med.* 2007;16:135-145.

5. Meredith A, Lord B. *BSAVA Manual of Rabbit Medicine*. Gloucester, UK: British Small Animal Veterinary Association; 2014.

6. Özkan C, Kaya A, Akgül Y. Normal values of haematological and some biochemical parameters in serum and urine of New Zealand white rabbits. *World Rabbit Sci.* 2012;20:253-259.

7. Varga M. *Textbook of Rabbit Medicine.* 2nd Ed. Philadelphia, PA: Elsevier; 2014.

8. Washington IM, van Hoosier G. Clinical biochemistry and hematology. In: Suckow M, Stevens K, Wilson R. *The Laboratory Rabbit, Guinea Pig, Hamster, and Other Rodents.* Boston, MA: Academic Press; 2011:57-116.

9. Eshar D, Wyre NR, Brown DC. Urine specific gravity values in clinically healthy young pet ferrets (*Mustela furo*). *J Small Anim Pract.* 2012;53:115-119.

10. Rishniw M, Bicalho R. Factors affecting urine specific gravity in apparently healthy cats presenting to first opinion practice for routine evaluation. *J Feline Med Surg.* 2015;17:329-337.

11. Van Vonderen IK, Kooistra HS, Rijnberk A. Intra- and interindividual variation in urine osmolality and urine specific gravity in healthy pet dogs of various ages. *J Vet Intern Med.* 1997;11:30-35.

12. Yeh H, Lin Y, Kuo C, et al. Urine osmolality in the US population: implications for environmental biomonitoring. *Environ Res.* 2015;136:482-490.

Liquid Gold: Review of Urinalysis in Exotic Companion Mammals from a Private Practice

Kaleigh Eichel, DVM, and
Alicia McLaughlin, DVM, CertAqVet

Session #3264

Affiliation: From Center for Bird and Exotic Animal Medicine, 11401 NE 195th St, Bothell, WA 98011, USA.

Note: Abstract format selected by author(s). See Front Matter for description of conference proceedings manuscripts submitted as scientific abstracts.

Abstract: Urinalysis is an inexpensive and rapid diagnostic test that can provide early detection of multiple diseases.[1-4] Over a 2-year period at a 24-hour general and emergency exotic animal private practice, a total of 307 small exotic mammals underwent urinalysis testing (sediment, dipstick, and/or culture) on emergency, urgent care, or wellness with the following species distribution: 167 rabbits, 79 guinea pigs, 44 ferrets, 6 chinchillas, 4 rats, 3 hamsters, 2 hedgehogs, 1 sugar glider, and 1 squirrel. Urinalysis was useful in the diagnosis and investigation of what was found to be commonly diagnosed conditions: urinary tract infections, bladder sludge, cystitis, urolithiasis, kidney disease (*E. cuniculi*, pyelonephritis), hepatitis, and reproductive diseases (prostatic cysts, cystic ovaries, pyometra). Best practices for urine collection, testing, interpretation, and normal values in common species in private practice are discussed based on clinical data.

References

1. Agundez MG, Porquet NC. Evaluation of urine dipstick for proteinuria assessment in pet rabbits. *Vet Rec.* 2021;188:e306. https://doi.org/10.1002/vetr.306.

2. Cernochova H, Hundakova A, Bardi E, Knotek Z. Biochemical profile or urine in guinea pigs (*Cavia procellus*). *Vet Med Czech.* 2020;65:445-450. https://doi.org/10.17221/32/2020-VETMED.

3. Doss GA, Mans C, Houseright RA, Webb JL. Urinalysis in chinchillas (*Chinchilla lanigera*). *J Am Vet Med Assoc.* 2016;248(8):901-7. https://doi.org/10.2460/javma.248.8.901.

4. Eschar D, Wyre NR, Brown DC. Urine specific gravity values in clinically healthy young pet ferrets (*Mustela furo*). *J Small Anim Pract.* 2012;53:115-119. https://doi.org/10.1111/j.1748-5827.2011.01173.x.

The Rat Preputial Gland in Health and Disease

Lauren V. Powers, DVM, DABVP (Avian Practice), DABVP (Exotic Companion Mammal Practice)

Session #3438

Affiliation: From Avian and Exotic Pet Service, Carolina Veterinary Specialists, 12117 Statesville Road, Huntersville, NC 28078, USA.

Abstract: The preputial glands are paired accessory sex glands found in both male and female rats. Clinically apparent disease of the preputial glands is relatively uncommon in pet rats but impaction, bacterial infection, abscessation, and neoplasia can be seen. Disease can be misdiagnosed by clinicians due to lack of knowledge about the preputial glands. Medical management options for impacted or infected glands include gland expression and/or irrigation under heavy sedation or anesthesia, antimicrobial therapy, and analgesic therapy. Surgical excision of the preputial gland(s) is relatively straightforward and is the treatment of choice for neoplastic conditions and for cases of impaction, inflammation, or abscessation unresponsive to medical management.

Introduction

Both male and female rats (*Rattus norvegicus*) possess paired preputial (clitoral) glands situated in the perineum on either side of the preputial opening. The glands can be a site for inflammation, infection, abscessation, impaction, and neoplasia. Although there are several published reports of non-neoplastic lesions and spontaneous and experimentally induced neoplastic lesions of the preputial glands in specific laboratory rat strains, reports are lacking that describe the medical and surgical management of clinical disease in individual rats.

The Rat Preputial Gland

Anatomy and function

The accessory sex glands in the male rat include the prostate gland, seminal vesicles (vesicular glands), coagulating glands (found in close association with the seminal vesicles), ampullary glands, bulbourethral glands (Cowper's glands), and the preputial glands.[1,2] The penis contains a bony process, the os penis. In the female rat the clitoris is situated inside a prepuce. The urethra opens at the urethral orifice located at the base of the clitoris. The urethra does not enter the vagina or vestibule as it does in other mammals.[3]

The preputial glands are elongated, flask-shaped, flattened organs of variable length of 5-20 millimeters.[4,5] The glands are located within a bed of adipose tissue in the perineum lateral to the prepuce. The preputial glands are similar in appearance and size in both sexes.[4,5] They are also referred to as clitoral glands in female rats. The excretory ducts open at the transition of the parietal (skin) layer of the prepuce in the male rat and near the tip of the clitoris in the female rat.[3-7]

The preputial glands are specialized sebaceous glands.[4,6,7] They are holocrine, branched, tubuloalveolar structures with acini that open into lateral ducts that combine into a wider central duct which becomes the solitary excretory duct.[4,6,7] The ducts are lined by stratified squamous cell epithelium that may be keratinized.[4,6]

The gland is surrounded by a fibrous connective tissue capsule.[4] The glands produce sebum prepuciale (smegma) which contains pheromones.[4,6] The preputial glands are androgen responsive and are important for courtship, mating, and pup survival.[5,7-9]

The size of the preputial glands generally does not decrease with advancing age; however, the amount of acinar tissue decreases and very little to no glandular parenchyma remains in very old rats.[4,6] Cystic and inflammatory changes are often present in the glands of older rats.[4] As rats age, the excretory duct lumen widens and the smaller, lateral ducts disappear.[4] The lumen of the excretory duct then tends to become ampulliform (flask-shaped) and serves more as a reservoir for smegma rather than as a draining glandular duct.[4] A caseous plug of amorphous debris that frequently contains lymphocytes and neutrophils may be present within the lumen of the excretory duct in older animals.[4] This plug differs from the normal sebum found in younger animals.[4] There is little difference in the histologic findings of the preputial glands between intact and castrated male rats with varying degrees of degenerative changes seen in aging castrated rats.[4]

Impaction, inflammation, infection, and abscessation

Apart from degenerative aging changes, inflammation is the most common non-neoplastic disease of the rat preputial gland.[3,6,10-12] In 9/32 (28.1%) of older male RccHan:WIST strain laboratory rats had inflammation or abscessation of the preputial gland.[12] Hematogenous spread of infection is also possible.[11] Ductal impaction, bacterial infection, and abscessation are occasionally diagnosed in pet rats. Clinical signs include evidence of localized pain or discomfort, dysuria or stranguria, hematuria, and the presence of a visible or palpable mass in the perineal area or a ruptured, draining preputial gland abscess. Antibiotics and medical management alone are generally considered ineffective and surgical excision of affected gland(s) is recommended for neoplasia and for chronic or recurrent cases of impaction or infection (Veterinary Information Network online consultation, September 2022).

Neoplasia

With the exception of the prostate and preputial glands, neoplasia of the accessory sex lands is uncommon in rats.[13] There are several published reports that describe neoplasms of the preputial gland in laboratory rats. Documented neoplasms include papillary adenomas, adenocarcinomas, basal cell carcinomas, squamous papillomas, squamous cell carcinomas, and other carcinomas.[6,10,12-16] A continuum of hyperplasia, adenoma, and carcinoma has been proposed for proliferative lesions and mixed patterns of cystic, papillary, or solid growth are often observed.[13] Certain drugs reportedly increase the incidence of tumors in carcinogenicity studies of laboratory rats such as ganciclovir, nalidixic acid, azoxymethane and nitrofurazone.[6,15] Carcinomas are often locally invasive and distant metastases is uncommon but can occur.[13]

Case Report

A case report is provided to illustrate the clinical management of preputial disease in the pet rat. A 2-year-old neutered male pet hairless rat (*Rattus norvegicus*) was treated for recurrent impaction and abscessation of both preputial glands incompletely responsive to medical management including gland expression and irrigation under general anesthesia, multiple antibiotics, and analgesics. Surgical excision of the right preputial gland was performed 2 months after initial presentation. After initial postoperative preputial and penile swelling with dysuria and stranguria that were responsive to gabapentin and prazocin, the surgical site eventually healed well; however, the rat died unexpectedly 3 months after initial presentation and had severe myocardial fibrosis and acute diffuse pulmonary congestion. Ductal impaction with low-grade inflammation and bacterial infection were present in the remaining left preputial gland.

Surgical description

The distal urethra was catheterized with a 3.5-French red rubber tube with the rat under general anesthesia. Catheterization was difficult and the catheter only advanced 1-2 centimeters. An elongated skin incision was made in the perineum to the right of the prepuce. The incision was extended caudolaterally over two palpable firm masses (empty vaginal tunic and abscessed preputial gland) and cranial to a small (~1 millimeter) ductal fistula in the skin. Blunt dissection was used to free the gland from its caudal attachment to the empty (castrated) right vaginal tunic caudally. The vaginal tunic was separated, ligated, and excised. The more cranial mass (abscessed preputial gland) was distinct as a firm, tan structure that was flattened dorsally to ventrally and tapered at its duct. A small bundle of blood vessels adjoining the gland laterally was ligated. Blunt dissection of the duct was continued, freeing the fistulated portion. The excretory duct was ligated as close to the prepuce as possible. Closure was routine.

Acknowledgments: The author expresses deep thanks to Leslie Jarrell JD, DVM, DACLAM, CertAqV (WAVMA); Thomas M Donnelley, BVSc, DACLAM, DABVP (Exotic Companion Mammal), DipECZM (Small Mammal); Ned Gentz, MS, DVM, Dipl ACZM; and Avery Bennett, DVM, MS, Dipl ACVS for clinical consultation through Veterinary Information Network. The author also expresses gratitude to Michael M Garner, DVM, DACVP for histopathology and consultation services.

Disclosure Statement: The author has no financial or commercial conflicts or interests to disclose.

References

1. Knoblaugh SE, True L, Tretiakova M, et al. Male reproductive system. In: Treuting PM, Dintzis SM, Montine KS, eds. *Comparative Anatomy and Histology: A Mouse, Rat, and Human Atlas*. 2nd Ed. London, UK: Elsevier; 2018:335-363.

2. Crisler R, Johnston NA, Sivula C, et al. Functional anatomy and physiology. In: Suckow MA, Hankenson FC, Wilson RP, et al, eds. *The Laboratory Rat.* 3rd Ed. London, UK: Elsevier; 2020:91-132.

3. Dixon D, Alison R, Bach U, et al. Nonproliferative and proliferative lesions of the rat and mouse female reproductive system. *J Toxicol Pathol.* 2014;27(3-4 Suppl):1S.

4. Montagna W, Noback CR. The histology of the preputial gland of the rat. *Anat Rec.* 1946;96(1):41-54.

5. Reznik G, Ward JM. Morphology of hyperplastic and neoplastic lesions in the clitoral and preputial gland of the F344 rat. *Vet Pathol.* 1981;18(2):228-238.

6. Greaves P. Male genital tract. In: *Histopathology of Preclinical Toxicity Studies: Interpretation and Relevance in Drug Safety Evaluation.* London, UK: Elsevier; 2012:615-666.

7. Dangelo JG, Munger BL. The ultrastructure of the rat preputial gland. *J Ultrastruct Res.* 1964;11(3-4):230-245.

8. Ponmanickam P, Palanivelu K, Govindaraj S, et al Identification of testosterone-dependent volatile compounds and proteins in the preputial gland of rat *Rattus norvegicus. General Comp Endocrinol.* 2010;167(1):35-43.

9. Natynczuk SE, Macdonald DW, Tattersall FH. Morphology and chemistry of brown rat, *Rattus norvegicus*, preputial and clitoral glands. *J Chem Ecol.* 1995;21:247-260.

10. Reznik G, Reznik-Schüller H. Pathology of the clitoral and prepucial glands in aging F344 rats. *Lab Anim Sci.* 1980;30(5):845-850.

11. Weisbroth SH, Kohn DF. Bacterial, mycoplasmal, and mycotic infections. In: Suckow MA, Hankenson FC, Wilson RP, et al, eds. *The Laboratory Rat.* 3rd Ed. London, UK: Elsevier; 2020:451-539.

12. Hojo M, Sakamoto Y, Maeno A, et al. A histopathological analysis of spontaneous neoplastic and non-neoplastic lesions in aged male RccHan:WIST rats. *J Toxicol Pathol.* 2020;33(1):47-55.

13. King-Herbert AP, Janardhan KS, Herbert RA. Neoplastic disease. In: Suckow MA, Hankenson FC, Wilson RP, et al, eds. *The Laboratory Rat.* 3rd Ed. London, UK: Elsevier; 2020:601-650.

14. Kaspareit-Rittinghausen J, Deerberg F, Nitsche B. Naturally-occurring neoplasms of the preputial and clitoral gland in Han:SPRD rats. *J Vet Med Series A.* 1988;35:596-603.

15. Ward JM. Dose response to a single injection of azoxymethane in rats: induction of tumors in the gastrointestinal tract, auditory sebaceous glands, kidney, liver and preputial gland. *Vet Pathol.* 1975;12(3):165-177.

16. Mitsumori K, Elwell MR. Proliferative lesions in the male reproductive system of F344 rats and B6C3F1 mice: incidence and classification. *Environ Health Perspect.* 1988;77:11-21.

The Trouble Down Under: Scrotal Ablation Surgery Techniques for Rabbits

Tracey K. Ritzman, DVM, DABVP (Avian Practice), DABVP (Exotic Companion Mammal Practice)

Session #3144

Affiliation: From Carolinas Veterinary Care Clinic, 10110 Northcross Center, Ste 100, Huntersville, NC 28078, USA, Carolina Raptor Center, 6000 Sample Road, Huntersville, NC 28078, USA.

Note: Abstract format selected by author(s). See Front Matter for description of conference proceedings manuscripts submitted as scientific abstracts.

Abstract: This presentation describes scrotal ablation and revision surgery technique in rabbits. Scrotal abnormalities such as elongation (congenital, traumatic), infection or abscess, trauma, testicular masses or neoplasia, and herniation are conditions that may require surgical treatment to either reduce scrotal size, revise the scrotal area such as with scrotal urethrostomy, or remove scrotal tissue.[1-6] Scrotal ablation surgery is described in humans, dogs, horses, goats, and marsupials like sugar gliders (a species that normally has a pendulous scrotum associated with a stalk). Scrotal ablation is defined as a surgical procedure to remove an elongated or pendulous scrotum or abnormal skin of the scrotum so that the skin sits flat in that area. Scrotal ablation surgery is usually associated with intact male patients. The surgery is inherently an open castration technique. In rabbits, consideration should be given to the normally open and large inguinal rings. A circumferential incision is made around the base of the scrotum and the tunic is incised and the spermatic cords are exposed. Each spermatic cord is double-ligated prior to being transected. After both testicles are removed the vascular pedicle on each side is traced cranially and the inguinal canal is identified. A single interrupted suture is placed across the inguinal canal to reduce its diameter to minimize the risk of future inguinal herniation. The subcutaneous tissues are sutured, and the dermis is closed with an intradermal pattern if possible, or with tissue adhesive or skin staples. Post operative care for scrotal ablation is like standard orchiectomy post-surgery care.

References

1. Bennet RA. Neutering companion exotic mammals. *DVM360*. 2009. Available at: https://www.dvm360.com/view/neutering-companion-exotic-mammals-proceedings-0. Accessed May 1, 2023.

2. Cusack L, Cutler D, Mayer J. The use of Ligasure™ device for scrotal ablation in marsupials. *J Zoo Wildl Med.* 2017;48(1):228-231.

3. Malbrue RA, Arsuaga Zorilla CB. Scrotal ablation and orchiectomy in the domestic laboratory goat. *Vet Anim Sci.* 2018;5:26-30.

4. Morges MA, Grant KR, MacPhail CM, Johnston MS. A novel technique for orchiectomy and scrotal ablation in the sugar glider (*Petaurus breviceps*). *J Zoo Wildl Med.* 2009;40(1):204-206.

5. Palmer SE, Passmore JL. Midline scrotal ablation technique for unilateral cryptorchid castration in horses. *J Am Vet Med Assoc.* 1987;190(3):283-285.

6. Stamper MA, Norton T, Spodnick G, et al. Hypospadias in a polar bear (*Ursus maritimus*). *J Zoo Wildl Med.* 1999;30(1):141-144.

Section 32

AEMV Mammal Medley Session

Carla Christman, DVM
Session Chair

Quantitative Stress Assessment in Guinea Pigs

Eva Stoffels, PhD

Session #3270

Affiliation: From Marumoto Veterinary Practice, Hoevenweg 18, 7722 PN Dalfsen, The Netherlands.

Abstract: Guinea pigs exposed to stressful stimuli react with elevated glucocorticoid levels in blood and other tissues. Chronic stress makes them prone to a number of diseases and contributes to a reduction of their life span. In this work, urinary cortisol has been proposed as a quantitative measure of the animal's stress load. Animals that are excessively handled and restrained by their owners have significantly higher cortisol levels and suffer from, among others conditions, reproductive disorders and cardiovascular diseases. Minimizing interactions and providing safe housing is essential to keep pet guinea pigs healthy.

Introduction

Pet guinea pigs are exposed to many stressors. Excessive handling, interactions with other pets, untreated diseases and inadequate nutrition contribute to their daily stress load. Stress greatly affects welfare and life expectancy of guinea pigs. Since guinea pig owners usually fail to recognize pain and stress signs in their animals, a quantitative method or a welfare meter is necessary.

Cortisol is the main glucocorticoid released from the zona fasciculata of the adrenal cortex. It participates in numerous physiologic processes, but its main function is the regulation of body responses to discomfort, pain and inflammation. Cortisol is widely considered as a biomarker for chronic disease, or for detection of diseases in their preclinical phases.[1] Measurement of serum, salivary and fecal cortisol were carried out in laboratory animals to study their responses to various stressors.[2-5] The normal levels of glucocorticoids in guinea pigs are much higher than in other species. This phenomenon is called cortisol resistance, and is caused by a mutation that greatly reduces the affinity of the glucocorticoid receptor.[6,7] The baseline levels in healthy animals were found to be 5-20 µg/dl in blood, and 5-10 times lower in saliva.[2,5] There was a good correlation between the two values. Blood and saliva cortisol levels were subject to significant diurnal fluctuations,[2,4] and increased drastically (> 100 µg/dl) during animals' activities and/or handling.[2,3] In contrast, fecal cortisol (60-100 µg/dl) provided a time-averaged value that reflected the stress load of the past hours/days.[2] To assess the animal's chronic exposure to stressors, this value may be more suitable than blood or salivary values. In this work, urinary cortisol has been proposed as a welfare meter.

Materials and Methods

Immunoturbidimetric measurements were carried out using the Solo analyzer (Eurolyser Diagnostica GmbH, Salzburg, Austria), with the measuring range of 0.8-24 µg/dl.[8] Saliva and/or urine samples were collected from 57 animals. Saliva samples were collected by inserting a cotton bud in the animal's cheek pouch, as described by Nemeth et al.[2] Urine samples were obtained by spontaneous micturition, or by applying gentle pressure to the animal's caudal abdomen. Urine samples (20 µl) were diluted by a factor of 20-50 using 0.9% NaCl, and 40 µl aliquots were used for the assay. For the urinary cortisol measurements, the control group (C-group, n = 12, seven females and five males, age 1-5 years) were the author's own clinically healthy guinea pigs, housed in groups of 20 animals (females) or in pairs (males). These animals were fed at regular intervals and only visually observed, without handling. The housing was as shown in Figure 1, with on average 0.5 m^2 space available per animal. For the test group (T-group, n = 12, seven females and five males, age 1-5 years) only the animals were selected which were frequently and intensively handled by their owners. The owners were asked to collect urine samples at home. The following conditions were present in the T-group: ovarian cysts (6/7 females), chronic cystitis (4), diabetes (1), cardiomyopathy (1), no apparent health issues (4). They were usually housed as shown in Figure 2 with at least 1 m^2 available per animal; the pens contained various shelters for the animals to hide.

Figure 1. A typical housing arrangement in the control (no handling, no health issues) group. Image on the left is a group of females. Image on the right is two males.

Figure 2. Typical housing conditions of the test group (extensively handled, with several health issues). Note the paucity of hay.

Results

Saliva measurements remained inconclusive. The values were poorly reproducible, so that no trends could be found. Presumably, this was caused by sample dilution with feed present in the animal's mouth. Besides, saliva samples were usually turbid, which may have introduced another source of error. In contrast, the urinary measurements provided reproducible values, ranging from 20 to 2100 μg/dl. In the C-group, urinary cortisol was 120 ± 80 (20-240) μg/dl. In the T-group it was 1000 ± 560 (530-2100) μg/dl. Based on one-way ANOVA, the difference between the groups was statistically significant ($P < 0.05$).

Discussion

Guinea pigs are rodents, and rodents are prey animals. In their natural environment, their survival depends on the continuous alertness to the presence of predators, and a rapid fight-or-flight (or solely flight) response. The natural behavior is mediated by catecholamines and glucocorticoids. For example, confrontations with new situations and unknown animals cause a significant rise in blood cortisol.[2,3] Rodents used for biomedical research learn to cope with stressful events (e.g., experimental procedures), but they do display somatic reactions when exposed to danger and pain.[9,10] Repeatability and predictability in their daily routine appear to be the key issue, as could be concluded from the studies of rats exposed to random versus predictable painful stimuli. Rats challenged in an unpredictable manner showed stress reactions which manifested themselves as gastric lesions.

Adaptive changes could be seen in animals exposed to noxious stimuli, as long as these occurred with a regularity, were preceded by a warning signal or could be deactivated by the animal itself.[9,10] This implies that being able to control the situation is important for being able to cope with it. Loss of control yields chronic stress.

In this sense, the current results in guinea pigs are not surprising. Guinea pigs perceive larger animals, including humans, as potential predators. Moreover, humans are unpredictable, because the guinea pig cannot see the whole human body and anticipate the movements. There is no manner for the guinea pig to control the situation. Therefore it is plausible that frequent and unexpected restraint evokes strong stress responses. Presumably, if guinea pigs could be conditioned by regularly occurring handling interactions with their owners, their stress responses would be tempered; however, this is usually not the case.

The consequences of chronic stress may be devastating for the animal. Since glucocorticoids affect both B- and T-cell dependent immunity,[1] stressed animals are prone to (opportunistic) infections and chronic stress and immunosuppression may be linked with neoplasia development and other illnesses. For example, stress predisposes females to bacterial cystitis, which is an opportunistic infection secondary to diet-induced urinary sludge.[11] Furthermore, stress is responsible for a spectrum of other ailments, such as reproductive disorders, diabetes and cardiovascular diseases in humans and other species.[1] All of the above are frequently observed in pet guinea pigs. Eminently, reproductive disorders (ovarian cysts) are very common in females that are excessively handled by their owners, or face other problems with their housing or husbandry. The regulation mechanisms of sex hormone secretion via the hypothalamus-pituitary-gonad axis provide an explanation (Fig 3). The inhibitory effect of glucocorticoids on the hypothalamus and pituitary hormone secretion[12] interferes with the physiologic estrogen-induced surge release of GnRH, FSH and LH. This impairs the aromatase production in the granulosa zone of the tertiary follicle.[13] Since androgens are continuously produced in the theca zone and adrenal glands, masculinization occurs when they are not converted to estrogens. This contributes to the well-known clinical condition of ovarian cysts, which can manifest itself in two ways. Type 1 results in typical flank alopecia (Fig 4) and/or sexually aggressive "male" behavior. This type is called "follicular" or "hormone producing/hormone responsive" and is relatively rare. Type 2 is characterized by accumulation of serous fluid in rete ovarii (Fig 5) and usually anestrus without alopecia or behavioral changes. Type 2 cysts may be present in more than 50% of sows and are usually an incidental finding during abdominal palpation or ultrasound examination. Both forms can coexist, or type 1 (follicular) can evolve into type 2. Although the majority of affected animals are around 3-4 years old, cysts can be found in sows as young as 6 months.

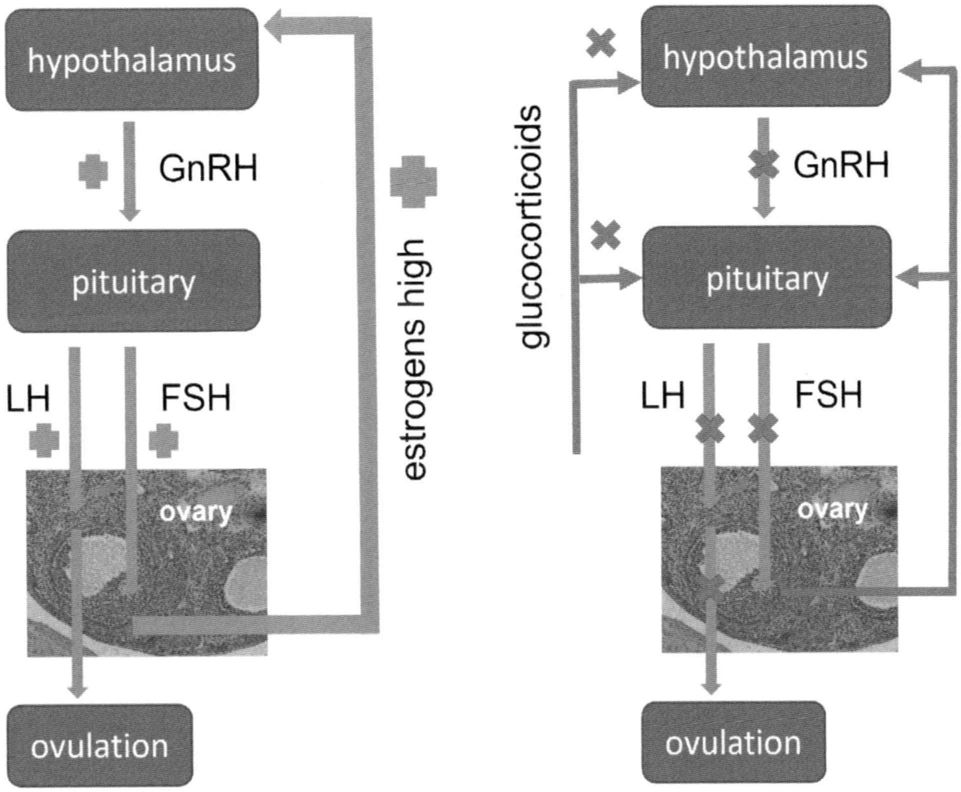

Figure 3. The hypothalamus-pituitary-gonad axis. Left: the physiologic situation where large amounts of estrogens from ripening follicles give positive feedback to the hypothalamus, resulting in a surge in GnRH release, followed by LH/FSH peaks and ovulation. Right: glucocorticoids attenuate the surge both at the hypothalamic and the pituitary levels. The low levels of estrogens exert negative feedback on the axis, preventing ovulation. The follicle persists and androgens are produced in its theca zone.

Figure 4. Left image is a female with type 1 (follicular) ovarian cysts. Right image is the same animal about 3 weeks after a single hCG injection.

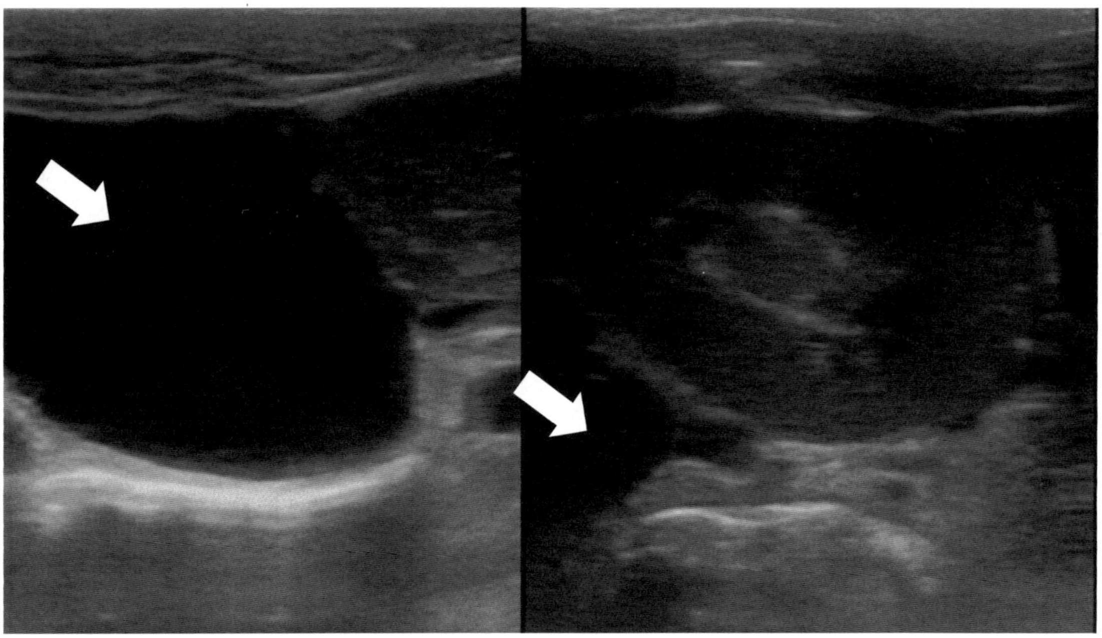

Figure 5. Ultrasound images of type 2 (rete ovarii) cysts showing accumulation of fluid in rete ovarii (arrows). Right image: note that this animal also suffered from kidney lesions.

Similar reproductive disorders (either with altered behavior or anestrus) are commonly seen in farm animals that undergo extreme industrial exploitation, such as dairy cattle and breeding pigs.[14] In contrast, ovarian cysts are uncommon in beef cattle because of completely different living conditions of these animals. The author does not advocate spaying guinea pig females with ovarian cysts. It is a painful procedure, while it does not provide a solution to the problem. A spayed female will be exposed to the same (or greater, if the perioperative stress is counted) stress load and will soon develop other stress-related disorders. Type 1 (follicular) cysts can be treated using GnRH (gonadorelin) or LH (hCG) analogues, but the recurrence is inevitable as long as the animal's living conditions are not improved. Moreover, the occurrence of ovarian cysts is transient; both type 1 and 2 cysts can resolve spontaneously.

The C-group with a low level of glucocorticoids was a random sample from the population of about 300 guinea pigs of both sexes. These animals follow steady daily patterns without unexpected events; handling and restraint are limited to what is necessary to evaluate the animal's health condition. Special care is taken to eliminate competition or aggression in the groups. Owing to a balanced diet,[11] typical diseases (urinary, dental) are scarce. The incidence of ovarian cysts in the female population (about 150 animals) is less than 1% (one animal). In the author's experience, these animals have a high life expectancy: the median life span is 5 years and 6 months, and many 7- or 8-year-old animals are present. The estrus cycle is used as a welfare meter: females that regularly (every 15-17 days) display agitation and lordosis reflex, can be considered cycling normally and free from major health issues.

Contrary to common judgements, the actual size of the pen plays a lesser role. While overcrowding of the enclosure results in serious health issues as shown in mice and rats,[15] the opposite - an excess of space may be detrimental, too. Prey animals fear open space, where they may be exposed to unknown and menacing stimuli (e.g., interactions with predators), so that they prefer to remain confined in the known area.[16] Obviously there is an optimum in the size of the enclosure. Both too large and too small (or too crowded) pens will result in loss of control and inherent stress responses. For example, guinea pigs housed as shown in Figure 2 are unable to control the whole area and will usually hide in small shelters. From the current experiment it can be tentatively concluded that the available surface area does not determine the animal's welfare, as long as the animals can consider the whole enclosure as their comfort zone.

Conclusions

Stress in the daily routine has severe impacts on the guinea pigs' health and welfare. While there is usually some level of misunderstanding on how to properly handle and work with their pet guinea pig, education of owners on just how their interactions could be contributing to stress and discomfort of their pet needs to be done. The urinary cortisol level provides an unbiased, quantitative measure that can be used to assess the full environment and life of the pet guinea pig. Excessive and unexpected interactions between the owner and the guinea pigs are not beneficial for the animals. More measurements are needed to establish a database and describe specific responses for various genders, age groups and types of interactions. For the sake of welfare and longevity of guinea pigs, it is imperative to reduce their stress load.

References

1. Lee DY, Kim E, Choi MH. Technical and clinical aspects of cortisol as a biochemical marker of chronic stress. *BMB Rep.* 2015;48:209-216.

2. Nemeth M, Pschernig E, Wallner B, et al. Non-invasive cortisol measurements as indicators of physiological stress responses in guinea pigs. *Peer J.* 2016;doi:10.7717/peerj.1590.

3. Hennessy MB, Zate R, Maken DS. Social buffering of the cortisol response of adult female guinea pigs. *Physiol Behav.* 2008;93:883-888.

4. Garris DR. Diurnal fluctuation of plasma cortisol levels in the guinea pig. *Acta Endocrinol.* 1979;90: 692-695.

5. Fenske M. Saliva cortisol and testosterone in the guinea pig: easures for the endocrine function of adrenals and testes? *J Exp Anim Sci.* 1996;38:13-19.

6. Keightley MC, Fuller PJ. Cortisol resistance and the guinea pig glucocorticoid receptor. *Steroids.* 1995;60:87-92.

7. Fuller PJ, Smith BJ, Rogerson FM. Cortisol resistance in the New World revisited. *Trends Endocrinol Metabol.* 2004;15:296-299.

8. Eurolyser Diagnostica GmbH. Available at: https://www.eurolyser.com/veterinary-diagnostics/point-of-care-tests/cortisol-test/. Accessed January 3, 2023.

9. Weiss JM. Effects of coping behavior with and without a feedback signal on stress pathology in rats. *J Comp Physiol Psychol.* 1971;77:22-30.

10. Tsuda A, Tanaka M, Hirai H. Effects of coping behavior on gastric lesions in rats as a function of predictability of shock. *Japan Psychol Res.* 1983;25:9-15.

11. Stoffels E. Lower urinary tract disease in guinea pigs as it relates to nutrition. *Proc ExoticsCon.* 2022; 267-283.

12. Calogero AE, Burrello N, Bosboom AM, et al. Glucocorticoids inhibit gonadotropin-releasing hormone by acting directly at the hypothalamic level. *J Endocrinol Invest.* 1999;22:666-670.

13. Erickson GF, Hsueh AJW, Stimulation of aromatase activity by follicle stimulating hormone in rat granulosa cells in vivo and in vitro. *Endocrinol.* 1978;102:1275-1282.

14. Bors SI, Bors A. Ovarian cysts, an anovulatory condition in dairy cattle. *J Vet Med Sci.* 2020;82:1515-1522.

15. Ramsden E, Adams J. Escaping the laboratory: the rodent experiments of John B. Calhoun & their cultural influence. *J Soc History.* 2009;42:761-797.

16. Ennaceur A, Michalikova S, Chazot PL. Models of anxiety: responses of rats to novelty in an open space and an enclosed space. *Behav Brain Res.* 2006;171:26-49.

Determining Amino Acid Profiles of Twelve Commercial Domestic Ferret Diets

Jake Decker, BA,
Jonathan Stockman, DVM, DACVIM (Nutrition),
Raphael Vanderstichel, DVM, PhD,
Shachar Malka, DVM, DABVP (Avian Practice)

Session #3215

Affiliation: From Long Island University's College of Veterinary Medicine, Brookville, NY 11548, USA (Decker, Stockman, Vanderstichel), Long Island Bird and Exotics Veterinary Clinic, Great Neck, New York 11021, USA (Malka).

Note: Abstract format selected by author(s). See Front Matter for description of conference proceedings manuscripts submitted as scientific abstracts.

Abstract: Commercial ferret diets have been implicated as contributors to increased prevalence of cystine urolithiasis in North American pet ferrets.[1,2] The aim of this study was to determine the amino acid profiles and selected minerals of 12 commercially available diets formulated for pet ferrets. Samples of each diet were submitted to Midwest Laboratories in Omaha, Nebraska for proximate analysis, mineral analysis by inductively coupled atomic plasma using a wet digest procedure and determination of their relative concentrations of glutamic acid, proline, glycine, alanine, cystine, valine, methionine, isoleucine, leucine, tyrosine, phenylalanine, lysine, histidine, arginine, and tryptophan. Five diets were also analyzed for taurine. The concentrations of these amino acids were compared to the requirements for adult non-reproducing cats outlined by the National Research Council (NRC) in 2006,[3] and protein requirements were compared to the requirements of adult minks at maintenance outlined in the NRC in 1982.[4] All analyzed diets met these guidelines.[3,4] The diets were then divided into four groups (grain-free, non-grain-free, legume-inclusive, and non-legume-inclusive) to compare their average protein content and methionine and cystine concentrations. The four grain-free diets were on average higher in protein, methionine, and cystine than the eight non-grain-free diets, but the four legume-inclusive diets were on average lower in protein, methionine, and cystine than the eight non-legume-inclusive diets. Nutritional diseases have been reported in ferrets.[5] Further research is needed to determine whether grain-free and non-legume-inclusive diets could be predisposing factors for the development of cystinuria and cystine urolithiasis in pet ferrets.

References

1. Pacheco RE. Cystine urolithiasis in ferrets. *Vet Clin North Am Exot Anim Pract.* 2020;23(2):309-319.

2. Nwaokorie EE, Osborne CA, Lulich JP, Albasan H. Epidemiological evaluation of cystine urolithiasis in domestic ferrets (*Mustela putorius furo*): 70 cases (1992-2009). *J Am Vet Med Assoc.* 2013;242(8): 1099-1103.

3. National Research Council. *Nutrient Requirements of Dogs and Cats.* Washington, D.C.: The National Academies Press; 2006.

4. National Research Council. *Nutrient Requirements of Mink and Foxes.* Washington, D.C.: National Academy Press; 1982.

5. Johnson-Delaney CA. Ferret nutrition. *Vet Clin North Am Exot Anim Pract.* 2014;17(3):449-470.

Presumptive Vaccine-Induced Distemper in Four Domestic Ferrets (*Mustela putorius furo*)

Hugo Gonzalez-Jassi, MVZ,
Julie Fithian, DVM,
Greta Doden, DVM,
Sunil More, BVSc & AH, MVSc, PhD, DACVP,
Akhilesh Ramachandran, BVSc, PhD, Dipl ACVM,
Brianne Taylor, DVM, MS, DACVP,
Giselle Cino Ozuna, DVM, PhD, DACVP,
Rebecca Wilkes, DVM, PhD, Dipl ACVM,
João Brandão, LMV, MS, DipECZM (Avian), Dipl ACZM

Session #3463

Affiliation: From the Department of Veterinary Clinical Sciences, College of Veterinary Medicine, Oklahoma State University, 2065 W. Farm Rd., Stillwater, OK 74078, USA (Gonzalez-Jassi, Fithian, Doden, Brandão), Department of Veterinary Pathobiology, College of Veterinary Medicine, Oklahoma State University, Stillwater, OK 74078, USA (More, Ramachandran, Taylor, Ozuna), Animal Disease Diagnostic Laboratory and Department of Comparative Pathobiology, College of Veterinary Medicine, Purdue University, West Lafayette, IN 47907, USA (Wilkes).

Note: Abstract format selected by author(s). See Front Matter for description of conference proceedings manuscripts submitted as scientific abstracts.

Abstract: Canine distemper virus (CDV) is a highly contagious virus that causes canine distemper (CD) in several mammalian species.[1-7] Naturally occurring, and vaccine-induced disease outbreaks have both been reported in mustelids.[8-9] The vaccine-induced disease occurs when modified live-virus (MLV) vaccines are used. Four juvenile ferrets were presented to the Oklahoma Animal Disease Diagnostic Laboratory for postmortem examination. All animals had clinical signs of CD and were suspected to originate from the same breeding facility. Allegedly, the ferrets were vaccinated with a commercially available MLV-CDV vaccine for mink (Distemink; United Vaccines Inc, Fitchburg, WI, USA). Gross and histopathologic findings were consistent with CD; lung tissue was positive for CDV by RT-PCR. Genetic sequencing of the matrix-fusion intergenic region was used for strain typing and the hemagglutinin (H) gene for phylogenetic analysis of the CDV from the animals. Sequencing of the matrix-fusion intergenic region obtained from three animals showed 100% identity with the Lederle strain, an America-1 lineage vaccine strain. Phylogenetic evaluation of the H gene of CDV from two ferrets grouped these sequences in a separate clade with 96-97% identity to Lederle. Sequencing results showed three of these animals were infected with a Lederle-like strain. Although Lederle strain is believed extinct in the wild, it has been used to produce MLV vaccines such as Distemink. These cases share similarities to previous reports of vaccine-induced CD. MLV vaccines should be avoided for immunization of ferret kits, especially in settings like breeding facilities and pet stores, where epidemiologic risks are amplified.[1,4]

Acknowledgments: The authors of this manuscript acknowledge the support of the Oklahoma State University Wayne and Debbie Bell Professorship in Veterinary Clinical Sciences for funding the molecular testing and the invaluable work of Dr. Alexa Hunter and Scott Mitchell.

References

1. Carpenter JW, Appel MJ, Erickson RC, Novilla MN. Fatal vaccine-induced canine distemper virus infection in black-footed ferrets. *J Am Vet Med Assoc*. 1976;169(9):961-964.

2. Durchfeld B, Baumgärtner W, Herbst W, Brahm R. Vaccine-associated canine distemper infection in a litter of African hunting dogs (*Lycaon pictus*). *Zentralbl Veterinarmed B*. 1990;37(3):203-212. doi:10.1111/j.1439-0450.1990.tb01048.x.

3. Ek-Kommonen C, Rudbäck E, Anttila M, Aho M, Huovilainen A. Canine distemper of vaccine origin in European mink, *Mustela lutreola*-a case report. *Vet Microbiol*. 2003;92(3):289-293. doi:10.1016/s0378-1135(02)00361-9.

4. Gill JM, Hartley WJ, Hodgkinson NL. An outbreak of post-vaccinal suspected distemper-like encephalitis in farmed ferrets (*Mustela putorius furo*). *NZ Vet J*. 1988;36(4):173-176. doi:10.1080/00480169.1988.35525.

5. Halbrooks RD, Swango LJ, Schnurrenberger PR, et al. Response of gray foxes to modified live-virus canine distemper vaccines. *J Am Vet Med Assoc*. 1981;179(11):1170-1174.

6. Perpiñán D, Ramis A, Tomás A, et al. Outbreak of canine distemper in domestic ferrets (*Mustela putorius furo*). *Vet Rec*. 2008;163(8):246-250. doi:10.1136/vr.163.8.246.

7. Tamukai K, Minami S, Kurihara R, et al. Molecular evidence for vaccine-induced canine distemper virus and canine adenovirus 2 coinfection in a fennec fox. *J Vet Diagn Invest*. 2020;32(4):598-603. doi:10.1177/1040638720934809.

8. Wimsatt J, Jay MT, Innes KE, et al. Serologic evaluation, efficacy, and safety of a commercial modified-live canine distemper vaccine in domestic ferrets. *Am J Vet Res*. 2001;62(5):736-740. doi:10.2460/ajvr.2001.62.736.

9. Quesenberry KE, de Matos R. Basic approach to veterinary care of ferrets. In: Quesenberry KE, Orcutt CJ, Mans C, Carpenter JW eds. *Ferrets, Rabbits, and Rodents: Clinical Medicine and Surgery.* 4th Ed. St. Louis, MO: Elsevier; 2021:13-26.

Outcomes and Management of Sugar Glider Self-Mutilation in a Veterinary Teaching Hospital: Thirteen Cases (2007-2022)

Lauren E. Thielen, DVM, DABVP (Avian Practice),
Melissa Hanson, DVM,
James K. Morrisey, DVM, DABVP (Avian Practice),
Ricardo De Matos, LMV, MSc, DABVP (Avian Practice), DipECZM (Avian), DipECZM (Small Mammal),
Nicola Di Girolamo, DVM, PhD, DipECZM (Herpetology), Dipl ACZM, Dipl ACVPM

Session #3489

Affiliation: From the Department of Clinical Sciences, College of Veterinary Medicine, Cornell University, 930 Campus Road, 14853, Ithaca, NY 14853, USA.

Note: Abstract format selected by author(s). See Front Matter for description of conference proceedings manuscripts submitted as scientific abstracts.

Abstract: Self-mutilation in sugar gliders is frequently mentioned in the literature; however, there are no retrospective case reports detailing the management and outcomes of self-mutilation in this species. The medical record database of a veterinary teaching hospital was searched for sugar gliders presented for trauma or mutilation. A total of 200 individual sugar glider cases were identified from June 2007 to November 2022. Of these, 13 sugar gliders (6.5%) presented for trauma or self-mutilation in various locations. Of these sugar gliders, seven presented for hind-limb mutilation, four for tail mutilation, and two for penile mutilation. All cases except two required surgery. One case of tail mutilation was medically managed, and one case was euthanized upon presentation for hind-limb mutilation. The case that was medically managed developed cellulitis of the hind leg and required prolonged wound care. Of the 11 cases undergoing surgery, five cases (45%) needed a repeat surgical procedure. Both penile ampullae amputation surgeries did not need a follow-up surgery; however, half of the limb mutilation cases did require surgical revision. Causes for repeated surgery were most commonly due to wound dehiscence or continued self-mutilation. Various post-operative medical management strategies, including placement of custom made and commercially available Elizabethan collars, were attempted to prevent post-operative mutilation. Self-mutilation is a relatively common occurrence in sugar gliders, and its surgical and post-operative management is complicated by several factors.

References

1. Miwa Y, Sladky KK. Small mammals: common surgical procedures of rodents, ferrets, hedgehogs, and sugar gliders. *Vet Clin North Am Exot Anim Pract.* 2016;19(1):205-244.

2. Tynes VV. Behavioral dermatopathies in small mammals. *Vet Clin North Am Exot Anim Pract.* 2013;16(3): 801-820.

Section 33

AEMV Posters

A Retrospective Study of Incidental Findings on Whole Body Computed Tomography Scans in Rabbits

Arianna Rizzi, BS,
Joanna Hedley, BVM&S DZooMed (Reptilian), DipECZM (Herpetology), MRCVS,
Vicki Baldrey, BVSc, BSc (Hons), DZooMed (Avian), FHEA, MRCVS

Session #3226

Affiliation: From the Royal Veterinary College Beaumont Sainsbury Animal Hospital, London NW1 0TU, UK.

Note: Abstract format selected by author(s). See Front Matter for description of conference proceedings manuscripts submitted as scientific abstracts.

Abstract: Computed tomography (CT) is an increasingly common imaging modality used in exotic pet medicine. This retrospective study investigated the frequency and significance of incidental findings (findings unrelated to the reason for the scan) in rabbits undergoing whole body CT between 2015-2022. During the study period, 176 rabbits underwent whole body CT scans: 80 females (45.5%) and 96 males (54.5%), ranging in age from 1 month to 12.8 years. A total of 366 incidental findings were documented in 85.8% (151/176) of scans. Older rabbits had more multiple findings when compared to younger rabbits. Rabbits euthanized immediately after CT scan (2.8%, 5/176) were excluded from further analysis. Of the remaining rabbits, 71.9% (123/171) had one or more incidental findings[1] that did not require follow-up. These included findings in the musculoskeletal (29.8%, 51/171), respiratory (28.7%, 49/171), urinary (23.4%, 40/171), aural (18.1%, 31/171), hepatic (16.4%, 28/171) dental (15.2%, 26/171), gastrointestinal (13.5%, 23/171), reproductive (9.94%, 17/171), ophthalmic (5.3%, 9/171), cardiovascular systems (5.3%, 9/171) and other systems (12.3%, 21/171). The remaining rabbits (28.1% 48/171) had further investigation, monitoring or treatment of an incidental finding recommended. The most common treatment recommendation (9.4%, 16/171) was ear cleaning. Of these, 62.5% (10/16) were lop breeds. The second most common recommendation (8.2%, 14/171) was analgesia for musculoskeletal changes (mostly osteoarthritis). Further investigations were recommended for 7.0% (12/171) of rabbits and 2.3% (4/171) had surgery recommended. This study found that incidental changes were commonly seen on CT scans of rabbits, with approximately a quarter leading to further investigation or treatment recommendations.

References

1. Berge P, Darsonval A, Nedelcu C, et al. Incidental findings on emergency CT scans: predictive factors and medico-economic impact. *Eur J Radiol.* 2020;129:10972. doi: 10.1016/j.ejrad.2020.109072.

Use of Micro-Computed Tomography in the Diagnosis of Chronic Skull Infection Originating from the Tympanic Bulla in a Guinea Pig

Yun Wen Peng, DVM, MVM,
Kuo Chen Hu, DVM, MVM,
Yen Tao Lee, DVM, MVM

Session #3246

Affiliation: From Brave Vet Exotic Animal Veterinary Hospital, No.227, Minquan W Rd, Datong District, Taipei City, Taiwan 103 (Peng, Lee), Department of Diagnostic Imaging, National Taiwan University Veterinary Hospital, No. 153, Section 3, Keelung Rd, Da'an District, Taipei City, Taiwan (Hu).

Note: Abstract format selected by author(s). See Front Matter for description of conference proceedings manuscripts submitted as scientific abstracts.

Abstract: A 2-year-old female guinea pig was presented with head tilt to the left and a frontal fistula associated with the left ear for 1 month without any improvement despite weeks of oral antibiotics and topical debridement. A large amount of pus and debris was found in both the left external ear canal and the fistula opening. A micro-computed tomography (micro-CT) scan of the skull revealed loss of left internal ear structures and bone lysis involving the left petrous part of temporal bone to the parietal bone. A total ear canal ablation and lateral bulla osteotomy (TECALBO), followed by marsupialization was performed. A piece of the otic bone was sent for bacterial culture and antibiotic sensitivity test. According to the results, there were no antibiotics suitable for use in guinea pigs, so enrofloxacin (10 mg/kg PO q12h, Bayer Baytril® Flavour Tablets 50 mg) was used only for the prevention of secondary gastrointestinal complications and surgical wound infection. The patient received wound care twice daily using honey (KRUUSE Manuka G, sterile, 15 g gel) and calcium alginate (Hartmann Sorbalgon® Ag) as dressing for 1 month along with supportive care. All symptoms except the neurologic signs resolved eventually. Pus from the frontal fistula recurred 3 months post-surgery. A repeated micro-CT scan found bone lysis of the left petrous part of temporal bone and the parietal bone, which showed progression. This time the owner opted for palliative care only and transfer to another hospital for Chinese herbal medicine treatment. The patient deteriorated and was euthanized 3 months after the recurrence. In this case, micro-CT was utilized to accurately diagnose and elucidate the details of the disease involving the skull, medial and inner ear in a guinea pig. The track of the infection from the inner ear to the frontal area of the skull was the same as what was described in I. Kunstyr's publication in 1979.[1] The ideal surgical treatment of otitis interna in guinea pigs remained inconclusive. Severe osteolysis of the temporal and parietal bone excluded our patient as a candidate for endoscopic myringotomy, and TECALBO was the only treatment option despite the high mortality rate described in published reports.[2,3]

References

1. Kunstyr I. Fistula channel from the ear to the frontal region in 2 guinea-pigs having streptococcal empyema of the middle and inner ear. *Lab Anim.* 1979;13(1):47-48.

2. Volait-Rosset L, Pignon C, Desprez I, et al. Development and validation of an endoscopic myringotomy technique to treat otitis media and interna in a case series of three guinea pigs (*Cavia porcellus*). *J Exotic Pet Med.* 2020;32:31-38.

3. Volait L, Curros A, Pignon C. Use of total ear canal ablation and lateral bulla osteotomy to treat otitis interna in guinea pigs. *Proc ExoticsCon.* 2015;381.

Section 34

ARAV Plenary Session

James E. Bogan, Jr., DVM, MVS (Conservation Medicine), DABVP (Canine/Feline), DABVP (Reptile/Amphibian), CertAqV,
ARAV President
Session Chair

Vehicular Trauma Patterns of Chelonians: Spatial Analysis and Identification of High-Risk Areas and Roads

Aswini Cherukuri,
Alexandra Sack, DVM, MPH, PhD,
Gregory Lewbart, MS, VMD, Dipl ACZM, DipECZM (Zoo Health Management)

Session #3150

Affiliation: From the Department of Clinical Sciences, College of Veterinary Medicine, North Carolina State University, Raleigh, NC 27607, USA (Cherukuri, Lewbart), Department of Biological Sciences, University of Notre Dame, Notre Dame, IN 46556, USA (Sack).

Note: Abstract format selected by author(s). See Front Matter for description of conference proceedings manuscripts submitted as scientific abstracts.

Abstract: With the rapid urbanization of turtle habitat, wildlife hospitals most commonly see turtles that have been injured by anthropogenic trauma.[1-3] Even low levels of mortality can significantly reduce the size and health of turtle populations.[4,5] Vehicular trauma is the most common cause of chelonians presenting to the Turtle Rescue Team (TRT) at North Carolina State University College of Veterinary Medicine.[2] This study uses retrospective analysis of turtles presenting for vehicular trauma from 2005-2019 to identify roads and areas that pose high risk for vehicular collisions. Statistical analysis of collision locations and descriptive analysis of high-risk roads and areas were conducted to determine factors associated with increased collisions and mortality. Geospatial factors and road attributes were evaluated for each collision location. Of the turtles that presented for vehicular trauma (n = 2553), 40.8% survived to be released and 53.8% died or were euthanized. Geocoded locations were obtained for 67.4% of all incidents (n = 1721). Logistic regression of geocoded locations within 60 kilometers of TRT (n = 1462) showed the relationship of dying or being euthanized from vehicular trauma was not significantly associated with the measured geospatial factors and road attributes. This indicated a more in-depth analysis of the roads was needed. Eleven high-risk roads and ten high-risk areas were identified based on the high spatial density of collisions. Through identification and analysis of high-risk roads and areas, management measures can be implemented. The high rate of vehicular caused mortality demonstrates the need for collision prevention as a conservation priority.

References

1. Schenk AN, Souza MJ. Major anthropogenic causes for and outcomes of wild animal presentation to a wildlife clinic in east Tennessee, USA, 2000-2011. *PLoS ONE.* 2014;9:e93517.

2. Sack A, Butler E, Cowen P, Lewbart GA. Morbidity and mortality of wild turtles at a North Carolina wildlife clinic: a 10-year retrospective. *J Zoo Wildl Med.* 2017;48:716-724.

3. Rayl JM, Adamovicz L, Stern AW, et al. Mortality investigation of monitored Eastern box turtles (*Terrapene carolina carolina*) in Central Illinois, USA, from 2016-18. *J Wildl Dis.* 2020;56:306-315.

4. Piczak ML, Markle CE, Chow-Fraser P. Decades of road mortality cause severe decline in a common snapping turtle (*Chelydra serpentina*) population from an urbanized wetland. *Chelonian Conserv Biol.* 2019;18:231-240.

5. Gibbs JP, Shriver WG. Estimating the effects of road mortality on turtle populations. *Conserv Biol.* 2002;16:1647-1652.

Clinical and Pathologic Findings in Bearded Dragons with Cardiovascular Disease (*Pogona vitticeps*): Fifty-Four Cases (2007-2022)

Sarah Ozawa, DVM, Dipl ACZM,
Kursten Pierce, DVM, DACVIM (Cardiology),
Amy Alexander, DVM, Dipl ACZM,
Krista A. Keller, DVM, Dipl ACZM,
Danielle Tarbert DVM, Dipl ACZM,
David Sanchez-Migallon Guzman, LV, MS, DipECZM (Avian),
DipECZM (Small Mammal), Dipl ACZM,
Miranda Sadar, DVM, Dipl ACZM,
Julie Sheldon, DVM, MS, Dipl ACZM,
Danielle Meritet, DVM, MS, DACVP

Session #3300

Affiliation: From the Department of Clinical Sciences, North Carolina State University, Raleigh, NC 27607, USA (Ozawa, Pierce), Department of Population Health and Pathobiology, North Carolina State University, Raleigh, NC 27606, USA (Meritet), Department of Comparative, Diagnostic and Population Medicine, College of Veterinary Medicine, University of Florida, Gainesville, FL 32608, USA (Alexander), Department of Veterinary Clinical Medicine, College of Veterinary Medicine, University of Illinois Champaign-Urbana, Urbana, IL 61802, USA (Keller), Department of Clinical Sciences, College of Veterinary Medicine, Colorado State University Fort Collins, CO 80521, USA (Sadar), Department of Medicine and Epidemiology (Sanchez-Migallon Guzman) and the William R Pritchard Veterinary Medical Teaching Hospital (Tarbert), School of Veterinary Medicine, University of California-Davis, Department of Small Animal Clinical Sciences, Davis, CA 95616, USA, College of Veterinary Medicine, University of Tennessee, Knoxville, TN 37996, USA (Sheldon).

Note: Abstract format selected by author(s). See Front Matter for description of conference proceedings manuscripts submitted as scientific abstracts.

Abstract: Descriptions of cardiovascular disease in bearded dragons are scarce and often limited to case reports and pathology descriptions.[1-4] The lack of information regarding cardiovascular disease may complicate the diagnosis and treatment in this species. The objective of this retrospective study was to describe the clinical and pathologic characteristics of cardiovascular disease in bearded dragons from seven institutions over a 15-year period. In total, 54 cases were diagnosed resulting in a total prevalence of 3.3% (range 0.7-5.4%). The median age was 5 years (range 0.6-11) and more females (n = 31) than males (n = 22) were included (1 unknown sex). The majority of cases (81.5%) were diagnosed on necropsy alone (n = 44). Four cases were diagnosed on echocardiography and necropsy, three on ultrasound and necropsy and three were diagnosed on diagnostic imaging only (echocardiogram or ultrasound). In total, 81 diagnoses of cardiovascular disease were found in 54 animals. The most common diagnoses were myocarditis (n = 14) followed by aneurysms (n = 11), atherosclerosis (n = 7), epicarditis (n = 7), myocardial degeneration/necrosis (n = 7), pericardial effusion (n = 7), neoplasia (n = 6), and vasculitis (n = 5). Overall, 18 animals died due to cardiovascular disease. Of the animals diagnosed with aneurysms, 91% died or were euthanized due to cardiovascular disease. Cardiovascular-specific treatment was only instituted in five cases (9.3%). The small number of cases diagnosed on antemortem diagnostics emphasizes the need for further research regarding the diagnosis of cardiovascular disease in bearded dragons as well as therapeutic options.

References

1. Crouch EE, McAloose D, McEntire MS, Morrisey JK, Miller AD. Pathology of the bearded dragon (*Pogona vitticeps*): a retrospective analysis of 36 cases. *J Comp Pathol.* 2021;186:51-61.

2. Furst N, Alexander A, Ossiboff RJ, et al. Tail base aneurysm in an inland bearded dragon (*Pogona vitticeps*). *J Exot Pet Med.* 2020;33:34-37.

3. Schilliger L, Paillusseau C, Gandar F, et al. Hypertensive heart disease and encephalopathy in a central bearded dragon (*Pogona vitticeps*) with severe atherosclerosis and first-degree atrioventricular block. *J Zoo Wildl Med.* 2019;50(2):482-486.

4. Schilliger L, Lemberger K, Chai N, et al. Atherosclerosis associated with pericardial effusion in a central bearded dragon (*Pogona vitticeps*). *J Vet Diagn Invest.* 2010;22(5):789-792.

Section 35

ARAV Novel Therapies and New Perspectives Session

La'Toya Latney, DVM, DECZM (ZHM), DABVP (Reptile and Amphibian Practice), CertAqV
Session Chair

Treatment of Three Parietal Aneurysms in Two Bearded Dragons (*Pogona vitticeps*) via Chemoembolization

Nicola Di Girolamo, DVM, PhD, DipECZM (Herpetology), Dipl ACZM, Dipl ACVPM,
Lawrence Santistevan, DVM,
Aaron Percival, DVM,
Christopher Tollefson, DVM, MS, Dipl ACVR

Session #3498

Affiliation: From Exotic Pet Service, Department of Clinical Sciences, College of Veterinary Medicine, Cornell University, 930 Campus Road, Ithaca, NY 14853, USA (Di Girolamo), Cardiology Service, Department of Clinical Sciences, College of Veterinary Medicine, Cornell University, 930 Campus Road, Ithaca, NY 14853, USA (Santistevan), Imaging Service, Department of Clinical Sciences, College of Veterinary Medicine, Cornell University, 930 Campus Road, Ithaca, NY 14853, USA (Percival, Tollefson).

Note: Abstract format selected by author(s). See Front Matter for description of conference proceedings manuscripts submitted as scientific abstracts.

Abstract: Currently, there is limited knowledge on aneurysms affecting bearded dragons.[1] There are occasional reports of aneurysms occurring in the dorsoventral portion of the neck, sometimes defined parietal or cervical aneurysms,[2] with anecdotal reports of recurrence after surgical correction of the disorder. Two bearded dragons presented for treatment of a parietal aneurysm to a veterinary teaching hospital. After angiographic computed tomography evaluation of the aneurysms, the bearded dragons were anesthetized, and chemoembolization of the temporo-orbital artery was performed under ultrasonographic guidance. In case 2, the procedure was repeated on the following day. Upon ultrasonographic recheck, both bearded dragons had ceased flow in the aneurysms. Both bearded dragons recovered uneventfully from the procedure. Both cases have been placed on amlodipine to lower systemic pressure. Systemic blood pressure obtained on case 1 over several days through a catheter in the coccygeal artery, showed a potential effectiveness of the amlodipine in lowering the blood pressure. Case 1 had a complete resolution of the aneurysm in the following 6 weeks, with no visible swelling up to 9 months after the procedure. Case 2 aneurysm was static for 2 months, but then suddenly started increasing in size. A further chemoembolization procedure was performed but following that, the patient was lost to follow-up. After 4.5 months from the initial appointment, case 1 presented with a similar aneurysm on the opposite side of the head, that was also occluded with multiple sessions of chemoembolization. The technique described herein showed potential to treat parietal aneurysms in bearded dragons, avoiding the need for surgery.

References

1. Furst N, Alexander A, Ossiboff RJ, et al. Tail base aneurysm in an inland bearded dragon (*Pogona vitticeps*). *J Exot Pet Med.* 2020;33:34-37.

2. Barten S, Wyneken J, Mader D, et al. Aneurysm in the dorsolateral neck of two bearded dragons (*Pogona vitticeps*). *Proc Annu Conf Assoc Reptile Amphib Vet.* 2006;43-44.

Determining the Effects of Furosemide on Male Red-Eared Sliders (*Trachemys scripta elegans*)

Katie Metcalf, DVM, MS,
Graciela Aguilar, DVM,
Shannon D. Dehghanpir, DVM, MS, DACVP (Clinical Pathology),
John Tuminello, BS,
Mark Acierno, DVM, MBA, Dipl ACVIM,
Mark A. Mitchell, DVM, MS, PhD, DipECZM (Herpetology)

Session #3408

Affiliation: From the Department of Veterinary Clinical Sciences, School of Veterinary Medicine, Louisiana State University, Baton Rouge, LA 70803, USA.

Note: Abstract format selected by author(s). See Front Matter for description of conference proceedings manuscripts submitted as scientific abstracts.

Abstract: Renal disease is a common ailment of captive reptiles that is often closely linked to chronic, subclinical states of dehydration. Currently, the diagnosis of renal disease in reptiles is poorly characterized. A potential outlet for the characterization of renal dysfunction in reptiles is the induction of dehydration by furosemide therapy. The utility of furosemide in reptiles is poorly characterized, however, both experimental and clinical utility of this medication has been documented in chelonians. The goals of this study were to confirm the utility of furosemide administration, to better characterize the diagnostic findings of dehydration or reduced renal function, and to determine the utility of urine for the diagnosis of dehydration in the red-eared slider. Ten adult, male, red-eared slider turtles (*Trachemys scripta elegans*) were used for this study. Turtles were randomly assigned to three different treatment groups in a complete-crossover study: 10 mg/kg furosemide q12h, four doses; 5 mg/kg furosemide q12h, four doses; and a control group. Baseline blood and urine samples were collected, and water and food were withheld for 48 hours. Blood and urine samples were re-collected, and they were given access to water and weighed over the next 48 hours. Urine and plasma osmolality measurements were also determined via freezing point osmometry for the red-eared sliders. The results of this study support the clinical utility of furosemide in the red-eared slider, as furosemide-treated turtles had evidence of prominent physical signs of dehydration, significantly increased urine production and osmolality, and disproportionately decreased chloride concentrations due to furosemide administration.

Incidence and Prognosis of Cloacal Prolapses in Reptiles: A Retrospective Study

Federica Fogli, DVM, MRCVS, and
Thomas Dutton, BVM&S, MRCVS, CertAVP (ZooMed) DipECZM (Avian)

Session #3139

Affiliation: From Great Western Exotics, Unit 10, Berkshire House, County Business Park, Shrivenham Rd, Swindon SN1 2NR, United Kingdom.

Note: Abstract format selected by author(s). See Front Matter for description of conference proceedings manuscripts submitted as scientific abstracts.

Abstract: Cloacal prolapses are a common problem affecting reptiles in private practice. They are described in several species, including lizards, chelonians and snakes. Prolapses are a medical emergency and prompt identification of the prolapsed tissue is fundamental to select adequate treatment. The objective of this study was to evaluate the types of cloacal prolapses in reptiles and analyze outcomes based on potential risk factors. Medical records of all reptile patients seen at Great Western Exotics between 2014 and 2022 were retrospectively evaluated. A total of 2228 reptiles were seen in the study period and prolapses were identified in 91 patients (4%). Overall, males appeared to be more prone to cloacal prolapses across all three species representing 53.8% of the cases. Also in 90.9% of cases, prolapses occurred in patients older than 1 year of age. Lizards presented more frequently (56%), compared to chelonians (34%) and snakes (9.8%). In lizards, true cloacal prolapses presented more commonly. Penile prolapses were the most identified prolapse in chelonians and hemipenile prolapses were the most occurring prolapse in snakes. In 60% of cases, underlying causes of cloacal prolapse were not investigated; however, surgery appeared to be the preferred form of treatment. These results suggest that cloacal prolapses are not a common occurrence but presented at a higher rate than previously reported in literature. This study also highlights that species and sex may impact the occurrence and type of cases seen, as well as the type of treatment chosen may affect the survival rate of the patient at presentation.

References

1. Alworth LC, Hernandez SM, Divers SJ. Laboratory reptile surgery: principles and techniques. *J Am Assoc Lab Anim Sci.* 2011;50(1):11-26. PMID: 21333158; PMCID: PMC3035398.

2. Bodri MS, Sadanaga KK. Circumcostal cloacapexy in a python. *J Am Vet Med Assoc.* 1991;198(2):297. PMID: 2004995.

3. Hedley J, Eatwell K. Cloacal prolapses in reptiles: a retrospective study of 56 cases. *J Small Anim Pract.* 2014;55(5):265-268. doi:10.1111/jsap.12199.

4. Hellebuyck T, Vilanova FS. The use of prefemoral endoscope-assisted surgery and transplastron coeliotomy in Chelonian reproductive disorders. *Animals.* 2022;12(23):3439. doi.org/10.3390/ani12233439.

5. Knotek Z, Cermakova E, Oliveri, M. Reproductive medicine in lizards. *Vet Clin North Am Exot Anim Pract.* 2017;20(2):411-438. doi:10.1016/j.cvex.2016.11.006.

6. Perry SM, Mitchell MA. Reproductive medicine in freshwater turtles and land tortoises. *Vet Clin North Am Exot Anim Pract.* 2017;20(2):371-389. doi:10.1016/j.cvex.2016.11.004.

Partial Hemimandibulectomy in a Bearded Dragon

Anna-Marie Ruoff, DVM

Session #3118

Affiliation: From Colorado Exotic Animal Hospital, 7120 E Hampden Ave, Ste 101, Denver, CO 80224, USA.

Note: Abstract format selected by author(s). See Front Matter for description of conference proceedings manuscripts submitted as scientific abstracts.

Abstract: A 9-year-old male bearded dragon (*Pogona vitticeps*) presented for hyporexia and a nodular gingival mass that bled during prehension and mastication. Excisional biopsy revealed an undifferentiated sarcoma most consistent with fibrosarcoma. The surgical margins were incomplete and the mass regrew within 2 weeks. Radiographs revealed lysis of the dentary bone in the region of the mass, so a partial hemimandibulectomy was performed and an esophagostomy tube was placed. Histopathology confirmed narrow but clear surgical margins and the client declined pursuing additional chemotherapy or radiation therapy options for management. The dragon recovered well from the procedure, began eating on his own, and maintained normal body condition. No regrowth of the mass was found at 2- and 8-week post-operative recheck examinations, the esophagostomy tube was removed, and the dragon continues to do well six months after the surgical procedure. To the author's knowledge, this is the first description of hemimandibulectomy in this species. Partial hemimandibulectomy may be considered for cases of mandibular neoplasia, fracture, or osteomyelitis that are refractory to medical and traditional surgical management.[1,2]

References

1. Goe AM, Heard DJ, Abbott JR, et al. Surgical management of an odontogenic tumor in a banded Gila monster (*Heloderma suspectum cinctum*) with a novel herpesvirus. *Vet Quart.* 2016;36(2):109-114.

2. Fossum TW. Surgery of the digestive system. *Small Animal Surgery.* 5th Ed. Philadelphia, PA: Elsevier; 2019.

Medicinal Uses of Honey and Beeswax in Turtle Shell Repair Techniques

Priya Patel, BVetMed, MRCVS

Session #3218

Affiliation: From the New England Wildlife Centers, 500 Columbian St, South Weymouth, MA 02190, USA.

Abstract: Manuka honey is frequently used in wound management protocols for its numerous wound healing properties. For chelonians that have sustained shearing injuries with missing fragments of shell, initial treatment with honey provides adequate debridement of the exposed tissue and then promotion of healthy granulation tissue. Honey bandages should be changed daily and kept dry. For terrestrial turtles these requirements pose no issue; however, for aquatic turtles, keeping them dry docked for extended periods can lead to secondary clinical issues such as inappetence and pressure sores. The use of beeswax as an alternative to honey provides similar wound healing benefits plus an added waterproof bonus due to its hydrophobic properties. The case studies presented in this lecture will demonstrate the successful treatment of three different species of aquatic turtles with significant shell defects using beeswax while maintaining all patients in aquatic environments during the entire treatment process.

Introduction

Honey and other bee products have been used for medical purposes for thousands of years dating back to Ancient Greek and Roman times.[1] Now due to several published research papers, honey is officially recognized for its wound healing properties and its antimicrobial, anti-inflammatory, and antioxidant effects.[2] In addition to aiding the debridement phase, honey also promotes faster wound healing by encouraging tissue regeneration and re-epithelization.[3] Honey's antimicrobial activity has been shown to be effective against many microbes including strains of antibiotic resistant bacteria.[2] The factors that are attributed to honey's antibacterial properties include its low pH, high sugar content, and presence of dilute hydrogen peroxide as it dissolves in the wound.[4] The medical quality of the honey does vary depending on the species of the bee and its diet.[2] Due to this variability many studies have focused on using Manuka honey, which in addition to all the beneficial properties all honey displays, also has advanced antibacterial properties due to the presence of a high level of methylglyoxal (MGO).[5,6] The level of MGO has been documented to be positively correlated with antibacterial activity.[7] Due to this finding, Manuka honey is the honey of choice used in this hospital and for the case studies described here.

When honey is used in wound treatment a non-adhesive bandage is applied to keep the wound clean and dry from the surrounding environment. A honey bandage is applied from varying times of duration from days to weeks depending on the degree of wound contamination. Typically, aquatic turtles are either completely dry docked or placed in shallow water during this process depending on the size of the wound and dimensions of the resulting honey bandage. This can pose physiologic and husbandry problems for aquatic and semi-aquatic turtles.[8] Dehydration, (even with administration of supplemental subcutaneous fluids) inappetence, and pressure sores have been observed by the author. Pressure sores are of particular concern with common snapping turtles due to their heavy body mass. The use of beeswax as an alternative to honey provides similar wound healing and antimicrobial benefits as well as an added waterproof bonus due to its hydrophobic properties.[9] This novel technique of using beeswax for treatment in cases with turtle shell defects reduced many of the deleterious effects of being dry docked. The following case studies demonstrate the successful treatment of three different species of aquatic turtles with significant shell defects using beeswax while maintaining all patients in aquatic environments during the entire treatment process.

Methods

For all cases described here yellow beeswax pastilles were used (Fig 1A). Pastilles were melted by microwaving them in thirty second increments until all beeswax beads were melted (Fig 1B). Beeswax was then stirred continuously as it cooled (Fig 1C). As soon as the beeswax was cool enough to the touch but still moldable it was pressed into the wound bed covering the entire field and overlapping the shell margins (Figs 1D and E). If at any point the beeswax cooled too fast and did not stick to the edges of the shell the beeswax was removed and the process was repeated. The beeswax had to be adequately affixed to the shell to prevent water from wicking underneath into the wound. Turtles were then placed into water once the beeswax was completely cooled and solidified, usually 6-8 hours after application (Fig 1F). Beeswax was replaced every 2-4 weeks depending on the severity of the initial wound. Once healthy granulation tissue was observed beeswax could be left in place for 6-8 weeks at a time.

Case Studies

Eastern painted turtle

An adult Eastern painted turtle (*Chryemys picta*) presented with a large shell defect to its carapace and bridge (Fig 2A). The coelomic cavity was exposed and internal organs were visualized. The turtle was bright, alert and responsive, so treatment was attempted. Due to the direct exposure of internal organs no topical ointments were used prior to placement of the beeswax. Beeswax was applied as previously described, and the turtle was placed in shallow water the next day (Fig 2B). Due to the concern for exposing the lungs and other organs to water a much more gradual introduction to a deep-water tank was used. Beeswax was replaced monthly for the first 4 months and then every 2 months for the remaining time in care. Systemic pain medications and antibiotics were also given until the coelomic cavity was sealed off by a healthy fibrous layer of tissue. After 11 months the coelomic cavity was closed off with a fibrous tissue layer (Fig 2C).

Diamondback terrapin

A female diamondback terrapin (*Malaclemys terrapin*) presented after a suspected boat strike with an almost complete amputation of its right bridge (Fig 3A). Surgical repair with wiring was attempted; however, due to reduced vascular supply the fragment devitalized and became necrotic. The coelomic membrane was still intact so beeswax was placed over the exposed membrane using the method previously described. The patient was then kept in water for the remainder of treatment (Fig 3B). The patient also received systemic pain medications and antibiotics until the appearance of healthy granulation tissue. After 10 months of treatment with beeswax a thick fibrous protective tissue layer was observed (Fig 3C).

Common snapping turtle

An adult common snapping turtle (*Chelydra serpentina*) weighing 11 kilograms presented with a full thickness shearing wound to its carapace. Both keratin and bone were removed, exposing underlying soft tissue (Fig 4A). Parts of the lung were visible at the center of the wound. Due to the level of contamination Manuka honey was applied to the wound for a week prior to applying beeswax. During this time the turtle was housed in shallow water. Once the wound was adequately debrided, beeswax was applied as previously described. The patient was placed in full water the day after application (Fig 4B). Due to the initial contamination of the wound the beeswax was replaced every 2 weeks for the first 2 months and then every month after that. The patient also received systemic pain medications and antibiotics until the appearance of healthy granulation tissue. After only 4 months, a thick fibrous layer of tissue had formed with shell pigmentation observed at the margins (Fig 4C).

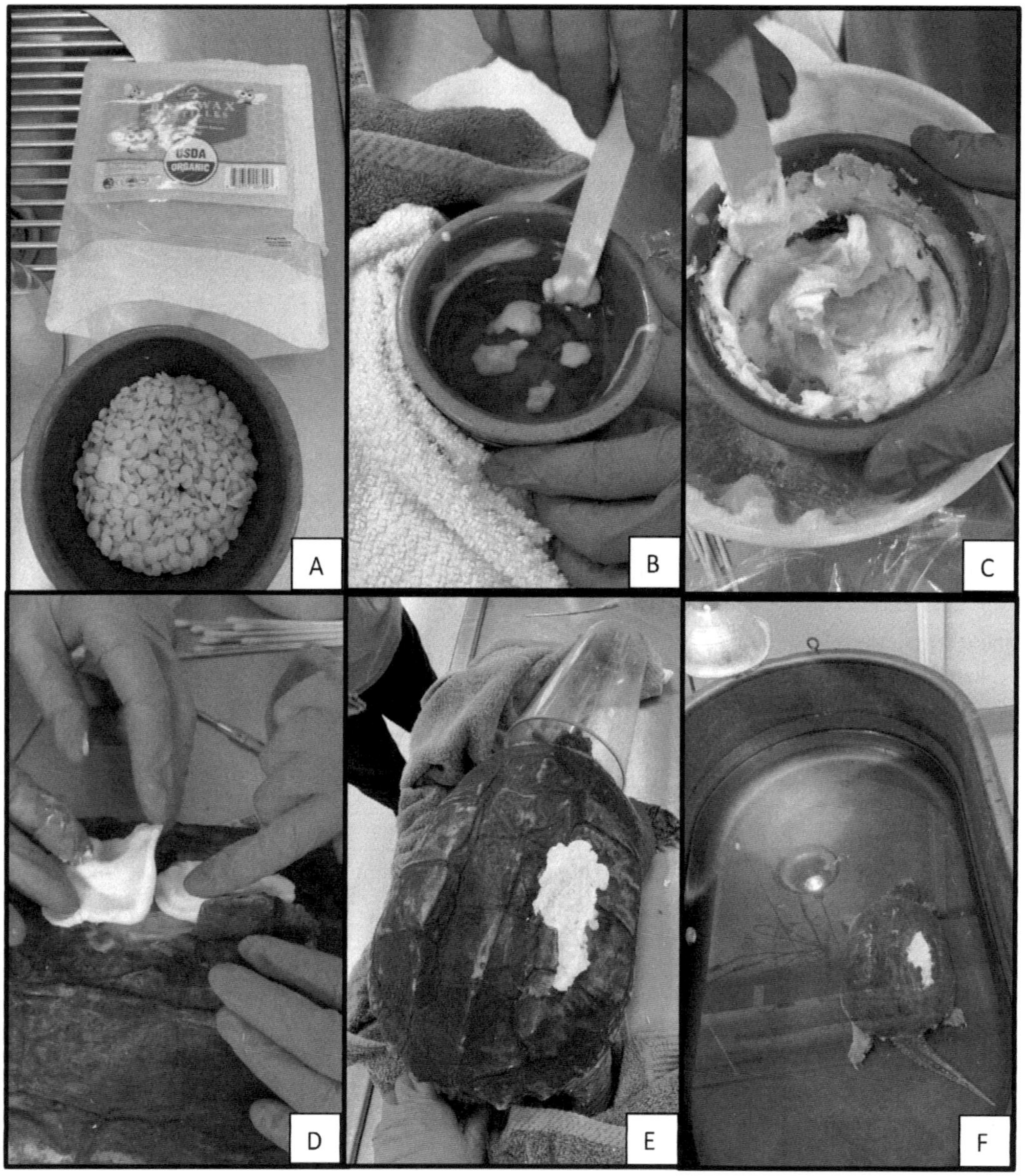

Figure 1. Step-by-step process of the application of beeswax on a turtle shell. (A) Yellow beeswax pastilles are placed in a microwave safe dish. (B) Pastilles were melted by microwaving them in thirty second increments until all beeswax beads were melted. (C) Beeswax was then stirred continuously as it cooled. (D) As soon as the beeswax was cool enough to the touch but still moldable it was pressed into the wound bed covering the entire field and overlapping the shell margins. (E) Beeswax adequately affixed to the shell. (F) Turtle placed into water once the beeswax was completely cooled and solidified.

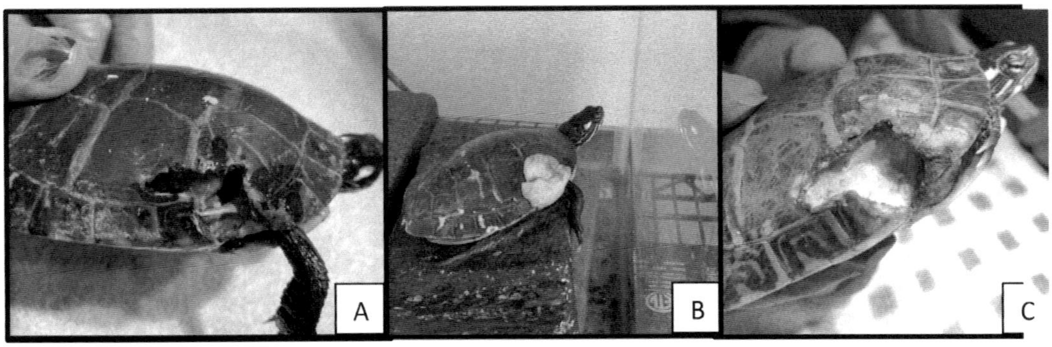

Figure 2. Eastern painted turtle (*Chryemys picta*) before and after results after using beeswax. (A) A large shell defect to the turtle's carapace and bridge with coelomic cavity and internal organ exposure was noted on initial exam. (B) Beeswax was applied and the turtle was placed in gradually deepening water throughout treatment. (C) After 11 months the coelomic cavity was sealed off with a fibrous tissue layer.

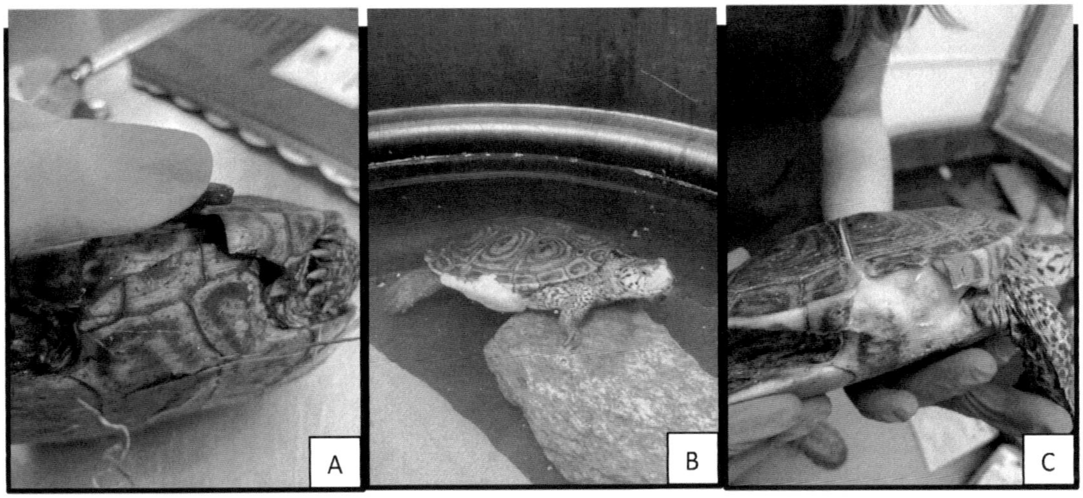

Figure 3. Diamondback terrapin (*Malaclemys terrapin*) before and after results after using beeswax. (A) Patient presented after a suspected boat strike with an almost complete amputation of its right bridge. Surgical repair with wiring was attempted; however, due to reduced vascular supply the fragment devitalized and became necrotic. (B) The coelomic membrane was still intact so beeswax was placed over the exposed membrane using the method previously described and the patient was then kept in water for the remainder of treatment. (C). After 10 months of treatment with beeswax a thick fibrous protective tissue layer was observed.

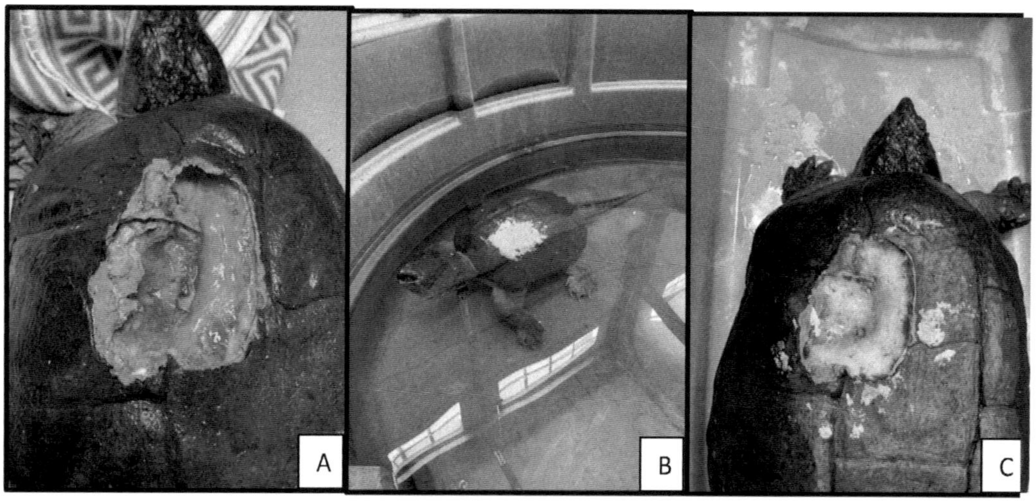

Figure 4. Common snapping turtle (*Chelydra serpentina*) before and after results after using beeswax. (A) Turtle presented with a full thickness shearing wound to its carapace with lung exposure. (B) Once the wound was adequately debrided, beeswax was applied and the patient was placed in full water the day after application. (C) After 4 months a thick fibrous layer of tissue had formed with shell pigmentation observed at the margins.

Summary

The use of honeybee products in wound healing techniques is gaining more recognition in the veterinary field. Manuka honey is already well-known for its numerous wound healing properties. Chelonian patients suffering from shell damage frequently spend months to a year in care due to their slow healing times.[10] This long treatment period affects quality of life for aquatic turtles that remain dry docked to keep injuries clean and dry. These case studies demonstrate a novel technique using beeswax that allows wound healing to occur while the patient is maintained in an aquatic environment thus negating some of the harmful secondary effects of being dry docked.

References

1. Meo SA, Al-Asiri SA, Mahesar AL, Ansari MJ. Role of honey in modern medicine. *Saudi J Biol Sci.* 2017;24(5):975-978.

2. Abeshu MA, Geleta B. Medicinal uses of honey. *Biol Med.* 2016;8:2.

3. Al-Waili N, Salmon K, Al-Ghamdi AA. Honey for wound healing, ulcers, and burns; data supporting its use in clinical practice. *Sci World J.* 2011;11:766-787.

4. Weston RJ. The contribution of catalase and other natural products to the antibacterial activity of honey: a review. *Food Chem.* 2000;71:235-239.

5. Adams CJ, Manley-Harris M, Molan PC. The origin of methylglyoxal in New Zealand Manuka (*Leptospermum scoparium*) honey. *Carbohyd Res.* 2009;344:1050-1053.

6. Atrott J, Henle T. Methylglyoxal in Manuka honey-correlation with antibacterial properties. *Czech J Food Sci.* 2009;27:S163-S165.

7. Johnston M, McBride M, Dahiya D, et al. Antibacterial activity of Manuka honey and its components: an overview. *AIMS Microbiol.* 2018;4(4):655-664.

8. Sypniewski LA, Hahn A, Murray J, et al. Novel shell wound care in the aquatic turtle. *J Exot Pet Med.* 2016;25(2):110-114.

9. Fratini F, Cilia G, Turchi B, et al. Beeswax: a minireview of its antimicrobial activity and its application in medicine. *Asian Pac J Med.* 2016;9(9):839-843.

10. Spadola F, Morici M. Treatment of turtle shell ulcerations using photopolymerizable nano-hybrid dental composite. *J Exot Pet Med.* 2016;25(4):288-294.

Pelvic Stabilization in the Painted Turtle: A Novel Surgical Technique

Allison Carter, DVM, CWR, and
Miranda Torkelson, DVM, CWR, CertAqV

Session #3318

Affiliation: From Wildlife Rehabilitation Center of Minnesota, 2530 Dale St N, Roseville, MN 55113, USA.

Abstract: This lecture discusses a novel surgical technique developed at the Wildlife Rehabilitation Center of Minnesota for stabilization of unilateral and bilateral ilial fractures in western painted turtles. We will discuss the development of the technique and follow the clinical progress of the nine patients that received this surgery in 2022. Discussion will include radiographic techniques, anesthesia for chelonian orthopedic procedures, post-operative care, physical therapy and post-operative monitoring.

Introduction

Painted turtles presenting for rehabilitation are frequently admitted for trauma, with vehicular trauma being most common. Although turtle shell repair methods have been published, there appears to be little knowledge of internal orthopedic repair in freshwater chelonians. At the Wildlife Rehabilitation Center of Minnesota, we find that almost 100% of painted turtles with caudal carapace fractures due to vehicular trauma sustain pelvic fractures as well. The majority of these fractures are unilateral or bilateral ilial fractures. The ilium serves multiple functions in both ambulating on land and swimming was well as forming part of the pelvic girdle through which eggs must pass in females. As such, inadequate healing or malalignment of an ilial fracture presents a potential barrier to release. The goal of this work was to develop and test the viability of a novel surgical technique for ilial fracture repair.

Radiographic Techniques

Two oblique views, a cranioventral caudodorsal 45 degree oblique and a caudoventral craniodorsal 45 degree oblique, were used to fully visualize and classify pelvic fractures. Understanding of radiographic anatomy and species variability was key for accurate interpretation of radiographs. Multiple intra-op and post-operative radiographs were taken to ensure proper pin placement.

Anesthesia

Orthopedic procedures were relatively long, so adequate anesthetic monitoring was important for success. Turtles were sedated with midazolam, hydromorphone, and alfaxalone intramuscularly. Patients were then intubated and ventilated with isoflurane. A jugular catheter was placed for intraoperative fluid support and a continuous rate infusion of Propofol. Heat support was provided with heating pads and a Bair Hugger. Patients were monitored with doppler, capnograph, and temperature probe.

Orthopedic Procedure

This novel procedure was developed and practiced many times on cadavers before being attempted in live patients. Authors are not advocating for the use of this procedure in live patients until outcomes can be better understood. No new procedure should ever be attempted on a live patient without adequate surgical experience and practice. The surgical window is the pre-femoral fossa. External-fixator pins were driven perpendicularly through the skin of the pre-femoral fossa into the fractured ilium. Pins were used to hold ilium in alignment. An epoxy puddy bar was formed from the carapace to the plastron holding pins in place.

Post-Operative Care

Patients all received meloxicam and tramadol for analgesia. Patients had e-tubes placed to provide supplemental nutrition post-operatively. Some patients received antibiotics post-operatively. Patients were transitioned into water housing as soon as possible.

Physical Therapy

Carts were made with 360 degree roller ball mini caster wheels adhered to a platform that could be taped to the turtle's plastron. This allowed the turtles to begin using the leg before they could bear full weight. This encouraged muscle strengthening and continued full range of motion of the limbs. As soon as possible, turtles were housed in water (even with implants in place). Swimming allowed turtles to use surgical legs without placing weight on the limb.

Post-Operative Monitoring

Although radiographs for post-operative monitoring of these cases were important, authors found that radiographically visible osseous callous often took months to develop. Accurate palpation was much more important for monitoring healing as fibrous callous was palpable within weeks. If computed tomography is available, it is recommended because it provides the best resolution image of fracture healing. Assessment of ambulation in turtles is difficult because the shell limits visibility of the entire limb. Authors find that there are a few tricks to more accurately assess ambulation in these species. Slow motion videos help show subtle changes in extension or lift. These videos should be taken from multiple angles. Allow the turtle to ambulate on multiple substrates since a slippery hospital floor may cause abnormal ambulation on its own. Listen to the turtle ambulating on a hard surface, the plastron should not tap on the ground. Check the plastron for ulcerations or abrasions which may indicate that the patient is not able to sufficiently lift its body.

Outcomes

Surgery was performed on nine painted turtles that presented for rehabilitation at the Wildlife Rehabilitation Center of Minnesota during the summer of 2022. Three patients died postoperatively. One was found to have coelomitis that was suspected to be secondary to initial trauma. The other two died of suspected internal trauma but it is unknown whether the trauma was due to the initial injury or secondary to surgery. Three patients successfully recovered from surgery but were later euthanized due to complications associated with the initial trauma. These included arthritic changes due to the proximity of the initial fracture to the joint and reproductive adhesions causing egg binding. Three turtles have recovered from surgery, have healed their fractures, and are expected to be released. Complications encountered during recovery of turtles from orthopedic procedures included infection, granulomatous reaction to implants, osteoarthritis, fixation failure, obstructive callous formation, nonunion, and soft tissue or nerve damage.

Next Steps

With a 30% success rate, this novel procedure has potential as a method of repair of ilial fractures in painted turtles but improvements should be made to increase success rates before it is widely used. The authors are considering various methods to better balance fracture biology and mechanics with the knowledge gained from these cases, including creating 3D printed models to assist with better fracture alignment, placing bone grafts, decreasing surgical trauma, and decreasing surgical time. The authors would also like to improve post-operative monitoring with force plate assessments of ambulation.

Expanding the Invertebrate Medicine Toolbox: Evaluation of Tonometry as a Novel Diagnostic Tool in Arachnids

Meghan Chung

Session #3141

Affiliation: From North Carolina State University College of Veterinary Medicine, Raleigh, NC 27607, USA.

Abstract: The goal of the current study was to investigate if tonometry could be used to gauge normal abdominal pressures in tarantulas. Five juvenile (yearling) female curlyhair tarantulas (*Tliltocatl albopilosus*) and one juvenile male curlyhair tarantula were used to track pressures over a period of time. An iCare TONOVET TV01 tonometer set to canine/feline species was used to measure intraabdominal pressures of spiders. Preliminary results show steady pressures when not in molt, with a significant drop in pressures post molt. These results indicate that abdominal pressures change throughout a molt cycle with predictability. If pressures prove to be consistent, tonometry could be used as a diagnostic indicator of health in arachnids.

Introduction

Diagnostics and health evaluation for invertebrates of many species is limited, specifically in the class Arachnida. These limitations are becoming more relevant as people seek care for their non-traditional pets. Specifically, in the order Araneae, trauma, dysecdysis, hair loss, and prolonged anorexia appear to be the most common medical problems seen in captivity. Recent methods on how to address trauma, amputation of limbs, leaking hemolymph, anesthetization of tarantulas and scorpions have been published, but healthy diagnostic indicators are relatively new to invertebrate medicine. Dysecdysis is a common reason for keepers and owners to seek veterinary care, and unless closely monitored from molt to molt, can be difficult to determine where an animal is within their molt cycle. Hydration status is an important variable for an animal to successfully molt. If dysecdysis is caught early, administration of intracardiac fluids has been shown to aid in the process. Keepers and owners have historically assessed hydration status based on body position and ability to move normally, as limb movement is dependent on hemolymph and hydraulics. A possible solution to monitoring an animal throughout their molt cycle is through the use of tonometry.

Materials and Methods

Five juvenile (yearling) female curlyhair tarantulas (*Tliltocatl albopilosus*) and one juvenile male curlyhair tarantula were used to track pressures over a period of time, as younger animals were more likely than adults to go through multiple molts in a 9-month period. These animals were sourced from a zoological collection and were captive born from the same clutch to maintain consistency. Spiders were housed in individual 14 × 10 × 14-inch (35.56 × 25.4 × 35.56-centimeter) plastic containers with holes drilled for ventilation. Within the containers was 2-3 inches of coco husk substrate, a black plastic hide, a small water dish, and a deeper dish filled with moistened paper towels. The containers were housed within an environmental chamber at a constant temperature of approximately 25 °C with humidity of 60-70% maintained via dishes of water placed within the chamber. Water dishes were filled with dechlorinated water daily. Animals were fed 0.75-inch killed crickets, increasing in amount accordingly as they grew. An iCare TONOVET TV01 tonometer set to canine/feline species was used to measure intraabdominal pressures of spiders. Pressures were evaluated once each week over a 9-month period; additional measurements were obtained 2 days following each observed episode of ecdysis. Pressures were measured from a site at the ventral opisthosoma, the thinnest area of cuticle throughout the body.

Results

Preliminary results show steady pressures when not in molt, with a significant drop in pressures post molt. Pressures of a tarantula that had self-amputated a limb showed a steady decrease throughout loss of hemolymph.

Discussion

Preliminary pressure data indicate that abdominal pressures change throughout a molt cycle with predictability. The goal of the current study was to investigate if tonometry could be used to gauge normal abdominal pressures in tarantulas. If pressures prove to be consistent, tonometry could be used as a diagnostic indicator of health in arachnids, where diagnostics are often limited to drawing hemolymph and necropsy. Confounding variables include variation in BCS, hydration status, nutritional quality and quantity, species, cetae interference, type of tonometer used, and site of measuring pressures. Additional areas to attempt reading pressures could be the joint spaces as they are another location of thin cuticle, as well as mapping tonometry pressures across the exoskeleton to determine cuticle thickness. Further research could determine how values compare when a female is gravid, if the values change before, during, and after sedation, and assessing pressures among other arachnid species, including scorpions. Experimental use of tonometry has shown that pressures of the pleural membrane in between tergite segments of the metasoma in Asian forest scorpions are relatively consistent, indicating a promising area for future research.

Clinical Features and Treatment of Fishhook Injuries in Freshwater Turtles: One-Hundred Twenty-Six Cases (1997-2022)

Lily M. Hale,
Sabrina L. Kapp,
James B. Robertson, MS,
Gregory A. Lewbart, MS, VMD, Dipl ACZM, DipECZM (Zoo Health Management),
Sarah M. Ozawa, DVM, Dipl ACZM

Session #3202

Affiliation: From College of Veterinary Medicine, North Carolina State University, Raleigh, NC 27607, USA (Hale, Kapp), Office of Research, North Carolina State University, Raleigh, NC 27695, USA (Roberston), Department of Clinical Sciences, North Carolina State University, Raleigh, NC 27607, USA (Lewbart, Ozawa).

Note: Abstract format selected by author(s). See Front Matter for description of conference proceedings manuscripts submitted as scientific abstracts.

Abstract: Fishhook injuries have been shown to impact a significant portion of freshwater turtle populations with the associated mortality being likely to cause population declines;[1,2] however, the main treatment and prognostic indicators reported for these injuries focus on sea turtle species.[3,4] The objective of this retrospective study was to describe the clinical features, treatment, and outcomes of wild freshwater turtles with fishing hook injuries. In total, 126 wild turtles with fishing hook injuries that were presented to a wildlife rescue clinic in North Carolina were reviewed. The most common species presenting for a fishhook injury was the yellow-bellied slider (*Trachemys scripta scripta*) (n = 69/126, 54.8%). The most common location identified was the oral cavity (n = 77/140, 55%) and the most common removal method was retrograde removal after cutting the barb off the hook (n = 76/120, 63.3%). Fish hooks embedded in the esophagus had a significantly higher chance of complications affecting recovery. There was no significant increase in mortality associated with the location of the injury, however there was a significant increase in mortality in patients that experienced complications ($P < 0.001$). The time in care ranged from 1-150 days (median 16 days). Of the turtles evaluated, 10.8% were euthanized or died after treatment and 89.2% were released. While no superior treatment was identified, considerations should be taken to provide patient comfort, decrease injury associated complications, and shorten recovery time by using minimally invasive techniques. Overall, freshwater turtles with fishhook injuries have a high release rate even when the injuries seem severe.

References

1. Steen DA, Hopkins BC, Van Dyke JU, et al. Prevalence of ingested fish hooks in freshwater turtles from five rivers in the southeastern United States. *PloS ONE* 2014;9:e91368.

2. Steen DA, Robinson OJ. Estimating freshwater turtle mortality rates and population declines following hook ingestion. *Conserv Biol.* 2017;31:1333-1339.

3. Valente ALS, Parga ML, Velarde R, et al. Fishhook lesions in loggerhead sea turtles. *J Wildl Dis.* 2007;43: 737-741.

4. Di Bello A, Valastro C, Freggi D, et al. Surgical treatment of injuries caused by fishing gear in the intra-coelomic digestive tract of sea turtles. *Dis Aquat Organ.* 2013;106:93-102.

Outcomes and Complications of Prefemoral Cystotomy for Urolith Removal in African Spurred Tortoises (*Centrochelys sulcata*): Eight Cases (2017-2022)

Nicola Di Girolamo, DVM, PhD, DipECZM (Herpetology), Dipl ACZM, Dipl ACVPM, and Christoph Mans, Dr med vet, Dipl ACZM

Session #3490

Affiliation: From Exotic Pet Service, Department of Clinical Sciences, College of Veterinary Medicine, Cornell University, 930 Campus Road, Ithaca, NY 14853, USA (Di Girolamo), Department of Surgical Sciences School of Veterinary Medicine, University of Wisconsin-Madison, 2015 Linden Drive, Madison, WI 53706, USA (Mans).

Note: Abstract format selected by author(s). See Front Matter for description of conference proceedings manuscripts submitted as scientific abstracts.

Abstract: Removal of uroliths in chelonians remains challenging. Therefore, the purpose of the present case series is to describe the outcomes and complications of a modified prefemoral surgical approach for urolith removal in chelonians. Between 2017 and 2022, a total of eight African-spurred tortoises underwent prefemoral cystotomy performed by a board-certified zoological medicine specialist and were included in this case series. Prefemoral cystotomy was achieved with the tortoises positioned in an oblique latero-dorsal recumbency,[1] which allowed the uroliths to gravitate toward the prefemoral incision. Access to the coelom via the prefemoral fossa was routine. The use of ring retractors and, for the last two individuals, circumferential retractors were necessary to provide and improve access and visibility. The uroliths were fragmented with a surgical drill or a rotary drilling tool and/or Allis or Babcock forceps. Partial urinary bladder amputation was needed in one case due to iatrogenic damage to the bladder during urolith removal. Closure of the body wall was performed in two or three layers. It was not always possible to close the coelomic membrane. Of the tortoises included, one animal died after the surgery. This animal was the first treated with this approach and demonstrated renal failure on pre-operative blood work, but the owner insisted on attempting surgical treatment. The remaining seven tortoises recovered from the surgery. Two tortoises developed dehiscence of the surgical wound but ultimately recovered, and they are currently healthy. The technique employed in this case series required some refinements but ultimately resulted in successful surgical outcomes.

References

1. Di Girolamo N, Studer K, Maranville R, et al. Feasibility, gross postmortem results and participant perception of oblique prefemoral ovariectomy in red eared sliders (*Trachemys scripta elegans*) during a teaching workshop. *J Herpetol Med Surg.* 2022;32(2):130-135.

Section 36

ARAV Nutrition and Diet Composition Session

Paul Gibbons, DVM, MS, DABVP (Reptile and Amphibian Practice)
Session Chair

Postprandial Uric Acid Level Increase Following Insect Ingestion in Inland Bearded Dragons (*Pogona vitticeps*)

Christoph Mans, Dr med vet, Dipl ACZM

Session #3384

Affiliation: From the School of Veterinary Medicine, University of Wisconsin-Madison, 2015 Linden Drive, Madison, WI 53706, USA.

Note: Abstract format selected by author(s). See Front Matter for description of conference proceedings manuscripts submitted as scientific abstracts.

Abstract: The postprandial increase in blood uric acid levels is a well-documented phenomenon in carnivorous and piscivorous avian species but has received limited attention in reptile species. Recently it has been documented that cricket ingestion results in a significant increase in uric acid levels for at least 24 hours in inland bearded dragons if fed at 1% body weight equivalent;[1] however, the effect of other feeder insects or different amounts of food ingested has not been previously evaluated. Therefore, in a randomized, blinded, complete crossover study, the effects of superworm (*Zophobas morio*) ingestion (1 versus 2% bodyweight equivalent) on blood uric acid levels were evaluated in healthy inland bearded dragons. Blood was collected before feeding (baseline) and at 12, 24, 48, 73, and 96 hours after food ingestions. Ingestion of 2% bodyweight equivalent of superworms resulted in higher peak uric acid levels at 12 hours compared to 1%. Plasma uric acid values returned to baseline within 24 hours after 1% insect ingestion. In contrast, plasma uric acid values remained significantly elevated for at least 48 hours following ingestion of 2% insects. The findings of this and a previous study show that feeder insect ingestion in bearded dragons will result in a significant increase in plasma uric acid levels, and the magnitude and duration of the increase in uric acid levels are dependent on the amount and type of feeder insects ingested. Therefore, inland bearded dragons should be fasted for at least 72 hours before blood collection for routine blood screening.

References

1. Parkinson L, Mans C. Investigation of the effects of cricket-meal ingestion on plasma uric acid concentration in inland bearded dragons (*Pogona vitticeps*). *J Am Vet Med Assoc.* 2020;257:933-936.

Systemic Metastatic Calcification in Chinese Stripe-Necked Turtle (*Mauremys sinensis*)

Chu Yung Weng, DVM, and
Yen-Chi Chang, DVM, MS

Session #3116

Affiliation: From No. 15, Ziqiang 6th St., Zhubei City, Hsinchu County 302044, Taiwan, R.O.C.

Abstract: A 4-year-old Chinese stripe-necked turtle was presented in the hospital and diagnosed with follicular stasis. Biochemistry revealed uric acid 13.0 mg/dl, calcium 19.3 mg/dl, and phosphorus 8.9 mg/dl. X-ray revealed pneumonia and ultrasound showed multiple follicles which were hyperechoic. After treatment, the value of uric acid was 1.8 mg/dl, calcium 14 mg/dl, phosphorus 7.7 mg/dl; however, stiffness of joints and skin were noted, the neck became gradually inflexible, and eventually the patient died. Histopathology revealed systemic calcification in heart, follicles, thyroid gland and liver. In addition, no gout tophi were found under polarizer observation and renal failure was diagnosed. Metabolic disruption by processes such as metabolic acidosis or dehydration can cause drastic shifts in calcium to its active ionized form, resulting in soft tissue mineralization.

Introduction

Pathologic calcification can be separated into dystrophic calcification and metastatic calcification. The former means that calcium salts deposit in damaged or nonviable tissues. The latter refers to deposition of calcium salts in healthy tissues secondary to hypercalcemia or abnormal calcium metabolism. In reptiles, reasons for pathologic calcification include hypervitaminosis D, renal diseases, dehydration and follicular stasis in female reptiles. On the other hand, improper captive diet may cause secondary hyperparathyroidism and lead to tissue mineralization.

Case Report

A 4-year-old intact female Chinese stripe-necked turtle (*Mauremys sinensis*) presented to Mystery Exotic Animal Hospital (MEAH) in June 2022 for dystocia, anorexia and respiratory problems. Egg retention was diagnosed since May 2022. X-ray revealed pneumonia and one egg still in the reproductive tract. The turtle was treated with oxytocin intramuscularly at 20 IU/kg once, calcium gluconate subcutaneously at 100 mg/kg once, Ringer's solution for reptiles that consists of one part Normol-R plus two parts 2.5% dextrose in 0.45% saline subcutaneously at 10 ml/kg once, and enrofloxacin (Baytril®, **Bayer)** intramuscularly at 8 mg/kg twice a day. The patient failed to lay the egg, but treatment of pneumonia continued for four rounds. On June 24, 2022 the patient was lethargic, blood tests revealed uric acid was 13.0 (reference range, 0.6-1.9 mg/dl), ionic phosphorus was 8.9 (reference range, 2.5-5.8 mg/dl) and the PCV was 23%. Follicular stasis was diagnosed by ultrasound. Enrofloxacin was replaced with ceftazidime intramuscularly at 20 mg/kg every 3 days and nebulization with 1:250 diluted F10.

On July 9, 2022, the patient started to eat pellets and the egg was laid when the patient was being treated with nebulization. Biochemistry revealed uric acid was 8.0 (reference range, 0.6-1.9 mg/dl), ionic phosphorus was 6.3 (reference range, 2.5-5.8 mg/dl) and the PCV was 20%. Oral medication with allopurinol 20 mg/kg was prescribed for 14 days.

On July 26, 2022, the owner reported that the appetite of the patient had decreased. Ultrasound revealed hyperechogenicity of follicles. Ceftazidime was administered intramuscularly at 40 mg/kg every 3 days.

On August 19, 2022, edema was noted but the patient started to eat pellets. Hard masses were detected on bilateral femoral fossa and biochemistry revealed LDH was over 4000 (reference range, 939-4356 mg/dl), uric acid was 3.6 (reference range, 0.6-1.9 mg/dl), ionic phosphorus was over 20 (reference range, 2.5-5.8 mg/dl) and the PCV was 33.3%. Oral medicine was given to the patient which was allopurinol 20 mg/kg and aluminum hydroxide 30 mg/kg.

The patient returned on September 6, 2022. Edema was resolved and the animal displayed normal activity and appetite. Biochemistry revealed LDH was over 4000 (reference range, 939-4356 mg/dl), uric acid was 1.8 (reference range, 0.6-1.9 mg/dl), ionic phosphorus was 7.7 (reference range, 2.5-5.8 mg/dl) and the PCV was 29.1%. No follicles were seen by ultrasound.

On September 24, 2022, the owner reported that the patient could not move its limbs and green feces were noted. All joints were swelling and hard. X-ray showed that several hyper opacity nodules accumulated around the joints and digits. Pneumonia also recurred. The opacity of the shell became more radiolucent than X-rays were taken before. Biochemistry revealed LDH was over 4000 (reference range, 939-4356 mg/dl), uric acid was 1.5 (reference range, 0.6-1.9 mg/dl), ionic phosphorus was 5.6 (reference range, 2.5-5.8 mg/dl) and the PCV was 24%.

The patient was treated with meloxicam 0.2 mg/kg intramuscularly at 20 mg/kg every 2 days, normal saline subcutaneously at 10 ml/kg every 2 days, nebulization with 1:250 diluted F10 and arthritis protocol of laser therapy (CTX, Companion) was applied to the patient every 2 days. Eventually, the patient died on October 24, 2022. The whole fresh body was sent to the Animal Disease Diagnostic Center College of Veterinary Medicine, National Chung Hsing University (NCHU) for necropsy.

Gross lesions

The extremities, head and neck were extremely swollen and stiff, and the skin was thickened and hard. At necropsy, a large amount of dark red fluid was seen in the coelomic cavity. The extremities' muscles were pale; there were yellow-white, pasty to chalky material accumulating between muscle bundles. White materials mentioned above were also seen in the serosa of all organs in coelomic cavity, subcutaneous adipose tissue and articular cavity. The thyroid gland was pale and hard. No obvious gross lesions were seen in cerebrum and cerebellum.

Cytologic examination

Under Diff-Quik stain, the imprint of coelomic fat revealed abundant polymorphic basophilic crystals and a few macrophages. The polarity of the unstained imprint was weak under polarized light microscopy.

Histologic findings

1. Right hindlimb: Abundant basophilic, variably shaped, angular, fragmented crystals deposited around articular surface and subcutaneous area with large amount of eosinophilic fibrinous exudate. The foci were infiltrated with a large number of macrophages and multinucleated giant cells. Occasionally, there were a few fibroblasts surrounding the lesion.

2. Thyroid gland: Colloid was absent in all follicles. The same crystals mentioned above were deposited in the basement membrane of all follicles. In the myocardium of multiple areas, the same calcium deposits as mentioned above can be seen. The muscle fiber gap in the lesion area is expanded, and there is light eosinophilic filamentous protein fluid in between. Spherical basophilic crystals with a diameter of about 1-2 μm can occasionally be seen, and the muscle fibers are homogeneous, fractured or dissolved, occasionally multiple nuclei of different sizes aggregated.

3. Heart: Multifocally, the crystals deposited within or between the cardiac muscle fibers which were hyalinized, fragmented or ruptured. The intercellular space was expanded by large amount of eosinophilic, filamentous, proteinaceous fluid. Multinucleated giant cells were occasionally seen.

4. Liver: The hepatic cords of zone II and zone III were effaced by moderate number of hemosiderin laden macrophages and heterophils. The endothelium of sinusoid was thickened due to basophilic minerals deposition.

5. Kidney: The walls of Bowman's capsule and mesangial cells in the cortex are thickened due to calcium salt deposits, and there are multiple areas of expanded interstitial spaces in both the cortex and medulla, with infiltration of inflammatory cells mainly composed of lymphocytes. Additionally, there are fragment-like crystals and psammoma bodies, which are characterized by concentric circular appearance.

6. Muscle of extremities: Calcium salts deposited over 75% area of muscle including bundles, perimysium and endomysium with widened interstitial spaces between muscle fibers in the affected areas. The muscle fibers exhibit homogenization, fragmentation, or dissolution. Multinucleated giant cells were occasionally seen.

7. Stomach: Calcium salts accumulated in lamina propria, muscularis mucosae and muscle layer.

8. Lung: Multiple lesions with accumulation of calcium salts were seen in mesenchymal tissue and submucosa.

9. Ovary duct: Multiple lesions with accumulation of calcium salts were seen in muscle layer and submucosa.

10. Pancreas: Over 90% adenyl follicles were replaced by calcium salts.

11. Duodenum: Heterophils infiltration and eosinophilic fibrinous fluid.

12. Adrenal gland: Calcium salts accumulated in multiple areas.

13. Coelomic fat: Fat cells were replaced by calcium salts.

Diagnosis

A diagnosis was made of systemic metastatic calcification. A discussion and review are presented based on this finding.

References

1. Garner MM, Jacobson ER, eds. *Noninfectious Diseases and Pathology of Reptiles: Color Atlas and Text, Diseases and Pathology of Reptiles.* 1st Ed. Vol 2. Boca Raton, FL: CRC Press; 2020. https://doi.org/10.1201/9780429154317.

2. Girling SJ, Raiti P. *BSAVA Manual of Reptiles.* Gloucester, UK: British Small Animal Veterinary Association; 2019.

3. Mader DR, Divers SJ, eds. *Current Therapy in Reptile Medicine and Surgery.* St. Louis, MO: Elsevier Saunders; 2014.

4. Divers SJ, Stahl S, eds. *Reptile Medicine and Surgery.* 3rd Ed. St. Louis, MO: Elsevier Saunders; 2016.

Hot Debate: Does Short-Term Artificial UVB Lighting Affect Plasma Vitamin D Concentrations in Leopard Geckos?

Amy Bitter,
Aspen Settle,
John Tuminello,
Mark A. Mitchell, DVM, MS, PhD, DipECZM (Herpetology)

Session #3133

Affiliation: From the Department of Veterinary Clinical Sciences, School of Veterinary Medicine, Louisiana State University, Baton Rouge, LA 70803, USA.

Note: Abstract format selected by author(s). See Front Matter for description of conference proceedings manuscripts submitted as scientific abstracts.

Abstract: Leopard geckos (*Eublepharis macularius*) are insectivorous lizards that are native to the rocky grassland and desert regions of the Middle East. This species is one of the most common reptiles kept in captivity, but there remains much we don't know about their basic husbandry. One of the most common diseases reported in captive leopard geckos is nutritional secondary hyperparathyroidism (a form of metabolic bone disease). This disease is preventable but is typically attributed to diets with low calcium/high phosphorus, low vitamin D, or a lack of ultraviolet B (UVB) exposure. Since leopard geckos are insectivorous, most commercially available feeder insects are low in vitamin D and calcium when not gut-loaded appropriately. Because they are a crepuscular species, it was long believed that they didn't require UVB lighting in their husbandry; however, recent work by our laboratory has shown that UVB exposure does increase plasma 25-hydroxyvitamn D_3 (25-OHD_3) concentrations in this species. Unfortunately, too much UVB exposure can lead to squamous cell carcinoma and cataracts. The purpose of this study was to determine the minimal amount of UVB exposure required to increase 25-OHD_3 concentrations in leopard geckos. Twenty-four leopard geckos were used for this study. The geckos were randomly assigned to three different UVB exposure groups: 15, 30, and 60 minute/daily UVB exposure. Each gecko served as its own control, and the geckos were exposed to the UVB for 28 days. The mean baseline 25-OHD_3 concentration in the geckos was 29 nmol/L. There was a significant difference in 25-OHD_3 over time ($P < 0.001$), with 25-OHD_3 concentrations being higher in all geckos at 28 days compared to baseline. No significant difference in 25-OHD_3 was found between groups. Because there were no differences between groups, reference intervals were established for baseline 25-OHD_3 concentrations in leopard geckos offered a low vitamin D diet and for being exposed to UVB for 15-60 minutes/day for 28 days.

Disclosure of support: Fluker Farms, Boehringer Ingelheim Summer Scholars.

Untargeted Primary Metabolomics for Biomarker Discovery in Bearded Dragons (*Pogona vitticeps*) with Hepatic Lipid Accumulation

Lisa Pacumio, BS,
Kevin Keel, DVM, PhD, DACVP,
Mélanie Ammersbach, DVM, DACVP,
Danielle Tarbert, DVM, Dipl ACZM,
Hugues Beaufrère, DVM, PhD, Dipl ACZM, DABVP (Avian Practice), DipECZM (Avian)

Session #3249

Affiliation: From the Department of Veterinary Medicine and Epidemiology (Pacumio, Beaufrère) and Department of Pathology, Microbiology and Immunology (Keel, Ammersbach), School of Veterinary Medicine, University of California-Davis, Davis, CA 95616, USA, Department of Small Animal Clinical Sciences, College of Veterinary Medicine, University of Tennessee, TN 37996, USA.

Note: Abstract format selected by author(s). See Front Matter for description of conference proceedings manuscripts submitted as scientific abstracts.

Abstract: Increased hepatic triacylglycerol and hepatic lipidosis are common in captive bearded dragons with an estimated prevalence of 27% of advanced lesions on necropsy.[1] Preliminary investigations on a small heterogeneous cohort of 14 bearded dragons, with varying degrees of hepatic lipid accumulation, identified several primary metabolites as candidate biomarkers such as β-hydroxybutyric acid (BHBA) and succinic acid.[2] The purpose of this study was to further screen primary metabolites for potential associations with hepatic lipid content. A cohort of 48 bearded dragons of similar age range was used. They were anesthetized with subcutaneous alfaxalone and a blood sample was taken from the caudal tail vein. Plasma was submitted for untargeted primary metabolomics screening for more than 200 metabolites using gas chromatography time-of-flight mass spectrometry (GCTOF-MS), a routine reptile biochemistry panel, and a lipoprotein panel determined by high-resolution polyacrylamide gel electrophoresis. Hepatic lipid content was quantified by liver attenuation measurements in Hounsfield units from computed tomography images of both liver lobes and digital image analysis of standardized histologic sections of the liver. Fibrosis was quantified by digital image analysis on Masson's trichrome stained histologic sections. Data filtering and normalization was performed prior to assessing statistical associations using serial linear models with a false discovery rate of 5% and multivariate statistics. Preliminary analysis found that several oligosaccharides (maltotriose, maltose, ribose) and ALP were significantly and linearly increased with hepatic lipid content (all $q < 0.05$). Succinic acid and BHBA were not significant biomarkers, invalidating results of a previous pilot study.

References

1. Barboza T, Susta L, Reavill D, Beaufrère H. Prevalence and risk factors of hepatic lipid changes in bearded dragons (*Pogona vitticeps*). *Vet Pathol.* 2023;60(1):123-132.

2. Barboza T, Susta L, zur Linden A, et al. Association of plasma metabolites and diagnostic imaging findings with hepatic lipidosis in bearded dragons (*Pogona vitticeps*) and effects of gemfibrozil therapy. *Plos ONE.* 2023;18(2):e0274060.

Measuring Effects of Diet and Ultraviolet B Exposure on 25-Hydroxyvitamin D Concentrations in Superworms (*Zophobas morio*)

Zoey Lex, BS,
Elise Barras DVM, MS,
John Tuminello, BS,
Graciela Aguilar, DVM,
Kimberly Boykin, DVM, MS,
Mark Mitchell, DVM, MS, PhD, DipECZM (Herpetology)

Session #3373

Affiliation: From the Department of Veterinary Clinical Sciences, School of Veterinary Medicine, Louisiana State University, Skip Bertman Drive, Baton Rouge, LA 70803, USA.

Note: Abstract format selected by author(s). See Front Matter for description of conference proceedings manuscripts submitted as scientific abstracts.

Abstract: Vitamin D is an important nutrient in vertebrates due to its role in calcium and phosphorus homeostasis. In vertebrates, vitamin D can be synthesized through ultraviolet B (UVB) exposure, obtained from the diet, or both. For most insects fed to vertebrates, vitamin D is introduced in the diet via gut-loading. To date, there has been limited study on the cutaneous synthesis of vitamin D in insects; however, recent work suggests some invertebrates, including mealworms (*Tenebrio molitor*), synthesize vitamin D following UVB exposure.[1] The purpose of this study was to determine if UVB exposure produces higher concentrations of 25-hydroxyvitamin D (25-OHD) in superworms (*Zophobas morio*). Our hypothesis, based on the mealworms, was that 25-OHD would be higher in superworms exposed to UVB compared to controls. Superworms (n = 500) were randomly and equally divided into four groups: a control group and three experimental groups. Experimental groups were exposed to UVB light for 8 hours per day. All groups were provided the same amount of substrate (Fluker's High Calcium Mealworm Diet, Port Allen, LA) daily. Superworms (15 g/sample) were collected from each group at baseline, 2, 4, and 6 days; worms were humanely euthanized; and 25-OHD concentrations were measured using liquid chromatography-mass spectrometry. There were no significant differences in 25-OHD concentrations between the control and treatment groups; however, there was a steady increase in 25-OHD concentrations in all groups until day 4, which was correlated to diet. These findings suggest there are differences among insects that should be considered when developing diets for vertebrates.

References

1. Oonincx DG, Van Keulen P, Finke MD, et al. Evidence of vitamin D synthesis in insects exposed to UVb light. *Sci Rep*. 2018;8:10807. doi:10.1038/s41598-018-29232-w.

Comparison of Shell Mineral Content Between Captive- and Wild-Bred Eastern Indigo Snakes (*Drymarchon couperi*)

Alexandra K. Mason, DVM, MS, CertAqV,
James Bogan, DVM, DABVP (Canine and Feline Practice), DABVP (Reptile and Amphibian Practice), CertAqV,
Nick Clark,
Mark A. Mitchell, DVM, MS, PhD, DipECZM (Herpetology)

Session #3250

Affiliation: From Central Florida Zoo & Botanical Gardens, 3755 W Seminole Blvd, Sanford, FL 32771, USA (Mason, Bogan), Central Florida Zoo's Orianne Center for Indigo Conservation, 30931 Brantley Branch Road, Eustis, FL 32726, USA (Mason, Bogan, Clark), Louisiana State University, School of Veterinary Medicine, Department of Veterinary Clinical Sciences, Skip Bertman Drive, Baton Rouge, LA 70803, USA (Mitchell).

Note: Abstract format selected by author(s). See Front Matter for description of conference proceedings manuscripts submitted as scientific abstracts.

Abstract: Eastern indigo snakes (*Drymarchon couperi*) (EIS) are large, nonvenomous, colubrids listed as threatened due to habitat loss, fragmentation and decline of gopher tortoise populations.[1-4] The Orianne Center for Indigo Conservation (OCIC) is the only captive breeding facility for the EIS that aims to reintroduce captive-bred snakes into the wild to re-establish them to their former range. Dystocia is well-documented in captive EIS, and differences in plasma biochemical parameters have been noted to be significantly different between successfully oviposited and dystocia snakes. The objectives of this study were to analyze the mineral content of EIS eggshells between four groups of snakes (wild-caught Florida snakes, wild-caught Georgia snakes, captive-bred snakes, and shells from dystocia clutches) for 13 minerals. Four clutches of dried, previously hatched or surgically removed and stored eggshells were randomly selected for evaluation from each EIS group, and three eggshells per each clutch were selected for analysis on a dry matter basis by spectroscopy at Michigan State Veterinary Diagnostic Laboratory (Lansing, MI). Preliminary findings are presented. Data are assessed for normality and if normally distributed, a one-way analysis of variance (ANOVA) determines if differences exist between mineral composition ± least significant difference tests if the ANOVA is significant. Non-normally distributed data are analyzed via Kruskal-Wallace test ± Mann Whitney U test to determine the differences in mineral composition by treatment group. A $P < 0.05$ is used to determine statistical significance.

References

1. Bogan JE, Hoffman M, Dickerson F, et al. A retrospective study of dystocia in Eastern indigo snakes (*Drymarchon couperi*). *J Zoo Wildl Med.* 2021;52(2):618-627.

2. Delaski KM, Gamble KC, Bernier D, et al. Shell mineral composition of artificially and naturally incubated eggs of ornate box turtles (*Terrapene ornata ornata*). *J Herpetol Med Surg.* 2014;24(3-4):101-104.

3. Osborne L, Thompson MB. Chemical composition and structure of the egg shell of three oviparous lizards. *Copeia.* 2005(3):683-692.

4. U.S. Fish and Wildlife Service. Recovery plan for the Eastern indigo snake first revision. Athens, GA: U.S. Fish and Wildlife Service, Georgia Ecological Services Field Office; 2019.

Hypervitaminosis D in Three Male Bearded Dragons (*Pogona vitticeps*)

Amanda D. Wong, DVM,
Maural Sowlat,
Danielle M. Lang, DVM,
Aubrey I. Engel, DVM,
Michelle C. Borsdorf, DVM, Dipl ACZM,
Krista A. Keller, DVM, Dipl ACZM

Session #3420

Affiliation: From the Department of Veterinary Clinical Medicine, College of Veterinary Medicine, University of Illinois at Urbana-Champaign, Urbana, IL 61802, USA.

Note: Abstract format selected by author(s). See Front Matter for description of conference proceedings manuscripts submitted as scientific abstracts.

Abstract: This case series describes hypervitaminosis D in three male bearded dragons. Two cases involve 9-year-old and 4-year-old bearded dragons primarily evaluated for malaise and hyporexia. The third case involves a 3-year-old bearded dragon referred for evaluation and treatment of cutaneous mycosis. In all cases, total hypercalcemia detected during plasma biochemical evaluation prompted evaluation of ionized calcium, 25-hydroxyvitamin D, and parathyroid hormone concentrations in the serum. All cases had serum 25-hydroxyvitamin D levels (920, 1150, 1160 nmol/L) that were elevated based on previously published experimental values in the species.[1,2] One bearded dragon had concurrent ionized hypercalcemia (1.87 mmol/L) based on published reference values extrapolated from green iguanas.[3] Sources of vitamin D in all cases included basking sites with appropriate ultraviolet B lighting and variable historic dietary supplementation with oral vitamin D_3. Dietary over supplementation is suspected to be the cause of excess vitamin D in all cases. All three animals are alive at the time of writing (5 months, 2 months, and 1 month from diagnosis) and under medical management to include removal of oral sources of vitamin D, reduction in UVB access, and supportive measures. Hypervitaminosis D is suspected to be underdiagnosed in companion bearded dragons as hypercalcemia diagnosed in females may be presumed to be due to a folliculogenic state. Submission of a vitamin D profile should be considered in a bearded dragon with total hypercalcemia.

References

1. Oonincx DG, Stevens Y, Van den Borne JJ, et al. Effects of vitamin D_3 supplementation and UVb exposure on the growth and plasma concentration of vitamin D_3 metabolites in juvenile bearded dragons (*Pogona vitticeps*). *Comp Biochem Physiol A Mol Integr Physiol.* 2010;15:122-128.

2. Oonincx DG, Van De Wal MD, Bosch G, et al. Blood vitamin D_3 metabolite concentrations of adult female bearded dragons (*Pogona vitticeps*) remain stable after ceasing UVb exposure. *Comp Biochem Physiol A Mol Integr Physiol.* 2013;165:196-200.

3. Dennis PM, Bennett RA, Harr KE, et al. Plasma concentration of ionized calcium in healthy iguanas. *J Am Vet Med Assoc.* 2001;219:326-328.

Effects of Three Different Dietary Calcium to Phosphorus Ratios on Bone Density in Ex Situ Critically Endangered Lemur Frogs (*Agalychnis lemur*)

Graciela Aguilar, DVM, and
Mark A. Mitchell, DVM, MS, PhD, DipECZM (Herpetology)

Session #3440

Affiliation: From Veterinary Clinical Science Department, School of Veterinary Medicine, Louisiana State University, Baton Rouge, LA 70802, USA.

Note: Abstract format selected by author(s). See Front Matter for description of conference proceedings manuscripts submitted as scientific abstracts.

Abstract: In the current Anthropocene era, amphibians are one of the most affected taxa, and *ex situ* conservation action plans are vital to preserving its most vulnerable species. The lemur frog (*Agalychnis lemur*), a native to Central America that used to be common in Costa Rica, is now listed by the International Union of Conservation of Nature as critically endangered due to these anthropogenic pressures. Managing these animals in captivity represents a challenge, and nutritional diseases, such as secondary nutritional hyperparathyroidism (SNHP), remain a common obstacle to our long-term success in managing this species *ex-situ*. SNHP in insectivores is primarily attributed to low dietary calcium in commercially available insects. House crickets (*Acheta domesticus*) typically have a 0.15:1 calcium to phosphorus (Ca:P) ratio. A positive ratio, preferentially a > 1.5 ratio, is required for vertebrates to ensure proper growth and reproductive success. To date, no study has evaluated how dietary Ca:P ratios impact bone development in lemur frogs or other arboreal frog species. This study aimed to determine how crickets gut-loaded with different Ca:P ratios (0.15:1[control], 2.5:1) diets affect bone density in 12 juvenile lemur frogs. The study's hypothesis was that the Hounsfield units (bone density) measured over time using computed tomography (CT) scans would be significantly lower in the animals fed crickets with a Ca:P ratio of 0.15:1 (control) compared to 2.5:1. CT scans were performed at baseline and 80 days after the animals received the diets. This study represents the first attempt to determine the impact of dietary Ca:P on bone density in critically endangered lemur frogs.

3D Printer and Polylactic Acid Use for Shell Repair in Turtles and Tortoises

Cristina Almeida,
J. Lozano,
Miguel Ramos

Session #3356

Affiliation: From Clínica de Aves e Exóticos de Lisboa, Rua Dom António Ribeiro n1 loja 2 1495-049 Algés Portugal (Almeida), Centre for Interdisciplinary Research in Animal Health, Faculty of Veterinary Medicine, Lisboa, Portugal (Lozano), Hospital Escolar Veterinário, Faculty of Veterinary Medicine, Lisboa, Portugal (Ramos).

Abstract: Shell injuries are a common presentation in captive turtles. Patients should be immediately assisted, and prognosis will depend upon the severity of injuries. There are many methods of fracture fixation and the decision for which method to use is based on the characteristics of the fracture, the veterinarian's familiarity with the method as well as the time and money available for the case. In this lecture we present the use of polylactic acid shell repair using 3D printing as an alternative method for both turtles and tortoises. The applications of 3D have been well known in medicine and dentistry since the 1990s. Some cases have been reported in cats and dogs making customized implants. In this case we present the steps and clinical outcome in injured tortoises and turtles with 3D implant repair as an alternative for the classical approaches.

Introduction

Shell injuries are a common presentation in turtles, and they should be provided with immediate analgesia and supportive care followed by fracture fixation if required. There are many methods of fracture fixation and the decision for which method to use is based on the characteristics of the fracture. In recent years, 3D printing products have been used for different purposes, thanks to methods that allow 3D image configuration such as computed tomography and magnetic resonance imaging, which are increasingly used in veterinary medicine. The 3D printing process can be accomplished using any of several different technologies. The choice of technology can depend on many factors including how the final product will be used and how easy the printer is to use.[1]

Materials such as polylactic acid (PLA), polyamide (PA, nylon), polycarbonates (PC), acrylonitrile butadiene styrene (ABS) can be used as raw materials in 3D printing products that will not be implanted into the organism.[2] Some biopolymers, bioglass and bioceramics, metals and alloys can be used as 3D printing raw materials in medical treatment applications.[3] It is possible to use more than one material and color. In addition, a wide variety of cellular materials can be used in the production of artificial tissues and organs as materials in bioprinters.[4]

While traditionally manufactured medical devices come in discrete sizes, patient-matched devices can be made in a continuous range of shapes. For instance, the specification may define a minimum and maximum wall thickness or how sharp a curve can be to maintain device performance for its intended use. Soares et al. (2018) achieved successful results in a cat with chronic oronasal fistula by applying a 3D printed patient specific soft tissue flap (mesh).[4] In another study, the bronchial structures of a Savanah monitor lizard and a Nile crocodile were 3D printed.[2,5] 3D printing was used to take advantage of advanced surgical planning and other advantages in the treatment of orbital and periorbital masses in three dogs.

Clinical Case

A clinical case is presented where a tortoise was injured with a bite wound. Complete blood work and x-ray studies were performed. Because there was an opened defect, the possibility of a 3D printed shell cover was investigated. With limited experience, the authors developed a local partnership with a dental medical practice. Measurements were taken and multiple models were provided. The principles were no secondary infection, better regeneration, and some aesthetical concerns. Also, the 3D model was removable making available the natural healing and the patient was monitored every week for several weeks and there were no secondary infections.

Conclusions

In conclusion, with 3D printing technology, products can be made in many areas and with various materials. This technology has been used in veterinary clinical sciences. Positive and effective results have been obtained, recommended, and are still developing with intensive study. This technology will be more common in the future.

Acknowledgments: Authors thank Maia, Clínica Dentária.

References

1. Deveci MZY, Altuğ M., İşler CT, et al. Three-dimensional printing applications in veterinary surgery. *J Adv VetBio Sci Tech.* 2022;7(1):130-142. doi.org/10.31797/vetbio. 983275.

2. Hespel A-M. 3D printers their clinical, experimental, and teaching uses. Master's Thesis: Auburn University. 2015.

3. Arslan N, Yaylacı B, Eyüpoğlu N, et al. Sağlıkta gelişen teknoloji: üç boyutlu yazıcılar. *Int J 3D Printing Technol Digital Ind.* 2018;2(2):99-110.

4. Soares CS, Barros LC, Saraiva V, et al. Bioengineered surgical repair of a chronic oronasal fistula in a cat using autologous platelet-rich fibrin and bone marrow with a tailored 3D printed implant. *J Feline Med Surg.* 2018;20(9):835-843.

5. Jamieson CD, Keenan P, Kirkwood DA, et al. A review of recent advances in 3D bioprinting with an eye on future regenerative therapies in veterinary medicine. *Front Vet Sci.* 2021;7:947. doi:10.3389/fvets.2020.584193.

6. Dorbandt DM, Joslyn SK, et al. Three-dimensional printing of orbital and peri-orbital masses in three dogs and its potential applications in veterinary ophthalmology. *Vet Ophthalmol.* 2017;20(1):58-64. doi:10.1111/vop.12352.

Section 37

ARAV Masterclasses Track 1

Paul Gibbons, DVM, MS, DABVP (Reptile and Amphibian Practice)
Session Chair

The Amphibian and Reptilian Liver: Normal Structure and Function

Jeanette Wyneken, PhD

Session #3235

Affiliation: From Florida Atlantic University, Boca Raton, FL 33431-0991, USA.

Abstract: The liver is an essential organ of the digestive system and is an integral part of the endocrine and hematopoietic systems. It is an intrinsic digestive gland in that it forms as a diverticulum off the developing intestine. The liver is a large visceral organ located in the caudal coelom; it is lobed in most taxa. It serves as a digestive gland, forming and releasing bile into the duodenal lumen. The hepatocytes and macrophages (including Kupffer cells) do much of the liver's work. Melanomacrophages provide pigmented protection from free radicals. The liver performs many other critical functions including forming red blood cells in embryos, storing iron, is the location of carbohydrate metabolism synthesizing glucose and storing sugars as glycogen. The liver is highly vascular receiving venous and arterial blood from hepatic portal vein, abdominal veins, and hepatic arteries and delivering processed blood via the postcava. Blood percolates through the hepatic sinusoids and cells filter blood and destroy damaged RBCs. Albumin, the most abundant circulating protein found in plasma, is formed in the liver, as are fibrinogen and the prothrombin group of clotting factors. The liver also detoxifies substances. Hepatocytes produce enzymes that combine toxins with amino acids so they can be excreted. During preparation for reproduction, the liver is responsible for fat mobilization for vitellogenesis. Bile acid metabolism in the form of biliverdin, in most reptilian and amphibian species, breaks down fats. Protein metabolism from the liver process amino acids and peptides to form proteins. Hormone metabolism occurs in the liver such that hormones (e.g., adrenalin, growth hormone, progesterone) are removed from circulation.

Introduction

This overview of amphibian and reptilian liver covers normal gross anatomy of the liver and gallbladder, hepatic products, and hepatic functions. The liver is one of two glands (the pancreas is the other) that form as direct out-pocketing from the developing digestive tube.[1] Fundamentally, it is an essential part of the digestive system. It also functions in hormone metabolism, protein metabolism, production or removal of some of bloods components, and in detoxification of intrinsic and extrinsic compounds.

Gross Anatomy of the Liver and Gall Bladder

Liver

The liver is the body's largest visceral organ and, along with the gallbladder and ducts, forms the hepatobiliary system. In tadpoles and hatchlings, the liver is yellowish orange and changes to reddish or reddish-brown as yolk stores are depleted and processed. It is reddish-brown to brownish-black in juveniles and adults. In amphibians the liver is lobed; the number of lobes varies from one to five depending on species with two lobes being a common morphology.[2-4] In most lizards, chelonians, and crocodilians the liver lobes are roughly pyramidal in shape, usually there are two lobes although exceptions are not uncommon. Some lizards have a single ventro-median lobe or right lobe only. Some chelonians have a small dorso-median third lobe that appears to be expansion to the dorsal bridge between lobes. The liver of snakes, particularly in boid snakes, is a single structure usually located along the left side of the body and adjacent to the right lung. In some species (e.g., garter snakes [*Thamnophis sirtalis*] and blind snakes [Xenotyphlopidae, *Typholops*, *Leptotyphlops*]), two asymmetric lobes form. The left lobe is large and long while the right lobe is small, short, and flattened.[2,5-7] The gallbladder collects bile from the liver and is present in all reptiles and anmphibians.[8] The gallbladder is a single structure, located between the liver's lobes in most amphibians and reptiles or embedded in the right lobe. In snakes, it is located at the distal end of the liver, often closely associated with the pancreas, and is in contact with the adjacent intestine.[9]

The liver is attached to the coelom's transverse septum via the coronary ligament. Ventrally, the liver attaches to the ventromedial coelomic wall (i.e., parietal peritoneum) via a short falciform ligament. The right lobe attaches to the right lung via a short hepatopulmonary ligament. The left lobe, when present, attaches to the stomach via the gastrohepatic ligament[10-12]

The liver is highly vascular. It receives the largest proportion of venous blood via the hepatic portal vein and the remainder from the hepatic artery.[1] The ventral abdominal vein joins the hepatic portal vein, usually in the abdomen, but may join it shortly after it enters the liver. Blood that has filtered through the parenchyma enters multiple hepatic veins that drain into the postcava.[9-11] The number of hepatic veins varies among taxa. Box turtles (*Terrapene carolina*) and the agamid forest lizards (*Calotes calotes*) have few short hepatic veins. In contrast, most snakes have a single long hepatic vein and Western fence lizards (*Sceloporus occidentalis*) have up to 60 from the right lobe and one from the left.[9]

Organization of liver parenchyma

The classic understanding of liver parenchymal organization based upon mammalian studies and older literature comparing amphibians and reptiles to mammals does not accurately reflect the hepatic architecture of amphibians and reptiles. Recent studies of amphibians and reptiles that draw upon modern imaging techniques allow for more effective comparisons with mammalian studies. The parenchyma is composed of hepatocytes, which function in digestion, macrophages that provide for immune function and detoxification, blood vessels that transport blood into and through the liver, and bile ducts that collect bile. Blood flow into the liver comes from the intestines via the hepatic portal vein, the ventral abdominal veins, and the hepatic artery and leaves via hepatic veins to the inferior vena cava.[1,4,9]

The organization of the parenchyma provide the functional relationships between the plates or trabecular cords of hepatocytes and blood flow within the liver. There are fewer studies with reptiles than with amphibians. Consequently, this area is ripe for functional anatomic studies. The liver of lower vertebrates is described as layered, plates or sheets of hepatocyte cords, net-like, trabecular, or tubular glands. A comparative study of parenchyma organization across amphibian orders found that anurans have two-cell-thick plate type and one-cell-thick plate organization of the sinusoids, caecilians (Gymnophiona) have one-cell-thick plate organization and salamanders and newts (Caudata) vary within and among taxa with two-cell-thick plate type and one-cell-thick plate organization of the sinusoids and others have several-cell-thick plate type and two-cell-thick plate type sinusoidal structure.[2] While reptilian hepatocytes cords and their sinusoidal architecture share some similarities with mammalian counterparts, they are organized as tubes within sheets or trabeculae, not lobules as in mammals.[2,4] Further, in addition to hepatocytes, there are at least three cell types and histologic levels of organization contributing to reptilian sinusoids that are not typical in mammals: (i) the sinusoid lumen is lined by endothelium; (ii) macrophages include Kupffer cells found within the lumen and attached to the endothelium plus melanomacrophages occur perisinusoidally; and (iii) stellate cells (Ito cells) and lipocytes, found between the endothelium and hepatocytes that sequester fat droplets that are most prevalent in chelonian and crocodilians.[4,9]

Gallbladder and ducts

The sac-like gallbladder is present in all amphibians and reptiles surveyed to date. In amphibians and reptiles, the gallbladder is located in, or along, the posteroventral aspect of the liver's right lobe.[13] In snakes and elongated-bodied amphibians and reptiles, the gallbladder tends to be located at the distal end of the liver and adjacent to the intestine.[2,13] It may be turgid in individuals that have not fed recently or flaccid in those that have recently had a meal.

In the African clawed frog (*Xenopus laevis*) the wall is formed columnar epithelium, fibromuscular layer, sub-serosa and serosa with three to four layers of blood vessels forming the vascular bed of the gallbladder wall. In the common toad (*Bufo americanus*), epithelial cells tend to be cuboidal and apical ends bulge associated with mucus production. The gallbladder wall has a thin fibrous lamina propria and blood vessels deep to the epithelium with smooth muscle fibers external to the lamina propria.[2]

The duct system that connects the gallbladder, liver, and intestine may vary. In addition, the literature is conflicting, likely due to casual usage of the terms hepatic duct and bile duct. Table 1 provides some definitions. In the bullfrog (*Rana catesbeianus* or *Lithobates catesbeianus*), a common bile duct passes through the pancreas and enters the intestine next to the pancreatic duct. In the painted turtle (*Chrysemys picta*), the hepatic ducts coalesce and directly join the gallbladder; bile from the gallbladder is transported to the duodenum directly by a cystic duct.[11,13] In the freshwater side-neck turtle (*Phrynops geoffroanus*), a common bile duct extends from, at least, the gallbladder which is embedded with the right lobe, to the duodenum.[14] In other reptiles, a common bile duct connects the cystic duct and the hepatic duct to the duodenum. In the green iguana (*Iguana iguana*), multiple bile ducts drain bile from the gallbladder to the duodenum.[2] The liver may be connected directly to the duodenum (as in *Varanus* spp.) via a hepatico-interior canal network along the surface of the gallbladder or from the live and gallbladder via common bile duct.[15] In crocodilians, a canal draining the gallbladder, crosses the pancreas and attaches to the intestine.[5]

Description of the arterial and venous circulation to and from the gallbladder are very limited. In anurans, one or two cystic arteries off the right hepatic artery supply the gallbladder and one or two cystic veins drain from the gallbladder to the abdominal vein.[16]

Table 1. Duct system descriptions.

Duct name	Duct landmarks (may vary with taxon)	Drainage
Bile duct[a]	Liver to cystic duct	Bile from liver lobes to cystic duct
Common bile duct	Bile ducts to duodenum	Bile from liver and gallbladder to intestine
Cystic duct	Common bile duct or bile ducts to gallbladder	Common bile duct
Hepatic duct	Ductules where bile is secreted between hepatocyte cords out flow coalescence	Bile from ductules to bile ducts

[a]Used by several authors as a general term for any duct draining bile from the liver to the gallbladder or to the intestine.

Functions of the Liver and Gall Bladder

Liver

The liver's functions in amphibians and reptiles are diverse. While the multiple hepatocyte functions have not been extensively documented in amphibians and reptiles, it is likely that the function in roughly similar ways as they do in mammals due to similar essential homeostatic processes in all vertebrates. As an exocrine organ, the liver is critical for digestion. Bile, produced by the hepatocytes, is delivered to the duodenum where it emulsifies fats as part of digestion.

Digestion, fats and sugars

Fats made available during digestion are taken up by stellate cells (i.e., Ito cells) and lipocytes. Fats and glycogen in the liver are high prior in species that hibernate just prior to undergoing this metabolic and behavioral torpor and deplete during hibernation.[9,10] The liver is responsible for storage vitamin A, D, E, K and B12. Vitamins A, D, E, and K are fat soluble. Consequently, bile secreted during digestion is essential for their absorption. If bile production or release is compromised, the proper absorption of these vitamins may be affected. Fat droplets in hepatocytes are stored as esters of preformed vitamin A. Stellate cells and lipocytes sequester fat droplets.[9] Vitamin A is critical for vision, serves as an antioxidant, functions as a developmental signaling cofactor, affects gene expression in skin, GI and respiratory linings, and more.

Key functions of liver include responsibility for glucose metabolism converting excess glucose to glycogen and as a glycogen storage organ (alongside muscle). The liver metabolizes glycogen to provide glucose when blood glucose is low.[9]

Bile

Bile salts are the major end metabolites of cholesterol and are also important in lipid and protein digestion, as well as shaping of the gut microflora.[13] Hoffman et al.[12] describe bile salts and define them as bile alcohol sulfates and conjugated bile acids. The liver drives the breakdown of cholesterol resulting in bile salts that are synthesized in the hepatocytes. Amphibians primarily produce bile alcohols and some bile acids. Reptiles primarily produce bile acids.[12,13] The path that bile salts take after secretion into the proximal intestine is described as a circulation. They are reabsorbed in the distal intestine and returned to the liver in blood, where they are removed and then resecreted by the hepatocytes as bile. Effectively the bile salt pool circulates between the intestine and the liver.[12,13]

Proteins and metabolites

Some proteins are processed by the liver and are removed or recycled. Breakdown of nucleic acids produces nitrogenous wastes as uric acid, urea, and ammonia. The liver plays a role in urea formation and uric acid which is then transported to the kidney's renal tubules where it is excreted. Seasonally upregulated hormones including estrogens, progesterone, and thyroid hormones are removed or recycled by the liver.[8,9]

The liver is the source of albumin which is responsible for transporting vitamins, hormones, and enzymes throughout the body. Albumin makes up approximately 50% of the proteins found in plasma.[1,2,9]

Blood elements and coagulation

Unlike mammals, blood element production (hematopoiesis) is tightly tied to the liver in the amphibia. Species differ in the extent of liver-based hematopoiesis; species such as axolotls (*Ambystoma mexicanum*) that retain regenerative capacity throughout life tend to have greater liver-based hematopoiesis and little is based in the limited bone marrow. Studies with axolotls and the African clawed frog (*Xenopus laevis*) found that peripheral liver is host to most erythroid lineage hematopoiesis and bone marrow tended to have greater myeloid hematopoietic transcription factors.[17,18] Anurans tend to have reduced hepatic blood element formation after metamorphosis. Reptiles have greater bone marrow-based hematopoiesis and less peripheral hepatic contribution. Key functional coagulation components, including fibrinogen, prothrombin, and factors V, VII, VIII, IX, (perhaps X), and XI (if present) are formed in the liver.[9] Recent surveys of vertebrates for the contact coagulation factors show the loss of genes for XI and XII in sea turtles and venomous snakes.[19]

Hormones

In female amphibians and reptiles, the liver is the site of vitellogenin synthesis in response to estrogens. Vitellogenin is transported in the blood to the ovaries where is forms lipovitellin and phosvitins (protein and lipid compounds) in yolks of the amphibian and reptile eggs.[20]

Immunology

Some immunologic cells are liver-associated. Melanomacrophage are abundant in, at least, chelonian and caudate livers. These cells function in the synthesis of melanin, phagocytosis, and neutralization of free radicals.[9,21] Comparisons of three amphibian orders showed similarities in immunologic function in Caudata (salamanders), Gymnophiona (caecilians) but not anurans (frogs and toads). In newt (*Notophthamus viridecens*) liver, lymphocytes form in the connective tissue region associated with hepatic portal drainage.[17,18]

Detoxification

Beyond detoxification of nitrogenous breakdown products (ammonia, urea, and uric acid), detoxification literature has, until relatively recently, focused on identification of abnormal end products. Often the literature simply states that the liver functions in detoxification and identifies biomarkers that indicate toxin exposure. For example, in amphibians, liver estrogen receptor assays can identify estrogen disruption factors.[21] The addition of molecular biological tools is beginning to increase understanding. A recent transcriptome analysis of chemical stress in an anuran highlighted the effectiveness of such an approach.[22].

Liver regeneration

There is limited evidence of hepatocyte regeneration in amphibians and no support for regeneration in reptiles.[23]

Gallbladder

The gallbladder functions as a reservoir and mechanical pump for bile from the liver. It is an organ of the liver that stores and concentrates bile via H_2O removal prior to delivery to the digestive tract. It also exchanges electrolytes, modifies, and actively secretes bile, and circulates bile through the enterohepatic circulation such that bile acids move from the liver to the small intestine and return to the liver.

Bile aids the digestive uptake of lipid-soluble compounds.[5]

References

1. Kardong KV. *Vertebrates: Comparative Anatomy, Function and Evolution.* 8th Ed. New York, NY: McGraw-Hill; 2018.

2. Jacobson E. *Infectious Diseases and Pathology of Reptiles: Color Atlas and Text.* Boca Raton, FL: CRC Press; 2007.

3. Reese AM. *The Alligator and Its Allies.* New York, NY: GP Putnam and Sons; 1915.

4. Akiyoshi H, Inoue AM. Comparative histological study of hepatic architecture in the three orders amphibian livers. *Comp Hepatol.* 2012;11(1):1-8.

5. Oldham-Ott CK, Gilloteaux J. Comparative morphology of the gallbladder and biliary tract in vertebrates: variation in structure, homology in function and gallstones. *Microsc Res Tech.* 1997;38(6):571-597.

6. Anderson GE, Secor SM. Ontogenetic shifts and spatial associations in organ positions for snakes. *Zoology.* 2015;118(6):403-412.

7. Perez D, Sheehy III CM, Lillywhite HB. Variation of organ position in snakes. *J Morphol.* 2019:280(12):1798-1807.

8. Schaffner F. The liver. In: Gans C, Gaunt A, eds. *Biology of the Reptilia.* Ithaca, NY: Society for the Study of Amphibians and Reptiles; 1998:485-531.

9. Diver SJ. Hepatology. In: Divers SJ, Stahl SS, eds. *Mader's Reptile and Amphibian Medicine and Surgery.* St. Louis, MO: Elsevier; 2019:649-668.

10. Wake MH. *Hyman's Comparative Vertebrate Anatomy.* Chicago, IL: University of Chicago Press; 1979.

11. Wyneken J. The anatomy of sea turtles. U.S. Department of Commerce National Oceanic and Atmospheric Administration Technical Memorandum NMFS-SEFSC-470, 2001. Available at: https://repository.library.noaa.gov/view/noaa/8502. Accessed June 1, 2023.

12. Hofmann AF, Hagey LR, Krasowski MD. Bile salts of vertebrates: structural variation and possible evolutionary significance. *J Lipid Res.* 2010;51(2):226-246.

13. Hagey LR, Møller PR, Hofmann AF, Krasowski MD. Diversity of bile salts in fish and amphibians: evolution of a complex biochemical pathway. *Physiol Biochem Zool.* 2010;83(2):308-321.

14. Moura LR, Santos AL, Belleti ME, et al. Morphological aspects of the liver of the freshwater turtle *Phrynops geoffroanus* Schweigger, 1812 (Testudines, Chelidae). *Braz J Morphol Sci.* 2017;16;26(3-4):129-134.

15. Srichairat N, Taksintum W, Chumnanpuen P. Histological and histochemical characteristics of the oral, pharyngeal and accessory digestive organs in the water monitor lizard (*Varanus salvator*) from Thailand. *Anatomia.* 2022;51:703-711.

16. Lametschwandtner A, Bartel H, Minnick B, et al. The microvasculature of the gallbladder in adult *Xenopus laevis* (Amphibia, Anura): a scanning electron microscope study of vascular corrosion casts and correlative light microscopy. *Ital J Zool.* 2011;78:1:27-34.

17. Goldblatt PJ, Hampton JA, DiDio LN, et al. Morphologic and histochemical analysis of the newt (*Notophthalmus viridescens*) liver. *Anat Rec.* 1987;217(4):328-338

18. Yaparla A, Reeves P, Grayfer L. Myelopoiesis of the amphibian *Xenopus laevis* is segregated to the bone marrow, away from their hematopoietic peripheral liver. *Front Immunol.* 2020;10:3015.

19. Mariz JP, Nery MF. Unraveling the molecular evolution of blood coagulation genes in fishes and cetaceans. *Front Mar Sci.* 2020;27(7):592383.

20. Wallace RA. Vitellogenesis and oocyte growth in nonmammalian vertebrates. In: Browder LW, ed. *Developmental Biology. Vol. 1. Oogenesis.* Boston, MA: Springer; 1985:127-177.

21. Venturino A, de D'Angelo AM. Biochemical targets of xenobiotics: biomarkers in amphibian ecotoxicology. *Appl Herpetol.* 2005;2(3):335-353.

22. Ma Y, Li B, Ke Y, et al. Transcriptome analysis of *Rana chensinensis* liver under trichlorfon stress. *Ecotoxicol Environ Saf.* 2018;147:487-493.

23. Delgado-Coello B. Liver regeneration observed across the different classes of vertebrates from an evolutionary perspective. *Heliyon.* 2021;7(3):e06449.

The Amphibian and Reptilian Liver: Disease Morphology and Pathogenesis

Michael M. Garner, DVM, DACVP

Session #3252

Affiliation: From Northwest ZooPath, 654 West Main Street, Monroe, WA, 98272, USA.

Abstract: This presentation is Part III of the Liver Masterclass series. This presentation focuses on the various infectious and noninfectious diseases that affect the liver. Topics include viral, bacterial, mycotic and parasitic infectious agents. Noninfectious diseases include degenerative, deposition, and neoplastic conditions.

Bacterial Diseases

Infectious disease processes involving the liver of reptiles and amphibians usually represent hepatic manifestations of systemic infections, embolic spread of primary infections from the gut, or ascending infections of the biliary tree. Culture and PCR findings should always be interpreted in corroboration with histologic findings to assess the true importance of the agent as a pathogen.

Salmonella spp.

Salmonella species and subspecies are gram-negative bacilli and normal components of the enteric flora of most wild and captive reptiles., Isolation rates are higher the longer the reptile is in human care. The most important aspect of *Salmonella* infections in reptiles is the potential for zoonotic disease, especially transmission from green iguanas and turtles to young children. Salmonellosis in humans contracted from reptiles can be especially virulent. As a disease-causing agent in reptiles *Salmonella*, especially *S. arizonae,* is perhaps most important for targeting the spinal column of snakes, where it causes osteomyelitis and eventual ankylosing spondylosis. Infections in the intestinal tract are usually secondary to other enteric problems such as amoebiasis, nematodiasis, obstructions or neoplasms. Hematogenous spread to the liver causes acute coagulative necrosis associated with emboli of bacteria in the sinusoids or phagocytized organisms in Kupffer cells, and if the animal survives long enough these lesions eventually develop into abscesses.[1]

Burkholderia pseudomallei

Burkholderia pseudmallei is a gram-negative bacillus and an important human pathogen that causes melioidosis. Infected humans can develop acute pneumonia and septicemia, or chronic abscesses. Recently there have been several reports of this agent causing dermal and hepatic abscesses and granulomas in green iguanas imported to the United States and Europe from Central America. Because the gross lesions are not specific for an etiologic agent, histologic examination, culture and/or PCR are important procedures to identify the agent and its potential for causing zoonotic disease.[1]

Mycobacterial spp.

Mycobacteria are gram-positive, acid-fast bacteria from the order Actinomycetales. *Mycobacterium* is the only genus within the family Mycobacteriaceae, the most common infectious actinomycete in reptiles and amphibians an infection can occur in all reptile and amphibian groups. Clinical disease may be due to chronic or acute manifestations of mycobacteriosis in reptiles, although all cases in amphibians appear to be chronic. Chronic lesions are characterized by formation of histiocytic nodules or discreet granulomas, sometimes with a necrotic or caseous core. Liver is a common site for formation of these lesions, but typically due to hematogenous spread from skin or gut. Acute manifestations of mycobacteriosis are characterized by widespread sepsis with bacterial emboli in multiple tissues, accompanied by vascular degeneration, thrombosis and perivascular tissue necrosis. An acid-fast stain is required to identify mycobacteria in granulomas, and to distinguish mycobacteria from other bacterial agents in cases of acute infection.

Mycobacterium marinum is the most common isolate from reptiles and amphibians with mycobacteriosis. *Mycobacterium hemophilus*, *m. fortuitum,* and *M. chelonae* are less frequently isolated from reptiles. Only a few cases of *M. avium* have been associated with infection, and most of these were in a green turtle (*Chelonia mydras*) farm. Advanced molecular diagnostic techniques and DNA sequencing have substantially improved the identification of mycobacteria in reptiles and amphibians and their role in disease or as normal enteric flora. Mycobacteriosis in reptiles and amphibians appears to have low potential for zoonotic disease in immunocompetent humans.[1]

Miscellaneous bacterial infections involving liver

There are numerous reports of bacterial infections in reptiles and amphibians including abscesses, granulomas or sepsis that involve liver. These are typically manifestations of sepsis, emboli from the gut, or ascending infections of the biliary tree. Typical isolates from reptiles include any of the Enterobacteriaceae, *Listeria monocytogenes*, *Chlamydia*, *Psuedomonas aerogunosa*, *Aeromonas hydrophila*, *Enterococcus* sp., and *Elizabethkingia meningoseptica.* Anaerobic bacteria, particularly *Clostridium* spp. are commonly isolated from blood of reptiles, but visceral disease associated with this phenomenon is apparently rare.[1] *Brucella*-associated abscesses are occasionally encountered in the skin and viscera of anurans.[2]

Mycotic Disease

There has been much scientific interest and expanded understanding of mycotic disease in recent years, across all animal Classes, including those of reptiles and amphibians; however, primary mycotic infection of the hepatobiliary system is rare. Hepatic mycoses typically occur when the organism ascends the biliary tree, spreads to the liver hematogenously, or extends directly from primary pulmonary infections. In amphibians, direct spread through the body wall to liver also is common. From a treatment perspective, histologic confirmation of infection is necessary to confirm the importance of culture and PCR results. Nonpigmented fungal infections are referred to as hyalohyphomycosis. Pigmented fungal infections are referred to as phaeohyphomycosis or chromomycosis. With few exceptions, identification of fungus to species is difficult histologically and PCR and/or culture are best for further characterizing the identity of the fungus. These procedures are best performed from fresh or frozen tissue, as sensitivity of PCR from paraffin blocks seems poor for fungi.[3]

Nannizziopsiaceae

Ophidiomyces, *Paranannizziopsis* and *Nannissiopsis* cause snake fungal disease, mycotic infections in aquatic snakes, and yellow fungus disease (formerly *Chrysosporium* anamorph of *Nannizziopsis vriesii*) in inland bearded dragons, respectively. A new family in the Onygenales, the Nannizziopsiaceae, was erected to accommodate these three genera. Although these agents are frequently associated with epidermal and deep dermal infection, they only rarely spread to viscera.[3]

Metarhizium spp.

Metarhizium granulomatis (formerly *Chamaeleomyces granulomatis*) and *Metarhizium viride* (formerly *Paecilomyces viridis*) are novel species in the genus *Metarhizium* that cause fulminant, disseminated granulomatous fungal disease in lizards, chelonians and crocodilians. Infection with *M. granulomatis* has been documented in veiled chameleons in zoological and private collections in Europe, with at least one case known from North America (Ladouceur E, pers. comm.). *Metarhizium viride* infection was first reported in newly imported carpet chameleons in France, but since has been described in veiled and panther chameleons and in inland bearded dragons in Europe. Cases of fatal nodular mycotic hepatitis, and from which an unspeciated *Penicillium* was repeatedly isolated, have been described in a group of captive Jackson's chameleons in New Zealand and these were likely misidentified cases of infection with *Metarhizium viridae*.[3]

Fusarium spp.

Fusarium solani is commonly isolated from lesions of mycotic pneumonia in marine turtles, crocodilians and snakes. Cases of pneumonia may extend to the liver. Several *Fusarium* spp., including *F. solani*, contribute to the severity of the cutaneous mycoses in Wyoming toads (*Bufo baxteri*) and Puerto Rican crested toads (*Peltophryne [Bufo] lemur*), diseases sometimes referred to as brown skin disease. These agents do not typically spread systemically or involve the liver.[4]

Beauveria bassiana

Beauveria bassiana is primarily an infectious agent of insects where it causes white muscarine, but this agent can infect reptiles. Fatal pneumonia and hepatic granulomas have been described in crocodilians and marine turtles. Experimental infection in box turtles revealed that the pathogenesis of this disease may be related to low environmental temperature, and cold stunned marine turtles also are more susceptible to infection.[3]

Acremonium spp.

Infection with an unspeciated *Acremonium* was documented in spectacled caimans (*Caiman* sp.) under the older appellation of cephalosporiosis. A fatal diffuse granulomatous mycotic pneumonia with focal necrotizing hepatitis was diagnosed in these three 6-month-old caimans. Slender conidiophores, 2- to 4-µm-wide hyphae, cigar-shaped conidia, and chlamydospores were present histologically, which were morphologically consistent with *Acremonium*, and an *Acremonium* sp. was cultured from swabs of the pulmonary nodules.[3]

Zygomycetes

Zygomycetous organisms causing disease in reptiles almost exclusively fall into two lineages: Phylum Entomophthoromycota (*Basidiobolus*, *Conidiobolus*, and *Schizangiella*), and subphylum Mucoromycotina, order Mucorales (*Mucor*, *Rhizopus*, *Rhizomucor*, *Absidia*, *Syncephalastrum*, and *Cunninghamella*). Hyphae are ribbon-like, of variable width, often exhibit bulbous swelling, branch at predominantly right angles, and are typically coenocytic, appearing aseptate or pauciseptate in tissue. These thin-walled hyphae collapse in tissue sections and often appear wrinkled and twisted or convoluted. Lesions are primarily cutaneous or enteric. *Schizangiella serpentis*, may be a serious pathogen of snakes, as it has been recognized within cutaneous and systemic lesions on multiple occasions. Although the unique, yeast-like microscopic appearance of this fungus is quite different from tissue forms of other fungi in this group, multi-locus phylogenetic analysis groups it with the *Basidiobolus* lineage.[3]

Oomycetes

Oomycetes are water molds in the kingdom Protista and are not considered true fungi due to the structural lack of chitin and ergosterol. Some important oomycetes are *Pythium* sp. and *Saprolegnia* sp. Infection in reptiles appears to be rare; however, *Saprolegnia* skin infection is common in all amphibian species, and usually associated with suboptimal environmental conditions or trauma. Infection can penetrate the gut wall and extend into the coelom. Antemortem infections can be difficult to distinguish from post-mortem colonization, as the agent tends to invoke little histologic reaction in antemortem cases, primarily edema and epidermal degeneration.[3]

Dematiaceous fungi

Dematiaceous fungi consist of a heterogeneous group of genera that are darkly pigmented due to the presence of melanin in the cell walls of hyphae, conidia, or both. The presence of melanin, with few exceptions, allows for easy identification of fungal elements in routine hematoxylin and eosin (H&E)-stained tissue sections. Most infections with these fungi occur following traumatic inoculation in the skin or subcutis. Infections with pigmented fungi are classified in three main categories: phaeohyphomycosis, chromoblastomycosis, and eumycotic mycetoma. Extension from skin lesions through the body wall to serosal surfaces of viscera is common in amphibians. The term phaeohyphomycosis has been broadened to include all mycotic infections characterized by the presence of pigmented hyphae, yeast cells, pseudohyphae-like elements, or any combination thereof, in tissue section. It is a histologic diagnosis based on the appearance of fungi in tissue, although some species are poorly pigmented and require additional staining (i.e., Fontana-Masson) to highlight or identify melanin. Many publications describing phaeohyphomycosis in reptiles concern chelonians, with box turtles (*Terrapene* spp.), various tortoise species (Testudines), and sea turtles representing the majority of reported cases. Most visceral infections in reptiles and amphibians are due to *Exophiala* spp. chromomycosis (or chromoblastomycosis) has a similar pathogenesis. Sclerotic bodies, also called muriform cells and fumagoid bodies, are the histologic hallmark of disease. They are round, 5- to 12-µm diameter, brown, thick-walled, structures with internal septation, and are believed to represent a vegetative form arrested between yeast and hyphal morphology. *Cladosporium* sp. are common isolates in reptile and amphibian lesions.[3]

Yeasts

Candidiasis, usually caused by *C. albicans*, is the most common form of yeast infection in reptiles, developing as cutaneous, pulmonary or enteric infections. Systemic spread from skin is rare, but extension from the lung or gut to liver is occasionally encountered. Oval, 3- to 6-µm budding cells with pseudohyphae, and occasionally hyphae, are seen in affected tissue. Less common yeast isolates from reptiles include *Cryptococcus*, *Coccidioides*, *Trichosporon*, *Torulopsis*, *Debaryomyces*, *Pseudozyma*, and *Rhodotorula*.[3]

Microsporidia

Microsporidia are considered to be closely related to the fungi, either as a basal branch or a sister group. This classification is based on molecular phylogenetics and physical characteristics. The life cycle of this organism is direct and may include vertical transmission. Microsporidiosis is occasionally encountered in reptiles and amphibians, especially young inland bearded dragons, where *Encephalitozoan pogonae* causes severe granulomatous disease, and appears to have a tropism for blood vessels. Infection occurs with rapid extrusion of the polar filament, which pierces the host cell membrane and allows transfer of cytoplasmic content (including the nucleus). The polar filament is a prominent feature allowing microsporidian species identification. Infection of a host cell is followed by merogony and sporogony, with eventual production of sporoblasts that mature into spores. Release of spores follows rupture of the host cell. These spores are typically ovoid or pyriform in shape and vary greatly in size, the smallest being 1-2 µm in diameter and the largest being a rod-shaped spore over 30 µm in length. The disease is diagnosed histologically, usually with the aid of special stains. Typically, mature spores are gram-positive, variably acid fast, have a PAS positive polar cap and stain bright pink with Luna.[3]

Myxozoa

Myxozoans are a group of microscopic, obligate endoparasitic cnidarians that have lost many genes over the course of their evolution from free-living ancestors. These organisms, primarily *Myxidium* spp., can occasionally infect turtles and anurans. Spore formation in the renal tubules and bile ducts and can be associated with substantial lesions.[5,6]

Viral Diseases

With few exceptions, hepatic lesions caused by viruses in reptiles represent manifestations of systemic or enteric viral infections. Targeted cells may be hepatocytes, endothelial cells, pigmented macrophages, stromal cells or biliary epithelium. Lesions of herpesvirus infection in the liver have not been formally documented in amphibians, although the author has seen herpesvirus associated hepatic lesions in Mexican tree frogs.

Herpesviruses

Herpesviridae (order *Herpesvirales*) is a diverse family of enveloped, double-stranded DNA viruses ranging from 150-200 nanometers in diameter with an icosahedral nucleocapsid. Replication is within the nucleus of host cells. In acute hepatic lesions, necrosis of hepatocytes is commonly encountered, and affected cells may form syncytia. During stages of replication, when examining H&E-stained tissue sections, eosinophilic- to amphophilic-staining intranuclear inclusions may be seen. Inclusions are more commonly observed during the acute or early stage of infection, whereas it is uncommon to observe them in the more advanced stages and they will not be seen when the virus is in a latent state. To date, published analyses of sequences indicate that all reptilian herpesviruses that have been analyzed belong to the subfamily *Alphaherpesvirinae*. Among reptiles, more herpesviruses have been detected in testudines than in any other taxa.[7]

Fibropapillomatosis

Fibropapillomatosis is a syndrome that affects marine turtles, particularly green turtles and loggerhead turtles. This disease is attributed to herpesvirus infection. Skin, corneal and esophageal lesions are typical fibropapillomas. Visceral lesions are fibromas, or less commonly fibrosarcomas, and can occur in any of the visceral tissues and in bone. In the liver the fibromas are associated with the stroma of the portal tracts.[7]

Herpesviruses in emydids

Herpesvirus infections seem most common in the emyids. These infections are typically systemic and frequently involve the liver, with associated inflammation, necrosis, syncytia and intranuclear inclusions. Affected species have included Pacific pond turtles (*Clemmys marmorata*), painted turtle (*Chrysemys picta*), map turtles (*Graptemys* spp.), and Eastern river cooter (*Pseudemys concinna concinna*) (Jungwirth et al., 2014). Herpesvirus infections are also common in tortoises, although the target sites for infection are primarily upper respiratory and upper alimentary tracts. Hepatitis associated with herpesvirus infection was reported in a Russian tortoise.[7]

Herpesviruses in squamates

In lizards, hepatic manifestations of herpesvirus infection with typical lesions have been documented in a green iguana, San Esteban chuckwalla (*Sauromalus varius*), red-headed agamas (*Agama agama*), and four juvenile monitor lizards (*Varanus* spp.). Hepatic infections in snakes include boa constrictors, and horned vipers (*Vipera ammodytes ammodytes*).[7]

Hepatic adenovirus infections in reptiles

Members of the family *Adenoviridae* are nonenveloped, linear, double-stranded DNA viruses with an icosahedral nucleocapsid and a diameter generally ranging from 80 to 110 nanometers. Replication is within the nucleus of host cells. The majority of known reptile adenoviruses are atadenoviruses and phylogenetic analyses support a squamate origin of this genus.[7]

In Testudines, atadenovirus-associated hepatic lesions with hepatocellular intranuclear inclusions have been documented in European pond turtles, and ornate box turtles. A novel siadenovirus caused high mortality in impressed tortoises and Sulawesi tortoises (*Indotestudo forsteni*), associated with intranuclear inclusions in endothelial cells of many tissues, including liver.[7]

Nile crocodiles (*Crocodylus niloticus*) from a crocodile farm in Zimbabwe had multifocal to diffuse hepatic necrosis with basophilic intranuclear inclusions, and viral particle morphology was consistent with an adenovirus.[7]

Adenovirus infections are common in inland bearded dragons (*Pogona vitticeps*) in which young animals may develop hepatic necrosis associated with intranuclear inclusions in hepatocytes. Additional documented cases are in coastal bearded dragon (*Pogona barbata* [*Amphibolurus barbatus*]), Rankin's dragon (*Pogona henrylawsoni*), a central netted dragon (*Ctenophorus nuchalis*), green striped tree dragons (*Diploderma* [*Japalura*] *splendidum*) and a Savannah monitor (*Varanus exanthematicus*).[7]

In snakes, adenovirus associated hepatic disease has been documented in boids, multiple species of colubrids and in an Indonesian pit viper (*Parias* [*Trimeresurus*] *hageni*).[7]

Iridoviruses

Iridoviruses are large (103 to 220 nanometers) double-stranded cytoplasmic DNA viruses. Iridoviruses can occur as enveloped and nonenveloped forms; both are infectious. Replication occurs in two stages: the first stage is in the nucleus and the second stage is in the cytoplasm. Inclusions can be difficult to identify and usually are in the cytoplasm of infected cells. Infections in reptiles most often involve the upper alimentary and respiratory tracts, conjunctiva and spleen. Hepatic iridovirus infection usually is a manifestation of systemic infection.

The family *Iridoviridae* is divided into two subfamilies: the *Alphairidovirinae*, which infect ectothermic vertebrates, and the *Betairidovirinae*, which mostly infect insects and crustaceans. The alphairidoviruses include three genera: *Ranavirus* (that infect fish, amphibians, and reptiles), *Megalocytivirus* (that infect fish), and *Lymphocystivirus* (that infect fish). The betairidoviruses also include three genera: *Iridovirus* (that infect a wide range of arthropods and which have also been found in amphibians and reptiles), *Chloriridovirus* (that infect various insect species), and *Decapodairidovirus* (that infect shrimp and crayfish). Many members of the genus *Ranavirus* have been shown to switch hosts. *Frog virus 3* (FV3), the type species for the genus *Ranavirus*, was first isolated in 1966 from a renal carcinoma in a leopard frog (*Lithobates* [*Rana*] *pipiens*). *Ranavirus* is a substantial contributor to mortality events in wild anurans. *Ranavirus* infections in reptiles are less common but have received much recent attention, due to their ability to cross classes and species, and potentially threaten wild populations of chelonians. In chelonians, *Ranavirus* infections have caused hepatic lesions (primarily necrosis and inflammation), in a spur-tailed (Hermann's) tortoise. An Eastern box turtle infected with an FV3-like *Ranavirus* also had intracytoplasmic inclusions in circulating leucocytes observed on routine examination of blood smears. Inclusions were also found in hepatocytes. Frequently, affected box turtles have systemic necrotizing vasculitis. Histologic examination of two green tree pythons (*Morelia* [*Chondropython*] *viridis*) that died after illegal importation into Australia from Irian Jaya, revealed ulceration of the nasal mucosa, necrotizing pharyngitis, and hepatic necrosis. In Germany, a *Ranavirus* was isolated from tissues of a deceased leaf-tailed gecko (*Uroplatus fimbriatus*) that died suddenly with ulcerative-necrotizing glossitis and hepatic necrosis.[7]

Paramyxoviridae

The family *Paramyxoviridae* consists of single-stranded negative-sense RNA viruses that range in diameter from 150 to 300 nanometers and can exist in spheroidal or filamentous forms. Mature virus undergoes envelopment at the cytoplasmic membrane. Inclusions consisting of nucleocapsid strands can sometimes be found in either the nucleus or cytoplasm. All known reptile paramyxoviruses that have been genetically characterized are part of the genus *Ferlavirus* (subfamily: *Orthoparamyxovirinae*) and can be divided into four genogroups: A, B, C, and tortoise. Another virus, *Reptile sunshinevirus 1*, found in pythons, was originally believed to be a paramyxovirus but has since been classified in a new family, *Sunviridae*.[7]

Ferlavirus infections have been described in squamates and chelonia. Infections associated with disease generally involve the lower respiratory tract and occasionally the brain. In snakes, additional manifestations may include involvement of the salivary glands, pancreatic ducts and bile ducts. Infected epithelium may have intranuclear or intracytoplasmic inclusions, although inclusions are uncommon. Ductular lesions include necrosis, dysplasia and squamous metaplasia.[7]

Arenaviridae

Viruses in the family *Arenaviridae* are sub-divided into four genera: one genus contains mammal-infecting arenaviruses (genus *Mammarenavirus*), one contains arenaviruses identified in fish (genus *Antennavirus*), and two contain arenaviruses that infect snakes (genera *Reptarenavirus* and *Hartmanivirus*). Arenavirus particles are enveloped, pleomorphic, 40-200 nm in diameter, and have surfaces decorated by glycoproteins. Arenaviruses have segmented single-stranded RNA genomes with two or three segments, each of which have one or two genes. Viral replication occurs in the cytoplasm. *Reptarenavirus* is the causative agent of inclusion body disease (IBD) in snakes. Inclusion body disease is primarily a disease of boid snakes but has been described in colubrids and palm vipers. The disease is associated with the formation of singular intracytoplasmic inclusions in the cytoplasm of all cell lines. In boas, inclusions are numerous in most tissues, including liver, and the disease may remain subclinical. In pythons, the disease primarily causes encephalitis, with few or no inclusions in any tissues. Prior to the identification of the causal agent, antemortem diagnosis was based on biopsy of the liver, or cytologic examination of gastric wash specimens stained with Wright-Giemsa or H&E. The identification of reptarenaviruses as the etiologic agents of IBD has facilitated specific serologic and molecular diagnostic tests. Reverse transcriptase-polymerase chain reaction (RT-PCR) can be used to detect viral RNA, and several antibodies have been developed that recognize reptarenavirus proteins. Good concordance between targeted testing and traditional detection based on the identification of inclusions in H&E-stained tissues has been reported. Antemortem tests in pythons are complicated by the fact that viral replication appears to be primarily restricted to the CNS.[7]

Parasitic Infections

As with most of the infectious processes in reptiles and amphibians, associated hepatic lesions usually are a manifestation of systemic or enteric infection.

Amoebiasis

Amoebiasis is common in tortoises and snakes. Historically, this disease has been accredited to *Entamoeba invadens*, based on the morphologic features of the trophs in tissue and the cysts in feces. It is likely that molecular diagnostic capabilities will greatly expand the range of pathogenic amoeba in reptiles and amphibians. Amoebiasis is primarily a disease of the alimentary tract, starting in the colon of chelonians, small intestine in snakes, and stomach in anurans. Amoeba infiltrate the gut wall, with subsequent necrosis, edema and inflammation. The amoeba spread hematogenously to the liver and other tissues, and subsequent necrosis and abscess formation is particularly severe in the liver. Secondary bacterial infections, including salmonellosis, may develop in the gut, and contribute to sepsis.[8] Amoebiasis is uncommon in lizards and crocodilians. Enteric amoebiasis is common in various anurans but spread to the liver is not a feature of the disease in these amphibians.[9]

Flagellates

Flagellates are primarily constituents of the normal enteric microbiota in reptiles and amphibians. Flagellate infection of the kidney, often attributed to *Hexamita* infection, is common in lizards and chelonians, and associated with severe tubulointerstitial nephitis. *Monocercomonas* sp. has been incriminated in outbreaks of systemic infection, and the latter may affect the gall bladder and bile ducts.[8]

Apicomplexans

Eimeria spp. are found in the gallbladder, bile ducts, and intestinal epithelium of snakes, lizards, and crocodilians. *Eimeria bitis* was reported to be pathogenic for several species of snakes including the puff adder (*Bitis arietans*), common garter snake (*Thamnophis sirtalis*), ribbon snake (*T. sauritus*), and smooth green snake (*Opheodrys vernalis*), where necrotizing cholecystitis was reported.[8]

Cryptosporidiosis is primarily a disease of the gastric mucosa of snakes; however, *Cryptosporidium* was also reported on the microvillous borders of the gallbladder and intra- and extrahepatic bile ducts in two of six corn snakes with gastric cryptosporidiosis.[8]

An unclassified intranuclear coccidian infection was reported in two radiated tortoises (*Astrochelys* [formerly *Geochelone*] *radiata*) that were captive bred and reared in the United States. Both tortoises were in renal failure and were euthanized. Light microscopic examination of multiple tissues revealed nephritis and pancreatitis, and an intranuclear protozoan was seen in renal epithelial cells, hepatocytes and ductular epithelium, pancreatic acinar cells, and enterocytes). multiple additional species of tortoises have subsequently been diagnosed with this disease. The precise identity of this apicomplexan coccidia has not been determined.[8]

Hemoparasitic apicomplexans

These coccidia include hemogregarines (*Haemogregarina*, *Hepatozoon*, *Karyolysus*, and *Haemolivia*), hemococcidia (*Schellackia* and *Lainsonia*), and *haemosporidians* (*Plasmodium*, *Haemocystidium*, *Fallisia*, *Saurocytozoon*, and *Haemoproteus*).[8] *Lankesterella* sp. are occasionally encountered in numerous cell types in amphibians.[10] Most of these agents have invertebrate intermediate hosts, and only a few have sexual and asexual stages in the definitive host. Reports of pathogenicity of these agents in reptiles and amphibians are scarce, and in most cases infection is incidental. Tissue phases of the haemosporidians are frequently found in pigmented macrophage centers of various organs, including the liver. One or two banana-shaped organisms may be seen within a single infected cell, which is often distorted in size and shape. Some species in snakes have a very narrow host range.[8]

Cestodes

The most important and unique cyclophyllideans in lizards and snakes are in the family Mescocestoididae, genus *Mesocestoides*. Members of this genus require three hosts to complete their life cycle, with mammals and birds serving as the final host. Arthropods (mites) serve as the first intermediate host, in which oncospheres develop into a cysticercoid. The larva that infects reptiles is called the *tetrathyridium* and development in reptiles occurs after they ingest an infected arthropod containing the cysticercoid.

In reptiles, tetrathyridia may be found within the intestines or various extraintestinal sites. Tetrathyridia of *Mesocestoides* have been found within white nodules in the liver, on the intestine, within the coelom, and within the mesentery of the island night lizard (*Xantusia riversiana*). There was minimal inflammatory response, with the parasites surrounded by a thin fibrous capsule. Mesocestoides also has been seen in the liver of a captive inland bearded dragon from California.[8]

Trematodes

Adult members of the family Spirorchiidae inhabit the circulatory system of susceptible reptiles, with turtles the most infected reptilian host. The most important lesions result from the release of eggs within the vascular compartment. Eggs ultimately become trapped within terminal vessels potentially at any site and invoke a granulomatous response. Cephalic and uveal lesions can be especially severe. *Rhytidodoides similis*, a member of the family Rhytidodidae, has been identified as a parasite of the gallbladder of green turtles, where papillomatous hyperplasia of the mucosa near the entrance to the cystic duct was seen. [8]

Nematodes

Nematode infections are primarily limited to the lungs and intestinal tract of reptiles and amphibians. Larval nematodes of another spirocercid, *Ascarops* sp., have been found in hepatic granulomas in the wild populations of Western fence lizard (*Sceloporus occidentalis*). In anurans, contaminated substrate can result in severe cutaneous larval migrans, with larvae embolizing to most visceral tissues, including liver.[8]

Perkinsus

Perkinsea is a microeukaryotic parasitic phylum composed of four main lineages of parasitic protists with broad host ranges. Three of the lineages primarily infect marine invertebrates and fish. The fourth lineage is a freshwater pathogen of tadpoles and has been associated with substantial losses in endangered Mississippi gopher frogs. Perkinsus is considered the third largest cause of mortality in wild anuran populations in North America. Vegetative forms infect the liver, and spores and meronts are released to the environment, where they are highly resistant.[11]

Neoplasia

There are four different cell lines in the normal reptilian and amphibian liver: hepatocytes, biliary epithelium, sinusoidal and vascular endothelium, pigmented macrophage centers, portal fibroblasts, and smooth muscle cells of the larger portal vessels. In amphibians, especially urodeles, hematopoietic cells are also normal constituents. Primary neoplastic processes can arise from any of these cell lines. Hepatocellular and biliary tumors, benign and malignant, are common. Neoplasms arriving from the other cell lines are uncommon to rare. Cysts arising from the biliary tree or endothelium are common. Ductal plate anomalies have recently been described in snakes. Historically referred to as biliary hamartomas, these anomalies present as nodules of hyperplastic or cystic bile ducts with associated fibrosis and are generally incidental findings. Metastatic or multicentric tumors are also frequently encountered in the liver. Some examples are lymphoma in many species, and gastric neuroendocrine tumors in bearded dragons.[12]

Deposition Disorders

Deposition disorders in general are common in reptiles and amphibians. These include Lipidosis, mineralization, gout, melanosis, xanthomatosis, hemosiderosis, lipofuscinosis, cholelithiasis and an amyloid-like disorder. Hepatic lipidosis is the only primary hepatic deposition disorder, although hepatic manifestations of mineralization, gout and melanosis can cause substantial hepatic lesions.[13]

Hepatic lipidosis

Hepatic lipidosis (HL) is the accumulation of excessive amounts of intracytoplasmic lipid in hepatocytes. HL can be associated with obesity, inanition, malnutrition, hypoxia, exposure to various drugs and toxins, metabolic derangements such as renal or endocrine disease, active reproductive status (in females), or combinations thereof. HL is potentially reversible; however, chronic manifestations of HL may lead to fibrosis, biliary hyperplasia and eventual cirrhosis. The latter is especially common in bearded dragons, where terminal hepatic disease may be encountered in relatively young adults (2-3 years). Some species seem to tolerate HL; for instance, HL is common in tortoises and possibly well tolerated, but is less common and may be associated with clinical hepatopathies in squamates, aquatic turtles, and anurans.[13]

Melanosis

Reptiles and amphibians begin to accumulate melanin and related pigments in pigmented macrophage centers (PMC) of the liver and to a lesser extent other tissues, starting at an early age. Over time, and likely also influenced by past or concurrent disease or inanition, the PMCs accumulate and enlarge. In some species the PMCs can occupy nearly the entire liver, displacing or effacing the normal parenchymal features, which may cause or contribute to hepatic insufficiency. This is especially common in old geckos and in old amphibians.[13]

Cholelithiasis

Cholelithiasis is common in reptiles and amphibians and in most cases is an incidental finding; however, obstruction of the biliary tree due to cholelithiasis can result in hepatic enlargement due to florid biliary hyperplasia, especially in green iguanas. When severe, this process can be misconstrued as biliary neoplasia. Patients with evidence of hepatobiliary disease should be carefully monitored for obstructive microcholeliths in the common bile duct and lumen of the gall bladder.[13]

Degenerative Disease

Degenerative hepatic disease of clinical importance in reptiles and amphibians is limited to cirrhosis. Cirrhosis is a condition in which there is excessive biliary hyperplasia and fibrosis in the portal tracts, that eventually bridges adjacent portal and centrilobular regions, and sequesters hepatocytes into nodules rather than organized plates. Hepatic lipidosis is also a common finding in cases of cirrhosis. The condition can be associated with aging, obstruction or ascending inflammation of the biliary tree, exposure to hepatic toxins, chronic hepatic lipidosis, or combinations thereof. Cirrhosis is not reversible.[14] A developmental anomaly, ductal plate malformation, has recently been described in snakes.[15]

References

1. Pasmans F, Martel A, Jacobson ER. Bacterial diseases of reptiles. In: Jacobson ER, Garner MM, eds. *Infectious Diseases and Pathology of Reptiles.* 2nd Ed. Boca Raton, FL: CRC Press; 2020.

2. Helmick KE, Garner MM, Rhyan J, et al. Clinicopathologic features of infection with novel *Brucella* organisms in captive waxy tree frogs (*Phyllomedusa sauvagii*) and Colorado River toads (*Incilus alvarius*). *J Zoo Wildl Med.* 2018;49(1):153-161.

3. Paré JA, Conley KJ. Mycotic diseases of reptiles. In: Jacobson ER, Garner MM, eds. *Infectious Diseases and Pathology of Reptiles.* 2nd Ed. Boca Raton, FL: CRC Press; 2020.

4. Crawshaw G, Pienkowski M, Lentini A, et al. Brown skin disease: a syndrome of dysecdysis in Puerto Rican crested toads (*Peltophryne lemur*). *Zoo Biol.* 2014;33(6):558-564.

5. Duncan AE, Garner MM, Bartholomew JL, et al. Renal myxosporidiosis in Asian horned frogs (*Megophrys nasuta*). *J Zoo Wildl Med.* 2004;35(3):381-386.

6. Garner MM, Bartholomew LJ, Whipps CM, et al. Renal myxozoanosis in crowned river turtles *Hardella thurjii*: description of the putative agent *Myxidium hardella* n. sp., by histopathology, electron microscopy, and DNA sequencing. *Vet Pathol.* 2005;42(5):589-595.

7. Marschang RE, Origgi FC, Stenglein MD, et al. Viruses and viral diseases of reptiles. In: Jacobson ER, Garner MM, eds. *Infectious Diseases and Pathology of Reptiles.* 2nd Ed. Boca Raton, FL: CRC Press; 2020.

8. Walden HDS, Greiner EC, Jacobson ER. Parasites and parasitic diseases of reptiles. In: Jacobson ER, Garner MM, eds. *Infectious Diseases and Pathology of Reptiles.* 2nd Ed. Boca Raton, FL: CRC Press; 2020.

9. Weisbrod TC, Jeon AB, Childress A, et al. Gastrointestinal entamoebiasis in captive anurans in North America. *Dis Aquat Organ.* 2021;143:109-118.

10. Gericota B, Garner MM, Barr B, et al. Morphologic, immunohistochemical, and molecular characterization of a novel *Lankesterella* protozoan in two White's tree frogs (*Litoria caerulea*). *J Zoo Wildl Med.* 2010;41(2):242-248.

11. Chambouvet A, Gower DJ, Jirků M, et al. Cryptic infection of a broad taxonomic and geographic diversity of tadpoles by *Perkinsea* protists. *Proc Natl Acad Sci.* 2015;112(34):E4743-51.

12. LaDouceur EEB. Reptile neoplasia. In: Garner MM, Jacobson ER, eds. *Noninfectious Diseases and Pathology of Reptiles.* 2nd Ed. Boca Raton, FL: CRC Press; 2020.

13. Graham EA, Burns RE, Ossiboff RJ. Deposition diseases. In: Garner MM, Jacobson ER, eds. *Noninfectious Diseases and Pathology of Reptiles.* 2nd Ed. Boca Raton, FL: CRC Press; 2020.

14. Stedman NL. Degenerative diseases. In: Garner MM, Jacobson ER, eds. *Noninfectious Diseases and Pathology of Reptiles.* 2nd Ed. Boca Raton, FL: CRC Press; 2020.

15. Patania OM, Troan BV, Cullen JM. Ductal plate malformations in captive snakes. *Vet Pathol.* 2021;58(4): 736-742.

Gross Lesion Recognition for the Reptile Clinician

Michael M. Garner, DVM, DACVP

Session #3255

Affiliation: From Northwest ZooPath, 654 West main Street, Monroe, WA 98272, USA.

Note: Reprinted with permission. Garner MM. Gross lesion recognition for the small mammal clinician. *Proc ExoticsCon.* 2022;385.

Abstract: Gross lesion recognition, like other imaging modalities, is a bit of an art form. Vast zones of grey may confront the investigator when considering what the lesion is, what it could be, and what to do with it. And nobody is good at all of it. A seasoned livestock pathologist could possibly clear an entire day's necropsies without the help of a single histology slide. A bee keeper would recognize immediately the maggot stages that parasitize his colony. A dog-cat pathologist in a busy private laboratory or a companion animal clinician may not have a clue about gross lesions in livestock or bees. Historically, our attempts at accurately diagnosing disease by gross lesion recognition have been a humbling experience, and that has been the impetus for all further diagnostic specialties. So now that we have put gross pathology in its proper place, why even bother with such an imperfect science? Obviously, gross lesions, or lack thereof, are the first visual indication of what may be wrong with the patient. These are the lumps, the effusions, the asymmetrical oddities, the discolorations, the odiferous clues to disease that (hopefully) stimulate a scientific thought process culminating in a list of differential diagnoses. The purpose of this masterclass is to present images of common and not-so-common reptile gross lesions in live and necropsy specimens, and in a participatory manner, establish a differential diagnosis and means for establishing a definitive diagnosis.

Recommended Reading

1. Campbell TW, Ellis C. *Avian and Exotic Animal Hematology and Cytology.* Ames, IA: Iowa State University Press; 2007.

2. Garner MM. An overview of biopsy, cytology and necropsy techniques for clinical veterinarians. In: Aguilar RF, Hernandez SM, Divers SJ, Perpinan D, eds. *Atlas de Medicina de Animales Exoticos.* 2 da Buenos Aires: Inter-Medica S.A. I.C.I.; 2010:437-443.

3. Garner MM. Overview of biopsy and necropsy techniques. In: Mader D, ed. *Reptile Medicine and Surgery.* 2nd Ed., St. Louis, MO: Elsevier Saunders; 2006:569-580.

4. Stacy BA. Reptile necropsy techniques. In: Jacobson ER, Garner MG, eds. *Reptile Diseases and Pathology.* Vol 1, 2nd Ed. London, UK: CRC Press; 2021:331-374.

Section 38

ARAV Masterclasses Track 2

Anna-Marie Ruoff, DVM
Session Chair

Untangling the Intricacies of Companion Reptile Nutrition, Associated Diseases, and Client Compliance

Kyra J. Berg, DVM, Dipl ACZM, and
Joanna K. Webb, DVM, MS

Session #3147

Affiliation: From VCA South Shore (Weymouth) Animal Hospital, 595 Columbian Street, South Weymouth, MA 02190, USA (Berg), the Chicago Zoological Society, 3300 Golf Rd, Brookfield, IL 60513, USA (Webb).

Abstract: Reptile nutrition is a labyrinth of data mining as a clinician looks to the species' natural history, life stage, and nutrient content of feed items. The importance of species-appropriate environmental temperature and humidity, as well as the influence of lighting, on micronutrient utilization must be emphasized. Considering the vast nature of the topic, this masterclass focuses on three essential components of nutrition and nutritional disease. The first focus reviews reptilian utilization of key micro and macronutrients, the vitamin D pathway, dietary recommendations for commonly managed species, gut-loading invertebrates, and supplement provision. The second component discusses diagnosis and management of nutrition-related diseases including but not limited to nutritional secondary hyperparathyroidism, metastatic calcification of soft tissues, obesity, periodontal disease in acrodonts, cloacal prolapses, aural abscesses, hepatopathies, renal disease, and neurologic disease. Finally, this information is assimilated in the third component into ways to maximize the nutrition discussion among client, staff member, and veterinarian to improve client compliance.

Introduction and Metabolism Basics

With over 10,000 currently identified reptile species, innate challenges in reptile nutrition include understanding the various needs of each individual species, such as temperature, humidity, microenvironments, physiology, and feeding strategy. Species specific temperatures are needed for appropriate physiologic processes including those relating to nutrition. There are also behavioral and seasonal changes in appetite; for example, both brumation and reproduction may lead to decreased appetite and thus decreased or increased nutritional requirements. These challenges are further compounded by the fact that these species are under human care, limiting the variety and nutritional content of food items. Poor nutrition is common and is responsible for 80% of the illnesses in managed reptiles, and the most common reason to seek veterinary care for reptiles is nutritional-related disease.[1,2] Therefore, a thorough husbandry anamnesis is essential, and a proactive approach is recommended over treating potentially preventable and life-threatening nutritionally associated conditions.

Reptiles typically have one-tenth the energy requirements of a comparably sized mammal. When sick, reptiles are often recommended to be kept at the upper range of the species' preferred optimum temperature zone (POTZ) to support immune function which subsequently increases metabolism and caloric needs. Thus, nutrition should address these increased requirements to avoid catabolic states. Specific dynamic action (SDA) refers to an increase in metabolism caused by ingestion of food, especially proteins.[3,4] This is most notably observed in *Python regius* and *P. molurus*, where amino acid ingestion independent of body temperature results in increased oxygen consumption, heart rate, and even growth of organs (e.g., the heart). Meal composition can affect SDA; for example, complete amino acid profiles (e.g., whole prey items) will maximize this effect. Fats and carbohydrates do not elicit SDA in reptiles.

Macro- and Micronutrients

Water

While reptiles may be able to withstand greater water losses than mammals, water is still considered the most important nutrient for all animals.[3] Regulation of water varies depending on the reptile's taxa or group. Most reptiles are uricotelic, meaning they produce large, insoluble uric acid to conserve water at the level of the renal tubule. Aquatic reptiles will produce urea or ammonia in greater percentages.[5] Insensitive cutaneous and respiratory water loss can be minimized in an environment with higher humidity.

Sources of water

Sources of water include dietary water (e.g., washed greens, moisture content within prey items), metabolic water formed from cellular metabolism, and environmental sources of water. Water should be always available to all reptiles. There are many clients under the impression that desert species do not need access to water sources. However, innate challenges of captivity to mimic accurate water provision and micro-environments of high humidity that would be present in the wild negate this impression.[1,3]

Water presentation is also important to encourage water consumption. Arboreal species such as vine snakes (*Ahaetulla nasuta*), tree boas (*Corallus* sp.), and Chameleonidae rarely come to the ground to drink, and some species (e.g., chameleons) are challenging to teach to drink out of bowl. Misting of the branches or a dripper is recommended, and, for some species, essential. The shape of the water container is also important, to avoid getting stuck while still allowing the reptile to drink or soak themselves.[1,3]

Soaking, whether self-induced or through the owner placing their reptile in a designated soaking bath, is another source of water provision. While controversial as to whether all or only some reptiles are capable of cloacal colonic absorption of fluids, supervised soaking in warm water poses little harm to health and potential health benefits. It is recommended that reptiles be soaked routinely to prevent clinical or subclinical dehydration.[1] Bearded dragons (*Pogona vitticeps*) administered furosemide showed active drinking behavior while soaked, and it was observed that both soaking and drinking resulted in significantly increased body weight.[6]

Calcium

Calcium is found in bones and the bloodstream and is needed on a regular basis. Nearly 99% of calcium in the body is bound within the inorganic matrix of bone.[2,7] The remainder, known as total calcium, is found in the bloodstream in three fractions: albumin- bound, complexed to other compounds, and the free ionized biologically active form.[8] Calcium should be supplemented independent of multivitamins, and should contain no phosphorus or vitamin D.[3] For accuracy of history and husbandry, ask owner to bring in the label or provide the information of supplementation on the phone.

Sources of calcium

Sources of calcium include those in metallic form, which are mixed into the food source; only a small percentage of calcium in this form is bioavailable to reptiles. Chelated calcium, which is found in greens, is highly absorbable and therefore the preferred source of calcium.[9] A study feeding hatchling ground snakes fed crickets loaded with turnip greens or calcium gel cubes showed that turnip green snakes developed normally while the other snakes developed bony abnormalities.[9] Calcium carbonate and calcium citrate are best absorbed, with 40% and 21% of elemental calcium available for absorption, respectively.[3]

Vitamin D

Vitamin D promotes increases of calcium and phosphorus in the blood through increased absorption in the gut, increased resorption from bone, decreased phosphorus excretion in kidneys, and increased calcium resorption in kidneys. It is a fat-soluble vitamin, therefore there is potential for toxicity.[2,3]

Vitamin D activation pathway

Vitamin D_3 (cholecalciferol) is produced by UVB and heat, but is inactive. Activation occurs when vitamin D_3 travels to the liver where hydroxylation occurs, and it is converted to calcidiol or 25-hydroxyvitamin D_3, the form measured in plasma. Calcidiol is then converted by the proximal tubules in the kidney to calcitriol or 1,25-dihydroxycholevitamin D_3, which is biologically active. The conversion of calcidiol to calcitriol is catalyzed by 1-α-hydroxylase. Parathyroid hormone (PTH) regulates this process in response to low calcium or increased phosphorus.[2]

Sources of vitamin D

Not all reptiles can utilize dietary vitamin D such as companion animal reptiles within the *Iguana* and *Cyclura* genera.[3] Green iguanas (*Iguana iguana*) do not appear to efficiently utilize dietary vitamin D without high doses, so they are dependent on UVB provision for vitamin D synthesis.[2] Adequate UVB light provision, regardless of species, results in less need for dietary supplementation, which can be challenging because recommended daily allowances in a majority of reptile species are unknown. While reptiles utilize vitamin D_3, many human products contain vitamin D_2, and reptiles cannot convert vitamin D_2 to vitamin D_3.[2] Having reptile-specific dietary supplementation for vitamin D once to twice monthly, with species-appropriate UVB light sources that are changed every six months, should be adequate for vitamin D contents in reptiles.

UVB Irradiation

Ultraviolet light remains essential for vitamin D and calcium homeostasis in nearly all reptiles studied, although this too is a controversial topic for certain groups of reptiles. Other functions of UVB include modulation of the immune system, disinfection, improvement of the skin barrier, and upregulation of melanin synthesis.

Nocturnal species can utilize UVB light.[10] While snakes may not require UVB, recent publications show that certain species increase circulating vitamin D after exposure to UVB.[10,11] Both appropriate UVB length and warmth are required for cutaneous vitamin D_3 synthesis. Cholesterol 7-dehydroxycholesterol in the skin is converted to previtamin D_3 when exposed to short-wavelength UVB (290-315 nm).[10]

Sources of UVB light

Glass, plastic, and mesh screens block UVB and some UVA, so it is important to ensure direct access to UVB. The spaces in mesh screens are the safest way to provide UVB through the small squares while preventing burn injury and escape of the captive reptile. Brands vary widely in UVB output and decline over time. The rule of thumb is to replace the bulb every six months. Alternatively, a UVB meter can be purchased to test lamps, which should be replaced when they reach below 70% of their original output. The distance and location (e.g., top vs side of the cage) can vary based on the bulb and directions by the company but is typically no more than 30-60 cm away from the reptile. If placed too closely, there may be an increased risk of developing keratitis and other keratopathies. Albino or hypomelanistic morphs may be at increased risk of skin damage or neoplasia secondary to UVB and may need lower UVB levels compared to normally pigmented conspecifics.[10]

Vitamin A

Vitamin A is important for epithelial cell health, immune system function, reproduction, embryonic development, bone metabolism, hematopoiesis, and vision. It is a fat-soluble vitamin, therefore there is potential for toxicity.[2] All studied herbivores appear to be able to synthesize vitamin A, and carnivores fed whole prey do not develop vitamin A deficiency because prey liver contains high amounts of vitamin A. Emydid turtles (e.g., box turtles and aquatic turtles) and insectivores such as Chameleonidae, leopard geckos (*Eublepharis macularius*), and fat-tailed geckos (*Hemitheconyx caudicinctus*) appear sensitive to hypovitaminosis A. This subset of reptiles needs vitamin A in their diet due to poor conversion of β carotene precursors (which are found in many plants) to vitamin A.[2] While there may be an indication that leopard geckos can utilize β carotene to create sufficient vitamin A quantities in the liver, this is still considered a controversial topic.[12,13,14]

Sources of vitamin A

For carnivores, vitamin A will be found in their prey's liver.[2] Dietary supplementations for insectivores and omnivorous chelonians should contain vitamin A, not β carotene, with an ideal ratio of vitamin A:D:E as 100:10:1. Insects should be gut-loaded with an appropriate multi-vitamin supplementation.[2] Additional supplementation with a multi-vitamin should be cautiously dusted on insects once to twice monthly because the vitamin A requirement for many reptiles remains unknown. Many supplements contain only carotenes rather than full vitamin A; reading the label for multi-vitamin supplements should ensure that vitamin A is included.[2]

Feeding Strategies

Feeding strategies in reptiles can vary throughout life, with different feeding habits of juveniles compared to adults. Considering the age of the reptilian patient remains an essential component of health assessment.

Insectivores

Many lizards, some turtles, and a few snakes are insectivores. Many obligate carnivore adults are insectivores as juveniles.[3] Regarding lizards, most members of Scincidae, Chamaeleonidae, Iguanidae, Agamidae, and Gekkonidae are insectivores.

Carnivores

All snakes are carnivorous, with most consuming rodents and rabbits but some also eat birds, fish, reptiles, and other vertebrates. Carnivorous lizards include many Varanidae species, primarily consuming small vertebrates, and Teiidae, which primarily consume insects and small vertebrates. Box turtles (*Terrepene* sp.) and hingeback tortoises (*Kinixys* sp.) are carnivorous; aquatic turtles are mainly carnivorous or scavengers. Carnivores require high levels of protein (25-60%) and fat (30-60%).[3]

Omnivores

Omnivores include lizard members of Scincidae, Chamaeleonidae, Iguanidae, Agamidae, and Gekkonidae. Many aquatic and Emydid turtles are also omnivores. Many omnivorous reptiles are more carnivorous as juveniles to aid in development. Omnivores typically have nutritional requirements of protein 15-40%, fat 5-40%, and carbohydrates 20-75%.[3]

Herbivores

Herbivores include Testudinidae, some Agamidae, some Iguanidae; and some specific species such as the Solomon Island prehensile-tailed skink (*Corucia zebrata*). Generally, these reptiles are browsers and grazers, but there is a great variety in their diet in the wild. Some species, like bearded dragons and Emydid turtles, become herbivores as adults after being insectivores or carnivores as juveniles. All herbivores require UVB light. These animals are hindgut fermenters with an enlarged cecum or colon containing symbiotic microbes that anaerobically ferment plant material. For tortoises, it is recommended to provide a majority hay or grass diet and good commercial pelleted diet for tortoises, with the remaining 20% as leafy greens. For iguanas, which are considered folivores, produce should be no more than 50% of the diet with the remainder a commercial diet. Herbivores typically have nutritional requirements of protein 15-35%, fat < 10%, and carbohydrates 50-75%, of which fiber should be 15-40%.[3]

Food Items

Vegetables

Herbivores should receive uncooked leafy greens as part of their normal diet. Oxalic acid, found in spinach, beets, and celery stalks, can interfere with normal calcium uptake, so should be limited. Edible flowers that can be offered as treats and enrichment include dandelions, hibiscus, and roses.[1]

Hay and plants

Most grass hays reach nutritional requirements for reptiles, but some are low in calcium. Alfalfa is higher in calcium but also high in protein, so should be offered sparingly to prevent obesity. Toxic plants exist and are to be avoided such as ramson (*Allium ursinum*).[15]

Vertebrate prey

Vertebrate prey should be completely thawed prior to feeding.[3] Muscle meat and viscera have low calcium, so provision of whole prey where bones are present is important for carnivores. To increase palatability, scenting prey with urine, exposing the intestines, exposing the brain, or tease feeding may be incorporated into feeding routines.

Live prey

Live prey is actively discouraged due to risk of injury to the reptile, the inhumanity of feeding live prey, and risk of parasite or other pathogen transmission. This practice is illegal in many countries. Live prey should be strictly limited to those reptiles that are being rehabilitated for release back into the wild to ensure hunting and survival skills are present.[1]

Mammals

The most common vertebrate mammals provided to reptiles are rats and mice. Consideration of what the prey are fed, much like with insects, is important to prevent obesity and ensure adequate nutritional content. Different life stages of the reptile mandate unique nutritional requirements, and different life stages of the mammalian prey provide distinct nutritional content. For example, juvenile rats and mice known as "fuzzies" have a positive calcium: phosphorus ratio, whereas adults are more likely to have increased fat content. Water content may change depending on the length of time that they have been frozen. Other prey items include rabbits, hamsters, gerbils, chicks, lizards, snakes, frogs, and toads.[3]

Fish

Fish are good sources of most nutrients except thiamin and vitamin E. Thiaminases, enzymes that destroy thiamin, are present in many marine animals, such as clams, herring, smelt, mackerel, Cyprinidae, bream, and dogfish. Therefore, thiamine supplementation is required (25-30 mg thiamine/kg of wet fish). Avoid goldfish due to risk of mycobacteriosis as well as high protein and fat content which can lead to gout.[3]

Fruit

Fruit can be provided in small amounts as nutrition, enrichment, and treats. Some reptiles are attracted to bright colors, such as reds and oranges. Some fruit may be low in calcium or have poor calcium: phosphorus ratios. Bananas and grapes contain tannins, which can interfere with protein metabolism. Avocados are considered toxic, although some wild species do consume them. Rhubarb and eggplant may be toxic.[3]

Insects

Types of insects

Commercially available prey items include crickets, wax moths, silk moths, cockroaches; mealworms, superworms, waxworms, silkworms, hornworms, earthworms (Canadian and European), red wriggler worms; houseflies, fruit flies, springtails, and black soldier flies and their larvae.[3]

Nutritional content

Monoculture insects have questionable nutritional value and alone are nutritionally incomplete. Larval insects have a high fat content, whereas adult insects have low carbohydrate, appropriate protein, appropriate essential amino acid, and high fiber content for reptiles. Generally, insects are calcium, iodine, and vitamin (B, D, A, E) deficient. Most insects have an inverted calcium: phosphorus ratio of 1:10, so supplementation is required to achieve the ideal calcium: phosphorus ratio of 1.5-2:1. Furthermore, dusting of calcium has variable results, may depend on the type of insect, and fastidious grooming by crickets, for example, can result in removal of the calcium dust before ingestion by the reptile occurs. Providing a variety of insects may help offset deficiencies, and gut-loading insects is essential, especially for captive crickets and mealworms. Species with better calcium: phosphorus ratios include Phoenix worms, wood lice, and earthworms.[3]

Gut-loading

Gut-loading involves providing a highly nutritious diet to insects for 24-48 hours prior to feeding the insect to the reptile, which gets incorporated into the insect. Insects will still need a water source during gut-loading. While more research is needed, recent studies evaluating duration and type of gut-loading diet to certain insect species has resulted in more accurate recommendations on how to gut-load for reptiles.[16,17] At this time, a combination of dusting and gut-loading insects is recommended, but it is important to be cognizant of the risk of overprovision of calcium and to monitor literature for new information in the field of reptile nutrition.[3]

Pelleted and manufactured diets

Some pelleted and manufactured diets have long-term feeding data, but many do not. While ideally containing all required nutrients, some omitted nutrients can result in severe deficiencies, and some reptiles will not eat these diets. Pelleted consumption is especially challenging in herbivorous reptiles. Reading product labels is important to ensure appropriate nutrition. For example, carnivorous animals require animal sources of protein.[1]

Nutritional Diseases

Nutritional secondary hyperparathyroidism

Nutritional secondary hyperparathyroidism (NSHP) is the most common nutritional disease of reptiles in captivity.[2,7] While there may be bias in what is presented to the hospital vs what is kept as pets, and their innate value to the owner, the most affected species include young tortoises, leopard geckos, and bearded dragons.

Etiology

Inadequate dietary calcium, unbalanced calcium: phosphorus ratio, lack of vitamin D_3, lack of UVB, or a combination thereof results in excess PTH, and subsequent NSHP. This is most often a lack of dietary vitamin D in non-herbivorous species and lack of UVB light in herbivorous species. Species eating whole prey are rarely if ever affected by NSHP due to the presence of calcium-rich bones.[2]

Pathophysiology

When ionized calcium decreases, PTH is released from the parathyroid gland, resulting in increased calcium resorption from bones, increased calcium reabsorption from kidneys, and increased absorption from the intestines. If the bones cannot replenish calcium, they become weaker. Furthermore, decreased ionized calcium can lead to clinical signs described below.[2]

Clinical signs

Clinical signs can range from nonspecific and vague to severe and life-threatening. Milder clinical signs include lethargy, weakness, anorexia, hyporexia, and lack of growth. Moderate clinicals signs include reduced ability to hold up the body, paresis, and plegia. Severe neurologic signs occur when blood ionized calcium levels cannot be maintained at adequate levels and include seizures, tetany, tremors, ataxia, flaccid paralysis, and death.[2]

Severe musculoskeletal abnormalities such as pathologic fractures, folding fractures, fibrous osteodystrophy, flattening of the ribs and scapula, bowing of the long bones, pliable jaws, kyphosis, scoliosis, and lordosis. In species with teeth, teeth may be translucent or glassy, deviated, or fall out. In chelonians, abnormalities of shell include deformities, pliable carapace or plastron, curving of the scutes, pyramiding, and black spot disease due to thinning and decreased ossification. Changes to the shell are permanent.[2]

Females with NSHP can suffer from egg retention, dystocia, follicular stasis, and soft-shelled eggs. Gastrointestinal signs include ileus, bloating, and colonic or cloacal prolapse. Gastrointestinal effects of NSHP need to be kept in mind during nutritional support for these critical animals.[2]

Diagnosis

Diagnosis can be achieved through multiple modalities. Strong suspicion can be gained from a thorough history of husbandry and physical examination findings.

Total and ionized calcium values are not necessarily correlated with each other, so having both can be helpful; ionized calcium has been established for some species. Calcium-deficient females may have elevated calcium due to egg production. Blood total and ionized calcium levels can be normal until skeletal reserves have been depleted. Thus, a gravid female with normal calcium levels may actually be calcium deficient. A calcium: phosphorus ratio below 1.5:1 is a concern. Plasma 25-hydroxycholecalciferol can be measured because calcitriol very unstable and has a very short half-life. While elevated PTH would be a definitive diagnosis, commercial mammalian assays are not accurate for reptiles.[2]

On radiographs, 40-50% of mineralization from the bones must be depleted before it is radiographically evident. Evidence of NSHP includes poor contrast from soft tissues, especially in the thinner bones, bowing of bones, absent metaphyseal trabecular markings, radiolucent metaphysis, folding fractures, or compression fractures of vertebrae. Bone density is best measured using quantitative CT or dual-energy X-ray absorptiometry.[2]

Treatment

Treatment depends on the severity of clinical signs and risk for new or more fractures to occur. Nutritional support is generally required for animals with this disease due to anorexia as well as the secondary gastrointestinal effects of NSHP.[2]

Exposure to sunlight and/or UVB light allows for cutaneous vitamin D synthesis as previously described. Administration of vitamin D_3 is not recommended because of the phosphorus increase associated with active vitamin D_3. While salmon calcitonin has been used to stimulate negative feedback on parathyroid gland to reduce PTH, this also results in the undesired effect of lowering calcium.[2]

Calcium gluconate at 50-100 mg/kg q12-24h can be administered IV, IO, or SC for moderate to severe cases of NSHP. Oral calcium supplementation may be started and the sole form of supplementation in mild cases of NSHP via calcium glubionate or calcium gluconate. Oral supplementation should be continued for 1-3 months until consistently eating calcium-rich foods, moving normally, appropriate husbandry is achieved, and with radiographic and biochemistry panel evidence of improvement. Calcium powder supplementation alone is not considered sufficient alone in treating any severity of NSHP when this disease is suspected.[2]

Some animals require cage and exercise restriction to prevent fractures or trauma, and to allow current fractures to heal. Fracture stabilization is best achieved with external coaptation; surgical reduction may result in progressive fracturing of these diseased bones. Analgesics are mandated for animals with fractures; injectable opioids can be used initially prior to transitioning to oral tramadol for at-home analgesia. Administration of NSAIDs should be considered as long as the patient is well hydrated and there is little to no concern for renal insufficiency.[2]

Prognosis depends on severity and pathologic changes present; multiple deficiencies are more challenging to recover from. Poor prognostic indicators include paralysis, urinary retention, and obstipation.[2]

Hypovitaminosis A

Pathophysiology

Retinoic acid differentiates keratinocytes into mature epidermal cells and maintains normal epithelium. Vitamin A deficiency leads to hyperkeratosis and squamous metaplasia, whereby cuboidal or columnar mucosal epithelium transform into keratinized stratified squamous epithelium. All portions of the body that contain epithelium can be affected. Deficiencies can take months to develop clinical signs because vitamin A is stored in the liver.[2]

Clinical signs

Given the extensive presence of epithelium, numerous body systems and clinical signs occur. These systemic clinical signs are often paired with weight loss, hyporexia, or anorexia, and lethargy. Ocular clinical signs include obstructed glands, xerophthalmia (dry eye), challenges in shedding the spectacle. Decreased tear production can lead to keratitis, corneal ulceration and/or erosion, keratomalacia, scarring, blindness. With chronicity, the eyes can swell shut and the yellow inflammatory exudate can form under the palpebrae, exacerbating damage to the cornea. In leopard geckos, periocular glands can distend and lead to cellulitis, resulting in abscesses periocularly or on the cranium.[2]

Lingual mucous glands can change, creating irregular thickening of the lips and mouth including along the oral commissures. Lingual nodules can be observed. Tympanic abscesses or peri-aural abscesses can occur in chelonians, chameleons, and leopard geckos. Pneumonia and upper respiratory tract infections are especially common in chameleons.[2]

Gout can occur due to renal tubular squamous metaplasia and hyperkeratosis. Recurrent hemipenal plugs can occur in males. Hypovitaminosis A often results in dysecdysis and dull coloration. Red-eared sliders will often lose the trademark red strip along their head and neck in chronic hypovitaminosis A.[2]

Diagnosis

Strong suspicion of this disease, like NSHP, can be gained from a thorough history of husbandry and physical examination findings, as well as response to therapy. Biopsy with histopathology showing squamous metaplasia is strongly suggestive of this disease, which has been weakly implicated in leopard geckos.[18] While vitamin A measurements are possible, this can be challenging because vitamin A rapidly degrades with light exposure, high temperatures, and autolysis. Plasma levels are not linear to intake or patient status, so liver levels provide the best information, and this is often via postmortem biopsy.[2]

Treatment

It is important to know that response is slow because epithelium must be replaced. Injectable vitamin A can be administered in severe cases, and oral supplementation is sufficient for most cases. Cellular debris in the eye must be carefully removed, with diligent flushing and lubrication of the eyes. Fluid and nutritional support must be provided as indicated.[2]

Prognosis depends on severity. Patients can recover in several weeks unless secondary or severe disease is present. Severe corneal ulcers and blindness worsen the prognosis because of the dependency on visual stimulus for the hunting response of many geckos.[2]

Toxicoses

Firefly intoxication

Fireflies (*Photinus* sp.) contain lucibufagin, a steroidal pyrone that is considered cardiotoxic. Clinical signs include focal facial seizures, violent head shaking, mouth gaping, protrusion or biting the tongue, nausea, vomiting, dyspnea, and black beard. Ingestion of one insect can be fatal in an adult bearded dragon, with clinical signs develop within 30-60 minutes of ingestion, and death within 90 minutes. No known effective therapy exists. Death has been reported in bearded dragons, a panther chameleon (*Furcifer pardalis*), and a Derjugin's lizard (*Darevskia derjugini*).[19]

Obesity

Obesity is unfortunately common in captive reptiles as a result of excess calorie intake and lack of exercise. Some breeders or producers push for rapid growth of reptiles and early breeding, which can contribute to obesity. Obesity is common in Varanidae, bearded dragons, red-eared sliders, and larger species of Boidae.[2]

Consequences of obesity include decreased lifespan, and predisposition to hepatic lipidosis, and yolk coelomitis and dystocia in females. In varanids and other lizards, the coelomic fat pads may constrict organs. Treatment involves gradual weight loss with caloric restriction, avoiding larval or neonatal foods that are high in fat, eliminating sugary items, monthly weighing, and exercise. Exercise can be promoted through larger cage setups, increasing climbing surfaces, live insects, hiding food to encourage foraging.[2]

Hepatic lipidosis

Normal fat storage sites in reptiles include the fat bodies and the liver. Hepatic fat stores are normal for certain reproductive states and seasons. In hepatic lipidosis (HL), normal lipid storage and metabolism is disrupted for multiple reasons, including obesity secondary to high carbohydrate or high fat diets, chronic hyporexia, certain vitamin or amino acid deficiencies that may impact liver function, and/or lack of exercise, hibernation, and reproductive activity.[2]

Pathophysiology

Hepatic lipidosis is the pathologic accumulation of lipid within hepatocytes. Starvation leads to depleting glucose and decreased insulin release. To compensate, peripheral lipolysis occurs to release free fatty acids into circulation, which travel to the liver and accumulate as triglycerides. Deficiencies in methionine, arginine, and carnitine, which play a role in lipid metabolism, may contribute to this disease process.[2]

Clinical signs

Clinical signs include hyporexia or anorexia, lethargy, weight loss, preovulatory stasis, elevated bile pigments and diarrhea. Rarely, neurologic signs can be observed, such as altered mentation, stupor, and ataxia.[2]

Diagnosis

Diagnosis is achieved through history and physical examination findings if the animal is obese or was reported by the owner to weigh significantly more prior to presentation. Changes to complete blood counts and plasma biochemistry can include anemia, hypoproteinemia, hypocalcemia, hypercholesterolemia, and hypertriglyceridemia; liver values are often normal.[2] Imaging (radiography, computed tomography [CT], ultrasonography) may show hepatomegaly, a hyperechogenic liver compared to the fat bodies, and/or negative Hounsfield units.[20] Definitive diagnosis is achieved through liver biopsy and histopathology;[21,22] placing the patient under sedation or anesthesia for these diagnostic biopsies should be balanced with the stability of the patient. Recent research on β-hydroxybutyric acid values in bearded dragons suggests this may be used as a supplemental diagnostic for HL for some individuals.[20]

Treatment

Nutritional support is paramount in this condition, whether through syringe, gavage, or esophagostomy tube feedings. Carnivore diets are often too high in fat, and supplements with amino acids can be provided.[2]

Prognosis is guarded. Many reports of HL in reptiles note long recovery periods, so clear communication with the client about at-home support and dedication to their pet is important.[2]

Periodontal disease

Periodontal disease occurs in lizards with acrodont dentition, namely Agamidae and Chameleonidae. This is thought to occur secondary trauma to or desiccation of the fragile gingival tissue along the maxilla and mandible, predisposing to periodontitis, stomatitis, and osteomyelitis in severe cases.[23] Acrodont teeth are attached to the biting edge of the jaws without sockets and are neither shed nor replaced.[24]

Gout and hyperuricemia

Gout is the deposition of uric acid in tissues, whereas hyperuricemia is elevated blood uric acid that can lead to gout.[25]

Etiology

Gout and hyperuricemia may be related to excessive uric acid production via excess dietary intake or reduced excretion secondary to renal insufficiency. High protein diets, dehydration, suboptimal temperatures, rapid growth and overfeeding, and nephrotoxic drugs can lead to gout. Renal disease in reptiles is often secondary to dehydration, which is important to educate the owner that this can be prevented with adequate at-home care. Hyperuricemia can lead to monosodium urate deposition in the joints (articular gout), around the joints (peri-articular gout), and in organs such as the spleen, liver, kidney, pericardium, and central nervous system (visceral gout). Clinical signs can be vague and include hyporexia and dehydration. Articular or peri-articular gout can lead to articular swelling.[25]

Diagnosis

Hyperuricemia is suggestive of gout, but not all patients with gout have hyperuricemia. Tophi, once mineralized, can be seen radiographically as increased areas of opacity. Aspirates of affected joints or organs yield monosodium urate crystals in fluid or tissue. On light microscopy, these tophi appear needle-like with strong negative birefringence. It is important to look at aspirates right away because pH and temperature affect solubility. Uric acid is removed by formalin and histologic processing, leaving characteristic holes with inflammation.[25]

Treatment

If uric acid is elevated, consider the use of allopurinol, which blocks xanthine oxidase to reduce the breakdown of xanthine to uric acid. This leads to excretion of xanthine and hypoxanthine, which requires more water for excretion; therefore, fluid therapy and soaking is needed. Gout is a painful condition, so analgesics are strongly recommended. Surgical removal of urate tophi from critical areas may be beneficial. Gout is not curable, but early cases may respond to medical therapy.[25]

Prognosis is poor in advanced stages. Visceral gout is difficult to treat and carries a guarded prognosis.[25]

Nutritional Support for Sick Reptiles

Regular diets can sometimes be used for assist feeding, whereby the reptile is hand- or tease-fed. If assist feeding is refused or not tolerated, gavage feeding is performed, or an esophagostomy tube may need to be placed. Prior to feeding, the patient needs to be of appropriate temperature and food warmed up as indicated. In cases of physiologic anorexia such as hibernation, torpor, or breeding, refeeding syndrome is not observed; however, refeeding syndrome can cause significant morbidity and mortality in sick, starved, and cachectic animals. Therefore, restarting to feed a reptile should occur very conservatively and slowly, and only after rehydration has occurred.[26]

Reptiles have larger intracellular fluid volume with presumed lower osmolality relative to mammals. Maintenance fluid rates are considered between 10-30 ml/kg/d. Rehydration usually occurs over 72-96 hours, and routes include soaking, subcutaneous, intravenous, and intraosseous. Subcutaneous administration of fluids has found to be similarly effective compared to intracoelomic administration with far less iatrogenic complications in bearded dragons.[27] Oral administration of fluids can be considered in mild dehydration, or with chelonians with e-tubes that are also receiving subcutaneous or intravenous fluids. Cloacal administration of fluids in chelonians can be performed with the intent of refilling the bladder, an important site of water storage.[26]

Feeding methods

Stomach tubing

Red rubber catheters or speculums may be used for stomach feeding. This is generally easy to do in lizards and snakes but is more challenging in chelonians. Stomach volume has not been determined in all species, but a rule of thumb is 5-25 ml/kg; start with smaller volumes first.[28]

Esophagostomy tubes

Esophagostomy tubes (E-tubes) are placed in chelonians when gavage feeding may cause too much stress and/or physical trauma, it is not possible to perform a gavage feeding due to patient temperament, and/or it is anticipated that the chelonian will need long-term nutritional support. Placement of e-tubes in chelonians is similar to felines, using 14- 24-Fr red rubber tubes or soft silicone tubes. Verification of correct placement is through contrast radiography or computed tomography (CT). These tubes can be left in place for prolonged periods of time.[29]

Nutritional support

Whole prey food items can be liquefied in a blender. For species that will tolerate syringe feedings, gavage feedings, or for individuals with an e-tube in place, several commercial diets can be used. These diets are nutrient-dense, easily digested, assimilated. Numerous commercial diets can be used; examples include Oxbow and Emeraid products.[26]

How to Increase Client Compliance

Mastering clinical reptilian husbandry is challenging enough for clinicians and staff; however, arguably the biggest roadblock is client compliance. Based on the authors' clinical experience, private owners commonly balk at the financial investment required to optimize husbandry and the time it requires to manage the reptile's enclosure and prey colony, if applicable. In addition, there are decades of information published in books and on the internet from well-intentioned hobbyists or breeders. This information is not always scientifically viable especially as more researchers invest in studying environmental and dietary factors related to this taxon.

As aforementioned, ensuring the owners submit or staff members collect a thorough husbandry history. This cannot be overemphasized. Many zoological companion animal diseases, including species of reptiles, get diagnosed with disorders that could have been prevented with proper routine husbandry. To save time during the appointment, have clients fill out an online form prior to the appointment. Incentivizing this behavior may improve client compliance. For example, the clinic can offer a free nail trim or beak trim (in applicable species), fecal wet mount, or a small discount on the appointment itself. Prioritize a full husbandry acquisition for new patients and annual exams.

In addition to the husbandry questionnaire, ask clients to bring or email pictures of their companion's enclosure, light labels or the bulb boxes, substrate bags, and formulated diet and supplement bottles. Considering how commonplace online shopping is in western society, clients may be able to quickly access previous orders on their preferred virtual marketplace.

Budgeting time within the appointment to allow the clinician or their staff to thoroughly review the importance of proper nutrition and overall husbandry is vital. Remember to show compassion; many of them 'know not what they do.' Explain the link between nutrition, lighting, temperature, hydration, and health. While the investment is high upfront, the family typically gets to spend more years with their companion. Another way to exhibit compassion is to include the financial investment pertaining to improvement of the patient's husbandry into the conversation surrounding the short-term and long-term treatment plan.

References

1. Rossi JV. General Husbandry and Management. In: Divers SJ, Stahl SJ, eds. *Mader's Reptile and Amphibian Medicine and Surgery.* 3rd Ed. St. Louis, MO: Elsevier; 2019:109-130.

2. Boyer TH, Scott PW. Nutritional diseases. In: Divers SJ, Stahl SJ, eds. *Mader's Reptile and Amphibian Medicine and Surgery.* 3rd Ed. St. Louis, MO: Elsevier; 2019:932-950.

3. Boyer TH, Scott PW. Nutrition. In: Divers SJ, Stahl SJ, eds. *Mader's Reptile and Amphibian Medicine and Surgery.* 3rd Ed. St. Louis, MO: Elsevier; 2019:201-223.

4. de Voe RS. Nutritional support of reptile patients. *Vet Clin North Am Exot Anim Pract.* 2014;17(2):249-261.

5. Holz PH. Anatomy and physiology of the reptile renal system. *Vet Clin North Am Exot Anim Pract.* 2020;23(1):103-114.

6. Parkinson L, Mans C. Effects of furosemide administration to water-deprived inland bearded dragons (*Pogona vitticeps*). *Am J Vet Res.* 2018;79(11):1204-1208.

7. Raiti P. Endocrinology. In: Divers SJ, Stahl SJ, eds. *Mader's Reptile and Amphibian Medicine and Surgery.* 3rd Ed. St. Louis, MO: Elsevier; 2019:835-848.

8. Berg KJ, Schexnayder M, Grasperge BJ, et al. Single time point reference intervals for complete blood counts and select biochemistries in juvenile red-footed tortoises (*Chelonoidis carbonaria*). *J Herpetol Med Surg.* 2021; 31(2): 124-131.

9. Rossi JV. General husbandry and management. In: Divers SJ, Stahl SJ, eds. *Mader's Reptile and Amphibian Medicine and Surgery.* 3rd Ed. St. Louis, MO: Elsevier; 2019:109-130.

10. Baines FM, Cusack LM. Environmental lighting. In: Divers SJ, Stahl SJ, eds. *Mader's Reptile and Amphibian Medicine and Surgery.* 3rd Ed. St. Louis, MO: Elsevier; 2019:131-138.

11. Bos JH, Klip FC, Oonincx DGAB. Artificial ultraviolet B radiation raises plasma 25-hydroxyvitamin D_3 concentrations in Burmese pythons (*Python bivittatus*). *J Zoo Wildl Med.* 2018;49(3):810-812.

12. Boykin KL, Carter RT, Butler-Perez K, et al. Digestibility of black soldier fly larvae (*Hermetia illucens*) fed to leopard geckos (*Eublepharis macularius*). *PLoS ONE.* 2020;15(5):e0232496.

13. Cojean O, Lair S, Vergneau-Grosset C. Evaluation of β-carotene assimilation in leopard geckos (*Eublepharis macularius*). *J Anim Physiol Anim Nutr (Berl).* 2018;102(5):1411-1418.

14. Scheelings TF, Hellebuyck T. Dermatology - skin. In: Divers SJ, Stahl SJ, eds. *Mader's Reptile and Amphibian Medicine and Surgery.* 3rd Ed. St. Louis, MO: Elsevier; 2019:699-711.

15. Hellebuyck T, Simard J, van de Velde N, Geerinckx L. Acute ramson (*Allium ursinum*) toxicosis in captive tortoises. *J Herpetol Med Surg.* 2019;29(1-2):34-39.

16. Boykin K, Bitter A, Mitchell MA. Using a commercial gut loading diet to create a positive calcium to phosphorus ratio in mealworms (*Tenebrio molitor*). *J Herpetol Med Surg.* 2012;31(4):302-306.

17. Latney LV, Toddes BD, Wyre NR, et al. Effects of various diets on the calcium and phosphorus composition of mealworms (*Tenebrio molitor* larvae) and superworms (*Zophobas morio* larvae). *Am J Vet Res.* 2017;78(2):178-185.

18. Wiggans KT, Guzman DSM, Reilly CM, et al. Diagnosis, treatment, and outcome of and risk factors for ophthalmic disease in leopard geckos (*Eublepharis macularius*) at a veterinary teaching hospital: 52 cases (1985-2013). *J Am Vet Med Assoc.* 2018;252(3):316-323.

19. Fizgerald KT, Martinez-Silvestre A. Toxicology. In: Divers SJ, Stahl SJ, eds. *Mader's Reptile and Amphibian Medicine and Surgery.* 3rd Ed. St. Louis, MO: Elsevier; 2019:977-991.

20. Barboza TK, Susta L, zur Linden A, et al. Association of plasma metabolites and diagnostic imaging findings with hepatic lipidosis in bearded dragons (*Pogona vitticeps*) and effects of gemfibrozil therapy. *PLoS ONE.* 2023;18(2):e0274060.

21. Barboza T, Beaufrère H, Reavill D, Susta L. Morphological features of hepatic lipid changes in bearded dragons (*Pogona vitticeps*), and a proposed grading system. *Vet Pathol.* 2023;60(1):123-132.

22. Barboza T, Beaufrère H, Reavill D, Susta L. Morphological features of hepatic lipid changes in bearded dragons (*Pogona vitticeps*), and a proposed grading system. *Vet Pathol.* 2023;60(1):133-138.

23. de Voe R. Gastroenterology - oral cavity, esophagus, and stomach. In: Divers SJ, Stahl SJ, eds. *Mader's Reptile and Amphibian Medicine and Surgery.* 3rd Ed. St. Louis, MO: Elsevier; 2019:752-760.

24. Barten S, Simpson S. Lizard taxonomy, anatomy, and physiology. In: Divers SJ, Stahl SJ, eds. *Mader's Reptile and Amphibian Medicine and Surgery.* 3rd Ed. St. Louis, MO: Elsevier; 2019:63-74.

25. Divers SJ, Innis CJ. Urology. In: Divers SJ, Stahl SJ, eds. *Mader's Reptile and Amphibian Medicine and Surgery.* 3rd Ed. St. Louis, MO: Elsevier; 2019:624-648.

26. Boyer TH, Scott PW. Nutritional therapy. In: Divers SJ, Stahl SJ, eds. *Mader's Reptile and Amphibian Medicine and Surgery.* 3rd Ed. St. Louis, MO: Elsevier; 2019:1173-1176.

27. Minor RL, Doss GA, Mans C. Evaluation of glucose absorption rates following intracoelomic or subcutaneous administration in experimentally dehydrated inland bearded dragons (*Pogona vitticeps*). *Am J Vet Res.* 2021;82(11):920-923.

28. Perry SM, Mitchell MA. Routes of administration. In: Divers SJ, Stahl SJ, eds. *Mader's Reptile and Amphibian Medicine and Surgery.* 3rd Ed. St. Louis, MO: Elsevier; 2019:1130-1137.

29. Cutler D, Divers SJ. Esophagostomy tube placement. In: Divers SJ, Stahl SJ, eds. *Mader's Reptile and Amphibian Medicine and Surgery.* 3rd Ed. St. Louis, MO: Elsevier; 2019:429-431.

Turtle Euthanasia Demystified

Renée Schott, DVM, CWR

Session #3442

Affiliation: From The Wildlife Rehabilitation Center of Minnesota, 2530 Dale St. N, Roseville, MN 55113, USA.

Abstract: Many veterinarians struggle with turtle euthanasia. Many have heard of euthanized turtles moving after being declared dead, which has led to odd turtle euthanasia practices in an attempt to prevent reanimation. There is abundant literature on turtle physiology and consciousness, yet it has not been applied it to euthanasia practices. This highly referenced and interesting master class reveals why turtles are so hard to euthanize and provides a protocol that will minimize turtle suffering during the euthanasia process and ensure non-reversible death.

Previously published in Schott R. Turtle euthanasia demystified. Proc ExoticsCon. 2021;483-489 and in Schott R. Turtle euthanasia demystified. Proc ExoticsCon 2022;428-435.

Introduction

Reptiles have all the anatomy necessary to feel pain and suffer and therefore should be treated with the same humane standards as other species, which includes providing them with proper analgesics.

The term euthanasia is derived from the Greek terms eu meaning well or good, and Thanatos meaning death. It is defined as ending the life of an animal in a way that eliminates or minimizes pain and distress.[1,2] The technique employed should result in rapid loss of consciousness followed by cardiac or respiratory arrest and, ultimately, a loss of brain function. Actually determining the pain/distress of a method of euthanasia can be very difficult because as humans we will never fully know/understand the subjective experience of the animal.[3] We use our best judgement; paddling, vocalizations, convulsions before apparent loss of consciousness is obviously suffering. We also use our knowledge of physiology and assume suffering in the absence of behaviors if a physiologic process theoretically leads to suffering. Pain, distress and suffering are subjective affective experiences that are perceived in the brain. Therefore, one must be conscious and alive to suffer. General rule: a gentle death that takes longer is preferable to a rapid but more distressing death and conversely, if all other methods are equally humane, the quickest method should be chosen.[4] Taking a life is abhorrent to many, we often forget this and need to keep it in mind.

Turtle Physiology Pertinent to Euthanasia Methods

Natural history

Turtles are poikilothermic (survive at a wide temperature range), and ectotherms (get their body heat from the outside world). Painted and snapping turtles have been seen brumating simply in the water (not in mud) but also in the mud, muskrat burrows/lodges, solo, together, seeps/springs, cattail stands, under banks, under logs, out in open water, etc, so many different environments.[5]

Anoxia tolerant brains

In commonly studied reptile species, anoxia tolerance appears to vary widely to opposite ends of the spectrum, although it is possible that if more species were investigated anoxia tolerance would be found as a relatively continuous spectrum.[5] Varying degrees of anoxia tolerance has been documented in chelonian species. These differences can be attributed in part to several metabolic factors, including the degree of metabolic depression, the effectiveness of lactic acid buffering, and the capacity of glycogen stores.[6] Electrical and biochemical brain activity is believed to persist in the anoxic chelonian brain for a considerable period of time.[7] Some species of freshwater turtles (*Trachemys* and *Chrysemys* genera) can survive many weeks without oxygen (commonly used as model animals for vertebrate anoxia tolerance).[8] Some freshwater turtle brains can survive anoxia for months (ie under the water during winter),[9] *Trachemys scripta* which can withstand complete anoxia for days at room temperature to weeks in winter hibernation.[5,6] Even at room temperature, 24 hours of anoxia and re-oxygenation results in no evident loss of neurons.[10]

During anoxic periods, these turtles may appear comatose due to brain and body down regulation: Turtles can down-regulate their brains to become virtually comatose.[9] One mechanism is to extend anoxic survival is entrance into a state of deep reversible hypo-metabolism.[11] Energy demand is reduced to meet the energy supplied by anaerobic glycolysis.[11] Energy demanding processes are greatly suppressed.[6,11] Ion channels, and neurotransmitters are downregulated by multiple intertwined mechanisms, suppression of action potentials, protein regulation, reviewed here.[11]

Neuroprotection (and why they don't die of reperfusion): A variety of protective mechanisms are activated at the molecular level in anoxic turtle brain, many of which may not only protect against damage under anoxic conditions, but also ameliorate oxidative stress when oxygen is restored.[11] These include increases in heat shock proteins, anti-apoptotic factors, the MAP kinases, antioxidants and modulation of the p53 pathway. Unlike the mammalian brain, which shows an overproduction of reactive oxygen species (ROS) following hypoxia or ischemia/reperfusion, the turtle brain appears to suppress ROS production upon re-oxygenation.[12]

Latitudes matter: In the northern parts of their ranges in North America, turtles may spend more than half of their lives in an overwintering state.[5] Ontario turtles tolerate anoxia much better vs Nebraska turtles.[13] The three northern subspecies are very anoxia-tolerant, and can remain responsive for more than 150 days in anoxic water, but the southern subspecies can survive only about half that time, having a more rapid fall in pH.[5] Northern turtles are better able to buffer accumulated lactate vs southern turtles, and thus survive longer in anoxic water.[14] Aquatic turtles with far northern ranges are more anoxia tolerant than southern animals even within a single species (probably because of the need to survive potentially long periods under ice or in hypoxic mud).[5]

Lactic acid buffering

Lactic acid can be both buffered and sequestered by bone. Chelonians may have particularly good capacity for buffering since there is so much well vascularized bone in the shell.[5,6]

Brain cell regeneration

It has been shown that after significant ischemic damage, within 3 weeks *T. scripta* has evidence of neuronal reproduction.[15] There are robust repair mechanisms present including neurogenesis of damaged brain cells.[12]

Ages in some species

Baby painted turtles hatch in fall, stay in ground, and live off yolk and come out in the spring.[13] Interestingly, hatchlings of many turtle species survive in the nest in their first winter through freeze tolerance or super-cooling;[16] however, in general hatchlings are far less able to tolerate anoxic submergence than adults. This could be due to their incomplete shell development, as the shell is important for lactate buffering.[5]a major portion of their life cycle has received relatively little attention until recently - namely what turtles are doing, and how they are doing it, during the winter. In the northern parts of their ranges in North America, turtles may spend more than half of their lives in an overwintering state. In this review, I emphasise the ecological aspects of overwintering among turtles, and consider how overwintering stresses affect the physiology, behaviour, distributions, and life histories of various species. Sea turtles are the only group of turtles that migrate extensively, and can therefore avoid northern winters. Nevertheless, each year a number of turtles, largely juveniles, are killed when trapped by cold fronts before they move to safer waters. Evidently this risk is an acceptable trade-off for the benefits to a population of inhabiting northern developmental habitats during the summer. Terrestrial turtles pass the winter underground, either in burrows that they excavate or that are preformed. These refugia must provide protection against desiccation and lethal freezing levels. Some burrows are extensive (tortoise genus Gopherus Snapping turtles emerge the same year they were laid as they cannot overwinter in the nest.

Unacceptable Euthanasia Methods

Hypothermia

Placing an animal in a freezer to cause death or dropping it into liquid nitrogen is unacceptable. These methods as standalone are not acceptable for euthanasia of reptiles, as ice crystals formed in the body are presumed to be painful, and the turtle might not die (consider physiology previously described).

Isoflurane

Because isoflurane[2] (and other gas anesthetics such as halothane or sevoflurane) have a noxious smell and can irritate respiratory epithelium, animals may hold their breath. In particular, turtles (because of the physiology mentioned above) can hold their breath for hours or days when exposed to isofurane.[17] For this reason, isoflurane as a euthanasia method is not recommended.[18]

Hyperthermia

Raising the temperature above the critical temperature of the species. This method is not acceptable for euthanasia of turtles.

Suffocation and drowning

Depriving animals of oxygen. This method is not acceptable for euthanasia of turtles.

Traditional Veterinary Euthanasia

Anesthesia

Anesthesia should first be induced in any turtle being euthanized. Whatever doses of whatever drugs you use for surgery in the species you are seeing, double it. For painted turtles and snapping turtles, this author uses alfaxalone (20 mg/kg) IM or SC and then 15 minutes later follows that up with 10-20 mg/kg propofol IV. Turtles are often too small to hit a vein awake so anesthetize with alfaxalone and then consider the occipital venous sinus.

Sodium pentobarbital: Sodium pentobarbital is a barbiturate developed in 1930s as an anesthetic[19] that has fallen out of use as an anesthetic. Currently, it is overdosed to cause death/euthanasia, and has been recommended as an effective and human method of euthanasia in reptiles.[3]

Sodium pentobarbital comes in two general formulations (with and without phenytoin), both of which are ~390 mg/ml and pHs are between 9.6-13. Routes of administration include:

- Intravenous
  - Ideally, pentobarbital is given intravenous (100 mg/kg).
  - May be difficult and likely stressful to try to hit a vein in an awake animal as most reptile pheblotomy is done blindly.
  - But the pH may cause venous endothelium cell damage on contact (painful).
- Intracardiac injection may only be used on a fully anaesthetized animal as this is very painful.[18]
- Intra-lung is presumed to be painful and should not be performed.
- Intracoelomic administration is described; however, all formulations of sodium pentobarbital are caustic, and some advise they not be given extravascularly as it is perceived to be painful.[20] In addition, it is slower acting.[18]
- Per os: administering sodium pentobarbital, typically via gavage tubing as it tastes bitter, has been used by some veterinarians but may be prohibited in certain states. Barbiturates have high oral bioavailability and can reach the brain quite quickly. A disadvantage is this method will be slower than IV so in a moribund animal may be ineffective.
- Intramuscular/subcutaneous should not be used as they are not effective and may cause pain.[18]

Sodium pentobarbital works in stages in mammals. These stages are:

Stage 1. Voluntary excitement

- Bloodstream → heart → cerebral cortex
- Begins to lose consciousness/coordination, hyperesthesia, loses voluntary motor, loss of superficial pain → disorientation, possible movement

Stage 2. Involuntary excitement

- Bloodstream → heart → cerebral cortex → cerebrum
- Loss of consciousness → possible uncontrolled motor activity (paddling, vocalizations)

Stage 3. Surgical anesthesia

- Bloodstream → heart → cerebral cortex → cerebrum → cerebellum (~5 sec)
- Cannot feel pain, doesn't respond to visual/auditory, reflexes begin to disappear
- Because of depressed metabolic rates in sick, dying, cold or otherwise depressed turtles, many individual turtles may not circulate the pentobarbital quickly enough to advance from this stage to Stage 4. If this happens, the turtle might wake up hours or days later.

Stage 4. Medullary paralysis

- Bloodstream → heart → cerebral cortex → cerebrum → cerebellum
- Depresses breathing, heartbeat, blood pressure→anoxia→brain cell death* (~40 sec) → possible heart fibrillation, agonal breaths, muscle spasms (unconscious, reflex)
- Twenty percent of the oxygen consumed by the body is used by the human brain[11] (making ATP to maintain membrane potentials). Thus, the brain is one of the first organs to fail in anoxia, interruption of the oxygen supply to the brain for more than a few minutes leads to irreversible neuronal death.[11]
- Even if a turtle is circulating the pentobarbital quickly enough to progress into Stage 4 and stop the heart, the brain cells do not die within a few minutes as in mammals. Turtle brains are so tolerant to hypoxia (see previous physiology described) that even with the heart stopped, brain cells could still be alive and may be interpreting pain.

Confirmation of Death

Confirmation of death in turtles can be difficult. The AVMA Guidelines state: "it is often difficult to ascertain that an amphibian or reptile is, in fact, dead." Although this statement has become a paradigm when dealing with euthanasia of herpetologic species, the truth is there are methods to ensure reptiles and amphibians are indeed dead, and it is our responsibility as veterinarians to educate ourselves and apply those methods in a manner that respects their welfare.

It seems remedial to discuss how we confirm death; however, with the unique physiology of turtles, it bears some extra attention. Traditionally, many interesting practices are performed to ensure death of turtles. These include the following.

After sodium pentobarbital, leave the turtle on a heating pad for 24 hours

While this method may help the pentobarbital circulate quick enough to bump the turtle from State 3 to Stage 4, it often is not enough to fully stop the heart. Even if it is, the heart stopping doesn't mean the brain cannot perceive pain, because brain cells are so tolerant to hypoxia. The brain could still be interpreting pain.

Check for a heartbeat with a doppler

Turtle hearts can beat so slow and still maintain life. During normal brumation, painted turtle hearts beat a little over 1 beat per 10 min. No veterinarian has placed a doppler probe on a turtle, listened for 10 minutes and then felt confident they were in the correct location enough to declare no heartbeat. Even if that did happen, again, lack of heartbeat does not mean the brain cell death. The brain could still be interpreting pain.

Declare death when the smell of decay is present

The smell of decay only means body cells are dying; it does not mean brain cells are dying. Review physiology above; turtle brains are so tolerant to hypoxia; they can survive past body cells surviving. The brain could still be interpreting pain.

Physically destroying the cerebrum (after anesthesia, sodium pentobarbital)

Pain is perceived in the cerebrum. This is where suffering happens. Death should always be confirmed by physical intervention, that is, destruction of the cerebrum. Death is judged to have occurred when recovery of brain activity is not possible.[7] Humane methods must achieve nonreversible damage or destruction of the brain tissue to ensure death.

Reflexes and other movements of limbs may remain intact after cerebrum destruction because of increased number of spinal reflexes in these species and/or tissues being tolerant to hypoxia. One must ask, if the cerebrum and the ability of the brain to perceive pain is destroyed, does it matter that the limbs are moving, or the heart is beating? The answer is no.

Author's Recommendation for Turtle Euthanasia

The entire sequence below is performed on every turtle euthanasia at WRCMN. It takes ~45-60 minutes time in total but ensures a complete death, no continued perception of pain, and prevents reanimation. After the below procedure, the body can immediately go into the freezer or be destroyed.

1. Alfaxalone double surgery dose (10-20 mg/kg) SC or IM into forelimbs. Subcutaneous is less reliable because may be inadvertently administered into a fat pad but intramuscular may be painful because of the large volume needed. Consider administering partial amounts into both forelimbs, or partial amount subcutaneously. In the author's experience, forelimb administration is more reliable vs rear limb administration, presumably due to the renal-portal system. Wait 20 minutes before proceeding to next step. If turtle is not anesthetized in 20 minutes, repeat this step.

2. Propofol 10-20 mg/kg IV. If the turtle is too small to hit a vein, utilize the occipital venous sinus. Wait 20 minutes before proceeding to next step.

3. Pentobarbital IV same volume as propofol. The occipital venous sinus is the most convenient venous access for this step in the author's experience. Wait 20 minutes before proceeding to next step. After this step, painted turtles' neck will extend dorsally; snapping turtles do not.

4. Destroy the cerebrum. The author prefers to destroy the cerebrum by pithing it with a long needle/pin. Measure from the back of the occipital protuberance to the eyes and ensure your needle/pin is long enough. Insert the needle/pin as you would for an occipital venous stick but advance it further. Scrape the inside of the cranium to confirm you are in the correct location and move the needle/pin around to destroy the cerebrum. Remember you need to destroy the cerebrum, not just the brainstem. Other physical methods may be used instead of pithing but are outside the scope of this class.

References

1. Canadian Council on Animal Care. Overview of acceptable euthanasia methods. *Canadian Council on Animal Care Guidelines on: Euthanasia of Animals Used in Science.* 2010:12-16.

2. Leary S, Underwood W, Anthony R, et al. *AVMA Guidelines for the Euthanasia of Animals: 2020 Edition.* 2020.

3. Meyer RE. Euthanasia and humane killing. In: Grimm KA, Lamont LA, Tranquilli WJ, et al, eds. *Veterinary Anesthesia and Analgesia.* 5th Ed. Ames, IA: Wiley Blackwell; 2015:130-146.

4. Hawkins P, Prescott MJ, Carbone L, et al. A good death? Report of the second Newcastle meeting on laboratory animal euthanasia. *Animals.* 2016;6(9):1-28. doi:10.3390/ani6090050.

5. Ultsch GR. The ecology of overwintering among turtles: where turtles overwinter and its consequences. *Biol Rev Camb Philos Soc.* 2006;81(May):339-367. doi:10.1017/S1464793106007032.

6. Jackson DC, Ultsch GR. Physiology of hibernation under the ice by turtles and frogs. *J Exp Zool. Part A Ecol Genet Physiol.* 2010;313 A(6):311-327. doi:10.1002/jez.603.

7. Pizzi R, McArthur S. Euthanasia technique for chelonians. *Vet Rec.* 2004;154(19):607-608.

8. Krivoruchko A, Storey KB. Turtle anoxia tolerance: Biochemistry and gene regulation. *Biochim Biophys Acta.* 2015;1850(6):1188-1196. doi:10.1016/j.bbagen.2015.02.001.

9. Nilsson GE, Lutz PL. Anoxia tolerant brains. *J Cereb Blood Flow Metab.* 2004;24(5):475-486. doi:10.1097/00004647-200405000-00001.

10. Kesaraju S, Schmidt-Kastner R, Prentice HM, Milton SL. Modulation of stress proteins and apoptotic regulators in the anoxia tolerant turtle brain. *J Neurochem.* 2009;109(5):1413-1426. doi:10.1111/j.1471-4159.2009.06068.x.

11. Larson J, Drew KL, Folkow LP, Milton SL, Park TJ. No oxygen? No problem! Intrinsic brain tolerance to hypoxia in vertebrates. *J Exp Biol.* 2014;217(Pt 7):1024-1039. doi:10.1242/jeb.085381.

12. Milton SL, Prentice HM. Beyond anoxia: the physiology of metabolic downregulation and recovery in the anoxia-tolerant turtle. *Comp Biochem Physiol A Mol Integr Physiol.* 2007;147(2 SPEC. ISS.):277-290. doi:10.1016/j.cbpa.2006.08.041.

13. Storey JM, Storey KB. Natural freezing survival in animals. *Annu Rev Ecol Syst.* 1996;27:365-386.

14. Reese SA, Stewart ER, Crocker CE, et al. Geographic variation of the physiological response to overwintering in the painted turtle (*Chrysemys picta*). *Physiol Biochem Zool.* 2004;77(4):619-630.

15. Kesaraju S, Milton SL. Preliminary evidence of neuronal regeneration in the anoxia tolerant vertebrate brain. *Exp Neurol.* 2009;215(2):401-403. doi:10.1016/j.expneurol.2008.10.017.

16. Packard GC, Packard MJ. Natural freeze-tolerance in hatchling painted turtles? *Comp Biochem Physiol Part A Mol Integr Physiol.* 2003;134(2):233-246. doi:10.1016/S1095-6433(02)00264-7.

17. Cooper JE. Humane euthanasia and post-mortem examination. In: Girling S, Raiti P, eds. *BSAVA Manual of Reptiles.* 2nd ed. Gloucester, UK: British Small Animal Veterinary Association; 2004:168-183.

18. Swiss Veterinary Office. Expert Panel. *Analysis on Humane Killing Methods for Reptiles in the Skin Trade.* 2013.

19. Tranquilli WJ, Grimm KA. Use, definitions, history, concepts, classification and considerations for anesthesia and analgesia. In: Grimm KA, Lamont LA, Tranquilli WJ, et al, eds. *Veterinary Anesthesia and Analgesia.* 5th Ed. Ames, IA: Wiley Blackwell; 2015:5.

20. Warren K. Reptile euthanasia – no easy solution? *Proc Annu Conf Austr New Zealand Council Care Anim Res Teach Thinking Outside the Cage: A Different Point of View.* 2012;111-113.

Section 39

ARAV Infectious Disease Session

Colin McDermott, VMD, DABVP (Reptile and Amphibian Practice), CertAqV
Session Chair

Brucellosis Caused by *Brucella* spp. Isolated from *Ceratophrys cranwelli* in Taiwan

Chu Yung Weng, DVM

Session #3350

Affiliation: From No. 15, Ziqiang 6th St., Zhubei City, Hsinchu County 302044 , Taiwan R.O.C.

Abstract: Atypical *Brucella* spp. causes brucellosis in amphibians and appears to be a potential zoonotic disease. Systemic disease caused by *Brucella* spp. may lead to death of amphibians. A 2-year-old intact female green Pacman frog (*Ceratophrys cranwelli*) presented to Mystery Exotic Animal Hospital. An abscess was noted on the right limb; red-leg syndrome was also noted. Paravertebral ganglia swelling was suspected in x-ray. *Brucella* spp. was detected by bacterial culture and sensitivity test of the abscess. The patient was treated with enrofloxacin according to the sensitivity test. After 5 months, remission of suspected paravertebral ganglia lesion in x-ray was noted and there was no recurrence of abscess.

Introduction

Brucella are gram-negative facultative intracellular bacteria that can infect many organs and soft tissues. Brucellosis is an important zoonotic disease worldwide caused by *Brucella* which was associated with warm-blooded animals in the past; however, the *Brucella* genus has been detected in amphibians and classified as an atypical strain. Several reports reveal that atypical *Brucella* strains have been isolated from several amphibians, including a big-eyed tree frog (*Leptopelis verniculatus*), several African bullfrogs (*Pyxicephalus Edulis*), White's tree frog (*Litoria Caerulea*), Pacman frog (*Ceratophrys ornata*), waxy tree frogs (*Phyllomedusa sauvagii*) and Colorado river toads (*Incilius alvarius*). Furthermore, there was one case report of human brucellosis caused by amphibian-type *Brucella* in 2021. The role of *Brucella* infection in amphibians is poorly understood, the potential of it being a novel emerging pathogen should be considered.

Case Report

A 2-year-old female green Pacman frog (*Ceratophrys cranwelli*) presented to Mystery Exotic Animal Hospital (MEAH) in November 2022 for an isolated 6 × 5 mm firm swelling on the right limb and lethargy; red leg syndrome was noted. Plain radiograph reveale periosteal reaction on the right femur and soft tissue opacity swelling caudal to the femur. In addition, there were several hyper-opacity nodules along the vertebrae which were suspected to be paravertebral ganglia. There were hyper-opacity feces in the intestine. Approximately 0.5 ml purulent material was aspirated from the swelling site and sent to the Animal Disease Diagnostic Center College of Veterinary Medicine, National Chung Hsing University (NCHU) for aerobic and anaerobic bacterial culture and sensitivity test.

The purulent material was inoculated on a blood agar plate (tryptic soy agar with 5% sheep blood) and MacConkey agar under 5% carbon dioxide environment for aerobic culture, anaerobic culture under anaerobic environment produced by MGC AnaeroPack and anaerobic indicator with anaerobic blood agar plate. All agars were stock at 35 °C ± 0.5 °C. The growth of gram-negative bacterial colonies was obtained on blood agar plate at 48 hours after inoculation. Identification of isolated bacterial strain was done by amplifying 16 rRNA and the closest species was *Brucella* sp. (IDENT 99.6%).

The frog was treated with enrofloxacin orally at 5 mg/kg once daily based on sensitivity test for 14 days and meloxicam orally at 0.1 mg/kg once daily for 14 days. The frog appeared clinically normal during the treatment period and repeat radiographs were taken every 14 days. Radiographs revealed suspected paravertebral ganglia swelling which regressed gradually. The medication was discontinued in March 2023, 5 months after presentation, and no recurrence has been reported.

References

1. Helmick KE, Garner MM, Rhyan J, Bradway D. Clinicopathologic features of infection with novel *Brucella* organisms in captive waxy tree frogs (*Phyllomedusa sauvagii*) and Colorado river toads (*Incilius alvarius*). *J Zoo Wildl Med.* 2018;49(1):153-161.

2. Latheef S, Keyburn A, Broz I, et al. Atypical *Brucella* sp in captive Australian green tree frogs (*Litoria caerulea*): clinical features, pathology, culture and molecular characterization. *Aust Vet J.* 2020;98(5): 216-221.

3. Mühldorfer K, Wibbelt G, Szentiks CA, et al. The role of 'atypical' *Brucella* in amphibians: are we facing novel emerging pathogens? *J Appl Microbiol.* 2017;122(1):40-53.

4. Rouzic N, Desmier L, Cariou M-E, et al. First case of brucellosis caused by an amphibian-type *Brucella*. *Clin Infect Dis.* 2021;72(9):e404-E407.

5. Soler-Lloréns PF, Quance CR, Lawhon SD, et al. A *Brucella* spp. isolate from a pac-man frog (*Ceratophrys ornata*) reveals characteristics departing from classical Brucellae. *Front Cell Infect Microbiol.* 2016;6:116.

Survey of *Nannizziopsis* spp. Infection in Wild Invasive Green Iguana (*Iguana iguana*) Populations in Taiwan

Xuan-Rui Liu, DVM, and
Pin-Huan Yu, DVM, PhD

Session #3464

Affiliation: From the Institute of Veterinary Clinical Science and National Taiwan University Veterinary Hospital, National Taiwan University, No. 153, Sec. 3, Keelung Road, Da'an District, Taipei 10617, Taiwan.

Note: Abstract format selected by author(s). See Front Matter for description of conference proceedings manuscripts submitted as scientific abstracts.

Abstract: Nannizziopsis spp. have been reported to infect both captive and wild lizards, leading to dermatomycosis, systemic infection, and even death. It will put native reptile species at risk if *Nannizziopsis* spp. is carried by alien species. Thus, we conducted a survey of *Nannizziopsis* spp. in wild green iguana (*Iguana iguana*) populations, a widely distributed invasive reptile species in Taiwan. The free-living green iguanas were captured as part of an eradication program conducted in Chiayi county and Tainan county. All skin lesions were biopsied aseptically for fungal culture and histopathological examinations. The fungal isolates underwent morphological and molecular identification once they grew on the agars. *Nannizziopsis arthrosporioides* was isolated from five out of 511 (0.98%) green iguanas, and the pathologic evidence of fungal infection was also present in these individuals. The gross lesions of *Nannizziopsis* spp. infection were non-specific, including granulomas, hyperkeratosis, scale defects, and crusts, and the green iguanas were generally in good condition. This study indicates that *Nannizziopsis* spp. exists in the wild in Taiwan and that the pathogen could possibly be spread by the exotic pet trade. In addition, this is the first report of *Nannizziopsis arthrosporioides* infection in green iguanas. The study could be an example of alien pathogen investigations while it is conducted along with an eradication program.

References

1. Paré JA, Sigler L. An overview of reptile fungal pathogens in the genera *Nannizziopsis*, *Paranannizziopsis*, and *Ophidiomyces*. *J Herpetol Med Surg*. 2016;26:46-53.

2. Peterson NR, Rose K, Shaw S, et al. Cross-continental emergence of *Nannizziopsis barbatae* disease may threaten wild Australian lizards. *Sci Rep*. 2020;10:1-12.

3. Lee KH, Chen TH, Shang G, et al. A check list and population trends of invasive amphibians and reptiles in Taiwan. *ZooKeys*. 2019;829:85-130.

4. Sigler L, Hambleton S, Paré JA. Molecular characterization of reptile pathogens currently known as members of the *Chrysosporium* anamorph of *Nannizziopsis vriesii* complex and relationship with some human-associated isolates. *J Clin Microbiol*. 2013;51:3338-3357.

Papillomatous-Like Lesions and Mortality Associated with *Nannizziopsis arthrosporioides* in a Shipment of Central American Boas (*Boa constrictor imperator*)

Cathy A. Johnson-Delaney, DVM, and
Denise Orleans, BS

Session #3251

Affiliation: From NW Zoological Supply, 3614 Smith Ave, Everett, WA 98201, USA.

Note: Abstract format selected by author(s). See Front Matter for description of conference proceedings manuscripts submitted as scientific abstracts.

Abstract: A shipment of Central American boas (*Boa constrictor imperator*) arrived infested with snake mites (*Ophionyssus natricis*) from a vendor in the United States. Within 2 weeks, proliferative lesions resembling papillomas began appearing on all surfaces on the majority of the snakes. Lips, eyes, and head lesions impaired the ability to eat and drink. Many began dying with clinical signs of dehydration and severe wasting. Samples were submitted from necropsy for histopathology as well as molecular testing for bacteria and fungal infection. Histopathology revealed keratinophilic lesions with underlying ulcerations and necrosis in the epidermis and dermis. Next-generation DNA sequencing on swabs and tissue samples revealed *Nannizziopsis arthrosporioides*.[1] Lesions differed in appearance from the *Nannizziopsis* sp. lesions commonly seen in lizards.[2] A program for treatment and prevention in the other reptile species in the facility was undertaken, as lesions were subsequently found in adjacent rooms to the holding room.[3] This program included culling of any snakes with lesions, increased cage and room disinfection, and topical dipping of all snakes in F10 1:250 (F10-SC, Health and Hygiene Pty Ltd, Roodepoort, South Africa). As snakes came from a vendor that had previously supplied various species to the facility, quarantine precautions were largely not implemented following the initial mite treatment. *N. arthrosporioides* has been isolated from lizards, but not previously from snakes. This case illustrates the need for strict quarantine procedures in any facility when receiving and holding reptiles from vendors where multiple imported species are housed.

Acknowledgments: Authors thank J. Krumbeck, MiDog; K.A. Keller, Univ IL Urbana-Champagne; M. Garner, NW Zoopath; and A. Orleans, NW Zoological Supply.

References

1. Stchigel AM, Sutton DA, Cano-lira JF, et al. Phylogeny of chrysosporia infecting reptiles: proposal of the new family Nannizziopsiaceae and five new species. *Persoonia.* 2013;31:86-100.

2. Pare JA, Sigler L. An overview of reptile fungal pathogens in the genera *Nannizziopsis*, *Paranannizziopsis*, and *Ophidiomyces*. *J Herpetol Med Surg.* 2016;26(1-2):46-52.

3. Chen YH, Chi MJ, Sun PL, et al. Histopathology, molecular identification and antifungal susceptibility testing of *Nannizziopsis arthrosporioides* from a captive Cuban rock iguana (*Cyclura nubila*). *Mycopathologia.* 2020;185(6):1005-1012. doi:10.1007/s11046-020-00481-6.

Current and Emerging Reptile Diseases, 2023

Robert J. Ossiboff, DVM, PhD, DACVP

Session #3344

Affiliation: From the Department of Comparative, Diagnostic, and Population Medicine, College of Veterinary Medicine, University of Florida, Gainesville, FL 32608, USA.

Abstract: This is the second of an ExoticsCon lecture series started in 2021 to distill and summarize important emerging or newly characterized diseases in free-ranging and captive reptiles for the reptile clinician. In this lecture, three important diseases of reptiles are covered, including chelonian cryptosporidiosis, invasive *Raillietiella orientalis* pentastomiasis in snakes and lizards, and *Austwickia chelonae* in chelonians and other reptiles. Cryptosporidiosis is an important parasitic disease of captive reptiles; while such infections may be commonly associated with squamates, chelonians are also susceptible to both gastric and intestinal cryptosporidiosis. Disease presentation and pathology are discussed. *Raillietiella orientalis* is a pentastome parasite native to Asia that was introduced into Florida in the early 1990's alongside the Burmese python, *Python bivittatus*. Since that time, the parasite has not only become established in southern Florida, but it has also spread north in the state successfully using native herpetofauna as intermediate hosts. The parasite has now been identified in the pet trade. The disease associated with *Raillietiella* pentastomiasis as well as implications for both herpetofaunal conservation in the Southwestern United States and captive reptile health also are discussed. Lastly, *Austwickia chelonae* is a filamentous, gram-positive bacterium that can infect a number of reptile hosts. Previously classified in the genus *Dermatophilus*, *A. chelonae* has been sporadically reported in the literature; however, the prevalence of *A. chelonae* cases in captive reptiles, particularly chelonians, appears to be increasing. The gross and microscopic lesions associated with this infection and potential implications to collections are discussed.

Introduction

The current state of knowledge of reptile medicine and disease is ever evolving. With busy clinic schedules and an ever-increasing pool of pertinent primary literature, it can be difficult for the reptile clinician to stay abreast of the current state of reptile disease discovery and characterization. This is the second installment of semiannual lecture series to distill and summarize important emerging or newly characterized diseases in free-ranging and captive reptiles for the reptile clinician. In this lecture, three important diseases of reptiles are covered, including chelonian cryptosporidiosis, invasive *Raillietiella* pentastomiasis, and *Austwickia chelonae* infections in reptiles.

Chelonian Cryptosporidiosis

Cryptosporidium species are important apicomplexan parasites of a variety of vertebrate hosts. These organisms can be a particularly important cause of morbidity and mortality in reptiles, and both *Cryptosporidium serpentis* and *C. varanii* of squamates are well characterized.[1,2] In chelonians much remains to be learned about the diversity and significance of *Cryptosporidium* infections. In a review by Adamovicz et al. (2020), reports of cryptosporidial infections were compiled representing 20 chelonian species in six countries.[3] Despite the variety of documented hosts and the relative frequency of reporting, cryptosporidiosis remains an infrequent clinical consideration for tortoises exhibiting weight loss, poor growth, and/or abnormal stool.

Three cryptosporidial species have been documented in chelonians, and most have been documented in tortoises. The best characterized, and arguably the most common and clinically significant *Cryptosporidium* in tortoises, is *C. ducismarci* (previously referred to as Genotype II).[4] This parasite has a tropism for intestinal enterocytes and can be clinically and pathologically similar to *C. varanii* infections in snakes and lizards. Tortoises with cryptosporidiosis can present clinically with weight loss despite appropriate appetite and diet, and may have chronic diarrhea with or without undigested food.[5] Microscopically, *C. ducismarci* can cause chronic, proliferative enteritis similar to the lesions associated with *C. varanii* infections. And as for *C. varanii* infections in squamates, *C. ducismarci* infections in chelonians are presumably associated with disruptions in nutrient absorption and water maintenance, and can be a cause of a chronic negative energy balance in infected tortoises.

Another chelonian *Cryptosporidium*, *C. testudinis* (formerly referred to as Genotype I) is less well characterized but should be considered an important chelonian pathogen. *C. testudinis* has a tropism for gastric epithelial cells, and can be considered to *C. serpentis* of squamates. Unlike *C. serpentis*, which causes chronic gastric inflammation, fibrosis, and proliferation of predominantly mucus cells, *C. testudinis* is associated with minimal to mild changes of the gastric mucosa in infected tortoises.[6] Despite the minimal associated microscopic alterations, *C. testudinis* appears to be associated with hatchling and juvenile stunting in the absence of other clear causes of poor growth, particularly in radiated tortoises (*Astrochelys radiata*). Concurrent *C. testudinis* and *C. ducismarci* infections also have been observed in stunted tortoises. Finally, the third cryptosporidial species documented in chelonians is currently referred to as *Cryptosporidium* sp. Genotype III. It is documented at the lowest frequency of chelonian cryptosporidia, and clear evidence of clinical disease has not been reported.[4]

Diagnosis of chelonian cryptosporidiosis is similar to that for other reptiles. Coproscopy, fecal PCR, and necropsy (if applicable) are all reliable methods of detection. Given the potential for multiple cryptosporidial species, even within individual animals, conventional PCR accompanied by sequencing is usually necessary for definitive identification.

Invasive *Raillietiella* Pentastomiasis

Pentastomes are crustacean parasites that can be found primarily in the respiratory tree of free-ranging reptiles. A number of genera are significant to reptiles, including *Porocephalus*, *Kiricephalus*, and *Armillifer* spp. in squamates, and *Sebekia* spp. in aquatic chelonians and crocodilians. Natural infections are usually rate-limiting and often are incidental. Because of the complex life-cycle of pentastomids that involves multiple intermediate hosts, infections are uncommon in captive born reptiles. The primary exception to this is pulmonary pentastomiasis (sebekiosis) in juvenile crocodilians fed wild-caught or commercially produced fish carrying pentastome nymphs.[7,8]

Pentastomes of the genus *Raillietiella* are naturally parasites of Asian herpetofauna; however, at least three species of *Raillietiella*, *R. frenatus*, *R. indica*, and *R. orientalis*, have been introduced as invasive pathogens in other parts of the world, including North America.[9-11] Of particular concern in the southeastern United States, is the introduction and spread of *R. orientalis* in native and invasive herpetofauna, and the effects the parasite may have on native reptile health. *R. orientalis* was first documented in southern Florida in 2017 by Miller et al. in both their natural, invasive host, the Burmese python (*Python bivittatus*) as well as spillover to native snakes.[11] Subsequent studies identified the parasite in a variety of native snakes, moving well beyond the invasive range of *P. bivittatus*.[12,13] In 2019, the parasite was documented in a free-ranging banded water snake (*Nerodia fasciata*) in north central Florida, nearly 340 kilometers north of the northernmost invasive range of *P. bivittatus* in Florida.[14] In 2022, a study by Palmisano et al. documented that common, native and invasive Florida lizards and frogs could serve as competent, intermediate hosts for *R. orientalis*, confirming that the invasive parasite could complete its lifecycle in herpetofauna native to the southeastern United States.[15]

The implications the invasive pentastome has for native snakes remains unclear, and is an area of active investigation. One particularly concerning finding as it relates to invasive *Raillietiella* pentastomiasis is that parasites can exhibit substantial phenotypic plasticity in different hosts,[16] and based on necropsy findings in both invasive pythons and native snakes with *R. orientalis*, the parasites can get considerably larger in non-native hosts and may be more clinically significant. Another concern is the potential spillover of *Raillietiella* into the pet trade. *R. orientalis* has been documented in a captive, pet banded water snake (*N. fasciata*) in Michigan, US[17] and although the animal was acquired as captive bred, it likely represented a wild-caught animal. Perhaps of greater concern is the wide-spread movement of wild-caught, invasive brown anoles nationwide from Florida as either pets or feeders, and what that might mean for other captive reptiles.

There are no efficacious pharmaceutical therapeutic options for pentastomes and they are an important differential to consider in captive snakes. Treatment may require surgical removal of the parasites from the lungs in definitive hosts. Diagnostics can include coproscopy and PCR.

Reptile *Austwickia chelonae* Infections

Austwickia chelonae is a gram-positive, acid-fast negative, facultative anaerobe that can be an important pathogen of certain reptiles. Historically classified in the genus *Dermatophilus* (dermatophilosis), *A. chelonae* was reassigned approximately in 2010.[18] First identified from a snapping turtle in Australia, the pathogen has been documented as a cause of typically granulomatous disease in turtles, lizards, and snakes.[19-21] *A. chelonae* has also been documented in free-ranging gopher tortoises (*G. polyphemus*) in Florida, US;[22] however, the bacterium seems to have the greatest disease potential in young, captive tortoises.[23]

Cases of austwickiosis in young captive tortoises seem to be commonplace in the American reptile hobby, particularly in sulcata tortoises (*Geochelone sulcata*), but can also be seen in other tortoise species, including leopard tortoises (*Stigmochelys pardalis*). In contrast to typical granulomatous disease as seen in other species, in these captive juvenile tortoises the infection generally manifests as crusts or plaques of tan-yellow material present on the skin of the limbs, axillary or inguinal areas, and particularly of the head, near the oral commissure, along the base of the mandible, and around the eyes. Microscopically, these crusts are composed of necrotic cellular debris, granulocytes, and dense aggregates of filamentous bacteria. The bacteria can be deeply invasive of the underlying tissue, where they can be associated with further necrosis, inflammation, and even bone loss. Secondary opportunistic bacterial infections are common, and septicemia due to bacterial translocation (usually the opportunistic bacteria) can occur. Although just based on the surface appearance of these lesions they may easily be written off by owners or keepers, the degree of change deeper to the crusts not evident grossly warrants appropriate characterization of the disease state and appropriate treatment.

Diagnosis is possible by one of several routine methods. Cytology of the inner surface of sloughed crusts will reveal mixed inflammation and the filamentous bacteria. Biopsy or (necropsy if appropriate) is also reliable for bacterial identification, and given the unique morphology of the bacterium, typically diagnostic. PCR (skin crusts, tissue samples, swabs of the lesion deep to crusts) can also identify the organism.

References

1. Bogan, JE. Gastric cryptosporidiosis in snakes, a review. *J Herpetol Med Surg.* 2019;29(3-4):71-86.

2. Pavlasek I, Ryan U. *Cryptosporidium varanii* takes precedence over *C. saurophilum. Exp Parasitol.* 2008;118(3):434-437.

3. Adamovicz L, Allender MC, Gibbons PM. Emerging infectious diseases of Chelonians: an update. *Vet Clin North Am Exot Anim Pract.* 2020;23(2):263-283.

4. Ježková J, Horčičková M, Hlásková L, et al. *Cryptosporidium testudinis* sp. n., *Cryptosporidium ducismarci* Traversa, 2010 and *Cryptosporidium* tortoise genotype III (Apicomplexa: Cryptosporidiidae) in tortoises. *Folia Parasitol.* 2016;63:035.

5. Richter B, Rasim R, Vrhovec MG, et al. Cryptosporidiosis outbreak in captive chelonians (*Testudo hermanni*) with identification of two *Cryptosporidium* genotypes. *J Vet Diagn Invest.* 2012;24(3):591-595.

6. Griffin C, Reavill DR, Stacy BA, et al. Cryptosporidiosis caused by two distinct species in Russian tortoises and a pancake tortoise. *Vet Parasitol.* 2010;170(1-2):14-19.

7. Boyce W, Cardeilhac P, Lane T, et al. Sebekiosis in captive alligator hatchlings. *J Am Vet Med Assoc.* 1984;185(11):1419-1420.

8. Adams L, Isaza R, Greiner E. Fatal pentastomiasis in captive African dwarf crocodile hatchlings (*Osteolaemus tetraspis*). *J Zoo Wildl Med.* 2001;32(4):500-502.

9. Kelehear C, Saltonstall K, Torchin ME. An introduced pentastomid parasite (*Raillietiella frenata*) infects native cane toads (*Rhinella marina*) in Panama. *Parasitology.* 2015;142(5):675-679.

10. Barton DP, Riley J. *Raillietiella indica* (Pentastomida) from the lungs of the giant toad, *Bufo marinus* (Amphibia), in Hawaii, U.S.A. *Comp Parasitol.* 2004;71(2):251-254.

11. Miller MA, Kinsella JM, Snow RW, et al. Parasite spillover: indirect effects of invasive Burmese pythons. *Ecol Evol.* 2017;8:830-840.

12. Farrell TM, Agugliaro J, Walden HDS, et al. Spillover of pentastome parasites from invasive Burmese pythons (*Python bivittatus*) to pygmy rattlesnakes (*Sistrurus miliarius*), extending parasite range in Florida, USA. *Herpetol Rev.* 2019;50(1):73-76.

13. Miller MA, Kinsella JM, Snow RW, et al. Highly competent native snake hosts extend the range of an introduced parasite beyond its invasive Burmese python host. *Ecosphere.* 2020;11(6):e03153.

14. Walden HDS, Iredale ME, Childress A, et al. Case report: invasive pentastomes, *Raillietiella orientalis* (Sambon, 1922), in a free-ranging banded water snake (*Nerodia fasciata*) in north central Florida, USA. *Front Vet Sci.* 2020;7:467.

15. Palmisano JN, Bockoven C, McPherson SM, et al. Infection experiments indicate that common Florida anurans and lizards may serve as intermediate hosts for the invasive pentastome parasite, *Raillietiella orientalis. J Herpetol.* 2022;56(3):355-361.

16. Westfall AK, Miller MA, Murray CM, et al. Host-specific phenotypic variation of a parasite co-introduced with invasive Burmese pythons. *PLoS ONE.* 2019;14(1):e0209252.

17. Farrell TM, Walden HDS, Ossiboff RJ. The invasive pentastome *Raillietiella orientalis* in a banded water snake from the pet trade. *J Vet Diagn Invest.* 2023;35(2):201-203.

18. Hamada M, Iino T, Iwami T, et al. *Mobilicoccus pelagius* gen. nov., sp. nov. and *Piscicoccus intestinalis* gen. nov., sp. nov., two new members of the family Dermatophilaceae, and reclassification of the *Dermatophilus chelonae* (Masters et al., 1995) as *Austwickia chelonae* gen. nov., comb. nov *J Gen Appl Microbiol.* 2010;56:427-736.

19. Wellehan JF, Turenne C, Heard DJ, et al. *Dermatophilus chelonae* in a king cobra (*Ophiophagus hannah*). *J Zoo Wildl Med.* 2004;35:553-556.

20. Tamukai K, Tokiwa T, Kobayashi H, et al. Ranavirus in an outbreak of dermatophilosis in captive inland bearded dragons (*Pogona vitticeps*). *Vet Dermatol.* 2016;27:99-e28.

21. Masters AM, Ellis TM, Carson JM, et al. *Dermatophilus chelonae* sp. nov., isolated from chelonids in Australia. *Int J Syst Evol.* 1995;45:50-56.

22. Liguori BL, Ossiboff RJ, Stacy NI, et al. Austwickia chelonae in a wild gopher tortoise (*Gopherus polyphemus*) and evidence of positive selection on the diphtheria-like toxin gene. *J Wildl Dis.* 2022;58(1):1-7.

23. Rostad SJ, Brandão J, Ramachandran A, et al. Austwickiosis in captive African spurred tortoises (*Geochelone sulcata*) co-infected with *Cryptosporidium ducismarci*. *J Comp Pathol.* 2019;173:1-7.

Meet the Emperor of the Forest: Eastern Indigo Snake (*Drymarchon couperi*), Natural History and Health

James E. Bogan, Jr., DVM, MVS (Conservation Medicine), DABVP (Canine and Feline Practice), DABVP (Reptile and Amphibian Practice), CertAqV, and Terry Norton, DVM, Dipl ACZM

Session #3702

Affiliation: From Central Florida Zoo & Botanical Gardens, 3755 West Seminole Blvd., Sanford, FL 32771, USA (Bogan), Georgia Sea Turtle Center, 214 Stable Rd., Jekyll Island, GA 31527, USA (Norton).

Abstract: The Eastern indigo snake (*Drymarchon couperi*) is the longest snake species native to North America. Eastern indigo snakes are considered a lynchpin species for the longleaf pine ecosystem as they are generalist predators, which helps regulate other animal species populations. The scientific name, *Drymarchon couperi*, roughly translates into Emperor of the Forest, a fitting name for this apex predator. This session reviews the natural history of the Eastern indigo snake and common heath issues seen in captivity.

Eastern Indigo Snake

The Eastern indigo snake (*Drymarchon couperi*, hereafter EIS), a nonvenomous colubrid, is the longest North American snake species with males reaching up to 2.6 meters in total length. The EIS was federally listed as Threatened in 1978 under the Endangered Species Act due to population declines attributed to habitat loss and fragmentation, persecution, over-collection for the pet trade, and dwindling gopher tortoise (*Gopherus polyphemus*) numbers. There is an intimate relationship between EIS and the longleaf pine (*Pinus palustris*) ecosystem. The longleaf pine ecosystem is one of the most imperiled ecosystems on the planet, with less than 3% of the original footprint. A keystone species for the longleaf pine ecosystem is the gopher tortoise, whose deep burrows serve as refugia for many other species including EIS. Historically, EIS occurred throughout Florida and the Coastal Plain of Georgia, with smaller populations present in Alabama and Mississippi. Today, the extant population species is limited to peninsular to Florida and southeastern Georgia.

Adult EIS display male-biased sexual size dimorphism with males having a larger body length up to 2.6 meters. Most adult males can also have small keels on one to five rows of scales along dorsal midline that are not present in females. Males require 2-3 years to reach sexual maturity, while females require 3-4 years. In southern Georgia, free-ranging adult EIS have been reported to live 8-12 years, whereas in captivity, they have been known to live over 25 years. As an active generalist predator, EIS feed on any prey they can overpower, including fish, amphibians, reptiles, birds, and small mammals. Studies have shown EIS to prefer snakes over other prey items, especially venomous species like rattlesnakes and southern copperheads (*Agkistrodon contotrix*). This generalist predatory nature regulates the populations of numerous species which denotes EIS as a lynchpin species.

A comprehensive population health study was conducted on EIS in southeastern Georgia during 2001 to 2005. Fifty-eight snakes were evaluated for baseline health, nutrition, toxicology, parasites, and disease. Skin biopsies with subsequent histopathology and bacterial and fungal cultures were obtained from live snakes with mild to severe diffuse dermatitis during the cooler months of the year. Selected prey items were collected from gastric contents and fresh dead on the road for nutritional analysis. Mean plasma total Ca concentrations were high (mean 50.6 mg/dl) when compared with baseline normal concentrations established for other free-ranging and captive snakes. There was no difference in Ca levels between males and females, but one individual female had total Ca 115.5mg/dl in April when they are developing eggs. Nutritional analysis of common prey items of the EIS (snakes, amphibians, juvenile gopher tortoises) revealed a higher protein and lower fat content, considerably higher concentrations of vitamins A and E, and variable mineral content (high Ca, P, Na; low Cu, Mn) compared to the diets of commercially reared and captive-fed rodents. Data suggest that diets for captive snakes may require modification to better duplicate natural food sources. Histopathologic evaluation of the skin biopsies revealed severe, subacute, heterophilic, hyperplastic, pustular, and ulcerative dermatitis in all snakes evaluated. A variety of bacterial and fungal species were isolated from these skin lesions indicative of an opportunistic infection. Recapture of some of these snakes during the warmer months of the year revealed that the dermatitis had resolved. The cause of the dermatitis is likely due to high humidity in the gopher tortoise burrow and immune suppression from cooler temperatures. Chiggers (*E. cinnabaris* and/or *P. gurneyi*) were found on six EIS in this study. Other ectoparasites reported to infest EIS (in Florida) are the chigger *Eutrombicula splendens* (Ewing) and the tick *Amblyomma dissimile*. The level of ectoparasitism recorded for EIS in this study was low and did not compromise the health of these individuals.

Necropsies were performed on 20 EIS found freshly dead on the road or in a few cases around gopher tortoise burrows. The most common finding other than trauma related changes was granulomatous inflammation in the gastrointestinal tract related to migrating larval parasites (acanthocephalans, cestodes, or nematodes). One snake recovering from radiotransmitter placement surgery developed a fatal colitis secondary to amoebiasis. Other significant findings identified were pulmonary parasitism with pentastomes or nematodes, focal dermatitis, and focal colitis. In two snakes, fungal infection had disseminated systemically from primary skin infection. Grossly skin lesions were consistent with snake fungal disease (*Ophidiomyces ophiodiicola*). In a third snake, fungal infection was in the respiratory tract and most likely an *Aspergillus* sp. based on oxalates found microscopically. Molecular diagnostics for *O. ophiodiicola* were performed on frozen skin and lung tissue from two cases documented to be fungal infections based on histopathologic findings and both were positive. One of the snakes was originally sampled in 2004, hence it is the earliest documentation of this disease in EIS.

In captivity, EIS are usually housed separately because they can be cannibalistic. At the author's facility, the snakes are housed either indoors or outdoors. Indoor snakes are housed in ARS rack systems with newspaper substrate, a cardboard hide box, and no supplemental lighting. Other facilities have recommended UV lighting when EIS are housed indoors. Outdoor snakes are housed in units with natural substrate and a simulated gopher tortoise burrow. Outdoor enclosures are best suited for pre-breeding male combat as well as breeding.

The breeding season is typically October-February. In captivity, the males are often temporarily paired to allow for combat, which may assist in getting them into breeding mode. After combat, the males are introduced to females for breeding. Oviposition occurs in the spring. Once eggs are laid, they are collected and artificially incubated for 90-100 days. The incubation temperatures are lower than many other snake species as the optimal incubation temperature for EIS is 26 °C.

A medical and disease survey of captive EIS housed in AZA accredited zoos reported common problems to include cataracts in older snakes, retained spectacles, skin lesions, impacted/infected hemipenal sacs, cryptosporidiosis, dystocia, and blood in feces of unknown origin which was often responsive to metronidazole. Miscellaneous problems reported included renal adenoma, fungal and bacterial abscesses/granulomas, stomatitis, pneumonia, cardiomyopathy, and proliferative osteopathy. Additional medical issues observed in captivity include dermatitis in outdoor housed animals especially after the winter, dystocia, pneumonia, breeding combat injuries, hyperviscosity syndrome, and gastric cryptosporidiosis.

The most common blood collection sites are ventral tail vein venipuncture and cardiocentesis. The ventral tail vein can be difficult to obtain large samples in EIS and lymph dilution is common. Manual restraint can be used safely for cardiocentesis but may require several individuals to restrain the snake. The jugular and palatine veins are other sites that are being used with some success. The authors prefer to use a sodium heparinized syringe; however, it has been documented that EDTA is also a suitable anticoagulant for EIS clinical pathology. Aseptic technique should be utilized when drawing blood.

As ectotherms, snakes exhibit dramatic variation in hematologic, nutritional, contaminant, and biochemical parameters depending on hydration status, season, metabolic rate, reproductive activity, and nutritional status. Captive EIS have unusually high concentrations of plasma calcium (Ca) and phosphorus (P) (mean Ca 158.9 mg/dl, mean P 35.3 mg/dl) when compared with other snake species (mean Ca 9 mg/dl, mean P 3 mg/dl). Ionized Ca concentration provides a clinical measurement of the physiologically active Ca in the circulation. In free ranging EIS, mean plasma ionized Ca concentration was 1.89 mM/L, SD 0.21, range 1.39-2.39, which is similar to other reptiles (mean ionized Ca concentration in green iguanas 1.47 mM/L). Phosphorus concentrations were lower in free ranging EIS (mean 9.4 mg/dl) when compared with captive EIS (mean 35.3 mg/dl); however, this was still elevated compared with healthy reptiles of other species (1-5 mg/dl). Plasma 25-hydroxy vitamin D (calcidiol) concentrations in both sexes of free-ranging EIS were high (median 150 ng/ml) when compared with other colubrid snakes (corn snakes) exposed to UVB light (mean 78.4, SD 6.69 ng/ml). Plasma calcidiol levels have been shown to be lower in captive EIS housed indoor (male mean 95, SD 45, female median 37) but similar in males housed outdoor (mean 152, SD 49) but elevated in females housed outdoor (mean 245, SD 47) compared to their wild counterparts, Plasma 1,25 dihydroxy vitamin D (calcitriol) in captive EIS was similar in males and females (mean 161 pM/L, SD 86), regardless if they were housed indoor or outdoor. It is still unclear why EIS have such significant elevations in total plasma Ca and 25-hydroxy-vitamin D concentrations, but it may be related to a uniquely high Ca diet. Elevated total protein and globulins, and a predominance of azurophils were found in free ranging snakes and may be evidence of chronic inflammation, possibly related to seasonal dermatitis and/or parasitism.

Depending on the procedure, there are several options that can be used for chemical restraint. Many minimally invasive procedures can be accomplished with sedation with or without local anesthesia. Mild to moderate sedation has been achieved in captive EIS with 2 mg/kg midazolam IM. Lidocaine (either 1% or 2%) can be used as a local, infiltrative anesthetic agent. If a deeper level of sedation is needed, 10-15 mg/kg alfaxalone IM divided over several injection sites can be effective. Administration of immobilization agents are more effective if administered in the cranial quarter of the body by avoiding the hepatic first-pass effect.

General anesthesia can be achieved by administration of 10 mg/kg propofol IV into the ventral tail vein. This regimen tends to be more reliable if the EIS is premedicated with 2 mg/kg midazolam IM 30 minutes prior to propofol administration. If vascular access to the tail vein proves difficult, another successful anesthetic regimen that has been used on a variety of snakes including EIS consists of a combination of ketamine at 5 mg/kg, dexmedetomidine at 50 micrograms/kg, and buprenorphine 0.03 mg/kg given IM. It takes about 15 to 20 minutes for the full effect. Buprenorphine is used for its sedative properties and not its analgesia properties as buprenorphine has been shown to not provide analgesia in reptiles.

Once the snake is adequately immobilized, it can be intubated with an appropriately sized uncuffed endotracheal tube. If the level of anesthesia is too light at the time of tracheal intubation, the snake can be manually ventilated with 5% isoflurane in oxygen until the desired plane of anesthesia is achieved. Once the EIS is at the desired plane of anesthesia, the isoflurane concentration is decreased to the lowest effective concentration, typically between 1-4%. Snakes at a surgical plane of anesthesia will typically be apneic and need to be ventilated (either mechanically or manually) through the procedure. To prevent respiratory alkalosis, the snakes should not be ventilated more than 2-3 breaths per minute.

The anesthetized snake should be maintained at its preferred optimal temperature zone around 77-80 °F. Overheating is more problematic than underheating in EIS. Many anesthetic events at OCIC have been done with an operating room temperature at 76-77 °F (24 °C) with no supplemental heat for the EIS. In this situation, the EIS can maintain a stable body temperature around 77 °F throughout anesthetic events lasting several hours. Anesthetic monitoring can be accomplished with a Doppler ultrasound with or without an ECG rhythm strip. Cloacal or esophageal temperatures should be monitored throughout the anesthetic procedure as well. Pulse oximetry, respiratory capnography, and oscillometric indirect blood pressure readings are variable and provide less consistent data.

The isoflurane should be turned off prior to the end of the procedure and oxygen is administered for a few minutes and then switched to an Ambu bag and room air. Atipamezole IM at the same volume as the dexmedetomidine. Midazolam can be reversed with 0.08 mg/kg flumazenil IM or SC. Additionally, administering 0.1 mg/kg epinephrine IM can hasten anesthetic recovery times if gas anesthetics are used.

Peri- and post-operative analgesia can be accomplished with NSAIDs and/or opioids. Commonly used NSAIDs are 0.2 mg/kg meloxicam SC or 2 mg/kg ketoprofen SC. Some studies suggest that a non-COX selective NSAID may be a better anti-inflammatory drug in reptiles. Commonly used effective opioids include 5 mg/kg tramadol SC or PO and fentanyl transdermal patches. Since the active metabolite of tramadol is produced after being processed by the liver, taking advantage of the hepatic first-effect may allow for a more rapid onset of analgesia if tramadol is injected in the caudal one-third of the body.

Common surgeries performed on EIS include egg removal, radiotransmitter placement and removal, and surgical repair of skin lacerations and other injuries. Surgical removal of eggs in dystocia cases is preferred over manually milking out the eggs as the surface of EIS eggs is rough. Endoscopic retrieval has been attempted on several occasions, but it tends not to work very well. The retained eggs may cause an intussusception-like telescoping in the oviduct caudal to the egg, preventing endoscopic removal. This telescoping corrects itself after the egg is surgically removed. One of the authors has waited for months to let eggs pass on their own, but that has not been successful and led to surgery anyway.

Surgery for dystocia can be approached from either side with standard ophidian celiotomy techniques. Using iodophor-impregnated adhesive drapes allows for easier maintenance of aseptic technique in EIS. The oviducts need to be identified and any visible pathology noted. Oviducts containing retained ova are exteriorized prior to salpingotomy. The oviduct mucosa should be swabbed for bacterial culture and the oviduct wall should be biopsied for histopathology. After the retained eggs have been removed the oviduct should be closed with a double layer Cushing pattern with 5-0 monofilament absorbable suture. The body wall is closed routinely with 3-0 monofilament absorbable suture. The skin can be closed with either 3-0 monofilament absorbable suture in a continuous intradermal pattern, 3-0 monofilament non-absorbable horizontal mattress pattern, or skin staples.

Although surgical intervention is the preferred treatment for dystocia in EIS, several cases of dystocia have been successfully managed at the OCIC with subcutaneous fluid therapy. One proposed cause of dystocia in EIS has been subclinical dehydration and subcutaneous fluid administration has supported the gravid EIS with suspected dystocia and prevented surgery in a few cases. The OCIC usually retires a snake if it has had three dystocia events. Additional reproductive disorders that have been identified in female EIS at the OCIC include yolk coelomitis, pre-ovulatory stasis, ectopic eggs, bacterial oophoritis, oviductal adenocarcinoma, and oviductal prolapse.

References

1. Bogan JE, Cray C, Rick M, Antonio F. Comparison of selected blood parameters of captive Eastern indigo snakes (*Drymarchon couperi*) housed indoor vs outdoor in central Florida, USA. J Herpetol Med Surg. 2020;30(3):165-172.

2. Drew M. Hypercalcemia and hyperphosphatemia in indigo snakes (*Drymarchon corais*) and serum biochemical reference values. *J Zoo Wildl Med.* 1994;25(1):48-52.

3. Knafo SE, Norton TM, Mitchell M, et al. Health and nutritional assessment of free-ranging Eastern indigo snakes (*Drymarchon couperi*) in Georgia, United States. *J Zoo Wildl Med.* 2016;47(4):1000-1012.

4. Norton TM, Bogan JE. Eastern indigo snake (*Drymarchon couperi*): natural history, conservation, and medicine. *Proc Conf Small Anim Exot North Am Vet.* 2021;814-818.

5. Wines MP, Johnson VM, Lock B, et al. Optimal husbandry of hatchling Eastern indigo snakes (*Drymarchon couperi*) during a captive head-start program. *Zoo Biol.* 2015;34(3): 230-238.

Subspectacular Abscesses in Snakes Associated with Concurrent Bacterial and Flagellated Protozoan Infection

Tom Hellebuyck, DVM, PhD, DipECZM (Herpetology),
Ferran Solanes Vilanova, DVM,
Michael Kotyk,
Ivan Čepička, PhD

Session #3352

Affiliation: From the Department of Pathobiology, Pharmacology and Zoological Medicine, Division of Poultry, Exotic Companion Animals, Wildlife and Experimental Animals, Salisburylaan 133, 9820 Merelbeke, Belgium (Hellebuyck, Solanes Vilanova), Department of Zoology, Faculty of Science, Charles University Prague, Vinicna 7, Prague 128 44, Czech Republic (Kotyk, Čepička).

Note: Abstract format selected by author(s). See Front Matter for description of conference proceedings manuscripts submitted as scientific abstracts.

Abstract: As the etiopathogenesis of subspectacular infections in snakes is poorly understood, these disorders can be challenging to manage. The present study describes the etiology, pathology and treatment of subspectacular abscesses in 14 snakes. Based on microbiologic examination of aspirates of the exudate collected from the subspectacular space, bacterial infection could be demonstrated in all snakes. In seven cases, *Pseudomonas aeruginosa* isolates that showed identical antimicrobial susceptibility patterns were cultured. Direct light microscopic examination demonstrated the simultaneous presence of flagellated protozoan parasites in subspectacular aspirates and oropharyngeal samples in eight snakes. Comparison of the SSU rDNA sequences revealed that the protozoan parasites represented either *Monocercomonas colubrorum* or *Hypotrichomonas acosta*. In all snakes, successful treatment was achieved following surgical debridement of the subspectacular space combined with antiparasitic and topical antimicrobial treatment. Bacteria as well as flagellates are considered to reach the subspectacular space through ascending infection from the oral cavity via the lacrimal duct. Although opportunistic bacterial infection presumably constitutes the predominant cause of subspectacular abscesses in snakes, the role of flagellate infection in the pathogenesis of this disorder should be further elucidated. Variation of the circuitous course of the lacrimal drainage may be the basis of a species predisposition towards the development of this disorder.

Adenoviruses and *Encephalitozoon pogonae* in Samples from Bearded Dragons (*Pogona* spp.) in Europe

Rachel E. Marschang, PD Dr med vet, DipECVM, DipECZM (Herpetology), and Lisa Schüler, vet

Session #3178

Affiliation: From Laboklin GmbH & Co. KG, Bad Kissingen, Germany.

Note: Abstract format selected by author(s). See Front Matter for description of conference proceedings manuscripts submitted as scientific abstracts.

Abstract: Both adenoviruses and the microsporidium *Encephalitozoon pogonae* have been repeatedly detected in bearded dragons (*Pogona* spp.). The majority of adenoviruses found in these animals have been characterized in the species *Lizard atadenovirus B* (LAdVB, a.k.a. agamid atadenovirus 1), although other adenoviruses also have been described sporadically. Infection with LAdVB has been hypothesized to increase susceptibility to other pathogens, and co-infections with microsporidia and adenoviruses have been described in individual cases. In a retrospective study, samples, including mostly cloacal swabs as well as feces and tissue samples, that had been submitted to a commercial laboratory (Laboklin GmbH & Co. KG, Bad Kissingen, Germany) for the detection of adenoviruses by PCR, were also screened for the presence of *E. pogonae*. Samples from 144 animals were included. The adenovirus PCR was positive for 35 (24.31%) samples, the *E. pogonae* PCR was positive for 28 (19.44%). Sequencing of the products from the adenovirus PCRs showed that 30 (85.71%) were clearly identifiable as belonging to LAdVB, 3 (8.57%) belonged to the species *Lizard atadenovirus A* (a.k.a. lizard atadenovirus 2), and 2 (5.71%) were not identifiable. Evaluation of coinfections showed no correlation between infection with an adenovirus and *E. pogonae* when all adenoviruses were included in the analysis ($P = 0.116849$), and the correlation remained insignificant when only LAdVB-like viruses were included ($P = 0.055707$). Both adenoviruses and microsporidia are commonly found in swabs and feces of captive bearded dragons in Europe. Understanding their clinical significance for these animals requires further study.

Section 40

ARAV Neoplasia Session

Kenneth Ford, DVM
Session Chair

Chemotherapy in Two Snakes as an Adjunctive Therapy to Surgical Treatment of Neoplasia

James E. Bogan, Jr., DVM, MVS (Conservation Medicine), DABVP (Canine and Feline Practice), DABVP (Reptile and Amphibian Practice), CertAqV,
Evan Sones, DVM, MS, DACVIM (Oncology),
Alexandra K. Mason, DVM, MS, CertAqV

Session #3157

Affiliation: From Central Florida Zoo & Botanical Gardens, 3755 West Seminole Blvd, Sanford, FL 32771, USA (Bogan, Mason), Animal Cancer Care Clinic, 106 N Dean Rd #120, Orlando, FL 32825, USA (Sones).

Note: Abstract format selected by author(s). See Front Matter for description of conference proceedings manuscripts submitted as scientific abstracts.

Abstract: Reptile chemotherapy is in its infancy and case reports are needed to help establish potential therapeutic studies.[1] In this session, two case reports are presented where chemotherapy was used as an adjunct to neoplasia treatment in a snake. The first case is an 11-year-old male cantil (*Agkistrodon bilineatus bilineatus*) that received 9.5 mg/kg cyclophosphamide PO q3wk for 10 doses following incomplete excision of a low-grade spindle-cell sarcoma involving the body wall. The animal tolerated the medication well, but the tumor regrew 16 months later. The animal underwent surgical debulking of the mass once again and 2.5 mg/kg carboplatin was applied intralesionally prior to skin closure. Unfortunately, the animal died immediately after surgery. The second case is a 12-year-old female Eastern indigo snake (*Drymarchon couperi*) that received 3.3 mg/kg carboplatin intracoelomically once monthly for 6 months following bilateral ovariectomy as treatment for malignant granulosa cell neoplasm in the ovaries. The animal was anorectic prior to surgery through the fifth dose of carboplatin and required regular tube feeding. After administration of 1 mg/kg mirtazapine transdermal ointment, the animal resumed eating. Two months after the last carboplatin dose, a mass was palpated in the mid coelom and the animal died before additional diagnostics could be performed. Complete necropsy revealed an unrelated intermediate grade spindle cell neoplasm of the oviduct.

References

1. Duke EG, Harrison SH, Moresco A, et al. A multi-institutional collaboration to understand neoplasia, treatment and survival of snakes. *Animals.* 2022;12:258.

Palliative Treatment of Renal Adenoma in an Aged Corn Snake

Angela M. Lennox, DVM, DABVP (Avian Practice), DABVP (Exotic Companion Mammal Practice), DipECZM (Small Mammal)

Session #3195

Affiliation: From Avian and Exotic Animal Clinic, 9330 Waldemar Road, Indianapolis, IN 46268, USA.

Abstract: Neoplasia is more frequently diagnosed in captive snakes. This finding is likely due to longer lifespans of snakes in captivity. In addition, the willingness of owners and caretakers to seek veterinary care may result in more frequent diagnosis of neoplasia in captive snakes.

Neoplasia in Snakes

Neoplasia is more frequently diagnosed in captive snakes, likely due to longer lifespans in captivity, and willingness of owners and caretakers to seek veterinary care. A 2017 survey of neoplasia in captive zoo snakes showed a diagnosis of neoplasia in 37 of 255 snakes (14.5%).[1] Snakes represented 27 species from five families. Of 37 snakes, 42 neoplasms of 18 cell types were diagnosed. 83.3% were diagnosed as malignant. Most common sites were skin and subcutis (12), bile duct and multi-organ (10), then liver (4) and kidney (4). Adenomas of the kidney were noted in a timber rattlesnake and a mole kingsnake (*Lampropeltis calligaster*), while adenocarcinoma was seen in a king brown snake *(Pseudechis australis)* and renal carcinoma in an African shield-nosed cobra (*Aspidelaps scutatus*). These snakes were 10-13 years old.

A multi-institutional review of 133 snakes (65 species) with confirmed neoplasia revealed 149 neoplasms, and 86.6% were malignant.[2] Sixty-seven snakes were female and 63 were male, and average age was 11.5 years. Of these neoplasms, 12 affected the kidney only, or kidney and multiple tissues. Malignant renal tumors including carcinoma in a kingsnake (*Lampropeltis* sp.), rat snake (*Elaphe obsoleta*), puff adder (*Bitis arietans*), and Sonoran lyresnake (*Trimorphodon lambda*) , renal carcinoma in a ringhals cobra (*Hemachatus haemachatus*), and renal adenocarcinoma in a radiated rat snake (*Coelognathus radiatus*), false water cobra (*Hydrodynastes gigas*), red spitting cobra (*Naja pallida*), puff adder, Baird's patchnose snake (*Salvadora bairdi*), rhinoceros viper (*Bitis nasicornis*), and western diamond-backed rattlesnake (*Crotalus atrox*). Benign renal tumors included renal adenoma in a kingsnake and renal cystadenoma in a terciopelo (*Bothrops asper*). Surgical excision of renal tumors was performed in two snakes (adenocarcinomas) with survival rates of 71 and 35 months each.[2]

Renal tumors appear to be uncommon in snakes (10% or less in two studies), and based on a few reviews, most are malignant. Three cases of renal neoplasms treated surgically (two renal adenocarcinomas and this case with adenoma) resulted in survival times of 3-6 years. Renal tumors in other species, including dogs, can carry a poor prognosis. Survival time for renal cell carcinoma varied from 0.5-3 years in dogs depending on tumor type.[3]

Neoplasia is thought to be more common in snakes than in other reptiles. Documented neoplasms in corn snakes include cardiac hemangioma, metastatic chondrosarcoma, and various cutaneous and subcutaneous neoplasms.[3] Renal neoplasia may be rare in corn snakes (*Pantherophis guttatus*). Renal anatomy is well described.[4] Snake kidneys are paired, elongated, and in the majority of snakes with multiple lobules, kidney length is about 10-15%

of the body length. The position of the kidney is challenging to pinpoint but has been calculated as a proportion of the distance between the snout and the cloaca; for corn snakes (elapids) the distance is 0.80-0.92. Urine flows from the kidneys to ureters and to the urodeum of the cloaca.[4]

Normal renal structures are generally not appreciated in snakes with radiography. Because there is no distinct renal pelvis, cortex and medulla, ultrasound identification of the kidney can be challenging; a 2012 study described ultrasonographic anatomy of coelomic organs in boid snakes.[5] Computed tomography evaluation of the snake urinary tract has not been described.

Renal Adenoma in a Corn Snake

An 11-year-old education facility corn snake presented for failure to defecate with stool retention proximal to a caudal coelomic mass in September of 2017. Surgical excision of the mass was performed, but biopsy and blood work were not performed at that time due to cost. The snake recovered well and was able to defecate normally post-surgery. Over a 6-year period the snake developed five additional masses with the same presentation, treatment (discrete renal lobectomy), and response. Additional diagnostic testing was permitted thanks to a successful fundraising appeal, with three separate biopsies indicating renal adenoma. At the sixth surgery in February 2023, biopsy showed renal adenoma with the addition of renal gout, and the development of an additional mass cranial to the renal masses. Histopathology again confirmed renal adenoma this time with urate tophi, with the additional mass a suspected adrenal cortical carcinoma with invasion of the oviduct. Due to loss of condition and rising uric acid (greater than 25 mg/dl), humane euthanasia was selected. The age of the snake at the time of euthanasia was 16 years, and survival time from initial diagnosis 5 years and 5 months, with excellent quality of life up until just prior to the final surgery.

References

1. Page-Karjian A, Hahne M, Leach K, et al. Neoplasia in snakes at zoo Atlanta during 1992-2012. *J Zoo Wildl Med.* 2017;48(20):521-524.

2. Duke E, Harrison G, Scott H, et al. A multi-institutional collaboration to understand neoplasia, treatment and survival of snakes. *Animals.* 2022;12(3):258.

3. Edmondson EF, Hess AM, Powers BE. Prognostic significance of histologic features in canine renal cell carcinomas: 70 nephrectomies. *Vet Pathol.* 2015;52(2):260-268.

4. Holz P. Anatomy and physiology of the reptile renal system. *Vet Clin North Am Exot Anim Pract.* 2020;23(1):103-114.

5. Banzato T, Russo E, Finotti L, et al. Ultrasonographic anatomy of the coelomic organs of boid snakes. *Am J Vet Res.* 2012;73(5):634-645.

Chromatophoroma in a Carpet Python (*Morelia spilota*)

Lindsey Elswick, DVM

Session #3313

Affiliation: From Cheat Lake Animal Hospital, 286 Fairchance Road, Morgantown, WV 26508, USA.

Note: Abstract format selected by author(s). See Front Matter for description of conference proceedings manuscripts submitted as scientific abstracts.

Abstract: A 19-year-old female carpet python (*Morelia spilota*) was presented for removal of a large subcutaneous mass growing on the dorsum. Prior fine needle aspirate of the mass was inconclusive. Survey radiographs showed mineralization within the mass, but no obvious invasion into the coelomic cavity or bony lysis was appreciated at that time. The snake was taken to surgery and an excisional biopsy was performed. Histologically, the tumor was composed of amorphous cells with abundant cytoplasm. Within the cytoplasm was a variable amount of brown-black pigmentation. Based on the evaluation of hematoxylin and eosin-stained sections, the tumor was confirmed to be a melanophoroma. The snake recovered well from surgery and no regrowth was noted approximately 6 weeks after the procedure. To the author's knowledge, the snake continues to do well at home.

Histologic and Immunohistochemical Characteristics of Squamous and Basal Cell Carcinomas in Reptiles

Ferran Solanes Vilanova, DVM,
Koen Chiers, DVM, PhD, DipECVP,
Tom Hellebuyck, DVM, PhD, DipECZM (Herpetology)

Session #3334

Affiliation: From the Department of Pathobiology, Pharmacology and Zoological Medicine, Division of Poultry, Exotic Companion Animals, Wildlife and Experimental Animals, Salisburylaan 133, 9820 Merelbeke, Belgium.

Note: Abstract format selected by author(s). See Front Matter for description of conference proceedings manuscripts submitted as scientific abstracts.

Abstract: This study re-evaluated the histologic characteristics of squamous cell carcinomas (SCCs) and basal cell carcinomas (BCCs) obtained from 22 squamates and 13 chelonians. The lesions were initially diagnosed as 29 SCCs and 6 BCCs following routine histologic examination. Five SCCs were reclassified as BCCs and five SCCs cases proved to be non-neoplastic lesions. While 18 SCCs and eight BCCs had a dermal origin, one SCC and three BCCs originated from the oral mucosa. SCCs were classified as poorly, moderately, or well-differentiated SCC. In addition, five dermal SCCs from lizards could be designated as keratoacanthomas, a not previously recognized histologic subtype of SCC. BCCs were classified as nodular, infiltrative, keratinizing, or basosquamous carcinoma histologic subtypes. BCCs have been previously documented in two Hermann's tortoises (*Testudo hermanni*)[1] and a Savannah monitor (*Varanus exanthematicus*).[2] The present study reports the occurrence of BCC in five additional reptile species for the first time. In contrast to what has been documented in humans, IHC staining with epithelial membrane antigen and Ber-EP4 does not allow the differentiation of SCCs from BCCs in reptiles, while COX-2 staining seems to have discriminative potential. Although the predilection sites and gross pathologic features of SCCs and BCCs are highly similar in reptiles, correct histologic identification and classification are highly important as the biological behavior seems to show considerable variation between SCCs and BCCs. Based on the results of this study, reptile BCCs are presumably often misdiagnosed as SCCs.

References

1. Hellebuyck T, Ducatelle R, Bosseler L, et al. Basal cell carcinoma in two Hermann's tortoises (*Testudo hermanni*). *J Vet Diagn Investig*. 2016;28(6):750-754. doi.org/10.1177/1040638716668560.

2. Kubiak M, Denk D, Stidworthy MF. Retrospective review of neoplasms of captive lizards in the United Kingdom. *Vet Rec*. 2020;186(1):28. doi.org/10.1136/vr.105308.

Gross, Histologic and Immunohistochemical Characteristics of Keratoacanthomas in Lizards

Ferran Solanes Vilanova, DVM,
Koen Chiers, DVM, PhD, DipECVP,
Marja Kik, DVM, PhD, DipECZM (Herpetology),
Tom Hellebuyck, DVM, PhD, DipECZM (Herpetology)

Session #3332

Affiliation: From the Department of Pathobiology, Pharmacology and Zoological Medicine, Division of Poultry, Exotic Companion Animals, Wildlife and Experimental Animals, Salisburylaan 133, 9820 Merelbeke, Belgium (Vilanova, Chiers, Hellebuyck), Utrecht University, Veterinary Medicine, Department of Biomedical Health Sciences, Pathology division. Pathology exotic animals and wildlife. Yalelaan 1, 3584 CL Utrecht, the Netherlands (Kik).

Note: Abstract format selected by author(s). See Front Matter for description of conference proceedings manuscripts submitted as scientific abstracts.

Abstract: The present study describes keratoacanthoma (KA), a not previously recognized neoplastic entity in lizards. While KAs presented as two dermal lesions in a bearded dragon (*Pogona vitticeps*), multicentric KAs were observed in three panther chameleons (*Furcifer pardalis*) and a veiled chameleon (*Chamaeleo calyptratus*). In all lizards, KAs were predominantly located at the dorsolateral body wall and KA of the eyelid was additionally observed in three out of four chameleons. KAs presented as crateriform skin tumors containing a central keratinous pearl and a diameter ranging from 0.1-1.5 centimeters with minimally infiltrating borders. A consistent immunohistochemical pattern was observed with the expression of cyclooxygenase-2, E-cadherin, and pan-cytokeratin. We describe KA as a low-grade, non-invasive but rapidly growing skin tumor that may show a multicentric appearance, especially in chameleons. Although dermal squamous cell carcinomas (SCC) in lizards show similar predilection sites and gross pathologic features, our results suggest that KA should be considered a histologic variant of SCC that represents a rather benign squamous proliferation in comparison to conventional SCCs. As previously postulated for dermal SCCs, artificial ultraviolet lighting may play an important role in the oncogenesis of KAs in lizards. Early diagnosis of KA and reliable discrimination from SCC are essential towards the prognosis of this neoplastic entity in lizards.

Curious Case of Cancer in a California Kingsnake (*Lampropeltis getula californiae*)

Krystal Woo, DVM, MVSc,
Eileen Henderson, DVM, MS, CertAqV, DACVP,
Javier Asin, DVM, PhD, DECVP

Session #3503

Affiliation: From Lindsay Wildlife Rehabilitation Hospital, 1931 1st Ave, Walnut Creek, CA 94597, USA.

Note: Abstract format selected by author(s). See Front Matter for description of conference proceedings manuscripts submitted as scientific abstracts.

Abstract: A 16-year-old, male California kingsnake (*Lampropeltis getula californiae*) was presented for a lifted scale. Fine needle aspirate had minimal yield. The lesion progressed over three months, so excisional biopsy was elected and histopathology reported a soft tissue sarcoma. Margins were incomplete, so another surgery was performed; margins were histologically clean and the patient recovered well. At month 18, intracoelomic masses were noted and barium study ruled out gastrointestinal involvement. Exploratory celiotomy revealed a fluid pocket and a renal mass; histopathology was also compatible with soft tissue sarcoma. The patient recovered well again, but new and recurrent masses were noted, so humane euthanasia was elected (28 months). On necropsy, there were multiple masses in the coelomic cavity, liver, and kidney. Histologically, all masses were consistent with a metastatic spindle cell neoplasm with rare, slightly pigmented cells suggesting a differential diagnosis of a poorly pigmented, undifferentiated melanocytic chromatophoroma. Neoplastic cells were PNL2 positive and S-100 negative via immunohistochemistry. Other findings included a hepatic biliary adenoma/cystadenoma, renal disease, hepatocellular atrophy, and bacterial granulomatous coelomitis. Soft tissue sarcomas are the most prevalent neoplasia in snakes with surgery being the most common treatment; median survival time with surgery alone is 5.5 months.[1-2] Reports of snakes having multiple neoplasms range from 10-25% of those with neoplasia.[3-5] This case report includes an interesting description of neoplasia in a California kingsnake with a relatively long survival time.

References

1. Duke EG, Harrison SH, Moresco A, et al. A multi-institutional collaboration to understand neoplasia, treatment and survival of snakes. *Animals.* 2022;12(3):258. doi.org/10.3390/ani12030258.

2. Garner MM, Hernandez-Divers SM, Raymond JT. Reptile neoplasia: a retrospective study of case submissions to a specialty diagnostic service. *Vet Clin North Am Exot Anim Pract.* 2004;7(3):653-671. doi.org/10.1016/j.cvex.2004.04.002.

3. Page-Karjian A, Hahne M, Leach K, et al. Neoplasia in snakes at Zoo Atlanta during 1992-2012. *J Zoo Wildl Med.* 2017;48(2):521-524. doi.org/10.1638/2016-0101R1.1.

4. Ramsay EC, Munson L, Lowenstine L, et al. A retrospective study of neoplasia in a collection of captive snakes. *J Zoo Wildl Med.* 1996;27(1):28-34.

5. Sykes JM, Trupkiewicz JG. Reptile neoplasia at the Philadelphia Zoological Garden, 1901-2002. *J Zoo Wildl Med.* 2006;37(1):11-19. doi.org/10.1638/04-112.1.

Clinical Presentation and Computed Tomography of Iridophoroma in a Lemon Frost Leopard Gecko (*Eublepharis macularius*)

Do Na Lee, DVM,
Minjong Ha, DVM, MS,
Seong Chan Yeon, DVM, MS, PhD

Session #3167

Affiliation: From Laboratory of Wildlife and Exotic Animal Medicine, College of Veterinary Medicine and the Research Institute for Veterinary Science, Seoul National University, Seoul 08826, Republic of Korea.

Note: Abstract format selected by author(s). See Front Matter for description of conference proceedings manuscripts submitted as scientific abstracts.

Abstract: Iridophoroma is a rare type of chromatophoroma characterized by the production of blue and white pigment granules and compromises only 3.3% of reptile neoplasms.[1-3] It has been reported in a few reptile species, including chameleons, bearded dragons (*Pogona vitticeps*), green iguanas (*Iguana iguana*), and snakes. Despite its rarity, hobbyists have noticed the prevalence of hereditary tumors in the lemon frost mutation of leopard gecko between the years 2015 to 2017. This was later confirmed to be iridophoroma through a private histopathologic examination in 2017. Since then, histopathology and genetic analysis have been reported in 2020 and 2021 respectively, but clinical presentation has yet to be discussed. This presentation focuses on the clinical finding of 11 lemon frost leopard geckos, nine of which had visible nodules. Routine clinical examinations were performed to assess metastasis, secondary infection and commodities, including hematoma, blindness, and anorexia. Fine needle aspiration of multiple lesions was performed and viewed under a polarized microscope for diagnosis. Two different computed tomography analyses were performed 8 days apart, while radiography and blood analysis were performed in critically affected individuals. Only prophylaxis treatment was attempted due to the hereditary nature of the disease, and patients were followed up for up to 2 years to track their progression. As the first discussion on the clinical presentation of iridophoroma in lemon frost leopard geckos, this lecture aims to guide diagnosis and treatment plans for the condition. It may also be used as reference for the treatment of other reptiles with iridophoroma.

References

1. Guo L, Bloom J, Sykes S, et al. Genetics of white color and iridophoroma in "lemon frost" leopard geckos. *PLoS Genet.* 2021;17(6): e1009580. doi.org/10.1371/journal.pgen.1009580.

2. Monahan CF, Garner MM, Kiupel M. Chromatophoromas in reptiles. *Vet Sci.* 2022;9(3):115. doi: 10.3390/vetsci9030115. PMID: 35324843; PMCID: PMC8955407.

3. Szydłowski P, Madej JP, Duda M. et al. Iridophoroma associated with the lemon frost colour morph of the leopard gecko (*Eublepharis macularius*). *Sci Rep.* 2020;10:5734. doi.org/10.1038/s41598-020-62828-9.

Seminoma Causing Liver Compression in a Black Greek Tortoise (*Testudo graeca*)

Marcie Logsdon, DVM

Session #3458

Affiliation: From Washington State University Veterinary Teaching Hospital, Exotics and Wildlife Department, 100 Ott Rd, Pullman, WA 99164, USA.

Abstract: A 15-year-old male black Greek tortoise (*Testudo graeca*) presented for evaluation of intermittent vomiting. Initially it was diagnosed with presumptive liver disease based on elevated liver enzymes (AST 704U/L, LDH 691 mg/dl, GLDH 1.85U/L) with normal creatinine kinase levels (CK 61U/L) and positive response to supportive care, antibiotic and liver supplement treatment. Rechecks at 3 and 6 months showed continued improvement and no recurrence of vomiting. The patient remained symptom free until 20 months post-initial presentation when it was examined for acute lethargy and depression. A computed tomography scan showed a large mass compressing the liver. Euthanasia was elected due to its inoperable nature. Necropsy and histopathology confirmed a seminoma. This clinical course was nearly identical to a seminoma in a *Testudo graeca* reported in 2015. These cases highlight the diagnostic value of advanced imaging and the importance of considering space occupying masses as a differential for elevated liver enzymes in tortoises.

Introduction

Testicular tumors are rarely reported in tortoises. A literature search for tortoise testicular tumor results in only two published reports.[1,2] Testicular tumor is not traditionally on the list of differentials for elevated liver enzymes in a tortoise. Given the unique nature of chelonian coelomic anatomy, namely that the shell is non-expansile and the testes are positioned internally in close proximity to the liver, it should not be surprising that liver compression and subsequent elevation of liver enzymes can be a response to a space-occupying testicular tumor. The following report outlines the progression of a case from initial presentation and clinical signs to eventual endpoint and definitive diagnosis. The hope is to show how advanced imaging can be of assistance in these cases, to point out early warning signs missed in this case, and to raise awareness of the atypical patient age at time of presentation. Earlier diagnosis of these tumors may result in viable surgical treatment options.

Case Report

A 15-year-old male black Greek tortoise (*Testudo graeca*) presented to the Washington State University Veterinary Teaching Hospital Exotics and Wildlife Department for evaluation of intermittent vomiting. On presentation, it had a slightly decreased body condition score (BSC) of 2/5 with some visible muscle wasting in the hind limbs. There were some mild chronic and static shell shape abnormalities (e.g., pyramiding of carapace). The patient passed grossly normal feces, urine, and urates in its travel carrier and the remainder of the physical exam was within normal limits. When offered, the tortoise immediately ate greens and blended vegetable topping without difficulty and with no signs of regurgitation. Initial radiographs were unremarkable (Fig 1). Initial CBC showed dehydration and occasional reactive lymphocytes but was otherwise unremarkable. VetScan Avian/Reptile chemistry panel showed elevated AST of 386 U/L with an error when reading the CK. A full liver panel was submitted to a reference laboratory 1 week later (Table 1).[3]

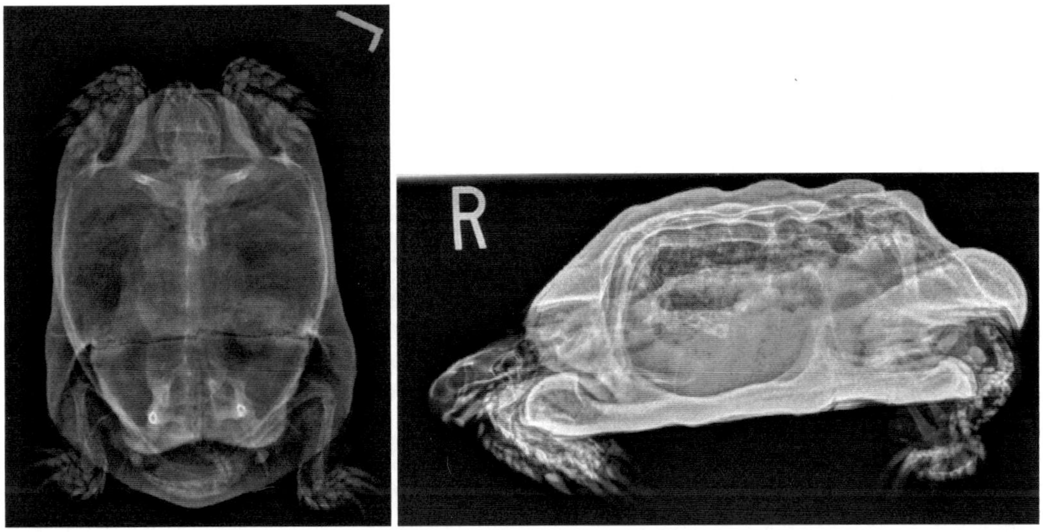

Figure 1. DV and cross-table left-to-right lateral of patient at time of initial presentation.

Table 1. Progressive blood chemistry levels of relevant analytes. Reference range adapted from Egyptian and Russian tortoises.

Analyte	11/20/20 VetScan	12/10/20 UF Miller Wildlife Lab	2/11/21 UF Miller Wildlife Lab	5/27/21 UF Miller Wildlife Lab	8/8/22 VetScan	Reference Range Egyptian Tortoise Russian Tortoise
AST (U/L)	386	704	511	498	527	(50)
GLDH (Umol/L)		1.85	5.57	5.94		0.6-1.5
CK (U/L)		61	101	27		6-344
LDH (mg/dl)		691	577	727		
GGT (mg/dl)		6	3	7		
TP (mg/dl)	5.1	4.1	4.8	5.2	5.1	(3.4)
BA (Umol/L)			3.2	2.5		
UA (mg/dl)	6.6				7.5	(3.8)
Glucose (mg/dl)	98				7	(84)

The patient continued eating normally and passing feces and urates while in the hospital. To rule out partial or intermittent obstruction, a barium study was performed. Radiographs taken 10 minutes and 24 hours after administration of 28 milliliters of barium confirmed passage through the GI tract (Fig 2). The initial radiology report focused only on the appropriate timed passage of barium through the GI tract and lack of any filling defect, but it did not comment on the mass effect that is apparent in retrospect.

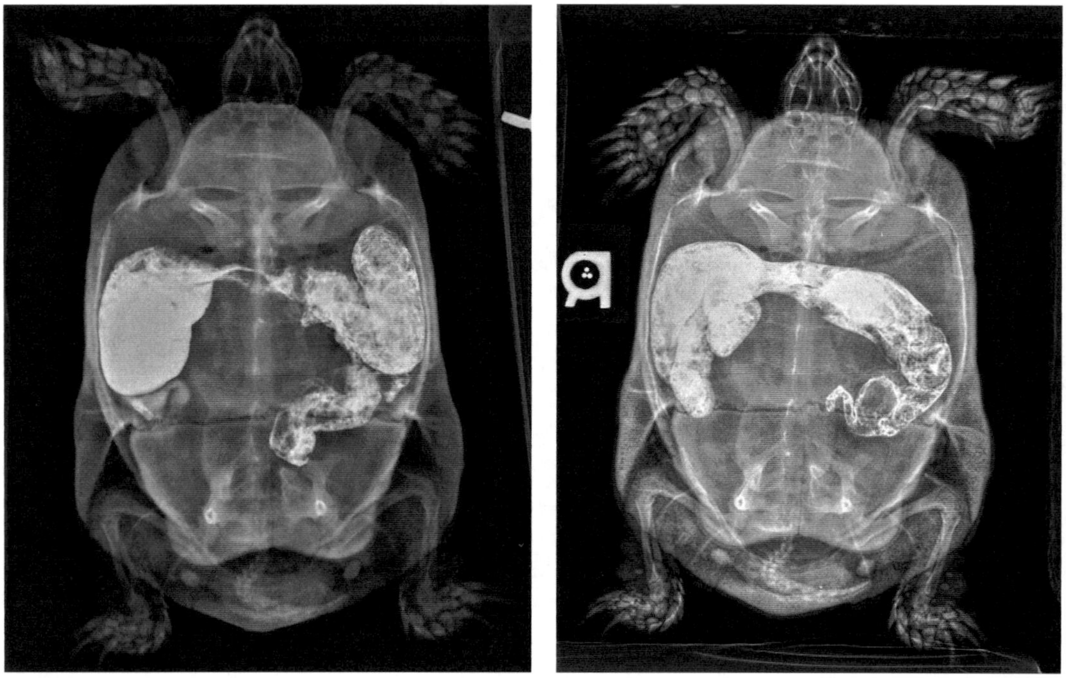

Figure 2. DV views 10 minutes (left) and 24 hours (right) post-barium administration.

Based on the patient's improved attitude and appetite the decision was made to treat for presumptive bacterial hepatitis and send home on antibiotics and liver supplements. Recheck liver panels at 2 and 5 months showed consistent liver enzyme elevations; however, the owner elected not to pursue additional treatments or diagnostics because the patient was acting normal at home (Table 1).

At 21 months post-initial presentation the patient presented for acute lethargy. On presentation, it was severely dehydrated, emaciated, and lethargic. Tableside chemistry analyzer showed an unchanged AST but severe hypoglycemia and elevated uric acid consistent with generalized state of debilitation. Emergency supportive care including intravenous fluids was initiated. The following day a computed tomography (CT) scan was performed. The CT scan revealed a large, space-occupying mass in the right cranial coelomic cavity compressing the liver and displacing the gastrointestinal tract (Fig 3).

Due to its inoperable nature and grave prognosis the owners elected euthanasia. Necropsy and histopathology revealed a large seminoma (Fig 4).

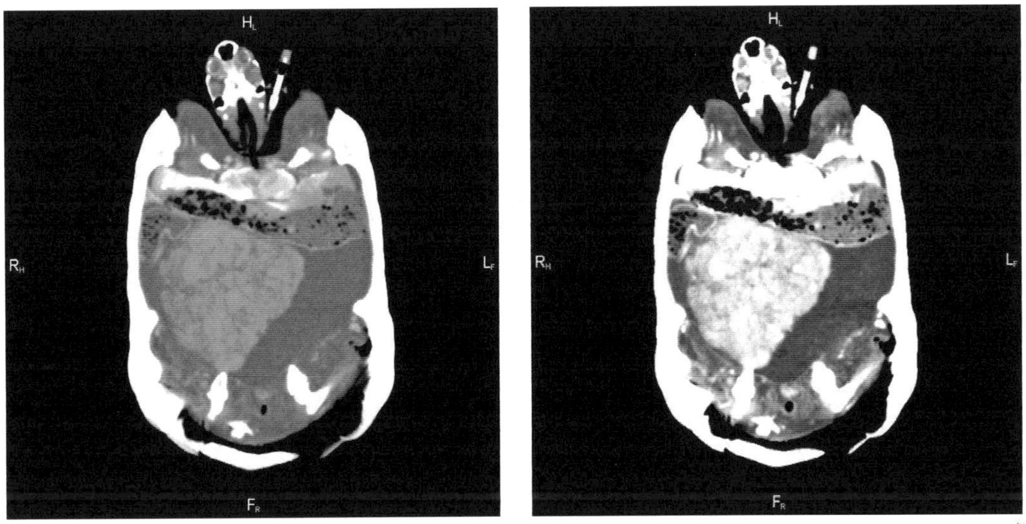

Figure 3. Computed tomography images pre- (left) and post- (right) contrast administration showing a contrast-enhancing mass in the cranial right coelomic cavity.

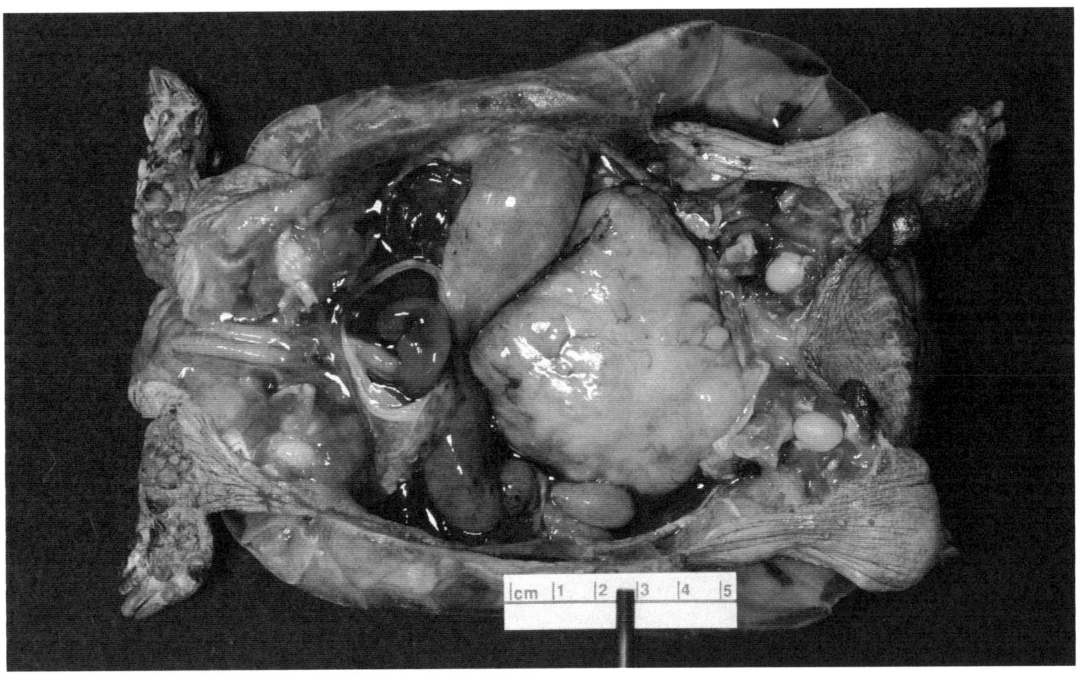

Figure 4. Necropsy of patient, plastron has been removed to allow for visualization of the coelomic cavity.

Discussion

Seminomas are rare, malignant tumors of the testes that are uncommonly reported in reptiles. After the final diagnosis was made, a literature search revealed a nearly identical case report in another black Greek tortoise (*Testudo graeco*) in 2015 (in this write-up the tortoise is referred to by its alternate common name of spur-thighed tortoise).[2] That tortoise also presented with severe liver enzyme elevations as the most prominent abnormality and advanced imaging (magnetic resonance imaging) was utilized to reach a diagnosis.

Combined, these two case reports show the benefits of advanced imaging in diagnosing abnormalities of the coelomic cavity in chelonians. The young age of both tortoises is concerning and serves as a reminder that neoplasia should always be on the list of differential diagnoses, regardless of patient age. Similarly, space-occupying mass should be a differential for elevated liver enzymes in chelonians.

Lastly, the author welcomes reports of similar encounters in practice. If more cases are encountered, there is interest in compiling and publishing a report on the occurrence of this tumor in Mediterranean tortoises.

References

1. Frye FL, Dybdal NO, Harshbarger JC. Testicular interstitial cell tumor in a desert tortoise (*Gopherus agassizi*). *J Zoo Anim Med.* 1988;19:55-58. doi.org/10.2307/20094856.

2. Pees M, Ludewig E, Plenz B, Schmidt V. Imaging diagnosis-seminoma causing liver compression in a spur-thighed tortoise (*Testudo graeca*). *Vet Radiol Ultrasound.* 2015;56(2):E21-4. doi:10.1111/vru.12145. Epub 2014 Feb 10. PMID: 24506145.

3. International Species Information System. 2013. Testudo_kleinmanni_No_selection_by_gender__All_ages_combined_Conventional_American_units__2013_CD.h tml. ISIS Physiological Reference Intervals for Captive Wildlife: A CD-ROM Resource. Eagan, MN: International Species Information System.

Nerve Sheath Tumor in a Green Iguana (*Iguana iguana*)

Jennilee Morrison, DVM, DABVP (Avian Practice)

Session #3475

Affiliation: From Avets Specialty & Emergency Trauma Center, 2674 Monroeville Blvd Monroeville, PA 15146, USA.

Note: Abstract format selected by author(s). See Front Matter for description of conference proceedings manuscripts submitted as scientific abstracts.

Abstract: A 30-year-old female green iguana (*Iguana iguana*) presented for further evaluation of a swelling of the left coelom along with worsening anorexia and aggression. The iguana had surgery previously at the referring veterinarian where most of her ovaries and uterus were removed due to chronic egg laying. An ultrasound, biopsy, and culture of the mass were digestive of chronic, sterile disease, likely a continuation of her previous issues. A computed tomography scan was performed along with baseline bloodwork. These were suggestive of chronic hypocalcemia, inflammation, and a large, space occupying mass. A coelomic exploratory surgery was performed and the mass was removed. Histopathology was submitted and special stains were required to show a peripheral nerve sheath tumor. Follow up showed significant clinical and bloodwork improvement in the patient. Husbandry as a predetermining factor is discussed.

Chromatophoromas and Regenerative Tail Amputation in an Axolotl (*Ambystoma mexicanum*) with Presumptive Metastasis

Kristina S. Vuong, Dr med vet,
Cristina Metral, Dr med vet,
Wesley Sheley, Dr med vet, PhD, DACVP,
Michelle Liu, Dr med vet, PhD, DACVP,
Ariel Carlson, Dr med vet, PhD, DACVP,
Michael Jones, Dr med vet, DABVP (Avian Practice)

Session #3322

Affiliation: From the Department of Small Animal Clinical Sciences (Vuong, Jones), Department of Biomedical and Diagnostic Sciences, University of Tennessee, College of Veterinary Medicine, 2407 River Drive, Knoxville, TN 37996, USA (Sheley, Liu, Carlson), BluePearl Pet Hospital, 3000 Busch Lake Blvd, Tampa, FL 33614, USA (Metral).

Abstract: Reported neoplasia in amphibians are rare with the most overrepresented species being axolotls (*Ambystoma mexicanum*) and African clawed frogs (*Xenopus laevis*).[1,2] Chromatophoromas are the most reported amphibian neoplasms and typically present as raised, pigmented masses.[1,3] The etiology is usually spontaneous, but water quality, ultraviolet radiation, and genetics have been implicated as predisposing risk factors.[3,4] The purpose of this presentation is to describe the diagnosis of spontaneous tail chromatophotomas in an axolotl with presumptive splenic and renal metastasis. This case highlights the surgical tail amputation technique and the regenerative capability of the axolotl. A 3-year-old intact female axolotl presented with a pigmented bilobed right proximal tail mass and a vascular left proximal tail mass. The patient was anesthetized with an alfaxalone immersion bath and subcutaneous alfaxalone and the tail was amputated proximal to the masses. The axolotl recovered uneventfully and was discharged with oral medications. Histopathology of the tail masses revealed incomplete excision of presumptive iridophoromas (a subset of chromatophoroma). Complete tail regeneration occurred after seven months with no gross reoccurrence of the masses. The patient presented nine months post-operatively with a 3-week history of decreased appetite. Coelomic ultrasound revealed a vascularized coelomic mass caudal to the liver. Euthanasia was performed and necropsy revealed presumptive metastasis of the chromotophoromas into the spleen and kidneys. Chromatophoromas with potential for metastasis should be considered in axolotls with cutaneous masses.

Introduction

The axolotl is an amphibian of the order Urodela that has become increasingly popular as pets and laboratory animals for cancer research and regenerative biology.[1,2] Reported neoplasia in amphibians is rare, with most neoplasms being described in axolotls (*Ambystoma mexicanum*) and African clawed frogs (*Xenopus laevis*).[1,2] Reported axolotl neoplasms in the literature include a proximal tail teratoma,[1] olfactory neuroblastomas,[5,6] chromatophoromas,[7] located on the dorsum,[8] base of the right forelimb,[8] face,[9] mandible,[10] and tail,[11] and mast cell tumors of the trunk.[12] The objective of this case report is to describe the diagnosis of spontaneous presumptive tail iridophoromas in an axolotl with suspected metastasis to the spleen and kidneys. This case also highlights the regenerative capability of the axolotl after tail amputation.

Clinical Report

A 3-year-old intact female axolotl (*Ambystoma mexicanum*) presented for two raised proximal tail masses and difficulty swimming. The axolotl initially had a raised, approximately 1 × 2 millimeter pigmented spot on the right proximal tail 18 months prior to presentation and this became progressively bilobed within the last 6 months. A left proximal tail mass was observed 6 months prior to presentation. The patient was housed in a 65-gallon aquarium with dual filters and a mixture of gravel and fine black sand substrate. The water temperature was maintained at 69 °F (20.6 °C) with a water cooler. The patient's diet consisted of shrimp pellets and occasional bloodworms.

On physical examination, a bilobed, pigmented right proximal tail mass and a vascular left proximal tail mass were observed approximately 1 centimeter caudal to the cloaca. The right cranial lobe was 1.5 × 1.0 × 1.5 centimeters and the right caudal lobe was 0.5 × 0.7 × 0.5 centimeters. Both right dermal lobes were firm and fixed with multifocal areas of brown and tan pigmentation on a light pink background. The 1.0 × 1.5 × 1.5 centimeter left dermal tail mass was soft, fixed, unpigmented on a light pink background and had prominent vascularization. The remainder of the physical examination was unremarkable. A tail amputation was recommended at this time. Prior to the scheduled tail amputation, the left mass ruptured resulting in significant blood loss. The patient presented with pale external gills, indicating anemia. Surgical amputation was elected to prevent further hemorrhage and potential infection with the understanding that the patient was a poor surgical candidate.

The axolotl was placed in an immersion bath of alfaxalone starting at 5 mg/L (Alfaxan 10 mg/ml; Jurox, MO, USA) based on a previous case report.[13] The immersion bath contained the patient's tank water with an air stone. The patient achieved only slight sedation despite increasing the concentration of the alfaxalone immersion bath by 5 mg/L increments to a total concentration of 20 mg/L over 60 minutes. Immobilization with loss of righting reflex was obtained after 10 minutes after administering subcutaneous alfaxalone at 5 mg/kg in new tank water. The patient was placed in dorsal recumbency on a sponge with the gills submerged in the tank water without additional water circulation (Fig 1a). The patient was monitored with Doppler (Model 811-B; Parks Medical Electronics, OR, USA) over the heart and by observing branchial motion. The tail was prepped with dilute povidone iodine and sterile saline and clamped with straight hemostats 0.5 centimeters cranial to the masses and 0.5 centimeters caudal to the cloaca. A ligature was placed over the crushed tissue as a tourniquet using monofilament 0 suture (PDS; Ethicon, NJ, USA). A circumferential skin incision was made 1 millimeter distal to the tourniquet using a scalpel (No. 15; McKesson, Medical-Surgical Inc, VA, USA). The underlying musculature was transected to the vertebrae using a scalpel and the vertebrae was transected with surgical scissors one vertebrae proximal to the skin incision. The mass was submitted for histopathology (Fig 1b). The tourniquet was removed, and hemorrhage was not observed. The surgical site was left open to heal by second intention (Fig 1c). The patient was moved into a recovery tank with an air stone where subcutaneous meloxicam 0.1 mg/kg (Metacam 5 mg/ml; Boehringer Ingelheim, GA, USA) and enrofloxacin 10 mg/kg (Baytril 22.7 mg/ml; Bayer, KS, USA; diluted 1:5 with sterile saline) were administered. Recovery was uneventful and the patient was discharged with oral meloxicam 0.2 mg/kg q24h for 7 days (Meloxicam 1.5 mg/ml; Boehringer Ingelheim, GA, USA) and oral enrofloxacin 10 mg/kg q24h for 21 days (Baytril 20.4 mg/ml; Bayer, KS, USA).

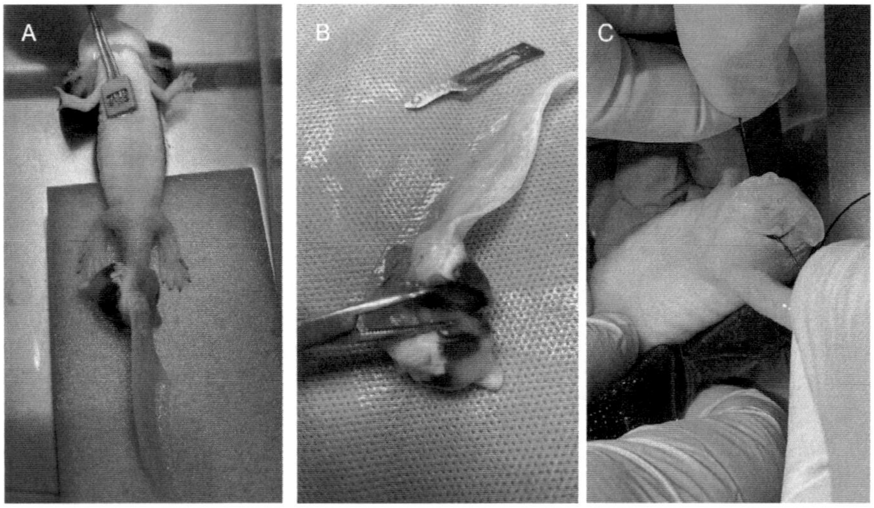

Figure 1. A 3-year-old female intact axolotl (*Ambystoma mexicanum*) presented for proximal tail amputation for two proximal tail masses. (A) Operative set-up with Doppler monitoring and the patient's gills submerged in an alfaxalone immersion bath. (B) Resected masses and distal tail. (C) Removal of the tourniquet monofilament 0 suture.

On histopathology, the tail masses were unencapsulated, multilobulated, densely cellular dermal neoplasms composed of spindle to polygonal cells forming loose, interlacing streams with invasion into the underlying skeletal muscle and vertebrae (Fig 2a). There was no interaction with the overlying epidermis and no evidence of vascular invasion in the sections examined. Neoplastic cells had ovoid to multilobulated nuclei with coarsely clumped chromatin and 1-3 prominent nucleoli. The neoplastic cells had marked (up to 8×) anisocytosis, marked anisokaryosis, and approximately 40% of the cells contained light brown intracytoplasmic pigment that was birefringent under polarized light (Fig 2a and 2b). There were no mitotic figures counted in 10 high power (2.37 mm^2) fields. Neoplastic cells extended to the surgical margin. The axolotl was diagnosed with presumptive iridophoromas on the proximal tail with incomplete excision. The incision healed 1-month post-operatively and the patient had dramatically improved energy and swimming ability. Regeneration of the tail was observed over 7 months with no gross evidence of mass reoccurrence (Fig 3a-c).

Nine months post operation, the patient presented with a 3-week history of decreased appetite. The patient had gained 70% of its body weight since the 1-month post-operative examination. A non-sedated coelomic ultrasound revealed a 2.09 × 3.71-centimeter vascularized mass caudal to the liver. The owners elected for humane euthanasia and necropsy was completed within 16 hours to minimize autolysis. A 1.5 × 3 × 4-centimeter friable, yellow-tan splenic mass replacing 65% of the spleen was identified (12.5% of total body weight; the mass occupied approximately one-third of the coelomic cavity) in addition to a 1.3-centimeter in diameter, soft, yellow-tan mass involving both kidneys (Fig 4). Histologic morphology of the splenic and renal masses was similar to that of the tail masses; however, no overt light brown birefringent pigment or melanin was found within these masses. Three immunohistochemical markers (Melan-A, S100, and PNL-2) from Kansas State Veterinary Diagnostic Laboratory were performed on sections of the tail, spleen, and renal masses. Neoplastic cells in the masses from all three sites as well as chromatophores within normal axolotl skin did not exhibit immunoreactivity. The axolotl was diagnosed with presumptive tail iridophoromas with suspected metastasis to the spleen and kidneys.

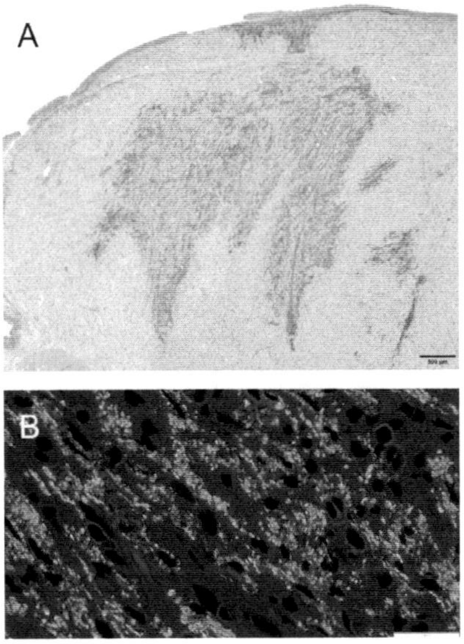

Figure 2. (A) Photomicrograph of a chromatophoroma (suspect iridophoroma) on the proximal right tail of an axolotl (*Ambystoma mexicanum*). Hematoxylin and eosin (H&E). 2x magnification. (B) Photomicrograph of the tail mass revealing intracytoplasmic birefringent pigment under polarized light. Hematoxylin and eosin (H&E). 40x magnification.

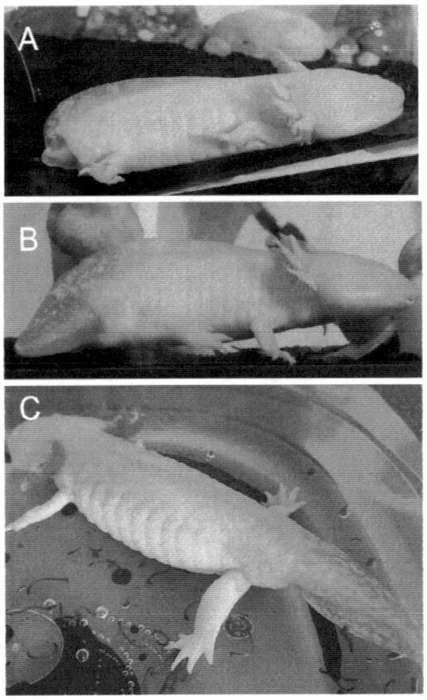

Figure 3. Tail regeneration in an axolotl (*Ambystoma mexicanum*) A) Two months post tail amputation. (B) Four months post tail amputation. (C) Seven months post tail amputation.

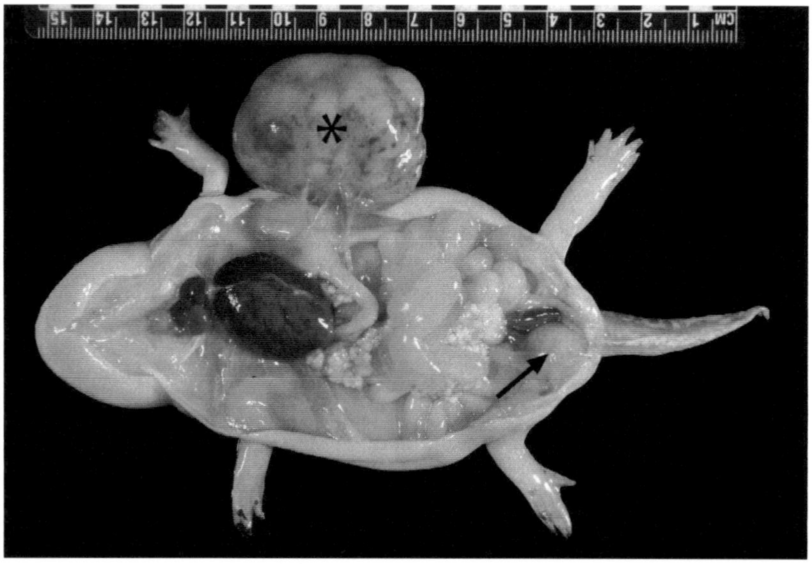

Figure 4. Gross necropsy of an axolotl (*Ambystoma mexicanum*) presenting with weight gain and decreased appetite nine-months after tail amputation for chromatophoromas (suspect iridophoromas). A large splenic mass (asterisk) filling approximately a third of the coelomic cavity and a similar smaller second mass involving both kidneys (black arrow) were both presumed to be metastases from the tail chromatophoromas.

Discussion

Chromatophoromas are composed of neoplastic dermal pigment cells in amphibians, reptiles, and fish and are categorized based on intracytoplasmic pigmentation/cell origin.[1,14-16] Chromatophoromas are the most reported amphibian neoplasms and typically present as raised, pigmented masses.[1,3] In reptiles, histopathology is the recommended diagnostic modality and masses typically consist of spindle to polygonal cells with variable intracytoplasmic pigment, mild-moderate nuclear atypia, and low mitotic counts.[15,16] Metastasis of chromatophoromas is rarely reported in amphibians and includes a report described in an axolotl.[1,3] Subtypes of chromatophoromas include iridophoromas that produce intracytoplasmic purine or pteridine crystals that are birefringent with polarized light, melanophoromas that produce intracytoplasmic melanin, xanthophoromas that produce intracytoplasmic red/orange/yellow carotenoid and pteridine, and mixed chromatophoromas.[15,16] The masses in this patient were most consistent with iridophoromas based on the birefringent intracytoplasmic pigmentation. Iridophoromas have been described to have light brown/amber intracytoplasmic pigment, as was found in the tail masses.[15,16] Immunohistochemistry (IHC) was conducted in an attempt to confirm neoplastic pigment cell origin; however, no immunoreactivity was observed. The use of these markers has not been validated in axolotls. The sensitivity of IHC to differentiate chromatophoromas in reptiles is low compared to mammals but electron microscopy has been described.[15,16] Aggressive surgical excision is the recommended treatment for chromatophoromas in amphibians and reptiles.[3,15,16] Complete excision is difficult due to the level of infiltration into surrounding tissue.[3,15,16]

Immersion anesthesia can be utilized in amphibians through transcutaneous and branchial absorption of drugs.[13] Anesthesia was attempted through an alfaxalone immersion bath based on a previous case report that achieved an hour of surgical anesthesia for endoscopy and exploratory celiotomy.[13] The anemia and the resulting poor perfusion of our patient may have contributed to the poor cutaneous absorption of alfaxalone. Alfaxalone is a $GABA_A$ (γ-aminobutyric acid) agonist[13] that may cause decreased respiratory rates,[17] increased heart rates,[17] and darkened coloration in amphibians.[18] No adverse effects were observed in this case. Other anesthetic immersion baths include tricaine methane sulphonate (MS-222), benzocaine, and eugenol (clove oil).[13,19,20] Due to the conservation of voltage-dependent sodium channels across vertebrates, lidocaine should be considered for future amphibian surgeries to provide regional anesthesia and multimodal analgesia.[21] In this report, a tail amputation was used to temporarily improve the quality of life in an axolotl with suspect metastatic tail chromatophoromas.

References

1. Hopewell E, Harrison SH, Posey R, et al. Analysis of published amphibian neoplasia case reports. *J Herp Med Surg.* 2020;30(3):148-155. doi:10.5818/19-09-212.1.

2. Brunst VV, Roque AL. A spontaneous teratoma in an axolotl (*Siredon mexicanum*). *Cancer Res.* 1969;29(1):223-229.

3. Stacy BA, Parker JM. Amphibian oncology. *Vet Clin North Am Exot Anim Pract.* 2004;7(3):673-695. doi:10.1016/j.cvex.2004.04.001.

4. Martinez-Silvestre A, Amat F, Bargallo F, et al. Incidence of pigmented skin tumors in a population of wild montseny brook newt (*Calotriton arnoldi*). *J Wildl Dis.* 2011;47(2):410-414.

5. Shioda C, Uchida K, Nakayama H. Pathological features of olfactory neuroblastoma in an axolotl (*Ambystoma mexicanum*). *J Vet Med Sci.* 2011;73(8):1109-1111. doi:10.1292/jvms.11-0105.

6. Modesto F, Nicolier A, Hurtrel C, et al. Excisional biopsy and radiotherapy for management of an olfactory neuroblastoma in an axolotl (*Ambystoma mexicanum*). J Am Vet Med Assoc. 2021;260(4):436-441. doi:10.2460/javma.20.09.0498.

7. Khudoley VV, Mizgireuv IV. On spontaneous skin tumours in amphibia. *Neoplasma.* 1980;27(3):289-299

8. Khudoley VV, Eliseiv VV. Multiple melanomas in the axolotl *Ambystoma mexicanum*. *J Natl Cancer Inst.* 1979;63(1):101-103.

9. Fernandes TL, Bernardino RI, Peleteiro C. Facial melanocytoma in an axolotl (*Ambystoma mexicanum*). *Proc Conference Dis Zoo Wild Anim.* 2014;28-31.

10. Menger B, Jahn S, Allmeling C, et al. Therapy of a melanosarcoma in an axolotl (*Ambystoma mexicanum*) by resection and skin grafting. *Kleintierpraxis.* 2009;54(3):149-154.

11. Sheremetieva EA. Spontaneous melanoma in regenerating tails of axolotls. *J Exp Zool.* 1965;158:101-121. doi:10.1002/jez.1401580110.

12. Harshbarger JC, Chang SC, DeLanney LE, et al. Cutaneous mastocytomas in the neotenic caudate amphibians *Ambystoma mexicanum* (axolotl) and *Ambystoma tigrinum* (tiger salamander). *J Cancer Res Clin Oncol.* 1999;125:187-192.

13. McMillan MW, Leece EA. Immersion and branchial/transcutaneous irrigation anesthesia with alfaxalone in a Mexican axolotl. *Vet Anaesth Analg.* 2011;38(6):619-623. doi:10.1111/j.1467-2995.2011.00660.

14. Muñoz-Gutiérrez JF, Garner MM, Kiupel M. Cutaneous chromatophoromas in captive snakes. *Vet Pathol.* 2016;53(6):1213-1219. doi:10.1177/0300985816644302.

15. Monahan CF, Meyer A, Garner MM, et al. Gross, histologic, and immunohistochemical characteristics of cutaneous chromatophoromas in captive bearded dragons. *J Vet Diagn Invest.* 2021;33(5):932-938.

16. Monahan CF, Garner MM, Kiupel M. Chromatophoromas in reptiles. *Vet Sci.* 2022;9(115):1-23.

17. Williams CJA, Alstrup AKO, Bertelsen MF, et al. Cardiovascular effects of alfaxalone and propofol in the bullfrog, *Lithobates catesceianus*. *J Zoo Wildl Med.* 2018;49(1):92-98.

18. Cermakova E, Oliveri M, Ceplecha V, et al. Anesthesia with intramuscular administration of alfaxalone in Spanish ribbed newt (*Pleurodeles waltl*). *J Exot Pet Med.* 2020;33(1):23-26.

19. Machin KL. Fish, amphibian, and reptile analgesia. *Vet Clin North Am Exot Anim Pract.* 2001;4(1):19-33. doi:10.1016/s1094-9194(17)30048-8.

20. Stetter M. Amphibians. In: West G, Heard D, Caulkett, N. *Zoo Animal and Wildlife Immobilization and Anesthesia.* 2nd Ed. Ames, IA: John Wiley & Sons; 2014:303-312.

21. DiGeronimo PM, Da Cunha AR. Local and regional anesthesia in zoological companion animal practice. *Vet Clin North Am Exot Anim Pract.* 2022;25:321-336.

Section 41

ARAV
ASSOCIATION OF REPTILE AND AMPHIBIAN VETERINARIANS

ARAV Clinical Pathology and Reproduction Session

Violaine Colon, DVM, MSc, DABVP (Reptile and Amphibian Practice)
Session Chair

Development of Ovotestis and Salpingitis in a Previously Ovariectomized Central Bearded Dragon (*Pogona vitticeps*)

Stacey Leonatti Wilkinson, DVM, DABVP (Reptile and Amphibian Practice),
Mary Cate Miller, DVM,
Victoria J. Bender, DVM,
Ethan Biswell, DVM, MS, DACVP (Anatomic Pathology)

Session #3364

Affiliation: From Avian and Exotic Animal Hospital of Georgia, 118 Pipemakers Circle Suite 110, Pooler, GA 31322, USA (Wilkinson), Biosphere Lab, 11700 Commonwealth Dr. Suite 200, Louisville, KY 40299, USA (Biswell).

Note: Abstract format selected by author(s). See Front Matter for description of conference proceedings manuscripts submitted as scientific abstracts.

Abstract: A 4-year-old ovariectomized, female central bearded dragon (*Pogona vitticeps*) presented with a 1-month history of blepharospasm and ocular discharge. Physical exam revealed conjunctivitis and significant weight loss. A CBC and chemistry panel demonstrated a leukocytosis with a lymphocytosis, heterophilic toxic change, and left shift. The patient was started on topical gentamicin ophthalmic solution 0.3% OU and systemic antibiotics (enrofloxacin 10 mg/kg PO q48h). A fecal exam was negative. Two weeks later, the patient's condition was improved but weight loss continued. A coelomic ultrasound revealed a large cystic mass of mixed echogenicity in the area of the previously removed left ovary. The patient's ovariectomy had been performed almost 4 years prior. Surgery was performed and the cystic mass and incorporated left oviduct were removed. In the area of the previous right ovariectomy, a 5- to 7-millimeter yellow nodule was present and removed, along with the right oviduct. All tissues were submitted for histopathology. The patient recovered routinely but over the next 48 hours remained quiet and regurgitated once. Seventy-two hours after surgery the patient regurgitated again, then rapidly declined over several hours and was euthanized. Histopathology revealed chronic salpingitis caused by a caseous necrotic follicle obstructing the lumen of the oviduct on the left, and the nodule on the right was an ovotestis. It was composed of mostly testicular tissue undergoing active spermatogenesis with remnants of ovarian and adrenal tissue. This is the first clinical report of ovotestis in this species, though it has been described in research on temperature versus genetic sex determination.[1-4]

References

1. Holleley CE, Meally DO, Sarre SD, et al. Sex reversal triggers the rapid transition from genetic to temperature-dependent sex. *Nature.* 2015;523:79-82.

2. Whiteley SL, Georges A, Weisbecker V, et al. Ovotestes suggest cryptic genetic influence in a reptile model for temperature-dependent sex determination. *Proc R Soc B.* 2020;288.

3. Whiteley SL, Holleley CE, Georges A. Developmental dynamics of sex reprogramming by high incubation temperatures in a dragon lizard. *BMC Genomics.* 2022;23:322-342.

4. Whiteley SL, Weisbecker V, Georges A, et al. Developmental asynchrony and antagonism of sex determination pathways in a lizard with temperature-induced sex reversal. *Nature Sci Rep.* 2018;8:1-9.

Pathology of Bearded Dragons (*Pogona vitticeps*)

Ethan Biswell, DVM, MS, DACVP (Anatomic Pathology)

Session #3751

Affiliation: From Biosphere Lab, Louisville, KY 40299, USA.

Abstract: This lecture reviews diseases commonly observed in bearded dragons. Information was obtained from cases submitted to a nationwide reference laboratory.

Non-Infectious Diseases

Non-infectious illnesses are common in bearded dragons (*Pogona vitticeps*) with poor diets and improper environments (e.g., lighting, temperature). The most common diseases include obesity, dysecdysis, and metabolic bone disease.[1] These diseases can result in decreased immunity leading to higher likelihood of contracting a secondary infectious agent.

Obesity is common in pet reptiles, including bearded dragons, due to excessive caloric intake and sedentary lifestyle. This often leads to hepatic lipidosis which can be severe and life threatening. It can result in metabolic derangement and increase the likelihood of additional illnesses. This must be interpreted carefully since the fat content of the liver can vary significantly based on the age, sex, and metabolic state of the animal. With hepatic lipidosis the liver grossly appears to be swollen and tan to yellow. Histologically, depending on the severity, hepatocytes are expanded by many small to one large lipid vacuole(s) within the cytoplasm.

Dysecdysis is an incomplete or abnormal shedding of the skin. This can be brought on by low humidity, lack of abrasive surfaces, nutritional deficiencies, and various infectious agents.

Metabolic bone disease, also called nutritional secondary hyperparathyroidism, is a complex illness that is most often observed in juvenile bearded dragons being fed an improper diet. This disease is caused by diets high in phosphorus and low in calcium or vitamin D_3. The low vitamin D_3 can also be secondary to a lack of exposure to UVB lighting which is required for synthesis of vitamin D in the skin. The most common gross lesions are swelling and softening of the jaw (typically referred to as rubber jaw), facial bones, and/or limbs. Histologically, these lesions are consistent with fibrous osteodystrophy featuring loss of trabecular bone and thinning of cortices with varying amounts of fibrosis.

Follicular stasis and egg binding are common in most captive reptiles. The pathogenesis is still being worked out but thought to be secondary to inappropriate environmental conditions. Obstructive dystocia is from any condition that physically prevents egg deposition which then leads to stasis of eggs or follicles within the coelomic cavity. This can lead to yolk coelomitis, which is caused by release of yolk material within the coelomic cavity resulting in a severe foreign body-type inflammatory reaction, mesothelial cell proliferation, and occasionally a secondary bacterial infection.

Renal, visceral, and articular gout can be caused by any process that affects the kidneys and leads to reduced uric acid excretion. The accumulated uric acid deposits are white and chalky, and most often on serosal surfaces (pericardium is typically most severe), and in and around joints. Within the kidneys gout is often identified as pin-point white foci with swelling of renal parenchyma. Histologically gouty tophi result in destruction of renal tubules with associated heterophils, macrophages, lymphocytes, plasma cells, and varying amounts of edema and fibrosis.

Infectious Diseases

Nannizziopsis and *Paranannizziopsis* spp., formerly classified as *Chrysosporium Anamorph of Nannizziopsis vriesii* (CANV), have been shown to cause yellow fungus disease. These agents are common infectious fungal agents observed within the skin of reptiles, with *N. guarroi* shown to be the most common in bearded dragons. Grossly, the scales become discolored and can break off. This fungal agent can invade into the dermis, skeletal muscle, and even bone resulting in severe trauma to the affected tissues. Secondary opportunistic bacterial infections can also occur.

Bacterial septicemia is a very common cause of death in reptiles, with lizards and snakes being more often affected. Gram-negative rods are frequently identified, though many types of primary and opportunistic bacteria can result in sepsis. Lesions are multifocal, can be in any organ, and depending on the chronicity the affected tissues are composed of necrotic areas infiltrated by varying amounts of heterophils, macrophages, lymphocytes, and fibrosis. Chronic lesions will have increased amounts of fibrosis and macrophages (commonly epithelioid) and formation of granulomas. Cutaneous infections can lead to sepsis and are typically associated with a suboptimal environment and diet.

Infectious stomatitis, also called mouth rot, is occasionally observed in bearded dragons and can be caused by opportunistic bacterial and/or fungal agents. Grossly, this can result in swollen gums and/or jaws, and commonly contains areas of ulceration and hemorrhage. Depending on the chronicity, the histologic lesions are typically necrosis, heterophilic to histiocytic inflammation, hemorrhage, and variable fibrosis with or without observation of the infectious agents.

Pneumonia is occasionally observed in lung samples submitted to this lab. This is typically seen in animals that have improper diets, are stressed, or live in cold, poor, and unclean environments. Bacterial pneumonias are the most common cause.

Adenovirus (*Agamid* adenovirus 1) is common in this species and is most often observed in young bearded dragons. Affected animals typically die acutely or become weak, don't eat, and fail to thrive which eventually leads to death. This virus results in hepatitis and gastroenteritis. Viral lesions are typically composed of varying amounts of necrosis admixed with mild to heavy amounts of heterophils and fewer lymphocytes, histiocytes, and plasma cells. Occasionally viral inclusions are observed.

Parasites are occasionally observed, with the most common that result in illness and lesions being *Coccidia* (*Isospora*), oxyurids, and *Cryptosporidium*. These are mainly gastrointestinal tract infections, though some *Cryptosporidium* species can be observed in other tissues (e.g., respiratory epithelium). Lesions can range from mild lymphohistiocytic to heterophilic inflammation to severe ulcers with associated necrosis and heavy heterophilic to histiocytic inflammation. Metazoan infections are rarely observed in our samples.

Neoplasia

Spindle cell sarcomas and squamous cell carcinomas are the most common tumors observed in this species. Gastric neuroendocrine carcinomas (GNTs) are uniquely found in higher occurrences in bearded dragons. These tumors tend to be highly malignant with invasive and metastatic potential.

References

1. Jacobson E, Garner M, eds. Infectious diseases and pathology of reptiles. Vol 1. In: Jacobsen E, Garner M, eds. *Diseases and Pathology of Reptiles: Color Atlas and Text.* Boca Raton: CRC Press; 2021.

Radiographic Demonstration of Ovoviviparity in an Eastern Diamondback Rattlesnake (*Crotalus adamanteus*)

Gordon Smok, DVM,
Ken Harkowitz, VMD,
Richard S. Funk, MA, DVM,
Chase Pirtle, AS

Session #3432

Affiliation: From the Department of Veterinary Technology, City College Gainesville, 2400 SW 13th St., Gainesville, FL 32608, USA (Smok), The Rattlesnake Conservancy, 7152 Lonestar Rd, Jacksonville, FL 32211, USA (Smok, Pirtle), Ashton Biological Preserve, 22215 SW 119th Ave, Archer, FL 32618, USA (Smok, Pirtle).

Note: Abstract format selected by author(s). See Front Matter for description of Conference proceedings manuscripts submitted as scientific abstracts.

Abstract: In reptiles we can see a variety of reproductive modalities and among those is ovoviviparous, seen in viperids as well as other species of snakes and lizards, such as in the Boidae family, garter snakes, skinks, European common lizards, and others.[1-12] Ovoviviparous is defined as producing eggs that develop within the maternal body (as of various fishes or reptiles) and hatch within or immediately after release from the parent.[8] This has been well documented and accepted in the veterinary profession; however, it has not been well represented in the literature or well documented though diagnostic modalities such as radiology. This study followed a captive Eastern diamondback rattlesnake (*Crotalus adamanteus*) through the process of ovoviviparity. This is documented through the digital radiography, computed radiography system, IDEXX PACS, illustrating the successive stages from egg formation to fetal skeletons and then through to parturition of offspring. In summary, all findings with regard to the series of images taken were confirmed by the authors listed and through IDEXX telemedicine consultants. The report from IDEXX stated "Impressions: Generally unremarkable gravid snake. The progression in size and development of mineralization of the age structures over time is a normal finding for the ovoviviparous nature of the species."

References

1. Banzato T, Russo E, Finotti L, et al. Ultrasonographic anatomy of the coelomic organs of boid snakes (*Boa constrictor imperator*, *Python regius*, *Python molurus molurus*, *and Python curtus*). *Am J Vet Res.* 2012;73(5):634-645.

2. Dreslik MJ, Hayes WK, Beaupre SJ, Mackessy SP, eds. *The Biology of Rattlesnakes.* Rodeo, NM: ECO Wear & Publishing; 2017.

3. Florida Museum. Florida Snake ID Guide, Eastern Diamond-backed Rattlesnake, *Crotalus adamanteus.* Available at: https://www.floridamuseum.ufl.edu/florida-snake-id/snake/eastern-diamond-backed-rattle-snake/. Accessed March 24, 2023.

4. Funk R. Lizard reproductive medicine and surgery. *Vet Clin North Am Exot Anim Pract.* 2002;5(3):579-613.

5. Gumpenberger M. Diagnostic imaging of reproductive tract disorders in reptiles. *Vet Clin North Am Exot Anim Pract*. 2017;20(2):327-343.

6. Laird MK, Thompson MB, Whittington CM. Facultative oviparity in a viviparous skink (*Saiphos equalis*). Biol Lett. 2019;15(4):20180827. doi: 10.1098/rsbl.2018.0827.

7. Mader D, Divers SJ, Stahl SJ, eds. *Mader's Reptile and Amphibian Medicine and Surgery*. 3rd Ed. St. Louis, MO: Elsevier; 2019.

8. Merriam-Webster, Incorporated. Ovoviviparous Definition and Meaning. 2022 Available at: https://www.merriam-webster.com/dictionary/ovoviviparous. Accessed March 24, 2023.

9. Rivera S. Health assessment of the reptilian reproductive tract, topics in medicine and surgery. *J Exot Pet Med*. 2008;17(4):259-266.

10. Sanath Krishna MK, Rajkumar K, Karthik M, et al. Standardization of ultrasonographic anatomy of coelomic organs in Indian spectacled cobra (*Naja naja*) and Russell's viper (*Daboia russelii*). *Proc Annu Conf Am Assoc Zoo Vet*. 2013.

11. Florida Fish and Wildlife Conservation Commission. Species Profile, Eastern Diamondback Rattlesnake, *Crotalus adamanteus*. Available at: https://myfwc.com/wildlifehabitats/profiles/reptiles/snakes/eastern-diamondback/. Accessed March 17, 2023.

12. Surget-Groba Y, Heulin B, Guillaume C-P, et al. Multiple origins of viviparity, or reversal from viviparity to oviparity? The European common lizard (*Zootoca vivipara*, Lacertidae) and the evolution of parity. *Biol J Linn*. 2006;87:1-11.

Effect of Methodology on Protein Electrophoresis in Eastern and Ornate Box Turtles (*Terrapene* spp.)

Laura Adamovicz, DVM, PhD

Session #3365

Affiliation: From the University of Illinois College of Veterinary Medicine Wildlife Epidemiology Laboratory and Veterinary Diagnostic Laboratory, 2001 S Lincoln Ave, Urbana, IL 61802, USA.

Note: Abstract format selected by author(s). See Front Matter for description of conference proceedings manuscripts submitted as scientific abstracts.

Abstract: Plasma protein electrophoresis (EPH) is useful for chelonian health assessment, though results can be method-dependent. Agarose gel electrophoresis (AGE) is the most common EPH method in reptiles, however, capillary zone electrophoresis (CZE) is a newer method that is easy to use, requires a smaller sample volume, and has superior fraction separation compared to AGE in multiple taxa, including chelonians. Despite the potential advantages of CZE, results do not always reliably correlate with AGE and methodological differences must be characterized prior to using CZE in new species. The purpose of this study was to assess analytical performance and agreement between AGE and CZE in wild Eastern (*Terrapene carolina carolina*; n = 46) and ornate (*Terrapene ornata*; n = 44) box turtles, to determine method-specific effects of biological predictors on EPH values, and to generate species and method-specific reference intervals. CZE had superior precision in both species evidenced by lower coefficients of variation for intra-assay (four replicates) and inter-assay (five replicates) analyses. Method-specific reference intervals are needed in box turtles due to poor-moderate correlation (Kendall's tau: -0.1 - 0.65) and frequent constant and proportional error (Passing-Bablok regression) between methods. Multiple statistical associations were identified between protein fractions and spatial, temporal, demographic, and health factors; however, these frequently differed between methods indicating that additional work is necessary to inform clinical interpretation of box turtle CZE. This study provides initial characterization of CZE in box turtles, identifies clinically relevant differences between CZE and AGE, and provides reference intervals for veterinary application.

Acknowledgments: Authors thank John Rucker and the turtle dogs for their assistance with turtle capture as well as the University of Illinois Turtle Team veterinary students for sample collection and processing. Authors also thank Carolyn Cray and the University of Miami Miller School of Medicine Acute Phase Protein Laboratory personnel for running the plasma protein electrophoresis assays.

Erythrocyte Sedimentation Rate in Eastern Indigo Snakes (*Drymarchon couperi*)

James E. Bogan, Jr., DVM, MVS (Conservation Medicine), DABVP (Canine and Feline Practice), DABVP (Reptile and Amphibian Practice), CertAqV

Session #3159

Affiliation: From Central Florida Zoo & Botanical Gardens, 3755 West Seminole Blvd, Sanford, FL 32771, USA.

Note: Abstract format selected by author(s). See Front Matter for description of conference proceedings manuscripts submitted as scientific abstracts.

Abstract: Erythrocyte sedimentation rate (ESR) is a hematologic test that can detect inflammatory activity within the body. Although not specific for any particular disease, ESR is often used as a screening "sickness indicator" due to its reliability and low cost. The Westergren method is a manual ESR technique commonly used but this method requires special graduated pipette and over 1 milliliter of whole blood, precluding its use in smaller patients where limited sample volumes can be obtained. A modified micro-ESR technique has been described using hematocrit capillary tubes but it used less commonly. ESR has been reported to be a useful inflammatory indicator in gopher tortoises (*Gopherus polyphemus*) and box turtles (*Terrapene* spp.) but not in Florida cottonmouth snakes (*Agkistrodon conanti*). Having an inexpensive screening test for inflammation can help guide medical decisions within conservation efforts of imperiled species. This study evaluated the correlation between these two methodologies in threatened Eastern indigo snakes (*Drymarchon couperi*, EIS) and found a very strong correlation ($r_s = 0.897$), without constant or proportional biases and a reference interval of 0 (90% CI -1-1) - 9 mm/h (90% CI 8-11) was defined. Additionally, a significant difference was found between healthy EIS and EIS in mid-ecdysis ($P = 0.006$) and EIS with gastric cryptosporidiosis ($P = 0.006$), indicating ESR as a useful inflammatory indicator in EIS.

References

1. Bogan JE. Analytical and clinical evaluation of two methods for measuring erythrocyte sedimentation rate in Eastern indigo snakes (*Drymarchon couperi*). *Animals.* 2023;13:464.

Evaluation of Contrast Radiography and Computed Tomography for Use in Sex Determination of Gila Monsters (*Heloderma suspectum*)

Caitlin Cunningham, DVM

Session #3453

Affiliation: From Arizona Exotic Animal Hospital 744 N Center St. Suite #101 Mesa, AZ 85201, USA.

Note: Abstract format selected by author(s). See Front Matter for description of conference proceedings manuscripts submitted as scientific abstracts.

Abstract: Sex determination in reptiles is pivotal for implementing appropriate husbandry, successful pairing for breeding programs, and reproductive medical care.[1-29] In reptiles that lack sexual dimorphism, accuracy is dependent upon techniques aside from physical observation. Sexing techniques that are quick, non-invasive, and reliable are preferred. Vast anatomic differences among reptiles make the efficacy of these techniques highly variable among species. This study evaluated hemipenis contrast radiography and computed tomography (CT) techniques previously described in other *Heloderma* species in their ability to reliably determine sex in 23 Gila monsters (*Heloderma suspectum*). The results were compared to the sex determined by coelomic ultrasound findings. Coelomic ultrasound showed identification of ovaries in eight animals and testes in 15 animals. Identification of hemipenes in males was shown in 1/15 (7%) by hemipenis contrast radiography and 9/15 (60%) by hemipenis contrast CT. These findings suggest that contrast radiography and CT are not reliable techniques to determine sex in Gila monsters and that coelomic ultrasound is still the most reliable, non-invasive technique for this species.

References

1. Beck DD. Ecology and behavior of the Gila monster in southwestern Utah. *J Herpetol.* 1990;24(1):54.

2. Beck DD, Martin BE, Lowe CH, Wiewandt TA. *Biology of Gila Monsters and Beaded Lizards.* Berkeley, CA: University of California Press; 2005.

3. Bertelsen MF, Sauer CD. Alfaxalone anaesthesia in the green iguana (*Iguana iguana*). *Vet Anaesth Analg.* 2011;38(5):461-466. doi:10.1111/j.1467-2995.2011.00640.x.

4. Davis JR, DeNardo DF. The urinary bladder as a physiological reservoir that moderates dehydration in a large desert lizard, the Gila monster *Heloderma suspectum*. *J Exp Biol.* 2007;210(8):1472-1480. doi:10.1242/jeb.003061.

5. DeFrancesco T, Royal K. A survey of point-of-care ultrasound use in veterinary general practice. *Educ Health Prof.* 2018;1:50-54.

6. Di Ianni F, Volta A, Pelizzone I, et al. Diagnostic sensitivity of ultrasound, radiography and computed tomography for gender determination in four species of lizards. *Vet Radiol Ultrasound.* 2014;56(1):40-45.

7. Divers SJ, Stahl SJ. *Mader's Reptile and Amphibian Medicine and Surgery.* St Louis, MO: Elsevier; 2019.

8. Enriquez KL, Lastica EA, Acorda JA. Ultrasonographic features of the reproductive organs of captive Asian reticulated pythons, *Python reticulatus* (Schneider, 1801) (Reptilia: Squamata: Pythonidae). *Philipp J Vet Med.* 2011;48(2):70-76.

9. Gienger CM, Beck DD. Heads or tails? sexual dimorphism in helodermatid lizards. *Can J Zool.* 2007;85(1): 92-98.

10. Gnudi G, Volta A, Di Ianni F, et al. Use of ultrasonography and contrast radiography for snake gender determination. *Vet Radiol Ultrasound.* 2009;50(3):309-311.

11. Goldberg SR, Lowe CH. Reproductive cycle of the Gila monster, *Heloderma suspectum*, in southern Arizona. *J Herpetol.* 1997;31(1):161. doi:10.2307/1565350.

12. Innis CJ. Endoscopy and endosurgery of the chelonian reproductive tract. *Vet Clin North Am Exot Anim Pract.* 2010;13(2):243-254.

13. Kuchling G. Endoscopic sex determination in juvenile freshwater turtles, *Erymnochelys madagascariensis*: Morphology of gonads and accessory ducts. *Chelonian Conserv Biol.* 2006;5(1):67.

14. Kuchling G, Griffiths O. Endoscopic imaging of gonads, sex ratios, and occurrence of intersexes in juvenile captive-bred Aldabra giant tortoises. *Chelonian Conserv Biol.* 2012;11(1):91-96.

15. Kuchling G, Goode E, Praschag P. Endoscopic imaging of gonads, sex ratio and temperature dependent sex determination in captive bred juvenile Burmese star tortoises, *Geochelone platynota. Asian Herpetol Res.* 2011;2(4):240-244.

16. Kuchling G, Goode E, Praschag P. Endoscopic imaging of gonads, sex ratio, and temperature-dependent sex determination in juvenile captive-bred radiated tortoises, *Astrochelys radiata.* In: Castellano CM, Rhodin AGJ, Ogle M, et al., eds. *Turtles on the Brink in Madagascar: Proceedings of Two Workshops on the Status, Conservation, and Biology of Malagasy Tortoises and Freshwater Turtles.* Lunenburg, MA: Chelonian Research Foundation; 2013:113-118.

17. Kuchling G, Kitimasak W. Endoscopic sexing of juvenile softshell turtles, *Amyda cartilaginea. Nat Hist J Chulalongkorn Univ.* 2009;9(1):91-93.

18. Martínez-Silvestre A, Bargalló F, Grífols J. Gender identification by cloacoscopy and cystoscopy in juvenile chelonians. *Vet Clin North Am Exot Anim Pract.* 2015;18(3):527-539.

19. Meza-Oviedo MA, Gómez Pizano DA, Carrillo D'Lacoste L, et al. Sex determination in Mexican-beaded lizard specimens (*Heloderma alvarezi* and *Heloderma horridum*) employing contrast radiography method. *J Exot Pet Med.* 2020;34:10-14.

20. Morris PJ, Alberts AC. Determination of sex in white-throated monitors (*Varanus albigularis*), Gila monsters (*Heloderma suspectum*), and beaded lizards (*H. horridum*) using two-dimensional ultrasound imaging. *J Zoo Wildl Med.* 1996;27(3):371-377.

21. Morris PJ, Henderson C. Gender determination in mature Gila monsters, *Heloderma suspectum*, and Mexican beaded lizards, *Heloderma horridum*, by ultrasound imaging of the ventral tail. *Bull Assoc Reptilian Amphib Vet.* 1998;8(4):4-5.

22. Pokorná MJ, Rovatsos M, Kratochvíl L. Sex chromosomes and karyotype of the (nearly) mythical creature, the Gila monster, *Heloderma suspectum* (Squamata: Helodermatidae). *PLoS ONE.* 2014;9(8).

23. Prades RB, Lastica EA, Acorda JA. Ultrasound features of the kidneys, urinary bladder, ovarian follicles and vaginal sacs of female water monitor lizard (*Varanus marmoratus*, Weigmann, 1834). *Philipp J Vet Anim Sci.* 2013;39(1):115-124.

24. Rizzi S, Faverzani S, Schnitzer P. Ultrasound sex determination in snakes: correlation between hemipenes or hemiclitorids and body morphological parameters. *Proc ICARE.* 2017;547-551.

25. Rovatsos M, Kratochvíl L. Molecular sexing applicable in 4000 species of lizards and snakes? From dream to real possibility. *Methods Ecol Evol.* 2017;(8):902-906.

26. Rovatsos M, Rehák I, Velenský P, Kratochvíl L. Shared ancient sex chromosomes in varanids, beaded lizards and alligator lizards. *Mol Biol Evol.* 2019;36(6):1113-1120.

27. Scheelings TF, Baker RT, Hammersley G, et al. A preliminary investigation into the chemical restraint with alfaxalone of selected Australian squamate species. *J Herpetol Med Surg.* 2011;21(2):63.

28. Schuler G, Feller S, Schwandt HJ. Determination of sex steroids in shed skin of the beaded lizard *Heloderma suspectum* (Gila monster). *Tierarztl Prax Ausg K Kleintiere Heimtiere.* 2020;48(6):410-419.

29. Webb JK, Keller KA, Chinnadurai SK, et al. Use of alfaxalone in bearded dragons (*Pogona vitticeps*): optimizing pharmacodynamics and evaluating cardiogenic effects via echocardiography. *J Am Vet Med Assoc.* 2023;261(1):126-131.

Hematology and Serum Protein in *Trachemys* ssp.: What Has Been Discovered, 2012-2023?

Gianluca Deli, MRCVS, DVM, CertAVP (ZM), GPCert (ExAP), PgCert (EAS), PgCert (HE), FHEA, VetGD Advisor,
Cristiana Manetti, DVM, GPCert (ExAP),
Daniele Petrini, DVM, MSc, GPCert (ExAP),
Stefano Cusaro, DVM, GPCert (ExAP),
Carlo Paoletti, DVM

Session #3291

Affiliation: From Medivet Faygate, Tower Road, Faygate, Horsham, RH12 4SD, UK (Deli), Clinica Veterinaria Cascina (Petrini).

Note: Abstract format selected by author(s). See Front Matter for description of conference proceedings manuscripts submitted as scientific abstracts.

Abstract: The purpose of this literature review was to identify updates over the last 11 years and compare them to the study that was performed in 2012. The aim of the prior study (2011-2012) was to characterize limited data on blood references in pond sliders (*Trachemys scripta* ssp.). This was performed by collecting blood from 101 pond sliders (*Trachemys scripta elegans* and *Trachemys scripta scripta*), analyzing blood cell counts, blood chemistry and serum protein electrophoresis, to obtain useful data for diagnostic purposes. For the assessment of biochemical analytes, three samples from three different time periods were analyzed: before hibernation in September 2011, in post hibernation in April 2012 and before the next hibernation period in September 2012. The subjects sampled came from a single semi natural pond. Results and values were compared using a Student's t-test. The data obtained in 2012 were comparable with the data available in the literature.[1-10] To the author's knowledge, this study reflects reference data evaluating the largest number of specimens (maintained under the same semi-captive conditions) ever reported.

References

1. Mader DR, Divers SJ. *Current Therapy in Reptile Medicine and Surgery.* St. Louis, MO: Elsevier; 2014.

2. Gimenez M, Saco Y, Pato R, et al. Plasma protein electrophoresis of *Trachemys scripta* and *Iguana iguana*. *Vet Clin Pathol.* 2010;39(2):227-235.

3. Carpenter JW. *Exotic Animal Formulary.* 6th Ed. Elsevier Saunders; 2022.

4. Deli G, Cusaro S, Petrini D. et al., Comparative study for the evaluation of hematological, blood chemistry and serum protein electrophoresis parameters in 101 specimens of *Trachemys scripta* ssp. *Proc ICARE.* 2013.

5. Leineweber C, Stöhr AC, Öfner S, et al. Plasma capillary zone electrophoresis and plasma chemistry analytes in tortoises (*Testudo hermanni, Testudo graeca*) and turtles (*Trachemys scripta elegans, Graptemys* spp.) in fall. *J Zoo Wildl Med.* 2021;51(4):915-925.

6. Andreani G, Carpenè E, Cannavacciuolo A, et al. Reference values for hematology and plasma biochemistry variables, and protein electrophoresis of healthy Hermann's tortoises (*Testudo hermanni* ssp.). *Vet Clin Pathol.* 2014;43(4):573-83. doi:10.1111/vcp.12203.

7. Musilová A, Knotková Z, Pinterová K, et al. Variations of plasma protein electrophoresis in healthy captive green iguanas (*Iguana iguana*). *Vet Clin Pathol.* 2015;44:243-248. doi.org/10.1111/vcp.12238.

8. Eshar D, Avni-Magen N, Kaufman E, et al. Effects of time and storage temperature on selected biochemical analytes in plasma of red-eared sliders (*Trachemys scripta elegans*). *Am J Vet Res.* 2018;79(8):852-857.

9. Gradela A, Souza VN, Queiroz MM, et al. Serum biochemistry of *Trachemys scripta elegans* and *Trachemys dorbignyi* (Testudines: Emydidae) bred in captivity in the Northeastern semiarid region of Brazil. *Vet World.* 2020;13(6):1083-1090.

10. Scope A, Schwendenwein I, Schauberger G. Characterization and quantification of the influence of season and gender on plasma chemistries of Hermann's tortoises (*Testudo hermanni, Gmelin* 1789). *Res Vet Sci.* 2013;95(1):59-68.

Axolotl Medicine

Rod Arad, DVM

Session #3414

Affiliation: From Raam Innovative Veterinary Solution, PC, 112 Main St, Unit 4, Pepperell, MA 01463, USA.

Abstract: Axolotls (*Ambystoma mexicanum*), also known as walking fish and Mexican salamander, are native to only two lakes in Mexico. Due to their limited environment and the encroachment of civilization, axolotls are currently a critically endangered species. While in the wild they are struggling to survive, in captivity axolotls are thriving and have become popular pets over the past few years. This has led to more clients seeking veterinary facilities to treat these unique patients. This presentation provides a brief overview of axolotl physiology, husbandry and common diseases.

Axolotl Characteristics

Axolotls have several unique traits, one of which is that they are neotenic. This means that they do not fully mature from the larval form retaining their gills and remaining aquatic. Axolotls have three gills on either side of their head which function similar to gills in other aquatic species. The gills absorb oxygen from the water and pass carbon monoxide back into the water. Visual inspection of the gills, their color and positioning, can give clues to underlying issues and disease. Axolotls have three other means of respiration including rudimentary lungs, cutaneous and buccal respiration. These additional respiratory mechanisms enable axolotls to maintain gas exchange even with compromised gills. Axolotls are also renowned for their regenerative abilities, which is one of the reasons they were taken from the wild to be studied. Damaged and lost limbs can be regrown in months, usually the process is faster in younger animals and slows or even stops as they get older. This ability to regenerate quickly is very helpful when they suffer external wounds due to the tank environment, axolotl bite wounds and surgical incisions.

Axolotls are part of the salamander (urodele) Family but they seem to have taken an evolutionary divergence and remained aquatic. Axolotls have six common colors: brown, melanoid, copper, albino, golden albino and leucistic. They are carnivorous, using their large mouths and grasping teeth to suck in their prey. They can measure anywhere between 18-40 centimeters. Females tend to be larger than the males. Sexual maturity is reached at about 18-24 months of age. Males can be differentiated from females due to the more pronounced cloacal bulge, the female body tends to be slightly rounder and sexually mature males will have black nails. Axolotls with proper husbandry and well controlled aquariums can live up to 15 years of age in captivity. Axolotls are poikilothermic, they prefer cooler temperatures in their environment, around 60-64 °F degrees. Their diet consists of worms, brine, mollusks and insect larva. As juveniles, axolotls are more interested in eating live feed which can make them more susceptible to parasitic infections from infected prey. Their strong predatory drive, large mouths and poor vision also makes them more prone to attacking anything smaller than themselves in the tank and prone to ingesting foreign bodies. This means that environmental substrate and aquarium mates need to be chosen carefully to prevent potential injury. If foreign material is ingested it may be grasped with long forceps and removed from the stomach; however, if it moves further down in the gastrointestinal tract surgical removal may be required.

Husbandry

As with all aquatic species, proper husbandry is crucial, fluctuations in the environment are frequently the cause of illness. Since axolotls are carnivorous, they will produce more nitrogenous wastes. The aquarium must be cleaned frequently, have water changes and water tests performed regularly. Frequent water testing would help detect alterations in pH, ammonia, nitrite and nitrate. Detecting issues with water parameters before major fluctuations occur can prevent stress and disease in Axolotls. Axolotls like most aquatic species are very sensitive; water temperature fluctuations can cause stress leading to disease susceptibility. Temperatures over 75 °F can lead to hyperthermia causing anorexia and complications with buoyancy. An aquarium with strong water currents can also lead to undo stress and increased vulnerability to illness. Like most aquatic patients we need to remind owners to make sure that dechlorinated water is used in their aquarium. The pH should be around 7.5 but they can tolerate anywhere from 6.5-8.0. Again, it is important to reiterate that any major or sudden shifts in the aquarium environment can lead to stress and increase susceptibility to disease. Failure to maintain a stable environment and feeding an improper diet leads to many of the diseases observed in captive bred axolotls.

Diseases

A list of some of the common diseases observed in axolotls includes bloating, bacterial infections, parasitic infection, fungal infections, gastrointestinal obstructions, hyperthermia, neoplasia and trauma.

The medical treatment of these conditions is often similar to those used in other amphibians. In cases of trauma and or limbs loss, it has been suspected that slowly lowering the temperature has increased the rate of healing. Conditions that require surgical intervention are treated similar to other amphibians. It is crucial to remember that we must minimize handling to prevent affecting the slime layer. Wearing non powder gloves that are wetted down with water will help to decrease irritation to the skin. Treating external infections can usually be accomplished with medicated dips. Parasitic, fungal and bacterial infections can be treated by applying treatment to the food. Sedation and anesthesia can be achieved using alfaxalone or MS-222.

Conclusions

Axolotls are unique, interesting creatures and make great pets. They do require a significant amount of upkeep to their aquarium. Many of the illnesses observed are due to inappropriate husbandry and lack of upkeep of the aquarium. Making gradual changes to the aquarium environment so that it coincides with the axolotl's ideal parameters will resolve many issues observed in practice. With more awareness being brought to these creatures, more doctors will be willing to take them on as patients.

References

1. Axolotl Central. Care Guide. Available at: https://www.axolotlcentral.com/axolotl-care-guide. Accessed April 22, 2023.

2. Divers SJ, Stahl SJ, eds. *Mader's Reptile and Amphibian Medicine and Surgery. E-Book*. Elsevier; 2018.

3. Lang E, ed. *Axolotls, Mexican Salamanders as Pets: Axolotls Care, Facts, Diet, Aquarium, Habitat, Breeding, Diseases and Where to Buy All Included.* IMB Publishing; 2013.

4. Mitchell M, Tully TN. *Current Therapy in Exotic Pet Practice*. St. Louis, MO: Elsevier; 2016.

5. Talbot B. The axolotl consult. *Proc Annu Conf Assoc Avian Vets Austr Unusual Pet Avian Vet.* 2018.

How Reliable Are Reptile Reference Ranges?

T. Franciscus Scheelings, BVSc, MVSc, PhD, MANZCVSc (Wildl Health),
DipECZM (Herpetology), and
Helen McCracken

Session #3238

Affiliation: From Melbourne Veterinary School, Faculty of Veterinary and Agricultural Sciences, Building 417, 250 Princes Hwy, Werribee The University of Melbourne, Victoria 3030 Australia.

Note: Abstract format selected by author(s). See Front Matter for description of conference proceedings manuscripts submitted as scientific abstracts.

Abstract: Reptile hematology and biochemistry reference ranges are often used by clinicians to assess the health of patients. A major issue with using these reference ranges is that often they are generated from small populations at a single point in time. Therefore, they may not be applicable to all patients with the potential for misinterpretation. A long-term study conducted over 8 years assessed impacts of location, sex, and time of year on hematologic values of a large free-ranging population of three species of chelonians within the Melbourne Zoo. In addition, impacts of location on hematology and biochemistry values in a population of free-ranging wild turtles within an urban setting were assessed. In both study populations, blood results were significantly influenced by year, season, reproductive status, and location. The results of this investigation clearly show that reptile hematology and biochemistry are heavily influenced by both extrinsic and intrinsic factors and that these need to be carefully considered when interpreting results.

Section 42

ARAV Surgery and Imaging Session

La'Toya Latney, DVM, DECZM (ZHM),
DABVP (Reptile and Amphibian Practice), CertAqV
Session Chair

Turtle Rehabilitation Cases: From the Wild Side

Renée Schott, DVM, CWR

Session #3439

Affiliation: From Wildlife Rehabilitation Center of Minnesota, 2530 Dale St. N, Roseville, MN 55113, USA.

Note: Abstract format selected by author(s). See Front Matter for description of conference proceedings manuscripts submitted as scientific abstracts.

Abstract: This masterclass is a case-based dive head-on into traumatic aquatic, freshwater turtle cases that include radiographs. Turtles have such interesting radiographs. The class is not just lung views and sideways swimming turtles; there are weird presentations and oddball cases that force participants to figure out answers to "How do we take a picture of that?" The class also covers how to treat these ailments.

Recommended Reading

1. Hernandez SM, Barron H, Miller E, et al., eds. *Medical Management of Wildlife Species*. 1st Ed. Hoboken, NJ: Wiley Blackwell; 2019. doi:10.1002/9781119036708.
2. Mullineaux E, Keeble E, eds. *BSAVA Manual of Wildlife Casualties*. 2nd Ed. Gloucester, UK: British Small Animal Veterinary Association; 2016.
3. Friend M, Franson JC, eds. *Field Manual of Wildlife Diseases: General Field Procedures and Diseases of Birds*. Madison, WI: U.S. Geological Survey; 1999.
4. Ruth I. *Wildlife Care Basics for Veterinary Hospitals: Before the Rehabilitator Arrives*. Davis, CA: Humane Society Veterinary Medical Association; 2007.
5. Gage L. *Hand-Rearing Wild and Domestic Mammals*. 1st Ed. Ames, IA: Blackwell Publishing; 2002.
6. Miller E, ed. *Minimum Standards for Wildlife Rehabilitation*. 4th Ed. St. Cloud, MN: National Wildlife Rehabilitators Association and International Wildlife Rehabilitation Council; 2012.
7. Stocker L. *Practical Wildlife Care*. 2nd Ed. Oxford, UK: Blackwell Publishing; 2005. doi:10.1638/1042-7260(2000)031[0586:br]2.0.co;2.
8. Meredith A, Keeble E. *Self-Assessment Colour Review: Wildlife Medicine and Rehabilitation*. 1st Ed. London, UK: Manson Publishing; 2011. doi:10.1201/b15226.
9. Schifino Valente AL. Diagnostic Imaging of the Loggerhead Sea Turtle (*Caretta caretta*). 2007. Available at: https://core.ac.uk/download/pdf/13288275.pdf. Accessed June 1, 2023.
10. Welsh M. *Diagnostic Imaging of Exotic Pets: Birds, Small Mammals, Reptiles*. Vol 89. Hanover, DE; 2011. doi:10.1111/j.1751-0813.2011.00824.x.

11. McArthur S, Wilkinson R, Meyer J, et al. *Medicine and Surgery of Tortoises and Turtles*. Oxford, UK: Blackwell Publishing; 2004.

12. Mader DR, ed. *Reptile Medicine and Surgery*. 2nd ed. St. Louis, MO: Elsevier; 2006.

Cystoscopic Sex Identification of Indonesian Blue-Tongued Skinks (*Tiliqua gigas*)

Giordano Nardini, DVM, PhD, DipECZM (Herpetology),
Nicola Di Girolamo, DVM, PhD, DipECZM (Herpetology), Dipl ACZM, Dipl ACVPM,
Alessandro Vetere, DVM,
Francesco Di Ianni, DVM, PhD

Session #3305

Affiliation: From Exotic Animal Section, Clinica Veterinaria Modena Sud, Piazza Dei Tintori 1, Spilamberto, (MO), Italy (Nardini, Vetere), Exotic Pet Service, Department of Clinical Sciences, College of Veterinary Medicine, Cornell University, 930 Campus Road, 14853, Ithaca, NY, USA (DiGirolamo), Department of Veterinary Science, University of Parma, Parma (PR), Italy (Di Ianni).

Note: Abstract format selected by author(s). See Front Matter for description of conference proceedings manuscripts submitted as scientific abstracts.

Abstract: This paper describes a non-surgical endoscopic technique for visualization of gonads and sex identification in Indonesian blue-tongued skinks (*Tiliqua gigas*). Forty-two, captive-bred, clinically healthy Indonesian blue-tongued skinks (*Tiliqua gigas*) between 6 months and 3 years of age, weighing between 22 and 550 grams, were included. With the skinks under general anesthesia, a cystoscopy was performed with a rigid endoscope. Gonads were then visualized through the transparency of the urinary bladder. Generalized linear models were built to identify significant associations between time required for cystoscopy and independent variables, including body weight, sex and order of execution. Cystoscopy was achieved in all skinks and allowed visualization of the gonads through the transparency of the urinary bladder in 41 (98%, 95% CI: 87 to 99%) of them. Based on the endoscopic appearance, 18 skinks were identified as males, and 23 were identified as females. Median procedure time was 60 (range 25-180) seconds and was not associated with procedure order (-0.69, 95% CI: -1.83 to 0.45), nor with the weight (0.02, 95% CI: -0.07 to 1.0) or the identified sex (11.7, 95% CI: -15.07 to 38.45) of the skink. All the skinks recovered from anesthesia and no clinical complications were observed prior to discharge. Considering the inherent difficulty in identifying the sex of blue-tongued skinks, cystoscopic gonadal identification could provide a significant improvement in the veterinary care of this species.

Novel Imaging Modalities for Antemortem Diagnosis of Jugular Thrombus in a Bearded Dragon (*Pogona vitticeps*)

Agnes (Aga) Galej, VMD,
Jessica Robertson, DVM, Dipl ACZM,
Yu Hung Hsieh, DVM, MS, Dipl ACVR

Session #3320

Affiliation: From the Department of Zoological Medicine and Surgery (Galej, Robertson) and the Department of Radiology (Hsieh), VCA West Los Angeles Animal Hospital, Los Angeles, CA 90025, USA.

Note: Abstract format selected by author(s). See Front Matter for description of conference proceedings manuscripts submitted as scientific abstracts.

Abstract: A 5-year-old male central bearded dragon (*Pogona vitticeps*) presented with a 3-day history of left forelimb lameness, minimal muscle tone, and absent pain sensation distal to the mid-humerus. Computed tomography revealed thickened and hypoattenuating soft tissues adjacent to the left scapulohumeral and left neck regions with no abnormal contrast enhancement. The suspected left external jugular vein in this region was ill-defined and small, with filling defects, consistent with the appearance of a venous thrombus. Focal ultrasound of the left side of the neck confirmed a suspected thrombus in this location, just cranial to a jugular venous valve. The patient was treated with supportive care and started on clopidogrel as an antithrombotic treatment. Clinical improvement was minimal following this visit and the patient presented again 5 months later after declining significantly. On follow-up imaging at this visit, a suspected cardiac mass was seen on echocardiography. A fine-needle aspirate of the mass at that time was suspicious for a sarcoma; however, the sample was non-diagnostic. The previously identified thrombus could not be detected on ultrasonography at this visit. The owner ultimately decided to pursue hospice care at home and no additional imaging or further workup was performed. This is the first report in reptiles describing a venous thrombus using diagnostic imaging.[1-12] Given the reported occurrence of aneurysms and thrombi in bearded dragons in clinical practice, this case demonstrates the use of an important tool for diagnosis of these conditions.

References

1. Bercker C, Hurtel C, Schilliger L. Spontaneous echo contrast in an asymptomatic McDowell's carpet python (*Morelia spilota mcdowelli*) with suspected restrictive cardiomyopathy. *J Herpetol Med Surg.* 2022;32: 113-115.

2. Buriko Y, Drobatz K, Silverstein DC. Establishment of normal reference intervals in dogs using a visco-elastic point-of-care coagulation monitor and its comparison with thromboelastography. *Vet Clin Pathol.* 2020;49:567-573.

3. Furst N, Alexander A, Ossiboff RJ, et al. Tail base aneurysm in an inland bearded dragon (*Pogona vitticeps*). *J Exot Pet Med.* 2020;33:34-37.

4. Hochleithner C, Holland M. Ultrasonography. In: Mader DR, Divers SJ, eds. *Current Therapy in Reptile Medicine and Surgery.* St. Louis, MO: Elsevier; 2014:107-127.

5. Mays E, Phillips K. Focused ultrasound of vascular system in dogs and cats - thromboembolic disease. *Vet Clin North Am Small Anim Pract.* 2021;51:1267-1282.

6. Raju S, Byrne J. External jugular vein thrombosis secondary to deep tissue neck massage. *J Vasc Surg Cases Innov Tech.* 2017;3:146-148.

7. Shilton CM. Necropsy. In: Doneley B, Monks D, Johnson R, Carmel B, eds. *Reptile Medicine and Surgery in Clinical Practice.* St. Louis, MO: Elsevier; 2017:409-424.

8. Stephens T, Rosenwax A. Diseases of the cardiovascular system. In: Doneley B, Monks D, Johnson R, Carmel B, eds. *Reptile Medicine and Surgery in Clinical Practice.* St. Louis, MO: Elsevier; 2017:287-298.

9. Wünschmann A, Armién AG, Childress AL, et al. Intrapericardial *Encephalitozoon pogonae*–associated arteritis with fatal hemopericardium in two juvenile central bearded dragons. *J Vet Diagn Invest.* 2019;31: 467-470.

10. Wyneken J. Computed tomography and magnetic resonance imaging. In: Mader DR, Divers SJ, eds. *Current Therapy in Reptile Medicine and Surgery.* St. Louis, MO: Elsevier; 2014:93-106.

11. Yarto E, Fajardo J, Morales M, et al. Bilateral atrial thrombosis in an African hedgehog (*Atelerix albiventris*) with cardiomyopathy, endometrial hyperplasia, and left cystic ovary. *J Exot Pet Med.* 2021;36:86-89.

12. Zellar AK, Klaphake E, Han S. Arteriosclerosis and arteritis in two colubrid snakes. *J Exot Pet Med.* 2019;28:173-179.

Comparison Between Cloacocystoscopic and Coeloscopic Gender Identification Using Two Anesthetic Protocols in Burmese Star Tortoises (*Geochelone platynota*)

Emanuele Lubian, DVM, PhD, EM, GPCert (ExAP),
Francesca Duse, DVM,
Gianluca Deli, MRCVS, DVM, CertAVP (ZM), CPCert (ExAP), PgCert (EAS), GPAdvCert (SMP), PGCert (HE), FHEA, VetGDAdvisor,
Filippo Spadola, PhD

Session #3158

Affiliation: From the Department of Veterinary Science, University of Messina, Polo Didattico Annunziata 98168 Messina, IT (Lubian, Spadola), MypetClinic, Viale Ranzoni 10, 20149, Milano, IT (Lubian, Duse), Ospedale Veterinario Universitario Piccoli Animali, Università degli Studi di Milano, Lodi, IT (Lubian, Duse), Medivet Faygate, Tower Road, Faygate, Horsham RH12 4SD, UK (Deli).

Note: Abstract format selected by author(s). See Front Matter for description of conference proceedings manuscripts submitted as scientific abstracts.

Abstract: Cloacoscopy, coelioscopy and cystoscopy are the three demonstrated endoscopic methods to identify gender in chelonians.[1,2] Few studies have been published regarding all three methods and there are not comparative studies. Until now there have been no studies to identify which method is better. Coelioscopy allows for the direct visualization of the gonads but the animals need to be anesthetized, contrary to cloacoscopy and cystoscopy, which can be performed without heavy sedation but gonads visualization is transvesical or trans cloacal. In this study cloaco-cystoscopy and coelioscopy were performed in 12 specimens of Burmese star tortoise between 61-151 grams, using two different anesthetic protocols. Cloacoscopy has been performed without sedation, in dorsal recumbency, to identify the health status of the animals before performing coelioscopy. Coelioscopy has been performed using two different anesthetic protocols (i.e., alfaxalone 5 mg/kg and local lidocaine 2 mg/kg or alfaxalone 5 mg/kg, dexmedetomidine 10 mg/kg, local lidocaine 2 mg/kg antagonized with atipamezole 50 mg/kg). Results show that gender identification in the animals studied was extremely hard using the cloacal cystoscopic method (due to presence of cloacal calculus, opacity of bladder wall and difficulty identifying genital papillae) compared with the coelioscopic method (in which gonads were identified in less than 10 minutes in all animals). Between the two anesthetic protocols, the one with dexmedetomidine provides good muscular relaxation, sedation and rapid and smooth awakening, which are optimal to perform the coelioscopic investigation.

References

1. Martínez-Silvestre A, Bargalló F, Grífols J. Gender identification by cloacoscopy and cystoscopy in juvenile chelonians. *Vet Clin North Am Exot Anim Pract.* 2015;18(3):527-539.

2. Hernandez-Divers SJ. Making the difference in exotic animal practice: the value of endoscopy. *Vet Clin North Am Exot Anim Pract.* 2015;18(3):351-357.

Cloacocystoscopy as a Diagnostic Tool for Emidids and Geoemidids

Emanuele Lubian, DVM, PhD, EM, GPCert (ExAP),
Niccolò Pagnini, DVM, MSc,
Gianluca Deli, MRCVS, DVM, CertAVP (ZM), CPCert (ExAP), PgCert (EAS), GPAdvCert (SMP), PGCert (HE), FHEA, VetGDAdvisor,
Filippo Spadola, PhD

Session #3303

Affiliation: From the Department of Veterinary Science, University of Messina, Polo Didattico Annunziata 98168 Messina, IT (Lubian, Spadola), MypetClinic, Viale Ranzoni 10, 20149, Milano, IT (Lubian), Ospedale Veterinario Universitario Piccoli Animali, Università degli Studi di Milano, Lodi, IT (Lubian), Ambulatorio Veterinario Galciana, Via Mannocci 32, 59100, Prato, IT (Pagnini), Medivet Faygate, Tower Road, Faygate, Horsham RH12 4SD, UK (Deli).

Note: Abstract format selected by author(s). See Front Matter for description of conference proceedings manuscripts submitted as scientific abstracts.

Abstract: The cloaca is a distinctive tract of avian and reptilian anatomy and has been subjected to numerous studies regarding morphology to identify species-specific differences and to develop better endoscopic techniques. Cloacoscopy and cystoscopy in chelonians has been described not only to diagnose and treat pathologies of the cloacal tract itself, but also to observe the viscera, since the bladder wall is characterized by great distensibility and transparency, especially regarding gonadal visualization in immature subjects. The present study evaluates the diagnostic potential of cloacocystoscopy in Emydidae and Geoemydidae. In this study 145 different cloaco-cystoscopic studies were taken into account, all of them were performed by the same operator, with the same tools (rigid telescope with 2.7-mm diameter, operating sheet of 4.8-mm diameter, xenon light source, endovideo, Ringer's lactate solution to irrigate the cloaca and the urinary bladder) on specimens weighing 120-970 grams. Turtles were not sedated and were kept in dorsal recumbency. Prevalence of visualization of the caudal margins of the liver, of the intestines and of the gonads was calculated, taking into account both the visualization obtained through the urinary bladder and through the accessory bladders. Results show better visualization of the liver and the intestine through the bladder wall, and better gonadal view through the accessory bladder. It also describes the possibility and the prevalence of entering in the distal part of the oviduct. Results demonstrate many differences with some past reports regarding these species.

References

1. Martínez-Silvestre A, Bargalló F, Grífols J. Gender identification by cloacoscopy and cystoscopy in juvenile chelonians. *Vet Clin North Am Exot Anim Pract.* 2015;18(3):527-539.

2. Hernandez-Divers SJ. Making the difference in exotic animal practice: the value of endoscopy. *Vet Clin North Am Exot Anim Pract.* 2015;18(3):351-357.

Celiotomy and Endoscopic Removal of Pentastomes in a Wild-Caught African Cape File Snake (*Limaformosa capensis*)

Alissa M. Tepedino, DVM, CertAqV

Session #3245

Affiliation: From Avian and Exotic Animal Care, 8711 Fidelity Boulevard, Raleigh, NC 27617, USA.

Note: Abstract format selected by author(s). See Front Matter for description of conference proceedings manuscripts submitted as scientific abstracts.

Abstract: An adult female wild-caught African Cape file snake (*Limaformosa capensis*) presented for chronic anorexia and lethargy. A fecal float cytology revealed larvated pentastome ova, and diagnosis of pentastomiasis was confirmed via tracheal wash cytology. Computed tomography located two worm-like adult stage pentastomes in the air sac of the patient. One month later the patient returned for surgery, and pre-operative radiographs showed two distinct areas with multiple adult parasites each. A celiotomy and incision into the air sac was performed to remove the cranial group of three female adult parasites. A rigid endoscope was then introduced through the air sac incision towards the caudal group and removed one female and one male adult parasite. Further endoscopy and post-operative radiographs confirmed removal of all adult parasites. The pentastomes were identified using PCR as a potentially novel *Kiricephalus* species with 99.23% homology to *K. coarctatus* and 98.98% homology to *K. pattoni*. The patient received medical treatment with ivermectin (0.2 mg/kg IM q14d) post-operatively. Resolution of the infection was confirmed with normal radiograph and fecal float cytology results after five treatments with ivermectin. This case report describes the successful outcome of surgical and medical treatment of pentastomiasis in a wild-caught snake.

Advances in Surgical Treatment of Urogenital Disorders of Chelonians

Nicola Di Girolamo, DMV, MSc (EBHC), PhD, DipECZM, Dipl ACZM, Dipl ACVPM

Session #3499

Affiliation: From the Department of Clinical Sciences, College of Veterinary Medicine, Cornell University, Ithaca, NY 14853, USA.

Abstract: Urogenital disorders such as urolithiasis, follicular stasis, ectopic eggs, salpingitis, oviductal prolapse and dystocia in chelonians are well documented and commonly require surgical intervention. Celiotomic access to the reproductive tract can be achieved through incision on the plastron (i.e., plastronotomy) or incision on the pre-femoral fossa (i.e., pre-femoral celiotomy). Plastronotomy is an invasive procedure, and prefemoral celiotomy in recent years has become the mainstay for treatment of urogenital disorders in chelonians. This presentation covers the most recent advances in urogenital surgery in chelonians that allow surgical procedures such as prefemoral cystotomy or prefemoral ovariosalpingectomy to be safely and effectively performed by the veterinary practitioner.

Introduction

Urogenital disorders such as urolithiasis, follicular stasis, ectopic eggs, salpingitis, oviductal prolapse and dystocia in chelonians are well documented and commonly require surgical intervention. Celiotomic access to the reproductive tract can be achieved through incision on the plastron (i.e., plastronotomy) or incision on the pre-femoral fossa (i.e., pre-femoral celiotomy). Plastronotomy is an invasive procedure, and prefemoral celiotomy in recent years has become the mainstay for treatment of urogenital disorders in chelonians. This presentation covers the most recent advances in urogenital surgery in chelonians that allow surgical procedures such as prefemoral cystotomy or prefemoral ovariosalpingectomy to be safely and effectively performed by the veterinary practitioner.[1-4]

Prefemoral Cystotomy

The use of prefemoral cystotomy and destruction of the stone with a mini-driver surgical drill has been successful in removing large and medium-sized stones. Prefemoral fossa celiotomy (syn. prefemoral coeliotomy) is currently considered the surgical access of choice for most urinary and reproductive surgeries. This approach avoids prolonged post-surgical dry-docking, which is necessary following plastron osteotomy and is particularly indicated in species with a relatively small plastron. In the author's experience prefemoral cystotomy is easier achieved with the chelonian positioned in oblique positioning with the lower side being the one in which the urinary bladder stone is more easily reached. This recumbency allows the stone to gravitate towards the prefemoral incision, making its approach sensibly easier. The skin, subcutaneous tissue, and the transverse and oblique abdominal muscles, and coelomic membrane are transected and the coelomic cavity accessed. The use of ring retractors is extremely useful to enhance access and visibility. Care should be taken to avoid trauma to the urinary bladder upon entering the coelom.

Cystocentesis should be performed if the bladder is distended and impairs surgical access to the coelom. The urinary bladder can be marsupialized to the skin. The stone is destroyed with a mini-driver or a rotary drilling tool and Babcock forceps. Depending on the size of the chelonian, urinary bladder thickness is significantly different. Thin urinary bladders are prone to rupture during suture and require extreme care during this process. Amputation of part of the urinary bladder may be the only solution if major damage has occurred during the procedure or because of the chronic presence of the stone. Closure of the body wall should be performed in two to three layers. It may not always be possible to close the coelomic membrane. Closure of the muscles and skin is routine. If substantial amounts of subcutaneous tissue are present, then additional subcutaneous sutures should be considered. The healing times after prefemoral coeliotomy (approximately 4 weeks) are substantially shorter than after a plastron osteotomy (approximately 1 to 2 years).

Percloacal Lithotomy

Cloacal stones are urinary stones that have been eliminated from the urinary bladder and get trapped in the caudal pelvis. Rarely they can reach the vent and therefore be visible externally. The following procedures should be performed under general and/or intrathecal anesthesia.

Depending on the species and size of the chelonian affected and, on the size and shape of the stone, different techniques can be employed for their removal. Currently, the easier and more effective way to remove these stones consists in a cloacal approach. The shape of caudal scutes of juvenile or adult tortoises of certain species (e.g., African spurred tortoises, *Centrochelys sulcata*) can limit the access to the tail and the cloaca. In those cases, removal of the caudal carapacial scutes increase visibility and accessibility to the stone. The scutes can be removed performing a straight-line incision with a rotary tool or an orthopedic driver. Care must be taken to prevent injury of the skin at the level of its attachment to the carapace. In species that have minimal development of caudal scutes or in young individuals where the caudal scutes are not yet fully developed to cover the tail and the cloaca, this procedure is not needed.

Once appropriate visualization of the tail and the cloaca is achieved, the stone can be visualized placing a retractor (e.g., Lone Star retractor) on the rims of the vent. Alternatively, an endoscope can be used to assist the procedure.[5] After the stone is properly visualized, it can be destroyed by means of a rotary drilling tool and Babcock forceps. Care must be taken to minimize damage to the cloacal mucosa.

Prefemoral Oophorectomy and Salpingectomy

A technique for exteriorization and excision of ovaries and oviducts of chelonians through the prefemoral fossa has been described.[6] We have successfully refined this technique by positioning chelonians in a dorso-oblique recumbency, rather than dorsal, ventral or lateral recumbencies.[2] The ovaries are identified and gently pulled out with cotton tip applicators. Care must be taken to avoid rupture of ovarian follicles. Once all ovarian follicles are exteriorized and the mesovarium is visible, the ovarian vasculature in the mesovarium is ligated and transected. Hemostasis and complete excision of the ovarian tissue should be confirmed with the endoscope. If the oviduct is diseased and surgical removal is intended, the ipsilateral oviduct is exteriorized, ligated and transected through the same prefemoral incision. The procedure is then repeated for the contralateral ovary and oviduct. Often the procedure may be performed for the contralateral ovary via the same prefemoral incision. Retained eggs in the oviduct and ectopic eggs free within the coelom or the urinary bladder can also be removed using this technique.

References

1. Di Girolamo N, Mans C. Reptile soft tissue surgery. *Vet Clin North Am Exot Anim Pract.* 2016;19(1): 97-131.

2. Di Girolamo N. Indications, outcomes and complications of prefemoral reproductive surgery in client-owned chelonians: 22 cases, 2017-2018. *Proc ExoticsCon.* 2019;560.

3. Studer KA, Di Girolamo N, Caron M, et al. Feasibility, gross postmortem results, and veterinarian opinions on oblique pre-femoral ovariosalpingectomy in red eared sliders (*Trachemys scripta elegans*) during a teaching workshop. *Proc ExoticsCon.* 2021.

4. Di Girolamo N. Techniques for minimally invasive removal of urinary stones in chelonians. *Proc ExoticsCon.* 2019;586-588.

5. Mans C, Sladky KK. Endoscopically guided removal of cloacal calculi in three African spurred tortoises (*Geochelone sulcata*). *J Am Vet Med Assoc.* 2012;240(7):869-875.

6. Innis CJ, Hernandez-Divers S, Martinez-Jimenez D. Coelioscopic-assisted prefemoral oophorectomy in chelonians. *J Am Vet Med Assoc.* 2007;230(7):1049-1052.

Section 43

ARAV Critical Care and Conservation Session

Kourtney Grimm, DVM
Session Chair

Suspected Bufotoxicosis Secondary to Toad (*Anaxyrus* spp.) Ingestion in an Inland Bearded Dragon (*Pogona vitticeps*)

Laura Whalen, DVM, CertAqV, and
Peter A. Sojka, DVM, DABVP (Avian Practice)

Session #3161

Affiliation: From Pieper Memorial Veterinary Hospital, 730 Randolph Road, Middletown, CT 06457, USA.

Note: Abstract format selected by author(s). See Front Matter for description of conference proceedings manuscripts submitted as scientific abstracts.

Abstract: A 5-year-old male inland bearded dragon (*Pogona vitticeps*) was presented for evaluation after toad (*Anaxyrus* spp.) ingestion. Onset of clinical signs, including writhing and pawing at the mouth, were noted within 2 minutes of ingestion and within 15 minutes seizure activity began. Concerns with ingestion of the toad included bufotoxicosis resulting in hyperkalemia and arrhythmias, seizures, and bradycardia. The ingestion of the toad itself poses risks for gastrointestinal obstruction. Supportive care including stabilization, decontamination, and treatment was ultimately unsuccessful. Decontamination with oral lavage and activated charcoal is recommended initially, and associated neurologic signs can be directly addressed with use of benzodiazepines. Heat support and continued and effective ECG monitoring are also recommended to increase the chance of recovery. In this case, obtaining bloodwork to monitor for electrolyte abnormalities, such as hyperkalemia, and access to digoxin-specific antibody fragments (digoxin immune fab) to counteract the digitalis-like effects of bufotoxins may have potentially changed the outcome. Digoxin immune fab use has been noted in a previous case with intraperitoneal administration (three doses at 6 mg/kg, fourth dose at 3 mg/kg), but was unavailable during hospitalization. Additional cases and treatments are described in the literature.[1-5] Implementation of the above monitoring and treatment methods may prove effective in future cases. Based on the case seen, it is suspected that bearded dragons are highly susceptible to the effects of bufotoxin; suspected exposure or ingestion should be treated as an emergency.   toad ingestion

References

1. Butler J. Toxicology case: successful treatment of *Bufo marinus* intoxication in a dog. *DVM 360.* Available at: https://www.dvm360.com/view/toxicology-case-successful-treatment-bufo-marinus-intoxication-dog. Accessed November 22, 2022.

2. Dobbs MR, Postma T. Neurotoxic animal poisons and venoms. In: Dobbs MR, ed. *Clinical Neurotoxicology Syndromes, Substances, Environments.* Philadelphia, PA: Elsevier Saunders; 2009:463-489.

3. Johnnides S, Green T, Eubig P. Toad intoxication in the dog by *Rhinella marina*: the clinical syndrome and current treatment recommendations. *J Am Anim Hosp Assoc.* 2016;52(4):205-211.

4. Fitzgerald KT, Newquist KL. Poisonings in reptiles. *Vet Clin North Am Exot Anim Pract.* 2008;11(2): 327-357.

5. Plumlee K, Roder JD. Biotoxins. In: Plumlee KH, ed. *Clinical Veterinary Toxicology.* St. Louis, MO: Mosby; 2004:98-116.

Health Assessment of Free-Ranging Eastern Indigo Snakes (*Drymarchon couperi*) from Hydrologic Restoration Construction Sites in South Florida

James E. Bogan, Jr., DVM, MVS (Conservation Medicine), DABVP (Canine and Feline Practice), DABVP (Reptile and Amphibian Practice), CertAqV,
Bradley M. O'Hanlon,
David Steen, PhD,
Terry Horan,
Robert Taylor,
Alexandra Mason, DVM, MS, CertAqV,
Timothy Breen,
Holly Andreotta,
Michael M. Garner, DVM, DACVP,
Bryan Cornelius,
April Childress,
Michele Elmore

Session #3156

Affiliation: From Central Florida Zoo's Orianne Center for Indigo Conservation, 30931 Brantley Branch Road, Eustis, FL 32736, USA (Bogan, Mason), Florida Fish and Wildlife Conservation Commission, 620 S. Meridian St., Tallahassee, FL 32399, USA (O'Hanlon), Fish and Wildlife Research Institute, Florida Fish and Wildlife Conservation Commission, 1105 SW Williston Road, Gainesville, FL 32601, USA (Steen), South Florida Water Management District, 3301 Gun Club Road, MSC 8120, West Palm Beach, FL 33406, USA (Horan, Taylor, Andreotta, Cornelius), United States Fish and Wildlife Service, Florida Ecological Services Field Office, 1339 20th Street, Vero Beach, FL 32960, USA (Breen), Northwest ZooPath, 654 West Main Street, Monroe, WA 98272, USA (Garner), Department of Comparative, Diagnostic, and Population Medicine, College of Veterinary Medicine, University of Florida, Gainesville, FL 32610, USA (Childress), United States Fish and Wildlife Service, Georgia Ecological Services, P.O. Box 52560, Fort Benning, GA 31995, USA (Elmore).

Note: Abstract format selected by author(s). See Front Matter for description of conference proceedings manuscripts submitted as scientific abstracts.

Abstract: There is a paucity of information regarding the health status of free-ranging Eastern indigo snakes (EIS; *Drymarchon couperi*) in heavily modified and developing landscapes. As a component of regional Florida Everglades restoration efforts, several areas occupied by EIS are being converted to reservoirs.[1] From 2020 to 2022, we opportunistically captured 28 EIS at two of these sites and brought them into captivity to join a captive breeding colony but 11 died within 5 months of capture. Health assessments were performed on 28 individuals and included hematology and plasma biochemistry analysis, as well as screening for pesticide contaminant levels, parasites, and other pathogens. Overall, the health of these EIS was poor compared to expectation and the presence of pathogens was relatively high (25/28 (89.4%) *Kalicephalus* sp., 12/28 (42.9%) *Raillietiella orientalis*, 11/28 (39.2%) *Ochetosoma validum*, 7/28 (25.0%) *Cryptosporidium serpentis*, 3/28 (10.7%) snake adenovirus 1, and 1/28 (3.6%) Ferlavirus genotype C), suggesting immunosuppression secondary to stress. Possible stressors include physical displacement, habitat modification, and noise pollution, in addition to the presence of harmful chemicals, and their impacts on this federally threatened species should be considered further.

References

1. Bogan JE, O'Hanlon BM, Steen DA, et al. Health assessment of free-ranging Eastern indigo snakes (*Drymarchon couperi*) from hydrologic restoration construction sites in south Florida. *J Wildl Dis*. 2023;59(1): *in press*. doi: 10.7589/JWD-D-22-00055.

Sea Turtle Rescue, Treatment, Rehabilitation and Release: The Challenge is Real, Caribbean Edition

Odette Doest, DVM

Session #3479

Affiliation: From Dierenartsenpraktijk Doest, St. Michielsweg 41, Julianadorp, Curacao.

Abstract: Sea turtles in the Caribbean Leeward Island territorial waters face many challenges. They are victims of ghost nets and boat strikes, they ingest fishhooks, suffer from poaching and many are infected with fibropapilloma. Their challenges are similar to those of other sea turtles around the world in tropical regions. The challenge is to treat them, without the availability of gas anesthesia and let them rehabilitate at the small animal clinic. The author manages to do surgery while using only injectable anesthesia and explains how to conduct the medical management, public education, and release of the sea turtles.

Health Issues

A total of 36 sea turtles have been treated at the facility from November 2018 to January 2023. In 2020 there was a peak in submitted sea turtles due an extreme amount of sargassum seaweed between January 2020 and March 2020, with a maximum of 13 sea turtles submitted at the same time. The most common species is the green sea turtle (*Chelonia midas*), followed by the loggerhead sea turtle (*Carettta caretta*) and the hawksbill sea turtle (*Eretmochelys imbricata*) which are all abundant in the waters surrounding Curaçao. Of the three species, the hawksbill sea turtles only came in as newly hatched juveniles.

Fishhooks and fish line entanglement

The NGO Sea Turtle Conservation Curacao is driven by volunteers and one paid manager to patrol the beaches during nesting season. They are the first responders for sea turtles that are entangled in fish lines or have fishhooks or lines attached to them. When they are unable to remove the fishhooks or lines at the beach itself, the sea turtles are transported to the clinic.

On radiographs more fishhooks are usually found than are apparent on physical examination. This situation requires surgery to remove the fishhooks from body orifices. Sometimes sedation-only protocols are used to remove hooks. Depending on the gravity of the procedure, turtles are released the next day or when their health is acceptable for release.

Boat strikes

Boat strike victims are brought in as soon as they can be caught and transported. Some of them have neurologic symptoms.

Immediate shell repair is attempted when feasible. If the damage is too severe on a scale from 1-5 with 5 being the most severe, turtles scoring 5 are either euthanized or die upon arrival.

Most boat strikes are reported from either the Spanish water which is a popular watersport area, or St. Joris Bay, which is the most popular kite surfing area.

Fibropapilloma victims

All fibropapilloma cases involved the green sea turtles (*Chelonia midas*) between estimated 3-30 years of age. The areas the papillomavirus sea turtles come from are two specific areas, the Caracasbaai and the north side of the island between Ascension and Wacawa. The most affected areas on the turtles were the eyes, front flippers, scholar and neck, and the underside of the carapace.

Fibropapilloma turtles come in severely emaciated, or because they were entangled in ghost nets with their tumors. Sometimes they are also the victims of a boat strike.

Some fibropapilloma sea turtles also have a variety of internal parasites. The treatment time for the turtles varies between 3 weeks to 1.5 years. In severe cases, surgical operations are performed in three to six sessions. Acyclovir has been used to try to reduce the virus load prior to operations. After recovering from the operations, they are on a treatment of analgesics and a broad-spectrum antibiotic.

Treatment

For certain procedures sea turtles are sedated. For shell repair surgery or fibropapilloma, they undergo general anesthesia. Antibiotic and antiviral medication as well as multivitamins are given via injection and also via gavage feeding. For gavage feeding formulas, a commercial piscivore formula is used that consists of grinded capelin, spinach, and a commercial rabbit feeding formula.

Rehabilitation and Public Education

Depending on received treatment, turtles are either dry docked or housed in small hard plastic basins of different sizes, depending on the carapace diameter, which are filled with seawater. Fibropapilloma sea turtles are treated and housed separately from the trauma sea turtles. When they can eat unassisted while on medication, they are transported to the private saltwater pool to regain their buoyancy and practice swimming.

When a sea turtle is ready for release, public press releases are held by the Sea Turtle Conservation. Announcements are made via radio and social media. Efforts are made to make sure local children will be present at the release. They are often asked to be a part of the sea turtle releases in the hope that such a special gesture will have a lasting impact on their lives and influence their decision-making in the future for the welfare of wildlife and wildlife conservation. The first release in 2023 was done by the Dutch Crown Princess Amalia van Oranje, in the presence of her parents the King and the Queen of the Netherlands.

Acknowledgments: The author thanks Dr. Marja JL Kik, DVM, DipECZM (Herpetology), Dr. Douglas Mader, MSc, Florida Keys Diplomate, DABVP (Canine and Feline Practice), DABVP (Reptile and Amphibian Practice), DipECZM (Herpetology).

References

1. Alfaro-Nuñez A, Bertelsen MF, Bojesen AM, et al. Global distribution of *Chelonia* fibropapilloma-associated herpesvirus among clinically healthy sea turtles. *BMC Evol Biol.* 2014;14:206.

2. Chatigny F, Kamunde C, Creighton CM, Stevens ED. Uses and doses of local anesthetics in fish, amphibians, and reptiles. *J Am Assoc Lab Anim Sci.* 2017;56(3):244-253. PMID: 28535859; PMCID: PMC5438917.

3. Nash A, Ryan EJ. Immunity in sea turtles: review of a host-pathogen arms race millions of years in the running. *Animals.* 2023;13:556. doi:10.3390/ani13040556. PMID: 36830343; PMCID: PMC9951749.

4. Work TM, Balazs GH, Summers TM, et al. Causes of mortality in green turtles from Hawaii and the insular Pacific exclusive of fibropapillomatosis. *Dis Aquat Organ.* 2015;115:103-110. doi.org/10.3354/dao02890.

5. Robinson NJ, Deguzman K, Bonacci-Sullivan L, et al. Rehabilitated sea turtles tend to resume typical migratory behaviors: satellite tracking juvenile loggerhead, green, and Kemp's ridley turtles in the northeastern USA. *Endang Species Res.* 2020;43:133-143. doi.org/10.3354/esr01065.

Evaluation of Histone Demethylase Inhibitor ML324 and Acyclovir Against Cyprinid Herpesvirus 3 Infection in Koi

Shelby Matsuoka,
Gloria Petri,
Kristen Larson,
Alexandra Behnke,
Xisheng Wang,
Muhui Peng,
Sean Spagnoli,
Christiane Lohr,
Ruth Milston-Clements,
Konstantin Divilov,
Ling Jin

Session #3162

Affiliation: From the Department of Biomedical Sciences, College of Veterinary Medicine, Oregon State University, Corvallis, OR 97331, USA (Matsuoka, Petri, Larson, Behnke, Wang, Peng, Spagnoli, Lohr, Jin), Department of Microbiology, College of Science, Oregon State University, Corvallis, OR 97331, USA (Milston-Clements, Jin), Department of Fisheries, Wildlife, and Conservation Sciences, Oregon State University, Newport, OR 97365, USA (Divilov).

Abstract: Cyprinid herpesvirus 3 (CyHV-3) can cause severe disease in koi and common carp (*Cyprinus carpio*), and no effective treatment is currently available.[1-57] Both LSD1 and JMJD2 are histone demethylases (HD) critical for immediate-early (IE) gene activation in herpesvirus replication. OG-L002 and ML324 are specific inhibitors to LSD1 and JMJD2, respectively. These inhibitors were compared with acyclovir (ACV) against CyHV-3 infection in vitro and in vivo. ML324, at 10-50 μM, completely blocked ~$1X10^3$ PFU CyHV-3 replication in vitro, while OG-L002 at 20 μM and 50 μM can produce 96% and 98% inhibition, respectively. ACV at 50μM produced 94% inhibition of ~$1X10^3$ PFU CyHV-3 replication in vitro. No detectable cytotoxicity was observed in KF-1 or CCB cells treated for 24 hours with 1 to 50 μM ML324. In 6-month-old infected koi, a significant reduction of CyHV-3 replication was observed with 20 μM of ML324 immersion bath for 3-4 hours on days 1, 3, and 5 post-infection compared to untreated and ACV-treated koi. Under heat stress, 50 to 70% of koi at 3-4 months old, survived CyHV-3 infection when they were treated daily in 20 μM of ML324 immersion bath for 3-4 hours within the first 5 dpi, compared to 11 to 19% in untreated or 22 to 27% in 20 μM ACV. In addition, ML324-treated koi had milder histologic pathology than untreated or ACV-treated koi on day 7 post-infection. Our study demonstrated that ML324 has the potential to be used against CyHV-3 infection in koi.

References

1. Gilad O, Yun S, Andree KB, et al. Initial characteristics of koi herpesvirus and development of a polymerase chain reaction assay to detect the virus in koi, *Cyprinus carpio koi*. *Dis Aquat Organ.* 2002;48:101-108. doi:10.3354/dao048101.

2. Hedrick RP, Gilad O, Yun S, et al. A herpesvirus associated with mass mortality of juvenile and adult koi, a strain of common carp. *J Aquat Anim Health.* 2000;12:44-57. doi:10.1577/1548-8667(2000)012<0044:AHAWMM>2.0.CO;2.

3. Cheng L, Chen CY, Tsai MA, et al. Koi herpesvirus epizootic in cultured carp and koi, *Cyprinus carpio* L., in Taiwan. *J Fish Dis.* 2011;34:547-554. doi:10.1111/j.1365-2761.2011.01266.x.

4. Gilad O, Yun S, Zagmutt-Vergara FJ, et al. Concentrations of a koi herpesvirus (KHV) in tissues of experimentally infected *Cyprinus carpio koi* as assessed by real-time TaqMan PCR. *Dis Aquat Organ.* 2004;60:179-187. doi:10.3354/dao060179.

5. Eide K, Miller-Morgan T, Heidel J, et al. Results of total DNA measurement in koi tissue by koi herpes virus real-time PCR. *J Virol Methods.* 2011;172:81-84. doi:10.1016/j.jviromet.2010.12.012.

6. Waltzek TB, Kelley GO, Stone DM, et al. Koi herpesvirus represents a third cyprinid herpesvirus (CyHV-3) in the family Herpesviridae. *J Gen Virol.* 2005;86:1659-1667. doi:10.1099/vir.0.80982-0.

7. St-Hilaire S, Beevers N, Joiner C, et al. Antibody response of two populations of common carp, *Cyprinus carpio* L., exposed to koi herpesvirus. *J Fish Dis.* 2009;32:311-320. doi:10.1111/j.1365-2761.2008.00993.x.

8. Eide KE, Miller-Morgan T, Heidel JR, et al. Investigation of koi herpesvirus latency in koi. *J Virol.* 2011;85:4954-4962. doi:10.1128/JVI.01384-10.

9. Xu JR, Bently J, Beck L, et al. Analysis of koi herpesvirus latency in wild common carp and ornamental koi in Oregon, USA. *J Virol Methods.* 2013;187:372-379. doi:10.1016/j.jviromet.2012.11.015.

10. Amin M, Adrianti DN, Lasmika NLA, Ali M. Detection of koi herpesvirus in healthy common carps, *Cyprinus carpio* L. *Virusdisease.* 2018;29:445-452. doi:10.1007/s13337-018-0488-7.

11. Lin L, Chen S, Russell DS, et al. Analysis of stress factors associated with KHV reactivation and pathological effects from KHV reactivation. *Virus Res.* 2017;240:200-206. doi:10.1016/j.virusres.2017.08.010.

12. De Clercq E. Antivirals for the treatment of herpesvirus infections. *J Antimicrob Chemother.* 1993;32(Suppl. A):121-132. doi:10.1093/jac/32.suppl_A.121.

13. Villarreal EC. Current and potential therapies for the treatment of herpes-virus infections. *Prog Drug Res.* 2003;60:263-307. doi:10.1007/978-3-0348-8012-1_8.

14. Troszok A, Kolek L, Szczygiel J, et al. Anti-CyHV-3 effect of fluorescent, tricyclic derivative of acyclovir 6-(4-MeOPh)-TACV in vitro. *J Vet Res.* 2019;63:513-518. doi:10.2478/jvetres-2019-0065.

15. Troszok A, Kolek L, Szczygiel J, et al. Acyclovir inhibits Cyprinid herpesvirus 3 multiplication in vitro. *J Fish Dis.* 2018;41:1709-1718. doi:10.1111/jfd.12880.

16. Quijano Carde EM, Yazdi Z, Yun S, et al. Pharmacokinetic and efficacy study of acyclovir against Cyprinid herpesvirus 3 in *Cyprinus carpio*. *Front Vet Sci.* 2020;7:587952. doi:10.3389/fvets.2020.587952.

17. Roizman B, Pellette PE. In: Knipe DM, Howley PM, ed. *The Family Herpesviridae: A Brief Introduction.* 4th Ed. Philadelphia, PA: Lippincott William & Wilkins; 2001:2381-2397.

18. Dulbecco R, Ginsberg H. Herpesviruses. In: *Virology.* Vol 2. Philadelphia, PA: J.B. Lippincott; 1988: 161-177.

19. Rice SA, Knipe DM. Gene-specific transactivation by herpes simplex virus type 1 alpha protein ICP27. *J Virol.* 1988;62:3814-3823. doi:10.1128/jvi.62.10.3814-3823.

20. Culhane JC, Cole PA. LSD1 and the chemistry of histone demethylation. *Curr Opin Chem Biol.* 2007;11: 561-568. doi:10.1016/j.cbpa.2007.07.014.

21. Shi Y, Lan F, Matson C, et al. Histone demethylation mediated by the nuclear amine oxidase homolog LSD1. *Cell.* 2004;119:941-953. doi:10.1016/j.cell.2004.12.012.

22. Klose RJ, Yamane K, Bae Y, et al. The transcriptional repressor JHDM3A demethylates trimethyl histone H3 lysine 9 and lysine 36. *Nature.* 2006;442:312-316. doi:10.1038/nature04853.

23. Whetstine JR, Nottke A, Lan F, et al. Reversal of histone lysine trimethylation by the JMJD2 family of histone demethylases. *Cell.* 2006;125:467-481. doi:10.1016/j.cell.2006.03.028.

24. Liang Y, Quenelle D, Vogel JL, et al. A novel selective LSD1/KDM1A inhibitor epigenetically blocks herpes simplex virus lytic replication and reactivation from latency. *mBio.* 2013;4:e00558-00512. doi:10.1128/mBio.00558-12.

25. Liang Y, Vogel JL, Arbuckle JH, et al. Targeting the JMJD2 histone demethylases to epigenetically control herpesvirus infection and reactivation from latency. *Sci Transl Med.* 2013;5:167ra165. doi:10.1126/scitranslmed.3005145.

26. Liang Y, Vogel JL, Narayanan A, et al. Inhibition of the histone demethylase LSD1 blocks alpha-herpesvirus lytic replication and reactivation from latency. *Nat Med.* 2009;15:1312-1317. doi:10.1038/nm.2051.

27. Nielsen LK, Smyth GK, Greenfield PF. Accuracy of the endpoint assay for virus titration. *Cytotechnology.* 1992;8:231-236. doi:10.1007/BF02522040.

28. St-Hilaire S, Beevers N, Way K, et al. Reactivation of koi herpesvirus infections in common carp *Cyprinus carpio*. *Dis Aquat Organ.* 2005;67:15-23. doi:10.3354/dao067015.

29. Zhou X, Ma H. Evolutionary history of histone demethylase families: distinct evolutionary patterns suggest functional divergence. *BMC Evol Biol.* 2008;8:294. doi:10.1186/1471-2148-8-294.

30. Ilouze M, Dishon A, Kotler M. Coordinated and sequential transcription of the cyprinid herpesvirus-3 annotated genes. *Virus Res.* 2012;169:98-106. doi:10.1016/j.virusres.2012.07.015.

31. Reed A, Lin L, Ostertag-Hill C, et al. Detection of ORF6 protein associated with latent KHV infection. *Virology.* 2017;500:82-90. doi:10.1016/j.virol.2016.09.030.

32. Omori R, Adams B. Disrupting seasonality to control disease outbreaks: the case of koi herpes virus. *J Theor Biol.* 2011;271:159-165. doi:10.1016/j.jtbi.2010.12.004.

33. Choi DL, Sohn SG, Bang JD, et al. Ultrastructural identification of a herpes-like virus infection in common carp *Cyprinus carpio* in Korea. *Dis Aquat Organ.* 2004;61:165-168. doi:10.3354/dao061165.

34. Minamoto T, Honjo MN, Uchii K, et al. Detection of cyprinid herpesvirus 3 DNA in river water during and after an outbreak. *Vet Microbiol.* 2009;135:261-266. doi:10.1016/j.vetmic.2008.09.081.

35. Boutier M., Gao Y., Donohoe O., Vanderplasschen A. Current knowledge and future prospects of vaccines against cyprinid herpesvirus 3 (CyHV-3) *Fish Shellfish Immunol.* 2019;93:531-541. doi:10.1016/j.fsi.2019.07.079.

36. Schroder L, Klafack S, Bergmann SM, et al. Generation of a potential koi herpesvirus live vaccine by simultaneous deletion of the viral thymidine kinase and dUTPase genes. *J Gen Virol.* 2019;100:642-655. doi:10.1099/jgv.0.001148.

37. Ma Y, Liu Z, Hao L, et al. Oral vaccination using *Artemia* coated with recombinant *Saccharomyces cerevisiae* expressing cyprinid herpesvirus-3 envelope antigen induces protective immunity in common carp (*Cyprinus carpio* var. Jian) larvae. *Res Vet Sci.* 2020;130:184-192. doi:10.1016/j.rvsc.2020.03.013.

38. Hu F, Li Y, Wang Q, et al. Carbon nanotube-based DNA vaccine against koi herpesvirus given by intramuscular injection. *Fish Shellfish Immunol.* 2020;98:810-818. doi:10.1016/j.fsi.2019.11.035.

39. Embregts CWE, Tadmor-Levi R, Vesely T, et al. Intra-muscular and oral vaccination using a koi herpesvirus ORF25 DNA vaccine does not confer protection in common carp (*Cyprinus carpio* L.). *Fish Shellfish Immunol.* 2019;85:90-98. doi:10.1016/j.fsi.2018.03.037.

40. Perelberg A, Ilouze M, Kotler M, Steinitz M. Antibody response and resistance of *Cyprinus carpio* immunized with cyprinid herpes virus 3 (CyHV-3). *Vaccine.* 2008;26:3750-3756. doi:10.1016/j.vaccine.2008.04.057.

41. Buchmann K. Evolution of innate immunity: clues from invertebrates via fish to mammals. *Front Immunol.* 2014;5:459. doi:10.3389/fimmu.2014.00459.

42. Boehm T, Iwanami N, Hess I. Evolution of the immune system in the lower vertebrates. *Annu Rev Genomics Hum Genet.* 2012;13:127-149. doi:10.1146/annurev-genom-090711-163747.

43. Cowen P, Li S, Guy JS, et al. Reactivation of latent pseudorabies virus infection in vaccinated commercial sows. *Am J Vet Res.* 1990;51:354-358.

44. Mengeling WL. Virus reactivation in pigs latently infected with a thymidine kinase negative vaccine strain of pseudorabies virus. *Arch Virol.* 1991;120:57-70. doi:10.1007/BF01310949.

45. Choi JY, Yoon SS, Kim SE, Ahn Jo S. KDM4B histone demethylase and G9a regulate expression of vascular adhesion proteins in cerebral microvessels. *Sci Rep.* 2017;7:45005. doi:10.1038/srep45005.

46. Li CH, Gao Y, Wang S, et al. Expression pattern of JMJD1C in oocytes and its impact on early embryonic development. *Genet Mol Res.* 2015;14:18249-18258. doi:10.4238/2015.December.23.12.

47. Carnesecchi J, Forcet C, Zhang L, et al. ERRalpha induces H3K9 demethylation by LSD1 to promote cell invasion. *Proc Natl Acad Sci.* 2017;114:3909-3914. doi:10.1073/pnas.1614664114.

48. Shi YG, Tsukada Y. The discovery of histone demethylases. *Cold Spring Harb Perspect Biol.* 2013;5:a017947. doi:10.1101/cshperspect.a017947.

49. Harancher MR, Packard JE, Cowan SP, et al. Antiviral properties of the LSD1 inhibitor SP-2509. *J Virol.* 2020;94:e00974-20. doi:10.1128/JVI.00974-20.

50. Tallmadge RL, Zygelyte E, Van de Walle GR, et al. Effect of a histone demethylase inhibitor on equine herpesvirus-1 activity in vitro. *Front Vet Sci.* 2018;5:34. doi:10.3389/fvets.2018.00034.

51. Hill JM, Quenelle DC, Cardin RD, et al. Inhibition of LSD1 reduces herpesvirus infection, shedding, and recurrence by promoting epigenetic suppression of viral genomes. *Sci Transl Med.* 2014;6:265ra169. doi:10.1126/scitranslmed.3010643.

52. Moerdyk-Schauwecker M, Stein DA, Eide K, et al. Inhibition of HSV-1 ocular infection with morpholino oligomers targeting ICP0 and ICP27. *Antivir Res.* 2009;84:131-141. doi:10.1016/j.antiviral.2009.07.020.

53. Rai G, Kawamura A, Tumber A, et al. Discovery of ML324, a JMJD2 demethylase inhibitor with demonstrated antiviral activity. *Probe Reports from the NIH Molecular Libraries Program.* 2010.

54. Hamada S, Suzuki T, Mino K, et al. Design, synthesis, enzyme-inhibitory activity, and effect on human cancer cells of a novel series of jumonji domain-containing protein 2 histone demethylase inhibitors. *J Med Chem.* 2010;53:5629-5638. doi:10.1021/jm1003655.

55. Mishra S, Van Rechem C, Pal S, et al. Cross-talk between lysine-modifying enzymes controls site-specific DNA amplifications. *Cell.* 2018;174:803-817:e816. doi:10.1016/j.cell.2018.06.018.

56. Fellous A, Earley RL, Silvestre F. The Kdm/Kmt gene families in the self-fertilizing mangrove rivulus fish, *Kryptolebias marmoratus*, suggest involvement of histone methylation machinery in development and reproduction. *Gene.* 2019;687:173-187. doi:10.1016/j.gene.2018.11.046.

57. Eide K, Moerdyk-Schauwecker M, Stein DA, et al. Reduction of herpes simplex virus type-2 replication in cell cultures and in rodent models with peptide-conjugated morpholino oligomers. *Antivir Ther.* 2010;15: 1141-1149. doi:10.3851/IMP1694.

Amphibian Fluid Solutions: Indications, Options and Storage

Alana Frum, LVTg, RVTg, BASVT

Session #3445

Affiliation: From Owings Mill, MD 21117, USA.

Abstract: Amphibian hydration processes differ significantly from other taxa, necessitating fluid therapy approaches appropriate to their specific physiologic needs. A variety of fluid solutions and formulas have been utilized for parental and immersion administration. While commercial amphibian fluid solutions exist, many practitioners prepare their solutions in-house. Common in-house prepared formulations include dilutions of dextrose in isotonic crystalloids and the measuring of individual analytes for dissolving in deionized water. It is imperative to measure the pH and osmolality of any solutions prepared in-house, aiming for a pH between 7.10-7.40 and an osmolality that is species and affliction appropriate. Amphibians seek out environments that are slightly hypo-osmolar to their interstitial fluid, allowing for their fluid and electrolyte transport mechanisms to maintain homeostasis with their surroundings[1] Fluid solutions can be prepared in-house to different osmolalities, to be hypotonic, isotonic or hypertonic.[2] Immersion fluid therapy is frequently utilized for ill amphibians due to decreased risks in comparison to parental administration, due to the required handling and potential of integumentary trauma or reactions. Commercially prepared amphibian fluid solutions provide information on their recommended storage conditions and expiration dates, but storage for in-house solutions is at the discretion of the practitioner. It is necessary to monitor the pH of any solutions prepared in-house, with refrigeration being recommended for most solutions to extend the shelf-life. Ideally, solutions should be prepared just prior to each administration to prevent changes in water quality.

Introduction

Amphibians are a class of vertebrate ectotherms, falling under the phylum Chordata with current estimations putting their number of species at over 5000.[3] The majority of species in the taxonomic class of Amphibia begin their lives in an aquatic setting and undergo a metamorphosis from their larval into their adult stages, in which they lose their gills and develop lungs for respiratory purposes. Amphibians are incredibly diverse, with species' habitats varying from entirely aquatic to completely terrestrial. In regard to approaching the treatment of amphibians in veterinary medicine, it is vital to have an understanding of their evolutionary and physiologic processes in order to best approach each species. Amphibian hydration processes differ significantly from that of the other taxa, thus requiring fluid therapy approaches appropriate to their specific physiology. Amphibians have a number of adaptations to their integumentary and renal systems that make their hydration processes unique.[4-6]

The integumentary system of some amphibians is involved in osmoregulation and respiration; skin disease can lead to massive electrolyte and fluid disturbances.[7] As amphibian skin is nonkeratinized and semipermeable to fluids and gases, these animals can dehydrate rapidly.[8] Amphibians are unable to alter evaporative water loss across the skin and rely on their mucus secretions to maintain homeostasis, providing a protective layer while also altering the electrolyte and water balance internally.[5,6]

The kidneys of amphibians filter blood and produce urine in the process of osmoregulation, removing waste products from the blood, reabsorbing nutrients and balancing water and ions.[4,5] The loops of Henle are reduced or absent in amphibians, with the closest related structure being an intermediate tubule connecting the proximal and distal tubules. In the proximal tubules, active transport mechanisms pump Na^+ and Cl^- for reabsorption into the bloodstream against the concentration gradient. Na^+ can also be absorbed across the bladder wall.[3,4,8] Amphibian renal system produces large volumes of dilute urine to excrete water while conserving ions due to the very high glomerular filtration rate. The amount in which water and ions are excreted is impacted by the environment, diet, hormones, blood volume and blood pressure of the animal. Amphibian kidneys do not have to work as hard to excrete extra water due to their lowered blood osmolality of around 250 mOsm, compared to the approximately 300 mOsm of mammals.[4] In aquatic amphibians, the blood plasma is hyperosmolar to freshwater.[7]

Most commonly, amphibian fluid therapy is accomplished via immersive baths. This route is less stress-inducing for the animal and easy to accomplish for the practitioner. In the event that dehydration is severe, the practitioner may administer sterile fluids via intracoelomic, intravenous or intraosseous route.[7,8] Fluids can also be administered into lymph sacs or orally in cases of mild to moderate dehydration.[7,9]

Indications

Hydrocoelom and edema

Ill amphibians commonly experience fluid retention and loss of electrolytes.[7] Fluid retention can lead to hydrocoelom, a collection of fluid in the coelomic cavity, which is commonly accompanied with lymphedema or subcutaneous edema.[10]

In cases of fluid overload, edema or ascites can also be observed. Hypertonic fluid solutions can help reduce fluid retention, but the correction of fluid overload is difficult and fluid therapy is typically more effective prophylactically in these cases.[7] Hypertonic solutions should not be used for more than 4 hours at a time and the animals should be closely monitored for adverse reactions.

Dehydration

The rate of evaporative loss through skin is higher for amphibians than other taxa. Dehydration is a common problem and can also be caused by vomiting, diarrhea or diuresis.[3,5,7] Clinical signs of dehydration in amphibians include tacky mucous membranes, ropy saliva, wrinkled skin, tightened skin on dorsum, oliguria, anuria, weight loss or sunken eyes.[7] Potassium-containing fluids should be initially avoided, with these animals typically benefiting from isotonic or hypotonic fluids.

Medication administration

Fluid therapy can also be used to deliver various medications to amphibians. Due to the permeable nature of their skin, medications dissolved in fluids can be used for immersive baths to deliver the medications transdermally. Rather than applying the medications directly to the skin, pH-balanced baths allow for transdermal administration.[8] This can also be used as a means of sedation, dissolving medications, commonly tricaine methanesulfonate (MS-222), in a fluid solution then buffering with sodium bicarbonate.

Options

Isotonic crystalloids with dextrose

An easy-to-make in-clinic option for fluid therapy is diluting dextrose with commercially prepared isotonic crystalloids.[8] Dextrose 50% can be added to solutions such as Normosol-R and Lactated Ringers to a 1.25% solution with an osmolality of approximately 240 mOsm, close to the blood osmolality of most amphibians. A 0.75% or 2.5% solution can be prepared for a more hypotonic or hypertonic solution, respectively. These solutions after preparation yield an acidic pH, warranting buffering with sodium bicarbonate until reaching a physiologically appropriate pH (Tables 1, 2). Dextrose 5% or higher can be used for short-term hypertonic baths.[2]

Table 1. pH and osmolality of Normosol-R with dextrose.

Dextrose (%)	0.75	1.25	2.5	5.0	7.5
pH	6.49	6.50	6.52	6.51	6.51
Osmolality	117	239	376	490	644

Table 2. pH and osmolality of Lactated Ringers solution with dextrose.

Dextrose (%)	0.75	1.25	2.5	5.0	7.5
pH	6.40	6.39	6.38	6.43	6.37
Osmolality	124	243	367	484	623

Analyte formulations

For practitioners with laboratory-grade salts maintained in their facilities, the option to create specific formulations exists. Salts commonly used include NaCl, KCl, $CaCl_2$, and $MgCl_2 7H_2O$[2,7,8] It is important to also have $NaHCO_3$ to serve as a physiologic buffer. Salts and sodium bicarbonate are dissolved in water free of chlorine, chloramines, ammonia, nitrites and nitrates; most commonly they are dissolved in distilled or reverse osmosis water. Formulations can be found in historic and current literature (Table 3).[2,8] The formulation commonly stated to be hypertonic found in Table 3 was found to only have an osmolality of around 240 following preparation (Table 4). If the practitioner is interested in providing a hypertonic bath, 2.5% dextrose in a commercial isotonic crystalloid may be considered for a higher osmolality. In addition, these formulations have been found to be basic following preparation, both before and after the addition of the sodium bicarbonate (Table 5). The pH should be checked prior to each use and buffering is warranted to bring the formulas to a physiologically appropriate pH (Table 4).[11]

Table 3. Analyte solution formulas dissolved in 1 liter of distilled water.

Salt	Isotonic[a]	Hypertonic[a]	Hypotonic[b]
NaCl (g)	6.6	7.3	3.46
KCl (g)	0.15	0.17	0.05
$CaCl_2$ (g)	0.15	0.17	0.1
$NaHCO_3$ (g)	0.2	0.22	0.2

[a]Wright KM, Whitaker BR. *Amphibian Medicine and Captive Husbandry.* Malabar, FL: Krieger Publishing; 2001.
[b]Hadfield CA, Whitaker BR. Amphibian emergency medicine and care. *Semin Avian Exot Pet Med.* 2005;14(2):79-89.

Table 4. pH, osmolality and salinity of analyte solutions after buffering with 12M HCl.[a]

Measure	Isotonic[b]	Hypertonic[b]	Hypotonic[c]
12M HCl (μl)	20	25	20
pH	7.17	7.26	7.27
Osmolality	217	240	112
Salinity	07.4	8.2	3.9

[a]Frum AS. Amphibian fluid solutions: formulation and storage. *Proc Annu Conf Assoc Zool Vet Tech.* 2022.
[b]Wright KM, Whitaker BR. *Amphibian Medicine and Captive Husbandry.* Malabar, FL: Krieger Publishing; 2001.
[c]Hadfield CA, Whitaker BR. Amphibian emergency medicine and care. *Semin Avian Exot Pet Med.* 2005;14(2):79-89.

Table 5. The pH of analyte solution formulas before and after the addition of $NaHCO_3$.[a]

pH	Isotonic[b]	Hypertonic[b]	Hypotonic[c]
pH without $NaHCO_3$	7.63	7.75	7.31
pH with $NaHCO_3$	8.07	7.95	8.05

[a]Frum AS. Amphibian fluid solutions: formulation and storage. *Proc Annu Conf Assoc Zool Vet Tech.* 2022.
[b]Wright KM, Whitaker BR. *Amphibian Medicine and Captive Husbandry.* Malabar, FL: Krieger Publishing; 2001.
[c]Hadfield CA, Whitaker BR. Amphibian emergency medicine and care. *Semin Avian Exot Pet Med.* 2005;14(2):79-89.

Commercially manufactured solutions

Isotonic amphibian ringer solutions can be purchased through commercial distributors, typically ranging in size from 500-milliliter to 1-liter bottles. As a rule, checking the pH before administering to an animal is the best practice. Ideally, the bottles should not be maintained in storage for extended periods. If possible, after opening a bottle the remainder of the solution after treatment should be wasted, using a fresh bottle for each treatment. Amphibian ringer solutions can be purchased online through various manufacturers (Electron Microscopy Sciences, Hatfield, Pennsylvania, United States; Ward's Science, Rochester, New York, United States; Carolina Biological, Burlington, North Carolina, United States).

Miscellaneous

Other solution options for amphibians are the use of saline and hetastarch. These include nine parts saline to one part sterile water and one part saline to two parts 5% dextrose.[8] Two parts 2.5% dextrose in 0.45% NaCl to one part Lactated Ringer's solution has also been reported for use in amphibians.[7] Hetastarch can be used as a short-term hypertonic bath.

Parental administration

In the event of parental administration of fluid therapy, sterile solutions are necessary. Liquid sterilization occurs by heating itself rather than the steam from steam sterilization in an autoclave. One-liter solutions can be autoclaved at 250 °F for 45 minutes for sterilization.

Subcutaneous administration, or administration into the lymphatic system, can be accomplished dorsally, laterally or in the proximal hind limbs. Generally, a rate of 10-20 ml/kg/day is used for this route.[7] Intracoelomic administration can be accomplished in dorsal recumbency at the lateral, caudal third of the coelom with the needle inserted parallel to the body wall and aspirated before administering.[7,8] Intravenous catheters are often not feasible for amphibians due to difficulty and tolerance, but bolus fluids may be administered at a rate of 5-10 ml/kg.[9] Intraosseous fluids are a potential for critically ill anurans, with a catheter placed in the tibiafibula normograde or in the femur retrograde.[7] The maximum volume for fluids is estimated to be at 25 ml/kg/day.[8]

Storage

Ideally, fluid solutions should be prepared fresh prior to each administration. Infiltrates and chemical changes have the potential to adversely affect solutions when stored for any period after opening. The pH of prepared analyte formations begins to increase after just one day in storage, leaving the range that is deemed physiologically appropriate for most amphibians in less than two days when stored in refrigeration.[11] These solutions have an accelerated rate of pH increase when stored at room temperature, indicating refrigeration to be ideal. Dilutions of dextrose in isotonic crystalloids should be stored in refrigeration, it is not recommended to store these solutions past 48 hours. For commercially manufactured products, it is recommended to follow storage and use instructions from the manufacturer for each specific product. It is important to warm solutions to an appropriate level prior to administering fluid therapy, ranging from 65-85 °F depending on species' needs.[2,6] Warming can be accomplished by leaving out at room temperature or immersing the fluid bottles in warm water.

References

1. Chinnadurai SK, Kane LP. Advances in amphibian clinical therapeutics. *J Exot Pet Med.* 2014;23(1):50-55.

2. Wright KM, Whitaker BR. *Amphibian Medicine and Captive Husbandry.* Malabar, FL: Krieger Publishing; 2001.

3. Cecil TR. Amphibian renal disease. *Vet Clin North Am Exot Anim Pract.* 2006;9(1):175-188.

4. Rhodes AE. The Dangers of Deicing: The Challenge of Osmoregulation in a Freshwater Environment Available at: https://static.nsta.org/case_study_docs/case_studies/deicing_flip.pdf. Accessed March 9, 2023

5. Parker-Graham C, Clayton LA, Mangus LM. Amphibian renal disease. *Vet Clin North Am Exot Anim Pract.* 2020;23(1):215-230.

6. Kanda I, Brandao J. *Exotic Animal Emergency and Critical Care Medicine.* Hoboken, NJ: John Wiley & Sons; 2021.

7. Pollock C. Fluid Administration in Amphibians. Available at: https://lafeber.com/vet/fluid-administration-in-amphibians/. Accessed March 9, 2023.

8. Hadfield CA, Whitaker BR. Amphibian emergency medicine and care. *Semin Avian Exot Pet Med.* 2005;14(2):79-89.

9. Whitaker BR, Yaw TJ. Emergency care of amphibians-exotic and laboratory animals. Available at: https://www.merckvetmanual.com/exotic-and-laboratory-animals/amphibians/emergency-care-of-amphibians. Accessed March 9, 2023.

10. Clancy MM, Clayton LA, Hadfield CA. Hydrocoelom and lymphedema in dendrobatid frogs at National Aquarium, Baltimore: 2003-2011. *J Zoo Wildl Med.* 2015;46(1):18-26.

11. Frum AS. Amphibian fluid solutions: formulation and storage. *Proc Annu Conf Assoc Zool Vet Tech.* 2022.

Successful Enterotomies in a Red Tegu (*Salvator rufescens*) and a Mali Uromastyx (*Uromastyx maliensis*)

Emily Thulson, DVM

Session #3429

Affiliation: From Arizona Exotic Animal Hospital, 744 N. Center St. Ste 101, Mesa, AZ 85201, USA.

Note: Abstract format selected by author(s). See Front Matter for description of conference proceedings manuscripts submitted as scientific abstracts.

Abstract: An adult male red tegu (*Salvator rufescens*) presented for straining, rolling, and not defecating in ten days. Blood chemistry suggested muscle trauma and stress. Radiographs showed mineral opacity round material in the caudal dorsal coelom. The tegu worsened despite supportive care including NSAIDs and subcutaneous fluids. The following day it was actively rolling on examination with a markedly distended coelom. A coelomic exploration was initiated and an enterotomy was performed, removing an enterolith from the colon. The patient recovered quickly and is still doing well 6 months post-operatively. In another case, an adult female Mali uromastyx (*Uromastyx maliensis*) presented for tissue prolapsing from the vent. The animal had laid eggs for the first time 2 months previously. Blood chemistry was unremarkable. Radiographs showed ingesta in the caudal coelom. Vent tacking sutures were placed under sedation. The lizard was given supportive care including antibiotics, NSAIDs, subcutaneous fluids, lactulose, and vibrational therapy. Its gastrointestinal tract became mildly more malleable on palpation over subsequent visits. The animal produced urates but no feces. It was elected to proceed with enterotomy 19 days post-prolapse. Normal-appearing feces were removed from the colon. The animal had various stages of follicles and free fluid in the coelom. Bacterial culture of the fluid yielded no growth. The lizard recovered well post-operatively but remained on low-dose lactulose for several months. Follow-up ultrasound showed quiescent follicles. These are both cases of successful enterotomies in lizards, few of which are published in the literature. Enterotomies should be considered when medical management is unsuccessful and when other modalities, such as cloacoscopy, are not achievable.

Evaluation of the Sensitivity of Plasma Biochemical and Blood Gas Parameters to Detect Renal Disease in Corn Snakes (*Pantherophis guttatus*)

Claire Vergneau-Grosset, DVM, IPSAV, Dipl ACZM,
Julie Pujol, DVM,
Jacobo R. Noriega, DVM
Shannon Ferrell, DVM, DABVP (Avian Practice), DACZM,
Stéphane Lair, DVM, DVSc, Dipl ACZM

Session #3168

Affiliation: From the Département de sciences cliniques, Faculté de médecine vétérinaire, Université de Montréal, 3200 rue Sicotte, Saint-Hyacinthe, QC, J2S 2M2, Canada.

Note: Abstract format selected by author(s). See Front Matter for description of conference proceedings manuscripts submitted as scientific abstracts.

Abstract: Despite the high incidence of renal disease in reptiles, the sensitivity of blood parameters are debated. The objective of this study was to evaluate the sensitivity of symmetric dimethylarginine (SDMA), N-acetyl-beta-D-glucosaminidase (NAG), gamma glutamyl transferase (GGT), alanine aminotransferase (ALT), aspartate aminotransferase (AST), total and ionized calcium concentrations, calcium-to-phosphorus ratio, lactate concentration and other blood parameters to detect impaired renal function in a model of experimentally induced acute renal tubular necrosis in colubrids. Six corn snakes (*P. guttatus*) were administered gentamicin at 50 mg/kg SC q24h for 11 days (IACUC approval 21-Rech-2134). Plasma biochemistry and blood gas analyses were obtained at day 0, 3, and 11. Parameters were compared between timepoints using paired Wilcoxon tests. Renal biopsies were collected prior to starting injections on all individuals, at days 3 and 11 in three individuals, and after euthanasia in all individuals. Snakes were euthanized due to anorexia, weight loss, and lethargy. Renal proximal and distal tubular necrosis was frequent among sampled individuals, with one case of gouty tophi, and the majority of snakes developed hepatic steatosis. Compared to baseline, an increase of uric acid, ALT, AST concentrations, and a decrease of ionized calcium, total calcium, calcium to phosphorus ratio, lactate, and sodium concentrations were noted. No changes of SDMA, NAG nor GGT were detected. Contrary to what has been proposed in tortoises, SDMA was not an indicator of renal lesions. Ionized calcium, total calcium, and lactate concentrations were the earliest indicators of renal lesions in this experimental model.

Acknowledgments: This project was funded by a grant from Natural Sciences and Engineering Research Council (NSERC) of Canada. Funding was also provided by the CHUV Fund, Faculty of Veterinary Medicine, Université de Montréal. The authors thank Hagen for the donation of equipment, and Amélie Aduriz and Bastien Rubin, Master students in Clinical Sciences, for their contribution.

Determining the Effects of Four Semen Extenders on Spermatozoa Motility in Corn Snakes (*Pantherophis guttatus*)

Aspen La-Riene Settle, BSc, and
Mark A. Mitchell, DVM, MS, PhD, DipECZM (Herpetology)

Session #3481

Affiliation: From Louisiana State University School of Veterinary Medicine, Baton Rouge, LA 70803, USA.

Note: Abstract format selected by author(s). See Front Matter for description of conference proceedings manuscripts submitted as scientific abstracts.

Abstract: More than 50% of all reptile species are anticipated to be lost to anthropogenic influences by the end of this century. To promote the survival of these animals it is vital to develop appropriate assisted reproductive methods, although this research has been limited in reptiles to date. While the use of semen extenders has been documented in a variety of mammalian species, strategies for preserving snake semen samples are largely restricted to simple refrigeration. This greatly hinders the ability to share genetic material between captive breeding programs for threatened and endangered species without incurring significant declines in spermatozoa motility after portable cooling systems have become ineffective (typically after 48 hours). This study evaluated four different semen extenders and crystalloids (INRA 96, Andro ProChill LT, Ham's F-10, and Hank's balanced salt solution) in extending the viability of corn snake (*Pantherophis guttatus*) semen. The hypothesis was that INRA 96 and Ham's F10 would maintain spermatozoa motility >50% for longer than 48 hours. Semen samples were manually collected from four corn snakes weekly in a complete cross-over study design. Semen samples were diluted directly into 1 milliliter of extender and a baseline motility was measured. Samples were refrigerated and evaluated at 24-hour intervals until spermatozoa motility reached 0%. The results of this study suggest that there is a high degree of individual variability in semen motility in corn snakes.

Evaluation of the Distribution and Viability of Chameleons in Zoological Institutions

Amélie Aduriz,
Johanna Staerk, PhD,
Dalia A. Conde, PhD,
Claire Vergneau-Grosset, DVM, IPSAV, CES, Dipl ACZM

Session #3148

Affiliation: From the Faculty of Veterinary Medicine, Clinical Sciences Department, Université de Montréal, 3200 Rue Sicotte, Saint-Hyacinthe, QC, J2S 2M2 Canada (Aduriz and Grosset), Species360 Conservation Science Alliance, Interdisciplinary Center on Population Dynamics CPop, Biology Department, Syddansk Universitet Campusvej 55, 5230 Odense, Denmark (Staerk and Conde).

Note: Abstract format selected by author(s). See Front Matter for description of conference proceedings manuscripts submitted as scientific abstracts.

Abstract: Of the 202 species of Chamaeleonidae, 38.6% are globally threatened. Currently, nearly 1000 individual chameleons are kept in zoological institutions worldwide. The objectives of this study were to assess species and sex distribution, reproductive status, life expectancy by sex, and main mortality causes of chameleons in zoos to aid research and support conservation efforts. Using Species360's Zoological Information Management Software (ZIMS), an analysis of the current species and sex distribution was conducted, and an estimate of life expectancy for seven species of chameleons was performed using Bayesian Survival Trajectory Analyses (BaSTA). An online survey was sent to 245 zoological institutions, and necropsy reports of the last 10 years were collected. Mortality causes and prevalence of renal lesions were recorded. Presently, 18% (36/202) of Chamaeleonidae species are represented in ZIMS, including 14% (11/78) of threatened species. Male and female chameleons are kept in equivalent numbers across 36 species, 203 zoos, and 6 continents. Among participating institutions (65/245), 51% reported viable chameleon births, including 12% within the last 12 months. Three species of chameleons (*Furcifer oustaleti*, *Trioceros jacksonii*, *Trioceros melleri*) showed no sex differences in life expectancy, whereas four species (*Calumma parsonii*, *Chamaeleo calyptratus*, *Chamaeleo chamaeleon*, *Furcifer pardalis*) displayed a significant male survival advantage. The main mortality causes in 14 species (n = 413) were infectious (48.4%), renal (12.3%), or reproductive (10.4%) diseases, and 34.6% of the studied individuals displayed renal lesions at necropsy. The results of this study warrant further investigation and suggest the need to develop a Species Survival Plan for Chamaeleonidae.

Evaluation of the Effects of Dexmedetomidine, Midazolam, Ketamine and Morphine Combination in *Aldabrachelys gigantea*: A Preliminary Study

Marco Masi, DVM, MS

Session #3366

Affiliation: From Centro Veterinario Specialistico, Via Sandro Giovannini, 51/53, 00137 Roma, Italy.

Note: Abstract format selected by author(s). See Front Matter for description of conference proceedings manuscripts submitted as scientific abstracts.

Abstract: In clinical practice, it is often difficult to carry out a complete clinical examination in Chelonians, as they tend to retract into their shell. In general, the larger the turtles are, the greater the difficulties encountered by the clinician. Injectable anesthesia is often used to assist in completing even the simplest procedures. This study assessed the effectiveness of a combination of dexmedetomidine, midazolam, ketamine, and morphine to provide adequate multimodal anesthesia to perform clinical examination and celioscopy. Twenty Aldabra giant tortoises (*Aldabrachelys gigantea*) were included in the study. The premedication was composed of dexmedetomidine (0.05 mg/kg), midazolam (1 mg/kg), ketamine (8 mg/kg), morphine (1 mg/kg), which was administrated into the right antebrachium. The type of administration (subcutaneously or intramuscular) was randomly assigned, and each turtle was anesthetized once. Heart rate and respiratory rate were recorded every 60 seconds, while body temperature (bT) was recorded continuously during the procedure. Other variables assessed included muscle tone of thoracic and pelvic limbs, neck retraction reflex and jaw tone every 15 minutes for up to 2 hours. When the subject was under anesthesia, characterized by the loss of the thoracic and pelvic limb retraction reflex and jaw tone, sex determination by celioscopy was performed via the left prefemoral fossa. At the end of the endoscopy, reversal agents were administered intramuscularly in the left antebrachium, atipamezole (0.5 mg/kg) and flumazenil (0.05 mg/kg). The combination of dexmedetomidine, midazolam, ketamine and morphine appeared effective to perform celioscopic sex determination in *Aldabrachelys gigantea*.

Comparison Between Cloaco-Cystoscopic Analysis Performed With and Without Intrathecal Anesthesia

Emanuele Lubian, DVM, PhD, EM, GPCert (ExAP),
Francesca Duse, DVM,
Gianluca Deli, MRCVS, DVM, CertAVP (ZM), CPCert (ExAP), PgCert (EAS),
GPAdvCert (SMP), PGCert (HE), FHEA, VetGDAdvisor,
Filippo Spadola, PhD

Session #3302

Affiliation: From the Department of Veterinary Science, University of Messina, Polo Didattico Annunziata 98168 Messina, IT (Lubian, Spadola), MypetClinic, Viale Ranzoni 10, 20149, Milano, IT (Lubian, Duse), Ospedale Veterinario Universitario Piccoli Animali, Università degli Studi di Milano, Lodi, IT (Lubian, Duse), Medivet Faygate, Tower Road, Faygate, Horsham RH12 4SD, UK (Deli).

Note: Abstract format selected by author(s). See Front Matter for description of conference proceedings manuscripts submitted as scientific abstracts.

Abstract: Cloacoscopy and cystoscopy are performed in chelonians for routinary diagnostic procedures. They can be performed without anesthesia, under sedation or under complete anesthesia. In the last 10 years a new anesthetic technique has been performed in surgical procedures of the caudal portion of the chelonian body; intrathecal anesthesia using lidocaine has been demonstrated to be useful for procedures of tail, cloaca, urinary bladder, genitalia and hindlimbs, allowing minor surgeries in the caudal body, such as phallectomy. This retrospective study compares the cloaco-cystoscopic procedure with and without intrathecal anaesthesia. Thirty-six animals between 247 grams and 1815 grams, both male and female, were included in this preliminary survey: 23 without anesthesia and 13 subjected to intrathecal anesthesia. Intrathecal anesthesia was performed using 2 mg/kg of lidocaine as reported in previous studies. Results demonstrate that intrathecal anesthesia reduces time of the entire procedure, reduces stress on the animal, and allows for better visualization of the coelomic organs and cloacal lumen because there are not contractions that empty the urinary bladder and because there is no fecal material inside cloaca. This study is pioneering for future cloacal and cystic investigation, because it demonstrates that using intrathecal anesthesia reduces animal stress and increases performance of the entire procedure.

References

1. Hernandez-Divers SJ, Stahl SJ, Camus A. Evaluation of diagnostic coelioscopy including liver and kidney biopsy in freshwater turtles (*Trachemys scripta*). *J Zoo Wildl Med.* 2010;41:677-687.

2. Mans C. Intrathecal drug administration in turtles and tortoises. *J Exot Pet Med.* 2014;23(1):67-70.

Section 44

ARAV Potpourri Session

Michael Corcoran, DVM, DABVP (Reptile and Amphibian Practice), CertAqV
Session Chair

“Alotl” Information about Axolotls

Alissa M. Tepedino, DVM, CertAqV

Session #3184

Affiliation: From Avian and Exotic Animal Care, 8711 Fidelity Boulevard, Raleigh, NC 27617, USA.

Abstract: The Mexican axolotl (*Ambystoma mexicanum*) is a species of neotenic aquatic salamanders native to central Mexico. It is one of the only species capable of fully regenerating tissues after injury or amputation throughout its life. When kept as household pets, axolotls are prone to a variety of illnesses relating to husbandry, infectious diseases, and other conditions requiring veterinary attention. Neoplasia in the axolotl is relatively uncommon. Anatomy and physiology, essential clinical techniques, anesthesia, surgery, and supportive care measures for the axolotl patient are presented in this session.

Introduction

Mexican axolotls (*Ambystoma mexicanum*) are an aquatic salamander species from the Urodela order of tailed amphibians. They have risen drastically in popularity as a household pet in recent years and are being seen in practice more now than ever before. The axolotl is neotenic and retains larval-stage features through its entire life. They are one of the only species capable of fully regenerating tissues after injury or amputation and have been extensively studied for their regenerative capacities in research laboratories. The purpose of this masterclass is to provide attendees with essential information regarding the natural history, husbandry, medicine, surgery, and supportive care of the axolotl patient.

Natural History

Axolotls are native to only two freshwater-to-slightly-brackish lakes in the Valley of Mexico, Lake Xochimilco and Lake Chalco. A 2019 assessment by the International Union for Conservation of Nature classified the axolotl as critically endangered with an estimated wild population of 50-1000 individuals.[1] Residential and commercial development in Mexico City, pollution, the draining of Lake Chalco as a flood control measure, and the introduction of non-native fish species to Lake Xochimilco have all contributed to their population decline. The vast majority of axolotls today are kept in home aquaria or research laboratories.

Unlike most other amphibian species, axolotls are neotenic and rarely undergo metamorphosis into a terrestrial adult stage. They maintain many of their larval-stage features into adulthood, including the external gills and dorsal fin. Axolotls are obligate carnivores, consuming a variety of worms, mollusks, arthropods, and small fish in the wild. They typically hunt at night and burrow into vegetation and mud during the daytime.

Anatomy and Physiology

The axolotl has a flat head, a tube-shaped body, primitive limbs, and a large, vertically oriented tail. The forelimbs have four digits and hindlimbs have five digits. The axolotl usually ambulates slowly along the bottom of their enclosure using their small limbs, but when startled they are capable of rapid swimming with the aid of their tail.

Adult axolotls are typically 23-25 centimeters in total length, although some individuals have reportedly reached 30-40 centimeters. Sexual maturity occurs around 10 months of age for males and 12-18 months of age for females. The average adult female weighs 180 grams and the average adult male weighs 130 grams.[2] Sexual maturity is indicated by the presence of a cloacal bulge in males and the development of black coloration to the nails in lightly colored axolotls. While axolotls are neotenic, exogenous thyroxine is known to cause metamorphosis into their terrestrial adult form.[3]

Wild-type axolotls are a black to mottled brown color, and selective breeding in captivity has led to many different color variations. The dermal layer is highly vascularized to allow cutaneous oxygen exchange across a thin epidermal layer. The epidermis is pseudo-stratified and does not contain the distinct stratum corneum seen in metamorphosed amphibians.[4] The skin is coated with a protective layer of mucus that is produced by Leydig cells in the epidermis.

Axolotls are capable of branchial, pulmonic, buccopharyngeal, and cutaneous methods of respiration. Branchial respiration through the external gills is the primary method of oxygen exchange. There are three projections, or rami, on each lateral aspect of the head and the caudal surface of each ramus is lined with gill filaments that facilitate oxygen exchange. The axolotl can flex and extend each set of rami to increase water flow across the gills, an action that increases with elevated water temperatures. The lungs are rudimentary and are primarily used to compensate for aquatic hypoxia. Little research has been completed on buccopharyngeal and pulmonic respiration in axolotls, but studies of the larval stage of the closely related tiger salamander (*Ambystoma tigrinum*) have demonstrated a likely pathway. The buccal cavity is expanded to draw air through the mouth while air is simultaneously exhaled from the lungs into the buccal cavity, where the fresh and expired air is mixed. The buccal cavity then compresses to pump a portion of the air mixture into the lungs, while the remainder is released rapidly through the mouth and gills.[5] Gaseous exchange occurs within the buccal cavity and pharynx as well, in a similar manner to cutaneous respiration.

The axolotl heart contains one ventricle, two atria separated by a partially perforated septum, and lacks coronary circulation.[6] Heart rate is influenced by environmental temperature. Both renal and hepatic portal vein systems are present, which can affect the efficacy of injectable medications administered in the caudal half of the axolotl. The spleen and liver are responsible for hematopoiesis.[7] Erythrocytes are nucleated and large in size. When injured, vasoconstriction and clot formation occur rapidly to prevent significant blood loss, even with major arteries.

As axolotls are carnivorous, they have a short, simple gastrointestinal tract. Their teeth are small and cone-shaped and are more suited to gripping than shredding food. They prehend prey items with a vacuum effect as they rapidly open the mouth. Their vision is poor but they have a strong sense of smell to aid in finding prey. Like fish and other aquatic amphibians, axolotls possess a lateral line system. This is used in detection of electric current using ampullary organs on the head, and detection of water movement using epithelial receptor organs along the lateral aspect of the body.[8]

The ability of axolotls to regenerate different tissues through their entire life has been extensively studied in research laboratories. They have been shown to be capable of regenerating limbs, cardiac tissue,[9] lungs,[10] and even parts of the spinal cord.[11] The following describes the tissue regeneration pathway following loss of a limb. After the initial injury, a layer of epidermal cells covers the wound site over a matter of days. Following this, a cone-shape group of cells called the blastema forms beneath the epidermis. The blastema cells are thought to be formed from the cells at the distal point of injury becoming progenitor cells that are capable of reforming new tissue. Blastema cells appear to stay consistent with their embryonic germ layer of origin[12] (i.e., blastema cells originating from muscle cells can only make new muscle cells, not epidermal or neural tissue). Over time blastema cells multiply and differentiate into diverse cell types to ultimately create a new limb.

Husbandry

The juvenile axolotl should be kept in a 10-gallon aquarium, while the single mature adult should be kept in a 20-gallon aquarium at minimum. They can be kept in a freshwater or slightly brackish environment. The aquarium water must be free of chlorine and chloramines, which are typically added to municipal tap water. Dechlorinating products that are commercially available for fish aquariums are safe for use for axolotls. The aquarium water system should undergo a complete nitrogen cycle and reach values of ammonia = 0 ppm, nitrites = 0 ppm, and nitrates < 10 ppm before an axolotl is introduced. An adequately sized mechanical filtration system should be used to remove nitrogenous waste from the water between water changes. Turbulent water flow can be stressful for the axolotl, so a low-power filtration system is recommended. Water changes of 20-30% of the total water volume are recommended when nitrates elevate above 10 ppm.

The author recommends weekly monitoring of pH, ammonia, nitrite, nitrate, KH, and GH levels using a commercially available water quality testing kit. The ideal water pH range is 7.4-7.6 while ranges from 6.5-8.0 are considered acceptable.[2] KH (carbonate hardness, also known as alkalinity) measures the buffering capacity of water and the ideal range for axolotls is 53-143 ppm. GH (general hardness) measures the amount of calcium and magnesium minerals in the water and the ideal range for axolotls is 125-250 ppm. Water temperature should be consistently monitored with an aquarium thermometer to maintain an ideal range of 60-64 °F, although a range of 55-68 °F is tolerated. An aquarium chiller is necessary for most keepers to maintain this temperature range.

Fine aquarium sand is the most highly recommended substrate choice for adult axolotls, as it provides a more textured surface for them to walk along and has a smaller risk of gastrointestinal impaction if accidentally ingested. Large river stones are another acceptable option, as long as they are at least twice the width of the axolotl's head to prevent accidental ingestion. Other options include tile or keeping them in a bare-bottom enclosure, but these provide less texture for walking and resting on. It is recommended to keep juvenile axolotls on a tile or bare-bottom enclosure until they are at least 6 inches in length. All axolotl enclosures should contain hides to help reduce stress. Since their natural environment contains low light, this should be recreated in captivity. For planted aquaria that requires a more intense light source, various hiding areas should be provided to allow them to retreat from the light. Aquarium decorations should lack sharp edges that could cause injury. All aquarium decorations should be larger than the width of the mouth to minimize risk of ingestion.

Juvenile axolotls should be kept individually, as they can be cannibalistic. Adult axolotls can be kept in pairs or groups if they are of similar size. For axolotls that are housed in pairs or groups, it is recommended to increase the aquarium size by 20 gallons per individual and to provide multiple different hides and feeding stations. Aggression between adults is uncommon and is often the result of insufficient space or competition for hides or food.

The diet should consist of a variety of feeder items. Juvenile axolotls should be fed daily or twice daily. Live brine shrimp and daphnia are recommended at this life stage. Adult axolotls should be fed two to three times per week. Common feeder items for adults include nightcrawlers (earthworms), blackworms, bloodworms, and sinking commercial salmon pellets. Obesity is a common issue for pet axolotls, so careful attention must be given to the amount and frequency of feedings. Only enough food that can be consumed in a 2- to 3-minute time period should be offered, and any uneaten food and waste should be removed after feeding to prevent negative effects on water quality.

Examination and Restraint

Routine wellness examinations should be completed annually or biannually for axolotls. The assessment should begin with an external visual examination of the skin, gills, eyes, oral cavity, mouth, limbs, and tail. The axolotl patient and some of the aquarium water can be placed in a clear, water-tight container to facilitate visualization from all angles. Endoscopic visualization can be used to examine the external gills and any identified external lesion in more detail without restraint. Axolotls should be handled only when necessary, as doing so can cause damage to the delicate gills or remove the protective mucous layer from the skin. When physical restraint is required, powderless nitrile gloves and/or a fine mesh net should be used. Chemical restraint is preferable to physical restraint for most diagnostic techniques.

Diagnostics

As many health conditions of axolotls are related to improper husbandry conditions, water quality testing should be completed with every physical examination. At minimum, pH, ammonia, nitrite, and nitrate values should be assessed to assure they are within the previously described normal ranges.

Diagnosis of ectoparasites can be completed by skin scrape, gill biopsy, or gill endoscopy. Skin scrape samples should be taken by using a cover slip to sample the mucous layer over a dermal lesion in a cranial-to-caudal direction. Gill biopsies are taken by removing the distal 2-3 millimeters of 3-5 gill filaments with iris scissors under sedation. Samples should be mounted on a microscope slide in the patient's source water, as abrupt changes in salinity or pH may cause changes to cellular structure of parasites that could make them less recognizable and could yield false negative results. Gill endoscopy using a small, handheld rigid endoscope has a higher diagnostic yield than gill biopsy due to increased visibility at the more proximal aspect of the gill filaments and rami. It is also well-tolerated by the axolotl patient without sedation.

Complete blood count and serum chemistry analysis can be completed by venipuncture of the ventral tail vein using a lateral approach or the femoral vein. A heparinized 1.0-cc syringe with a 27-gauge needle should be used. Transillumination may be helpful in aiding visualization of the vessel. Sedation or anesthesia is recommended to minimize handling and reduce patient stress.

Radiographic imaging is useful in evaluation of the skeletal system, gas in the lungs or coelomic cavity, and in cases of suspected foreign body ingestion. Taking radiographs of the patient out of the water by placing them in a clear plastic container greatly increases image detail compared to radiographs taken of the patient in a container or bag of water. As axolotls are capable of several methods of respiration, they tolerate short periods of time out of the water well as long as their skin is kept moist. If possible, lateral images should be taken using a horizontal beam. Limitations of radiography include poor coelomic detail and difficulty evaluating patients of smaller size. As amphibians have the largest red blood cells of any taxa, they are more susceptible to radiation damage, so repeated imaging should be avoided if possible.

Ultrasound is useful for evaluation of coelomic organs, monitoring of heart rate during general anesthesia, diagnosis of some foreign bodies, and in cases of ascites for fluid sampling. While ultrasound imaging is often successful with the conscious axolotl patient, sedation may yield better quality images and minimize patient stress on a case-by-case basis. Ultrasonographic images of the entire coelom can be obtained by holding the probe head slightly offset from the patient beneath the water surface and scanning cranial to caudal on each lateral aspect and the dorsal aspect of the axolotl. Ventral images can be obtained by applying ultrasound gel to the probe and scanning through the bottom of a clear container, or by direct contact with the sedated patient.

Common Conditions

While axolotls are generally regarded as a hardy species to keep in captivity, they can become stressed and succumb to illnesses requiring veterinary care. Many health conditions are a result of improper husbandry conditions, so a thorough husbandry review and water quality testing should accompany the physical exam. Signs of illness are generally non-specific and include lethargy, hyporexia, ascites, skin and ocular lesions, degeneration or blunting of the gills, and floating abnormally.

Conditions of water quality

Ammonia toxicity (new tank syndrome)

Ammonia is excreted as waste from the axolotl through the gills and urinary system, and can further be elevated by the breakdown of protein and uneaten food in the aquarium. Ammonia spikes can occur from damage to the biofilter or source of nitrifying bacteria, increasing the stocking density, or increasing the feeding rate. At ammonia levels as low as 0.05 ppm, thickening and hemorrhage of gill lamellae occurs, inhibiting oxygen exchange. At 2 ppm, neurologic signs, internal organ damage, and sudden death can occur. Ammonia exists in equilibrium in the water between the severely toxic unionized ammonia (NH_3) and the less toxic ammonium ion (NH_4^+), the sum of which is called total ammonia nitrogen (TAN). The proportion of NH_3 in solution elevates with increasing pH, increasing temperature, and decreasing salinity.[13] Commercial water quality test kits usually measure TAN, and determining toxic levels of NH_3 can be completed using conversion charts. Treatment for ammonia toxicity includes daily water changes of 25-50%, adding nitrifying bacteria to the aquarium, and minimizing feeding.

Nitrite toxicity (brown blood disease)

Ammonia is converted into nitrite by nitrifying bacteria in the biofilter, and thus nitrite spikes often follow ammonia spikes. The species of nitrifying bacteria that convert nitrite into nitrate are less adherent to filter media, so cleaning media too thoroughly can cause spikes in nitrite without spikes in ammonia. Nitrite oxidizes hemoglobin into methemoglobin, which cannot transport oxygen and makes blood appear brown in color. Increasing salinity by up to 2 mg/L will aid in treatment, as chloride will competitively inhibit nitrite uptake by the gills. Treatment with methylene blue baths at 1-2 mg/L helps reverse methemoglobin formation. Supplemental oxygenation should also be provided with an air stone.

Nitrate toxicity (old tank syndrome)

Nitrite is converted by nitrifying bacteria into nitrate. Nitrate is removed from the aquarium system through water changes and through uptake by aquarium plants. Nitrate levels > 50 ppm indicates chronic poor husbandry and can lead to exophthalmia and corneal opacities in axolotls. Extremely high nitrate levels correspond with a pH crash, as nitrate-rich water loses its buffering capacity (decrease in KH) while decomposing waste in the aquarium contributes additional hydrogen ions. Small daily water changes of < 25% are recommended to slowly increase the pH and restore buffering capacity to the aquarium. Old tank syndrome is preventable through routine water quality testing and aquarium maintenance.

Chlorine or chloramine toxicity

Chlorine and chloramines are chemical disinfectants commonly added to municipal water systems at 0.5-2.0 mg/L. While levels of these compounds are safe for human consumption, levels over 0.02 mg/L causes toxicosis in amphibians. Dechlorinating products that are commercially available for fish are safe for use in axolotl aquaria.

Toxicosis can be seen when dechlorinating products are not added to tap water prior to water changes or when larger-scale chlorination occurs at municipal water supply centers. Clinical signs include skin erythema, gill tissue damage and necrosis, neurologic signs, difficulty maintaining buoyancy, or sudden death without any clinical signs. Once clinical signs occur, prognosis is very poor. Adding sodium thiosulfate at 7.4 mg/L for every 1 mg/L chlorine/chloramine present in the water will neutralize chlorine.[13] This reaction will consume dissolved oxygen and lower pH, so aeration with an air stone and buffering the water with sodium bicarbonate is also necessary.

Hyperthermia

As with all poikilotherms, axolotls are sensitive to environmental temperature changes. Clinical signs of hyperthermia may be seen at prolonged temperatures over 75 °F and include inappetence, ascites, and abnormal floating. The water temperature of the affected patient should be slowly decreased until they return to their preferred optimal temperature zone. Antibiotic therapy may be needed to prevent secondary bacterial infections. Elevations in water temperatures also shift ammonia towards its more toxic unionized form, so management of new tank syndrome may also be necessary. Hyperthermia can be prevented by maintaining water temperatures of the axolotl aquarium with a water chiller controlled by a thermostat.

Infectious diseases

Bacterial infections

In early stages of bacterial infections, clinical signs may include dermatitis, inappetence and lethargy. As progression towards septicemia occurs, hydrocoelom may develop. Diagnosis can be made from skin scrapes of superficial lesions, complete blood count analysis, and/or coelomocentesis for fluid analysis and culture. Bacterial infections are often secondary to husbandry or water quality abnormalities, leading to chronic stress and immunosuppression, which must be corrected as part of treatment. Commonly isolated pathogens include *Pseudomonas aeruginosa*, *Aeromonas hydrophila*, and other gram-negative bacteria. Antibiotic options include enrofloxacin, gentamicin, and amikacin (5 mg/kg IM, ICe), ceftazidime (20 mg/kg IM, SC), and trimethoprim sulfonamines (15-20 mg/kg IM, PO).[14] Tetracyclines are contraindicated in axolotls. Maintaining the patient at a lower temperature range of 41-59 °F during treatment may aid in recovery.[15] There are no differentiating clinical signs between bacterial septicemia and mycobacteriosis. Chronic stress, improper quarantine measures when mixing populations, and offering live prey foods may predispose axolotls to infection. At necropsy, multiple coelomic granulomas are indicative of mycobacteriosis, and cytology or histology of the lesions showing acid-fast bacilli is diagnostic. There is no treatment for mycobacteriosis, and it does pose risk of zoonosis.

Fungal infections

In a similar manner to bacterial infections, fungal infections can be seen secondary to poor husbandry and chronic stress and immunosuppression. The most common fungal infections of axolotls seen in clinical practice are caused by *Saprolegnia* spp., a ubiquitous water mold that causes white to grey cotton-like patches in areas of pre-existing damage to the skin or gills. Addressing the underlying cause is usually curative, but treatment with salt baths or itraconazole may be indicated for more severe infections. Infections with *Batrachochytrium dendrobatidis* (chytridiomycosis) have been reported in axolotls. Clinical signs may include erythema, increased shedding or sloughing of skin, or sudden fatalities without previous clinical signs. Itraconazole baths at 0.002-0.0025% for 5 minutes daily for 10 days have been effective in treatment of chytridiomycosis in axolotls.[16] Proper quarantine and biosecurity measures are essential for prevention of infection.

Parasitic infections

Axolotls that are fed a diet of live fish have increased risk of exposure to gastrointestinal parasites. Commonly seen parasites include the flagellated protozoan *Hexamita* and ciliated *Opalina*, both of which can be diagnosed through fecal cytology and can be treated with metronidazole at 5 mg/g food for 3-4 days.[15] Infections with ectoparasites are often the result of improper quarantine and biosecurity measures. Commonly seen ectoparasitic organisms include *Trichodina* and *Ichthyodobo* and can be diagnosed by skin scrape and wet mount cytology. Hypersalinity soaks via addition of salt at 6-10g/L for up to 1 hour[14] are effective for treatment. Infection with the macroscopic copepod crustacean *Lernaea cyprinacea* (anchor worm) can be diagnosed via visual or endoscopic examination. A recent case report describes manual removal of the parasites followed by treatment with lufenuron 0.1 mg/L every 7 days for 5 treatments for effective treatment of a severe anchor worm infection.[17]

Gastrointestinal foreign bodies

Axolotls housed on substrate that is smaller than the width of the mouth, such as gravel, small stones, or glass aquarium marbles are at increased risk of gastrointestinal obstruction from foreign body ingestion. Smaller ingested foreign material may be regurgitated. Cases involving large singular foreign bodies or ingestion of multiple items may necessitate manual removal. After anesthetizing the axolotl, the foreign body may be removed through the mouth using lubricated forceps or an endoscope passed into the stomach. Enterotomy surgery may be necessary in cases of ingestion of items with sharp edges or if the foreign body moves aboral to the stomach.

Trauma

Aggression may be seen between axolotls under conditions of overcrowding, keeping animals of different sizes together, or keeping animals in mixed-gender aquaria during mating season. Minor wounds should be debrided and treated with local antiseptics to prevent secondary infection and left to heal by second intention. Larger wounds affecting the tail, limbs, or digits may necessitate amputation. In these cases, the wound should still be left to heal by second intention, as suturing the site can inhibit blastema formation and regeneration. In cases of predation by household cats, diagnostic imaging should be performed to assess for internal trauma and the patient should be treated with systemic antibiotics.

Neoplasia

Cases of neoplasia in amphibians are relatively uncommon. Cutaneous tumors in axolotls are typically benign and originate from pigment cells,[18] although one case of malignant chromatophoroma has been reported.[19] Surgical removal of cutaneous masses is recommended when they interfere with normal mobility or put the patient at risk of injury. Olfactory neuroblastoma has been diagnosed in two reported cases of intraoral masses.[20,21] Treatment with surgical excision and radiotherapy was curative in one case report.

Anesthesia and Analgesia

Sedation and anesthesia are necessary for surgical procedures and certain methods of diagnostic testing in the axolotl. Tricaine methanesulfonate (MS-222) is administered by immersion, making it ideal for use in aquatic amphibians. A wide dose range for amphibians of 100 mg/L to 3 g/L are reported in the literature,[14] but the author has used immersion baths of 1.0-1.5 g/L for diagnostic testing and short surgical procedures in axolotls with success. As MS-222 is acidic, it must be buffered with sodium bicarbonate (baking soda) or sodium carbonate (soda ash) until the solution reaches a pH equal to the pH of the axolotl's source water. Alfaxalone can be used for anesthesia by injection (15-20 mg/kg IM) or through immersion (5 mg/L induction)[14] Isoflurane is effective for anesthesia when delivered topically or through aeration into the patient's water, but both methods are difficult to deliver safely for veterinary staff involved.

Depth of anesthesia is measured by loss of righting reflex and loss of reaction to nociceptive stimuli. After induction of anesthesia through immersion, the patient can be removed from the water and kept on a moist surface for the duration of the procedure. Irrigating the gills with the induction solution after removal aids in maintenance of anesthesia for longer procedures. Anesthetic monitoring should include monitoring heart rate monitoring with a Doppler or ultrasound and respiratory rate monitoring, although both factors are dependent on environmental temperature. Reversal of anesthesia is completed by returning the patient to their original environment, where the anesthetic is expelled through the gills.

There are few studies that exist that have evaluated pain responses in the axolotl, however, analgesia should be provided peri-operatively for surgical cases and where pain is suspected. Buprenorphine (50 mg/kg ICe) and butorphanol (0.5 mg/L and 0.75 mg/L) had no significant effect on pain in a behavioral assessment, and buprenorphine caused severe adverse gastrointestinal side effects in two of six treated axolotls.[22] Non-steroidal anti-inflammatory medications such as meloxicam (0.2-0.4 mg/kg IM) and ketoprofen (1 mg/kg SC) have been evaluated in other amphibian species[14] and the practitioner may consider extrapolating dosages for use in the axolotl patient. Local anesthetics such as lidocaine and bupivacaine are another option that may provide effective analgesia. As with MS-222, lidocaine and bupivacaine should be buffered with sodium bicarbonate to minimize pain at the injection site. The neurokinin-1 receptor antagonist maropitant citrate (Cerenia) has yet to be studied in the axolotl, but anecdotally has shown encouraging results in fish and amphibian patients at dosages of 1-2 mg/kg SC.

Surgery

Enterotomy

Celiotomy with enterotomy is necessary in some cases of gastrointestinal foreign body obstruction. The anesthetized patient should be placed in dorsal recumbency, and the planned surgical field should be cleaned with sterile saline. The mucous layer should be kept intact if possible, to aid in post-operative healing. A paramedian incision is made through the skin and body wall to avoid damage to the ventral midline abdominal vein. The location of the foreign body should be identified and isolated by packing the surrounding field with sterile saline-soaked surgical gauze. Enterotomy and foreign body removal follows procedural guidelines described for other species. An absorbable synthetic monofilament suture should be used for simple continuous closure of the enterotomy site and coelomic wall incision. A non-absorbable synthetic monofilament suture should be used for closure of the skin in a horizontal mattress pattern. Sutures can be removed 4 weeks post operation.[23]

Amputation

Indications for amputation include severe trauma or infection involving a limb or the tail. After anesthetizing the patient, the affected area can be amputated using a scalpel blade. Hemostasis is very rapid in axolotls, so blood vessels typically do not require ligation. The surgical field should be left open to heal by the previously described blastema regeneration. Suturing the site should be avoided, as this can inhibit regeneration.

Cutaneous biopsy

Indications for biopsy include diagnosis of cutaneous masses or non-resolving skin lesions. Excisional biopsy of masses can be performed with sharp dissection. The skin can be sutured or allowed to heal by second intention if closure is not possible, as axolotl skin is poorly elastic. Topical antiseptics can be used for minimizing risk of secondary infection at the surgical site post-operatively.

Supportive Care

The sick or post-operative axolotl patient should be separated from any tankmates in a setup with normal water quality parameters. Ideally the client and the practitioner would have a cycled quarantine tank available. If this is not available, the patient may be kept in an aquarium or tub setup with daily full water changes. In either setup, the patient should have adequate hides available to minimize stress. Severely ill patients may be kept in a tub in a refrigerator at 44-46 °F. This creates an environment that is inhospitable for most infectious pathogens and decreases the basal metabolic rate to inhibit most nonvital processes to induce artificial torpor.[24] Most infections that can be resolved in this manner should show improvement within a week, but the patient can be kept in this state for over a month if necessary.

Illness and injury to axolotls and other amphibians results in osmotic stress through alteration of the pathways they use to maintain fluid homeostasis. Treatment with sodium chloride baths will help minimize this stress by minimizing loss of chloride ions and decreasing the osmotic pressure gradient, which in turn decreases absorption of nitrogenous toxins from the environment. Addition of non-iodized aquarium salt at 2-5 g/L is well-tolerated for long treatment periods. A solution created by axolotl researcher, Johannes Holtfreter, contains a mixture of several different salts that helps to maintain normal electrolyte levels. Holtfreter's solution does not contain a dechlorinating agent, so if used it should be made with deionized or distilled water or tap water that has been treated with a dechlorinating product.

References

1. International Union for Conservation of Nature. Available at: https://www.iucnredlist.org/species/1095/53947343. Accessed March 17, 2023.

2. LafeberVet: Basic Information Sheet: Axolotl. Available at: https://lafeber.com/vet/basic-information-sheet-axolotl/. Accessed March 17, 2023.

3. Page RB, Voss SR. Induction of metamorphosis in axolotls (*Ambystoma mexicanum*). Cold Spring Harb Protoc. 2009.

4. Seifert AW, Monaghan JR, Voss SR, Maden M. Skin regeneration in adult axolotls: a blueprint for scarfree healing in vertebrates. *PLoS ONE.* 2012;7(4):1-19.

5. Brainerd EL. Mechanics of lung ventilation in a larval salamander *Ambystoma tigrinum. J Exp Biol.* 1998;201:2891-2901.

6. Sanches PG, Veld RC, de Graaf W, et al. Novel axolotl cardiac function analysis method using magnetic resonance imaging. *PLoS ONE.* 2017;12(8):e0183446.

7. Lopez D, Lin L, Monaghan JR, et al. Mapping hematopoiesis in a fully regenerative vertebrate: the axolotl. *Blood.* 2014;124(8):1232-1241.

8. Farkas JE, Monaghan JR. Housing and maintenance of *Ambystoma mexicanum*, the Mexican axolotl. *Methods Mol Biol.* 2015;1290:27-46.

9. Cano-Martinez A, Vargas-Gonzalez A, Guarner-Lans A, et al. Functional and structural regeneration in the axolotl heart (*Ambystoma mexicanum*) after partial ventricular amputation. *Arch Cardiol Mex.* 2010;80(2): 79-86.

10. Jensen TB, Giunta P, Shultz NG, et al. Lung injury in axolotl salamanders induces an organ-wide proliferation response. *Dev Dynam.* 2021;250(6):866-879.

11. Freitas PD, Yandulskaya AS, Monaghan HR. Spinal cord regeneration in amphibians: a historical perspective. *Dev Neurobiol.* 2019;79(5):437-452.

12. Kragl M, Knapp D, Nacu E, et al. Cells keep a memory of their tissue origin during axolotl limb regeneration. *Nature.* 2009;460(2):60-67.

13. Loh R. Water quality explained: how it can affect your axolotl's health. *Proc World Small Anim Vet Assoc World Congr.* 2015.

14. Carpenter JW, Harms CA. *Carpenter's Exotic Animal Formulary.* 6th Ed. St. Louis, MO: Elsevier; 2023.

15. Loh R. Common disease conditions in axolotls. *Proc World Small Anim Vet Assoc World Congr.* 2015.

16. Del Valle JM, Eisthen HL. Treatment of chytridiomycosis in laboratory axolotls (*Ambystoma mexicanum*) and rough-skinned newts (*Taricha granulosa*). *Comp Med.* 2019;69(3):204-211.

17. Stanley J, MacHale J, Hedley J. Successful treatment of anchor worm (*Lernaea cyprinacea*) using lufenuron in the Mexican axolotl (*Ambystoma mexicanum*). *J Herpetol Med Surg.* 2021;31(2):107-110.

18. Hopewell E, Harrison SH, Posey R, et al. Analysis of published amphibian neoplasia case reports. *J Herpetol Med Surg.* 2020;30(3):148-155.

19. Vuong KS, Metral C, Sheley W, et al. Chromatophoromas on the tail of a pet axolotl (*Ambystoma mexicanum*) with presumptive metastasis. *J Exot Pet Med.* 2023;45:8-13.

20. Shioda C, Uchida K, Nakayama H. Pathological features of olfactory neuroblastoma in an axolotl (*Ambystoma mexicanum*). *J Vet Med Sci.* 2011;73(8):1109-1111.

21. Modesto F, Nicolier A, Hurtrel C, et al. Excisional biopsy and radiotherapy for management of an olfactory neuroblastoma in an axolotl (*Ambystoma mexicanum*). *J Am Vet Med Assoc.* 2021;260(4):436-441.

22. Llaniguez JT, Szczepaniak MA, Rickman BH, et al. Quantitative and qualitative behavioral measurements to assess pain in axolotls (*Ambystoma mexicanum*). *J Am Assoc Lab Anim Sci.* 2020;59(2):186-196.

23. Lord B. Approaches to axolotl gastrotomy. Available at: https://www.vettimes.co.uk/app/uploads/wp-post-to-pdf-enhanced-cache/1/approaches-to-axolotl-gastrotomy.pdf. Accessed March 20, 2023.

24. Gresens J. An introduction to the Mexican axolotl (*Ambystoma mexicanum*). *Lab Anim.* 2004;33(9):41-46.

Agreement of Temporoorbital Pulse Rate and Heart Rate in Ball Pythons (*Python regius*)

Mio Ito, BS,
Alexander A. Levitskiy, BS,
Nicola Di Girolamo, DVM, PhD, DipECZM (Herpetology), Dipl ACZM, Dipl ACVPM

Session #3477

Affiliation: From Exotic Pet Service, Department of Clinical Sciences, College of Veterinary Medicine, Cornell University, 930 Campus Road, Ithaca, NY 14853, USA.

Note: Abstract format selected by author(s). See Front Matter for description of conference proceedings manuscripts submitted as scientific abstracts.

Abstract: Currently, there are no described techniques to measure pulse rate in snakes. The purpose of the present study was to establish feasibility of measurement of temporoorbital pulse rate in ball pythons of different size and to evaluate its agreement with heart rate. Forty ball pythons (*Python regius*) ranging in body weight from 50 grams to 2.2 kilograms were included in the study. Heart rate and pulse rate were recorded by two operators, one immediately following the other. A random sequence was generated to randomize the measurement that was obtained first for each animal. The second operator was blinded to the results of the first operator. Beats were counted for 15 seconds using a standard veterinary ultrasonic Doppler flow detector. Heart rate was obtained over the apex of the heart. Pulse rate was obtained on the dorsal aspect of the head, slightly off midline at approximately the level of the parietal bone. In five additional snakes, heart and pulse rate were obtained simultaneously with two Dopplers by two operators to verify whether pulse rate was synchronous to heart rate. Both pulse rate and heart rate were successfully obtained in all snakes tested. Mean pulse rate was 63.4 ± 11.8 and heart rate was 62.2 ± 10.9. Mean difference between pulse and heart rate was 1.2 beats (95% limits of agreement: -10.6 to 13.0). When pulse and heart rate were measured at the same time, they were synchronous. Temporoorbital pulse rate measurement in ball pythons has good agreement with heart rate and is feasible in specimens of any size.

Acknowledgments: Authors thank Hudson Valley Herpetological for letting us obtain the measurements.

An Effective Location to Obtain Pulse Rate in Chelonians, Snakes and Lizards of Various Sizes

Nicola Di Girolamo, DVM, PhD, DipECZM (Herpetology), Dipl ACZM, Dipl ACVPM

Session #3469

Affiliation: From Exotic Pet Service, Department of Clinical Sciences, College of Veterinary Medicine, Cornell University, 930 Campus Road, Ithaca, NY 14853, USA.

Note: Abstract format selected by author(s). See Front Matter for description of conference proceedings manuscripts submitted as scientific abstracts.

Abstract: Techniques to obtain a pulse rate in reptiles are scantly reported in the literature. The purpose of this presentation is to describe a novel anatomic location that has been effectively used to obtain a pulse rate in chelonians, snakes, and lizards using a standard veterinary ultrasonic Doppler flow detector in a veterinary teaching hospital. We initially identified the presence of a superficial arterial location in bearded dragons (*Pogona vitticeps*) affected from parietal aneurysms. This artery was identified as the temporoorbital artery. Cardiovascularly healthy bearded dragons were found to also have a pulse rate in the location where affected bearded dragons developed aneurysms. Subsequently, we found that the temporoorbital artery could be consistently located also in snakes and chelonians. Anatomically, the temporoorbital artery lies in the laterotemporal fossa, an area of soft tissue in the head between the parietal and the postorbital bone. Clinically, the pulse of the temporoorbital artery can be easily identified applying a standard veterinary ultrasonic Doppler flow detector with a flat probe on the dorso-caudal aspect of the head, slightly off midline. Using this technique, we have been able to measure pulse rate in almost every chelonian, snake and lizard examined so far, with body weights ranging from 50 grams to 7.8 kilograms. The opportunity to reliably measure pulse rate also in reptiles opens up to a large number of future investigations and expands the range of diagnostic tools during physical examination or monitoring options under general anesthesia.

Performance of Three Portable Blood Glucose Monitors in Inland Bearded Dragons (*Pogona vitticeps*)

Danielle K. Tarbert, DVM, Dipl ACZM,
Lisa Pacumio, BS,
Melanie Ammersbach, DVM, DACVP,
Hugues Beaufrere, DVM, PhD, Dipl ACZM,
DABVP (Avian Practice), DipECZM (Avian)

Session #3431

Affiliation: From the William R. Pritchard Veterinary Medical Teaching Hospital (Tarbert, Pacumio), the Department of Pathology, Immunology, and Microbiology (Ammersbach), and the Department of Medicine and Epidemiology (Beaufrere), School of Veterinary Medicine, University of California-Davis, Davis, CA 95616, USA. Present address: Department of Small Animal Clinical Sciences, College of Veterinary Medicine, University of Tennessee, 2407 River Drive, Knoxville, TN 37996, USA (Tarbert).

Note: Abstract format selected by author(s). See Front Matter for description of conference proceedings manuscripts submitted as scientific abstracts.

Abstract: Blood glucose measurement is essential in bearded dragon disease diagnosis and management.[1,2] Point-of-care (POC) analyzers can be helpful in small patients as they use 0.5 μl blood or less; however, in other species, significant differences exist between POC analyzers and traditional benchtop methods resulting in recommendations to generate device-specific reference intervals, use correction formulas, or avoid use altogether.[3] In this study, samples from 47 bearded dragons *(Pogona vitticeps)* were used to compare three POCs (AlphaTrak II, canine and feline settings; AccuCheck Guide; LDX) to a benchtop analyzer (Cobas C 501). Correlation was evaluated using Spearman correlation coefficients.[4] Bias was investigated using difference plots and Passing-Bablok regression.[4] Clinical and analytical agreement was assessed using published mammalian total allowable error values and bearded dragon-specific analyzer coefficients of variation, respectively.[4,5] A multiple linear regression model was used to investigate potential effect of hematocrit, total solids, and lipemia. Good precision and strong correlation with the reference analyzer was found for all POC; however, significant proportional bias was present for all POC and significant constant bias was present for the AlphaTrak. The AlphaTrak overestimated while the AccuCheck and LDX underestimated blood glucose. No POC was in analytical agreement with the reference analyzer. Only the LDX was in clinical agreement; however, when corrective formulas were applied, all analyzers were within clinical decision limits. Higher PCV was associated with negative increase in bias. This study supports the use of POCs for blood glucose measurement in bearded dragons provided species-specific corrective formulas or method-specific reference intervals are utilized.

Acknowledgments: This study was funded by the Reptile Research Fund, UC Davis, School of Veterinary Medicine.

References

1. Minor RL, Doss GA, Mans C. Evaluation of glucose absorption rates following intracoelomic or subcutaneous administration in experimentally dehydrated inland bearded dragons (*Pogona vitticeps*). *Am J Vet Res*. 2021;82:920-923.

2. Ritter JM, Garner MM, Chilton JA, et al. Gastric neuroendocrine carcinomas in bearded dragons (*Pogona vitticeps*). *Vet Pathol.* 2009;46:1109-1116.

3. Cutler DC, Koenig A, Di Girolamo N, Mayer J. Investigation for correction formulas on the basis of packed cell volume for blood glucose concentration measurements obtained with portable glucometers when used in rabbits. *Am J Vet Res*. 2020;81:642-650.

4. Jensen AL, Kjelgaard-Hansen M. Method comparison in the clinical laboratory. *Vet Clin Pathol.* 2006;35:276-286.

5. Harr KE, Flatland B, Nabity M, et al. ASVCP guidelines: allowable total error guidelines for biochemistry. *Vet Clin Pathol.* 2013;42:424-436.

Co-Infections with Respiratory Pathogens in Snakes

Christoph Leineweber, Dr med vet,
Elisabeth Müller, Dr med vet, DipECVM,
Rachel E. Marschang, PD Dr med vet, DipECVM, DipECZM (Herpetology)

Session #3177

Affiliation: From Laboklin GmbH & Co. KG, Bad Kissingen, Germany.

Note: Abstract format selected by author(s). See Front Matter for description of conference proceedings manuscripts submitted as scientific abstracts.

Abstract: Many viruses have been shown to cause respiratory disease in snakes. These include serpentoviruses (order *Nidovirales*), paramyxoviruses, and reoviruses. Adenoviruses and reptarenaviruses, the cause of inclusion body disease (IDB) in boas and pythons, have also been associated with respiratory disease. Mycoplasma has been increasingly found in snakes, with a possible association with respiratory disease. In this retrospective study, results of diagnostic testing of samples from various snake species by PCR for serpento-, paramyxo-, reo-, adeno-, and reptarenaviruses and mycoplasma were evaluated for the likelihood of detection of multiple pathogens in a given sample. Results from 1656 to 1685 animals were evaluated for the presence of viruses and 602 for mycoplasma. Serpentoviruses were found in 243 of 1682 samples (14.45%), paramyxoviruses in 15 of 1678 samples (0.89%), reoviruses in 39 of 1656 samples (2.36%), adenoviruses were found in 58 of 1662 samples (3.49%), reptarenaviruses in 84 of 1674 samples (5.02%), and mycoplasma in 259 of 602 samples (43.02%). Significant correlations were found for detection of adenoviruses and paramyxoviruses ($P = 0.0371$), reptarenaviruses and reoviruses ($P = 0.0002$), serpento- and paramyxoviruses ($P = 0.0046$), serpentoviruses and mycoplasma ($P < 0.0001$), and for paramyxo- and reoviruses ($P < 0.0001$). Coinfections between many of these pathogens have been described before, but not for this number of samples and this range of pathogens. Additional analysis to understand the role of host species in these infections is ongoing. The role of individual pathogens in disease development is not yet understood and requires further study.

Evaluating Effectiveness of Nasal Flush Technique in *Mycoplasma* PCR- or DNA-Sequencing-Positive Tortoises

Marianne Caron, DVM,
Ian Kanda, RVT, VTS (Exotics),
Thomas Boyer, DVM, DABVP (Reptile and Amphibian Practice),
Mark Mitchell, DVM, MS, PhD, DipECZM (Herpetology)

Session #3452

Affiliation: From Pet Hospital of Penasquitos, 9888 Carmel Mountain Rd Ste F, San Diego, CA 92129, USA (Caron, Kanda, Boyer), Louisiana State University, Department of Veterinary Clinical Sciences, 1026 Louisiana Animal Disease Diagnostic Laboratory, Baton Rouge, LA 70803, USA (Mitchell).

Note: Abstract format selected by author(s). See Front Matter for description of conference proceedings manuscripts submitted as scientific abstracts.

Abstract: Tortoise mycoplasmosis is one of the most extensively characterized infectious diseases of chelonians; however, there has been little study on treatment of tortoise mycoplasmosis.[1,2] The purpose of this study was to compare positive and negative *Mycoplasma* results prior to and after a modified Jarchow nasal flush protocol, as well as comparing positive and negative results between MiDOG, a next-generation DNA sequencing test, and PCR. Of the 13 tortoises in this study, 12/13 (92.31%) tortoises with nasal discharge were *Mycoplasma* sp. positive on at least one diagnostic test. Only 1/11 (9.10%) remained positive 1 month after the last nasal flush, potentially a case of re-infection (i.e., was placed with a cage mate that previously had nasal discharge after its treatment) rather than a lack of treatment success. Twelve of 12 tortoises (100%) had a resolution of nasal discharge 1 month after the last nasal flush. These results indicate that the nasal flush protocol outlined is effective at decreasing the likelihood of testing positive for *Mycoplasma* sp. by both PCR and DNA sequencing, as well as resolving nasal discharge, 1 month after the last nasal flush.

References

1. Jarchow J. Treatment protocol for upper respiratory tract disease of desert tortoises. *Proc 7th Int Symp Path Med Reptile Amphib.* 2004.

2. Rettenmund CL, Boyer DM, Orrico WJ, et al. Long-term oral Clarithromycin administration in chelonians with subclinical *Mycoplasma* spp. infection. *J Herpetol Med Surg.* 2017;27:58-61.

3. Brown MB, McLaughlin GS, Klein PA, et al. Upper respiratory tract disease in the gopher tortoise is caused by *Mycoplasma agassizii*. *J Clin Microbiol.* 1999;37(7):2262-2269.

4. Jacobson ER, Gaskin JM, Brown MB, et al. Chronic upper respiratory tract disease of free-ranging desert tortoises (*Xerobates agassizii*). *J Wildl Dis.* 1991;27(2):296-316.

5. Jacobson ER, Brown MB, Wendland LD, et al. Mycoplasmosis and upper respiratory tract disease of tortoises: a review and update. *Vet J.* 2014;201(3):257-264.

In Vitro Characterization of Ophidian Serpentoviruses

Robert J. Ossiboff, DVM, PhD, DACVP,
Steven B. Tillis,
Tracey D. Logan, MSc

Session #3337

Affiliation: From the Department of Comparative, Diagnostic, and Population Medicine, College of Veterinary Medicine, University of Florida, Gainesville, FL 32608, USA.

Note: Abstract format selected by author(s). See Front Matter for description of conference proceedings manuscripts submitted as scientific abstracts.

Abstract: Ophidian serpentoviruses, positive-sense RNA viruses commonly referred to as nidoviruses, are increasingly being identified as important infectious agents in both captive and free-ranging snake populations. The clinical significance of these viruses can be variable, but they have the potential to cause severe and sometimes fatal disease in captive snakes. While much has been learned about the diversity and disease potential of these viruses, little is known about the properties of these viruses, including their environmental stability, susceptibility to common disinfectants, and their kinetics of growth. To address this, serpentoviruses were isolated in culture from three PCR positive snakes representing three genera of pythons. The viruses were used to establish a median tissue culture infectious dose ($TCID_{50}$) to determine changes in infectious titer in samples to assess viral stability and growth. All three isolated viruses showed limited environmental stability, losing all detectable infectivity after 10-12 days at room temperature. All viruses likewise showed a wide susceptibility to sanitizers, with 10% bleach and 70% ethanol rendering virus inert in one minute and 7% peroxide, 2% chlorhexidine, and quaternary ammonium solution within 3 minutes. Although all viruses successfully grew at 32 °C with varied rates across three different snake cell lines, all viruses failed to replicate at 35 °C. These results not only increase our fundamental understanding of these emerging reptile viruses, they also provide the basis for mitigating the effects of viral outbreaks in captive snake colonies as well as potentially offering a non-pharmacologic method to eliminate ophidian serpentoviral infections.

References

1. Ossiboff RJ, Origgi FC, Stacy NI. Health and disease in free-ranging and captive wildlife. *Front Vet Sci.* 2020;1056.

Tortoise Anatomy Project

M. Scott Echols, DVM, DABVP (Avian Practice), and
Heather F. Smith, PhD

Session #3750

Affiliation: From Medical Center for Birds, Oakley, CA 94561, USA (Echols), Echols Veterinary Services, Sandy, UT 84093, USA (Echols), Scarlet Imaging, Murray, UT 84157, USA (Echols), Midwestern University, Glendale, AZ 85308, USA (Smith).

Note: Abstract format selected by author(s). See Front Matter for description of conference proceedings manuscripts submitted as scientific abstracts.

Abstract: Veterinary care of captive tortoises, including proper diagnosis and treatment, is often hindered by a paucity of detailed clinical anatomic and imaging references. Captive tortoises frequently present with upper and lower respiratory tract infections. Joint luxations, such as the hip and shoulder may result in lameness, and often lack known etiology. Shell trauma, especially resulting from vehicular impacts, is common in chelonians. Additionally, the tortoise vascular system is poorly understood, which complicates procedures such as venipuncture, catheter placement and surgical exploration. Several tortoise species are also critically endangered, and there is a pressing need to document their anatomy to provide proper veterinary care. The primary objective of this research is to improve care of captive tortoises by: 1) generating detailed maps of the arterial and venous systems to improve catheter placement, blood collection, and identification of vascular anomalies and insufficiencies; 2) improving the interpretation and treatment of subluxated and luxated joints and other orthopedic disorders by providing detailed anatomical documentation and descriptions of associated osteological landmarks and muscle attachment sites; 3) informing the treatment of upper and lower respiratory tract infections by improving the understanding of underlying respiratory anatomy; 4) identifying divisions and soft tissue and bony attachments to the shell (carapace and plastron) in effort to better guide diagnosis and appropriate treatment of shell injuries; and 5) generating high-quality anatomic, tomographic, MRI and radiographic atlases of tortoise anatomy for improved surgical precision and other medical procedures.

Section 45

ARAV Posters

Shell Fracture Repair with Reconstruction PAX Locking Plate: Case Report

Rita Sousa, LMV

Session #3203

Affiliation: From Hospital Veterinário do Oeste, EN 247 Condomínio Valpraia Lj B, Vale de Adares, 2530-086 Lourinhã, Portugal.

Note: Abstract format selected by author(s). See Front Matter for description of conference proceedings manuscripts submitted as scientific abstracts.

Abstract: A 29-year-old red-eared slider (*Trachemys scripta elegans*) was brought to Hospital Veterinário do Oeste after falling from a balcony during the owner's vacation. The turtle had a broken shell, and a computed tomography scan was advised to plan the surgery. The surgery consisted of placing a PAX 2.4-mm titanium locking reconstruction plate (Securos Surgical, USA), cut into four pieces, using six 2.0-mm screws. During the recovering period, the turtle was medicated with meloxicam, vitamin A, vitamin D_3, vitamin E complex, ceftazidime, and tramadol. As soon as the turtle was returned to its normal enclosure, it started eating on its own. Locking bone plating systems are increasingly being utilized for fracture repair in veterinary patients. These systems have a greater flexibility of applications, which is ideal for exotic pet orthopedics. Advantages include reservation of periosteal blood supply, improved fracture site stability (as in this case), reduced risk of altering fracture reduction during screw placement, and a lack of precise anatomical plate contouring required. The PAX plate hole and screw head is a self-locking, poly-axial, angle stable system and screws do not need to be inserted at a fixed angle within the plate for coupling to occur. Locked monocortical screws will not toggle within the bone and require a smaller number of cortices to be engaged above and below the fracture. Stability at the fracture site is solely dependent on the stiffness of the locking construct, not on the screw bone interface nor the compression or friction between the plate and bone.

Comparison of Portable Uric Acid Analyzers for Measurement of Uric Acid in Red-Eared Slider Turtles

Alma Paulina Acosta-Nieves, MVZ, EMCV (FS),
Marina Guadarrama-Olhovich, MVZ, MMVZ,
Carlos Gustavo Vázquez-Peláez, MSc, PhD,
Juliana Leal-Nieves, MVZ, EMCV (FS)

Session #3311

Affiliation: From Faculty of Veterinary Medicine, National Autonomous University of Mexico (Acosta-Nieves), Pathology Department, Facultad de Medicina Veterinaria y Zootecnia, Universidad Nacional Autónoma de México (Guadarrama-Olhovich), Genetics and Biostatistics Department, Facultad de Medicina Veterinaria y Zootecnia, Universidad Nacional Autónoma de México (Vázquez-Peláez), Ethology, Wildlife and Laboratory Animal Department, Facultad de Medicina Veterinaria y Zootecnia, Universidad Nacional Autónoma de México (Leal-Nieves).

Abstract: Portable electrochemical meters and spectrophotometry are used to determine uric acid levels in humans. The aim of this study was to validate the effectiveness of two portable meters for human use to determine the serum concentration of uric acid in red-eared slider turtles (*Trachemys scripta elegans*). The study was performed on thirty female turtles in captivity. The samples were taken from the jugular and coccygeal vein of the turtles and were analyzed using two portable uric acid meters and a spectrophotometry method. The Bland-Altam procedure was used to analyze the uric acid response. The results suggest that neither of the two portable meters used were equivalent in measuring uric acid levels when compared to spectrophotometry. Uric acid measurements using portable electrochemical meters are not precise. The meters were 40% and 27% different for Xper E1 and UASure, respectively, and were found out of the boundaries. Therefore, the value determined by spectrophotometry cannot be replaced. Portable uric acid meters and spectrophotometry cannot be used indistinctly in veterinary clinical practice in this species. Further studies must consider the validation of these portable uric acid meters on different reptile and bird species.

Introduction

Uric acid is the end product of the catabolism of purines, the constituent nitrogenous bases of nucleic acid in most mammals and reptiles. The endogenous production of uric acid occurs in the liver, intestines, and other tissues like muscles, kidneys, and vascular endothelium. The kidneys excrete two-thirds of the total daily uric acid production; the rest is metabolized by intestinal microbiota and eliminated through feces. Diminished excretion of uric acid leads to hyperuricemia associated with kidney disease, dehydration, or high protein intake.[1-3]

Chronic hyperuricemia may cause uratosis: a deposit of monosodium crystals in articulations.[4] In reptiles, visceral, articular, and periarticular uratosis may be present when uric acid levels exceed 2,210 μmol/L (25 mg/dl). Early detection of high serum levels of uric acid is decisive to avoid clinical complications given the decrease in renal excretion.[1]

In humans, portable meters are widely described in measuring and monitoring uric acid levels. These devices have shown efficacy compared to standardized spectrophotometry methods; however, its functionality in animals is unknown.[5] Based on the above, an observational, cross-sectional clinical study was designed to evaluate the reliability of the measurements of uric acid levels in red-eared slider turtles (*Trachemys scripta elegans*).

Two different portable electrochemical meters were used to determine uric acid levels and were correlated with the results from the standardized spectrophotometry method.

Materials and Methods

The study included thirty adult female red-eared slider turtles (*Trachemys scripta elegans*), which belong to the herpetary of the Zacango Ecological Park located in Calimaya, Estado de México. Individuals weighed approximately 1.350 kilograms and physical examination showed they were clinically healthy. The turtles lived in a 2 × 2 meter plastic tub with 30 centimeters of water. Each tub contained five turtles that were fed every third day with 20 grams of pellets. A Calorex® heater maintained the temperature between 24 and 26 °C; this was monitored using a digital thermometer. Before obtaining the samples, the room temperature was determined at 18 °C (64.4 °F) with 20% humidity, measured by Combometer® Exo terra digital thermo-hygrometer for indoor use.

Portable Uric Acid Meter Mesurements

The portable uric acid meters used for this study were UASure® (Apex Biotechnology Corp., Hsinchu, Taiwán) and Xper technology E1® (Technology Corp, Taiwán). The measurement was initiated by placing a capillary drop of blood on the reaction zone on the reactive strip. Both meters are based on amperometric electrochemistry and use a non-enzymatic, printed, two-layered, carbon-pasted, dry-electrode strip; at 300mV and pH = 7.0-10.0, the electrons are transferred to an underlying electrode through the major electron mediator ($K_3Fe(CN)_6$, potassium ferricyanide). The current detected by the electrode on the test strip is in proportion to the uric acid concentration in the blood. Both meters were used according to the manufacturer's instructions. Using a single badge of reactive strips, the analytical performance was determined. Both portable meters had a dynamic range for uric acid detection between 265-1768 μmol/L (3-20 mg/dl). Results above or below displayed Hi or Lo.

Spectrophotometry Uric Acid Determination

Samples for the spectrophotometry comparisons were processed in a Dirui® CS-T240 Auto-Chemistry Analyzer. A SEKURE® uric acid-SL reagent was used: buffer solution containing 1.8 mmol/l DHBS, 0.5 mmol/l 4-aminoantipyrine, more than 3500 u/l peroxidase (botanical), more than 200 u/l of uricase (microbial), stabilizing agents and preservatives.

Bioethical Committee

This study was evaluated and approved by the Internal Committee for the Care and Use of Animals (CICUA) de la Facultad de Medicina Veterinaria y Zootécnia de la Universidad Nacional Autónoma de México (FMVZ-UNAM).

Sample Collection

Blood samples were collected from each of the dorsal coccygeal and jugular veins with syringes impregnated with sodium heparin (Hinepar®1000 UI/ml, Laboratorios PiSA). The physical containment of the specimens was performed using the conventional restriction technique described for this species.[6] To assess the precision of the method, and to verify variations, three readings per sample were made. One drop of blood (4.0 to 6.0 µl) was added to the UASure® device, giving reads every 30 seconds between each strip. The Xper E1® gave the readings every 15 seconds between each strip. The same technician performed all readings, and each measurement required a different strip. The remaining blood was placed in a Vacutainer® tube without anticoagulant. The samples were stored in a cooler flask and were taken to the FMVZ-UNAM pathology department. Samples were centrifuged and plasma was separated.

Statistical Analysis

The descriptive statistics determined included: 95% confidence interval, mean, and standard deviation. The Bland-Altam procedure and the MedCalc software was used to analyze the response of two portable electrochemical meters for human use (Xper E1 and UASure) to determine uric acid in red-eared slider turtles (*Trachemys scripta elegans*).

Results

Thirty adult female red-eared slider turtle blood samples were analyzed using two portable uric acid meters and a spectrophotometer. In 84% of the samples, no numeric values were obtained. Meaning, that uric acid was low or below 265 µmmol/L (3 mg/dl). These unidentified levels were excluded from the statistical analysis. Table 1 provides values of the three replicates from each portable meter and the spectrophotometry measures.

Descriptive statistics and 95% confidence interval estimated for uric acid spectrophotometric measurements are shown in Table 2. There was an important variation among the samples that were above 70%, with a confidence interval between 142-260 µmmol/L (1.6-2.9 mg/dl). Nonetheless, the sample size was adequate with a 95% confidence throughout the whole sample. Because many values could not be quantified due to Lo markings, only five observations per group were used for comparison among methods. Because the three replicates (strips) were taken at the same time in each portable uric acid meter (Xper and UASure) and only one sample from the spectrophotometer each strip was compare with the spectrophotometer value within individual, therefore 15 observations were use in the analysis.

Mean uric acid descriptive statistics and Bland-Alman plot of the differences of spectrophotometry estimation compared with the Xper E1 are provided in Figure 1. Mean uric acid descriptive statistics and Bland-Alman plot of the differences of spectrophotometry estimation compared with the and the UASure method are presented in Figure 2.

Table 1. Uric acid level (µmmol/L) of three replicates from each portable meter and the spectrophotometer.

N	Xper strip #1	Xper strip #2	Xper strip #3	UASure strip #1	UASure strip #2	UASure strip #3	Dirui®
1	283	300	283	300	327	300	238
2	265	265	267	344	346	344	298
3	Lo	Lo	Lo	Lo	Lo	Lo	150
4	Lo	Lo	Lo	Lo	Lo	Lo	194
5	Lo	Lo	Lo	Lo	Lo	Lo	205
6	Lo	Lo	Lo	Lo	Lo	Lo	190
7	Lo	Lo	Lo	Lo	Lo	Lo	154
8	Lo	Lo	Lo	Lo	Lo	Lo	157
9	Lo	Lo	Lo	Lo	Lo	Lo	140
10	Lo	Lo	Lo	Lo	Lo	Lo	264
11	Lo	Lo	Lo	Lo	Lo	Lo	124
12	265	274	265	274	274	265	218
13	716	707	716	618	689	689	612
14	Lo	Lo	Lo	Lo	Lo	Lo	133
15	539	592	539	901	1255	901	857
16	Lo	Lo	Lo	Lo	Lo	Lo	150
17	Lo	Lo	Lo	Lo	Lo	Lo	178
18	Lo	Lo	Lo	Lo	Lo	Lo	185
19	Lo	Lo	Lo	Lo	Lo	Lo	158
20	Lo	Lo	Lo	Lo	Lo	Lo	129
21	Lo	Lo	Lo	Lo	Lo	Lo	60
22	Lo	Lo	Lo	Lo	Lo	Lo	105
23	Lo	Lo	Lo	Lo	Lo	Lo	130
24	Lo	Lo	Lo	Lo	Lo	Lo	94
25	Lo	Lo	Lo	Lo	Lo	Lo	196
26	Lo	Lo	Lo	Lo	Lo	Lo	174
27	Lo	Lo	Lo	Lo	Lo	Lo	189
28	Lo	Lo	Lo	Lo	Lo	Lo	76
29	Lo	Lo	Lo	Lo	Lo	Lo	79
30	Lo	Lo	Lo	Lo	Lo	Lo	187

Table 2. Mean, standard deviation, standard error, and coefficient of variation for portable electrochemical meters for human use (Xper E1 and UASure) and spectrophotometry used to determine uric acid in red-eared slider turtles (*Trachemys scripta elegans*).

Method	Mean	Standard deviation	Standard error	Coefficient of variation
Xper E1	418.40	190.53	49.20	48%
UASure	498.47	255.04	65.85	21%
Spectrophotometry	444.60	259.24	66.93	58%

Method A spectrophotometry and Method B Xper E1	
Differences	
Sample size	15
Arithmetic mean	26.2
95% CI	-56.39 to 108.79
Standard deviation	149.1
Lower limit	-266.1247
95% CI	-410.55 to -121.69
Upper limit	318.5247
95% CI	174.09 to 462.96

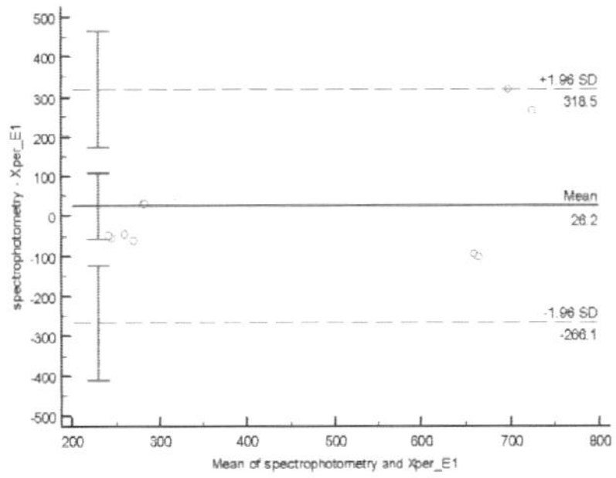

Figure 1. Left: Mean (± SD) uric acid of the spectrophotometry estimation method was 444.6 (± 259.24) compared with 418.4 (± 190.53) for the Xper E1 method. The basis of the Xper E1was 26.2 (95% CI -56.4, 108.8), precision was 149.1 (95% confidence limit = -266.1, 318.5), and the percentage error was 89.55%. Right: Bland-Altman plot of differences (Xper E1 - spectrophotometry) to determine uric acid in red-eared slider turtles (*Trachemys scripta elegans*), with mean (bias) and intervals confidence 95%.

Method A spectrophotometry and Method B UASure	
Differences	
Sample size	15
Arithmetic mean	-53.8667
95% CI	-64.5104 to -43.2230
Standard deviation	19.2
Lower limit	-91.5379
95% CI	-110.1507 to -72.9251
Upper limit	-16.1954
95% CI	-34.8082 to 2.4174

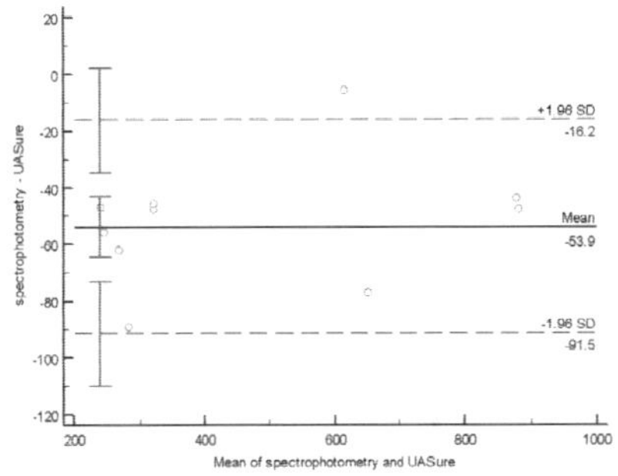

Figure 2. Left: Mean (± SD) uric acid of the spectrophotometry estimation method was 444.6 (± 259.24) compared with 498.47 (± 255.04) for the UASure method. The basis of the UASure was -53.9 (95% CI -64.5, -43.2), precision was 19.2 (95% confidence limit = -110.2, -34.8), and the percentage error was 23.36%. Right: Bland-Altman plot of differences (UASure - spectrophotometry) to determine uric acid in red-eared slider turtles (*Trachemys scripta elegans*), with mean (bias) and intervals confidence 95%.

Discussion

Uric acid levels of red-eared slider turtles, obtained by two portable meters for human use (UASure® and Xper E1®), do not correspond to the values obtained by the Dirui® spectrophotometer method. Neither of the portable meters had good estimation compared to values obtained with the spectrophotometry; suggesting that the meters have a wide variability when measuring red-eared slider turtle blood samples.

Kuo et al.[5] described the use of the portable meter UASure® in humans and determined that the device could be used as a reliable test for the measurement of uric acid. Based on this study, UASure® and Xper E1® showed 89% and 23% errors when compared to the results obtained from Dirui® spectrophotometry. These findings could not be considered reliable determinations compared to the gold standard (spectrophotometry).

Paraskos et al.[7] conducted a study and evaluated four portable uric acid meters for human use and found that UASure® had the most variations in the results. When comparing it to this study, major ranges of variation were found. Our results show that portable meters did not detect 84% of uric acid samples. Lo markings were registered; meaning that the portable meters did not register uric acid below 265 µmmol/L (3 mg/dl). This was later confirmed by the results found in Dirui® spectrophotometry.

While comparing the results from the strips on the portable meters with the values found on the spectrophotometry, in some cases, both portable meters overestimated the concentrations of uric acid. This is why some samples showed high concentrations of uric acid; however, this was dismissed by the results found on the Dirui® spectrophotometry.

There are different reasons why turtles showed low uric acid concentrations. Eatwell et al.[8] mention that room temperature has a significant effect on the different metabolic routes on reptiles; therefore, the parameters of some biochemical analytes may vary according to temperature. Klaphake[9] claims that the optimal temperature preferred by the red-eared slider turtles (*Trachemys scripta elegans*) is between 22 and 30 °C; turtles sampled in this study, lived in 18 °C, a suboptimal temperature. Levels of uric acid vary throughout the day depending on body temperature, diet, and the hydration of the patient.

Paraskos et al.[7] described that, in humans, the UASure® portable meter required numerous tries to obtain readable results. They associated the intangible readings to the inadequate filling of the strips. In the present study, the majority of the readings showed Lo which means values for uric acid were below 265 µmmol/L (3 mg/dl). It is important to report that all strips tested were filled with adequate amounts of blood, therefore, intangible readings were not a limitation in this study.

In addition, Kuo et al.[5] reported interference in the quantities of uric acid with elevated metabolic endogenous like ascorbic acid and cholesterol. Haque[10] studied the relationship between serum levels of uric acid and glucose on healthy humans, prediabetics, and diabetics, showing that individuals with hyperglycemia had lower levels of uric acid in comparison to healthy individuals. Also, Eatwell et al.[8] referred that levels of glucose in reptiles could vary notably among each species and can be influenced by nutrition, environmental conditions, and the season. In our study, blood samples were taken with 10- to 15-minute intervals; due to the difficulty presented in accessing the veins, the time-lapse of the sample collection was prolonged, possibly generating transitory hyperglycemia. We were not able to give a complete correlation for this finding because glucose was not quantified.

Finally, Sykes[6] described that erythrocytes in reptiles measure between 13-25 µm and have a nucleus whereas human erythrocytes measure 6-8 µm and have no nucleus. Speculations on the size and form of turtle erythrocytes could interfere with the precision determination because the portable meters are meant for human blood. Also, Gulbrandsen et al.[11] mentioned that the number of erythrocytes and the state of hydration could interfere with the results because hematocrits above 50% and below 35% may cause incorrect uric acid readings.

Conclusions and Clinical Relevance

The present study shows that portable meters for human use are not a precise method to obtain concentrations of uric acid on red-eared slider turtles (*Trachemys scripta elegans*) and thus, they cannot replace the values obtained with spectrophotometry. Even though our sample size was small, the results suggest that there is an important variation between uric acid levels obtained by the portable meters when compared to the results obtained from the spectrophotometry. In this study, the usefulness of portable meters for uric acid levels in red-eared slider turtles is questionable. Further studies on different reptile and bird species are needed.

Disclosure Statement: This research did not receive any specific grant from funding agencies in the public, commercial, or not-for-profit sectors.

References

1. Maixner JM, Ramsay EC, Arp LH. Effects of feeding on serum uric acid in captive reptiles. *J Zoo Anim Med.* 1987;18(2/3):62. doi:10.2307/20460240.

2. Reavill DR, Schmidt RE. Urinary tract diseases of reptiles. *J Exot Pet Med.* 2010;19(4):280-289. doi.org/10.1053/j.jepm.2010.10.007.

3. Selleri P, Hernandez-Divers SJ. Renal diseases of reptiles. *Vet Clin North Am Exot Anim Pract.* 2006;9(1):161-174. doi.org/10.1016/j.cvex.2019.08.006.

4. Bardin T, Richette P. Definition of hyperuricemia and gouty conditions. *Curr Opin Rheumatol.* 2014;26(2):186-191. doi:10.1097/BOR.0000000000000028.

5. Kuo CS, Hwu CM, Lin YH, et al. Portable electrochemical blood uric acid meter. *J Clin Lab Anal.* 2002;16(2):109-114. doi:10.1002/jcla.10030.

6. Sykes JM, Klaphake E. Reptile hematology. *Clin Lab Med.* 2015;35(3):661-680. doi.org/10.1016/j.cll.2015.05.014.

7. Paraskos J, Berke Z, Cook J, et al. An analytical comparison between point-of-care uric acid testing meters. *Expert Rev Mol Diagn.* 2016;16(3):373-382. doi:10.1586/14737159.2016.1134326.

8. Eatwell K, Hedley J, Barron R. Reptile haematology and biochemistry. *In Pract.* 2014;36(1):34-42. doi:10.1136/inp.f7488.

9. Klaphake E, Gibbons PM, Sladky KK, Carpenter JW. Reptiles. In: Carpenter JW, ed. *Exotic Animal Formulary.* 5th Ed. St. Louis, MO: Elsevier; 2018:81-166.

10. Haque T, Rahman S, Islam S, et al. Assessment of the relationship between serum uric acid and glucose levels in healthy, prediabetic and diabetic individuals. *Diabetol Metab Syndr.* 2019;11(1):1-8. doi.org/10.1186/s13098-019-0446-6.

11. Gulbrandsen CL, Morton NE, Rao DC, et al. Determinants of plasma uric acid. *Hum Genet.* 1979;50(3):307-312. doi:10.1007/BF00399397.

Successful Treatment of Spectacle Avulsion in a Schultze's Pit Viper (*Trimeresurus schultzei*) and a Burmese Python (*Python bivittatus*)

Tom Hellebuyck, DVM, PhD, DipECZM (Herpetology), and
Ferran Solanes Vilanova, DVM

Session #3354

Affiliation: From the Department of Pathobiology, Pharmacology and Zoological Medicine, Division of Poultry, Exotic Companion Animals, Wildlife and Experimental Animals, Salisburylaan 133, 9820 Merelbeke, Belgium.

Note: Abstract format selected by author(s). See Front Matter for description of conference proceedings manuscripts submitted as scientific abstracts.

Abstract: A 5-year-old female Schultze's pit viper (*Trimeresurus schultzei*) and a 3-year-old male Burmese python (*Python bivittatus*) were presented with a chronic, unilateral subspectacular abscess. Partial spectaculectomy was attempted to sample and debride the exudate from the subspectacular space, but near-complete avulsion of the spectacle occurred because of the severe, chronic inflammatory changes in both cases, and a complete spectaculectomy needed to be performed. *Pseudomonas aeruginosa* was isolated following microbiologic examination of the exudate in both snakes. Different techniques that rely on the use of grafts were considered but deemed inappropriate, mainly because of the relatively small size of the spectacle. Ofloxacin ophthalmic ointment was applied every six to eight hours during a 2-week period, followed by the application of vitamin A ophthalmic ointment until the first shed occurred. On average, 3 weeks after the start of the treatment, centripetal migration of the germinal epithelium of the spectacle was noticed, re-establishing a germinal center to produce new spectacle material. Following the first shed, the regenerated spectacle showed a relatively normal appearance. The regeneration of a functional spectacle and subspectacular space was confirmed by the infiltration of fluorescein into the subspectacular space and assured patency of the drainage system. While follow-up in the viper was lost, the spectacle regained its normal appearance after a second shedding cycle in the python. This report demonstrates that complete regeneration of the spectacle following avulsion in snakes is possible, if the transition zone to the spectacle is still intact.

Section 46

Hybrid Session 1

Andrea Winkel, DVM, MPH
Session Chair

How to Incorporate Exotic Animals into General Practice

Jesse A. Fallon, DVM, PhD

Session #3316

Affiliation: From Cheat Lake Animal Hospital, Morgantown, WV 26508, USA.

Abstract: Veterinarians love their exotic animal patients. Whether furred, feathered, or scaly, these patients are what get them excited about going to work; however, exotic animal medicine poses challenges when trying to incorporate this much-needed care into general practice. From scheduling problems to adequate staff training, to financial productivity, to appropriate medical and surgical equipment, there are many hurdles to successfully building an exotic case load. This presentation establishes a clear path for the general practitioner to establish themselves as an exotic animal veterinarian for the public, their staff, and their administrators.

Introduction

The practice of exotic animal medicine has improved markedly over the past several decades, and each year, after this ExoticsCon conference, veterinarians feel energized, ready to take on new knowledge and apply it to patients. Then, as they get back to clinical practice, they encounter numerous hurdles that prevent them from practicing at the level of efficiency and confidence desired. There have been many presentations on what equipment to purchase to make care of exotic animal patients easier and more effective. More common than equipment inadequacies are challenges of scheduling, productivity, and staff training. This presentation focuses on overcoming these more nuanced obstacles.

Scheduling

One of the biggest challenges that veterinarians face occurs well before they meet exotic animal patients. What strategies can you employ to help fill your schedule with exotic animal patients? How do you deal with no-shows and late appointments? How do you address a wellness appointment with a patient that really has significant deficits in husbandry or has a serious illness? These questions must be addressed with a structured approach directed to relatively uneducated front desk staff that want to make the right decisions about when and how to schedule, but need the right training and tools to succeed.

The first step when training receptionists is to help them understand that the initial approach to exotic animal patients is very similar to dog and cat patients. Get the patient's name, species, sex, status, age, and presenting complaint. Teach them to be specific (e.g., 17-year-old, male, blue and gold macaw presenting for regurgitation and decreased appetite rather than parrot: ain't doin' right). Staff are capable of this, and are typically eager to learn. This training can occur one-on-one in a small practice or may occur at routine reception meetings in larger practices. If it is not already a routine, encourage the practice manager or front desk manager to have routine receptionist meetings. Ask to attend one of these meetings, or ideally attend them semi-annually, as a reminder for experienced staff or to engage with new staff members.

Remember, developing a full but not overloaded schedule starts with the receptionist. The front desk needs to know which patients can wait until the next available appointment (e.g., chronic feather destructive behavior) and which need to be addressed quickly (e.g., a rabbit that isn't eating or passing stool). These descriptions can be outlined in a simple, bulleted document or dichotomous key that is laminated and displayed next to each reception workstation. Categorize common presenting complaints into emergencies, priority cases, and cases that can wait until the first available appointment. Unless staff are well trained and there is low turnover (unlikely), these documents should be in language that can be easily understood by the layperson. This will help prevent downtime and inappropriately scheduled sick patients. To help provide optimum care for their patients, many clinicians choose to have most of their appointments open for the receptionist to schedule any animal but keep one or two reserved each day for urgent care cases that remained blocked until a predetermined time before the appointment (e.g., 24 hours). This will help avoid a really complicated, sick patient from being squeezed into an inadequate time slot.

No-shows

Appointments that do not show up can be extremely frustrating. The downtime associated with these no-shows is detrimental to staff morale and reduces your production. To help reduce the number of no-shows, call ahead, text, and email reminders can be helpful. It is important not to rely on a single form of communication. Some clients are very responsive to text messages, while others need a phone call reminder. Most digital practice software can automatically text or email clients at a predetermined time. If a client does not show up for an appointment, have a staff member call them when they are 15 minutes late. Phrase this conversation in a way that is not accusatory, but gets the message across with a phone call saying something like "We noticed that you are late for your 3:30 pm appointment with Polly; we just wanted to check and make sure everything was okay."

Even when everything had been done to remind the client of their upcoming appointment, no-shows are still inevitable. As a clinician, there are essentially only two different strategies to deal with no-show exotic appointments, which seem to be more common than no-show dog and cat appointments. The first is to expect that there will be the occasional no-show, and structure your appointments with this in mind. This might mean scheduling more cases than would be comfortable during each day, with the expectation that one or two will not show up. The benefit to this approach is that it is relatively easy to accomplish – just book more appointments. The downside of this approach is the feeling of being overloaded if everyone shows up. The second approach is to preempt the no-show appointment by: 1. Have the client pre-pay for the exam, which will serve both to increase incidence of clients showing up for the appointment and if they don't show up for the scheduled appointment some financial compensation for the prescheduled time is attained. 2. Charge a no-show fee, which is a typical response for other professions (e.g., dentists). If this concept is new to the practice, it may generate resistance from the practice manager or owner. These strategies are not as negatively received by clients as some might think, particularly a pre-pay policy.

With any approach chosen, a full appointment calendar is the goal. While at work, it makes sense to be seeing patients!

Productivity

Let's face it, it can be hard to be a high-producing associate when a large percentage of exotic animal patients is being seen. Owners and practice managers that value exotic pets recognize this, and hopefully want to support efforts to see more exotic patients. How can you both see exotics routinely and be productive?

First, you must charge for your time. Exotic animal exams, even wellness exams, often take more time than dog and cat patients. These patients are harder to restrain and often require more lengthy husbandry and diet discussions. Pre-empt these challenges by having your exotic animal client owners fill out a simple diet and husbandry form. With the acceptance of digital communication, this can be done before the client shows up at the office. Take the initiative to make a husbandry and diet form for the species or classes of animals seen (e.g., pet birds, backyard poultry, reptiles, amphibians, mammals) and make it a routine for receptionists to email these forms the week of the client's appointment. Then, train technicians or assistants to review these forms prior to presenting the case to the veterinarian so they already know what areas need further discussion and which are already adequately addressed. Congratulate them in the areas where they are succeeding early during the appointment. This will help strengthen relationships and will make diagnostic and treatment recommendations more palatable. Remember, veterinarians are working with clients as a team to help care for their pet. Finally, charge more for exotic animal examinations than for dog and cat visits. Again, these patients require more time and expertise than domestic species, so it is perfectly reasonable to charge more.

After (or before if it is better for a particular patient) the examination and initial discussion, don't make judgements on what the client will or won't be willing to do diagnostically for their pet. Recommend the appropriate diagnostic, offer an estimate if needed, then give them time to decide how they want to proceed. If sedation would benefit the patient, explain to the owner why this is the safest, most efficient, and thorough method for obtaining the diagnosis or performing a complete examination. Again, charge appropriately for these services. They require expertise that is not likely captured on a per volume of drug basis.

Finally, schedule rechecks as appropriate for the patient and condition, and do it before the client leaves the exam room. Let the client know that their pet should be seen again in 3-4 weeks and have the technician schedule them while in the exam room. This will ensure that your patients are well cared for and that the appointment calendar remains full.

Staffing

One of the areas that most veterinarians can do better is training and utilizing staff appropriately. This is particularly true for exotic animal patients. Find a technician or assistant that can appropriately discuss husbandry and diet, safely restrain, monitor sedation, and perform diagnostics on patients. Consider how a dog or cat is treated; the technician collects the blood, takes x-rays, and educates the client. There is no reason that the same can't be done for exotic animal patients; however, to get to this level of efficiency and competency requires training.

Once an individual with aptitude and ability has been identified, encourage them to improve their knowledge and skills. Take videos of proper restraint, blood collection, and client discussions and have staff review them. Ask the practice manager if they will support sending the technician to ExoticsCon or another conference where they can learn from other veterinarians and technicians. Many times, these staff members are excited to learn how to help care for these patients. Encourage them, provide verbal reinforcement, and show enthusiasm! If exotic-animal-capable staff are only available 2 days per week, use those days as your exotic-animal-heavy appointment days.

Equipment and Medications

This topic has been thoroughly covered in other lectures. There are some necessities that are worth mentioning. Examination and treatment equipment that you cannot do without include:

- Gram scale. At least two, with appropriately sized perches for birds.
- Magnification and light. These are critical in small patients and can be as simple as an otoscope light with magnification or hobby loupes.
- Drugs for species-appropriate sedation (e.g., midazolam, alfaxalone, butorphanol).
- Liquid antibiotics, NSAIDs, gabapentin, and parasiticides.
- Small climate-controlled oxygen cage.

Conclusions

Exotic animal medicine is exciting, and staff will respond to a veterinarian's enthusiasm for it. Following the suggestions outlined here will make the time seeing exotic animal patients more efficient, more productive, and more enjoyable!

How to Convince Your Local Emergency Practice to See Avian and Exotic Animal Patients

Megan Turner, DVM, and
Bruce Nixon, DVM

Session #3389

Affiliation: From Animal Emergency Hospital of North Texas, Grapevine, TX 76051, USA (Turner), Texas Avian and Exotic Hospital, Grapevine, TX 76051, USA (Nixon).

Abstract: A challenge for any practice that treats avian and exotic animal patients is how to provide after-hours and emergency care. Traditionally, this responsibility has fallen on primary care veterinarians meaning they are forced to take emergency calls at night and on the weekends. This leads to a poor work-life balance and burnout over time as these veterinarians never get a break from work. Ultimately, this results in poorer outcomes for these patients when those tasked with providing them with care are exhausted and overwhelmed. The first part of this presentation addresses the best strategy that primary care veterinarians can employ to convince their local emergency practice to start seeing avian and exotic patients in addition to dogs and cats. Tips like identifying and persuading the key decision maker of the emergency practice and encouraging buy-in from the entire emergency practice staff—not just a couple of veterinarians that might be willing to see these patients when they are on shift—are discussed in detail. The second part of the lecture provides ideas for how to support and train emergency practice staff once they have agreed to start treating avian and exotic patients.

How to Convince Your Local Emergency Practice to See Avian and Exotic Patients

Below is an outline of information covered in this lecture.

Introduction

How many of you see emergencies yourself?

How many use intern(s)?

- Foreshadow a unique way to use interns later in the lecture.

Just don't see them?

- State ethics requirements.

Other methods?

Identify Neo- s/he is "The One"

Every emergency practice has a person, usually singular, who has the ability to make decisions.

Positions.

- Medical or Hospital Director (especially corporate practices)
- Chief of Staff (especially shareholder-owned practices)
- Owner and privately owned practices
- Area or Regional Administrator (corporate practices)
- Highly motivated associate

The Pitch

Determine what motivates "The One."

- Money
  - Show actual financial data.
  - New income source.
  - Avian and exotic animal owners are also dog and cat owners.
- Expanding their referral base
  - If you are not already exclusively referring to a particular emergency practice, then promise to start doing so.
- Institution building
  - This is especially valuable for 24/7 emergency practices associated with a referral center.

Close the Deal

Train and support their staff.

If at all possible, send them personnel.

- ERs everywhere are desperate for staff: doctors, technicians and receptionists
- Obviously a part-time only arrangement
  - Enter agreement about poaching staff in either direction.
  - This involves weekends mostly but also day and night weekday work for 24/7 facilities.
- This is good for your staff too
  - Emergency practices tend to pay extremely well.
  - It also helps them sharpen their emergency skills for those patients who present to you in critical shape.

- Offer for their staff to do part-time shifts at your practice
    - Having several staff members who work both practices strongly bonds their commitment to seeing avian and exotic animal patients.

Unique solutions

- Sharing an internship with the ER

How to Support and Train Your Local Emergency Practice to See Avian and Exotic Animal Patients

Megan Turner, DVM, and
Bruce Nixon, DVM

Session #3391

Affiliation: From Animal Emergency Hospital of North Texas, Grapevine, TX 76051, USA (Turner), Texas Avian and Exotic Hospital, Grapevine, TX 76051, USA (Nixon).

Abstract: In the first session, ideas for how to convince your local emergency practice to start treating avian and exotic patients were addressed. This session dives into how to support and train the emergency practice staff. Many veterinarians are intimidated by avian and exotic animal patients. They received minimal, if any, training with them during veterinary school and they do not know where to start. Furthermore, they do not feel qualified to treat these unique and special patients. It is important to provide support to the emergency practice staff as they gain confidence. Simple strategies like providing onsite training, recommending textbooks and other resources they should have on hand, providing constructive feedback on cases, and making yourself available for phone consultations can be an extremely efficient way at bringing emergency practice staff up to speed. Remember, all levels of staff will need training including receptionists, technicians and veterinarians. You are the best person to teach the staff how to care for your patients. By putting in the time up front and creating open lines of communication between your hospital and the emergency practice, you will ensure that your clients and patients are receiving the same quality of care they would from you. In turn, you will get your free time back to enjoy life. Finally, you will need to have a strategic plan to maintain your emergency practice's commitment to these patients as these facilities typically have high staff turnover and are a frequent target of buy-outs and mergers.

How to Support and Train Your Local Emergency Practice to See Avian and Exotic Animal Patients

Below is an outline of information covered in this lecture.

Now you have them committed, so it is time for you to commit to them

Training

Reference materials

Phone access

Follow-up

Training

Explicitly state the goal of the emergency practice

- We are not avian and exotic animal experts, but we will try to help your pet to the best of our ability.
- Stress that the emergency practice doctors are not specialists.
- Provide phone scripts for receptionists and others.
- Help them create a list of patients they are not willing to see as client owned patients (e.g., primates, state protected wildlife, venomous snakes).

Focus on stabilization and support, not definitive diagnosis and treatment

- Help them understand that all species are approached the same way during evaluation
- Based on history and PE, create a problem list and then figure out where in the DAMNIT scheme the patient falls and develop your plan from there.
- Stress that it is not important that emergency practice veterinarians know every single illness that can befall every single species

If there is interest, then you can provide them with pathways for diagnostics and more intensive treatment

Stress that there is no reason to be intimidated

- Remind them that most owners are grateful that somebody is willing to help their pet

Handling labs

- Physical examinations for veterinarians and handling of the patients for technicians
- Can do on the same day, but might be best to split into morning for veterinarians and afternoon for technicians

Species

- At least two birds: one large and one small
- Exotic mammals
- Rabbit
- Ferret
- GP (possibly a rat as well)
- Hedgehogs

- Reptiles
- Lizard
- Snake
- Tortoise

Triage and supportive care

- Emphasize that these unique patients are even better at hiding illness than dogs and cats when they get stressed and have an adrenaline dump
  - Do not underestimate their degree of illness because they look pretty good at admission.
  - Listen to the history. If owners say the animal is lethargic and hardly moving at home, believe them.
- All birds should immediately receive heat and oxygen; it will never hurt them can help a lot

Top five problems for each species group

Reference materials

Your own personal, laminated cheat sheet

Books

The top five recommended texts are a starting place. They are listed below.

- Carpenter JW, ed. *Exotic Animal Formulary.* 5th Ed. St. Louis, MO: Elsevier; 2018. (Have more than one copy and always have the latest edition on hand.)
- Graham J, Doss G, Beaufrere H. *Exotic Animal Emergency and Critical Care Medicine.* Hoboken, NJ: Wiley-Blackwell; 2021.
- Quesenberry KE, Orcutt CJ, Mans C, Carpenter JW, eds. *Ferrets, Rabbits, and Rodents: Clinical Medicine and Surgery.* 4th Ed. St. Louis, MO: Elsevier; 2021.
- Morishita TY, Greenacre CB, eds. *Backyard Poultry Medicine and Surgery: A Guide for Veterinary Practitioners.* 2nd Ed. Hoboken, NJ: Wiley-Blackwell; 2021.
- Speer B, ed. *Current Therapy in Avian Medicine and Surgery.* St. Louis, MO: Elsevier; 2016.

Online support

- Veterinary Information Network (vin.com)

Phone support

- Phone support is essential. Someone on your staff must always be available for a phone consultation. Failure to do this two or three times will undermine all your work. Failure to do this means you will likely have to start over with another emergency practice.

Follow-up

- Quarterly personal visits (with food)
- Feedback on individual patients
- Positive and constructive criticism; nobody only wants to hear what they did wrong

Compounding Considerations and Making Modified Pharmaceuticals Work for Patients

Julia K. Whittington, DVM, and
Lauren R. Forsythe, PharmD, DICVP

Session #3411

Affiliation: From University of Illinois College of Veterinary Medicine, 1008 W. Hazelwood Dr., Urbana, IL 61802, USA.

Abstract: Commercial pharmaceutical products can feel like one-size-fits-no-one as professionals try to figure out how to treat patients, especially when trying to get an accurate dosage and struggling to the dosage into the patient reliably. Compounded medications are options that are made specifically to suit a particular patient's needs. Compounded medications are not without risk, which requires consideration of the risk-benefit ratio of utilizing these medications. This session identifies pros and cons of commonly compounded dosage forms, as well as the risks to be aware of when opting for compounded medications. The session covers the points that should be considered when reaching for a compounded product as well as considerations to ensure the product is the best option to meet the specific needs of the individual patient.

Veterinary Pharmaceutical Regulations

The U.S. Food and Drug Administration (FDA) oversees the development and approval of pharmaceutical agents in animals. The FDA approval process is where safety and efficacy are proven before a drug can be marketed. Two pieces of legislation provide the framework for this oversight: The Federal Food, Drug, and Cosmetic Act (21 CFR 9) and the Animal Medicinal Drug Use Clarification Act (AMDUCA) (21 CFR 530). The Federal Food, Drug, and Cosmetic Act defines what a drug is and gives authority to the FDA to approve and regulate drugs for humans and animals including safety standards, proper labeling, and purity of foods, drugs, vaccines, devises, and cosmetics. AMDUCA was passed into law in 1994 and permits veterinarians to prescribe or use certain approved animal and human medications in a manner that is outside of the conditions described on the product's label. This off-label or extra-label use is defined as the actual or intended use of a drug in an animal in a manner that is not in accordance with the approved labeling and includes such situations as use in a non-target species, use for a condition for which the drug was not intended, and deviations in dosing, frequency or route of administration, as well as subsequent withdrawal times. There are few drugs that are approved for use in zoological species, making extra-label drug use common. AMDUCA requires that any drug used in this manner must be administered under the oversight of a veterinarian, and that a valid veterinarian-client-patient relationship (VCPR) exist. Furthermore, extra-label drug use is limited to cases where the health of the animal is threatened, and significant suffering or death of the animal will result if it is not treated. If a drug is available that is labeled for the species or condition, the extra-label use of an alternate drug is not allowed. AMDUCA also regulates the compounding of approved animal and human drugs. Specifically, veterinarians and pharmacists may compound drugs for use in animals but only if there is not an FDA-approved product that can be used as labeled or in an extra-label manner.

Compounding Medication

Compounding medication has vastly expanded the availability of some medications that were previously impractical due to dose and/or dosage form for use in veterinary patients. Compounding is the process of mixing, combining, or altering ingredients to create a medication that meets the unique needs of a specific patient. Compounding includes the combining of two or more drugs. Compounded medications are not FDA approved.[1] Diluting or reconstituting medication in accordance with the approved labeling is not considered compounding. The United States Pharmacopeia (USP) is an independent, scientific nonprofit organization that collaborates with the FDA to set standards for safe, quality medications. The USP compounding standards are USP chapter <795> for compounding nonsterile preparations, USP chapter <797> for compounding sterile preparations, and USP chapter <800> for compounding hazardous drugs.[2] With the wide variety of species and the size variety within each species, it is not possible to have approved products available to meet the needs of every, or even most, patients. Compounded medications are relied on to fill the gaps where manufactured products fail to meet the needs. In veterinary medicine, compounded medications are used to treat animals when an FDA-approved drug is not available in the required dosage form, strength, or flavor. FDA-approved drugs can be used for compounding as long as all criteria for extra-label drug use are met. Compounding using drugs or materials such as bulk powder that are not FDA-approved is only allowed in certain circumstances.[3] Compounded products can improve patient care in several ways, but they also have inherent risks. Therefore, it is important to limit compounded medication prescribing to instances where manufactured products are not a feasible option and consider the risk-benefit analysis of using a compounded product.

Benefits Of Compounded Medications

There are five main benefits of compounding medication for veterinary species: dosage form, flavoring, strength, availability, and elimination of toxic ingredients. Modifying the dosage form may facilitate medication administration. Veterinary patients, and specifically zoological species, aren't always willing to take medications and forced administration may be challenging. Therefore, compounding to modify the dosage form to facilitate administration is warranted. Some strategies that are facilitated by compounding include a more desirable dosage form such as a liquid instead of a tablet or a transdermal instead of an oral medication, a medication flavored to the patient's preference, or a treat formulation that encourages consumption of the drug.

In addition to dosage form, medications are compounded because of the need for customized dosage strength. For example, enrofloxacin is commonly used in rabbits but is not commercially available as an oral liquid and available tablets are not practical. Compounding medications is also indicated when a medication is needed and is unavailable commercially or no longer being manufactured. Cisapride is a commonly used medication in veterinary patients; however, it was pulled from the human market in 2000 due to its cardiac side effect profile in humans. Therefore, compounding is the only way to obtain this medication for use in veterinary patients. Similarly, with the ever-present supply chain issues, medications that are unavailable for long stretches of time may need to be compounded. Compounding provides a mechanism to continue therapy until the manufactured product is available.

Risk of Compounded Medications

While there are benefits and indications for compounding medications, it is not without risk. One risk is that the compound contains either more or less active ingredient than stated on the label. The compounding process has several points where ingredient loss can occur to the point that even though the correct amount is weighed or measured, the final product does not contain enough active ingredient. Errors during the compounding can also occur, with a decimal point transposed or a scale read incorrectly, leading to too much active ingredient. Errors also include the wrong medication being put into the compound. Many compounding ingredients are white powders. A stringent verification process is necessary to prevent an incorrect ingredient or an incorrect strength of an active ingredient from accidentally being used.

Compounded medications may be unstable. If medication compounding is done in practice, the best way to minimize the risk of an adverse event is to take the simplistic approach.[4] Minimal changes to the product, such as diluting a solution or adding a flavor, are less likely to have a negative impact on the final product. Trissel's Stability of Compounded Formulations[5] is a resource that compiles all currently available stability information on drugs in compounded oral, enteral, topical, ophthalmic, and other specialized preparations. The information contained is based on human medicine, but it has useful data on compound stability for the anticipated duration of use and proper storage of compounds. If a compounded medication is unstable, the active ingredient degrades before the beyond-use date (BUD) is reached. The USP standards provide default BUDs that are generally short enough that drug degradation and microbial contamination will not occur during the use period of the product; however, that is not guaranteed. There is also concern about whether extended dates are based on true stability-indicating assays or if they are based on potency over time. The latter is not sufficient for determining extended BUDs.

Efficacy and safety are two concerns one must consider when weighing the risks of compounding medication. The compound may be prepared correctly but not produce the intended effect in the patient. One cause may be that the bioavailability (amount absorbed) is too low. Likewise, the compound may be prepared correctly but there is something that makes it unsafe for the patient. This could be an inactive ingredient, dose or route that has a high potential of causing harm to a patient.[6]

Non-sterile compounds are exactly as the name implies; they are not sterile. They still need to have adequate preservatives or short-enough dating that excessive microbial growth does not occur while the medication is in use. Sterile products, on the other hand, must be sterile. While a sterile environment that complies with USP chapter <797> greatly increases the probability that the final product will be sterile, there is still the potential for contamination to occur resulting in a final product that is not actually sterile. Even if batch testing is done and a certain percentage of the batch is determined to be sterile, there is still the potential that there is contamination in an item that was not tested as contamination is not uniformly distributed.

Controlling the risks of medication compounding

AMDUCA stipulates that the first choice of pharmaceutical for any patient be an FDA-approved drug proven to be safe and effective and used as indicated on the label. In veterinary medicine, the first choice is often not feasible. The second choice is to use an approved drug, but in an extra-label manner. When the use of an approved drug is not an option, that is when AMDUCA allows for compounding medication using a commercial product. AMDUCA does not address the use of compounding using bulk drug. By legally requiring the use of approved products if possible, AMDUCA prevents unnecessary compounding while protecting the integrity of the drug approval process. Compounding may be performed by a licensed pharmacist in a state-licensed pharmacy, federal facility, or by a physician or veterinarian. In 2013, the Drug Quality and Security Act established the provision that compounding can be performed in outsourcing facilities if under the direct supervision of a licensed pharmacist and the facility is registered with the FDA.[1]

While AMDUCA legalizes compounding by manipulating approved products, it does not address compounding from bulk chemical. Bulk chemical is necessary in some cases such as when the approved product is on backorder, or when the active ingredient does not exist as an approved product. There are also cases where a high-quality formulation cannot be prepared from a commercial product, such as when the only approved product is a chewable tablet. Therefore, although bulk chemical compounding is technically illegal, it is necessary under certain circumstances. Because of this, the FDA has released GFI #256 to describe its expectations and when it plans to use regulatory discretion in the enforcement of bulk chemical compounding. While a full discussion of GFI #256 is beyond the scope of this session, Figure 1 provides a flow chart of what to expect regarding clinical impact.[7]

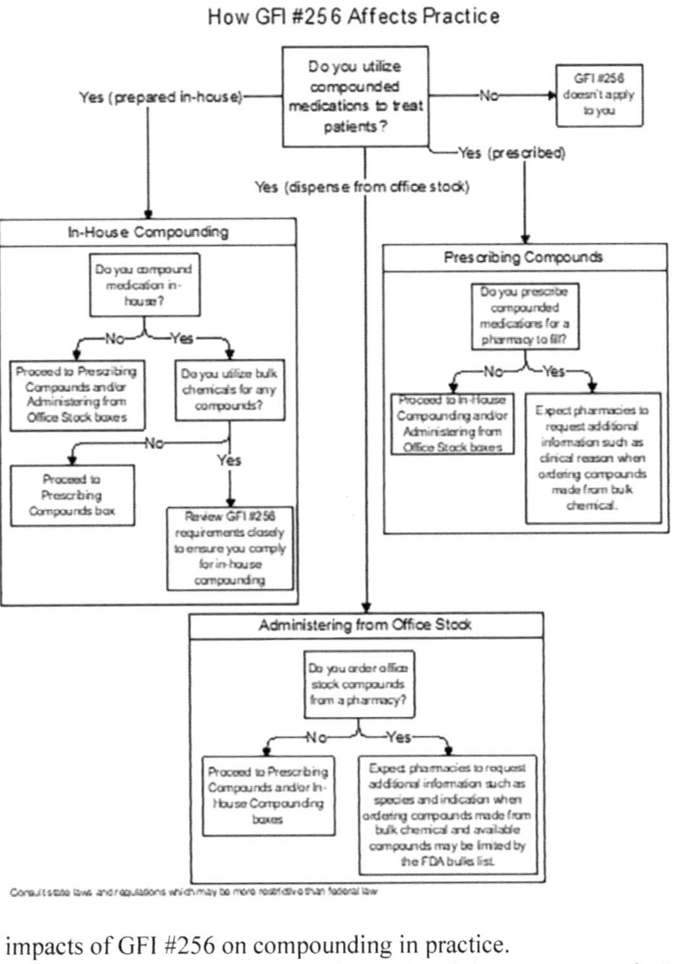

Figure 1. Expected impacts of GFI #256 on compounding in practice.
Source: DVM360. Forsythe L. What is GFI #256 and how does it impact my practice? Available at: https://www.dvm360.com/view/what-is-gfi-256-and-how-does-it-impact-my-practice-. Accessed April 3, 2023.

The risks associated with compounded medications are partially dependent on whether the compound is prepared in a 503a (traditional pharmacy) or 503b (outsourcing) facility. Traditional pharmacies should follow USP standards when compounding medications which are designed for small scale preparation of compounded products. These pharmacies are not held to the current good manufacturing practice (CGMP) requirements and are subject to less USP standards as they are enforced by the state(s) regulating them. This results in a higher risk of an adverse event when using compounded medication from these facilities. Conversely, compounds prepared in an outsourcing facility are held to CGMP requirements which are the same requirements that drug manufacturing facilities must adhere to. The final testing that is required by CGMP requires testing a percentage of each batch of compounded drug prior to release. This testing verifies the correct identity and potency of the active ingredient, as well as sterility if the product is marketed as sterile. There are also requirements that the outsourcing facility has proven the preparation process yields a consistently uniform batch, as well as that the formulation has sufficient stability for the expiration date given. For sterility, there is potential that the tested products are sterile but other products in the batch are contaminated. The large-scale manufacturing process designed to ensure consistency and repeatability between batches decreases the risks of an adverse event when using compounded medications from these facilities; however, because 503b outsourcing facilities are manufacturing compounds and not FDA-approved drugs, there is still a risk of lack of safety and/or efficacy. Safety and efficacy proof is required for the FDA-approval process, and not required for 503b compounds.

Compounding in Practice

Despite the risks, the benefits of compounding medications to meet the needs of veterinary patients often outweigh the risks. Knowledge of both state and federal compounding laws is essential for any veterinarian intending to compound medications. Most veterinarians receive little or no training in medication compounding as part of their veterinary training,[3] and veterinary practitioners report interacting with dispensing and compounding pharmacists more than any other professional collaboration.[8] A survey of zoo veterinarians found that only 25% were comfortable compounding medications for their patients. The majority of respondents stated that the reasons for not compounding medications included a lack of training, lack of access to ingredients, and the availability of commercial products. In situations where a compounded medication was needed for patient care, these veterinarians sought a pharmacist or compounding pharmacy to fill this need.[9] Another survey of over 1500 veterinary practitioners indicated that most used compounded medications in the treatment of their patients. Over 60% of the respondents indicated that they compounded medications in their practice, with the remainder utilizing compounding pharmacies. Less than 25% of the medications dispensed in primarily dog/cat practices were compounds, as compared to 50-95% of the medications used by practitioners treating companion zoologic species. Despite the significant number of practitioners engaged in medication compounding, relatively few understood their state laws regarding this practice.[3] State veterinary boards have regulatory oversight of veterinary practitioners through enforcement of their state practice act(s). Many states do not have specific compounding regulations for veterinarians.

Unlike in human health care, veterinarians can dispense medications directly to a client for a patient without a pharmacy prescription. Therefore, most veterinary prescriptions that are written are for compounded medications or for products that the veterinarian chooses not to stock. Any pharmacist can dispense a prescription medicine at the direction of a veterinarian if there is a valid prescription. The prescription becomes the legal documentation of the medical order. If the drug proves to be unsafe, or is not efficacious, both the pharmacist and veterinarian share the responsibility when it comes to compounded medications.[4] Therefore, it is important for the veterinarian to maintain documentation regarding the clinical need for a compounded medication and that the client was made aware of the risks and benefits of the compounded medication.

A veterinary pharmacist focuses on animal health in collaboration with the veterinarian, and veterinary pharmacists have a deeper understanding of animal health which translates to better patient care.[9] In many instances, utilizing a veterinary pharmacist to compound medications for veterinary patients is preferred because of their access to cost-effective drugs and advanced training in compounding. This has the added advantage of reducing preparation errors, contamination and having access to formulations that require specialized equipment to make.[3]

When partnering with a compounding pharmacy, take the time to investigate the facility and speak to the pharmacist. Ensure that they are experienced with veterinary compounds, that they understand animal toxicities, and that they utilize veterinary specific formulary information.[10] Compounding pharmacies must be licensed or registered to practice in the resident state and the state of dispensing. In addition, the pharmacy must follow USP standards as outlined in applicable state pharmacy regulations and adhere to criteria for compounding set by AMDUCA. Active pharmaceutical ingredients must be used in accordance with GFI #256, should be obtained from FDA-registered and DEA-registered suppliers, should be substances of USP grade, and should be accompanied by a valid certificate of analysis. Ideally, the pharmacy should be accredited by the Pharmacy Compounding Accreditation Board.[11] The key is to consider all options and have an open discussion with the client so they are aware of the benefits and risks of compounded medications and can participate in considering if a compounded medication is the best option in that particular case.

Resources described in Table 1 can be used for additional information on compounded medications.

Table 1. References for information on medication compounding in practice.

Reference	Web link	Information available
USP website	www.usp.org	Background information on USP. Current expert committee members. Updates on chapter updates in progress and key dates for comment submission. USP publications available for purchase.
USP Compounding Compendia	Purchase at https://www.usp.org/products/usp-compounding-compendium	Compounding chapters. Compounded preparation monographs.
State Pharmacy Boards	List of all websites found at: https://nabp.pharmacy/about/boards-of-pharmacy/	State regulations.
State Veterinary Boards	List of all websites found at: https://www.aavsb.org/public-resources/find-regulatory-board-information	State regulations.
AVMA	www.avma.org	News and position statements on compounding legislation. Summary of state office use laws. AMDUCA ELDU requirements.
FDA	www.fda.gov	GFIs and CPGs. Warning letters issued to compounding pharmacies. Current and resolved drug shortages. New drug approvals. Approved animal drugs and marketing status. Approved human drugs and marketing status. Appropriate drug disposal. Indexed drugs for Minor Use/Minor Species. Adverse event reporting. Compounding risk alerts.
DailyMed	https://dailymed.nlm.nih.gov/dailymed/	Package inserts for FDA-approved human and veterinary medications. Note: Package inserts different depending on the manufacturer.
AMDUCA	https://www.ecfr.gov/cgi-bin/text-idx?SID=054808d261de27898e02fb175b7c9ff9&node=21:6.0.1.1.16&rgn=div5	ELDU requirements.
DQSA	https://www.fda.gov/drugs/human-drug-compounding/text-compounding-quality-act	503a versus 503b pharmacies.
cGMP	https://www.accessdata.fda.gov/scripts/cdrh/cfdocs/cfcfr/CFRSearch.cfm?CFRPart=211	Requirements for medications prepared by 503b pharmacies and for manufacturing approved medications.
GFI #256	https://www.fda.gov/regulatory-information/search-fda-guidance-documents/cvm-gfi-256-compounding-animal-drugs-bulk-drug-substances	Current FDA guidance on when compounding from bulk chemicals is appropriate.
Non-sterile and sterile compounding for the veterinary practitioner	Textbook only; currently available for preorder through Barnes and Noble	Guide to all aspects of prescribing compounds as well as considerations for preparing simple compounding in house.

References

1. U.S. Food and Drug Administration. Compounding and the FDA: Questions and Answers. Available at: https://www.fda.gov/drugs/human-drug-compounding/compounding-and-fda-questions-and-answers. Accessed March 18, 2023.

2. United States Pharmacopeia. USP Compounding Compendium. Available at: https://www.usp.org/compounding. Accessed March 19, 2023.

3. Gochenauer AE, Rishniw M, Holmes ER, Forsythe LR. The current landscape of veterinary compounding in the veterinary setting. *Int J Pharm Compd.* 2019;23(5):428-433.

4. Boothe DM. Veterinary compounding in small animals: a clinical pharmacologist's perspective. *Vet Clin North Am Small Anim Pract.* 2006;36:1129-1173.

5. Trissel LA, Ashworth LD, Ashworth J. *Trissel's Stability of Compounded Formulations.* 6th Ed. Washington, DC: The American Pharmacists Association; 2018.

6. Davidson G. Veterinary compounding: regulation, challenges, and resources. *Pharmaceutics.* 2017;9(1):5.

7. DVM360. Forsythe L. What is GFI #256 and how does it impact my practice? Available at: https://www.dvm360.com/view/what-is-gfi-256-and-how-does-it-impact-my-practice-. Accessed April 3, 2023.

8. Kustritz MVR, Molgaard LK, Tegzes JH. Frequency of interactions between veterinarians and other professionals to guide interprofessional education. *J Vet Med Educ.* 2013;40(4):370-377.

9. Besley N, Browne P, Park M, et al. Pharmacists in zoos? A qualitative study investigating the potential for pharmacist involvement in wildlife health care in Aotearoa New Zealand. *J Am Pharm Assoc.* 2023; 63(3):825-831. doi:10.1016/j.japh.2023.01.008.

10. Anderson M. Compounding: myths and facts. *VMX.* 2021;38-40.

11. Powers LV, Davidson G. Compounding and extralabel use of drugs in exotic animal medicine. *Vet Clin North Am Exot Anim Pract.* 2018;21:195-214.

Recession Proof Your Exotic Animals Practice: For Owners and Associates

Len Donato, VMD, DABVP (Avian Practice)

Session #3294

Affiliation: From Radnor Veterinary Hospital Wayne, PA 19087, USA, the Department of Clinical Sciences and Advanced Medicine, The University of Pennsylvania School of Veterinary Medicine, Philadelphia, PA 19104, USA.

Abstract: During the COVID-19 pandemic, clinical veterinary medicine did very well financially. As time moves on from the pandemic, veterinary medicine is seeing that client visits are beginning to drop to pre-pandemic levels. When the number of visits decreases, a decrease in gross revenue is typically observed. Veterinary medicine has not experienced that yet because inflation caused fees to increase significantly, thereby increasing the average client transaction (ACT). Raising fees alone will not sustain meaningful growth, and in fact contributes to practice stagnation and regression. Understanding a variety of strategies to promote more visits and increase ACT will benefit the financial health of the hospital overall. Exotic animal practitioners have to be savvier than dog and cat counterparts as a majority of the patients do not get vaccines, flea and tick medication, and heartworm prevention. Those products and services help gross revenue by increasing transactions and ACT. Strategies affecting the number of transactions and the ACT at the same time will benefit the practice owner, the non-owner veterinarians, and staff. Increasing the number of transactions involves veterinarians maximizing time management, proper follow up and reminders, along with a comprehensive internal and external marketing campaign. ACT is addressed by appropriate fee increases, managing lost fees, and expanding services and skills. Knowing these strategies and how they affect gross revenue, veterinarian salary, and the health of the clinical environment will minimize the effects of a recession and will likely result in growth financially and professionally.

Introduction

Recessions are defined by the National Bureau of Economic Research (NBER) as a significant decline in economic activity that is spread across the economy and last for more than a few months.[1] Historical data demonstrates the most recent prolonged recessions were The Gulf War Recession (7/90-3/91), The Dot Com Recession (3/01-11/01), The Great Recession (12/07-6/09), and The COVID Recession, which is on-going.[2] When recessions are prolonged they are defined as depressions, but there is debate as to definition.[1,2]

It is widely thought that veterinary medicine is recession-proof, particularly with the most recent experience. In reality there have been industry-wide effects from the beginning of the COVID-19 pandemic. Food-animal supply chains were dramatically affected,[3] zoo income was reduced by lack of visitors and donations,[4] and a large number of veterinarians retiring from the profession or going part time has even made the everyday news.[5] The Great Recession in 2008 resulted in decreased gross revenue for veterinary practices and a significant decrease in net that was previously 35-45% and fell to 10% or below for some practices. The effects of that are still felt today as net revenue for a non-specialty practice reached 15%.[6] Non-owner veterinarians also feel recessions as numbers of clients typically decrease as well as their number of visits (N) and how much they spend at each visit (average client transaction, ACT). Pro-sal is a common method to pay associates which is based on revenue generated by the associate. Since N and ACT decrease, that affects additional pay over base salary. This makes Pro-sal a bit more risky for the non-owner veterinarian during a recession. Also, if a practice has a decrease in net revenue, there are fewer resources that could go to additional staff, equipment, or capital improvements for the hospital building, all of which make for a better-quality hospital and work environment.

Everything feels good financially right now because of the COVID/post COVID bump, but data show the number of visits decreased by 4% in 2022.[7] As seen from the recessions listed above, it is not if a recession occurs. The questions are: how many recessions will be experienced in a career, and how severe will they be? What can be done to maintain income during those times? There are strategies to make sure a similar drop-off is not seen as has been seen with other recessions. It is better to implement and practice the strategies now, rather than when the economic environment is not healthy.

Understanding Profitability

Veterinary practices should be managed to be profitable. This is a function of gross revenue and expenses to run the practice. While this simplifies what a practice needs to achieve, executing is a big challenge. Focusing on gross revenue, but not ignoring the expenses, will make for a profitable practice. Expenses need to be monitored as most will be within a percentage range of gross revenue. As an example, as the number of staff increases, so does payroll. The goal behind the increased staff is to accommodate more patients, which increases gross revenue. In this example, as the expense of staff increases so does revenue. There are two exceptions to expenses where a difference can be made; these are discussed later. The focus is on revenue generation.

Managing gross revenue has the greatest impact on profitability. To understand income there is a simple formula:

$$\#\text{ of transactions (N)} \times \text{average client transaction (ACT)} = \text{gross revenue}$$ [8]

It is important to understand the components that make up both factors in the equation, and the best ways they can be affected to produce higher income or gross revenue. Increasing ACT has a greater impact on gross revenue than N. Both are important to manage. This is why specialists typically generate more than general practitioners as one transaction could be in the thousands regularly.

One of the easiest ways to increase revenue through ACT is to raise prices. This has occurred in the last few years because of the extreme increase in N during the pandemic. It was meant to handle overall increases in payroll, increases due to inflation, and to become more profitable as a practice. The continued price increases during a recession eventually catch up and surpass what owners are willing to pay.[7] A precipitous drop in N occurs because the high prices keep clients from scheduling. This was observed in the Great Recession as the years before were quite prosperous. This isn't to say prices should not be raised. It should be a thoughtful process understanding the value of services, inventory costs, lab fees, and other expenses. Raising fees to weed out clients will result in having a hard time during a recession.

How to properly raise ACT

Yes, raise fees

As outlined above, make this a thoughtful process with good intentions.

- Exams and surgery, generally value based.
- Lab fees, generally 2 × cost ± a technician fee. Monitor monthly as fees from reference lab or materials for in-house equipment may increase suddenly.
- Inventory, generally 2 × cost ± a pharmacy fee. Monitor invoices as they arrive for fee increases.

Practice quality medicine

This sounds obvious. This is the greatest part of our careers, as not only does that satisfy our desire to be an exotics veterinarian, but it also increases ACT. It allows us to offer a variety of services. Consider adding a new service this coming year.

- Preventive care.
- In-hospital diagnostics.
- Advanced imaging, radiology, ultrasound, CT, and endoscopy.
- Soft tissue surgery and orthopedics.
- Behavior consultations.
- Senior wellness programs.
- Many others; you know what you like!

Address missed fees

This refers to fees not charged by the veterinary practice for services that were performed. This is a long-standing problem that can lose a practice 5-10% of gross revenue a year.[9] A practice that generates $2,000,000 in gross revenue can be missing up to $200,000/year.

- Identify areas where fees may be missed and address.
- Institute a checks-and-balance system that works for the doctors and staff.

How to properly raise N

Time management

Managing time sounds like it is obvious and intuitive. It is not. It must be practiced like sports or playing an instrument. Having more time allows you to see more cases or at least offer more services, which can affect ACT too. So how should time be managed?

- Delegate. This is an art and a skill. The art is you must be able to trust others. The skill is that you have to follow up. Even though you delegate a task, does not mean you are not to check on said task and provide feedback. It is the most forgotten step of delegating. As a veterinarian one should perform exams and surgery and follow up with the clients on patient care. No one else can do those tasks. Everything else can be delegated, but you still must check. Take a class on delegating if needed.
- Have the right people. In the practice have proper staffing type, numbers, and skills. Your external team is just as important. These are the payroll company, accountant, attorneys, and financial planner. Performing these financial and legal tasks on your own will end just as badly as if you tried to be the only one handling all the patients.

- Evaluate scheduling. This becomes a topic of debate. There are many scheduling strategies ranging from straight booking of timed appointments to high density schedule with catch-up slots. There are even schedules with overlapping appointments considering support staff contact and doctor contact for each patient. Scheduling considers the veterinarians and staff as a whole entity. Evaluate your schedule to determine if it is efficient and fits the practice philosophy.

- Be paperless or paper light. Time savings looking for files alone is worth every penny spent on an electronic medical record system. This also aids in addressing missed fees.

Client reminders

These are not for just annual examinations and vaccines. Set reminders for blood work when monitoring chronic patients, behavior visits for young birds to make sure clients get into good habits early, examination and husbandry reviews for reptiles before winter, and anything important to the specific patient.

Recall system

Call clients far more than if they were in for a sick visit. Call clients or have staff call for weight checks, diet conversions, recheck reminders, blood work monitoring, setting spay/neuter, post surgery check in, and whatever you think is remotely important. It helps with owner compliance, checks on the pet, increases visits, and builds trust, which results in return visits.

Marketing

Managing a successful veterinary practice involves strategic planning within and between several functional areas of the business; just a few of those areas include finance, human resources, inventory, operations, and marketing. The hospital's marketing strategy should provide a clear depiction of the specific elements that set your practice apart from the crowd. The marketing strategy's key elements ideally include defined internal and external marketing components, brand messaging, print and digital-based marketing tools, word-of-mouth communications, and social media. All these marketing mediums work together to reach and engage different types of consumers that would benefit from your services. Having a general understanding of the different types of marketing tools and their costs will help a practice owner determine specific marketing mediums to focus on, relative to their overall return on investment.

Expenses

As previously mentioned, expenses are not the focus when evaluating profitability. They cannot be ignored and must be managed, but your profitability will not dramatically change by continually switching companies for electricity. An office manager or hospital administrator will make sure those expenses are kept in check, but not a lot of resources should be placed on expenses with a couple of exceptions.

Payroll

The cost of labor should be calculated as a percentage of gross revenue. It will be close to 20% for veterinary salaries and 20% for non-veterinary salaries, totaling 40% of gross revenue.[10] The office manager or hospital administrator should monitor overtime. That is the one area payroll can get out of hand. If overtime is managed and payroll is above 40%, review the amount and type of staff to find where costs are going.

Inventory management

Inventory is an expense that can get way out of control. There are a few ways this can happen. The first is that there are no controls over how much is ordered or fear buying. There can be a lot of anxiety of running out of a product. Use ordering history to determine how much is needed over a given period. While taking longer than it has in the past, products still arrive in a fairly timely manner. Learn the patterns of delivery from your distributors. The second is that an owner of a practice may just want what they want. This can be challenging to overcome as everything and anything gets ordered. Cost of goods should be about 17.5% of revenue.[11] If it is above that, look for excess inventory, duplication of medications, lack of use of an online pharmacy, and even theft.

Conclusions

These outlined concepts are not challenging to understand. Implementation is the most challenging part. The strategies and tactics above will prevent having a challenging time during a recession. They require practice while growth and gross revenue is high so when it is not, you are prepared.

References

1. National Bureau of Economic Research. Frequently Asked Questions. Available at: https://www-nber-org.proxy.library.upenn.edu/research/business-cycle-dating/business-cycle-dating-procedure-frequently-asked-questions. Accessed March 18,2023.

2. Rodeck D. What is a recession? *Forbes.* 2022;7.

3. Hayes DJ, Schultz LL, Hart CE, KL Jacobs. A descriptive analysis of the COVID-19 impacts on U.S. pork, turkey, and egg markets. *Agribusiness.* 2021;37(1):122-141.

4. Fine L, Barnes C, Niedbalski A, Deem S. Staff perceptions of COVID-19 impacts on wildlife conservation at a zoological institution. *Zoo Biol.* 2022;41(3):234-243.

5. Columbia Broadcasting System (CBS). Available at: https://www.news9.com/story/62900baa45556307273f185c/future-of-veterinarian-care-uncertain-as-many-begin-to-retire. Accessed March 18, 2023.

6. Felsted KE. What is your veterinary clinic worth? *DVM360.* 2021;53(3):54.

7. Zirkle K. Reading the signs of healthy veterinary practices. *Trends.* 2022;15(1):26-28.

8. Donato L. Avian practice management: the art of delegating. *Proc Annu Conf Assoc Avian Vet.* 2008.

9. Fuhrman J. Don't misfire on missed charges. *Today's Vet Bus.* 2021;5(1):46-47.

10. O'Neil T. 5 Ways to better manage veterinary hospital labor. *Today's Vet Bus. 2019;2(6):12-16.*

11. American Veterinary Medical Association. The Economic State of the Veterinary Profession 2019. Available at: https://ebusiness-avma-org.proxy.library.upenn.edu/files/productdownloads/HR-Secured-ECO-StateofProfessionReport-20190927.pdf. Accessed March 18, 2023.

The Opioid Epidemic and Why a Human Health Crisis Matters in Veterinary Medicine

Julia K. Whittington, DVM, and
Lauren R. Forsythe, PharmD, DICVP

Session #3410

Affiliation: From University of Illinois College of Veterinary Medicine, 1008 W. Hazelwood Dr., Urbana, IL 61802, USA.

Abstract: The complexity of controlled substance regulations as well as a limited understanding about the prevalence of drug diversion can make compliance difficult. This session covers background information about the opioid epidemic to provide perspective on why the regulations are in place. It also includes information about prevalence and signs of in-hospital drug diversion. Human statistics are tied to veterinary medicine with specific case examples, and considerations when utilizing controlled substances to treat animal patients. Appropriate precautions to avoid supplying controlled substances to a drug-seeking client are also covered.

The Opioid Crisis

The opioid crisis in the United States is not a new problem. In fact, it has been brewing since the introduction of morphine in the late 1800s. Throughout history, there are examples of addiction resulting from the popularity of opioids, sedatives, and stimulants, and this popularity has been largely driven by pharmaceutical companies that embrace the popularity to enhance their profits. Along with addiction and the associated problems, an evolution of regulatory oversight culminating in our current tiered structure of controls and oversight has occurred. While many prescribers were content to decrease their prescribing of these addictive medications, some resisted the restrictions and continued as they always had. As a result, history has shown that regulations still leave gaps and are only as effective as their enforcement. One example is the case of Dr C. who was a physician with an illustrious career practicing in Madison, WI, near the University of Wisconsin from the 1960s until his retirement in 1979. He became known for readily prescribing Quaalude (methaqualone), to tense and restless college students during a time when other practitioners had largely ceased prescribing this drug due to the addiction concerns. Every college student that wanted a sedative or a stimulant (or both) learned to go to Dr. C and say that they had problems sleeping. They would then receive prescriptions for whatever they wanted with directions to one of the three pharmacies in town that would readily supply the medications. While this case was dramatic, it is only one example in the history of addictive drugs where a respected doctor served as a legal source of controlled substances.[1]

From a more recent perspective, the current opioid crisis has been progressively worsening since the 1990s (Fig 1). While veterinarians were not the original source for obtaining controlled substances, as controlled substance restrictions have continued to tighten, those addicted to a prescription medication have had to seek alternative options. Veterinary medicine presented a possible alternative. There are a variety of reasons for this; however, it is important to understand the significance of the current opioid crisis first.

Concerning statistics

Between 1999 and 2020, more than 932,000 Americans died of a drug overdose. While this is concerning on its own, it is also important to note that the problem is getting worse each year (Fig 1). Overdose deaths in the United States were six times higher in 2017 compared to 1999, with two-thirds of the 2017 deaths attributed to opioids. In 2020, overdose deaths increased 38% over 2019 deaths, with two-thirds again contributed to opioids. While this problem is not specific to the United States, the United States does account for 80% of the world's opioid consumption and 99% of the world's hydrocodone consumption despite only being 4.4% of the world's population. The result is 187 Americans dying each day from opioid overdoses (Fig 2).[3]

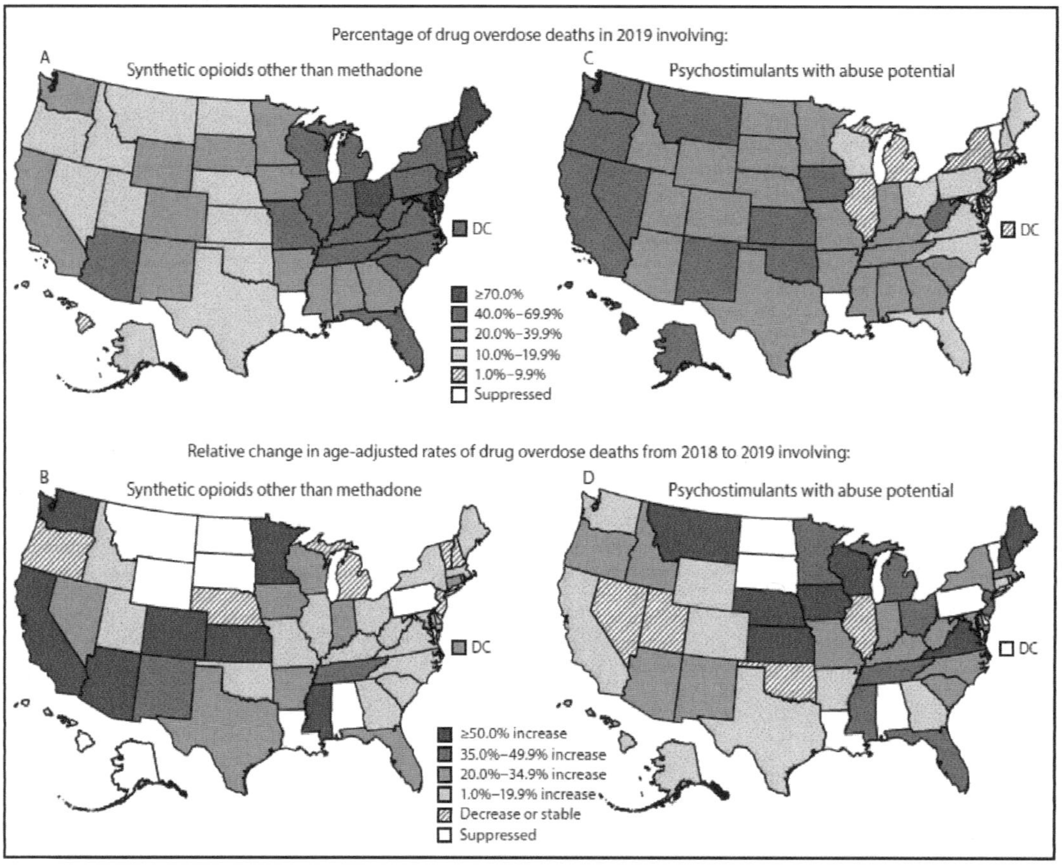

Figure 1. Drug overdose deaths by state and drug class in 2019.
Source: Mattson CL, Tanz LJ, Quinn K, et al. Trends and geographic patterns in drug and synthetic opioid overdose deaths-United States, 2013-2019. *Morb Mortal Wkly Rep.* 2021;70(6):202-207.

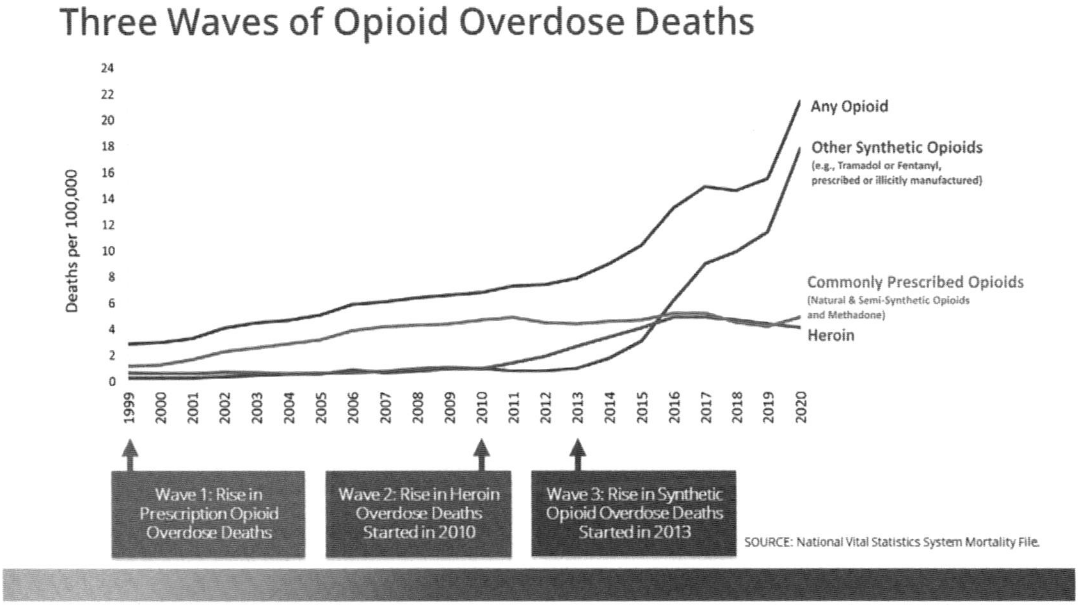

Figure 2. The three waves of the opioid crisis.
Source: Centers for Disease Control and Prevention. Understanding the Opioid Overdose Epidemic. Opioids. Available at: https://www.cdc.gov/opioids/basics/epidemic.html. Accessed March 26, 2023.

There are several studies looking at the prevalence of drug diversion in human healthcare. Based on the knowledge that the incidence of drug diversion in healthcare reflects that of the general population but with increased access, the authors assert that 10% of healthcare workers are abusing drugs at any given time. During their career, 15% of pharmacists, 10% of nurses and 8% of physicians will abuse drugs at some point. Despite knowing this is a problem and having stringent regulations designed to address it, 80-93% of diversion goes undetected.[4]

A 2021 study in human healthcare found that 96% of providers are concerned about hospital drug diversion, with 73% indicating that they believe most diversion goes undetected. Of the executives surveyed, 82% indicated that they personally know someone who has diverted drugs; however, 94% are confident or very confident in the effectiveness of the drug diversion program at their facility.[5]

While it is common knowledge in human healthcare that drug diversion in healthcare professionals is a significant problem, there is a disconnect between healthcare professionals believing it's a problem in general and believing that it could occur at their facility. In a 2019 study of over 650 hospital executives and providers, 85% of providers were concerned about hospital drug diversion, and 50% reported observing suspicious activity; however, despite the level of concern and observation of suspicious activity, only 20% believed that drug diversion was a problem at their facility.[6]

Xylazine, a new player

Recently, a new threat emerged that is worsening the opioid crisis. The presence of xylazine in illicit fentanyl is quickly becoming more prevalent and poses significant problems in humans. Xylazine is being added to fentanyl by drug dealers to boost the effects of fentanyl and extend the associated high. Since the effects of fentanyl are short-lived, the ability to extend the effects is desirable in the illicit drugs market; however, xylazine is not safe for humans. In the 1960s, xylazine was evaluated for potential use in humans, but proved inappropriate due to severe CNS depression.[7] Naloxone is not an effective reversal agent for xylazine since it is not an opioid, and the veterinary-approved reversal agents are not available for use in humans. In addition, xylazine can potentiate the effects of opioids such as sedation and respiratory depression, making naloxone reversal of opioid overdoses less effective. This occurs through the CNS depression from xylazine blunting the response to hypoxia associated with an opioid overdose. In addition to the systemic effects of xylazine, eschars commonly occur and can lead to amputation.[8] The result is xylazine making the opioid epidemic worse through increased mortality and addiction to xylazine in addition to opioids.[7]

Given the immediate public health crisis resulting from xylazine, the federal government is moving forward with scheduling this drug. Permanent scheduling is a long process where the Attorney General, Drug Enforcement Administration (DEA), Department of Health and Human Services (DHHS) and the Food and Drug Administration (FDA) place substances into the controlled substance schedules based on eight factors, which are collectively termed the 8-Factor Analysis. These eight factors are the following:[8]

1. Actual or relative potential for abuse;
2. Known scientific evidence of pharmacologic effects;
3. Current scientific knowledge of the substance;
4. History and current pattern of abuse;
5. Scope, duration, and significance of abuse;
6. Risk to public health;
7. Psychic or physiologic dependence liability; and
8. Whether the substance is an immediate precursor to an already-scheduled substance.

There is also the option for emergency scheduling of a substance into Schedule I; however, since this path is only available for Schedule I, and xylazine has accepted use in veterinary medicine, it is unlikely that this route to scheduling will be taken.[8] As of February 2023, it was reported that the DEA has conducted the 8-Factor Analysis described above for xylazine and that the scheduling process has been in the works since October 2021, with supplemental information provided to HHS in September 2022 to accelerate the scheduling process.[9] On February 28, 2023, the FDA announced that it is taking action to restrict unlawful entry of xylazine in both finished dosage form and as bulk chemical into the United States. This includes reviewing all xylazine shipments to ensure that they are headed to appropriate places such as veterinarians.[10] While this will not solve the problem, it is the first step giving the FDA the authority to seize shipments that are not directed to an appropriate channel. In addition, legislation can be passed through Congress and signed by the President to schedule a substance. At the time of writing, the American Veterinary Medical Association (AVMA) has endorsed the Combating Illicit Xylazine Act, which was introduced into Congress on March 28, 2023. If passed, this act would make xylazine a Schedule III controlled substance.[11]

The introduction of xylazine into the illicit drug supply likely has origins in Puerto Rico, with the earliest report of xylazine contaminated drugs being from the early 2000s and a rapid increase in 2019. A recent report issued by the DEA and Department of Justice noted that identification of xylazine as an illicit drug is increasing throughout the country with 193% and 112% increases in a 2-year period in the South and West respectively. While the xylazine found initially may have been raw chemical illegally imported, the xylazine on the streets now is not purely illicit and has likely been obtained through diversion. Empty bottles of pharmaceutical grade xylazine with VetOne (distributed by MWI Animal Heath) labeling are being found in areas where dealers are bagging drugs.[7,9] Diversion has likely been going undetected due to the uncontrolled nature of xylazine and limited inventory systems in veterinary medicine; however, a 2020 break-in at Peace River Veterinary Clinic located in Alberta, Canada involved theft of three sedatives which included xylazine, ketamine, and detomidine.[12]

Applying the human healthcare data to veterinary medicine

The information up to this point has focused on the data from the perspective of the impact to human healthcare. Afterall, in veterinary medicine, our patients are not the ones we are worried about diverting drugs. It is the humans attached to them that regulations are designed to protect. Compared to human healthcare, the data on drug diversion in veterinary medicine is limited; however, there is evidence that this issue is equally prevalent in veterinary medicine based on reports of DEA fines and court cases against veterinarians. The following are a few examples of these instances of drug diversion occurring in veterinary practices.

- In Wyoming, a veterinary receptionist was accused of stealing tramadol and fraudulently calling in prescriptions.[13]

- In Alabama, a veterinary technician was arrested on 20 counts of controlled substance possession in connection with theft from Auburn.[13]

- In Florida, a veterinary technician was arrested for stealing controlled drugs from two veterinary clinics where they were employed.[13]

- In Massachusetts, a veterinary technician was charged with stealing 7,800 opioid pills and changing clinic records to cover.[13]

- In Virginia, a veterinarian was charged with diverting hydromorphone by injecting himself and replacing the volume with saline or butorphanol.[14]

A 2018 survey of 189 Colorado veterinarians, with an average time in practice of 22 years, found that 44% were conscious of opioid misuse by either a client or staff member, 12% were aware of opioid misuse and diversion, and 13% were aware of a pet owner intentionally making their pet appear ill or injured to obtain opioids.[15]

History has shown that as controlled substance sources tighten up, diverters will find alternatives, and veterinary medicine presents an alternative to human healthcare. Some possible reasons include lack of inventory systems, need to transport controlled substances, large volumes required for some species, and historically less emphasis on diversion prevention in veterinary medicine compared to human healthcare. There is also limited use of Prescription Drug Monitoring Programs (PDMPs) due to challenges in applying these reporting systems to veterinary prescribing.

Who is harmed by drug diversion in healthcare?

While it can be tempting to assume that drug diversion affects the person diverting the drugs more than anyone else, there are significant risks to non-diverting healthcare workers, the organization, and the patients. The following is a partial list of potential risks to each group (Table 1).

Table 1. Summary of risks associated with drug diversion.

Risks to the person diverting drugs	Risks to other healthcare workers
Medication side effects or overdose. Infection from non-sterile drugs or supplies. Loss of license. Criminal prosecution.	Shared patient care responsibilities may lead to shared medical-legal liability. Potential victim of manipulations by the diverter. Risk of disciplinary action for innocent colleagues. Poor-quality work environment.
Risks to the organization	**Risks to patients**
Loss of revenue. Civil liability and investigations. Reduced employee morale. Reduced trust of clients.	Not receiving prescribed drug or dosage. Poor-quality care. Decreased quality of life. Increased morbidity and mortality.

Understanding Drug Diversion in Healthcare

The Joint Commission defines drug diversion as "the transfer of any legally prescribed substance from the individual for whom it was prescribed to another person."[16] This definition includes both controlled and non-controlled drugs. Although the focus here is on controlled substances, xylazine provides a prime example of a non-controlled drug that may be diverted, thus considered for controlled substance scheduling after being marketed as uncontrolled for many years.

Reasons for diversion vary and include an individual with a substance use disorder diverting drugs for their own use, a desire to sell the medications for a profit in the illicit drug market, or to relieve physical pain that is being inadequately managed. The most diverted drug class is opioids with fentanyl being the most common drug.[17] Other commonly diverted drugs include benzodiazepines, ketamine, propofol, pregabalin, and gabapentin.

It is important to consider that drug diversion can happen at any point during the life of a drug. Opportunities for diversion in practice include the following:

- Procurement: Ordering or inventorying incorrectly and then diverting.
- Preparation and dispensing: Replacing the drug with a product similar in appearance.
- Prescribing: Altering prescriptions and/or prescribing unnecessary drugs for personal pets.
- Administration: Documented as given to a patient but not actually given.
- Wasting: Logged as wasted but taken by healthcare worker.

Identifying diversion in practice

Controlled substance regulations state "A healthcare organization and its employees are mandated to report any incident of drug diversion within the organization."[18] To report drug diversion, it must first be identified. Signs of drug diversion are generally non-specific and can look like healthcare professionals that are overworked, short on time, stressed, tired, or generally careless. While those are problems on their own, they are not illegal. Therefore, to determine if drug diversion suspicions are justified, it's important to consider the combination of signs and trends. Below are some potential signs that may indicate a colleague is diverting controlled substances.[17]

- Chronic absenteeism
- Frequent breaks or disappearances from the workplace
- Excessive amount of time near the drug supply
- Unreliability in keeping appointments and deadlines
- Variable work performance
- Confusion, memory loss, difficulty concentrating
- Sloppy recordkeeping
- Excessive drug wastage
- Personality changes
- Removing controlled substances without an order
- Frequent association with compromised product containers
- Frequent association with volume missing from premixed infusion
- Missing multi-dose vial overfill
- Late documentation of certain medications
- Lack of effective patient analgesia despite what should be adequate pain control
- Frequent/unusual efforts to help administer pain medication to patients
- Controlled substance waste tampering

Diversion Prevention in Practice

The approach throughout history to address the problems associated with potentially addictive drugs has been to restrict access without complete prohibition so that those that need the drugs have access and those that do not are protected from their addictive potential. This is based on the 1800s mentality that white market (i.e., physician involvement) use of addictive drugs such as opioids, sedatives, and stimulants is necessary in many cases, but that all illicit use is unmoral and therefore should be punished; however, this creates a situation where white market addiction is viewed as uninformed physicians and patients accidentally being harmed, thus necessitating additional protection and education. This is in direct contrast with those having less access to healthcare being viewed as criminals for obtaining more dangerous versions of the drugs through unregulated supply chains. Regardless of your personal opinions on the merits of this structure, this mentality has led us to our current situation where regulations are used to protect the potentially naive white market consumers while punishing those obtaining drugs through other channels. The result is increasingly stringent regulations with increasing oversight of compliance.

When regulations are applied correctly and completely, they are designed to make it difficult to divert drugs and quickly identify any diversion that does get through the system. When regulations are not fully implemented, they can result in a false sense of security as if there is a safety net when an effective one does not actually exist. DEA fines are more than $15,000 per violation[19] and the DEA does not schedule visits at your convenience. It is critical for veterinarians to take controlled substance security and recordkeeping seriously. Following the regulations is a solid start, and there are practice-specific considerations that need to be addressed to find the most effective and feasible way to stay in compliance.

To make sense of the complicated landscape of controlled substance regulations, a global view is needed. The federal- and state-controlled substance acts provide minimum requirements for handling controlled substances; however, these are minimums, and additional precautions may be warranted in some situations. The DEA monitors and tracks all controlled substances from synthesis or import through the time you purchase it. They expect that this level of tracking will continue through the time the drug is administered or disposed of. To track these drugs effectively and report diversion, there needs to be a certain level of security and documentation. Therefore, as a DEA license holder, you are responsible for ensuring that you maintain your controlled substances with adequate security and documentation. The problem is that what meets that requirement is a matter of professional judgement and what a DEA agent thinks is acceptable on inspection will likely vary from one agent to the next.

While the ambiguity of adequate controlled substance security cannot be completely simplified, the following steps will help to critically evaluate current processes to identify potential gaps that need to be addressed.

- Own the process. If you are the DEA registrant, you are ultimately responsible for all controlled substances obtained under your license. You are allowed to delegate management of this process as you deem appropriate, but when the DEA comes walking through the door, it will be you that is responsible for whatever has been occurring under your license.

- Review the Federal Controlled Substance Act as well as your state Controlled Substance Act to ensure that you are compliant with all stated requirements. Remember, in the case that federal and state laws differ, you must follow whichever law is most strict. You can find the Federal Controlled Substance Act at https://www.ecfr.gov/current/title-21/chapter-II.

- Think like a diverter. Evaluate all of the steps that occur in the lifecycle of a controlled drug in your practice and look for opportunities where drug could be pocketed without anyone likely to notice. As stated earlier, the Controlled Substance Acts provide the minimum requirements, but you will likely need additional checks and balances that fit your unique practice situation. Keep in mind that the possible opportunities are likely different for injectable drugs only administered in house versus medications that may be dispensed for a client to administer at home.

- Rank the potential diversion opportunities identified in order of likelihood of diversion not being noticed and address the most problematic areas first.

- Review the 2022 American Society of Hospital Pharmacists (ASHP) Guidelines on Preventing Diversion of Controlled Substances. These are written for human healthcare and in some cases refer to technology that isn't readily available in veterinary medicine. They still provide valuable insight into areas of high concern and a lot of the recommendations are adaptable to veterinary medicine.

- Consider reaching out to the local DEA field agents with questions. Each inspector will be looking for different things and will often interpret compliance differently. They want you to be successful, so they are willing to answer questions proactively. Your successful controlled substance management contributes to their goal of keeping the controlled drug supply secure. You can also contact your state veterinary medical association. Many associations have a DEA resource and/or are willing to contact the DEA for you in the event you are hesitant to ask a question.

Human healthcare is ahead of veterinary medicine in drug diversion prevention. That means there is a lot of valuable data and literature on problems, diversion potential, and compliance that will help guide actions and policies in veterinary medicine. In veterinary medicine, we need to do our part to help curb the opioid crisis, and effectively managing our access to controlled substances is the way to accomplish that goal.

References

1. Herzberg D. *White Market Drugs: Big Pharma and the Hidden History of Addiction in America.* Chicago, IL: The University of Chicago Press; 2020.

2. Mattson CL, Tanz LJ, Quinn K, et al. Trends and geographic patterns in drug and synthetic opioid overdose deaths-United States, 2013-2019. *Morb Mortal Wkly Rep.* 2021;70(6):202-207.

3. Centers for Disease Control and Prevention. Understanding the Opioid Overdose Epidemic. Opioids. Available at: https://www.cdc.gov/opioids/basics/epidemic.html. Accessed March 26, 2023.

4. Baldisseri MR. Impaired healthcare professional. *Crit Care Med.* 2007;35(Suppl):S106-S116.

5. Porter Research Survey 2021: Drug Diversion in Healthcare. Available at: https://invistics.com/2021-porter-survey/. Accessed March 26, 2023.

6. BD Institute for Medication Management Excellence. Health Care's Hidden Epidemic. Available at: https://go.bd.com/BD-Institute-for-Medication-Management-Excellence-Drug-Diversion.html. Accessed March 26, 2023.

7. MedPage Today. What Doctors Should Know About Xylazine in Fentanyl. Available at: https://www.medpagetoday.com/special-reports/features/102915. Accessed March 26, 2023.

8. Congressional Research Service. Xylazine: considerations for federal control. *Insight.* 2023;IN12086.

9. MedPage Today. Will Xylazine Become a Controlled Substance? Available at: https://www.medpagetoday.com/special-reports/exclusives/103244. Accessed March 26, 2023.

10. Food and Drug Administration. FDA Takes Action to Restrict Unlawful Import of Xylazine. Available at: https://www.fda.gov/news-events/press-announcements/fda-takes-action-restrict-unlawful-import-xylazine. Accessed March 26, 2023.

11. American Veterinary Medical Association. AVMA endorses federal legislation to address illicit xylazine. Available at: https://www.avma.org/blog/avma-endorses-federal-legislation-address-illicit-xylazine. Accessed March 30, 2023.

12. Global News Canada. Veterinary Drugs, Equipment Stolen from Peace River Clinic - Edmonton. Available at: https://globalnews.ca/news/7387857/peace-river-veterinary-clinic-theft/. Accessed March 26, 2023.

13. Today's Veterinary Business. Dirty little secrets. Available at: https://todaysveterinarybusiness.com/dirty-little-secrets/. Accessed March 26, 2023.

14. U.S. Department of Justice. Amherst County Veterinarian Sentenced for Opioid Theft. Western District of Virginia. Available at: https://www.justice.gov/usao-wdva/pr/amherst-county-veterinarian-sentenced-opioid-theft. Accessed March 26, 2023.

15. Mason DS, Tenney L, Hellyer PW, et al. Prescription opioid epidemic: do veterinarians have a dog in the fight? *Am J Public Health.* 2018;108(9):1162-1163.

16. The Bulletin. Joint Commission focuses on strategies to detect, prevent drug diversion. Available at: https://bulletin.facs.org/2019/06/joint-commission-focuses-on-strategies-to-detect-prevent-drug-diversion/. Accessed March 26, 2023.

17. American Nurse. Drug diversion in healthcare - prevention and detection for nurses. Available at: https://www.myamericannurse.com/drug-diversion-in-healthcare/. Accessed March 26, 2023.

18. Code of Federal Regulations. 21 CFR 1301.91. Employee responsibility to report drug diversion. Available at: https://www.govinfo.gov/app/details/CFR-2011-title21-vol9/CFR-2011-title21-vol9-sec1301-91. Accessed April 2, 2023.

19. DVM360. Commonly ignored controlled substance regulations that can increase your risk and cost you money when audited by the DEA. Available at: https://www.dvm360.com/view/commonly-ignored-controlled-substance-regulations-that-can-increase-your-risk-and-cost-you-money-when-audited-by-the-dea. Accessed March 26, 2023.

The Unkind Truth About Conflict Avoidance and Managing Difficult Conversations

Julia K. Whittington, DVM

Session #3407

Affiliation: From University of Illinois College of Veterinary Medicine, 1008 W. Hazelwood Dr., Urbana, IL 61802, USA.

Abstract: Conflicts among personnel and with clients can be incredibly disruptive to effective functioning of the veterinary care team. Disagreements are compounded by the stress and challenges facing veterinary medicine leading to client dissatisfaction and staff turnover. Most people are hard-wired to avoid conflict, which tends to cause these individuals to harbor on-going resentment and feelings of disgruntlement because the root cause of the conflict is not addressed. Whether you are the practice manager, a staff member, or the client, addressing conflict through implementation of effective communication and de-escalation techniques will help in managing difficult conversations. Strategies for managing difficult conversations will be discussed.

Communication

Communication is essential for effective interaction between individuals of any species. As communities and societies developed in human culture, unique language and forms of communication developed to meet the needs of those communities. People communicate for several different reasons and the form of communication varies from spoken word to images, body language and actions, writings, pictures, symbols, and music. The way that humans communicate build relationships, which is essential to our humanness.

Components of communication

Written and spoken words have the capacity to be direct and have clear meaning; however, the way that written or spoken words are delivered to the recipient may impact the message that is received. An example is the use of an exclamation point at the end of a sentence when trying to express excitement. Another example is changing your tone of voice to ask a question. Non-verbal communication involves body language which includes facial expressions, body movements, posture, and gestures that have specific meanings. Non-verbal communication can be used alone or with verbal communication. When used alone, non-verbal communication is often used to relay how the person is feeling, but this carries the risk of being misunderstood. When used with verbal communication, non-verbal communication can add further clarity to the meaning of the words and has less risk of being misunderstood because there is more information being provided to the receiver. Verbal communication is rule-bound since it is through established words and all the associated grammar. Non-verbal communication is rule-bound by social norms and a general understanding of meaning; however, communication is constantly evolving because society and cultures change.

Humans tend to privilege nonverbal communication to verbal communication when our nonverbal messages conflict with our verbal messages. This means that nonverbal communication has a greater impact on our emotions and feelings when compared to verbal communication. The reason for this may be because a primitive survival trait is to react to feelings and sensory stimulation. It is the fight or flight response when the body just takes over for the survival of the individual. In times like that, people are not really thinking clearly. So, especially during times of high stress or emotion, visual and sensory cues are more impactful than words that require thinking to understand. It is difficult to choose if there is a reason that humans would privilege nonverbal communication given our culture. Both types of communication are used every day. Verbal communication, delivered in a clear and polite way, is very effective. Nonverbal communication is open to interpretation under the influence of the receiver's perception; however, people are emotional beings and social interactions are very important. Sometimes, words take on more meaning when combined with body language. The two together make a bigger impact and are most effective. Indirect routes of communication, such as emails, memos, and social media posts, lose the benefit of non-verbal communication.

Communication stereotypes

In general, women are more indirect, use more words, and are more emotional in their communication than are men. Women historically are the caregivers in the family unit. Historically, women are nurturers and the 'weaker' sex. Men are the breadwinners of the family. They are the disciplinarian and set the rules. Obviously, this is not true in modern day and many families are breaking stereotypes. Communication style is individual, even though stereotypes exist in modern day culture. There are men who are emotional and freely show their feelings. Unfortunately, sometimes these individuals are viewed as being weak and are dismissed as being unmanly. There are also women that are very direct and don't display their feelings. These individuals are sometimes viewed as heartless and uncaring. Neither of these assumptions are true. Sometimes people are not honest in their communication to hide their true self or to fit in. Breaking stereotypes and avoiding actions based on unintended biases requires that people stop and think before they act and speak. Being open-minded and viewing an individual as a person, not as a male or female, will allow communication to be more meaningful. Breaking stereotypes requires an open mind and empathy. Empathy is the ability to share or understand what another person may be feeling in a given situation. It requires an understanding of the reality a person is facing and putting yourself into their shoes. If that is done, then communication will become less stereotypical and more effective.

Communication filters

Perception is how an individual becomes aware or views things by using their senses. A person's perception can change their experiences and experiences can change a person's perception. The adage "a glass is either half full or half empty" is an example of perception. If a person views a situation as being negative, then they will likely have a negative experience. The same is true for the opposite, with a person's positive view resulting in a positive experience. Conversely, a person may have a negative experience that then causes their perception to be negative about future similar experiences.

Perception relates to identity and self-esteem. Identity is how a person views themselves in comparison to social standards. These comparisons result in a person having the feeling of accomplishment or failure, inclusion or being-marginalized, joy or sadness, fear or confidence, etc., which are all based on perception. Self-esteem is the value a person places on themselves and their usefulness in society. It is impacted by the community and the individual's place in it. Perception of the community by the individual, as well as the community's perception of the individual and their role, directly impact a person's sense of self-esteem. Perception has the capacity to induce a wide spectrum of feelings in the recipient of the communication. These feelings are unique to an individual, so can't be assumed. These feelings may be the result of some unmet need which is also highly variable between individuals because it is based on their unique experiences and perception. For example, three people may hear the same story and one will feel happy, one will feel anger, and one will feel sadness.

Managing perceptions that are based on fact and actual experiences are easier to manage that those that are based on preconceived ideas and those that are developed using the attribution theory. Preconceived ideas are those that are based on opinions made without facts. These are usually emotional and can be biased. Sometimes, perceptions that are preconceived cause an individual to think or do something in a way that reinforces their perception. This results in a self-fulfilling prophecy which increases the chance that the perception will remain. Another perception pitfall is described by the Attribution Theory,[1] and occurs when a cause-and-effect relationship is ascribed to a person's behavior or a random event when no such relationship exists. For instance, a difficult client may be a "bad person," or an untoward event may have occurred because of "carelessness" on the part of an individual. Human nature induces us to try to make sense of the social world, and as scientists, veterinarians are trained to look for causes and effects. The danger in having this perception is that people will be incorrectly judged, especially during first impressions. One way to combat this is with the Rule of 6.[2] For every story that we tell ourselves, we need to come up with five other possible stories. Maybe our initial story is true, but maybe it's a variety of other things. This creates a conscious effort to be curious and not judgmental. Everyone delivers and receives information through their filter of perceptions and their own sense of self. In modern society, we are taught from a very young age to think in terms of right and wrong, to judge and demand, and to compete. This results in inevitable conflict.[3] Being self-aware and understanding how the filters of another might impact communication is important to effective communication.

Communication in Healthcare

Providing quality patient care and client service in veterinary medicine requires positive and collaborative working relationships among team members. Efficient and effective communication among personal, and with clients, is paramount to relationship building. This is becoming more important in present day veterinary medicine in which personnel are experiencing high levels of stress and burnout. Veterinarians are often trained to focus on problems and to identify solutions. Indeed, advances in veterinary medical knowledge and biotechnology makes it impractical to didactically deliver to veterinary students the sheer expanse of information available in veterinary medicine. Because of this, many curricula have adopted problem-based learning to teach students to be life-long learners, considering basic sciences and clinical knowledge to problem solve.[4] There is relatively little content in veterinary education regarding communication or managing conflict. Therefore, veterinary professionals tend to approach conversations, and often conflict, with a find-it-and-fix-it attitude like their approach to the diagnosis and treatment of their patients;[5] however, not everyone wants their problems to be fixed. Sometimes what a person is really seeking is an empathetic statement. In some situations, it may be worth asking the person if they want empathy or a solution. That sets you up to successfully address their need.

Communication within the veterinary healthcare team

Promoting effective and efficient communication within the veterinary healthcare team has been shown to increase team satisfaction, increase client satisfaction, reduce errors and complaints, and improve patient care.[5] Communication is most effective when team members trust each other; trust is developed through vulnerability, shared experiences, time, and a belief in mutual good will. Often, trust is promoted through shared goals and a commitment of individuals to their respective roles and responsibilities. Though each person's roles and responsibilities differ, each contributes to the delivery of patient care. Respectful engagement to hear opinions and provide feedback enhances the performance of all members of the team.

Conflict Management

Conflicts happen and can occur within an individual, an organization, or between persons or groups. A conflict is a disagreement that either has the potential to cause harm or causes harm; however, not all disagreements result in conflict. Multiple factors contribute to conflict development including differences in perspectives, priorities, goals, values, and even organizational hierarchy.[6] Not all conflict is disruptive. Conflicts that are managed well can result in sharing of ideas, problem-solving, and improved efficiency of the team. However, conflicts that go unresolved are often disruptive resulting in reduced efficiency, and decreased confidence, self-esteem, self-reliance, and morale in team members. The result is usually increased staff turnover and increased incidence of adverse patient care.[7] There are four main reasons underlying most conflicts: 1) differences in values that are real or imagined, 2) competing individual goals, 3) poor communication, and 4) personalizing a generic or organizational issue.

The anatomy of a conflict

Conflicts evolve through four phases.[6,7] A disagreement occurs, and the individuals or groups involved become frustrated. This is the first phase of a conflict. Frustration is a strong emotion that can be overwhelming and broadly directed. When frustration is experienced, the natural inclination is finding the cause. This is the second phase, and it allows us to rapidly identify a plan to make this unpleasant emotion go away. Unfortunately, the identified cause is often inaccurate or more complicated than we can appreciate while feeling frustrated. Regardless, we start implementing our plan of action directed at addressing the "cause" in the third phase. This leads to a series of destructive outcomes in phase four as the conflict becomes ingrained or escalates. Conflicts may escalate for a variety of reasons including differences in roles and responsibilities, with individuals feeling unsupported or unheard, and also personal feelings of insecurity and discomfort in the individual (e.g., introvert, imposter syndrome).

Managing conflicts

Conflict is common in healthcare teams, and is often unavoidable, however, the impact of the conflict can be minimized through effective management. Management of conflict must not diminish the importance of the existing interactions. There is an inherent power struggle in a conflict with competing interests: assertiveness in the extent the individual (or group) will go to meet their goals, and cooperativeness in the extent the individual (or group) will go to meet the goals of the other individual (or group) with whom they have the conflict. Five conflict management strategies are defined. These are Avoiding, Accommodating, Compromising, Competing, and Collaborating.[6,7] In each strategy, the balance between self-interest and the interests of others defines the outcome (Fig 1). Collaboration is the most time consuming and can be costly in emotional energy as well as in resources. It is also the most likely to result in sustainable change for the betterment of all. When there is long-standing conflict or mistrust among team members, it may be difficult to get to the point where collaboration is possible.

Unfortunately, avoiding is often the conflict management strategy selected by individuals. In this case, neither party "wins" and the conflict becomes entrenched in culture, and possibly escalates, with individuals or groups divided over negativity, and team productivity and satisfaction being diminished.

The goal of conflict management is resolution, and ideally that resolution will include unification of the parties involved and a recommitment to the common goals of the team. It is important to consider those goals and understand that goals immediately post-conflict need to be short-term and flexible. Try to understand the 'root cause' of the conflict and realize that it may not even be well understood by the conflicting parties. The environment for conflict discussions needs to be safe and neutral for all parties. Whenever possible, conflict management conversations should not be relegated to a third party (e.g., an arbitrator or mediator) as this is a form of conflict avoidance. Allow individuals to express themselves openly but focus the conversation on the problem. Personal attacks and disrespectful interjections that interrupt another person who is talking are disruptive to the process. Identify the common ground and expand on it to build consensus. Most importantly, stay calm. To move toward a productive solution, both parties need to identify their unmet needs that are central to the conflict. Once the unmet needs are identified, then both sides can work together to identify potential solutions to meet these needs.

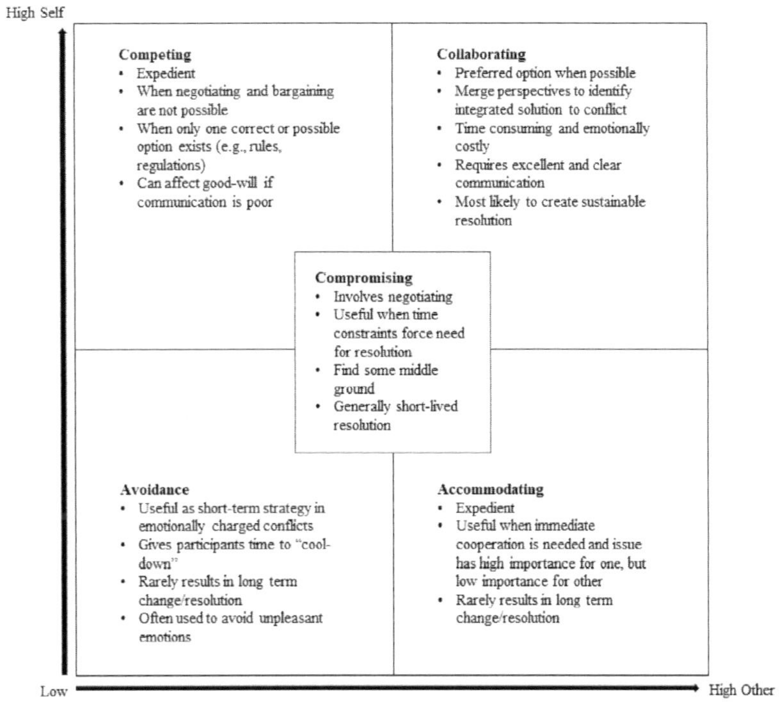

Figure 1. Conflict management strategies that address needs of self and other to varying degrees.

Communication Toolbox

Whether managing conflict, providing feedback, or having a difficult conversation, good communication skills are the foundation of a positive and successful interaction. The key to a positive interaction includes establishing a connection with the individual. Assume they have positive intent and thank them for their participation. Be curious and not judgmental. Be accountable for your role in any conflict, conceding or apologizing when appropriate. Be prepared to deliver facts clearly. When describing a specific situation, explain only the facts and avoid applying judgment through your filter. For example, say "You raised your voice and slammed a book when talking to your colleague. Tell me what was going on in that situation." instead of "Why did you get angry when talking to your colleague?" Be specific and avoid vague generalizations. Clear is kind, especially when delivering bad news or providing corrective feedback.[8] Communicate any pre-determined goals and the subsequent action plan that results from the interaction. When possible, make the action plan a collaborative approach by asking how you can support the other person.

Core skills

Communication skills are like any other skill in veterinary medicine and require practice to achieve and retain proficiency.[9] Communication in healthcare settings historically have the veterinarian, or physician in the case of human medicine, in the 'guardian' role. The veterinarian is viewed as the expert who provides the recommendations that the client is expected to follow with little shared decision making. With advances in veterinary medicine, more and more options have become available for our patients and veterinarians are needing to engage clients in decision making. In this manner, the veterinarian and client become collaborators and the result is relationship-centered care.[10] As in conflict resolution, effective communication and collaboration requires skill.

Making a connection

The first communication goal in any engagement is making a connection so that ideas may be shared clearly and respectfully. This goal may be hampered by factors including the nature of the interaction (e.g., emergency patient visit, providing negative feedback), but minimally the participants should acknowledge that each person has a valuable perspective and important information to share. A personal statement of introduction is a good first step. Using the person's name helps to make the individual feel seen.

Open-ended questions are those that cannot be answered with an affirmative or negative. They begin with when, what, how, or where, and allow the individual to share information openly. People want to share their ideas and information and often provide more detail when answering an open-ended question than if asked as series of yes-or-no questions. It is tempting to interrupt for clarification or to ask a related question when an individual is answering an open-ended question but avoid doing so as it may limit shared information at best, or negatively impact the collaborative interaction. Remember to pause before asking another question so that individuals have time to organize their thoughts.

Once information has been gathered, confirm that you have a clear and correct understanding of it by using reflective listening. By summarizing your understanding of the information relayed, you can correct misconceptions while making the individual feel heard.

Demonstrating empathy

To be empathetic is to be concerned about the feelings and values that another person has. Situational empathy is being able to feel and appreciate what another person is going through. Both are useful communication skills for promoting positive interactions. Empathy statements acknowledge, validate, and normalize the client's perceptions and emotions.[11] Fortunately, even persons who do not have innate feelings of empathy are able to make empathy statements which can help an individual feel supported and understood. People don't want to be judged and giving them the gift of empathy can go a long way towards collaborative agreements and conflict resolution. Beware of non-verbal cues that contradict your words.

Ensuring understanding

Through the sharing of information, a plan emerges. During the planning and ultimate execution of the plan, it is important to ensure that all people are on-board and understand what their responsibility will be. This can be done through a process called chunk-n-check.[11] This is especially helpful when working with clients who may feel overwhelmed with information, options, or simply the situation. By sharing information in manageable pieces, the receiver has time to absorb the news and the communicator can check-in to ensure understanding before moving on to the next piece of information. Another skill is the use of a signpost transition when moving between topics. In this way, the current information is summarized and then a verbal cue is given to indicate that the conversation is transitioning to a new topic.

Setting the stage

Anyone who has visited the Seattle, Washington, Pike's Place Market has likely enjoyed the energy exhibited by the employees of the World-Famous Pike Place Fish Market. This business is the focus of Fish! A Remarkable Way to Boost Morale and Improve Results, a book by Stephen C. Lundin, Harry Paul, and John Christensen.[12] In telling the story, the authors promote four key lessons to improve workplace morale. They are simple, common-sense lessons that can have a profound impact on promoting effective communication and conflict resolution.

Choose your attitude

You may not have a choice about what work you do, but you have a choice about the way you do it. If you have a good attitude, you can have a great day. If something bad happens to you or if someone makes you mad, you can choose to let it get to you or you can choose to let it go. If you have a grouchy attitude, you can have a bad day and make everyone else's day bad as well. By choosing your attitude you are being proactive, and you show personal accountability.

Play

You can be serious about your work, but you can have fun in the way you do it. Having fun is energizing.

Make their day

This is done by including others in your work and giving customers and co-workers a good experience, which leads to positive energy.

Be present

Look for opportunities to be fully engaged in your work and focus on other people. Be aware of customer needs so that you serve them well. Give people your full attention when engaging them, and really listen. Serving others brings personal satisfaction.

Conclusions

Communication style is as individualized as the individuals communicating, and every person approaches communication filtered through their personality, their perceptions, their experiences, their biases, and their sense of self. Communication skills are not a one-size-fits-all and not every tool is used in every interaction. Finding the tools that work best for you and practicing is the key to being an effective communicator. Take time to reflect on your own communication style and conflict resolution preferences. While not the only options, assessments such as Meyer-Briggs for personality type and the Thomas-Kilmann Instrument (TKI) for conflict resolution preferences can provide keys insights into your preferred processes. By knowing what you prefer, you can build off that to recognize your biases and hone your skills to be an effective communicator.

References

1. Mcleod S. Attribution Theory-Situational vs Dispositional. 2023. Available at: https://simplypsychology.org/attribution-theory.html. Accessed April 2, 2023.

2. Brown J. The rule of six. In: *The Art and Spirit of Leadership.* Bloomington, IN: Trafford Publishing; 2012. Available at: https://www.judysorumbrown.com/rule-of-six. Accessed April 2, 2023.

3. Bayer communication project-clear the air: tools for managing team conflict and enhancing team collaboration, module 15 syllabus. *Inst Healthcare Comm.* 2015;1-75.

4. Lane EA. Problem-based learning in veterinary education. *J Vet Med Educ.* 2008;35(4):631-636.

5. Bayer communication project-easy for you to say: communication within the veterinary healthcare team, module 12 syllabus. *Inst Healthcare Comm.* 2015;1-53.

6. Saltman DC, O'Dea NA, Kidd MR. Conflict management: a primer for doctors in training. *Postgrad Med J.* 2006;82:9-12.

7. Piryani RM, Piryani S. Conflict management in healthcare. *J Nepal Health Res Counc.* 2018;16(4):481-482.

8. Brown B. The call to courage. In: *Dare to Lead: Brave Work. Tough Conversations. Whole Hearts.* New York: Random House; 2018:45-69.

9. Coleman NL, Berlin A, Fischkoff K, et al. Annual structured communication skills training for surgery residents. *J Surg Res.* 2023;281:314-320.

10. Cornell KK, Kopcha M. Client-veterinarian communication: skills for client centered dialogue and shared decision making. *Vet Clin North Am Small Anim Pract.* 2007;37(1):37-47.

11. Shaw J. Guide to FRANK communication skills. *Frank Communication Workshop.* Ft. Collins, CO; 2014.

12. Lundin SC, Paul H, Christensen J. *Fish! A Remarkable Way to Boost Morale and Improve Results.* New York, NY: Hyperion; 2000.

Ten Tips for Supervising and Strengthening Teams When You Are Out of Bandwidth

Renée Schott, DVM, CWR

Session #3446

Affiliation: From Wildlife Rehabilitation Center of Minnesota, 2530 Dale St. N, Roseville, MN 55113, USA.

Abstract: Fortunately, or unfortunately, many veterinarians fall into formal or informal supervisory roles. Even if not a formal manager, veterinarians lead teams. Building a strong, cohesive team is not something taught well in many veterinary schools. The top 10 tips to help manage people and teams are provided in 15 minutes.

Tip 1

Show interest in each person's personal life. This is an easy way to foster connection.

Tip 2

Show vulnerability. If you cannot be vulnerable with your staff, they will feel they cannot be vulnerable with you, and will bottle up more emotions.

Tip 3

Let others know how you are feeling. It sets expectations and demonstrates vulnerability.

Tip 4

Ditch the formal reviews and instead schedule regular one-on-one meetings with your direct reports. It may seem laborious, but having a dedicated time and space for a regularly scheduled one-on-one meeting will naturally create the space for you and your report to connect on a human level. Meet weekly or a few times per month. The agenda could look something like: How are you? How is the work? How are we?

Tip 5

Know yourself. You need to dive into your own self-development to be able to be self-aware. Your inner world affects your outer world. Sometimes simple self-inquiry can make a difference in how you show up to work. Some self-inquiry questions may include: What's going on? What am I feeling? What is the dialogue inside my head? What are my feelings in my body? What is another way to see the situation?

Tip 6

Avoid blame (of yourself and others). Encourage accountability.

Tip 7

Ditch the golden rule; switch to the platinum rule. Focus on fairness (not equal) based in individual needs.

Tip 8

Recognition goes much further than praise!

Tip 9

Do not ask why. It implies blame. Instead ask what and how questions because they invite accountability.

Tip 10

Address conflict as soon as both parties are back to a grounded state. Discuss conflict at meetings and have staff create rules around how conflict should be repaired.

Bonus Tip

A great way to deliver feedback is to use the COIN method.

C: Connect and/or give context.

O: Observation. Describe the observable behavior.

I: Impact. Describe how it impacted the team negatively.

N: Next steps. Make sure to agree on the next steps.

There are multiple resources available to help build strong teams.[1-10]

References

1. Brown B. *Daring Greatly.* 2013. New York, NY: Random House; 2013.

2. Rock D. SCARF: A brain-based model for collaborating with and influencing others. *NeuroLeadership J.* 2018;1.

3. Brown B. *Dare to Lead: Brave Work. Tough Conversations. Whole Hearts.* New York, NY: Random House; 2018.

4. Wilson J, Gislason M. *Coaching Skills for Nonprofit Managers and Leaders: Developing People to Achieve Your Mission.* San Francisco, CA: Jossey-Bass; 2010.

5. Scott K. *Radical Candor: Be a Kick-Ass Boss Without Losing Your Humanity.* New York, NY: St. Martin's Press; 2017.

6. Rø, A, Fridjhon M. *Creating Intelligent Teams: Leading with Relationship Systems Intelligence.* Bryanston, JHB, South Africa: KR Publishing; 2016.

7. Stanfield RB. *The Art of Focused Conversation: 100 Ways to Access Group Wisdom in the Workplace.* BC, Canada: New Society Publishers; 2000.

8. Patterson K, Grenny J, McMillan R, Switzler A. *Crucial Conversations Tools for Talking When Stakes Are High.* New York, NY: McGraw-Hill; 2011.

9. Lencioni P. *5 Dysfunctions of a Team.* San Francisco, CA: Jossey-Bass; 2002.

10. Eoyang G, Holladay R. *Adaptive Action: Leveraging Uncertainty in Your Organization.* Palo Alto, CA: Stanford University Press; 2013.

How to Avoid a Board Complaint and What to Do if You Receive One

Jesse A. Fallon, DVM, PhD

Session #3315

Affiliation: From West Virginia State Board of Veterinary Medicine, Cross Lanes, WV 25313, USA, Cheat Lake Animal Hospital, Morgantown, WV 26508, USA, Avian Conservation Center of Appalachia, Morgantown, WV 26508, USA.

Abstract: Veterinarians work hard to provide the best medical care for their patients and good service for their clients. Unfortunately, there are occasions when a client may be unsatisfied with the service they receive. Most of the time, we can recover from these negative client interactions and regain their trust, or clients may simply choose to take their pet elsewhere; however, some very unhappy clients may choose to submit a complaint to the state Board of Veterinary Medicine, either because of a real or perceived medical error, or simply because of inadequate communication. How do you best respond to a board complaint? What steps can you take to ensure that a board complaint is resolved with no probable cause? What can you do in your day-to-day practice to help prevent a board complaint in the first place? This presentation will walk you through the steps of how to best respond to a client-driven board complaint and what you can do to help prevent a complaint from ever being filed.

Introduction

One of the greatest fears of a licensed veterinarian or veterinary technician is the dreaded board complaint. Whether it is the outcome of a medical error or simply an outlet for a disgruntled client, a complaint against you to your state's Board of Veterinary Medicine causes an enormous amount of worry, feelings of disappointment or shame, and concern for your license. Perhaps there was an undesirable outcome from surgery, a misdiagnosis, a transcription error on a medication label, a patient died under anesthesia, a rabbit fractured its vertebrae during restraint, or simply a misunderstanding in communication. Whatever the sequence of events that lead up to a complaint being filed, it is important to understand how to respond to a complaint, what you can do to ensure a finding of "no probable cause" from the board, and how you can avoid a complaint altogether. To accomplish these goals, you must first understand how your state board functions, identify the key components of a medical record that can help keep protect you, and recognize steps you can take to prevent a complaint from being filed in the first place.

What is the Board of Veterinary Medicine?

Each state has its own Board of Veterinary Medicine, and the composition, precise duties, and appointment process for board members varies only slightly from state to state. Generally, the mission of your state board is to protect the health, safety, and welfare of animals and the general public by enforcing the licensing and regulation of veterinarians, veterinary technicians, veterinary premises, certified euthanasia technicians, and animal crematories. The board has two major areas of regulatory authority. First, it determines which veterinarians, technicians, premises, and animal crematories are qualified to be licensed. Second, when a consumer (pet owner) complaint has been received and an investigation has been initiated, the board serves to determine whether a licensee's actions have fallen below the standard of care. In summary, the board provides qualified individuals the guidelines for acquiring and maintaining licensure in their state. To fulfill this duty, the board is made up of a predetermined number and composition of members.

The board's members are charged with ensuring that the licensees have completed and maintain the legal right to practice in their state. To ensure objective evaluation of a complaint, your board members follow an individual state practice act that defines the roles and requirements of licensees. The practice act, as well as the meetings and actions of the board, are accessible by the public. The practice act is written to withstand legal scrutiny and provide maximum public and animal protection. You should be familiar with your state practice act, as it is the playbook by which the board makes decisions about licensees.

The makeup of the board differs from state to state, but most include a majority of licensed veterinarians from different regions in the state, in some cases include a registered veterinary technician, and finally a minority of members who are not licensed to practice veterinary medicine. This diversity helps maintain a balanced perspective and attempts to adequately represent the profession and the public. Additionally, your board will have paid staff including at least an executive director, an attorney, and in most cases, one or more staff members. The full-time staff members do not have voting authority, but rather oversee the day-to-day administrative duties of the board (e.g., license renewals). The individual board members have specific responsibilities including a board chair and committee assignments.

How is a Complaint Filed and How does the Board Respond?

Any pet owner or representative (e.g., an attorney) can file a complaint to the state board. After a complaint is filed, the board sends you (i.e., the licensee) a certified letter stating that a complaint has been filed against you, requests the medical records, includes the letter from the complainant, and requests that your respond to the complaint. You can author the response yourself or you can have an attorney that is retained by your professional liability insurance (e.g., AVMA PLIT) author the letter. If you elect to use an attorney to help author your response, you provide the liability insurance company with a description of the incident presented within the complaint. Once you have provided your response and other required documents, the board members receive all of these redacted documents (no names, facility identification etc.) in advance of the upcoming board meeting. Prior to the board meeting, the complaint committee reviews the complaint letter, your response, and the medical records. At the prescheduled board meeting, the committee summarizes and reviews the case with the board, and typically makes a recommendation as to whether or not there is probable cause of deviation of standard of care, and the board then votes on the case. They may elect to: 1. Find no probable cause of malpractice; 2. Find probable cause of malpractice; or 3. Request that the complaint committee acquire more information (e.g., incomplete medical records, missing details that are important to the case, etc.)

What Should You Do if a Complaint is Filed Against You?

What do you do when or if you receive a letter from the board stating that they have received a complaint? First, don't panic. Review the medical record and the case details, and then immediately contact your PLIT representative. The liability company will ask you to provide them a description of the case that triggered the complaint. Be honest, open, and describe all details surrounding the case. Do not withhold any information about the case, client, or animal that is in question. Their job is to help defend you from the complaint.

After you provide details on the case, you and your liability company work together to author your response letter. This response is extremely important, and should not be accusatory, inflammatory, contain typos, or emotional overtones. It should include a description of your role in the case using clear and concise language, the diagnostic progression of the case, your recommended and accepted treatments, the client communication, and the outcome of the case. Consider the following when composing your response:

- If a medical error was made and the complaint was clearly justified and represents malpractice, often direct admission is the best approach.

- If the complaint was inaccurate or does not represent an error in judgement or decision making, say so. Describe the motivation and thought process you used in the clinical decision making and include supportive evidence, referring both to the medical record, peer reviewed publication, or expert opinion.

- Explain why the complainant is incorrect in their assessment, and state clearly that you did not make a mistake and that, while you have empathy for the client, they are incorrect in their assessment. It is okay and may be helpful to cite or include statements from your state practice act. You may also elect to include supplementary material in your response. This may include phone or text records, pictures or videos, or notarized statements from technicians, receptionists, or other doctors.

Receiving a complaint is stressful, and you will not be thinking objectively in this situation. You will be angry, frustrated, hurt, scared, and anxious. It is okay to feel this way. You may be inclined to take immediate action to rectify or address the complaint, but you should avoid any of the following:

- Falsify or alter the medical record

- Reach out to the client personally

- Post anything on social media

How Can You Ensure Appropriate Medical Records?

One of the most common problems that we see at the level of the state board is incomplete or inadequate medical records. Common gaps include incomplete or inadequate documentation of physical exam, inadequate or lack of anesthetic monitoring, and not documenting communications. Often there are major discrepancies in how the client perceived versus how the veterinarian remembers the sequence of events surrounding the incident that triggered the complaint. From the board's perspective, the most reliable source of information is the documentation in the medical record. Remember, if it is not documented in the medical record, there is no way that the board can determine whose side of the story is true. Your record does not need to contain copious amounts of writing. In fact, a succinct (but complete) medical record is often easier for the board to come to a decision. Be sure that your medical records contain the following:

1. Complete physical exam (including vital parameters)

2. Adequate anesthetic monitoring

3. Clearly stated treatment plan

4. Thorough communication details including discussion of expectations, risks, and complications

5. Thoroughly documented phone or email communication

6. Signed documents as appropriate (e.g., anesthetic consent, euthanasia consent)

Your thorough, complete, yet succinct medical record makes the job of the board's complaint committee easy. If standard of care is met and the record is complete, the outcome of your complaint will be no probable cause.

What Happens if the Board Finds Probable Cause?

If probable cause is found, the board will respond with a letter stating their findings and offer you a consent agreement. This describes the nature of the malpractice and the adverse action (i.e., punishment) which may include specific CE requirements, financial reimbursement for the board's time and expenses, probation, or revocation (very uncommon) of the license. They also provide a timeline for you to complete the required actions. At this point, you can choose to sign the consent agreement and accept the board's adverse action, or if you feel that the finding is unjustified, you can elect to decline to sign. This will most likely result in a court hearing, where you and your attorney will have to defend your position in front of a judge. If you are found at fault, you will be responsible for attorney and court fees. For more details on this, you should rely on your professional liability attorney.

How Can You Avoid a Complaint in the First Place?

There is no foolproof way to prevent a client from filing a complaint; however, one of the consistent themes in complaint cases is that the client often feels that the veterinarian or veterinary staff did not adequately communicate some aspect of their pets' care. This may include feelings that there was inadequate discussion of risks or expectations of a particular surgery, too slow or inadequate response to phone calls with concerns following the discharge of a patient, or a misunderstanding about a diagnosis or treatment course. A large percentage of the complaint cases that I have worked on could have been prevented with a mutual understanding of the pet's current condition, the diagnostic options, and the treatment plan. Ideally, you should have these conversations in person, ask the client if they understand the plan, and then document this reported understanding. If the owners decline to follow your treatment or diagnostic advice, document this in the medical record.

Conclusions

Keep in mind that your State Board of Veterinary Medicine is made up of people that are sympathetic to your situation. They are not your enemy; however, they have been charged by the governor or state legislature to protect the public, and they work hard to be objective in their assessment of complaint cases. Therefore, be sure to maintain a thorough medical record, have clear and consistent communication directly with your clients, and be diligent to uphold current standard of care for whatever species you are working on. Finally, if you do receive a complaint, don't panic. Many excellent veterinarians have had the misfortune of a complaint against them. You can overcome this challenge and continue to serve your patients and clients.

Section 47

Hybrid Potpourri Session

Renee Schott, DVM, CWR
Session Chair

Pradofloxacin Pharmacokinetics in Two Exotic Pet Species: Rabbit and Eastern Long-Necked Turtle

Shangzhe Xie, BSc/BVM, MVS, PhD, DABVP (Avian Practice), DACCM

Session #3103

Affiliation: From Mandai Wildlife Group, 80 Mandai Lake Road, Singapore 729826.

Abstract: The pharmacokinetics of some fluoroquinolones in rabbits and chelonians are well described but not for pradofloxacin, a new broad-spectrum veterinary fluoroquinolone available as an oral suspension for cats and dogs. In two separate studies, the single dose pharmacokinetic profile of pradofloxacin oral suspension at 7.5 mg/kg was investigated in four rabbits (*Oryctolagus cuniculus*) and 10 Eastern long-necked turtles (*Chelodina longicollis*). Plasma concentrations were measured using high-performance liquid chromatography on numerous samples obtained over time. Rabbit pharmacokinetic results indicated that oral absorption of pradofloxacin was slower and elimination faster compared to other fluoroquinolones in healthy rabbits, as well as relative to cats and dogs. Following treatment with pradofloxacin, fecal microbiota profiling showed some compositional differences between treated and control rabbits. This was the result of a significant decrease in the abundance of Proteobacteria, in particular bacteria belonging to the *Pseudomonas*, *Atopostipes* and *Parabacteroides* genera. Despite these microbiota changes, no adverse effects were observed in the treated rabbits in this study, but further studies utilizing multiple dose protocols (i.e., 7 days) are needed to confirm safety. Similarly, no adverse effects were observed in Eastern long-necked turtles administered 7.5 mg/kg q24hr for 7 days; however, using mammalian MIC data, the AUC0-24/MIC ratios for a range of bacterial isolates suggest that this dose of pradofloxacin is unlikely to be effective against many bacterial pathogens in Eastern long-necked turtles. Further studies are required for both species to establish clinically relevant dosages, but these two studies provide a platform for the understanding of this fluoroquinolone.

Introduction

Pradofloxacin is a third-generation fluoroquinolone developed specifically for veterinary use. This fluoroquinolone has good efficacy against gram-negative, gram-positive and anaerobic bacteria in dogs and cats.[1,2] Its main inhibitory effect on bacterial pathogens is through dual targeting of bacterial DNA gyrase and topoisomerase IV enzymes, and its bactericidal activity is maintained during stationary or slow growing phases of microbial replication.[3] Pradofloxacin has been shown to be effective in treating infections when the pathogen showed reduced susceptibility to other fluoroquinolones.[2] It also has lower potential for inducing antimicrobial resistance than other fluoroquinolones due to its low mutant prevention concentration.[4] Pharmacokinetic investigations in dogs and cats have shown that the maximum plasma concentrations are achieved within 3 hours of dosing and that the half-life is 5 to 10 hours.[5,6] Prior to these studies, there were no published reports of pradofloxacin use in rabbits and reptiles. There were also no previous pharmacokinetic studies for any drug in Eastern long-necked turtles (*Chelodina longicollis*) or any other Australian freshwater turtle (Chelidae, Carettochelydae).

Methods and Materials

In the first study, four female mature New Zealand white rabbits (body weight median 3.1 kg; range: 2.9-3.4 kg), were housed indoors in separate stainless steel cages with free access to feed (i.e., commercial pelleted diet) and water via automated dispensers, including the days of pradofloxacin administration and sample collection.[7] Rabbits were clinically examined daily and weight, food consumption, feces production and presence of any abnormal signs (i.e., ruffled coat, poor posture, abdominal pain, lung sounds, gut sounds, dehydration, diarrhea or lethargy) noted on the clinical record sheet. Each rabbit was administered pradofloxacin (25 mg/ml) oral suspension (Veraflox Oral Solution for Cats, Bayer Australia, Pymble, NSW), at a dose of 7.5 mg/kg via oral syringe.[7] To quantify plasma pradofloxacin concentrations, blood samples (0.5 ml per time point) were collected using a 25-gauge needle via the ear or lateral saphenous vein into lithium heparin tubes at t = 0, and hourly there after until t = 10 hours.[7]

In the second study, 10 captive, adult, Eastern long-necked turtles (*Chelodina longicollis*, body weight 450-1500 g, sex undetermined) were housed in two outdoor partially shaded enclosures.[8] Five turtles were housed in each enclosure, assigned to optimize the similarity of turtle weights between enclosures. Each enclosure comprised a 700-L pool, water temperature was maintained between 26-30 °C, with a small dry haul-out area with leaf-litter substrate and a basking light. A gelatin based pudding made in house with whitebait, beef mince, and Wombaroo Reptile Supplement (Wombaroo Food Products, Glen Osmond, South Australia) was fed once daily. Eight turtles were administered 7.5 mg/kg pradofloxacin oral suspension (Veraflox, 25 mg/ml, Bayer Australia, Pymble, NSW) delivered by syringe into the esophagus under manual restraint prior to feeding. Turtles were held for 1 minute to monitor for regurgitation of drug and then returned to the enclosure. Food was offered 1-2 hours after drug administration. One additional turtle in each tank received no pradofloxacin to evaluate the possibility of environmental drug exposure. Following iodine disinfection over the jugular vein, blood samples (1 milliliter) were collected using a 25-gauge needle and placed into lithium heparin tubes at t = 0, 1, 3, 6, 9, 12, 24, 48, 72, 96, 120, 144 and 168 hours relative to pradofloxacin administration.[8] Total blood volume over the 7-day sampling period from each turtle was limited to 1% body weight. All turtles were sampled at t = 0. Only one turtle was of sufficient size for all 13 blood samples to be collected; the remaining seven treated turtles had their respective allowable number of 1-milliliter samples distributed such that there were seven samples at 1 hour reducing to four samples at 72 hours. Aliquots of pool water from each enclosure were also collected at each time point.[8]

Plasma and water samples from both studies were stored at -80 °C prior to determining pradofloxacin concentration by high pressure liquid chromatography.

Results and Conclusions

In rabbits, the pradofloxacin median (range) T_{max} was 4.50 (2.00-5.00) hours, C_{max} 600.66 (395.85-886.72) ng/ml and $T_{1/2}$ was 1.27 (0.12-1.39) hours.[7] These results indicate that oral absorption of pradofloxacin was slower, and elimination faster compared with other fluoroquinolones in healthy rabbits, as well as relative to cats and dogs.[7] Following treatment with pradofloxacin, fecal microbiota profiling showed some compositional differences between treated and control animals. This was the result of a significant decrease in the abundance of Proteobacteria, in particular bacteria belonging to the *Pseudomonas*, *Atopostipes* and *Parabacteroides* genera.[7]

In Eastern long-necked turtles, pradofloxacin was quantifiable in plasma and tank water samples for up to 48 hours after drug administration.[8] The T_{max} (9.0 hours) and $T_{1/2}$ to 48 hours (13.16 hours) were longer, and the C_{max} (0.2 µg/ml) and AUC0-24 (2.2 hr*µg/ml) lower, than previously reported in cats and dogs.8 No adverse effects were observed in six turtles administered 7.5 mg/kg q24h for 7 days.[8] Using mammalian MIC data, the AUC_{0-24}/MIC ratios for a range of bacterial isolates suggest that this dose of pradofloxacin in turtles is unlikely to be effective against many bacterial pathogens.[8]

References

1. Silley P, Stephan B, Greife HA, Pridmore A. Comparative activity of pradofloxacin against anaerobic bacteria isolated from dogs and cats. *J Antimicrob Chemother.* 2007;60(5):999-1003.

2. Lees P. Pharmacokinetics, pharmacodynamics and therapeutics of pradofloxacin in the dog and cat. *J Vet Pharmacol Ther.* 2013;36:209-221.

3. Silley P, Stephan B, Greife HA, Pridmore A. Bactericidal properties of pradofloxacin against veterinary pathogens. *Vet Microbiol.* 2012;157:106-111.

4. Wetzstein H-G. Comparative mutant prevention concentrations of pradofloxacin and other veterinary fluoroquinolones indicate differing potentials in preventing selection of resistance. *Antimicrob Agents Chemother.* 2005;49(10):4166-4173.

5. Sykes JE, Blondeau JM. Pradofloxacin: a novel veterinary fluoroquinolone for treatment of bacterial infections in cats. *Vet J.* 2014;201(2):207-214.

6. Hauschild G, Rohn K, Engelhardt E, et al. Pharmacokinetic study on pradofloxacin in the dog-comparison of serum analysis, ultrafiltration and tissue sampling after oral administration. *BMC Vet Res.* 2013;9(1):1-8.

7. Xie S, Trott DJ, Saputra S, et al. Pharmacokinetic profile and effect on the faecal microbiome of a single dose of pradofloxacin oral suspension in the rabbit (*Oryctolagus cuniculus*). *J Vet Pharmacol Ther.* 2022;45(2):203-212.

8. Taylor E, Trott DJ, Kimble B, et al. Pharmacokinetic profile of a single dose of an oral pradofloxacin suspension administered to Eastern long-necked turtles (*Chelodina longicollis*). *J Vet Pharmacol Ther.* 2021;44:503-509.

VetGyver: How to Apply a Normal Thing to Exotic Pet Equipment

Chaowaphan Yinharnmingmongkol, DVM, Dipl,TBVM

Session #3243

Affiliation: From AnimalSpace Exotic Pet Hospital, 99/30 Puttamonthon Sai2, Taweewuttana, Salatummasop, Bangkok, 10170, Thailand.

Abstract: Exotic pets have unique anatomy and behaviors, which means that when working with multiple species, veterinarians must adapt to each animal's needs. Unfortunately, most veterinary equipment is designed for use with humans, dogs, and cats, making it challenging to find suitable equipment for exotic pets. By using standard equipment or everyday items creatively, veterinarians can often find cost-effective solutions. For example, intubating a rabbit typically requires an endoscope, but a USB ear scope can be used instead, saving thousands of dollars. A 1.5-millimeter silicone endotracheal tube, designed for small avians or mammals, can also be used as a feeding tube for tiny animals. Clear plastic surgical drapes are useful for visualizing small mammals during surgery, but an oven bag can serve the same purpose. Transparent hooks can be used in many scenarios, including hanging an intravenous line, and can be covered with an electronic cord cover to protect against animal biting. A saline bottle can be modified to function as an oxygen or gas supply mask and can even be adjusted to fit an Elizabethan collar for a prairie dog. In some cases, LEGO® toys can be repurposed in certain situations to prevent feces and urine from contaminating the surgical site following a plastronotomy procedure on a tortoise. By thinking outside the box and utilizing a range of everyday items, exotic pet veterinarians can save money while still providing quality care for their patients.

Ecotourism in the Caribbean: The Good, the Bad and the Ugly

Odette Doest, DVM

Session #3493

Affiliation: From Dierenartsenpraktijk Doest, St. Michielsweg 41, Julianadorp, Curacao.

Note: Abstract format selected by author(s). See Front Matter for description of conference proceedings manuscripts submitted as scientific abstracts.

Abstract: The main economic income for most Caribbean destinations is tourism. Exploring nature, hiking, camping, and ecotourism is booming. This phenomenon increases the possibility of conflict between nature conservation and nature exploitation. Ecotourism can be done in various ways and animal encounters are increasingly popular. We compare the different practices in Trinidad and Tobago, Suriname, the Dominican Republic, Mexico and the Dutch Caribbean and the way it affects wildlife.

Definition of Ecotourism

According to the website of the international ecotourism society, ecotourism is now defined as "responsible travel to natural areas that conserves the environment, sustains the well-being of the local people, and involves interpretation and education (The International Ecotourism Society, 2015). Education is meant to be inclusive of both staff and guests."[1]

Impact of Ecotourism

Wildlife observation excursions often are part the Caribbean ecotourism experience and involve tours to flamingo breeding and foraging sites, sea turtle nesting areas by night, and adventurous hikes into bat caves and ecosystems. Feeding wild animals often is a way to entice people to choose for certain tours that "guarantee" wildlife encounters. The feeding of wildlife can alter behaviors and has long-term impacts on animal diets and animal health.[2] Wildlife species are impacted by the presence of human visitors in the area be it on foot, boat or by car.[3] This presentation demonstrates the impact of ecotourists' actions and the importance of educating ourselves and our clientele about how to interact with wildlife species. Veterinarians often participate in wildlife excursions during vacations and should be aware of how to avoid potential harm. Veterinarians in an ecotourism area should know how to guide clients in responsible behavior.

References

1. The International Ecotourism Society. What is Ecotourism? The Definition. Available at: http://www.ecotourism.org. Accessed April 1, 2023.

2. Smulders FOH, O'Shea OR, Christianen MJA. Animal-borne video reveals atypical behaviours in provisioned green turtles: a global perspective of a widespread tourist activity. *Global Ecol Conserv.* 2012;25:e01417.

3. Ouboter DA, Kadosoe VS, Ouboter PE. Impact of ecotourism on abundance, diversity and activity patterns of medium-large terrestrial mammals at Brownsberg Nature Park, Suriname. *PLoS One.* 2021;16(6):e0250390. doi: 10.1371/journal.pone.0250390. PMID: 34077471; PMCID: PMC8171955.

Use of Alfaxalone in Exotic Animal Practice

Elizabeth A. Martinez, DVM, Dipl ACVAA

Session #3224

Affiliation: From Jurox Inc, now part of Zoetis, 1321 Burlington St. Suite D, N Kansas City MO 64116, USA.

Abstract: Alfaxalone is a sedative and anesthetic with a formulation that is indexed by the U.S. Food and Drug Administration for use in multiple minor species commonly encountered in exotic animal practice, including several species of small mammals, reptiles, amphibians, and birds. Following a brief summary of the pharmacology of alfaxalone and the indexing process, discussion of clinical case examples provides information to help veterinary professionals achieve optimal results and patient outcomes.

Introduction

It is often necessary to sedate or anesthetize exotic patients for physical examinations, laboratory sample collection, and diagnostic or surgical procedures. The ideal drug would be safe and versatile, have multiple routes of administration options, and provide a smooth sedation, anesthesia, and recovery process. Alfaxalone has been studied in numerous exotic species and has been shown to be safe and effective with a wide margin of safety.[1,2]

Clinical Pharmacology

Alfaxalone is a synthetic neuroactive steroid with the properties of a general anesthetic. The primary mechanism of action is via positive allosteric modulation of the gamma-aminobutyric (GABA) subtype A receptor ($GABA_A$) in the central nervous system (CNS). The $GABA_A$ receptor is a transmembrane ion channel. When alfaxalone binds to the $GABA_A$ receptor, the effect of endogenous GABA is potentiated, with movement of chloride ions intracellularly and hyperpolarization of the neuron and inhibition of action potential propagation.[3] It is important to remember that alfaxalone has no analgesic properties and that an appropriate analgesic plan is instituted. Other drugs that have a similar mechanism of action include benzodiazepines, propofol, etomidate, and barbiturates.

Indexing Process

Indexing is the process to legally market unapproved new animal drugs for non-food producing minor species. To be placed on the Index, the drug must have its safety and effectiveness affirmed through an alternative U.S. Food and Drug Administration process through the Minor Use Minor Species (MUMS) group. A drug that is Indexed is not registered. To have a product on the Index is especially helpful to individuals treating animals or classes of animals representing markets too small to support the costs of the traditional drug approval process.

A group of Expert Panelists reviewed over 100 peer and non-peer reviewed publications, abstracts and personal communications regarding alfaxalone's use in various species. All the information was compiled into a database and reviewed by the MUMS group for determination of eligibility of alfaxalone to be listed on the Index.

Alfaxalone (Alfaxan® Multidose IDX, Jurox Inc, N Kansas City, MO, USA) was indexed in February 2020 and joins a handful of other medications that can be found on the FDA's website, along with links to the Freedom of Information Summary (FOI) and Product Labeling.[4] To date there are only 15 other products on the FDA's Index.

Clinical Use of Alfaxalone

Exotic companion mammals

Rodents, ferrets (*Mustela furo*), guinea pigs (*Cavia porcellus*), and rabbits (*Oryctolagus*) are some of the most common exotic companion mammals. The use of alfaxalone in these species has been reported. Most commonly, alfaxalone is included in a multimodal sedation or anesthetic plan and administered intravenously (IV) or intramuscularly (IM).

Rodents

In mice (*Mus musculus*), alfaxalone has been administered IV, IM, and subcutaneously (SC).[5-7] Intravenous dosing without premedication was between 15-20 mg/kg producing successful anesthesia and was maintained with a CRI at 0.25-0.75 mg/kg/min.[5] Many may find IV access in a mouse challenging to achieve and the intraperitoneal (IP) route of administration of 80 mg/kg with 10 mg/kg of xylazine produced anesthesia with longer sleep times observed in female versus male mice.[6] In rats, alfaxalone has been administered IV, across a large dosing range, for both induction and maintenance of anesthesia, as well as IP.[2,5,8,9]

Ferrets

Because physical restraint may cause undue stress and discomfort on the animal, ferrets may require sedation to facilitate routine noninvasive medical procedures. Alfaxalone has been administered IM for sedation with butorphanol,[10] or butorphanol and midazolam.[11] Alfaxalone has also been administered IV for both induction and maintenance of anesthesia.[12,13]

Guinea pigs

The intramuscular administration of 5 mg/kg in guinea pigs produced successful sedation when administered alone.[14] Onset of sedation was rapid (2-3 minutes) and physiological variables were maintained within normal ranges. Duration of sedation was approximately 30 minutes. Alternatively, 20 mg/kg of alfaxalone alone has been used for non-painful procedures for less than 60 minutes.[15] The authors also concluded that 15 mg/kg alfaxalone with dexmedetomidine at 0.25 mg/kg and buprenorphine at 0.05 mg/kg subcutaneously would be adequate for non-painful procedures requiring consistent immobility for a minimum of 90 minutes.[15]

Rabbits

Administration of alfaxalone in rabbits is considered off-label since this species is considered a food-producing minor species. Numerous publications are available demonstrating that alfaxalone is a safe and effective drug for sedation, induction, and maintenance of anesthesia.[16-19]

Reptiles

Snakes

Alfaxalone may be administered IV, to effect, for induction of anesthesia in snakes. If IV access is not possible, alfaxalone may also be administered IM, SC, or intracelomic (ICe). In the ball python (*Python regius*), alfaxalone has been shown to provide a rapid smooth induction when administered IM, when injected caudally at 30 mg/kg and cranially at 20 mg/kg.[20] Cranial injection resulted in more successful intubation and significantly greater depth of anesthesia, likely due to hepatic shunting and first-pass metabolism via the renal portal system. In corn snakes (*Pantherophis guttatus*), SC administration of alfaxalone, 15 mg/kg, resulted in consistent anesthesia and is recommended for intubation and short procedures.[21] In the common garter snake (*Thamnophis sirtalis*), ICe administration of alfaxalone, 30 mg/kg, resulted in reliable loss of righting reflex (LRR). The anesthetic depth was greatest between 10 and 30 minutes after LRR.[22]

Lizards

Alfaxalone has been evaluated in numerous lizard species and with various routes of administration. A wide dose range has been published, that is likely due to whether the desired effect was sedation or anesthesia, the route of administration, the anatomical site of injection, and the effects of concurrent drugs before or simultaneous with alfaxalone. An IM injection of alfaxalone, 10 mg/kg, in either the triceps or quadriceps muscles provided reliable chemical restraint in bearded dragons (*Pogona vitticeps*) for short, nonpainful, diagnostic procedures.[23] A combination of alfaxalone, 15 mg/kg, and midazolam, 1 mg/kg, administered SC craniodorsal to the thoracic limb provided appropriate sedation to allow for physical examination and venipuncture in leopard geckos (*Eublepharis macularius*).[24] In veiled chameleons (*Chameleo calyptratus*), alfaxalone, 5 mg/kg IV following butorphanol premedication, provided light anesthesia and allowed for intubation; alfaxalone, 15 mg/kg IV, provided light anesthesia suitable for electroejaculation.[25]

Chelonians

Induction of anesthesia in chelonians with alfaxalone IV or IM has been described. In terrapins and tortoise species, 5 mg/kg IV was suitable for intubation, with full recovery in 30 minutes.[26] In loggerhead sea turtles (*Carreta caretta*), 3, 5, and 10 mg/kg alfaxalone IV was suitable for intubation, with dose-dependent duration of anesthesia and respiratory depression.[27] Alfaxalone, 10 or 20 mg/kg, IM in red-eared sliders (*Trachemys scripta* elegans) resulted in a smooth, rapid induction and uneventful recovery.[28] Depth and duration of sedation was body temperature dependent.

Amphibians

Studies of alfaxalone have investigated its administration IM and via immersion. In American bullfrogs (*Lithobates catesbeiana*), doses of 10-17.5 mg/kg IM resulted in immobilization with similar duration of effect, with higher doses having a shorter time to recumbency.[29] However, no dose prevented responsiveness to noxious stimuli. Immersion in alfaxalone at 2 g/L for 30 min did not produce immobilization or anesthesia.[29] In Oriental fire-bellied toads (*Bombina orientalis*) immersion in alfaxalone at 200 mg/L for 14 minutes resulted in sedation but remained responsive to noxious stimuli and is only suitable for non-invasive short procedures.[30] Immersion with a combination of alfaxalone, 30 mg/L and morphine, 50 mg/L or butorphanol, 25 mg/L improved antinociception.[31] Several case reports have been published describing anesthesia via immersion in axolotls (*Ambystoma mexicanum*) for surgical procedures and radiotherapy.[32-34]

Birds

Psittaciformes

Dose-dependent sedation was seen following IM alfaxalone in quaker parrots (*Myiopsitta monachus*);[35] however, muscle tremors and hyperexcitation was observed. Premedication with midazolam IM 10 minutes prior to alfaxalone produced a smoother induction but prolonged the recovery time compared to alfaxalone alone. The combination of intranasal midazolam and butorphanol and IM alfaxalone was found to be a safe and effective protocol for sedating quaker parrots.[36] Budgerigars (*Melopsittacus undulatus*) administered 5 mg/kg or 10 mg/kg IM achieved sedation adequate for handling nonpainful procedures. Some birds did show excitement during induction and recovery.[37]

Passeriformes

In the Bengalese finch (*Lonchura domestica*), dose-dependent effects on induction time, duration, and depth of anesthesia were observed with SC alfaxalone administration.[38] The addition of midazolam, 0.7 mg/kg, or butorphanol, 1 mg/kg, to 30 mg/kg alfaxalone improved muscle relaxation and increased depth of anesthesia. In black-cheeked lovebirds (*Agapornis nigrigenis*), alfaxalone, 12.5 mg/kg SC, provided nearly 1 hour of excellent sedation appropriate for noninvasive procedures. Rough induction and recoveries, along with minor movements and twitching were observed.

Waterfowl

Rose flamingos (*Phoenicopterus roseus*) undergoing orthopedic surgery, alfaxalone, 2 mg/kg, IV produced a smooth and rapid induction of anesthesia.[40] Alfaxalone, 10 mg/kg IV, was found to be a suitable induction drug for use in mute swans (*Cygnus olor*), producing a smooth a rapid induction adequate for intubation.[41] Agitation and ataxia was observed during recovery, requiring manual restraint in some birds.

Disclosure Statement: The author, Dr. Elizabeth A Martinez, is an employee of Jurox Inc, now part of Zoetis.

References

1. Full Prescribing Information Alfaxan® Multidose IDX. Available at: https://jurox.com/us/product/alfaxan-multidose-idx. Accessed March 28, 2023.

2. Lau C, Ranasinghe MG, Shiels I, et al. Plasma pharmacokinetics of alfaxalone after a single intraperitoneal or intravenous injection of Alfaxan(®) in rats. *J Vet Pharmacol Ther.* 2013;36(5):516-520.

3. Warne LN, Beths T, Whittem T, et al. A review of the pharmacology and clinical application of alfaxalone in cats. *Vet J.* 2015;203(2):141-148.

4. The Index of Legally Marketed Unapproved New Animal Drugs for Minor Species. Available at: https://www.fda.gov/animal-veterinary/minor-useminor-species/index-legally-marketed-unapproved-new-animal-drugs-minor-species. Accessed March 28, 2023.

5. Tremoleda JL, Kerton A, Gsell W. Anaesthesia and physiological monitoring during in vivo imaging of laboratory rodents: considerations on experimental outcomes and animal welfare. *EJNMMI Res.* 2012;2(44):123.

6. Siriarchavatana P, Ayers JD, Kendall LV. Anesthetic activity of alfaxalone compared with ketamine in mice. *J Am Assoc Lab Anim Sci.* 2016;55(4):426430.

7. Higuchi S, Yamada R, Hashimoto A, et al. Evaluation of a combination of alfaxalone with medetomidine and butorphanol for inducing surgical anesthesia in laboratory mice. *Jpn J Vet Res.* 2016;64(2):131139.

8. White KL, Paine S, Harris J. A clinical evaluation of the pharmacokinetics and pharmacodynamics of intravenous alfaxalone in cyclodextrin in male and female rats following a loading dose and constant rate infusion. *Vet Anaesth Analg.* 2017;44(4):865875.

9. White K, Aldurdunji M, Harris J, et al. Alfaxalone population pharmacokinetics in the rat: model application for pharmacokinetic and pharmacodynamic design in inbred and outbred strains and sexes. *Pharmacol Res Perspect.* 2022;10(6):e01031.

10. Milloway MC, Posner LP, Balko JA. Sedative and cardiorespiratory effects of intramuscular alfaxalone and butorphanol at two dosages in ferrets (*Musela putorius furo*). *J Zoo Wildl Med.* 2021;51(4):841-847.

11. Bercier M, Langlois I, Dunn M, et al. Cytological analysis of bronchoalveolar lavage fluid acquired by bronchoscopy in healthy ferrets: A pilot study. *Can J Vet Res.* 2016;80(1):7480.

12. Thas I. Acquired salivary mucoceles in two domestic ferrets (*Mustela putorius furo*). *Vet Rec Case Rep.* 2014;2:18.

13. Giral M, GarcíaOlmo DC, GómezJuárez M, Gómez de Segura IA. Anaesthetic effects in the ferret of alfaxalone alone and in combination with medetomidine or tramadol: a pilot study. *Lab Anim.* 2014;48(4):313320.

14. d'Ovidio D, Marino F, Noviello E, et al. Sedative effects of intramuscular alfaxalone in pet guinea pigs (*Cavia porcellus*). *Vet Anaesth Analg.* 2018;45(2):183189.

15. Doerning CM, Bradley MP, Lester PA, Nowland MH. Effects of subcutaneous alfaxalone alone and in combination with dexmedetomidine and buprenorphine in guinea pigs (*Cavia porcellus*). *Vet Anaesth Analg.* 2018;45(5):658666.

16. Costa RS, Ciotti-McClallen M, Tilley R, et al. Intramuscular alfaxalone with or without buprenorphine or hydromorphone provides sedation with minimal adverse effects in healthy rabbits (*Oryctolagus cuniculus*) in a randomized blinded controlled trial. *J Am Vet Med Assoc.* 2022;261(2):223-228.

17. Knutson KA, Petritz OA, Thomson AE, Balko JA. Intramuscular alfaxalone-butorphanol-midazolam compared with ketamine-butorphanol-midazolam in New Zealand white rabbits. *J Am Assoc Lab Anim Sci.* 2022;61(5):475-481.

18. Bradley MP, Doerning CM, Nowland MH, et al. Evaluation of alfaxalone total intravenous anesthesia in rabbits (*Oryctolagus cuniculus*) premedicated with dexmedetomidine or dexmedetomidine and buprenorphine. *Vet Anaesth Analg.* 2022;49(3):308-312.

19. Reabel SN, Queiroz-Williams P, Cremer J, Hampton CE, et al. Assessment of intramuscular administration of three doses of alfaxalone combined with hydromorphone and dexmedetomidine for endoscopic-guided orotracheal intubation in domestic rabbits (*Oryctolagus cuniculus*). *J Am Vet Med Assoc.* 2021;259(10): 1148-1153.

20. Yaw TJ, Mans C, Johnson SM, et al. Effect of injection site on alfaxalone-induced sedation in ball pythons (*Python regius*). *J Small Anim Pract.* 2018;59(12):747-751.

21. Rockwell K, Boykin K, Padlo J, et al. Evaluating the efficacy of alfaxalone in corn snakes (*Pantherophis guttatus*). *Vet Anaesth Analg.* 2021;48(3):364-371.

22. Strahl-Heldreth DE, Clark-Price SC, Keating SCJ, et al. Effect of intracoelomic administration of alfaxalone on the righting reflex and tactile stimulus response of common garter snakes (*Thamnophis sirtalis*). *Am J Vet Res.* 2019;80(2):144-151.

23. Shippy S, Allgood H, Messenger K, et al. Pharmacokinetics and pharmacodynamics of intramuscular alfaxalone in central bearded dragons (*Pogona vitticeps*): effect of injection site. *Vet Anaesth Analg.* 2023:S1467-2987(23)00044-2.

24. Doss GA, Fink DM, Sladky KK, Mans C. Comparison of subcutaneous dexmedetomidine-midazolam versus alfaxalone-midazolam sedation in leopard geckos (*Eublepharis macularius*). *Vet Anaesth Analg.* 2017;44(5):1175-1183.

25. Ferreira TH, Mans C. Sedation and anesthesia of lizards. *Vet Clin North Am Exot Anim Pract.* 2022;25(1):73-95.

26. Knotek Z. Alfaxalone as an induction agent for anaesthesia in terrapins and tortoises. *Vet Rec.* 2014;175(13):327.

27. Phillips BE, Posner LP, Lewbart GA, et al. Effects of alfaxalone administered intravenously to healthy yearling loggerhead sea turtles (*Caretta caretta*) at three different doses. *J Am Vet Med Assoc.* 2017;250(8):909-917.

28. Kischinovsky M, Duse A, Wang T, Bertelsen MF. Intramuscular administration of alfaxalone in red-eared sliders (*Trachemys scripta elegans*)-effects of dose and body temperature. *Vet Anaesth Analg.* 2013;40(1):13-20.

29. Posner LP, Bailey KM, Richardson EY, et al. Alfaxalone anesthesia in bullfrogs (*Lithobates catesbeiana*) by injection or immersion. *J Zoo Wildl Med.* 2013;44(4):965-971.

30. Adami C, Spadavecchia C, Angeli G, d'Ovidio D. Alfaxalone anesthesia by immersion in oriental fire-bellied toads (*Bombina orientalis*). *Vet Anaesth Analg.* 2015;42(5):547-551.

31. Adami C, d'Ovidio D, Casoni D. Alfaxalone-butorphanol versus alfaxalone-morphine combination for immersion anaesthesia in oriental fire-bellied toads (*Bombina orientalis*). *Lab Anim.* 2016;50(3):204-211.

32. McMillan MW, Leece EA. Immersion and branchial/transcutaneous irrigation anaesthesia with alfaxalone in a Mexican axolotl. *Vet Anaesth Analg.* 2011;38(6):619-623.

33. Modesto F, Nicolier A, Hurtrel C, Benoît J. Excisional biopsy and radiotherapy for management of an olfactory neuroblastoma in an axolotl (*Ambystoma mexicanum*). *J Am Vet Med Assoc.* 2021;260(4):436-441.

34. Vuong KS, Sheley W, Liu M, et al. Chromatophoromas on the tail of a pet axolotl (*Ambystoma mexicanum*) with presumptive metastasis. *J Exot Pet Med.* 2023;45:8-13.

35. Whitehead MC, Hoppes SM, Musser JMB, et al. The use of alfaxalone in quaker parrots (*Myiopsitta monachus*). *J Avian Med Surg.* 2019;33(4):340-348.

36. Conner CM, Hoppes SM, Stevens BJ, Simon BT. Intranasal butorphanol and midazolam administered prior to intramuscular alfaxalone provides safe and effective sedation in quaker parrots (*Myiopsitta monachus*). *Am J Vet Res.* 2022;83(12):ajvr.22.08.0140.

37. Balko JA, Lindemann DA, Allender MC, Chinnadurai SK. Evaluation of the anesthetic and cardiorespiratory effects of intramuscular alfaxalone administration and isoflurane in budgerigars (*Melopsittacus undulatus*) and comparison with manual restraint. *J Am Vet Med Assoc.* 2019;254(12):1427-1435.

38. Perrin KL, Nielsen JB, Thomsen AF, Bertelsen MF. Alfaxalone anesthesia in the Bengalese finch (*Lonchura domestica*). *J Zoo Wildl Med.* 2017;48(4):1146-1153.

39. Greunz EM, Limón D, Bertelsen MF. Alfaxalone sedation in black-cheeked lovebirds (*Agapornis nigrigenis*) for non-invasive procedures. *J Avian Med Surg.* 2021;35(2):161-166.

40. Villaverde-Morcillo S, Benito J, García-Sánchez R, et al. Comparison of isoflurane and alfaxalone (Alfaxan) for the induction of anesthesia in flamingos (*Phoenicopterus roseus*) undergoing orthopedic surgery. *J Zoo Wildl Med.* 2014;45(2):361-366.

41. Baldrey V, Stanford M, Bacon H. A comparison of the pharmacodynamic effects of intravenous ketamine-xylazine with alfaxalone in mute swans (*Cygnus olor*) presenting at a wildlife veterinary hospital. *Vet Anaesth Analg.* 2021;48(5):716-724.

Ooo That Smell! Maggot-Infested Wounds in Wild Mammals

Renée Schott, DVM, CWR

Session #3385

Affiliation: From Wildlife Rehabilitation Center of Minnesota, 2530 Dale St. N, Roseville, MN 55113, USA.

Abstract: Young (and some adult) mammals often present during the summer months with maggot-infested wounds. This gross, but fun, talk covers the basics on what the crud that bug is doing in the wound and what that means for the patient. Although wild mammals are used as case studies, the information in this masterclass can be applied to domestic species.

Introduction to Maggots

Maggots are the larval form of true flies, which are in the order Diptera.[1-5] These true flies have two main wings and two greatly reduced wings and include over 150,000 species. This presentation focuses on maggots in wounds and *Cuterebra*.

Generalized Life Cycle of Flies

Because there are so many species of true flies, and the timing of the life cycle varies greatly depending on ambient temperature, tissue temperature, and a variety of other factors, a very generalized fly life cycle is discussed here.

Egg (Day 0)

Most females lay approximately 2-millimeter-long eggs in clumps up to 300. Some fly species lay eggs on decaying matter, some need to be laid on flesh wounds, and some on intact flesh. Some flies need a living host and others don't. In around a day the egg will hatch into the larva or maggot.

Larva, maggot (Days 1-15)

The larva, called the first instar, initially feeds on fluid exuded from the body and then migrates into the body or wound. The maggot will double in size to approximately 5 millimeters and molt into the second instar. The second instar will double in size again (to 10 millimeters) and molt. This process is repeated via the third instar, which can be up to 20 millimeters long. The third instar molts into a pre-pupa and migrates away from the body, usually burrowing into the soil. There it remains in the soil for approximately 4 days, not feeding, until it develops into the pupa.

Instars (maggots) have hooks to shred and rake in decaying flesh. They have a muscular, segmented body, a simple intestine and a pair of very large salivary glands. These salivary glands secrete digestive enzymes which help to create their soup environment they eat. They have anterior and posterior spiracles to breathe, so they can feed and breathe 24 hours a day.

Pupa

The pupa does not feed and remains in the soil until it emerges as the fly (approximately 10 days).

Fly

Upon emerging from the pupa, the fly mates. Flies feed on protein from body fluids and lay eggs in approximately 2 days.

Cuterebra Life Cycle

Although *Cuterebra* life cycle follows the above generalized fly life cycle, a few differences are worth mentioning. Eggs are laid on grass stems, wood chips, or bark that is near the opening to a rodent or rabbit burrow. The eggs hatch and the larva then move onto the fur of passing rodents and rabbits. Larva enters the body through any body opening (not through the skin). The larva then migrates to the subcutaneous tissue, creating a subcutaneous cyst, it creates a breathing pore and grows. In 3-6 weeks, it drops to the ground once it has developed into a pupa. The pupa burrows into the ground and the adult fly emerges several months to years later.

Types of Myiasis Based on How it Infects

Myiasis can be generally grouped into five categories. Each category is described below.

Cutaneous

Fly eggs are laid on an animal or other object, hatching into the first instar. As an animal brushes against the object, the instar hitches a ride onto the animal and burrows into skin or enters through a body orifice. These larvae produce lesions that look like boils, pimples or furnucular lesions.

Wounds

Maggots are in an open wound that they either helped create, or infested a wound that was already present.

Creeping or larva migrans

Larva migrate around the body. *Baylisascaris* in non-hosts is an example. This group is not addressed here.

Myiasis of body cavities

Myiasis of body cavities is maggot infestation in the eye, nasal passes, ear canal or mouth.

Accidental myiasis

Accidental myiasis happens when eggs or maggots are accidentally eaten by an animal, such as an animal eating carrion with a maggot infestation. Usually this causes only gastrointestinal symptoms. The eggs or larvae may or may not be defecated whole. This grouping is not covered here.

Types of Myiasis Based on What It Eats

Facultative

These eat only dead or decaying organic matter or living matter.

Obligatory

These require a living host for development.

Medicinal Maggots

Medicinal maggots are generally the species *Lucilia sericata*, which is a species that only eats dead tissue. In a laboratory, they are propagated in a sterile environment and fed a strict diet. Modern companies now put these maggots into a porous, sterile bag that goes directly into the wounds. The enzymes the maggots secrete exit the bag to digest the dead tissue and they can suck up the soup through the bag as well. The enzymes this species secretes have antibacterial, antifungal and sometimes antiviral properties. Each bag is left in the wound for approximately 4 days and each would generally need 1-12 sessions.

Cases

Cases from the Wildlife Rehabilitation Center of Minnesota (WRCMN) are used to demonstrate some treatment options when dealing with maggot infested wounds. Prognosis is discussed using release proportion data from WRCMN.

References

1. Gerhardt RR, Hribar LJ. Flies (Diptera). In: Mullen GR, Durden LA, eds. *Medical and Veterinary Entomology.* 3rd Ed. London, UK: Academic Press; 2019. doi:10.1016/B978-0-12-814043-7.00011-X.

2. Ahmadnejad M, Rezazadeh F. Maggot therapy for snakebite necrotic wound in a horse. *Iran J Vet Surg.* 2021;16:156-160.

3. Harvey ML, Dadour IR, Gasz NE. Maggot therapy in chronic wounds: new approaches to historical practices. *Ann Entomol Soc Am.* 2021;114:415-424.

4. Stadler F. The maggot therapy supply chain: a review of the literature and practice. *Med Vet Entomol.* 2020;34:1-9.

5. Cousquer G. Veterinary care of rabbits with myiasis. *In Pract.* 2006;28:342-349.

Spatial Analysis of Wildlife Rehabilitation Data for Optimization of Outreach Efforts

Katherine Y. Choo, SVM,
Krystal M.T. Woo, DVM, MVSC,
Peter Flowers, RVT,
Carlos L. de la Rosa, PhD

Session #3427

Affiliation: From University of California, Davis, School of Veterinary Medicine, 944 Garrod Drive, Davis, CA 94566, USA (Choo), Lindsay Wildlife Experience, 1931 First Avenue, Walnut Creek, CA 94597, USA (Woo, Flowers), Center for Plant Conservation, 15600 San Pasqual Valley Rd., Escondido, CA 92027, USA (de la Rosa).

Note: Abstract format selected by author(s). See Front Matter for description of conference proceedings manuscripts submitted as scientific abstracts.

Abstract: Over the past decade, studies at wildlife hospitals have worked to highlight leading causes of wildlife injuries, providing information to improve local conservation strategies.[1-4] Identifying spatial patterns of the incidence of these injuries for application in outreach and rehabilitation has previously been difficult due to the constraints of wildlife tracking and monitoring. This study visualized and analyzed data from a wildlife hospital using geographic information systems software to evaluate the types of injuries that occur in cities in the California San Francisco Bay area. Of approximately 62,000 cases evaluated, the primary reasons for intake in 60% of cases were identified using specific search features, word processing programs, and manual assessment. Among the categorized cases, leading causes included domestic attack by dogs and cats (25.3%), abandonment/abduction (21.4%), fall trauma (15.0%), window collision (6.6%), and vehicular collision (6.2%). Upon spatial analysis of these causes, the top ten cities contributing to wildlife patients at Lindsay Wildlife Rehabilitation Hospital (LW) were found to have statistically significant differences in their proportions of common reasons for presentation when compared to those of the overall population of cities serviced by LW and to each other. Domestic attacks by dogs and cats ranged from 18.1% to 30.9% of categorized cases from these cities, and window collisions from 3.2% to 13.0%. The results indicate that cities with similar numbers of wildlife patients can differ in the primary types of injuries presented. While linear regression did not reveal significant correlations between the proportions of individual injury categories and city characteristics as defined by distance from LW, city population, and city area, this study demonstrates potential for future studies to determine other combinations of predictors to identify risk probability of these injuries in a city of interest. By identifying these patterns, we aim to focus and optimize prevention efforts and future outreach approaches to better inform local communities.

References

1. Long RB, Krumlauf K, Young AM. Characterizing trends in human-wildlife conflicts in the American Midwest using wildlife rehabilitation records. *PLoS ONE* 2020;15(9):e0238805.

2. Romero F, Espinoza A, Sallaberry-Pincheira N, et al. A five-year retrospective study on patterns of casuistry and insights on the current status of wildlife rescue and rehabilitation centers in Chile. *Revista Chilena de Historia Natural.* 2019;92:6.

3. Schenk AN, Souza MJ. Major anthropogenic causes for and outcomes of wild animal presentation to a wildlife clinic in east Tennessee, USA, 2000-2011. *PLoS ONE* 2014;9(3):e93517.

4. Baker TM. Exploring public opinion of urban wildlife and effective use of urban wildlife education in Tucson, Arizona. Master's Dissertation: Prescott College in Environmental Studies, 2017.

Secret Tips for Intubation in Tiny Exotic Pets (Hamsters, Sugar Gliders, Chinchillas, Hedgehogs, Squirrels)

Chaowaphan Yinharnmingmongkol, DVM, Dipl, TBVM, and
Rapon Kanokpornwasin, DVM

Session #3244

Affiliation: From AnimalSpace Exotic Pet Hospital, 99/30 Puttamonthon Sai2, Taweewuttana, Salatummasop, Bangkok, 10170, Thailand.

Abstract: One of the most critical aspects of small mammal anesthesia is successful intubation. The risk of anesthesia-related complications is higher in small exotic pets, such as guinea pigs, chinchillas, hamsters, hedgehogs, and sugar gliders, and intubation may help facilitate a safer anesthesia.[1,2] During anesthesia, hypoxemia and apnea are common problems that can be alleviated through proper intubation. Endotracheal intubation in small exotic patients can be challenging. It is recommended to secure the airway during anesthesia. While using a reinforcement silicone tube works well for the blind intubation technique in rabbits, small exotic pets may prefer a blind method using a feeding tube. To intubate small exotic pets, the animal should be positioned in sternal recumbency with hyperextension of the neck. The mouth can be opened using a small thread to hook the incisor, and the feeding tube is inserted into the mouth, same as the line technique in rabbits. To ensure success, it is important to cut the feeding tube at a 45-degree angle and reduce the sharpness using a flame. The feeding tube method can be used to intubate hamsters, hedgehogs, sugar gliders, chinchillas, and other small exotic animals. The feeding tube is a cheaper and more flexible alternative to a silicone endotracheal tube, and its head can be connected to an endotracheal adapter. Overall, intubation is a critical aspect of anesthesia in small exotic pets, and proper technique can reduce the risk of complications. By using a feeding tube and blind technique, veterinarians can successfully intubate a range of small exotic pets and provide quality care while minimizing costs. These techniques are demonstrated during the presentation.

References

1. Bennett K, Lewis K. Sedation and anesthesia in rodents. *Vet Clin North Am Exot Anim Pract.* 2022;25(1): 211-255.

2. Gardhouse S, Sanchez A. Rabbit sedation and anesthesia. *Vet Clin North Am Exot Anim Pract.* 2022;25(1): 181-210.

Roles of Veterinarians in Rewilding Projects in Victoria, Australia

Patricia Macwhirter, BVSc (Hons), MA, PhD, FANZCVS, and
Madeleine Rowe, DVM, MANZCVS

Session #3163

Affiliation: From Melbourne Bird Veterinary Hospital, 1 George St, Scoresby, Victoria, Australia, 3179 and Healesville to Phillip Island Naturelink Inc.

Abstract: Human-caused species extinctions and global collapse of biodiversity are critical issues affecting our planet. Rewilding is a comprehensive conservation effort focused on restoring sustainable biodiversity and ecosystem health by protecting core wild areas, providing connectivity between such areas, and protecting or re-introducing keystone and other locally indigenous species. A key element of this work is re-establishing local ecological vegetation communities and their associated ecological webs including indigenous flora and fauna species, both vertebrate and invertebrate. Veterinarians have attributes, alongside their veterinary skills, that make them valuable team members for groups working to halt biodiversity collapse. These include organizational abilities, entrepreneurship, influence within their community, local knowledge and broad networks. Roles complementing traditional services where veterinarians can assist rewilding projects include advocating for landscape-wide rewilding, communicating global and local biodiversity goals, research into conservation assets and threatening processes, engaging in rewilding on private or public land and helping to navigate biodiversity incentives. In this talk the development of biolinks in the face of urban expansion near Western Port, a bay south-east of Melbourne, Victoria, Australia, is used as a case study.

Introduction

Species extinctions and global collapse of biodiversity are critical issues affecting our planet, the magnitude of which has not been seen since the extinction of the dinosaurs 65 million years ago.

- The United Nations (UN) most recent Climate Change 2023: Synthesis Report attributed, with high or very high confidence, changes in ecosystem structure, species ranges and seasonal timing to climate change caused by human activity.[1]

- World Wildlife Fund's Living Planet Report (2022) estimated an average decline of 69% in monitored wildlife populations since 1970.[2]

- The UNs' IPBES Global Assessment Report on Biodiversity and Ecosystems, 2019, noted the rate of global change in nature is unprecedented and direct drivers with the largest global impact have been changes in land and sea use; direct exploitation of organisms; climate change; pollution; and invasion of alien species.[3]

- Following European colonization, Australia has had one of the worst extinction records in the world, including the worst record for mammal extinctions, with greater than 10% of endemic terrestrial mammal species now lost.[4] Alien species, human activity and changed fire regimes, partly induced by climate change, have been key factors in local species declines.

Like other people, veterinarians have contributed to these existential problems. While we are not ecologists, understanding disrupted ecologies is centered in biological sciences and based on the same principles as epidemiology. These are areas where veterinarians are well trained and experienced. Given the magnitude of the problem, there is a compelling 'why' veterinarians should be part of teams finding and applying solutions to halt global species declines.

Rewilding is a comprehensive conservation effort focused on restoring sustainable biodiversity and ecosystem health by protecting core wild areas, providing connectivity between such areas, and protecting or re-introducing keystone and other locally indigenous species.[5] It is a paradigm shift in the way people interact with nature and seeks to address the direct drivers of biodiversity and ecosystem decline identified in the multiple reports noted above. For success, it needs scientific and local knowledge, community support and a focus on restoration of dynamic ecological webs.

Roles for Veterinarians

Veterinarians in private practices and wildlife facilities see injuries and disease in native species caused by human activities, as well as the close inter-species connections that humans form with pets and wildlife. We appreciate the benefits, and challenges, these connections have for people, wildlife, and the broader community. In the Australian context, cats, and dogs, whose ownership our profession encourages, have contributed to biodiversity collapse through predation and as feral populations. So, too, has farming and farm animals that have become feral (including pigs, goats, horses, and camel), and alien species that have acclimatized such as European rabbits, hare, deer, rats, and mice.

Many veterinarians help sick and injured wildlife by providing traditional veterinary care. While this is useful and rewarding on a micro-scale, it generally has negligible impact on global biodiversity collapse. However, veterinarians have additional attributes, alongside veterinary skills, that make them valuable team members for groups working to halt species extinctions. These include organizational abilities; entrepreneurship; influence within their community; local knowledge and broad networks. Roles complementing traditional services where veterinarians can assist include advocating for landscape-wide rewilding projects, communicating global and local biodiversity goals; researching conservation assets and threatening processes; engaging in hands-on habitat repair and helping to navigate biodiversity incentives. Conservation begins where you stand and, in this presentation, biolinks (wildlife corridors that connect larger habitat cores) being developed in the face of urban expansion near Western Port,[6] a bay south-east of Melbourne, Victoria, Australia, close to where the authors work and live, is used as a case study of roles veterinarians can play in rewilding projects.

Advocating For Landscape-Wide Rewilding

Western Port has a remarkable diversity of flora and fauna, many of which are iconic, threatened, or endangered species. It is home to breeding colonies of the little penguin (*Eudyptula minor*), koala (*Phascolarctos cinereus*), platypus (*Ornithorhynchus anatinus*), echidna (*Tachyglossus aculeatus*), southern brown bandicoot (*Isoodon obesulus*), Australian fur seal (*Arctocephalus pusillus*), hooded plover (*Thinornis rubricollis*), short-tailed shearwater (*Puffinus tenurostris*), growling grass frogs (*Litoria raniformis*), migratory shorebirds and many more. In the 1970s a strong environmental movement emerged in Western Port in response to threats of major industrialization, including plans for a nuclear power plant. Strong activism, with veterinarians being minor players, led to a reversal of government policy. Western Port became recognized as a Ramsar Site of international significance[7] for migratory waders and, in 2002, it was designated the Mornington Peninsula and Western Port UNESCO Biosphere Reserve,[8] part of a global network of 738 reserves in 134 countries world-wide and one of only five Biosphere Reserves in Australia.

Compared with other jurisdictions, citizen advocacy has kept government support for conservation at the forefront in Victoria and, as a result, any direct exploitation of native wildlife is tightly controlled. However, populations are far from secure. Habitat loss from human expansion, development and extractive industries, climate change, pollution, and invasion of alien species (especially European red foxes and cats) are ongoing challenges. These themes are echoed in many places around the world.

Communicating Global and Local Biodiversity Goals

With species decline continuing unabated world-wide, acting both locally and globally is essential if the current extinction trends are to be halted. Online resources, including reports referenced earlier, are useful in identifying why, where, and how conservation efforts might best be prioritized, as well as giving global context for local conservation projects or species recovery plans.

The United Nations (UN) Sustainable Development Goals[9] are a collection of 17 interlinked objectives designed to serve as a shared blueprint for peace and prosperity for people and the planet now and into the future. They emphasize the interconnected environmental, social, and economic aspects of sustainability. There is a plethora of worthy opportunities across the UN Goals where veterinarians, with agile minds and interests in birds, animals, and science, can play positive roles.

Rewilding as a means of safeguarding global biodiversity falls primarily within the Life on Land and Climate Change UN Sustainability Goals. One of the key findings in the 2022 U.N. Environment Program report,[10] reviewing progress with these goals, was that environmental emergencies and human well-being need to be addressed together to achieve sustainability. The report also advocates that everyone has a role to play in ensuring that human knowledge, ingenuity, technology, and cooperation are redeployed from transforming nature to transforming humankind's relationship with nature. For some veterinarians the 'why' for local rewilding will not be as good a match as other worthy causes, but there may be synergies elsewhere on the planet where they might assist. For instance, a U.S. veterinarian helped crowd fund financial aid following the 2020 Victorian bushfires (Driggers T, Facebook and email communications, 2020) while Australian veterinarians have helped Indonesian colleagues in the Friends of the National Parks Foundation in Bali with their work with wildlife conservation.[11]

Successful rewilding projects have proven useful in addressing species declines, but there are significant challenges. These include starvation and native vegetation loss when herbivore populations are not effectively controlled or community backlash if predators dangerous to humans expand from the rewilding site. Based on difficulties with the Oostvaardersplassen project in the Netherlands,[12] which started in 1983 and aimed to recreate natural Pleistocene-like conditions, Kopina et al 2019 listed golden rules of rewilding as Cores, Carnivores, Corridors and Compassion, with an emphasis on putting the welfare of the animals used in rewilding experiments at the forefront.[13] Multiple sites in Australia where predator-proof fences or island introductions have excluded exotic predators such as cats and foxes (*Vulpes vulpes*) but not re-introduced indigenous predators such as dingos (*Canis lupus dingo*) or quolls (*Dasyurus* spp.) have seen population explosions and mass starvation, for instance with koalas on Western Port islands causing eucalyptus die back, or macropod overgrazing causing grassland ecosystem collapse.[14] The International Union for the Conservation of Nature discusses these and other challenges in its publication, Guiding Principles for Rewilding.[15]

Indigenous peoples' knowledge and perspectives, which often co-evolved with local flora and fauna, can offer valuable insights for rewilding projects, for instance First Nations Australians traditionally used fire-stick farming in shaping the Australian landscape. The cessation of regular burning after European colonization has contributed to the increased ferocity of Australian bush fires. Fuel-reduction burns in line with indigenous cultural practice are now commonplace. Managing the risk of fire is a complex challenge in rewilding projects, especially where connectivity is lacking, as fires are both dangerous to human lives and assets and can destroy critical wildlife populations.

Researching Conservation Assets and Threatening Processes

Understanding the unique flora, fauna and other conservation assets of a region, as well as threatening processes that are impacting upon them, is important both for advocacy and in undertaking rewilding projects. Town planning provides for roads and other built infrastructure, but it has not traditionally included provisions for conservation hubs and biolink connectivity for wildlife. This is an area where veterinarians can join with others to press for critical research and change.

In the context of Western Port, urban growth corridors eliminated known wildlife corridors and splintered habitats into isolated reserves. As a result, wildlife was over-exposed to dogs, cats, traffic, and foxes, and some species, such as bandicoots and koalas, became locally extinct and unable to repopulate. This was a catalyst that led a coalition of environmental groups initiating projects to identify possible biolinks (which often corresponded to drains and waterways) across the landscape. Counter-balancing developers, a citizen advocacy group called Healesville to Phillip Island Nature Link Inc, emerged to promote naturelinks across both private and public land so current and future generations will be able to live sustainably with Victoria's unique flora and fauna, mitigate climate change and enjoy positive benefits of connecting with nature.[16]

A landscape-wide, intergenerational, well-connected, citizen advocacy approach has been helpful in encouraging local government to adopt their own biolink plans. The Victorian state government has also supported this, for instance by developing and promoting NatureKit, a publicly accessible online mapping tool for biodiversity data integration and decision support.[17] Using NatureKit and the Victorian Biodiversity Atlas[18] it is possible to generate lists of local flora, fauna and ecological vegetation classes that land owners and managers can use in rewilding projects. Indigenous plant nurseries can supply plants on these lists to recreate historic ecological vegetation communities from farmland or degraded sites. As a result, landowners, including veterinarians, can embark upon rewilding projects even if their own knowledge of indigenous vegetation or ecological webs is limited.

While these are good starting points, knowledge of locally indigenous ecological vegetation classes, including their web of flora and invertebrate and vertebrate fauna, will always be limited. There are many simple opportunities to contribute to greater understanding, such as survey work using traditional or innovative methods (e.g., environmental DNA) to establish baseline information and information regarding response following rewilding efforts. Investigation of potential wildlife pathogens, including ones with zoonotic potential such as *Chlamydia psittaci*, is an area where veterinarians are well trained. Combining skills, it is possible to use both veterinary and ecological approaches to solve complex biodiversity challenges, for instance the discovery of chytrid fugus as the cause of global die-offs in frog populations by the team now based at Melbourne University.[19]

Hands-On Habitat Repair

Rewilding can involve protecting and 'restoring' indigenous ecological vegetation classes and their associated fauna communities or landscape remediation (e.g., of farmland or industrial sites) to re-establish such communities. Both provide healthy opportunities to get away from computers and re-connect with the natural world. Rewilding and fox control over the past 20 years on the author's (PM's) property Harewood has seen the expansion of three endangered species (southern brown bandicoot, growling grass frogs and swamp skink *Lissolepis coventryi*) and eleven indigenous ecological vegetation classes, six of them endangered. Environmental DNA studies are currently underway to better understand how ecological webs function in these different, threatened, communities.

Combining Endangered Species Recovery and Rewilding Programs

While endangered species recovery programs focus on individual species, the focus in rewilding programs is on re-establishing broad, functional ecological webs. These are complementary. Captive breeding programs need suitable habitats for species release if successful re-introduction is to occur and rewilding projects need captive breeding projects to support their work.

From a public relations perspective, it is often easier to promote saving an iconic bird or animal than it is saving a complex ecosystem. Joining forces with local artists, educators, and farmers to promote the plight of multiple species of endangered parrots kept as pets has been used by bird owners and veterinarians in developed countries in trying to protect these species in the birds' home ranges, for instance in the Caribbean and South and Central America where the global extinction crisis is particularly acute. Such efforts need to be backed up with in-depth understanding of the species habitat needs and human social context if protection or re-introduction efforts are to be successful.

In Victoria, veterinarians at Healesville Sanctuary, a state government native wildlife and ecotourism/education facility with a state-of-the-art wildlife hospital, have served in traditional roles as well as fostering, alongside other team members, core habitat areas and biolink connectivity for captive-bred endangered species release. Programs include those for helmeted honeyeaters (*Lichenostomus melanops cassidix*) and orange-bellied parrots (*Neophema chrysogaster*). A privately owned facility, Moonlit Sanctuary, with team support from private veterinarians, has also had success in breeding orange-bellied parrots. These parrots migrate from their natural breeding grounds on the west coast of Tasmania to coastal wintering grounds on the mainland, including Western Port. This species feeds on salt marsh, habitat that has been greatly reduced because coastal areas are prime real estate for human development. Cirocovirus has also been a problem in wild populations. Riparian forested areas near Melbourne, Helmeted Honeyeater's former stronghold, have also been impacted by forest clearing and urban expansion with invasion of more aggressive native and exotic species including noisy miners (*Manorina melanocephala*) and common mynahs (*Acridotheres tristis*). Re-establishing large enough areas of habitat for these endangered species, controlling predators and aggressive competitors, and addressing disease challenges are some of the multiple issues with which recovery teams, including veterinarians, are reckoning.

While these recovery efforts have meet with some success, status of other local species (including swift parrots, *Lathamus discolor,* and blue-winged parrots, *Neophema chrysostoma,* both of whom also migrate across the Bass Strait) have been in decline with re-classifications moving from near-threatened towards endangered. This demonstrates why it is critical to invest effort in near-threatened species before their status deteriorates and to review lists of regionally threatened species in a timely manner. When trying to preserve habitats, a stronger case can often be made when survey data is available, and a site is known to support multiple threatened species.

In the Western Port context, the Victorian Wader Study Group, which includes veterinarians amongst its diverse membership, has surveyed shorebirds since the 1970s. The citizen science data they have produced has been key in encouraging government to protect key areas for conservation.

Helping to Navigate Incentives for Biodiversity Projects

In December 2022, 188 countries adopted the Kunming-Montreal Global Biodiversity Framework (GBF) which set new targets aimed at halting extinction of known threatened species. It offers aid for biodiversity initiatives with financial incentives increasing, both in Australia and globally, through a GBF Fund and other avenues.[20] As one of 17 megadiverse countries that occupy less than 10% of the world's area but have 70% of biodiversity, Australia was a key player and donor.[21] Domestically, Australia committed to zero new extinctions and establishment of a voluntary Nature Repair Market to support landholders to restore and protect nature, which will operate alongside existing sources such as offsets funding provided by developers.

GBF initiatives will only succeed if communities, including veterinarians, take ownership and work to protect their indigenous flora and fauna through projects such as rewilding. This presentation has suggested possible roles to help individuals tweak their own imaginations on how they might work to reverse current extinction trends. It will need a great deal of creative and collective effort by many people around the world if future generations are to experience the beautiful natural areas and extraordinary biodiversity our generation has been so privileged to enjoy.

Acknowledgments: Thank you to Graeme Phipps from the R.A.R.E.S. Foundation for his review of this paper.

References

1. Intergovernmental Panel on Climate Change. Climate Change 2023: Synthesis Report. p. 7. Available at: https://www.ipcc.ch/report/ar6/syr/. Accessed June 1, 2023.

2. World Wildlife Fund. The Living Planet Report. 2022. Available at: https://livingplanet.panda.org/. Accessed June 1, 2023.

3. Brondizio ES, Settele J, Díaz S, Ngo HT, eds. *Global Assessment Report on Biodiversity and Ecosystem Services of the Intergovernmental Science-Policy Platform on Biodiversity and Ecosystem Services.* Bonn, DE: Intergovernmental Science-Policy Platform on Biodiversity and Ecosystem Services; 2019. doi.org/10.5281/zenodo.6417333.

4. Woinarski J, Burbridge A, Harrison P. Ongoing unravelling of a continental fauna: decline and extinction of Australian mammals since European settlement. *Proc Natl Acad Sci.* 2015;112(15):4531-4540. doi.org/10.1073/pnas.1417301112.

5. International Union for Conservation of Nature. Issues Brief: The Benefits and Risks of Rewilding. Available at: https://www.iucn.org/resources/issues-brief/benefits-and-risks-rewilding. Accessed June 1, 2023.

6. Wikipedia. Western Port. Available at: https://en.wikipedia.org/wiki/Western_Port . Accessed June 1, 2023.

7. The Ramsar Convention Secretariat. Wetlands of International Important. Available at: https://www.ramsar.org/about/wetlands-of-international-importance-ramsar-sites. Accessed June 1, 2023.

8. Western Port Biosphere. Mornington Peninsula and Western Port UNESCO Biosphere Reserve. Available at: https://www.biosphere.org.au/. Accessed June 1, 2023.

9. United Nations. Sustainable Development Goals. Available at: https://www.undp.org/sustainable-development-goals. Accessed June 1, 2023.

10. U.N. Environment Programme. Annual Report 2022. Available at: https://www.unep.org/resources/annual-report-2022. Accessed June 1, 2023.

11. Friends of the National Parks Foundation, Bali. Available at: https://www.fnpf.org. Accessed June 1, 2023.

12. Schwartz D. European Experiments in Rewilding: Oostvaardersplassen. Available at: https://rewilding.org/european-experiments-in-rewilding-oostvaardersplassen/. Accessed June 1, 2023.

13. British Association of Nature Conservationists. The golden rules of rewilding - examining the case of Oostvaardersplassen. *ECOS.* 2019;40(6).

14. Whisson D, Ashman K. When an iconic native animal is overabundant: the koala in southern Australia. *Conserv Sci Pract.* 2020;2:5e188. doi:full/10.1111/csp2.188.

15. International Union for the Conservation of Nature. Rewilding Thematic Group, Global Definition and Guiding Principles for Rewilding. Available at: https://www.iucn.org/our-union/commissions/group/iucn-cem-rewilding-thematic-group. Accessed June 1, 2023.

16. Healesville to Phillip Island Nature Link Inc. Available at: https://www.hpnl.org.au/. Accessed June 1, 2023.

17. State Government of Victoria. Energy, Environment and Climate Action. NatureKit. Available at: https://www.environment.vic.gov.au/biodiversity/naturekit. Accessed June 1, 2023.

18. State Government of Victoria. Energy, Environment and Climate Action. Victorian Biodiversity Atlas. Available at: https://www.environment.vic.gov.au/biodiversity/victorian-biodiversity-atlas. Accessed June 1, 2023.

19. Winthrope S. The invasive fungus threatening earth's biodiversity. Pursuit. 2019. Available at: https://pursuit.unimelb.edu.au/articles/the-invasive-fungus-threatening-earth-s-biodiversity. Accessed June 1, 2023.

20. Convention on Biodiversity. COP15: Final Text of Kunming-Montreal Global Biodiversity Framework. Available at: https://www.cbd.int/article/cop15-final-text-kunming-montreal-gbf-221222. Accessed June 1, 2023.

21. The High Ambition Coalition for Nature and People. Available at: https://www.hacfornatureandpeople.org/home. Accessed June 1, 2023.

How Non-Traditional Companion Animals Impact Climate Change

Gianluca Deli, MRCVS, DVM, CertAVP (ZM), GPCert (ExAP), PgCert (EAS), GpAdvCert (SMP), PgCert (HE), FHEA, VetGD Advisor, and Emanuele Lubian, DVM, GPCert (ExAP)

Session #3292

Affiliation: From Medivet Faygate, Tower road, Faygate, Horsham, RH12 4SD, UK (Deli), Clinica MyPet, Viale D. Ranzoni, 10 20149 Milano, IT (Lubian).

Note: Abstract format selected by author(s). See Front Matter for description of conference proceedings manuscripts submitted as scientific abstracts.

Abstract: This presentation will go through a wide-ranging bibliographic review and a reflection on how patients, non-traditional companion animals (NTCAs), can directly and indirectly affect climate change. The intent is to show how the maintenance and care of these animals can significantly affect the balance of our planet, much more than one might expect. The problem also is seen from the human point-of-view, where various factors come into play, such as natural disasters, epidemic, war, which further aggravates the problem. This is to underline how responsible management can also bring benefits and positive changes in the long term: preferential exploitation of local resources, reduction in the movement of goods (production of food in developing countries), breeding of mutations and their potential liberation (species allochthonous), reduction of the impoverishment of flora and fauna. Veterinary surgeons can play an important role both in daily clinical activity and in guiding and educating clients towards a more sustainable management of NTCAs.

References

1. United Nations. UN News: 5 things you should know about the greenhouse gases warming the planet. Available at: https://news.un.org/en/story/2022/01/1109322?gclid=Cj0KCQjwk7ugBhDIARIsAGuvgPY86dtb4iALK66YWyBiuQIke-C3kXQxpEewJQOLdMpMdniY0DUKBv8aAp7bEALw_wcB. Accessed June 1, 2023.

2. Protopopova A, Ly LH, Eagan BH, Brown KM. Climate change and companion animals: identifying links and opportunities for mitigation and adaptation strategies. *Integr Comp Biol.* 2021;61(1):166-181.

3. Benessaiah K, Sengupta R. How is shrimp aquaculture transforming coastal livelihoods and lagoons in Estero Real, Nicaragua? The need to integrate social-ecological research and ecosystem-based approaches. *Environ Manage.* 2014;54(2):162-179. doi:10.1007/s00267-014-0295-x. PMID: 24912580.

4. Nemecek T, Kägi T. Life Cycle Inventories of Agricultural Production Systems. 2007. Available at: https://db.ecoinvent.org/reports/15_Agriculture.pdf. Accessed June 1, 2023.

5. World Association of Zoos and Aquariums. A Guide. Reducing, Measuring, and Offsetting Carbon at Your Zoo or Aquarium. Available at: https://www.waza.org/wp-content/uploads/2022/09/WAZA-Carbon-Guide_WEB.pdf. Accessed June 1, 2023.

Section 48

Hybrid Special Session

Use of 3D Printed Surgical Models to Aid Exotic Animal Surgeries

Angela M. Lennox, DVM, DABVP (Avian Practice), DABVP (Exotic Companion Mammal), DipECZM (Small Mammal), and
Sean Bellefeuille

Session #3191

Affiliation: From Avian and Exotic Animal Clinic, 9330 Waldemar Road Indianapolis, IN 46268, USA (Lennox), Med Dimensions, LLC, Ontario, NY 14519, USA (Bellefeuille).

Abstract: The use of 3D printed anatomic and surgical models are useful for specific skill learning and can be used for complicated surgeries where a clear understanding of anatomy is critical for surgical success. A systematic review and meta-analysis of medical students learning with models showed increased satisfaction and equal or improved test scores than students learning from cadavers.[1] 3D print models are used successfully and being investigated in human surgery, for example for planning patient specific liver resection where disease and anatomic variation make surgical decisions challenging.[2] In veterinary medicine, a study showed similar learning results for veterinary students using horse limb bones of an actual specimen versus a 3D printed model.[3] Use of 3D model for teaching purposes and for surgical management of several cases are presented.

Models

Models were printed from computed tomography (CT) scans by Med Dimensions (med-dimensions.com). Med Dimensions uses patient medical images (CT, magnetic resonance imaging [MRI]) to create models, tools or implants to improve surgical accuracy and clinician confidence. Volumetric data from CT or MRI scans are used to extract anatomy of interest and create 3D reconstructions that can be analyzed and manipulated digitally. The 3D model is used to create either a 3D print for surgical preparation/rehearsal or a model in a computer-aided design software to create patient-specific tools. These tools can be cutting/drilling guides or can be custom implants created to fit the patient perfectly. The tools/implants are then 3D printed as well, typically in a different material than the models.

Models are typically 3D printed in plastic using an additive manufacturing method called fused deposition modeling, commonly known as extrusion based printing or filament printing. The materials can be multicolored to point out regions of interest or specific lesions like a tumor. Intraoperative tools like guides or implants are typically printed using more advanced methods such as stereolithography and they are typically made out of resins, metals or polymers. Most of these materials can be autoclaved which is very important to bring them into the operating room safely.

Applications

The applications for this technology are very broad. Orthopedics, neurosurgery or oncology come to mind for surgical preparation and execution. Educational models have been increasingly popular and in demand. With 3D printing and other associated technologies it has become easier to create models that are anatomically accurate and cost-effective. Recently, Med Dimensions has been exploring the combination of silicone molding with 3D printed parts to create multi material (multiple tissue type) models that reflect the anatomy and feel more realistic. An example of this application would be intubation models that have a realistic silicone airway.

References

1. Zhen Y, Aisher D, Hanming J, et al. The role of 3D printed models in the teaching of human anatomy: a systematic review and meta-analysis. *BMC Med Educ.* 2020;20(1):335.

2. Huber T, Hanke LI, Boedecker C, et al. Patient-individualized resection planning in liver surgery using 3D print and virtual reality (i-LiVR)-a study protocol for a prospective randomized controlled trial. *Trials.* 2022;23(1):403.

3. De Alcantara Leite dos Reis D, Gouveia BLR, et al. Comparative assessment of anatomical details of thoracic limb bones of a horse to that of models produced via scanning and 3D printing. *3D Print Med.* 2019;5(1):10-13.

Section 49

Hybrid Marla Lichtenberger Emergency and Critical Care Session

Salina Locke, DVM, DABVP (Avian Practice)
Session Chair

Stayin' Alive 3. You Can't Always Get What You Want: Brass-Tacks versus Gold-Standard Diagnostic and Treatment Options

Meera Music, BVM&S, MRCVS,
Alicia McLaughlin, DVM, CertAqVet,
Anneliese Strunk, DVM, DABVP (Avian Practice)

Session #3268

Affiliation: From Center for Bird and Exotic Animal Medicine, 11401 NE 195th St. Bothell, WA 98011, USA.

Abstract: Exotic animals commonly present to veterinary clinics on an emergency basis. Both acute and chronic diseases may manifest with an acute onset of symptoms due to the tendency of exotic species to hide signs of illness until it is advanced. Accurate diagnosis of the underlying cause for illness is essential to provide appropriate treatment, but not all owners can do all the recommended testing. In this presentation, authors discuss how to optimize the clinical approach to the critically ill exotic animal regardless of budget provided, and review common emergency presentations and recommended approaches for exotic mammals, reptiles, and birds.

Introduction

This presentation is intended to highlight the broad spectrum that exists within exotic emergency medicine between the brass-tacks approach needed to establish prognosis and guide treatment plans, and the gold-standard approach that often allows for a deeper understanding of a patient's disease process and improved outcomes for clients and patients. There is no one-size-fits-all approach that will work in every case, and it is possible to practice evidence-based medicine on a budget.[1-11] Flexibility and adaptability are important.

Pet owners who are dealing with financial hardship are often very connected to their pet(s). While they may not have monetary resources, they may be able to commit a significant amount of time and home management to their pets' medical care. When faced with a pet's illness, some of these pet owners may elect euthanasia preemptively to try to prevent pain and suffering, or from a lack of understanding of what long-term care may entail, even if there are inexpensive management options available. Clinicians who have less experience dealing with complicated health issues/diagnostic approaches on a budget may also jump to recommending euthanasia prematurely. While quality of life concerns must always be assessed carefully, and explicitly, it is the veterinarian's duty to advocate for the animal when appropriate. This includes discussing multiple management options and not making assumptions about what treatments owners are willing to consider. Clinicians have access to a wealth of information from the detailed history/husbandry assessment and physical examination alone. Thoughtfully interpreting these findings based on current scientific literature will allow the clinician to prioritize the case approach based on evidence.

Euthanasia

There are some situations where it is reasonable to consider euthanasia without diagnostics or attempting treatment. These can include the following situations.

- Severe disease or injury resulting in unrelenting pain and suffering, regardless of an owner's financial capability.
- When pet owners are unable or unwilling to commit the time and resources for even the bare minimum of diagnostics or ongoing care.
- Animals with multiple concurrent diseases identified on physical exam that are showing signs of declining quality of life.

Many diseases can be managed adequately with less than gold-standard care as long as a considerate approach is taken. While euthanasia may be a consideration in a budget case, it is possible for many exotic pets to survive even catastrophic disease processes with inexpensive case management, if the owner is dedicated to providing appropriate home care and follow ups as recommended. For example, an animal with traumatic exophthalmia or ocular rupture may be able to heal with medical management and develop a phthisical or comfortable eye without extensive surgical care.

Managing Expectations

A long-term plan for monitoring and treatment must be discussed from the outset. For example, dental disease in rabbits and rodents typically requires lifelong management with ongoing anesthetized occlusal adjustments and if owners are not properly prepared for this, it can result in decreased quality of life for the pet, significant disease progression, and client dissatisfaction. From the moment of diagnosis, owners should be provided with details about the need for and costs of the long-term management required for the rest of the patient's life.

Embarking on treatment for diseases that require significant home management (e.g., wounds and metabolic diseases like chronic renal failure and insulinoma) should include a full discussion about what home management and ongoing bare minimum monitoring entails. In some cases, rechecking blood work may be a vital part of disease management (e.g., monitoring glucose levels in a ferret with insulinoma). A stable cardiac patient with disease that has been historically well managed may be able to skip monitoring blood work every 6 months, even if this is the usual standard of care. This is dependent on the species involved, medical history of the individual patient, and the disease being treated. It is important to clarify to the client that management plans may need to be modified depending on disease progression.

Minimizing Stress

Minimizing stress should continue to be a part of baseline case management, including the use of minimal stress handling and medicating techniques and sedation when appropriate. Fear-free methods can be utilized by staff in hospital and be a part of owner home care. An owner with the ability to commit time towards training can make a huge difference in the quality of life for a pet requiring ongoing home care. For example, there are some bird owners who will not medicate their birds directly and use inadequate measures such as putting the medication on food items, which will likely be incompletely consumed, or lead to consumption of excessive treat foods. Prescribing training for owners to learn how to teach their birds to accept medication willingly allows for consistent and low-stress management of diseases.

Antibiotic Usage

Exotic animal medicine on a budget should not entail indiscriminate antibiotic usage. Every veterinary clinic should develop established guidelines for antibiotic use.[1,2] Although empirical medications may be necessary in some cases, boundaries and expectations must be set for the owner at the initiation of empirically used medications, to promote responsible use. If an infection is suspected, and culture or DNA sequencing of potential pathogens are out of budgetary range, cytology should be performed of lesions as a low-cost, high-yield diagnostic test to help guide antimicrobial usage, if any.

Management Considerations

When a client expresses concern about finances impacting medical care of an animal, the following steps can be very helpful.

- Get a budget from the client to assess whether it is reasonable to attempt management of the disease. Find out if there is any flexibility for financial input later or long term for the pet.

- Be sure to obtain clear, detailed histories including going through husbandry with a fine-toothed comb. Little things like variations in husbandry or timing of medication administration, among others, can make a big impact on treatment outcomes.

- Review the scientific literature, talk to colleagues, and search VIN and other online exotic forums for advice on managing the medical condition inexpensively.

- When making decisions about prescribing medications, think about the timing around starting or stopping administration of medications so you can easily evaluate if the treatment is helping the pet (e.g., therapeutic trials).

- Thoughtfully time follow-up visits based on the ability to have maximum impact for the visit; plan the recheck around an expected time to recheck diagnostic testing, and/or before the course of medications has been completed so that you can decide whether treatment should continue. Always provide written communication to clients. Consider video demonstrations of medication administration or other home treatments so that clients can refer to these and minimize the risk of mistakes in implementation.

- Provide the client with clear, objective measurement methods (e.g., weight, respiratory rate, quality of life assessments) to help keep track of how the pet is doing at home.

- Don't stop recommending and discussing high-quality medicine, even if a client expresses financial constraints. The goal is not to pressure the client, but to make sure that they are educated about possible negative outcomes associated with missing a concurrent or primary diagnosis due to the current case limitations. This will minimize surprises for them along the way, improve satisfaction in the communication process, and give the client the option to change their mind over time about the level of medicine they desire to pursue. Document all recommendations thoroughly in the medical record.

Different diagnostics can range significantly in significance of information gathered to treatment decisions, cost to the client, and possible risk to the patient (e.g., sedation or anesthesia needed, phlebotomy). Being able to convey clearly and unambiguously the pros and cons of different diagnostics is a vital skill for any clinician. This will help an owner to prioritize their finances in the face of a significant illness, as well as helping the clinician prioritize order of testing. Table 1 describes considerations for diagnostic tests that can be useful in management of budget cases.

Table 1. Diagnostic pros and cons.

Test	Cost[a]	Risk to patient	Time to take diagnostics, scheduling	Need for sedation, anesthesia	Potential uses	Other considerations
Single lateral radiograph (including entire caudal aspect of the patient)	$	Minimal	Immediate	Likely not needed.	Mammals straining to urinate, defecate or with hematuria.	May be stressful if patient is dyspneic (orthopneic posture exhibited). Lack of ability to triangulate (i.e., in the case of a ureteral stone that overlies the bladder).
Single ventrodorsal/ dorsoventral radiograph / Single "bird in a box" or "plop" radiograph	$	Minimal	Immediate	Likely not needed.	Useful to identify presence of mineralized, metallic structures or get assessment of respiratory tract, cardiac size. Helps minimize stress on dyspneic animals because it does not require handling.	Without orthogonal view, definitive interpretation of superimposed tissues/bone is not possible (e.g., stone in caudal urinary tract may be difficult to identify).
Orthogonal radiographs (2-3 view)	$$	Moderate, depending on species and stability of patient.	Immediate	May be needed.	Important to prevent misdiagnosis of a variety of disorders.	May be stressful if patient is dyspneic.
In-house ultrasound (ranging from AFAST/TFAST/FLASH or single organ scans to full abdominal or thoracic scans depending on clinician skill, comfort	$	Minimal	Immediate	Likely not needed.	Rabbits with anemia/elevated ALT; animals with urinary signs; coelomic or abdominal distention; increased respiratory effort.	Useful to identify source of mass effects, presence or absence of blood flow in an organ or mass, bladder or renal abnormalities, cystic or follicular structures, or free fluid within the thorax or abdomen.

Table 1 Continued. Diagnostic pros and cons.

Test	Cost[a]	Risk to patient	Time to take diagnostics, scheduling	Need for sedation, anesthesia	Potential uses	Other considerations
Echocardiogram, cardiologist consult	$$$	Moderate, depending on species and stability of patient.	Can take a week or more to coordinate with specialist.	May be needed.	Help confirm diagnosis and guide treatment of a variety of cardiac disorders.	May be difficult to arrange, depending on specialist availability. Specialists may require clinic to facilitate sedation/anesthesia for any procedures performed.
Full abdominal or thoracic ultrasound (by a radiologist or with consultation with a radiologist)	$$$	Moderate, depending on species and stability of patient.	Can take a week or more to coordinate with specialist.	May be needed.	Useful when searching for subtle lesions.	May be difficult to arrange, depending on specialist availability. Specialists may require clinic to facilitate sedation/anesthesia for any procedures performed.
Computed tomography (CT)	$$$	Moderate, depending on species and stability of patient.	Immediate if machine in house. Can take a week or more to coordinate with another facility.	Likely needed (aside from most non-contrast chelonian studies).	Useful for diagnosing a wide range of subtle disorders in patients of all sizes.	External facilities may require clinic to facilitate sedation/anesthesia for any procedures performed.
Cytology (in-house)	$	Minimal	Immediate	May be needed to obtain sample.	Can help guide antimicrobial therapy and inform prognosis.	Skilled staff needed, or doctor clinician may need to read themselves.
Fecal occult blood	$	None	Immediate	None needed.	Can confirm presence of GI bleeding in non-carnivores.	There may be interference from carnivore diets.
Urinalysis	$	Minimal-moderate, depending on if sedation or anesthesia is needed to obtain sample.	Immediate	May be needed.	Useful for assessing presence of pyuria, crystalluria, renal function, ketoacidosis.	None

Table 1 Continued. Diagnostic pros and cons.

Test	Cost[a]	Risk to patient	Time to take diagnostics, scheduling	Need for sedation, anesthesia	Potential uses	Other considerations
Cultures, next-generation DNA sequencing (MiDog)	$$$	Minimal-moderate, depending on if sedation or anesthesia is needed to obtain sample.	Prolonged time until results (7-10 business days in most cases).	May be needed to obtain sample for submission.	May be necessary to achieve good outcomes in cases with resistant infections.	MiDog provides resistance patterns based on genetic sequencing rather than sensitivity testing. Both culture and sensitivity and MiDog should be performed if a resistant or atypical infection is suspected.
In-house blood smear evaluation	$	Minimal-moderate, depending on if sedation or anesthesia is needed to obtain sample.	Immediate	Does not typically require sedation, depending on phlebotomy site.	Useful as a patient-side diagnostic test to determine if antibiotic therapy should be considered.	Skilled staff needed, or doctor clinician may need to read themselves. Very small sample needed in most cases (< 100 μl).
CBC (send out)	$	Minimal-moderate, depending on if sedation or anesthesia is needed to obtain sample.	Immediate with in-house point-of-care machine. 1-3 days for mammals, 3-5 business days for reptiles and avian patients.	Does not typically require sedation, depending on phlebotomy site. Very small sample needed in most cases (< 100 μl).	May be more accurate than in-house analysis depending on skill level of staff reading in-house labs. Useful as a second opinion.	Does not typically require large volumes of blood.
Point-of-care analyzer (iStat 1, Abbott Point of Care Inc. Princeton, NJ, USA)	$$	Minimal-moderate, depending on if sedation or anesthesia is needed to obtain sample.	Immediate	May require sedation, depending on phlebotomy site.	Useful for assessing specialized analytes immediately (glucose, ionized calcium, blood gasses).	Very small sample needed (< 100 μl) but must be unclotted.

Table 1 Continued. Diagnostic pros and cons.

Test	Cost[a]	Risk to patient	Time to take diagnostics, scheduling	Need for sedation, anesthesia	Potential uses	Other considerations
Glucometer	$	Minimal	Immediate	May require sedation, depending on phlebotomy site.	Rapid assessment possible.	Significant variations in precision and accuracy depending on species and machine.
Chemistry (in-house)	$$	Minimal-moderate, depending on if sedation or anesthesia is needed to obtain sample.	Immediate with in-house point-of-care machine.	May require sedation, depending on phlebotomy site.	Excellent way to assess prognosis quickly for some patients on emergency.	Will not include some analytes, depending on which rotors/test slides/cartridges are stocked/methodolog y use. May be challenging to get enough blood to run in house depending on patient size and machine in use.
Chemistry (send out)	$$	Minimal-moderate, depending on if sedation or anesthesia is needed to obtain sample.	1-3 days in most cases.	May require sedation, depending on phlebotomy site.	Same uses as in-house chemistry, but delayed results.	May be challenging to get enough blood to run in house depending on patient size. Volume of blood obtained will determine number of analytes able to be tested.
Surgery (mass removal, exploratory surgery with biopsies	$$$	High	Soon after need is identified, depending on availability of on-site surgical services.	Yes	May be the only way to diagnose and treat the underlying disease.	Underutilized tool. Consider surgical intervention as primary diagnostic in all-or-nothing cases. Risk may outweigh benefits in some patients.

[a]$ = < $100; $$ = $100-350; $$$ = > $350.

References

1. McLaughlin A, Music M, Strunk A. Stayin' alive: current topics in exotic emergency medicine. *Proc ExoticsCon.* 2019;20-33.

2. Music M, McLaughlin A, Strunk A. Stayin' alive 2: triage, medicine, and surgery. *Proc ExoticsCon.* 2022;489-499.

3. Music M, Strunk A. Reptile critical care and common emergencies. *Vet Clin North Am Exot Anim Pract.* 2016.

4. McLaughlin A, Strunk A. Common emergencies in small rodents, hedgehogs, and sugar gliders. *Vet Clin North Am Exot Anim Pract.* 2016;19(2):465-499.

5. Graham JE, Doss GA, Beaufrere H. *Exotic Animal Emergency and Critical Care Medicine.* 1st Ed. Ames, IA: Wiley-Blackwell; 2021.

6. Block G. Doing the most for our patients with the least: evidence-based veterinary care on the cheap. Available at: https://hsvma.memberclicks.net/assets/pdfs/veterinarycareonthecheap_handout.pdf. Accessed June 1, 2023.

7. Enzler MJ, Berbari E, Osmon DR. Antimicrobial prophylaxis in adults. Mayo Clin Proc 2011;86(7): 686-701.

8. Buisson C. Getting comfortable with hospice and palliative medicine. *Proc Southwest Vet Symp.* 2021.

9. Lichtenberger M, Lennox A. Critical care of the exotic companion mammal (with a focus on herbivorous species): the first twenty-four hours. *J Exotic Pet Med.* 2012;21:284-282.

10. Costello M. Principles of cardiopulmonary cerebral resuscitation in special species. *Semin Avian Exot Pet Med.* 2004;13(2):132-141.

11. Lichtenberger M. Shock and cardiopulmonary-cerebral resuscitation in small mammals and birds. *Vet Clin North Am Exot Anim Pract.* 2007;10:275-291.

12. American Veterinary Association. Antimicrobial use in veterinary practice. Available at: https://www.avma.org/resources-tools/one-health/antimicrobial-use-and-antimicrobial-resistance/antimicrobial-use-veterinary-practice/. Accessed March 24, 2023.

13. American Veterinary Association. AAFP/AAHA antimicrobial stewardship guidelines. Available at: https://www.avma.org/resources-tools/avma-policies/aafpaaha-antimicrobial-stewardship-guidelines/. Accessed March 24, 2023.

Containing Contagion and Understanding Zoonotic Disease Potential of Wildlife Species

Julia K. Whittington, DVM

Session #3404

Affiliation: From University of Illinois College of Veterinary Medicine, 1008 W. Hazelwood Dr., Urbana, IL 61802, USA.

Abstract: Avian influenza, salmonellosis, plague, mange, tularemia...and the list goes on. Exposure to infectious diseases, some of which are zoonotic, is likely in wildlife rehabilitation and care as many wild animals are affected by or carry these pathogens. Understanding disease transmission and preparing for the introduction of a potential pathogen into a health care facility is of utmost importance for biosafety and biosecurity. This presentation introduces important concepts in preventing disease transmission during wildlife care.

Wildlife, Infectious Diseases, and Zoonoses

Over 60% of known human pathogens and 75% of emerging diseases that affect humans are zoonotic. More than 50 zoonotic diseases are found in the United States alone, and this list is published as an appendix to the Compendium of Veterinary Standard Precautions for Zoonotic Disease Prevention in Veterinary Personnel.[1] Wildlife species are more likely to be associated with zoonotic diseases than domestic animals and over 80% of the zoonotic diseases found in North America are also found in wildlife.[1,2] Exposure to zoonotic diseases is an occupational hazard of veterinary professionals, and those working with live or dead wildlife species are at greater risk of coming into contact with potentially zoonotic pathogens.[3] Zoonotic diseases commonly associated with North American wildlife include hantaviruses, plague, rabies, and tularemia.[3]

Transmission of infectious pathogens requires a source of infection, a susceptible host, an effective route of transmission and a point of entry (Table 1).

Table 1. Route, point of infection, and high-risk activitie3s associated with transmission of zoonotic diseases of wildlife.

Transmission route	Point of infection	High-risk activities	Examples of disease exposure
Direct contact	Ingestion, contamination of mucous membranes, contact through non-intact skin	Handling animal without proper restraint or use of personal protective equipment (PPE)	Brucellosis, plague, rabies, *Salmonella* spp., tularemia, scabies, avian influenza
Indirect contact	Ingestion, contamination of mucous membranes, contact through non-intact skin	Handling contaminated equipment or material without proper PPE	Brucellosis, plague, rabies, *Salmonella* spp., tularemia, scabies, avian influenza
Aerosol	Inhaled particles or droplets deposited on mucous membranes	Power washing contaminated surface without PPE, handling animals with infectious respiratory disease without proper PPE	Hantaviruses, highly pathogenic influenza, plague
Vector-borne	Bites from an infected arthropod	Working with animals or in an environment without appropriate vector control	Lyme disease, plague, tularemia, Rocky Mountain spotted fever, West Nile virus

Source: Taylor T, Buttke D. Safe work practice for working with wildlife. In: Richgels KLD, Gibbs SEJ, Wild MA, eds. *Field Manual of Wildlife Diseases.* Reston, VA: U.S. Geological Survey; 2020:1-26.

Wildlife is susceptible to a myriad of infectious pathogens. In some cases, large-scale die-offs may be experienced during an outbreak event, and in other cases, individual animals may experience clinical signs while most of the population remains unaffected. Asymptomatic carriers of infectious disease are common in wildlife species. Therefore, it is prudent to maintain the mindset that a wildlife patient has the potential to be at risk for transmitting a zoonotic agent. This index of suspicion is the driving factor during outbreak investigations where the initial steps involve determining if the problem is an individual incident or a population problem. Early identification and quick implementation of a planned intervention can reduce the risk and severity of the incident or outbreak. During this process of case definition and classification, the veterinarian must define the cause of illness in the animal, confirming the diagnosis, while also looking for trends that pertain to the animal type, timing of the presentation, and location, that may indicate that the problem is at the population scale.[4] It is important to be familiar with the diseases of wildlife found in the region and any signs or symptoms seen in animals and humans. It is also prudent to stay abreast of local wildlife trends, including unusual mortality events. Maintaining open communication with wildlife care professionals in the area, as well as with local natural resources managers, will allow for networking and more rapid detection of an emerging issue.[3]

An incomplete summary of zoonotic agents of concern in wild animals is found in Table 2.

Table 2. Incomplete list of zoonotic diseases associated with wildlife of North America. Causative agent, host, route of transmission and associated symptoms are listed.

Disease or agent	Host	Transmission, incubation	Signs and symptoms of disease in humans
Campylobacteriosis *C. jejuni*	Multiple	Fecal-oral, 1-10 days	Diarrhea, ± mucous and blood, nausea, fever
Cryptosporidiosis *Cryptosporidium* spp.	Multiple	Fecal-oral, 3-7 days	Abdominal pain and cramping, profuse water diarrhea, anorexia
Dermatomycosis *M. canis, Trichophyton* spp.	Multiple	Direct, indirect contact, 4-14 days	Focal, flat, spreading circular skin lesions with peripheral crusted, scaly redness
Erysipiloidiosis *E. rhusiopathiae*	Rodents	Direct contact, 1-7 days	Local raised, reddened skin lesions, progressive, painful and itchy
Giardiasis *Giardia* spp.	Multiple	Fecal-oral, 1-4 weeks	Abdominal cramps, bloating, anorexia, nausea, intermittent diarrhea
Hantavirus Bunyaviridae	Rodents	Aerosolized, direct contact, 12-16 days	Severe pulmonary syndrome, fever, weakness, cough
Histoplasmosis *H. capsulatum*	Bats, birds	Aerosolized, 3-17 days	Asymptomatic, acute respiratory disease with chronic pulmonary pathology
Leptospirosis *L. interrogans*	Multiple	Direct contact, 7-12 days	Fever, headache, weakness, red and watery eyes
Lyme disease *B. burgdorferi*	Multiple	Vector-borne, 1-2 weeks	Bull's-eye rash, fever, muscle and joint pain
Lymphocytic choriomeningitis virus (LCMV) Arenaviridae	Rodents	Direct contact, Aerosolized, vector-borne, 1-3 weeks	Fever, weakness, headache, lymphadenopathy
Orf Poxvirus	Wild hoof stock	Direct, indirect contact, 3-6 days	Solitary skin lesion, discolored or pustular, progressing to weeping proliferative
Pasteurellosis *Pasteurella* spp.	Multiple	Direct, indirect contact, aerosolized, variable incubation	Local redness, swelling, pain, fever, can lead to septicemia
Plague *Y. pestis*	Rodents	Direct contact, vector-borne, 2-6 days	Fever, lymphadenopathy, septicemia
Psittacosis *C. psittaci*	Birds	Direct contact, aerosolized, 4-15 days	Fever, headache, weakness, respiratory disease, ± hepatitis
Q fever *C. burnetii*	Wild hoof stock	Direct contact, aerosolized, 2-4 weeks	Fever, headache, muscle aches, sweating
Rabies *Lyssavirus*	Mammals	Direct contact, 3-8 weeks	Headache, fever, malaise, sensory changes, excitability, behavior changes, hydrophobia
Salmonellosis *Salmonella* spp.	Multiple	Fecal-oral, 12-36 hours	Abdominal pain, diarrhea, headache, vomiting
Tularemia *F. tularensis*	Multiple, lagomorphs, rodents	Direct contact, aerosolized, 3-5 days	Dependent on point of entry, fever, headache, weakness, vomiting, diarrhea, pneumonia
Tuberculosis *Mycobacterium* spp.	Multiple	Aerosolized, direct contact, 4-12 weeks	Cough, weight loss, fatigue, fever

Biosecurity

Biosecurity refers to both management practices and physical measures taken to reduce the risk of introduction, establishment and spread of disease-causing biological agents, either from natural or intentional exposures, to humans, animals, plants, and the environment.[5] Biosecurity is based on the principles of preparation, prevention, response, and recovery. Basic to any biosecurity protocol are two main tenets: containment and prevention. These are achieved through identification of potential infectious agents, prevention of their introduction into populations or facilities, and control of their transmission through isolation, treatment, quarantine, or removal of affected subjects. Wildlife care facilities are presented with a unique challenge. Caregivers must protect hospitalized patients and staff from potential infectious disease exposure while providing treatment for injured, ill, or orphaned wild animals that may be the source of exposure.

Facility preparedness for dealing with the risk of infectious disease or zoonotic disease exposure will depend on its biosecurity, the monitoring of disease trends in wildlife populations, and open communication with staff, diagnostic laboratories, wildlife rehabilitators, wildlife biologists, and government agencies.

Standard Precautions

Standard precautions are a set of practices put in place to prevent the transmission of infectious diseases through blood or bodily fluids. These standards are universal in that they typically refer to basic measures needed to prevent human-to-human transmission of infection; however, the standards are considered best practice in the clinical veterinary setting as well. These include the following actions.

- Use protective barriers such as gloves and protective outerwear.
- No eating, drinking or smoking when working with animals or in animal care areas.
- Practice good hand hygiene.
- Avoid activities that increase risk of bites, scratches, needle sticks or other injuries.
- Provide good ventilation of the work environment.
- Disinfect soiled and contaminated surfaces and equipment.
- Handle tissues and biological samples safely.
- Seek medical attention when ill and advise care provider of potential infectious disease exposure.

Minimum standards for biosecurity are available; however, individual facilities must undergo their own risk assessment to personalize procedures. It is important to note that no biosecurity measure is 100% effective and each measure, if adhered to, adds layers of protection to the total plan (Fig 1). Modification of the physical design of an environment, although difficult and costly, is the most effective means of biosecurity enforcement as it does not rely on human behavior or compliance. Control measures used to mitigate the potential for zoonotic disease transmission vary in effectiveness and are cumulative. Those that remove the risk or replace the risk with a less risky option are preferred to those that simply manage the risk. Prevention strategies that utilize control measures to interrupt potential pathogen transmission at multiple points along the continuum of source, susceptible host, effective route, and point of entry are preferred.[1]

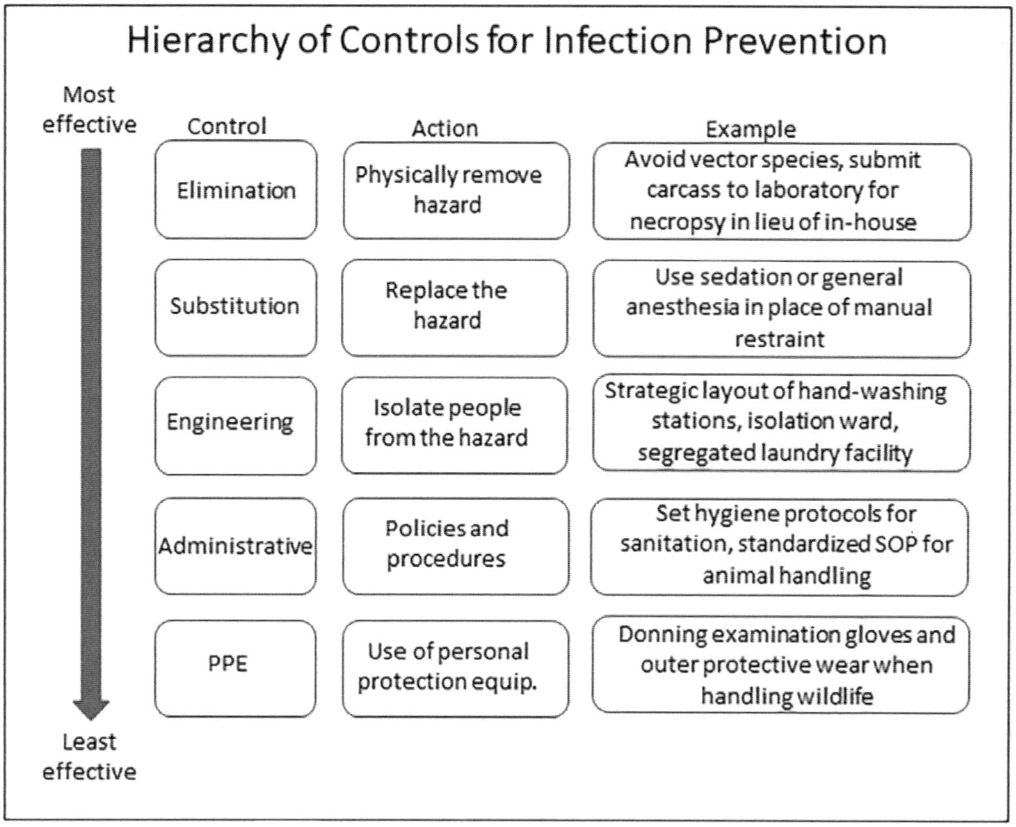

Figure 1. The hierarchy of controls for infection prevention indicating most effective to least effective control measure. Control measures are additive in effectiveness and should be implemented as needed to mitigate risks of exposure to infectious pathogens.

Hand hygiene

Hand hygiene is the single most important veterinary standard precaution in the prevention of disease transmission. Hand hygiene includes hand washing with soap and water or the use of alcohol-based hand rubs. Hand hygiene is important when handling potentially infectious animals or being in contact with contaminated materials or surfaces. It is important to practice good hand hygiene between handling individual animals or groups (e.g., litter or clutch).

Hand washing

Most zoonotic pathogens are transient flora and reside in the upper layers of the stratum corneum of the skin. The mechanical abrasion experienced during hand washing in conjunction with the detergent properties of soap effectively remove or deactivate most potential pathogens present on the hands. Commercial soap is appropriate for hand-washing stations, but bacterial overgrowth is possible with refillable containers that are not emptied and cleaned well prior to refilling. The temperature of the water used during hand washing is only important in promoting good compliance with the technique but has no effect on actual microbe removal. Moisturizing soaps are a reasonable option that may help hands from becoming chaffed. Nails should be kept short and clean. Artificial nails and hand jewelry that can harbor potential pathogens should be avoided when working with wildlife. The correct hand washing procedure involves wetting the hands with running water, placing soap in the palms, and rubbing the hands together to make lather. All surfaces should be engaged for a total of 20-30 seconds. The hands are then rinsed with running water and dried using a disposable towel. The towel can be used to turn off the faucet if there is concern that the faucet handle is contaminated.

Alcohol rubs

When soap and a water source is unavailable, alcohol-based hand rubs may be used for hand hygiene. Alcohol rubs should be between 60-95% ethyl or isopropyl alcohol to allow for fast-acting denaturing of the microbial protein. Alcohol rubs have broad microbial spectrum and are well tolerated due to the presence of emollients in the products. Organic debris and visible soiling on the hands should be removed prior to applying the alcohol rub. The correct procedure for using an alcohol hand rub is to remove all gross debris, then place the alcohol-based hand rub in the palms. Apply the rub to all surfaces of the hands and rub the hands together until dry.

Personal protective equipment

Even though most veterinarians have knowledge and concern about risks associated with zoonotic pathogens, a large number do not engage in the use of appropriate personal protective equipment (PPE), even when working with animals displaying signs of infectious disease.[6] No PPE is 100% effective. Protective measures must be augmented with good behavioral practices including appropriate hand hygiene. The effectiveness of using PPE depends on its consistent use and accessibility.[1] If the risk of zoonotic pathogen exposure exists, then employers are required to provide equipment to mitigate that risk and training on the appropriate use of that equipment. Training should include what PPE to use and when, how to use it including putting it on and taking it off, how to maintain and dispose of the PPE, and what limitations are inherent in the use of PPE. Non-disposable PPE must be disinfected before its storage or reuse.

Gloves

Hand protection should be appropriate for the activities. Disposable gloves made of lightweight nitrile or vinyl offer protection against infectious disease while heavy gloves are more appropriate for protection from injury. Ideally, exam gloves should be worn inside any outer glove being used for restraint. Gloves are barrier protection between the potential pathogen and the hands of the handler. They should be used when handling biological specimens, when the handler has lost skin integrity, and when performing procedures that have a higher risk for contacting potential infectious organisms. These include dental procedures, obstetric procedures, resuscitation, and necropsy. Gloves should also be worn when cleaning cages, litter boxes, contaminated equipment or surfaces, and handling soiled laundry. Unlike in human medicine, veterinary practitioners very often do not wear gloves when examining healthy animal patients unless the index of suspicion for the presence of a zoonotic disease is high. This should not be the case for wildlife; gloves should be worn whenever handling a wild animal. Additionally, gloves should be changed between examination of animals, between performing 'dirty' and 'clean' procedures, and when torn. Care must be taken to avoid contact between contaminated gloves and mucous membranes or non-contaminated surfaces. Glove removal is performed by pulling the glove off each hand inside-out, avoiding contact with the exterior of the glove. Hand washing must still be performed after removing the gloves.

Outerwear

If possible, protective clothing such as lab coats or smocks should be worn over street clothes to limit the transfer of potential pathogens via direct or indirect contact. Avoid wearing protective outerwear outside of the clinical environment and change or clean the garment when it becomes soiled. Disposable gowns should be used any time infectious disease is suspected and the garment should be used solely for that individual animal to avoid indirect transmission to another. Foot protection is important in a comprehensive biosecurity plan and should be suitable for the specific environment. Disposable foot covers add a layer of protection over street shoes and should be removed when leaving the contaminated environment.

Facial and respiratory protection

Eye protection in the form of safety glasses or goggles protects eyes and conjunctiva from trauma as well as exposure to potential pathogens. It should be used when splashes or sprays are likely. Respirators and face shields are recommended for all persons potentially exposed to air-borne pathogens. The National Institute of Occupational Safety and Health has approved disposable particulate respirators (N95, N99, or N100) as a minimum standard of respiratory protection against viral pathogens such as avian influenza. These require individual health assessment, training, and fit testing to ensure effective protection. Surgical face masks do not offer adequate protection against particulate air-borne pathogens.

Animal management

A fundamental concern for wildlife care facilities is cross-contamination or co-mingling of species. The potential for transmission of zoonotic diseases between wild and domestic or zoo species increases with their proximity and interaction. Human traffic and common treatment areas raise the risk for pathogen dissemination. Patient intake and movement are key points of control regarding biosecurity. If the safe and effective handling of a wild animal in the clinical care setting is impractical, while also maintaining biosecurity for other patients and humans, do not attempt to provide the care and refer the patient to a more appropriate facility.

Patient intake

Ideally, wildlife and domestic animals should be segregated at all times. Wildlife being admitted should bypass the waiting room and be moved directly to a quarantine area within the clinical space dedicated to wildlife. Classic signs that the animal is suffering from a potential infectious disease include neurologic signs, diarrhea, respiratory tract signs, fever, wounds or chronic infections, and irritated or inflamed skin. Assume there is a potential for an infectious etiology even if clinical signs are not immediately evident. If a common clinical space is used for triage or exam, do not use the space for a different patient until it has been thoroughly disinfected. Limit contact of animals with suspected zoonotic disease to essential personnel.

Patient screening

Establishment of a minimum health screening protocol for admitted patients is essential to identify infected animals and disease trends. Diagnostic procedures should include physical examination, baseline bloodwork, fecal evaluation, and survey radiographs. Further testing should be based on individual animal presentation and species. Consideration of diseases endemic to the facilities' region must be considered when developing specific screening protocols. Screening does not preclude the need for quarantine as incubation periods for infectious diseases vary. Medical records for patients should include information such as the age, sex, and species of the animal as well as the location the animal originated to facilitate disease monitoring.[7]

Most reported injuries among veterinary personnel occur during animal handling or treatment.[8] When performing animal handling, be prepared. If manual restraint is risky, consider sedation or anesthetic restraint of the patient. If working with large wildlife species, always have an escape plan ready. Preventing animal-related injuries requires pre-planning, appropriate equipment, and good communication among the clinical care team. Have a post-exposure/injury management plan established for use in an emergency.

Necropsy and handling biological samples

Useful information can be learned from a necropsy, but necropsies can be a high-risk procedure due to the possibility of exposure to infectious agents in body tissue, fluids, and aerosols, as well as the possibility of injury.

In-house necropsy

Only essential personnel should participate in a necropsy to limit exposure. Minimally, protective gloves, facial protection, and impermeable protective outerwear should be worn while performing the necropsy. When there is a concern for aerosolization of particles, eye wear and respiratory PPE should be used. The area in which the necropsy is to be performed should be well ventilated and have good lighting.

Submission to outside laboratory

If the necropsy is performed at an outside laboratory, or if biological samples are being submitted to an outside lab for processing, a thorough history with the submission is of critical importance. Be sure to clearly label all samples. Assume that infectious disease is a differential diagnosis for wildlife and wear gloves when handling samples. If there is information that raises the index of suspicion for an infectious and/or zoonotic etiology, clearly note that in the submission information. Underreporting of zoonotic disease potential significantly increases the risk of exposure for laboratory personnel.[9]

The U.S. Department of Health and Human Services, in partnership with the Public Health Service, Centers for Disease Control and Prevention (CDC), and National Institutes of Health publishes the Biosafety in Microbiological and Biomedical Laboratories (BMBL) as an advisory document recommending best practices for the safe conduct of work in biomedical and clinical laboratories from a biosafety perspective. The document defines four primary biosafety levels for laboratories working with animal pathogens and details facility design features and safety equipment, facility practices and procedures, and personal protective equipment to mitigate risks.[10] Since these standards are already in place for diagnostic laboratories, it is prudent to submit the carcass of an animal suspected of having a zoonotic disease to a commercial laboratory for post-mortem evaluation. In addition, the CDC, in conjunction with the U.S. Department of Agriculture, administer the Federal Select Agent Program that determines which agents and toxins have the potential to pose a severe threat to both human and animal health.[11] Listed among these are several pathogens known to cause zoonotic disease in wildlife including *Francisella tularensis* and *Yersinia pestis*, both Tier 1 agents of gravest concern, as well as *Coxiella burnetii*, Eastern equine encephalitis virus, avian influenza virus, Newcastle disease virus, *Brucella abortis*, and *B. melitensis*. If any of these infectious agents are suspected or identified, maintain biosecurity and contact your state veterinary and/or public health officials.

Facility management

Isolation, quarantine, and segregation

Isolation is a critical aspect of biosecurity that can contain the spread of infectious diseases in wildlife care facilities. To protect hospitalized patients, animals suspected of infectious disease should be placed in an isolation room with appropriate precautions and signage. Equipment and PPE should be exclusive to this area, and protocols should be clearly posted. Ideally, isolation rooms have negative air pressure in relation to the adjacent rooms and should exhaust air directly outside or have the air recirculated using a HEPA filter. If a true isolation area is not available, a room dedicated to treating infectious animals can be used. The purpose of quarantine is to segregate an animal that may be incubating a disease from the main population while monitoring it for signs of clinical disease. Clinical signs of infectious disease are often vague and non-specific. All animals should be considered potentially infectious and placed in quarantine before grouped with other patients. Wildlife should be kept segregated from domestic/captive patients. Whenever possible, dedicated handling equipment should be used for wildlife and those suspected of infectious diseases. Equipment that can't be dedicated for use (e.g., ultrasound unit) should be disinfected between patients.

Ventilation

Controls to prevent the spread of airborne contaminants focus on proper ventilation of clinical space including the dilution and removal of contaminants, the pressurization of wildlife clinical space in relation to other spaces, and the pressurization of the building to the outdoors.[11] Despite the importance of proper ventilation, most veterinarians are unfamiliar with the ventilation in their facility. Clinic design should ensure that air contaminants are being filtered or exhausted and not introduced to surrounding clinic space. To maximize the dilution and removal of contaminants, air changes per hour should equal or exceed twelve.[11] Similar to isolation areas described previously, air pressure should be negative with relation to the rest of the clinic.

Sanitation

Surface contamination with some pathogens may persist for months. Disinfectants effective against a broad range of pathogens are ideal for veterinary facilities that care for wildlife. Common disinfectants with good spectrum of activity for veterinary use include quaternary ammonium compounds, hypochlorite (e.g., household bleach), and hydrogen peroxide-based oxidizing agents.[1,13] Each of these should be stored, diluted, and used according to recommended standards. The CDC maintains a list of all disinfectants labeled to be efficacious against potential pathogens and should be consulted when preparing a biosecurity plan.[14] Frequent routine cleaning and removal of gross contamination from surfaces and equipment, regardless of the disinfectant used, is essential for optimal sanitation. Be sure to include high-touch non-animal surfaces such as doorknobs, cage door handles, and faucet handles. It is prudent to create a facility sanitation plan and schedule. If a contaminated surface is porous or the material is organic (e.g., soil or bark for burrowing, branches for perching, etc.), it is best to discard the item and replace it.

Foot baths

Disinfectant footbaths and footpads, using peroxygen-based disinfectant, have been shown to reduce bacterial numbers on footwear. Both mechanical removal of organic debris and the antimicrobial activity of the disinfectant help achieve this effect. Adhesive mats offer an alternative and remove contaminants through their adhesive and antimicrobial properties. These measures do not sterilize footwear and should not be relied upon solely to prevent pathogen transmission. Bioaccumulation will require frequent changes of footwear disinfectants. In wildlife care facilities where prevention and containment are equally important, the use of disposable shoe coverings may be useful. Personnel training to ensure proper removal and disposal is essential to avoid hand contact with contaminated shoe covers.

Medical waste

The disposal of medical waste is regulated at the state level by multiple agencies including the Environmental Protection Agency, which categorizes medical waste as infectious, hazardous, radioactive, and other. Veterinary hospitals, including wildlife care facilities, most often generate infectious or hazardous waste including sharps, tissues, contaminated materials, and dead animals. Dedicated biohazard receptacles, including sharps containers, should be made available and used routinely in wildlife care facilities. Standard operating procedures to prevent needle sticks should be followed and include proper restraint of the animal, keeping needles capped until in use, never placing needles or syringes in the mouth, using the scoop method to recap needles, and removing uncapped needles from a syringe using an instrument.

Laundry

Soiled laundry may also pose a hazard for exposure to potential pathogens and proper handling is required to minimize the risk of disease transmission. Separate clean and dirty laundry bins should be used, and gloves should be worn when handling soiled laundry. Machine washing with standard laundry detergent followed by drying is effective in removing contaminants. Wash personal laundry separately from laundry soiled during animal care.

Vector control

Rodents and insects are important vectors for transmission of many diseases. Patient triage often includes implementation of routine or as needed ectoparasite control. Additionally, insect vectors such as flies and gnats must be controlled. Effective prevention and control of rodents is dependent on limiting access to food and nesting material. Food stores should be stored in air-tight containers and access to paper, clutter, or hay should be minimized. Food for animals should be stored and prepared separately from food intended for human consumption.

Personnel management

It is essential that worker education is a priority. Occupational health and safety plans, based on identification of facility hazards, should be developed. All personnel must be trained regarding zoonosis awareness, infectious disease prevention and control practices, safe animal handling and behavioral cues of animals they will be handling, the proper use and disposal of PPE, safe handling of drugs and equipment, proper handling and storage of biological tissues, basic first aid, and emergency protocols. In addition, employees must understand the potential risk in working with wildlife and the risk posed to other animals that they interact with (e.g., domestic patients, personal pets, production animals). Providing baseline training for new employees, as well as regular refresher training for current personnel, is critical. Training completion should be documented, and the employee provided a written copy of all relevant material.[1]

Vaccination against rabies, tetanus, and seasonal influenza are often recommended or required for persons working with wildlife. Additionally, veterinary practitioners who are at high risk for unrecognized exposure to rabies should have their serum anti-rabies virus antibody titer checked every two years. The Advisory Committee on Immunization Policies recognizes a titer of ≥ 0.5 IU/ml as correlating with protection for rabies. Those with potential exposure that is recognized should pursue post-exposure prophylaxis.[15] Vaccination against hepatitis B is recommended for those working with nonhuman primates, and individuals working with marine mammals and nonhuman primates should speak with their healthcare provider about annual tuberculosis testing. Immunocompromised and pregnant personnel should be encouraged to speak with their healthcare provider to understand the risks and necessary measures to mitigate those risks when working with wildlife.

Creating and Implementing an Infection Control Plan

Providing care for wildlife patients is rewarding on both professional and personal levels. However, there is risk to the individuals working with these animals, and to animals that may come into direct or indirect contact with pathogens of wildlife patients. Wildlife care facilities are in a unique position to identify emerging disease or bioterrorism through care of wild animal sentinels.[8] Preparedness planning for infectious disease outbreaks and developing biosecurity protocols will help ensure the safety of personnel while continuing to care for hospitalized wildlife.

The Occupational Safety and Health Administration (OSHA) sets specific standards for employer responsibility to provide PPE and respiratory protection for workers who are potentially exposed to infectious agents.[16] Some states also set standards (e.g., California, General Industry Safety orders exclusively for veterinarians). Infection control plans that address these requirements should be developed prior to being needed and must be reviewed regularly to ensure that the plan is meeting evolving needs. Each plan should be tailored to the individual practice and should provide clear, thorough guidance for how risks associated with infectious disease exposure will be mitigated. Templates for such plans are available (Williams) and often follow standardized veterinary precautions as discussed above. The plan must indicate which personnel are responsible for each area discussed and the activities assigned. It will be necessary to identify essential personnel responsible for oversight of facility operation in the event of emergencies and these people will need to be fitted with appropriate PPE. Contact and resource information for staff to use if questions arise should be provided. Every staff member should certify that they have received and understand the training. Once a plan is established, there should be a non-punitive mechanism for reporting untoward events that resulted from the plan not being followed or failing. This allows for each team member to have ownership in the plan's success and for trends to be identified that indicate the plan needs to be modified. Integrating risk management processes into daily clinic operation promotes a positive culture of safety.

References

1. Williams CJ, Scheftel JM, Elchos BL, et al. Compendium of veterinary standard precautions for zoonotic disease prevention in veterinary personnel: National Association of State Public Health Veterinarians: Veterinary Infection Control Committee 2015. *J Am Vet Med Assoc.* 2015;247(11):1252-1277.

2. Jones KE, Patel NG, Levy MA, et al. Global trends in emerging infectious diseases. *Nature.* 2008;451: 990-993.

3. Taylor T, Buttke D. Safe work practice for working with wildlife. In: Richgels KLD, Gibbs SEJ, Wild MA, eds. *Field Manual of Wildlife Diseases.* Reston, VA: U.S. Geological Survey; 2020:1-26.

4. Lim ZJK, Xie S. Application of epidemiology and principles of herd/flock health or the exotic animal veterinarian. *Vet Clin North Am Exot Anim Pract.* 2021;24:495-450.

5. Terrestrial Animal Health Code. World Organisation for Animal Health (OIE). 2022. Available at: https://www.woah.org/en/what-we-do/standards/codes-and-manuals/terrestrial-code-online-access/. Accessed April 2, 2023.

6. Wright JG, Jung S, Holman RC, et al. Infection control practices and zoonotic disease risks among veterinarians in the United States. *J Am Vet Med Assoc.* 2008;232(12):1863-1872.

7. Miller E, ed. *Minimum Standards for Wildlife Rehabilitation.* 3rd Ed. St. Cloud, MN: National Wildlife Rehabilitators Association; 2000.

8. Nordgren LD, Gerberich SG, Alexander H, et al. Evaluation of factors associated with work-related injuries to veterinary technicians certified in Minnesota. *J Am Vet Med Assoc.* 2014;245:425-433.

9. Morse S, Henkel R. *Francisella tularensis*: understanding reported occupational exposure and laboratory methods used for the identification of *Francisella tularensis*. *Appl Biosafety.* 2018;23(1):11-18.

10. Centers for Disease Control and Prevention. Biosafety in microbiological and biomedical laboratories (BMBL), 6th Ed. CDC Laboratories. Strengthening Lab Safety. Available at: https://www.cdc.gov/labs/BMBL.html. Accessed April 3, 2023.

11. Centers for Disease Control and Prevention. Select Agents and Toxin List. Available at: https://www.selectagents.gov/sat/list.htm. Accessed April 2, 2023.

12. Centers for Disease Control and Prevnetion. Background C. Air. Environmental Infection Control Guidelines. Infection Control. Available at: https://www.cdc.gov/infectioncontrol/guidelines/environmental/background/air.html. Accessed April 3, 2023.

13. The Center for Food Security& Public Health. Iowa State University. The Antimicrobial Spectrum of Disinfectants. 2018. Available at: https://www.cfsph.iastate.edu/pdf/antimicrobial-spectrum-of-disinfectants. Accessed April 2, 2023.

14. Centers for Disease Control and Prevention. Chemical Disinfectants. Infection Control. Available at: https://www.cdc.gov/infectioncontrol/guidelines/disinfection/disinfection-methods/chemical.html. Accessed April 2, 2023.

15. Rao AK, Briggs D, Moore SM, et al. Use of a modified preexposure prophylaxis vaccination schedule to prevent human rabies: recommendations of the Advisory Committee on Immunization Practices – United States, 2022. *Morb Mortality Wkly Rep.* 2022;71(18):619-627.

16. U.S. Department of Labor Occupational Safety and Health Administration. Safety and Health Topics. Healthcare. Standards. Available at: https://www.osha.gov/healthcare/standards. Accessed April 3, 2023.

Wildlife: Coping with Common to Crazy Emergencies

Heather Wilson Barron, DVM, DABVP (Avian Practice), CertAqV

Session #3221

Affiliation: From Loggerhead Marinelife Center, 14200 US Hwy 1, Juno Beach, FL 33408, USA.

Abstract: Wildlife rehabilitation centers and veterinary hospitals that commonly admit wildlife patients often encounter a wide variety of wildlife from a large geographic area. As a result, these entities may be well-suited to participate in disease surveillance and One Health objectives. Obstacles to participation may include permit restrictions, financial disincentives, staff safety, absence of mandates to collect relevant data, and lack of contact between wildlife and public health agencies. Practitioners may be greatly rewarded for investment of time and other resources by gaining more knowledge about wildlife diseases that impact humans or companion and food animals.

Introduction

Wildlife may present to veterinary hospitals for a number of emergent conditions. Although trauma is the primary reason that wildlife species are admitted, a variety of pathogens and toxins should also be considered. Additionally, the health of wildlife species has received growing attention both from a conservation perspective and from the global 'One Health' initiative, which integrates human, animal, and ecosystem health. Aside from their intrinsic value, wildlife can be important as sentinel species for both the health of the environment and people. Of the 1,461 infectious diseases now recognized in humans, approximately 60% are due to multi-host pathogens characterized by their movement across species lines. Over the last three decades, approximately 75% of new emerging human infectious diseases have been zoonotic; however, routine surveillance for pathogens in free-ranging wildlife presents financial, logistical, welfare, and other constraints. Veterinary hospitals and wildlife rehabilitative centers may present an alternative source to facilitate the monitoring of wildlife populations for a wide variety of pathogens. Data collected this way may be biased, but if the ecology of the tested species and the specific temporal, spatial and age group distribution of admissions are considered, then the information can be a useful window into the health of regional wildlife.

Wildlife Triage

Triage is the process of prioritizing patients based on presenting condition and the resources available to deal with that condition. Therefore, triage of wildlife may necessarily vary to some degree based on individual philosophy, case load, local, state, and federal regulations, and the presenting situation (i.e., oil spill affecting hundreds of animals versus the normal daily load of injury and illness). In any case, the primary objective to keep in mind during the triaging process is eventual return of the patient to the wild. If triaging wildlife in a private practice setting, there are some considerations that may help set triage policy. Guidelines about what the practice is willing to accept should be made clear to staff and to the public. Recognize when the animal should be returned to the wild, as is often the case with abducted neonates. Also be aware of reasons why a wild animal may not be releasable. The goal of wildlife medicine is always eventual release and the possibility of euthanasia needs to be kept in mind during triage if chances of release are remote. Realistic resource allocation is especially important in a for-profit clinic that is underwriting wildlife treatment.

Pathogens Under Surveillance

Surveillance is critical for identifying and monitoring pathogen activity in wildlife populations, and although there are some limitations, wildlife and veterinary hospitals have been shown to be useful in monitoring for a variety of pathogens, including West Nile virus, Newcastle disease virus (NDV), avian influenza, bacteria such as *Escherichia coli* and *Salmonella* spp., and the following pathogens listed here.[1-9]

Ranavirus

Ranavirus typically exists at low prevalence (less than 5 percent of individuals infected in a population), and then rapidly emerges, with mortality often exceeding 90 percent in multiple species. With one in three amphibian species and over 40 percent of turtles at risk of extinction, *Ranavirus* represents a significant threat to herpetofaunal biodiversity. Ranaviral disease has been likened to Ebola virus for amphibians because their bodies swell and bleed profusely. The reasons for *Ranavirus* emergence at a site vary, but often are related to stressors that can be natural or anthropogenic (human-related) in origin. For example, outbreaks might signal that local ponds and wetlands are becoming more susceptible to disease under the stressors of climate change, pollution and development. A *Ranavirus* epizootic event was documented involving several Florida box turtles (*Terrapene carolina bauri*) from Sanibel and Captiva islands. Clinical signs included facial swelling, stomatitis, open-mouth breathing, mucopurulent nasal discharge, bilateral chemosis, dehydration, emaciation, and skin sloughing over the head and neck. The animals were either dead or euthanized upon presentation to the clinic. Postmortem findings in all carcasses examined included buphthalmos, oronasal discharge, fibrinonecrotizing stomatitis, glossitis, and red-tinged fluid and blood within the coelomic cavity. Histologic findings included fibrinonecrotizing and heterophilic stomatitis and glossitis, as well as marked skeletal muscle degeneration. *Ranavirus*-specific polymerase chain reaction (PCR) was done on blood samples and oral swabs when available, from all turtles ($n = 8$) examined. Sequencing and phylogenetic analysis confirmed FV3 in 4/7 oral swabs and 3/8 blood samples. Immunohistochemical staining for *Ranavirus* sp. confirmed the presence of antigen in the oral mucosa and peri-oral skin (three turtles) and in the kidney (one turtle). Diagnosing *Ranavirus* infection in free-ranging box turtles can be challenging in chronic stages of disease. Box turtles may live for a hundred years or more, but their populations are increasingly threatened by disease, habitat loss, vehicle strikes, and illegal collection. Reptiles can be good indicators of ecological health because of their reliance on the environment for appropriate physiologic functions, such as immunity, digestion, and thermoregulation.

Newcastle disease virus

Paramyxoviruses can be widespread in avian populations, but do not always cause illness. NDV was previously synonymous with avian paramyxovirus type 1 (APMV-1); however, due to changes in taxonomy is now referred to as avian avulavirus. The virus strains can be differentiated as lentogenic, mesogenic, or velogenic based on a mean death time in chicken embryos. Currently, velogenic strains of NDV are not endemic in U.S. domestic poultry; however, these strains are present in other countries and are occasionally detected in wild birds in the United States, which could be potentially devastating for the poultry industry, making wildlife reservoirs a concern. This disease is listed by the World Health Organisation for Animals (OIE) as vitally important for avian species and products. Virus detection in a specific geographical location often leads to trade restrictions and embargos and thus virulent NDV is listed as a Tier 1 U.S. Department of Agriculture Select Agent as it is a pathogen of national concern and a significant threat to animal agriculture in the United States.

Double-crested cormorants (*Phalacrocorax auritus*) (DCCO) are the wild avian species most associated with NDV illness; however white pelicans and ring-billed gulls have also been affected during outbreaks. Depending upon the strain, signs of infection can vary from no signs to dehydration, emaciation, weakness, diarrhea, and neurologic signs such as depression, wing or leg paralysis, clenching of toes, torticollis, ataxia, and head tremors. NDV can spread from wild avian populations to domestic poultry populations; however, thanks to good biosecurity, this has been documented very rarely and only in association with DCCO virus outbreaks. NDV is not considered a risk to human health, with signs of conjunctivitis and mild flu-like symptoms being the most severe clinical signs demonstrated.

North American river otter parasites of clinical concern

North American river otters (*Lontra canadensis*) (LOCA) may have a variety of parasites that can have an impact on health, especially in a rehabilitation setting or in relocation efforts. *Dirofilaria lutrae* and *Dirofilaria immitis* have been reported in LOCA with the former species considered otter-specific. In-house ELISA tests for heartworm disease may be positive with both infections because female antigens are conserved across species. Confirmation of a positive result, either by an echocardiogram or examination of blood for microfilaria or a Knott's test, is needed. Adult *Di. lutrae* worms are found subcutaneously and the microfilaria in the blood are much larger than those of *D. immitis*. The subcutaneous abscesses that can develop due to *Di. lutrae*, may be lanced and the worm gently removed. The wound should be flushed, and the animal treated with doxycycline and ivermectin. Otters in an outdoor rehabilitative setting should ideally be placed on monthly heartworm preventative. Another differential for worms found subcutaneously in NA otters is *Dracunculus lutrae* (or Guinea worm). When these worms or nodules are lanced, larvae with long taping tails will be detected vs. the microfilaria that would be present with *Di. lutrae* infections. Also, no *Dr. lutrae* larvae enter circulation. Local hair loss and skin damage due to self-inflicted trauma (scratching) may occur with *Dr. lutrae*. There is no treatment other than physical removal and because *Dracunculus* have no evidence of *Wolbachia* endosymbionts, treatment with doxycycline would probably not be indicated. Animals with evidence of pneumonia should be checked for an eosinophilia on CBC. If present, this may indicate a high probability for *Crenosoma goblei* (lungworm). Although *C. goblei* larvae may be found on a fecal exam, bronchoscopy or bronchoalveolar lavage may be more sensitive. In the author's hands, treatment with fenbendazole (50 mg/kg PO q24h for 3 days) appeared to be more effective than ivermectin (0.3 mg/kg PO) in treating several individuals.

Anticoagulant Rodenticides

In some populations, more than 80% of individuals test positive for anticoagulant rodenticides (AR), pointing to the potential for ecosystem-wide effects. This secondary exposure can be directly lethal to some mammalian and avian predators, including endangered species. These poisons are mostly derived from coumarin and function primarily by interrupting the vitamin K cycle leading to failure to synthesize clotting factors I, II, VII, IX, and X in the liver. Today, most commercially available rodenticides are second-generation anticoagulant rodenticides (SGARs). Relay toxicosis is a problem in many different species of wildlife and may manifest clinically in a variety of ways.

For example, SAGRs have been shown to contribute to immunosuppression and mange in bobcats and mountain lions in California.[9,10] Bobcats, like domestic cats, seem tolerant to the vitamin K antagonistic effects of ARs, as indicated by normal clotting times in AR-exposed bobcats; however, this tolerance may lead to chronic sublethal exposure that influences other physiological parameters, as recently indicated in one study looking at genome-wide expression of multiple systemic effects.[11] Other documented non-target affected wildlife commonly include deer, raccoons, foxes, squirrels, weasels, voles, and skunks.[12]

Symptoms in raptors may include weakness, depression, hemorrhage, pale mucous membranes, peripheral vascular collapse, and shock. Normal thrombocyte counts in the face of prolonged bleeding or clotting times and petechial hemorrhages are highly suggestive of anticoagulant rodenticide intoxication. Assessment of coagulation times in avian patients is an area that has only been established in select species. Commercial coagulation tests are unavailable for avian patients, and those currently used have been modified from available mammalian products. In birds, the intrinsic pathway is thought to play only a minor role in coagulation; therefore, the prothrombin time would be the most clinically useful parameter in diagnosing a coagulopathy but has limited clinical availability. Activated coagulation (aka clotting) time (ACT) is a test that has long been used in mammalian patients as a method to evaluate the intrinsic and common pathways of the coagulation cascade. ACT is a more economical and easily performed method in any hospital, and in cases of wildlife patients can be used to assess for coagulopathies affecting the intrinsic and common pathways. Anticoagulant rodenticide screening tests are available and practical for use in some species, but the risk of encouraging further hemorrhage from venipuncture must be considered.

Treatment consists of intravenous fluids, iron dextran (10 mg/kg IM), vitamin B12 (0.5 mg/kg IM), with oxygen and heat support. The mainstay of treatment used by the author is administration of vitamin K1 1.0 mg/kg IM or SC q8-12h until stable (usually 1-2 days), then q24h PO for 7-30 days. This author has not found it necessary to go longer than that for most cases; birds make their own vitamin K and usually appear to go back to clotting normally quickly. The injectable vitamin K1 can be given orally in patients too small for the oral capsules/tablets; there is also a pediatric oral solution that is 120 **µ**g/0.8 ml). Blood transfusions are given in more severe cases (this author uses a PCV < 15% as a guideline, but this should always be assessed in light of the clinical picture; patients too BAR to tolerate an intravenous catheter are probably not going to get a blood transfusion).

Lead

Lead is a persistent inorganic environmental pollutant that affects humans and animals worldwide. Virginia opossums (*Didelphis virginiana*) were found to have elevated lead levels relative to other mesomammals in the southeastern study area.[13] Median blood lead level was 7.3 **µ**g/dL (n=104) in free-ranging raccoons in New York City.[14] Recent studies suggest that lead poisoning is estimated to kill a million wildfowl a year and cause sub-lethal poisoning in another ≥3 million in Europe alone. Modelling and correlative studies have supported the potential for population-level effects and lead clearly inhibits reproductive capabilities in adult birds as well as altering growth and development of hatchlings.[15] Interestingly, even sub-lethal exposure has been shown to significantly affect behavior leading to increased mortality risk in eagles.[16] Chronic low levels of exposure can cause immunosuppression as well as build up in the tissues and eventually manifest similarly to that of an acute exposure, especially in birds in the breeding season as hens mobilize calcium from their bones for their eggs and free stored lead along with it.

Lead toxicosis in wildlife may be diagnosed by sending whole blood to a diagnostic lab or in house with something like the LeadCare II system, which has been shown to be highly accurate in domestic animals and is used frequently in wildlife centers.[17] Heparinized whole blood concentrations greater than 20 µg/dl are suggestive. Levels greater than 50 µg/dl are diagnostic. A recent study in bald eagles has shown that histologic lesions compatible with lead toxicity occurred within the heart (76/93 birds), brain (59/93 birds), and eyes (24/87 birds) caused by ischemic necrosis of the small to medium arteries in these organs and that lead levels of 4 ppm or greater are more likely to lead to lesions in the heart.[18]

Treatment involves using chelation therapy. The author prefers starting with calcium EDTA (Calcium disodium versenate, 3M Pharmaceuticals) as the initial chelator for lead. It is important to maintain hydration as the bound molecules are large and may cause renal insult if the patient is dehydrated. This drug should not be given orally as it may increase lead absorption from GI tract. [19] The usual dose used by the author is 30-35 mg/kg IM or

SC q12h for 3-5 days, off 3-5 days, then repeat as needed. In extremely depressed patients where there may be a concern about efficacy of subcutaneous or intramuscular absorption, it can be used intravenously short term (48 hours) at 20-35 mg/kg diluted from 200 mg/ml to 60 mg/ml in saline as per the VIN formulary when given to mammals. It is not known if this is necessary in birds or reptiles but the author has used it safely for the past 30 years in many different species. Dimercaptosuccinic acid (DMSA) is often used as a after initial treatment with CaEDTA but can also be used in conjunction with it at a dose of 25-35 mg/kg PO q24h for as many days as needed to resolve elevated blood lead levels. Removal of any lead particles from the gastrointestinal tract can be an important part of treatment and can be managed in a variety of creative ways.

Euthanasia and Assessing Life

Veterinarians, veterinary staff, and rehabilitators care for wildlife for a number of professional and personal reasons. Many wild animals are fully releasable once recovered from their illness or injury. Individual animals may be ecologically important in areas of declining wildlife populations. Members of the general public expect veterinarians and rehabilitators to save the wild animals they have found. Medical and surgical skills developed from working on wildlife can be adapted and used to help treat companion animals; however, for those animals with severe injuries and illnesses that negate ultimate release and where permanent placement is not available, practical, or humane, euthanasia must be considered. Ultimately, veterinarians and rehabilitators are advocates for wild animal welfare and well-being. Euthanasia should not be considered a failure.

Euthanasia should be conducted without undue patient stress or restraint. Premedication and a multi-modal drug delivery approach to euthanasia should strongly be considered. For example, mammals can be premedicated with drugs such as dexmedetomidine and midazolam prior to induction with a gas anesthetic before administration of sodium pentobarbital. If sodium pentobarbital or other barbiturates are used, proper carcass disposal must be practiced, as fatal secondary barbiturate poisonings have recently been documented in species including golden eagles (*Aquila chrysaetos).* Veterinarians have been prosecuted for such poisonings under the federal Eagle Protection Act.

Conclusions

Wildlife rehabilitation centers and veterinary hospitals that commonly admit wildlife patients often encounter a wide variety of wildlife from a large geographic area and so may be ideally suited to participate in disease surveillance and One Health objectives. Obstacles to participation may include permit restrictions, financial disincentives, staff safety, no mandate to collect relevant data, and lack of contact between wildlife and public health agencies; however, practitioners may be greatly rewarded for investment of time and other resources by being more knowledgeable about wildlife diseases that may impact humans or companion and food animals.

References

1. Randall NJ, Blitvich BJ, Blanchong JA. Efficacy of wildlife rehabilitation centers in surveillance and monitoring of pathogen activity: a case study with West Nile virus. *J Wildl Dis.* 2012;48(3):646-653.

2. Camacho M, Hernández JM, Lima-Barbero JF, Höfle U. Use of wildlife rehabilitation centers in pathogen surveillance: A case study in white storks (*Ciconia ciconia*). *Prev Vet Med.* 2016;130:106-111.

3. Ana A, Perez Andrés M, Julia P, et al. Syndromic surveillance for West Nile virus using raptors in rehabilitation. *BMC Vet Res.* 2017;13(1):368.

4. Willette MM, Ponder JB, McRuer DL, Clark EE Jr. Wildlife health monitoring systems in North America: from sentinel species to public policy. In: Aguirre AA, ed. *New Directions in Conservation Medicine: Applied Cases of Ecological Health.* New York, NY: Oxford Press; 2012:552-562.

5. Sleeman JM, Clark EE, Jr. Clinical wildlife medicine: a new paradigm for a new century. *J Avian Med Surg.* 2003;17:33-37.

6. Farley JM, Romero CH, Spalding MG, et al. Newcastle disease virus in double-crested cormorants in Alabama, Florida, and Mississippi. *J Wildl Dis.* 2001;37(4):808-812.

7. Sleeman JM. Use of wildlife rehabilitation centers as monitors of ecosystem health. In: Fowler ME, Miller RE, eds. *Zoo and Wild Animal Medicine: Current Therapy.* Vol 6. St. Louis, MO: Elsevier Saunders; 2008.

8. Brown JD, Sleeman JM. Morbidity and mortality of reptiles admitted to the Wildlife Center of Virginia, 1991-2000. *J Wildl Dis.* 2002;38:699-705.

9. Serieys LEK, Lea AJ, Epeldegui M, et al. Urbanization and anticoagulant poisons promote immune dysfunction in bobcats. *Proc Biol Sci.* 2018;31;285.

10. Riley SPD, Bromley C, Poppenga R, et al. Anticoagulant exposure and notoedric mange in bobcats and mountain lions in urban southern California. *J Wildl Manag.* 2007;71:1874-1884.

11. Fraser D, Mouton A, Serieys LEK, et al. Genome-wide expression reveals multiple systemic effects associated with detection of anticoagulant poisons in bobcats (*Lynx rufus*). *Mol Ecol.* 2018;27(5):1170-1187.

12. Stone WB, Okoniewski JC, Stedelin JR. Poisoning of wildlife with anticoagulant rodenticides in New York. *J Wildl Dis.* 1999;35(2):187-193.

13. Lockhart JM, Siddiqui S, Loughry WJ, Bielmyer-Fraser GK. Metal accumulation in wild-caught opossum. *Environ Monit Assess.* 2016;188(6):317.

14. Rainwater KL, Marchese K, Slavinski S, et al. Health survey of free-ranging raccoons (*Procyon lotor*) in Central Park, New York, New York, USA: implications for human and domestic animal health. *J Wildl Dis.* 2017;53(2):272-284.

15. Pain DJ, Mateo R, Green RE. Effects of lead from ammunition on birds and other wildlife: A review and update. *Ambio.* 2019;48(9):935-953.

16. Ecke F, Singh NJ, Arnemo JM, et al. Sublethal Lead Exposure Alters Movement Behavior in Free-Ranging Golden Eagles. *Environ Sci Technol.* 2017;51(10):5729-5736.

17. Bischoff K, Gaskill C, Erb HN, et al. Comparison of two methods for blood lead analysis in cattle: graphite-furnace atomic absorption spectrometry and LeadCare(R) II system. *J Vet Diagn Invest.* 2010;22(5):729-733.

18. Manning LK, Wünschmann A, Armién AG, et al. Lead intoxication in free-ranging bald eagles (*Haliaeetus leucocephalus*). *Vet Pathol.* 2019;56(2):289-299.

19. Bradberry S, Vale A. A comparison of sodium calcium edetate (edetate calcium disodium) and succimer (DMSA) in the treatment of inorganic lead poisoning. *Clin Toxicol (Phila).* 2009;47(9):841-58. doi: 10.3109/15563650903321064. PMID: 19852620.

Addressing Dogma in Exotic Animal Practice

Nicolle Lofgren, VMD, DACEPM

Session #3394

Affiliation: From Mountain West Veterinary Specialists, Layton, UT 84041, USA.

Abstract: Exotic animal care suffers from reliance on dogma likely due to a paucity of exotic-exclusive clinics and veterinarians. This requires veterinarians to rely on handed down information and treatment protocols, which may be out of date or no longer clinically relevant. Treatment of exotic animal patients is just as complex as any species, with no single treatment fitting all presenting patients in a single species. This masterclass addresses some of the commonly held dogmas in exotic animal practice to help provide more effective diagnostics and treatments, as well as more accurate information for clients.

Introduction

Dogma is common in exotic animal practice, with many treatment and diagnostic recommendations being based on long-held beliefs with little or no scientific backing, potentially setting a dangerous precedent. These dogmas range from the assumption that feather plucking is caused by boredom to treating all rabbit infections as if they are *Pasteurella*. Dogmas addressed in this lecture include peas as a causative agent for bladder stones in ferrets, treatment of feather plucking as a behavioral disease, calcium restriction as bladder stone prevention in guinea pigs, gastrointestinal (GI) stasis as a disease rather than a symptom, and others. Recent literature on each topic is summarized to help break each dogma. This lecture serves as an update to exotic pet medicine to help practitioners make better-informed treatment and diagnostic decisions and better guide their clients.

Breaking Down Common Dogma

Bladder stones in guinea pigs are caused by high-calcium foods, and should be prevented with calcium restricted diets

Bladder stones in guinea pigs are classically thought of as a problem of genetics and calcium metabolism. Calcium carbonate stones are the most common stone identified in guinea pigs[1] and owners are frequently advised to restrict calcium and oxalates in their guinea pig's diet to prevent calciuria. This calcium restriction can increase calciuria by triggering secondary calcium mobilization by the parathyroids in response to dietary restriction. More recent studies in humans have shown nitrate ingestion and excess sodium are more likely the culprits for calciuria and secondary stones.[2] Excessive ascorbic acid supplementation may also contribute as the breakdown of NO2- into NO is catalyzed by ascorbic acid, and NO can then inhibit sodium transport in the proximal convoluted tubule, reducing calcium reabsorption. Reducing foods high in nitrates such as spinach[3,4] and advising against calcium restriction is likely key to stone prevention in guinea pigs.

Ferret cystine bladder stones are caused by peas

Ferret urinary obstruction due to cystine urolithiasis has become a common presenting emergency in the last decade.[5] Ferret owners are often guided to avoid grain free diets due to grain replacers such as peas, however recent studies have not shown any correlation between peas or specific legumes and cystine stone formation.[6,7]

As noted in one review, not all ferrets fed a grain free diet form cystine stones, and the disease is likely more complex than a simple ingredient. The ideal diet for a cystine stone former is an alkalinizing, low protein, high water content and low sodium diet. Unfortunately, ferrets require a high protein level and as carnivores have acidic urine normally. Although we are not yet sure what component of certain diets is causing this stone formation in ferrets, the assumption that peas and legumes are the culprit is unfounded and misleading. Because we cannot screen for risk, grain free foods should be discouraged, but owners should be made aware that research has not yet elucidated the specific components of grain free foods and low-quality dog/cat kibble causing stone formation.[8] Until further information is available, the safest preventative option is to maintain ferrets on a non-grain free diet formulated specifically for ferrets, and non-grain free treats labelled for ferrets. Additionally it is recommended not to mix multiple brands of food to limit risk.

GI stasis is a disease

Rabbits frequently present on emergency for having GI stasis. Many GI stasis protocols commonly aim to encourage gut motility, without taking into account that often gut motility is affected by elevated cortisol levels in response to stressors[9,10] and treating GI stasis as a disease rather than a symptom can reduce the efficacy of treatment for the animal. Framing GI stasis as a disease which can be treated with motility agents may also encourage owners to try to treat at home rather than ruling out potential causes for anorexia in their pet. Finding the underlying disease process is imperative to treating the animal's discomfort. Veterinarians are encouraged to reframe these symptoms to owners as a sign of distress, and to narrow down a differential list via diagnostic testing prior to administering a medication protocol.

Differentials vary based on an animal's age, husbandry, recent introduction to new pets, vaccine status, spay/neuter status and other medical history, but generally will start with a CBC and chemistry and radiographs to rule out renal failure, liver lobe torsion, anemia, sepsis, foreign body ingestion and gas distension of the GI indicating enteritis. Further work up can include *Encephalitozoon cuniculi* testing, advanced imaging for thorough evaluation of dental structures, joints and internal anatomy, and clostridium toxin testing, etc. Commonly, GI stasis protocols will include metoclopramide or other gut motility agents and meloxicam, without evaluation for underlying cause or normal renal function. Practitioners treating rabbits are encouraged to consider their medications similarly to an older cat when evaluated for anorexia, judging medications based on safety and chances of improving comfort.

Respiratory rabbits have snuffles, which can be assumed to be *Pasteurella* infection

Rabbits presenting with signs of respiratory disease are frequently diagnosed with snuffles, a layman's term assumed to be *Pasteurella* infection, without culture. As with any potential bacterial infection, empirical antibiotic use without culture and susceptibility testing goes against antibiotic stewardship and should be avoided when able. Additionally, assuming the identity of an infection may lead to frustration if not responding to empirical antibiotic choice. Terming any respiratory signs snuffles additionally simplifies cases which are potentially more complex than primary infection. Any diagnosis in a respiratory rabbit should take into account signalment and potential source of infection- what is the patient's dental disease risk based on age, breed, and diet, and is a dental abscess possibly the source of infection? Is this pneumonia or localized to the upper respiratory tract? Are the patient's naso-lacrimal ducts patent and what are the implications for long-term symptoms and treatment? These considerations may be sidelined by the assumptions made when diagnosing snuffles.

Retrospective reports of cultures in cases of oral and respiratory disease in rabbits show that infections are often mixed, composed of aerobic and anaerobic bacteria. Most studies find infections to mainly be composed of *Fusobacterium*, *Bacteroides*, *Pseudomonas*, *Prevotella*, *Pasteurella*, *Moraxella*, *Mycoplasma*, *Actinomyces*, *Staphylococcus* and *Streptococcus*.[11-13] Culture and susceptibility often indicate the necessity of multi-modal antibiotic therapy. Veterinarians are encouraged to work up any infection based on presenting symptoms and to the letter of antibiotic stewardship, including imaging, culture, and blood work to fully evaluate origin and extent of infection when able.

In this author's search, few studies in the literature include anaerobic culture,[14,15] potentially, causing over-reporting of aerobic and facultatively anaerobic species such as *Pasteurella*. Because rabbits form caseous pus and encapsulated abscesses, and due to the normally anaerobic inclusive flora of the rabbit oral cavity,[13] anaerobic infection should be considered in all respiratory rabbits and either aerobic/anaerobic culture, or next-gen DNA sequencing culture are recommended.

Feather plucking is caused by boredom

Feather destructive behavior (FDB) is classically considered a sign of boredom in avian species; however, this is a diagnosis of exclusion. Reducing stressors ultimately reduces FDB in many cases, however underlying causes are extensive and should be evaluated as FDB can indicate underlying pathology. Considerations such as age, recent introduction of new birds, boarding facility stay, access to household toxins and diet should be evaluated and a differential list should be formed. Primary differentials include infectious disease pathology (e.g., avian ganglioneuritis caused by bornavirus, chlamydiosis, circovirus, polyomavirus, parasitism, and *Aspergillus*), toxins such as heavy metals or chronic exposure to aerosolized toxins such as smoke, heart disease, hormonal and reproductive causes and causes for pain such as arthritis or gout. In animals presenting for or with FDB, a workup is recommended with priority based on signalment and history. In reproductive cases with no other indicators of disease or negative testing presumed to be behavioral this author will sometimes employ a Lupron trial. A single dose of Leuprolide acetate (Lupron Depot, AbbVie, Chicago, IL, USA) is administered and the owner is instructed that if signs improve after the initial week a deslorelin implant (Suprelorin, Virbac Animal Health, Fort Worth, TX, USA) may be considered. In cases responsive to hormonal control, testing for ganglioneuritis caused by bornavirus is still recommended as reproductive/ hormonal activity can be a trigger for bornavirus symptoms.[16]

Furosemide is not effective in reptiles due to the lack of a loop of Henle

As reptiles become more common pets and our husbandry improves, reptiles are living longer lifespans and we are beginning to see diseases of age. Heart failure in reptiles is occasionally reported in the literature, and medical management is rarely pursued. Although reptiles lack a loop of Henle, furosemide has been used successfully to cause experimental dehydration in bearded dragons, as well as having successful therapeutic use in one reported case.[17,18] Despite lacking a loop of Henle, furosemide is thought to work in the Na-K-Cl transporters in the distal tubules of the reptile nephron, or possibly during post-renal processing in the colon.[18,19] Regardless of mechanism, medical management of heart failure in reptiles can be offered in a clinical setting, with similar monitoring recommendations to mammals.

References

1. Hawkins MG, Ruby AL, Drazenovich TL, Westropp JL. Composition and characteristics of urinary calculi from guinea pigs. *J Am Vet Med Assoc*. 2009;234(2):214-220.

2. Damasio PC, Amaro CR, Cunha NB, et al. The role of salt abuse on risk for hypercalciuria. *Nutr J*. 2011;10:1-4.

3. Nuñez de González MT, Osburn WN, Hardin MD, et al. A survey of nitrate and nitrite concentrations in conventional and organic-labeled raw vegetables at retail. *J Food Sci.* 2015;80(5):C942-C949.

4. Brkić D, Bošnir J, Bevardi M, et al. Nitrate in leafy green vegetables and estimated intake. *Afr J Tradit Complement Altern Med.* 2017;14(3):31-41.

5. Hanak EH, Di Girolamo N, DeSilva U. Composition of ferret uroliths in North America and Europe: 1055 Cases (2010-2018). *Proc ExoticsCon.* 2019;403-404.

6. Pacheco RE. Cystine urolithiasis in ferrets. *Vet Clin North Am Exot Anim Pract.* 2020;23(2):309-319.

7. Lamglait B, Brieger A, Rainville MP, et al. Retrospective case control study of pet ferrets with cystine urolithiasis in Quebec, Canada: epidemiological and clinical features. *J Vet Med Surg.* 2021;5(1):31.

8. Stockman J, Malka S, Lofgren N, et al. Determining the urine cystine and amino acid concentrations in domestic ferrets (*Mustela putorius furo*). *J Exot Pet Med.* 2023.

9. Jang SJ, Kang SS, Son SJ, et al. Cortisol levels and gastrointestinal disorders after stressful surgery in rabbits. *In Vivo.* 2017;31(4):637-640.

10. Schoemaker N. Recurrent gut stasis: advanced therapeutic options and how to diagnose and treat the underlying causes. *Proc BSAVA Congr.* 2016;148-149.

11. Gardhouse S, Sanchez-Migallon Guzman D, Paul-Murphy J, et al. Bacterial isolates and antimicrobial susceptibilities from odontogenic abscesses in rabbits: 48 cases. *Vet Rec.* 2017;181(20):538.

12. Taylor WM, Beaufrère H, Mans C, Smith DA. Long-term outcome of treatment of dental abscesses with a wound-packing technique in pet rabbits: 13 cases (1998–2007). *J Am Vet Med Assoc.* 2010;237(12): 1444-1449.

13. Tyrrell KL, Citron DM, Jenkins JR, Goldstein EJ. Periodontal bacteria in rabbit mandibular and maxillary abscesses. *J Clin Microbiol.* 2002;40(3):1044-1047.

14. Abd-Algawad R, Mowafy R, Salem SM, et al. Molecular characterization of snuffle-causing bacterial pathogens in rabbits in terms of prevalence and pathogenicity. *Egypt J Agr Res.* 2021;99(4):475-485.

15. Pellett S, Harcourt-Brown F, Pinborough M. Bacterial and fungal culture and sensitivity from rabbits presenting with nasal discharge. *Proc BSAVA Congr.* 2016;553-553.

16. Rossi G, Dahlhausen R, Orosz S. Avian ganglioneuritis in clinical practice. *Vet Clin North Am Exot Anim Pract.* 2018;21:33-67.

17. Parkinson, Lily A, Mans C. Effects of furosemide administration to water-deprived inland bearded dragons (*Pogona vitticeps*). *Am J Vet Res.* 2018;79(11):1204-1208.

18. Lofgren N, Malka S, Vecere G. Therapeutic use of furosemide in a bearded dragon with pericardial effusion. *Proc ExoticsCon.* 2021;475.

19. Goncharevskaya OA, Monin Y, Natochin Y. The influence of furosemide and Co 2+ on electrolyte and water transport in newt distal tubule and frog skin. *Pflügers Arch.* 1986;406(6):557-562.

Section 50

Hybrid Session 2

Dan Johnson, DVM, DABVP (Exotic Companion Mammal Practice)
Session Chair

No Patient is Too Small to Help: Microsurgery for Exotic Pets

Zoltan Szabo, Dr med vet, Dipl ACZM, DABVP (Exotic Companion Mammal Practice), DABVP (Avian Practice), GpCert(ExAP), MRCVS

Session #3175

Affiliation: From Concordia Pet Care, No. 5-7, Blue Pool Road, Happy Valley, Hong Kong.

Abstract: Similar to the increasing popularity of small mammals as pets, owners' demands for state-of-the-art surgical treatments are also increasing. Guinea pigs, chinchillas, hedgehogs, and hamsters are the smallest mammals that most veterinarians see; however, the small size of these patients has some unique surgical challenges. Microsurgical techniques require the surgeon to work under magnification with delicate tissues in a limited space. The masterclass discusses the recommended surgical instruments, magnification methods, and surgical techniques necessary to perform successful soft tissue and orthopedic surgeries on the smallest of exotic pets.

Introduction

The anesthetic risk of rodents is higher than that of dogs and cats. These small mammals are especially prone to complications such as hypovolemia from blood loss, hypothermia, and respiratory compromise. The owners' lack of knowledge and experience with the species can also lead to unrealistic expectations and complications. Some clinical signs, like anorexia, are mild in dogs but can indicate a potentially life-threatening problem in rodents. Common diseases, like abscesses or intestinal obstruction, can be straightforward in dogs and cats, but the prognosis is usually worse in rodents. Postoperative complications are also more frequent in these species.[1]

Preoperative Period

Ideally, every patient should undergo preoperative diagnostic testing, including hematology and blood biochemistry, radiography, or contrast-enhanced computer tomography. Subclinical renal disease is common in rabbits, and respiratory disease in guinea pigs. Perioperative antibiotic therapy for routine, sterile surgeries (e.g., neutering and skin mass removals) is unnecessary.[2]

Preoperative fasting

Rabbits and rodents cannot vomit, so fasting before surgery is unnecessary; however, the syringe feeding of anorectic patients should be ceased for 1-2 hours before anesthesia to ensure the oral cavity is empty. If the mouth is full of food, that makes endotracheal intubation in rabbits difficult, and it could lead to aspiration pneumonia in guinea pigs or chinchillas. To prevent this complication the oral cavity should be cleaned at them time of the premedication.[3]

Fluid therapy

Parenteral fluid administration is recommended for every surgical procedure and should be initiated before surgery. Intravenous catheter placement is surprisingly easy in rabbits and hedgehogs because their skin is relatively soft. Guinea pigs are the opposite; the thick skin on their limbs makes catheter placement challenging. The standard recommendation for intravenous fluid support during anesthesia and surgery is 10 ml/kg per hour; however, IV catheter placement could be impossible in smaller patients, and intraosseous catheters are rarely used in clinical practice because most clinicians don't find them practical. Therefore, subcutaneous fluid administration is the norm. If maintaining vascular access is possible after surgery, continue administering intravenous fluids until the patient is discharged from the hospital.[4]

Thermal support

Smaller patients have a relatively larger body surface than larger ones. Therefore, hypothermia during anesthesia and surgery is very common in small rodents. Therefore, every anesthetized patient's core temperature should be monitored continuously, and warming devices must be used during every surgery.[4]

Hemodynamic support

The total circulating blood volume of these tiny patients is minimal; therefore, seemingly small amounts of blood loss can have disastrous effects. In addition, blood from a conspecific for transfusion is usually not available. Therefore, strict attention to hemostasis and prevention of blood loss is critical. During surgery, bipolar electrosurgical forceps should be used to seal any bleeding blood vessels immediately. In addition, metal hemostatic clips can be used for bigger blood vessels and in areas that are difficult to reach with conventional ligatures.[4]

Analgesia and Anesthesia

Because of the fragile nature of rodents, sedation prior to any stressful intervention is essential. In the perioperative period, multimodal anesthesia, using a combination of systemic and local anesthetic drugs that target different steps of the body's pain transmission pathway, is essential. Gas anesthesia is usually delivered via tight-fitting facemasks, which is the preferred method to maintain anesthesia in small rodents. While endotracheal intubation is technically possible, it is rarely done in clinical practice. The physiologic parameters (heart rate, respiratory rate, oxygen saturation, body temperature) should be monitored and maintained during the surgery.[5]

Surgical Preparation

The skin is aseptically prepared prior to surgery. The skin of most exotic mammals is thin, and the fur is dense; therefore, clipping should be performed carefully to prevent skin damage. Gentle and atraumatic skin handling reduces postsurgical pain and self-inflicted wound trauma. Water-based disinfectants should be avoided to reduce heat loss, and alcohol-based ones should be used instead. The organs should be handled gently and examined in situ during abdominal exploration without unnecessary manipulation. Regular lavage with warm, sterile saline, as well as the use of moistened gauze squares, helps prevent drying of the abdominal organ surface of the organs. Wound infection is usually due to contamination or self-inflicted wound trauma. It can be prevented with gentle surgical site preparation, atraumatic tissue handling, wound closure by apposing edges correctly and with no tension, minimally reactive suture materials, and perioperative local anesthesia.

Surgical Instruments

Most instruments used for eye surgeries are unsuitable for small rodents' microsurgery. The ophthalmic instruments are short, making them challenging to handle and manipulate tissues within a body cavity. Full-length, dedicated microsurgical instruments are designed so that only their tip is miniaturized. The instruments' tips extend beyond the surgeon's fingers far enough to use within the patient's body cavity. Rounded handles are important for the needle holders, where a curved needle must be moved through the tissue by rolling the needle holder with the fingers.

A microsurgical kit minimally should consist of a microsurgical needle holder, micro scissor, and two thumb forceps. Ring-tipped microsurgical forceps are of a standard length, counterbalanced, and delicate, with a tip that has a ring to distribute the pressure over a wider surface, making them less traumatic but allowing the surgeon to hold the delicate strand of suture for knot tying.

Sterile cotton-tipped applicators are helpful for gentle tissue dissection and manipulation, and they can also be used to absorb blood or fluids. Moisten the applicators for tissue manipulation and dissection and use them dry to absorb fluids. An absorbable gelatin sponge and other absorbable hemostatic materials help to control bleeding.

Magnification

Magnification is recommended because the surgeon's inability to see small structures limits dexterity and manipulation. In addition, with magnification, small amounts of hemorrhage appear more important, and individual vessels are more easily identified. Binocular loupes are available in different styles, with cheap hobby loupes being the simplest but having several disadvantages, including a set focal distance, no focal light source, and the inability to look above the lenses when magnification is not needed. In general, the higher the magnification, the closer the lenses must be to the object to be in focus. Anything outside of the specific focal distance of the lenses will not be in focus, resulting in a blurry image.

Surgical loupes eliminate most of the problems. They are designed to be used with a good neck position, the lenses do not cover the whole visual field so the surgeon can look above the lenses, they have a focal LED light source, and the lenses have a wider focal range rather than a set focal distance, which generally allows the entire patient to be in focus during surgery.[6]

Sutures

Sutures and needles used in the average dog and cat practice are too large for use in small mammals. Suture sizes 4-0 to 7-0 are commonly used in small rodent species. Skin closure is best accomplished with an intradermal and subcuticular technique to reduce the chance of wound chewing.

References

1. Lee HW, Machin H, Adami C. Peri-anaesthetic mortality and nonfatal gastrointestinal complications in pet rabbits: a retrospective study on 210 cases. *Vet Anaesth Analg.* 2018;45(4):520-528.

2. Szabo Z. Soft tissue surgery: rodents *In:* Quesenberry KE, Orcutt CJ, Mans C, Carpenter JW, eds. *Ferrets, Rabbits, and Rodents: Clinical Medicine and Surgery.* 4th Ed. St. Louis, MO: Elsevier; 2021:467-482.

3. Szabo Z, Bradley K, Cahalane AK. Rabbit soft tissue surgery. *Vet Clin North Am Exot Anim Pract.* 2016;19(1):159-188.

4. Bennett RA. Rabbit Soft Tissue Surgery. In: Bennett RA, Pye GW, eds. *Surgery of Exotic Animals.* Ames, IA: Wiley-Blackwell; 2021:240-276.

5. Bennett K, Lewis K. Sedation and anesthesia in rodents. *Vet Clin North Am Exot Anim Pract.* 2022;25(1): 211-255.

6. Phillips H. Magnification surgery. In: Bennett RA, Pye GW, eds. *Surgery of Exotic Animals.* Ames, IA: Wiley-Blackwell; 2021:23-34.

How to Plan and Write a Case Series: Tips from the Editor of the Journal of Exotic Pet Medicine

Nicola Di Girolamo, DMV, PhD, DipECZM (Herpetology), Dipl ACZM, Dipl ACVPM

Session #3501

Affiliation: From Exotic Pet Service, Department of Clinical Sciences, College of Veterinary Medicine, Cornell University, 930 Campus Road, Ithaca, NY 14853, USA.

Abstract: Proper development and reporting of original research are fundamental to avoid research waste and ensure maximum effectiveness of resources invested in research. Case series can be defined as original research articles in which information from individual patients that share something in common is collected. This lecture covers how to plan, write, and interpret a case series.

Introduction

Proper development and reporting of original research are fundamental to avoid research waste and ensure maximum effectiveness of resources invested in research. Case series can be defined as original research articles in which information from individual patients that share something in common is collected.[1] Although case series are considered studies lower on the overall evidence pyramid,[2] they may provide important information on new or poorly described disorders.[1] Considering the paucity of research on clinical diseases in exotic patients, case series are still among the most common study designs used for clinical articles in exotic pet medicine. Typical topics suited for a cases series, include a novel or poorly described condition, or diagnostic imaging findings with a specific diagnostic (perhaps new) technique in a specific clinical condition, or outcomes of a specific intervention for a specific clinical condition.[1] Case series cannot establish effectively the effectiveness of a treatment, or the diagnostic accuracy of a technique, and as such, investigators should be careful about the conclusions drawn when interpreting the results.[1] Case series can be either prospective (i.e., when the collection of the information from the patients is planned at a certain time and is then performed over a period of time) or retrospective (i.e., when collection of information from the patients is planned at a certain time when the information is already available in a database or similar). Retrospective case series are by far the most common design, and this presentation focuses primarily on this specific study design.

How to Set Up a Retrospective Case Series

The first step in setting up a case series is typically to identify a knowledge gap. A knowledge gap is a specific piece of clinical information that is currently unknown. A knowledge gap could be something as simple as the rectal temperature of rabbits presented to a hospital, or as complex as the change in a specific analyte after a specific surgery. In exotic pet medicine, many simple knowledge gaps are currently present, and research should be tailored to answer them.

The second step consists of drafting a protocol (i.e., a document that contains background information) that will become the introduction of the final manuscript and the methods. While drafting the protocol and working on the background information, a comprehensive literature search should be performed. At this stage, it is not uncommon to find out that your knowledge gap already has been answered in another study. Previous studies should be carefully evaluated and the need for research should be reconsidered or aims may need to be slightly edited to provide useful information. For example, if interested in describing the rectal temperature of rabbits presented to a hospital, and a previous research article already reported similar data, one might either decide to avoid performing the study and employ the information found from the other study in clinical activity, or look to a certain detail (e.g., the rectal temperature during a certain season or during certain clinical conditions). It is also important to describe and record the literature search(es) performed. A common template that is advised for use is the following:

"The following databases (XX, XX and XX) have been searched with the following keywords XX XX on [date; e.g., 01/01/2020]; the following textbooks have been consulted (References). No other reports of XX have been found doing these searches."

The second part of the protocol typically includes in detail the methods that are going to be used to carry on the case series. Important methodologic aspects that need to be considered when performing a case series can be summarized in those that pertain to inclusion of patients, such as population of interest, inclusion criteria, medical record search, and those that pertain to extraction of information from the records, such as what data were extracted and how data were extracted. It is critical at this stage to define exactly how and when the medical record database will be searched, to evaluate the need for multiple assessors of the medical records, in case extraction of certain information may be debatable, and to create definition for data that will be extracted. For example, something as simple as the presence of fever in a patient may be extracted in many ways from a medical records and appropriate definition is important to avoid imprecisions.

Reporting a Case Series

Introduction

The introduction explains the context of the case series and should clarify (1) what is the current scientific knowledge on the topic and (2) why the topic deserves further investigation. Keep the introduction brief and focus only on previous original research articles. Typically two to three paragraphs summarizing relevant research are sufficient. The final paragraph of the introduction presents the specific objectives. The terminology "To our knowledge…" is not acceptable anymore; instead to replace it with a thorough literature search.

Materials and methods

The following headings (or a variation thereof) are expected to be included when considering a case series for publication.

- Study design
- Population (inclusion and exclusion criteria, recruitment)
- Medical record search
- Data extracted
- Variables of interest and outcomes
- Data reporting ± statistical analysis

Results

The following headings (or a variation thereof) are expected to be included when considering a case series for publication.

- Population
- Signalment and history
- Presenting complaints
- Physical examination
- Diagnostics
- Treatments
- Outcome

These headings should be adapted to the topic of the case series. For a case series focusing on diagnostic imaging findings in a specific condition, it would be natural to expect slightly different headings like the following:

- Population
- Signalment, complaints & PE
- Diagnostics
- Radiographs
- Ultrasound
- Computed tomography
- Treatments and outcome

Discussion

In the discussion the results of the case series should be reported in the context of existing evidence. Cautiously interpret the results (and methods) compared to existing literature. In terms of paragraphs, a typical discussion could be composed as follows:

(1) three to four paragraphs that summarize the main three or four results considering the study objectives;

(2) limitations of the study, including sources of potential bias and imprecision; and

(3) generalizability (external validity) of the results.

One of the most common mistakes made when writing the discussion and conclusions section of a case series is to reach conclusions based on the results of the study without considering the lack of a control group. For example, if a case series on appendicitis in rabbits found that 90% of the appendicitis occurred during summer and fall, this does not automatically mean that appendicitis as a disease is more common during those seasons. Since the higher percentages of cases reported during those seasons could be a mere consequence of overall increased caseload in that season, or even something more trivial, such as data extraction that for some reason excluded cases from other seasons. As such, authors of case series should understand that only limited conclusions can be drawn from this uncontrolled type of research.

References

1. Albrecht J, Meves A, Bigby M. Case reports and case series from Lancet had significant impact on medical literature. *J Clin Epidemiol.* 2005;58(12):1227-1232.

2. Murad MH, Asi N, Alsawas M, Alahdab F. New evidence pyramid. *Evid Based Med.* 2016;21(4):125-127.

Section 51

Hybrid Session 3

Byron de la Navarre, DVM
Session Chair

Electrochemotherapy in Exotic Animals: A Review of the Literature

Brynn McCleery, DVM, DABVP (Avian Practice)

Session #3135

Affiliation: From Red Bank Veterinary Hospital, 197 Hance Avenue, Tinton Falls, NJ 07722, USA.

Abstract: Electrochemotherapy (ECT) is an ablative therapy for tumors that combines electroporation with the use of hydrophilic chemotherapeutics, commonly bleomycin or cisplatin, either locally or systemically. ECT allows the tumor cells to become more permeable to the chemotherapeutic agents resulting in high intracellular concentrations and, subsequently, increased antitumor activity. Use of ECT in human medicine is well established, while this therapy is still novel in veterinary medicine. Preclinical and clinical veterinary studies have shown ECT to be an effective and safe therapeutic option for multiple cutaneous and subcutaneous tumor types in dogs, cats, and horses. This therapy has also been utilized in multiple zoological species. This master class serves as a review of ECT and its use in exotic animal species.

Introduction

To reach their biological targets, chemotherapy drugs must overcome a major obstacle: the cell membrane of tumor cells.[1,2] Electroporation (i.e., electropermeabilization) is a method utilizing high-voltage electric field pulses to induce an electrically mediated reorganization of the plasma membrane of cells. This leads to formation of aqueous pathways (i.e., electropores or transient permeant defects), and, therefore, leads to targeted permeabilization permitting the cross flow of material.[1,3,4] Electrochemotherapy (ECT) combines electroporation of tumors with local or systemic administration of chemotherapeutic drugs that have poor membrane permeability on their own and greatly increases uptake of drugs by the tumor cells, leading to more effective cellular death.[1,3,4]

ECT has been added to the armamentarium available for local and systemic cancer management, along with surgery, radiation therapy, and chemotherapy. In veterinary oncology, late referrals and/or rapid tumor growth often prevent achievement of local control with surgery alone. A multimodality approach involving adjuvant therapies is often required.[1] ECT, along with chemotherapy and radiotherapy, can help achieve better control in these cases. Chemotherapy can be used to reduce tumor volume permitting wide excision, reduce the probability of local recurrence after surgical excision of some tumor types, and reduce the chances of distant dissemination (its major application).[1] The clinician must balance attaining sufficiently high levels of chemotherapeutic drug in the tumor (through high dose intensity and dosing frequency) to achieve cytotoxicity of tumor cells without causing systemic toxic effects, such as bone marrow suppression or gastrointestinal symptoms.[2,4] This problem can be avoided with ECT, which utilizes lower doses of chemotherapy drugs and is associated with greatly reduced side effects. Radiation therapy is utilized for the treatment of solid tumors as an adjuvant therapy or for treatment of inoperable neoplasms. Compared to radiation therapy, ECT provides more targeted tumoricidal therapy and spares healthy tissue with less side effects to collateral tissues.[2]

Overall, ECT can be used as a curative, palliative, adjuvant, neoadjuvant, or concomitant therapy for local control of solid neoplasms.[1,5]

Mechanism of Action

The principal mechanism of action of ECT is electropermeabilization of cells in tumors allowing chemotherapeutic drugs to reach their intracellular targets (i.e., increased drug permeability).[3] Cells have two possible fates after electroporation: (1) reversal of the process and return to the previous steady state or (2) impossibility to reverse the ion fluxes, activation of the caspase apoptotic or necrotic pathways, and, ultimately, cell death.[1] Reversible electroporation is performed most commonly to introduce drugs or genetic material into the cells without affecting cell viability.[6] Membrane sealing in vitro typically occurs within minutes, depending on the electrical parameters used. In a mouse model, 63% of membrane resealing occurred within 9 minutes of the applied electrical pulses.[2]

Other mechanisms exist in addition to electropermeabilization that increase the effectiveness of ECT. These include the following mechanisms.

Vascular lock phenomenon

A transient but reversible reduced blood flow occurs immediately following electrical pulse application leading to drug entrapment in the tissue for several hours.[3,4] The reduced blood flow occurs due to reflex vasoconstriction of the afferent arterioles and is medicated by the sympathetic nervous system.[2] The hypoperfusion is transient in normal tissue but lasts up to 12 hours with electroporation only and 5 days with electrochemotherapy.[3] This mechanism traps the chemotherapy drug in the treated area, providing more time for the drug to act, and also prevents bleeding from the tissue, which is an important feature when treating hemorrhagic tumors.[2,3]

Vascular disrupting action

Cytotoxic damage extends beyond just the tumor cells and affects stromal cells, including endothelial cells of tumor blood vessels, resulting in cell death. This leads to disruption of tumor blood flow and death of tumor cells surrounding the vessel.[3,4]

Massive tumor antigen shedding

After ECT, tumor antigen shedding leads to induction and upregulation of systemic immunity to neoplastic cells via biological response modifiers including interleukins (ILs) 2 and 12, granulocyte-macrophage colony stimulating factor, and tumor necrosis factor alpha.[3,4] The immune response appears to play a crucial role in the efficacy of ECT. Immunocompromised laboratory mice had a lower response to treatment compared to those with normal immune systems.[5]

Development

ECT has received considerable attention over the past 30 years. The term was first coined in 1991 in a study evaluation the modality in nude mice.[7] Further in vivo, preclinical studies in various laboratory animal models (including mice, rats, hamsters, rabbits) showed antitumor effectiveness of ECT on subcutaneous tumors grown in muscle, brain, or liver and of different histologic type (including sarcomas, carcinomas, gliomas, and melanomas). These studies showed minimal toxicity and good safety of ECT.[3] Based on the preclinical studies, clinical studies were started in human patients. Clinical studies were initiated in veterinary medicine soon after the first human clinical trials were underway, and the first veterinary clinical trials involved treatment of soft tissue sarcomas in feline patients.[1,3] The modality became more routine in 2006 after the European Standard Operating Procedures for Electrochemotherapy (ESOPE) published a standard operating procedure for ECT in human patients and a medical grade electroporator was approved (Cliniporator).[5,7]

Effectiveness

Clinical studies in humans have demonstrated a pronounced antitumor effectiveness of ECT. In a prospective, nonrandomized, multi-institutional study, an objective response rate of 85% (74% complete and 11% partial responses) of cutaneous and subcutaneous tumors was reported, regardless of tumor type (histologic origin), drug utilized, route of drug administration, or type of electrodes employed for tumors 3cm in diameter or smaller.[2,3,8] Additionally, a good response was noted in tumors that were previously resistant to chemotherapy or radiotherapy.[2]

Clinical studies in veterinary patients (i.e., dogs, cats, horses) have also shown high efficacy of ECT in the treatment of cutaneous and subcutaneous tumors of different histologic types with an approximately 80% long-lasting objective response rate.[3] Types of tumors that have been successfully treated include mesenchymal tumors (soft tissue sarcomas, fibromatosis), epithelial tumors (perianal and anal sac tumors, carcinomas, ganglioneuroblastoma, trichoblastoma), round cell tumors (melanoma, mast cell tumors, lymphoma, Sticker's sarcoma), and equine sarcoids.[1]

Indications

ECT is indicated for the following cases[5]:

- Primary or metastatic cutaneous or subcutaneous tumors that cannot be sufficiently treated with their respective first-line treatments
- Tumors affecting quality of life due to bleeding, ulceration, and/or pain
- Oral or nasal tumors as a solo treatment or in combination with surgery
- Incompletely resected tumors or as a method to extend the safety margins during surgery
- As a primary therapy if first-line therapies are possible but rejected by the owner
- To reduce cancer burden in primary or metastatic tumors that are accessible to the electrodes (superficial or via surgery)
    - Patients showing a poor response to systemic therapy
    - Large tumors prior to surgery to improve relapse-free survival
    - Palliative intent if no other treatment options exist
- Curative intent for tumors up to 3cm[3] to avoid post-operative complications or if surgery would be difficult based on location

Chemotherapeutic Drugs Utilized

Drugs used with ECT cannot cross the cell membrane on their own due to chemical properties or lack of a transport mechanism (nonpermeant or poorly permeant).[2,3] Increased permeability of the tumor cells through electroporation allows the use of drugs that would otherwise require high systemic doses to be effective, but the doses utilized in ECT are so low as to have no systemic toxicity.[3,4] Drug doses with minimal or no antitumor effectiveness produced nearly an 80% complete response when delivered by ECT.[3]

Bleomycin

Bleomycin in normally transported across the cell plasma membrane by protein receptors and cannot diffuse across the membrane due to its larger size.[1,2,4] Carrier proteins internalize the drug via an endocytotic pathway.[4] Transport is therefore limited by a low number of carrier proteins exposed at the cell surface and withdrawal of the proteins from the membrane during endocytosis.[2,4] Due to this required transport system, uptake in tumor cells is slow and restricted under normal conditions. Furthermore, loss of membrane receptors is a main mechanism of chemoresistance again bleomycin. These obstacles are bypassed by electropermeabilization, which allows almost free diffusion of bleomycin into cells for as long as they remain permeable (up to 1 hour).[1,2] Once inside the cell, bleomycin's mechanism of action is through fragmentation of DNA resulting in G2-M phase arrest of the neoplastic cell cycle.[1,5] The number of DNA lesions dramatically increases leading to apoptosis when bleomycin is used in combination with ECT.[1]

Bleomycin can be administered intravenous (IV) or intratumorally. ECT with bleomycin has a comparable effect despite the route of drug administration used.[3] Typical doses for bleomycin are: 15-30 mg/m^2 (15,000-30,000IU/m^2).[1,3] Electropulsation of cells increases the cytotoxicity of bleomycin 100- to 5000-fold in vitro, depending on the study.[2,3,9] The author could not find any pharmacokinetic studies for bleomycin in avian or reptile species; a few PK studies exist in laboratory rodents.

Cisplatin

Cisplatin has poor cellular permeability with only approximately 50% of drug transported through the plasma membrane by passive diffusion.[2,4] The platinum complexes in cisplatin bind DNA bases leading to cross-linking of DNA and subsequent cell death.[1,2] Electroporation enhances the transmembrane passage of cisplatin by a factor of 2-8 times, and, therefore, increases the number of cross-links noted in tumor cells.[1,4] Although typically not tolerated well by cats systemically, cisplatin used intratumorally with ECT is well tolerated.[1] Electropulsation increases the cytotoxicity of cisplatin 8- to 80-fold in vitro.[2-4] Cisplatin is only administered intratumorally when used for ECT in veterinary patients.[3] It is considered an ideal drug for intratumoral injection as it does not cause necrosis.[10] However, necrosis was seen in a case of ECT used in a cockatiel with a facial fibroma and has also been reported in canine patients. In the cockatiel, necrosis occurred approximately 2-3 days after the first treatment and sloughing followed.[11] The pharmacokinetics of cisplatin have been evaluated in an avian species (sulphur-crested cockatoos), rabbits, and rodents.[12-16]

Combination

Bleomycin and cisplatin can also be combined during a single ECT session, typically with bleomycin administered by bolus IV and cisplatin administered intratumorally.[1,9]

Other Drugs Reported

Doxorubicin

Doxorubicin belongs to the anthracycline class. Multiple mechanisms of action are reported including interactions with tumor DNA (base intercalation, DNA strand breakage, and topoisomerase II inhibition), DNA polymerase activity inhibition, altered regulation of gene expression, and release of free radicals. Preclinical studies suggest that ECT may improve the efficacy of doxorubicin, especially in the treatment of mammary cancer in veterinary patients.[1]

Mitoxantrone

Mitoxantrone is a synthetic anthracycline analogue with a similar mechanism of action as doxorubicin but less cardiotoxicity than other members of this drug family. Reports of improved efficacy of mitoxantrone when used with ECT exist for both human and veterinary patients. Dosing of 5 mg/mm^2 IV has been reported.[1]

Benefits

ECT is a highly effective treatment of various primary tumors and metastases of different histologic types across species. It can be used with curative intent for solitary or multiple cutaneous or subcutaneous tumors or as an adjunct treatment to surgery or radiotherapy to prevent localized recurrence by removing occult micrometastases or as a neoadjuvant therapy (facilitating other modalities by reducing the size of the mass prior to surgical resection or radiotherapy).[1,3,4] Results of treatment are comparable to other standard therapeutic approaches with a lower cost and relative ease of application.[1,3] Electroporation is conducted in seconds to minutes. Along with preparation and post-treatment observation, ECT is a quick therapy that can be performed on an out-patient basis in most cases.[2]

ECT has a high specificity for targeting cancer cells (sparing to healthy tissue) and high degree of localization of treatment effect. It also preserves innate immune response and structure of the extracellular matrix.[5,8] Compared to surgery, ECT is often less invasive.[7] Furthermore, ECT utilizes lower drug doses and is a nonthermal ablation modality.[1-6] These characteristics lead to lower morbidity and side effects. Since tumor cells are not dying secondary to application of physical energy (as with irreversible electroporation, radiotherapy, and many other ablative therapies), the cells die off in a more controllable way from apoptosis or other programmed types of cell death. This results in a slower shrinkage without the development of severe necrosis and reduces the risk of infection.[7] ECT can also be repeated without the development of chemoresistance.[7]

Adverse Effects and Contraindications

Overall, side effects of ECT are limited.[1] The most common local adverse side effect is contraction of muscles underlying the treatment area. The contractions are only present during application of the electrical pulses and resolve at the completion of treatment.[3,4] However, this modality causes discomfort, and heavy sedation or general anesthesia is required during treatment of veterinary patients.[1,3,5] In humans, ECT is typically performed under local anesthetic.[1]

Other reported side effects are mostly local and transient including erythema and edema of the skin.[6] Alopecia and pigment change have also been reported.[5] For post-operative patients, delayed healing of surgical beds and wound dehiscence are reported side effects.[17] Formation of a fistula, extensive necrosis and anatomic defects may occur if the tumor compromises the whole thickness of the tissue.[5] Severe radiation recall has also been reported in a cat.[17] If repeated or used in conjunction with standard chemotherapy, cumulative effects of bleomycin are possible and may lead to lung fibrosis. Maximum doses should be calculated and monitored (200,000 IU/m^2 in dogs, unknown in cats and exotics).[5]

ECT is contraindicated in pregnant and lactating animals.[5]

Treatment Protocol

Electropermeabilization protocols vary in terms of electrical shapes, voltage, amperage, frequency, and interpulse duration.[1] Only very brief electrical pulses are required for electroporation.[4] Protocols usually use trains of 8 square or biphasic wave electrical pulses with amplitudes (voltage to distance ratios) of approximately 1000-1300 V/cm and a pulse duration of up 100 μs (50+50 μs for biphasic pulses, interpulse interval of 300 μs to 1 ms) with a frequency of 1-5000 Hz.[1-3,5] Square wave pulses offer the advantage of controlling the pulse amplitude and length.[2,5]

Electrodes

Multiple types of electrodes exist for electroporation. These include the following electrodes.

Parallel plate

Effective for surface tumors of different sizes with limited depth.[3,5] Plate electrodes provide a more uniform electrical field distribution, but plates typically have a maximum treatment depth of < 3 mm.[3,5]

Needles

Allow for deeper application of pulses in tissues but a more heterogenous electrical field distribution. Needle probes are more invasive and are difficult to use on tough skin (e.g., horses).[3] One group of authors recommends use of needle electrodes for most tumors as needles can help surpass the electrical impedance from thicker skin and allow a more adequate distribution of the electric field to the tumor.[5]

Wire

Wire electrodes are easy to bring into contact with shaved skin and can be moved with ease in different orientations. However, a limited amount of tissue is affected by field discharge, and successive treatments are usually required for eradication of tumors.[3]

Time interval after injection

The time interval between drug injection and application of electrical pulses to the tumor is very important. Sufficient drug must be in the tumor at the time of electroporation.[3,4] During intravenous administration of cisplatin (5 mg/kg) or bleomycin (0.5 mg/kg) to laboratory rabbits and rodents, an interval of only a few minutes is required to reach maximum drug concentration in the tumor. After intratumoral administration of cisplatin (2 mg/cm^3) or bleomycin (3 mg/cm^3), the interval is even shorter, and electrical pulses should follow administration of the drug immediately (within 1 minute).[3] In larger animals, it is recommended to deliver electrical pulses to the tumor site during the pharmacokinetic peak of the medication if delivered IV; this is typically 8-28 minutes later in humans.[4] After intratumoral application, electroporation is performed 1-10 minutes later in humans, dogs, cats, and horses.[4]

Protocol details

Pre-treatment

Blood work, evaluation of anesthetic risk, cytology or histopathology confirming tumor type, and oncological staging is recommended for patients prior to ECT.[5]

Anesthesia or heavy sedation

Similar protocols should be used as for surgery in the same area.

Chemotherapy injection

IV bolus of bleomycin (over 30-45 seconds) is recommended in dogs and cats as this route almost always achieves adequate distribution and concentration of the drug in the tumor and its margins when timing of electroporation is appropriate, is safer due to less leakage after injection, avoids areas of tumor with insufficient drug concentrations secondary to errors in intratumoral injection technique, and is an immune system activator that contributes to a local immune response.[5] However, the intratumoral route is preferred in exotics due to lack of information on dosing and effects of systemic bleomycin.[5]

Time interval

Electrical pulses are applied 5-8 minutes after IV bleomycin or immediately to 5 minutes later after intratumoral injection of bleomycin or cisplatin in most species.[5] As noted above, timing in laboratory rabbits and rodents was shortened with peak drug concentrations in the tumor occurring within minutes after IV dosing and < 1 minutes for intratumoral application.[3]

Electroporation pulses

Electroconductive gel improves contact of the external electrodes to lesions.[1] ECT only affects cells directly around the electrode, so electroporation should be repeated several times until the clinician is satisfied that the full surface of the tumor and tumor margins have been treated, taking care to avoid treatment overlap.[4]

Post-treatment

Monitor patients after therapy for a couple of hours.[2] Post-treatment NSAIDs have been recommended by some clinicians.[5]

Repeat

ECT is often performed at 2-week intervals for two sessions for post-operative adjunctive therapy or until complete response/tumor regression if for direct ablation of gross disease.[1] A minimum treatment interval of 4 weeks had been recommended for ECT with IV bleomycin.[5] Maximum therapeutic effect may occur 6-8 weeks later or longer as tumors can enter a quiescent state and shrink again after 2-3 months once cells try to divide again (as bleomycin's mechanism of action affects dividing cells). Based on this delay, some clinicians recommend monitoring and only treating again if the tumor is not sufficiently treated in the first session, starts to grow, or new lesions develop to prevent overtreating and development of necrosis.[5] Stagnant lesions can be biopsied to evaluate for persistent tumor versus scar tissue. Frequent monitoring is recommended after treatment, often at 2 weeks, monthly for 3 months, then every 3 months (or a similar variation).[5]

Evolutions

Four evolutions have been reported for tumors after treatment with ECT. Understanding the evolutions can help determine if repeated treatment is needed for a particular neoplasm.[5]

1. *Common evolution:* The tumor shrinks after initial swelling and continues to shrink to the final response.

2. *Two times evolution:* The tumor shrinks, becomes quiescent, then shrinks again.

3. *No change evolution:* No shrinking is seen due to a high population of quiescent cells.

4. *Tumoral escape evolution:* After an initial regression, the tumor starts growing again. These masses should be retreated.

Future Application

Newer research on ECT involves treatment of new tumor types (e.g., feline clear cell thymoma, nasal cell tumors, rectal neoplasms), use of ECT as an adjunct to novel therapies (e.g., cold plasma in a mouse melanoma model), evaluation of new or nonvalidated chemotherapy drugs (e.g., doxorubicin, 5-fluorouracil), and development of new electrodes (e.g. grid, single needle, endoscopic). There are also newer studies evaluating the use of electroporation with gene-based products (electrogene therapy), immunomodulatory agents, and other candidate compounds, such as calcium, oligonucleotides, plasmids, and small proteins.[1,2,4,7]

New electrodes have been developed to allow ECT therapy to be used on visceral and intracavitary tumors using endoscopic guidance or an intraoperative approach.[1,2,4] The endoscopic electrodes attach to standard endoscopes and allow treatment of surgically inoperable neoplasms and cancers recalcitrant to radiation therapy.[2] One of the endoscopic electrodes in development has a vacuum feature which sucks tissue into contact with the electrodes during treatment.[2]

Cases in the Literature

A summary of published cases of ECT in exotic animals is provided in Table 1.

Table 1. Published reports of electrochemotherapy in exotic animals.

Species (Age, sex)	Tumor	Treatment (Intratumoral)	Outcome
Central bearded dragon[a] *Pogona vitticeps* (7.8-year-old male)	Fibrosarcoma (oral mucocutaneous) 10 × 20 mm	Cisplatin (10 mg/ml) 0.1 mg/cm^3 0.5- to 1-second intervals Six treatments q2wk	Reduced size 90% after sixth treatment, resolved at 18 months No recurrence at 2 years
Yellow bellied slider[b] *Trachemys scripta scripta* 5-year-old (female)	Squamous cell carcinoma (cutaneous on neck, well differentiated incompletely excised) 2 × 1.5 cm	Bleomycin (1 mg/ml) 1 mg, 1 ml 5 minutes after injection, 1200V/cm, 1Hz, 100µsec (50 + 50 µsec) Two treatments q2wk	Treated surgical scar and 1-cm margin No recurrence at 1 year
Green sea turtles[c] *Chelonia mydas* n = 2	Fibropapillomas (herpesvirus) a. 1.43 × 1.4 cm carapace b. 1.52 × 1.12 cm left anterior flipper	Bleomycin 1U/cm^3 4 minutes after injection, needle electrode, 1000V Two treatments, q33d	Response starting at 4 weeks, complete regression at 10 months No recurrence at 1 year Note: local anesthetic (lidocaine) only
Cockatiel[d] *Nymphicus hollandicus* 12-year-old female)	Squamous cell carcinoma (uropygial, ulcerated) 9 × 7 × 4 mm	Cisplatin (2 mg/ml) 1 mg/kg (first) Bleomycin 0.5 mg/kg (second, third[d]) q1mo Bleomycin 1 mg/kg (fourth) 3 months later 1-2 minutes after injection, plate (first) and needle electrodes, trains of eight pulses, 1300V/cm, 5 Hz, 100 µsec	No change in tumor size with cisplatin Mass necrotic and regressed with two treatments of bleomycin (90% reduced at 4 weeks, 100% after third treatment) Recurrence 3 months later- fourth treatment No recurrence at 18 months
Cockatiel[e] *Nymphicus hollandicus* (11-year-old male)	Fibroma (cutaneous, left face) 10 mm diameter	Cisplatin 1 mg/kg, 0.05 ml 1-2 minutes after injection, 1300 V/cm, 5 Hz, 100 seconds Two treatments, q3mo	Regressed to 7 mm diameter after first treatment. Regressed to 4 mm after second Unchanged after 3 years
Budgerigar[f] *Melopsittacus undulatus* (10-year-old male)	Epithelioid hemangioendothelioma (subcutaneous, low-grade, left wing, incompletely excised) 2 × 1 cm	Bleomycin (2.5 mg/ml) 0.25 mg, 0.1 ml 5 minutes after injection, 1200 V/cm, 1 Hz, 50+50 µsec Two treatments q2wk	Complete regression. No recurrence at 12 months
African pygmy hedgehog[g] *Atelerix albiventris* (5 years old)	Squamous cell carcinoma (oral, ulcerated) 1 × 1.5 cm	Bleomycin (1.5 mg/ml) 1.5 mg total, 5 minutes after injection, 1200 V/cm, 1Hz, 50+50 µsec Five treatments, q1mo	Partial remission (25% regression) 5 months until tumor progression and euthanasia

Table 1 Continued. Published reports of electrochemotherapy in exotic animals.

Species (Age, sex)	Tumor	Treatment (Intratumoral)	Outcome
African pygmy hedgehog[h] *Atelerix albiventris* (5.5-year-old male)	Hemangiosarcoma (cutaneous, left scapula/forelimb, incompletely excised) 3 × 3.3 cm	Bleomycin 1U (first) and 0.7U (second) 5 minutes after injection, 1200 V/cm Two treatments, q2wk	Euthanized 2 months later for oral/cervical masses (poorly differentiated sarcomas, thyroid carcinoma), no recurrence of HSA on necropsy
Ferret[i] *Mustela putorius furo* (5-year-old castrated male)	Mast cell tumor x2 (cutaneous, tail base 0.8 × 0.8 × 0.5 cm; lateral right hock 1.2 × 1 × 0.7 cm)	Bleomycin (3 mg/ml) 0.5 mg, 0.1 ml 1-2 minutes after injection, 1300 V/cm, 5 kHz, 100 µsec One treatment	Complete regression (day 30 first, day 67 second), no recurrence at 15 months for first tumor and 11 months for second
Ferret[i] *Mustela putorius furo* (7-year-old spayed female)	Sebaceous adenoma (cutaneous, lateral left hock) 0.7 × 0.7 × 0.5 cm	Bleomycin (3 mg/ml) 0.5 mg, 0.1 ml 1-2 minutes after injection, 1300 V/cm, 5 kHz, 100 µsec One treatment	Complete regression (day 70), no recurrence at 8 months
Ferret[i] *Mustela putorius furo* (7-year-old spayed female)	Squamous papilloma (cutaneous, dorsolateral left thorax) 0.4 × 0.4 × 0.2 cm	Bleomycin (3 mg/ml) 0.5 mg, 0.1 ml 1-2 minutes after injection, 1300 V/cm, freq 5 kHz, 100 µsec One treatment	Complete regression, no recurrence at 8 months
Rats[j] *Rattus norvegicus* 18 months old, 24 months old n = 2	Mammary carcinoma (axillary, incompletely excised) 3 cm, 3.5 cm (max diam)	Cisplatin (0.5 mg/ml), 1 mg, 2 ml 5 min after injection, 1200 V/cm, freq 1 Hz, 50 + 50 µsec Two treatments q2wk	Treated surgical scar and 1-cm margin Disease free at 10 months (older rat, no tumor at necropsy) and 14 months (younger rat)

Source: [a]Doneley RJT, Sprohnle-Barrera CH, Jones EE, et al. The use of electrochemotherapy to treat an oral mucocutaneous fibrosarcoma in a central bearded dragon (*Pogona vitticeps*). *J Exot Pet Med.* 2022;41:38-41.
[b]Lanza A, Baldi A, Spugnini EP. Surgery and electrochemotherapy for the treatment of cutaneous squamous cell carcinoma in a yellow-bellied slider (*Trachemys scripta scripta*). *J Am Vet Med Assoc.* 2015;246(4):455-457.
[c]Brunner CHM, Dutra G, Silva CB, et al. Electrochemotherapy for the treatment of fibropapillomas in *Chelonia mydas*. *J Zoo Wildl Med.* 2014;45(2):213-218.
[d]Racnik J, Svara T, Zadravec M, et al. Electrochemotherapy with bleomycin in the treatment of squamous cell carcinoma of the uropygial gland in a cockatiel (*Nymphicus hollandicus*). *J Exot Pet Med.* 2019;29:217-221.
[e]Racnik J, Svara T, Zadravec M, et al. Electrochemotherapy with cisplatin for the treatment of a non-operable cutaneous fibroma in a cockatiel (*Nymphicus hollandicus*). *N Z Vet J.* 2019;67(3):155-158.
[f]Lanza A, Baldi A, Rossi G, Spugnini EP. Electrochemotherapy for the treatment of an incompletely excised subcutaneous low-grade epithelioid hemangioendothelioma in a budgerigar parakeet (*Melopsittacus undulatus*). *Open Vet J.* 2019;9(3):269-272.
[g]Spugnini EP, Lanza A, Sebasti S, et al. Electrochemotherapy palliation of an oral squamous cell carcinoma in an African pygmy hedgehog (*Atelerix albiventris*). *Vet Res Forum.* 2018;9(4):379-381.
[h]McCleery B, Morges M. Electrochemotherapy for the treatment of an incompletely excised hemangiosarcoma in an African pygmy hedgehog. *Proc ExoticsCon.* 2021;2788.
[i]Racnik J, Svara T, Zadravec M, et al. Electrochemotherapy with bleomycin of different types of cutaneous tumours in a ferret (*Mustela putorius furo*). *Radiol Oncol.* 2018;52(1):98-104.
[j]Lanza A, Pettorali M, Baldi A, Spugnini EP. Surgery and electrochemotherapy treatment of incompletely excised mammary carcinoma in two male pet rats (*Rattus norvegicus*). *J Vet Med Sci.* 2017;79(3):623-625.

Reptiles

ECT has been reported in multiple reptile species. Use of ECT may be due to difficulties associated with vascular access for chemotherapy and relative radiation resistance in these species, making traditional therapies more difficult or less effective. Treatment has been reported in a bearded dragon with an ulcerated, low-grade (based on a canine grading scheme) fibrosarcoma of the rostral maxillary mucocutaneous region of several months duration; yellow-bellied slider with an incompletely excised, cutaneous, well-differentiated squamous cell carcinoma; and two green sea turtles with herpesvirus associated fibropapillomas on the carapace and flipper.[4,8,18] The only side effects reported in these cases was muscle contraction during electric pulse application. The green sea turtles received ECT under local anesthetic only due to concerns for systemic debilitation.[18]

Birds

Various tumors have been treated with ECT in avian patients including an ulcerated, rapidly growing squamous cell carcinoma of the uropygial gland of a cockatiel; non-operable cutaneous fibroma in a cockatiel; and an incompletely excised, subcutaneous, low-grade epithelioid hemangioendothelioma on the left wing of a budgerigar.[11,19,20] Not surprisingly, all of the reports of ECT in avian patients are in small birds where complete surgical excision may not be possible, ease of administration of other therapies may be more difficult, and risks associated with traditional therapies may not be acceptable to owners.

Exotic companion mammals

ECT has been reported in two hedgehogs. The first was diagnosed with oral squamous cell carcinoma of 1-month duration and was treated with palliative ECT using bleomycin.[21] The second hedgehog received ECT with bleomycin for an incompletely excised hemangiosarcoma on the left scapula and mid-brachium.[22] Other reports of ECT in exotic companion mammals include two ferrets each treated for two cutaneous tumors (two mast cell tumors, a squamous papilloma, and a sebaceous adenoma) and two rats treated post-operatively for incompletely excised mammary carcinomas in the axillary region.[17,23] All four skin tumors in the two ferrets treated with ECT became necrotic after treatment (between days 6-15 after therapy) but regressed completely (within 30-70 days from treatment).[23]

Additional reports in zoological species:

- Asian elephant: Treated for a malignant mammary mass at the El Paso Zoo.
- Asian small-clawed otter (*Aonyx cinereus*):[24] A 7-year-old female was diagnosed with oral squamous cell carcinoma. The mass was nonresectable on CT scan and involved the tongue, ventral pharynx, and hyoid bones. The patient received three ECT treatments every 2-4 weeks using bleomycin (20 mg/m^2) IV 10 minutes prior to intralesional cisplatin (0.2 mg/kg). Reversible and irreversible electroporation was performed with biphasic electrical pulses. The rostral portion of the tumor regressed with poor response caudally. The procedure was reported to provide palliation with good quality of life (follow up or 7 months after diagnosis).
- Binturong (*Arctictis binturong*):[25] A 3-year-old male binturong at the Naples Zoo was diagnosed with squamous cell carcinoma of the left jaw (5 cm diameter) that was invasive on CT scan. ECT was performed with bleomycin (10 mg/kg IV). Repeat CT showed metastasis to the lungs and spleen, and the patient was euthanized.

- Vietnamese pot-bellied pig (*Sus scrofa*):[26] A 9-year-old FS pot-bellied pig was treated for multiple cutaneous squamous cell carcinoma lesions and a history of pruritus. The patient had raised, hemorrhagic crusts and ulcerated lesions on the flanks (bilaterally), dorsum, ventral abdomen, pelvic limbs, and caudal to the ears. ECT was performed eight times over the course of 1 year (recommended q2wk but up to 4-month delays due to travel constraints of the owners) using bleomycin IV (15 IU/m^2) 5 minutes prior to electroporation (1200-1300V/cm). Partial remission and marked improvement in pruritus was sustained for 18 months post-diagnosis. A new lesion in an untreated location was present at 2 months. Side effects of the ECT included mild erythema 1 week after treatment, mucopurulent discharge from the lesions 3-5 days after treatment, and transient weight loss.

One review[5] of ECT in veterinary medicine mentioned use of the modality in fishes and snakes; however, no references were provided and no published reports could be found on literature review.

Conclusions

Electrochemotherapy increases the internalization of nonpermeant or poorly permeant chemotherapy drugs into the cytosols of target cells.[2] ECT has a good cost-benefit ratio and can be used alone or as a neoadjuvant, adjuvant, or concomitant treatment with standard therapies for the curative or palliative treatment of primary or metastatic cutaneous and subcutaneous tumors of various histologic types in many species with low cost and side effects.[5,7] In exotic animals, ECT is an alternative option to other ablative locoregional therapies for tumors in sensitive or difficult areas to treat, when other therapies are considered too invasive or risky by the owner, where mutilation may occur, or when cost of standard therapies is too high.[7]

References

1. Spugnini EP, Baldi A. Electrochemotherapy in veterinary oncology: state-of-the-art and perspectives. *Vet Clin North Am Small Anim Pract.* 2019;49(5):967-979.

2. Impellizeri J, Aurisicchio L, Forde P, et al. Electroporation in veterinary oncology. *Vet J.* 2016;217:18-25.

3. Cemazar M, Tamzali Y, Sersa G, et al. Electrochemotherapy in veterinary oncology. *J Vet Intern Med.* 2008;22:826-831.

4. Doneley RJT, Sprohnle-Barrera CH, Jones EE, et al. The use of electrochemotherapy to treat an oral mucocutaneous fibrosarcoma in a central bearded dragon (*Pogona vitticeps*). *J Exot Pet Med.* 2022;41:38-41.

5. Tallado M, Mir LM, Maglietti F. Veterinary guidelines for electrochemotherapy for superficial tumors. Front Vet Sci. 2022;9:868989.

6. Fiorentzis M, Viestenz A, Siebolts U, et al. The potential use of electrochemotherapy in the treatment of uveal melanoma: in vitro results in 3D tumor cultures and in vivo results in a chick embryo model. *Cancers.* 2019;11:1344.

7. Cemazar M, Sersa G. Recent advances in electrochemotherapy. *Bioelectricity.* 2019;1(4):204-213.

8. Lanza A, Baldi A, Spugnini EP. Surgery and electrochemotherapy for the treatment of cutaneous squamous cell carcinoma in a yellow-bellied slider (*Trachemys scripta scripta*). *J Am Vet Med Assoc.* 2015;246(4): 455-457.

9. Torrigiani F, Pierini A, Lowe R, et al. Soft tissue sarcoma in dogs: a treatment review and a novel approach using electrochemotherapy in a case series. *Vet Comp Oncol.* 2019;17:234-241.

10. Graham JE, Kent MS, Theon A. Current therapies in exotic animal oncology. *Vet Clin North Am Exot Anim Pract.* 2004;7:757–81.

11. Racnik J, Svara T, Zadravec M, et al. Electrochemotherapy with cisplatin for the treatment of a non-operable cutaneous fibroma in a cockatiel (*Nymphicus hollandicus*). *N Z Vet J.* 2019;67(3):155-158.

12. Najjar TAO, Saad SY. Cisplatin pharmacokinetics and its nephrotoxicity in diabetic rabbits. *Chemotherapy.* 2001;47l;128-135.

13. Siddik ZH, Newell DR, Boxall FE, Harrap KR. The comparative pharmacokinetics of carboplatin and cisplatin in mice and rats. *Biochem Pharmacol.* 1987;36(12):1925-1932

14. Filippich LJ, Charles BG, Sutton RH, Bucher AM. Carboplatin pharmacokinetics following a single-dose infusion in sulphur-crested cockatoos (*Cacatua galerita*). *Aust Vet J.* 2004;82(6):366-369.

15. Filippich LJ, Bucher AM, Charles BG, Sutton RH. Intravenous cisplatin administration in sulphur-crested cockatoos (*Cacatua galerita*): clinical and pathologic observations. *J Avian Med Surg.* 2001;15(1):23-390.

16. Zehnder A, Graham J, Reavill DR, McLaughlin A. Neoplastic diseases in avian species. In: Speer BL, ed. *Current Therapy in Avian Medicine and Surgery.* St. Louis, MO: Elsevier; 2016;107-141.

17. Lanza A, Pettorali M, Baldi A, Spugnini EP. Surgery and electrochemotherapy treatment of incompletely excised mammary carcinoma in two male pet rats (*Rattus norvegicus*). *J Vet Med Sci.* 2017;79(3):623-625.

18. Brunner CHM, Dutra G, Silva CB, et al. Electrochemotherapy for the treatment of fibropapillomas in *Chelonia mydas*. *J Zoo Wildl Med.* 2014;45(2):213-218.

19. Racnik J, Svara T, Zadravec M, et al. Electrochemotherapy with bleomycin in the treatment of squamous cell carcinoma of the uropygial gland in a cockatiel (*Nymphicus hollandicus*). *J Exot Pet Med.* 2019;29:217-221.

20. Lanza A, Baldi A, Rossi G, Spugnini EP. Electrochemotherapy for the treatment of an incompletely excised subcutaneous low-grade epithelioid hemangioendothelioma in a budgerigar parakeet (*Melopsittacus undulatus*). *Open Vet J.* 2019;9(3):269-272.

21. Spugnini EP, Lanza A, Sebasti S, et al. Electrochemotherapy palliation of an oral squamous cell carcinoma in an African pygmy hedgehog (*Atelerix albiventris*). *Vet Res Forum.* 2018;9(4):379-381.

22. McCleery B, Morges M. Electrochemotherapy for the treatment of an incompletely excised hemangiosarcoma in an African pygmy hedgehog. *Proc ExoticsCon.* 2021;2788.

23. Racnik J, Svara T, Zadravec M, et al. Electrochemotherapy with bleomycin of different types of cutaneous tumours in a ferret (*Mustela putorius furo*). *Radiol Oncol.* 2018;52(1):98-104.

24. Anderson KB, Siegal-Willott J, Merrick CH, et al. Electrochemotherapy for the treatment of oral squamous cell carcinoma in an Asian small-clawed ottern (*Aonyx cinereus*). *Proc Annu Conf Am Assoc Zoo Vet.* 2019.

25. Silvestre P, Prisco F, Capasso M, et al. Oral squamous cell carcinoma in a captive-born bearcat (*Arctictis binturong*). *Proc Zoo Wildl Health.* 2019;153.

26. Weissman M, Donnelly LL, Branson K, Glahn A, Nagy D, Havis B, Johnson G, Flesner B. Electrochemotherapy for a cutaneous squamous cell carcinoma in a Vietnamese pot-bellied pig (*Sus scrofa*). *J Exot Pet Med.* 2020;30:37-43.

Introduction to Beekeeping

Elizabeth B. Mackey, DVM

Session #3368

Affiliation: From Mackey Exotic Animal Clinic, 1760 Antioch Church Road, Watkinsville, GA 30677, USA.

Abstract: This masterclass is designed to introduce veterinarians to the science of beekeeping and their role in working with the beekeepers. The unique aspects of their eusocial behavior are discussed as well as common diseases and parasites. Equipment used in the colony is also presented. The goal is to have the attendees comfortable with developing a veterinary-client-patient relationship with keepers.

Introduction

There are over 20,000 species of bees worldwide,[1] but it is *Apis millifera*, the western honey bee, that is the predominant species in beekeeping. Native to Africa, the Middle East and northern Europe, these bees are often managed so we can take advantage of their products, such as beeswax, honey, pollen and propolis. We also use their behavior of foraging for nectar, which results in pollination, as a way of boosting the yield of our commercial crops, resulting in an estimated increased yield of $15 billion.[2]

As with other species, there are bacterial, viral and fungal diseases of the honey bee. In 2017 FDA required Veterinary Feed Directives for food producing animals, and in June of 2013 over-the-counter antibiotics will no longer be available.[3] While involvement of veterinarians and their understanding of appropriate use of antibiotics is a positive change, many of us do not understand the basics of hive management and the terminology used by beekeepers. As with any other animal we treat, we need to understand the husbandry of the patient, as well as the anatomy, common diseases and common parasites.

Become familiar with your state's regulations for honey bees and how they define a veterinary-client-patient relationship. The focus with food animals tends to be having a relationship with the owner and familiarity with their apiary, whether a few hives or hundreds or more. Diseases that are reportable, as well as options to treat, may have state variations. It is also important to understand the beekeeper's experience level. Individuals may have an interest and purchase impulsively but have no understanding of the basic health and husbandry requirements. They may have poor frame handling technique and be unable to evaluate brood patterns. They may also be uncomfortable with alcohol washes that sacrifice bees for diagnostic purposes. Commercial beekeepers, on the other hand, have vast knowledge and experience. Do not be offended if they will not allow you to remove frames and insist on handling the bees themselves. The health of their queen is everything, and if they don't want to risk you rolling the queen, they can remove the frames and present them to you for assessment.

Your examination with the hive begins as with any other species you see. Questions about feeding and treatments for parasites are important, as well as details on how the colony originated. Take time to observe the bees and assess their behaviors. Are they showing normal levels of activity? Are there normal behaviors being observed, such as orientation flights, trophallaxis and waggle dances? Look closely at the frames. Are there adequate stores of nectar, pollen and honey? Is the brood pattern sufficient for raising new worker bees? Is there evidence of a failing queen, as may be seen with poor brood patterns or the presence of supercedure cells?

A key difference that must be remembered with honey bees is that the colony is the patient, not the individual bee. Honey bees, along with termites and ants, are eusocial, with generations that overlap and care for the young of others. Often referred to as a superorganism,[1] they work collectively for survival and reproduction of the colony. Understanding their natural behaviors is essential for beekeepers, who can then develop a management plan to work with, or against these behaviors.

Although the colony comprises many bees working together, it is easier to understand them if you think of them as a single organism. The colony begins small at approximately 3-5 pounds of bees. It grows during times of plenty, with the goal of storing enough food to survive the winter, as well as growing large enough to be able to reproduce. An adult colony may weigh as much as 20 pounds and be comprised of 50-60,000 individual bees. Reproduction involves the colony dividing, with the queen and approximately two-thirds of the bees leaving with her to a new location of their choice. Although this swarming process is natural for the bees, we as beekeepers do not want to lose our bees. Beekeeping requires managing.

The Bees

As an insect, the honey bee has a life cycle of complete metamorphosis. It begins as an egg, hatches into a larva, pupates, and then emerges as an adult bee. The details of that are variable, though, as within the colony not all bees are created equally! The cell size, diet, and days to reach maturity are different for queens, workers and drones.

The worker bee begins as an egg that was laid by the queen. The diploid egg (16 pair of chromosomes) is laid at the bottom of a cell of comb. The egg hatches at day three and the larva eats royal jelly for the next 3 days. This high protein substance is produced by the mixing of pollen and the secretions from the mandibular and hypopharyngeal gland of the nurse bee.[4] She is then transitioned to brood food, which is a combination of pollen and nectar that is slightly fermented to allow for longer storage. The cell is then capped and she continues to develop until ready to emerge as a full-sized adult around day 21.

Worker bees have numerous jobs and as they age they rotate through these various positions.[4] They start as being nurse bees, cleaning up the cell from which they just emerged, then taking care of feeding other developing bees. These bees are all developing in the same region of the hive, and this is collectively referred to as the brood. It is the worker bees who care for the queen and guard the entrance of the hive to protect from pests and robbers. It is the workers who communicate with the foragers (older workers) to let them know what supplies are needed and then relieve them of those loads on their return. These supplies must then be properly stored. It is the workers who determine when the colony is ready to swarm, agitating the queen until she is ready to leave. Workers remove the dead from the hive and remove the males from the colony when they are no longer needed. They protect the hive, stinging an invader if necessary. The stinging will only occur once, though, as the barbed stinger will pull the venom sac from the abdomen of the bee, resulting in her death. The last job a worker will have is that of a forager. This is a high-risk job as they must leave the protection of the hive and may travel several miles in search of nectar, exposing themselves to predators and the elements. The average lifespan for these workers is ~ 4-6 weeks. When the temperatures drop in the fall, brood production will begin to decrease, and then longer-lived workers bees will remain through the winter months. Now their focus is to keep the colony warm and survive until they can begin the cycle again in the spring.

Drones are the male bees that develop from unfertilized eggs (haploid) and they have no stinger, but do have a reproductive endophallus.[4] They are much larger than the worker bees and are readily distinguishable by their size and larger eyes. They have only one job- mate with a queen. In the spring when drone brood emerges (24 days from egg), the males from multiple hives will join together in drone congregations. Here they spend their time waiting for a queen. If one arrives they will attempt to mate with her. This is their last act as the process will remove their endophallus and cause death. If they do not find a queen they will return to the hive for the night and then go back to the congregation area the next day. When the days turn cool and the focus of the colony is winter survival, the drones are no longer necessary and are cast from the hive.

The queen has one responsibility: lay eggs. Within a few days of emerging from her cell (16 days from egg) she will go on her maiden flight in search of a drone congregation. She will mate with 15-20 drones and then return home. If she doesn't find a congregation or not many drones are present she will seek them out the next day. This mating process provides all of the sperm she will ever have, which she will store in her spermotheca.[4] The queen is recognized by her elongated abdomen. She is usually found in the vicinity of the brood as she is focused on laying egg, producing as many as 1500 in a day. She is capable of stinging multiple times, as the barbs are minimal and it is not a fatal process for her. She does not engage in hive protection, so her stinging is directed at rival queens, even if they have not emerged yet.

The Hive

The hive is the man-made physical structure we put colonies in. There are numerous types but the most common is the Langstroth hive. This hive was designed by Lorenzo Langstroth and provided removeable frames and, extremely important, the preservation of bee space.[5] This is the critical distance of one-fourth to three-eighths of an inch between structures in a hive through which a bee navigates. If the space is too small, it is not navigable, and the bees will fill the gap with propolis. If the space is wider, then there is likelihood of the bees using the space for comb. The traditional Langstroth hive is placed on a stand or bricks to elevate it above the ground. This helps to protect against insects and mice as well as being more comfortable for the beekeeper. The base is called a bottom board and can be screened (additional ventilation) or solid. On this is placed the hive body, or deep super. The structure of the bottom board results in a gap between it and the hive body that is placed on it, and this then becomes the entrance to the hive. Additional hive bodies/supers are added on top and ultimately there is a lid to protect from the elements.

Within the hive bodies are frames. These are placed in the hive body like index cards in a file box. Hive bodies can be different depths (deep, medium, shallow) and can hold different numbers of frames (e.g., 10 frames, 8 frames, 5 frames). Correct bee space is maintained regardless of the size of the box. A frame holds the wax honeycomb that is the foundation on which a colony lives. It is here that they store nectar and pollen, create and store honey, and raise the brood. Most cells are the same size, but workers can build other sizes and types as needed. If the queen detects a larger cell, she will lay an unfertilized egg which will develop into the larger drone bee. Swarm cells are designed to raise new queens. When the original queen leaves with many of the workers to start a colony elsewhere, the new queen from the swarm cell will emerge, go on her maiden flight, and return to re-build the colony. Supercedure cells, another type of queen cell, are built when the workers perceive a new queen is needed. It may be that the original one has died, or run out of sperm, or is ill. For whatever reason, they have determined that the old queen needs to be replaced.

Feeding of the colony is often provided in the form of syrup (sugar and water in a ratio of 1:1 to 2:1), which can be in containers placed inside the hive or outside the hive, or even as a feeding station well away from the hive. This may be done with a new hive that doesn't have much comb. Building the wax comb is very energy intensive, so keepers can provide support with syrup when the nectar flow isn't strong enough. Feeding might also be necessary at other times if the colony doesn't have enough reserves. Starvation is a leading cause of death in colonies, and part of beekeeping is monitoring to assure they have adequate reserves or sources of their protein and carbohydrate needs.

Protective gear is recommended when working with bees. This should always include a veil to protect the face. It is natural for bees to go for the face of an intruder. Other protective gear may include full body suits or just jackets, gloves (heavy or lightweight disposable) and a smoker. The smoke is used to block the pheromone communication and is used just enough to provide a distraction but not so much the bees fear there is a fire and abscond.

References

1. Seeley TD. *Honeybee Democracy.* Princeton, NJ: Princeton University Press. 2010;20.

2. U.S. Food and Drug Administration. Helping Agriculture's Helpful Honey Bees. 2018. Available at: https://www.fda.gov/animal-veterinary/animal-health-literacy/helping-agricultures-helpful-honey-bees. Accessed June 1, 2023.

3. U.S. Food and Drug Administration. Using Medically Important Antimicrobial in Bees. 2023. Available at: https://www.fda.gov/animal-veterinary/development-approval-process/using-medically-important-antimicrobials-bees-questions-and-answers. Accessed June 1, 2023.

4. Winston ML. *The Biology of the Honey Bee.* Cambridge, MA: Harvard University Press; 1987.

5. Delaplane KS. *First Lessons in Beekeeping.* Hamilton, IL: Dadant and Sons; 2007:29-39.

Section 52

Technician Session 1

Christina Miller, RVT, BSc
Session Chair

Gimme Shelter (and UVB and Tube Feeding): Caring for Hospitalized Reptiles

Jovanna Berls, CVA, LVT, FFCP (Veterinary), FFCP (Avian)

Session #3152

Affiliation: From ABC Animal & Bird Clinic, 11930 Highway 6 South, Sugar Land, TX 77498, USA.

Abstract: Hospitalized reptiles in some ways are similar to other species commonly seen in veterinary clinics; like all patients, they need appropriate food, caging, and medication. But they do have some specialized needs, such as ultraviolet lighting and particular attention to thermal support. This presentation reviews methods for enhancing comfort and welfare for reptile patients during a hospital stay, with attention to the different requirements for commonly seen species of lizards, snakes, and chelonians. Techniques for providing heat, light, nutrition, and hydration will be covered, along with tips for the care and feeding of live food items.

Introduction

Like any other hospitalized patient, reptiles need housing, food, water, and medication; however, there are nearly 12,000 known species of reptiles,[1] originating from across the world, with a wide range of natural histories, feeding strategies, and environmental requirements. It is simply not possible for even the most expert of veterinary technicians to know from memory what is needed for every species, but a basic grounding in commonly seen pet species will go a long way toward preparing technicians to deal with the myriad of possible reptile patients. This presentation focuses on frequently seen species of lizards (bearded dragons [*Pogona vitticeps*], leopard geckos [*Eublepharis macularius*], and veiled/Yemen chameleons [*Chamaeleo calyptratus*]), snakes (ball pythons [*Python regius*], corn snakes [*Pantherophis guttatus*]), boa constrictors [*Boa constrictor*]), and chelonians (red-eared sliders [*Trachemys scripta elegans*], box turtles [*Terrapene carolina*], and Russian tortoises [*Testudo horsfieldii*]).

These patients have specialized requirements that set them apart from other species commonly seen in veterinary hospitals. As ectotherms, they often need thermal support and specific humidity levels; many need provision of ultraviolet light; they may also require live foods. These things are necessary to maintain homeostasis, a complex process that requires a balance of appropriate nutrition, hydration, temperatures, and activity. Homeostasis is essential for normal animals to remain healthy, and ill or injured animals are less likely to recover well without them.

Effects of Stress on Healing and Recovery

For many exotic species that are essentially wild animals, captivity alone can be a source of stress, making appropriate husbandry techniques especially critical for their health, reproductive success, and even survival. If an exotic pet has been stressed in its usual home, it is likely to be more so in a hospital environment in which it will be handled by unfamiliar people and subjected to sometimes unpleasant procedures, such as venipuncture or administration of medications.[2]

Stress can be defined as a response to anything that disturbs homeostasis and the body's response as it tries to adapt and restore homeostasis.[3] It is a physiologic response related to release of catecholamine and, in reptiles, corticosterone from the adrenal glands. This chemical release has many effects, some helpful and some not, such as elevated heart and respiratory rates, reduced digestive function, and redirection of blood flow from visceral organs to skeletal muscle.[4] In addition, elevated levels of corticosterone can alter metabolism and suppress the immune system.[5] Altered metabolism, in turn, may affect the absorption, distribution, and elimination of pharmaceutical agents used to treat the patient.[3,5]

Detrimental effects of stress may present acutely, as in a rabbit that collapses suddenly as a result of stress-induced cardiomyopathy,[4] or they may appear very slowly, as in a reptile that has suffered low-level, chronic thermal stress before developing an infection.[6] It is important to remember that corticosterone release can occur very rapidly. Recent research shows that cockatiels can experience elevated levels of plasma corticosterone in as little as 3 minutes after a stressful event.[7]

Thus, maintaining an environment that is as comfortable and stress-free as possible is extremely important for any hospitalized patient. When dealing with an unfamiliar species, always check the patient's natural history to determine the least stressful conditions in which to house it.

Also keep in mind that one of the most stressful events to which patients will be exposed is capture and restraint, whether for administration of medication or simply moving the animal for cleaning of its enclosure. Research shows that, for many animals, handling can be far more stressful than an actual medical procedure.[8] Thus, minimizing restraint and using low-stress techniques can significantly reduce the negative effects of stress; this is especially important for shy species, such as chameleons and ball pythons.

General Principles of Reptile Hospital Care

Housing and substrate

Reptiles should be kept in a secure enclosure to prevent escape. Small reptiles can be wily and squeeze through surprisingly tiny openings. Tearing a ward apart in an effort to find them is not a pleasant way to spend a day! Spartan caging, defined as caging without decorative furnishings or a particulate substrate, is generally preferred, as it facilitates monitoring of the patient. A particulate substrate, such as coconut fiber, may be needed if encouraging egg laying or other natural behaviors, such as burrowing behavior in fossorial species. A paper substrate, such as a paper towel, may be preferable for monitoring purposes, but bear in mind that it can interfere with a small reptile's ability to find and ingest prey by allowing food items to hide underneath it. Remove the paper and leave out the substrate if needed.[9]

Aquatic or semi-aquatic patients will require at least part of the enclosure to be filled with water. Some chelonian patients will only feed, urinate, and defecate in water, so it is important to carefully monitor water quality.[9]

Temperature

As mentioned above, all reptiles are ectotherms, dependent on the environment for the temperature necessary to support their physiological functions, such as heart rate, respiratory rate, digestion, and elimination. Every species has what is called a preferred optimal temperature zone (POTZ), and it is necessary for a reptile patient to be maintained within that range for adequate physiological function and healing. These ranges can differ markedly even between similar species. For example, the daytime POTZ for a veiled chameleon is 22.2-37.8 °C (72-100 °F),[10] with a night-time drop of about 12.2 °C (10 °F), while a Jackson's chameleon *(Trioceros jacksonii)* must be kept in a lower daytime POTZ of 20-28.3 °C (68-83 °F), with a drop to 10-15 °C (50-59 °F) at night, and it should not be kept above 28.9 °C (84 °F) for long periods of time.[11]

The ideal situation is to provide a temperature gradient, with a warm area and a cooler area available in the same enclosure, so the patient can choose the temperature that feels most comfortable; however, that may not be practical in smaller enclosures, and a sick reptile may not be willing or able to move around. Providing a steady temperature at the upper end of the POTZ will provide adequate care.[9]

A wide variety of lamps and other heating equipment is available, and it may take some research to determine which of them will work best in an individual hospital situation. Keep in mind that it can be even more dangerous to overheat a patient than to keep him/her too cool, so good control of the temperature is essential.

Hydration and humidity

Because reptiles originate from a wide variety of climates from tropical to desert, it is critical to provide an appropriate amount of humidity for proper hydration and ecdysis. Humidity should be monitored regularly, and depending on the species, a humid hide and/or water bowl large enough for the animal to soak in may be needed. Appropriate humidity for the macrohabitat of many desert species is approximately 30%, while for tropical species, 60% or higher humidity will be preferred.[12] Note that even desert species will need access to water to remain adequately hydrated and often will take advantage of a more humid microhabitat, such as a humid hide (in captivity).

A variety of misting and drip systems are available or can be made from inexpensive materials; humid hides can be fashioned from small plastic boxes with moist paper towels or moss placed inside.[9] Providing a shallow warm-water bath for 20-30 minutes once or twice a day can assist with hydration, as well as stimulating digestion and metabolic activity. Take care to ensure the water is neither too hot nor too cold. What feels barely warm to humans, whose normal body temperature is 37 °C (98.6 °F), may feel too hot for an animal accustomed to a normal temperature of 29.4 °C (85 °F).[12]

Some reptiles, especially chameleons, typically drink water clinging to leaves, rather than lapping water from a bowl. These patients will need leaves, natural or artificial, that can be sprayed with water in order to remain hydrated.[10] Bearded dragons sometimes can be persuaded to drink by dripping water on their heads. Many species, especially crepuscular or nocturnal species such as leopard geckos, will appreciate provision of a humid hide, for maintaining hydration and/or because it is natural for them to stay concealed during the day.

Ultraviolet light (UVA and UVB)

In the wild, reptiles use every part of the spectrum of light provided by the sun, from infrared to ultraviolet, even if they are crepuscular or nocturnal.[13] When kept indoors, however, that spectrum is unavailable, at least in the same form. Many reptiles who are housed indoors must be provided with artificial ultraviolet light in the range of 290-320 nanometers, as well as warmth, to produce the vitamin D necessary for processing calcium. Without calcium supplementation and ultraviolet light (whether natural or artificial), these reptiles are highly susceptible to nutritional secondary hyperparathyroidism, also known as NSHP or metabolic bone disease. Even reptiles that do not spend much time in the sun during the day and carnivorous species that ingest adequate amounts of calcium from their diet, such as ball pythons, can benefit from provision of ultraviolet light. Hospital accommodation for these patients should provide ultraviolet light whenever possible.

Most reptiles see colors, and UVA light is a crucial component of color vision. Light in this spectrum also can reach nonvisual receptors in the reptile retina, pineal body, and parietal eye (if present), which transmit information to a complex neuroendocrine network that controls daily and seasonal behaviors. Thus, red and blue daylight or night light bulbs should not be used at night because they are visible to the reptile and can disrupt the animal's circadian rhythms.[13]

The length of time to which an animal should be exposed to ultraviolet light also varies by species. Animals originating from regions near the equator generally need approximately 12 hours of light and 12 hours of darkness, while those whose natural habitat is in more temperate latitudes may do better with a more variable light cycle, from 9-14 hours, depending on the season.[14] Guidelines on the appropriate amount of exposure are purely experimental, however.[13] Never leave the lights on 24 hours a day, as that can be a significant source of stress.[14] Provide a hide in the enclosure so that the patient can get away from the light if it prefers to do so.

A wide variety of ultraviolet lamps are available, and some research will be necessary to determine the best set up for an individual hospital's situation. Keep in mind that most such lamps only emit ultraviolet light for about six months, regardless of what the label says, and that any screening or other material between the lamp and the animal may reduce the amount of light being received. In addition, the depth to which the ultraviolet light will penetrate varies greatly from one lamp to another, so the size of the enclosure may impact the effectiveness of the lamp.

Feeding

Reptiles in the wild have a variety of feeding strategies; the reptiles that are the focus of this presentation are primarily herbivores (Russian tortoises), insectivores (leopard geckos, veiled chameleons, juvenile bearded dragons), omnivores (adult bearded dragons, adult box turtles, adult red-eared sliders), and carnivores (ball pythons, boa constrictors, corn snakes). Some insectivorous species, such as bearded dragons, eat primarily animal protein (e.g., insects and worms) when they are young, then become more omnivorous as they age, eating various greens and fruit. This is also true of box turtles and red-eared sliders. In addition, some mostly insectivorous species, such as veiled chameleons, occasionally eat some plant material, so if live plants are placed in the enclosure, care should be taken to ensure they are non-toxic. Leaves and branches of ficus (*Ficus benjamina*) or hibiscus (*Hibiscus* spp.) are appropriate. It is generally preferable to give frozen and thawed small mammal prey to species that will eat them, such as snakes, as opposed to live mice or rats that may fight back against attacks and injure the reptile. For insectivorous lizards, such as chameleons, leopard geckos, or omnivorous chelonians, it is acceptable to feed live insects and insect larvae; however, attention must be paid to the nutrients provided, especially calcium, when feeding insects to reptiles that are housed indoors. Nearly all common feeder insects have a negative calcium:phosphorus ratio, and reptiles that consume them will need calcium supplementation to compensate.[15]

For reptiles that are not self-feeding while in the hospital, force feeding may be indicated. Keep in mind that force or tube feeding can be extremely stressful for the patient, so it should not be undertaken unless it is truly needed. Reptiles' feeding often is influenced by physiological factors such as gravidity, season, brumation, or ecdysis. Thus, there often is no rush to ensure the reptile is fed, especially every day, and it is essential to ensure the patient is well-hydrated and warmed to its POTZ before attempting to feed. A useful rule of thumb is to feed if the patient has experienced a weight reduction of 10% or more.[16]

The rate for maintenance fluid replacement in reptiles has been estimated at 10-15 ml/kg/day, given over 72 to 96 hours.[17] Fluids can be given via intracoelomic (ICo), intraosseous (IO), intravenous (IV), or subcutaneous (SC) routes and should be administered in the caudal one-third of the body. Position the patient in dorsal recumbency to reduce the likelihood of introducing fluids into the lungs.[5,17]

In general, daily energy needs are estimated at $SMR = 32\ (BW^{0.75})$ for a reptile kept at 30 °C (86 °F), where SMR represents standard metabolic rate in kcal/day and BW represents weight in kg. Reptiles are subject to "refeeding syndrome," in which potentially life-threatening hypophosphatemia and hypokalemia can occur. Start by feeding 40-75% of daily energy requirements for the first two to four days, then gradually increase to the full calculated energy requirements. Note that the stomach capacity of most adult reptiles is approximately 2-5% of body weight.[16]

Either a metal gavage tube or a red rubber catheter can be used to tube feed most snakes and lizards. Make sure the food is warm before it is given. Cold food can result in decomposition rather than digestion of food in the gastrointestinal tract. Measure the distance from the mouth to the stomach, which generally is about mid-body on the left side. Be gentle when opening the mouth, especially in small reptiles or if there is any suspicion of NSHP, as it is possible to fracture the jaw. Lizards' teeth and the keratin surfaces of chelonians' beaks also can be fragile, so exercise caution with metal gavage tubes and/or specula used to hold the mouth open. It can be helpful to hold the patient upright for a minute or two after feeding to minimize the risk of regurgitation, particularly if the tube used was too short to reach the stomach. Oral medications can be administered through the gavage tube along with food if desired.[17]

Feeding chelonians can be considerably more challenging, especially if the patient is large. Even a small, sick turtle or tortoise can be very strong and have defenses such as withdrawing firmly into the shell that are not easily overcome. If the patient must be gavage fed repeatedly, it may be more advantageous to place a pharyngostomy or esophagostomy tube, especially if the patient is aggressive.[19]

As with other species, changing a reptile patient's diet while in hospital is not advisable, even if the diet fed at home is deficient, because doing so runs the risk of triggering anorexia. Patients who have been eating solely live food are unlikely to accept frozen-thawed mice or rats immediately. If the patient is accustomed to exclusively eating mealworms, it may refuse crickets or vegetables, and vice versa. That said, it may be possible to make minor adjustments that would have minimal or no effect on the appearance or taste of the food, such as providing appropriately gut-loaded insects or adding a powdered calcium supplement to greens or vegetables.

Medications

The renal portal system in reptiles directs blood from the caudal region of the body into the kidneys, especially at times of water conservation. Research indicates that a valve in the abdominal vein closes, redirecting blood through the iliac veins and into the kidneys. This can have an impact on the absorption and distribution of injected medications in the caudal half of a reptile patient. The extent of that impact is a subject of some controversy, but it is generally agreed to be safer and more effective to administer parenteral medications in the cranial half of the body.[19]

Reptile morphology differs significantly from one species to another, which can affect where and how medications are administered. Intravenous access can be difficult, especially in very small reptiles, and giving medications per os can be challenging in some species (chelonians). Research indicates oral administration may be of little value because the reptile gastrointestinal tract is quiescent between meals, resulting in little or no absorption of the drug.[20] The intramuscular (IM) route is generally preferred in most species, although there are some limitations on the volume that can be administered in a single site. Preferred sites for IM injections in chelonians are the upper forelimb (deltoid or triceps muscles); in lizards, the forelimbs (avoid the cranial surface of the limb to minimize the risk of damage to the radial nerve); for snakes, epaxial muscles in the cranial half of the body.[21]

Subcutaneous space is available for injections, but it is generally smaller and less well vascularized than in mammals. It is not necessary to tent the skin to give SC injections. Preferred sites are the axillary and inguinal spaces in chelonians; for lizards, the lateral body wall and the inguinal space; for snakes, the lateral body wall.[20]

Intracoelomic injections can be used but limit the amount of fluids being introduced to the coelomic cavity, as excessive amounts can cause lung compression. This route also requires taking care not to penetrate visceral structures with the needle, which has prompted some experts to recommend avoiding it unless absolutely necessary.[19]

When IV access is available, preferred sites in lizards are the tail vein and the ventral abdominal vein; for chelonians, the right jugular vein; in snakes, the tail vein is most commonly used.[20]

Monitoring

Hospitalized reptiles require at least daily monitoring, and that includes monitoring of the environmental conditions in the enclosure to ensure they remain appropriate. Changes in a reptile's condition can be subtle and easily missed without careful and regular monitoring. Minimum monitoring parameters include weight, mentation, appetite, and eliminations. More critical patients may require more frequent monitoring.[9]

Body weight is an important indicator, as continued weight loss is a poor clinical sign; however, continuing weight gain is also a concern if it does not correlate well with the amount of food and water being consumed or improvement in hydration. Increasing weight gain may indicate retention of fluid or constipation/obstipation or both, which can ultimately result in compression of other visceral organs, including the lungs. Production of feces may not occur every day, and that is not necessarily abnormal.[12]

Species-Specific Data

Bearded dragons[10]

- *Temperature in hospital incubator:* 31.1 °C (88 °F)[5]
- *Feeding strategy:* Insectivore (juvenile); omnivore (adult). For juveniles, offer live insects or worms with a body length no more than the space between the lizard's eyes.[10] Salad greens can be offered even to young dragons so they become accustomed to eating vegetable matter early. Adult dragons can eat larger insects/larvae and larger amounts of vegetables. Feed in the morning.
- *Supplementation:* Gut load insects/larvae; powdered calcium and multivitamin supplements can be provided on salad.
- *Humidity:*30-50%. Provide a water feature large enough for the dragon to soak itself in if possible.
- *UVB and UVA light:* Yes.
- *Diurnal*

Veiled or Yemen chameleons[10]

- *Temperature in hospital incubator:* 29.4 °C (85 °F)[5]
- *Feeding strategy:* Insectivore but will eat some plant material. Feed appropriately gut-loaded live insects in the morning. Additional supplementation may be needed, depending on the nutrients provided by the gut-loading formula used.
- *Humidity:* 40-60%. Spray leaves in the enclosure at least q12h. Enclosure should have no loose substrate.
- *UVB and UVA light:* Yes.
- *Diurnal*

Leopard geckos[10]

- *Temperature in hospital incubator:* 29.4 °C (85.0 °F)
- *Feeding strategy:* Insectivore. Feed appropriately gut-loaded live insects no greater in length than the length of the lizards' head.[10] Additional supplementation may be needed, depending on the nutrients provided by the gut-loading formula used. Feed at night.
- *Humidity:* 40-60%, with a humid hide and a water feature provided.
- *UVB and UVA light:* Optional but may do better with it, depending on the patient, such as NSHP patients.
- *Crepuscular/nocturnal*

Ball pythons[10]

- *Temperature in hospital incubator:* 30.5 °C (87 °F)
- *Feeding strategy:* Carnivore. Feed frozen-thawed rodents no wider than the widest part of the snake's body.[10] Feed at night. Acceptance may be improved by ensuring the feeder rodent is warm.
- *Humidity:* 60-80%; provide a water feature large enough for the patient to soak in it if possible.
- *UVB and UVA light:* Optional but may do better with it, depending on the patient.
- *Nocturnal*

Corn snakes[10]

- *Temperature in hospital incubator:* 29.4 °C (85 °F)
- *Feeding strategy:* Carnivore. Feed frozen-thawed rodents no wider than the widest part of the snake's body.[10] Feed at night. Acceptance may be improved by ensuring the feeder rodent is warm.
- *Humidity:*40%-60%. Provide a water feature large enough for the patient to soak in it if possible.
- *UVB and UVA light:* Optional but may do better with it, depending on the patient.
- *Diurnal*

Boa constrictors[10]

- *Temperature in hospital incubator:* 30.5 °C (87 °F)
- *Feeding strategy:* Carnivore. Feed frozen-thawed rodents no wider than the widest part of the snake's body.[10] Feed at night. Acceptance may be improved by ensuring the feeder rodent is warm.
- *Humidity:* 60-80%; provide a water feature large enough for the snake to coil its entire body in it if possible.
- *UVB and UVA light:* Optional but may do better with it, depending on the patient.
- *Nocturnal*

Red-eared sliders[10]

- *Temperature in hospital incubator (dry-docked):* 28.8 °C (84 °F)
- *Water temperature:* 23.8-30 °C (75-86 °F)
- *Feeding strategy:* Insectivore and or carnivore (juvenile, up to 40% protein), omnivore (adult, 25-50% may be commercially available turtle pellets). Appropriate sources of protein include earthworms, mealworms, and feeder fish. Appropriate plant matter includes finely chopped fresh dark leafy greens that are high in calcium or mixed vegetables.
- *Humidity:* Semi-aquatic, needs water feature unless dry-docked because of condition.
- *UVB and UVA light:* Yes.

Box turtles[10]

- *Temperature in hospital incubator:* 30.0 °C (86 °F)
- *Feeding strategy:* Omnivore. Appetite is stimulated by color and odor/ripeness of fruits and vegetables, and movement of prey items. Recommended foods include dandelion leaves, sweet potatoes, flowers such as hibiscus and geraniums (*Pelargonium* spp.), earthworms, slugs, and snails. Ensure food is obtained from a pesticide-free source. Feed when most active, early morning and late evening.
- *Humidity:* 50-70%; provide a water feature large enough for the patient to soak in it.
- *UVB and UVA light:* Yes.

Russian tortoise[10]

- *Temperature in hospital incubator:* 27-30 °C (80-85 °F)
- *Feeding strategy:* Herbivore. Dry leafy grass such as timothy hay should be available at all times. Fresh greens and vegetables also may be offered, but do not feed fruit. Dust greens with calcium and multivitamin supplement.
- *Humidity:* 30-50%. Provide fresh, clean water at all times.
- *UVB and UVA light:* Yes.
- Diurnal

Tips on Husbandry for Live Feeder Worms and Insects

Feeding insectivores can pose particular challenges because most insects and larvae contain little calcium and are low in fat-soluble vitamins, both of which are essential for maintaining health in these species.[22,23] The recommended solution for this is to gut load the insects. In other words, feed the insects with a nutrient-rich diet that then is transferred to the reptile when the insect or larva is eaten.

Preferred gut loading diets are those that provide a positive calcium:phosphorus ratio (calcium higher than phosphorus). For most vertebrate animals, a ratio of 2:1 or 3:1 is generally considered adequate, but it is unknown whether that is sufficient for health in rapidly growing juvenile reptiles.[22]

The most common insects used are house crickets (*Acheta domesticus*), mealworms (*Tenebrio molitor*), black soldier fly larvae (*Hermetia illucens*), dubia roaches, sometimes sold as tropical spotted roaches (*Blaptica dubia*), and tomato hornworms (*Manduca sexta*).

Not every veterinary hospital will keep its own colonies of feeder insects, but it is usually not necessary to gut load feeder larvae/insects for more than about 24 hours, so it is possible to keep small numbers of them or ask clients to bring food for their reptiles and gut load as needed. It is recommended to fast the insects for 24 hours before gut loading to encourage them to eat the offered diet. Several different gut-loading formulas are available, but it is important to select one that provides at least 8% calcium to improve the calcium:phosphorus ratio to a beneficial level.[22] Gut-loading diets that provide fat-soluble vitamins are also available. Note that some insectivorous species, such as chameleons and leopard geckos, are unable to convert beta-carotene to vitamin A, so will need insects fed a diet that includes preformed vitamin A.[15]

Crickets

Research has shown that crickets are a good source of moisture, protein, and fat but are deficient in calcium and fat-soluble vitamins. They should be gut loaded with a high-calcium diet (8% or more) and fed to reptiles within 6-12 hours (no more than 24 hours) of gut loading. Beyond that time, the crickets may stop eating the gut-loading formula because of its high mineral content, at which point the calcium:phosphorus ratio will revert to an undesirable level.[22]

Mealworms

These feeder insects tend to be significantly higher in fat than crickets; if fed excessively, they can lead to obesity in reptiles. They should be gut loaded for about 24 hours before being fed to patients.[22]

Black soldier fly larvae

Sometimes appearing in pet stores as Calci Worms or Phoenix Worms®, these feeder larvae have a naturally high calcium:phosphorus ratio; however, they have poor digestibility and may pass through the reptile's digestive tract without the calcium being absorbed. They can be fed the same gut-loading formula as mealworms, which will increase provision of fat-soluble vitamins to the reptile, but they are unlikely to eat the formula unless provided with a water source.[22]

Dubia roaches

Similar to crickets, these insects are available in some locations in various sizes and provide adequate amounts of protein, fat, and moisture. They can be fed a gut-loading formula used for crickets and/or mealworms for a period of about 48 hours to achieve a beneficial calcium:phosphorus ratio.[22]

Tomato hornworms

Little has been reported about the nutritional content of these worms, and they also have an inverse calcium:phosphorus ratio. Research has shown that gut loading with the only commercially available species-specific diet for them does not provide an adequate level of calcium; however, their moisture content is approximately 85%, so they may be beneficial if there is concern about the reptile's hydration.[22]

Conclusions

Watching a reptile patient recover and leave the hospital after some time in your care is a wonderful experience. Ensuring those patients' special needs are met during their hospital stay is the key to creating those positive outcomes.

References

1. Uetz P. Species Numbers (as of December 2022). The Reptile Database. Available at: http://www.reptile-database.org/db-info/SpeciesStat.html. Accessed March 2, 2023.

2. Chitty J. Hospitalization of birds and reptiles. *J Exot Pet Med.* 2011;20:102-103.

3. Hunt C. Stress and welfare. In: Divers S, Stahl S, eds. *Mader's Reptile and Amphibian Medicine and Surgery.* 3rd Ed. Philadelphia, PA: Elsevier; 2019:380.

4. Minh H, Boyeaux A, Pignon C. Assessment and care of the clinically ill rabbit. *Vet Clin North Am Exot Anim Pract.* 2016: 19(2): 105.

5. Mitchell M. Managing the reptile patient in the veterinary hospital: establishing a standards of care model for nontraditional species. *J Exot Pet Med.* 2010: 19(1);56-72.

6. Rossi J. General husbandry and management. In: Divers S, Stahl S, eds. *Mader's Reptile and Amphibian Medicine and Surgery.* 3rd Ed. Philadelphia, PA: Elsevier; 2019:114-115.

7. Turpen K, Welle K, Trail J, et al. Establishing stress behaviors in response to manual restraint in cockatiels (*Nymphicus hollandicus*). *J Avian Med Surg.* 2019;33:41.

8. Heidenreich B. Reducing stress for avian capture and restraint. *Proc Annu Conf Assoc Avian Vet.* 2014;367.

9. Cutler D, Divers S. Hospitalization. In: Divers S, Stahl S, eds. *Mader's Reptile and Amphibian Medicine and Surgery.* 3rd Ed. Philadelphia, PA: Elsevier; 2019:432-436.

10. Association of Amphibian and Reptile Veterinarians. Companion Reptile Care Series. Available at: https://arav.site-ym.com/page/CompanionCareGuides. Accessed March 2, 2023.

11. Healey M. Lighting and Temperature Requirements for Jackson's Chameleons. Available at: https://reptifiles.com/jacksons-chameleon-care/jacksons-chameleon-lighting-temperatures/. Accessed March 2, 2023.

12. Pees M, Girling S. Emergency care. In: Girling S, Raiti P, eds. *BSAVA Manual of Reptiles.*,3rd Ed. Quedgeley, UK: British Small Animal Veterinary Association; 2019:110.

13. Baines F, Cusack L. Environmental lighting. In: Divers S, Stahl S, eds. *Reptile and Amphibian Medicine and Surgery.* 3rd Ed. Philadelphia, PA: Elsevier; 2019:131-138.

14. Varga M. Captive maintenance. In: Girling S, Raiti P. *BSAVA Manual of Reptiles.* 3rd Ed. Quedgeley, UK: British Small Animal Veterinary Association; 2019:43-45.

15. Boyer T, Scott P. Nutrition. In: Divers S, Stahl S, eds. *Mader's Reptile and Amphibian Medicine and Surgery.* 3rd Ed. Philadelphia, PA: Elsevier; 2019:210-215.

16. Pollock C. Feeding the Hospitalized Lizard. LafeberVet. Available at: https://lafeber.com/vet/feeding-the-hospitalized-lizard/. Accessed March 2, 2023.

17. Petritz O, Son T. Emergency and critical care. In: Divers S, Stahl S, eds. *Reptile and Amphibian Medicine and Surgery.* 3rd Ed. Philadelphia, PA: Elsevier; 2019:971.

18. Pollock C. Nutritional Support in Reptiles. LafeberVet. Available at: https://lafeber.com/vet/how-to-guide-nutritional-support-in-reptiles/. Accessed March 2, 2023.

19. Pollock C. Feeding the Hospitalized Turtle or Tortoise. LafeberVet. Available at: https://lafeber.com/vet/feeding-the-hospitalized-turtle-or-tortoise/. Accessed March 2, 2023.

20. Pollock C. Administration of Medications in Reptiles. Lafebervet. Available at: https://lafeber.com/vet/administration-of-medication-in-reptiles-2/. Accessed March 2, 2023.

21. Perry S, Mitchell M. Routes of administration. In: Divers S, Stahl S, eds. *Mader's Reptile and Amphibian Medicine and Surgery.* 3rd Ed. Philadelphia, PA: Elsevier; 2019:1130-1138.

22. Boykin K, Mitchell M. Serving up a healthy plate of evidence-based nutrition ensures good digestion for reptiles. *Proc ExoticsCon.* 2022;364-370.

23. Boyer T, Scott P. Nutritional diseases. In: Divers S, Stahl S, eds. *Mader's Reptile and Amphibian Medicine and Surgery.* 3rd Ed. Philadelphia, PA: Elsevier; 2019:939.

Rabbit Anesthesia 101 for the Veterinary Technician

Stefanie Kotschwar, RVT, BScH (Zoology, Nutrition), MSc (Zoo Conservation), AAS (Vet Tech), Diploma (Web Design and Equine Science)

Session #3234

Affiliation: From Lynwood Animal Hospital, 107-30 Stafford Rd, Ottawa, ON, K2H 1W8, Canada.

Abstract: Many veterinary technicians find rabbit anesthesia overwhelming and scary. The purpose of this talk is to go over all the steps from intake to discharge of rabbit patients presenting to the clinic for surgery. Topics covered include history, pre-surgical exam, anesthetic drugs, anesthetic equipment, monitoring equipment, monitoring patient and assessing anesthetic depth, monitoring during recovery, post-surgical in hospital care, post-surgical home care client education There is emphasis on areas of risk and how reduce them along the way. The focus is on what veterinary technicians can do at each step of the way for a successful outcome.

Introduction

Rabbits have become wildly popular as pets in North America and more are brought to the veterinary clinic for regular check-ups.[1] The veterinary team might encounter these patients on a regular basis for procedures that require sedation or general anesthesia. Many pet owners and even veterinary professionals may be under the impression that rabbits have a high anesthetic risk and procedures should be avoided. While it is true that they have a higher anesthetic risk than cats or dogs, they can do well under general anesthesia as well. One published article went over studies that have shown the anesthetic risk in rabbits can range anywhere from 1.39-4.8%.[1] These studies have also shown that deaths not only occur during the surgical procedure but up to 48-hours post anesthesia.[1] Successful anesthesia rates require understanding of rabbit anatomy and husbandry, all components of an anesthetic procedure, our limitations, and staff training. While there are no general anesthetic guidelines published for exotic companion animals, the veterinary team can refer to the 2020 American Animal Hospital Association (AAHA) Anesthesia Monitoring Guidelines for Dogs and Cats as a basic outline and apply it to not-so-typical patients. Any strategies used to keep dogs and cats safe under general anesthesia can be applied to rabbits with some modifications.[2] Individual components for each procedure are described below.

Patient History

The rabbit owner should have ideally been counseled on appropriate care and nutrition of their pet at an exam prior to the day of anesthesia. Nevertheless, the admitting support staff personnel should obtain a brief history on the patient's behavior, appetite, and stool production, as this can help identify any changes in health (or husbandry) that may influence anesthetic morbidity.[2,3] The focus should be on questions regarding alertness, eating and defecating habits. Any changes should be investigated further. Provide stabilization first, if required.

Pre-Anesthetic Examination

Prior to handling the rabbit for a pre-anesthetic examination, observe it in the carrier for any signs of distress. Rabbits are prey species that try to mask clinical signs of illness when not feeling well. An examination prior to sedation is important to ensure a proper anesthetic plan can be made and adjusted based on the rabbit's health status. Not only should vital signs be taken including heart rate, respiratory rate, and temperature, but also check the color of mucus membranes and capillary refill time.[2,3] Perform an abdominal auscultation to detect borborygmi, as the absence of these sounds suggests that the patient may be stressed or painful. Table 1 highlights normal vital signs for rabbit patients.[1-3]

Table 1. Normal vital signs for rabbits.

Vital sign (unit)	Normal range
Heart rate (beats per minute)	200-300
Respiration rate (breaths per minute)	32-60 (higher rates are not uncommon in stressed patients)
Temperature (°C)	38.5-39.5
Weight (kilograms)	1-10

Hospital Cage Set-Up

Rabbits are a prey species so keeping them in an area specifically for small exotic animals away from any predatory species can help reduce patient stress.[2,3] If no separate ward is available, provide hiding shelters as well as litter boxes to make the cage more comfortable and place them as far away from predator species as possible.

Anxiolytics such as gabapentin are commonly used in cats and dogs to be given at home prior to the veterinary visit or surgical procedures to ensure their fear, anxiety and stress are reduced.[1] One study has investigated the efficiency of a one dose use of gabapentin in rabbits and has had some positive outcomes. It seems that even one dose of gabapentin can have relaxing effects on rabbits, but further research in this area is needed.[4]

Pre-Anesthetic Blood Work

Ideally a complete blood cell count, plasma biochemistry along with a blood smear evaluation should be performed to rule out any underlying issues prior to any anesthetic events, especially on senior patients. If full blood work is not possible, performing a packed cell volume and total protein at minimum should be considered.[1]

Pre-Anesthetic Fasting

Fasting is only recommended 30 minutes to 1 hour prior to sedation to reduce the incidence of ingesta remaining in the oropharyngeal space that could be accidentally introduced into the respiratory tract, particularly if a blind intubation technique is used. This fasting time can also help reduce the volume of gastrointestinal contents which could put pressure on the diaphragm which may result in the stomach applying pressure to the diaphragm, affecting the patient's ability to ventilate under anesthesia. Longer fasting times are not recommended as withholding food from rabbits can result in an increased risk of gastrointestinal stasis, or ileus. Rabbits cannot vomit or eructate due to a muscular cardiac sphincter.[1,3]

Procedure and Duration

Consider what type of surgical procedure is being performed, anticipated length, type of heat support, fluid support and how painful it could be.[2]

American Society of Anesthesiologists Risk Status

The American Society of Anesthesiologists created a physical classification status that has been widely used not only in human medicine but also in veterinary medicine. This system is useful in assessing the patient and any comorbidities that could contribute to anesthetic risks.[5] Table 2 reviews the risk assessment grades.

Table 2. American Society of Anesthesiologists (ASA) physical status classifications.

ASA classification	Definition
ASA I	A normal healthy patient
ASA II	A patient with mild systemic disease
ASA III	A patient with severe systemic disease
ASA IV	A patient with severe systemic disease that is a constant threat to life
ASA V	A moribund patient who is not expected to survive without the procedure

Source: American Society of Anesthesiologists. ASA Physical Classification Status. Available at: https://www.asahq.org/standards-and-guidelines/asa-physical-status-classification-system. Accessed June 1, 2023. Cripwell D, Dugdale A, Michou J. ASA Physical Classification Status: A Guide for Veterinarians. Available at: https://www.thinkanesthesia.education/sites/default/files/2021-03/US%20-%20ASA%20physical%20Status%20Classification%20Chart%20-%20%20V2.pdf.

Anesthetic Drugs

A multimodal anesthesia approach is always preferred as a combination of drugs is used to maximize their positive effect while minimizing any negative effects.

Pre-anesthetic drugs

The goal is to sedate patients and to provide preemptive analgesia. The use of premedication facilitates catheter placement, decreases the amount of induction agent needed and decreases minimum alveolar concentration (MAC) during the procedure. The protocol use should be tailored to each individual patient based on the type of procedure, their age, breed, and any existing comorbidities.[1] Having a preset plan that can be used as a starting base and modified as needed is appropriate.[2]

Midazolam is a commonly used sedative and anxiolytic belonging to the benzodiazepine class. It is a short acting drug that is water soluble and can be given intramuscularly (IM), subcutaneously (SC), intravenously (IV), or intraosseously (IO). This drug can be used alone for minor, noninvasive and non-painful procedures, and in combination with ketamine, opioid, alpha-2 agonists for extensive procedures.[1]

Alpha-2 agonists include drugs such as xylazine, detomidine, medetomidine and dexmedetomidine and they provide excellent sedation and muscle relaxation. They can have significant cardiovascular effects that include vasoconstriction, decreased cardiac output, second degree heart block and bradyarrhythmia.[1,2]

Ketamine is an NMDA antagonist and provides analgesia and dissociative anesthesia. It is considered a good adjunct in premedication but can cause muscle rigidity and should be used with muscle relaxants. It can also be used as induction agent alongside midazolam.[1]

Anticholinergic drugs may be used to minimize salivary and bronchial secretion and reduction of vagally-induced bradyarrhythmia. They can increase the viscosity of airway secretions. Their use has been questioned in human medicine and there are no controlled studies in rabbits. About 61%[1] of rabbits have significant atropine esterase activity which necessitates higher doses of atropine compared to other mammalian species. Glycopyrrolate has a duration of effect of 60 minutes, but due to lack or proven efficacy and effect on gastrointestinal motility routine use is not recommended.[1,3]

Opioids

Opioids can provide potent analgesia. Their receptor affinities can be used to classify this group of drugs.

Pure μ agonists

Pure μ agonists are preferred for procedures where moderate to severe pain is expected (e.g., limb amputation, laparotomy). These drugs include hydromorphone, fentanyl, morphine, and methadone.[1]

Partial opioid agonists, agonists and antagonists

This subgroup of opioids can provide sedation (and potentiate the effect of other sedatives) but is not highly effective for analgesia. Butorphanol, for example, is not appropriate for painful procedures, but can be used in conjunction with other drugs (example, midazolam) to improve sedation for minor procedures such as skull radiographs that require adequate patient immobilization. Butorphanol in particular has a notable ceiling effect to its analgesic and sedative properties at higher doses, but minimal MAC sparing effects are observed. This kind of drug is only appropriate for mild pain and as a sedative for minor diagnostic procedures such as skull radiographs. It can cause significant respiratory depression when combined with other sedatives. It can be given IM or IV.[2]

Buprenorphine is a partial μ agonist and effective analgesic to treat mild to moderate pain. When used alone it only provides mild sedation. It does not have any cardiovascular effects but causes mild respiratory depression. If combined with midazolam the sedative effects can last up to 90-120 minutes.[1]

Local anesthetics

Local anesthetics such as lidocaine and bupivacaine are valuable tools in the anesthetic protocol as they reversibly block transmission of nociceptive signals from reaching the central nervous system. Local anesthetics are applied via topical, infiltration, intraarticularly or epidural route.[1,7]

Epidural provides preemptive analgesia by injecting the drug into the epidural space that surrounds the spine. It allows for anesthesia without systemic effect, reduced recovery time, as well as decreased use of gas anesthesia. The lumbosacral junction site is the most used site with a technique like in dogs and cats. This type of anesthesia is beneficial in abdominal surgeries and orthopedic procedures. A study using morphine and lidocaine proved effective analgesia for limb orthopedic surgery. Potential side effects include transient paresis or paralysis.[1,7]

Induction

Anesthetic induction can be achieved either via injectable or inhalant drugs. Inhalant or volatile anesthetics are noxious and irritating to the airway and may cause laryngospasm. If used for induction, adequate pre-anesthetic medication is necessary to provide moderate to deep sedation to avoid producing an excitement phase.

Induction agents are ideally administered with an IV catheter in place. Propofol produces rapid induction without an excitement phase, however apnea is common so preoxygenate the patient and ensure that a visual intubation technique is possible. Side effects are dose dependent, short lived and might decrease cardiac contractility. Side effects can be exacerbated by rapid administration.[1]

Alfaxalone is effective as an IV induction agent. It has a rapid onset and short duration of action. The cardiovascular effects of this drug has not been well studied in rabbits. Alfaxalone can also be used via IM or SC route for sedation. Like with propofol, apnea can occur when administered IV, so it is important to pre oxygenate and visual intubation is needed.

Etomidate is an ultrashort-acting non-barbiturate hypnotic anesthetic agent. It does not provide analgesia. It is reported to cause hemolysis, thrombophlebitis, and pain on injection.[1]

Ketamine and midazolam can be given together to effect for a smooth induction. Apnea can occur but it is not common.

Equipment

All equipment needed should be prepared prior to sedating a patient to ensure readiness.[2]

Vascular access supplies

Intravenous access should be obtained in all anesthetic patients and should be considered for patients undergoing sedation. Applying a topical local anesthetic cream at least 20 minutes prior can reduce discomfort when placing an IV catheter. Sites that can be used for IV catheter placement include cephalic vein, marginal ear vein and the lateral saphenous. The IV catheter sizes are selected based on patient size and can range from 22-26 gauge. Necrosis of the ear veins can occur with irritating drug administration. Some drugs will require to be diluted prior to be given IV as they can cause irritation.[1]

The use of an IO catheter is an option, especially for sick patients but may not be needed for stable surgical patients.

Airway access supplies

Ideally any rabbits undergoing any surgical procedures, especially lengthy ones, should have airway access available to support them as needed. There are several methods for securing airway access. It is important to remember that rabbits are obligate nasal breathers.[1,3] Once the rabbit has been sedated, oxygen via a facemask should be provided to improve oxygen saturation.[3]

Oral endotracheal intubation

This method provides multiple challenges due to the rabbit's narrow oral cavity, long tongue, narrow gape and potential for laryngospasm. These anatomic features make placement difficult and could lead to laryngeal trauma and tube displacement. There are different methods that can be used to place an endotracheal tube in a rabbit: Blind, direct visualization and with the use of an endoscope. Being proficient in more than way is advantageous. Use the largest possible endotracheal tube size to reduce airway resistance. Appropriate tube placement can be confirmed as in other mammals: Fogging of the tube observed on inspiration, positive capnography readings, and auscultating the lungs for breath sounds with intermittent positive-pressure ventilation. Complications that can arise post extubating include tracheal injury, ulceration, post-intubation tracheal stricture.[1]

Supraglottic airway devices

Another available option is supraglottic airway devices which are fast to place (14-38 seconds[1]), cause minimal trauma and allow for intermittent positive pressure if needed. Whenever this airway device is used a capnograph to measure $ETCO_2$ is a must. They are also not recommended for extended procedures due to potential lingual edema and cyanosis, gastric tympany and potential dislodgement.[1]

Incubators

With smaller patients, hypothermia is a major concern due to the large surface-area-to-volume ratio. Thermoregulation can be further depressed due to the different anesthetic drug use, so consider placing patient in an incubator once sedated to reduced heat loss.[1] The patient should also be placed in the incubator during recovery.

Anesthetic machine

Most rabbit patients weigh under 10 kilograms, so the use of the non-rebreathing system would be most appropriate. Select a reservoir bag that is appropriate for the size of rabbit undergoing anesthesia. Check the system for any leaks and ensure the pop off valve is open. Ensure there is enough inhalant anesthetic in the machine.

Fluid Therapy

Gut hydration is needed to ensure proper functioning and motility, otherwise ileus can develop, which can be fatal in some cases. Daily water intake ranges from 50-150 ml/kg of BW. This high requirement makes fluid therapy especially important.[1]

Monitoring

A quick summary of the available monitoring tools that should be used will be reviewed below. The best monitoring tool is a highly trained and observant technician.

Positioning

Always monitor the patient's head position and ensure it is straight and there are no kinks in the endotracheal tube. The chest should also be slightly elevated to reduce the pressure from the gastrointestinal tract on the diaphragm.[1,3]

Anesthetic depth

Anesthetic depth is an indicator to determine if the patient is at an appropriate anesthetic plane and should be monitored by a trained veterinary technician at least every 5 minutes. Anesthetic depth can be monitored by assessing reflexes including righting, palpebral, corneal, toe withdrawal and pinnal reflex.[1,7] The author finds the jaw tone to be a reliable indicator of light anesthetic depth.

Thoracic auscultation

Evaluate the heart rate and rhythm with a stethoscope of even with the use of a Doppler flow detector.[3]

Pulse strength

Peripheral pulses can be palpated in rabbits at the central auricular and metatarsal arteries. The femoral pulse can also be palpated, but in many surgical procedures this region may not be accessible to the anesthetist.[3]

Mucous membrane color and capillary refill time

The color of the mucous membranes should always be pink and any change could indicate an airway obstruction or other issues like decreased perfusion or internal bleeding.[3]

Blood pressure

While direct blood pressure measuring is the most accurate method, it is not always feasible in a general practice environment. Direct blood pressure is an invasive method that requires special equipment, arterial catheter placement and maintenance that can also be time consuming. Indirect blood pressure measurements with a Doppler is preferred over non-invasive oscillometry inaccurate with rapid heart rates. The use of the forelimb preferred.[1,8]

Electrocardiogram

The electrocardiogram (ECG) allows for continuous monitoring of the heart rate and rhythm of the patient. Atraumatic clips are preferred for rabbits to avoid iatrogenic injury.[1,3,10]

Pulse oximeter

This is a very useful tool but can be unreliable in some animals due to their thick fur. The probe can be attached to ears, tongues, feet and tails.[3,11]

Capnograph

This is an extremely helpful monitoring tool to ensure the endotracheal tube is properly placed as well as an early indication that there is a potential obstruction. Normocapnia in small mammals is associated with $ETCO_2$ values of 35-45 mmHg.[1,9]

Temperature

Rectal or esophageal thermometers are appropriate to use in surgery. Use of various heating sources, IV fluid warmer, socks and bubble wrap to maintain body temperature is recommended.[3,12]

Post-Surgical In-Hospital Care

Many anesthetic mortalities occur in the post-anesthesia period, so close monitoring is required. The patient should be maintained on oxygen until it is ready to be extubated and movement occurs. Once extubated, keep the patient in a stress free and warm recovery cage or even incubator. Ongoing monitoring of temperature, heart rate and respiration rate is recommended. The IV catheter should be maintained in case of an emergency until close to discharge time. Provide food and water as soon as possible to avoid hypoglycemia and to ensure proper gastrointestinal motility. Monitor fecal output closely. Provide syringe feeding for any patient that is not eating hay after a surgical procedure.[3,11,13]

Pain Assessment

Rabbits tend to be alert and active in their normal environment. As they are prey species, they tend to hide signs of disease and discomfort for as long as possible, so it is imperative to monitor them closely for any signs of pain. Changes in their normal behavior such as lack of activity, decreased food consumption, hunched posture, hiding, signs of aggression, bruxism, and lack of grooming can be indicators of pain.[13] A validated resource that can help all staff members identify whether a rabbit is experiencing pain is the grimace scale from N3CRs.[14] The grimace scale is based off easily identifiable facial expressions which provide a score that can indicate pain level.[15]

Post-Surgical Home Care Client Education

As studies have shown, post-anesthetic complications can occur days after the procedure, so it is important to thoroughly go over the home care with owners. As with dog and cat patients, explaining how the procedure went and highlighting the importance of reduced exercise and daily incision checks is key. Next ensure the owner knows how to measure out medication and knows how to give it to their rabbit. It can be beneficial to break down the home care into different sections. Focus on the first 24-48 hours where the priority is pain control, return of appetite and fecal production. Once the owner is comfortable with the care for the first 2 days, go over the details for the next 1-2 weeks of home care. As needed show the owner and provide tips and tricks on how to feed and medicate the rabbit at home. If possible, show the owner what the incision looks like on discharge and even have them take a picture, so they have a reference.

Conclusions

While rabbits are known to have higher anesthetic-related deaths compared to cats and dogs, there are many tools that can be implemented to ensure a positive outcome. It is important as veterinary technicians to be skilled in using and implementing as many tools as possible to ensure rabbit patients do well under anesthesia and to keep our knowledge up to date.

References

1. Gardhouse S, Sanchez A. Rabbit sedation and anesthesia. *Vet Clin North Am Exot Anim Pract.* 2021;25: 181-210.

2. Grubb T, Sager J, Gaynor JS, et al. AAHA anesthesia and monitoring guidelines for dogs and cats. *J Am Anim Hosp Assoc.* 2020;56(2):1-24.

3. Longley L. Rabbit anesthesia. In: Longly L, ed. *Anaesthesia of Exotic Pets.* Philadelphia, PA: Elsevier Saunders; 2008:27-35.

4. Burton M, Conway R, Mishkin N. et al. Pharmacokinetics of gabapentin after single, oral administration in domestic rabbits (*Oryctolagus cuniculus*). *J Exot Pet Med.* 2023;45:1-5.

5. American Society of Anesthesiologists. ASA Physical Classification Status. Available at: https://www.asahq.org/standards-and-guidelines/asa-physical-status-classification-system. Accessed June 1, 2023.

6. Cripwell D, Dugdale A, Michou J. ASA Physical Classification Status: A Guide for Veterinarians. Available at: https://www.thinkanesthesia.education/sites/default/files/2021-03/US%20-%20ASA%20physical%20Status%20Classification%20Chart%20-%20%20V2.pdf.

7. Pollock C, Lichtenberger M, Echols MS. Epidural anesthesia in small mammals. LafeberVet Website. 2013. Available at https://lafeber.com/vet/epidural-anesthesia-in-small-mammals/. Accessed June 1, 2023.

8. Lafferty K, Pollock CG. Blood pressure monitoring in exotic animal species. LafeberVet Website. 2018. Available at https://lafeber.com/vet/blood-pressure/. Accessed June 1, 2023.

9. Lafferty K, Pollock CG. Capnometry in exotic animal species. LafeberVet Website. 2018. Available at https://lafeber.com/vet/capnometry/. Accessed June 1, 2023.

10. Lafferty K, Pollock CG. Electrocardiography in exotic animal species. LafeberVet Website. 2018. Available at https://lafeber.com/vet/electrocardiography/. Accessed June 1, 2023.

11. Lafferty K, Pollock CG. Pulse oximetry in exotic animal species. LafeberVet Website. 2018. Available at https://lafeber.com/vet/pulse-oximetry/. Accessed June 1, 2023.

12. Lafferty K, Pollock CG. Monitoring vital signs in exotic animal species. LafeberVet Website. 2018. Available at https://lafeber.com/vet/monitoring-vital-signs-in-exotic-animal-species/. Accessed June 1, 2023.

13. van Zeeland YRA, Schoemaker NJ. Analgesia, anesthesia, and monitoring. In: Graham JE, Doss GA, Beaufrère H, eds. *Exotic Animal Emergency and Critical Care Medicine.* Hoboken, NJ: Wiley-Blackwell; 2021:95-108.

14. National Centre for the Replacement Refinement and Reduction of Animals in Research. The Rabbit Grimace Scale. Available at: https://www.nc3rs.org.uk/3rs-resources/grimace-scales/grimace-scale-rabbit. Accessed June 1, 2023.

15. Keating SCJ, Thomas AA, Flecknell PA, et al. Evaluation of EMLA cream for preventing pain during tattooing of rabbits: changes in physiological, behavioural and facial expression responses. *PLoS ONE.* 2012;7(9):1-11.

Section 53

Technician Session 2

Brittany Skinner, RVT
Session Chair

Introduction to Zoology for Veterinary Technicians

Christina Miller, RVT, BSc

Session #3510

Affiliation: From Clayton Park Veterinary Hospital, 278 Lacewood Dr, Halifax, NS, B3M 3N8, Canada.

Abstract: Veterinary hospitals working with exotic species encounter a wonderful diversity of animals compared to the standard canine and feline patients in the average small animal practice. Consider how differently dog and cat patients are treated: drug doses, behavioral principles, vital signs, test values, and common diseases differ from these two ubiquitous companion animals but how can technicians prepare to care for an even greater range of unique and distinctive species? Understanding some key concepts in zoology will increase understanding of the patient's nature and needs.

Introduction

Zoology is the scientific study of the behavior, structure, physiology, classification, and distribution of animals.[1] While medicine is a key focus in the day-to-day operations of technicians and nurses, zoology undeniably has a place in our work. Understanding the natural history, phylogenetic relationships, anatomy, and physiology of patients makes us better able to provide appropriate care for them.

A few other terms that are important to define are as follows:

- Animal: Organisms that belong to kingdom Animalia (also called Metazoa), that are multicellular and heterotrophic organisms.[2]
- Heterotroph: An organism deriving its nutritional requirements from complex organic substances[1] (e.g., from other organisms). They lack the ability to chemically produce their own food from inorganic molecules.
- Taxonomy: The branch of science concerned with classification, especially of organisms.[1]
- Species: A group of living organisms consisting of similar individuals capable of exchanging genes or interbreeding. The species is the principal natural taxonomic unit, ranking below a genus and denoted by a Latin binomial.[1]
- Phylogenetics: The study of organisms' shared evolutionary history.[3]
- Evolution: Descent with inherited modification, including changes in gene frequency in a given population over time.[4]
- Evolutionary fitness: The effectiveness of a particular genotype in producing successful offspring in its environment.[4]

Classifying Animals

Organisms are currently classified based on *shared ancestry* and not *shared characteristics*. Classically, Linnaean taxonomy was used to organize animal life based on the shared characteristics of animals, meaning "like" was grouped with "like." This was based on physician, botanist, and zoologist Carl Linnaeus' (as well as a few other early biologists) work, *Systema Naturæ*, first published in 1735. With the advent of molecular biology and greater accessibility to means of analyzing the DNA of animals and other life forms, organisms are now classified based on shared ancestry to better illustrate their evolutionary relationships.

Animal organisms are named following the International Code of Zoological Nomenclature, which is published by the International Commission on Zoological Nomenclature (ICZN). The code regulates a uniform system that maintains both universality as well as continuity in naming animal organisms so that scientific communication regarding these organisms can remain clear across all languages. The rules include how names are established with appropriate binomial nomenclature, which name will be officially used if there are conflicts in the literature, and how scientific literature must cite animal names. Note that the naming of plants is dictated by a different authority, and overlap exists between plants and animals (for example, there are both plant and animal genera *Abronia* and *Dracaena*).

Perhaps one of the most fluid concepts in animal classification is the designation of species. While the textbook definition of species that is listed above often applies, there are many examples of gray areas that are more challenging to definitively decide upon. For example, the presence of animal hybrids, such as fertile offspring of the tiger (*Panthera tigris*) and lion (*Panthera leo*), breaks this rule as it suggests that these two very distinct species would "technically" be the same species. It has instead been suggested that "species" should refer only to naturally occurring interbreeding animals.

Examples exist of naturally occurring hybridization, such as Algonquin or Eastern wolf (*Canis lupus lycaon*) naturally hybridizing with their close relative, the coyote (*Canis latrans*). Where the Eastern wolf is a protected species, but coyotes are not (and are often considered a pest animal in many regions), where is the line drawn with legal protection of wolf-coyote hybrids?[5] It has even been determined that many of these "coywolf" hybrids not only have both wolf and coyote genes, but a 2016 study found domestic dog (*Canis familiaris,* or *Canis lupus familiaris*) genes in these populations. It has been suggested that this hybridization be treated as a speciation event, and be recognized as a new species (*Canis oriens*).[6]

Zoogeography and Phylogenetics

Zoogeography is the science of the geographic distribution of animals,[1] and is best interpreted when paired with phylogeny. Whereas zoogeography aims to answer "what lives where, and why does it live there," phylogenetics adds a temporal scale to these spatial questions to provide a better resolution of the answers.

An example of questions that can be answered by these sciences, is "Why are there so many marsupials in Australia?" Nearly 70% of extant marsupial species live in Australia; the remaining 30% live in the Americas, and they are not distributed on any other continent.[7] The inquisitive naturalist may ask, why is this the case? Are the marsupials in Australia closely related to the marsupials in the Americas? Have they lived anywhere else, and if so, why has that distribution changed?

Much of our understanding of modern phylogenetic relationships comes from the theory that Earth during the Paleozoic and Mesozoic eras was made up of one mostly fused supercontinent, Pangea. There is significant geological evidence supporting this theory. Evolutionary relationships between taxa are supported by this theory, as well[8]: Zoogeographic regions with spatial patterns of animal distributions have been recognized, and the patterns demonstrated by these regions vastly match the geological evidence for the supercontinent theory.[9]

Why Relationships Matter

Understanding that shared characteristics do not necessarily equate to shared ancestry is a key concept in modern zoology. While many species look related (i.e., have shared characteristics), this can be the result of convergent evolution. This can be the result of similar environmental pressures and adaptive mechanisms that allow for better survival (or better evolutionary fitness), or simply random chance.

A complex example is the evolution of the camera eye in both vertebrates and cephalopods, which are invertebrates. These eyes have both evolved an iris around a circular lens, a light-sensitive retina with photoreceptors, as well as anterior and posterior chambers containing humours.[10] The cephalopod eye differs in that it develops embryonically as an invagination of the body surface instead of from brain tissue, the cornea is congruent with the epidermis instead of compromising the surface of the globe, and the cephalopod eye also lacks a retinal blind spot.[11]

Understanding relationships between species can be immensely helpful in veterinary practice when data are deficient on a particular patient that the clinician and technician or nurse are treating. While cautious judgment must always be exercised, it can be far more relevant to extrapolate data from a closely related species, compared to a very distant relative. Normal physiologic parameters, clinical chemistry and hematologic values, and even drug doses can be relevant in many scenarios.

Conversely, some species are closely related, but inhabit differing ecological niches and have demonstrable differences in their environmental needs. In the absence of published care guidelines, examining other species within similar taxa that have comparable natural behaviors and habitats can yield life-saving information in the hospital.

Where Do Patients Fit on The Tree of Life?

A cladogram can be used to illustrate the phylogenetic relationships between organisms. This helps to determine the significance of shared characters, which may be anatomic, behavioral, or genetic.

Synapomorphies are shared derived characters, thought to be evolved in an ancestral species to a group and then shared with the descendants. Symplesiomorphies are shared characters between distant, related groups. Homoplasies are characters that are a result of convergent evolution, or a regression to an ancestral or primitive version of a character.[1]

Conclusions

The exotic animal veterinary technician or nurse will be exposed to a plethora of different species. Understanding relationships between species helps to deepen the appreciation for this wonderful diversity of animals and provide better care for unusual patients.

References

1. Oxford University Press. Oxford English Dictionary. Available at: https://www.oed.com. Accessed February 2, 2023.

2. Animal Diversity Web. Animalia (animals). Available at: https://animaldiversity.org/accounts/Animalia/. Accessed February 2, 2023.

3. Stanford Encyclopedia of Philosophy. Phylogenetic Inference. Available at: https://plato.stanford.edu/entries/phylogenetic-inference/. Accessed February 2, 2023.

4. University of California, Berkeley. Understanding Evolution. Available at: https://evolution.berkeley.edu/evolution-101/mechanisms-the-processes-of-evolution/evolutionary-fitness/. Accessed February 2, 2023.

5. Zimmer C. What is a species? *Scientific American*. 2008. Available at: https://www.scientificamerican.com/article/what-is-a-species/ Accessed February 2, 2023.

6. Way JG, WS Lynn. Northeastern coyote/coywolf taxonomy and admixture: a meta-analysis. *Canid Biol Cons*. 2016;19(1):1-7.

7. Wikipedia. Marsupial. Available at: https://en.wikipedia.org/wiki/Marsupial. Accessed February 4, 2023.

8. Cox CB, Moore PD, Ladle RJ. *Biogeography: An Ecological and Evolutionary Approach.* Hoboken, NJ: Wiley-Blackwell; 2016.

9. Holt BG, Lessard J-P, Borregaard MK, et al. An update of Wallace's zoogeographic regions of the world. *Science*. 2013;339(6115):74-78. doi:10.1126/science.1228282.

10. Serb JM. Toward developing models to study the disease, ecology, and evolution of the eye in Mollusca. *Am Malac Bull*. 2008;26:3-18.

11. Hanke FD, Kelber A. The eye of the common octopus (*Octopus vulgaris*). *Front Physiol*. 2019;10:1637.

Getting a Bird's Eye View on Avian Anesthesia and Analgesia

Liz Vetrano, BS, CVT, VTS (Exotic Companion Animals)

Session #3263

Affiliation: From Mount Laurel Animal Hospital, Mount Laurel, NJ 08054, USA.

Abstract: Anesthesia has always had a reputation for being an intimidating and difficult aspect of avian medicine. With appropriate knowledge and training regarding anatomy, multimodal analgesia, and anesthetic monitoring, these cases can be both challenging and successful. This presentation discusses the process of avian anesthesia from start to finish. This discussion includes several types of common avian procedures and how they may make a difference in surgical protocols including multimodal pain management. Preoperative management is also discussed to create the best outcome for the patient. It is essential to evaluate the case in collaboration with a clinician to determine stability and appropriate timing of any anesthetic event. This evaluation also relates to determining the most accurate American Society of Anesthesiologists' status and preparing for potential complications both during and post operation. During the presentation a procedure is followed while reviewing correct set up, sedation, pain management (including constant rate infusions), preparation (operating room set up and pre-operating table), anesthetic monitoring, and management, including use of a ventilator, and recovery. All these details can simultaneously contribute to setting up the patient for a successful anesthetic event and overall recovery.

Introduction

Avian anesthesia is often viewed as a challenging aspect of veterinary medicine instilling hesitation in most support staff, and occasionally clinicians. With the appropriate knowledge and skills set, any avian anesthetic event can have a successful outcome with a favorable prognosis and good quality of life. Evaluating and performing appropriate diagnostics prior to sedation or anesthesia is the proper way to fully evaluate the patient's current condition and any other possible underlying conditions that may affect the overall anesthetic protocol and treatment plan. Once the patient is determined to be a stable anesthetic candidate, an appropriate protocol should be deliberated between the clinician and technician, including suitable analgesia. Analgesia can vary depending on species, type of procedure, patient status, or availability of drugs. Once the protocol has been solidified, it is the technician's responsibility to be knowledgeable about surgery preparation including catheter placement, preparing the surgical site, and equipment/instruments required. During anesthetic monitoring, it is important to be familiar with normal vital ranges, injectable sedation, gas anesthesia, monitoring equipment, and how to manage the patient if there are anesthetic complications. Recovery can be considered one of the most important steps in this whole process as it has been thought that most patients decline during this time frame during the surgical case. Observing the patient and providing appropriate care can determine the outcome. Overall, it is important for the entire team to collaborate and communicate to determine the most appropriate plan, for both analgesia and anesthesia, for the individual case that will provide a positive prognosis.

Pre-Surgery

Surgical preparation for avian cases can be very diverse. A young, stable chicken could be presenting for a simple laceration repair, or a critical, egg-bound, 30-year-old Moluccan cockatoo could be presenting for an emergency salpingohysterectomy (and don't forget the patient's history of atherosclerosis). Each of these cases should be evaluated based on their species, status, and history. An important part of having a successful anesthetic event is evaluating the patient's current condition. Typically, radiographs and blood work are baseline diagnostics required for clinicians to make a reasonable list of differentials and determine if any other treatments need to be completed prior to the procedure. In addition to these, sometimes the clinician will determine that the patient would benefit from an ultrasound or computed tomography scan. This can help then solidify a more detailed surgical approach depending on the procedure (e.g., mass removal, exploratory coeliotomy, endoscopy). Depending on the patient's status, starting treatments prior to the procedure can be an imperative part of ensuring a successful anesthetic event. For example, an egg-bound patient typically benefits from fluid therapy, antibiotics, and very importantly, pain management, prior to surgery. Fluid therapy is a valuable treatment used to maintain blood pressure and increase hydration. Administering a warmed crystalloid fluid, such as Normosol R® or Lactated Ringers®, subcutaneously can correct fluid deficits and improve perfusion. Antibiotics starting pre-operatively or intra-operatively can also be valuable if an infection is suspected due to their current condition or reported white blood cell count. Getting ahead of the pain, otherwise known as avoiding the wind up, is also a key feature in preparing for any procedure. Starting pain management as quickly as possible before a procedure can not only improve the patient's level of comfort, but also the anesthesia protocol itself, and the recovery process. Specific analgesia and protocols will be discussed separately.

On occasion with the critical patient, checking a packed cell volume/total solids (PCV/TS) in addition to other baseline blood work can determine if the patient requires a blood transfusion. For most species, a normal PCV range can be between 40-55%[1]. If a patient presents with a PCV of 15% or lower, or if the patient is clinically weak from anemia, a whole blood transfusion may be required.[1] Birds can tolerate lower PCVs if the condition is chronic. Since birds do not possess specific blood types, whole blood from other species may be used if that is all that is available; however, homologous whole blood has a much longer half-life. The author has used chicken blood for psittacine patients requiring emergency surgery and experiencing hemorrhage. Ten percent of the donor's body weight may be taken and administered to receiver over 1-4 hours.[1]

Analgesia and Sedation Plan

Providing appropriate analgesia for any exotic species typically has had many challenges, including lack of pain recognition and appropriate pain medications capable of providing appropriate therapeutic levels. With an increase in the number of studies performed supporting the suspicion that all of these exotic companion animals experience pain as much as our canine and feline patients it is imperative to provide the appropriate analgesic protocols. Several factors play a role in determining the correct protocol, including species, procedure, co-morbidities, age, and medications. One of the first steps in determining a protocol is evaluating each individual case and referring to a pain scale. There have been several scales developed for exotics species which show specific signs indicating levels of pain. For birds, these signs may include but are not limited to orbital tightening, clenched feet, increased respiratory effort, fluffed plumage, not perching, among other signs. Some of the more common types of analgesics and how they can be applied within the anesthetic protocol before, during, and after the procedure are described below.

Opioids are a common type of analgesic used regularly in the medical field. They can be used for cases experiencing mild to moderate amounts of pain. It is important to refer to pain scales and determine which specific opioid is most suitable for the signs of pain the patient is experiencing, as well as the species. Buprenorphine is a partial mu-agonist which has proven to be a suitable choice for some avian species including American kestrels (*Falco sparverius*) and Columbiformes.[2] This is acceptable for mild pain in these species as these animals initially present to the hospital for a painful condition or as a surgical pre-medication. Butorphanol is typically the preferred opioid of choice for most avian species due to its affinity for kappa receptors. Most species have been demonstrated to have predominantly kappa receptors, so butorphanol provides excellent pain management. This drug can be administered either intramuscularly into the pectoral muscle, or intranasally (which is less invasive and avoids bruising). The only disadvantage of butorphanol in some cases is the frequency of dosing, which can be anywhere from every 2-4 hours to maintain appropriate therapeutic levels.[2] This means more handling, stress, and risk of iatrogenic injury which can be detrimental for a critical patient. One of the major advantages is that butorphanol can be used as a constant rate infusion during hospitalization and/or procedures. Not only will it provide adequate pain relief, but also help reduce the amount of gas anesthesia required to keep the patient at an appropriate plane of anesthesia. A loading dose is typically delivered in conjunction with a sedative, in order to improve the amount of sedation, and the CRI is started shortly afterwards. Tramadol is another opioid with an affinity for mu receptors. Although it does not target kappa receptors, this provides mild pain management for birds, simply because of its ability as a serotonin reuptake inhibitor.[3] This can be administered both by mouth or by intramuscular injection for mild pain.

Midazolam is a water-soluble benzodiazepine typically used in conjunction with butorphanol, or by itself for avian sedation. It does not provide any analgesia when used by itself. It is important to note that just like any other medication, its effects can vary between species and individuals. It can be delivered intramuscularly or intranasally and is an essential part of the avian pre-medication protocol. Doses can be anywhere from 0.1-2 mg/kg IM or IV and 2 mg/kg IN (species may vary).[2] Flumazenil is the antagonist for midazolam which is typically administered after the procedure at 0.02-0.1 mg/kg IM/IV for most species.[2] For very non-invasive or brief procedures (i.e., prolapse repair, diagnostics), a dose of 1-2 mg/kg can be paired with 2 mg/kg of butorphanol and can be a great alternative for gas anesthesia.[2]

Alfaxalone is a neuroactive steroid used for sedation in many species; however, there has been very little literature on its use in avian patients. Due to the lack of peer-based reviews and evidence-based recommendations. The author has not utilized this medication in avian patients.

Ketamine is a dissociative anesthetic that when used in collaboration with a sedative, such as midazolam, it not only can provide significant sedation, but some pain management as well, as it has some affinity for the NMDA receptor. Although there are some clinicians that utilize ketamine in avian patients, the author has no experience using this medication in them.

Gabapentin is a GABA analogue used to treat human neuropathic pain.[2] This can be a useful analgesic both pre and post any surgical event. Studies have been completed involving Senegal parrots

Meloxicam is a popular non-steroidal anti-inflammatory drug (NSAID), typically administered post-operatively in a multi-modal analgesic plan. It preferentially inhibits cyclooxygenase-2 and is administered after gas anesthesia in case of renal dysfunction or hypovolemia. The NSAIDs can play a major role in limiting the amount of inflammation that can occur from any traumatic procedure. Celecoxib is another NSAID that can be useful post-operatively; however, it can only be administered by mouth. Like meloxicam, it is a selective COX-2 inhibitor and helps reduce inflammation.

Surgery Preparation and Operating Room Setup

It is imperative to not only prepare for what is to be expected, but also for the unexpected during a surgical procedure. Prior to the procedure, it is necessary to collaborate with the clinician to determine what specialized equipment or instruments they need to streamline that specific procedure. A procedure such as a cloacal prolapse repair would require different instruments than a salpingohysterectomy, so it is important to familiarize yourself with common tools used in the more frequent surgeries being performed. Smaller instruments are typically more desirable for any avian procedure. This includes hemostats, towel clamps, needle holders, forceps, and scissors. It is important to note that certain instruments are preferred for avian surgeries due to their extremely thin and friable tissue. This includes Brown-Adson forceps (not rat-toothed) and suture with a tapered needle. Cautery is also a very useful piece of equipment for cutting and cauterizing the friable tissue throughout all stages of the procedure. Providing the clinician with bipolar cautery and very small, thin cautery forceps are essential to streamline cutting and provide coagulation as to limit hemorrhaging. Hemostatic clips (e.g., Weck® Hemo-clip® Metal Ligation System, Teleflex, Morrisville, NC) are another essential tool for hemostasis during an avian procedure. There are small, medium, and large options depending on the size of the patient. Right-angled hemostatic clips are also available and sometimes more useful for the clinician on smaller patients. Statinsky Debakey vascular clamps are also useful in these patients, as they are designed for occluding large, deep vessels. Hemostatic sponges (Vetspon®, Elanco Animal Health, Greenfield, IN) are another useful item to have available in case of excess bleeding as they are easily placed onto the site to give the clinician some time to evaluate the source of bleeding.

In addition to discussing specific instruments and equipment needed, it is also important to discuss positioning. Positioning of the patient is an essential part of the surgery itself and setting the surgeon up for success. For example, salpingohysterectomies should always be placed in right lateral recumbency since the reproductive tract is located on the left side of the patient. Different surgical procedures require different positions which can vary greatly. Knowing how the surgeon would like the patient to be placed on the table ahead of time can streamline the preparation process and limit the patient's anesthetic time.

The surgical prep area also plays a valuable role in streamlining the surgical process, prior to moving the patient into the operating room. Sedation, induction, intubation, intravenous catheter placement, and plucking (if necessary) typically take place during this phase. All equipment and items should be set up prior to the patient being sedated. This includes heat support. Heat support is one of the most essential pieces of equipment in any bird procedure. Birds have very naturally high body temperatures ranging from 38-45 °C (100.4-113.0 °F). Failing to provide heat support or even administering room-temperature fluids can result in a compromising decrease in body temperature and potentially shock. Providing adequate heat support, such as a forced air warmer (3M™ Bair Hugger™ System, Saint Paul, Minnesota) during all aspects of the procedure is a requirement.

Streamlining surgical preparation once the patient has been sedated is clearly one of the major players in ensuring a shorter, and more efficient anesthetic time. In addition to heat support, having equipment prepared for intubation and intravenous or intraosseous (IO) catheterization is also important. This includes various IVC size options, hypodermic needle options (if IO access is a possibility), local anesthetic if needed, aseptic scrub, suture, tape, bite block, and different Cole endotracheal (ET) tube sizes. Once the patient has been sedated with premedications, received pre-oxygenation, and placed under gas anesthesia, securing an airway is the number one priority. Birds possess a trachea with fully encircled cartilaginous rings. This puts them at a much higher risk for trauma and strictures due to intubation. Only uncuffed tubes should be used. Cole tubes are the most commonly used endotracheal tubes in birds. They allow a good sealing of the trachea, and they limit the dispersion of air with no risk to the tracheal epithelium.[4] To ensure appropriate placement, a capnograph should be utilized to detect an appropriate end-tidal CO_2 ($ETCO_2$) reading. A bite block is also necessary to prevent the patient from biting through the ET tube. The author prefers to make these from tongue depressors or cotton-tipped applicators wrapped in cohesive bandage based on the size of the patient. Bite blocks also provide additional support for securing the ET tube to the patient. Once an airway has been secured, vascular access should also be secured.

Like in any other species, vascular access is essential for any procedure to allow for fluid therapy and/or emergency medications. The size and status of the patient are both factors that could help determine whether an IV or IO catheter is preferred. For medium to larger-sized patients, an IV catheter is preferred as they are less invasive, painful, and easier to place. Potential IV sites include the basilic (ulnar or wing) vein, jugular vein (preferred in more critical patients as this site can put the patient at risk for hemorrhage), medial metatarsal vein, or pedal vein. Any vascular access in the caudal half of the patient's body may pass through the renal portal system before entering systemic circulation. For smaller patients, an IO catheter may be the only option. The preferred site for this is the ulnar bone as this is easily accessible, is not a pneumatic bone, and can be wrapped to keep comfortably in place post-surgery. Once vascular access has been achieved, fluid therapy should be started as soon as possible. Again, it is essential that these are warmed and never at room temperature since it could cause them to go into shock.

For site preparation, it is necessary to pluck the feathers since feather follicles are still intrusive to the surgical site. It is not always necessary to pluck for certain procedures, such a cloacal prolapse repair, but it is for any invasive procedure involving entering the coelomic cavity (e.g., salpingohysterectomies, mass removals, biopsies via endoscopy). Plucking can either be achieved simply by hand or with hemostats. The author recommends using hemostats for feathers with larger shafts. It is also important to hold the skin taught with the other hand while plucking since their skin is very thin and at risk for tearing. Confirm with the surgeon that the area is appropriately prepped prior to moving the patient in the OR.

Surgery and Monitoring

After moving the patient into the operating room, ensure that the patient is in the optimal position prior to taping and securing them to the table. Be mindful and gentle where possible when taping the patient down, especially when positioning the wings. Another technician should simultaneously be attaching the patient to the anesthetic machine. The author prefers sevoflurane as an inhalant anesthetic over isoflurane for avian patients due its high potency, lack of odor, and fast-acting effects. Depending on the procedure and the size of the patient, a ventilator is a great option for avian procedures. Due to their advanced respiratory systems and their inability to perform gas exchange adequately under anesthesia, it is essential to provide intermittent positive pressure ventilation every 6-12 seconds.[5] It is important to also remember the differences in species such as Anseriformes (waterfowl) that possess a dive reflex allowing them in instinctively hold their breath. Intermittent positive pressure ventilation bypasses their ability to shunt. The ventilator is extremely convenient and allows technicians to perform other tasks such as monitor or circulate the operating room. Ventilators also allow for tight control of the delivered tidal volume with long and deep breaths allowing for beneficial gas exchange and better planes of anesthesia. A ventilator is typically not recommended for procedures that require the clinician to enter the air sac. Once the air sac is punctured, the pressure that the ventilator requires to operate is no longer possible, and the ventilator will fail to function. If a ventilator is not available (or if an air sac needs to be punctured), a non-rebreathing circuit is also acceptable. This requires a single dedicated staff member who will provide IPPV during the entire procedure.

Monitoring smaller exotic patients has always been a challenge due to the lack of usable equipment available and their small size. Thankfully, more equipment has been developed to allow for more precise monitoring options other than just a Doppler probe. For respiration, attaching a side-stream capnograph, such as an Airmate®, (Sentier, Waukesha, WI, USA) is necessary to not only monitor respiratory rate but $ETCO_2$, as well. This will also ensure appropriate gas exchange, confirm tube placement throughout the procedure, and that the breaths being delivered are adequate. A study on Congo African grey parrots (*Psittacus erithacus*) demonstrated that a normal $ETCO_2$ range is 30-40 mmHg.[6] Although this is a good reference point, it is important to remember that this can vary amongst species. If a clinician is required to enter an air sac, the $ETCO_2$ readings are invalid until the air sac is closed.

Evaluating an accurate electrocardiogram (ECG) can also be difficult in these species due to their small size and high heart rates. Despite these challenges, Sentier has developed a very small, light, and mobile machine, called a Vetcorder® (Sentier, Waukesha, WI, USA) that can accommodate these patients and potentially produce an ECG. Not only does it come with small metal clips for mammals, but they also provide electrode needles suitable for avian patients. These are placed gently through the skin. An ECG reading can be obtained by placing the right atrial lead on the right side of the cranial sternum and the left atrial lead on the left side of the caudal sternum. Both atrial leads may also be placed on the caudal sternum in cases where it can disrupt the surgical site. Once the needles are in place, small amounts of isopropyl alcohol 70% can be used to help with the reading. Be mindful to only use small amounts as too much can lower the body temperature since alcohol is very volatile and this may contribute to significant evaporative heat loss. The normal ECG in the avian patient can have the appearance of ventricular tachycardia, primarily because of the large negative S wave. This QRS morphology is caused by the high density of sympathetic and parasympathetic nerve fibers in the avian ventricles and atria when compared to mammals.

A Vetcorder® (Sentier, Waukesha, WI, USA) is also capable of providing SPO_2 values. The pulse oximeter is used to measure the oxygen saturation of hemoglobin in the bloodstream. Like any other species, values should be anywhere between 96-100% for normal perfusion. Pulse oximetry also provides a heart rate reading. Normal heart rates can have a wide range, depending on the patient's body size (the smaller the patient, the higher the heart rate. See Table 1).[7] Just like with any other vital sign, it is always important to gather them prior to the procedure to have a general understanding of what is normal for that individual or to ensure their values are normal prior to an anesthetic event. The pulse oximeter may be placed on various sites. The most common is the foot or toes depending on the patient size. Other options for the pulse oximeter probe are the basilic (wing vein) or the medial metatarsal vessel. Pulse oximetry uses infrared lighting to shine through and determine the amount of oxygen traveling in the bloodstream. Because of this, it can have difficulty providing an accurate reading on patients that have darker skin pigmentation. In these cases, a Doppler probe is always great to have as a backup to evaluate heart rate.

Table 1. Normal avian heart rates based on weight.

Weight (grams)	Heart rate, resting (bpm)
25	380
100	329
200	245
300	226
400	213
500	203
1000	175
2000	152

Source: Speer B. *Current Therapy in Avian Medicine and Surgery*. St. Louis, MO: Elsevier; 2015.

In addition to ECG and SPO_2, a digital temperature probe is a necessity. This is a great way to evaluate the patient's internal temperature and ensure appropriate heat support is being provided throughout the procedure. Note that a patient's temperature is more likely to drop when there is a large, open incision versus no incision required. Regardless, heat support is always a necessity when the patient is sedated. Hypothermia can be one of the most common causes of death for avian patients under anesthesia. The author prefers forced air warmers for avian patients as they provide a large amount of consistent heat support without risk of burning. Heat lamps should be available to provide additional heat support if needed. Providing warmed IV fluids is also necessary to maintain normal body temperature. There are small fluid warmers available specifically for IV fluid administration and should be placed as close to the patient as possible.

Blood pressure is another value that is typically measured during anesthetic procedures. This is probably the most challenging value to obtain due to the limited small cuff sizes available that could gather an accurate reading. The pressure cuff can be placed on the distal humerus, femur, or tibiotarsus. Normal pressures range between 90-180 mmHg in conscious psittacine birds.[8] It is important to recognize that normal values can vary based on species. Unfortunately, certain studies have shown that, in birds less than 2 kilograms, direct blood pressure values do not tend to correlate with indirect values.[8] As a result, indirect blood pressure is typically an unreliable value, especially on smaller species. Since it may be possible to still obtain values, it is still valuable to monitor trends.[8]

In addition to all the monitoring equipment, it can be argued that a technician with a stethoscope is the most valuable implement in the operating room. Sometimes the equipment can be faulty or unable to read. It is imperative to learn how to troubleshoot all the equipment, including the anesthesia machine, while simultaneously evaluating the patient. In the event of a true emergency such as respiratory or cardiac arrest, it is important to be prepared if the patient has been labeled as a cardiopulmonary resuscitation. Having emergency drugs calculated and already drawn up can streamline any code.

Recovery

Recovery can be the most vulnerable time for an avian patient after an anesthetic event. This is the time that hypothermia can occur, or they can seemingly suddenly go into cardiac or respiratory arrest. During closing, it is important to communicate with the surgeon and if appropriate, begin to taper the gas anesthesia. Once completed and the gas turned off, disconnect the patient, flush the machine (to rid the remaining anesthetic gas from the line), and re-hook the patient to the machine so that they can continue to receive oxygenation while waking up. Continuing fluid therapy (with warmed fluid) and heat support is necessary. Reversal agents should be given at this time (in regard to sedation, not analgesics). Birds have the reputation of waking up very quickly. It is important to be prepared with a towel for restraint (and to prevent injury) and to extubate efficiently when appropriate. The bite block should provide protection if the bird wakes up faster than the technician can extubate. The tube should be inspected for any blood. Sometimes a small amount of trauma, simply from movement, can be a possibility. Once the patient is ambulatory, they should be placed in an enclosure that can provide both heat support and oxygen therapy. Depending on the invasiveness of the procedure, it is sometimes recommended that they remain hospitalized and under observation for an additional 24 hours. Fluid therapy, gavage feedings, heat support, and other treatments may be necessary to ensure a positive prognosis.

Conclusions

With the appropriate skills and knowledge, avian anesthesia does not have to bear the stigma of resulting in death. As a team, it is imperative to collaborate from start to finish. This includes determining an appropriate ASA status, surgical protocol, preparation plan, and recovery plan. Like any other treatment plan, communication between all parties is essential throughout each phase. Ensure that everyone is playing a role. With teamwork, practice, continuing education, and communication, bird procedures can have a positive outcome with a good prognosis. It is important to remember that all animals, including birds, deserve gold-standard medicine to the best of our ability. In veterinary medicine, it is our job to challenge ourselves enough to be able to provide these services.

References

1. Graham JE, Doss GE, Hugues B, eds. *Exotic Animal Emergency and Critical Care Medicine.* Hoboken, NJ: Wiley-Blackwell; 2021.

2. Plumb DC. *Plumb's Veterinary Drug Handbook.* 6th Ed. Ames, IA: Blackwell Publishing; 2008.

3. Carpenter JW, ed. *Exotic Animal Formulary.* 5th Ed. St. Louis, MO: Elsevier; 2018.

4. Crosta L. LafeberVet. Avian anesthesia webinar. Available at: https://lafeber.com/vet/avian-anesthesia-webinar/. Accessed June 1, 2023.

5. Longley LA. *Anaesthesia of Exotic Pets.* Philadelphia, PA: Elsevier Saunders; 2008.

6. Edling TM, Degernes LA, Flammer K, et al. Capnographic monitoring of anesthetized African grey parrots receiving intermittent positive pressure ventilation. *J Am Med Vet Assoc.* 2001;219(12):1714-1718.

7. Speer B. *Current Therapy in Avian Medicine and Surgery.* St. Louis, MO: Elsevier; 2015.

8. Lafferty KA, Pollock C, Ambros B, Nugent-Deal J. Blood pressure monitoring in exotic animal species. LafeberVet Website. Available at: https://lafeber.com/vet/blood-pressure/. Accessed June 1, 2023.

Section 54

Technician Tutor Session

Lorelei D'Avolio, LVT, VTS (Clinical Practice- Exotics), CVPM
Session Chair

Avian Pain Assessment for Veterinary Technicians

Jovanna Berls, CVA, LVT, FFCP (Veterinary), FFCP (Avian)

Session #3151

Affiliation: From ABC Animal & Bird Clinic, 11930 Highway 6 South, Sugar Land, TX 77498, USA.

Abstract: Pain control is an essential part of nursing care for all species of veterinary patients, but managing pain appropriately depends on being able to determine whether a patient is experiencing discomfort. Validated methods of assessing pain have been developed for some types of animals, including some exotic species, but so far, no such pain scale has been devised for birds. As in other species, evaluation of pain in avian patients relies heavily on behavioral cues such as self-grooming and social interactions. These behaviors vary from one species of bird to another, and birds are very adept at using masking behaviors to avoid being perceived as ill or weak. This presentation reviews behaviors likely to be observed in birds suffering discomfort and discusses some of the ways technicians can act as advocates for their patients' healing and welfare.

Introduction

There is ample evidence that birds have the physiologic and anatomic mechanisms to experience pain, but validated methods of measuring it have not yet been developed. Assessing the severity of pain in avian species is challenging for several reasons. Some assessments that have proved helpful for mammals, such as grimace scales available for rabbits and rodents, rely on facial expressions that are much more limited in avian species. Physiological indicators, such as changes in heart rate, respiratory rate, and blood pressure, have been investigated, but some of these parameters are more reliable than others, and measuring them can itself induce stresses that can mimic a pain response. Hypothermia has been reported as interfering with accurate pain assessment.[1,2]

Thus, as in other species of exotic pets, pain assessment relies heavily on behavioral cues, and there have been some efforts to develop a pain scale based on such signs that would be applicable to birds. These behavioral cues vary from one species of bird to another: A scale that works for chickens may not be accurate for parrots, for example. In addition, birds are adept at using masking behaviors to avoid exhibiting signs of discomfort.

Behavioral Cues

Most research into behavioral indicators of pain in birds has focused on poultry, not the psittacine species that often are the bulk of the exotic veterinary technician's caseload. A recently published integrative literature review of the subject showed that, out of 132 total articles on pain-related behaviors in birds, only 5 articles (3.8%) involved parrots, with 73 articles (55.4%) devoted to chickens and turkeys.[3]

Before attempting to evaluate whether behavioral cues indicate pain, it is important to be familiar with the normal behavior of the species being studied. For example, *Pionus* spp., when excited, frequently make a wheezing or panting sound, which should not be mistaken for a sign of pain or illness. In addition, because of the limited usefulness of facial expressions, it is useful to assess the individual as a whole, noting posture, position of the tail and the head crest (in species that have them), and how the wings are carried. Keep in mind also that stress and the presence of an observer can alter behavior regardless of whether pain is present.[1]

In general, birds tend to respond to noxious stimuli with a fight-or-flight response (e.g., attempts to escape, vocalization, thrashing) or with conservation and withdrawal reactions (e.g., making no effort to escape, immobility, no vocalization). Pet birds who are in pain may allow interactions with humans that they normally would not permit.[4]

Some behaviors that may indicate pain include struggling, vocalizing, aggressive behaviors (especially in a normally tolerant bird); reduction of normal behaviors (such as eating, perching, ambulating, preening); changes in interactions with conspecifics or human handlers; new abnormal behaviors; and conservation or withdrawal responses (e.g., hiding, decreased self-grooming, hyporexia). In studies of musculoskeletal pain, Hispaniolan Amazons (*Amazona ventralis*) and green-cheeked conures (*Pyrrhura molinae*) with experimentally induced arthritis were found to ambulate less on the affected limb, groom the affected limb more than needed for general feather maintenance, and have a ruffled or unkempt appearance to the feathers.[1] African grey parrots (*Psittacus erythacus*) subjected to thermal or electrical stimuli were seen to vigorously flinch their wings when the stimulus was perceived to be noxious.[3]

Behavioral cues that have been recommended for consideration when evaluating pain in birds include:

- Dull mentation
- Eyes closed or squinting in a novel environment
- Rapid or labored breathing
- Change in environmental interaction
- Hiding
- Decreased or abnormal perching
- Increased or decreased maintenance behaviors
- Decreased locomotion
- Focal preening of a specific area
- Fluffed appearance or unkempt feathers
- Abnormal hunched body posture
- Decreased appetite or change in food choice.[1,5]

It is important to observe the avian patient both before and after administration of any analgesic drugs.[2] After administration of analgesics, a marked, easily noticed change toward more normal posture and behaviors should be observed; if it is not, pain management should be re-evaluated.[4] It is interesting to note that chickens with limb pain were observed to self-select food containing an analgesic medication when offered a choice of foods.[1]

Types of Pain: Acute, Chronic, Visceral, Neuropathic

Behavioral signs of acute pain, such as trauma, may differ from those exhibited in cases of chronic pain, such as osteoarthritis. Acute pain may result in restlessness, hunched body posture, change in appetite and physical activity, and/or lameness. Also, in cases of severe, acute pain, changes in the eyes may be noted, including pupil size, squinting, and abnormal tear production.[1]

For chronic pain, the bird may engage in increased guarding of a painful area, avoiding behaviors that the bird may perceive as exacerbating pain, self-isolation, feather destructive behavior and/or self-mutilation, inappetence, and weight loss.[1] Chronic pain also may be manifested by a reduction in social grooming behaviors, such as decreased dust bathing in chickens, and in a reduction in the performance of comfort and self-maintenance behaviors.[2] By contrast, chronic pain also may result in overgrooming or feather destructive behavior or both.[4]

Visceral pain may be more difficult to notice, as it may involve more nonspecific signs such as anorexia, lethargy, and weight loss. Neuropathic pain has been reported less often but may result in dull mentation and overgrooming at the site of the pain.[1]

Severity of Pain

In the absence of numerical pain scores for avian patients, pain can be characterized as mild, moderate, or severe.[1] It can be useful to keep in mind how you might feel if you were experiencing a similar medical condition.[2]

There have been some efforts to develop actual pain scores that may be useful as well, such as one developed in 2012 to evaluate fracture pain in pigeons (*Columba livia*). This scale set numerical values from 0-5, with 0 indicating no pain and 5 indicating extreme pain, with values assigned to four criteria: Pigeon's attitude in the presence of the observer, fracture limb position in the presence of the observer, observer's evaluation of pigeon's degree of pain (i.e., overall assessment), and pigeon's motor activity during 10 minutes of video recording.[6]

Keep in mind that a scale validated for pigeons may not be accurate for psittacine birds or other species.

The lack of scientifically validated methods of pain assessment for avian patients increases the importance of frequent and thoughtful monitoring. Always keep an eye out for subtle signs of abnormal behavior and remain aware that the patient may expend considerable effort to prevent observation of discomfort. If all else fails, the familiar standard applies to birds as much as it does to other veterinary patients: If it would be painful for you, assume it is painful for your patient.

References

1. Mikoni N, Sanchez-Migallon Guzman D, Paul-Murphy J. Pain recognition and assessment in birds. *Vet Clin North Am Exot Anim Pract.* 2023;26:65-81.

2. Speer B. Basic anaesthesia. In: Chitty J, Monks D, eds. *BSAVA Manual of Avian Practice: A Foundation Manual.* Quedgeley, UK: British Small Animal Veterinary Association; 2018:234-235.

3. Mikoni N, Sanchez-Migallon Guzman D, Fausak E. Recognition and assessment of pain-related behaviors in avian species: an integrative review. *J Avian Med Surg.* 2022;36(2):153-172.

4. Hawkins M, Paul-Murphy J, Sanchez-Migallon Guzman D. Recognition, assessment, and management of pain in birds. In: Speer B, ed. *Current Therapy in Avian Medicine and Surgery.* St. Louis, MO: Elsevier; 2016:616.

5. Turner T. Pain recognition in exotic species: a training guide for veterinary technicians. *Proc Annu Conf Assoc Avian Vet.* 2009:103-106.

Caudal Intravenous Catheter Placement in Squamate Reptiles

Christina Miller, RVT, BSc

Session #3511

Affiliation: From Clayton Park Veterinary Hospital, 278 Lacewood Dr, Halifax, NS, B3M 3N8, Canada.

Abstract: Vascular access in reptile patients can be challenging. Small patient size, absence of obvious peripheral veins in many animals, and the presence of thick skin all make identifying suitable vessels for either venipuncture or intravenous (IV) catheterization difficult. The caudal vein (ventral coccygeal vein) is a suitable site for IV catheter in some patients. Appropriate patient candidate selection, placement technique, as well as catheter maintenance will be reviewed.

Introduction

Reptiles are reputed as being challenging patients to obtain vascular access due to a lack of obvious peripheral veins in many species, the presence of thick and often scaled skin, and sometimes diminutive patient size. Recognizing techniques that are effective in a given species and appropriate for a given patient is a critical skill in treating these animals. Caudal vein (ventral coccygeal vein) catheterization has been discussed in the literature since as early as 2004.[1]

Who Are the Squamate Reptiles?

Order Squamata (the scaled reptiles) is the most speciose of class Reptilia; as of December, 2022 the total number of reported squamate reptiles is 11,549.[2] Squamata traditionally contained three sub-orders that were grouped morphologically: Serpentes (snakes), Sauria (lizards), and Amphisbaenia (worm-lizards); these classifications, although convenient, are now considered defunct in light of genetic data that has provided better resolution to the squamate reptile phylogeny.[3-5] In spite of this, squamate reptiles that are considered snakes and lizards are discussed as these animals can be readily identified as such in the veterinary practice.

What Sites Exist for Vascular Access?

Sites for catheter placement are relatively limited in these species compared to more traditional companion animals, particularly mammals:

- Cephalic: Accessible in lizards, this requires a surgical cut-down. Sedation with or without local anesthesia is recommended in patients except perhaps those that are moribund.

- Jugular: Accessible to varying degrees in lizards and snakes, this site also requires a surgical cut-down. General anesthesia is required for placement in snakes.

- Intracardiac: An emergency technique in snakes, a suitably-sized catheter can be introduced directly into the ventricle of the heart.[6,7,8] These can reportedly be left in place for up to 24 hours,[7] although long-term consequences of placement do not appear to have been studied.
- Intraosseous (IO): Several sites exist in most lizards, however there are no suitable IO catheter sites present in snakes or legless lizards species.

The caudal or ventral coccygeal vein is present in both lizards and snakes, and is found along the length of the tail. The vein is located immediately ventral to the caudal vertebrae.

Selecting Suitable Candidates

Vascular access is advantageous when treating critically ill and/or anesthetic patients, not only for rapid parenteral administration for emergency drugs, but also for providing fluid support. According to the American Animal Hospital Association, intravenous (or intraosseous) fluids are advantageous for hospitalized patients that are not taking in food or water voluntarily, patients undergoing anesthesia, and patients that require relatively large and/or rapid fluid bolus administration.[9] Reptile patients that are over 8% dehydrated would ideally have intravenous or intraosseous access for fluid administration, as absorption from other routes may be ineffective or increase intracoelomic pressure (subcutaneous and intracoelomic, respectively).[10]

When is intraosseous NOT an option?

While intraosseous catheters are a viable option for many patients, a few contraindications exist:[11,12]

- Pyoderma at the insertion site
- Pre-existing fracture
- Pre-existing osteomyelitis
- Nutritional secondary hyperparathyroidism
- Recent (within 48h) IO catheterization of the same site

Caudal autotomy

Many lizard species (and a few snakes) can perform caudal autotomy (tail drop), an anti-predator behavior that involves self-amputation of the tail at pre existing fracture planes. The amputated tail will continue to twitch and writhe, which is suspected to be a distraction for the would-be predator while the animal escapes. A ring of muscle constricts around the remaining caudal tissue, limiting bleeding. Some species can regenerate the tail in the form of a rod of cartilage in lieu of new vertebrae; the new tail often has differing colouration and scalation compared to the original appendage.[13] Caudal autotomy occurs in most lizard taxa, excluding true chameleons (Chamaeleonidae), most agamids (Agamidae), beaded lizards (Helodermatidae), monitor lizards (Varanoidea: Varanidae and Lanthanotidae), and the knob-scaled lizards (Xenosauridae). In some lizard species, autotomy and caudal regeneration is only present in juveniles, such as some members of Iguanidae (notably, *Ctenosaura* sp., *Cyclura* sp., *Iguana* sp., and *Sauromalus* sp.) and some members of Scincidae (*Tiliqua* sp.).[13]

Tail manipulations (including catheter placement) in these species should be limited to avoid inducing autotomy. Tail loss is not without consequences, as many species rely on the tail for balance when locomoting, social signaling, and energy storage.[14] It has been demonstrated that tail regrowth is energetically costly,[15] which is an additional stressor for already sick patients.

The renal portal system

Reptiles exhibit a circulatory mechanism absent in mammals, the renal portal system. Blood circulating in the caudal half of the body, including blood from the caudal, iliac, or hypogastric veins, may circulate through the renal parenchyma and be excreted via the renal tubules prior to transiting through general circulation.[16,17] This is hypothesized to be a compensatory mechanism for dehydration, sequestering blood to the kidneys could prevent ischemic injury.[17]

A few studies have demonstrated variable circulation through the renal portal system.[18-21] It is thought that reptiles have valves present that control blood flow into the renal portal veins as has been demonstrated in birds, but this has not been thoroughly evaluated in reptiles.[22,23]

The presence of the renal portal system in reptiles creates a potential complication: Drugs injected into the caudal half of the body could be concentrated in the kidneys prior to entering general venous circulation. It is considered a reasonable precaution to avoid administering potentially nephrotoxic or drugs excreted through the renal tubules (but not the glomeruli) in the caudal half of the body, including via the caudal vein.[16]

Technique and Maintenance

The need for sedation depends on the patient's demeanor. The author finds that in patients where a lateral approach to the tail is used, mild sedation may be needed. For patients where a ventral approach is taken, deep sedation or anesthesia is advised to eliminate the righting reflex. Over-the-needle catheters are suitable for this technique. Patients should be warmed to their preferred optimum temperature zone prior to placement to avoid peripheral vasoconstriction.

Site selection

A distance of 25-50% of the tail length from the cloaca is preferred to avoid accidental trauma of the hemipenes, hemiclitori, or scent glands.[24]

The author prefers a ventral approach in species with a dorsoventrally compressed body shape, and the lateral approach in species with a laterally compressed tail. Species with a round cross-section of the tail are amenable to either approach.

Ventrally-placed caudal catheters can be difficult to maintain in active patients due to the drag that this catheter will experience with patient movement. The site can also easily become contaminated with fecal and urate matter. The author typically reserves this site for anesthetic patients where the catheter is unlikely to require long-term maintenance (e.g., beyond the inpatient procedure), or in inactive patients where the lateral approach is less appealing. This catheter is not manageable in snakes that are active in any capacity.

Site preparation

Appropriate hand hygiene is performed,[25] followed by skin preparation of the insertion site. The site is prepared aseptically; the author routinely uses a combination of chlorhexidine 2% scrub to remove all obvious organic debris, followed by a rinse of isopropyl alcohol 70%. The scrub phase should be quite vigorous to clean underneath the scales, and depending on the ruggedness of the scales, a scrub brush may be needed instead of gauze. Preoperative surgical scrub brushes can be very effective. Hand hygiene is performed again prior to catheter placement, and sterile or clean gloves should be donned.

Catheter insertion, ventral approach

The ventral approach is performed on the midline of the ventral aspect of the tail. Gentle digital pressure may be placed at the base of the tail by the restrainer to temporarily impede venous return. The catheter is inserted at approximately a 30° angle, and slowly advanced until either a flash of blood is observed in the catheter hub, or the presumed anatomical location of the vein is reached. Because of the relatively low blood pressure and slow heart rate of reptiles, catheter filling may be very slow, or effectively nonexistent in volume-depleted patients. The catheter is threaded into the vein, and no resistance should be encountered. Physiologic saline can be used to flush the catheter: There should be no resistance, and no backflow around the catheter hub. It can be difficult or impossible to palpate a small fluid bolus proximal to the placement site, unlike in mammals. Aspirating blood from the catheter should be performed slowly with very gentle negative pressure to avoid causing venous collapse. A PRN port or pediatric infusion port (such as a "T-port") can be attached to the catheter hub. Both of these implements should be primed with sterile physiologic saline to avoid causing air emboli during catheter use.

Catheter insertion, lateral approach

The lateral approach is performed immediately ventral to the midline on the lateral aspect of the tail. Gentle digital pressure may be placed at the base of the tail, as in the ventral approach. The transverse processes of the caudal vertebrae may be palpated as a guide, and catheter insertion should be planned ventral to the level of the processes. Some species, like the green iguana (*Iguana iguana*) have a divet between the dorsal and ventral muscle groups of the tail, and this can be used as a placement guide. The tail may be slightly bent laterally away from the operator to better expose the vein.[8,12] The catheter is inserted as per the technique described above.

Securing the catheter

The catheter can be secured using cloth tape as per typical technique (the author finds that paper tape and transpire tape are both ineffective on reptile skin). In very active patients where a catheter is projected to be maintained for several days, the author has had success placing simple dermal stay sutures (using non-absorbable suture) on a tape butterfly around the catheter hub. For anesthetic patients that will be immobile, tape alone is usually effective in securing the catheter. For catheters maintained in awake patients, a layer of gauze bandage followed by self-adhesive bandage can be applied, with the partial circumference of the cranial margin of the bandage secured to the skin with tape or other adhesive bandage.

Alternatively, transparent film bandage can be used to secure catheters for anesthesia. Padding between the catheter hub and PRN or extension port and the tail can be used if desired, such as a piece of gauze or cotton ball.

Maintaining the catheter

The catheter insertion site should be checked at least daily for signs of infection, phlebitis, or extravasation. When flushed to verify patency, the animal's total fluid intake should be considered, as reptiles have a much lower daily fluid requirement compared to mammals and birds. Other possible complications include hematoma and thrombosis. If any complications are observed, the catheter should be promptly removed.[8]

Conclusions

The caudal vein is an acceptable site for intravenous catheter placement in many lizard and snake patients, particularly where intraosseous catheters are not possible. This site is best avoided in patients that perform caudal autotomy as an anti-predator behavior. Catheter maintenance is similar to other intravenous catheters.

References

1. Wellehan JFX, Lafortune M, Gunkel C, et al. Coccygeal vascular catheterization in lizards and crocodilians. *J Herpetol Med Surg.* 2004;14(2):26-28.

2. Reptile Database. Species Statistics Aug 2019 [sic]. Available at: http://www.reptile-database.org/db-info/SpeciesStat.html. Accessed March 5, 2023.

3. Vidal N, Blair Hedges S. The molecular evolutionary tree of lizards, snakes, and amphisbaenians. *Comptes Rendus Biologies.* 2009;332(2-3):129-139. doi:10.1016/j.crvi.2008.07.010.

4. Pyron RA, Burbrink FT, Wiens JJ. A phylogeny and revised classification of Squamata, including 4161 species of lizards and snakes. *BMC Evol Biol.* 2013;13:93. doi:10.1186/1471-2148-13-93.

5. Wiens JJ, Hutter CR, Mulcahy DG, et al. Resolving the phylogeny of lizards and snakes (Squamata) with extensive sampling of genes and species. *Biol Lett.* 2012;8(6):1043-1046. doi:10.1098/rsbl.2012.0703.

6. Mader DM. DVM 360: Critical care techniques in reptiles. Available at: https://www.dvm360.com/view/critical-care-techniques-reptile-patients-proceedings. Accessed March 5, 2023.

7. Divers SJ. Clinical procedures for reptiles. In: Winter AL, ed. *Merck Veterinary Manual.* Available at: https://www.merckvetmanual.com/exotic-and-laboratory-animals/reptiles/clinical-procedures-for-reptiles. Accessed March 5, 2023.

8. Pardo MA, Divers SJ. Catheter placement. In: Divers SJ, Stahl SJ, eds. *Mader's Reptile and Amphibian Medicine and Surgery.* 3rd Ed. St. Louis, MO: Elsevier; 2019:422-428.

9. American Animal Hospital Association. AAHA/AAFP Fluid Therapy Guidelines for Dogs and Cats: Implementation Toolkit. Available at https://www.aaha.org/globalassets/02-guidelines/fluid-therapy/fluid-therapy_guidlines_toolkit.pdf. Accessed March 5, 2023.

10. Herrin KV. Clinical techniques and supportive care. In: Doneley B, Monks D, Johnson R, Carmel B, eds. *Reptile Medicine and Surgery in Clinical Practice.* Hoboken, NJ: Wiley-Blackwell; 2018:159-173.

11. Lee JA, Guieu LS, Bussières G, Smith CK. Advanced vascular access in small animal emergency and critical care. *Front Vet Sci.* 2021;8:703595. doi:10.3389/fvets.2021.703595.

12. Kanda I, Brandão J. Herptiles: catheterization and venipuncture. In: Graham JE, Doss GA, Beaufrère H, eds. *Exotic Animal Emergency and Critical Care Medicine.* Hoboken, NJ: Wiley-Blackwell; 2021:722-731.

13. Funk RS, Johnson R, Divers SJ. Tail abnormalities. In: Divers SJ, Stahl SJ, eds. *Mader's Reptile and Amphibian Medicine and Surgery.* 3rd Ed. St. Louis, MO: Elsevier; 2019:1347-1348.

14. Durso AM, JC Maerz. Natural behavior. In: Divers SJ, Stahl SJ, eds. *Mader's Reptile and Amphibian Medicine and Surgery.* 3rd Ed. St. Louis, MO: Elsevier; 2019:90-99.

15. Naya DE, Veloso C, Muñoz JL, Bozinovic F. Some vaguely explored (but not trivial) costs of tail autotomy in lizards. *Comp Biochem Physiol A Mol Integr Physiol.* 2007;146(2):189-193. doi:10.1016/j.cbpa.2006.10.014.

16. Girling S. Vascular, hematopoietic, and immune systems. In: Divers SJ, Stahl SJ, eds. *Mader's Reptile and Amphibian Medicine and Surgery.* 3rd Ed. St. Louis, MO: Elsevier; 2019:917-921.

17. Holz PH. The reptilian renal portal system - a review. *Bull Assoc Reptilian Amphib Vet.* 1999;9(1):4-9.

18. Kummrow MS, Tseng F, Hesse L, et al. Pharmacokinetics of buprenorphine after single-dose subcutaneous administration in red-eared sliders (*Trachemys scripta elegans*). *J Zoo Wildl Med.* 2008;39:590-595.

19. Holz P, Barker IK, Crawshaw GJ, Dobson H. The anatomy and perfusion of the renal portal system in the red-eared slider, *Trachemys scripta elegans*. *J Zoo Wildl Med.* 1997;28:378-385.

20. Holz PH, Barker IK, Burger JP, Crawshaw GJ. The effect of the renal portal system on pharmacokinetic parameters in the red-eared slider (*Trachemys scripta elegans*). *J Zoo Wildl Med.* 1998;28(4):386-393.

21. Benson KG, Forrest L. Characterization of the renal portal system of the common green iguana (*Iguana iguana*) by digital subtraction imaging. *J Zoo Wildl Med.* 1999;30:235.

22. O'Malley B. *Clinical Anatomy and Physiology of Exotic Species.* Edinburgh, UK: Elsevier Saunders; 2005.

23. Stephens T, Rosenwax A. Disease of the cardiovascular system. In: Doneley B, Monks D, Johnson R, Carmel B, eds. *Reptile Medicine and Surgery in Clinical Practice.* Hoboken, NJ: Wiley-Blackwell; 2018:287-297.

24. Olsson A, Simpson M. Analgesia and anaesthesia. In: Doneley B, Monks D, Johnson R, Carmel B, eds. *Reptile Medicine and Surgery in Clinical Practice.* Hoboken, NJ: Wiley-Blackwell; 2018:369-381.

25. American Animal Hospital Association. Protocol for hand hygiene using soap and water, or alcohol-based hand sanitizer Available at: https://www.aaha.org/aaha-guidelines/infection-control-configuration/protocols/hand-hygiene2/. Accessed March 5, 2023.

26. Barten S, Simpson S. Lizard taxonomy, anatomy, and physiology. In: Divers SJ, Stahl SJ, eds. *Mader's Reptile and Amphibian Medicine and Surgery.* 3rd Ed. St. Louis, MO: Elsevier; 2019:63-74.

Technique Tutor: Intravenous Catheter Placement in Small Mammals

Stefanie Kotschwar, RVT, BScH (Zoology, Nutrition), MSc (Zoo Conservation), AAS (Vet Tech), Diploma (Web Design and Equine Science)

Session #3335

Affiliation: From Lynwood Animal Hospital, 107-30 Stafford Rd, Ottawa, ON, K2H 1W8, Canada.

Abstract: Small exotic mammals might require intravenous (IV) access while being in the hospital to receive fluids while sick and dehydrated, administration of medications, induce anesthesia or provide support during surgical procedures. In this lecture we review various locations where an IV catheter can be placed, the supplies needed and go over the steps on how to place it with some tips and tricks along the way.

Introduction

Intravenous (IV) catheters are commonly placed in small exotic mammals that require fluid support while sick, in emergency situations or while undergoing a surgical procedure. Any small exotic mammal having a surgical procedure under general anesthesia should have IV access. IV access should even be considered for patients undergoing a procedure under sedation.

Catheterization techniques like those used in cats and dogs are applied to small exotic mammals, the main difference being the patient size. Another factor to keep in mind is that most small exotic mammals are considered prey species and can therefore experience higher levels of stress from handling.[1] If possible, consider using some level of sedation prior to catheterization. Ensure the treatment area used has no predators and keep the noise level as low as possible.

Applying a topical local anesthetic cream at least 20 minutes prior can provide comfort when placing an IV catheter. Sites that can be used for IV catheter placement include cephalic vein, marginal ear vein and the lateral saphenous. The IV catheter sizes are selected based on patient size and can range from 22-26 gauge. Some drugs will require dilution prior to IV administration to avoid vasculitis.[1-5]

Supplies

Supplies needed for intravenous catheter placement are listed below.

- Clippers
- Indwelling IV catheters of varying sizes: 22-26 gauge
- 25-18 gauge hypodermic needle
- Intermittent infusion plug

- White porous tape
- Topical anesthetic cream like EMLA® (lidocaine 2.5% and prilocaine 2.5%; Astra Pharmaceuticals LP, Wayne, PA)
- Tourniquet
- Gauze squares
- Sterile sodium chloride 0.9%
- Chlorhexidine gluconate 0.5% solution
- Self-adherent flexible bandage
- Clean gloves

IV Catheter Placement Sites

Placement sites for IV catheters in multiple small mammal species are described below.

Rabbits

- *Marginal ear vein*
- *Cephalic*
- *Saphenous*

Guinea Pigs

- *Cephalic*
- *Saphenous*

Hedgehogs

- *Cephalic*
- *Saphenous*

Ferrets

- *Cephalic*
- *Saphenous*

Rats

- *Cephalic*
- *Saphenous*
- *Tail*

Other rodents

- *Cephalic*
- *Saphenous*
- *Consider IO*

General IV Catheter Steps

Steps for placing a catheter are outlined below.

- Place the patient on a non-slip surface.
- Maintain the patient in sternal recumbency.
- Shave catheter site – rabbit fur can be difficult to shave!
- Place topical anesthetic cream if available.
- For small mammals, the use of a tourniquet is recommended to help raise the vein. It can be difficult for an assistant to hold off the vein while restraining the patient.
- Prepare the site aseptically.
- Make a small cut down hole with a hypodermic needle: Size will depend on patient size, but a 21-gauge is a good rule of thumb.
- Preload your catheter with sterile 0.9% sodium chloride as this will help you visualize a small blood flush easier and avoid inadvertent air embolism.
- If the IV catheter does not have wings, place some of the white tape across to make tape butterflies as this will make the taping easier. Experiment with this step as not everyone likes IV catheters with wings, but it is the author's preference.
- Hold the IV catheter at an about 20-degree angle to the site. Insert IV catheter slowly until you see a small flash of blood in the catheter. Advance another 1 millimeter, before advancing the IV catheter.
- Place at least one strip of tape to stabilize the catheter, especially if working with a non-sedated patient.

- Prime the intermittent infusion plug with sterile 0.9% sodium chloride and then attach it.
- Continue securing the IV catheter with at least one or two more strips of white tape.
- Flush the IV catheter with sterile 0.9% sodium chloride to confirm placement.
- Wrap the site with Self-adherent flexible bandage. A layer of cling gauze is optional.

Considerations for IV Catheter Placement in Other Sites

Multiple considerations for placing IV catheters in other sites are discussed below.

Saphenous vein

This vein can be a little harder to stabilize, try stretching the skin to stabilize the vein.

Marginal ear vein, rabbits

Ensure you are using the marginal ear vein and not the central auricular artery! The central auricular artery runs along the midline of the ear.

Reports or ear necrosis when using this vein have been reported. Keep that in mind if you are injecting any medication that could be irritating if injected perivascular.

This site does not require a cut down prior to placing the IV catheter.

Roll the pinna around a roll of gauze or rolled up gauze squares to help tape and stabilize the IV catheter.

IV Catheter Maintenance

If the catheter is placed on a hospitalized patient, check it regularly for patency and check the site for erythema, inflammation, and discharge. If there are any concerns, remove the IV catheter and place a new one in a different site.[2-5]

Conclusions

While placing an IV catheter in small exotic mammals can prove to be challenging, there are some tips that can facilitate the process. In addition, continual practice is beneficial.

References

1. Gardhouse S, Sanchez A. Rabbit sedation and anesthesia. *Vet Clin North Am Exot Anim Pract.* 2022;25: 181-210.

2. Nugent-Deal J. Intravenous Catheter Placement in Rabbits. LafeberVet Website. 2021 Available at: https://lafeber.com/vet/intravenous-catheter-placement-in-rabbits/. Accessed March 1, 2023.

3. Echols MS, Lichtenberger M, Pollock C. Intravenous catheter placement in small mammals. LafeberVet website. May 7, 2013. Available at: https://lafeber.com/vet/intravenous-catheter-placement-in-small-mammals/. Accessed March 1, 2023.

4. Nugent-Deal J. Exotic anesthesia and analgesia. In: Ballard B, Cheek R, eds. *Exotic Animal Medicine for the Veterinary Technician.* 3rd Ed. Hoboken, NJ: John Wiley & Sons; 2017:11-30.

5. van Zeeland YRA, Shoemaker NJ. Catheterization and venipuncture. In: Graham JE, Doss JA, Beaufrère H, eds. *Exotic Animal Emergency and Critical Care Medicine.* 1st Ed. Hoboken, NJ: John Wiley & Sons; 2021:51-69.

Technique Tutor: Feeding the Hospitalized Small Mammal

Stefanie Kotschwar, RVT, BScH (Zoology, Nutrition), MSc (Zoo Conservation), AAS (Vet Tech), Diploma (Web Design and Equine Science)

Session #3336

Affiliation: From Lynwood Animal Hospital, 107-30 Stafford Rd, Ottawa, ON, K2H 1W8, Canada

Abstract: This session reviews nutritional support for hospitalized small exotic mammals, with a focus on rabbits. It includes a brief review of digestive anatomy, available feeding formulas, supplies required and tips and tricks on how to feed patients with the least amount of stress.

Introduction

Nutritional support is required for any small mammal presenting to the clinic that is anorexic or hyporexic, due to disease or recovering from a surgical procedure. These animals have a high metabolic rate and a decrease in appetite can very quickly lead to glycogen store depletion and hypoglycemia. In addition, sick patients have an increased energy requirement that can lead to catabolism of fat and muscle if there is decreased caloric intake.[1]

Gastrointestinal Anatomy of Small Exotic Mammals

There is a wide variety of species encompassing small exotic mammals with a variety of feeding strategies. Thus, there are variations in gastrointestinal anatomy among small exotic mammals.

Most small exotic mammals encountered in veterinary practice have herbivorous diets and are monogastric hindgut fermenters. They have a single stomach and a large cecum that provides for breakdown of fiber from their diet, and these species engage in cecotrophy or coprophagy. Species such as rabbits, guinea pigs and chinchillas are hindgut fermenters.[2,3]

Smaller rodents have different adaptations. For example, hamsters have a compartmentalized stomach with fermentation taking place in one of those compartments, the forestomach.[2,3]

Ferrets are obligate carnivores and this is reflected in their gastrointestinal anatomy. They have a simple digestive tract that allows food to pass quickly. They have a rapid gastrointestinal transit time, so they should not be fasted for extended periods of time.[2,4]

Feeding Requirements

While individual nutrient requirements vary among species, daily nutrient requirements can be calculated. Basal metabolic rate (BMR = 70x $BW^{0.75}$kcal/kg/d where BW is the patient's body weight in kilograms) can be used to calculate energy requirements for small mammals. This can then be multiplied by a factor of 1.2-2 depending on their health status.[1]

Feeding Formulas

There are many products available that can be used to provide nutritional support to small mammals. Diet selection depends on the preference of the patient and its requirements. Below are examples of products available from several different companies.

Oxbow Animal Health (Papillion, NE)

- Critical Care® for herbivores comes in apple/banana or anise flavor and is considered a complete maintenance diet. Not appropriate for nasogastric tube feeding as it will clog the tube due to its high fiber content.
- Critical Care® for herbivores fine grind is available for nasogastric feeding tubes.
- Critical Care® for carnivores is a complete maintenance diet.
- Critical Care® for omnivores is a complete maintenance diet.

Lafeber Company (Cornell, IL)

- Emeraid® IC intensive care that is intended for debilitated patients, and animals with digestive disorders that require highly digestible ingredients. It should not be used long term as the fiber content is limited.
- Emeraid® sustain for herbivores is a good maintenance diet.
- A maintenance diet for carnivores and omnivores is also available.
- Supreme™ Pet Foods (Hadleigh, Ipswitch, United Kingdom).
- Recovery for small herbivores.
- Recovery plus has a higher level of fibers and contains pre- and probiotics.

Sherwood® Pet Health (Ephraim, UT)

- Line of recovery foods to support sick animals until appetite has been restored.

Feeding Routes

How a patient is fed depends on its status and compliance. While several options exist, including use of an orogastric tube, nasogastric tube and esophagostomy tube, the focus of this session is syringe feeding.

Supplies

- Bowl
- Feeding formula
- Water, unsweetened applesauce or baby food
- Appropriately sized syringe for the patient

Feeding Tips and Tricks

Once the food is prepared as indicated on the package, there are things to keep in mind and to try when feeding the small exotic mammal.

- Ensure that you are knowledgeable in the husbandry, behavior, and restraint of the species you are working with.
- Always approach slowly and talk in a gentle manner.
- Preparing the food with unsweetened applesauce or baby food adds a little bit of flavor that could entice the patient to eat.
- Always offer food in a bowl first to determine if they will lick it off directly.
- If using a syringe, place the tip of the syringe in the diastema of herbivores. For carnivores, place the tip of the syringe in the corner of their mouth.
- Give little bits of food at a time.
- Try feeding the patient inside the hospital cage first. Most feel safer in that space.
- Use the least amount of restraint necessary. Only use a towel for restraint and do the feeding on a treatment table if all else fails.
- Use a lot of appeasing and stress-reducing techniques like a body massage or even a face rub.
- Be patient and try different techniques.
- Talk to the patient! Bunnies especially like to be told they are a king or queen and that they are the best bunny.
- Giving the patient a face rub and then lifting the lip slightly to get the syringe into the mouth and giving a little bit of food works most of the time.
- Letting the patient move around the cage in between the feedings will give them the opportunity to slowly chew and swallow their food.

- If you have an uncooperative patient that needs to be swaddled in a towel, try keeping them in the cage or feeding them on the floor. Use the treatment table as the last resort as herbivores do not like to be picked up and carried around.
- Be flexible! Some well socialized rodents like rats like to be fed while perching on your shoulder.
- Clean their faces after feeding. No one likes a dirty face and if any food residue is left on their face, it will dry up and make them uncomfortable.
- Sometimes you must let someone else try; do not take it personally.

Conclusions

There are many reasons why small exotic mammals need to be syringe-fed while in the hospital. There is not one way of doing this because each animal is an individual. There are many techniques that can be employed to find the best way to feed each patient to ensure their comfort and low stress. A lot of patience and creativity is required by the veterinary technician to accomplish this process.

References

1. van Zeeland YRA, Schoemaker NJ. Nutrition and fluid therapy. *Exotic Animal Emergency and Critical Care Medicine.* Hoboken, NJ: John Wiley & Sons. 2021;109-122.

2. Bullen LE. Nutrition for pocket pets (ferrets, rabbits, and rodents). *Vet Clin North Am Exot Anim Pract.* 2021;15:583-604.

3. Grant K. Rodent nutrition: digestive comparisons of 4 common rodent species. *Vet Clin North Am Exot Anim Pract.* 2014;17:471-483.

Section 55

Technician Emergencies Session

Debbie Sykes, CWR, VT, FFCP
Session Chair

Exotic Animal Emergencies for Technicians, Part 1: Evaluation of the Emergent Exotic Mammal Patient

Christina Miller, RVT, BSc

Session #3367

Affiliation: From Clayton Park Veterinary Hospital, 278 Lacewood Dr, Halifax, NS, B3M 3N8, Canada.

Abstract: Exotic companion mammals commonly present to the veterinary hospital in a state of marked illness. This masterclass reviews emergency presentations in common exotic companion mammal species. It covers the technician's role in the primary survey, obtaining vitals efficiently and safely, obtaining airway and vascular access, and proceeding with cardiopulmonary and cerebral resuscitation.

Introduction

Exotic pets present some unique challenges in both general practice and the emergency room. Most exotic animals encountered in the veterinary practice exhibit some degree of masking phenomenon, either because they are not truly domesticated, they are prey animals, or both. Masking phenomenon is an instinct to hide signs of illness with the goal of preventing predation, as predators may exhibit a preference towards old, weak, or sick-looking prey since they may be more efficient to capture and subdue (this theory is partially supported by ecological literature; however, predator-prey interactions can be complex and dependent on the species as well as the environment).[1-3]

Regardless, there are some presentations that customer service staff can generally assume are at a high triage level, including:

- Recumbent or non-responsive patient
- Bloated abdomen or coelom, especially if acute
- Dyspnea
- Seizures, muscle tremors
- Toxic substance ingestion or exposure
- Stranguria, hematuria
- Trauma (blunt, crushing, falling, attacked by predator)
- Active bleeding
- Hypothermia, hyperthermia
- Exophthalmos, proptosis, other ocular trauma

- Large open wounds, evisceration
- Ataxia

Unfortunately, acute presentations of chronic conditions are common in exotic pets. This may be due to a combination of masking phenomenon (e.g., the patient is hiding early signs of illness) and poor pet owner knowledge of how health problems present in exotic taxa. Captive exotic species are also very susceptible to husbandry-related disease, which may lend to a slow development of some health problems that reinforces client perception that a pet's behavior or anatomy is normal.

Because of all of these factors, a general rule has been suggested that if a client is calling with a concern about their exotic pet, that the pet should be examined (ideally promptly).[4,5] This is operating on the assumption that if the client has appreciated something wrong or abnormal, that the disease process may be quite advanced.

The Exotic Mammal Primary Survey

The primary survey starts with an initial general impression of the patient, which can be observed in the pet's carrier, or the pet may be allowed to explore a secure exam room (particularly rabbits, ferrets). Try to observe the pet covertly: What is their mentation like? Most exotic mammals should be bright, alert, and responsive to your presence. Rabbits and guinea pigs will adopt a freeze-and-assess response to your presence. This behavior is reflected in wild prey animals, who upon detecting a possible threat, will freeze (to potentially evade predator detection) and assess whether hiding, fight, or flight is appropriate. Sugar Gliders are quite strictly nocturnal and may be asleep or drowsy, but should be rousable. Ferrets may similarly be sleepy or drowsy, as they typically sleep between 14-18 hours per day, but should likewise be rousable. If the patient is dull, dyspneic, or non-responsive to touch, proceed to the ABCDE protocol. The ABCDE protocol is a systematic approach to patient assessment in order of the most acutely life-threatening problems to least life-threatening problems.

The author has personally observed numerous small mammal patients initially appear to be stable during the initial impression observation, then decompensate during history-taking (e.g., patient becomes dyspneic when they initially presented with a normal respiratory pattern). This highlights the amazing ability of these animals to mask their illness.

A: Airway

Airway and breathing assessment are effectively performed simultaneously; however, signs of airway obstruction should be considered immediately, including: Paradoxical thoracic and abdominal movements, increased respiratory effort (especially on inspiration), cyanosis, stertor (low-pitched sound, suggests nasal or pharyngeal obstruction) or stridor (high-pitched sound, suggests lower airway obstruction), and open-mouth breathing in obligate nasal breathers. If there are signs of airway obstruction, oxygen therapy should be implemented immediately.[6]

Anxiolysis and sedation may be considered to help lower the patient's oxygen demand and help to lower their stress. If it is possible to localize and remove the obstruction, it should be done as quickly as possible to hopefully alleviate the dyspnea. Rodents and lagomorphs are obligate nasal breathers, and nasal secretions can cause significant obstruction. Mild suction using a nasal bulb syringe intended for human infants can be very effective; however, if the secretions are dried, the humectant action of humidified oxygen may be needed prior to removal. Oral foreign bodies can be challenging to identify and retrieve due to the long and narrow oral cavity with an infolding of the cheek tissue, the inflexa pellita. Use of an otoscope, bivalve speculum, or endoscope can be used to visualize the oral cavity. Guinea pigs should have ingesta in their mouths, which can make visualization difficult.[6]

If respiratory arrest occurs or the obstruction cannot be identified and relieved, airway access must be secured. If there is jaw tone present, use of sedation and/or induction agents may be necessary. Airway access will be discussed below.

Oxygen therapy

There are several techniques for providing oxygen therapy that can be considered in these animals:

- Flow-by oxygen, oxygen flow rate = 2-3 L/min: This method may be used during initial triage, but it is wasteful to use for longer-term therapy since the flow rate needs to be very high. The oxygen outflow must be held within 2 centimeters of the patient's nares to be effective.[7] The handling required to maintain this technique is inherently stressful to many small mammal patients.

- Tight-fitting face mask, oxygen flow rate => 100 ml/kg/min. FiO_2 is better with a tight-fitting facemask compared to a loose mask or mask lacking a diaphragm, or an oxygen cage. This can be very stressful in some animals, as rodents and lagomorphs have a very sensitive face. Restraint is required to keep the mask on in all but the most compliant of patients, further adding to the patient's stress. If the seal on the mask is very tight, there is less CO_2 escape (less so on a non rebreathing anesthetic circuit), and prolonged use may contribute to rebreathing of waste gas and subsequent hypercapnia, as well as accumulation of heat.[8]

- Loose face mask or oxygen cage, oxygen flow rate => 300 ml/kg/min. The FiO_2 can reach up to 40% using an appropriate cage.[9] Commercial units that allow for temperature and humidity control, and monitor FiO_2 and CO_2 are ideal. House-made oxygen chambers (e.g., using modified steel kennel doors, or smaller containers with an oxygen inlet and passive scavenging ventilation) can still be useful. If using a more rudimentary oxygen chamber, ensure that a bubbler is used to maintain a humid atmosphere (dry oxygen can desiccate tiny airways very quickly) and monitor environmental temperature closely (overheating can occur rapidly with some setups and species like chinchillas and ferrets).[10]

B: Breathing

The quality and rate of the patient's respirations should be assessed, and oxygen therapy initiated immediately if abnormal. Abnormal respiratory depth and rate (e.g., tachypnea, bradypnea, hyperpnea, hypopnea) and patterns (e.g., Cheyne-Stokes, Kussmaul, Biot, ataxic) should be noted. Normal respiratory rates for a selection of species is available in Table 1. Accessory respiratory movements may be appreciated. Marked abdominal effort resulting in paradoxical abdominal and thoracic movements can occur with markedly increased effort, and this is often paired with orthopneic posturing (e.g., head pointed up, elbows out). Obligate nasal breathers may exhibit nasal flaring, which may precede open-mouth breathing, which is a sign of severe dyspnea in these species.[6]

The lung fields may be assessed next via auscultation. Referred upper airway sounds can be normal in obligate nasal breathers. The presence and location of any abnormal sounds (e.g., rales, crackles, wheezing, stridor) should be noted. Thoracic percussion is not typically helpful due to the small size of most exotic mammal patients encountered in practice.[6]

Table 1. Normal vital signs in exotic companion mammals.[6]

Species	Rectal temperature °C	Rectal temperature °F	Heart rate (bpm)	Respiratory rate (rpm)
African pygmy hedgehog (*Atelerix albiventris*)	35.4-37.0	95.7-98.6	180-280	25-50
Chinchilla (*Chinchilla lanigera*)	37.0-38.0	98.6-100.4	100-300	40-80
Domestic ferret (*Mustela putorious furo*)	37.8-39.2	100-102.6	180-250	30-60
Domestic rabbit (*Oryctolagus cuniculus*)	37.8-39.2	100-102.6	150-300	30-60
Fancy rat (*Rattus norvegicus domestica*)	37.2-39.2	99-102.6	280-500	60-120
Gerbil (*Meriones unguiculatus*)	37.4-38.2	99.3-100.8	260-600	70-120
Guinea pig (*Cavia porcellus*)	37.2-39.5	99-103.1	220-300	40-120
Hamsters (Cricetidae)	36.1-38.9	97-102	280-420	100-250
Mouse (*Mus musculus*)	37.1	98.8	500-725	80-230
Sugar glider (*Petaurus breviceps*)	32.0 (cloacal)	89.6	200-300	16-40

C: Circulation

Pulse quality, rate, and regularity can be appreciated at the femoral artery, which is palpable in the inguinal region where the limb meets the ventrum, in the dip between the muscle bellies. The author finds that with a gentle touch and using your pinky fingers, this pulse can be palpated in animals as small as guinea pigs and sometimes rats (obesity makes palpation challenging in any species). The central auricular artery can be palpated in most rabbits (exceptions being particularly diminutive dwarf breeds with very small pinnae). Pulse deficits can be very difficult to appreciate in animals with a rapid heart beat (over 200 bpm in the author's experience).[6]

The mucous membranes can be assessed at the conjunctiva, oral mucosa, and vulva or prepuce in most species. The mucous membranes should be medium pink and moist, as in other companion mammals. Capillary refill time can be assessed at the mucous membranes, or on unpigmented footpads, or the inside of the pinnae in rabbits. The extremities should be relatively warm to the touch.[6]

Cardiac auscultation may be performed simultaneously with pulmonary auscultation. Pediatric or infant stethoscopes are essential in exotic mammal practice for more precise localization of heart and lung sounds.

The apex beat can be located at the left hemithorax at about the 5th intercostal space. The exception to this is mustelid species such as the domestic ferret, where the apex beat is located more caudally (6-8th intercostal space). Heart rate, rhythm, and the presence of abnormal sounds should be characterized.[6] Normal heart rates are described in Table 1. Localizing heart murmurs to specific valves is challenging if not impossible in small exotics; however, left-sided versus right-sided can sometimes be identified. Ferasin and associates[11] describe iatrogenic murmurs caused by excessive pressure placed on the thorax of kittens with the stethoscope diaphragm, due to their relatively compliant thorax compared to adult cats. While this phenomenon has not been described in the literature in exotic mammals, the author recommends remaining cautious with the pressure placed on the small mammal thorax.

Both arrhythmias and murmurs can be difficult to appreciate with high heart rates. Electrocardiogram evaluation can be helpful.

If hemorrhage is present, bleeding can be controlled in a similar manner to other species. Non-adherent dressing pads can be used to place pressure. Alternatively, sterile cotton-tipped swabs can be applied to bleeding vessels on very small patients, although removal may dislodge a clot.

D: Disability

Patient mentation and reflexes are examined next. Corneal or palpebral reflexes should be present. A study evaluating neurologic examination in healthy rabbits found that pupillary light reflex (PLR) was slow and weak, and consensual PLR was often absent.[12] Menace reflex is often absent in rabbits and rodents. Dazzle reflex may be absent in normal chinchillas.[13]

Common neurologic deficits in small mammal herbivores (e.g., rabbits, chinchillas, guinea pigs) as well as myomorph rodents (e.g., mice, rats) include head tilt, also referred to as wryneck or torticollis, circling and rolling, ataxia, and nystagmus. The causes for these signs are varied, but an effort should be made to reduce potential trauma by offering a padded environment, ideally in a low light and quiet region of the hospital. Rabbits with these clinical signs may be infected with *Encephalitozoon cuniculi*, which may be zoonotic, so appropriate biosecurity measures are advised until the patient's serologic status for this pathogen is known.[14]

E: Exposure of environment

Body temperature is ideally taken from the rectum, or cloaca where applicable, such as in marsupials. Rectal thermometers intended for companion mammal use are effective. Male guinea pigs have prominent rectal folds, and improper placement in this cloacal region results in a lower temperature reading. Ensure that the thermometer is directed dorsally in these folds to enter the anus (the author finds that gently pulling dorsally on the skin dorsal to the coccyx helps to position the thermometer). Tympanic or auricular thermometers are not considered reliable in most exotic mammal species because the ears are a thermoregulatory surface in rabbits and chinchillas, and the ears of most other exotic mammals are too small for their use. A 2021 study[15] suggests that a tympanic thermometer can be used in the interscapular region in ferrets; however, this has not been validated in patients that are not normothermic.

A low rectal temperature is a strong predictor of a poor prognosis in several species, even if the patient is bright, alert, and responsive. Rabbits that present with a rectal temperature below 36.7 °C[16] and guinea pigs that present with a rectal temperature below 37.9 °C[17] should be treated as if they are in shock.

Survey the body for skin lesions, fractures, bruises, and other injuries.

Airway Access

Airway access should be secured in apneic patients. Intubation of the trachea is ideal, as this gives staff the most control over oxygenation and ventilation, but unfortunately not always possible depending on the species and equipment at your disposal. The author recommends reinforcing these skills with stable patients (i.e. during routine surgeries) so that your staff builds confidence and competence for emergent patients.

With any patient that is not apneic, support staff can attempt preoxygenation to improve/increase the patient's oxygen saturation, in the event apnea does occur, with the goal of preventing hypoxemia. The relatively high or fast metabolic rate of these species results in a greater need for oxygen per unit of body weight compared to larger mammals, so hypoxemic events can rapidly result in ischemic injury.[18] In any dyspneic patient, assume that apnea can occur, and prepare intubation equipment as per the species' needs. The methods of providing oxygen support are described in section A, above.

Ferrets

Ferrets have similar anatomy to canine or feline patients, and orotracheal intubation is comparable to intubating a pediatric cat or kitten. The glottis can be directly visualized with a laryngoscope. Ferrets have incomplete tracheal rings, and a cuffed or uncuffed endotracheal (ET) tube is acceptable. Most ferrets will accommodate an ET tube between 2.0-3.5 millimeters in diameter. The use of a stylet is a personal preference.

The tube length is measured from the rostrum to the thoracic inlet, and lubricated with a water-soluble lubricant. Local anesthetic (e.g., lidocaine) can be applied to the arytenoid cartilages, and then allowing 15-30 seconds for the anesthetic to take effect will help to prevent laryngospasm during intubation attempts (and reduce the risk of traumatic iatrogenic laryngeal edema). The ET tube can be secured using cling gauze tie around the back and arms in a figure-of-8 pattern, as ferrets have a very floppy neck and a simple tie around the back of the skull is ineffective. Take note that ferrets will maintain a jaw tone until a moderate plane of surgical anesthesia is achieved, and may retain the jaw tone for a couple of minutes once they reach this plane. Take care not to get bitten while intubating these patients.

Hedgehogs and sugar gliders

Orotracheal intubation technique in these species is, once again, similar to in canine and feline patients, albeit much smaller in size. A pediatric blade on a laryngoscope (e.g., Miller size 0 blade), an otoscope, or small rigid endoscope may be used to aid in visualizing the glottis. Small uncuffed ET tubes with a diameter of 1.0-2.0 millimeters are appropriately-sized.

The tube length is once again measured from the rostrum to the thoracic inlet, lubricated, and local anesthetic may be applied to the arytenoids (take care with using small volumes to avoid accidental overdose). The tube may again be secured with a cling gauze tie around the back of the head and arms in a figure-of-8 pattern.

Van Zeeland and Schoemaker[6] describe a technique using the V-Gel® supraglottic device (Docsinnovent, Hemel Hempstead, United Kingdom) intended for use in rabbits: Size R1 may fit appropriately in a hedgehog's oral cavity. Capnography must be used to ensure placement.

Rats and mice

Small myomorph rodents follow a similar protocol as above; however, their anatomy is considerably smaller in size. Rats can accommodate an uncuffed ET tube with a diameter of 1.0-1.5 millimeters. Alternatively, an over-the-needle IV catheter with the stylet removed can be modified using a scalpel blade to create a Murphy eye. Size 14-16 gauge can fit comfortably in a rat patient. The catheter hub will fit snugly on a small ET tube adapter. Likewise, the mouse glottis can accommodate a 20-gauge IV catheter.

Rabbits

Rabbits have a long and narrow oropharyngeal cavity, the glottis is hidden by the soft palate in a natural posture, and visualizing the glottis without specialized equipment is challenging if not impossible. Several methods of airway access will be discussed.

Orotracheal intubation

A small uncuffed ET tube (size 2.0-3.5 millimeter will be appropriate for a 2-4 kg rabbit, and a PVC tube is preferred for reasons discussed below) is measured from the rostrum to the thoracic inlet, and lubricated. Local anesthetic should be applied to the arytenoids. The tube may be secured with a cling gauze tie or rubber band around the back of the head. Oxygenation via the nares before and during intubation is ideal.

- Blind method: This is a technically difficult intubation method. There is no visualization of the glottis, and relies on breath sounds, the ET tube fogging (PVC tubes work best), or a capnograph for placement. This technique is not effectively possible during apnea or arrest. The head is positioned in dorsoflexion and the tube is fed along the roof of the mouth, and directed towards the glottis. Local anesthetic can be applied to the glottis via the ET tube once a visual (tube fogging) or audio (breath sounds) cue is obtained. The tube is then introduced into the glottis during a breath. Heavy sedation is required as the animal's swallowing reflex and any jaw movement must be absent for this technique.[19] Unsuccessful intubation attempts leads to traumatic edema of the pharyngeal region, which can contribute to post-extubation dyspnea and respiratory arrest.[20] This technique also does not allow for inspection of the oral cavity for ingesta, which may accidentally be pushed into the glottis.[19]

- Endoscopic: The use of an endoscope allows for visualization of the glottis but necessitates specialized equipment. An oral gag should be used to prevent damage to the scope. A rigid endoscope may be used for a side-by-side technique, where the glottis is visualized, then the ET tube (or a stylet, such as a polypropylene catheter) is introduced into the oropharyngeal space parallel to the scope, then into the glottis. A flexible endoscope may be passed through an ET tube and used effectively as a stylus with direct visualization of the glottis for intubation. The latter technique is limited with very small rabbits.[19]

- Laryngoscope or otoscope: The device is introduced into the side of the patient's mouth behind the incisors, and used to gently move the soft palate dorsally so that the glottis may be visualized. The ET tube or stylus is then directed into the glottis. A neonatal or pediatric straight laryngoscope blade (such as a Miller size 0) is needed for many smaller patients if using the laryngoscopic technique.

Supraglottic airway devices (SAD)

These are species-specific devices that are designed to form a seal around the larynx while occupying space in the oral cavity to prevent accidental esophageal intubation. The V-Gel® (Docsinnovent, Hemel Hempstead, United Kingdom) is designed for rabbit use, and at the time of preparing this presentation, there are six sizes available for rabbits weighing between 0.6 to over 4.5 kg.[21] These devices are best used with a capnograph to confirm placement, as the device may shift if the patient is moved, or if it is not appropriately sized. Docsinnovent has excellent training materials available on their website (https://docsinnovent.com/products/v-gel-rabbit/). The SAD may be used in patients who present in respiratory arrest, as it does not require patient respirations for placement, and may be used to administer positive pressure ventilation.[19] The SAD is relatively easy to use and placement is generally atraumatic.[22] Note that gastric bloat/tympany can occur with IPPV, and the resulting increased abdominal pressure can impede diaphragmatic expansion.

Nasotracheal intubation

This method can be used to secure an airway without obstructing the oral cavity, or if a prolonged recovery is predicted as the tube cannot be chewed on once jaw tone returns. A 1.0- 2.5-millimeter diameter flexible tube (ET tube, or other product suitable as a nasoesophageal tube) can be used in most rabbits. The tube is measured from rostrum to thoracic inlet. Lidocaine (0.1-0.2 milliliters) is instilled into the nostril to be used (maximum

2 mg/kg dose). The head is positioned in dorsoflexion, and the lubricated tube is directed ventromedially. The narrow airway may provide some drag on the tube; however, there should not be any significant resistance. A crunchy sensation suggests that the tube has been directed into the turbinates. Both of these events can result in edema and bleeding of the nasal passage. It is recommended to only attempt nasotracheal intubation of one nostril, since rabbits are obligate nasal breathers. Contraindications include the presence of an upper respiratory tract infection (as this technique may carry pathogens into the lower airway; Devalle[23] demonstrated that this may not be a concern in healthy rabbits) or pre-existing edema or stenosis of the nasal passages (e.g., patients with rhinitis, or maxillary dental disease).[24]

Guinea pigs, chinchillas, and degus

These species present a significant obstacle to orotracheal intubation: The palatal ostium. This structure is a well-vascularized, thin, membrane of tissue from the soft palate, the palatoglossal arches laterally, and the base of the tongue. There is a very small laryngeal opening. The palatal ostium may very easily be traumatized, and significant edema and hemorrhage can occur with unsuccessful intubation attempts.[25,26] Guinea Pigs typically have ingesta in their oral cavity, further making visualization challenging (in awake and stable patients, syringe-feeding a 3- to 5-milliliter bolus of water can help clear the oral cavity; cotton-tipped swabs can otherwise be used).

The use of a dental rack or board with an endoscope (or potentially otoscope) is ideal to attempt intubation. Guinea pigs and chinchillas can accommodate a 2.0- to 2.5-millimeter diameter uncuffed ET tube, or an 8-Fr red rubber catheter (length cut to reduce dead space, and a Murphy eye should be added).[25]

Vascular Access: Intravenous

Vascular access should be obtained in any patient presenting in an urgent situation. Unless the patient is moribund, anxiolysis, sedation or anesthesia is typically required to safely place an intravenous or intraosseous catheter. The exception to this could be patients such as well-socialized rabbits or ferrets; however, the benefits of anxiolytics should not be neglected as even sociable animals are likely stressed when presenting for an emergency.

Local anesthesia is very helpful in obtaining vascular access. EMLA cream ® (Aspen Pharmacare Canada, Oakville, ON, Canada) is a lidocaine 2.5% and prilocaine 2.5% topical anesthetic that can be applied prior to IV catheter or even venipuncture attempts.

See Table 2 for a list of equipment needed for IV catheter placement. Placement sites vary between species.

Table 2. Equipment needed for intravenous catheter placement in mammals.

Equipment
Clippers with #40 blade
Rubber tourniquet, paper clip (marginal auricular vein in rabbits), optional
Aseptic preparation for skin: Chlorhexidine or iodophor, isopropyl alcohol or sterile saline
Gauze sponges
Intermittent infusion/PRN plug, or pediatric T-port (slip-tip/Luer-slip preferred for T-port)
Sterile saline ± heparin for flushing
Indwelling over-the-needle catheter (20-26 gauge, 0.75-1")
Porous tape, roll cling gauze 0.5-1", self-adhesive wrap 0.5-1"
Topical anesthetic (e.g., E,MLA cream®), optional

Ferrets

Cephalic and lateral saphenous vein sites are similar in location to canine patients, and 22- to 26-gauge indwelling catheter is typically effective. Jugular placement requires a surgical cutdown and can accommodate a 20- to 24-gauge catheter, but can be difficult to maintain in a patient that is not anesthetized or moribund.

Hedgehogs

Cephalic catheters can only be maintained if the patient cannot/does not curl/ball, and this vein can typically accommodate a 24- to 26-gauge catheter.

Sugar gliders, mice, and hamsters

There are no great sites for IV catheter placement in this species as while peripheral veins may be accessible, catheters are typically too long to feed appropriately. Opt for an intraosseous catheter in these patients, instead.

Rabbits

Cephalic and lateral saphenous sites are predictable; however, both of these sites may have significant positional occlusion in awake patients. Since rabbits vary in body size significantly compared to other companion mammals, appropriate catheter sizes can vary between 20-26 gauge. The marginal auricular vein is an excellent, accessible site in most rabbits except dwarf breeds with very small pinnae. The caudal marginal vein is usually more prominent, but the author has found that the cranial vein can be the better option in some individuals. This vein can typically accommodate 22- to 26-gauge catheters. Occluding the vein naturally makes it more easily visualized; the restrainer may gently occlude using the index finger and middle finger in a scissor fashion, or a paperclip may be placed at the proximal end of the vein temporarily.

Guinea pigs

Guinea pigs tend to have very tough skin, and a cut-down using a hypodermic needle can be helpful to avoid burring of the catheter at insertion. The cephalic, accessory cephalic (located more laterally on the forelimb), and lateral saphenous veins are effective sites. A 24- to 26-gauge catheter is typically suitable.

Chinchillas and degus

The cephalic and lateral saphenous sites are often accessible. The cephalic veins are very superficial. Because of their small size, 26-gauge catheters are typically suitable.

Rats

The lateral saphenous veins as well as the lateral tail vein are accessible for catheter placement, and typically accommodate a 24- to 26-gauge catheter. The jugular requires a surgical cutdown but can accommodate a larger catheter, 22- to 26-gauge.

Vascular Access: Intraosseous

Intraosseous access is advised in patients that are too small to place an IV catheter, if the patient is experiencing peripheral venous collapse and the typically accessible veins are too small, or if the accessible veins are no longer usable due to prior catheter placement or injury.

Contraindications to IO catheter placement include the presence of a skin infection at the insertion site, pre-existing fracture of the target bone, osteomalacia (typically due to nutritional secondary hyperparathyroidism, common in sugar gliders), or pre-existing osteomyelitis of the target bone.

Local anesthesia in the form of a lidocaine block in the subcutaneous tissue is advised, with the goal of blocking the periosteal tissue. Lidocaine has a quick onset of action, as fast as 2-5 minutes after injection.[27] Sedation is also necessary unless the patient is moribund. The placement of IO catheters is suspected to be mildly to moderately painful; however, there is evidence that infusion of drugs and/or fluids is quite painful. An excellent summary of the evidence is presented by Cummings (2023).[28]

See Table 3 for a list of equipment needed for IO catheter placement. Placement sites vary between species. Needle selection is based on the following measurements:

- The internal needle diameter should equal ⅓-⅔ the medullary cavity's thinnest diameter of the diaphysis.
- The ideal needle length should equal ⅓-½ the length of the medullary cavity.

Orthogonal view radiographs are needed to confirm placement. Aspirating bone marrow contents and smooth infusion of flush solution are not as sensitive as radiographs to confirm placement.

Table 3. Equipment needed for intraosseous catheter placement in mammals.

Equipment
Adequate sedation if patient is not moribund
Clippers with #40 blade
Rubber tourniquet, paper clip (marginal auricular vein in rabbits), optional
Aseptic preparation for skin: Chlorhexidine or iodophor, isopropyl alcohol or sterile saline
Gauze sponges
Intermittent infusion/PRN plug, or pediatric T-port (slip-tip/Luer-slip preferred for T-port)
Radiology suite to confirm placement
Sterile gloves
Sterile saline ± heparin for flushing
Hypodermic or spinal needle (18-25 gauge)
Local anesthetic and needle, syringe for infusion
Porous tape, non-absorbable sutures (size 3-0 to 4-0), needle drivers, tissue forceps, optional
Scalpel, #11 or #15, optional
Cerclage wire optional as a stylet in hypodermic needle, optional

The procedure for placement of an IO catheter is as follows (consider that the patient is either adequately sedated, or is moribund):

- Locate the anatomic reference points.
- Perform an aseptic preparation of the site (shave fur, scrub) and instill local anesthetic.
- Stabilize the limb with your non-dominant hand.
- A stab incision with a scalpel may be made over the site if desired.
- Using your dominant hand, a firm, rotating motion is used to insert the needle. A slight pop will be felt when the medullary cavity is breached.
- Flush with saline (heparinized or non-heparinized).
- Take orthogonal-view radiographs to confirm.
- The catheter may be secured with a tape butterfly and simple interrupted or horizontal mattress sutures, if desired.

Intraosseous catheters can be maintained in place for up to 72 hours. The catheter insertion site should be inspected regularly for infection and extravasation of fluids.[29]

There are three commonly used IO catheter placement sites in mammals: The proximal femur, the proximal tibia, and the proximal humerus.

Proximal femur

The patient is placed in lateral recumbency. The non-recumbent limb is where the IO catheter will be placed. The target limb is slightly adducted, and the proximal limb is palpated to identify either the greater trochanter (rabbits) or trochanteric fossa (ferrets and rodents). The musculature here can be thick, making palpation a bit more challenging. The needle is inserted into the site as described above, in a proximodistal direction.

Proximal tibia

The patient is placed in lateral recumbency. The non-recumbent limb is where the IO catheter will be placed. The target limb is flexed at the stifle, and the tibial tuberosity/tibial crest is palpated (the insertion point of the patellar ligament), this is the insertion site. The needle is inserted into the site as described above, in a proximodistal direction.

Proximal humerus

The patient is placed in lateral recumbency. The non-recumbent limb is where the IO catheter will be placed. The shoulder of the target limb is flexed, and the greater tubercle of the humerus is palpated. This is the insertion point. The needle is inserted in a proximodistal direction.

Cardiopulmonary Cerebral Resuscitation

The goal of cardiopulmonary cerebral resuscitation (CPCR) in exotic mammals is the same as in other species: Return of spontaneous circulation (ROSC) and ventilation. The RECOVER Initiative (www.recoverinitiative.org) is a foundational resource for canine and feline emergency medicine, and the principles of modern CPCR in these companion species will be used as a basis for discussing exotic mammal resuscitation.

It is important to recognize that there is a significant knowledge gap for exotic species, as few studies have been performed to evaluate the effectiveness of CPCR protocols in these animals. There are no specific exotic animal guidelines from the RECOVER Initiative, at this time.

Basic life support versus advanced life support

Basic life support (BLS) is the recognition of cardiopulmonary arrest (CPA) and the immediate response to this event. It includes airway management and ventilation, as well as chest compressions.

Advanced life support (ALS) includes the basic interventions of BLS, as well as rescue drugs (e.g., anticholinergics, positive inotropes, vasopressors), correction of electrolyte imbalances, severe anemia, and/or volume deficits, and defibrillation where appropriate.

It is important to establish a hospital protocol to differentiate between BLS, ALS, and do not resuscitate (DNR) orders, to direct resuscitation efforts appropriately. In the author's experience, a client agreement for CPCR incorporates both BLS and ALS, versus a DNR order.

Some important statistics

Out-of-hospital CPA has a low success rate of CPCR for any species. CPA that occurs during general anesthesia has the highest success rate of CPCR for any species. Overall, CPA in exotic mammal species has a low success rate: Onuma and associates[30] found a survival rate of 0-5.6% and a discharge rate of 0-2.8% for exotic mammal species, including rabbits, hamsters, ferrets, and some other small mammal species.

There are many challenges with small exotic mammal species that could explain these poor survival and discharge rates. Some of those challenges are described below.

- Due to their relatively high metabolic rate, the time between the onset of CPA and the development of severe neurologic damage is likely shorter than in dogs or cats.
- If a facemask is used for ventilation, this can cause gastric bloat or tympany, which impedes diaphragmatic movements.
- Doses of rescue drugs are often extrapolated from other species, and efficacy has not been verified for all species.
- Animals with a small body size present various challenges, and simultaneous interventions are not always possible.

Ventilation

Always remember to ensure correct tube placement during ventilation, which can be reinforced by observing chest movements during intermittent positive pressure ventilation (IPPV), auscultating the lungs, and finally, capnography when ROSC is achieved.

Most patients can be ventilated at 10-20 breaths/minute, which equals to IPPV every 3-6 seconds (with a one second inspiratory time). If a ventilator is being used, a tidal volume of 10 ml/kg of body weight can be calculated. Respect a maximum airway pressure during IPPV of 10-20 mmHg to avoid barotrauma. High peak inspiratory pressure will also compromise venous return. If the $ETCO_2$ is greater than 45 mmHg (e.g., if spontaneous respirations have returned but are ineffective, or when ROSC is achieved), adjust the IPPV rate, inspiratory time, and tidal volume to better ventilate the patient.[31]

Chest compressions

Most exotic mammal patients encountered will be small enough to create a thoracic pump effect with one of two techniques: (1) place fingers and thumb on opposing sides of the thorax at the widest diameter; or (2) use the fingers of both hands on either side of the thorax, held straight, compressing using the full length of your fingers at the widest diameter of the thorax. The compression should make up half of the total compression and release cycle to allow for complete recoiling of the thorax.[31]

Electrocardiography

Electrocardiograms are helpful to identify arrhythmias, which helps direct treatment. The leads may be applied to the patient using atraumatic clips. Alternatively, hypodermic needles may be inserted intradermally in the lead placement locations, and then the leads attached to the needles.

In small mammals, there are several common rhythms that can be identified. They are described below.

- Sinus bradycardia: The sinus node signals appropriate cardiac muscle depolarization; however, the heart rate is slow.

- Asystole: Cessation of electrical and mechanical activity of the heart.

- Pulseless electrical activity (PEA): There is sufficient electrical activity in the heart; however, no pulse is detected (therefore, no perfusion occurring).

- Ventricular tachycardia: A fast heart rate that originates in the ventricles.

Doppler probe

A Doppler probe may be applied over the patient's heart, jugular/carotid, or limb arteries to attempt to detect a pulse in-between compression cycles.

Capnometry

The use of a capnograph is considered the best indicator of ROSC.[31,32] Side-stream capnography will produce a delay in $ETCO_2$ readings, and the sample volume of some units can surpass the tidal volume of very small patients.[33]

When monitoring the capnograph during a CPA event, consider the following:[31]

- $ETCO_2$ < 10 mmHg: If there is cardiac output, this can indicate hyperventilation (in response, the rate of IPPV should be reduced). This may also occur if the patient is not correctly intubated, or if CPCR is ineffective.

- $ETCO_2$ > 45 mmHg: A transient increase over 45 mmHg may occur at the onset of ROSC. This may also indicate hypoventilation if there is cardiac output.

- $ETCO_2$ 12-18 mmHg: These values indicate sustained ROSC.

CPCR algorithm

The following CPCR algorithms (Figs 1 and 2) have been adapted from Lichtenberger and Lennox,[34] and van Zeeland and Schoemaker.[31]

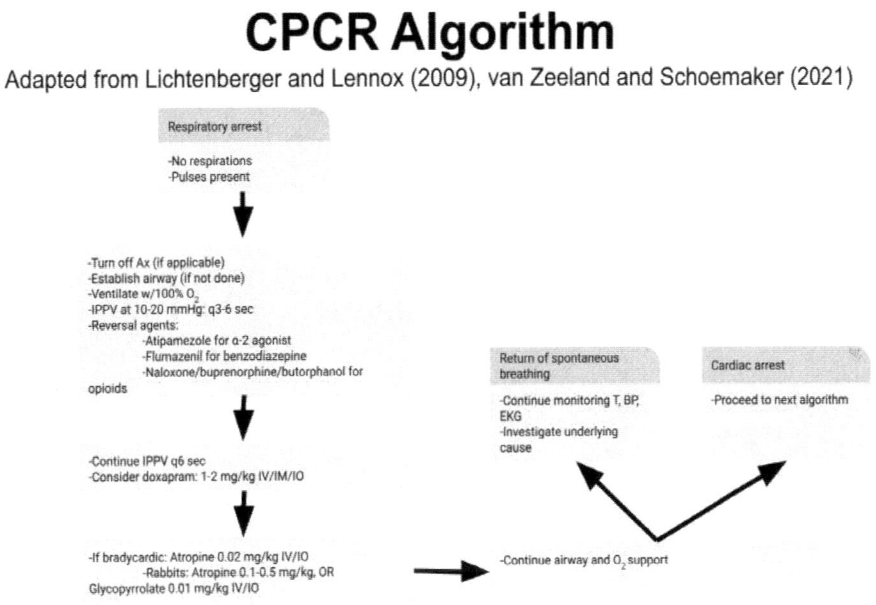

Figure 1. Respiratory arrest cardiopulmonary cerebral resuscitation (CPCR) algorithm.

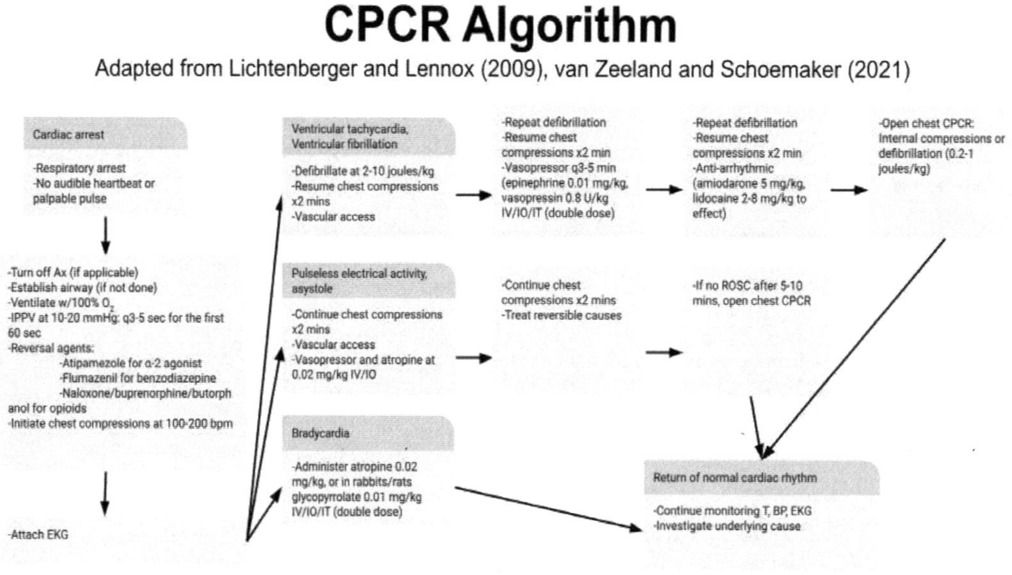

Figure 2. Cardiac arrest cardiopulmonary cerebral resuscitation (CPCR) algorithm.

Other preparations

Being prepared for emergencies will give your patients the best chance at survival. Aside from regular review of emergency protocols with staff and undertaking continuing education to be aware of current CPCR guidelines, what else can be done?

The crash cart should be kept stocked with adequate supplies, including drugs that are not out-of-date. See Table 4 for a list of supplies that are kept in the crash cart, which is similar to the canine and feline crash cart.

Anesthetic patients should have emergency drugs and fluid bolus volumes calculated in advance in the event of an anesthetic CPA. For patients with an American Society of Anesthesiologists (ASA) score over 3/5 (i.e., moderate to severe risk of death under anesthesia), emergency drugs can be pre-drawn to save vital seconds during a CPA event.

Generic drug charts with pre-calculated drug doses based on common body weight ranges can be prepared for commonly seen species. This can be kept with any CPCR guidelines or algorithm charts for quick access.

Table 4. Crash cart recommended supply list.

Supplies to be kept in the crash cart	Supplies to be kept near the crash cart
Emergency drugs, including epinephrine, atropine, glycopyrrolate, vasopressin, doxapram, lidocaine	CPCR protocols; quick drug reference charts
Various sizes of endotracheal tubes, supraglottic airway devices, oxygen masks	Defibrillator
1-ml, 3-ml syringes, ideally with needles attached	Syringe driver/pump
U-100 insulin syringes, ⅓ ml volume	Warming equipment for patient
Needles of various sizes (20-26 gauge)	Bag valve mask
Sterile physiologic saline flush	Crystalloids (e.g., Lactated Ringer's solution), colloids
IV catheters of various sizes (20-26 gauge)	Infusion sets and appropriate extensions
Spinal needles of various sizes (18-22 gauge)	Indirected blood pressure monitoring equipment
Fenestrated red rubber catheters (5- to 12-Fr)	Pulse oximeter

Conclusions

Many principles of exotic mammal emergency medicine are similar to canine and feline medicine; however, like with all exotic taxa there are many idiosyncrasies with these species that present unique challenges. Preparation of the treatment and surgical suites as well as staff members is critical for success.

References

1. Genovart M, Negre N, Tavecchia G, et al. The young, the weak and the sick: evidence of natural selection by predation. *PLoS ONE.* 2010;5(3):e9774. doi:10.1371/journal.pone.0009774.

2. Krumm CE, Conner MM, Hobbs NT, et al. Mountain lions prey selectively on prion-infected mule deer. *Biol Lett.* 2010;6:209-211. doi:10.1098/rsbl.2009.0742.

3. Mukherjee S, Heithaus M. Dangerous prey and daring predators: a review. *Biol Rev Camb Philos Soc.* 2013; 88:550-563. doi:10.1111/brv.12014.

4. McLaughlin A, Strunk A. Common emergencies in small rodents, hedgehogs, and sugar gliders. *Vet Clin North Am Exot Anim Pract.* 2016;19(2):465-99. doi:10.1016/j.cvex.2016.01.008. PMID: 27131160.

5. Covey E. How to triage. *Vet Nurse.* 2018;9(5). doi.org/10.12968/vetn.2018.9.5.262.

6. van Zeeland YRA, Schoemaker NK. Mammals: history and clinical rxamination. In: Graham JE, Doss GA, Beaufrère H, eds. *Exotic Animal Emergency and Critical Care Medicine.* Hoboken, NJ: Wiley-Blackwell; 2021:5-23.

7. Mazzaferro EM. Oxygen therapy. In: Silverstein DC, Hopper K, eds. *Small Animal Critical Care Medicine.* 2nd Ed. St. Louis, MO: Elsevier Saunders; 2015:77-80.

8. Boysen SR, Mathews KA. Respiratory emergencies. In: Mathews KA, ed. *Veterinary Emergency and Critical Care Manual.* 3rd Ed. Guelph, ON: Lifelearn, Inc; 2017:37-84.

9. Lee-Fowler T. Feline lower airway disease. In: Drobatz KJ, Hopper K, Rozanski E, Silverstein DC, eds. *Textbook of Small Animal Emergency Medicine.* Hoboken, NJ: Wiley; 2018:220-224.

10. Gardhouse S. Oxygen therapy. In: Graham JE, Doss GA, Beaufrère H, eds. *Exotic Animal Emergency and Critical Care Medicine.* Hoboken, NJ: Wiley-Blackwell; 2021:38-50.

11. Ferasin L, Ferasin H, Kilkenny E. Heart murmurs in apparently healthy cats caused by iatrogenic dynamic right ventricular outflow tract obstruction. *J Vet Int Med.* 2020;34(3):1102-1107. https://doi.org/10.1111/jvim.15774.

12. Warnefors E, Rueløkke ML, Gredal H. Results of a modified neurological examination in 26 healthy rabbits. *J Exot Pet Med.* 2019;30:54-59.

13. van der Woerdt A. Ophthalmologic diseases of small mammals. In: Quesenberry KE, Orcutt CJ, Mans C, Carpenter JW, eds. *Ferrets, Rabbits, and Rodents: Clinical Medicine and Surgery.* 4th Ed. St. Louis, MO: Elsevier; 2021:583-594.

14. Didier ES, Maddry JA, Brindley PJ, et al. Therapeutic strategies for human microsporidia infections. *Expert Rev Anti Infect Ther.* 2005;3(3):419-434.

15. Keeney CMH, Hung CS, Harrison TM. Comparison of body temperature using digital, infrared, and tympanic thermometry in healthy ferrets (*Mustela putorius furo*). *J Exot Pet Med.* 2021;36:16-21.

16. Di Girolamo N, Toth G, Selleri P. Prognostic value of rectal temperature at hospital admission in client-owned rabbits. *J Am Vet Med Assoc.* 2016;248(3):288-297. doi:10.2460/javma.248.3.288.

17. Levy IH, Di Girolamo N, Keller KA. Rectal temperature is a prognostic indicator in client-owned guinea pigs. *J Small Anim Pract.* 2021;62(10):861-865.

18. Heard D. Rodents. In: West G, Heard D, Caulkett N, eds. *Zoo Animal and Wildlife Immobilization and Anesthesia.* 2nd Ed. Ames, IA: Wiley-Blackwell; 2014:893-903.

19. Varga M. Airway management in the rabbit. *J Exot Pet Med.* 2017;26:29-35.

20. Thompson KL, Meier TR, Scholz JA. Endotracheal intubation of rabbits using a polypropylene guide catheter. *J Vis Exp.* 2017:1-8.

21. Docsinnovent. V-gel ADVANCED for rabbits. Available at: https://docsinnovent.com/products/v-gel-rabbit/. Accessed March 6, 2023.

22. Engbers S, Larkin A, Rousset A, et al. Comparison of a supraglottic airway device (V-gel ®) with blind orotracheal intubation in rabbits. *Front Vet Sci.* 2017;4:49.

23. Devalle JMS. Successful management of rabbit anesthesia through the use of nasotracheal intubation. *J Am Assoc Lab Anim Sci.* 2009;48:166-170.

24. Lafferty K. A guide to nasotracheal intubation in the rabbit. Available at: https://lafeber.com/vet/a-guide-to-nasotracheal-intubation-in-rabbits/. Accessed March 6, 2023.

25. Johnson DH. Endoscopic intubation of exotic companion mammals. *Vet Clin North Am Exot Anim Pract.* 2010;13:273-289.

26. Turner MA, Thomas P, Sheridan DJ. An improved method for direct laryngeal intubation in the guinea pig. *Lab Anim.* 1992;26:25-28.

27. Cousins MJ, Bridenbaugh PO. *Neural Blockade in Clinical Anesthesia and Management of Pain.* 4th Ed. Philadelphia, PA: Lippincott Williams & Wilkins; 2008.

28. Cummings CO. Are we providing adequate analgesia for intraosseous infusion in exotic animal practice? *J Exot Pet Med.* 2023;44:27.

29. van Zeeland YRA, Schoemaker NK. Mammals: Catheterization and Venipuncture. In: Graham JE, Doss GA, Beaufrère H, eds. *Exotic Animal Emergency and Critical Caré Medicine.* Hoboken, NJ: Wiley-Blackwell; 2021b:51-69.

30. Onuma M, Kondo H, Ono S, et al. Retrospective investigation of cardiopulmonary resuscitation outcome in 146 exotic animals. *J Vet Med Sci.* 2017;79(9):1611-1614. doi:10.1292/jvms.16-0360.

31. van Zeeland YRA, Schoemaker NK. CPR and euthanasia. In: Graham JE, Doss GA, Beaufrère H, eds. *Exotic Animal Emergency and Critical Care Medicine.* Hoboken, NJ: Wiley-Blackwell; 2021:80-94.

32. Brainard BM, Boller M, Fletcher DJ. RECOVER evidence and knowledge gap analysis on veterinary CPR. Part 5: monitoring. *J Vet Emerg Crit Care.* 2012;22(1):S65-84.

33. Lafferty K, Pollock CG. Capnometry in exotic animal species. Available at: https://lafeber.com/vet/capnometry/. Accessed March 20, 2023.

34. Lichtenberger M, Lennox AM. Exotic companion mammal emergency medicine and critical care. *Proc Annu Conf Assoc Exot Mamm Vet.* 2009.

Exotic Animal Emergencies for Technicians, Part 2: Evaluation of the Emergent Reptile Patient

Christina Miller, RVT, BSc

Session #3508

Affiliation: From Clayton Park Veterinary Hospital, 278 Lacewood Dr, Halifax, NS, B3M 3N8, Canada.

Abstract: Reptiles commonly present to the veterinary hospital in a state of severe illness. This masterclass reviews emergency presentations in non-crocodilian and non-avian reptiles (e.g., squamates and chelonians). It covers the technician's role in the primary survey, obtaining vitals efficiently and safely, obtaining airway and vascular access, and proceeding with cardiopulmonary and cerebral resuscitation.

Introduction

The basic principles of triage and masking phenomenon are covered in "Exotic Animal Emergencies for Technicians, Part 1: Evaluation of the Emergent Exotic Mammal Patient."

Reptiles present unique challenges as emergent patients due to their anatomy and ectothermic nature. Ectotherms are animals which rely on external heat sources to maintain their body temperature within a preferred optimum temperature zone (POTZ), that is species-specific.[1,2] They are sometimes described as poikilotherms; however, this term relates to thermoconformers, which does not characterize the entire daily habits of many reptiles (i.e., some reptiles are always thermoconformers, whereas others only become thermoconformers when they are inactive and are thermoregulators when active). The ectotherm's mentation, metabolism, immune function, drug metabolism, among innumerable other physiological processes rely on the animals' access to an appropriate thermal gradient to reach their preferred body temperature (PBT) within the POTZ.[1,2]

A husbandry review and client education of how to correct deficiencies is critical for success once the patient is discharged from the hospital.[1]

The Reptile Primary Survey

At the initial assessment, the patient may not be within their POTZ, so their mentation and vitals may be altered. They may have dulled responsiveness, neurological deficits, bradycardia, and bradypnea.[1] Many lizard species adopt a darker overall body colour.[3] If the patient is cool to the touch on arrival, active warming should be initiated. Cloacal temperature is not a useful assessment except for detecting hyperthermia and monitoring warming progress with regards to environmental temperature. Most rectal thermometers intended for use in mammals will not read below 32-34 °C (89.6-93.2 °F), and many species will present with a body temperature below these values. A digital probe thermometer intended for use in reptile enclosures can instead be used; ensure that you are using a quality unit by verifying temperatures against another thermometer.

If the species' POTZ is unknown, provide thermal support at 28-29.4 °C (82-85 °F).[1,4] Ideally, the patient should be warmed slowly, over 4-6 hours.[1] Once the patient is warmed to their POTZ, the initial assessment should be performed. Latent clinical signs may be appreciated once the animal has been warmed within their POTZ.

Masking phenomenon may not be as prominent in these animals as other exotic taxa; however, many pet owners simply do not recognize signs of illness in reptiles. This may either be due to unfamiliarity with normal behaviours and anatomy, or because some illnesses develop very slowly in reptiles.

Reptiles are morphologically and behaviorally diverse, and activity cycle as well as posture varies with the species being presented. Generally, reptiles should be bright, alert, and responsive to your presence and/or touch. Squamates will generally exhibit tongue-flicking behavior. Most lizard species should be able to raise their body off and above the surface they are resting on, with the exception of some stocky, low-bodied taxa such as Scincidae, Gerrhosauridae, and a few others. Chameleons should perch when offered a branch or limb, with their body lifted. Snakes may coil into a defensive position, Ball Pythons (*Python regius*) most famously will curl into a ball with their head hidden in the center. Many snakes will instinctively wrap around a hand or arm when being handled, and their grip should be substantial. Chelonians should be bright and alert, monitoring their environment, but will recede into the shell if threatened.[5]

If the patient is dull, dyspneic, or non-responsive to touch, proceed to the ABCDE protocol. The ABCDE protocol is a systematic approach to patient assessment in order of the most acutely life-threatening problems to least life-threatening problems.

A: Airway

Airway and breathing assessment is effectively performed simultaneously; however, signs of airway obstruction should be considered immediately, including: Increased respiratory effort (especially on inspiration), cyanosis (rare in reptiles; some species have dark blue or purple oral mucous membranes naturally),[6] stertor (low-pitched sound, suggests nasal or pharyngeal obstruction) or stridor (high-pitched sound, suggests lower airway obstruction).[6,7] Upper airway obstructions can be easily identified as the glottis is relatively rostrally located in squamates and chelonians, at the base of the tongue. There is no epiglottis. Reptiles are nasal breathers, and open-mouth breathing (aside from threat displays, vocalizations, or thermoregulation, e.g., increasing evaporative water loss in lizards[2,6]) is considered abnormal.[7] Orthopneic posture may also be exhibited by reptiles: Snakes may elevate the cranial ¼-⅓ of the body against a wall or object in their enclosure, and lizards may adopt a similar posture with or without support of an object. Chelonians may extend limbs and the neck to increase the limited space for lung expansion due to the presence of the rigid, boney shell.[6]

Reptiles have a relatively low metabolic rate compared to endotherms, and during anaerobic metabolism they exhibit a substantial capacity to buffer increased concentrations of lactic acid and hydrogen ions.[8] These characteristics allow reptiles to compensate effectively for partial airway obstruction or chronic disease and may only present when they decompensate from chronic disease.[6]

Oxygen therapy:

Oxygen therapy can be counterproductive in some patients, as the reptile's respiratory drive is stimulated by hypoxia.[6,7] Oxygen therapy is indicated if there are signs of hypoxia: Dyspnea, tachycardia, and cyanosis (which is rare). A positive response to oxygen therapy may be considered confirmatory of hypoxia.[9] Exposure to airborne irritants and toxins, evidence of drowning, and decompensating heart disease are all indications for oxygen therapy in reptiles.[6,9]

Note that pulse oximetry and blood gas analysis are not validated in reptiles.[6,9] Pulse oximetry may be used to evaluate trends in a patient; however, the value obtained may not be representative.

Flow-by oxygen may not suppress the respiratory rate: The FiO_2 provided by flow-by oxygen is generally between 25-40% at a flow rate of 2-3 L/min,[10] and the oxygen should be humidified to prevent airway desiccation.[6] Intermittent positive pressure ventilation (IPPV) may be necessary in debilitated individuals if apnea occurs from a high FiO_2 of oxygen.[9]

If respiratory arrest occurs or the obstruction cannot be identified and relieved, airway access must be secured. If there is jaw tone present, use of sedation and/or induction agents may be necessary. Airway access will be discussed below. Apneic patients can be ventilated on room air using a bag valve mask unless there is an indication (as discussed above) for oxygen therapy.

B: Breathing

The quality and rate of the patient's respirations should be assessed. Rate correlates to various physiologic parameters: Hypoxia results in an increased respiratory rate, and while normoxia in theory results in eupnea, supplemental oxygen results in respiratory suppression. Tachypnea can occur with hyperthermia and bradypnea occurs when the patient is below their POTZ. Hypercapnia and acidosis result in increased tidal and minute volumes whereas hypocapnia and alkalosis result in a decrease in these volumes.[7]

Note that some species will exhibit buccal pumping; therefore, thoracic respirations should be counted in these animals.

Lung fields may be assessed with auscultation, but this is difficult due to the presence of scales or a shell. A moistened gauze placed between the patient's skin or shell and the stethoscope diaphragm can be helpful in dampening the interfering sounds.[6]

C: Circulation

Circulation is challenging to evaluate. Peripheral pulses are typically not palpable. Capillary refill time and mucous membrane color are not used to evaluate perfusion,[11] although mucous membrane pallor can be suggestive of anemia. Heart rate is currently the best indicator of cardiovascular function.[11]

Noninvasive and indirect blood pressure (NIBP) is considered inaccurate but may be used to observe trends in a patient. A Doppler with a sphygmomanometer or an oscillometric unit can be used.[12,13] Direct arterial blood pressure monitoring is not practical for clinical applications.[1]

The heart is located within the pectoral girdle in most lizard and chelonian species, although some lizard taxa (Chamaeleonidae, Varanidae, Teiidae, and a few others) have a more caudally positioned heart. In snakes, heart position varies with the species' ecological niche: Arboreal species' heart is located approximately 15% of the body length from the rostrum; in snakes that are not arboreal or aquatic, the heart is located approximately 25% of the body length from the rostrum; in marine and freshwater species, the heart is located at nearly 50% of the body length.[14-16] Heart rate can be evaluated using a Doppler either over the heart, or at major vessels such as the carotid and brachial arteries. Like with pulmonary auscultation, cardiac auscultation is possible with a dampened gauze but may not be rewarding, especially in species with the bony pectoral girdle overlying the heart. Murmurs may be auscultated in larger species where the heart is located more caudally using a pressure-sensitive acoustic stethoscope.[17]

D: Disability

As previously noted, patients should be within their POTZ in order to accurately evaluate mentation and neurological function.[1] When below the POTZ, ectotherms such as reptiles exhibit reduced nerve conduction. This results in apparent paresis, lethargy, abnormal behaviours, or even neurological deficits.[18]

A description of normal mentation and posture in some common species was described in the previous section, The Reptile Primary Survey.

In addition to observing mentation and posture, gait may be evaluated on a surface with reasonable traction, such as a towel or yoga mat (reptiles may have difficulty ambulating on smooth surfaces such as a metal exam table or tile floor). The presence of ataxia, paresis, and lameness may all be observed during a gait evaluation.[19] Aquatic and semiaquatic species benefit from evaluation in water deep enough for them to submerge;[19] the water should be warmed to be within the species' preferred temperature, and close attention must be paid to the patient in case of deficits to avoid aspiration of water or potential drowning.

Proprioception can be evaluated in some larger lizards but cannot be evaluated in snakes and is difficult to interpret in chelonians.[19] Righting reflex can be evaluated in snakes.[19]

The menace response is described in some texts;[19] however, it is observed to be unreliable or absent in some species.[20-22] Palpebral and corneal reflex cannot be assessed in species with a spectacle (fused, transparent eyelid).[19] Pupillary light response is unreliable in reptiles due to their ability to override the reflex.[19]

E: Exposure of environment

At this point in the triage evaluation, patient temperature has been evaluated already. Survey the body for skin lesions, fractures, bruises, and other injuries.

Airway Access

Airway access should be obtained in apneic patients. Unlike in mammals and birds, pre-oxygenation may not be beneficial, depending on the patient's presentation as discussed above. Commercially available uncuffed tubes can be used for many patients; however, smaller purpose-made endotracheal (ET) tubes made from items such as red rubber catheters and polypropylene IV catheters are commonplace. Many of these house-made ET tubes will attach to the anesthetic circuit using the smallest adapter for commercial 2-3 mm internal diameter tubes. These can be cut to length and a Murphy eye can be added using a scalpel blade; a brief pass over a flame can dull the edges.[6] Squamate reptiles have dorsally incomplete tracheal rings, and chelonians have complete tracheal rings;[2] therefore, cuffed tubes must be avoided with the latter, as inflating the cuff has a great risk of tracheal tears and/or necrosis.[6] Tubes should ideally be clear when possible to allow for observation of spontaneous breathing (fogging), since capnometry is often ineffective in small reptiles due to their small tidal volume.[23]

Snakes

The mouth is opened with a soft speculum, or gentle manual traction on the mandible. The glottis in snake species is mobile to accommodate ventilation during ingestion of large prey items.[2,24] It is located caudodorsal to the tongue sheath, and a muscular valve remains closed until inspiration. Genus *Pituophis* has an epiglottal keel across the transverse aspect of the glottis, thought to amplify their defensive hissing sound; care should be taken to avoid damaging this structure during intubation.[24,25] The ET tube is premeasured to the level of the heart. Most snakes have only a single lung on the right side; in boid snakes (boas and pythons) a smaller left lung exists.[24] The ET tube should be lubricated, and no resistance should be encountered when intubating.

Lizards

The mouth is opened using a soft speculum; the author avoids opening the mouth in lizards with manual traction on the mandible in case of osteomalacia, as iatrogenic fracture may occur. The glottis is located caudodorsal to the tongue sheath and has a muscular valve as in snakes. Agamidae and Chamaeleonidae possess a very fleshy tongue and expandable gular region, so the glottis may be more difficult to visualize in these taxa (using a finger in the gular region to gently prop the tissue dorsally can help the operator visualize and stabilize the glottis). The ET tube is measured to the level of tracheal bifurcation, in many species this coincides with the level of the heart at the pectoral girdle. In Chamaeleonidae, the trachea bifurcates cranial to the thoracic inlet.[6] The ET tube should be lubricated, and no resistance should be encountered when intubating.

Chelonians

The mouth is opened with a speculum. The glottis is located at the base of the tongue and may be obscured by the fleshy tongue. Gentle rostral traction of the tongue can be helpful in propping up the glottis for better visualization. Members of suborder Cryptodira have a variable trachea length that either bifurcates in the mid-cervical region,[6,26-28] or immediately caudal to the head in members of the genus *Testudo*.[29] In suborder Pleurodira, the trachea length may be 85% of the carapace length, or longer.[29] The ET tube should be measured appropriately as per the species, lubricated, and no resistance should be encountered during intubation.

Vascular Access: Intravenous

Vascular access should be obtained in any patient presenting in an urgent situation. Reptiles do not have obvious peripheral veins, and the presence of thick and often scaled skin or the presence of a shell makes palpation of vessels difficult. Small patient size also contributes to the difficulty (or impossibility) of placing intravenous (IV) catheters in some patients. Furthermore, reptiles have a relatively low blood pressure compared to similarly sized birds or mammals, and in patients with cardiovascular or perfusion deficits, the flash of blood that helps to confirm IV catheter placement may not occur. Sites for IV catheter placement will be discussed.

Jugular vein

The right side is typically larger than the left. In lizards and snakes, a cutdown is typically needed for this site. Note that in snakes this requires opening the coelomic cavity as the vein is located quite deep and runs parallel to the trachea.[30] Patients as small as leopard geckos (*Eublepharis macularius*) can accommodate a jugular catheter.[31] In chelonians, the jugular can be superficial and easily accessed.[31]

Cephalic vein

This site may be accessed in lizards via a surgical cutdown in a transverse direction across the antebrachium.[30] An ultrasound-guided technique has been described in chelonians under 15 kg in body weight.[32]

Ventral caudal coccygeal vein

This vein may be accessed via a lateral or ventral approach in squamate reptiles, at 25-50% of the tail length to avoid accidental trauma of the hemipenes, hemiclitori, or scent glands.[33] This site should be avoided in species that perform caudal autotomy.

Dorsal caudal coccygeal vein

This site can be accessed in chelonians; it is difficult to maintain in active animals.[30]

Cardiac, ventricular

This site may be used in snakes where the catheter can be introduced directly into the ventricle of the heart with the patient in dorsal recumbency; this catheter may only be maintained in moribund or anesthetized patients.[30,34] Long-term consequences of catheter placement do not appear to have been studied.

Vascular Access: Intraosseous

Venous access may not be possible in very small, dehydrated, or hypotensive patients, and intraosseous (IO) catheters are an acceptable alternative in many animals.

Contraindications to IO catheter placement include the presence of a skin infection at the insertion site, pre-existing fracture of the target bone, osteomalacia (typically due to nutritional or renal secondary hyperparathyroidism), or pre-existing osteomyelitis of the target bone.[30] Local anesthesia and sedation for IO catheter placement is preferred unless the patient is moribund. Ongoing pain management is advised for IO catheters used for fluid and/or drug infusion.[35]

Placement sites vary between species. Needle selection is based on the following measurements: the internal needle diameter should equal ⅓-⅔ the medullary cavity's thinnest diameter of the diaphysis. The ideal needle length should equal ⅓-½ the length of the medullary cavity.[30,31]

Proximal tibia

This site is accessible in lizards and may be easier to maintain than other sites. The stifle is flexed to expose the tibial crest, which is the insertion site.[30,31]

Distal femur

This site is accessible in lizards and chelonians, and it is typically easy to maintain in lizards. The stifle is flexed to expose the diaphysis of the distal femur, which is the insertion site.[30,31]

Distal humerus

This site is accessible in chelonians and lizards and is easy to maintain in lizards. The elbow is flexed to expose the great tubercle of the humerus, and the needle introduced at the craniolateral aspect.[30,31]

Gular plastron and plastrocarapacial junction

These sites are only present in chelonians and may be easier to maintain in awake patients compared to IO catheters placed in the limbs. A drill (e.g., EZ-IO® Intraosseous Vascular Access System, Teleflex, Morrisville, NC, USA) is generally needed for placement in larger specimens.[30,31] A 2012 study found that fluid uptake from these sites is less effective compared to IO catheters placed in limbs.[36]

Cardiopulmonary Cerebral Resuscitation

The goal of cardiopulmonary cerebral resuscitation (CPCR) in reptiles is the same as in other species: Return of spontaneous circulation (ROSC) and ventilation. The knowledge gap between reptiles and even other exotic taxa is significant, and reptiles present many challenges during CPCR. The following discussion is informed by the RECOVER Initiative (Reassessment Campaign on Veterinary Resuscitation), found at https://recoverinitiative.org.

As discussed in "Exotic Animal Emergencies for Technicians, Part 1: Evaluation of the Emergent Exotic Mammal Patient," it is important to operate with protocols for basic life support (BLS), advanced life support (ALS), and do not resuscitate (DNR) orders for individual patients. A basic algorithm for CPCR is found in Part 1, and deviations from this algorithm will be discussed below.

Ventilation

Ventilation with 100% oxygen is advised until ROSC is achieved. At this time, the patient may be transitioned to room air (or alternatively, 30-40% oxygen)[37] as hypoxemia is the primary stimulus for spontaneous respirations in reptiles.[6,7] Ventilating the patient with a maximum circuit pressure of 12 cmH_2O and 2-6 breaths per minute has been advised.[38]

Chest compressions

Chest compressions may be less effective in reptiles compared to other taxa. The heart is located within the pectoral girdle in many lizards, or protected by the plastron in chelonians: Both of these features limit the effectiveness of compressions; a Doppler probe may be placed over a major artery (heart, carotid, brachial; spectacle in applicable squamates) to determine the effectiveness of compressions.[38] Compressions in snakes can be very effective, recall that the location of the heart varies with the species.[14] There appear to be no published recommended rates for evidence-based CPCR in reptiles; the author typically aims for 60-100 bpm.

Electrocardiography

Electrocardiography can be challenging to interpret in reptiles,[39] and arrhythmias are not well described in the literature. Arrhythmias that can be identified are treated as in other species.

Capnometry

Capnometry may not provide accurate readings in reptiles. In small patients (< 100 g body weight), side-stream capnography may have a sampling volume greater than the patient's tidal volume, resulting in the sample becoming diluted with fresh gas.[40] End tidal CO_2 ($ETCO_2$) values may not be representative of $PaCO_2$ due to pulmonary shunting.[41] Values rarely surpass 25 mmHg, and a return to normal values observed during anesthesia (e.g., 11-25 mmHg) may indicate ROSC.[38]

Conclusions

Reptiles are challenging patients due to their unique anatomy and physiology. There is a lack of evidence-based recommendations for CPCR in these taxa.

References

1. Petritz OA, Son TT. Emergency and critical care. In: Divers SJ, Stahl SJ, eds. *Mader's Reptile and Amphibian Medicine and Surgery.* 3rd Ed. St. Louis, MO; Elsevier; 2019:63-74.

2. Vitt LJ, Caldwell JP. *Herpetology: An Introductory Biology of Reptiles and Amphibians.* 4th Ed. St. Louis, MO; Elsevier; 2014.

3. Barten S, Simpson S. Lizard taxonomy, anatomy, and physiology. In: Divers SJ, Stahl SJ, eds. *Mader's Reptile and Amphibian Medicine and Surgery.* 3rd Ed. St. Louis, MO; Elsevier; 2019:967-976.

4. Pollock, C. Lafebervet: Exotic ICU: Nursing Care for Reptiles. Available at: https://lafeber.com/vet/exotic-icu-nursing-care-for-reptiles/. Accessed March 5, 2023.

5. Durso AM, JC Maerz. Natural behavior. In: Divers SJ, Stahl SJ, eds. *Mader's Reptile and Amphibian Medicine and Surgery.* 3rd Ed. St. Louis, MO: Elsevier; 2019;90-99.

6. Kanda I, Brandão J. Herptiles: oxygen therapy. In: Graham JE, Doss GA, Beaufrère H, eds. *Exotic Animal Emergency and Critical Care Medicine.* Hoboken, NJ: Wiley-Blackwell; 2021:716-721.

7. Knotek Z, Divers SJ. Pulmonology. In: Divers SJ, Stahl SJ, eds. *Mader's Reptile and Amphibian Medicine and Surgery.* 3rd Ed. St. Louis, MO; Elsevier; 2019:786-804.

8. Bickler PE, Buck LT. Hypoxia tolerances in reptiles, amphibians, and fishes: life with variable oxygen availability. *Annu Rev Physiol.* 2007;69:145-170.

9. Schumacher J. Respiratory medicine of reptiles. *Vet Clin North Am Exot Anim Pract.* 2011;14:207-224.

10. Mazzaferro EM. Oxygen therapy. In: Silverstein DC, Hopper K, eds. *Small Animal Critical Care Medicine.* 2nd Ed. St. Louis, MO: Elsevier Saunders; 2015:77-80.

11. Schumacher J, Mans C. Anesthesia. In: Mader DR, Diver SJ, eds. *Current Therapy in Reptile Medicine and Surgery.* St. Louis: Elsevier; 2014:145-148.

12. Chinnadurai SK, Wrenn A, DeVoe RS. Evaluation of noninvasive oscillometric blood pressure monitoring in anesthetized boid snakes. *J Am Vet Med Assoc.* 2009;234:625-630.

13. Chinnadurai SK, DeVoe R, Koenig A, et al. Comparison of an implantable telemetry device and an oscillometric monitor for measurement of blood pressure in anaesthetized and unrestrained green iguanas (*Iguana iguana*). *Vet Anaesth Analg.* 2010;37:434-439.

14. Farrell AP, Gamperl AK, Francis ETB. Comparative aspects of heart morphology. In: Gans C, Gaunt A, eds. *Biology of the Reptilia.* Vol. 19. Morphology G. Visceral Organs. Ithaca, NY: Society for the Study of Amphibians and Reptiles; 1998:375-424.

15. Wyneken J. Normal reptile heart morphology and function. *Vet Clin North Am Exot Anim Pract.* 2009;12: 51-63, vi.

16. McCracken HE, Fowler ME, Miller RE. Organ location in snakes for diagnostic and surgical evaluation. In: Miller RE, Fowler ME, eds. *Zoo and Wild Animal Medicine: Current Therapy.* 4th Ed. Philadelphia: WB Saunders; 1999:243-248.

17. Schilliger L, Girling S. Cardiology. In: Divers SJ, Stahl SJ, eds. *Mader's Reptile and Amphibian Medicine and Surgery.* 3rd Ed. St. Louis, MO: Elsevier; 2019:669-698.

18. Long SY. Approach to reptile emergency medicine. *Vet Clin North Am Exot Anim Pract.* 2016;19:567-590.

19. Platt SR. Neurology. In: Divers SJ, Stahl SJ, eds. *Mader's Reptile and Amphibian Medicine and Surgery.* 3rd Ed. St. Louis, MO: Elsevier; 2019:805-826.

20. Tarbert DK, Murthy VD, Sanchez-Migallon Guzman D. Neurological examination in healthy adult inland bearded dragons (*Pogona vitticeps*). *J Am Vet Med Assoc.* 2022;260(9).

21. Hedley J, MacHale J, Rendle M, Crawford A. Neurological examinations in healthy juvenile bearded dragons (*Pogona vitticeps*) and adult leopard geckos (*Eublepharis macularius*). *J Herpetol Med Surg.* 2021;31(2):141-146.

22. Monchaux M. Neurology in exotic animals. *Proc Int Conf VIII FAUNA.* 2019.

23. Briscoe LA, Syring R. Techniques for emergency airway and vascular access in special species. *Semin Avian Exot Pet Med.* 2004; 13(3): 118-131.

24. Funk RS, Bogan JE, Jr. Snake taxonomy, anatomy, and physiology. In: Divers SJ, Stahl SJ, eds. *Mader's Reptile and Amphibian Medicine and Surgery.* 3rd ed. St. Louis, MO; Elsevier; 2019:50-62.

25. Young BA, Sheft S, Yost W. Sound production in *Pituophis melanoleucus* (Serpentes: Colubridae) with the first description of a vocal cord in snakes. *J Exp Zool.* 1995;273(6):472-481.

26. Boyer TH, Innis CJ. Chelonian taxonomy, anatomy, and physiology. In: Divers SJ, Stahl SJ, eds. *Mader's Reptile and Amphibian Medicine and Surgery.* 3rd Ed. St. Louis, MO; Elsevier; 2019:31-49.

27. Zehtabvar O, Tootian Z, Vajhi A, et al. Computed tomographic anatomy and topography of the lower respiratory system of the European pond turtle (*Emys orbicularis*). *Int J Vet Sci.* 2014;9(2):9-16.

28. Lambertz M, Bohme W, Perry SF. The anatomy of the respiratory system in *Platysternon megacephalum* Gray, 1831 (Testudines: Cryptodira) and related species, and its phylogenetic implications. *Comp Biochem Physiol A: Mol Integr Physiol.* 2010;156(3):330-336.

29. Habova M, Pyszko M, Horak O, et al. Differences in the anatomy of the lower respiratory tract in selected species of the order Testudines. *Vet Med-Czech.* 2022;67(2):78-86.

30. Pardo MA, Divers SJ. Catheter placement. In: Divers SJ, Stahl SJ, eds. *Mader's Reptile and Amphibian Medicine and Surgery.* 3rd Ed. St. Louis, MO: Elsevier; 2019:422-428.

31. Kanda I, Brandão J. Herptiles: catheterization and venipuncture. In: Graham JE, Doss GA, Beaufrère H, eds. *Exotic Animal Emergency and Critical Care Medicine.* Hoboken, NJ: Wiley-Blackwell; 2021:722-731.

32. Di Bello A, Valastro C, Freggi D, et al. Ultrasound-guided vascular catheterization in loggerhead sea turtles (*Caretta caretta*). *J Zoo Wildl Med.* 2010;41(3):516-518.

33. Olsson A, Simpson M. Analgesia and anaesthesia. In: Doneley B, Monks D, Johnson R, Carmel B, eds. *Reptile Medicine and Surgery in Clinical Practice.* Hoboken, NJ: Wiley-Blackwell; 2018:369-381.

34. Divers SJ. Merck Veterinary Manual: Clinical Procedures for Reptiles. Available at: https://www.merck-vetmanual.com/exotic-and-laboratory-animals/reptiles/clinical-procedures-for-reptiles. Accessed March 5, 2023.

35. Cummings CO. Are we providing adequate analgesia for intraosseous infusion in exotic animal practice? *J Exot Pet Med.* 2023;44:27.

36. Young BD, Stegeman N, Norby B, Heatley JJ. Comparison of intraosseous and peripheral venous fluid dynamics in the desert tortoise (*Gopherus agassizii*). *J Zoo Wildl Med.* 2012;43(1):59-66. doi:10.1638/2011-0058.1.

37. Costello MF. Principles of cardiopulmonary cerebral resuscitation in special species. *Semin Avian Exot Pet Med.* 2004;13(3):132-141.

38. Doss GA, Sladky KK. Herptiles: CPR and euthanasia. In: Graham JE, Doss GA, Beaufrère H, eds. *Exotic Animal Emergency and Critical Care Medicine.* Hoboken, NJ: Wiley-Blackwell; 2021:738-745.

39. Doss GA, Mans C, Sladky KK. Herptiles: analgesia, anesthesia, and monitoring. In: Graham JE, Doss GA, Beaufrère H, eds. *Exotic Animal Emergency and Critical Care Medicine.* Hoboken, NJ: Wiley-Blackwell; 2021:746-757.

40. Ozeki L, Caulkett N. Monitoring. In: West G, Heard D, Caulkett N, eds. *Zoo Animal and Wildlife Immobilization and Anesthesia.* 2nd Ed. Ames, IA: Wiley-Blackwell; 2015:149-165.

41. Bertelsen MF. Squamates (snakes and lizards). In: West G, Heard D, Caulkett N, eds. *Zoo Animal and Wildlife Immobilization and Anesthesia.* 2nd Ed. Ames, IA: Wiley-Blackwell; 2015:657-659.

Exotic Animal Emergencies for Technicians, Part 3: Evaluation of the Emergent Avian Patient

Christina Miller, RVT, BSc

Session #3509

Affiliation: From Clayton Park Veterinary Hospital, 278 Lacewood Dr, Halifax, NS, B3M 3N8, Canada.

Abstract: Birds commonly present to the veterinary hospital in a state of severe illness. This masterclass reviews emergency presentations in avian patients (focusing on Psittaciformes, Galliformes, and Columbiformes), the technician's role in the primary survey, obtaining vitals efficiently and safely, obtaining airway and vascular access, and proceeding with cardiopulmonary and cerebral resuscitation.

Introduction

The basic principles of triage and masking phenomenon are covered in "Exotic Animal Emergencies for Technicians, Part 1: Evaluation of the Emergent Exotic Mammal Patient."

Birds are challenging patients due to their propensity to mask clinical signs from all but the most observant caretakers, only appearing obviously ill when the patient has become relatively unstable: What pet owners perceive as acute illness is often a chronic illness that has gone unrecognized.[1,2] Masking phenomenon and a failure to recognize early, subtle signs of illness on the part of the pet owner has resulted in birds having a reputation for being fragile patients.

The Avian Primary Survey

When initially triaging a bird, the patient should first be observed covertly. If the patient is dull, dyspneic, or non-responsive to touch, proceed to the ABCDE protocol. The ABCDE protocol is a systematic approach to patient assessment in order of the most acutely life-threatening problems to least life-threatening problems.

At presentation, birds should always be bright and alert, and able to perch. Psittacine birds will especially be exhibiting exploratory behavior. Birds should always notice the veterinary staff and respond to your presence. The author uses a brief mental algorithm to determine if an avian patient that does not qualify for immediate assessment with the ABCDE protocol is safe to handle, see Table 1.

Immediate supportive care in the form of a low-stress environment, heat support, and oxygen (if indicated), is paramount for birds that appear overtly ill.[1]

Table 1. "Is this bird stable?" algorithm.

Response when the patient is approached	Assessment and recommended support
It fluffs up, but still watches you. It reacts to your movements. It exhibits an increased respiratory rate, but the effort is normal.	This patient may be stable. Proceed with exam; however, monitor for signs of distress. If handling causes signs of distress, cease the exam and provide supportive care ± anxiolysis.
It is fluffed up and ignores you. It exhibits an increased respiratory rate and effort or is dyspneic. It appears weak: Unable to perch, abnormal splay-legged posture, resting at the bottom of the cage.	This patient is unstable and may deteriorate with handling. Supportive care ± anxiolysis is strongly recommended before proceeding with hands-on exam.

A: Airway

Airway and breathing assessment are effectively performed simultaneously; however, signs of airway obstruction should be considered immediately, including: Collapse or weakness, dyspnea (open-mouth breathing, exaggerated keel movements with respirations, tail bobbing with respirations).[1] Upper airway obstructions can be easily identified as the glottis is relatively rostrally located, at the base of the tongue. There is no epiglottis. The nares should also be surveyed for obstruction. Birds are nasal breathers, and open-mouth breathing (aside from threat displays, vocalizations, or thermoregulation) is considered abnormal.[1,2] Orthopneic posture is abnormal and may also be adopted by birds: The wings may be held away from the body, the neck outstretched.

Oxygen therapy

Birds have the highest metabolic rates among vertebrates, and therefore, the highest rates of oxygen consumption per unit of body mass. Any avian patient presenting for illness, especially those with an obviously increased respiratory effort or dyspnea, could benefit from oxygen supplementation.[3]

Oxygen provided by face mask may be very stressful for conscious bird patients due to the restraint necessary to maintain positioning, or they may find the face mask aversive.[3] FiO_2 of 35-60% can be achieved by using a face mask.[4] An oxygen chamber provides support that does not necessitate ongoing patient handling, resulting in less stress for the bird. The FiO_2 in an oxygen chamber can vary from 30-60%,[3] which is comparable to a face mask. Commercial incubators that allow for monitoring and control of O_2, CO_2, temperature, and humidity are ideal.

B: Breathing

The quality and rate of the patient's respirations should be assessed. Rate and effort correlates to various physiologic parameters, including $PaCO_2$, PaO_2, intrapulmonary PCO_2, and pH.[3] Rates correlate to patient body weight: Birds weighing < 300 g typically have a normal respiratory rate of 40-60 rpm, and birds weighing between 400-1000 g typically range between 13-30 rpm.[5] Respiratory rate and effort will increase will handling, so a hands-off assessment of these signs should always be performed first during the primary survey. After handling, respiratory rate and effort should normalize within 3-4 minutes in a healthy bird.[5]

Birds lack a diaphragm and rely on the bellows-like system of air sacs and some accessory muscles to ventilate the lungs. An increased respiratory effort may at first be subtle, involving accessory muscles (the tail depressor and suprapubic muscles) that create a "tail bob" movement.[6] Open-mouth breathing and an orthopneic posture suggest severe dyspnea.

While breathing should be assessed first as a hands-off observation, air sac and lung auscultation should be reserved for a more stable patient (or, moribund patients may not perceive this handling as stressful). The presence and location of abnormal sounds should be noted. Pediatric or infant stethoscopes are invaluable for evaluating small avian patients.

C: Circulation

Circulation is challenging to evaluate in birds presenting in distress, as the handling required to examine peripheral vessels, mucosal surfaces, or even performing a cardiac auscultation can be very stressful.

The heart may be auscultated at either side of the cranial portion of the keel bone, or immediately caudal to the scapulae on the patient's dorsum.[1] See Table 2 for a summary of normal heart rates in a variety of avian species.

Table 2. Normal heart rates in select avian species.

Species	Heart rate (bpm)
African grey parrot, Congo (*Psittacus erithacus*)	340-600
Amazon parrot (*Amazona* spp.)	340-600
Canary (*Serinus* spp.)	~274
Cockatiel (*Nymphicus hollandicus*)	~206
Eclectus parrot (*Eclectus roratus*)	340-600
Finch (Estrildidae and Fringillidae)	~274
Lovebird (*Agapornis* spp.)	206-274
Macaw, large (*Ara* spp. and others)	~275
Macaw, small (*Diospittaca* sp. and others)	~389
Rock dove (*Columba livia*)	160-300

Source: Lafferty K, Pollock CG. Monitoring vital signs in exotic animal species. May 17, 2018. LafeberVet Website. Available at: https://lafeber.com/vet/monitoring-vital-signs-in-exotic-animal-species/. Accessed March 6, 2023.

Few mucosal surfaces are easy to access on an alert bird. The cloacal (proctodeal) mucosa can be gently everted using a sterile, lubricated cotton-tipped applicator to assess mucous membrane color and capillary refill time. Likewise, the conjunctiva can be gently assessed for color. The choana can also be assessed for color but requires an oral exam,[1] which may not be tolerated by a distressed bird. Some species have dark pigmentation of the choanal mucosa, so pallor cannot be appreciated in this location.

Capillary refill time can likewise be assessed at the ulnar or basilic vein, or the dorsal metatarsal vein.[1]

D: Disability

As prey species, birds are excellent at exhibiting masking phenomenon and in the unfamiliar exam room will put maximum effort into appearing healthy. Thus, birds that are presenting with an abnormal mentation are likely severely compromised. Flighted birds prefer perching, so an inability to perch (e.g., the patient presents at the bottom of a cage or carrier where a perch is provided) is very concerning.

Typical reflexes should be present such as the corneal reflex and menace response. The pupillary light reflex is present in birds although it is non-consensual, and may be overridden by voluntary muscular control of the iris.[8]

E: Exposure of environment

The patient can be surveyed for injury such as obvious skin lesions, fractures, bruises, etc.

Cloacal temperature is not routinely taken in birds due to their propensity to exhibit elevated body temperature due to the stress of transport, illness, handling, among other things. A low cloacal temperature (hypothermia) is significant in that it suggests severe hemodynamic compromise.[9] The author will typically obtain a cloacal temperature on an avian patient that is moribund or during anesthetic monitoring.

Note that birds with feather loss due to either feather destructive behaviors or avian circovirus, and birds with a poor body condition score and/or muscle condition score, are at a higher risk of becoming hypothermic when compromised due to the lack of insulation or heat-generating muscle mass. These patients should be provided with heat support upon presentation unless there is a contraindication (e.g., presenting for heat stress).

Airway Access

Airway access should be obtained in apneic patients. Dyspneic birds with a possible lower airway obstruction may require air sac cannulation to secure an airway; however, this technique will not be discussed here.

Orotracheal intubation is relatively straightforward in birds. Birds lack an epiglottis, and the glottis is located at the base of the tongue. The fleshy tongue of psittacine birds may obscure visualization of the glottis, so the tongue may be gently retracted to better expose the tracheal opening. Birds have complete tracheal rings, so uncuffed tubes are preferred to avoid tracheal trauma. If a cuffed tube is used, the cuff should not be inflated. The trachea narrows a short distance from the glottis in many species,[2,10] so the width of the glottis is not a perfect predictor for the width of the lumen of the caudal trachea (i.e., the endotracheal tube selected should generally be narrower than the width of the glottis[10]). Cole tubes may reduce potential trauma.[2] Take note that the smaller the diameter of the tube, the greater the airway resistance encountered as per Poiseuille's law.[11] Because of this, it could be advantageous in very small patients (< 100 g body weight) to switch to a face mask once spontaneous respirations return.

Vascular Access: Intravenous

Anxiolysis, sedation, and/or anesthesia is typically required to obtain vascular access in birds, unless the patient is moribund.[12] Avian skin appears thin and translucent; however, it is also tough: A dermal cut down may be necessary immediately parallel to the vein to help prevent the intravenous (IV) catheter from burring.[13] Any site selected should ideally be plucked of feathers (if present) to allow for aseptic preparation.

It is important to note that birds can exsanguinate themselves if an IV catheter is displaced or removed by the patient, either due to excessive movement, or the patient's curious nature. Intraosseous (IO) catheters may be more suitable for some patients for these reasons. IV catheters should be appropriately secured and patients monitored closely if conscious.

Basilic or ulnar vein

This vein is located on the ventral surface of the wing, immediately distal to the elbow joint. The catheter is inserted in a proximal direction, and secured with either a transparent adhesive dressing or a tape butterfly around the hub of the catheter with simple dermal sutures. The wing can be stabilized with a figure-of-8 bandage, and distractor tabs (tape) added for conscious patients.[9,13]

Medial metatarsal vein

This vein is located on the medial surface of the tibiotarsus, immediately proximal to the tibiotarsus-tarsometatarsal joint, at the junction of scaled to scaleless skin. It is an excellent choice in birds with a body weight over 300 g. It can be secured with porous tape.[9,13]

Jugular vein

This vein is accessible in birds as small as 75 g in body weight. The right side is typically much larger than the left side and is generally preferred for catheterization (or venipuncture) attempts. The jugular can be found in a naturally featherless tract of skin (apteryla) in most species.[9,13] The jugular is somewhat mobile in the soft tissue, and the cervical air sac is sometimes precariously positioned immediately medial to the vein: A gentle digital manipulation of the area can improve its placement.

Vascular Access: Intraosseous

Venous access may not be possible in very small, dehydrated, or hypotensive patients, and intraosseous (IO) catheters are an acceptable alternative in many animals. Confirmation of placement requires orthogonal radiographic views. IO catheters can be maintained in place for up to 72 hours.[13]

Contraindications to IO catheter placement include the presence of a skin infection at the insertion site, pre-existing fracture of the target bone, osteomalacia (typically due to nutritional or renal secondary hyperparathyroidism), or pre-existing osteomyelitis of the target bone.[13] Local anesthesia and deep sedation or anesthesia for IO catheter placement is preferred unless the patient is moribund due to the stress and pain associated with placement. Ongoing pain management is advised for IO catheters used for fluid and/or drug infusion.[14]

Placement sites vary between species: The humerus, and in some species the femur, are pneumatic bones and should not be used for IO catheter placement. Infusing drugs or fluid into pneumatic bones can result in pneumonia, air sacculitis, or asphyxiation.[13] Needle selection is based on the following measurements: The internal needle diameter should equal ⅓-⅔ the medullary cavity's thinnest diameter of the diaphysis. The ideal needle length should equal ⅓-½ the length of the medullary cavity.[13]

Distal ulna

The dorsal tubercle of the distal ulna is localized, and the metacarpals slightly pronated to expose the insertion site. Once placed, flushing the catheter should result in blanching of the ulnar vein. A figure-of-eight bandage of the wing helps to stabilize this catheter site for prolonged use.[13]

Proximal tibiotarsus

The leg is flexed at the knee to expose the cnemial crest of the tibiotarsus. The needle is introduced in a disto-medial direction immediately posterior to the patellar tendon insertion site.[13] This site may be tolerated poorly in conscious birds, as standing and/or perching on the catheterized limb is likely uncomfortable.

Cardiopulmonary Cerebral Resuscitation

The goal of cardiopulmonary cerebral resuscitation (CPCR) in birds is the same as in other species: Return of spontaneous circulation (ROSC) and ventilation. The knowledge gap between birds and even other exotic taxa is significant, and these animals present many challenges during CPCR: Their metabolic rate and thus oxygen demands are higher than other taxa, drug doses are mostly extrapolated from other species, the small body size of many patients make interventions challenging, and the presence of the keel renders chest compressions less effective.[15] The following discussion is informed by the RECOVER Initiative (Reassessment Campaign on Veterinary Resuscitation), found at https://recoverinitiative.org.

As discussed in "Exotic Animal Emergencies for Technicians, Part 1: Evaluation of the Emergent Exotic Mammal Patient," it is important to operate with protocols for basic life support (BLS), advanced life support (ALS), and do not resuscitate (DNR) orders for individual patients. A basic algorithm for CPCR is found in Part 1, and deviations from this algorithm will be discussed below.

Ventilation

Ventilation with 100% oxygen is advised, at a rate of 10-20 breaths/min. Ventilation should be held for one second at a maximum airway pressure of 15-20 cmH_2O.[15]

Chest compressions

The large sternum shields the heart and is non-pliable. Chest compressions in birds function on the "thoracic pump" principle, where increasing intrathoracic pressure via compressions is presumed to move blood from the heart to emulate circulation.[16] Compressions over the lateral or dorsal body (approximating the anatomic location of the heart) may be more effective than ventral compressions over the keel.[15] Compressions at a rate of 120 compressions/min, or approximating a normal heart rate for the species, is advised.[15]

Electrocardiography

Electrocardiogram (ECG) leads may be placed at the wing and thigh skin adjacent to the body wall.[15] Avian taxa exhibit some differences in the normal electrocardiogram compared to mammals:

- The mean electrical axis is negative in most species.
- The S wave is very prominent, except in meat poultry (e.g., *Gallus gallus domesticus*).
- In smaller bird species with a relatively high heart rate, P on T phenomenon is common, and ST segment appears short (i.e., ST slurring).[15]

In birds, there are several common rhythms that can be identified as in mammals:

- Sinus bradycardia: The sinus node signals appropriate cardiac muscle depolarization; however, the heart rate is slow.
- Asystole: Cessation of electrical and mechanical activity of the heart.
- Pulseless electrical activity (PEA): There is sufficient electrical activity in the heart; however, no pulse is detected (therefore, no perfusion occurring).
- Ventricular tachycardia: A fast heart rate that originates in the ventricles.

Capnometry

In small patients (< 100 g body weight), side-stream capnography may have a sampling volume greater than the patient's tidal volume, resulting in the sample becoming diluted with fresh gas, although monitoring trends can still be helpful for patient assessment.[17] A return to normal $ETCO_2$ values observed during anesthesia may indicate ROSC (30-45 mmHg).[18]

Conclusions

Birds are challenging patients due to their unique anatomy and physiology. There is a lack of evidence-based recommendations for CPCR for avian species.

References

1. Doss GA, Mans C. Birds: history and clinical exam. In: Graham JE, Doss GA, Beaufrère H, eds. *Exotic Animal Emergency and Critical Care Medicine.* Hoboken, NJ: Wiley-Blackwell; 2021;435-444.

2. Doneley B. The clinical examination. In: Samour J, ed. *Avian Medicine.* 3rd Ed. St. Louis, MO: Elsevier; 2016:49-72.

3. Beaufrère H. Birds: oxygen therapy. In: Graham JE, Doss GA, Beaufrère H, eds. *Exotic Animal Emergency and Critical Care Medicine.* Hoboken, NJ: Wiley-Blackwell; 2021:450-456.

4. Manning AM. Oxygen therapy and toxicity. *Vet Clin Small Anim Pract.* 2002;32(5):1005-1020.

5. Hoppes SM. Merck Veterinary Manual: Management of Pet Birds - Exotic and Laboratory Animals. Available: at https://www.merckvetmanual.com/exotic-and-laboratory-animals/pet-birds/management-of-pet-birds. Accessed March 6, 2023.

6. Baumel J, Wilson J, Bergren D. The ventilatory movements of the avian pelvis and tail: function of the muscles of the tail region of the pigeon (*Columba livia*). *J Exp Biol.* 1990;151(1):263-277.

7. Lafferty K, Pollock CG. Monitoring vital signs in exotic animal species. May 17, 2018. LafeberVet Website. Available at: https://lafeber.com/vet/monitoring-vital-signs-in-exotic-animal-species/. Accessed March 6, 2023.

8. Williams D. Ophthalmology. In: Harrison GJ, Lightfoot TL, eds. *Clinical Avian Medicine.* Palm Springs, FL: Spix Publishing, Inc.; 2011:673-694.

9. Beaufrère H. Medical, nursing, and cosmetic procedures. In: Samour J, ed. *Avian Medicine.* 3rd Ed. St. Louis, MO: Elsevier; 2016:204-245.

10. Doneley B. *Avian Medicine and Surgery in Practice: Companion and Aviary Birds.* 2nd Ed. Boca Raton, FL: CRC Press; 2016.

11. Li Y, Lu Y, Tan X, Zhang Z, et al. Effect of endotracheal tube size on airway resistance and dynamic lung compliance. *Med (Baltimore).* 2022;101(43):e31410. doi:10.1097/MD.0000000000031410.

12. Dubé C, Dubois I, Struthers J. Intravenous and intraosseous fluid therapy in critically ill birds of prey. *J Exot Pet Med.* 2011;20(1):21-26.

13. Schnellbacher R, Beaufrère H. Birds: catheterization and venipuncture. In: Graham JE, Doss GA, Beaufrère H, eds. *Exotic Animal Emergency and Critical Care Medicine.* Hoboken, NJ: Wiley-Blackwell; 2021:457-467.

14. Cummings CO. Are we providing adequate analgesia for intraosseous infusion in exotic animal practice? *J Exot Pet Med.* 2023;44:27.

15. Kabakchiev C, Beaufrère H. Birds: catheterization and venipuncture. In: Graham JE, Doss GA, Beaufrère H, eds. *Exotic Animal Emergency and Critical Care Medicine.* Hoboken, NJ: Wiley-Blackwell; 2021: 477-487.

16. Costello MF. Principles of cardiopulmonary cerebral resuscitation in special species. *Semin Avian Exot Pet Med.* 2004;13(2):132-141.

17. Ozeki L, Caulkett N. Monitoring. In: West G, Heard D, Caulkett N, eds. *Zoo Animal and Wildlife Immobilization and Anesthesia.* 2nd Ed. Ames, IA: Wiley-Blackwell; 2015:149-165.

18. Sanchez Migallon-Guzman D, Beaufrère H. Birds: avian pain management and anesthesia. In: Graham JE, Doss GA, Beaufrère H, eds. *Exotic Animal Emergency and Critical Care Medicine.* Hoboken, NJ: Wiley-Blackwell; 2021:488-502.

Index

D

E

F

G

H

I

J

K

L

M

N

O

P

Q

R

S

U

V

W

X

Z